Crystallography in Modern Chemistry

A RESOURCE BOOK OF CRYSTAL STRUCTURES

現代化學的晶體學

結構資料述評

麥松威
周公度 編著

Crystallography in Modern Chemistry

A RESOURCE BOOK OF CRYSTAL STRUCTURES

Thomas C. W. Mak

Department of Chemistry
The Chinese University of Hong Kong
Shatin, New Territories, Hong Kong

Gong-Du Zhou

Department of Chemistry
Peking University
Beijing, China

A Wiley-Interscience Publication

JOHN WILEY & SONS, INC.

New York   Chichester   Brisbane   Toronto   Singapore

Copyright © 1992 by John Wiley & Sons, Inc.

All rights reserved. Published simultaneously in Canada.

Reproduction or translation of any part of this work beyond that permitted
by Section 107 or 108 of the 1976 United States Copyright Act without the
permission of the copyright owner is unlawful. Requests for permission
or further information should be addressed to the Permissions Department,
John Wiley & Sons, Inc.

Library of Congress Cataloging in Publication Data

Mak, Thomas C. W., 1936–
 Crystallography in modern chemistry : a resource book of crystal
structures / Thomas C.W. Mak and Gong-Du Zhou.
  p.   cm.
"A Wiley–interscience publication."
Includes bibliographical references and index.
ISBN 0–471–54702–6
 1. X-ray crystallography.   I. Chou, Kung-tu.   II. Title.
QD945.M244   1992
548—dc20   91–23395
                                                           CIP

Printed in the United States of America

10 9 8 7 6 5 4 3 2 1

Preface

This book is based on lecture notes used by the authors in teaching a variety of courses (General Chemistry, Chemical Bonding, Structure of Matter, Inorganic Chemistry, and Chemical Crystallography) at *The Chinese University of Hong Kong*, *The University of Western Ontario* (1965-1969), and *Peking University* in the past quarter-century, a period that has witnessed the emergence of X-ray crystallography as an important tool, and often the method of choice, for structure elucidation in all branches of chemistry. Our purpose is to expand the original scope and present a broadly organized survey of the wealth of structural information on the underlying theme that, historically, crystallography and modern chemistry have undergone parallel development in a complementary manner. It is hoped that this book will provide budding scholars with the necessary background to comprehend the global nature and richness of structural chemistry and motivate them to make their own contributions.

The text is organized in seven **chapters** and opens with an essay which traces the birth and development of X-ray crystallography and its enormous influence on modern science. Chapter 2 covers **fundamental crystal structures** such as sodium chloride, graphite, ice, benzene and urea, which generally fall within the purview of the undergraduate chemistry curriculum. Chapter 3 portrays the structural chemistry of the **main-group elements**, and chapter 4 that of the **transition metals**. The next two chapters deal with **organic compounds** and **organometallic compounds**, respectively. **Inclusion compounds** constitute the subject matter of chapter 7, and the book closes with an appendix listing some reference books and an index. The vast and exciting field of **macromolecular crystallography**, owing to limitation of space and the authors' non-expertise, has been accorded only cursory treatment in chapter 1.

All chapters, except the first, each consists of a number of **sections** labelled in the format *x.yy* where *x* and *yy* are, respectively, the chapter number and the section number. Each section features a **principal crystal structure** chosen on the basis of its historical significance and/or novel structural features. The description is then followed by a number of **remarks** devoted to relevant topics of interest, thus making it possible for a wide range of related compounds to be covered. The unit-cell parameters and atomic coordinates are listed for each principal structure so that students can explore various structural details with the aid of an interactive molecular graphics program; experience has shown that this is the most effective way of learning about the intricacies of a three-dimensional structure, especially in the case of the layer, channel, or framework type of crystal lattice.

Most of the molecular and crystal structures discussed in the text are accompanied by **original** figures, the sources of which are duly acknowledged as a fitting tribute to their authors. Although the book is not intended to be comprehensive, an attempt is made to bring the readers to the forefront of active research by including many topics of current interest with relevant references for further reading. In many instances literature citations are updated to the end of 1991. The authors are apprehensive that a volume of this nature and size, prepared in camera-ready form and without direct access to crystallographic data bases, is susceptible to a plethora of errors. An appeal is made to knowledgeable readers for their constructive criticisms and suggestions, especially with regard to important structures which we have inadvertently neglected.

Acknowledgments

We are indebted to our colleagues and friends who rendered their help in reading the manuscript: chapter 2 by Dr. Choi-Nang Lam (林才能), CUHK; chapter 3 by Dr. Seik Weng Ng (伍錫榮), University of Malaya; chapter 4 by Dr. T.S. Andy Hor (賀子森), National University of Singapore, Dr. Wai-Kee Li (李偉基), CUHK, and Dr. Kin-Shing Chan (陳建成), CUHK; chapter 5 by Dr. Tze-Lock Chan (陳子樂), CUHK; and chapter 6 by Dr. Kevin Wing-Por Leung (梁永波), CUHK, Dr. Kwan-Yu Hui (許均如), CUHK, and Dr. T.S. Andy Hor. We thank Miss Ronney Sau-Ling Chim (詹秀玲) and Miss Shuk-Wah Chan (陳淑華) for their skill and perseverance in typing several versions of the manuscript, Prof. Zhou Zhong-yuan (周忠遠), Chengdu Center of Analysis and Testing, Academia Sinica, and Prof. Wang Ru-ji (王如驥), Nankai University, for preparing some of the structure plots, and Miss Shek-Lin Huen (禤錫蓮), Miss Kwok Sik (郭 適) and Dr. T.S. Andy Hor for checking some of the literature references. Our warm appreciation is due to Prof. Ambrose Yeo-Chi King (金耀基), CUHK, whose calligraphy graced the Chinese title of the book, and to Dr. Ma Pui-Han (馬佩嫻醫生) who provided generous financial support which enabled the second author to spend five months in Hong Kong. Last but most of all, we gratefully dedicate this volume to our wives, Mrs. Gloria Sau-Hing Mak (麥葉秀卿) and Prof. Hao Run-rong (郝潤蓉), for their sacrifice, encouragement and support over an untold number of years.

Zhou Gong-du (周公度)   Thomas Chung-Wai Mak (麥松威)
Beijing, China   *Shatin, N.T., Hong Kong*

Foreword

Modern research in many aspects of the biological, chemical and physical sciences without an intimate knowledge of the atomic architecture of the molecules involved is like chemistry before Dalton. For this reason, crystal structure analysis has become an essential component of these scientific disciplines. The method is unique in its capability to provide atomic structural information on crystalline materials of molecular weights ranging from ten to several million daltons, although precision is difficult to achieve at both extremes.

The intellectual and technical advances in crystallography of the past twenty-five years have changed X-ray crystal structure analysis from a painfully tedious, although mentally challenging, methodology to a fast analytical procedure. Not only has there been a dramatic reduction in the time required to elucidate a crystal structure, from months to days or even hours in the most favourable cases, but the level of precision and range of complexity have greatly increased. Modern crystallography provides the principal tools for studying the detailed electron distribution in chemical bonds, the thermal motion of atoms in crystals, and the atomic architecture of macromolecules of biological interest. The science of X-ray crystallography was invented and developed by physicists and has been applied by biologists, chemists, material scientists and mineralogists. Although the title of this book refers to chemistry, it contains subject matter of relevance to all these related fields. Most modern chemistry departments now supplement their service NMR facility with service crystal structure analysis in support of research in synthesis and reaction mechanisms. As with spectroscopy, users tend to forget the contributions of the pioneers that made this possible.

This book is based on experience gained in teaching a variety of lecture courses. As always, this is the best prerequisite for a text that combines both educational and reference material for a broad readership. The book addresses the historical development of crystallography in the first chapter, then goes on to provide quite detailed descriptions of crystal structures in increasing levels of complexity, ranging from metals, inorganics, organics, organometallics to inclusion complexes. As the number of known crystal structures already exceeds 100,000 and further increases by about 5,000 every year, the examples used to illustrate various topics have been carefully selected to reflect the major advances of recent years.

George Alan Jeffrey
Pittsburgh, PA, U.S.A.

Acknowledgments

The authors are most grateful to the following scientific publishers for granting us permission to reproduce original figures from their copyrighted materials (free of charge unless otherwise marked with an asterisk; a double asterisk indicates concession of a much reduced permissions fee). We are particularly indebted to The American Chemical Society, The Royal Society of Chemistry, VCH Verlagsgesellschaft mbH, Pergamon Press, and The International Union of Crystallography in view of the large number of figures taken from their research journals.

Most publishers responded sympathetically to our plea to limit the size of the book by skipping the standard credit line for each original figure. The source of the figure is indicated in the form "After ref. *n*." at the end of the legend, and a full citation of reference [*n*] appears under **References** at the end of the **Section**. In each instance the copyright holder is either the publisher of a book or indentified indirectly from the journal title.

American Association for the
  Advancement of Science*
Acta Chemical Scandinavia
The American Chemical Society**
American Institute of Physics
Annual Reviews, Inc.
Academic Press, Inc.*
American Society of Metals
Bell and Hyman Ltd.
Butterworth-Heinemann
Cornell University Press
Cold Spring Harbour Laboratory
The Chemical Society of Japan
Chinese University Press
Elsevier Science Publishers B.V.*
Hutchinson Ross Publishing
  Company
International Union of
  Crystallography
International Union of Pure
  & Applied Chemistry
Johann Ambrosius Barth Verlag
Journal of Chemical Education
Kluwer Academic Publishers
Macmillan Magazines Ltd.

Marcel Dekker, Inc.
National Research Council of
  Canada
R. Odenbourg Verlag GmbH
Oak Ridge National Laboratory
Oxford University Press
Pergamon Press**
Plenum Publishing Corporation
The Royal Society of Chemistry
Science Press, Beijing
Springer Verlag
Taylor & Francis Ltd.
VCH Verlagsgesellschaft mbH
John Wiley & Sons, Inc.
Walter de Gruyter Publishers
W.H. Freeman & Company
WE. Saarback Gbmh
Fujian Institute of Research
  on the Structure of Matter
Koninklijke Nederlandse Akademie
  van Wetenschappen
Verlag der Zeitschrift für
  Naturforschung
Bayerische Akademie der
  Wissenschaften

Contents

Chapter 1 Development of X-Ray Crystallography 1

Chapter 2 Fundamental Structures 21

2.1. Copper Cu 22

2.2. Magnesium Mg 26

2.3. α-Iron Fe 30

2.4. Diamond C 37

2.5. Graphite C 43

2.6. Iodine I_2 57

2.7. Sodium Chloride (Rock Salt, Common Salt, Halite) NaCl 64

2.8. Cesium Chloride CsCl 69

2.9. Zinc Blende (Cubic Zinc Sulfide, Sphalerite) ZnS 72

2.10. Wurtzite (Hexagonal Zinc Sulfide) ZnS 77

2.11. Nickel Arsenide NiAs 81

2.12. Sodium Thallide NaTl 86

2.13. Fluorite (Fluospar) CaF_2 89

2.14. Rutile (Titanium Dioxide) TiO_2 93

2.15. α-Quartz SiO_2 97

2.16. Ice Ih (Hexagonal Ice) H_2O 104

2.17. Cadmium Iodide CdI_2 111

2.18. Spinel $MgAl_2O_4$ 114

2.19. Perovskite $CaTiO_3$ 119

2.20. Potassium Hexachloroplatinate(IV) K_2PtCl_6 129

2.21. Copper Sulfate Pentahydrate $CuSO_4 \cdot 5H_2O$ 134

2.22. Diopside $CaMg(SiO_3)_2$ 139

2.23. Calcite (Calcium Carbonate) $CaCO_3$ 146

2.24. Hexamethylenetetramine $(CH_2)_6N_4$ 151

2.25. Hexamethylbenzene $C_6(CH_3)_6$ 158

2.26. Naphthalene $C_{10}H_8$ 169

2.27. Urea $(NH_2)_2CO$ 175

Addenda to Chapters 1 and 2 182

Chapter 3   *Inorganic Compounds (Main Groups)*

3.1.	High Pressure Ices H_2O	**185**
3.2.	Potassium Hydrogen Fluoride (Potassium Bifluoride) KHF_2	**196**
3.3.	Hydrochloric Acid Dihydrate $HCl \cdot 2H_2O$ ($H_5O_2{}^+Cl^-$)	**204**
3.4.	Complex Containing Trinuclear Molybdenum Clusters Bridged by the Hydrogen Oxide Ligand $H_3O_2{}^-$ {[$Mo_3O_2(C_2H_5COO)_6(H_2O)_2]_2(H_3O_2)$}$Br_3 \cdot 6H_2O$	**210**
3.5.	Lithium Nitride Li_3N	**214**
3.6.	Complex Containing a Cryptated Sodium Cation and a Natride Anion $Na^+(C_{18}H_{36}N_2O_6) \cdot Na^-$	**224**
3.7.	Cesium Suboxide Cs_7O	**232**
3.8.	Basic Beryllium Acetate $Be_4O(CH_3COO)_6$	**237**
3.9.	α-Rhombohedral Boron (α-Boron, R12 Boron) B	**241**
3.10.	Decaborane(14) $B_{10}H_{14}$	**249**
3.11.	π-Cyclopentadienyl-π-(1)-2,3-dicarbollyliron(III), a Metallocarborane $C_5H_5FeB_9C_2H_{11}$	**256**
3.12.	Borax $Na_2B_4O_5(OH)_4 \cdot 8H_2O$	**265**
3.13.	Sodium β-Alumina $Na_2O \cdot 11 Al_2O_3$	**272**
3.14.	Gallium(I) Tetrachlorogallate(III) ("Gallium Dichloride") $Ga(GaCl_4)$	**280**
3.15.	Calcium Carbide CaC_2	**285**
3.16.	α-Oxalic Acid Dihydrate $H_2C_2O_4 \cdot 2H_2O$	**291**
3.17.	Forsterite (Olivine) Mg_2SiO_4	**297**
3.18.	Beryl $Be_3Al_2[Si_6O_{18}]$	**301**
3.19.	Muscovite (Mica) $KAl_2[AlSi_3O_{10}](OH)_2$	**304**
3.20.	Low-Albite (Feldspar) $Na[AlSi_3O_8]$	**309**
3.21.	Dehydrated Zeolite 4A (Linde Molecular Sieve Type 4A) $Na_{12}[Al_{12}Si_{12}O_{48}]$	**313**
3.22.	Phenyl-(2,2′,2″-Nitrilotriethoxy)silane (Silatrane) $(C_6H_5)\overline{Si(OCH_2CH_2)_3N}$	**323**
3.23.	Complex Containing Nonanuclear Germanium Anions $[K(crypt)^+]_6Ge_9{}^{2-}Ge_9{}^{4-} \cdot 2\frac{1}{2}en$	**330**
3.24.	Tin(II) Chloride Dihydrate $[SnCl_2(H_2O)] \cdot H_2O$	
3.25.	Hexanuclear Basic Lead(II) Perchlorate Hydrate $[Pb_6O(OH)_6](ClO_4)_4 \cdot H_2O$	**339**
3.26.	Nitrogen Triiodide–Ammonia (1/1) $NI_3 \cdot NH_3$	**346**
3.27.	Orthorhombic Black Phosphorus P	**351**
3.28.	Hexachlorocyclophosphazene $(PNCl_2)_3$	**364**
3.29.	Tetraphosphorus Trisulfide P_4S_3	**372**
3.30.	Potassium Antimony Tartrate Trihydrate (Tartar Emetic) $K_2[Sb_2(d\text{-}C_4H_2O_6)_2] \cdot 3H_2O$	**378**
3.31.	Bismuth Subchloride $Bi_{24}Cl_{28}$ or $(Bi_9{}^{5+})_2(BiCl_5{}^{2-})_4(Bi_2Cl_8{}^{2-})$	**382**
3.32.	Hydrogen Peroxide H_2O_2	**388**
3.33.	Orthorhombic α-Sulfur S_8	**394**
3.34.	Tetrasulfur Tetranitride S_4N_4	**402**
3.35.	Polythiazyl (Polymeric Sulfur Nitride) $(SN)_x$	**409**
3.36.	Decaselenium Hexafluoroantimonate $Se_{10}(SbF_6)_2$	**414**

3.37. Hexatellurium Tetrakis(hexafluoroarsenate)–Arsenic Trifluoride (1/2)
Te$_6$(AsS$_6$)$_4$•2AsF$_3$ **423**

3.38. Xenon Difluoride XeF$_2$ **430**

3.39. Gamma Brass Cu$_5$Zn$_6$ **440**

3.40. α-Silver Iodide AgI **448**

3.41. Silver Nitrate Oxide Ag(Ag$_6$O$_8$)NO$_3$ **455**

3.42. Tri-iodoheptakis(tri-*p*-fluorophenylphosphine)undecagold Au$_{11}$I$_3$[P(*p*-FC$_6$H$_4$)$_3$]$_7$ **463**

3.43. Calomel (Mercury(I) Chloride, Merchlite) Hg$_2$Cl$_2$ **473**

Chapter 4 *Inorganic Compounds (Transition Elements)*

4.1. Vanadyl Bisacetylacetonate VO(C$_5$H$_7$O$_2$)$_2$ **484**

4.2. Potassium Octacyanomolybdate(IV) Dihydrate K$_4$Mo(CN)$_8$•2H$_2$O **493**

4.3. Cesium Iron Fluoride Cs$_3$Fe$_2$F$_9$ **501**

4.4. Sodium Tris(ethylenediamine)cobalt Chloride Hexahydrate
(+)$_{589}$[Co(en)$_3$]$_2$Cl$_6$•NaCl•6H$_2$O **509**

4.5. (1,3,6,8,10,13,16,19-Octaazabicyclo[6.6.6]eicosane)cobalt(II) Dithionate Monohydrate
(Cobalt Sepulchrate Complex) (+)$_{490}$[Co(C$_{12}$H$_{30}$N$_8$)]S$_2$O$_6$•H$_2$O **517**

4.6. Chlorotris(triphenylphosphine)rhodium(I) (Wilkinson's Catalyst) RhCl[P(C$_6$H$_5$)$_3$]$_3$ **534**

4.7. Potassium Rhenium Hydride K$_2$ReH$_9$ **543**

4.8. Trinuclear Molybdenum–Sulfur Cluster Complex
(NH$_4$)$_4$[Mo$_3$S$_2$(NO)$_3$(S$_2$)$_4$(S$_3$NO)]•3H$_2$O **559**

4.9. Pentakis(triphenylphosphine)hexakis(4-chlorobenzenethiolato)hexasilver(I)
Toluene (1/2) Solvate Ag$_6$(SC$_6$H$_4$Cl)$_6$(PPh$_3$)$_5$•2C$_6$H$_5$CH$_3$ **569**

4.10. Iron(II) Phthalocyanine (FePc) C$_{32}$H$_{16}$N$_8$Fe **578**

4.11. Copper–Oxygen Based High-Temperature Superconductor Tl$_2$Ba$_2$Ca$_2$Cu$_3$O$_{10}$ **589**

4.12. Ternary Molybdenum Oxide System K$_2$Mo$_8$O$_{16}$ **600**

4.13. Potassium Tetracyanoplatinate (1.75/1) Sesquihydrate K$_{1.75}$[Pt(CN)$_4$]•1.5H$_2$O **610**

4.14. Dodecatungstophosphoric Acid Hexahydrate (Keggin Structure) (H$_5$O$_2$)$_3$[PW$_{12}$O$_{40}$] **618**

4.15. Nitrogenpentammineruthenium(II) Dichloride [Ru(NH$_3$)$_5$(N$_2$)]Cl$_2$ **633**

4.16. Iron–Sulfur Tetranuclear Cluster [(CH$_3$)$_4$N]$_2$[Fe$_4$S$_4$(SC$_6$H$_5$)$_4$] **644**

4.17. Dioxygen Adduct of Chlorocarbonylbis(triphenylphosphine)iridium
(Vaska's Compound) IrO$_2$Cl(CO)[P(C$_6$H$_5$)$_3$]$_2$ **659**

4.18. (*meso*-Tetraphenylporphinato)iron(II) [Fe(TPP)] C$_{44}$H$_{28}$N$_4$Fe **669**

4.19. Dioxygen Adduct of (2-Methylimidazole)-*meso*-tetra-
(α,α,α,α–*o*-pivalamidophenyl)porphyrinatoiron(II)-Ethanol C$_{70}$H$_{76}$N$_{10}$O$_7$Fe **686**

4.20. *cis*-Dichlorodiammineplatinum(II) (cisplatin, *cis*-DDP) PtCl$_2$(NH$_3$)$_2$ **694**

4.21. Cobalt Inosine 5′-Phosphate Heptahydrate C$_{10}$H$_{11}$O$_8$N$_4$PCo•7H$_2$O **702**

4.22. Dipotassium Octachlorodirhenate(III) Dihydrate K$_2$[Re$_2$Cl$_8$]•2H$_2$O **718**

4.23. Benzyltriphenylphosphonium Dodeca-μ-carbonyl-dodecacarbonyl-
dihydrido-*polyhedro*-tridecarhodate [P(CH$_2$Ph)Ph$_3$]$_3$[Rh$_{13}$H$_2$(CO)$_{24}$] **730**

Addendum to Chapter 4 **742**

Chapter 5   *Organic Compounds*

5.1.   Prostaglandin PGF$_{1\beta}$ (Tri-*p*-bromobenzoate Methyl Ester)   C$_{42}$H$_{47}$O$_8$Br$_3$ **744**

5.2. α-Glycine $^+$H$_3$N–CH$_2$–CO$_2^-$ **757**

5.3.   Perdeuterio-α-Glycylglycine $^+$D$_3$N–CD$_2$–CO–ND–CD$_2$–CO$_2^-$ **767**

5.4.   Potassium Dihydrogen Isocitrate   K(C$_6$H$_7$O$_7$) **786**

5.5.   Methyl *p*-Bromocinnamate   BrC$_6$H$_4$CH=CHCOOCH$_3$ **794**

5.6.   Cyclooctatetraene (COT)   C$_8$H$_8$ **799**

5.7.   Polyethylene Adipate (PEA)   [–CO(CH$_2$)$_4$CO•O(CH$_2$)$_2$O–]$_n$ **811**

5.8.   8,8-Dichlorotricyclo[3.2.1.0^{1,5}]octane   C$_8$H$_{10}$Cl$_2$ **821**

5.9.   Cubane   C$_8$H$_8$ **830**

5.10. 1′,8′ : 3,5-Naphtho[5.2.2]propella-3,8,10-triene   C$_{18}$H$_{14}$ **839**

5.11. [2.2]Paracyclophane   C$_{16}$H$_{16}$ **849**

5.12. 1,1′-Binaphthyl   C$_{20}$H$_{14}$ **862**

5.13. Kekulene (Cyclo[*d.e.d.e.d.e.d.e.d.e.d.e.*]dodecakisbenzene)   C$_{48}$H$_{24}$ **873**

5.14. s-Triazine   C$_3$H$_3$N$_3$ **881**

5.15. Di(2,3,6,7-tetramethyl-1,4,5,8-tetraselenafulvalenium) Hexafluorophosphate
       (C$_{10}$H$_{12}$Se$_4$)$_2$PF$_6$ **890**

5.16. Silenes (Silaethenes)   I Me$_2$Si=C(SiMe$_3$)(SiBu$^t_2$Me)•C$_4$H$_8$O
                               II Me$_2$Si=C(SiMe$_3$)(SiBu$^t_2$Me) **902**

5.17. Sucrose   C$_{12}$H$_{22}$O$_{11}$ **911**

5.18. (+)-8-Bromocamphor   C$_{10}$H$_{15}$OBr **923**

5.19. Cholesterol Hemiethanolate   C$_{27}$H$_{46}$O•½C$_2$H$_5$OH **932**

5.20. Codeine Hydrobromide Dihydrate   C$_{18}$H$_{21}$O$_3$N•HBr•2H$_2$O **947**

5.21. Sodium Benzylpenicillin (Sodium Penicillin G)   C$_{16}$H$_{17}$N$_2$O$_4$SNa **956**

5.22. Vitamin B$_{12}$ Coenzyme   C$_{72}$H$_{100}$CoN$_{18}$O$_{17}$P•17H$_2$O **968**

5.23. Valinomycin   C$_{54}$H$_{90}$N$_6$O$_{18}$•3(CH$_3$)$_2$SO **984**

Chapter 6   *Organometallic Compounds*

6.1.   Methyllithium   (CH$_3$Li)$_4$ **998**

6.2.   Ethylmagnesium Bromide Dietherate (Grignard Reagent)   C$_2$H$_5$MgBr•2(C$_2$H$_5$)$_2$O **1012**

6.3.   Trimethylaluminum   Al$_2$(CH$_3$)$_6$ **1023**

6.4.   Zeise's Salt   K[(H$_2$C=CH$_2$)PtCl$_3$]•H$_2$O **1032**

6.5.   Bis(triphenylphosphine)hexafluorobut-2-yneplatinum(0)   (Ph$_3$P)$_2$(CF$_3$C≡CCF$_3$)Pt **1038**

6.6.   [1,2-Bis(dimethylphosphino)ethane](neopentylidyne)-
       (neopentylidene)(neopentyl)tungsten(IV)   W(≡CCMe$_3$)(=CHCMe$_3$)(CH$_2$CMe$_3$)(dmpe) **1051**

6.7.   Dimanganese Decacarbonyl   Mn$_2$(CO)$_{10}$ **1061**

6.8.   Bis(tetraethyl)ammonium μ$_6$-Carbido-penta-μ-carbonyl-octacarbonyl-
       octahedro-hexacobaltate   [NEt$_4$]$_2$[Co$_6$C(CO)$_{13}$] **1074**

6.9.   Bis(cyclopentadienyl)iron (Ferrocene)   Fe(C$_5$H$_5$)$_2$ **1092**

6.10. Bis(cyclopentadienyl)beryllium (Beryllocene)   Be(C$_5$H$_5$)$_2$ **1106**

6.11. μ-Dinitrogen-bis[bis(pentamethylcyclopentadienyl)titanium]   {(η^5-C$_5$Me$_5$)$_2$Ti}$_2$N$_2$ **1119**

6.12. Tetraphenylcyclobutadiene Iron Tricarbonyl Fe(CO)$_3$(C$_4$Ph$_4$) 1131

6.13. Bis(cyclooctatetraenyl)uranium(IV) (Uranocene) U(η^8-C$_8$H$_8$)$_2$ 1139

6.14. Tribenzo[*b,e,h*][1,4,7]trimercuronin (*o*-Phenylenemercury Trimer) (C$_6$H$_4$Hg)$_3$ 1146

6.15. Trimethyltin(IV) Chloride (CH$_3$)$_3$SnCl 1152

6.16. (*R*-1-Cyanoethyl)(3-methylpyridine)cobaloxime C$_{17}$H$_{25}$N$_6$O$_4$Co 1164

Chapter 7 Inclusion Compounds

7.1. Ethylene Oxide Deuterohydrate 6(CH$_2$)$_2$O•46D$_2$O 1174

7.2. Tetra-*n*-Butyl Ammonium Benzoate Clathrate Hydrate
(*n*-C$_4$H$_9$)$_4$N$^+$C$_6$H$_5$COO$^-$•39½H$_2$O 1191

7.3. β-Hydroquinone Hydrogen Sulfide Clathrate 3C$_6$H$_4$(OH)$_2$•*x*H$_2$S (*x* = 0.874) 1207

7.4. 4-*p*-Hydroxyphenyl-*cis*-2,4-dimethylchroman Carbon Tetrachloride Clathrate
6[C$_{17}$H$_{18}$O$_2$]•CCl$_4$ 1217

7.5. Hexakis(benzylthiomethyl)benzene–Dioxan 1:1 clathrate
C$_6$(CH$_2$SCH$_2$C$_6$H$_5$)$_6$•C$_4$H$_8$O$_2$ 1228

7.6. Tri-*o*-thymotide Chlorocyclohexane Clathrate
2(C$_{33}$H$_{36}$O$_6$)•C$_6$H$_{11}$Cl 1238

7.7. *trans*-Diamminemanganese *catena*-Tetra-μ-cyanonickelate-Benzene (1/2)
(Hofman-type Clathrate) Mn(NH$_3$)$_2$Ni(CN)$_4$•2C$_6$H$_6$ 1248

7.8. (α-Cyclodextrin)$_2$•Cd$_{0.5}$•I$_5$•27H$_2$O (C$_{36}$H$_{60}$O$_{30}$)$_2$•Cd$_{0.5}$•I$_5$•27H$_2$O 1258

7.9. 1:1 Paraquat Bisparaphenylene-34-crown-10 Inclusion Compound
[C$_{12}$H$_{14}$N$_2$•C$_{28}$H$_{40}$O$_{12}$][PF$_6$]$_2$•2Me$_2$CO 1278

Bibliography

1295

Index

1303

Crystallography in Modern Chemistry

A RESOURCE BOOK OF CRYSTAL STRUCTURES

Chapter 1

Development of X-Ray Crystallography

Of all the current methods for structural characterization of chemical compounds, X-ray crystallography (complemented by neutron diffraction) is the only one capable of providing detailed information on interatomic distances, bond angles, molecular architecture, absolute configuration, thermal vibration parameters, crystal packing, as well as possible order-disorder and/or non-stoichiometry from the same experiment. The only requirement is that the sample be available in the form of a single crystal (linear dimension *ca.* 0.3 mm), which is not destroyed during data collection. For compounds of simple structures, the analysis of powder samples can often yield useful information. The experience of "seeing" a complex structure unravelled directly (*via* an electron-density map or a molecular plot) from the observed reflection intensities is exhilarating to the crystallographer and seldom fails to impress his science colleagues. The present essay briefly traces the historical development of crystallography and cites a number of crystal structure determinations which profoundly influenced the course of chemistry and related fields such as biochemistry and molecular biology.

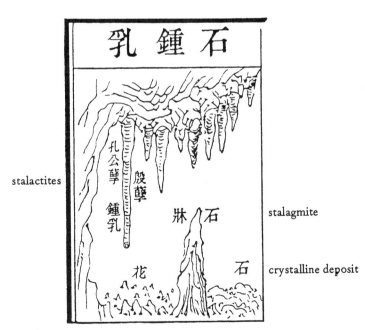

Fig. 1-1 Stalactites, stalagmites, and crystalline deposits of calcite ($CaCO_3$). From Li Shih-Chen's（李時珍）*Pên Tshao Kang Mu*（本草綱目）of A.D. 1596.

1

Early Developments

The word *crystal* is derived from the Greek name for "clear ice", a term applied to transparent quartz in a mistaken reference about its origin. The first encounter of our ancestors with crystals most likely took place in a cave, where deposits of calcite often occur as stalactites and stalagmites (Fig. 1-1). As natural objects of great beauty, crystals in the form of minerals and precious stones have aroused human curiosity and imagination since the dawn of civilization, and from time to time spectacular finds are unearthed in archaeological excavations. Crystallization as a means of purifying chemical substances was discovered long ago by the ancients, and a wide variety of crystalline compounds was investigated and sometimes synthesized, for example, by Chinese alchemists searching for the drug of immortality and pursuing the making of gold (Table 1-1).

Table 1.1 Some crystalline substances in Chinese alchemy texts.

cinnabar	HgS	朱砂，辰砂，丹砂，仙砂
realgar	As_4S_4	雄黃
orpiment	As_2O_3	雌黃
arsenolite	As_4O_6	礜石，秋石
stalactitic calcite	$CaCO_3$	石鐘乳
potash alum	$KAl(SO_4)_2 \cdot 12H_2O$	白礬
quartz (transparent)	SiO_2	水玉，水精
cassiterite	SnO_2	石桂
mosaic gold	SnS_2	豆金

Around A.D. 300, the Chinese alchemist Ko Hung (葛洪) gave a recipe for the preparation of "mosaic gold"——actually tiny stannic sulphide crystals displaying a golden lustre. In modern times Joseph Needham and his co-workers successfully repeated this recipe and identified "cold salt" (寒鹽) as ammonium chloride.[1] This and other similar experimental verifications have led to their statement that **'Chinese alchemical texts should be taken very seriously until proved to be erroneous.'**

$$6Sn + 2KAl(SO_4)_2 \xrightarrow[\text{prolonged heating}]{\text{"cold salt"}} (3/2)SnS_2 + (9/2)SnO_2 + Al_2O_3 + K_2SO_4 + \text{elemental S}$$

Crystals generally occur as polyhedral figures, although individual specimens may often be small and imperfectly formed. Scientific studies of the crystalline state probably began in the early seventeenth century, when Johannes Kepler (discoverer of the three planetary laws of motion) attempted to explain the hexagonal symmetry of snowflakes with illustrations showing the

packing of spherical particles.[2]   In 1669 Steno demonstrated that although
specimens of the same crystalline substance may vary greatly in size and
shape, the angles between corresponding faces are constant.[3]   In 1784 Haüy
discovered the law of rational indices. According to this fundamental law of
crystallography, every face of a single crystal may be described by a set of
three *small* integers (the Miller indices).[4]   In 1830 Hessel showed that
crystals, considered as idealized, finite geometrical figures, can be
classified into thirty-two distinct "symmetry classes".[5]

  The preceding studies based on the external shapes of crystals
inevitably led to the fundamental idea of a "lattice", namely a regular,
translational repetition of identical structural units in the crystalline
state. In 1850 Bravais showed that there exist only fourteen types of space
lattices.[6]   To simplify matters, one's attention should be focused on the
"unit cell" (Fig. 1-2) since the entire lattice can be regenerated by the
stacking of identical unit cells.

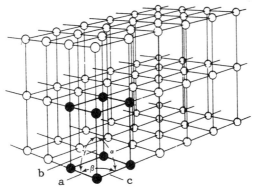

Fig. 1-2 The crystal as an infinitely extended three-dimensional lattice.
Each lattice point represents a structural unit (an atom, a group of atoms, a
molecule, or a group of molecules). A unit cell is outlined in black.

  The general problem of enumerating all possible ways of symmetrical
repetition of an asymmetric object in space was solved independently by
Fedorov,[7] Schönflies,[8] and Barlow[9] towards the end of the last century.
Their investigations showed that every crystal belongs to one of the two
hundred and thirty possible symmetry patterns, called "space groups", in the
same way that a given type of wallpaper belongs to one of the seventeen two-
dimensional "plane groups".

  The English amateur crystallographer William Barlow was a school teacher
who worked out the theory of space groups in his spare time. Lord Kelvin had
considered the closest packing of identical spheres and concluded that there
existed only one possible arrangement: cubic closest packing. But Barlow,

with his command of space group theory, came up with a second: hexagonal closest packing. He went on to discuss the probable internal structures of a number of simple crystalline compounds (Table 1-2), and got most of them right,[10] some thirty years before their determination by X-ray analysis!

Table 1-2 Symmetry and structure.

Stoichiometry External Symmetry	Space Group Theory	Probable Internal Structure of Crystals
Crystal Structure		Assignment by Barlow, 1883-1887
cubic closest packing		Cu, Ag, Au
hexagonal closest packing		Mg, Zn, Cd
halite (rock salt)		NaCl, KCl
cesium chloride		CsCl
zinc blende (sphalerite)		ZnS (cubic form)
wurtzite		ZnS (hexagonal form)
diamond		---

Birth of X-ray Crystallography

It is generally agreed that Röntgen's discovery of X-rays in 1895[11] marked the beginning of modern science. Another milestone was reached in 1912 with Laue's discovery of the diffraction of X-rays by crystals.[12] At that time, the University of Munich was a major research center of physics in Europe. Röntgen had moved there as Director of the Physical Institute in 1900 and received the first Nobel Prize in Physics the following year. Arnold Sommerfeld, the premier theoretical physicist, was appointed Director of the newly organized Institute of Theoretical Physics in 1906 with the blessing of Röntgen. Among Sommerfeld's flock of students was the young Paul Ewald, who in 1912 was writing his doctoral dissertation on the propagation of light waves through a lattice of resonating oscillators (scattering atoms). Some results of his mathematical analysis seemed puzzling to him, and he decided to consult Max von Laue, a Privat-Dozent who had just written an article on optics in the authoritative *Handbuch der Physik*. Laue graciously agreed and invited Ewald to dinner at his house. As they were walking from the Institute towards their destination, Ewald began to explain his treatment of the problem. Totally unaware of the lattice concept, Laue was astonished to learn that crystals were generally assumed to possess such internal regularity. He asked Ewald about the distance between the resonators. Ewald replied that this was very small compared with the wavelength of visible light, but no exact value could be given since the internal structures of crystals were not known. In Laue's mind flashed the question: "What would happen if an X-ray beam, instead of visible light, were to pass through the crystal?"

   While Ewald was putting the finishing touches to his thesis and preparing for his oral defence, Laue treated the crystal as a three-dimensional diffraction grating and arrived at the famous Laue equations:

$$\mathbf{a}_i.\mathbf{S} = h_i$$

where $\mathbf{a}_i$ ($i = 1,2,3$) are the primitive translations of the lattice, $\mathbf{S}$ is the scattering vector of magnitude $2\sin\theta/\lambda$, and h_i are the diffraction indices which take integral values. When he presented his ideas in a seminar, Röntgen and Sommerfeld both expressed skepticism. With tactful diplomacy Laue persuaded Walther Friedrich (Sommerfeld's assistant) and Paul Knipping (a student who assisted Friedrich) to test his ideas experimentally. The resulting paper made history by showing the first recorded X-ray diffraction patterns and their interpretation (Fig. 1-3).[12] Recognition was quick and universal, and in his congratulatory postcard to Laue,[13] Albert Einstein

INTERFERENZ-ERSCHEINUNGEN BEI RÖNTGENSTRAHLEN
(*Interference Phenomena for X-rays*)
von

W. Friedrich, P. Knipping und M. Laue
Vorgelegt von A. Sommerfeld in der Sitzung am 8. Juni 1912

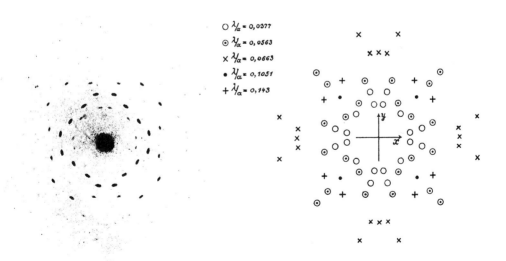

Cubic ZnS, 10 x 10 x 0.5 mm; X-ray beam perpendicular to (100) face.

Fig. 1-3 The first recorded X-ray diffraction pattern and its interpretation. (After ref. 12).

Table 1-3 Nobel prizes --- the X-ray connection.

Physics	1901	Wilhelm Conrad Röntgen	discovery of X-rays
	1914	Max von Laue	diffraction of X-rays by crystals
	1915	William Henry Bragg* William Lawrence Bragg*	X-ray analysis of crystal structures
	1917	Charles G. Barkla	characteristic X-ray spectra of the elements
	1924	Manne Siegbahn	X-ray spectroscopy
	1927	Arthur Holly Compton (Charles T.R. Wilson)	wavelength change in diffused X-rays (the "Compton effect")
Chemistry	1936	Peter Debye	dipole moments, X-ray and electron diffraction in liquids and gases
	1954	Linus Carl Pauling*	nature of the chemical bond
	1962	John Charles Kendrew* Max Ferdinand Perutz*	X-ray structure determination of globular hemoproteins (myoglobin and hemoglobin)
Physiology- Medicine	1962	Francis H.C. Crick* James D. Watson Maurice H.F. Wilkins*	structure of deoxyribonucleic acid
Chemistry	1963	Giulio Natta (Karl Ziegler)	structural chemistry and technology of high polymers
	1964	Dorothy Crowfoot Hodgkin*	X-ray analysis of biologically important compounds (e.g. penicillin and vitamin B_{12})
	1969	Odd Hassel* (Derek H.R. Barton)	conformational analysis
	1976	William Nunn Lipscomb*	structure and bonding in boron hydrides
Physics	1981	Kai M. Siegbahn (Nicolaas Bloembergen, Arthur L. Schawlow)	electron spectroscopy for chemical analysis (ESCA)
Chemistry	1982	Aaron Klug*	reconstruction of three-dimensional images of complex molecular aggregates (e.g. viruses, membranes, chromosomes and muscle fibres) from electron micrographs
Chemistry	1985	Herbert Hauptman* Jerome Karle*	solution of the phase problem in crystallography by direct methods
Chemistry	1987	Donald J. Cram Jean-Marie Lehn Charles J. Pedersen	host-guest chemistry and structure-specific interactions
Chemistry	1988	Robert Huber* Johann Deisenhofer* Hartmut Michel	isolation and X-ray structural determination of a membrane-bound photosynthesizing protein cluster

Enclosed in parentheses are the names of Nobel Laureates whose awards were unrelated to X-rays. The names of crystallographers are marked by asterisks. It is of interest to note that Yuan T. Lee, one of the 1986 Nobel Laureates in Chemistry, carried out research in crystallography while putting in a year of "military service" after obtaining his M.Sc. degree from National Tsing Hua University in Taiwan, which resulted in a paper on the structure of tris(cyclopentadienyl)samarium(III) [Acta Crystallogr., Sect. B 25, 2580 (1969)].

remarked that his experiment 'ranks among the greatest in physics'. Einstein's assessment was well-founded since on the one hand, Laue's discovery established the nature of X-rays and initiated the field of X-ray spectroscopy, while on the other it provided a powerful new tool for investigating the structure of crystalline matter on an atomic scale. As stated by the Braggs:[14] 'In physics, chemistry, biochemistry, metallurgy, mineralogy and other sciences new points of view have been adopted as a result of this more intimate knowledge of the ultimate structure of matter.'

The unique contributions of Röntgen and Laue in the development of modern science may be appreciated from a tabulation of twenty-eight Nobel Prize winners whose awards had something to do with X-rays (Table 1-3).[15] The awards were initially in physics and gradually shifted to chemistry and molecular biology. Remarkably but not fortuitously, fifteen *bona fide* crystallographers have their names (marked by asterisks) on this celebrated list.

Basic Principles

The applications of X-rays in modern science and technology are summarized in Fig. 1-4. X-Ray crystallography is used extensively nowadays in the elucidation of crystal and molecular structures. In light microscopy, the image of a magnified object may be seen directly or recorded on film because scattered visible light can be focused by optical lenses. However, scattered X-rays cannot be focused, and only the intensities of the diffraction pattern are recorded in the experiment. The crystallographer's job is to supply the missing phases for the computation of a Fourier map in which atoms show up as

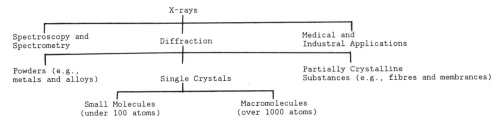

Fig. 1-4 Applications of X-rays in modern science and technology.

electron-density maxima. The nature of this "phase problem" can be understood by reference to Fig. 1-5. The structure factor, $F(hk\ell)$, represents the wave scattered by the contents of the unit cell in a direction specified by the diffraction indices $hk\ell$. It is in general a complex quantity composed of an amplitude $|F(hk\ell)|$ and a phase $\alpha(hk\ell)$. The measured intensity $I(hk\ell)$ is proportional to the square of the structure amplitude. Thus $|F(hk\ell)|$ can be derived from experiment, but the phases $\alpha(hk\ell)$ as required in the electron-density expression are not immediately available. Solution of the crystal structure therefore amounts to finding an approximately correct set of phases. Unlike the electron density $\rho(xyz)$, the Patterson function $P(uvw)$ can be computed directly from the measured intensities. As Patterson showed in his pioneering paper of 1934,[16] this function represents a superposition of all interatomic vectors in the crystal and can lead to a solution of the phase problem in favourable cases. If the structure contains a heavy atom, namely

$$F(hk\ell) = \sum_j f_j e^{i2\pi(hx_j+ky_j+\ell z_j)} = |F(hk\ell)|e^{i\alpha(hk\ell)}$$

structure
factor   =   amplitude x phase factor

$$\rho(xyz) = \frac{1}{V}\sum_h \sum_k \sum |F(hk\ell)|\cos[2\alpha(hx+ky+\ell z) - \alpha(hk\ell)]$$

electron
density

$I(hk\ell) \longrightarrow |F(hk\ell)|^2 \longrightarrow |F(hk\ell)| \dashrightarrow \rho(xyz)$

experimental
intensity   structure

$P(uvw)$

Patterson
function

Fig. 1-5 The phase problem. The missing phases $\alpha(hk\ell)$ must be supplied in the last step as indicated by the broken line.

an atom of dominant scattering power, its position can be readily located from a Patterson map. The calculated phase α_H based on the heavy atom may be taken

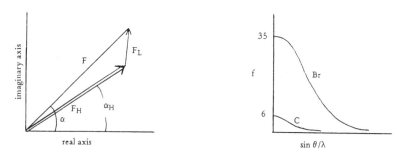

Fig. 1-6 Principle of the heavy-atom method. $F = F_H + F_L$ (H = heavy atom, L = other light atoms). Note the relative scattering powers of Br and C atoms. (i) Locate H with $P(uvw)$ and calculate α_H. Set $\alpha \simeq \alpha_H$. (ii) Locate some or all light atoms from ρ map computed with $|F|$ and α_H.

as an approximation to the true phase α (Fig. 1-6), and an electron-density map computed with $|F|$ and α_H then reveals the positions of part or all of the remaining light atoms in the unit cell. The "heavy-atom method", together with the related method of "isomorphous replacement", have since been used to solve crystal structures of great complexity.

Attempts to derive useful phase information directly from the observed intensities began in the late 1940s and evolved into the powerful "direct methods" of today.[17,18] The physical basis of direct methods is the non-negativity of the electron-density function in a crystal. The foundations of a mathematical approach based on determinantal inequalities were laid in 1950 by Karle and Hauptman,[19] culminating in a monograph which listed a set of probabilistic formulae and measures for direct phase determination of centrosymmetric crystals.[20] The much more difficult task of solving a non-centrosymmetric structure by the direct method was first accomplished in 1964[21] using the "tangent formula".[22]

To gain some idea of the flavour of direct methods, consider the general expression derived from the Sayre equation[23] for a centrosymmetric crystal (Fig. 1-7), which shows that the sign of reflection $hk\ell$ is related to those of many others in the set. The probability that $S(hk\ell)$ is positive can be estimated by a formula involving a hyperbolic tangent function,[24] and in practice phases are assigned and modified with the aid of probability formulae until self-consistency is attained.

S means "sign of"; $\sim$ means "is probably equal to"

$$|E(hk\ell)| = |F(hk\ell)|/(\epsilon\Sigma f_j^2)^{\frac{1}{2}} \quad \text{(normalized structure factors)}$$

From the Sayre equation

$$S(hk\ell) \sim S(h'k'\ell')S(h-h',k-k',\ell-\ell')$$

or $S(hk\ell)S(h'k'\ell')S(h-h',k-k',\ell-\ell') \sim +1$

if $|E(hk\ell)|$, $|E(h'k'\ell')|$ and $|E(h-h',k-k',\ell-\ell')|$ are all large, i.e., all three reflections are strong.

The probability that this relation holds is

$$P = 1/2 + (1/2)\tanh(\sigma_3/\sigma_2^{3/2}|E(hk\ell)E(h'k'\ell')E(h-h',k-k',\ell-\ell')|)$$

where $\sigma_3 = \Sigma n_j^3$, $\sigma_2 = \Sigma n_j^2$, and $n_j = f_j/\Sigma f_j$.

Fig. 1-7 Direct methods (centric case).

Structures of Fundamental Importance

Soon after Laue's discovery, structure analysis by means of X-rays quickly developed into a major discipline through the pioneering researches of William H. and W. Lawrence Bragg (father and son). The first structure

The Structure of Some Crystals as Indicated by their Diffraction
of X-rays.

By W. L. Bragg, B.A.

(Communicated by Prof. W. H. Bragg, F.R.S.   Received June 21,—Read
June 26, 1913.)

Proc. Roy. Soc. (London) **A89**:248–277 (1913)

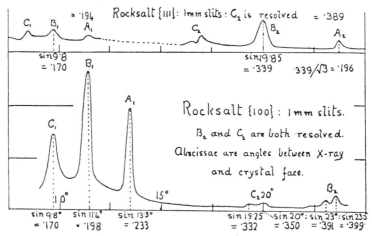

Fig. 1-8 X-Ray diffraction analysis of sodium chloride. (After ref. 25).

determined was that of sodium chloride by Lawrence Bragg (Fig. 1-8).[25]   For the first time, concrete evidence was obtained for an ionic compound which defied the Daltonian hypothesis that 'atoms combine to form molecules'. This 1913 paper also contains the famous Bragg equation, $n\lambda = 2d\sin\theta$, derived from a re-interpretation of the Laue photographs. The structure of diamond, solved by the Braggs in the same year, aroused great interest since it confirmed the tetrahedral arrangement of valencies around a saturated carbon atom as proposed by Butlerov, van't Hoff and Le Bel nearly half a century earlier, and furthermore provided the first measure of 1.52 Å for the C-C single bond.[26] Ten years later, Dickinson and Raymond determined the structure of hexamethylenetetramine,[27] a highly symmetric cage compound obtained in nearly quantitative yield from the condensation of ammonia and formaldehyde. This remarkable achievement represents the earliest complete determination of a structure in the whole organic field. An equally outstanding contribution was made by Kathleen Lonsdale in 1928 in her analysis of hexamethyl-benzene.[28]   This work established the hexagonal geometry of the benzene ring and clearly differentiated the aromatic and aliphatic C-C bonds. In 1929 Linus Pauling, who learnt his crystallography from Dickinson and had determined the first alloy structure, Mg_2Sn,[29] at the age of twenty-three, formulated a set of principles governing the structure and stability of

complex ionic crystals.[30] The all-important "electrostatic valence rule"
looks deceptively simple, yet in practice it quickly eliminates implausible
models and furnishes clues to the correct structure of an ionic crystal.[31]
This was later generalized by Brown to give the "valence sum rule", which
provides a quantitative description of the chemical connectivity in complex
crystals and liquids and constitutes the basis of bond valence--bond length
correlations for about two thousand atom pairs, including hydrogen bonds.[32]

In 1936 the power of the heavy-atom and isomorphous replacement methods
was demonstrated by J. Monteath Robertson in his elegant solution of the
structure of phthalocyanine,[33] a centrosymmetric, planar $C_{32}N_8H_{18}$ macrocycle
whose two inner protons can be replaced by a divalent metal ion with minimal
disturbance of the resulting crystal structure. Another milestone was reached
in 1951, when Bijvoet[34] used the anomalous dispersion of X-rays to establish
that the absolute configuration of natural dextrorotatory tartaric acid does
agree with the arbitrary Emil Fischer convention. Starting in the late 1940s,
Dorothy Crowfoot (later Hodgkin) advanced X-ray analysis to the next pinnacle
of complexity. Structure determination of the first important antibiotic,
penicillin, was considered a masterpiece in 1949,[35], only to be surpassed by
the solution of the vitamin B_{12} enigma six years later.[36]

Meanwhile, dramatic advances were made in biological chemistry
concerning the structures of proteins and nucleic acids. Drawing upon
structural data gathered from their earlier studies of amino acids and simple
peptides, Pauling and Corey constructed a model for the α-helix in 1951.[37]
The key to their success was the realization that the amide group in
polypeptides must be planar due to resonance. In 1954, Max Perutz achieved a
breakthrough by showing that the isomorphous replacement method could, in
principle, lead to the structure determination of macromolecules.[38] The
first protein to have its structure solved was myoglobin, with a molecular
weight of about 18000, by Kendrew in 1958.[39] This analysis showed that
large portions of the myoglobin molecule consist of right-handed α-helices.
The next protein structure to be worked out was that of an enzyme, namely
lysozyme, by Phillips,[40] and soon afterwards Perutz completed his detailed
analysis of hemoglobin,[41] which has a molecular weight approximately four
times that of myoglobin. Several hundred proteins have since been subjected
to X-ray analysis, generating a data bank[42] which underlies current
understanding of structure-function relationships and enzymatic catalysis.

Pauling's success with the α-helix prompted Watson and Crick to race him
in building an acceptable model for deoxyribonucleic acid, the chemical of
life. The result is the famous double-helix paper of 1953,[43] which

initiated new and exciting research fields in biochemistry, molecular biology, and genetic engineering. Two decades were to pass before the "L-shaped" backbone structure (Fig. 1-9) of phenylalanine yeast transfer ribonucleic acid (yeast tRNA[Phe]) was unravelled in the first three-dimensional X-ray analysis of a nucleic acid.[44]

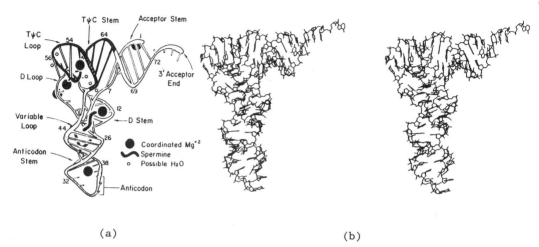

(a)   (b)

Fig. 1-9 Structure of yeast tRNA[Phe]. (a) Schematic diagram showing the backbone as a coiled tube. Cross rungs are used to represent base pairs and shorter rungs indicate bases. (b) Stereoview of the molecule. (After ref. 45).

In the late 1930s Bernal and Fankuchen pioneered the structural studies of viruses and obtained the first X-ray diffraction patterns from oriented gels of tobacco mosaic virus (TMV)[46] and single crystals of tomato bushy stunt virus (TBSV).[47] Success in tackling these enormously difficult problems was realized after nearly four decades, when structural details of the assembly of the viral coat protein subunits in TMV[48] and TBSV[49] were determined.

Small Molecules

Crystallographic studies of small molecules continue to be an impetus in the development of chemistry. In the mid 1950's, Lipscomb used low-temperature techniques to determine the structures of several boron hydrides and rationalized their geometries on the basis of a topological theory involving "open" and "closed" three-center, two-electron bonds.[50] This work greatly enriched our understanding of chemical bonding and paved the way for later studies of electron-deficient compounds. An important field of modern inorganic research is concerned with metal-metal bonds. Brosset in 1935[51]

actually showed that the metal-metal distance in $[W_2Cl_9]^{3-}$ is shorter than
that in metallic tungsten, but then his paper drew virtually no attention.
The first structural determination of a transition metal carbonyl, $Fe_2(CO)_9$,
by Powell in 1939,[52] also strongly suggested the presence of a covalent bond
between the iron atoms, yet the field remained largely stagnant until the
first metal-metal double bond was found in the complex $[Re_3Cl_{12}]^{3-}$.[53] This
was quickly followed by Cotton's discovery of a quadruple bond in
$[Re_2Cl_8]^{2-}$,[54] and the field of metal-metal multiple bonds has been pursued
with vigour ever since. Another major area of modern chemical research is
organometallic chemistry. An X-ray study of the metal-olefin interaction in
Zeise's salt[55] led to the important notion of "back bonding", which greatly
stimulated subsequent experimental and theoretical researches on metal-π
complexes. In a similar vein, the determination of the molecular structure of
ferrocene[56] triggered an explosive growth in the chemistry of sandwich
compounds.

Making use of X-ray crystallography to elucidate the three-dimensional
architecture of crystalline "molecular compounds", Powell opened up the
fertile frontier of molecular inclusion phenomena in 1947. In the β-
hydroquinone clathrate of sulfur dioxide, Powell showed that the SO_2 molecules
occupy cages in a host lattice formed by hydrogen bonding between the
hydroquinone molecules.[57] In 1967 Pedersen reported his discovery of the
crown ethers,[58] which heralded the beginning of a new *era* of host-guest
chemistry, in particular the study of transport of ions across membranes.
From 1969 onwards, Lehn and his associates have synthesized various
macropolycyclic cryptands and crypto-spherands to explore the subtle
structure-function relationship in supramolecular systems.[59] In the early
1970's Cram entered the field through the synthesis of chiral crown ethers,
and has since expounded the general principles of host-guest complexation,
preorganization, and complementarity through the synthetic design of a series
of spherands, hemispherands, cavitands (molecular vessels), and carcerands
(molecular cells) as biomimetic systems.[60] The concepts of molecular
recognition and receptor/substrate association, long cherished by biologists
but lacking in geometrical detail, were placed on a sound footing through the
rational design of a bewildering variety of receptor molecules.[61,62] For
their contributions to molecular inclusion phenomena Pedersen,[63] Lehn,[64]
and Cram[65] shared the Nobel Prize in Chemistry for 1987.

Modern Techniques and Macromolecular Crystallography

The development of low-temperature, high-temperature, high-pressure, and

microgravity (performed on a space satellite or shuttle)[66] techniques for
growing and handling single crystals have greatly extended the range of
chemical substances that can be investigated by diffraction methods. Rapid
advances in instrumentation (synchrotron X-ray sources permitting the use of
weakly-diffracting and unstable crystals; area detectors for simultaneous
recording of many reflections) and methodology (rotational and translational
Patterson searches using a known molecular fragment;[67] new direct methods
incorporating isomorphous replacement and anomalous dispersion data;[68] rapid
and robust least-squares refinement procedures; interactive molecular graphics
modeling systems) underlie the spectacular achievements of macromolecular
(protein) crystallography and drug design in the last decade. A recent
crowning achievement is the isolation and structure determination (Fig. 1-10)
of a membrane protein, the photosynthetic reaction center from
Rhodopseudomonas viridis (with a molecular weight of 145,000 in the asymmetric

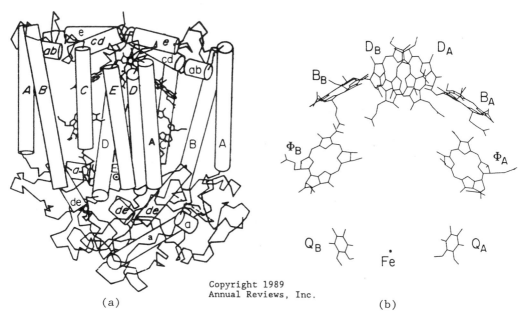

(a)   (b)

Fig. 1-10 (a) The RC (reaction center) structure with protein subunits
and cofactors. The α-helixes have been approximated by straight
cylinders. Helices of the L subunit are lettered in plain type, while
helices of the M subunit are lettered in italic type. H subunit helices
(**A** and **a**) are in bold font. The phytyl and isoprenoid tails of the
cofactors have been truncated. (b) Cofactor structure of the RC from
Rb. sphaeroides R-26. Phytyl and isoprenoid tails of the cofactors have
been omitted for clarity. The cofactors are displayed in the same
orientation as (a). (After ref. 72).

unit) to 2.3Å resolution ($R = 0.193$ for 95,762 unique reflections and 10,288 non-hydrogen atoms, including 201 ordered water molecules) by Deisenhofer, Michel and Huber, who received honours as the 1988 Nobel Laureates in Chemistry.[69-71]

The structural elucidation of large macromolecular complexes arising from nucleic acid-protein interactions was fascilitated by three-dimensional image reconstruction methods developed by Aaron Klug (1982 Nobel Laureate in Chemistry), who combined X-ray diffraction and electron microscopy data to derive the structures of biological assemblies such as tobacco mosaic virus (TMV), T4 bacteriophage, nucleosomes and chromatin.[73]

Spherical viruses are nucleoprotein particles in which the nucleic acid is packaged in a "capsid" composed of quasi-equivalent protein subunits organized in the form of an icosahedron (Fig. 1-11).[74] Modern advances in macromolecular crystallography have greatly increased the number of spherical viruses determined to high precision (Table 1-8).[75,76] In addition, the

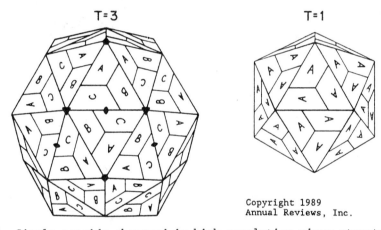

T=3   T=1

Fig. 1-11 Simple capsids observed in high-resolution virus structures. The shells have exact icosahedral symmetry. The $T = 1$ shell contains 60 subunits related by icosahedral symmetry. Each subunit is represented by a trapezoid that is the approximate shape of the β-barrel when viewed from the top. All subunits in the $T = 1$ capsid are identical and labeled A for comparison with the $T = 3$ capsid. The icosahedral asymmetric unit of $T = 1$ viruses is one subunit and the threefold symmetry in the central triangle is exact. The asymmetric unit of the $T = 3$ capsid is the central triangle containing subunits A, B, and C. The subunits labeled A, B, and C have the same amino acid sequence but are in slightly different environments. While similar to the triangle in the $T = 1$ structure, the threefold axis relating A, B, and C is not exact. This quasi-threefold axis relates the quasi-sixfold axes (left and right vertexes of the triangle) to a fivefold axis (top vertex). Like the $T = 1$ structure, the $T = 3$ structures are formed by identical subunits with the β-barrel fold. (After ref. 75).

Table 1-8   Spherical viruses determined to high resolution.[*]

Plant RNA viruses:		Capsid symmetry[**]	Resolution	Animal RNA viruses:		Capsid symmetry	Resolution
Tombus group	TBSV	T = 3	2.9Å	Rhino	HRV14	P = 3	3.0Å
	TCV	T = 3	3.2Å		HRV1A	P = 3	
Sobeamo group	SBMV	T = 3	2.9Å	Entero	Polio Mahoney 1	P = 3	2.9Å
Como group	CPMV	P = 3	3.0Å		Polio Sabin 3	P = 3	2.4Å
	BPMV	P = 3		Cardio	Mengo	P = 3	3.0Å
	STNV	T = 1	2.5Å	Aphtho	FMDV 01K	P = 3	2.9Å
				Insect RNA viruses:			
					BBV	T = 3	3.0Å

[*] Abbreviations used: BBV, black beetle virus; BPMV, beanpod mottle virus; CPMV, cowpea mosaic virus; FMDV, foot-and-mouth disease virus; HRV, human rhinovirus; SBMV, southern bean mosaic virus; STNV, satellite tobacco necrosis virus; TBSV, tomato bushy stunt virus; TCV, turnip crinkle virus.

[**] T = 1,3,4,7,... relates to the triangulation number. P = 3 implies a pseudo T = 3 surface lattice arrangement where there are three nonidentical but similarly folded subunits arranged as in a T = 3 lattice.

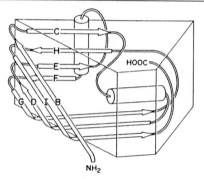

Diagrammatic representation of the polypeptide fold of one subunit of poliovirus found also in the shell-forming portion of all viral subunit structures determined to date. Shown also is the nomenclature for the secondary structural elements βB, βC,...,βI.

structure of TMV, a helical rod-shaped plant virus, has been refined to 2.9Å resolution; there is a loss of information, amounting to a factor of about 2.5, owing to cylindrical averaging of the X-ray fiber diffraction data.[77]

It seems appropriate to end the present discourse with a landmark study providing intimate details of the mutual recognition between a tRNA and its specific synthetase, an enzyme which binds it to the correct amino acid. Crystal-structure analysis of *Escherichia coli* glutaminyl-tRNA synthetase (GlnRS) complexed with its cognate glutaminyl transfer RNA (tRNA^{Gln}) and adenosine triphosphate (ATP) at 2.8Å resolution has shown that the tip of the long arm (anticodon loop and stem) of the tRNA "L" fits snugly into a deep pocket in the protein, whereas the tip of the short leg (acceptor stem and 3' end) is inserted into a gaping cavern that contains the binding site for the other two substrates, ATP and glutamine (Fig. 1-12).[78]

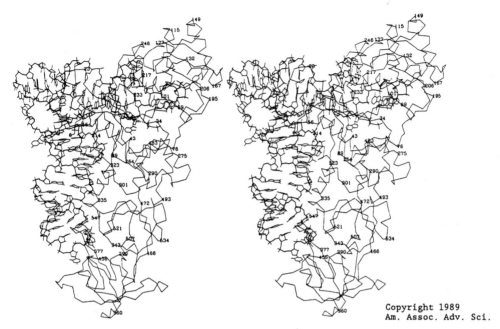

Fig. 1-12 Stereoview showing the α-carbon backbone of GlnRS and all atoms of tRNA^{Gln} and ATP. The first five residues of the amino terminus and the last six of the carboxyl terminus are disordered and not shown. (After ref. 78).

Judging by the rapid advances in macromolecular crystallography and related fields in the last two decades, we are certain that the design and synthesis of new proteins, by cloning or peptide coupling, will most likely be achieved sometime in the next century. The present status of our knowledge of the principles and patterns of protein conformation and computational methods for the prediction of protein structure has been reviewed recently.[79]

References

[1] A.R. Butler, C. Glidewell, J. Needham and S. Pritchard, *Chem. Brit.*, 132 (1983).

[2] J. Kepler, *Strena seu de Nive Sexangula*, G. Tampach, Frankfurt, 1611.

[3] N. Steno, *De solido intro solidum naturaliter contento (dissertationis prodromus)*, Florentiae, 1669.

[4] R.J. Haüy, *Essai d'une théorie sur la structures des crystaux appliquée a plusieurs genres de substances crystallisées*, Paris, 1784.

[5] J.F.C. Hessel, *Kristallometrie oder Krystallonomie und Kristallographie*, Leipzig, 1831.

[6] A. Bravais, *Mémoire sur les systèmes formés par des points distribués regulièrement sur un plan ou dans l'espace*, J. l'École Polytech. (Paris) **19**, 1 (1850).

[7] E.S. Fedorov, *Trans. Russ. Min. Soc.* 21, 1 (1885); **25**, 1 (1888); *Symmetry of Regular Systems of Figures*, 1890. [In Russian.]

[8] A. Schönflies, *Krystallsysteme und Krystallstruktur*, Leipzig, 1891.

[9] W. Barlow, *Z. Kryst. Miner.* 23, 1 (1894).

[10] W. Barlow, *Nature (London)* 29, 186, 205 (1883); 29, 404 (1884).

[11] W.C. Röntgen, *Über eine neue Art von Strahlen*, Sitzungsberichte der Würzburger Physikalisch-Medizinischen Gesellschaft, 132 (1895).

[12] W. Friedrich, P. Knipping and M. Laue, *Interferenz-Erscheinungen bei Röntgenstrahlen*, Sitzungber. (Kgl.) Bayerische Akad. Wiss., 303 (1912).

[13] E. Streller, R. Winau and A. Hermann, *Wilhelm Conrad Röntgen 1845-1923* (translated from the German by D. Thompson), Inter Nationes, Bonn-Bad Godesberg, 1973, p. 60.

[14] W.H. Bragg and W.L. Bragg, *Curr. Sci.*, 9 (1937).

[15] T.C.W. Mak, *Chinese University Bulletin*, Supplement 11, The Chinese University of Hong Kong, 1983, p. 1.

[16] A.L. Patterson, *Phys. Rev.* 46, 372 (1934); J.P. Glusker, B.K. Patterson and M. Rossi (eds.), *Patterson and Pattersons*, Oxford University Press, Oxford, 1987.

[17] H.A. Hauptman, *Physic Today* 42, No. 11, 24 (1989); *Angew. Chem. Int. Ed. Engl.* 25, 603 (1986). [Nobel Lecture.]

[18] J. Karle, *Angew. Chem. Int. Ed. Engl.* 25, 614 (1986). [Nobel Lecture.]

[19] J. Karle and H.A. Hauptman, *Acta Crystallogr.* 3, 181 (1950).

[20] H.A. Hauptman and J. Karle, *Solution of the Phase Problem I. The Centrosymmetric Crystal*, Am. Crystallogr. Assoc., Washington, 1953.

[21] I.L. Karle and J. Karle, *Acta Crystallogr.* 17, 835 (1964).

[22] H.A. Hauptman and J. Karle, *Acta Crystallogr.* 9, 635 (1956).

[23] D. Sayre, *Acta Crystallogr.* 5, 60 (1952).

[24] W. Cochran and M.M. Woolfson, *Acta Crystallogr.* 8, 1 (1955).

[25] W.L. Bragg, *Proc. Roy. Soc. London* A89, 248 (1913); M.F. Perutz, *Acta Crystallogr.*, *Sect. A* 46, 633 (1990). [Anecdotic account of the scientific life of Bragg commemorating the centenary of his birth.]

[26] W.H. Bragg and W.L. Bragg, *Nature (London)* 91, 557 (1913).

[27] R.G. Dickinson and A.L. Raymond, *J. Am. Chem. Soc.* 45, 22 (1923).

[28] K. Lonsdale, *Nature (London)* 122, 810 (1928).

[29] L. Pauling, *J. Am. Chem. Soc.* 45, 2777 (1923).

[30] L. Pauling, *J. Am. Chem. Soc.* 51, 1010 (1929).

[31] L. Pauling, *The Nature of the Chemical Bond*, 3rd ed., Cornell University Press, Ithaca, 1960, p. 547.

[32] I.D. Brown in M. O'Keeffe and A. Navrotsky (eds.), *Structure and Bonding in Crystals*, Vol. 2, Academic Press, New York, p. 1; D. Altermatt and I.D. Brown, *Acta Crystallogr.*, *Sect. B* 41, 240 (1985).

[33] J.M. Robertson, *J. Chem. Soc.*, 1195 (1936).

[34] J.M. Bijvoet, A.F. Peerdeman and A.J. van Bommel, *Nature (London)* 168, 271 (1951).

[35] D. Crowfoot, C.W. Bunn, B.W. Rogers-Low and A. Turner-Jones, *The Chemistry of Penicillin*, Princeton Universit Press, New Jersey, 1949, p. 310.

[36] D. Crowfoot Hodgkin, J. Pickworth, J.H. Robertson, K.N. Trueblood, R.J. Prosen and J.G. White, *Nature (London)* 176, 325 (1955).

[37] L. Pauling, R.B. Corey and H.R. Branson, *Proc. Natl. Acad. Sci. USA* 37, 205 (1951).

[38] D.W. Green, V.M. Ingram and M.F. Perutz, *Proc. Roy. Soc. London* **A225**, 287 (1954).

[39] J.C. Kendrew, G. Bodo, H.M. Dintzis, R.G. Parrish, H.W. Wyckoff and D.C. Phillips, *Nature (London)* **181**, 662 (1958); J.C. Kendrew, R.E. Dickerson, B.E. Strandberg, R.G. Hart, D.R. Davis, D.C. Phillips and V.C. Shore, *Nature (London)* **185**, 422 (1960).

[40] C.C.F. Blake, D.F. König, G.A. Mair, A.C.T. North, D.C. Phillips and V.R. Sarma, *Nature (London)* **206**, 757 (1965); D.C. Phillips, *Sci. Am.* **215**, 78 (1966).

[41] A.F. Cullis, H. Muirhead, M.F. Perutz and M.G. Rossmann, *Proc. Roy. Soc. London* **A265**, 161 (1962); M.F. Perutz, H. Muirhead, J.M. Cox, L.C.G. Goaman, F.S. Mathews, E.L. McGandy and L.E. Webb, *Nature (London)* **219**, 29 (1968).

[42] F.C. Bernstein, T.F. Koetzle, G.J.B. Williams, E.F. Meyer Jr., M.D. Brice, J.R. Rodgers, O. Kennard, T. Shimanouchi and M. Tasumi, *J. Mol. Biol.* **112**, 535 (1977). [The Brookhaven Protein Data Bank.]

[43] J.D. Watson and F.H.C. Crick, *Nature (London)* **171**, 737 (1953).

[44] S.-H. Kim, G.J. Quigley, F.L. Suddath, A. McPherson, D. Sneden, J.J. Kim, J. Weinzierl and A. Rich, *Science (Washington)* **179**, 285 (1973); S.-H. Kim in P.R. Schimmel, D. Söll and J.N. Abelson (eds.), *Transfer RNA: Structure, Properties, and Recognition*, Cold Spring Harbor Laboratory, New York, 1979, p. 83.

[45] A. Rich, G.J. Quigley, M.M. Teeter, A. Ducruix and N. Woo in P.R. Schimmel, D. Söll and J.N. Abelson (eds.), *Transfer RNA: Structure, Properties, and Recognition*, Cold Spring Habor Laboratory, New York, 1979, p. 103; M. Sundaralingam, *ibid.*, p. 115.

[46] F.C. Bawden, N.W. Pirie, J.D. Bernal and I. Fankuchen, *Nature (London)* **138**, 1051 (1936); J.D. Bernal and I. Fankuchen, *J. Gen. Physiol.* **25**, 111 (1941).

[47] J.D. Bernal and I. Fankuchen, *Nature (London)* **142**, 1075 (1938).

[48] G. Stubbs, S. Warren and K. Holmes, *Nature (London)* **267**, 216 (1977).

[49] S.C. Harrison, A.J. Olson, C.E. Schutt, F.K. Winkler and G. Bricogne, *Nature (London)* **276**, 368 (1978).

[50] W.H. Eberhardt, B. Crawford Jr. and W.N. Lipscomb, *J. Chem. Phys.* **22**, 989 (1954).

[51] C. Brosset. *Arkiv Kemi, Mineral. Geol.* **12A**, No. 4 (1935); W.H. Watson Jr. and J. Waser, *Acta Crystallogr.* **11**, 689 (1958).

[52] H.M. Powell and R.V.G. Ewens, *J. Chem. Soc.*, 286 (1939); F.A. Cotton and J.M. Troup, *J. Chem. Soc., Dalton Trans.*, 800 (1974).

[53] J.A. Bertrand, F.A. Cotton and W.A. Dollase, *J. Am. Chem. Soc.* **85**, 1349 (1963); W.T. Robinson, J.E. Fergusson and B.R. Penfold, *Proc. Chem. Soc. (London)*, 116 (1963).

[54] F.A. Cotton, N.F. Curtis, C.B. Harris, B.F.G. Johnson, S.J. Lippard, J.T. Mague, W.R. Robinson and J.S. Wood, *Science (Washington)* **145**, 1305 (1964); F.A. Cotton and C.B. Harris, *Inorg. Chem.* **4**, 330 (1965); F.A. Cotton, *Inorg. Chem.* **4**, 334 (1965).

[55] J.A. Wunderlich and D.P. Mellor, *Acta Crystallogr.* **7**, 130 (1954); J.A.J. Jarvis, B.T. Kilbourn and P.G. Owston, *Acta Crystallogr., Sect. B* **26**, 876 (1970); **27**, 366 (1971); R.A. Love, T.F. Koetzle, G.J.B. Williams, L.C. Andrews and R. Bau, *Inorg. Chem.* **14**, 2653 (1975).

[56] J.D. Dunitz, L.E. Orgel and A. Rich, *Acta Crystallogr.* **9**, 373 (1956); P. Seiler and J.D. Dunitz, *Acta Crystallogr., Sect. B* **35**, 1068 (1979).

[57] D.E. Palin and H.M. Powell, *J. Chem. Soc.*, 208 (1947).

[58] C.J. Pedersen, *J. Am. Chem. Soc.* **89**, 2495, 7017 (1967).

[59] B. Dietrich, J.-M. Lehn and J.-P. Sauvage, *Tetrahedron Lett.*, 2885 (1969).

[60] D.J. Cram and J.M. Cram, *Science (Washington)* **183**, 803 (1974).

[61] F. Vögtle (ed.), *Host Guest Chemistry*, *Top. Curr. Chem.* **98** (1981), **101** (1982); F. Vögtle and E. Weber (eds.), *ibid.* **121** (1984).

[62] E. Weber, J.L. Toner, I. Goldberg, F. Vögtle, D.A. Laidler, J.F. Stoddart, R.A. Bartsch and C.L. Liotta, *Crown Ethers and Analogs*, Wiley, New York, 1989.

[63] C.J. Pedersen, *J. Incl. Phenom.* **6**, 337 (1988). [Nobel Lecture.]

[64] J.-M. Lehn, *J. Incl. Phenom.* **6**, 351 (1988). [Nobel Lecture.]

[65] D.J. Cram, *J. Incl. Phenom.* **6**, 397 (1988). [Nobel Lecture.]

[66] L.J. DeLucas, C.D. Smith, H.W. Smith, S. Vijay-Kumar, S.E. Senadhi, S.E. Ealick, D.C. Carter, R.S. Snyder, P.C. Weber, F.R. Salemme, D.H. Ohlendorf, H.M. Einspahr, L.L. Clancy, M.A. Navia, B.M. McKeever, T.L. Nagabhushan, G. Nelson, A. McPherson, S. Koszelak, G. Taylor, D. Stammers, K. Powell, G. Darby and C.E. Bugg, *Science (Washington)* **246**, 651 (1989).

[67] M.G. Rossmann, *The Molecular Replacement Method*, Gordon and Breach, New York, 1972.

[68] J. Karle, *Acta Crystallogr., Sect. A* **45**, 765 (1989).

[69] J. Deisenhofer, R. Huber and H. Michel in G.D. Fasman (ed.), *Prediction of Protein Structure and the Principles of Protein Conformation*, Plenum Press, New York, 1989, p. 99.

[70] J. Deisenhofer and H. Michel, *Angew. Chem. Int. Ed. Engl.* **28**, 829 (1989). [Nobel Lecture.]

[71] R. Huber, *Angew. Chem. Int. Ed. Engl.* **28**, 848 (1989). [Nobel Lecture.]

[72] D.C. Rees, H. Komiya, T.O. Yeates, J.P. Allen and G. Feher, *Annu. Rev. Biochem.* **58**, 607 (1989).

[73] A. Klug, *Angew. Chem. Int. Ed. Engl.* **22**, 565 (1983). [Nobel Lecture.]

[74] D.L.D. Caspar and A. Klug, *Cold Spring Harbor Symp. Quant. Biol.* **27**, 1 (1962).

[75] M.G. Rossmann and J.E. Johnson, *Annu. Rev. Biochem.* **58**, 533 (1989); J.-P. Wery and J.E. Johnson, *Anal. Chem.* **61**, 1341A (1989).

[76] G. Stubbs in G.D. Fasman (ed.), *Prediction of Protein Structure and the Principles of Protein Conformation*, Plenum Press, New York, 1989, p. 117.

[77] K. Namba and G. Stubbs, *Science (Washington)* **231**, 1401 (1986); G. Stubbs, K. Namba and L. Makowski, *Biophys. J.* **49**, 58 (1986).

[78] M.A. Rould, J.J. Perona, D. Söll and T.A. Steitz, *Science (Washington)* **246**, 1135 (1989).

[79] G.D. Fasman (ed.), *Prediction of Protein Structure and the Principles of Protein Conformation*, Plenum Press, New York, 1989.

Note The Nobel Lectures delivered by the prizewinners in Chemistry during the period 1971-1990 are collected together in two new volumes that also include their updated biographical data, portraits and presentation speeches: S. Forsén (ed.), *Nobel Lectures in Chemistry 1971-1980*; B. Malmstrom (ed.), *Nobel Lectures in Chemistry 1981-1990*; World Scientific, Singapore, 1992.

Additional references are listed in p. 182.

Chapter 2

Fundamental Structures

2.1. Copper   Cu   22

2.2. Magnesium   Mg   26

2.3. α-Iron   Fe   30

2.4. Diamond   C   37

2.5. Graphite   C   43

2.6. Iodine   I₂   57

2.7. Sodium Chloride (Rock Salt, Common Salt, Halite)   NaCl   64

2.8. Cesium Chloride   CsCl   69

2.9. Zinc Blende (Cubic Zinc Sulfide, Sphalerite)   ZnS   72

2.10. Wurtzite (Hexagonal Zinc Sulfide)   ZnS   77

2.11. Nickel Arsenide   NiAs   81

2.12. Sodium Thallide   NaTl   86

2.13. Fluorite (Fluospar)   CaF₂   89

2.14. Rutile (Titanium Dioxide)   TiO₂   93

2.15. α-Quartz   SiO₂   97

2.16. Ice Ih (Hexagonal Ice)   H₂O   104

2.17. Cadmium Iodide   CdI₂   111

2.18. Spinel   MgAl₂O₄   114

2.19. Perovskite   CaTiO₃   119

2.20. Potassium Hexachloroplatinate (IV)   K₂PtCl₆   129

2.21. Copper Sulfate Pentahydrate   CuSO₄ · 5H₂O   134

2.22. Diopside   CaMg(SiO₃)₂   139

2.23. Calcite (Calcium Carbonate)   CaCO₃   146

2.24. Hexamethylenetetramine   (CH₂)₆N₄   151

2.25. Hexamethylbenzene   C₆(CH₃)₆   158

2.26. Naphthalene   C₁₀H₈   169

2.27. Urea   (NH₂)₂CO   175

Addenda to Chapters 1 and 2   182

2.1 Copper

Cu

Crystal Data

Cubic, space group $Fm3m$ (No. 225)

$a = 3.61529(7)$Å (25°C), $Z = 4$

Atom	Position	x	y	z
Cu	4(a)	0	0	0

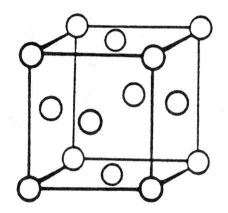

Fig. 2.1-1 Unit cell of copper, a typical cubic closest packed (ccp) structure.

Crystal Structure

The crystal structure of copper (Fig. 2.1-1) was predicted correctly by W. Barlow[1] in the 19th century and later determined by W.L. Bragg[2]. The structure of copper can be considered as the stacking of layers of spherical atoms in a closest packed manner.

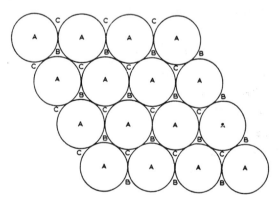

Fig. 2.1-2 Closest packed layer of identical spheres.

In two-dimensional closest packing of identical spheres (Fig. 2.1-2),
each sphere is in contact with six neighbours at the corners of a regular
hexagon. The layer contains twice as many interstices (holes) as the spheres.
As shown in Fig. 2.1-2, the centers of the spheres are labelled A, one set of
interstices B, and the other set C. If we denote the initial layer by A, the
spheres in the second layer are placed over the interstices of B. Each sphere
in this second layer is now in contact with three neighbours in the first
layer. The upper surface of the second layer displays two sets of dimples
about A and C. Stacking the third layer over the interstices C, and repeating
the fourth layer like the first, generates the infinitely extended sequence
ABCABC... as illustrated in Fig. 2.1-3. This arrangement possesses cubic
symmetry, and is therefore termed cubic closest packing (ccp). The structure
is conveniently described in terms of a face-centered cubic (fcc) unit cell
(Fig. 2.1-3), with atoms occupying the centers of the faces of the cube as
well as the corners. A view of the unit cell along any of its four body
diagonals (namely the <111> directions) shows the closest packed layers
clearly.

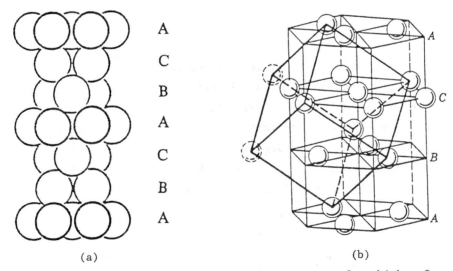

(a)   (b)

Fig. 2.1-3 Sequence of layers in the ccp structure, for which a fcc
unit cell is outlined with thicker lines.

Remarks

1. Coordination and proportion of space occupied

In the ccp structure, each sphere has twelve neighbours: six in its own
layer, three in the layer below, and three in the layer above. The twelve
neighbouring atoms form a coordination polyhedron (Fig. 2.1-4) in the shape of

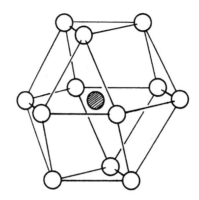

Fig. 2.1-4 The coordination polyhedron (a cubo-octahedron) in the ccp
structure. (After ref. 4).

a cubo-octahedron, which is derived from a cube with its eight corners cut
off.

In the ccp structure, if the radius of each sphere is R, the fcc unit
cell is defined by the parameter $a = 2\sqrt{2}R$. The proportion of space occupied
by the four spheres is $4(4\pi R^3/3)/(2\sqrt{2}R)^3 = 0.7405$, and the effective volume
per atom is $5.66R^3$ [$= (4\pi R^3/3)/0.7405$]. These values are the same for the
hexagonal closest packed (hcp) structure described in section 2.2.

2. Interstices

There are interstices (holes) of two types in the ccp structure:
tetrahedral and octahedral holes. In three-dimensional closest packing, the
number of tetrahedral holes is twice that of the spheres, while the number of
octahedral holes is equal to that of the spheres. The holes may be filled
with smaller spheres contacting the principal spheres of radius R. The radius
of such a small sphere in a tetrahedral hole is equal to $(\sqrt{3}/\sqrt{2} - 1)R =$
$0.225R$, and in an octahedral hole $(\sqrt{2} - 1)R = 0.414R$. (Fig. 2.1-5).

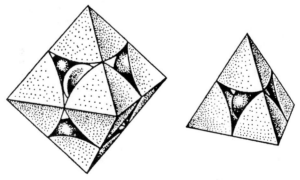

Fig. 2.1-5 Octahedral and tetrahedral holes in a closest packed
structure with a small sphere inserted in each. (After ref. 5).

3. Elements with the ccp structure

The ccp structure occurs commonly in many elements[3]; some relevant data are listed in Table 2.1-1.

Table 2.1-1 Elements with the cubic closest packed structure.

Element	Cell edge a/Å	mp /°C	Density /g cm^{-3}
Ac	5.311	1600	10.07
Ag	4.0862	960.5	10.49
Al	4.050	660	2.70
Am	4.894	1170	13.78
Ar (4.2°K)	5.256	-184	1.82
Au	4.078	1063	19.30
α-Ca	5.582	850	1.55
γ-Ce (-10 to 730°C)	5.1604	795	6.771
β-Co	4.548	1490	8.70
β-Cr	3.68	1550	7.20
Cu	3.615	1083	8.92
γ-Fe	3.591	1535	7.87
Ir	3.839	2454	22.50
Kr (58°K)	5.721	-157	2.97
β-La (310 to 868°C)	5.303	920	6.186
γ-Mn (1100°C)	3.855	1244	6.368
Ne(4.2°K)	4.429	-249	1.54
Ni	3.524	1455	8.90
Pb	4.951	327	11.34
Pd	3.890	1554	12.00
Pt	3.923	1774	21.45
δ-Pu (320°C)	4.637	640	15.92
Rh	3.8031	1966	12.44
β-Sc (1000°C to mp)	4.541	1539	3.20
α-Sr	6.085	770	2.60
α-Th	5.0843	1800	11.50
Xe (58°K)	6.197	-112	3.66
α-Yb (to 798°C)	5.481	824	6.977

References

[1] W. Barlow, *Nature (London)* **29**, 186, 205, 404 (1883, 1884).

[2] W.L. Bragg, *Phil. Mag.* **28**, 355 (1914).

[3] R.W.G. Wyckoff, *Crystal Structures*, 2nd ed., Vol. 1, Wiley-Interscience, New York, 1963, p. 7.

[4] D.M. Adams, *Inorganic Solids*, Wiley, New York, 1974.

[5] B.K. Vainshtein, V.M. Fridkin and V.L. Indenbom, *Modern Crystallography II, Structure of Crystals*, Springer-Verlag, Berlin, 1982.

Note A recent account of the structure of the pure metals in given by H.W. King in R.W. Cahn and P. Haasen (eds.), *Physical Metallurgy*, 3rd ed., Part 1, North-Holland, Amsterdam, 1983, chap. 2.

2.2 Magnesium
Mg

Crystal Data

Hexagonal, space group $P6_3/mmc$ (No. 194)

a = 3.20927(5), c = 5.21033(6)Å (25°C), Z = 2

Atom	Position*	x	y	z
Mg	2(c)	0	0	0

*Origin at $\bar{6}$m2

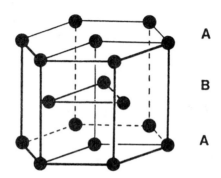

Fig. 2.2-1 Unit cell of magnesium, a typical example of the hexagonal closest packed (hcp) structure.

Crystal Structure

The crystal structure of magnesium (Fig. 2.2-1), first suggested by Barlow[1] and subsequently determined by Hull,[2] is built of closest packed layers of spheres stacked in the infinitely extended sequence A̲B̲ABAB......
(Fig. 2.2-2). This arrangement possesses hexagonal symmetry and is therefore termed hexagonal closest packing (hcp).

There are two atoms in the unit cell: one is located at the corner, and the other at (1/3, 2/3, 1/2). In the three-dimensional crystal structure each sphere has twelve nearest neighbours: six lie in its own layer, three in the layer below, and three in the layer above. These twelve spheres form a coordination polyhedron as shown in Fig. 2.2-3.

Remarks

1. Comparison of the hcp and ccp structures

The hcp structure has interstices (holes) of the same types as the ccp structure. In ccp neighbouring octahedral interstices share edges, and the tetrahedral interstices behave likewise. In hcp neighbouring octahedral interstices share faces, and a pair of tetrahedral interstices share a common

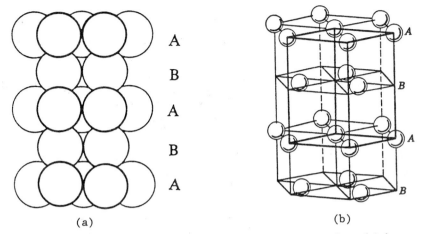

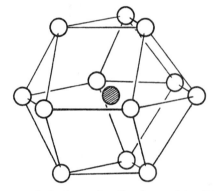

Fig. 2.2-2 Sequence of layers in the hcp structure, for which a primitive hexagonal unit cell is outlined with thicker lines.

Fig. 2.2-3 The coordination polyhedron in the hcp structure. (After ref. 4).

face to form a trigonal bipyramid. Furthermore, the hcp and ccp structures differ in an important geometrical respect, namely the disposition of the closest packed layers. In hcp there is only one direction normal to which the spheres are arranged in individually closest packed layers, whereas in ccp such layers occur in four directions normal to the cubic diagonals. This distinction gives rise to important differences between the mechanical properties associated with the two structures. The ccp metals tend to be more ductile and easier to work with mechanically than the hcp metals. Owing to the higher symmetry, the ccp structure contains more possible slip directions than the hcp structure.

Some metals occur in both the ccp and hcp structures. The energy associated with the two is substantially the same, and there is little tendency to convert from one to the other. The hexagonal form of calcium is obtained above 450°C, and then only if the metal is especially pure. For cobalt and nickel the hexagonal form seems to be the more stable at room temperature, although the cubic form occurs more commonly. The cubic form of nickel can be made hexagonal by annealing for several days at 170°C. Cobalt prepared by reduction above 450°C is cubic, and remains so on cooling, but grinding brings about a change to the hexagonal form.

2. Elements with the hcp structure

Table 2.2-1   Elements with the hexagonal closest-packed structure.

Element	Cell edge/Å		mp/°C	Density g cm^{-3}
	a	c		
Be	2.287	3.583	1280	1.86
β-Ca(450°C to mp)	3.98	6.52	842	1.55
Cd	2.979	5.618	3209	8.65
α-Co	2.507	4.069	1490	8.9
γ-Cr	2.722	4.427	1550	8.56
Dy	3.5925	5.6545	1407	8.536
Er	3.5590	5.592	1497	9.051
Gd	3.6315	5.777	1312	7.895
He	3.57	5.83	-272	0.126
α-Hf	3.197	5.058	1700	11.4
Ho	3.5761	5.6174	1461	8.803
α-La	3.770	12.159	920	6.162
Li(78°K)	3.111	5.093	186	0.53
Lu	3.5050	5.5486	1652	9.842
Mg	3.209	5.210	650	1.74
Na(5°K)	3.657	5.902	97.5	0.97
α-Nd	3.658	11.799	1024	7.003
Ni	2.65	4.33	1455	8.90
Os	2.735	4.319	2700	22.5
α-Pr	3.673	11.835	935	6.769
Re	2.761	4.458	3170	20
Ru	2.704	4.282	2500	12.2
α-Sc(to 1000°C)	3.3080	5.2653	1539	2.992
α-Sm	3.621	26.25	1072	7.54
β-Sr(248°C)	4.32	7.06	757	2.6
Tb	3.5990	5.696	1356	8.272
α-Ti	2.950	4.686	1820	4.54
α-Tl	3.456	5.525	300	9.35
Tm	3.5372	5.5619	1545	9.332
α-Y(to 1490°C)	3.6451	5.7305	1509	4.478
Zn	2.665	4.947	419	7.13
α-Zr	3.232	5.147	1750	6.49

A list of elements which crystallize in the hcp structure[3] is given in Table 2.2-1.

In almost all hcp metals, the axial ratio c/a is slightly less than the "ideal" value of $2\sqrt{2}/\sqrt{3} = 1.633$, lying between 1.63 and *ca.* 1.57. Thus atoms in the same layer are usually a little farther apart than those in

neighbouring layers. With beryllium as the only apparent exception, the axial
ratios of these elements increase with temperature, so that the action of heat
tends to equalize the interatomic distances. Zinc and cadmium are notable
exceptions to the general trend as their axial ratios are far greater than
1.63, being about 1.86, and this makes atomic separations in a layer more than
10% less than those between adjacent layers. Their axial ratios also increase
with temperature, but the effect of this is to exaggerate the departure from
spherical closest packing.

3. Double hexagonal close-packed structure

Iron hydride formed at high pressure above 3.5 gigapascals (GPa) has a
double hexagonal closed-packed (dhcp) structure [Fig. 2.2-4(b)] in which the
layers follow the pattern ABACABAC..., with a c axis doubling that of the hcp
ABAB... stacking in pure iron [Fig. 2.2-4(a)].[5] The interstitial hydrogen
atoms occupy octahedral sites, resulting in a stoichiometry of $FeH_{0.94}$ at 62
GPa and an increase in unit-cell ($a = 2.666$ and $c = 8.74$Å) volume by 17% with
respect to hcp iron.

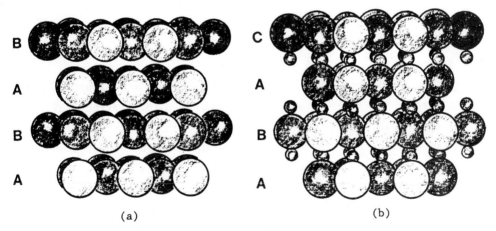

(a)   (b)

Fig. 2.2-4 Comparison of the crystal structures of (a) hcp iron
and (b) dhcp iron hydride. (Adapted from ref. 5).

References
[1] W. Barlow, *Nature (London)* **29**, 186, 205, 404 (1883,1884).

[2] A.W. Hull, *Proc. Natl. Acad. Sci. USA* **3**, 470 (1917).

[3] R.W.G. Wyckoff, *Crystal Structure*, 2nd ed., Vol. 1, Wiley-Interscience,
    New York, 1963, p. 8.

[4] D.M. Adams, *Inorganic Solids*, Wiley, New York, 1974.

[5] J.V. Badding, R.J. Hemley and H.K. Mao, *Science (Washington)* **253**, 421
    (1991).

2.3 α-Iron

Fe

Crystal Data

Cubic, space group *Im3m* (No. 229)

$a = 2.86645(1)$Å (20°C), $Z = 2$

Atom	Position	x	y	z
Fe	2(a)	0	0	0

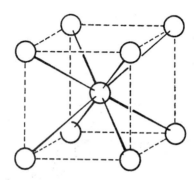

Fig. 2.3-1   Unit cell of α-iron, a typical example of the body-centered cubic (bcc) structure.   (After ref. 7).

Crystal Structure

The structural modification of pure iron at room temperature is called α-iron or ferrite.   α-Iron has a body-centered cubic (bcc) structure (Fig. 2.3-1).[1]   There are two atoms per unit cell: one at the corner, and the other at the center of the cube.   Each atom has eight equidistant nearest neighbours at a distance of $(\sqrt{3}/2)a = 0.866a$.   The bcc structure is *not* a closest packed structure.   If the radius of each sphere is R, a is equal to $4R/\sqrt{3}$, and the proportion of space occupied by the spheres in the unit cell is $2[(4/3)\pi R^3]/(4R/\sqrt{3})^3 = \pi\sqrt{3}/8 = 0.6802$.

Remarks

1. Interstices in the bcc structure

In the bcc structure there are two types of interstices to be considered.   The larger of the two is tetrahedral: at (1/2, 1/4, 0) and its equivalent positions (Fig. 2.3-2).   These interstices may be filled with smaller spheres in contact with the principal spheres of radius R; the radius of such a smaller sphere is equal to $[(\sqrt{5}/\sqrt{3}) - 1]R = 0.291R$.   In α-Fe, these tetrahedral interstices can accommodate an atom of radius 0.36Å.   The smaller interstices of the other type are found at the midpoints of the edges

(0, 0, 1/2; etc.) and at the centers of the faces (1/2, 1/2, 0; etc.), where
they are surrounded by six atoms at the corners of a compressed octahedron
(Fig. 2.3-2). These interstices could be filled with smaller spheres of
radius $(2\sqrt{3} - 1)R = 0.154R$. In α-Fe there is room only for spheres of 0.19Å
radius.

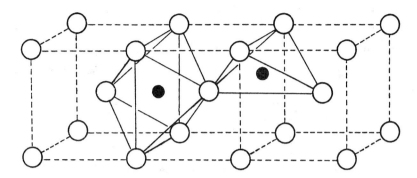

Fig. 2.3-2 Octahedral and tetrahedral interstices in the bcc
structure. (After ref. 7).

2. Phase transition of iron and the ferrite and austenite phases

The room-temperature bcc form of iron (α-Fe) changes to the ccp
structure (γ-Fe) on heating to above 910°C, and reverts to the bcc structure
(δ-Fe) again at about 1390°C. Iron is a ferromagnetic material under 767°C,
and above this temperature it becomes paramagnetic.

α-Fe bcc	γ-Fe ccp	δ-Fe bcc	liquid-Fe
767°C 910°C	1390°C	1535°C	
ferromagnetic paramagnetic			

Ferrite and austenite are the most important components in the iron-
carbon system, which is of enormous practical importance in industry.

Ferrite, the solid solution of carbon in bcc iron (α-Fe), has a maximum
solubility of 0.025 weight percent C at the eutectoid temperature. The fact
that ferrite dissolves far less carbon as does α-Fe is accounted for by the
smaller size of the interstices. When heated above 910°C, α-Fe changes to
γ-Fe (ccp structure), which provides its octahedral interstices for
accommodating the carbon atoms, thus giving rise to austenite, an interstitial
solid solution of carbon in γ-Fe. Austenite has a maximum solubility of 2.0
weight percent C at 1130°C. On cooling, austenite decomposes and converts to
other phases.

3. Structure of cementite

Cementite, Fe_3C, which is an important phase in carbon steel, has been the subject of a large number of publications bearing upon its phase and structural relationships to iron and the other iron carbides. Lipson and Petch first determined its structure by the powder method.[3] Fasiska and Jeffrey prepared the single crystal of $(Fe_{2.7}Mn_{0.3})C$ and refined its structure using three-dimensional data.[4]

Cementite belongs to the orthorhombic system, space group *Pnma* (No. 62), with $a = 5.0598$, $b = 6.7462$, $c = 4.5074$Å, and $Z = 4$. Its atomic coordinates are as follows:

Atom	Position	x	y	z
Fe(1)	4(c)	.0367	1/4	.8402
Fe(2)	8(d)	.1816	.0666	.3374
C	4(c)	-.123	1/4	.44

The arrangement of the iron atoms of the cementite structure is best described by starting with the hcp structure, then compressing it in a $[10\bar{1}0]$ direction so that the basal hexagonal layers fold at the lines of intersection of the $(2\bar{1}\bar{1}0)$ planes to form regularly pleated layers. The folding is alternately up and down and is symmetrical, so that the $(2\bar{1}\bar{1}0)$ planes retain their mirror symmetry property. The $(2\bar{1}\bar{1}0)$ mirror planes in the hcp structure become the mirror planes at $y = 1/4$, $3/4$ in the space group *Pnma*. The iron atoms which lie in these planes are those in the $4(c)$ positions. The characteristic ABAB... arrangement of successive layers occurs in the $[c]_H$ direction of the hcp structure, so that the atoms of one pleated layer lie over the intersitices of those in the layer below.

At the line of intersection of the mirror planes and the pleated layers, i.e. on the fold-lines, the iron atoms are separated by two sets of interstices, X and Y, as shown in Fig. 2.3-3. The carbon atoms occupy these X and Y sites in an ordered manner, alternating at X and Y between successive pleated layers. There are two other theoretically possible carbon atom arrangements making use of these 4-fold type interstices which are consistent with the Fe_3C stoichiometry. These are (i), X and Y sites both occupied and both vacant between alternate layers, and (ii), all X sites (or all Y sites) occupied between successive layers. The actual arrangement is that in which the adjacent carbon atoms are farthest apart, as shown in Fig. 2.3-3.

In cementite, $(Fe_{2.7}Mn_{0.3})C$, the intermetallic distances range from 2.475 to 2.682Å, i.e. from values equal to that in a-iron to 10% greater. The atoms in general positions have eleven metal atom neighbours, and those in the special positions have twelve. Each carbon atom lies inside a trigonal prism of iron atoms, and have Fe-C distances from 1.96 to 2.04Å, as shown in Fig. 2.3-4. Cementite is very hard and brittle and is ferromagnetic; its melting point is about 1550°C and its Curie point about 210°C.

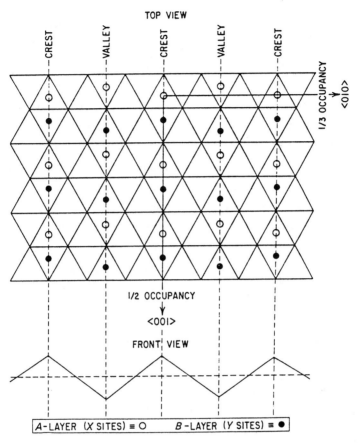

Fig. 2.3-3 Carbon atom positions on A and B pleated layers. (After ref. 4).

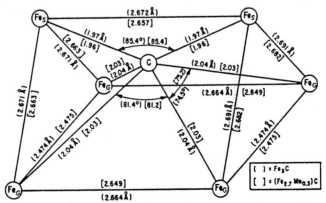

Fig. 2.3-4 Geometrical parameters of a triangular prism enclosing a C atom, calculated from the coordinates with both the $(Fe_{2.7}Mn_{0.3})C$ and the Fe_3C lattice dimensions. (After ref. 4).

4. Carbon steels[2,5]

Precise knowledge of the iron-carbon system is of basic importance in metallography and in metallurgy. All kinds of hammerable iron alloys are nowadays called steel. The irons used in practice - except for very special cases - always contain manganese, silicon, sulfur and phosphorus besides carbon. As long as the amount of these components is small (not more than 2-3% for Mn and Si, 0.1% for S and P), the corresponding alloy is simply called steel or more precisely carbon steel.

Iron containing carbon melts at much lower temperatures than pure iron. This fact is technically very important, since iron containing carbon can easily be melted and cast. In the liquid state cast iron or pig iron contains at least 1.7% dissolved carbon, in most cases considerably more. Grey pig iron contains about 3.8% of dissolved carbon; its melting temperature is about 1200°C. On cooling it down iron and cementite or graphite is formed, and the grey colour is due to this latter constituent. White pig iron contains even more carbon, which is present in the form of cementite provided the sample has been cooled down rapidly. Such castings are hard and brittle.

Austenite is the most important component of quenchable iron. During the course of quenching, i.e. rapid cooling down after heating, martensite crystal needles are formed which impart hardness and elasticity to steel. The stability of the martensite structure can be increased by introducing alloying components such as vanadium, chromium, manganese, cobalt, nickel and tungsten, which widen the γ-phase region. Above 150°C martensite is decomposed to ferrite and cementite if it is not stabilized by these alloying metals. At higher temperatures, above 721°C, austenite is decomposed to perlite, which consists of roughly parallel layers of ferrite and cementite (eutectoid). Another important function of the alloying metals is to control the undesirable effects of impurities in the iron such as sulfur and phosphorus.

Tempering, i.e. heating up to 250-300°C for a short time, results in the decomposition of martensite into ferrite and cementite. The tempered steel becomes soft as its internal tensions are eliminated.

Martensite has a tetragonal structure, which can be regarded as being built of a bcc iron lattice with carbon atoms (only 10-12% occupancy) at the centers of the basal faces and of the c edges: (1/2, 1/2, 0); (0, 0, 1/2). The unit cell is correspondingly slightly distorted, and the axial ratio $c/a >$ 1 (e.g. 1.066) depends on the carbon content. The carbon atoms cause strong tensions in the crystal, enhancing its hardness and strength. Martensite is unstable at higher temperatures, so that it is not represented in the iron-carbon phase diagram.

The various phases encountered in carbon steels and their crystal structures have always been of major importance in studies of transformations in the solid state as well as in industrial metallurgy. The reason for this is that the mechanical and other physical properties of steel not only depend on the number and nature of the phases present but are also very sensitive to the microstructure of the alloy, particularly in respect of the sizes of the crystal grains present and their mutual orientation. It is because all these factors can be varied independently to a large extent, by the appropriate control of composition, heat treatment, and mechanical working that steels can be prepared with an almost infinite variety of desirable physical properties.

5. Elements with the bcc structure

The bcc structure is common among metallic elements; some examples are given in Table 2.3-1.[2,6]

Table 2.3-1 Elements with the body-centered cubic structure.

Element	Cell edge a/Å	mp /°C	Density /g cm^{-3}
Ba	5.025	704	3.59
γ-Ca (500°C)	4.38	850	1.55
δ-Ce (730°C to mp)	4.12	795	6.67
α-Cr	2.884	1550	7.19
Cs (78°K)	6.067	28.5	1.87
Eu	4.578	826	5.259
α-Fe	2.8606	1535	7.89
K (78°K)	5.247	62.3	0.86
γ-La (868°C to mp)	4.26	920	5.98
Li	3.5093	186	0.53
δ-Mn	3.075	1244	7.43
Mo	3.147	2625	10.2
Na	4.291	97.5	0.97
Nb	3.300	2415	8.57
β-Nd (868°C to mp)	4.13	1024	6.80
γ-Np (~600°C)	3.52	640	20.45
β-Pr (798°C to mp)	4.13	935	6.64
ε-Pu (500°C)	3.638	640	19.86
Rb (75°K)	5.605	38.5	1.53
β-Sn (917°C to mp)	4.07	1072	7.40
γ-Sr (614°C)	4.85	757	2.6
Ta	3.306	2996	16.6
β-Th (1450°C)	4.11	1800	11.5
β-Ti (900°C)	3.307	1820	4.43
β-Tl	3.882	300	11.85

(Table 2.3-1 continued)

γ-U	3.474	1130	18.06
V	3.024	1735	6.0
W	3.165	3410	19.26
β-Y (1490°C to mp)	4.11	1509	4.25
β-Yb (798°C to mp)	4.44	824	6.54
β-Zr (850°C)	3.62	1750	6.49

References

[1] A.W. Hull, *Phys. Rev.* **9**, 84 (1917).

[2] C.S. Barrett and T.B. Massalski, *Structure of Metals*, 3rd ed., Pergamon Press, Oxford, 1980.

[3] H. Lipson and N.J. Petch, *J. Iron Steel Inst.* **142**, 95 (1940).

[4] E.J. Fasiska and G.A. Jeffrey, *Acta Crystallogr.* **19**, 463 (1965).

[5] I. Náray-Szabó, *Inorganic Crystal Chemistry*, Akadémiai Kiadó, Budapest, 1969.

[6] F.S. Galasso, *Structure and Properties of Inorganic Solids*, Pergamon Press, Oxford, 1970.

[7] D.M. Adams, *Inorganic Solids*, Wiley, New York, 1974.

2.4   Diamond
C

Crystal Data

Cubic, space group $Fd3m$ (No. 227)

$a = 3.56679(2)$Å (20°C), $Z = 8$

Atom	Position*	x	y	z
C	8(a)	0	0	0

*Origin at $\overline{4}3m$; inversion center at (1/8,1/8,1/8).

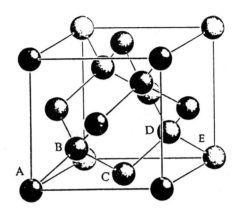

Fig. 2.4-1 Unit cell of diamond.

Crystal Structure

The diamond structure was, in fact, one of the very first to be determined by the Braggs[1], who wrote:

"We have applied the new methods of investigation involving the use of X-rays to the case of the diamond, and have arrived at a result which seems of considerable interest. The structure is extremely simple. Every carbon atom has four neighbours at equal distances from it, and in directions symmetrically related to each other. The directions are perpendicular to the four cleavage or (111) planes of the diamond; parallel, therefore, to the four lines which join the center of a given regular tetrahedron to the four corners. The elements of the whole structure are four directions and one length, the latter being, in fact, 1.52×10^{-8} cm. There is no acute angle in the figure. These facts supply enough information for the construction of a model which is easier to understand than a written description."

The diamond structure (Fig. 2.4-1) can be considered to comprise two interpenetrating fcc sublattices displaced with respect to each other by the vector from (0,0,0) to (1/4,1/4,1/4). The midpoint of each C-C bond [best value presently known is $(\sqrt{3}/4)a/4 = 1.5445Å$] is located at a center of symmetry so that the adjacent carbon atoms conform to the staggered arrangement.

The diamond structure is quite open, as the proportion of space occupied by the carbon atoms is only $\pi\sqrt{3}/16 = 0.3401$. On the other hand, the strong covalent bonding throughout the entire diamond lattice readily accounts for its extreme hardness and stability.

Remarks

1. The tetrahedral covalency of carbon

The determination of the diamond structure provided concrete evidence for the tetrahedral covalency of a saturated carbon atom, first proposed by van't Hoff and Le Bel in 1874, as well as a direct measurement of the length of a C-C single bond.[2,3] This work laid the foundation of the chemistry of aliphatic and alicyclic compounds by begetting such basic concepts as covalent radius, bond length, bond angle, molecular shape (symmetry), configuration and conformation. For instance, the carbon skeleton of an alkane or a cycloalkane can be picked out from a portion of the diamond structure.

2. Electron density and C-C σ-bonding[4]

In the diamond structure, the carbon atoms are bonded covalently by electron pairs which occupy localized molecular orbitals formed by overlapping neighbouring sp^3 hybrids. Fig.2.4-2(a) shows the electron density in a plane passing through the series of atoms A, B, C, D and E in the unit cell. The corresponding difference deformation density map displays positive peaks of

(a)   (b)

Fig 2.4-2 Experimental electron density maps of diamond: (a) cross section through the centers of C atoms (contours in $eÅ^{-3}$); (b) the corresponding difference deformation density. (Adapted from ref. 4).

height 0.51 $e\text{Å}^{-3}$ midway between the carbon atoms, indicating a concentration of electron density at the centers of the σ bonds.

3. Hexagonal diamond

A metastable hexagonal form of diamond, which has been found in aerolite, has been prepared from graphite at 130 kbar above 1000°C. Hexagonal diamond crystallizes in space group $P6_3/mmc$ with $a = 2.51$, $c = 4.12$Å, and $Z = 4$.[5] The bond type and the C-C bond distance are the same as in cubic diamond. Fig. 2.4-3 shows the carbon skeleton of hexagonal diamond.

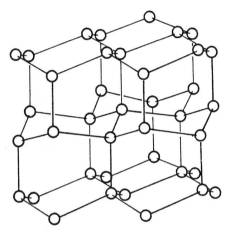

Fig. 2.4-3 A portion of the hexagonal diamond lattice. (After ref. 5).

4. Elements with the diamond structure

The group IV elements Si, Ge and Sn also conform to the cubic diamond structure type (Table 2.4-1).

Table 2.4-1 The structural parameters of group IV elements.[6]

Element	a/Å	Temperature /°C	Interatomic distance/Å	Density /g cm^{-3}
C	3.56679	20	1.5445	3.5150
Si	5.43086	20	2.3517	2.3283
Ge	5.65735	25	2.4498	5.3234
Sn	6.4892	20	2.8099	5.765

5. "Diamond molecules"

Adamantane, whose highly symmetric carbocyclic skeleton is a fragment of the diamond lattice, may be viewed as the methane analogue in a homologous series in which diamantane ("congressane", congress emblem of the International Union of Pure and Applied Chemistry, I)[7] and triamantane (II) are analogues of ethane and propane, while *anti*-tetramantane (**III**) and skew-tetramantane (**IV**) are related to *n*-butane conformers and isotetramantane (**V**) matches isobutane (Fig. 2.4-4). The analogy has led to diamond being termed the "infinite adamantylogue of adamantane".[8]

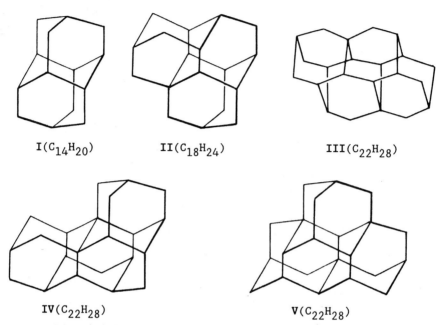

$I(C_{14}H_{20})$ $II(C_{18}H_{24})$ $III(C_{22}H_{28})$

$IV(C_{22}H_{28})$ $V(C_{22}H_{28})$

Fig. 2.4-4   Higher homologues of adamantane.

X-Ray analysis of *anti*-tetramantane has revealed an interesting bond-length progression: $CH-CH_2 = 1.524$, $C-CH_2 = 1.528$, $CH-CH = 1.537$, and $C-CH = 1.542$Å, approaching toward the limit of 1.5445Å in diamond as the number of bonded H atoms decreases.[9]

6. Conformations of medium-size and large cycloalkanes[10]

The preferred conformations of cycloalkanes C_nH_{2n} should correspond to low strain energies. The rings with $n = 4i+2$ have $2/m$ symmetry; the resulting arrangements of carbon atoms can be traced out of the diamond structure, as shown in Fig. 2.4-5(a), and should, therefore, be virtually strain-free. On the other hand, the rings with $n = 4i$ possessing 422 symmetry cannot be fitted exactly into the diamond structure, as shown in Fig. 2.4-5(b).

For fourteen-membered ring systems, structure analyses of 1,8-diazacyclotetradecane dihydrobromide[11] and 1,8-diazacyclotetradecane-1,8-diol[12] have provided experimental verification that the stable ring conformation is indeed based on the diamond structure. In the large cyclic paraffin cyclo-tetratriacontane, $[CH_2]_{34}$, the molecular ring is basically in

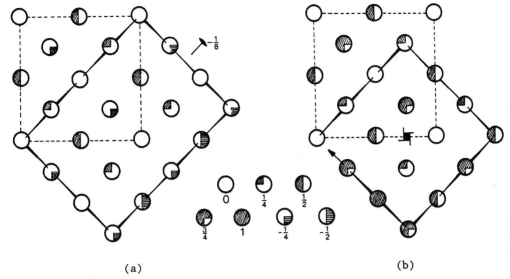

(a)   (b)

Fig. 2.4-5 (a) Showing how a fourteen-membered ring with 2/m symmetry can be traced out of the diamond structure. All rings with $n = 4i+2$ can be derived in a similar fashion. (b) The corresponding twelve-membered ring with 422 symmetry cannot be traced out of the diamond structure, which does not possess four-fold rotation axes. The four-fold screw axis gives rise to a helix. (After ref. 10).

the form of two roughly parallel zigzag chains of 15 carbon atoms, linked at each end by two closure atoms, and conforming to a path traceable to the diamond structure.[13]

The diamond structure has also been used for deriving stable conformations of alkanes and strain-free helical conformations of polymeric chains.

References

[1]   W.H. Bragg and W.L. Bragg, *Nature (London)* **91**, 557 (1913); *Proc. Roy. Soc. London* **89A**, 277 (1913).

[2]   J.H. van't Hoff, *Naturelles* **9**, 445 (1874).

[3]   J.A. Le Bel, *Bull. Chim. Soc. Fr.* **22**, 337 (1874).

[4]   R. Brill, *Solid State Phys.* **20**, 1 (1967).

[5]   J. Donohue, *The Structures of the Elements*, Wiley, New York, 1974.

[6]   I. Náray-Szabó, *Inorganic Crystal Chemistry*, Akadémiai Kiadó, Budapest, 1969.

[7]   I.L. Karle and J. Karle, *J. Am. Chem. Soc.* **87**, 918 (1965).

[8] D. Lenoir and P. von R. Schleyer, *J. Chem. Soc., Chem. Commun.*, 941 (1970).

[9] P.J. Roberts and G. Ferguson, *Acta Crystallogr., Sect. B* **33**, 2335 (1977).

[10] J.D. Dunitz in J.D. Dunitz and J.A. Ibers (eds.), *Perspectives in Structural Chemistry*, Vol. II, Wiley, New York, 1968 p. 1.

[11] J.D. Dunitz and E.F. Meyer, *Helv. Chim. Acta* **48**, 1441 (1965).

[12] C.J. Brown, *J. Chem. Soc. (C)*, 1108 (1966).

[13] H.F. Kay and B.A. Newman, *Acta Crystallogr., Sect. B* **24**, 615 (1968).

Note Diamond (including diamond films) edged out buckminsterfullerene, C_{60}, as "molecule of the year 1990". See R.L. Guyer and D.E. Koshland, *Science (Washington)* **250**, 1640 (1990).

Theoretical studies of clusters based on the diamond lattice, ranging from cyclohexane to the unknown superadamantane-5, $C_{35}H_{36}$, are discussed in M. Shen, H.F. Schaefer III, C. Liang, J.-H. Lii, N.L. Allinger and P.v.R. Schleyler, *J. Am. Chem. Soc.* **114**, 497 (1992).

Diamond-like ("diamondoid") three-dimensional networks can be contructed from tetrafuntional organic molecules containing tetrahedrally directed substituents. The carboxyl groups of adamantane-1,3,5,7-tetracarboxylic acid are interlinked by pairwise intermolecular hydrogen bonds to generate a hollow diamondoid network leaving very large cavities. These interstices are filled in such a way that the crystal structure consists of five translationally equivalent, interpenetrating, but unconnected diamondoid networks. Likewise the "threefold diamondoid" structure of 3,3-bis(carboxyymethyl)glutaric acid ("methanetetraacetic acid") is built of three interlaced diamondoid networks. See O. Ermer, *J. Am. Chem. Soc.* **110**, 3747 (1988); O. Ermer and A. Eling, *Angew. Chem. Int. Ed. Engl.* **27**, 829 (1988).

2.5 Graphite
C

Crystal Data

Hexagonal, space group $P6_3/mmc$ (No. 194)

$a = 2.4562(1)$, $c = 6.6943(7)$Å (15°C), $Z = 4$

Atom	Position	x	y	z
C(1)	2(b)	0	0	1/4
C(2)	2(d)	2/3	1/3	1/4

Origin at center ($\bar{3}$m)

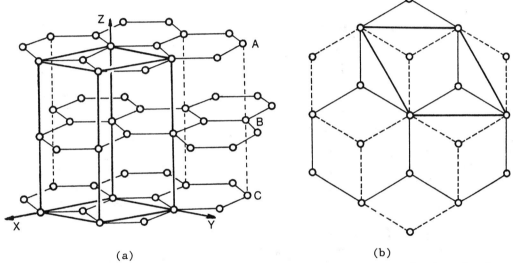

(a)   (b)

Fig. 2.5-1 Structure of hexagonal graphite. (a) General view showing unit cell. (b) Viewed in projection down the c axis: solid lines, layer at $z = 0$; dashed lines, layer at $z = 1/2$; thin lines, outline of unit cell. (After ref. 3).

Crystal Structure

In the same way that diamond proves to be the prototype for the carbon skeleton in the aliphatic and alicyclic compounds of organic chemistry, graphite typifies the grouping present in aromatic compounds. Although graphite was studied at an early stage, on account of the difficulty in obtaining good single crystals, details of the atomic arrangement remained in doubt until the investigations of Bernal[1] and Hassel and Mark[2] in 1924.

Graphite in its common form has the hexagonal structure (Fig. 2.5-1), which comprises a stacking of parallel layers, in each of which the carbon atoms are arranged at the corners of a network of planar regular hexagons.

The C-C distance is $a/\sqrt{3} = 1.418$Å within each layer, and the distance between the layers is $c/2 = 3.347$Å.

Two adjacent graphite layers can be stacked in only one way, namely with three alternate atoms of one ring above ring centres in the lower layer. If the third layer lies above the first, the stacking sequence is ...ABABAB..., as found in hexagonal graphite. If the third layer is displaced with respect to the first two, the resulting sequence ...ABCABCABC... gives rise to rhombohedral graphite (Fig. 2.5-2). The unit cell of rhombohedral graphite has $a = 3.642$Å, $\alpha = 39.5°$, space group $R\bar{3}m$ (No. 166), and $Z = 2$; for the triply-primitive hexagonal unit cell a_H and $c_H = 2.461$ and 10.064Å, respectively.

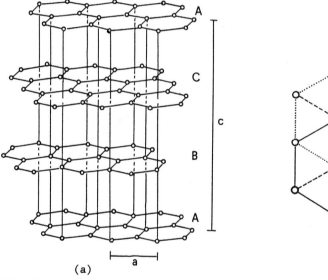
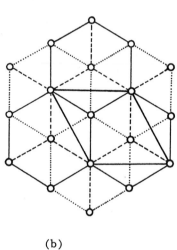

(a)   (b)

Fig. 2.5-2 The structure of rhombohedral graphite: (a) general view, (b) viewed along the hexagonal c_H axis. Solid lines, layer at $z = 0$; dashed lines, layer at $z = 1/3$; dotted lines, layer at $z = 2/3$; thin lines, outline of unit cell. (After refs. 3 and 4).

The two forms of graphite are interconvertible by grinding (hexagonal → rhombohedral) or heating above 1025 °C (rhombohedral → hexagonal). Partial conversion leads to an increase in the average spacing between layers; this reaches a maximum of 3.44Å for turbostratic graphite in which the stacking sequence of the parallel layers is completely random. The enthalpy difference between hexagonal and rhombohedral graphite is only 0.59 ± 0.17 kJ mol^{-1}.[3]

Remarks

1. Graphite and diamond

In graphite, intralayer bonds are formed by sp^2 hybrids as in benzene.

The p orbitals normal to the plane overlap to form π-MO's covering the entire layer, giving rise to a partly filled conduction band. The mobility of the π electrons accounts for both the colour (black) and the metal-like electrical conductivity within the layers of graphite. The layers are widely separated (3.347Å) and bound by van der Waals forces only; electrical conductivity normal to them is therefore extremely small. Owing to the weakness of interlayer bonding, graphite is readily cleaved; this same property is also associated with its use as a heavy-duty lubricant, although the mechanism of its lubrication is far from simple.

The structural differences between graphite and diamond are reflected in their differing physical and chemical properties (Table 2.5-1).

Table 2.5-1 Some properties of graphite and diamond.[5]

Property	Graphite	Diamond
Density/g cm^{-3}	2.266 (ideal)	3.514
Hardness/Mohs	< 1	10
mp/°K	4100 ± 100 (at 9 kbar)	4100 ± 200 (at 125 kbar)
Refractive index n(at 546 nm)	2.15 ($\parallel$ layer) 1.81 ($\perp$ layer)	2.41
Band gap E_g/kJ mol^{-1}	-	580
ρ/ohm cm	(0.4-5.0) x 10^{-4}($\parallel$ layer) 0.2-1.0 ($\perp$ layer)	10^{14}-10^{16}
$\Delta H_f°$/kJ mol^{-1}	0.00 (standard state)	1.90

2. Graphite intercalation compounds

The large interlayer spacing in graphite enables a wide range of substances to occupy the intervening space, yielding graphite intercalates such as alkali metal graphite compounds (C_8K, $C_{24}K$, $C_{36}K$, $C_{48}K$, $C_{60}K$, C_6Li, $C_{12}Li$, $C_{18}Li$, $C_{40}Li$, $C_{32}Na$, $C_{64}Na$, $C_{120}Na$...), graphite salts ($C_{24}{}^+HSO_4{}^-$.$2H_2SO_4$, $C_{24}{}^+NO_3{}^-$.$2HNO_3$, $C_{48}{}^+NO_3{}^-$.$3HNO_3$...), and other graphite intercalation compounds of halides. Halides which have been reported to form intercalates include the following[5]:

HF, ClF_3, BrF_3, IF_5; XeF_6, $XeOF_4$; CrO_2F_2, SbF_3Cl_2, TiF_4, MF_5 (M = As, Sb, Nb, Ta), UF_6

MCl_2: M = Be, Mn, Co, Ni, Cu, Zn, Cd, Hg

MCl_3: M = B, Al, Ga, In, Tl; Y, Sm, Eu, Gd, Tb, Dy; Cr, Fe, Co; Ru, Rh, Au

MCl_4: M = Zr, Hf; Re, Ir; Pd, Pt

MCl_5: M = Sb; Mo; U

MCl_6: M = W, U

Mixtures of $AlCl_3$ plus Br_2, I_2, ICl_3, $FeCl_3$, WCl_6

Bromides: $CuBr_2$, $AlBr_3$, $GaBr_3$, $AuBr_3$

Fig. 2.5-3 shows the structure of C_8K.[6] Between each pair of carbon layers
is a potassium layer. The position of the graphite layers remains the same
since in this way each K atom can be surrounded both above and below by carbon
hexagons. C_8K is a metallic conductor which shows a negative temperature
coefficient. In the direction of the layers the conductivity is about ten
times as large as in pure graphite, while perpendicular to the layers it is

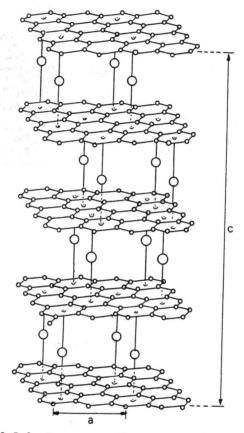

Fig. 2.5-3 Structure of C_8K. (After ref. 4).

small although still a hundred times greater than in the corresponding
direction in graphite itself. By the addition of K atoms the number of $p\pi$
electrons in the conduction band is appreciably increased.

3. Novel graphite-like materials

A novel graphite-like material of composition BC_3 has been prepared by
the reaction of benzene with boron trichloride at 800°C. The probable sheet
structure of BC_3, as deduced from electron microscopy (interlayer spacing
3.4Å), electron diffraction ($c/a = 2.76$) and X-ray powder diffraction, is
illustrated in Fig. 2.5-4. A similar layer-form material of approximate
compositions C_5N (electron diffraction patterns similar to those of BC_3) was

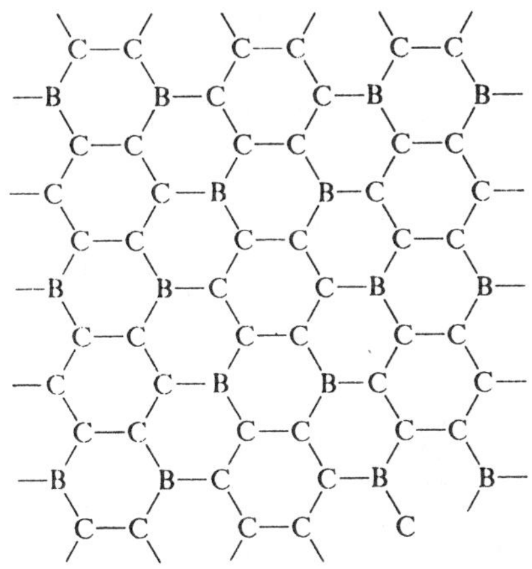

Fig. 2.5-4 The probable atomic arrangement in a layer of the boron-carbon hybrid BC_3. (After ref. 7).

obtained by the interaction of gaseous pyridine and chlorine at ~800°C. A sample of C_5N heated to 1400°C, at 4000 lb in^{-2} in a graphite die, has a composition $C_{57}N$; the interlayer spacing of 3.42Å and a = 2.44Å are derived from X-ray powder diffraction data.

4. The hexagonal form of boron nitride

Hexagonal BN, prepared industrially by the fusion of urea with $B(OH)_3$ in an atmosphere of NH_3 at 500-950°C, is often described as "graphite-like". Actually it is a simple layer structure (made up of six-membered rings of alternating B and N atoms) in which the B atoms in one layer is laid directly over the N atoms in the next layer and *vice versa*.[8] The crystallographic data for hexagonal BN are: space group $P6_3/mmc$, a = 2.504, c = 6.661Å, and Z = 2(BN). The intralayer B-N distance of 1.446Å is similar to the value 1.44Å in borazine but much less than twice the single-bond covalent radius (0.79Å) of boron.

Two other modifications of BN are known: a cubic form known as borazone (charge transfer of 0.46e from B to N)[9] with the zinc blende structure (Section 2.9) and a rarer form obtained by fusion, which has the structure of rhombohedral graphite.

5. Carbon fibers

Nowadays carbon fibers (often misnamed graphite fibers) are used in a matrix of another substance to form a durable composite material which can be three times stronger but four times lighter than steel.[10]

Carbon fibers are composed of "graphene" sheets, which are hexagonal layers of carbon atoms similar to the arrangement in crystalline graphite but

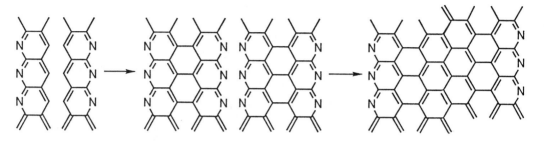

400-600°C Dehydrogenation   600-1300°C Denitrogenation

Fig. 2.5-5 Generation of a graphene sheet from polyacrylonitrile (PAN) by heat treatment. The long open chains of PAN first close their borders to form hexagonal rings and then link edges to form a sheet. (After ref. 10).

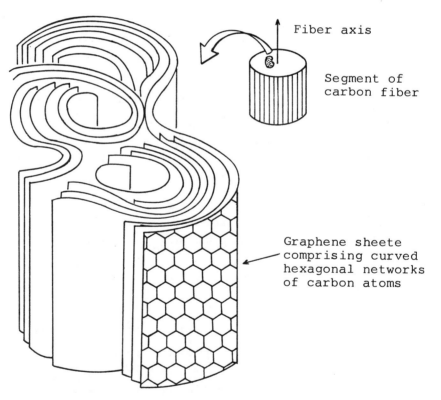

Fig. 2.5-6 Structure of a segment of a carbon fiber. (After ref. 10).

not constrained to rigid planarity. The extraordinary strength of a carbon
fiber is derived from the parallel alignment of the graphene sheets with
respect to its longitudinal axis, thus taking full advantage of the intralayer covalent bonding. A well-developed commercial process for the
production of carbon fibers employs polyacrylonitrile (PAN) as the starting
material. The PAN filaments undergo cyclization at 200-300 °C in an oxidizing
atmosphere (usually dry air) while being stretched, forming linear chains of
edge-sharing hexagonal heterocyclic rings. The polymer chains are then
condensed by dehydrogenation and denitrogenation that eventually produces
graphene sheets.

6. The quest for all-carbon molecules

    Fascination with the possible existence of discrete, stable pure-carbon
molecules C_n, also known as polycarbons, has led to a number of recent
experimental and theoretical studies.[11,12] For the molecules C_2 through C_9,
both linear and cyclic structures have been discussed. The theoretical
consensus is that the larger molecules C_{10} through C_{29} should have monocyclic
structures with special stability for those comprising (4n+2) atoms.
Molecules in the range C_{30} to C_{40} are expected to be very reactive. Above
C_{40}, more stable even-numbered molecules with closed spheroidal structures
begin to dominate, as highlighted by the exceptional stability of the
buckminsterfullerene C_{60} cluster.

    For the cyclo[18]carbon target molecule, C_{18}, *ab initio* calculations
predicted a relatively stable, cyclic D_{9h} ground state geometry with
alternating single (1.362Å) and triple (1.199Å) bonds [Fig. 2.5-7(a)].[11]
The compound $C_{60}H_{30}$ possessing D_{3h} geometry [Fig. 2.5-7(b)], a direct
precursor to C_{18}, has been characterized by X-ray crystallography.[11] The

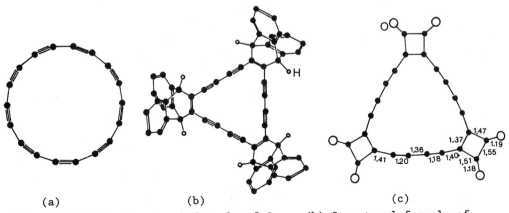

| (a) | (b) | (c) |

Fig. 2.5-7 (a) Structural formula of C_{18}. (b) Structural formula of
$C_{60}H_{30}$. (c) Molecular geometry of $C_{24}O_6$. (After refs. 11 and 13).

analysis of laser flash heating experiments on this precursor by time-of-flight mass spectrometry shows a sequence of retro-Diels-Alder reactions leading to C_{18} as the predominant fragmentation product.

The higher oxides of carbon $C_{8n}O_{2n}$ ($n = 3$-5) are stable deep red solids that serve as ideal precursors to the cyclo[n]carbons C_{18}, C_{24} and C_{30}.[13] X-Ray analysis has established that the $C_{24}O_6$ molecule has C_3 symmetry with a planar C_{18} ring and bent diyne moieties. The cyclobutenedione unit is planar and twisted by 2.2Å with respect to the C_{18} ring skeleton [Fig. 2.5-7(c)]. The laser desorption Fourier transform mass spectra of $C_{24}O_6$, $C_{32}O_8$ and $C_{40}O_{10}$ show successive losses of two CO fragments to give cyclo[n]carbon ions C_n^+ ($n = 18,24$) and C_n^- ($n = 18,24,30$) in both positive and negative modes.

7. Buckminsterfullerene, a spheroidal C_{60} naked cluster

Experimental[14] and theoretical[15] studies in the mid 1980s provided convincing evidence for the existence of a stable C_{60} species which approximates to a spherical shell of graphite. The vaporization of graphite by laser irradiation produced even-numbered carbon clusters of 38-120 atoms, which were expanded in a supersonic molecular beam, photoionized using an excimer laser, and detected by time-of-flight mass spectrometry.

The C_{60} peak was the largest in the resulting gaussian distribution of maxima but not completely dominant. By increasing the time between vaporization and expansion the C_{60} peak was made about 40 times larger than neighbouring clusters. The most plausible model of this stable C_{60} cluster is a truncated icosahedron, a polygon with 60 vertices and 32 faces, 12 of which are pentagons and 20 are hexagons (Fig. 2.5-8). For such a unique structure convenient names such as "buckminsterfullerene" (after Buckminster Fuller who popularized this form for the construction of geodesic domes), "footballene", "soccerene", "buckyball", and "fullerene-60" have been suggested. The diameter of this C_{60} cage molecule is ~7Å, and mass spectral evidence for $C_{60}^{\bullet+}$ (and also $C_{70}^{\bullet+}$) containing an encaged helium atom in its inner cavity has been obtained.[16] The complexes $C_{60}La$, $C_{60}U$ and $C_{60}H_{36}$ are also known.

In buckminsterfullene, each carbon atom is sp^2 hybridized and bonded to three other carbon atoms. Each carbon atom is equivalent to every other carbon atom, being the common vertex of a pentagon and two hexagons. The inner and outer surfaces of the molecule are thus covered with a sea of π electrons. Simple Hückel theory gives 33.1616β for the delocalization energy of buckminsterfullerene, or 0.5527β per carbon atom.[15] This is to be compared with the value of 2β for benzene, or 0.3333β per carbon atom. The energetically stable, highly conjugated cluster model of buckminsterfullerene

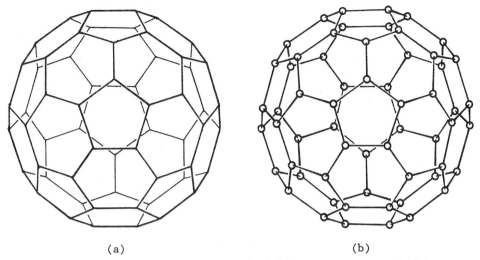

(a)   (b)

Fig. 2.5-8 The structure of buckminsterfullerene represented (a) as a
polyhedron and (b) as an assembly of bonded carbon atoms. (After ref. 3).

has been substantiated by MNDO calculations[17] and spectroscopic
evidence.[18]

     Corroborative experimental evidence of the buckminsterfullerene C_{60}
structure is provided by the existence of saucer-like [5.6.1]corannulene [Fig.
2.5-9(a)], whose carbon skeleton constitutes one-third of that of C_{60}[19] and
by the considerable bending of the "naphthalene bow" in
[2](2,6)naphthalino[2]paracyclophane-1,11-diene [Fig. 2.5-9(b)].[20]

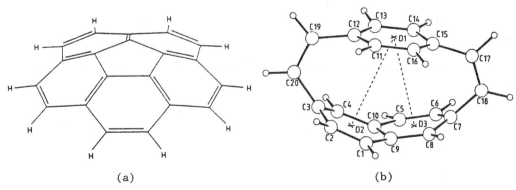

(a)   (b)

Fig. 2.5-9 (a) Structural formula of [5.6.1]corannulene. (b) Molecular
structure of [2](2,6)naphthalino[2]paracyclophane-1,11-diene. (After
ref. 20).

8. Icospiral graphite particles and soot[21]

     For a small sheet of graphitic net, energetics favours closure to a
hollow cage (with simultaneous generation of pentagonal faces) owing to the
bond energy released by eliminating the reactive benzyne-like edges. An

elegant "icospiral particle nucleation scheme" has been proposed to account for the spontaneous creation of C_{60} and other larger quasi-crystalline carbon particles from a chaotic plasma.[22] The initial hypothetical steps involve highly reactive open spiral shell (nautilus-like) embryos (Fig. 2.5-10). Small accreting carbon fragments are mopped up by adsorption on the surface of such shells and rapidly knitted into the advancing edge. Once the trailing edge has been bypassed [Fig. 2.5-10(c)], closure becomes impossible (unless annealing takes place) and new network will form under epitaxial control so that the skin lies 3.4Å (the graphite interlayer spacing) above the inner layer [Fig. 2.5-10(d)]. The formation of C_{60} and related spheroidal C_n clusters, the "fullerenes", is readily explained by occasional, statistical closure during nucleation of a network with the correct disposition of the required twelve pentagons.

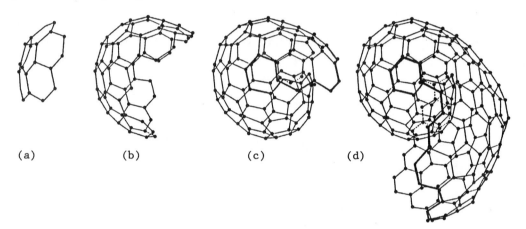

(a)   (b)   (c)   (d)

Fig. 2.5-10 The icospiral nucleation process starts with a reactive saucer-shaped C_{20} (corannulene skeleton) (a) and grows by accretion *via* the half-shell (b) to (c), where edge bypass has occurred, and then on to (d), where the second growth shell is forming under epitaxial control. Statistical closure at stage (c) is proposed as the explanation for fullerene formation. (Adapted from ref. 21).

The scheme predicts C_{60} to be a byproduct of soot formation, and C_{60}^{+} has been established as a dominant ion in a sooting flame. The presence of much hydrogen in soot is accounted for by internal C-H bond formation at the pentagonal faces. A soot particle forms under conditions where the balance between the C-C network and C-H bond formation is continuously oscillating, and gaps which develop as an icospiral grows are rapidly covered over, trapping C-H and dangling bonds in a three-dimensional spiral with a smoother and more spheroidal shape.

9. The third form of carbon: fullerenes C_{60} and C_{70}

Pure samples of C_{60} and C_{70} (fullerene-70) have been isolated from soot by an amazingly simple technique.[23] When a high electric current was passed through a graphite electrode in a vacuum chamber filled with Ar gas at 50-100 mbar, a black soot-like solid material deposited on surfaces inside the chamber. Solvent extraction of batches of this material with benzene gave plum-coloured solutions which yielded, after solvent removal and chromatographic separation (alumina, hexane), pure crystalline mustard coloured C_{60} and reddish-brown C_{70} in a ratio of approximately 5:1.[24]

A study of C_{60} using X-ray powder diffraction established a somewhat disordered hcp structure with $a = 10.02$ and $c = 16.39$Å, confirming the size and shape of the spherical molecule in a close-packed structure.[23] The ^{13}C NMR spectrum of C_{60} consists of a single line, as required, at 142.68 ppm and unaltered by proton decoupling.[24] Fig. 2.5-11(a) shows the ^{13}C NMR spectrum for C_{70}, which exhibits five lines at 150.07, 147.52, 146.82, 144.77 and 130.28 ppm, also unaltered by proton decoupling. The data provide compelling evidence for the prolate-spheroidal fullerene-70 structure [Fig. 2.5-11(b)] that possesses five sets of inequivalent carbon atoms with a $n_a:n_b:n_c:n_d:n_e$ ratio of 10:10:20:20:10.[24]

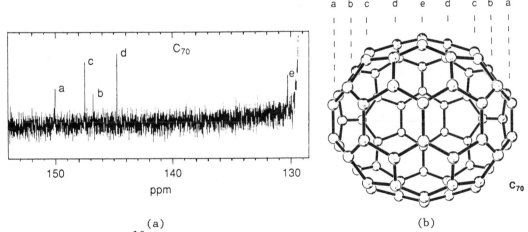

(a)   (b)

Fig. 2.5-11 (a) ^{13}C NMR spectrum of C_{70} (fullerene-70). (b) Molecular structure of C_{70} showing sets of equivalent carbon nuclei. (After ref. 24).

The fullerenes C_{60} and C_{70} can be regarded as a new form of carbon in addition to diamond and graphite. The possibility of attaching other atoms or groups (e.g. hydrogen atoms or an aliphatic chain) to the closed cages may open up an entirely new field of carbon chemistry. Other relatively stable fullerenes C_n, as deduced from the mass spectrum of carbon smoke from discharge processed graphite, correspond to $n = 62, 64, 66, 68$, etc.[24]

The carbon framework of buckminsterfullerene remains intact when heteroatoms are attached to it. Selective osmylation yields 1:1 and 1:2 adducts of C_{60} with OsO_4L_2 (L = pyridine ligand) depending on the reaction conditions.[25] X-Ray analysis of $C_{60}(OsO_4)(4\text{-}t\text{-butylpyridine})_2$ has confirmed that the osmyl unit serves to break the pseudo-spherical symmetry of C_{60} to give an ordered crystal structure.[26] The O(1)-Os-O(2) fragment bridges across the edge [1.62(4)Å] involved in *six-six ring fusion* [Fig. 2.5-12(a)]. The tetracoordinate oxygen-bonded carbon atoms lie 3.81(3)Å from the center of the cluster fragment comprising the remaining 58 carbon atoms (average C-C bond lengths are 1.388(9)Å for six-six and 1.432(5)Å for six-five ring fusions), which all lie within a shell of mean radius 3.512(3)Å. The bonding pattern in $(Ph_3P)_2Pt(\eta^2\text{-}C_{60})\cdot C_4H_8O$ bears a striking similarity to the dihapto alkene-platinum interaction known for $(Ph_3P)_2Pt(\eta^2\text{-}C_2H_4)$ [Fig. 2.5-12(b)].[27]

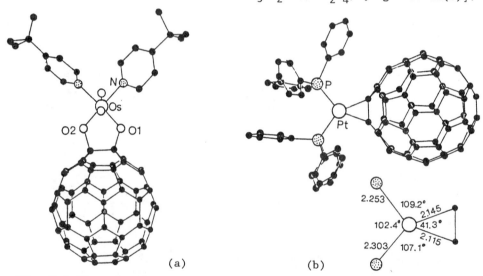

(a)   (b)

Fig. 2.5-12 Molecular structure of (a) $C_{60}(OsO_4)(4\text{-}t\text{-butylpyridine})_2$ and (b) $(Ph_3P)_2Pt(\eta^2\text{-}C_{60})$. (Adapted from refs. 26 and 27).

References

[1] J.D. Bernal, *Proc. Roy. Soc. London* **106A**, 749 (1924).

[2] O. Hassel and H. Mark, *Z. Phys.* **25**, 317 (1924).

[3] J. Donohue, *The Structures of the Elements*, Wiley, New York, 1974.

[4] H. Krebs, *Fundamentals of Inorganic Crystal Chemistry*, (translated into English by P.H.L. Walter), McGraw-Hill, London, 1968.

[5] N.N. Greenwood, *Chemistry of the Elements*, Pergamon Press, Oxford, 1984.

[6] W. Rüdorff and E. Schulze, *Z. Anorg. Allg. Chem.* **277**, 156 (1954).

[7] J. Kouvetakis, R.B. Kaner, M.L. Sattler and N. Bartlett, *J. Chem. Soc., Chem. Commun.*, 1758 (1986).

[8]   R.W.G. Wyckoff, *Crystal Structures*, 2nd ed., Vol. 1, Wiley-Interscience, New York, 1963, p. 184.

[9]   G. Will, A. Kirfel and B. Josten, *J. Less-Common Met.* **117**, 61 (1986).

[10] J. FitzGerald and G. Taylor, *New Scientist* **20 May**, 48 (1989).

[11] F. Diederich, Y. Rubin, C.B. Knobler, R.L. Whetten, K.E. Schriver, K.N. Houk and Y. Li, *Science (Washington)* **245**, 1088 (1989).

[12] W. Weltner Jr. and R.J. Van Zee, *Chem. Rev.* **89**, 1713 (1989); H.W. Kroto, A.W. Allaf and S.P. Balm, *Chem. Rev.* **91**, 1213 (1991).

[13] Y. Rubin, M. Kahr, C.B. Knobler, F. Diederich and C.L. Wilkins, *J. Am. Chem. Soc.* **113**, 495 (1991).

[14] H.W. Kroto, J.R. Heath, S.C. O'Brien, R.F. Curl and R.E. Smalley, *Nature (London)* **318**, 162 (1985).

[15] A.D.J. Haymet, *J. Am. Chem. Soc.* **108**, 319 (1986).

[16] T. Weiske, D.K. Böhme, J. Hrušàk, W. Krätschmer and H. Schwarz, *Angew. Chem. Int. Ed. Engl.* **30**, 884 (1991).

[17] M.D. Newton and R.E. Stanton, *J. Am. Chem. Soc.* **108**, 2469 (1986).

[18] W. Krätschmer, K. Fostiropoulos and D.R. Huffman, *Chem. Phys. Lett.* **170**, 167 (1990), and references cited therein.

[19] M. Randić and N. Trinajstić, *J. Am. Chem. Soc.* **106**, 4428 (1984).

[20] N.E. Blank, M.W. Haenel, C. Krüger, Y.-H. Tsay and H. Wientges, *Angew. Chem. Int. Ed. Engl.* **27**, 1064 (1988).

[21] H.W. Kroto, *Science (Washington)* **242**, 1139 (1988).

[22] H.W. Kroto and K.G. McKay, *Nature (London)* **331**, 328 (1988).

[23] W. Krätschmer, L.D. Lamb, K. Fostiropoulos and D.R. Huffman, *Nature (London)* **347**, 354 (1990).

[24] R. Taylor, J.P. Hare, A.K. Abdul-Sada and H.W. Kroto, *J. Chem. Soc., Chem Commun.*, 1423 (1990).

[25] J.M. Hawkins, T.A. Lewis, S.D. Loren, A. Meyer, J.R. Heath, Y. Shibato and R.J. Saykally, *J. Org. Chem.* **55**, 6250 (1990).

[26] J.M. Hawkins, A. Meyer, T.A. Lewis, S. Loren and F.J. Hollander, *Science (Washington)* **252**, 312 (1991).

[27] P.J.Fagan, J.C. Calabrese and B. Malone, *Science (Washington)* **252**, 1160 (1991).

Note The discovery of the fullerenes is recounted in D.R. Huffman, *Physics Today* **44**, 22 (1991); G. Taubes, *Science (Washington)* **253**, 1476 (1991).

As "Molecule of the Year 1991", C_{60} made the front covers of two science journals: E. Culotta and D.E. Koshland, Jr., *Science (Washington)* **254**, 1706 (1991); R.F. Curl and R.E. Smalley, *Scientific American* **265**, 32 (October, 1991). The latter article contains several strikingly beautiful illustrations in colour.

A series of five articles on C_{60} covering its likely astronomical significance, Hückel energy levels, Kekulé structures, and molecular vibrations are published in I. Hargittai (ed.), *Quasicrystals, Networks, and Molecules of Fivefold Symmetry*, VCH Publishers, New York, 1990, chap. 14-18.

A neutron powder diffraction study of ordered C_{60} at 5K [space group $Pa\bar{3}$, $a = 14.0408(1)$Å, $Z = 4$] showed that each cluster has six electron-rich inter-pentagon 6:6 bonds (1.391Å) and six electron-poor pentagons (6:5 bonds of length 1.466Å as edges) facing its twelve C_{60} neighbours; the "non-facing" pentagons have edges of 1.444Å. See W.I.F David, R.M. Ibberson, J.C. Matthewman, K. Prassides, T.J.S. Dennis, J.P. Hare, H.W. Kroto, R. Taylor and D.R.M. Walton, *Nature (London)* **353**, 147 (1991).

An electron diffraction study of C_{60} yielded r_g values of 1.401(10) and 1.458(6)Å for the 6:6 and 6:5 bonds, respectively. See K. Hedberg, L. Hedberg, D.S. Bethune, C.A. Brown, H.C. Dorn, R.D. Johnson and M. de Vries, *Science (Washington)* **254**, 410 (1991).

Crystalline cyclohexane solvates of C_{60} and mixed C_{60}/C_{70} are reported in S.M. Gorun, K.M. Creegan, R.D. Sherwood, D.M. Cox, V.W. day, C.S. Day, R.M. Upton and C.E. Briant, *J. Chem. Soc., Chem. Commun.*, 1556 (1991).

Analytical rules have been formulated for the derivation of three homologous series of fullerenes with closed-shell electronic structures:

$$C_{60+6k}; \quad k = 0, 2, 3, 4, \ldots\ldots$$
$$C_{70+30k}; \quad k = 0, 1, 2, 3, 4, \ldots\ldots$$
$$C_{84+36k}; \quad k = 0, 1, 2, 3, 4, \ldots\ldots$$

See P.W. Fowler and J.I. Steer, *J. Chem. Soc., Chem. Commun.*, 1403 (1987); P.W. Fowler, J.E. Cremona and J.I. Steer, *Theor. Chim. Acta* **73**, 1 (1988).; P.W. Fowler, *J. Chem. Soc., Faraday Trans.* **86**, 2073 (1990). Proposed structures for the higher fullerenes (C_{76}, C_{78} and C_{84}), fullerides, and fullerenium ions are discussed in P.W. Fowler, *J. Chem. Soc., Faraday Trans.* **87**, 1945 (1991); D.E. Manolopoulos, *ibid.*, 2861; P.W. Fowler, R.C. Batten and D.E. Manolopoulos, *ibid.*, 3103 ; P.W. Fowler and D.E. Manolopoulos, *Nature (London)* **355**, 428 (1992).

Isolation and characterization of the higher fullerenes C_{76}, C_{84}, C_{90} and C_{94} are announced in F. Diederich, R. Ettl, Y. Rubin, R.L. Whetten, R. Beck, M. Alvarez, S. Anz, D. Sensharma, F. Wudl, K.C. Khemani and A. Koch, *Science (Washington)* **252**, 548 (1991).

The chiral nature of C_{76} (D_2 point group) is established in R. Ettl, I. Chao, F. Diederich and R.L. Whetten, *Nature (London)* **353**, 149 (1991).

Isolation of the major C_{2v} and minor chiral D_3 isomers of C_{78} are reported in F. Diederich, R.L. Whetten, C. Thilgen, R. Ettl, I. Chao and M.M. Alvarez, *Science (Washington)* **254**, 1768 (1991).

The geometric and electronic structures of hypothetical $C_{60}H_{60}$, $C_{60}F_{60}$ and $C_{60}H_{36}$ (T_h symmetry) are discussed in B.I. Dunlap, D.W. Brenner, J.W. Mintmire, R.C. Mowery and C.T. White, *J. Phys. Chem.* **95**, 5763 (1991).

The hexa-substitued derivative, $[(Et_3P)_2Pt]_6C_{60}$, has crystallogrphic $\bar{1}$ and nearly ideal T_h molecular symmetry with an octahedral array of metal atoms each bound to the C_{60} unit in a dihapto manner. See P.J. Fagan, J.C. Calabrese and B. Malone, *J. Am. Chem. Soc.* **113**, 9408 (1991).

The latest advances in fullerene research are covered in G.S. Hammond and V.J. Kuck (eds.), *Fullerenes: Synthesis, Properties, and Chemistry of Large Carbon Clusters*, American Chemistry Society, Washington, DC, 1992.

The energetics of two hypothetical carbon networks of *negative* curvature (named "schwarzites" in honour of the mathematician H.A. Schwarz) derived from the presence of *seven-membered* rings are discussed in T. Lenosky, X. Gonze, M. Teter and V. Elser, *Nature (London)* **355**, 333 (1992).

Additional references are listed in p. 182.

2.6 Iodine

I_2

Crystal Data

Orthorhombic, space group *Cmca* (No. 64)

$a = 7.136(10)$, $b = 4.686(7)$, $c = 9.784(15)$Å (110K), $Z = 4$

Atom	Position	x	y	z
I	8(f)	0	.15434	.11741

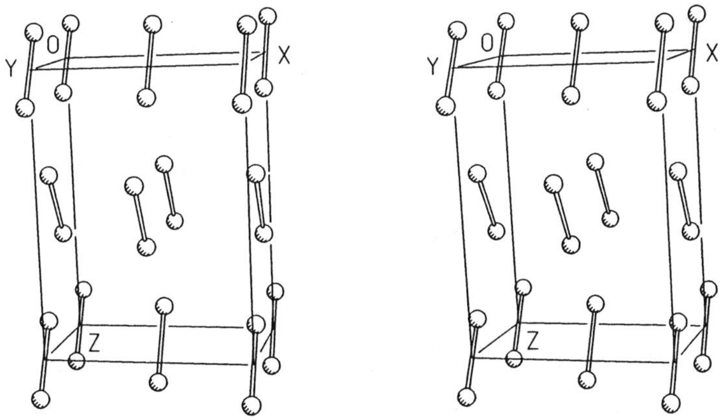

Fig. 2.6-1 Stereoview of the crystal structure of iodine.

Crystal Structure

The crystal structure of iodine was first determined in 1928,[1] and was later refined in 1953[2] and 1967.[3] Solid iodine consists of diatomic I_2 molecules, which are arranged as shown in Fig.2.6-1. The molecules lie essentially in layers, which run parallel to the *bc* plane. Fig. 2.6-2 shows the structure projected along the *a*-axis.

The low-temperature data yielded the following bond lengths:

intramolecular	I(1)-I(2)	2.715(6)Å
intermolecular		
intralayer	I(1)-I(3)	3.496(6)Å
	I(1)-I(2')	3.972(7)Å
interlayer	I(1)-I(5)	4.269(6)Å
	I(1)-I(6)	4.337(7)Å
	I(1)-I(7)	4.412(7)Å

In the solid state, the intramolecular I-I bond length 2.715Å is significantly longer than that in the gaseous molecule, 2.662Å.[4] An elongation of the bond may be expected on theoretical grounds from the short intermolecular contacts between iodine molecules in the same layer.

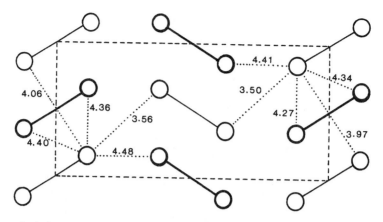

Fig. 2.6-2 Crystal structure of iodine projected along the *a* axis. Molecules at $x = 0$ (light lines) and $x = 1/2$ (heavy lines) are shown. Intermolecular distances at room temperature are given on the left, and those at -163°C on the right. (Adapted from ref. 10).

Remarks

1. Layer structure and properties

     Fig. 2.6-3 shows that owing to the very short intermolecular contacts (shorter than the expected van der Waals separation of 4.30Å by ~0.8Å), a two dimensional network is formed of nearly linear chains $(.I...I-I...I.)_\infty$, each

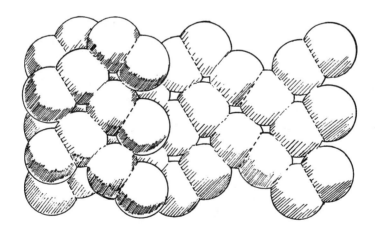

Fig. 2.6-3 Packing of space-filling models showing the strong intermolecular interaction between I_2 molecules in each layer.

iodine atom being involved in two nearly perpendicular chains. The resulting checkerboard packing arrangement suggests that the bonds in intersecting chains are formed by orthogonal iodine 5p orbitals. For each of the chains a description of the bonding as 4-center, 6 electron bonding may be assumed,

resulting in a bond order smaller than unity for the I-I bond,[5] in agreement
with the observed elongation from the structural data. The remaining short
intermolecular distances of 3.97Å are expected to have relatively little
influence on the I-I bond length.

Judging from the crystal structure, one would expect that within each
layer some delocalization of electrons takes place, and that the electrical
mobility transverse to the layers should be much smaller than that parallel to
the bc plane. Solid iodine is thus an anisotropic material. The specific
conductivity, k, of single crystals of iodine at room temperature are:[6]

5×10^{-12} ohm^{-1} cm^{-1}, normal to the bc plane, and

1.7×10^{-8} ohm^{-1} cm^{-1}, parallel to the bc plane.

2. Crystal structure of the halogens[7]

Halogens have high volatility and relatively low enthalpy of
vaporization, which reflect their diatomic molecular structure. In the solid
state the molecules align to give a layer structure: F_2 has two modifications:
a low temperature α form and a higher-temperature β form, neither of which
resembles the orthorhombic layer structure of the isostructural crystalline
Cl_2, Br_2, and I_2. Table 2.6-1 gives the interatomic distances in crystalline
halogens, from which two further features of interest emerge: (a) the
intralayer intermolecular distances Cl...Cl and Br...Br are almost identical,
and (b) the ratio of the intra-layer X...X distance to the X-X bond distance
decreases with an increase in atomic number. Fluorine is not directly
comparable because it has a different stucture.

Table 2.6-1 Interatomic distances in crystalline halogens (Å).

X	X-X	X...X		Ratio $\dfrac{X...X}{X-X}$
		within layer	between layers	
F	1.49	3.24	2.84	(1.91)
Cl	1.98	3.32	3.74	1.68
Br	2.27	3.31	3.99	1.46
I	2.72	3.50	4.27	1.29

3. Polyiodide anions[7-9]

The additional weak bonding between molecules in the crystalline iodine
is very evident also in the structures of polyiodides. It invites comparison
with the "charge-transfer" bond in the numerous molecular species formed by
iodine with iodides, in which this I...I bond is collinear with the I-I
molecule and much shorter than twice the van der Waals radius of iodine. The
I_3^- ion occurs in many crystals. In $KI_3 \cdot H_2O$, I_3^- is symmetrical with a bond
length of 2.93Å. In NH_4I_3, I_3^- is an unsymmetrical ion with bond lengths 2.79

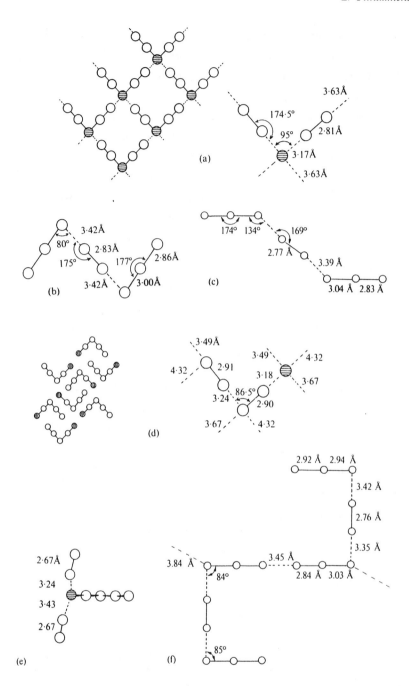

Fig. 2.6-4 Structures of some higher polyiodide anions: (a) I_5^- in $[N(CH_3)_4]I_5$; (b)I_8^{2-} in Cs_2I_8; (c) I_8^{2-} in $[(CH_2)_6N_4CH_3]_2I_8$; (d) the I_5^- layer in $[N(CH_3)_4]I_9$; (e) the environment of I^- (shaded circle) in $[N(CH_3)_4]I_9$; (f) I_{16}^{4-} in (theobromine)$_2$H$_2$I$_8$. (After ref. 11).

and 3.11Å, indicating the formulation $(I-I...I)^-$. Crystalline $[Cu(NH_3)_4]I_4$
contains linear I_4^{2-} ions; the bond lengths are 3.34, 2.80 and 3.34Å, and the
next shortest I...I distance is 4.65Å, indicating the formulation $(I...I-$
$I...I)^{2-}$. The structures of some higher polyiodide anions are shown in Fig.
2.6-4.

4. Polyiodide cations[12]

The structures of $I_2Sb_2F_{11}$,[13] I_3AsF_6,[14] $I_4(AsF_6)_2$,[15] I_5AsF_6,[16]
and $I_{15}AsF_6$[17] have been determined by X-ray crystallography. In these
compounds, homopolyatomic cations of iodine occur. $I_2Sb_2F_{11}$ contains a
chemically isolated paramagnetic I_2^+ cation, in which the I-I bond length of
2.557(4)Å is shorter than that in I_2 [2.715(6)Å] and has bond order 1.5.
$I_4(AsF_6)_2$ contains an I_4^{2+} cation, the dimer of I_2^+, which has the shape of a
planar rectangle with I-I bond lengths of 2.587(3) and 3.256(3)Å. The
structures of other high polyiodide cations are shown in Fig. 2.6-5.

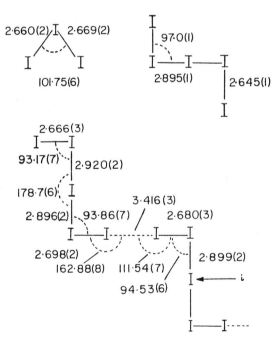

Fig. 2.6-5 Structures of polyiodide cations I_3^+, I_5^+, and I_{15}^+. (After
ref. 12).

5. The intercalation compound $(Te_2)_2(I_2)_x$

Reinvestigation of the metastable subiodide of tellurium, formulated as
Te_3I, revealed a variable composition ranging from Te_2I to $Te_2I_{0.42}$. X-Ray
analysis of the shiny, black-metallic, air-stable iodine-rich phase showed
that it is an intercalate with the idealized structural formula

$(Te_2)_2(I_2)$.[18] As shown in Fig. 2.6-6, the crystal structure [space group
Cmmm, $a = 4.464(4)$, $b = 4.924(5)$, $c = 9.037(16)$Å, and $Z = 1$] contains planar
double layers of Te_2 units arranged parallel to the c axis. Each Te atom thus
has a distorted tetragonal-pyramidal coordination with one short and four
longer Te-Te bonds, of lengths 2.713(7) and 2.835(2)Å, respectively.

The I_2 molecules form planar layers located in between the tellurium
double layers; the iodine atoms are two-fold disordered, corresponding to
alternating orientation of the I_2 molecules in successive layers. Within each
iodine layer the arrangement of the I_2 molecules matches that in the (100)
plane of elemental iodine (Fig. 2.6-2); in the present structure, however, the
I-I bond length of 2.866(12)Å is longer and the intermolecular I...I contacts,
3.324(14) and 3.841(16)Å, are shorter.

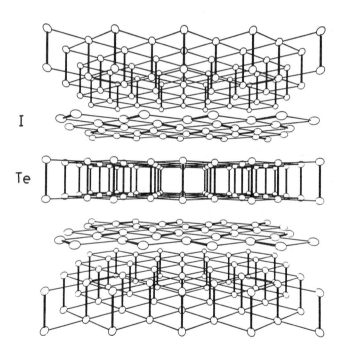

Fig. 2.6-6 Perspective representation of the layer sequence [001] in
$(Te_2)_2(I_2)$. The Te_2 pairs and I_2 molecules are accentuated by bolder
lines. (After ref. 18).

The crystal used in the structure analysis showed an occupation of the
iodine layers of only 88%, which is attributable to both vacant sites within
the iodine layers and to partial substitution of I_2 by HI in the hydrothermal
synthesis of the intercalate.

References

[1] P.M. Harris, E. Mack and F.C. Black, *J. Am. Chem. Soc.* **50**, 1583 (1928).

[2] A.I. Kitaigorodskii, T.L. Khotsynova and Yu. T. Struchkov, *Zh. Fiz. Khim.* **27**, 780 (1953).

[3] F. van Bolhuis, P.B. Koster and T. Migchelsen, *Acta Crystallogr.* **23**, 90 (1967).

[4] I.L. Karle, *J. Chem. Phys.* **23**, 1739 (1955).

[5] E.E. Havinga and E.H. Wiebenga, *Recl. Trav. Chim. Pays-bas* **78**, 724 (1959).

[6] N. Nakamura and H. Chihara, *J. Phys. Soc. Japan* **22**, 201 (1967).

[7] P.K. Hon, T.C.W. Mak and J. Trotter, *Inorg. Chem.* **18**, 2916 (1979).

[8] K.F. Tebbe, in A.L. Rheingold (ed.), *Homoatomic Rigs, chains and Macromolecules of Main Group Elements*, Elsevier, New York, 1977.

[9] N.N. Greenwood and A. Earnshaw, *Chemistry of the Elements*, Pergamon Press, Oxford, 1986.

[10] J. Donohue, *The Structures of the Elements*, Wiley, New York, 1974.

[11] A.F. Wells, *Structural Inorganic Chemistry*, 5th ed., Clarendon Press, Oxford, 1984.

[12] N. Burford, J. Passmore and J.C.P. Sanders in J.F. Liebman and A. Greenberg (eds.), *From Atoms to Polymers: Isoelectronic Analogies*, VCH Publishers, New York, 1989, p. 53.

[13] C.G. Davies, R.J. Gillespie, P.R. Ireland and J.M. Sowa, *Can. J. Chem.* **52**, 2048 (1974).

[14] J. Passmore, G. Sutherland and P.S. White, *Inorg. Chem.* **20**, 2169 (1981).

[15] R.J. Gillespie, R. Kapoor, R. Faggiani, C.J.L. Lock, M. Murchie and J. Passmore, *J. Chem. Soc., Chem. Commun.*, 8 (1983).

[16] A. Apblett, F. Grein, J.P. Johnson, J. Passmore and P.S. White, *Inorg. Chem.* **25**, 422 (1986).

[17] J. Passmore, P. Taylor, T.K. Whidden and P.S. White, *Can. J. Chem.* **57**, 968 (1979).

[18] R. Kniep and H.-J. Beister, *Angew. Chem. Int. Ed. Engl.* **24**, 393 (1985).

Note The cation in $Br_5^+MF_6^-$ (M = As,Sb) has a planar Z-shape, as reported in H. Hartl, J. Nowicki and R. Minkwitz, *ibid.* **30**, 328 (1991).

2.7   Sodium Chloride (Rock salt, Common salt, Halite)
NaCl

Crystal Data

Cubic, space group $Fm3m$ (No. 225)

$a = 5.64006(6)$Å (20°C), $Z = 4$

Atom	Position	x	y	z
Na	4(a)	0	0	0
Cl	4(b)	1/2	1/2	1/2

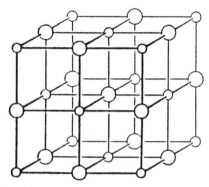

Fig. 2.7-1   Unit cell of NaCl.

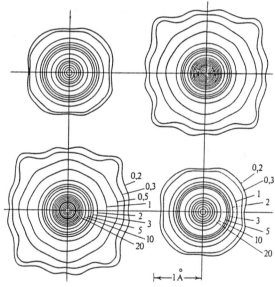

Fig. 2.7-2   Electron-density section in NaCl through the plane $xy0$.
Numerals indicate $e\text{Å}^{-3}$ for the contours.   (After ref. 2).

Crystal Structure

The crystal structure of NaCl can be described as a fcc lattice of Na^+
ions with Cl^- ions occupying all the octahedral holes, or *vice versa* (Fig.

2.7-1).[1] The crystal is hard and has a high melting point of 801°C. The
ions cannot easily move in the crystal lattice, and accordingly the ionic
solid is an electrical insulator. When NaCl is dissolved in water or is
molten, the ions become mobile and the aqueous solution or the melt becomes
conducting.

     Fig. 2.7-2 shows the electron-density distribution $\rho(xyz)$ in crystalline
NaCl.[2] The electron density around the atomic nuclei is spherical, falling
to zero in the region in between. The number of electrons associated with
each atomic nucleus, estimated by integration of $\rho(xyz)$ over a reasonable
volume centered about it, is 10.1 for Na and 17.8 for Cl. This provides
direct experimental evidence that NaCl is composed of Na^+ and Cl^- ions, which
may be considered as hard spheres that are very slightly polarized by
neighbouring ions of opposite charge.

Remarks

1. Compounds with the NaCl structure
     The rock-salt structure is the most common one for MX compounds; more
than half of the over 400 compounds so far investigated adopt this structure.
It occurs principally among the alkali metal halides and hydrides, the
alkaline-earth oxides and chalcogenides, the oxides of divalent early first-
row transition metals, and the chalcogenides of divalent lanthanides and
actinides. It is also found among less ionic compounds such as the nitrides,
phosphides, arsenides, and bismuthides of the lanthanide and actinide
elements, the silver halides (except the iodide), and the tin and lead
chalcogenides. Many interstitial alloys of appropriate composition such as
carbides and nitrides also adopt the NaCl structure. Some examples are listed
in Table 2.7-1.

     A recent deformation density study has shown that the electron transfer
from cation to anion is 0.54(1) for NaF and 0.13(1) for LiF.[3]

2. Historical significance
     The structure of NaCl had been predicted by Barlow as early as 1883.[4]
In 1913, W.L. Bragg[1] first determined the crystal structure and showed that
the interionic distance is 2.8Å. He used the X-ray spectrometer designed by
his father, W.H. Bragg, to measure the intensities of the various diffracted
beams, and derived the symmetry of the diffraction pattern from the systematic
absences. W.L. Bragg considered diffraction as reflection from sets of
parallel planes of equal spacing d in the crystal. For a beam diffracted at
an angle 2θ from the incident beam, he deduced that

$$2d\sin\theta = n\lambda,$$

Table 2.7-1   Compounds with the rock-salt structure.

MX	M
MH	Li, Na, K, Rb, Cs
MF	Li, Na, K, Rb, Cs, Ag
MCl	Li, Na, K, Rb, Cs, Ag
MBr	Li, Na, K, Rb, Ag
MI	Li, Na, K, Rb
MO	Mg, Ca, Sr, Ba, Ti, Zr, Hf, V, Nb, Ta
	Mn, Fe, Co, Ni, Pa, Pu, Pd, Cd, Eu, Am
MS	Mg, Ca, Sr, Ba, Mn, Pb, Ce, Sm, Eu, Th, U, Pu
MSe	Mg, Ca, Sr, Ba, Mn, Pb, Sn, Th, U
MTe	Ca, Sr, Ba, Pb, Sn, U, Pu
MC	Ti, Zr, Hf, V, Nb, Ta, Th
MN	Sc, Y, Ti, Zr, V, Nb, Cr, Np, Pu, Th, U
others	InP, InAs, SnP, SnAs, ThAs, ThSb, UP, UAs, USb, UBi

where n represents the order of diffraction, and λ the X-ray wavelength.   This
is the famous Bragg's Law.   In 1915, W.H. Bragg and W.L. Bragg shared the
Nobel Prize in Physics for their pioneering researches on the determination of
crystal structures by X-ray diffraction.   The formula weight FW (in atomic-
weight units) of a crystalline compound, the measured density D_x (in g cm^{-3}),
the unit-cell volume V (in $\mathring{A}^3$), the number of formula units per unit cell Z,
and Avogadro's number N_o are interrelated by the formula

$$D_x = (FW \times Z)/(N_o \times V \times 10^{-24}) = (FW \times Z)/(0.6022 \times V),$$

which was used in providing an early but accurate estimation of Avogadro's
number.

Using a single crystal as a three-dimensional diffraction grating for X-
rays, H.G.I. Moseley in the years 1913-1914 measured the wavelengths of the
characteristic K and L lines in the emission spectra of a range of metallic
elements.   The resulting Moseley's Law, $\nu^2 \alpha (Z-\sigma)^2$ where Z stands for the
atomic number of the element, placed Mendeleev's Periodic Table on a sound
theoretical footing and played a central role in the evolution of modern
atomic physics.[5]

3. Incorporation of organic molecules

Simple ionic solids such as NaCl continue to be objects of considerable
interest in current research.   For example, when crystalline NaCl is heated in
an atmosphere of sodium vapour, it develops "color centers" as a result of
diffusion of a small quantity of Na atoms into the ionic crystal.[6]

Michl and co-workers have prepared stable materials containing 0.1-1% of

a volatile nonionic guest "salted" in an alkali halide matrix by condensing
organic vapors with excess alkali halide vapor on a 77°K surface.[7] Volatile
compounds (such as benzene, anthracene and phenols) have been trapped inside
the microscopic cavities of inorganic salts (such as NaCl) at low temperature
to yield new materials that are stable above room temperature. The guest
molecules that are aggregated within microcavities can react with one another
under ultraviolet irradiation. This new development opens up the possibility
of trapping individual covalent molecules (such as iron pentacarbonyl,
guadricyclane and norbornadiene) inside an ionic crystal.[8]

Cesium fluoride absorbs bromine or iodine to form 1:1 and 1:2 phases.
X-Ray analysis of CsF.Br$_2$ has shown that it is an intercalation compound with
Br$_2$ guests stacked between ionic lattice planes (Fig. 2.7-3).[9] The Cs$^+$...F$^-$
distance of 2.940Å is significantly reduced from that of 3.001Å in NaCl-type
CsF, but virtually identitical to that of 2.936Å in the high-pressure, CsCl-
type CsF modification (48 kbar).

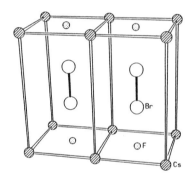

Fig. 2.7-3 Two adjacent unit cells in the crystal structure of CsF.Br$_2$
(P4/mmm, a = 4.177(2), c = 7.364(2)Å, Z = 1, z parameter of Br = .3422).
(Adapted from ref. 9).

4. Clusters containing "ionic crystal fragments"

Phenyllithium reacts with silver bromide in ether solution to form the
salt [Li$_6$Br$_4$(Et$_2$O)$_{10}$]$^{2+}$[Ag$_3$Li$_2$Ph$_6$$^-$]$_2$.[10] Fig. 2.7-4 shows two views of the
structure of the centrosymmetric cation with the ether ligands (two each on
Li3 and Li4; one on Li5) omitted for clarity. The Li$_6$Br$_4$ skeleton of the
cation bears a striking resemblance to the corresponding ten-atom fragment of
crystalline LiBr which has the rock-salt structure. The average Li-Br
distance of 2.58(2)Å in this ion is comparable to 2.75Å in the LiBr crystal.
Other related "salt clusters" which have been structurally characterized
include the [Li$_4$Cl$_2$(Et$_2$O)$_{10}$]$^{2+}$ cation consisting of a planar Li$_2$Cl$_2$ core with
a terminal Li attached to each Cl,[11] and neutral [Li$_4$Cl$_4$(HMPA)$_4$], where HMPA
= (Me$_2$N)$_3$PO, which possesses a cubane-like Li$_4$Cl$_4$ core.[12] It is reasonable

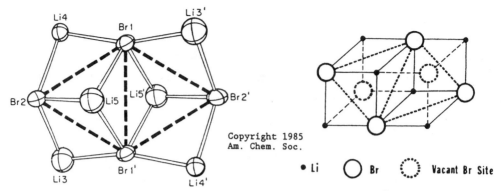

Fig. 2.7-4 Molecular skeleton of the $[Li_6Br_4(Et_2O)_{10}]^{2+}$ cation (left) as a fragment of crystalline LiBr (right). (After ref. 10).

to speculate that these salt clusters, especially the ether-solvated ones, may represent the early stages of nucleation and growth of a lithium halide crystal as it crystallizes from ether.

References

[1] W.L. Bragg, *Proc. Roy. Soc. London* **A89**, 248 (1913); M.F. Perutz, *Acta Crystallogr., Sect. A* **46**, 633 (1990). [Anecdotic account of the scientific life of Bragg given in celebration of the centenary of his birth.]

[2] H. Witte and E. Wölfel, *Z. Phys. Chem. (Frankfurt)* **3**, 296 (1955).

[3] E.N. Maslen, *Acta Crystallogr., Sect. A* **43**, Supplement, C102 (1987).

[4] W. Barlow, *Nature (London)* **29**, 186, 205, 404 (1883, 1884).

[5] F.K. Richtmeyer, E.H. Kennard and T. Lauritsen, *Introduction to Modern Physics*, 5th ed., McGraw-Hill, New York, 1955. Chapter 8 contains a concise and lucid discussion of X-rays.

[6] N.N. Greenwood, *Ionic Crystals, Lattice Defects, and Nonstoichiometry*, Butterworths, London, 1968, p. 170.

[7] E. Kirkor, J. Gebicki, D.R. Philips, and J. Michl, *J. Am. Chem. Soc.* **108**, 7106 (1986).

[8] E.S. Kirkor, D.E. David and J. Michl, *J. Am. Chem. Soc.* **112**, 139 (1990); E.S. Kirkor, V.M. Maloney and J. Michl, *J. Am. Chem. Soc.* **112**, 148 (1990).

[9] D.D. DesMarteau, T. Grelbig, S.-H. Hwang and K. Seppelt, *Angew. Chem. Int. Ed. Engl.* **29**, 1448 (1990).

[10] M.Y. Chang, E. Böhlen and R. Bau, *J. Am. Chem. Soc.* **107**, 1679 (1985).

[11] H. Hope, D. Oram and P.P. Power, *J. Am. Chem. Soc.* **106**, 1149 (1984).

[12] D. Barr, W. Clegg, R.E. Mulvey and R. Snaith, *J. Chem. Soc., Chem. Commun.*, 79 (1984).

2.8   Cesium Chloride
CsCl

Crystal Data

Cubic, space group $P\bar{4}3m$ (No. 215)

$a = 4.121(3)$Å, $Z = 1$

Atom	Position	x	y	z
Cs	1(a)	0	0	0
Cl	1(b)	1/2	1/2	1/2

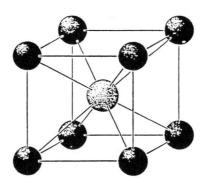

Fig. 2.8-1   Unit cell of CsCl.

Crystal Structure

The CsCl structural type for AB compounds had been suggested by Barlow[1] prior to its actual determination by Davey and Wick in 1921.[2]

The unit cell of CsCl is shown in Fig. 2.8-1. It is primitive cubic, with Cl$^-$ ions at the corners and a Cs$^+$ ion at the body center, or vice versa. The coordination number of either ion is eight, and the interionic separation is $(\sqrt{3}/2)a$.

Remarks

1. Compounds with the CsCl structure

Two principal kinds of compounds crystallize with the CsCl structure. In one group are halides of the largest univalent ions, and in the other are intermetallic compounds.[3] Some examples are listed in Table 2.8-1.

2. Prevalence of the CsCl structure among intermetallics

Why is the CsCl structure so widely adopted by intermetallic compounds, but is rare for ionic ones? A simple rationalization proceeds as follows. If atoms or ions are treated as hard spheres with definite radii r_A and r_B, the

Table 2.8-1   Compounds with the CsCl structure.

Compounds	
Halides	M
MCl	Cs, Rb, Tl, NH_4^+
MBr	Cs, Tl, NH_4^+
MI	Cs, Tl
Intermetallics	M
MAg	Cd, Ce, La, Li, Mg, Nd, Y, Yb, Zn
MAl	Co, Fe, Nd, Ni, Pd, Sc
MAu	Cd, Mg, Mn, Yb, Zn
MBa	Cd, Hg
MBe	Co, Cu, Ni, Pd
MCa	Ti
MCd	Ce, Eu, La, Pr, Sr
MCu	Eu, Pd, Y, Zn
MGa	Ni, Rh
MHg	Li, Mg, Mn, Nd, Pr, Sr, Ce
MIn	La, Pd, Pr, Tm, Yb
MMg	Pr, Sc, Sr, Ti, Ce
MZn	Pr, La, En, Ce
Others	X
CsX	CN, NH_2, SH, SeH

fractional volume occupied by them in a crystal is

$$\phi = 4\pi R n_i r_i^3 / 3V,$$

where n_i is the number of atoms of type i in a unit cell of volume V. The
curves of ϕ vs r_A/r_B for the CsCl, NaCl and ZnS (both wurtzite and zinc
blende) structures are shown in Fig. 2.8-2 with proper scaling.

It is seen that in the r_A/r_B range of 0.59-1.70, the CsCl structure
makes more efficient use of space than the NaCl and ZnS structures. For most
intermetallic compounds of composition AB, the r_A/r_B values lie in this range,
so the CsCl structure might be expected to be favoured if space filling were
the predominant criterion.

For ionic compounds, the inflections at 0.225(ZnS), 0.414(NaCl) and
0.732(CsCl) reflect the known critical radius ratios below which each
structure becomes unstable owing to anion-anion (B-B) contact. In ionic
crystals of composition AB, the r_A/r_B ratios generally fall within the range
0.225-0.732, and very seldomly exceed 0.732. Furthermore, some covalent
interaction is always present even in the most ionic materials, and the

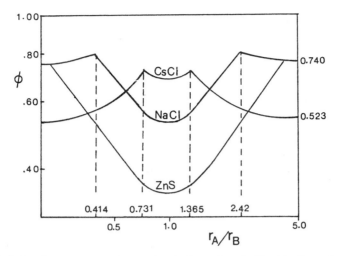

Fig. 2.8-2 ϕ versus r_A/r_B curves for the CsCl, NaCl and ZnS structures.[4] (After ref. 5).

tetrahedral (ZnS type) and octahedral (NaCl type) coordination environments are better suited to the requirement of bond directionality. These factors combine to make the CsCl structure rare among ionic compounds.

References

[1] W. Balow, *Nature (London)* **29**, 186, 205, 404 (1883,1884).

[2] W.P. Davey and F.G. Wick, *Phys. Rev.* **17**, 403 (1921).

[3] W.H. Pearson, *A Handbook of Lattice Spaces and Structures of Metals and Alloys*, Pergamon Press, Oxford, 1958.

[4] E. Parthé, *Z. Kristallogr.* **115**, 52 (1961).

[5] D.M. Adams, *Inorganic Solids*, Wiley, New York, 1974.

2.9   Zinc Blende (Cubic Zinc Sulfide, Sphalerite)

ZnS

Crystal Data

Cubic, space group $F\bar{4}3m$ (No. 216)

$a = 5.4093(4)$Å (26°C), $Z = 4$

Atom	Position	x	y	z
Zn	4(a)	0	0	0
S	4(c)	1/4	1/4	1/4

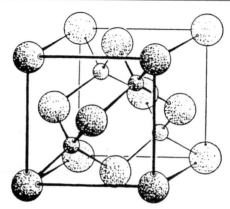

Fig. 2.9-1 Unit cell of cubic zinc sulfide.

Crystal Structure

In the zinc blende structure, each atom is symmetrically surrounded by four atoms of the other sort disposed at the corners of a regular tetrahedron (Fig. 2.9-1). If all the atoms were alike, the structure would be the same as that of diamond. When the structure of diamond was determined by W.H. Bragg and W.L. Bragg,[1] they indicated that zinc blende must have a very similar structure.

In the zinc blende structure, the sulfur atoms are arranged in cubic closest packing (fcc unit cell), in which one-half of the tetrahedral sites are filled with zinc atoms; the arrangement of the filled sites is such that the coordination numbers of sulfur and zinc are both four. The structure can also be considered as being composed of two interpenetrating fcc sublattices of Zn^{2+} and S^{2-} ions displaced by the vector $(a + b + c)/4$.

Remarks

1. Binary compounds with the zinc blende structure

Compounds with the zinc blende and other tetrahedral structures are commonly formed when one element belongs to the Nth and the other to the

(8-N)th B sub-groups. In these structures each atom is bound to four neighbours by covalent bonds and it is mandatory that the number of valence electrons available for the formation of these bonds should average four per atom. It is not necessary that these be contributed equally by the two types of atoms, and the condition is therefore satisfied if these atoms are related in the manner indicated. Some binary compounds which adopt the zinc blende structure are listed in Table 2.9-1.

Table 2.9-1   Some binary compounds with the zinc blende structure. The *a* values are in Å.[2,3]

CuF	4.255	BeSe	5.139	CdTe	6.481	AlAs	5.656
CuCl	5.4057	BeTe	5.626	HgS	5.8517	AlSb	6.1355
γ-CuBr	5.6905	ZnS	5.4093	HgSe	6.085	GaP	5.450
γ-CuI	6.051	ZnSe	5.667	HgTe	6.453	GaAs	5.6534
γ-AgI	6.495	ZnTe	6.1026	BN	3.615	GaSb	6.096
β-MnS	5.600	b-SiC	4.358	BP	4.538	InP	5.869
β-MnSe	5.83	b-CdS	5.818	BAs	4.777	InAs	6.058
BeS	4.8624	CdSe	6.077	AlP	5.451	InSb	6.4788

2. Multicomponent compounds with the zinc blende structure[4]

Many ternary, quaternary, and multicomponent compounds also exhibit tetrahedral coordination of atoms and possess structures similar to those indicated in Fig. 2.9-2. Their formulae can be obtained from those of binary compounds by horizontal and diagonal substitution in the Periodic Table, such that the condition (number of valence electrons/number of atoms) = 4 is satisfied. For instance, replacing In in indium arsenide by Cd and Sn gives $CdSnAs_2$, and replacing Cd in cadmium selenide by Ag and In, $AgInSe_2$. Fig. 2.9-2 shows the structural relationship of chalcopyrite, $CuFeS_2$(b) and stannite, Cu_2FeSnS_4(c) to zinc blende, ZnS(a).

3. The cubic form of BN

Borazone, a cubic form of boron nitride converted from the hexagonal form at 1800°C and 85 kbar in the presence of an alkali or alkaline-earth catalyst, is an air-stable (though slowly hydrolyzed by water) and extremely hard material which scratches even diamond. Though often described as "diamond-like" by analogy of the isoelectronic C-C and B-N units, cubic BN is actually isostructural with zinc blende. Electron density distribution studies using accurate X-ray diffraction data have revealed the bonding features in cubic BN. Fig. 2.9-3 compares the dynamic deformation density map of BN (multiple refinements employing rigid pseudoatoms) in the (110) plane with the corresponding map of diamond (conventional high-order structure

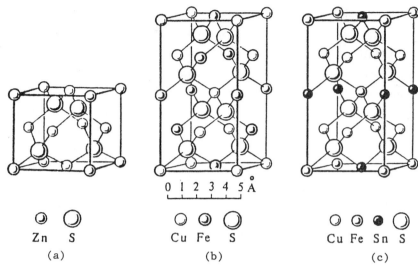

Zn S   Cu Fe S   Cu Fe Sn S
(a)   (b)   (c)

Fig. 2.9-2 The structure of (a) zinc blende, ZnS; (b) chalcopyrite,
CuFeS$_2$; (c) stannite, Cu$_2$FeSnS$_4$. (After ref. 4).

refinement).[5] The striking difference is the lack of a well-developed bond
peak in BN, and considerable excess charge is accumulated at the N atom.
Charge integrations about the atomic sites led to a formal transfer of 0.44 e
from B to N, in agreement with their electronegativity difference of about 1.

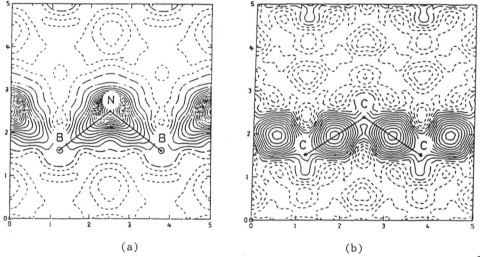

(a)   (b)

Fig. 2.9-3 Dynamic deformation density in the (110) plane of (a) BN (X-X^{mult})
and (b) diamond (X-X^{HO}). Contours are at 0.05 eÅ^{-3}. (After ref. 5).

4. Anomalous dispersion and polarity sense of a zinc blende crystal

        "Anomalous dispersion" is an effect on the scattering of X-rays at
wavelengths near those for which the radiation is strongly absorbed by an atom
in exciting its inner electrons to higher energy levels. The phase change due

to one or more strongly absorbing atoms, or "anomalous scatters", is in general different from that caused by the other atoms in the structure. In normal circumstances "anomalous scattering" is negligible, and the intensities $I(hkl)$ and $I(\bar{h}\bar{k}\bar{l})$ of reflections (hkl) and $(\bar{h}\bar{k}\bar{l})$ are the same regardless of the crystal class, which is the essence of Friedel's Law.[6] When anomalous scattering becomes considerable, Friedel's Law no longer holds and $I(hkl) = I(\bar{h}\bar{k}\bar{l})$.

The zinc blends structure may be considered as being composed of an alternating sequence of unequally-spaced, polar double layers perpendicular to the [111] direction (Fig. 2.9-4).[7] The polarity sense of a single crystal of zinc blende was determined by Coster, Knol and Prins, who first demonstrated the breakdown of Friedel's Law under conditions of anomalous scattering.[8] They used Au$L\alpha$ radiation consisting of a doublet, Au$L\alpha_1$ ($\lambda = 1.276$Å) and Au$L\alpha_2$ ($\lambda = 1.288$Å), which brackets the K absorption edge of Zn ($\lambda_K = 1.283$Å). The two opposite {111} faces of a ZnS crystal are often visually distinguishable. One face tends to be well developed with a brilliant, shiny appearance; the other is poorly developed and dull. For the pair of the first order reflections from these opposite faces, the intensity reflected from the dull face was less than that reflected from the shiny face under anomalous scattering conditions. For normal scattering the intensities were roughly equal.

Anomalous scattering introduces a phase change into a given atomic scattering factor, which becomes a complex quantity:

$$f = f_o + \Delta f' + i\Delta f'' = f' + i\Delta f''.$$

The real correction term $\Delta f'$ to the normal scattering factor f_o is usually negative. The imaginary component $\Delta f''$ is $\pi/2$ ahead of the phase of the real part $(f_o + \Delta f')$ in the complex plane, and this advance in phase is independent of the incident beam direction and of the atomic positions in the structure. For the layer arrangement shown in Fig. 2.9-4, the equations

$$F(A) = f_S + i(f'_{Zn} + i\Delta f''_{Zn}) = f_S - \Delta f''_{Zn} + if'_{Zn}$$

$$F(B) = f_S - i(f'_{Zn} + i\Delta f''_{Zn}) = f_S + \Delta f''_{Zn} - if'_{Zn}$$

should hold for the first-order reflection, leading to $F^2(A) < F^2(B)$ if f''_{Zn} is positive. For the third-order reflection, the relation is reversed. Hence A (reflection from the side that terminates in a layer of Zn atoms) must be identified with reflection from the dull face, B with reflection from the shiny face, and the polarity sense of the crystal is established.

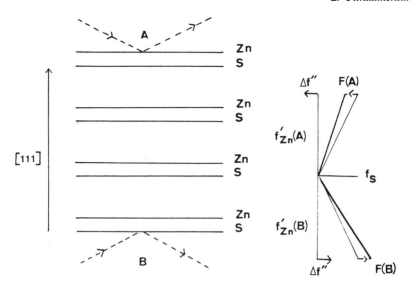

Fig. 2.9-4 Double layers of Zn and S atoms along [111] direction of ZnS (S layers at 0, 4/12, 8/12, ...; Zn layers at 1/12, 5/12, 9/12,...). Phase diagram for first-order reflection from opposite **A** and **B** faces, assuming $f_{Zn} = f'_{Zn} + i\Delta f'_{Zn}$ with $\Delta f''_{Zn}$ positive. (After ref. 7).

Anomalous dispersion techniques are used extensively to establish the absolute configuration of molecules (Section **5.4**) and as an aid in the solution of the phase problem.[9]

References

[1] W.H. Bragg and W.L. Bragg, *Nature (London)* **91**, 557 (1913); *Proc. Roy. Soc. London* **89A**, 277 (1913).

[2] R.W.G. Wyckoff, *Crystal Structures*, 2nd ed., Vol. **1**, Wiley-Interscience, New York, 1971.

[3] F.S. Galasso, *Structure and Properties of Inorganic Solids*, Pergamon Press, Oxford, 1970.

[4] B.K. Vainshtein, V.M. Fridkin and V.L. Indenbom, *Modern Crystallography II, Structure of Crystals*, Springer-Verlag, Berlin, 1982.

[5] G. Will and A. Kirfel in D. Emin, T. Aselage, C.L. Beckel, I.A. Howard and C. Wood (eds.), *Boron-Rich Solids*, American Institute of Physics, New York, 1986, p. 89. Refinements of the cubic BN (borazone) structure using AgKα and synchrotron radiations, respectively, are reported in G. Will, A. Kirfel and B. Josten, *J. Less Common Met.* **117**, 61 (1986) and K. Eichhorn, A. Kirfel, J. Grochowski and P. Serda, *Acta Crystallogr., Sect. B* **47**, 843 (1991).

[6] G. Friedel, *Compt. Rend.* **157**, 1533 (1913).

[7] J.D. Dunitz, *X-Ray Analysis and the Structure of Organic Molecules*, Cornell University Press, Ithaca, 1979.

[8] D. Coster, K.S. Knol and J.A. Prins, *Z. Phys.* **63**, 345 (1930).

[9] S. Ramaseshan and S.C. Abrahams (eds.), *Anomalous Scattering*, Munksgaard, Copenhagen, 1975.

2.10 Wurtzite (Hexagonal Zinc Sulfide)
ZnS

Crystal Data

Hexagonal, space group $P6_3mc$ (No. 186)

$a = 3.811(4)$, $c = 6.234(6)$Å, $Z = 2$

Atom	Position	x	y	z
Zn	2(b)	1/3	2/3	0
S	2(b)	1/3	2/3	5/8

Origin at $6_3(3m)$.

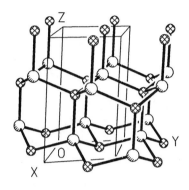 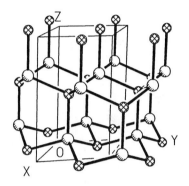

Fig. 2.10-1 Stereoview of the hexagonal zinc sulfide structure. Larger spheres represent S atoms.

Crystal Structure

The structure of wurtzite was first determined in 1923 by Aminoff.[1] The crystal structure is hexagonal and, as in zinc blende, each atom is surrounded by four neighbours at the corners of a tetrahedron. The sulfur atoms are arranged in hexagonal closest packing, in which one-half of the tetrahedral sites are filled with zinc atoms. The Zn^{2+} and S^{2-} sublattices are equivalent, being displaced by the length of the Zn-S bond in the c direction.

Remarks

1. Polytypes of ZnS

The cubic zinc sulfide (zinc blende) and hexagonal zinc sulfide (wurtzite) structures are closely related. The two sets of unit cell parameters are nearly exactly interconvertible by $a_H \approx (1/\sqrt{2})a_C$ and $c_H \approx (2/\sqrt{3})a_C$. The atomic arrangements are very similar and the environment of zinc or sulfur atoms in both structures is nearly the same out to the second-nearest

neighbours. In wurtzite each Zn atom is tetrahedrally surrounded by four
nearest S neighbours at virtually the same distance (2.342 versus 2.338Å) as
in zinc blende. There are also twelve next-nearest neighbours at the same
distance in the two structures and the only difference in coordination is the
substitution of a trigonal prism in wurtzite for an antiprism in zinc blende.
Neglecting the slight deviations from spherical closest packing, in wurtzite
the arrangement of S atoms along the [00.1] direction is in layers according
to the hcp sequence ...ABAB... [Fig. 2.10-2(a)]; in zinc blende, the
arrangement of S atoms along the [111] direction follows the ccp sequence
...ABCABC... [Fig. 2.10-2(b)].

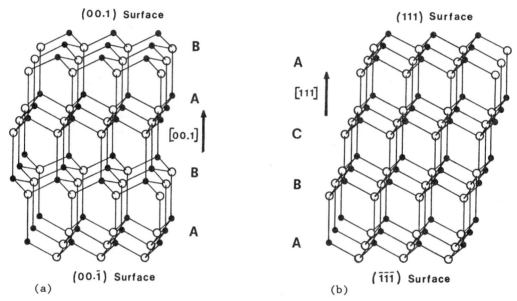

Fig. 2.10-2 Spherical closest packing in (a) hexagonal zinc sulfide
(wurtzite) and (b) cubic zinc sulfide (zinc blende). (Adapted from ref. 2).

"Polymorphs" refer to different crystalline modifications of a
substance and the term "polytype" is reserved for a class of polymorphs for
which there is a special structural relationship between the contents of the
unit cells. As a consequence of the small difference in energy between the
wurtzite and zinc blende arrangements, other close packed sequences with
larger repeat distances can occur. The stacking sequences and crystal-
structure parameters are known for at least ten polytypes of ZnS. Fig. 2.10-3
shows the structures of some polytypes, in which H, C and R represent the
hexagonal, cubic and rhombohedral crystal systems, respectively; 2,3,4,6...
are the numbers of stacking layer in a period, and the stacking sequence of S
atoms is given for each polytype.

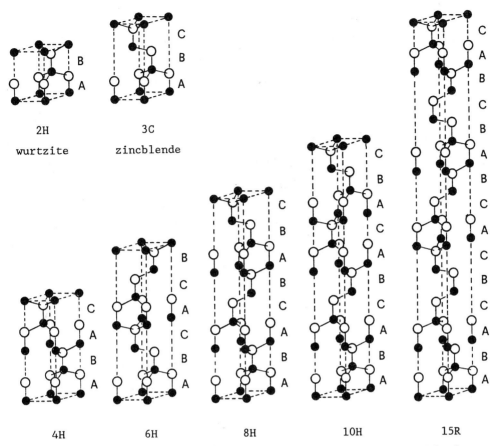

4H   6H   8H   10H   15R

Fig. 2.10-3 Polytypes of ZnS; the S and Zn atoms are represented by large open circles and small filled circles, respectively. (Adapted from ref. 2).

2. Compounds with the wurtzite structure

Some compounds which exhibit the wurtzite structure are listed in Table 2.10-1 together with their hexagonal unit cell parameters.

Table 2.10-1 Some compounds with the wurtzite structure.
The a and c values are in Å.[3,4]

Compound	a	c	Compound	a	c
ZnO	3.2495	5.2069	MnSe	4.12	6.72
ZnS	3.811	6.234	AgI	4.580	7.494
ZnSe	3.98	6.53	AlN	3.111	4.978
ZnTe	4.27	6.99	GaN	3.180	5.166
BeO	2.698	4.379	InN	3.533	5.693
CdS	4.1348	6.7490	CuH	2.893	4.614
CdSe	4.309	7.021	NH_4F	4.39	7.02
MnS(red)	3.976	6.432	α-SiC	3.076	5.048

3. Compositions and properties

Compounds with the zinc blende, wurtzite or other tetrahedral structures have received much attention in materials science, since useful correlations can be derived for these compounds in regard to the chemical composition, the crystal structure, and the electronic structure.[5] A very large number of derivatives exists in this group of related phases, as it is possible to form defect tetrahedral structures by omitting atoms from certain tetrahedral vertices, or to form filled tetrahedral structures by stuffing extra atoms into available voids.

The semiconducting properties of industrially important materials such as ZnO (zincite) and SiC have resulted in an intensive search for new compounds with the wurtzite, zinc blende and other tetrahedral structures. Many such compounds contain nonmetallic atoms as partners and show no obvious metallic properties. The boundary between quasi-metallic and nonmetallic behavior then becomes difficult to define. An example is the series Ge-GaAs-ZnSe-CuBr; the compounds are related only by the similarity in their tetrahedral structures.[6]

Normally ZnS is a n-type semiconductor, but p-type conduction has been observed with ZnS+Cu, prepared under sulfurizing conditions. ZnS doped with foreign donors (Al^{3+}, Cl^-) and prepared under medium sulfur pressure (H_2S) shows a blue luminescence band which is usually referred to as the "self-activition" band. The luminescence of a ZnS phosphor prepared with the addition of a small quantity of silver is strongest if an equal amount of aluminum is also present; this suggests that both Ag and Al retain their normal valencies at normal lattice sites, forming a solid solution (Zn,Ag,Al)S and that the enhanced luminescence is due to the presence of silver.

References

[1] G. Aminoff, *Z. Kristallogr.* **58**, 203 (1923).

[2] W.L. Roth in M. Aven and J.S. Prener (eds.), *Physics and Chemistry of II-VI Compounds*, North-Holland, Amsterdam, 1967, chap. 3.

[3] R.W.G. Wyckoff, *Crystal Structures*, 2nd ed., Vol. 1, Wiley-Interscience, New York, 1971.

[4] F.S. Galasso, *Structure and Properties of Inorganic Solids*, Pergamon Press, Oxford, 1970.

[5] E. Parthé, *Crystal Chemistry of Tetrahedral Structures*, Gordon and Breach, New York, 1964.

[6] C.S. Barrett and T.B. Massalski, *Structure of Metals*, 3rd ed., Pergamon Press, Oxford, 1980.

2.11   Nickel Arsenide
NiAs

Crystal Data

Hexagonal, space group $P6_3mc$ (No. 186)

a = 3.619(2), c = 5.034(2)Å, Z = 2

Atom	Position	x	y	z
Ni	2(a)	0	0	0
As	2(b)	1/3	2/3	1/4

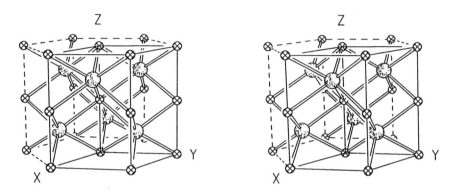

Fig. 2.11-1 Stereoview of the crystal structure of NiAs.   Larger spheres represent As atoms.

Crystal Structure

The crystal structure of NiAs was determined in 1923 by Aminoff.[1]   The As atoms are in hexagonal closest packing with all available octahedral holes occupied by the Ni atoms (Fig. 2.11-1).   In an alternative description, the Ni atoms are arranged in hexagonal layers which exactly eclipse one another, and only half of the large six-coordinate interstices are filled by the As atoms (Fig. 2.11-1).   An important feature of this structure is that the Ni and As atoms have different coordination environments.   Each As atom is surrounded by six equidistant Ni atoms situated at the corners of a right trigonal prism, as shown in Fig. 2.11-1.   Each Ni atom, on the other hand, has eight close neighbours, six of which are As atoms arranged octahedrally about it, while the other two are those Ni atoms immediately above and below at $z = \pm c/2$. Along each chain of Ni atoms parallel to the c axis, the extra Ni-Ni bond strength compensates for the compression.   The axial ratio (c/a) of NiAs is 1.39, which suggests that with a decrease in ionic character, metal atoms are less likely to repel each other, and more likely to form metal-metal bonds.

Remarks

1. Compounds with the NiAs structure

The flexibility of the NiAs structure makes it particularly suitable for compounds with bond types intermediate between ionic and intermetallic. It is thus only found among transition-metal compounds. Besides being adopted by many phosphides, arsenides and germanides, it is also adopted by intermetallic compounds and in a related but distorted form by sulfides, selenides and tellurides. Some examples are listed in Table 2.11-1.[2,4]

Table 2.11-1   Compounds with the NiAs Structure.

Compounds	a/Å	c/Å	c/a
Arsenides			
MnAs	3.724	5.706	1.53
NiAs	3.602	5.009	1.39
β-TiAs	3.64	6.15	1.69
Antimonides			
CrSb	4.13	5.51	1.33
CuSb	3.874	5.193	1.34
FeSb	4.072	5.140	1.26
IrSb	3.987	5.521	1.38
MnSb	4.15	5.78	1.39
NiSb	3.942	5.155	1.31
PdSb	4.078	5.593	1.37
PtSb	4.13	5.483	1.33
Bismuthides			
MnBi	4.27	6.15	1.44
NiBi	4.070	5.35	1.31
PtBi	4.315	5.490	1.27
RhBi	4.075	5.669	1.39
Stannides			
AuSn	4.323	5.523	1.28
CuSn	4.198	5.096	1.32
IrSn	3.988	5.567	1.40
NiSn	4.048	5.123	1.27
PdSn	4.11	5.44	1.32
PtSn	4.111	5.439	1.32
RhSn	4.340	5.553	1.28
Sulfides			
CoS	3.374	5.187	1.54
NbS	3.32	6.46	1.95
β-NiS	3.4392	5.3484	1.55
TiS	3.299	6.380	1.93
VS	3.33	5.82	1.75
Selendies			
CrSe	3.71	6.03	1.63
FeSe	3.617	5.88	1.63
β-NiSe	3.6613	5.3562	1.46
TiSe	3.5722	6.205	1.74
VSe	3.66	5.95	1.63
Tellurides			
CrTe	3.93	6.15	1.56
MnTe	4.087	6.701	1.64
NiTe	3.98	5.38	1.35
PdTe	4.152	5.672	1.37
RhTe	3.99	5.66	1.42
ScTe	4.120	6.748	1.64
VTe	3.942	6.126	1.55
ZrTe	3.953	6.647	1.68

2. Deficient and enriched NiAs structures

A clear understanding of the NiAs structure is basic to a more complete appreciation of the factors influencing the occurrence of defect materials.[3] The NiAs structure is certainly the most enigmatic of the simple AB_n

structures, and the reasons for its adoption are by no means self-evident. It is an uniquely flexible structure. It has the ability to absorb internal stresses by changing the axial ratio with retention of the symmetry of the structure, rather than by a phase change. A wide range of composition can also be accommodated, either on the metal deficient side of MX, or by accommodating extra metal atoms up to the limit M_2X. In the latter case, such

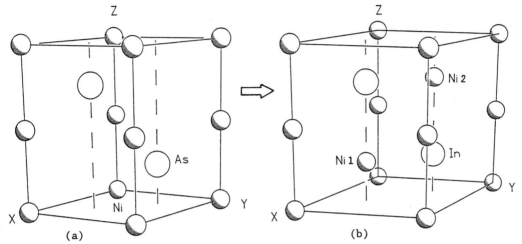

Fig. 2.11-2 The relationship between the NiAs (a) and Ni_2In (b) structures. (After ref. 4).

as Ni_2In, the additional Ni atoms enter into the trigonal-bipyramidal holes of the hcp structure. This is a typically metallic structure with a *c/a* ratio near 1.22. Fig. 2.11-2 shows the relationship between the Ni_2In and NiAs structures.

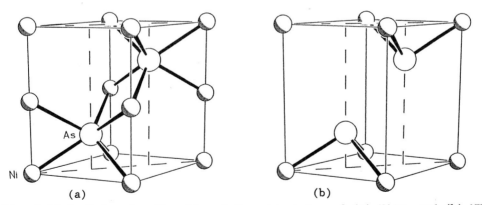

Fig. 2.11-3 Related unit cells of the structures of (a) NiAs, and (b) MX_2 with (0,0,1/2) sites all vacant (CdI_2 structure). (After ref. 5).

In a number of metal-deficient phases, vacant metal sites in the NiAs structure can arise in two simple ways:

(a) Random vacancies in (0,0,0) and (0,0,1/2) [see Fig. 2.11-3(a)].

(b) Sites (0,0,0) fully occupied but (0,0,1/2) only partly filled. The further alternatives are then (i) random, or (ii) ordered vacancies in alternate metal layers. The latter forms superstructures of the NiAs type with various arrangements of the vacancies. When the (0,0,1/2) sites are all vacant, the composition changes to MX_2 [see Fig. 2.11-3(b)].

In the Co-Te system, the phase with the NiAs structure is homogeneous over the range 50-66.7 atomic per cent Te, and at the latter limit the composition corresponds to the formula $CoTe_2$. Over this whole range the cell dimensions vary continuously but by only a very small amount:

	a/Å	c/Å
CoTe	3.882	5.367
$CoTe_2$	3.784	5.403

The explanation of this unusual phenomenon, a continuous change from CoTe to $CoTe_2$, is as follows. The quenched $CoTe_2$ compound crystallizes at high temperatures with the CdI_2 structure [Fig. 2.11-3(b)], which is very simply related to the NiAs structure of CoTe [Fig. 2.11-3(a)]. In both the extreme structures, the Te atoms are arranged in hcp. At the composition CoTe all the octahedral holes are occupied by Co. As the proportion of Co decreases some of these positions become vacant, and finally, at the composition $CoTe_2$, only one-half are occupied.

In the Fe-S system, the NiAs structure is stable over only a small range of composition corresponding approximately to the formulae Fe_7S_8, Fe_9S_{10}, $Fe_{10}S_{11}$ and $Fe_{11}S_{12}$, thus providing many examples of super-structures of the NiAs type with various arrangements of the vacancies. For example, the supercell of Fe_7S_8 is composed of 12 NiAs-type subcells, and the vacant sites for Fe atoms are distributed in every second layer, as shown in Fig. 2.11-4, so that they are as far as possible from one another. All the Fe atoms are octahedrally coordinated with S atoms and the octahedra are not appreciably distorted.[6]

Phases with low c/a ratios, as listed in Table 2.11-1, are largely metallic in bond type. Lowering of the c/a value brings two next-nearest metal neighbours close to any one metal in an octahedral site, thereby giving it a coordination number of eight. At the same time, this close approach allows d-d overlap parallel to the c axis.

The octahedral and trigonal-bipyramidal sites of the hcp B lattice share a common face, making movement from one type of site to the other especially easy and accounting, at least in part, for some of the experimental

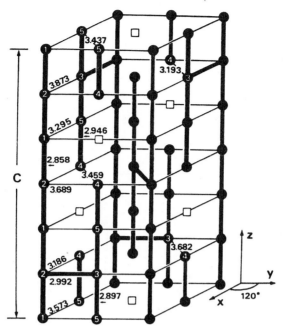

Fig. 2.11-4 Structure of Fe_7S_8 (space group $P3_12_11$, with $a = 6.8652$, $c = 17.046$Å, and $Z = 3$). Only Fe atoms are shown; crystallographically independent atoms are indicated by numbers in solid circles. The Fe-Fe distances are given by the numbers near the lines connecting the atoms. The thick lines represent Fe-Fe distances under 3.0Å. (After ref. 6).

difficulties encountered in trying to synthesize ordered materials. Commonly it proves impossible to obtain a compound with the NiAs structure satisfying an exact 1:1 stoichiometry.

With the exception of some alloy phases, the NiAs structure is essentially restricted to high temperatures. It is avoided in the equilibrium state at low temperatures by one of three means: (i) a change of the stacking sequence, (ii) a structural deformation, and (iii) the absence of any homogeneous phase at the ideal composition.

References

[1] G. Aminoff, Z. Kristallogr. **58**, 203 (1923).

[2] C.S. Barrett and T.B. Massalski, Structure of Metals, 3rd ed., Pergamon Press, Oxford, 1980, p. 245.

[3] D.M. Adams, Inorganic Solids, Wiley, Chichester, 1974, pp. 55, 312.

[4] F.S. Galasso, Structure and Properties of Inorganic Solids, Pergamon Press, Oxford, 1970, p. 120.

[5] A.F. Wells, Structural Inorganic Chemistry, 5th ed., Clarendon Press, Oxford, 1984, p. 754.

[6] A. Nakano, M. Tokonami and N. Morimoto, Acta Crystallogr., Sect. B **35**, 722 (1979).

2.12   Sodium Thallide
NaTl

Crystal Data

Cubic, space group $Fd3m$ (No. 227)

$a = 7.488(8)$Å, $Z = 8$

Atom	Position	x	y	z
Tl	8(a)	0	0	0
Na	8(b)	1/2	1/2	1/2

Origin at $\bar{4}3m$.

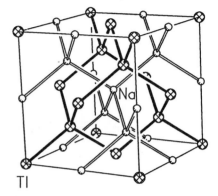

Fig. 2.12-1 Stereoview of the NaTl structure. The diamond-like Tl and Na sublattices are indicated by solid and open lines, respectively.

Crystal Structure

The crystal structure of this intermetallic compound (Fig. 2.12-1) was first determined by Zintl.[1] Each Na (or Tl) atom possesses four equivalent Na (or Tl) and four different Tl (or Na) neighbours at the same distance. Thus every atom is in a cubic coordination environment analogous to that in a body-centered cubic structure or a CsCl-type structure. The atoms of each type form a diamond-like sublattice that is usually favoured by tetravalent atoms. The structure of NaTl can also be understood as a diamond-like lattice of Tl atoms whose vacant sites are completely filled with Na atoms. According to the valence bond picture the covalent bonds in the NaTl crystal are formed by saturated sp^3-like hybrids, leading to semiconducting behaviour.

Remarks

1. Nature of chemical bonding in NaTl

The Tl atom has three valence electrons, which are insufficient for the construction of a stable diamond lattice. This deficit can be partially compensated by the introduction of Na atoms. Charge transfer from the Na atoms to the Tl atoms enables the latter to form a diamond-like structure, in which the effective radius of the Na atom is considerably smaller than that in pure metallic sodium. Therefore, the chemical bonding in NaTl is expected to be a mixture of covalent, ionic, and metallic contributions. The available experimental data lend support to this picture.[2]

2. The Zintl phases

According to Laves, Zintl phases are those intermetallic compounds which crystallize in typical "non-metal" crystal structures.[3] The NaTl type structure is the prototype for Zintl phases.

Solid compounds with the NaTl-type structure are formed by alkali or alkaline-earth metals with metallic or semimetallic elements of the fourth, fifth and third (only partly) groups of the Periodic Table.[4] Table 2.12-1 lists some binary NaTl-type compounds and their parameters; for AB compounds, B is more electronegative than A, r is one-half of the distance between nearest neighbours, while r_A and r_B denote the atomic radii in the corresponding purely metallic systems.

As can be seen from Table 2.12-1, r and r_B differ by a few percent. The ratio r/r_A is significantly smaller for the sodium compounds than for the

Table 2.12-1 Parameters of some binary NaTl-type compounds.

AB compound	$a/\text{Å}$	$r/\text{Å}$	$r_A/\text{Å}$	$r_B/\text{Å}$	r/r_A	r/r_B
LiZn	6.209	1.34	1.51	1.33	0.89	1.01
LiCd	6.687	1.45	1.51	1.48	0.96	0.98
LiAl	6.360	1.38	1.51	1.36	0.91	1.01
LiGa	6.195	1.34	1.51	1.39	0.89	0.96
LiIn	6.786	1.47	1.51	1.52	0.97	0.97
NaIn	7.297	1.58	1.86	1.52	0.85	1.03
NaTl	7.473	1.62	1.86	1.66	0.87	0.98

lithium compounds. In those cases where r and r_A are very different, the interatomic separations involving the larger B atoms, B-B and A-B, must either be compressed, or a charge transfer to the more electronegative non-alkali element and/or strengthening of the covalent bonds within the non-alkali sublattice will take place, thereby reducing the effective atomic radii in Zintl phases as compared to pure metallic bonding.

Theoretical studies of NaTl-type compounds show that the overall charge transfer from the alkali to the non-alkali sublattice is less than -0.1e per atom for the Li-containing compounds, and less than -0.2e per atom for the Na-containing compounds.

From the crystal structure, physical measurements, and theoretical calculations, the nature of the chemical bond in the NaTl type compounds can be deduced as belows:

    a)   Strong covalent bonds exist between the atoms of the B atom (Zn, Cd, Hg, Al, Ga, In, Tl) sublattice.

    b)   The alkali atoms are not in bonding contact.

    c)   The chemical bond between the two sublattices is metallic with a small ionic component.

    d)   For the upper valence/conduction electron states a partial metal-like charge distribution can be identified.

3. Physical properties of the NaTl-type Zintl phases

The NaTl-type Zintl phases have been of special interest because of their unusual properties. Many experimental investigations have been reported for binary (AB) and ternary ($AB_{1-x}C_x$) NaTl-type compounds. Besides the crystal structures, the thermodynamic behavior, electrical conductivity, magnetic susceptibility, NMR spectra, elastic constants and optical properties have been studied. Additionally, for LiAl electrochemical investigations have been performed in view of the recent interest in fast ionic conductors.

As metallic systems the NaTl-type compounds $A_{1-x}B_x$ possess a distinct phase width in the range of approximately $0.45 \leq x \leq 0.55$. Furthermore they exhibit metallic conductivity and as in metals the electrical resistivity increases with temperature. From the thermal and elastic behaviour of these phases it was concluded that strong covalent bond contributions are present in NaTl-type phases.[5] The magnetic measurements show deviations from the typical paramagnetism of metals. The chemical shifts in the NMR signals of the AB (B are Group IIIB elements) compounds are much smaller than in the pure parent metals, and at least for NaTl the valence electrons show a large diamagnetic contribution to the magnetic susceptibility. Some Zintl phases are colored and in ternary systems the color changes continuously as a function of the composition.

References

[1] E. Zintl and W. Dullenkopf, *Z. Phys. Chem. B* **16**, 195 (1932).

[2] P.C. Schmidt, *Structure and Bonding* **65**, 91 (1987).

[3] F. Laves, *Naturwissenschaften* **29**, 244 (1941).

[4] H. Schäfer, B. Eisenmann and W. Müller, *Angew. Chem. Int. Ed. Engl.* **12**, 694 (1973).

[5] H.G. von Schnering, *Angew. Chem. Int. Ed. Engl.* **20**, 33 (1981).

[6] H. Krebs, *Fundamentals of Inorganic Crystal Chemistry*, (translated into English by P.H.L. Walter), McGraw-Hill, London, 1968.

2.13   Fluorite (Fluospar)
CaF$_2$

Crystal Data

Cubic, space group $Fm3m$ (No. 225)

a = 5.462(3)Å, Z = 4

Atom	Position	x	y	z
Ca	4(a)	0	0	0
F	8(c)	1/4	1/4	1/4

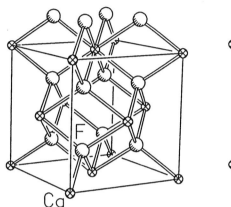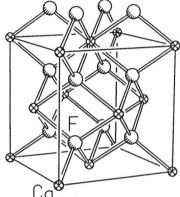

Fig. 2.13-1 Stereoview of the crystal structure of fluorite.

Crystal Structure

The structure of fluorite was determined by Bragg in 1914.[1] It may be regarded as being composed of cubic closest packed calcium cations with all the tetrahedral interstices occupied by fluoride anions (Fig. 2.13-1).

The antifluorite structure, which is adopted by a large number of oxides and other chalcogenides of the alkali metals, has the inverse structure in which the roles of the cations and anions are interchanged.

Remarks

1. Fluorite-like crystalline compounds

The many known fluorite-like crystals are of four kinds:[2,3]

a) All halides of the larger divalent cations except two fluorides in this class.

b) Oxides, sulfides, etc. of univalent ions, mostly group IA, that have the antifluorite structure with a radius ratio greater than unity.

c) Oxides of large quadrivalent cations.

d) Intermetallic compounds.

Some compounds with the fluorite and antifluorite structures are listed in Table 2.13-1.[2]

Table 2.13-1 Some compounds with the fluorite and antifluorite structures.

Fluorite type			Antifluorite type		
CeO_2	AmO_2	CuF_2	Li_2O	Na_2Te	Mg_2Si
ThO_2	ZrO_2	CdF_2	Li_2S	K_2O	Mg_2Sn
PrO_2	HfO_2	HgF_2	Li_2Se	K_2S	Mg_2Pb
PaO_2	TbO_2	PbF_2	Li_2Te	K_2Se	Cu_2S
UO_2	CaF_2	$SrCl_2$	Na_2O	K_2Te	Cu_2Se
NpO_2	SrF_2	$BaCl_2$	Na_2S	Rb_2O	
PuO_2	BaF_2		Na_2Se	Rb_2S	

2. The first alloy structure

The crystal structure of magnesium stannide was determined by Pauling in 1923.[4] Mg_2Sn, an antifluorite structure with $a = 6.765$Å, was the first intermetallic compound to be completely characterized by X-ray diffraction.

3. Superionic solids[5,6]

Superionic solids are ionic materials with high electrical conductivity comparable with those of liquid electrolytes. These materials are also termed "solid electrolytes" or "fast ion conductors". Typically a superionic solid exhibits the following characteristics:

a)   the crystal lattice is consolidated by ionic bonding;

b)   electrical conductivity is high, *ca.* 10^{-1}-10^{-4} $ohm^{-1}cm^{-1}$;

c)   the ionic transference number is almost equal to 1, since the principal charge carriers are ions; and

d)   the electronic conductivity is small; generally materials with electronic transference number less than 10^{-4} are considered satisfactory superionic solids.

From the crystallographic point of view a perfect crystal of an ionic compound would be an insulator. The presence of defects or disorder is a necessity to sustain significant ionic transport. It is natural to classify the ionic solids according to the types of defect or disorder responsible for ionic transport. These are principally of two types:

a)   Point defect type. In this type of solid, the transport is through Frenkel or Schottky defect pairs which are thermally generated. As a result, the number of defects, and hence charge carriers, is a function of temperature. The activation energy is generally high, ~1eV or more.

b)   Molten sublattice type. In this type of solid, the number of ions of a particular type is less than the number of sites available for them in its sublattice. As a result, all these ions can "hop" or "move like a free ion" from one position to another. Since all these ions are available for transport, the conductivity is large and the activation energy is low. These materials/ions possess an "average structure" rather than a rigid structure.

Besides ZrO_2 (remark 4), Li_3N (Section 3.5), β-aluminas (Section 3.13) and α-AgI (Section 3.40) are the most extensively studied superionic solids. The highest conductivity at room temperature obtained so far is for $RbAg_4I_5$ (Section 3.40), which is 0.27 $ohm^{-1}cm^{-1}$.

4. Solid electrolytes of the fluorite type

Some fluorite-type oxides such as ZrO_2 and CeO_2 exhibit properties characteristic of solid electrolytes. In this type of MO_2 structure, the O^{2-} ions which occupy the tetrahedral holes can easily migrate to the neighbouring empty octahedral holes. Doping of ZrO_2 or CeO_2 with rare-earth or alkaline-earth oxides stabilizes the fluorite lattice and introduces vacant O^{2-} sites. The resulting doped crystalline material shows higher electrical conductivity when the temperature is increased.

$ZrO_2(Y_2O_3)$ is a good solid electrolyte of the anionic (O^{2-}) type which can be utilized in devices for measuring the amount of dissolved oxygen in molten steel, and as an electrolyte in fuel cells.

5. Inter-alkali metal chalcogenides

Investigation of the Na_2O/K_2O phase diagram led to the discovery of the ternary oxide KNaO in 1982,[7] and since then many new inter-alkali metal chalcogenides have been synthesized and characterized by X-ray crystallography. Although the binary alkali metal chalcogenides are all of the antifluorite type (Table 2.13-1), the ternary chalcogenides adopt entirely different structures (Table 2.13-2). RbLiO has a structure related to the anti-PbFCl-type, but the Li atom has only three O atoms as its nearest neighbours. KLiO has a distinctly different structure with Li also surrounded

Table 2.13-2   Crystallographic data of inter-alkali metal chalcogenides.[8]

Compound	Space group	a/Å	b/Å	c/Å	Z	Structure	Ref.
KNaO	P4/nmm	4.002		6.214	2	anti-PbFCl	[9]
RbNaO	P4/nmm	4.093		6.531	2	anti-PbFCl	[9]
NaLiS	P4/nmm	4.026		6.495	2	anti-PbFCl	[9]
KLiS	P4/nmm	4.318		6.962	2	anti-PbFCl	[9]
RbLiS	P4/nmm	4.429		7.236	2	anti-PbFCl	[9]
RbNaS	P4/nmm	4.711		7.560	2	anti-PbFCl	[9]
KNaS	Pnma	7.815	4.597	8.329	4	anti-$PbCl_2$	[9]
RbKS	Pnma	8.222	5.043	9.452	4	anti-$PbCl_2$	[9]
KNaSe	Pnma	7.884	4.706	8.704	4	anti-$PbCl_2$	[9]
RbLiO	Pnma	6.568	3.518	8.888	4		
KLiO	Cmca	8.618	6.403	6.417	8	Fig. 2.13-2	

by three oxygens (Fig. 2.13-2).[10]   The Rb_2O/K_2O system only yielded a mixed crystal $Rb_{1-x}K_xO$ with the antifluorite structure.

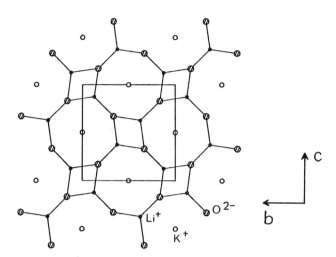

Fig. 2.13-2 Structure of KLiO showing a condensed planar [LiO⁻]-net at $z = 1/2$. Crystal data: $a = 8.618(4)$, $b = 6.403(2)$, $c = 6.417(2)$Å, $Z = 8$; K in 8(d) with $x = -0.1777(1)$, Li in 8(f) with $y = 0.1260(6)$ and $z = 0.3746(7)$, O in 8(f) with $y = 0.1692(3)$ and $z = 0.6668(3)$. (After ref. 10).

References

[1]   W.L. Bragg, *Proc. Roy. Soc. London* **A89**, 468 (1914).

[2]   A.R. West, *Solid State Chemistry and its Applications*, Wiley, Chichester, 1984.

[3]   R.W.G. Wyckoff, *Crystal Structures*, 2nd ed., Vol. 1, Wiley-Interscience, New York, 1963.

[4]   L. Pauling, *J. Am. Chem. Soc.* **45**, 2777 (1923).

[5]   S. Chandra, *Superionic Solids*, North-Holland, Amsterdam, 1981.

[6]   P. Hagenmuller and W. van Gool, *Solid Electrolytes*, Academic Press, New York, 1978.

[7]   H. Sabrowsky and V. Schröer, *Z. Naturforsch.* **37b**, 818 (1982).

[8]   H. Sabrowsky, P. Vogt and A. Thimm, *Acta Crystallogr.*, *Sect. A* **43**, Supplement, C145 (1987), and references cited therein.

[9]   A.F. Wells, *Structural Inorganic Chemistry*, 5th ed., Clarendon Press, Oxford, 1984, pp. 273, 487.

[10] H. Sabrowsky, P. Mertens and A. Thimm, *Z. Kristallogr.* **171**, 1 (1985).

Note The question of bending of the *monomeric* alkaline earth dihalides, together with *ab initio* structural and vibration spectroscopic parameters, is discussed in M. Kaupp, P.v.R. Schleyer, H. Stoll and H. Preuss, *J. Am. Chem. Soc.* **113**, 6012 (1991).

2.14   Rutile (Titanium Dioxide)
TiO$_2$

Crystal Data

Tetragonal, space group $P4_2/mnm$ (No. 136)

a = 4.593659(18), c = 2.958682(8)Å, (298K) Z = 2[1]

Atom	Position	x	y	z
Ti	2(a)	0	0	0
O	4(f)	.30479(10)	.30479(10)	0

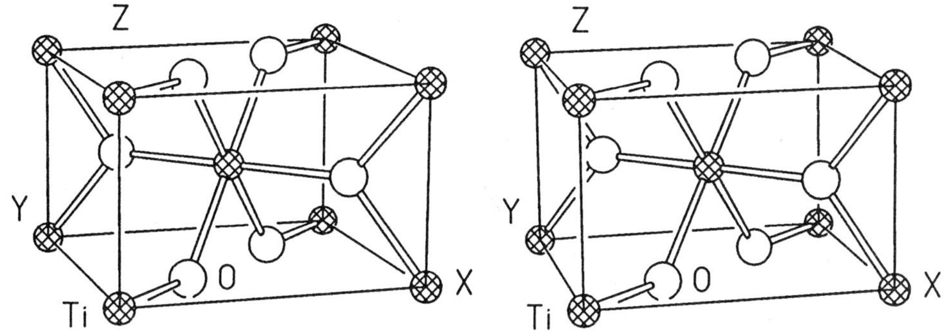

Fig. 2.14-1 Stereoview of the crystal structure of rutile.

Crystal Structure

The crystal structure of rutile was first determined by Vegard in 1916.[2] In this structure of 6:3 coordination, every titanium atom is surrounded by six oxygen atoms approximately at the corners of a regular octahedron, and every oxygen atom by three titanium atoms approximately at the

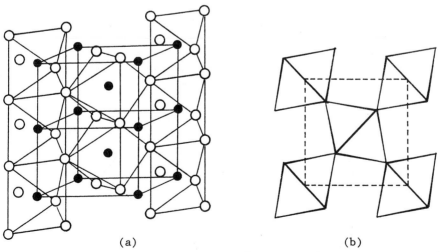

(a)   (b)

Fig. 2.14-2 (a) The arrangement of TiO$_6$ octahedra in chains (After ref. 3); (b) the structure viewed along the c axis (After ref. 4).

corners of an equilateral triangle (Fig. 2.14-1). From Fig. 2.14-2, it is seen that the coordination of Ti is octahedral with two long apical Ti-O bonds of length 1.9800(9)Å and four short equatorial Ti-O bonds of length 1.9485(5)Å. For Ti at (0,0,0), the apical bonds are along the twofold axis ±[110] and the equatorial ones in the mirror plane (110). The TiO_6 octahedra share equatorial edges to form ribbons parallel to [001]. The ribbons are connected through apical corners, resulting in a planar coordination of O by three Ti atoms, with two Ti-O-Ti angles of 130.60(4)° and one Ti-O-Ti angle of 98.79(4)°. Fig. 3.14-2(a) shows the arrangement of the octahedral chains, and Fig. 3.14-2(b) the structure viewed along the c axis.

Remarks

1. Polymorphs of TiO_2

Anatase, brookite and rutile are three naturally occurring forms of TiO_2.[5] They are all composed of octahedral groups of oxygen atoms around titanium, displaying the typical six-coordination of this element, but the way in which the groups are linked together is different for each form. In rutile, each coordination octahedron shares two edges with other octahedra and the two shared edges are slightly shorter than the remaining ten. In brookite three and in anatase four of the twelve edges are shared (Fig. 2.14-3). One would thus expect rutile to be the most stable and anatase the least stable of the three polymorphs.

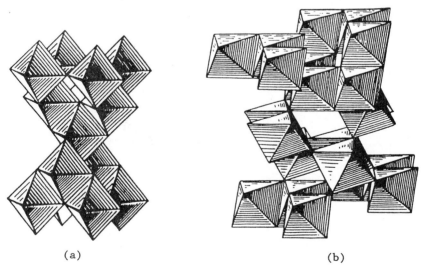

(a)   (b)

Fig. 2.14-3 Crystal structure of (a) anatase and (b) brookite, represented by linked polyhedra. (After ref. 6).

A new polymorph known as $TiO_2(B)$ has been prepared by Tournaux and co-workers[7] from the precursor phase $K_2Ti_4O_9$. Hydrolysis and dehydration lead to the effective removal of K_2O from $K_2Ti_4O_9$, leaving the remainder of the structure intact. The structure of $TiO_2(B)$ is built of TiO_6 octahedra but these are linked up in a different way in relation to the structures of the other three TiO_2 polymorphs.

2. Compounds with the rutile structure

Two main groups of compounds exhibit the rutile structure: oxides of some tetravalent metal ions and fluorides of small divalent metal ions (Table 2.14-1). In both cases, these M^{4+} and M^{2+} ions are too small to form the fluorite structure with O^{2-} and F^-, respectively. The rutile structure may be regarded as essentially ionic.

Table 2.14-1   Compounds with the rutile structure.

TiO_2	OsO_2	CoF_2
CrO_2	PbO_2	FeF_2
GeO_2	RuO_2	MgF_2
IrO_2	SnO_2	MnF_2
$\beta\text{-}MnO_2$	TaO_2	NiF_2
MoO_2	VO_2	PdF_2
NbO_2	WO_2	ZnF_2

3. The charge density in rutile

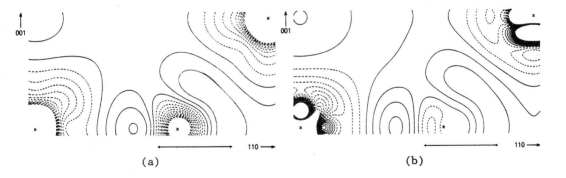

Fig. 2.14-4   Static model deformation maps at infinite resolution in the plane (110), 295K. Interval 0.1 $e\text{Å}^{-3}$, zero and positive contours full lines, negative contours broken. E.s.d.'s at distances larger than 0.5Å from the atomic centers are 0.05 $e\text{Å}^{-3}$; near the atoms, they are very model dependent, i.e. large for steep features and small for flat features. The orientations of the Ti atoms at (000)(lower left) and (1/2,1/2,1/2)(upper right) differ by a rotation of 90° around [001]; a mirror line runs parallel to the long Ti-O bond through Ti at (000) and O at (xx0); the short bonds are between (xx0) and (1/2,1/2,±1/2). (a) Refinement 1ILHE, (b) refinement 1ILGM. (After ref. 8).

Three high-resolution X-ray data sets of rutile have been measured with three different crystals from two sources at two temperatures (two synthetic specimens at 295K and a natural specimen at 100K). In addition, five second-order Bragg intensities have been measured at 295K on an absolute scale with X-ray diffraction. Numerous deformation density refinements have been carried out in order to study the effects of data-reduction procedures, extinction corrections, anharmonicity, and radial parts of the multipolar deformation functions. Invariant features of the resulting static model deformation maps (Fig. 2.14-4) are (a) residual density maxima of about 0.3 $e\text{Å}^{-3}$ on both types of Ti-O bonds, and (b) a region of negative density at Ti indicating an electron transfer from Ti to O.[8] Symmetry-inequivalent bonds of different lengths show nearly identical features.

References

[1] S.C. Abrahams and J.L. Bernstein, *J. Chem. Phys.* **55**, 3206 (1971).

[2] L. Vegard, *Phil. Mag.* **32**, 65 (1916).

[3] A.F. Wells, *Structural Inorganic Chemistry*, 5th ed., Clarendon Press, Oxford, 1984.

[4] D.M. Adams, *Inorganic Solids*, Wiley, London, 1974.

[5] W.L. Bragg and G.F. Claringbull, *Crystal Structures of Minerals*, Bell, London, 1965.

[6] A.S. Povarennykh, *Crystal Chemical Classification of Minerals*, Vols. 1 and 2, Plenum Press, New York, 1972.

[7] A.R. West, *Solid State Chemistry and its Applications*, Wiley, Chichester, 1984.

[8] R. Restori, D. Schwarzenbach and J.R. Schneider, *Acta Crystallogr., Sect. B* **43**, 251 (1987).

Note The phase transition sequence: rutile ⟶ columbite-type (α-PbO_2) ⟶ baddeleyite-type (ambient-presure ZrO_2 and HfO_2; metal atom coordinated by seven oxygen atoms) occurs as pressure is increased. See H. Sato, M. Sugiyama, T. Kikegawa, O. Shimomura and K. Kusaba, *Science (Washington)* **251**, 786 (1991). The structures of orthorhombic α-PbO_2 and monoclinic ZrO_2 are described in R.W.G. Wyckoff, *Crystal Structures*, 2nd ed., Vol. 1, Wiley, New York, 1963, pp. 259, 244.

Very accurate structural and thermal perameters for rutile and anatase from neutron powder diffraction are reported in C.J. Howard, T.M. Sabine and F. Dickson, *Acta Crystallogr., Sect. B* **47**, 462 (1991).

Energy minimization methods are able to model accurately the crystal structure of rutile starting from the experimental unit-cell dimensions and random atomic distributions, as reported in C.M. Freeman and C.R.A. Catlow, *J. Chem. Soc., Chem. Commun.*, 89 (1992).

2.15   α-Quartz
SiO₂

Crystal Data

Trigonal, space group $P3_121$ (No. 152)

a = 4.9134(2), c = 5.4052(3)Å (25°C), Z = 3

Atom	Position	x	y	z
Si	3(a)	.4701	0	1/3
O	6(c)	.4136	.2676	.2142

From the neutron diffraction study of ref. 8; the coordinate parameters are transformed

to agree with the standard setting of the International Tables for X-ray Crystallography.

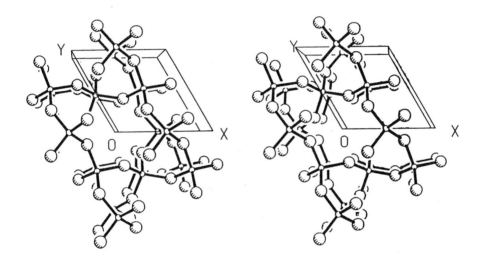

Fig. 2.15-1 Stereoview of the crystal structure of α-quartz. The larger spheres represent the O atoms.

Crystal Structure

At room temperature α-quartz is thermodynamically the most stable form which consists of infinite arrays of corner-shared [SiO₄] tetrahedra (Fig. 2.15-1). The tetrahedra form interlinked helical chains. There are two slightly different Si-O distances: 1.608 and 1.610Å, and the angle Si-O-Si is 143.6°. The helices in any one crystal can be either right-handed or left-handed, so that individual crystals as non-superimposable mirror images can be readily separated by hand. The absolute configuration has been determined by showing that for a levorotatory crystal the space group is $P3_121$ rather than $P3_221$. The specific rotation of α-quartz with the Na-D line is 27.71°/mm.

At 573°C α-quartz transforms into β-quartz, which has the same general structure but is somewhat less distorted (angle Si-O-Si = 155°). Only slight displacements of the atoms are required, so the transition is readily reversible on cooling and the "handedness" of the crystal is preserved throughout.

The first complete determinations of the simpler β-quartz structure were made by W.H. Bragg and Gibbs[1] in 1925, and by Wyckoff[2] in 1926. Fig. 2.15-2 illustrates the relationship between the α- and β-forms according to Gibbs.[3] The structures of α-quartz[4-8] and β-quartz[8] have been refined to high precision.

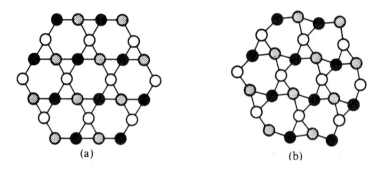

(a)   (b)

Fig. 2.15-2 The relation between (a) β-quartz and (b) α-quartz. Both structures are viewed in the same projection; only the Si atoms are shown, and heights are differentiated by shading. (After ref. 9).

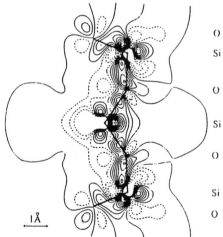

Fig. 2.15-3 Deformation difference electron density map of α-quartz: a section through a chain of bonded atoms; contour interval 0.1 $e\text{Å}^{-3}$, and broken lines represent negative densities. (After ref. 5).

Remarks

1. The nature of the Si-O bond in α-quartz

The nature of the Si-O bond in α-quartz has been clarified through its deformation difference electron density map (Fig. 2.15-3), which shows that

the charges in quartz are +1.22 at Si and -0.61 at O. In the crystal an
electron density corresponding to the covalent fraction of the bond is
observed. Lone-pair densities of O atoms are displaced from the bonding
plane, and oxygen appears to be sp^3 hybridized.[10]

2. The modifications of SiO₂

Silica (SiO₂) crystallizes commonly in three low-pressure forms as
quartz, tridymite, and cristobalite, each having an α-form (low-temperature)
and a β-form (high-temperature). They are all built of tetrahedral groups of
oxygen atoms surrounding a central silicon atom, and each oxygen atom is
shared between two silicon atoms, i.e., the tetrahedral groups are linked by
having common corners. Since a silicon atom has a half share of each of its
four surrounding oxygen atoms the silicon-oxygen ratio in SiO₂ is obeyed.

α-Quartz changes into β-quartz at 573°C, α-tridymite into β-tridymite
between 120°C and 160°C, and α-cristobalite into β-cristobalite between 200°C
and 275°C. The high-temperature forms have higher symmetry or smaller unit
cell than the low-temperature forms: α-quartz has a trigonal axis, β-quartz a
hexagonal axis (Fig. 2.15-2); α-tridymite (several forms known) a larger cell
of lower symmetry, β-tridymite (two forms known) a small cell of higher
symmetry; α-cristobalite is tetragonal whereas β-cristobalite is cubic.

Quartz is the stable form up to 870°C, tridymite between 870°C and
1470°C, and cristobalite from this temperature to the melting point (1720°C).
The changes from one type to the other are extremely slow, and tridymite and
cristobalite can remain in the metastable state indefinitely at ordinary
temperatures. On the other hand the α-β transformation of each modification
takes place rapidly and is reversible. These transitions are summarized
below:

$$
\beta\text{-quartz} \underset{\text{slow}}{\overset{870°}{\rightleftharpoons}} \beta\text{-tridymite} \underset{\text{slow}}{\overset{1470°}{\rightleftharpoons}} \beta\text{-cristobalite} \underset{\text{sluggish}}{\overset{1720°}{\rightleftharpoons}} \text{liq. SiO}_2
$$

fast ↕ 573°   fast ↕ 120-160°   fast ↕ 200-275°

α-quartz   α-tridymite   α-cristobalite

At high-pressure and high-temperature, silica forms denser
modifications: coesite and stishovite. Monoclinic coesite was shown by Zoltai
and Buerger to be built of silicon-oxygen tetrahedra in a new arrangement
somewhat resembling the alumina-silica framwork of the feldspars.[11]
Stishovite has been obtained under laboratory conditions at 1200-1400°C and
160 kbars; it has the rutile structure with nearly octahedral coordination

about silicon.[12]   Fig. 2.15-4 shows the structures of several modifications of silica.

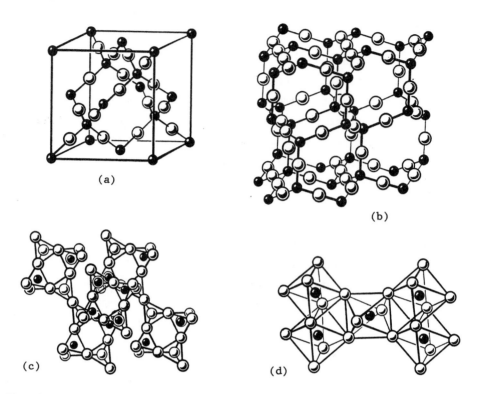

(a)

(b)

(c)

(d)

Fig. 2.15-4 Structures of some modifications of SiO_2.   (After ref. 5).
(a) β-cristobalite   (b) α-tridymite
(c) coesite   (d) stishovite

At still higher pressure (40-120 kbar, 380-585°C) keatite is formed under hydrothermal conditions from amorphous silica and dilute alkali. In this crystal, the [SiO_4] tetrahedra are connected into 5-, 7- and 8-membered rings as in high pressure ice III (Section 3.1). Fig. 2.15-5 shows the structure of keatite.[14]

A modification known as "fibrous silica" has been obtained by heating "SiO" at 1200-1400°C.[15]   Other compounds which are known to have this structure are SiS_2, $SiSe_2$, and $BeCl_2$. The crystal data of the modifications of silica are summarized in Table 2.15-1.

3. Clathrate inclusion compounds of SiO_2

Some clathrate inclusion compounds of SiO_2, known as "clathrasils", have been prepared, In the dodecasils, for example, [SiO_4] tetrahedra are linked to form pentagonal dodecahedra, which are in turn linked to form a neutral three-dimensional framework of SiO_2.[16]   The clathrasils contain cages for the

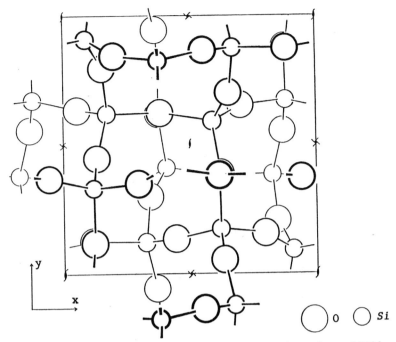

Fig. 2.15-5 The structure of keatite in projection along [001]. (Adapted from ref. 14).

Table 2.15-1 Crystal data of the modificaions of silica.

Modification	Space group	Z	Cell parameters/Å	$\rho/g\ cm^{-3}$
vitreous	-		-	2.196
"fibrous"	Ibam	4	a = 8.36, b = 5.16, c = 4.72	1.960
α-quartz	P3₁21	3	a = 4.9134, c = 5.4052 (25°C)	2.649
β-quartz	P6₂22	3	a = 4.9977, c = 5.4601Å (590°C)	2.533
α-tridymite	P6₃/mmc[17]	864	a = 30.08, c = 49.08	2.241
	Cc[18]	48	a = 18.494, b = 4.991, c = 23.758, β = 105.79°	2.269
	F1[19]	320	a = 9.932, b = 17.216 c = 81.864, α = β = γ = 90°	2.281
β-tridymite	P6₃/mmc[20]	4	a = 5.052, c = 8.270 (460°C)	2.183
	C222₁[21]	8	a = 8.730, b = 5.000 c = 8.201 (170°C)	2.217
	P2₁2₁2₁[22]	24	a = 26.171, b = 4.986, c = 8.196 (155°C)	2.239
α-cristobalite[23]	P4₁2₁2	4	a = 4.9570, c = 6.8907 (10°K)	2.357
			a = 4.9709, c = 6.9278 (296°K)	2.331
			a = 4.9877, c = 6.9697 (473°K)	2.302
β-cristobalite	Fd3m	8	a = 7.3Å	2.052
coesite	C2/c	16	a = 7.135, b = 12.372 c = 7.174, β = 120.36°	2.921
stishovite	P4₂/mnm	2	a = 4.1790, c = 2.6649	4.288
keatite	P4₃2₁2	12	a = 7.46, c = 8.61	2.499

accommodation of guest molecules of appropriate sizes (Fig. 2.15-6). The general structural formula of dodecasil-3C is $136SiO_2.16X.8Y$ [X = N_2, Y = $N(CH_3)_3$, $HN(CH_3)_2$]. The clathrasils are further discussed in Section 7.1.

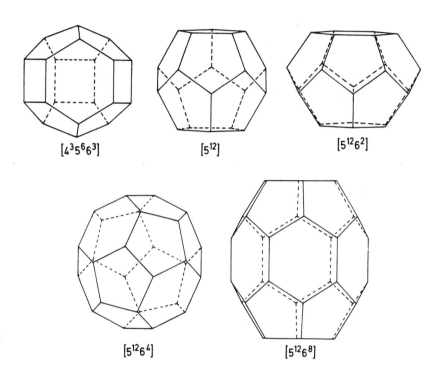

$$[4^3 5^6 6^3] \qquad [5^{12}] \qquad [5^{12} 6^2]$$

$$[5^{12} 6^4] \qquad [5^{12} 6^8]$$

Fig. 2.15-6 The polyhedral cavities observed in clathrasils: silicon atoms are located at the corners of the polyhedra, and oxygen atoms near the centers of the lines connecting the corners. (After ref. 13).

References

[1] W.H. Bragg and R.E. Gibbs, *Proc. Roy. Soc. London* **A109**, 405 (1925).

[2] R.W.G. Wyckoff, *Amer. J. Sci.* **11**, 101 (1926).

[3] R.E. Gibbs, *Proc. Roy. Soc. London* **A110**, 443 (1926).

[4] W.H. Zachariasen and H.A. Plettinger, *Acta Crystallogr.* **18**, 710 (1965).

[5] B.K. Vainshtein, V.M. Fridkin and V.L. Indenbom, *Modern Crystallography II, Structure of Crystals*, Springer-Verlag, Berlin, 1982.

[6] L. Levien, C.T. Prewett and D.J. Weidner, *Amer. Min.* **65**, 920 (1980).

[7] Y. LePage, L.D. Calvert and E.J. Gabe, *J. Phys. Chem. Solids* **41**, 721 (1980).

[8] A.F. Wright and M.S. Lehmann, *J. Solid State Chem.* **36**, 371 (1981).

[9] A.F. Wells, *Structural Inorganic Chemistry*, 5th ed., Clarendon Press, Oxford, 1984.

[10] N. Thong and D. Schwarzenbach, *5th European Crystallographic Meeting, Abstracts*, Copenhagen, 1979, p. 348.

[11] T. Zoltai and M.J. Buerger, *Z. Kristallogr.* **111**, 129 (1959); L. Levien and C.T. Prewitt, *Amer. Min.* **66**, 324 (1981).

[12] E.C.T. Chao, J.J. Fahey, J. Littler and D.J. Milton, *J. Geophys. Res.* **67**, 419 (1962); W.H. Baur and A.A. Khan, *Acta Crystallogr., Sect. B* **27**, 2133 (1977).

[13] F. Liebau, *Structural Chemistry of Silicates*, Springer-Verlag, Berlin, 1985.

[14] J. Shropshire, P.P. Keat and P.A. Vaughan, *Z. Kristallogr.* **112**, 409 (1959).

[15] A. Weiss and A. Weiss, *Naturwissenschaften* **41**, 12 (1954); *Z. Anorg. Allg. Chem.* **276**, 95 (1954).

[16] H. Gies, F. Liebau and H. Gerke, *Angew. Chem. Int. Ed. Engl.* **21**, 206 (1982).

[17] J.E. Fleming and H. Lynton, *Phys. Chem. Glasses* **1**, 148 (1960).

[18] W.H. Baur, *Acta Crystallogr., Sect. B* **33**, 2615 (1977).

[19] J.H. Konnert and D.E. Appleman, *Acta Crystallogr., Sect. B* **34**, 391 (1978).

[20] K. Kihara, *Z. Kristallogr.* **148**, 237 (1979); K. Kihara, *Z. Kristallogr.* **157**, 93 (1981) [Structure refined in space group *Cc2m* with $a = 8.75$, $b = 5.052$ and $c = 8.27$Å].

[21] K. Kihara, T. Matsumoto and M. Iwamura, *Z. Kristallogr.* **177**, 27, 39 (1986).

[22] K. Kihara, *Z. Kristallogr.* **146**, 185 (1977).

[23] J.J. Pluth, J.V. Smith and J. Faber, *J. Appl. Phys.* **57**, 1045 (1985).

2.16   Ice Ih (Hexagonal Ice)

H_2O

Crystal Data

Hexagonal, space group $P6_3/mmc$ (No. 194)

$a = 4.5135(14)$, $c = 7.3521(12)$Å, (H_2O, 0°C).

$a = 4.5085(20)$, $c = 7.3380(35)$Å, (H_2O, -66°C).

$a = 4.5165(14)$, $c = 7.3537(12)$Å, (D_2O, 0°C).

$a = 4.5055(20)$, $c = 7.3380(35)$Å, (D_2O, -66°C).

$Z = 4$

Atom	Position	x	y	z	Occupancy
O	4(f)	1/3	2/3	.0629	1
D(1)	4(f)	1/3	2/3	.1989	1/2
D(2)	12(k)	.4551	.9102	.0182	1/2

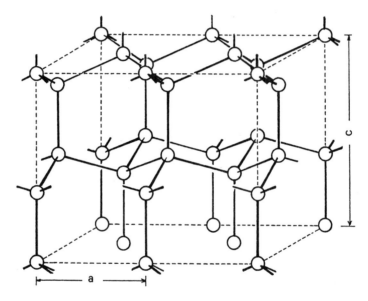

Fig. 2.16-1 Structure of ice Ih. The oxygen atom of each water molecule is shown by a sphere, and hydrogen bonds by thick lines. The hydrogen atoms are omitted; they lie essentially on the hydrogen bond lines. The outlined rectangular (orthohexagonal) cell has a volume that is four times as large as that of the true hexagonal unit cell. The hexagonal *c* axis and one of the equivalent *a* axes are labeled.

Crystal Structure

When liquid water is frozen under normal atmospheric conditions or water vapour is deposited at temperatures below 0°C and above about -80°C, the water molecules are arranged to form hexagonal ice, ice Ih. The deposition of water vapour onto a surface at between -80°C and -130°C produces cubic ice, ice Ic. When water vapour is deposited onto a surface at a temperature below about -140°C, the deposit appears to be either noncrystalline or to consist of very small crystals and is often referred to as vitreous or amorphous ice. In addition, at least eight high-pressure crystalline forms of the water substance, ice II to IX, have been identified (see Section 3.1).

The symmetry, unit cell and oxygen atom arrangement of ice Ih were determined in an X-ray diffraction study by Barnes[1] in 1929. In 1957, Peterson and Levy[2] used neutron diffraction to investigate a single crystal of D_2O at -50°C and -150°C. The arrangement of the atoms in the unit cell is shown in Fig. 2.16-1. Each oxygen atom is tetrahedrally surrounded by four other oxygen atoms, forming O-D...O hydrogen bonds, one of which measures 2.752Å (parallel to the c-axis) and the other three 2.765Å. In the ice crystal, the hydrogen atoms of one H_2O molecule are directed towards the lone electron pairs of the O atoms of the other H_2O molecules, and the H_2O molecules further interconnect by hydrogen bonds to form a three-dimensional framework. Several important features of the structure of ice Ih can be seen from Fig. 2.16-1.

a)   It is apparent that the tetrahedral coordination of the oxygen atoms gives rise to a crystal structure possessing hexagonal symmetry. The hexagonal shape exhibited by many ice crystals (snowflakes, frost, ice) is clearly related to the symmetry of the internal molecular arrangement.

b)   The structure of an isolated water molecule is shown in Fig. 2.16-2. From the data of the D_2O ice Ih structure, the D-O-D angle is the same as that in an isolated water molecule, so that each deuteron is about 0.04Å off the O...O line.[3]

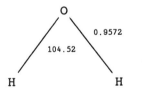   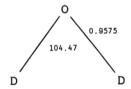

Fig. 2.16-2 The bond angle (degree) and bond distance (Å) of the isolated molecules of H_2O and D_2O.

c)   Ice Ih has a very open channel structure stabilized by hydrogen bonding, as shown in Fig. 2.16-3, and this is reflected in its low density compared to that of liquid water. At 0°C, the density of ice Ih is 0.9168 g cm^{-3}, and that of water is 1.0000 g cm^{-3}.

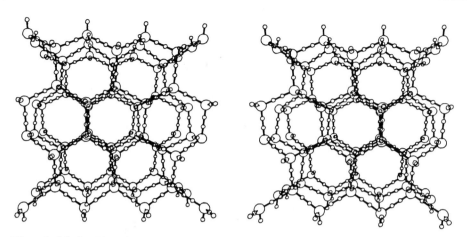

Fig. 2.16-3 The structure of Ice Ih viewed down the hexagonal *c* axis. The large circles are oxygen atoms; the small circles represent the half-hydrogens in the disordered model. Note the large voids in the structure. This is a stereo pair and may be viewed by use of a small hand-held stereoscope. (From ref. 8).

Remarks

1. The distribution of H atoms in ice Ih

Each oxygen atom in ice Ih must have two hydrogen atoms associated with it, and the question now arises as to how these hydrogen atoms are distributed in the structure. The fact that the infrared spectra of ice Ih and liquid water are essentially the same as that of water vapour suggests that the structure of the water molecule is similar in all three phases. This led Bernal and Fowler[4] to suggest that in ice Ih a hydrogen atom is situated on the line joining each pair of oxygen atoms but is approximately 1Å from one oxygen atom and 1.76Å from the other. In 1935, Pauling proposed[5] a disordered model based upon the assumption that ice Ih can exist in any one of a large number of configurations [the number is $(3/2)^{N_o}$ per mole of ice crystal], each corresponding to a unique arrangement of the orientations of the water molecules. This leads to the theoretical value $k\ln(3/2)^{N_o} = R\ln(3/2) = 0.806$ cal deg^{-1} mol^{-1} for the residual entropy of ice Ih, in excellent agreement with the experimental values of 0.82 cal deg^{-1} mol^{-1} for ordinary ice and 0.77 cal deg^{-1} mol^{-1} for D_2O ice.

The results obtained by neutron diffraction of D_2O ice Ih[2] are in complete agreement with Pauling's disordered model. In Fig. 2.16-4 the deuteron associated with each O...O bond is shown as two semicircles at the two equilibrium positions which it may occupy. It can be seen that the geometrical arrangement about the central O atom is very close to that of a perfect tetrahedron.

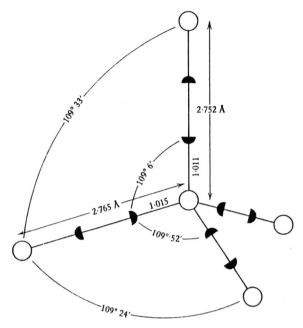

Fig. 2.16-4 Average environment of a molecule in D₂O ice Ih at -50°C as determined by neutron diffraction. (After ref. 2).

2. The structure of ice Ic

    In ice Ic (Fig. 2.16-5), the oxygen atoms are arranged in a diamond-type structure with lattice constant a = 6.350(8)Å (-130°C), giving an O...O distance of 2.750Å.[6] Electron diffraction has shown that the hydrogen atoms are disordered.[7]

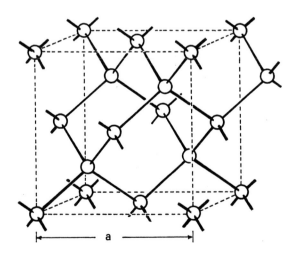

Fig. 2.16-5 Structure of ice Ic. The oxygen atom of each water molecule is shown by a sphere, and hydrogen bonds by thick lines. The hydrogen atoms are omitted; they lie essentially on the hydrogen bond lines. The unit cell is outlined with dashed lines.

3. The ordered phase of ice I

Following the discovery of a first-order phase transition in annealed KOH-doped ice at 72K, which was identified as the order-disorder transition associated with the proton positions, a structural study was carried out using neutron powder diffraction data.[9]

The crystal of ordered ice I at 5K is orthorhombic, space group $Cmc2_1$, with a = 4.5019(5), b = 7.7978(8), c = 7.3280(2)Å and Z = 12. The coordinate parameters are listed in Table 2.16-1, and the crystal structure is shown in Fig. 2.16-6.

Table 2.16-1 The coordinate parameters of ordered ice I.

Atom	x	y	z
O(1)	0	.6648	.0631
O(2)	1/2	.8255	-.0631
D(1)	0	.6636	.1963
D(2)	0	.5363	.0183
D(3)	.6766	-.2252	-.0183

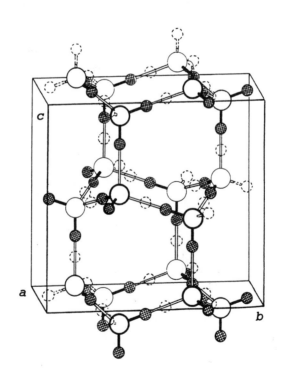

Fig. 2.16-6 The structure of ordered ice I, in which the deuterium atoms (hatched circles) have unit site occupancy. The fully disordered ice Ih structure is obtained when "half-deuterium" atoms occupy the sites represented by the hatched and dashed circles. (After ref. 9).

4. Solid phases of hydrogen sulfide

Hydrogen sulfide solidifies below 187.6 K and phase transitions occur at 126.2 K and 103.5 K; the corresponding melting point and phase transition temperatures of deuterium sulfide are 187.1, 132.8 and 107.8 K, respectively.[10] The structural data for the three solid phases of D_2S as determined from a recent powder neutron diffraction study are summarized in Table 2.16-2.[10]

Table 2.16-2   Structural data for the known solid phases of deuterium sulfide.

Phase	Crystal data	Structural characteristic
I	Fm3m, a = 5.8486(8)Å (160 K), Z = 4	12-fold orientational disorder
II	Pa3, a = 5.7647(6)Å (120 K), Z = 4	6-fold orientational disorder
III	Pbcm, a = 4.0760(1), b = 13.3801(5), c = 6.7215(3)Å (1.5 K), Z = 8	fully-ordered

The coordinate parameters of the fully-ordered lowest-temperature phase III are listed in Table 2.16-3. The structure can be described in terms of a hcp arrangement of two kinds of crystallographically distinct D_2S molecules [Fig. 2.16-7(a)], which is distorted by accommodation of the deuterium atoms. The close-packed layers lie normal to the c axis, with the molecular dipoles

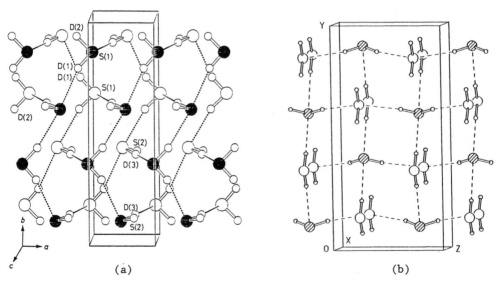

(a)   (b)

Fig. 2.16-7 (a) Perspective view of $D_2S(III)$ showing the hcp layers; the shaded and unshaded S atoms are at z = 3/4 and 1/4, respectively. The potential D...S "hydrogen-bonding" interactions are indicated by dotted lines and thin solid lines. (b) The pseudo-tetragonal structure as seen along the a axis. (After ref. 10).

aligned in the +**a** and -**a** directions in an antiferroelectric array. Fig. 2.16-7(b) shows the pseudo-tetragonal arrangement of the S atoms.

Table 2.16-3 The coordiate parameters of phase III D_2S.

Atom	Site symmetry	x	y	z
S(1)	m	-.042	.1449	1/4
S(2)	m	.460	.4017	1/4
D(1)	m	.173	.2168	1/4
D(2)	m	.194	.0736	1/4
D(3)	1	.6724	.3797	.1057

In the crystal structure there are three S-D...S interactions that implicate hydrogen bonding, with S...S distances of 4.001(9), 4.027(9) and 3.985(5)Å involving D(1), D(2) and D(3), respectively. These are however much longer than either the shortest *direct* S...S contact of 3.672(9)Å in the structure or twice the van der Waals radius for sulfur (1.85Å), so that hydrogen bonding (if it exists at all) plays no significant part in consolidating the crystal structure.

The strongest S-H...S hydrogen bonds known occur in dithiophosphinic acids, which have S...S distances of 3.75-3.84Å and S-H...S angles in the range 159-173°.[11] The first H_2S complex characterized by X-ray diffraction, [Ru(SH$_2$)(PPh$_3$)(L)].THF (L = 2,2'-(ethylenedithio)bis(thiophenolate)), contains S-H...O(THF) and S-H...S(L) bridge bonds with S...S distances of 3.33 and 3.69Å, respectively.[12]

References

[1] W.H. Barnes, *Proc. Roy. Soc. London* **125A**, 670 (1929).

[2] S.W. Peterson and H.A. Levy, *Acta Crystallogr.* **10**, 70 (1957).

[3] P.V. Hobbs, *Ice Physics*, Clarendon Press, Oxford, 1974.

[4] J.D. Bernal and R.H. Fowler, *J. Chem. Phys.* **1**, 515 (1933).

[5] L. Pauling, *J. Am. Chem. Soc.* **57**, 2680 (1935).

[6] H. König, *Z. Kristallogr.* **105**, 279 (1944).

[7] G. Honjo and K. Shimaoka, *Acta Crystallogr.* **10**, 710 (1957).

[8] A. Rich and N. Davidson (eds.), *Structural Chemistry and Molecular Biology*, Freeman, San Francisco, 1968, pp. 471, 509.

[9] A.J. Leadbetter, R.C. Ward, J.W. Clark, P.A. Tucker, T. Matsuo and H. Suga, *J. Chem. Phys.* **82**, 425 (1985).

[10] A.N. Fitch and J.K. Cockcroft, *J. Chem. Soc., Chem. Commun.*, 515 (1990); J.K. Cockcroft and A.N. Fitch, *Z. Kristallogr.* **193**, 1 (1990).

[11] B. Krebs and G. Henkel, *Z. Kristallogr.* **179**, 373 (1987).

[12] D. Sellman, P. Lechner, F. Knoch and M. Moll, *Angew. Chem. Int. Ed. Engl.* **30**, 552 (1991).

Note A recent review of the properties of the H_2O (also H_2S) molecule in its ground electronic state and the electrostatic nature of the hydrogen bond is given in A.D. Buckingham, *J. Mol. Struct.* **250**, 111 (1991).

2.17 Cadmium Iodide

CdI_2

Crystal Data

Trigonal, space group $P\bar{3}m$ (No. 164)

$a = 4.244(10)$, $c = 6.835(14)$Å, $Z = 1$

Atom	Position	x	y	z
Cd	1(a)	0	0	0
I	2(d)	1/3	2/3	.2492

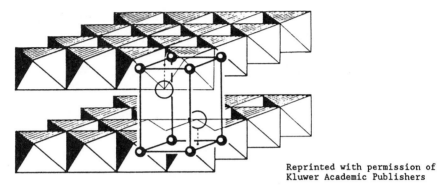

Reprinted with permission of
Kluwer Academic Publishers

Fig. 2.17-1 Crystal structure of CdI_2. (After ref. 4).

Crystal Structure

The layer structure of CdI_2 was first determined by Bozorth in 1922.[1] It may be described as a hexagonal closest packing of I⁻ anions, in which half the octahedral interstices are occupied by Cd^{2+} ions. The manner of occupancy of the octahedral interstices in such that entire layers of octahedral interstices are occupied and these alternate with layers of empty interstices. The layer stacking sequence along the c axis of the unit cell in CdI_2 is shown schematically in Fig. 2.17-1: the I⁻ layers form a ...ABABAB... sequence, and the Cd^{2+} ions occupy octahedral interstices which may be regarded as the C positions relative to the A and B positions for I⁻ ions. Hence the CdI_2 structure may be described as the sequence ...<u>AcBAcBAcB</u>.... The composite CdI_2 layers are held together by weak van der Waals forces between the layers of I⁻ ions. In this respect, CdI_2 bears certain similarities to molecular crystals.

In the CdI_2 crystal, the Cd-I (2.99Å) and I...I (4.21-4.24Å) distances are about equal to the sums of the relevant ionic radii.

Remarks

1. Compounds with the CdI_2 structure

More than 50 compounds have been found to have the CdI_2 structure (Table 2.17-1). They are for the most part hydroxides, bromides, and iodides of divalent metals, and sulfides, selenides, and tellurides of tetravalent metals.

Table 2.17-1   Some compounds with the CdI_2 structure.[2]

CdI_2	TmI_2	$Ca(OH)_2$	HfS_2	$CoTe_2$
CaI_2	YI_2	$Cd(OH)_2$	PtS_2	$IrTe_2$
CoI_2	YbI_2	$Co(OH)_2$	SnS_2	$NiTe_2$
FeI_2	ZnI_2	$Fe(OH)_2$	TaS_2	$PdTe_2$
GeI_2	$CoBr_2$	$Mg(OH)_2$	ZrS_2	$PtTe_2$
MgI_2	$FeBr_2$	$Mn(OH)_2$	$HfSe_2$	$RhTe_2$
MnI_2	$MgBr_2$	$Ni(OH)_2$	$PtSe_2$	$SiTe_2$
PbI_2	$MnBr_2$		$SnSe_2$	$TiTe_2$
ThI_2	$TiBr_2$		$TiSe_2$	$ZrTe_2$
TiI_2	VBr_2		$ZrSe_2$	

2. Crystal structure of $CdCl_2$

The crystal structure of $CdCl_2$ (Fig. 2.17-2) is closely related to that of CdI_2. Both originate from the same type of XMX sandwiches, differing only

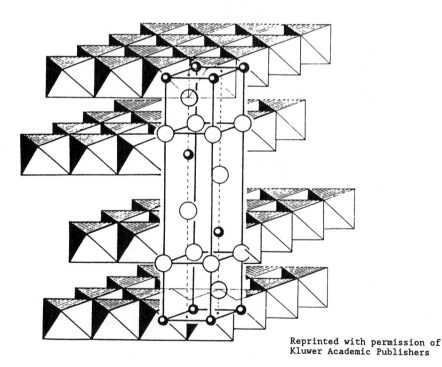

Reprinted with permission of
Kluwer Academic Publishers

Fig. 2.17-2 Crystal structure of $CdCl_2$. (After ref. 4).

in the way of stacking. The CdI_2 structure is based on hcp packing, whereas the $CdCl_2$ structure is based on ccp packing.

The $CdCl_2$ structure may be represented by a hexagonal unit cell, although a smaller rhombohedral cell can also be chosen. The base of the hexagonal cell is of similar size and shape to that in CdI_2, but the c axis of $CdCl_2$ is three times as long as the c axis in CdI_2. This is because in $CdCl_2$ the $[CdCl_6]$ octahedra are staggered along c, giving rise to a three-layer repeat for Cd^{2+} (CBA) and a six-layer repeat for Cl^- ions (ABCABC), as shown in Fig. 2.17-2. In contrast, in CdI_2 the $[CdI_6]$ octahedra are stacked on top of each other and the c repeat contains only two I^- layers (AB) and one Cd^{2+} layer (C). $CdCl_2$ structure may be described as the sequence ...AcBCbABaCAcBCbABaCA...

3. Polytypes of CdI_2

According to the manner of stacking of composite XMX layers, many other stacking arrangements, the so called polytypes, have been found. Nearly 200 polytypes have been reported for CdI_2. These are all built of I-Cd-I layers but differ in the closest packing layer sequences. The three simplest closest packing sequences of iodide layers are: AB..., ABC..., and ABAC... . The AB... sequence as found in CdI_2 is only the next common type after ABAC... . All other polytypes of CdI_2 are of rather rare occurrence.[3]

References

[1] R.M. Bozorth, *J. Am. Chem. Soc.* **44**, 2232 (1922).

[2] R.W.G. Wyckoff, *Crystal Structures*, 2nd ed , Vol. **1**, Wiley-Interscience, New York, 1963.

[3] F. Hulliger, *Structural Chemistry of Layer-type Phases*, D. Reidel, Dordrecht, 1976.

[4] F. Lévy, *Crystallography and Crystal Chemistry of Materials with Layered Structures*, D. Reidel, Dordrecht, 1976.

2.18   Spinel
$MgAl_2O_4$

Crystal Data

Cubic, space group *Fd3m* (No. 227)

$a = 8.0800(4)$Å, $Z = 8$

Atom	Position	x	y	z
Mg	8(a)	0	0	0
Al	16(d)	5/8	5/8	5/8
O	32(e)	.387	.387	.387

Origin at $\overline{4}3m$.

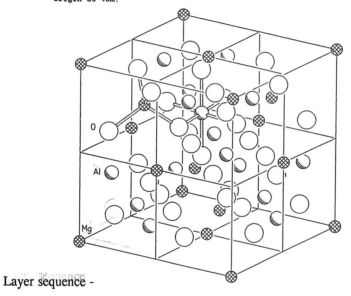

Layer sequence -

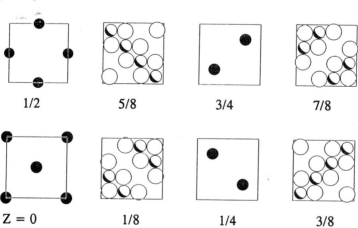

| 1/2 | 5/8 | 3/4 | 7/8 |

| Z = 0 | 1/8 | 1/4 | 3/8 |

Fig. 2.18-1 Crystal structure of spinel. (After ref. 3).

Crystal Structure

The structure of spinel was first analyzed by W.H. Bragg[1] and by S. Nishikawa [2] in 1915. It has an essentially cubic close packed array of oxide ions with Mg^{2+} and Al^{3+} ions in tetrahedral and octahedral interstices, respectively. Fig. 2.18-1 shows the structure of spinel, the unit cell being divided into octants. The A tetrahedra and the B octahedra are arranged alternately, so that every oxide ion is the common vertex of one tetrahedron and three octahedra.

Remarks

1. Compounds with the spinel structure

   Well over a hundred compounds with the spinel structure have been reported to date: mostly oxides, some sulfides, selenides and tellurides, and only a few halides. Many different cations may be introduced into the spinel structure and several different charge combinations are possible, viz.:

2,3	as in $MgAl_2O_4$,
4,2	as in $TiMg_2O_4$,
1,3,4	as in $LiAlTiO_4$
1,2,5	as in $LiNiVO_4$, and
6,1	as in WNa_2O_4.

Similar cation combinations occur with sulfides, e.g. 2,3: $ZnAl_2S_4$ and 4,2: $SnCu_2S_4$. With halide spinels, the cations are limited to charges of 1 and 2 in order to give an overall cation:anion ratio of 3:4, e.g. $NiLi_2Cl_4$.

2. Normal spinels and inverse spinels

   A complication factor in the spinel structure is the cation distribution over the 8(a) and 16(d) sites. Examples of normal spinels are $MgAl_2O_4$ and $MgTi_2O_4$, in which the oxidation state of Ti is +3; their respective structural formulae are $[Mg^{2+}]_t[Al^{3+}_2]_oO_4$ and $[Mg^{2+}]_t[Ti^{3+}_2]_oO_4$, the subscripts t and o indicating tetrahedral and octahedral sites, respectively.

   In inverse spinels, half of the B ions are in tetrahedral 8(a) sites, leaving the remaining B ions and the A ions to fill the 16(d) sites. Usually, the occupancy of these 16(d) sites is disordered. Examples of inverse spinels are $MgFe_2O_4$ and Fe_3O_4; their structural formulae are $[Fe^{3+}]_t[Mg^{2+}Fe^{3+}]_oO_4$ and $[Fe^{3+}]_t[Fe^{2+}Fe^{3+}]_oO_4$, respectively.

   In intermediate spinels any combination of cation arrangement between the extremes of normal and inverse spinels is possible. Sometimes the cation distribution varies with temperature. The cation distribution may be quantified in a simple way by using a "degree of inversion" parameter, γ, which corresponds to the fraction of A ions at the octahedral sites, such as:

Normal: $[A]_t[B_2]_oO_4$ $\gamma = 0$

Inverse: $[B]_t[AB]_oO_4$ $\gamma = 1$

Random: $[A_{1-\gamma}B_\gamma]_t[A_\gamma B_{2-\gamma}]_oO_4$ $0 < \gamma < 1$

A representative selection of normal and inverse spinels, with their unit-cell and coordinate parameters, is listed in Table 2.18-1.

Table 2.18-1 Crystallographic data for some spinels.[4]

Crystal	Oxidation states of cations	a(Å)	x(=y=z)	Structure
$MgAl_2O_4$	2,3	8.0800	0.387	normal
$CoAl_2O_4$	2,3	8.1068	0.39	normal
$CuCr_2S_4$	2,3	9.629	0.381	normal
$CuCu_2Se_4$	2,3	10.357	0.380	normal
$CuCr_2Te_4$	2,3	11.051	0.379	normal
$GeCo_2O_4$	4,2	8.318	–	normal
WNa_2O_4	6,1	8.99	–	normal
$MgFe_2O_4$	2,3	8.389	0.382	inverse
$MgIn_2O_4$	2,3	8.81	0.372	inverse
$MgIn_2S_4$	2,3	10.708	0.384	inverse
Fe_3O_4	2,3	8.39	–	inverse
$TiMg_2O_4$	4,2	8.44	0.39	inverse
$SnZn_2O_4$	4,2	8.70	0.39	inverse
$TiZn_2O_4$	4,2	8.467	0.380	inverse
$LiAlTiO_4$	1,3,4	8.34	–	mixed, Li in 8(a)
$LiCoSbO_4$	1,2,5	8.56	–	mixed, Li in 8(a)

3. LFSE effect

The cation distribution in spinels and the degree of inversion, γ, have been studied in considerable detail. Several factors influence γ, including the site preferences of ions in terms of size, covalent bonding effects, and ligand field stabilization energies (LFSE). The actual γ value in any particular spinel is governed by the net effect of these various factors taken together.

Most transition metal ions prefer octahedral coordination or a distorted variant, and an important factor is their large LFSE in octahedral sites. This can be estimated as follows. In octahedral coordination, each t_{2g} electron experiences a stabilization of $0.4\Delta_o$, and each e_g electron a destabilization of $0.6\Delta_o$; thus $Cr^{3+}(d^3)$ has a LFSE of $1.2\Delta_o$ whereas $Cu^{2+}(d^9)$ has a LFSE of $0.6\Delta_o$. In tetrahedral coordination, each e_g electron is stabilized by $0.6\Delta_t$, and each t_2 electron destabilized by $0.4\Delta_t$. More accurate values may be obtained spectroscopically and are given in Table 2.18-2 for some oxides of transition metal ions. It can be seen that high-spin d^5 ions, as well as d^0 and d^{10} ions, have no particular preference for octahedral or tetrahedral sites insofar as ligand field effects are concerned. Ions such as Cr^{3+}, Ni^{2+} and Mn^{3+} show the strongest preference for octahedral coordination, and hence tetrahedral coordination is quite rare for Ni^{2+}.

Table 2.18-2   Estimated LFSE (kJ mol^{-1}) of transition metal oxides.[5,6]

Ion	Electronic configuration	Octahedral stabilization	Tetrahedral stabilization	Excess octahedral stabilization
Ti^{3+}	d^1	87.4	58.5	28.9
V^{3+}	d^2	160.1	106.6	53.5
Cr^{3+}	d^3	224.5	66.9	157.6
Mn^{3+}	d^4	135.4	40.1	95.3
Fe^{3+}	d^5	0	0	0
Mn^{2+}	d^5	0	0	0
Fe^{2+}	d^6	49.7	33.0	16.7
Co^{2+}	d^7	92.8	61.9	30.9
Ni^{2+}	d^8	122.1	35.9	86.2
Cu^{2+}	d^9	90.3	26.8	63.5

The coordination preferences of ions are shown by the type of spinel structure that they adopt. Lattice energy calculations show that, in the absence of LFSE, spinels of oxidation state 2,3 (i.e. A = M^{2+}, B = M^{3+}; e.g. MgAl₂O₄) tend to be normal, whereas spinels of oxidation state 4,2 (i.e. A = M^{4+}, B = M^{2+}; e.g. TiMg₂O₄) tend to be inverse. However, these preferences may be changed by the intervention of LFSE effects, as shown by the γ parameters of some 2,3 spinels in Table 2.18-3. Some examples are:

a) All chromate spinels are normal. This is consistent with the very large LFSE of Cr^{3+}.

b) Most 2,3 Mg^{2+} spinels are normal apart from MgFe₂O₄ which is essentially inverse. This reflects the lack of any LFSE for Fe^{3+}.

c) Co₃O₄ is normal, because low spin Co^{3+} gains more LFSE by going into the octahedral site than Co^{2+} loses by occupying the tetrahedral site, so its structural formula is $[Co^{2+}]_t[Co^{3+}_2]_oO_4$. Mn₃O₄ is also normal. Magnetite, Fe₃O₄, is inverse because whereas Fe^{3+} has no LFSE in either tetrahedral or octahedral coordination, Fe^{2+} has a preference for octahedral sites; its structural formula is accordingly $[Fe^{3+}]_t[Fe^{2+}Fe^{3+}]_oO_4$.

4. Magnetic spinels

The commerically important magnetic spinels, known as ferrites, are those of the type MFe₂O₄, where M is a divalent cation such as Fe^{2+}, Ni^{2+}, Cu^{2+} and Mg^{2+}. They are all inverse, either partially or completely. This is probably because Fe^{3+}, being a d^5 ion, has no LFSE in an octahedral site;

Table 2.18-3 The γ parameters of some spinels.[5,6]

M^{3+} \\ M^{2+}	Mg^{2+}	Mn^{2+}	Fe^{2+}	Co^{2+}	Ni^{2+}	Cu^{2+}	Zn^{2+}
Al^{3+}	0	0.3	0	0	0.75	0.4	0
Cr^{3+}	0	0	0	0	0	0	0
Fe^{3+}	0.9	0.2	1	1	1	1	0
Mn^{3+}	0	0	0.67	0	1	0	0
Co^{3+}	-	-	-	0	-	-	0

hence, the larger divalent ions go preferentially into octahedral sites and Fe^{3+} is distributed over both tetrahedral and octahedral sites.

These ferrite phases have interesting magnetic structures and are all either antiferromagnetic or ferrimagnetic. This is because the ions in the tetrahedral 8(a) sites have magnetic spins that are antiparallel to those of the ions in the 16(d) octahedral sites. There are many factors which influence the magnetic moments of ferrites, such as the composition of M^{2+}, the degree of inversion, and the thermal history.

Rererences

[1] W.H. Bragg, *Phil. Mag.* **30**, 305 (1915).

[2] S. Nishikawa, *Proc. Tokyo Math. Phys. Soc.* **8**, 199 (1915).

[3] F.S. Galasso, *Structure and Properties of Inorganic Solids*, Pergamon Press, Oxford, 1970, p. 211.

[4] A.R. West, *Solid State Chemistry and its Applications*, Wiley, Chichester, 1984, p. 569.

[5] J.D. Dunitz and L.E. Orgel, *Adv. Inorg. Chem. Radiochem.* 2, 1 (1960).

[6] N.N. Greenwood, *Ionic Crystals, Lattice Defects and Nonstoichiometry*, Butterworths, London, 1968.

2.19 Perovskite
CaTiO$_3$

Crystal Data

Cubic, space group *Pm3m* (No. 221)

a = 3.853Å, Z = 1

Atom	Position	x	y	z
Ti	1(a)	0	0	0
Ca	1(b)	1/2	1/2	1/2
O	3(d)	0	0	1/2

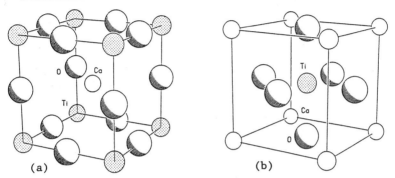

(a)   (b)

Fig. 2.19-1   (a) A-Type and (b) B-type unit cell of perovskite.

Crystal Structure

The crystal structure of perovskite is a very simple one. The structure in its idealized form is cubic, with Ti^{4+} ions at the corners of the unit cell, a Ca^{2+} ion at the body center, and O^{2-} ions at the mid-points of the edges; this A-type cell is shown in Fig. 2.19-1(a). When the origin of the cubic unit cell is taken at the Ca^{2+} ion, then the Ti^{4+} ion occupies the body center and the O^{2-} ions the face centers; this B-type unit cell is shown in Fig. 2.19-1(b).

Each Ca^{2+} is thus 12-coordinated and each Ti^{4+} 6-coordinated by oxygen neighbours, while each O^{2-} is linked to four Ca^{2+} and two Ti^{4+} ions. As expected, it is the larger metal ion which occupies the site of higher coordination. It is noted that geometrically the structure can be regarded as a close-packed array of (O^{2-} + Ca^{2+}) ions, with the Ti^{4+} ions occupying 1/4 of the octahedral interstices.[1]

Remarks

1. Compounds with the CaTiO$_3$ structure

Many ABO$_3$ crystals give exceedingly simple diffraction patterns that can be accounted for in terms of a cubic cell containing one formula unit like

$CaTiO_3$. Some of these have true cubic symmetry, whereas some owing to strain or to small departures from perfect cubic symmetry have appreciably distorted atomic arrangements. It is clear that in such a symmetrical structure a simple relationship must exist between the radii of the component ions. Ideally this relationship is

$$r_A + r_O = \sqrt{2}(r_B + r_O),$$

where A is the larger cation, but it is found that in practice the structure appears whenever the condition

$$r_A + r_O = t\sqrt{2}(r_B + r_O)$$

helds. Here t is a "factor of tolerance" which may lie within the approximate limiting range 0.7-1.0. If t lies outside this range, other structures are obtained. Table 2.19-1 lists some compounds with the perovskite structure.

Table 2.19-1   Some compounds with the perovskite structure.[1]

$CaTiO_3$	$SrHfO_3$	$LaCrO_3$	$LaNiO_3$	$KMgF_3$
$SrTiO_3$	$BaHfO_3$	$NaWO_3$	$CaSnO_3$	$PbMgF_3$
$BaTiO_3$	$NdVO_3$	$LiWO_3$	$SrSnO_3$	$KNiF_3$
$CdTiO_3$	$LaVO_3$	$LaMnO_3$	$BaSnO_3$	$KZnF_3$
$PbTiO_3$	$NaNbO_3$	$LaFeO_3$	$CaCeO_3$	$KFeF_3$
$CaZrO_3$	$KNbO_3$	$SrFeO_3$	$SrCeO_3$	$KCoF_3$
$BaZrO_3$	$NaTaO_3$	$YFeO_3$	$BaCeO_3$	$TlCoF_3$
$PbZrO_3$	$KTaO_3$	$BaFeO_3$	$CdCeO_3$	$LiBaF_3$
$SrZrO_3$	$YCrO_3$	$LaCoO_3$	$BaPrO_3$	$RbCaF_3$

Several points of importance emerge from a consideration of this list, as discussed by Evans.[2]

a) The sizes of ions A and B

In all the compounds the A ions are large (e.g. K, Ca, Sr, Ba) and comparable in size to the oxygen or fluoride ion, as is to be expected, since the A and O (or F) ions together form a close-packed array. Similarly, the B ions are small, since they must have a radius appropriate for 6-coordination by oxygen (or fluorine). These conditions are, of course, merely another expression of the fact that the radii satisfy the relation given above with a tolerance factor t within the range quoted. Quite generally, for oxides (and fluorides) the radii of the A and B ions must lie within the ranges 1.0-1.4 and 0.45-0.75Å, respectively.

b) The valencies of ions A and B

Among oxides the perovskite structure is not exclusively restricted to those compounds in which the A and B ions are divalent and quadrivalent, respectively, as is shown by the fact that $KNbO_3$ and $LaCrO_3$ also have this structure. It thus appears that the valencies of individual cations in the structure are of only secondary importance, and that any pair of ions are compatible provided that they have radii appropriate to the coordination and an aggregate valency of 6 to confer electrical neutrality on the structure as a whole; among the oxides listed in Table 2.19-1 are compounds with pairs of cations of valencies 1 and 5, 2 and 4, and 3 and 3. This point is made even more clearly by the fact that the perovskite structure is also found in a number of oxides in which the A and/or B sites are not all occupied by atoms of the same kind. Thus $(K_{0.5}La_{0.5})TiO_3$ has the perovskite structure with the A ions replaced by equal numbers of ions of K and La, while in $Sr(Ga_{0.5}Nb_{0.5})O_3$ the B ions are replaced by equal numbers of ions of Ga and Nb. In $(Ba_{0.5}K_{0.5})(Ti_{0.5}Nb_{0.5})O_3$ the same structure is again found, with (Ba+K) in place of A and (Ti+Nb) in place of B.

A still more extreme example shows that the perovskite structure can even occur with some of the A sites unoccupied. Sodium tungsten bronze has the ideal composition $NaWO_3$, with the perovskite structure, but this compound shows very variable composition and colour, and is better represented by the formula Na_xWO_3 with $1 > x > 0$. In the sodium-poor varieties the structure remains essentially unaltered but some of the sites normally occupied by sodium are vacant. To preserve neutrality one tungsten ion is converted from W^{5+} to W^{6+} for every site so unoccupied, and this change in oxidation state gives rise to the characteristic alteration in colour and explains its association with the sodium content.

c) Perovskite as a complex oxide

Among compounds with the perovskite structure are many "titanates", "niobates", "stannates", etc., which would normally be regarded as inorganic salts. Structurally, there is no justification for this view. In the true salts of inorganic acids finite complex anions have a discrete existence in the crystal structure: in calcium carbonate, for example, CO_3^{2-} anions are clearly recognizable and the structure as a whole is built up of these anions and of Ca^{2+} cations arranged in a manner very similar to that of the ions in sodium chloride. In "calcium titanate" on the other hand, each titanium ion is coordinated symmetrically by six oxygen neighbours and no TiO_3^{2-} complex ion can be discerned. Thus in spite of the resemblance between the empirical formulae $CaCO_3$ and $CaTiO_3$, the compounds are structurally entirely distinct, and while the former is a salt the latter should more properly be regarded as a complex oxide.

2. Distortion of the CaTiO$_3$ structure

The "ideal" highly symmetric cubic perovskite structure is found in only a limited number of the compounds given in Table 2.19-1. At high temperature, or when the tolerance factor t is very close to unity, this simple structure does indeed often occur, but in many compounds the actual structure is a pseudosymmetric variant of the ideal arrangement, distorted from it by small displacements of the atoms. In some cases, these displacements result in a slight distortion of the unit cell, the symmetry of which is accordingly reduced, and in others the deformation is such that adjacent cells are no longer precisely identical so that the true unit cell comprises more than one of the smaller ideal units. The number of these pseudosymmetric structures is enormous, but it is important to stress that in many cases the degree of departure from the ideal arrangement is only very slight.

a) Crystal structure of BaTiO$_3$

The departure from the ideal structure is of profound importance, for it is from this that the ferroelectric properties of many of these oxides originate. Ferroelectricity is not compatible with the high symmetry of the ideal structure, and the property is manifested only in those members of the perovskite family which have structures of lower symmetry. BaTiO$_3$ has a tetragonal unit cell with axial ratio $c/a = 1.01$ derived from the B-type cubic cell as shown in Fig. 2.19-2. When the distortion is eliminated by heating BaTiO$_3$ to 120°C, it no longer exhibits ferroelectric properties.

b) Crystal structure of K$_2$NiF$_4$

The K$_2$NiF$_4$ structure can be derived from one B-type unit cell of the perovskite structure in combination with two A-type unit cells, each with one

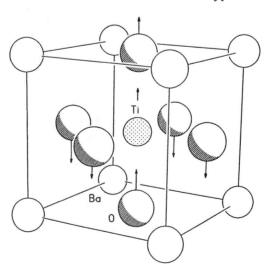

Fig. 2.19-2   Crystal structure of BaTiO$_3$.   (After ref. 3).

layer removed, as shown in Fig. 2.19-3. In this manner a tetragonal unit cell
with $c\sim3a$, where a is the length of the unit cell edge of perovskite, is
generated. The resulting structure in space group *I4/mmm* (No. 139) is
described by the following atomic positional parameters:

K :	4(e),	0, 0, 0.35;
Ni :	2(a),	0, 0, 0;
F(1):	4(c),	0, 1/2, 0;
F(2):	4(e),	0, 0, 0.15.

In this structure, the Ni atom is in an octahedral environment like the
Ti atom in perovskite, but the K atom has only nine F atoms around it.

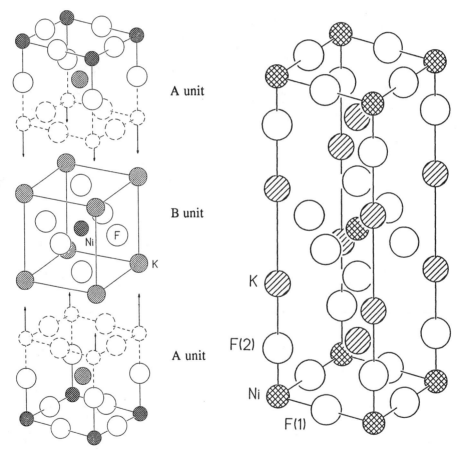

Fig. 2.19-3 Structure of K_2NiF_4. (After ref. 3).

3.   Frameworks derived from the ideal perovskite structure[4]

The basic and elegant perovskite structure ABX_3 forms the parent or
"aristotype" for a wide range of other structures related to it by
combinations of topological distortions, substitution of the A, B and X ions,
and intergrowth with other structure types. These compounds exhibit a range

of magnetic, electrical, optical and catalytic properties that make them attractive to solid state physics, chemistry and material science.

Fig. 2.19-4 illustrates some structural frameworks generated from the perfect perovskite structure with various degrees of oxygen deficiency. Fig. 2.19-4(a) shows structures made up of corner-sharing BO_6 octahedra and BO_5 square pyramids in the stoichiometry range between ABO_3 and $ABO_{2.5}$. Several phases with ordered vacancies and therefore with definit stoichiometries have been determined by X-ray diffraction and electron microscopy in the systems Ca-Mn-O and Sr-Mn-O. The frameworks of the metal cations are retained intact in the oxygen-deficient phases. Fig. 2.19-4(b) represents structures made up

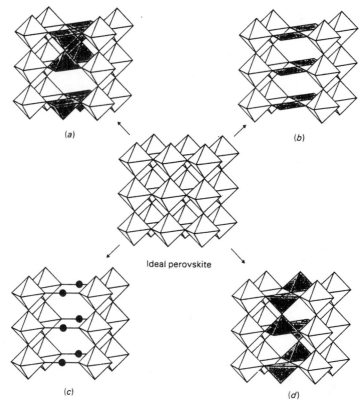

(a)

(b)

Ideal perovskite

(c)

(d)

Fig. 2.19-4 Different structural frameworks derived from the ideal perovskite structure (center) by loss of oxygen. (After ref. 4).

of interlinking layers of corner-sharing BO_6 octahedra and BO_4 planar squares, as found in $LaNiO_{2.5}$. Fig. 2.19-4(c) illustrates structures made up of layers of corner-sharing BO_6 octahedra interlinked by linearly coordinated B-metal cations. Fig. 2.19-4(d) shows structures made up of interlinked layers of corner-sharing BO_6 octahedra and BO_4 tetrahedra, as found in the mineral brownmillerite as well as in phases such as $CaFeO_{3-x}$ and $LaCoO_{2.5}$.

Fragments of the perovskite structure can be combined with many different structural elements. This adaptability opens up an enormous potential of new and interesting frameworks. The structure of K_2NiF_4 can be regard as being composed of one perovskite layer and one rock salt layer, leading to an overall composition of A_2BO_4.

The insertion of double $[Bi_2O_4]_\infty$ layers (with Bi located at the apices of $[BiO_4]$ square pyramids) between perovskitic elements leads to the "Aurivillius phases", where the number of perovskitic layers can be varied between one and four; Fig. 2.19-5(a) shows the first member of the series. The Aurivillius phases often display modulated structures because of mismatch between the bismuth oxide layers and the perovskite layers. The bismuth-copper-oxide superconductors exhibit similar layered structures. A homologous series is based on variation in the perovskite-like layer thickness, i.e. one, two or three CuO layers [Fig. 2.19-5(b) and (c)]. There are superconducting composite copper oxide phases made up of a double perovskitic copper oxide layer alternating with a lead oxide/copper/lead oxide sandwich, $Pb_2Sr_2LnCu_3O_{8+\delta}$, where Ln is a lanthanide or a mixture of a lanthanide with Sr or Ca [Fig. 2.19-5(d)]. Alkaline earth metal cations such as Ca and Sr and rare earth elements such as Y and Nd occupy the perovskitic A-sites. As in the bismuth-copper oxides, the perovskitic copper atoms adopts a square-pyramidal oxygen coordination.

4. High-temperature superconductors

The year 1987 marked an exciting epoch of breakthroughs and triumphs in the field of superconductivity. Studies of pervoskite-like systems of the general formula $A_xB_yCu_zO_w$ (A = Ba, Sr, ...; B = La, Y,...) showed that they exhibit superconducting properties above the melting point of liquid nitrogen,[5] thus rendering the practical applicability of superconducting materials in everyday life a foreseeable goal. The ensuing world-wide research activity was unprecedented, as best exemplified by a timely international meeting on this hot topic fondly acclaimed as the "Woodstock of Physics".[6] Most appropriately, Bednorz and Müller, who started it all by their announcement of the discovery of the prototype "high-temperature superconductor" in 1986,[7] were awarded the 1987 Nobel Prize in Physics.

Complex oxides of the type $A_xB_yCu_zO_w$ are notable for their variability in chemical composition and structure, depending on the starting materials and method of preparation. An isostructural Sr analogue of the original Bednorz-Müller La-Ba-Cu-O (LBCO) high-T_c superconductor has been subjected to single-crystal X-ray analysis.[8] The tetragonal black shiny platelet crystals of

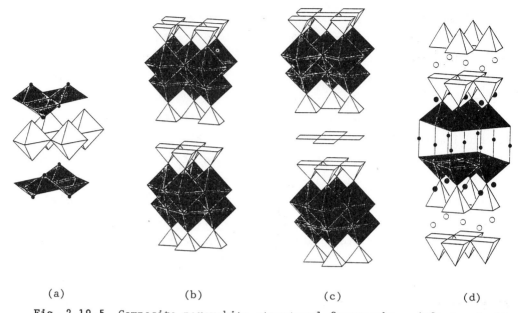

| | (a) | (b) | (c) | (d) |

Fig. 2.19-5 Composite perovskite structural frameworks. (After ref. 4)

$La_{1.85}Sr_{0.15}CuO_4$ belong to space group $I4/mmm$, with $a = 3.7873(1)$, $c = 13.2883(3)$Å, $Z = 2$, and the atomic coordinates are:

Atom	x	y	z
La(Sr)	0	0	.36094
Cu	0	0	0
O(1)	0	1/2	0
O(2)	0	0	.1825

The X-ray study confirmed that $La_{1.85}Sr_{0.15}CuO_4$ is of the K_2NiF_4 layer-perovskite type, in agreement with the earlier powder diffraction data (X-ray and neutron). As seen in Fig. 2.19-6(a), the structure contains layers of corner-sharing elongated octahedra. The lanthanum ions occupy cavities formed primarily by the terminal oxygen atoms, O(2). The coordination geometry around the copper atom is a tetragonally elongated octahedron, with Cu-O(1) = 1.898(1)Å and Cu-O(2) = 2.406(4)Å. The lanthanum (and strontium) site has a coordination number of 9 (C_{4v} symmetry) with bond distances La-O(2) = 2.354(4)Å (1 bond) and 2.745(1)Å (4 bonds), and La-O(1) = 2.639(1)Å (4 bonds).

Powder diffraction studies revealed a stability range of $6.0 \leq x \leq 7.0$ for high T_c superconducting materials of the $A_xB_yCu_zO_w$ class. The crystal data for the end members are: $Ba_2YCu_3O_7$, orthorhombic, space group $Pmmm$, $a = 3.8198(1)$, $b = 3.8849(1)$, $c = 11.6762(3)$Å and $Z = 1$; $Ba_2YCu_3O_6$, tetragonal, space group $P4/mmm$, $a = 3.8570(1)$, $c = 11.8194(3)$Å and $Z = 1$.[9] As shown in

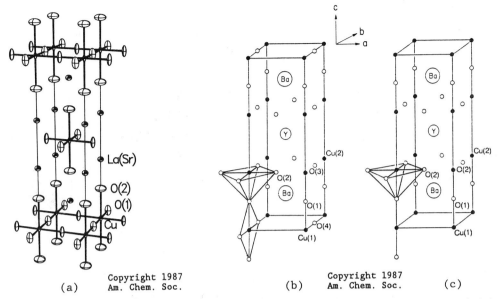

(a)   (b)   (c)

Fig. 2.19-6 Unit cells of (a) $La_{1.85}Sr_{0.15}CuO_4$, (b) $Ba_2YCu_3O_7$ and (c) $Ba_2YCu_3O_6$. (After refs. 8 and 9).

Fig. 2.19-6(b) and (c), these structures are conveniently described as oxygen-deficient perovskites with triple A-type unit cells owing to Ba-Y ordering along the c axis. In $Ba_2YCu_3O_7$ the Cu(1) atoms form linear chains of corner-

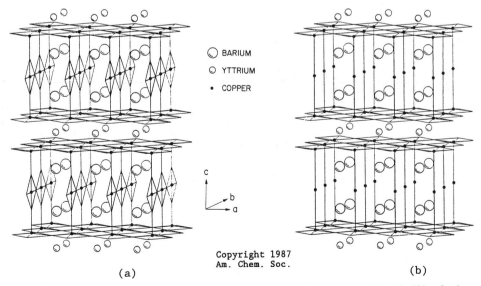

○ BARIUM
○ YTTRIUM
• COPPER

(a)   (b)

Fig. 2.19-7 (a) Perspective view of $Ba_2YCu_3O_7$ emphasizing the Cu(1) chains and Cu(2) layers. (b) Perspective view of $Ba_2YCu_3O_6$ showing linear two-coordination about Cu(1) and the Cu(2) layers. (After ref. 9).

linked squares orientated along the *b* axis, and the Cu(2) atoms form two-dimensional layers of corner-shared square pyramids [copper-oxygen bond distances: Cu(1)-O(1) 1.846, Cu(1)-O(4) 1.943, Cu(2)-O(1) 2.295, Cu(2)-O(2) and Cu(2)-O(3) 1.930Å].

The structure of $Ba_2YCu_3O_6$ is derived from that of $Ba_2YCu_3O_7$ by removal of the O(4) atoms. The coordination geometry at Cu(1) then becomes linear two-coordinate, but the square-pyramidal environment of Cu(2) remains unchanged [Cu(1)-O(1) 1.795, Cu(2)-O(1) 2.469, Cu(2)-O(2) 1.941Å].

Comparison of the crystal structures of the end members of the $Ba_2YCu_3O_x$ series leads to the conclusion that Cu(2) is Cu^{2+}, and that the phase change from $x = 7$ to $x = 6$ arises primarily from a reduction of Cu^{3+} to Cu^+ at the Cu(1) sites. Superconductivity is optimal for $x \sim 7$, and for $x \sim 6$ the material is a semiconductor. The available structural evidence strongly suggests that $-(Cu-O)_n-$ chains are required for superconductivity.

Metal-oxide based high-temperature superconductors are further discussed in Section 4.11.

References

[1] R.W.G. Wyckoff, *Crystal Structures*, 2nd ed. Vol. 2, Wiley-Interscience, New York, 1964.

[2] R.C. Evans, *An Introduction to Crystal Chemistry*, 2nd ed., Cambridge University Press, London, 1976.

[3] F.S. Galasso, *Structure and Properties of Inorganic Solids*, Pergamon Press, Oxford, 1970, p. 162.

[4] A. Reller and T. Williams, *Chem. Brit.* **25**, 1227 (1989).

[5] M.K. Wu, J.R. Ashburn, C.J. Torng, P.H. Hor, R.L. Meng, L. Gao, Z.J. Huang, Y.Q. Wang and C.W. Chu, *Phys. Rev. Lett.* **58**, 908 (1987).

[6] A. Khurana, *Physics Today* **41** 21 (1988).

[7] J.G. Bednorz and K.A. Müller, *Z. Phys.* **B64**, 189 (1986).

[8] H.H. Wang, U. Geiser, R.J. Thorn, K.D. Carlson, M.A. Beno, M.R. Monaghan, T.J. Allen, R.B. Proksch, D.L. Stupka, W.K. Kwok, G.W. Crabtree and J.M. Williams, *Inorg. Chem.* **26**, 1190 (1987).

[9] D.W. Murphy, S.A. Sunshine, P.K. Gallagher, H.M. O'Bryan, R.J. Cava, B. Batlogg, R.B. van Dover, L.F. Schneemeyer and S.M. Zahurak in D.L. Nelson, M.S. Whittingham and T.F. George (eds.), *Chemistry of High-Temperature Superconductors*, American Chemical Society, Washington, DC, 1987, chap. 18, and references cited therein.

Note Na_3NO_3 is not an ortho salt but an oxide nitrite, $(NO_2)ONa_3$, with the antiperovskite structure in which the NO_2^- anion exhibits dynamic orientational disorder. See the detailed review in M. Jansen, *Angew. Chem. Int. Ed. Engl.* **30**, 1547 (1991).

2.20 Potassium Hexachloroplatinate(IV)
K_2PtCl_6

Crystal Data

Cubic, space group $Fm3m$ (No. 225)

$a = 9.6862(3)$Å (MoKα_1, 120K), $Z = 4$

Atom	Position	x	y	z
Pt	4(a)	0	0	0
Cl	24(e)	.23902	0	0
K	8(c)	1/4	1/4	1/4

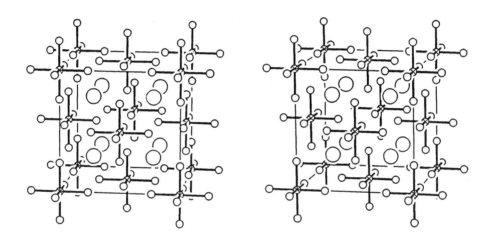

Fig. 2.20-1 Stereoview of the crystal structure of K_2PtCl_6.

Crystal Structure

The crystal structure of K_2PtCl_6, first determined by Ewing and Pauling in 1928,[2], has been accurately refined by Saito and his co-workers in 1984 using room-temperature MoKα data[3] and in 1990 using 120K AgKα and MoKα data.[1] Fig. 2.20-1 shows a unit cell of K_2PtCl_6. The Pt(IV) atom is surrounded octahedrally by six Cl⁻ ions with Pt-Cl = 2.315(1)Å. The K⁺ ion is coordinated by twelve Cl⁻ ions with K...Cl = 3.426(2)Å.

Remarks

1. Historical significance

The structural determination of $(NH_4)_2PtCl_6$ by Wyckoff and Posnjak[4] in 1921 provided the first direct verification of Alfred Werner's postulate of octahedral metal complexes. The existence of square-planar coordination compounds was also substantiated by Dickinson's study of K_2PtCl_4 and $(NH_4)_2PdCl_4$ in the same year.[5]

2. d-Electron distribution of the Pt atom in K_2PtCl_6

The aspherical d-electron distribution in 3d or 4d transition-metal complexes have been examined,[6-8] but 5d metals have been seldom studied because of the strong absorption effect and a small valence/total electron ratio. K_2PtCl_6 is one of the rare examples.

From the difference electron density map,[1] a section of the (110) plane through the Pt nucleus is shown in Fig. 2.20-2. A positive peak of

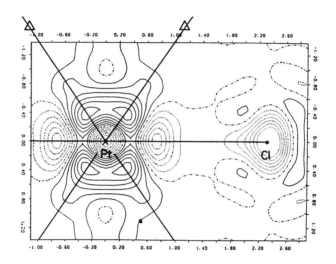

Fig. 2.20-2 A section of the X-X synthesis through the Pt nucleus and two threefold axes. Contours are drawn at intervals of 0.2 $e/Å^{-3}$. Negative contours are dotted, and zero contours chain-dotted. (After ref. 1).

1.3(3) $e/Å^{-3}$ is located on the threefold axis 0.47Å from the Pt nucleus, whereas a negative trough is observed on the Pt-Cl bond axis. The charge asphericity is caused by the non-bonding 5d electrons of Pt(IV) which occupy the t_{2g} set of orbitals in an octahedral ligand field, with an increase in the electron density in the [111] direction relative to that in the [100] direction. The present work confirmed that the charge asphericity around a 5d transition-metal atom can be detected on a deformation density map even though the (d/total electron) ratio is as low as 6/78.

3. Compounds with the K_2PtCl_6 structure

A large number of compounds with the general formula A_2BX_6, where A is NH_4^+, K^+, Rb^+, Cs^+, Tl^+ and B is a metal in oxidation state +IV, adopt this structure. Some of the compounds are listed in Table 2.20-1.

Table 2.20-1 Compounds with the K_2PtCl_6 structure.

Compound	a/Å	Compound	a/Å
Cs_2CoF_6	8.91	K_2MnCl_6	9.0445
Cs_2CrF_6	9.02	K_2SnCl_6	10.002
Cs_2GeCl_6	10.21	K_2ReCl_6	9.795 (120K)
Cs_2MnCl_6	10.17	$(NH_4)_2PtCl_6$	9.858
Cs_2PdBr_6	10.62	Rb_2MnCl_6	9.82
Cs_2SnCl_6	10.38	Rb_2PbCl_6	10.215
K_2SiF_6	8.174	Rb_2SnBr_6	10.85
K_2NiF_6	8.124	Rb_2ZrCl_6	10.198
K_2OsCl_6	9.720 (120K)	Tl_2PtCl_6	9.775

The aspherical 5d-electron density distribution in isomorphous crystals of $K_2[MCl_6]$ (M = Re, Os, Pt) has also been clearly observed by the X-ray diffraction method using AgKα radiation at 120K.[1] In all three compounds the appearance of positive peaks on the threefold axes and negative troughs on the metal-Cl bond axes indicates an excess 5d electron population in the t_{2g} orbitals and a deficiency in the e_g orbitals of the metal atom. The peak heights and their distances from the metal center are listed in Table 2.20-2, along with similar data for the K_2PdCl_6 and $K_2[MCl_4]$ (M = Pd, Pt) complexes.

Table 2.20-2 Peak heights ($e\text{Å}^{-3}$) and their distances (Å) from the central atom in deformation density maps of $K_2[MCl_6]$ and $K_2[MCl_4]$ complexes.

Compound	Radiation	Peak location	Peak height	Distance	Ref.
K_2ReCl_6	AgKα	on threefold axis	1.2(3)	0.38	[1]
K_2OsCl_6	AgKα	on threefold axis	0.8(2)	0.58	[1]
K_2PtCl_6	AgKα	on threefold axis	0.9(3)	0.46	[1]
	MoKα	on threefold axis	1.3(3)	0.47	[1]
K_2PtCl_4	AgKα	on fourfold axis	1.8(4)	0.52	[1]
	AgKα	on bisector of Cl-Pt-Cl	1.6(4)	0.50	[1]
	MoKα	on fourfold axis	2.8(3)	0.47	[1]
	MoKα	on bisector of Cl-Pt-Cl	0.7(3)	0.54	[1]
K_2PdCl_6	MoKα	on threefold axis	1.4(3)	0.51	[8]
K_2PdCl_4	MoKα	on fourfold axis	2.3(3)	0.47	[8]
	MoKα	on bisector of Cl-Pd-Cl	0.8(2)	0.5	[8]

4.   d-Electron distribution of the metal atom in $K_2[MCl_4]$ (M = Pt, Pd)

K_2PtCl_4 crystallizes in space group $P4/mmm$ with a = 6.9813(3), c = 4.1048(3)Å (MoKα₁ radiation, 120K) and Z = 1.[1] The atomic coordinates are: Pt at (0,0,0), Cl at (x, x, 0) with x = 0.23377(4), and K at (0, 1/2, 1/2).

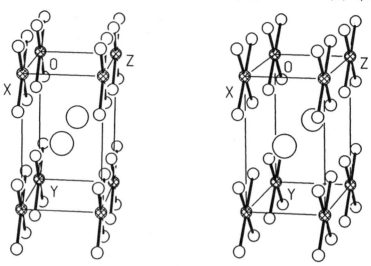

Fig. 2.20-3   Stereoview of the crystal structure of K_2PtCl_4.

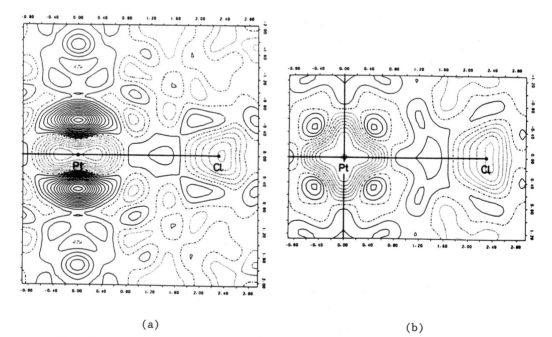

(a)   (b)

Fig. 2.20-4   Deformation densities on (a) the (100) and (b) (001) planes of K_2PtCl_4 (MoKα data at 120K); $[(sin\theta/\lambda)]_{max}$ = 1.0Å⁻¹, contour intervals at 0.2eÅ⁻³.   (After ref. 1).

The crystal structure is illustrated in Fig. 2.20-3; the Pt-Cl bond length is 2.308(1)Å, and the K^+ ion has eight nearest Cl^- neighbours at 3.214(2)Å. Deformation-density sections on the (100) and (001) planes are shown in Fig. 2.20-4. Two positive peaks appear above and below the Pt nucleus, and in the plane of the square-planar anion the deformation density is higher in the direction of the bisector of the Cl-Pt-Cl bond angle than that on the Pt-Cl bond axis. These observations indicate an excess 5d electron population in the $a_{1g}(d_z2)$ and $b_{2g}(d_{xy})$ orbitals and a population deficiency in the $b_{1g}(d_{x^2-y^2})$ orbital. Analogous results are obtained for the 4d electron-density distribution in K_2PdCl_4.[8]

References

[1] H. Takazawa, S. Ohba, Y. Saito and M. Sano, *Acta Crystallogr., Sect. B* **46**, 166 (1990).

[2] F.J. Ewing and L. Pauling, *Z. Kristallogr.* **68**, 223 (1928).

[3] S. Ohba and Y. Saito, *Acta Crystallogr.* **C40**, 1639 (1984).

[4] R.W.G. Wyckoff and E. Posnjak, *J. Am. Chem. Soc.* **43**, 2292 (1921).

[5] R.G. Dickinson, *J. Am. Chem. Soc.* **44**, 2292 (1921).

[6] P. Coppens and M.B. Hall (eds.), *Electron Distributions and the Chemical Bonds*, Plenum Press, New York, 1982.

[7] K. Toriumi and Y. Saito, *Adv. Inorg. Chem. Radiochem.* **27**, 28 (1983); Y. Saito, *Int. Rev. Phys. Chem.* **8**, 235 (1989).

[8] H. Takazawa, S. Ohba and Y. Saito, *Acta Crystallogr., Sect. B* **44**, 580 (1988).

2.21   Copper Sulfate Pentahydrate
$CuSO_4 \cdot 5H_2O$

Crystal Data

Triclinic, space group $P\bar{1}$ (No. 2).

$a = 6.1224(4)$, $b = 10.7223(4)$, $c = 5.9681(4)$Å

$\alpha = 82.35(2)$, $\beta = 107.33(2)$, $\gamma = 102.60(4)°$

$Z = 2$

Atom	x	y	z	Atom	x	y	z
Cu(1)	0.00000	0.00000	0.00000	O(9)	0.43430	0.12430	0.62801
Cu(2)	0.50000	0.50000	0.00000	H(5a)	-0.1146	0.1213	0.2365
S(1)	0.01317	0.28636	0.62528	H(5b)	-0.2716	0.0196	0.2070
O(1)	-0.09284	0.15180	0.67315	H(6a)	0.3015	0.1876	0.0802
O(2)	0.24421	0.31741	0.79641	H(6b)	0.3241	0.1215	0.2832
O(3)	-0.14025	0.37249	0.63681	H(7a)	0.5726	0.4008	0.4031
O(4)	0.04314	0.30144	0.38429	H(7b)	0.3479	0.3829	0.3269
O(5)	-0.18256	0.07339	0.15133	H(8a)	0.7927	0.4020	-0.0812
O(6)	0.28886	0.11724	0.14836	H(8b)	0.8375	0.3898	0.1340
O(7)	0.46547	0.40643	0.29664	H(9a)	0.5737	0.1339	0.6521
O(8)	0.75479	0.41609	0.01918	H(9b)	0.4175	0.1852	0.6816

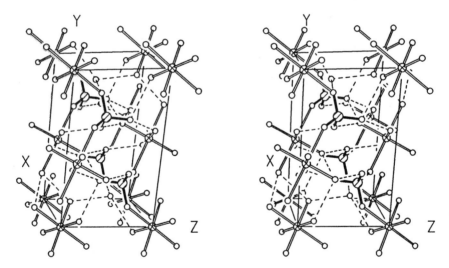

Fig. 2.21-1 Stereoview of the crystal structure of $CuSO_4 \cdot 5D_2O$.
Hydrogen bonds are indicated by broken lines.

Crystal Structure[1]

In the unit cell (Fig. 2.21-1), each of the two non-equivalent Cu atoms has four H_2O molecules and two *trans* sulfato O atoms as ligands, arranged in the form of an elongated octahedron.   The residual water molecule O(9) plays

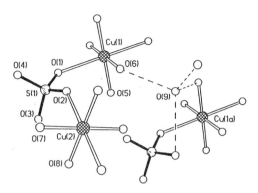

Fig. 2.21-2 Atom numbering and coordination modes in $CuSO_4.5H_2O$. The hydrogen-bonds formed by water molecule O(9) are represented as dotted lines.

a space-filling role and links the two independent octahedra and the sulfato groups through hydrogen bonding (Fig. 2.21-2). Since the water molecules are of two different kinds, the compound is more appropriately formulated as $Cu(H_2O)_4SO_4.H_2O$. The bond lengths and angles of the coordination octahedra about the Cu atoms and the sulfato group are shown in Fig. 2.21-3.

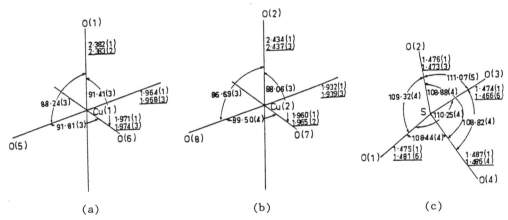

(a)	(b)	(c)

Fig. 2.21-3 Bond lengths (Å) and angles (°) involving (a) Cu(1), (b) Cu(2), and (c) the sulfato group. Neutron diffraction bond lengths, calculated from the coordinates of Bacon and Titterton (1975) with the X-ray cell dimensions, are underlined. (After ref. 1).

Remarks

Copper sulfate pentahydrate is unique in its historical significance, and studies of its crystal structure over a span of 73 years reflect the development of diffraction methods and chemical bonding theories and their mutual interaction.

1. The first diffraction pattern

    Copper sulfate pentahydrate has the distinction of being the first crystal used to record an X-ray diffraction pattern.[2]

    In 1895, when Röntgen discovered X-rays, scientists could not decide whether it consisted of waves or of streams of particles, since the wavelength is small, of the order of 10^{-8} cm, and early attempts to detect diffraction effects using a slit were unsuccessful. The idea came to von Laue, as a result of a discussion with Ewald on a theoretical problem, that a crystal might act as a three-dimensional diffraction grating for X-rays.

    Friedrich and Knipping, who were research students, selected a copper sulfate pentahydrate crystal to record the first X-ray diffraction pattern, as shown in Fig. 2.21-4. This experimental finding clarified the nature of X-rays and marked the beginning of the science of X-ray crystallography.

Fig. 2.21-4   The first X-ray diffraction pattern.   (After ref. 2).

2. Locating the non-hydrogen atoms by X-ray diffraction

    In 1934, Beevers and Lipson determined the crystal structure of copper sulfate pentahydrate.[3]   Their work showed that the compound is composed of $[Cu(H_2O)_4]^{2+}$ and SO_4^{2-} ions and isolated H_2O molecules. Each Cu atom is coordinated octahedrally by four H_2O molecules and two more distant oxygens from the SO_4^{2-} groups. The H_2O molecules in this crystal have two different functions, ligating and space-filling. The results obtained in the investigation thus directly elucidated the relation between the chemical and physical properties of copper sulfate pentahydrate and its structure.

3. Locating the protons by neutron diffraction

    In 1962, Bacon and Curry located the protons of copper sulfate pentahydrate by neutron diffraction,[4] and later in 1975, refined the structural parameters using more precise data.[5]   The accurate O-H...O, O-H and H...O bond lengths are listed in Table 2.21-1. The angles at the O atoms of the water molecules are listed in Table 2.21-2. The H-O-H angles of the ligating water molecules are larger than that of the space-filling water molecule. This is consistent with the fact that the ligating water molecules form part of a positively charged complex cation.

Table 2.21-1 Distances (Å) in hydrogen bonds.[1]

	O...O	O-H	H...O
O(5)-H(5a)...O(4)	2.852(4)	0.962(6)	1.911(6)
O(5)-H(5b)...O(9)	2.778(4)	0.958(5)	1.832(6)
O(6)-H(6a)...O(2)	2.793(4)	0.964(8)	1.894(6)
O(6)-H(6b)...O(9)	2.742(4)	0.980(4)	1.768(4)
O(7)-H(7b)...O(4)	2.749(3)	0.956(4)	1.804(4)
O(7)-H(7a)...O(3)	2.709(3)	0.970(3)	1.742(3)
O(8)-H(8a)...O(3)	2.670(4)	0.965(6)	1.706(6)
O(8)-H(8b)...O(4)	2.704(3)	0.962(4)	1.756(4)
O(9)-H(9a)...O(1)	2.778(3)	0.976(4)	1.816(4)
O(9)-H(9b)...O(2)	2.971(6)	0.912(12)	2.092(10)

Table 2.21-2 Angles (°) at O atoms of water molecules.

	O...O...O	H-O-H
O(4), H(5a)-O(5)-H(5b), O(9)	120.1(1)	108.5(5)
O(2), H(6a)-O(6)-H(6b), O(9)	129.9(2)	109.2(6)
O(4), H(7b)-O(7)-H(7a), O(3)	118.3(1)	112.9(3)
O(3), H(8a)-O(8)-H(8b), O(4)	105.4(1)	110.2(5)
O(1), H(9a)-O(9)-H(9b), O(2)	122.4(2)	107.2(6)

4. The Jahn-Teller distortion

For the Cu^{2+} ion in an ideal octahedral ligand field the 3d wave-functions split into a lower triplet (t_{2g}) which is completely filled and an upper doublet (e_g) which accommodates the remaining three electrons. The state functions for the latter are $(3d_{z^2})^2(3d_{x^2-y^2})^1$ and $(3d_{z^2})^1(3d_{x^2-y^2})^2$. It is this degeneracy which is removed by distortion according to the Jahn-Teller Theorem, the four water O atoms defining the *xy* plane moving towards Cu, and the two sulphato O atoms along *z* moving away. The e_g doublet splits into two levels, d_{z^2} being stabilized and the uppermore $d_{x^2-y^2}$ being destabilized and hence only half occupied. The remaining t_{2g} orbitals undergo similar splitting, but this should not affect their occupancy.

In 1985, the structure of copper sulfate pentahydrate was subjected to a detailed study based on parameters derived from accurate data collected for a charge density analysis.[1] The integrated difference density within a distance of 1.1Å of the Cu nuclei was evaluated for six equal parts of a sphere, described by bisecting idealized angles between the Cu-O bonds. The results listed in Table 2.21-3 are qualitatively consistent with the Jahn-Teller prediction. $P_{x^2-y^2}$ is the net electron count in the segments containing the Cu-O(water) vectors, and P_{z^2} the corresponding figure for the Cu-O(sulphato) vectors. P_{tot}, the total electron count for the whole sphere, is close to the expected value of -1 in each case.

Table 2.21-3   Integrated electron populations within a sphere of radius 1.1Å centered on the Cu atoms.

	$P_{x^2-y^2}$	P_{z^2}	P_{tot}
Cu(1)	-0.72(1)	-0.29(1)	-1.01(2)
Cu(2)	-0.84(1)	-0.12(1)	-0.96(2)

$CrSO_4 \cdot 5H_2O$[6] [space group $P\bar{1}$, a = 6.188(1), b = 10.929(2), c = 6.039(1)Å, α = 82.40(2), β = 107.77(1), γ = 102.71(2)°] is to a first approximation isomorphous with $CuSO_4 \cdot 5H_2O$, and its bonding features have been explored in an electron-density study using an accurate set of X-ray data. The anisotropy of the deformation density $\Delta\rho$ around the two independent Cr nuclei is related to unequal occupancy of the 3d metal orbitals, as expected in view of the Jahn-Teller distortion in the structure. The electron density along the Cr-O bonds is depleted less heavily than that along the Cu-O vectors in the nearly isostructural $CuSO_4 \cdot 5H_2O$ structure.

References

[1] J.N. Varghese and E.N. Maslen, *Acta Crystallogr., Sect. B* **41**, 184 (1985).

[2] W. Friedrich, P. Knipping and M. Laue, *Sitzungber. (Kgl.) Bayerische Akad. Wiss.*, 303-322 (1912). [English translation by J.J. Stezowski in J.P. Glusker (ed.), *Structural Crystallography in Chemistry and Biology*, Hutchinson Ross, Stroudsburg, 1981, pp. 23-39; distributed by Academic Press.]

[3] C.H. Beevers and H. Lipson, *Proc. Roy. Soc. London* **A146**, 570 (1934).

[4] G.E. Bacon and N.A. Curry, *Proc. Roy. Soc. London* **A266**, 95 (1962).

[5] G.E. Bacon and D.H. Titterton, *Z. Kristallogr.* **141**, 330 (1975).

[6] T.P. Vaalstra and E.N. Maslen, *Acta Crystallogr. Sect. B* **43**, 448 (1987).

Note Electron deformation densities and chemical bonding are discussed by A.A. Low and M.B. Hall in Maksic (ed.), *Theoretical Models of Chemical Bonding*, Part 2: *The Concept of the Chemical Bond*, Springer-Verlag, Berlin, 1990, p.543.

2.22 Diopside
CaMg(SiO₃)₂

Crystal Data

    Monoclinic, space group $C2/c$ (No. 15)
    $a = 9.7456(7)$, $b = 8.9198(8)$, $c = 5.2516(5)$Å, $\beta = 105.86(1)°$[7]
    $Z = 4$

Atom	x	y	z
Ca	0	.30144(3)	1/4
Mg	0	.90814(5)	1/4
Si	.28627(3)	.09330(3)	.22936(5)
O(1)	.11550(7)	.08728(7)	.1422(1)
O(2)	.36136(7)	.25013(8)	.3183(1)
O(3)	.35083(7)	.01759(8)	.9953(1)

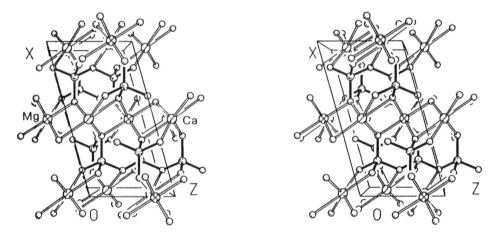

Fig. 2.22-1 Crystal structure of diopside. Pairs of one-dimensional silicate chains overlap each other in this view down the *b* axis. The Si-O bonds are represented by solid black lines, and the Mg and Ca atoms differentiated by different shading.

Crystal Structure

    Diopside is a member of the well-known rock-forming pyroxene group of minerals, which are unbranched single chain silicates. Its structure was first determined by Warren and Bragg[1] and refined in 1969[2], 1981[7] and 1982.[8] In the structure (Fig. 2.22-1), the [SiO₃]ₙ chains run parallel to the *c* axis. Pairs of chains are arranged back to back without displacement in the *b* direction, but staggered in the *c* direction so that monoclinic symmetry is attained. The mean bridging Si-O distance is 1.676Å and the mean non-bridging Si-O distance is 1.594Å. The chains are held together by the Ca and Mg ions. Each Mg ion is octahedrally surrounded by six "active" oxygens, i.e.

oxygens with only one link to silicon. Each "active" oxygen is linked to one
Ca ion and one Mg ion. The Ca ion is in an irregular eight-fold coordination
site, having four short Ca-O distances with a mean of 2.36Å and four longer
ones with a mean of 2.64Å.

Remarks

1. First application of two-dimensional Fourier synthesis in the solution of a
   crystal structure[3]

   W.H. Bragg proposed in 1915 that the Fourier method should be applicable
to X-ray analysis. The crystal may be considered to be built up of stacks of
strata, each being parallel to a set of planes in the crystal lattice. If the
amplitudes and phases for all the sets of crystal planes are known, they can
be combined, criss-crossing each other in all directions, and the result will
be a picture of the electron density $\rho(xyz)$ everywhere in the crystal.

   Putting this in formal mathematical language:

$$\rho(xyz) = \frac{1}{V} \overset{hk\ell}{\Sigma\Sigma\Sigma} \left| F(hk\ell) \right| \cos[2\pi hx/a + 2\pi ky/b + 2\pi \ell z/c + \alpha(hk\ell)],$$

where the structure amplitude $\left| F(hk\ell) \right|$ is measured in electron units, and
the cosine function gives the sinusoidal variation at fractional intervals of
a/h, b/k, and c/ℓ along the unit cell axes modified by a phase $\alpha(hk\ell)$. When
scaled by the factor $1/V$, where V is the volume of the unit cell, the
resulting electron density $\rho(xyz)$ is measured in number of electrons per $Å^3$.
The summation of the series gives the electron density at each point (x,y,z),
and the maxima correspond to the positions of the atoms.

   Diopside was the first crystal structure to be solved using a two-
dimensional Fourier series. Fig. 2.22-2 shows the $\rho(xz)$ map matching a
projection of the diopside structure upon (010).

2. Single and double chains in silicates[9,10]

   Chain silicates formed by corner-sharing of $[SiO_4]$ tetrahedra are
particularly prevalent in nature and many important minerals have this basic
structural unit. Despite the apparent simplicity of their motif and
stoichiometry, considerable structural diversity is encountered because of the
differing conformations that can be adopted by the linked tetrahedra. As a
result, the repeat distance along the chain axis can correspond to (1), 2, 3,
4, 5, 6, 7, 9, 12... tetrahedra, as illustrated schematically for some cases
in Fig. 2.22-3. The most common conformation for the single chain silicates
is a repeat after every second tetrahedron with the chains stacked parallel so
as to provide sites of 6- or 8-coordination for the cations; e.g. the pyroxene

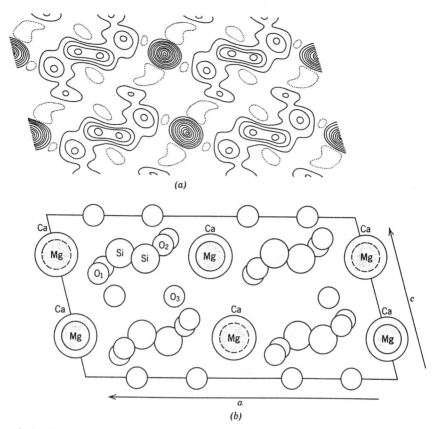

Fig. 2.22-2 The $\rho(xz)$ map and the projection of the diopside structure
upon (010). (After ref. 3).

minerals enstatite $Mg_2(SiO_3)_2$, diopside $CaMg(SiO_3)_2$, jadeite $NaAl(SiO_3)_2$, and
spodumene $LiAl(SiO_3)_2$. The synthetic silicates Li_2SiO_3 and Na_2SiO_3 are
similar. Other examples of unbranched single chains found in silicates
illustrated in Fig. 2.22-3(a) are: wollastonite $Ca_3(SiO_3)_3$, synthetic silicate
$Na_2Cu_3(SiO_3)_4$, rhodonite $(Mn,Ca)_5(SiO_3)_5$, and pyroxfemoite $(Fe,Ca)_7(SiO_3)_7$.

The single $(SiO_3)_\infty$ chains can link laterally to form double chains or
ribbons whose stoichiometry depends on the repeat unit of the single chain.
Some examples of double chains of $[SiO_4]$ tetrahedra are shown in Fig. 2.22-
3(b). The double chains based on the stoichoimetry $[Si_2O_5]^{2-}$ are found in the
aluminosilicate sillimanite $Al(AlSiO_5)$. By far the most numerous are the
amphiboles or asbestos minerals which adopt the $[Si_4O_{11}]^{6-}$ double chain, e.g.
tremolite $Ca_2Mg_5(Si_4O_{11})_2(OH)_2$, the structure of which is very similar to that
of diopside. The double chain $[Si_6O_{17}]^{10-}$ occurs in the mineral xonotlite
$Ca_6(Si_6O_{17})(OH)_2$.

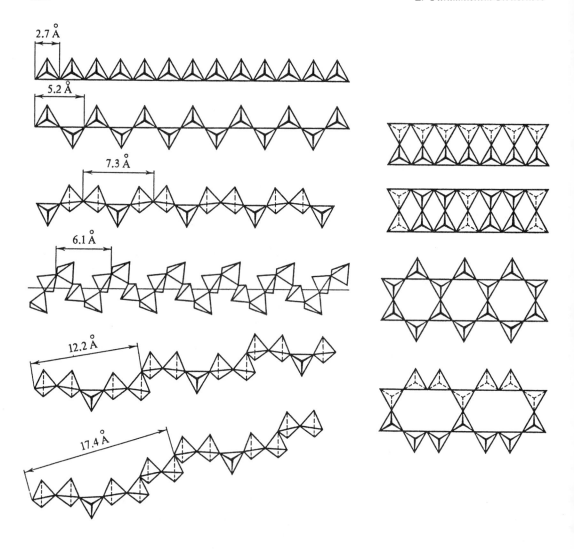

(a)   (b)

Fig. 2.22-3 Various types of chains in silicates, (a) single chains, (b) double chains. (After ref. 9).

3. The pyroxenes and amphiboles

   The pyroxenes and amphiboles represent a well-defined group of minerals varying widely in composition but having certain close relationships in their optical properties, form and cleavage, which have led mineralogists to classify them together. Crystal structure determination has afforded ample verification of the soundness of the mineralogical classification. The

pyroxenes may all be referred to a general formula of the type $M[SiO_3]$, with
in some cases a partial substitution of Si by Al and M by M'. The amphiboles
are similar in composition; for example, the formula of tremolite, which used
to be given as $CaMg_3[SiO_3]_4$, is actually $Ca_2Mg_5Si_8O_{22}(OH)_2$. The essential
difference between the pyroxene and amphibole structures is that the former is
based on single chains of linked tetrahedral groups, whereas the latter is
based on double chains, as shown in Fig. 2.22-4.[4]

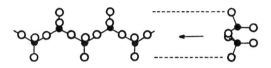

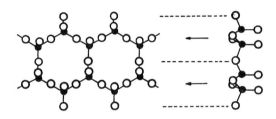

Fig. 2.22-4 Silicon-oxygen chains in pyroxenes (above) and amphiboles
(below). (Adapted from ref. 4).

The characteristic difference between pyroxenes and amphiboles lies in
their cleavage. The angle between the cleavage cracks is 93° in a pyroxene,
and 56° in an amphibole. Fig. 2.22-5 shows the relation between the structure
and cleavage of pyroxene and amphibole.

4. An example of electron transfer in the Si-O bond

In orthoenstatite, $Mg_2[SiO_3]_2$, a member of the pyroxenes, the structure
of its single chain of tetrahedral silicate anions is similar to that in
diopside. The distribution of the valence electrons of $Mg_2[SiO_3]_2$ has been
determined from very accurately measured X-ray and neutron diffraction
data.[5] The charge density distributions demostrate that there is an
accumulation of electron density between the silicon and the oxygen atoms,
suggesting that the Si-O bond has a significant covalent component. The
O-Si-O bond angles show only small deviations from the tetrahedral value
109.47°. The directional character of the covalent Si-O bond is more
convincingly displayed by the fact that the majority of Si-O-Si angles are
found in a rather small range near 140°.

The transfer of electrons from the Mg cations to oxygen permits an

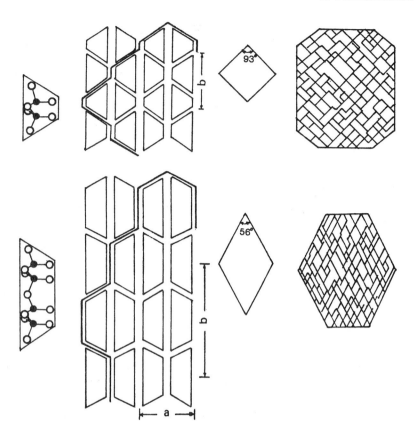

Fig. 2.22-5 The relation between the structure and cleavage of pyroxene and amphibole. (Adapted from ref. 4).

explanation of the positions of the residual electron density maxima in the map obtained for orthoenstatite (Fig. 2.22-6).[5,6] The average Si-to-residual maximum distance is 0.71Å for $Si-O_t$ bonds and 1.14Å for the $Si-O_b$ bonds, whereas the corresponding Si-O distances are 1.602Å and 1.666Å, respectively. The electrons in the covalent component of the bonds are nearer to the silicon atom than to the terminal oxygen atom, but are farther from the silicon atom than from the bridging oxygen.

In the bond system $Si-O_b-Si-O_t...Mg$, Mg transfers more electrons to its neighboring oxygen atom than does Si, owing to its lower electronegativity ($x_{Mg} = 1.23$, $x_{Si} = 1.74$). As a consequence, the slightly more negative O_t atom transfers part of its electron density to the adjacent silicon atom, thus shifting the electron density maximum further towards Si than does the less negative oxygen atom O_b. In agreement with this model, slightly higher charges have been found for the terminal than the bridging oxygen atom.

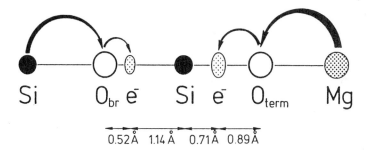

Fig. 2.22-6 Location of the residual electron density maxima obtained from difference Fourier maps of orthoenstatite, and suggested electron transfer in the bond system Si-O_b-Si-O_t...Mg. (After ref. 6).

The bond between silicon and oxygen is expected to be partly ionic and partly covalent. From the ionic model, the mean atomic charges averaged over the crystallographically non-equivalent atoms in $Mg_2[SiO_3]_2$ are: Mg +1.82, Si +2.28, O_t -1.42, and O_b -1.27. According to the covalent model, each lobe of the Si sp^3 hybrid set can overlap "head on" with a 2p orbital of an oxygen atom to form a σ bond, and there is also some π overlap of the remaining 2p orbitals of oxygen with the d orbitals of silicon, imparting some double bond character to the Si-O bond.

References

[1] B.E. Warren and W.L. Bragg, *Z. Kristallogr.* **69**, 168 (1928).

[2] J.R. Clark, D.E. Appleman and J.J. Papike, *Min. Soc. Am. Spec. Pap.* **2**, 31 (1969).

[3] W.L. Bragg, *The Development of X-ray Analysis*, Bell, London, 1975; W.L. Bragg, *Z. Kristallogr.* **70**, 475 (1929).

[4] W.L. Bragg and G.F. Claringbull, *Crystal Structures of Minerals*, Bell, London, 1965.

[5] S. Sasaki, Y. Takéuchi, K. Fujino and S.I. Akimoto, *Z. Kristallogr.* **158**, 279 (1982).

[6] F. Liebau, *Structural Chemistry of Silicates*, Springer-Verlag, Berlin, 1985.

[7] L. Levien and C.T. Prewitt, *Amer. Min.* **66**, 315 (1981).

[8] G. Rossi, S. Ghose and W.R. Busing, *Geol. Soc. Am. Abstr.* **14**, 603 (1982).

[9] B.K. Vainshtein, V.M. Fridkin and V.L. Indenbom, *Modern Crystallography II, Structure of Crystals*, Springer-Verlag, Berlin, 1982.

[10] F. Liebau, *Acta Crystallogr.* **12**, 180 (1959).

2.23 Calcite (Calcium Carbonate)
CaCO₃

Crystal Data

Trigonal, space group $R\bar{3}c$ (No. 167)

Hexagonal cell: $a = 4.9898(3)$, $c = 17.060(5)$Å, (300K), $Z = 6$

Atom	Position	x	y	z
Ca	6(b)	0	0	0
C	6(a)	0	0	1/4
O	18(e)	.25706	0	1/4

Rhombohedral cell: $a = 6.3748(5)$Å, $\alpha = 46.08(4)°$, $Z = 2$

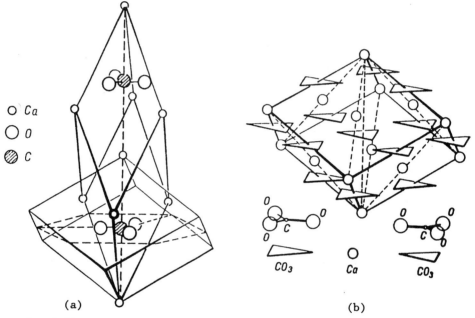

(a)   (b)

Fig. 2.23-1 Crystal structure of calcite: (a) true unit cell, (b) as a
distorted NaCl structure. (After ref. 8).

Crystal Structure

The structure of calcite was one of the earliest to be determined by
W.L. Bragg,[1] and was later refined by Chessin *et al*.[2] Calcite has a very
perfect cleavage which outlines crystal blocks of the well known rhombohedral
form. It is customary to choose the morphological axes of calcite in such a
way that the cleavage rhomb may be described as having the faces {100}, but
this does not correspond to the smallest unit cell. The true unit cell has a
much more elongated form with a rhombohedral angle of 46.1°, whereas the
cleavage rhomb has a rhombohedral angle of 101.9° (Fig. 2.23-1).

The CO_3^{2-} ion is a planar equilateral triangle with carbon at the center and C-O = 1.283Å; the whole group lies in a plane at right angles to the three-fold axis. The Ca^{2+} ions occupy the corner of the cell, each being surrounded by six oxygen atoms of six different CO_3^{2+} ions, with Ca-O = 2.36Å; each oxygen atom is bound to two Ca^{2+} ions.

Remarks

1. Polymorphs of calcium carbonate

    Calcium carbonate crystallizes naturally with either the calcite or aragonite structure. Calcite is the thermodynamically stable phase at atmospheric pressure and room temperature, while aragonite is considered the stable high-pressure phase. Bridgman[5] discovered two new room-temperature transitions at 14 and 17 kbar (now revised to 15 and 22 kbars) based on volumetric compression data. Both of the phases CaCO₃(II) and CaCO₃(III) occur within the commonly accepted stability field of aragonite.

2. Crystal structure of aragonite

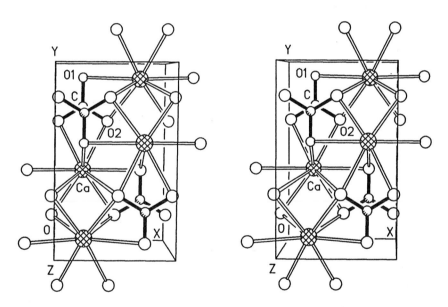

Fig. 2.23-2 Stereoview of the crystal structure of aragonite.

Aragonite is one of three high-pressure phases of CaCO₃, the form stable up to 15 kbar at ordinary temperatures.[3] Aragonite is orthorhombic, space group *Pmcn*, with a tetramolecular unit of dimensions: a = 4.9614, b = 7.9671, and c = 5.7404Å. The Ca^{2+} ions in calcite and those in aragonite are arranged in approximate cubic and hexagonal close packing, respectively. In each case the CO_3^{2-} group occupy a position between six Ca^{2+} ions, but there is a very important difference. In calcite the CO_3^{2-} group is so placed that each oxygen touches two Ca^{2+}, and is just half way between the upper and lower groups of three Ca^{2+}. In aragonite, each oxygen atom touches three Ca^{2+}; the CO_3^{2-} group consequently lies nearer the upper than the lower groups of oxygens, and Ca^{2+} is nine coordinated by oxygen atoms (Fig. 2.23-2).

3. Compounds with the calcite and aragonite structures

The calcite and aragonite structures are found in a large number of ABO_3 compounds, including the nitrates, carbonates and borates shown in Table 2.23-1. The choice of structure is determined by the size of the cation. When the cation is small, the calcite structure would be favored, and when it is large, the aragonite structure would be preferred, the transition taking place at a cation radius of about $0.7r_0 = 0.98$Å.[4] Since the radius of the Ca^{2+} ion is so close to this critical value, calcium carbonate occurs in both structures. When the radius of the cation exceeds about 1.45Å, the aragonite structure in turn becomes unstable, and compounds like $RbNO_3$ and $CsNO_3$ adopt other structures.

Table 2.23-1 Some compounds with the calcite and aragonite structure.[4]

	Calcite structure			Aragonite structure	
Compound	Cation radius/Å	Compound	Cation radius/Å	Compound	Cation radius/Å
$LiNO_3$	0.60	$ScBO_3$	0.81	$CaCO_3$	0.99
$MgCO_3$	0.65	$InBO_3$	0.81	$SrCO_3$	1.13
$CoCO_3$	0.72	YBO_3	0.93	$LaBO_3$	1.15
$ZnCO_3$	0.74	$NaNO_3$	0.95	$PbCO_3$	1.21
$MnCO_3$	0.80	$CdCO_3$	0.97	KNO_3	1.33
$FeCO_3$	0.80	$CaCO_3$	0.99	$BaCO_3$	1.35

4. The refractivity of calcite and aragonite

The birefringence of calcite and of aragonite is so large that these crystals provide an excellent example of the relation between refractivity and crystal structure.[6]

The refractive indices of calcite and aragonite for the sodium D line are as follows (electric vector **E**):

Calcite $\epsilon = 1.486$ (**E** parallel to threefold axis)

                  $\omega = 1.658$ (**E** perpendicular to threefold axis)

Aragonite $\alpha = 1.530$ (**E** parallel to *c* axis)

                  $\beta = 1.681$ (**E** parallel to *a* axis)

                  $\gamma = 1.686$ (**E** parallel to *b* axis)

Acute bisectrix, *c* axis; optic axial plane, (100).

The density of calcite is 2.75, and that of aragonite 2.94 g cm^{-3}. Allowing for this difference in density, it is noted that ϵ for calcite corresponds to α for aragonite, and ω for calcite to β and γ for aragonite. The cause of this correspondence is that the CO_3^{2-} group is responsible for the birefringence in both cases. The CO_3^{2-} groups are perpendicular to the threefold axis of calcite, and very nearly perpendicular to the c-axis of aragonite. A high refractive index implies a large degree of polarization of the atoms in the crystal as the light waves sweep through it. In both crystals the refractive index is high when **E** of the light wave is parallel to the plane of the CO_3^{2-} group and low when **E** is perpendicular to the plane of the CO_3^{2-} group. It is clear that the CO_3^{2-} group, which has delocalized π bonding over the four atoms, is more easily polarizable when the electric field is parallel to the molecular plane.

5. Electron density distribution in the carbonate group

Magnesite (MgCO₃) is an isomorph of calcite. A study of the electron-density distribution in magnesite[7] has yielded the absolute electron density [Fig. 2.23-3(a)] and the deformation density [Fig. 2.23-3(b)] through the plane of the CO_3^{2-} anion.

From these maps neither deviation from planarity nor disorder was observed in the CO_3^{2-} group. In the absolute density map, ρ has a maximum of 2.55 eÅ^{-3} at the midpoint of the C-O bond and 0.18 eÅ^{-3} between the Mg and O atoms. The deformation density map shows a maximum of 0.50 eÅ^{-3} at the midpoint of the C-O bond and another two maxima of 0.52 eÅ^{-3}, related by a

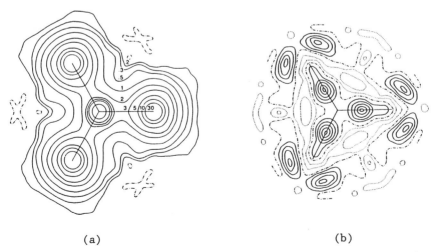

(a)   (b)

Fig. 2.23-3 Absolute electron density (a) and deformation density ($\rho_o - \rho_c$) (b) through the plane of the CO_3^{2-} anion in MgCO₃. (After ref. 7).

twofold axis, in the lone-pair region of the O atom, forming with it an angle of 101(1)°. One of these maxima is directed towards the Mg atom although shifted from the Mg-O bond line, the distance to the Mg and O atoms being 1.90(5) and 0.55(5)Å, respectively. It can also be observed that the maximum at the C-O bond is elongated in the z direction as a consequence of the π bonding in the $CO_3{}^{2-}$ group. These results elucidated the nature of the covalent bond in the $CO_3{}^{2-}$ anion.

References

[1] W.L. Bragg, *Proc. Roy. Soc. London* **A89**, 468 (1914).

[2] H. Chessin, W.C. Hamilton, B. Post, *Acta Crystallogr.* **18**, 689 (1965).

[3] J.P.R. de Villiers, *Amer. Min.* **56**, 758, (1971); A. Dal Negro and L. Ungaretti, *Amer. Min.* **56**, 768 (1971).

[4] R.C. Evans, *An Introduction to Crystal Chemistry*, 2nd ed., Cambridge University Press, London, 1976.

[5] P.W. Bridgman, *Am. J. Sci.* **237**, 7 (1939).

[6] W.L. Bragg and G.F. Claringbull, *Crystal Structures of Minerals*, Bell, London, 1965.

[7] S. Göttlicher and A. Vegas, *Acta Crystallogr., Sect. B* **44**, 362 (1988).

[8] A.S. Povarennyhk, *Crystal Chemical Classification of Minerals*, Vol. I, (translated from Russian by J.E.S. Bradley), Plenum Press, New York, 1972.

2.24 Hexamethylenetetramine
(1,3,5,7-Tetraazaadamantane, Urotropine)
$(CH_2)_6N_4$

Crystal Data

Cubic, space group $I\bar{4}3m$ (No. 217)

$a = 7.021(9)$Å, $Z = 2$

Atom	Position	x	y	z
C	12(e)	.2377	0	0
N	8(c)	.1235	.1235	.1235
H	24(g)	.0885	.0885	-.3275

Values are from ref. 5; 298K, MoKα data.

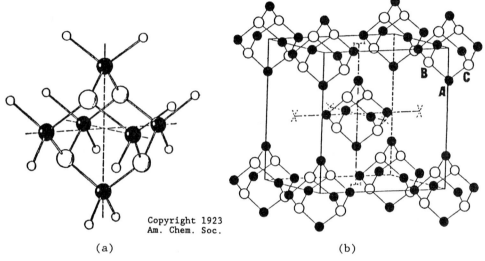

Copyright 1923
Am. Chem. Soc.

(a)   (b)

Fig. 2.24-1 (a) Molecular structure of $(CH_2)_6)N_4$, with H atoms represented by small open circles. (b) Unit cell of $(CH_2)_6N_4$; the H atoms are not shown. (After ref. 1).

Crystal Structure

The cage geometry of hexamethylenetetramine, $(CH_2)_6N_4$, may be described as consisting of four nitrogen atoms tetrahedrally, and six CH_2 groups octahedrally, grouped about a common center [Fig. 2.24-1(a)]. The ideal molecular symmetry of $T_d(\bar{4}3m)$ is fully retained in the crystal, which consists of a body-centered packing of $(CH_2)_6N_4$ molecules in the unit cell as shown in Fig. 2.24-1(b). Each molecule is surrounded by eight neighbours along <111> and six others at <100>; the former contacts allow a nitrogen of one molecule to fit snugly into a recess formed by three hydrogen atoms of a neighbouring molecule, while the latter contacts allow each CH_2 group to nestle at 90°

across a neighbouring CH_2 group belonging to a molecule in an adjacent unit cell.

The crystal structure was first determined by Dickinson and Raymond in 1923[1] and refined in the 1930s.[2,3] In 1957, Andresen[4] used single-crystal neutron diffraction to determine the coordinates, particularly those of hydrogen, and to study the atomic vibrations. In 1963, Becka and Cruickshank[5] refined the structure with Cu and Mo X-ray data collected at 298, 100 and 34K, and analysed the anisotropic atomic-vibration amplitudes in terms of a lattice model. The corrected weighted-mean dimensions of the $(CH_2)_6N_4$ molecule at 298K were derived: C-N 1.476(2)Å, N-C-N = 113.6(2)°, C-N-C = 107.2(1)°, and C-H = 1.088(11)Å.

Making use of X-ray and neutron diffraction data, Duckworth, Willis and Pawley in 1970 computed an X-N deformation density map which clearly revealed the nitrogen lone pairs.[6]

Remarks

1. Historical significance

Hexamethylenetetramine, in view of its high symmetry and cage geometry, holds the distinction of being the first organic structure to be completely determined by X-ray analysis. This elegant work by Dickinson and Raymond[1] is a classic example of the application of space group theory to molecular crystals.

2. Hexamethylenetetramine as a reagent in organic synthesis[7]

Hexamethylenetetramine (263°C, sublimation) is formed in almost quantitative yield from the condensation of formaldehyde and ammonia:

$$6HCHO + 4NH_3 \underset{H^+}{\overset{NH_3}{\rightleftharpoons}} (CH_2)_6N_4 + 6H_2O.$$

The polycyclic tertiary amine is a monoacidic base, and its quaternary salts with alkyl halides are used in the well known Delépine reaction for the synthesis of primary amines and the Sommelet reaction for the synthesis of aldehydes.

Besides serving as a source of formaldehyde and ammonia, hexamethylene-tetramine has been used for the synthesis of triaza- and tetraaza systems and for ring-closure to form five-, six-, or seven-membered heterocycles.

3. Structural chemistry of hexamethylenetetramine

Hexamethylenetetramine forms a bewildering number of molecular adducts with various organic compounds and coordination compounds with many inorganic salts.[8] Its structural chemistry is dominated mainly by the nitrogen lone pair, as summarized in Table 2.24-1.

Table 2.24-1 Structural chemistry of $(CH_2)_6N_4$.

+ HX (weak)	$\longrightarrow$	$\equiv$N....H-X	hydrogen-bonded molecular adduct
+ HX (strong)	$\longrightarrow$	$\equiv$N-H....X⁻	salt formation involving proton transfer
+ RX	$\longrightarrow$	$\equiv\overset{+}{N}$-R X⁻	quaternary ammonium salt
+ M^{n+}	$\longrightarrow$	$\equiv$N→M^{n+}	coordination compound
+ H$_2$O$_2$	$\longrightarrow$	$\equiv$N→O + H$_2$O	tertiary amine oxide

When one of its four lone pairs coordinates to a metal center as in $(CH_2)_6N_4 \cdot Mo(CO)_5$, or engages a proton as in the $[(CH_2)_6N_4H]^+$ cation, or partakes in quaternization of its parent N atom as in the $[(CH_2)_6N_4CH_3]^+$ and $[(CH_2)_6N_4CH_2C_6H_5]^+$ cations, the $(CH_2)_6N_4 \cdot BH_3$ donor-acceptor adduct, and the $(CH_2)_6N_4O$ amine oxide, the cage symmetry is reduced to $C_{3v}(3m)$.[9,10] Furthermore, moving away from the quaternary N atom, the three sets of C-N bonds vary in the order long, short, and normal relative to the standard bond length of 1.476(2)Å in crystalline $(CH_2)_6N_4$; as a typical example the measured molecular dimensions of $(CH_2)_6N_4O$ are given in Fig. 2.24-2.[11]

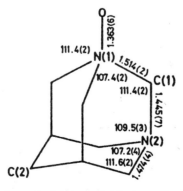

Fig. 2.24-2 Bond lengths (Å) and bond angles (°) of $(CH_2)_6N_4O$. (After ref. 11).

The extent of distortion of the $(CH_2)_6N_4$ cage system by various means of quaternizing one bridge-head nitrogen atom is compared in Table 2.24-2.[9] The data show that cage distortion decreases in the order methylation > borine adduct formation $\approx$ *N*-oxide formation $\approx$ protonation > metal coordination and that, in the last instance, variation of the formal oxidation state of the metal atom produces virtually no effect.

Table 2.24-2   Distortion of the $(CH_2)_6N_4$ cage system through quaternization of one nitrogen atom.[a]

Type	Compound	$C_\alpha-N_q$	$C_\alpha-N$	C-N	$C_\alpha N_q-C_\alpha$
	$(CH_2)_6N_4$	1.476(2)	1.476(2)	1.476(2)	107.2(1)
Methylation	$[(CH_2)_6N_4CH_3]Br.H_2O$	1.535(15)	1.430(5)	1.469(9)	107.8(5)
Borine adduct formation	$(CH_2)_6N_4.BH_3$[b]	1.527(5)	1.475(6)	1.475(6)	107.4(3)
N-oxide formation	$(CH_2)_6N_4O$	1.514(2)	1.445(7)	1.474(4)	107.4(2)
Protonation	$[(CH_2)_6N_4H]Cl$	1.526(7)	1.453(10)	1.469(5)	109.3(4)
	$[(CH_2)_6N_4H]Br$	1.512(8)	1.458(7)	1.465(5)	109.7(3)
Coordination	$[(CH_2)_6N_4]_4OCu_4Cl_6$	1.514(6)	1.453(10)	1.468(12)	107.0(9)[c]
	$(CH_2)_6N_4.Mo(CO)_5$	1.504(5)	1.455(4)	1.466(6)	106.5(4)

[a] The quaternary nitrogen is indicated by subscript q and a carbon atom bonded to it by subscript α.

[b] The measured $C_\alpha-N$ bond in this compound is abnormally long and does not fit the general pattern of C-N bond-length variation.

[c] Value not given in the original paper but calculated from the reported atomic coordinates.

Table 2.24-3   Hydrogen-bonded molecular adducts of hexamethylenetetramine.

Number of nitrogen lone pairs utilized	Example	Structure
1	$(CH_2)_6N_4.C_6H_5CH_2COOH$	monomeric
2	$(CH_2)_6N_4.2m-CH_3C_6H_4OH$	monomeric, butterfly-like
	$(CH_2)_6N_4.p-C_6H_4(OH)_2$	infinite zigzag chain
3	$(CH_2)_6N_4.3C_6H_5OH$	monomeric, propeller-like
	$(CH_2)_6N_4.6H_2O$	clathrate hydrate
4	$(CH_2)_6N_4.2(NH_2)_2CS$	corrugated layer

All four nitrogen lone pairs of $(CH_2)_6N_4$ can form acceptor hydrogen bonds with a variety of donor molecules (Table 2.24-3), and in all cases there is no significant distortion of its cage geometry.[10]

As a ligand in metal complexes $(CH_2)_6N_4$ tends to generate polymeric crystal structures with variable coordination geometries about metal ions of the same kind; for example, in $(CH_2)_6N_4 \cdot 4AgCl$ octahedral $AgCl_5N$, tetrahedral $AgCl_3N$ and tetragonal pyramidal $AgCl_4N$ coordination polyhedra interconnect to form a three-dimensional network.[12]

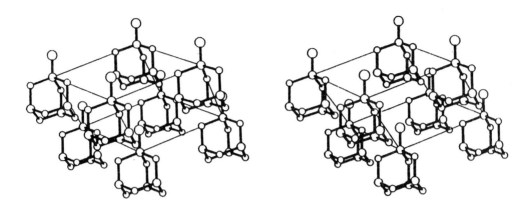

Fig. 2.24-3 Stereo plot showing the packing of $(CH_2)_6N_4O$ molecules in a rhombohedral unit cell. (After ref. 11).

4. Structural analogy of hexamethylenetetramine and some quaternized derivatives

A stereo view of the crystal structure of $(CH_2)_6N_4O$, which crystallizes in space group $R3m$ (No. 160) with $Z = 1$ in the rhombohedral unit cell, is shown in Fig. 2.24-3. The molecular packing is related to that of $(CH_2)_6N_4$ (space group $I\bar{4}3m$, $Z = 2$) in a simple way. Starting with a cubic unit cell of $(CH_2)_6N_4$, one can arrive at the $(CH_2)N_4$ structure by first removing the $(CH_2)_6N_4$ molecule at the body-center, then introducing the N→O function in the [111] direction to each remaining molecule, and finally compressing the resulting primitive lattice along [111] until the oxygen atom of each $(CH_2)_6N_4O$ molecule nests comfortably in a recess formed by three hydrogen atoms of a neighbouring molecule.

The borine adduct $(CH_2)_6N_4 \cdot BH_3$ and the hydrohalides $[(CH_2)_6N_4H]^+X^-$ (X = Cl, Br) are "isostructural" with $(CH_2)_6N_4O$; Table 2.24-4 shows the relationship of this series of compounds and the relative bulk of the different "substituents".

5. Crystal structure of adamantane

Adamantane, $(CH_2)_6(CH)_4$, crystallizes at room temperature in space group $Fm3m$ (No. 225) with $a = 9.445$Å and $Z = 4$.[13] The fcc arrangement is favored over the bcc structure of $(CH_2)_6N_4$ because the $(CH_2)_6(CH)_4$ molecule is

Table 2.24-4   Structural relationship of $(CH_2)_6N_4$ and some derivatives.

Compound	Space group	Z	α_R (Å)	α_R (°)	V_R (Å^3)/Z
$(CH_2)_6N_4$	$I\bar{4}3m$	2	7.021	90	173.0
$(CH_2)_6N_4O$	R3m	1	5.914	106.3	175.5
$[(CH_2)_6N_4H]Cl$	R3m	1	5.947	97.11	205.0
$(CH_2)_6N_4 \cdot BH_3$	R3m	1	6.15	103.2	210.6
$[(CH_2)_6N_4H]Br$	R3m	1	6.040	96.13	216.2

approximately spherical. The structure is disordered, *not* ordered in space
group $F\bar{4}3m$ as previously reported.[13] The model which gives the best fit
with the intensity data consists of hindered reorientations of a rigid
molecular skeleton with the CH groups aligned in the [111] directions.[14]

Below -65°C disordered cubic adamantane transforms to an ordered
tetragonal form [molecular symmetry $\bar{4}$, space group $P\bar{4}2_1c$ (No. 114), $a = 6.60$,
$c = 8.81$Å (at -110°C), $Z = 2$].[13] The measured C-C bond lengths have a mean
of 1.536(11)Å. The structure of the low-temperature phase (Fig. 2.24-4) is
related to that at room temperature in that a_T lies along the *ab* face diagonal
of the cubic cell ($a_T \sim a_C/\sqrt{2}$) and c_T corresponds to c_C.[15]

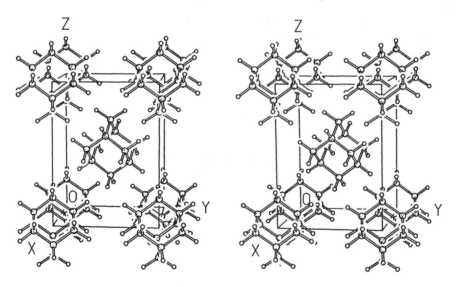

Fig. 2.24-4   Stereoview of the crystal structure of adamantane at -110°C.

6. Compounds possessing the adamantane skeleton

Many compounds with the adamantane (and also cubane) skeleton can be
formally derived by the exchange of "equivalent" atoms or groups.[16] For
instance, replacing all four "pseudonitrogen" CH groups of adamantane,
$(CH)_4(CH_2)_6$, gives rise to hexamethylenetetramine, $N_4(CH_2)_6$. Some examples of
adamantane-like cage compounds containing "paraelements"[17] are E_4O_6 (E = P,

As, Sb), $(MeC)_4(AlMe)_6$, $(HC)_4(BMe)_6$, $P_4(SnR_2)_6$ (R = Me, Bu, Ph), $(HSi)_4E_6$ (E = S, Se), $(MeE)_4S_6$ (E = Si, Sn, Ge), $(EP)_4S_6$ (E = O, S), $(EP)_4(NMe)_6$ (E = O, S, Se), and $(F_3Ta)_4O_6^{4-}$.

A general discussion of the concept of "element displacement" as a guiding principle in p-block element chemistry is to be found in ref. 17.

References

[1] R.G. Dickinson and A.L. Raymond, *J. Am. Chem. Soc.* **45**, 22 (1923).

[2] R.W.G. Wyckoff and R.B. Corey, *Z. Kristallogr.* **89**, 462 (1934).

[3] R. Brill, H.G. Grimm, C. Hermann and C. Peters, *Ann. Phys. Leipzig.* **34**, 393 (1939).

[4] A.F. Andresen, *Acta Crystallogr.* **10**, 107 (1957).

[5] L.N. Becka and D.W.J. Cruickshank, *Proc. Roy. Soc. London* **A273**, 435, 455 (1963).

[6] J.A.K. Duckworth, B.T.M. Willis and G.S. Pawley, *Acta Crystallogr., Sect. A* **26**, 263 (1970).

[7] N. Blazevic, D. Kolbah, B. Belin, V. Sunjic and F. Kajfez, *Synthesis* 161 (1979).

[8] J. Altpeter, *Das Hexamethylenetetramin und seine Vervendung*, Knapp, Halle, 1931.

[9] K.Y. Hui, P.C. Chan and T.C.W. Mak, *Inorg. Chim. Acta* **84**, 25 (1984), and references cited therein.

[10] T.C.W. Mak, X. Chen, K. Shi, J. Yao and C. Zheng, *J. Cryst. Spectrosc. Res.* **116**, 639 (1986), and references cited therein.

[11] T.C.W. Mak, M.F.C. Ladd and D.C. Povey, *J. Chem. Soc., Perkin Trans. II*, 593 (1979).

[12] T.C.W. Mak, *Inorg. Chim. Acta* **84**, 19 (1984); **90**, 153 (1984), and references cited therein.

[13] C.E. Nordman and D.L. Schmitkons, *Acta Crystallogr.* **18**, 764 (1965).

[14] J.P. Amoureux, M. Bee and J.C. Damien, *Acta Crystallogr., Sect. B* **36**, 2633 (1980).

[15] J. Donohue and S.H. Goodman, *Acta Crystallogr.* **22**, 352 (1965).

[16] T.C.W. Mak and F.C. Mok, *J. Cryst. Mol. Struct.* **8**, 183 (1978).

[17] A. Haas, *Adv. Inorg. Chem. Radiochem.* **28**, 167 (1984).

Note $(t$-BuGe$)_4S_6$ has a "double-decker" (rather than an adamantane-like) structure whose C_2 skeleton comprises two planar Ge_2S_2 rings bridged by a pair of S atoms. See W. Ando, T. Kadowaki, Y. Kabe and M. Ishii, *Angew. Chem. Int. Ed. Engl.* **31**, 59 (1992).

2.25   Hexamethylbenzene

$$C_6(CH_3)_6$$

Crystal Data

Triclinic, space group $P\bar{1}$ (No. 2)

$a = 8.92(2)$, $b = 8.86(2)$, $c = 5.30(1)$Å, $\alpha = 44.45°$,

$\beta = 116.72°$, $\gamma = 119.57°$, $Z = 1$

Atom	x	y	z
C(1)	.372	.234	-.016
C(2)	.144	.383	-.008
C(3)	-.228	.146	.008
C(4)	.177	.111	-.007
C(5)	.069	.180	-.003
C(6)	-.108	.069	.004

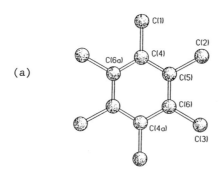

(a)

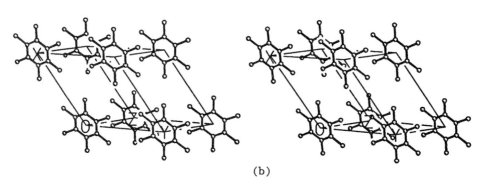

(b)

Fig. 2.25-1   (a) Atom numbering and (b) stereoview of the crystal structure of hexamethylbenzene.

Crystal Structure

Hexamethylbenzene crystals are triclinic, and there is only one molecule of $C_6(CH_3)_6$ in the unit cell.[1]   Fourier synthesis gave a bond length of 1.39Å within the benzene ring and 1.53Å between the methyl carbon and the benzene carbon.[2]   This result is in agreement with the expectation that a methyl group attached to a fully substituted benzene ring should have about the same C-C bond length as in saturated aliphatic hydrocarbons.   Fig. 2.25-1 shows the arrangement of hexamethylbenzene molecules in the unit cell (Fig. 2.25-1). The closest intermolecular C...C contacts in a single (001) layer all exceed 3.9Å. The molecular planes are actually tilted very slightly out of the (001) crystal plane, by just over 1°, permitting somewhat greater separations between intermolecular methyl hydrogen atoms.

Remarks

1. First structural determination of an aromatic system

Kekulé, in 1856, proposed the celebrated planar hexagonal geometry of the benzene ring.[3] Lonsdale's work on hexamethylbenzene showed that the aromatic ring is "almost if not quite *flat*; that is, it resembles the rings of six carbon atoms existing in graphite rather than those in diamond." Subsequent crystallographic studies on condensed-ring hydrocarbons by Robertson[4] and his students provided a wealth of fundamental structural information for comparison with the predictions of newly developed bonding theories.

2. Crystal structure of benzene

Benzene crystallizes in the orthorhombic system, with space group *Pbca*, $a = 7.46$, $b = 9.66$, $c = 7.03$Å, and four molecules in the unit cell.[5] Early studies of the crystal structure were conducted at -3°C, near its melting point, and the thermal motions of the atoms precluded precise determination of the coordinate parameters. An X-ray diffraction analysis in 1958 gave an average C-C bond length of 1.392Å. In 1964 single-crystal neutron diffraction measurements were made at both 218K and 138K.[6] A two-dimensional plot of the scattering density (Fig. 2.25-2) offers a striking experimental verification of Kekulé's concept of the benzene molecule. This is a classic example of the power of neutron diffraction as the work was carried out with a flux density of only about 10^{12} neutrons $cm^{-2}s^{-1}$. Analysis of the results showed that the molecule executes an oscillatory motion in its own plane, with angular displacements of 7.9° and 4.9° respectively, at the two temperatures. The more accurately measured carbon and hydrogen coordinates resulted in C-C = 1.398(8)Å and C-H = 1.077Å at 218K.

3. Symmetry of the benzene molecule from vibrational spectra

The infrared and Raman spectra of C_6H_6 and C_6D_6 in the vapour and liquid phases have been investigated by many different groups of workers.[7,8] The representations for the 30 fundamental vibrations of the planar hexagonal D_{6h} model of benzene are $\Gamma_{vib} = 2A_{1g} + A_{2g} + A_{2u} + 2B_{1u} + 2B_{2g} + 2B_{2u} + E_{1g} + 3E_{1u} + 4E_{2g} + 2E_{2u}$. Confirmation of the rule of mutual exclusion and the number of theoretically allowed infrared bands (four; A_{2u} and $3E_{2u}$) and Raman

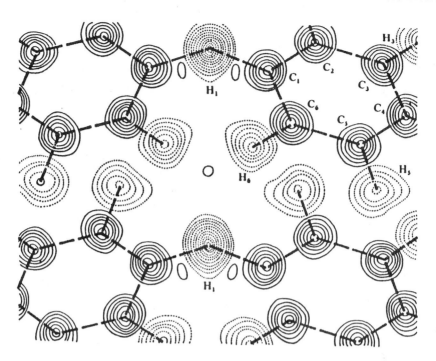

Fig. 2.25-2 A projection of the molecular structure of solid benzene as determined from neutron diffraction. Broken-line contours for the hydrogen atoms indicate their negative scattering amplitude. Hydrogen atoms from two neighbouring molecules overlap at positions such as H_1. (After ref. 6).

lines (seven; $2A_{1g} + E_{1g} + 4E_{2g}$) may be regarded both as experimental substantiation of the D_{6h} structure of benzene and also as a demonstration of the power of group theory. Further detailed discussions of the vibrational spectra of benzene are to be found in the classical treatise of Herzberg.[9]

4. The true equilibrium geometry of benzene

Ermer has emphasized that there is as yet no irrefutable *experimental proof* of the D_{6h} equilibrium geometry of benzene by any physical method.[10]

"The atomic positions originating from the X-ray (and neutron) diffraction analysis of crystalline benzene are averaged over space and time and are compatible not only with a crystallographically ordered D_{6h} model of benzene, but also with a disordered D_{3h} model corresponding to a superposition of benzene molecules rotated with respect to each other by 60° around the three-fold axis. The disorder may either be static or dynamic. In the first case, we would deal with a statistical distribution of both molecular orientations disregarding the crystallographic symmetry requirements, and in the second instance with a rapid interconversion (relative to the measuring time of usually many hours) of D_{3h} forms via the D_{6h} form corresponding to a symmetric-double minimum potential. Dynamic disorder could also be due to a hindered rotation of the benzene molecules around the three-fold axis."

Assuming a difference of 0.10Å between the C-C and C=C bond lengths in the D_{3h} model, the superimposed carbon atoms are calculated to be only 0.05/sin60° = 0.058Å apart [Fig. 2.25-3]. This small separation is far below the resolving power of current diffraction measurements.

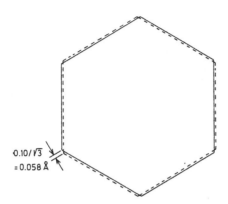

Fig. 2.25-3 Disorder model of benzene with localized single and double bonds. Both D_{3h} Kekulé structures are rotated with respect to each other by 60° around the three-fold axis: C-C and C=C bond lengths of 1.45 and 1.35Å, respectively, are assumed. (After ref. 10).

5. Hexasubstituted benzenes

The crystal structure of hexachlorobenzene (space group $P2_1/c$, a = 8.08, b = 3.87, c = 16.65Å, β = 117.0° and Z = 2)[11] [Fig. 2.25-4(a)] is a good example of the "projection-to-hollow" principle of packing in molecular crystals.[12] The molecules are closely packed with each Cl atom fitted into the void between two vicinal Cl substituents of a neighbouring C_6Cl_6 molecule.

The crystal structure of hexanitrobenzene (space group $I2/c$, a = 13.32, b = 9.13, c = 9.68Å, β = 95.5° and Z = 4) is built up from molecular layers parallel to the (10$\bar{1}$) plane; the molecules are arranged in hexagonal close packing in each layer [Fig. 2.25-4(b)].[13] The $C_6(NO_2)_6$ molecule occupies a site of symmetry 2, the plane of the nitro group making an angle of 53° with the aromatic ring. As the benzene rings are mutually screened by the nitro groups, the interlayer spacing $d(10\bar{1})/2$ = 4.00Å is large compared to the usual value of 3.4-3.6Å for crystalline aromatic hydrocarbons.

Hexaaminobenzene crystallizes in the unusual cubic space group $Pa3$ (No. 205) with a = 9.192Å and Z = 4. The molecule has approximate S_6 symmetry

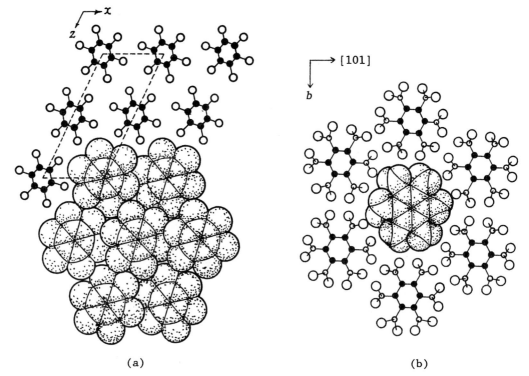

(a)   (b)

Fig. 2.25-4 (a) Crystal structure of hexachlorobenzene projected on the *ac* face. (b) Closest-packed layer of hexanitrobenzene molecules. (Adapted from ref. 12).

with six identical CNH_2 moieties of pyramidal configuration. The measured C-C and C-N bond distances are 1.383(6) and 1.432(8)Å, respectively.[14]

Recent studies have resulted in the synthesis of hexaazaoctadecahydro-coronene, HAOC, and isolation of its mono-, di-, tri- and tetracations as crystalline salts.[15] While the highly symmetric benzene derivative HAOC [Fig. 2.25-5(a)] conforms to D_{3d} symmetry, the C_{2v} dication $HAOC^{2+}$ exhibits a Jahn-Teller distorted structure corresponding to coupled cyanine fragments with distinctly different short (1.395Å) and long (1.471Å) central-ring C-C and radial C-N (1.337 and 1.405Å) bond distances in its BF_4^- and PF_6^- salts.[15] Further oxidation of HAOC to its tetracation $HAOC^{4+}$ as the salt [HAOC][SbF_6]$_4$.MeCN has led to the first structural determination of a benzene derivative in the +4 oxidation state.[16]

The tetracation occupies a 2/m site in the crystal, and its structure deviates only slightly from idealized D_{3d} symmetry, with a significantly puckered $(C_B)_6N_6$ central skeleton [Fig. 2.25-5(b)] in contrast to the approximate planarity of this moiety in HAOC.[16]

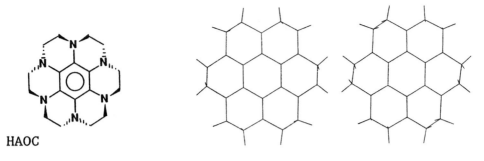

HAOC

(a)   (b)

Fig. 2.25-5 (a) Structural formula of hexaazaoctadecahydrocoronene, HAOC. (b) Stereoview down the approximate C_3 axis of the tetracation in [HAOC][SbF$_6$]$_4$.MeCN. (After ref. 16).

The [HAOC]$^+$ and [HAOC]$^{3+}$ radical cations also show distortions though not to the same extent as the dication. The average bond distances in the series [HAOC]^n (n = 0, 1+ 2+, 3+, 4+) are compared with calculated (*ab initio*) values in Table 2.25-1.

Table 2.25-1   Observed and calculated bond lengths (Å) for [HAOC]^n systems (n = 0, 1+, 2+, 3+, 4+).[*]

Compound	Obs/Calc	a	b	c	d	e	f	Δ
HAOC	obs	1.397	1.397	1.416	1.416	1.442	1.492	±0.065
HAOC (D_{3d})	calc	1.401 (1.37)	1.401 (1.37)	1.453 (1.02)	1.453 (1.02)	1.474	1.537	
[HAOC][BF$_4$]	obs	1.382	1.444	1.371	1.422	1.449	1.487	±0.014
[HAOC]$^{.+}$	calc	1.404	1.463	1.392	1.459	1.478	1.536	
[HAOC][BF$_4$]$_2$	obs	1.397	1.476	1.333	1.425	1.462	1.507	±0.038
[HAOC][PF$_6$]$_2$	obs	1.390	1.478	1.332	1.411	1.459	1.494	±0.040
[HAOC]$^{2+}$ (C_{2h})	calc	1.401	1.522	1.349	1.458	1.489	1.537	
[HAOC]$^{2+}$ (S_2)	calc	1.491	1.353	1.423	1.318	1.482	1.536	
^{3}[HAOC]$^{:2+}$	calc	1.429 (1.23)	1.429 (1.23)	1.383 (1.22)	1.383 (1.22)	1.480	1.539	
[HAOC][PF$_6$]$_3$	obs	1.417	1.439	1.326	1.339	1.463	1.512	±0.111
[HAOC]$^{.3+}$	calc	1.44	1.49	1.33	1.38	1.49	1.53	
[HAOC][SbF$_6$]$_4$.MeCN	obs	1.436	1.436	1.318	1.318	1.472	1.518	±0.149
[HAOC]$^{4+}$	calc	1.475 (1.11)	1.475 (1.11)	1.330 (1.52)	1.330 (1.52)	1.506	1.542	

[*] Calculated total bond order enclosed in parentheses. Δ denotes the average deviation of the $(C_B)_6N_6$ core atoms from their least-squares plane.

6. Inorganic analogues of benzene

Borazine (1), the prototype inorganic analogue of benzene, was first synthesized by Stock and Pohland in 1926.[17] Although its regular hexagonal molecular structure (B-N 1.44, B-H 1.20, N-H 1.02Å) and physical properties closely resemble those of benzene, there is little evidence of aromatic character in the conventional sense as the ring is easily disrupted by oxidation or solvolysis.[18] Hexaethylborazine is more stable and reacts with $Cr(CO)_3(MeCN)_3$ to give a "half-sandwich" complex $Cr(CO)_3(\eta^6\text{-}B_3N_3Et_6)$[19] which closely resembles the corresponding hexamethylbenzene analogue $Cr(CO)_3(\eta^6\text{-}C_6Me_6)$.

1

2   X = mesityl
    Y = cyclohexyl

3   X = methyl
    Y = 2,6-di(*iso*-
        propyl)phenyl

Recently the first examples of boron-phosphorus and aluminum-nitrogen analogues of borazine, $(2,4,6\text{-}Me_3C_6H_2BPC_6H_{11})_3$ (2)[20] and [MeAlN(2,6-$iPr_2C_6H_3)]_3$ (3),[21] respectively, were synthesized and structurally characterized by X-ray crystallography and NMR (^{31}P, ^{11}B; ^{1}H, ^{27}Al) spectroscopy. Compound 2 possesses a planar central B_3P_3 ring (P-B-P 115°) in which all the B-P bonds are essentially equal at 1.84Å. Compound 3 has a similar "alumazene" Al_3N_3 ring (N-Al-N 115°), the much greater ionic character of the Al-N bond being reflected by its average length of 1.78Å [Fig. 2.25-6(a)]. In both compounds the C atoms bonded to the ring (B-C 1.58, P-C 1.84; Al-C 1.98, N-C 1.44Å) are coplanar with it.

The "germanazene" [GeN(2,6-$iPr_2C_6H_3$)]_3 is the first stable example of an unsaturated six-membered ring involving a heavier main group 4 element as an integral part of the delocalized, multiply bonded system.[22] X-Ray analysis has revealed a planar skeleton [Fig. 2.25-6(b)] with large internal angular distortions from idealized hexagonal geometry [average G-N 1.859(2), N-C 1.452(5)Å; N-Ge-N 101.8(1), Ge-N-Ge 138.0(2)°]. The samll bond angle at Ge is attributable to the increased reluctance of the heavy main-group elements to engage in hybridization. The shorter Ge-N distance as compared to the value ~1.89Å found in the acyclic Ge(II) amides, $Ge[N(SiMe_3)_2]_2$ and $Ge[NCMe_2(CH_2)_3CMe_2]_2$, and the fact that the planar conformation is preferred over a less strained "twist" one, strongly suggest that there is enhanced overlap between the empty Ge 4p and filled N 2p orbitals.

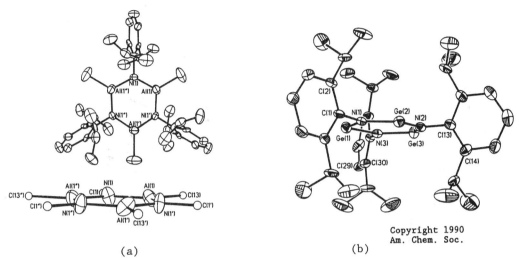

    (a)   (b)

Fig. 2.25-6 (a) Two views of the molecular structure of [MeAlN(2,6-$iPr_2C_6H_3$)]₃. (b) Molecular structure of [GeN(2,6-$iPr_2C_6H_3$)]₃; the average dihedral angle between the Ge₃N₃ plane and the phenyl rings is 86.2°. (After refs. 21 and 22).

Pyridine-like heterobenzenes C_5H_5E are known for E = P, As, Sb and Bi, the stability decreasing rapidly with increasing atomic number. The structure of a fully substituted phosphabenzene is shown in Fig. 2.25-7.[23] Steric overcrowding in the *vic*-tri-*tert*-butyl-substituted part of the molecule causes the aromatic ring to adopt a rather twisted boat conformation.

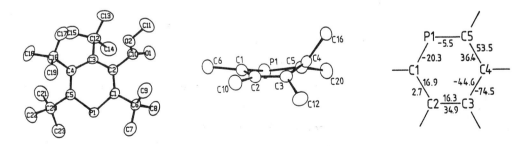

Fig. 2.25-7 Molecular structure of 2,4,5,6-tetra-*tert*-butyl-3-methoxycarbonylphosphabenzene showing, from left to right, a perspective view with atom numbering, the non-planar boat conformation, and torsion angles (°) around the C₅P ring. (After ref. 23).

The sandwich-type borabenzene complex shown in Fig. 2.25-8 has C_2 symmetry with the boron atoms in a *syn* arrangement in the staggered conformation (Fig. 2.25-8).[24]   The measured bond lengths are: Fe-B 2.370(2), Fe-C(1) 2.236(2), Fe-C(2) 2.121(2), Fe-C(3) 2.147(2), Fe-C(4) 2.087(2), and Fe-C(5) 2.128(2)Å.   The borabenzene ring is planar, and the torsion angle B-centroid(B,C1-C5)-centroid(B',C1'-C5')-B' has the value 39.3°.

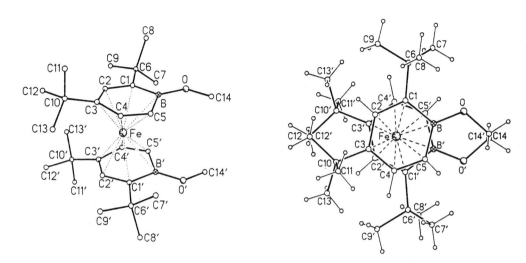

Fig. 2.25-8 Molecular structure of bis(2,4-di-*tert*-butyl-1-methoxyborabenzene)iron: (left) perspective view and (right) projected along an axis which passes through the centroids of the two C_5B rings. (After ref. 24).

Reddish-black, air-sensitive bis(η^6-arsabenzene)chromium has been synthesized by metal atom-ligand co-condensation:[25]

$$Cr(g) \ + \ 2C_5H_5As \ \xrightarrow[\text{2.} \quad 30 \ °C]{\text{1.} \ -196 \ °C} \ (C_5H_5As)_2Cr$$

X-Ray analysis has revealed that the crystals are twinned along (001) and highly disordered. In the model described, the As atom equally populates positions 1 and 4 of the η^6 ring, and the structure is consistent with either the *cis*- or *trans*-eclipsed form. The Cr atom is located 0.25Å from a crystallographic inversion center, so that the two alternative orientations of the complex are slightly displaced from each other.

Recent studies have reported the synthesis and characterization of *cyclo*-P_6 (hexaphosphabenzene) and *cyclo*-As_6 (hexaarsenabenzene) as the stabilized middle deck of triple-decker transition metal complexes (see Section 6.9).[26-28]   X-Ray analyses of the series of complexes $[\{(\eta^5\text{-}C_5Me_4R)M\}_2(\mu,\eta^6\text{-}E_6)]$ (R = Me, M = Mo, E = P;[26] R = Me, M = W,

E = P;[27] R = Et, M = V, E = P;[27] and R = Et, M = Mo, E = As[28]) yielded
a planar hexagonal middle deck of edge length 2.17, 2.17, 2.13, and 2.35Å,
respectively.

SCF theoretical calculations suggest that the hypothetical D_{6h}
"hexaazabenzene" molecule is a classic aromatic system with a N-N bond length
of 1.288Å.[29] Stabilization of this "hexazine", *cyclo*-N_6, species by
coordination to transition metals is likely to be a more attainable goal in
view of the successful synthesis of the hexaphospha- and hexaarsena-analogues
in the triple-decker sandwich compounds.

References

[1] K. Lonsdale, *Nature (London)* **122**, 810 (1928).

[2] L.O. Brockway and J.M. Robertson, *J. Chem. Soc.*, 1324 (1939).

[3] A. Kekulé, *Bull. Soc. Chim. Fr.* **3**, 98 (1865).

[4] J.M. Robertson, *Organic Crystals and Molecules*, Cornell University Press,
     Ithaca, 1953.

[5] E.G. Cox, D.W.J. Cruickshank and J.A.S. Smith, *Proc. Roy. Soc. London*
     A247, 1 (1958).

[6] G.E. Bacon, N.A. Curry and S.A. Wilson, *Proc. Roy. Soc. London* **A279**, 98
     (1964).

[7] N. Herzfeld, C.K. Ingold and H.G. Poole, *J. Chem. Soc.*, 316 (1946). [The
     last of a series of 21 publications].

[8] P.C. Painter and J.L. König, *Spectrochim. Acta* **A33**, 1003, 1019 (1977).

[9] G. Herzberg, *Molecular Spectra and Molecular Structure II. Infrared and
     Raman Spectra of Polyatomic Molecules*, D. Van Nostrand, Princeton, 1945,
     p. 362.

[10] O. Ermer, *Angew. Chem. Int. Ed. Engl.* **26**, 782 (1987).

[11] I.N. Streltsova and Yu.T. Struchkov, *Zh. Strukt. Khim.* **2**, 312 (1961).

[12] A.I. Kitaigorodsky, *Molecular Crystals and Molecules*, Academic Press, New
     York, 1973.

[13] P. Vaughan and J. Donohue, *Acta Crystallogr.* **5**, 530 (1952).

[14] D.A. Dixon, J.C. Calabrese and J.S. Miller, *Angew. Chem. Int. Ed. Engl.*
     28, 90 (1989).

[15] J.S. Miller, D.A. Dixon, J.C. Calabrese, C. Vazquez, P.J. Krusic, M.D.
     Ward, E. Wasserman and R.J. Harlon, *J. Am. Chem. Soc.* **112**, 381 (1990).

[16] D.A. Dixon, J.C. Calabrese, R.L. Harlow and J.S. Miller, *Angew. Chem.
     Int. Ed. Engl.* **28**, 92 (1989).

[17] A. Stock and E. Pohland, *Chem. Ber.* **59**, 2215 (1926).

[18] G. Huttner and B. Krieg, *Angew. Chem. Int. Ed. Engl.* **10**, 512 (1971); *Chem. Ber.* **105**, 3437 (1972).

[19] E.L. Mutterties (ed.), *The Chemistry of Boron and its Compounds*, Wiley, New York, 1967, p. 411.

[20] H.V.R. Dias and P.P. Power, *Angew. Chem. Int. Ed. Engl.* **26**, 1270 (1987); *J. Am. Chem. Soc.* **111**, 144 (1989).

[21] K.M. Waggoner, H. Hope and P.P. Power, *Angew. Chem. Int. Ed. Engl.* **27**, 1699 (1988).

[22] R.A. Bartlett and P.P. Power, *J. Am. Chem. Soc.* **112**, 3660 (1990).

[23] G. Mass, J. Fink, H. Wingert, K. Blatter and M. Regitz, *Chem. Ber.* **120**, 819 (1987).

[24] G. Maier, H.-J. Wolf and R. Boese, *Chem. Ber.* **123**, 505 (1990).

[25] C. Elschenbroich, J. Kroker, W. Massa, M. Wünsch and A.J. Ashe III, *Angew. Chem. Int. Ed. Engl.* **25**, 571 (1986).

[26] O.J. Scherer, H. Sitzmann and G. Wolmerschäuer, *Angew. Chem. Int. Ed. Engl.* **24**, 351 (1985).

[27] O.J. Scherer, J. Schwalb, H. Sawarowsky, G. Wolmerschäuser, W. Kaim and R. Gloss, *Chem. Ber.* **121**, 443 (1988).

[28] O.J. Scherer, H. Sitzmann and G. Wolmershäuser, *Angew. Chem. Int. Ed. Engl.* **28**, 212 (1989).

[29] P. Saxe and H.F. Schaefer III, *J. Am. Chem. Soc.* **105**, 1760 (1983).

Note Geometrical parameters describing the ring deformation in monosubstituted benzene derivatives are discussed in A. Domenicano and A. Viciago, *Acta Crystallogr., Sect. B* **35**, 1382 (1979); A. Domenicano, P. Murray-Rust and A. Viciago, *Acta Crystallogr., Sect. B* **39**, 457 (1983).

$[Li(THF)]_2[(Me_3Si)_6C_6]$ contains the first reported nonconjugated benzene dianion (benzenide) with both lithium atoms on the same side of the boat-shaped ring (two normal bonds of average length 1.393Å and four long ones of 1.512Å). $[Li(dimethoxyethane)]_2[1,2,4,5-(Me_3Si)_4C_6H_2]$ contains a novel $6C$-8π antiaromatic benzene dianion having a nearly planar D_{2h} ring (two long bonds of 1.553Å and four normal ones of 1.409Å) located symmetrically between two lithium atoms. See A. Sekiguchi, K. Ebata, C. Kabuto and H. Sakurai, *J. Am. Chem. Soc.* **113**, 1464, 7081 (1991).

Multiple bonding, π interaction and aromatic character in isoelectronic B-P, B-As, Al-N and Zn-S compounds are discussed in P.P. Power, A. Moezzi, D.C. Pestana, M.A. Petrie, S.C. Shoner and K.M. Waggoner, *Pure Appl. Chem.* **63**, 859 (1991).

2.26 Naphthalene
$C_{10}H_8$

Crystal Data

Monoclinic, space group $P2_1/a$ (No. 14)

$a = 8.108(5)$, $b = 5.940(2)$, $c = 8.647(5)$Å, $\beta = 124.38(4)°$ at (92K)[1]

$Z = 2$

Atom	x	y	z	Atom	x	y	z
C(1)	.0824	.0189	.3288	H(1)	.1229	.0572	.4471
C(2)	.1132	.1638	.2233	H(2)	.1798	.3067	.2726
C(3)	.0479	.1055	.0371	H(4)	.1412	.3883	-.0253
C(4)	.0770	.2521	-.0756	H(5)	.0323	.2916	-.3289
C(5)	.0134	.1906	-.2545				

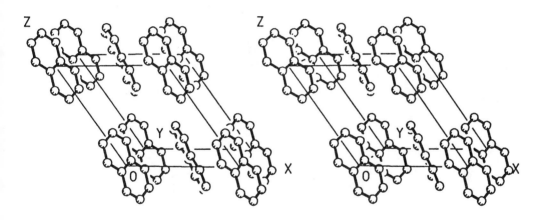

Fig. 2.26-1 Stereoview of the crystal structure of naphthalene.

Crystal Structure

The first definitive study of the crystal and molecular structures of naphthalene and anthracene was completed by Robertson.[2] The results established the strictly planar form of the molecules to within narrow limits and gave their orientation in the crystal to within about 1°. The molecular arrangement in the unit cell is shown for naphthalene in Fig. 2.26-1. The long axes of the molecules are not exactly coincident with the crystal *c* axis, as at first supposed, the deviation of naphthalene being about 14° and of anthracene about 9°. In both cases the molecular planes are steeply inclined, at an angle of about 64°, to the (010) plane.

A more comprehensive investigation based on three-dimensional photographic data was undertaken around 1950 for naphthalene[3] and anthracene.[4]   In both cases the molecules were found to be coplanar to within 0.01Å, and the bond angles all 120° to within 2°.   The evaluation of the electron density function in the molecular plane [Fig. 2.26-2] was an enormous task at that time, but it has the advantage of giving an aesthetically appealing picture of the molecular structure, on which the bond lengths and angles can be measured directly.[5]   Owing to the high resolution achieved, the hydrogen electron clouds become visible, although distorted, at the 0.5 $eÅ^{-3}$ level.   In anthracene, which has somewhat lower atomic vibrational amplitudes, the hydrogen resolution is even better.

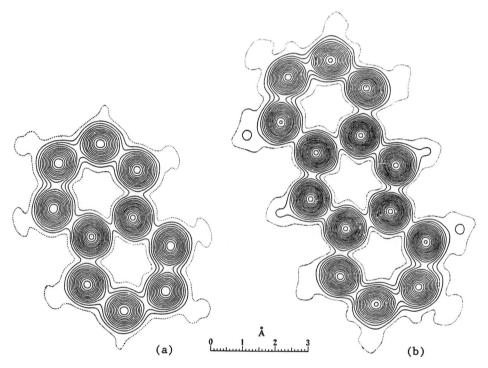

(a)   Å
0   1   2   3
(b)

Fig. 2.26-2   (a) Section through the plane of the naphthalene molecule.   (b) Section through the plane of the anthracene molecule. Contour levels are drawn at intervals of 1 $eÅ^{-3}$, the 0.5 $eÅ^{-3}$ line being dotted.   (After ref. 5).

A systematic study of the temperature dependence of the experimental thermal vibrational parameters in naphthalene (at five temperatures between 90 and 240K)[1] and anthracene (at six temperatures from 94 to 295K)[6] has been completed.   There is also a neutron diffraction study of perdeuteroanthracene at 16K.[7]

The averaged bond lengths (Å) and bond angles (°) in naphthalene[1] and anthracene[6] are shown in Fig. 2.26-3.

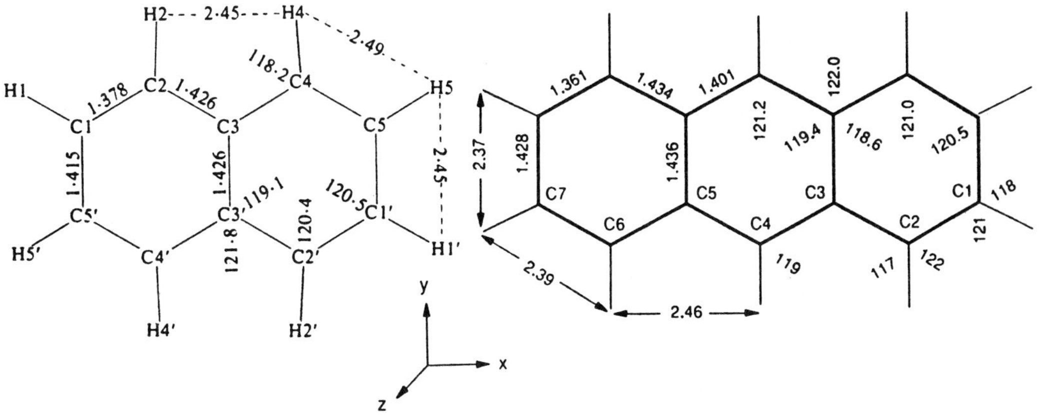

Fig. 2.26-3 Averaged molecular dimensions of naphthalene and anthracene. (After refs. 1 and 6).

1. Unit cell dimensions and molecular size

In 1921, W.H. Bragg published the preliminary results of his X-ray examination of a number of fairly complex organic crystals, including naphthalene and anthracene.[8] In this work he introduced the important idea that certain units of structure, like the benzene or naphthalene ring that possesses a definite size and form, might be preserved with little or no alteration in passing from one crystalline organic derivative to another. Furthermore, the size and shape of these units could presumably be deduced from the known structures of diamond and graphite and so compared with X-ray measurements of the unit cell dimensions of the organic crystals.

Bragg's approach worked for naphthalene and anthracene because the long axes of the molecules happen to be nearly coincident with the c crystal axes. Fig. 2.26-4 shows the unit cells of the two crystals and the relative positions of the two molecules within each. While a, b and β remain almost the same in the two unit cells, the c axis increases by about 2.5Å in passing from naphthalene to anthracene. Bragg attributed this increase to the extra ring and concluded that both crystals have similar structures with aromatic molecules lying end-to-end along the c axis.

2. Electron-density maps and bond lengths

Electron-density maps such as Fig. 2.26-2 did much to convince skeptics that a rigorous determination of the molecular geometry of organic crystals can be achieved by X-ray diffraction methods. The measured molecular

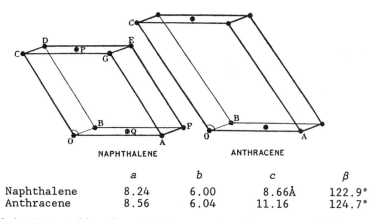

	a	b	c	β
Naphthalene	8.24	6.00	8.66Å	122.9°
Anthracene	8.56	6.04	11.16	124.7°

Fig. 2.26-4 Unit cells of naphthalene and anthracene. (After ref. 5).

dimensions of the condensed six-membered ring aromatic hydrocarbons provided
much valuable data for testing the predictions of budding theories of chemical
bonding.[5]

X-Ray analysis showed that, in contrast to benzene, all C-C bonds in
naphthalene are not the same; in particular, the C(1)-C(2) bond (1.378Å) is
considerably shorter than the C(2)-C(3) bond (1.425Å). Simple valence bond
theory advances three structure formulae for naphthalene:

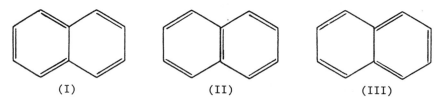

If structures (I), (II), and (III) are equally weighted, one would expect the
C(1)-C(2) bond to have more double-bond character than the C(2)-C(3) bond,
which is in agreement with the observed data. Often naphthalene is written as
the single structure (IV), in which the circles stand for partially
overlapping aromatic sextets. Although representation (IV) suggests a greater
symmetry for naphthalene than the real situation, it has the advantage of
emphasizing the delocalized nature of the π system. A quantum chemical
computation has yielded the SCF bond orders for the C-C bonds in naphthalene,
as shown in (V),[9] which give a better fit to the measured bond lengths.

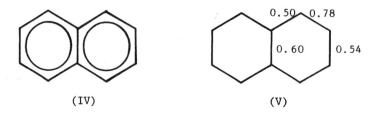

3. Structure of tetracene and pentacene

The higher homologues tetracene[10] and pentacene[11] are closely similar to the two preceding members of the benzenoid series. Although there are definite variations in orientation in the crystals, the general arrangements of the tetracene and pentacene molecules in the unit cell are like that of anthracene. The two molecules are completely planar within the limits of experimental error.

The measured bond lengths and bond angles suggest that the symmetry of the molecules does not differ significantly from *mmm*, and the mean values are shown in Fig. 2.26-5.

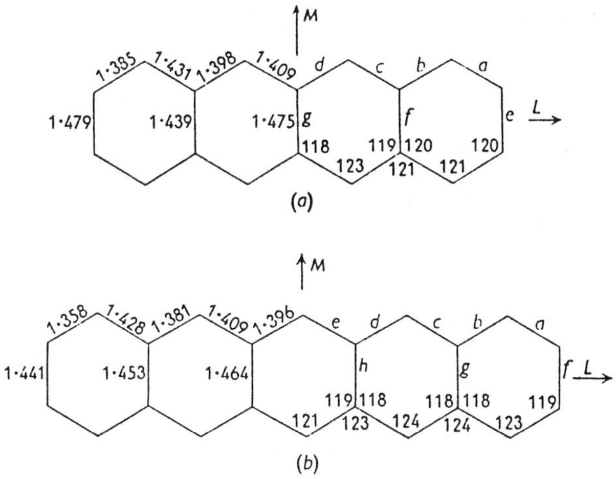

Fig. 2.26-5 Mean bond lengths and bond angles in (a) tetracene and (b) pentacene. (After refs. 10 and 11).

Table 2.26-1 Measured and calculated bond distances (in Å).

Tetracene			Pentacene		
Bond [Fig. 2.26-5(a)]	Measured	Calculated	Bond [Fig. 2.26-5(b)]	Measured	Calculated
a	1.385	1.355	a	1.358	1.351
b	1.431	1.450	b	1.428	1.458
c	1.398	1.378	c	1.381	1.370
d	1.409	1.409	d	1.409	1.420
e	1.479	1.450	e	1.396	1.392
f	1.439	1.450	f	1.441	1.458
g	1.475	1.450	g	1.453	1.458
			h	1.464	1.458

A calculation of the bond lengths for comparison with the measured distances was made from the non-excited valence bond (Kekulé) structures for the two molecules. The percentage double-bond character was derived for each bond, and the corresponding bond length from a correlation curve drawn through the points (0,1.50Å), (0.33,1.42Å), (0.5,1.392Å), (1.0,1.337Å) (Table 2.26-1).

References

[1] C.P. Block and J.D. Dunitz, *Acta Crystallogr., Sect. B* **38**, 2218 (1982).

[2] J.M. Robertson, *Proc. Roy. Soc. London* **A140**, 79 (1933); **A142**, 674 (1933).

[3] C. Abrahams, J.M. Robertson and J.G. White, *Acta Crystallogr.* **2**, 233, 238 (1949).

[4] A. McL. Mathieson, J.M. Robertson and V.C. Sinclair, *Acta Crystallogr.* **3**, 245, 251 (1950).

[5] J.M. Robertson, *Organic Crystals and Molecules*, Cornell University Press, Ithaca, 1953.

[6] C.P. Block and J.D. Dunitz, *Acta Crystallogr., Sect. B* **46**, 795 (1990).

[7] S.L. Chaplot, N. Lehner and G.S. Pawley, *Acta Crystallogr., Sect. B* **38**, 483 (1982).

[8] W.H. Bragg, *Proc. Roy. Soc. London* **34**, 33 (1921); **35**, 167 (1922).

[9] J.A. Pople, *Trans. Faraday Soc.* **49**, 1375 (1953).

[10] J.M. Robertson, V.C. Sinclair and J. Trotter, *Acta Crystallogr.* **14**, 697 (1961).

[11] R.B. Campbell, J.M. Robertson and J. Trotter, *Acta Crystallogr.* **14**, 705 (1961).

Note The latest crystallographic refinement of the structure of naphthalene and anthracene is reported in C.P. Block, J.D. Dunitz and F.L. Hirshfeld, *Acta Crystallogr., Sect. B* **47**, 789 (1991).

Examples of naphthalene ring deformation caused by crowded substituents include the 1,4,5,8-tetramethyl and 1,3,6,8-tetra-*tert*-butyl derivatives. See F. Imashiro, K. Takegoshi, A. Saika, Z. Taira and Y. Asahi, *J. Am. Chem. Soc.* **107**, 2341 (1985); J. Handal, J.G. White, R.W. Franck, Y.H. Yuh and N.J. Allinger, *ibid*. **99**, 3345 (1977). Several highly substituted naphthalenes with nonplanar skeletons are described in J.B. Bremner, L.M. Engelhardt, A.H. White and K.N. Winzenberg, *ibid*., **107**, 3910 (1985).

1,8-Bis(dimethylamino)naphthalene is the prototype "proton sponge", a term applied to certain aromatic diamines that exhibit exceptionally high basicities owing to the close proximity of their nitrogen lone pairs. *Mono*protonation of such a diamine leads to the formation of a strong intermolecular N...H...N hydrogen bond with a considerable relaxation of the steric strain. For a recent review on proton sponges, see H.A. Staab and T. Saupe, *Angew. Chem. Int. Ed. Engl.* **27**, 865 (1988).

X-Ray structure analysis has established that the pronounced twisting of 18.8° about the central C(4a)-C(8a) bond in the "double proton sponge" 1,4,5,8-tetrakis(dimethylamino)naphthalene is almost completely suppressed by the formation of two strong N...H...N hydrogen bridges in its dihydrobromide pentahydrate. See T. Barth, C. Krieger, F.A. Neugebauer and H.A. Staab, *Angew. Chem. Int. Ed. Engl.* **30**, 1028 (1991).

2.27   Urea

$(NH_2)_2CO$

Crystal Data

Tetragonal, space group $P\bar{4}2_1m$ (No. 113)

T/K*	12	30	60	123	150	173	293
a/Å	5.565	5.565	5.570	5.584	5.590	5.598	5.645
c/Å	4.684	4.685	4.688	4.689	4.692	4.694	4.704

Z = 2

Atom	Position	Temperature/K*	x	y	z
C	2(c)	12	0	1/2	.3260
		123	0	1/2	.3280
		293	0	1/2	.3328
O	2(c)	12	0	1/2	.5953
		123	0	1/2	.5976
		293	0	1/2	.5962
N	4(e)	12	.1459	.6459	.1766
		123	.1447	.6447	.1785
		293	.1418	.6418	.1830
H(1)	4(e)	12	.2575	.7575	.2827
		123	.2557	.7557	.2841
		293	.2527	.7527	.2839
H(2)	4(e)	12	.1441	.6441	-.0380
		123	.1431	.6431	-.0348
		293	.1389	.6389	-.0306

* Values at 293K are from ref. 2; others are from ref. 1. The standard deviations for the lattice parameters are all 0.001Å

Crystal Structure

Urea was the first organic compound to be synthesized in the laboratory, a feat accomplished by Wöhler in 1828. It is a very important product of the chemical and agricultural industries.

Urea was subjected to X-ray analysis at an early stage by Hendricks[3] and Wyckoff,[4] but these studies were not sufficiently accurate to determine all atomic parameters. The structure was later refined several times using improved X-ray and neutron diffraction data.[5-9] The urea molecule is planar, and its measured bond distances and angles are listed in Table 2.27-1.

In the crystal, urea occupies a special position where its molecular symmetry (mm2) is fully utilized. Every hydrogen atom is involved in a N-H...O hydrogen bond, and this accounts for the fact that urea occurs as a solid at room temperature. A feature of chemical interest in the urea

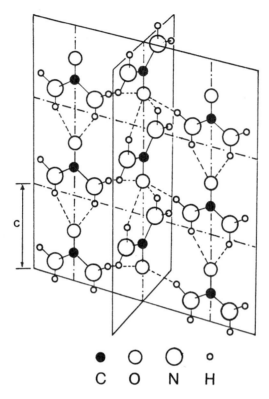

Fig. 2.27-1 Crystal structure of urea. (Adapted from ref. 27).

Table 2.27-1 Bond distances and angles of urea.[*]

	12K	123K	293K
Bond distances (Å)			
C-O	1.265(1)	1.264(1)	1.246(2)
C-N	1.349(1)	1.351(1)	1.333(1)
N-H(1)	1.022(3)	1.051(3)	1.004(2)
N-H(2)	1.018(3)	1.005(3)	1.005(3)
O...H(1)	1.992(2)	2.009(2)	2.052(2)
O...H(2)	2.058(2)	2.067(2)	2.071(3)
N-H(1)...O	2.985(1)	2.998(1)	3.040(3)
N-H(2)...O	2.955(1)	2.960(1)	2.977(2)
Bond angles (°)			
O-C-N	121.4(1)	121.4(1)	121.9(1)
N-C-N	117.2(1)	117.2(1)	116.2(1)
C-N-H(1)	119.1(1)	119.2(1)	119.9(2)
C-N-H(2)	120.5(1)	120.7(1)	120.6(1)
H(1)-N-H(2)	120.4(2)	120.1(2)	119.5(2)
N-H(1)...O	167.2(2)	166.8(2)	167.6(3)
N-H(2)...O	147.4(2)	147.6(2)	149.0(2)

[*] The values are all from neutron diffraction: at 12K and 123K the values from ref. 1 are corrected for harmonic thermal motion; values at 293K are from ref. 2.

structure is that it appears to provide the only instance of a carbonyl O atom which forms four acceptor N-H...O hydrogen bonds (Fig. 2.27-2).

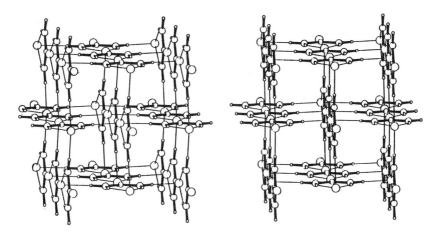

Fig. 2.27-2 The hydrogen-bond framework in crystalline urea viewed approximately along the *c* axis. (After ref. 1).

Remarks

1. Acid adducts of urea

The urea molecule can act as a monoacidic base as the partially negatively charged oxygen atom is capable of coordinating with one proton. In aqueous solution urea and many acids form 1:1 adducts such as $(NH_2)_2CO.HNO_3$, $(NH_2)_2CO.HCl$, $(NH_2)_2CO.H_2SO_4$, and $(NH_2)_2CO.H_3PO_4$.[10] X-Ray analysis has shown that in most cases the acidic proton is attached to the carbonyl oxygen atom to give a uronium salt, for example $[(NH_2)_2COH]^+.NO_3^-$.[11] In the short hydrogen bond which links the urea and phosphoric acid moieties, the O...O separation is 2.421Å, the angle at H is 169.9°, and the acid proton is essentially centered at distances of 1.223 and 1.207Å respectively from the O atoms, so the compound is neither a true adduct $(NH_2)_2CO.H_3PO_4$ nor a true uronium salt $(NH_2)_2COH^+.H_2PO_4^-$, but of an intermediate nature.[12-14]

$$
\begin{array}{c}
\quad\quad\quad 1.22 \quad\quad 1.21 \\
\quad\quad\quad O \text{----} H \text{----} O \quad\quad OH \\
H_2N-C \quad\quad\quad\quad\quad\quad\quad P \\
\quad\quad\quad 1.00 \quad\quad 1.99 \quad\quad OH \\
\quad\quad\quad N \text{----} H \text{-----} O \\
\quad\quad\quad H
\end{array}
$$

In methanol solution, urea and H_2SiF_6 form a 4:1 adduct which has the structural formula $[(NH_2)_2CO...H...OC(NH_2)_2]_2SiF_6$.[15] The two independent cations in the asymmetric unit are each stabilized by a symmetrical hydrogen bond (of length 2.424 and 2.443Å, respectively). H_2SiF_6 finds agricultural use in the prevention of the staleness of wheat, but it easily hydrolyzes to release HF. Its adduct with urea serves the same purpose with the obvious advantage of being a stable solid.

The acid adducts of urea have also been investigated by infrared spectroscopy, which indicates that urea coordinates with the proton through an $O-H^+$ bond, forming the uronium cation in the company of a wide variety of anions, such as $PtCl_6^{2-}$, $SnCl_6^{2-}$, $SbCl_6^-$, ClO_4^-, and $C_2O_4^{2-}$. [10]

2. Metal-urea complexes

Urea has three coordination sites: the carbonyl oxygen and the two nitrogen atoms. Usually it acts a monodentate O-bonded ligand as in $Zn[(NH_2)_2CO]_6^{2+}$ [16] [Fig. 2.27-3(a)]; sometimes the bidentate N,O-coordination mode of urea is found in its metal complexes, for example $Co[(NH_2)_2CO]_4^{2+}$ [17] as shown in Fig. 2.27-3(b).

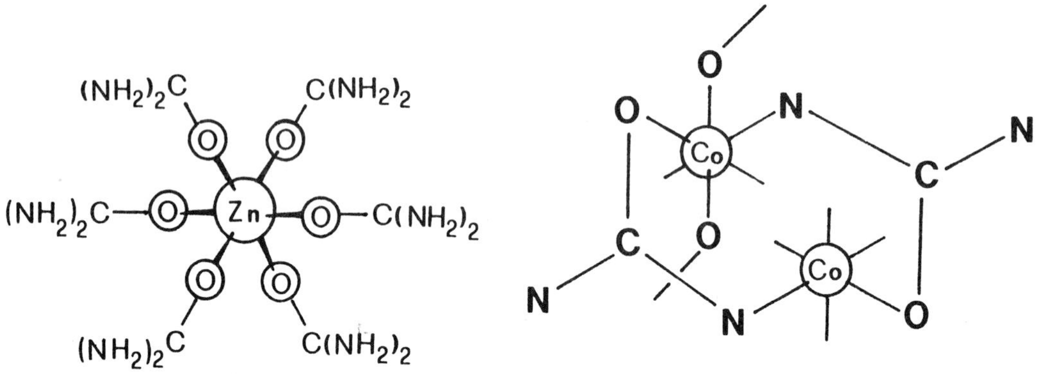

(a)   (b)

Fig. 2.27-3 The coordination modes of urea: as (a) a monodentate ligand in $Zn[(NH_2)_2CO]_6^{2+}$; and (b) a bidentate ligand in $Co[(NH_2)_2CO]_4^{2+}$.

Metal-urea complexes are an interesting group of compounds which has been systematically investigated in regard to the metal-oxygen and metal-nitrogen chemical bonds and the effect of metal interaction on the structure and properties of urea. Some of these complexes have antimicrobial and antivirial properties and are potentially effective drugs. The fact that vitamine B_{12} is rich in $-CONH_2$ groups is a reminder of the biological importance of urea in the design of complicated molecules in life.

3. Inclusion compounds of urea [18,19]

Urea forms an isomorphous series of crystalline, non-stoichiometric inclusion compounds with n-alkanes and their derivatives (including alcohols, esters, ethers, aldehydes, ketones, carboxylic acids, amines, nitriles, thioalcohols and thioethers) provided that their main chain consists of *six* or more carbon atoms. The unit cell is hexagonal, space group $P6_12$ or $P6_52$ with $a = 8.230(4)$ and $c = 11.005(5)$Å, and contains six urea molecules. [20,21]

In the urea-n-hydrocarbon complex, the urea molecules are arranged in three interpenetrating helical spirals, which are linked by hydrogen bonds to

form the walls of each hexagonal channel in the crystal lattice (Fig. 2.27-4).
The n-hydrocarbon molecules occupy the hollow channels in an extended planar
zigzag configuration, generally with positional disorder of the atoms. The
packing arrangement viewed along c is shown in Fig. 2.27-5, in which the cross
section of the channel (nearly circular with a diameter of about 5.25Å) is
compared with those of several potential guest molecules. It is seen that
while n-octane and 3-methylheptane can fit into the available space, 2,2,4-

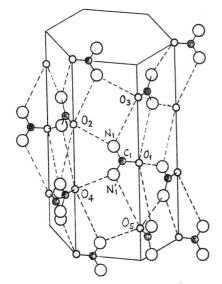

The shorter hydrogen bonds
N_1-H...O_2 and N_1'-H...O_4 are
about 2.93Å in length; the
longer bonds N_1-H...O_3 and
N_1'-H...O_5 are about 3.04Å.

Fig. 2.27-4 Hydrogen bonding in a urea channel adduct.

trimethylpentane cannot be so accommodated. Although benzene and other small
ring compounds might be expected from size and shape considerations to be
capable of forming urea inclusions compounds, none has yet been found. If
however a long chain is attached to the aromatic system, e.g.
octadecylbenzene, then a channel adduct can be obtained. The selectivity
dictated by channel size has been exploited in the industrial separation of n-
alkanes from branched-chain and cyclic compounds.

The composition of the non-stoichiometric urea-n-hydrocarbon adduct can
be calculated from a formula given by Smith:[22]

Urea/hydrocarbon mole ratio = 0.684(N-1) + 2.175, where N is the number
of carbon atoms in the extended zigzag chain (by taking C-C = 1.54Å, C-C-C =
109.5°, and radius of CH_3 group = 2.0Å).

The hexagonal host lattice of urea inclusion compounds occurs in two
enantiomorphic forms, depending on the handedness of the spirals. This
asymmetry can be used for optical resolution by crystallizing part of a
racemic mixture as a guest in the urea complex.

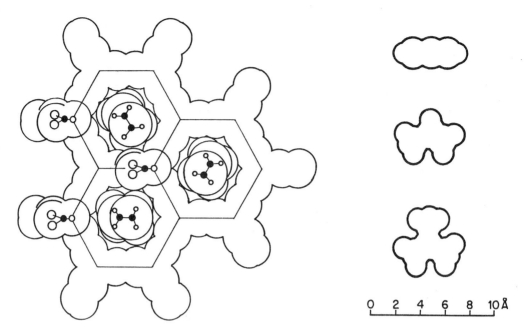

Fig. 2.27-5 Cross section of a urea channel compared with those of n-octane (inside channel), benzene (top right), 3-methylheptane (middle right) and 2,2,4-trimethylpentane (bottom right). (After refs. 20 and 21).

In the disordered crystal structure of the urea inclusion compounds $(NH_2)_2CO + C_nH_{2n+2}$ (n = 12, 16), the orientation of the included paraffin chain was deduced and the hydrogen (deuterium) positions determined for the hexadecane complex (C16) using neutron diffiaction data.[23] The atomic coordinates of the urea molecule from the C16 Mo-$K\alpha$ X-ray refinement are:

	O1	C1	N1	H1	H2
x	0.3205(0)	0.4085(1)	0.4772(1)	0.5344(11)	0.4512(11)
y	0.6410(1)	0.8171(1)	0.9136(1)	0.0348(15)	0.8555(11)
z	0.2500	0.2500	0.3527(2)	0.3569(5)	0.4036(7)

Thiourea[24,25] and selenourea[26] both form crystalline inclusion compounds with rhombohedral host lattices which bear some similarity to those of urea. With a channel diameter of about 6.1Å, the thiourea host lattice can enclose highly branched alkanes and their derivatives, haloalkanes, 5-, 6- and 8-membered ring compounds, condensed ring systems, and even ferrocene and other metallocenes. On the other hand, n-alkanes do not form thiourea complexes owing to their loose fit in the large channel, and this property can be used to separate benzene and cyclohexane from n-heptane.

References

[1]   S. Swaminathan, B.M. Craven and R.K. McMullan, *Acta Crystallogr., Sect. B* **40**, 300 (1984).

[2] H. Guth, G. Heger, S. Klein, W. Treutmann and C. Scheringer, *Z. Kristallogr.* **153**, 237 (1980).

[3] S.B. Hendricks, *J. Am. Chem. Soc.* **50**, 2455 (1928).

[4] R.W.G. Wyckoff, *Z. Kristallogr.* **75**, 529 (1930).

[5] J.E. Worsham, H.A. Levy and S.W. Peterson, *Acta Crystallogr.* **10**, 319 (1957).

[6] N. Sklar, M. Senko and B. Post, *Acta Crystallogr.* **14**, 716 (1961).

[7] A. Caron and J. Donohue, *Acta Crystallogr., Sect. B* **25**, 404 (1969).

[8] A. Pryor and P.L. Sanger, *Acta Crystallogr., Sect. A* **26**, 543 (1970).

[9] D. Mullen and E. Hellner, *Acta Crystallogr., Sect. B* **34**, 1624 (1978).

[10] T. Theophanides and P.D. Harvey, *Coord. Chem. Rev.* **76**, 237 (1987).

[11] J.E. Worsham and W.R. Busing, *Acta Crystallogr., Sect. B* **25**, 572 (1969).

[12] E.C. Kostansek and W.R. Busing, *Acta Crystallogr., Sect. B* **28**, 2454 (1972).

[13] D. Mootz and K.R. Albrand, *Acta Crystallogr., Sect. B* **28**, 2459 (1972).

[14] H.F.J. Savage, R.H. Blessing and H. Wunderlich, *Trans. Am. Crystallogr. Assoc.* **23**, 97 (1987).

[15] Z. Zhang, M. Shao, X. Xu, Y. Tang and Y. Tu, *Kexue Tongbao*, 658 (1982). [In Chinese.]

[16] W. van de Giesen and C.H. Stam, *Cryst. Struct. Commun.*, 1257 (1972).

[17] P.S. Gentile, J. White and S. Haddad, *Inorg. Chim. Acta* **8**, 97 (1974).

[18] L.C. Fetterly in L. Mandelcorn (ed.), *Non-stoichiometric Compounds*, Academic Press, New York, 1964, chap. 8.

[19] K. Takemoto and N. Sonoda in J.L. Atwood, J.E.D. Davis and D.D. MacNicol (eds.), *Inclusion Compounds*, Academic Press, London, 1984, Vol. 2, chap. 2.

[20] A.E. Smith, *Acta Crystallogr.* **5**, 224 (1952).

[21] W. Schlenk, *Liebigs Ann. Chem.* **565**, 204 (1949).

[22] A.E. Smith, *J. Chem. Phys.* **18**, 150 (1950).

[23] R. Forst, H. Jagodzinski, H. Boysen and F. Frey, *Acta Crystallogr., Sect. B* **46**, 70 (1990).

[24] A.E. Smith, *J. Chem. Soc.*, 2416 (1951).

[25] W. Schlenk, *Liebigs Ann. Chem.* **573**, 142 (1951).

[26] H. van Bekkum, J.D. Remijnse and B.M. Wepster, *J. Chem. Soc., Chem. Commun.*, 67 (1967).

[27] A.I. Kitaigorodsky, *Molecular Crystals and Molecules*, Academic Press, New York, 1973.

Note The crystal structure of the $3(NH_2)_2CO.CCl_4$ adduct at 170K is reported in J.F. Fait, A. Fitzgerald, C.N. Coughlan and F.P. McCandless, *Acta Crystallogr., Sect. C* **47**, 332 (1991).

Additional references are listed in p. 182.

Addenda to Chapter 1 and Chapter 2

Chapter 1

An interesting discussion of the pursuits of chemists in the next twenty
years is given in G.M. Whitesides, *Angew. Chem. Int. Ed. Engl.* **29**, 1209
(1990).

The important developments in organic synthesis in the past twenty-five
years and future projections are sketched in D. Seebach, *Angew. Chem.
Int. Ed. Engl.* **29**, 1320 (1990).

Chapter 2

Section 2.5

For the X-ray structures of $(\eta^2\text{-}C_n)\text{Ir(CO)Cl(PPh}_3)_2 \cdot mC_6H_6$ ($n = 60$, $m = 5$;
$n = 70$, $m = 2.5$), see A.L. Balch, V.J. Catalano and J.W. Lee, *Inorg.
Chem.* **30**, 3980 (1991); A.L. Balch, V.J. Catalano, J.W. Lee, M.M. Olmstead
and S.R. Parkin, *J. Am. Chem. Soc.* **113**, 8953 (1991).

The experimental $\Delta H_f°(c)$ for C_{60} is 545 kcal mol^{-1}. See H.-D. Beckhaus,
C. Rüchardt, M. Kao, F. Diederich and C.S. Foote, *Angew. Chem. Int. Ed.
Engl.* **31**, 63 (1992).

Air-stable films containing the endohedral fullerene complexes La@C_n
($n = 60$, 70, 74, 82; the symbol "@" indicates encagement) have been
made. Evidence has also been obtained for the successful production of
K@C_{60}, C_{59}B and K@C_{59}B where a boron atom subsitutes for a carbon vertex.
See Y. Chai, T. Guo, C. Jin, R.E. Haufler, L.P.F. Chibante, J. Fure,
L. Wang, J.M. Alford and R.E. Smalley, *J. Phys. Chem.* **95**, 7564 (1991).

Superconductivity occurs in K_3C_{60} ($T_c = 18$ K) and Rb_3C_{60} ($T_c = 30$ K).
See A.F. Hebard, M.J. Rosseinsky, R.C. Haddon, D.W. Murphy, S.H. Glarum,
T.T.M. Palstra, A.P. Ramirez and A.R. Kortan, *Nature (London)* **350**, 600
(1991); K. Holczer, O. Klein, S.-M. Huang, R.B. Kaner, K.-J. Fu, R.L.
Whetten and F. Diederich, *Science (Washington)* **252**, 1154 (1991).

The structure of single-phase superconducting K_3C_{60} is described in
P.W. Stephens, L. Mihaly, P.L. Lee, R.L. Whetten, S.-M. Huang, R. Kaner,
F. Diederich and K. Holczer, *Nature (London)* **351**, 632 (1991).

Preparation and structure of the alkali-metal fulleride A_4C_{60} is reported
in R.M. Fleming, M.J. Rosseinsky, A.P. Ramirez, D.W. Murphy, J.C. Tully,
R.C. Haddon, T. Siegrist, R. Tycko, S.H. Glarum, P.Marsh, G. Dabbagh,
S.M. Zahurak, A.V. Makhija and C. Hampton, *Nature (London)* **352**, 701
(1991); erratum, *ibid.* **353**, 868 (1991).

Carbon needles ranging from 4-30 nm in diameter and up to 1 μm in length
have been prepared; electron microscopy reveals that each needle
comprises coaxial tubes of 2-50 graphitic sheets separated by a spacing
of 3.4Å, and in each tube the carbon hexagons are arranged helically
about the needle axis. See S. Iijima, *Nature (London)* **354**, 56 (1991).

The entire March 1992 issue of *Accounts of Chemical Research* consists of
11 articles devoted to buckminsterfullerenes.

Section 2.27

Crystalline inclusion complexes comprising urea-water-halide/pseudohalide
and thiourea-halide lattices that feature a ribbon-like arrangement of
hydrogen-bonded urea and thiourea molecules, respectively, are reported
in T.C.W. Mak and R.K. McMullan, *J. Incl. Phenom.* **6**, 473 (1988) and
T.C.W. Mak, *ibid.* **8**, 199 (1990).

Chapter 3

Inorganic Compounds (Main Groups)

3.1. *High Pressure Ices* H_2O 185

3.2. *Potassium Hydrogen Fluoride (Potassium Bifluoride)* KHF_2 196

3.3. *Hydrochloric Acid Dihydrate* $HCl \cdot 2H_2O$ $(H_5O_2^+Cl^-)$ 204

3.4. *Complex Containing Trinuclear Molybdenum Clusters Bridged by the Hydrogen Oxide Ligand* $H_3O_2^-$
$\{[Mo_3O_2(C_2H_5COO)_6(H_2O)_2]_2(H_3O_2)\}Br_3 \cdot 6H_2O$ 210

3.5. *Lithium Nitride* Li_3N 214

3.6. *Complex Containing a Cryptated Sodium Cation and a Natride Anion* $Na^+(C_{18}H_{36}N_2O_6) \cdot Na^-$ 224

3.7. *Cesium Suboxide* Cs_7O 232

3.8. *Basic Beryllium Acetate* $Be_4O(CH_3COO)_6$ 237

3.9. *α-Rhombohedral Boron (α-Boron, R12 Boron)* B 241

3.10. *Decaborane(14)* $B_{10}H_{14}$ 249

3.11. *π-Cyclopentadienyl-π-(1)-2,3-dicarbollyliron(III), a Metallocarborane* $C_5H_5FeB_9C_2H_{11}$ 256

3.12. *Borax* $Na_2B_4O_5(OH)_4 \cdot 8H_2O$ 265

3.13. *Sodium β-Alumina* $Na_2O \cdot 11Al_2O_3$ 272

3.14. *Gallium(I) Tetrachlorogallate(III) ("Gallium Dichloride")* $Ga(GaCl_4)$ 280

3.15. *Calcium Carbide* CaC_2 285

3.16. *α-Oxalic Acid Dihydrate* $H_2C_2O_4 \cdot 2H_2O$ 291

3.17. *Forsterite (Olivine)* Mg_2SiO_4 297

3.18. *Beryl* $Be_3Al_2[Si_6O_{18}]$ 301

3.19. *Muscovite (Mica)* $KAl_2[AlSi_3O_{10}](OH)_2$ 304

3.20. *Low-Albite (Feldspar)* $Na[AlSi_3O_8]$ 309

3.21. *Dehydrated Zeolite 4A (Linde Molecular Sieve Type 4A)* $Na_{12}[Al_{12}Si_{12}O_{48}]$ 313

3.22. *Pheynl-(2,2',2"-nitrilotriethoxy)silane (Silatrane)* $(C_6H_5)Si(OCH_2CH_2)_3N$ 323

3.23. *Complex Containing Nonanuclear Germanium Anions* $[K(crypt)^+]_6Ge_9^{2-}Ge_9^{4-} \cdot 2\frac{1}{2}en$ 330

3.24. *Tin(II) Chloride Dihydrate* $[SnCl_2(H_2O)] \cdot H_2O$ 335

3.25. Hexanuclear Basic Lead(II) Perchlorate Hydrate
   [Pb$_6$O(OH)$_6$](ClO$_4$)$_4$·H$_2$O   **339**

3.26. Nitrogen Triiodide–Ammonia (1/1)   NI$_3$·NH$_3$ **346**

3.27. Orthorhombic Black Phosphorus   P   **351**

3.28. Hexachlorocyclophosphazene   (PNCl$_2$)$_3$ **364**

3.29. Tetraphosphorus Trisulfide   P$_4$S$_3$ **372**

3.30. Potassium Antimony Tartrate Trihydrate (Tartar Emetic)
   K$_2$[Sb$_2$(d-C$_4$H$_2$O$_6$)$_2$]·3H$_2$O   **378**

3.31. Bismuth Subchloride   Bi$_{24}$Cl$_{28}$, or (Bi$_9^{5+}$)$_2$(BiCl$_5^{2-}$)$_4$(Bi$_2$Cl$_8^{2-}$)   **382**

3.32. Hydrogen Peroxide   H$_2$O$_2$ **388**

3.33. Orthorhombic α-Sulfur   S$_8$ **394**

3.34. Tetrasulfur Tetranitride   S$_4$N$_4$ **402**

3.35. Polythiazyl (Polymeric Sulfur Nitride)   (SN)$_x$ **409**

3.36. Decaselenium Hexafluoroantimonate   Se$_{10}$(SbF$_6$)$_2$ **414**

3.37. Hexatellurium Tetrakis(hexafluoroarsenate)–Arsenic Trifluoride (1/2)
   Te$_6$(AsF$_6$)$_4$·2AsF$_3$ **423**

3.38. Xenon Difluoride   XeF$_2$ **430**

3.39. Gamma Brass   Cu$_5$Zn$_6$ **440**

3.40. α-Silver Iodide   AgI   **448**

3.41. Silver Nitrate Oxide   Ag(Ag$_6$O$_8$)NO$_3$ **455**

3.42.   Tri-iodoheptakis(tri-p-fluorophenylphosphine)undecagold
   Au$_{11}$I$_3$[P(p-FC$_6$H$_4$)$_3$]$_7$ **463**

3.43. Calomel (Mercury(I) Chloride, Merchlite)   Hg$_2$Cl$_2$ **473**

3.1 High Pressure Ices
H$_2$O

Crystal Data

Polymorph	Space group	Unit cell parameters*	Diffraction study**	Ref.
ice II	$R\bar{3}$ (No. 148)	a=7.79(1)Å, α=113.1(2)°, Z=12	X, N	[1]
ice III	$P4_12_12$ (No. 92)	a=6.73(1), c=6.83(1)Å, Z=12	X, N	[2]
ice IV	$R\bar{3}c$ (No. 167)	a=7.60(1)Å, α=70.1(2)°, Z=16	X	[3]
ice V	$A2/a$ (No. 15)	a=9.22(2), b=7.54(1), c=10.35(2)Å β=109.2(2)° Z=28	X	[4]
ice VI	$P4_2/nmc$ (No. 137)	a=6.27, c=5.79Å Z=10	X, N	[5]
ice VII	$Pn3m$ (No. 224)	a=3.43Å, Z=2	X	[6,9]
ice VIII	$I4_1/amd$ (No. 141)	a=4.6779(5), c=6.8029(10)Å Z=8	N	[7]
ice IX	$P4_12_12$ (No. 92)	a=6.73(1), c=6.83(1)Å, Z=12	X, N	[8]

* Reduced to values at 1 atm and -163°C [10].
** X and N stand for X-ray and neutron diffraction studies, respectively.

Crystal Structure and Remarks

1. Solid phases of water at high pressure

   Much research work, mainly by Kamb, has been done on the phase transitions and the structures of the ice polymorphs, and several general reviews have been published.[10-14]

   At high pressure and low temperature water can exist in at least eight solid forms, the structure adopted in each case being dependent upon the conditions and method of preparation. In all high pressure polymorphs of ice the water molecule remains intact; consequently they exhibit many of the unique properties of ice Ih.

   The experimentally determined phase diagram for the solid phases of water as it is presently known is shown in Fig. 3.1-1, in which the solid and long-dashed lines represent directly measured stable and metastable boundary lines respectively, and short-dashed and dotted lines represent boundaries that are extrapolated or estimated as accurately as possible from the available data. The only naturally occurring form is ice Ih. Ice Ic and vitreous ice are not indicated. Ices II, III, V, VI, VII and VIII can exist as stable phases at suitable temperatures and pressures, whereas ice IV and IX cannot.

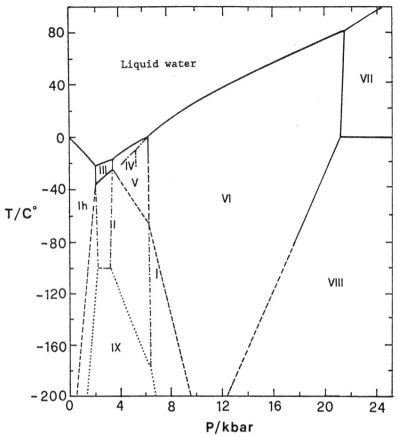

Fig. 3.1-1 Phase diagram of the solid phases of water. Ice IV is
metastable in the region of stability of ice V; ice Ic and vitreous ice
are not indicated. (After ref. 10).

2. Ice II

On the basis of an X-ray diffraction study conducted in 1964, a single
crystal neutron diffraction investigation of D_2O ice II was carried out to
locate the protons in the structure. The atomic coordinates in D_2O ice II are
listed in Table 3.1-1. The study confirmed the ordered structure proposed in
the X-ray work.

In D_2O ice II, two crystallographically-independent types of water
molecules form a tetrahedrally-linked network of hydrogen bonds, with each
deuteron lying near but not on the O...O line. The orientation of the
molecules is nearly, but not exactly, symmetrical with respect to the donor
O...O...O angles. The D-O-D angles (average 105.4°) do not differ
significantly from the angle in D_2O vapour (104.5°), although the
corresponding O...O...O angles are smaller. The structure of ice II is shown
in Fig. 3.1-2.

Table 3.1-1 Atomic coordinates of ice II.

Atom	Position	x	y	z
O(1)	6(f)	.272	.026	-.147
O(2)	6(f)	.480	.757	.339
D(1)	6(f)	.728	.404	.403
D(2)	6(f)	.149	.041	-.202
D(3)	6(f)	.742	.198	.371
D(4)	6(f)	.423	.195	-.016

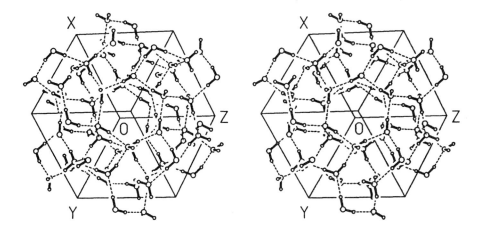

Fig. 3.1-2 Structure of ice II viewed in the [111] (hexagonal *c*)
direction. Hydrogen bonds between the water molecules are shown as dashed
lines. Puckered hexagonal rings of the two non-equivalent types, stacked
alternately one above the other, are linked together into columns by
hydrogen bonds in the same way as in ice Ih. Adjacent columns are linked
together in a more tightly fitting way than in ice Ih.

The basic structure of rhombohedral ice II in relation to ice Ih may be
described in the following way. The hexagonal columns of ice Ih, each
containing puckered layers of water molecules, are detached from one another,
moved relatively up and down parallel to the *c* axis so as to give a
rhombohedral stacking sequence, and then rotated 30° arround their *c* axis so
that they can be re-linked in a more compact way to give ice II. The manner
in which the hexagonal columns are linked in ice II causes the six-membered
rings in each column to twist relative to one another through an angle of
about 15°. This twist, plus a considerable flattening of the puckered six-
membered rings, causes the *c* axis to decrease from the value of 7.36Å in ice
Ih to 6.25Å in ice II. Each oxygen atom is hydrogen bonded to four nearest
neighbours at 2.77-2.84Å, and the O...O...O angles range from 80 to 128°.

3. Ice III and ice IX

Single-crystal X-ray diffraction has shown that ice III has a proton
disordered structure. Single-crystal neutron diffraction has revealed that
ice IX has an almost completely proton ordered structure, in which the ordered
component comprises two types of water molecules: type (1) in a general
position, and type (2) on a two-fold axis, each forming four hydrogen bonds in
a three-dimensional framework. Table 3.1-2 lists the atomic coordinates of
ice III and ice IX.

Table 3.1-2 Atomic coordinates of ice III and ice IX.

Atom	Position	Ice III (H_2O)				Ice IX (D_2O)			
		x	y	z	Occu-pancy	x	y	z	Occu-pancy
O(1)	8(b)	.1063	.2993	.2865	1	.1092	.3015	.2858	1.0000
O(2)	4(a)	.3895	.3895	0	1	.3926	.3926	0	1.0000
H or D(1)	8(b)	-.018	.335	.233	1/2	-.0124	.3321	.2137	.949
H or D(2)	8(b)	.129	.166	.301	1/2	.1138	.1577	.2968	.966
H or D(3)	8(b)	.319	.350	.086	1/2	.3002	.3594	.1057	.966
H or D(4)	8(b)	.217	.341	.206	1/2	.2181	.3289	.1809	.034
H or D(5)	8(b)	.154	.395	.341	1/2	.1344	.3930	.3866	.051
H or D(6)	8(b)	.374	.524	-.009	1/2	.3687	.5494	-.0216	.034

The positions of the oxygen atoms in ice III and ice IX appear to be
same. In these structures each oxygen atom is surrounded in an approximately
tetrahedral manner by four others at distances of 2.76 to 2.80Å. The
arrangement is such that the tetrahedral character of the water molecule is
preserved, and no hydrogen bond deviates by more than 17° from its idealized
direction. The density of ice III is 1.160 g cm^{-3} at -175°C and 1 atm. The
increase in density relative to ice Ih is accomplished by distortions from
ideal tetrahedral coordination which allow two or three nonbonded neighbours
to approach each water molecule to distances of about 3.6Å, as compared with
4.5Å in ice Ih. It is noteworthy that in the transition of ice Ih to ice II,
half of the hexagonal rings are destroyed, but when ice Ih transforms
to ice III all the hexagonal rings are destroyed and replaced by pentagonal
rings.

The hydrogen atoms in ice III are disordered. When ice III changes to
ice IX the hydrogen atoms become ordered, but some measurements of neutron
diffraction intensities for ice IX showed that the proton ordering is not
quite complete. The structure of ice IX is shown in Fig. 3.1-3.

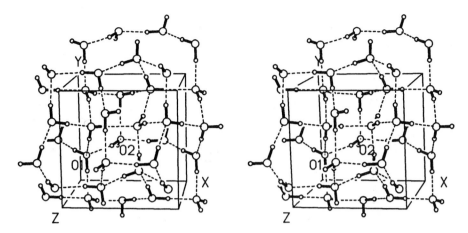

Fig. 3.1-3 Stereoview of the crystal structure of ice IX. Water molecules of type (1) form helical chains around the 4_1 axes parallel to the *c* axis. These helices are linked together laterally by molecules of type (2). Ice III has the same structure except that the protons are disordered.

4. Ice IV

The crystal structure of metastable high-pressure phase ice IV, which forms in the region of stability of ice V, has been determined by X-ray diffraction.[3] The atomic coordinates are listed in Table 3.1-3.

Table 3.1-3   Atomic coordinates of ice IV.

Atom	Position	x	y	z	Occupancy
O(1)	12(f)	.3804	-.1109	-.2396	1
O(2)	4(c)	.0855	.0855	.0855	1
H(1)	4(c)	.03	.03	.03	1/2
H(2)	12(f)	.14	.18	.01	1/2
H(3)	12(f)	.21	.33	-.10	1/2
H(4)	12(f)	.30	-.01	-.28	1/2
H(5)	12(f)	.22	.11	-.34	1/2
H(6)	12(f)	.28	.41	-.02	1/2

Fig. 3.1-4 shows the structure of ice IV. The calculated density at 110 K and 1 atm is 1.272 g cm^{-3}. Every molecule is linked by asymmetric hydrogen bonds to four others, giving rise to a new type of tetrahedrally connected network. Molecules of type (1) are linked by O(1)...O(1′) bonds into puckered six-membered rings of 3 symmetry, through the center of each of which passes an O(2)...O(2′) bond between a pair of type (2) molecules, along the threefold axis. The six-membered rings are linked laterally by type (2) molecules to form puckered sheets that are topologically similar to such sheets in ice Ih,

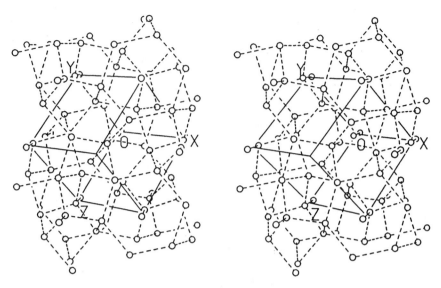

Fig. 3.1-4 Stereoview of the crystal structure of ice IV.

but are connected to one another in a very different and novel way. One
quarter of the intersheet bonds connect not directly between adjacent sheets
but remotely from one sheet to the second nearest sheet, through holes in the
intervening sheet. These remote connections are the O(2)...O(2') bonds,
passing through the O(1)-type six-membered rings. The sheets are stacked in a
sequence based on ice Ic, modified by reversal of the puckering to form the
remote connections and by internal distortion of the sheets to complete the
remaining intersheet bonds. Of the four nonequivalent hydrogen bonded O...O
distances in the structure, two (2.79 and 2.81Å) are only moderately
lengthened relative to the bonds in ice Ih (2.76Å), whereas the O(1)...O(1')
bond (2.88Å) and O(2)...O(2') bond (2.92Å) are unusually long. This is caused
by repulsion between O(1) and O(2) at nonbonded distances of 3.14 and 3.29Å in
the molecular cluster consisting of the O(1)-type six-membered ring threaded
by the O(2)...O(2') bond.

5. Ice V

The structure of ice V was determined from X-ray diffraction data.
Table 3.1-4 lists the atomic coordinates of the oxygen atoms of ice V. The
crystal structure is illustrated in Fig. 3.1-5.

The water molecules in ice V are hydrogen bonded to their four nearest
neighbours at distances of 2.76 to 2.87Å (average 2.80Å), and the shortest
next nearest neighbours distance is 3.28Å. The distortion from an ideal
tetrahedral coordination is quite large, the bond angle at the oxygen atoms
ranging from 84 to 128° with a root mean square deviation of 18° from 109.5°.

Table 3.1-4 Atomic coordinates of oxygen atoms of ice V.

Atom	Position	x	y	z
O(1)	4(e)	1/4	-.1847	0
O(2)	8(f)	.4629	.0565	.1544
O(3)	8(f)	.2751	-.3475	.2477
O(4)	8(f)	.3993	.3596	-.0146

The hydrogen atoms in ice V are disordered, but neutron diffraction measurements on D_2O ice V at -163°C indicate significant proton ordering.[11] It remains uncertain as to whether the partial ordering occurs only when ice V is quenched at low temperatures.

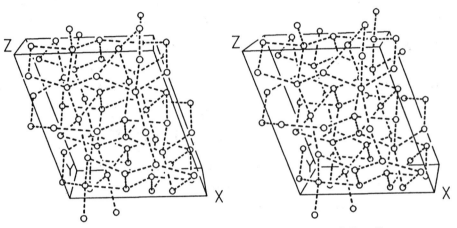

Fig. 3.1-5 Stereoview of the crystal structure of Ice V.

6. Ice VI

The atomic coordinates of the oxygen atoms in ice VI are listed in Table 3.1-5, and its structure is shown in Fig. 3.1-6.

Table 3.1-5 Atomic coordinates of oxygen atom in ice VI.

Atom	Position	x	y	z
O(1)	8(g)	1/4	.026	.632
O(2)	2(a)	1/4	-1/4	1/4

Origin at $\bar{1}$.

The water molecules are placed in two separate but interpenetrating frameworks. The molecules in each of these frameworks are hydrogen bonded and four-coordinated and they occupy the voids in the other framework. The four nearest neighbours to each water molecule are at distances of 2.80 to 2.82Å, forming a rather highly distorted tetrahedron in which the angle between the hydrogen bonds varies from 76° to 128°. This structural feature is a natural way to achieve high density in tetrahedrally linked framework structures.

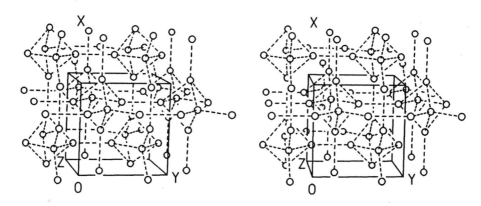

Fig. 3.1-6 Stereoview of the crystal structure of ice VI. Note that the two hydrogen bonded frameworks interpenetrate but do not interconnect.

The space group $P4_2/nmc$ requires a completely proton disordered structure. If any proton ordering occurs, the symmetry of the crystal must be degraded. Such a degradation is observed at -163°C in both X-ray and neutron diffraction studies, which corresponds to a violation of the c glide plane and the 4_2 axis. The space group symmetry is therefore reduced from tetragonal $P4_2/nmc$ to at least orthorhombic $Pmmn$. The phase which occurs at -163°C with space group $Pmmn$ may be referred to as ice VI'. However, in view of the fact that the distinction between these two phases is much better than that between ice III and ice IX, ice VI' warrants being called ice X.

7. Ice VII and ice VIII

Cubic ice VII, which was discovered by Bridgman,[9] is produced when ice VI is subjected at room temperature to a pressure ≥ 21.4 kbar.

In ice VII, each oxygen atom has eight nearest neighbours, four of which are located at the vertices of a regular tetrahedron and are hydrogen bonded to the central molecule (Fig. 3.1-7). The structure consists of two interpenetrating but not interconnecting frameworks, each of which is of the diamond type (like ice Ic). The density of ice VII at 25 kbar (1.66 g cm^{-3}) would be exactly twice that of ice I except that the hydrogen bond length in ice VII is 2.96Å as compared with 2.76Å in ice I at 1 atm. The increase is caused by four repulsive contacts, which have the same O...O distance as the four bonds. In ice VII, the protons appear to be disordered.

Ice VII becomes ordered at temperatures below about 0°C to form ice VIII. The crystal structure of D_2O ice VIII (at 28 kbar and -4°C) from neutron diffraction is shown in Fig. 3.1-8; its atomic coordinates are given in Table 3.1-6.

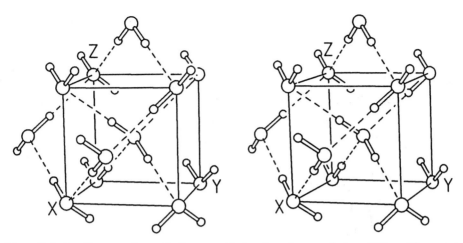

Fig. 3.1-7 Stereoview of the crystal structure of ice VII. The protons occupying a set of plausible positions are included for aesthetic reasons.

Table 3.1-6   Atomic coordinates of D_2O ice VIII.

Atom	Position	x	y	z
O	8(e)	0	1/4	0.1049
D	16(h)	0	.4137	0.1932

Origin at centre (2/m).

The tetragonal ice VIII and cubic ice VII unit cell parameters are related by $a_{VIII} = \sqrt{2}a_{VII}$, and $c_{VIII} = 2a_{VII}$. The water molecules are all oriented along the +[001] direction in one sublattice, and along the -[001] direction in the other.

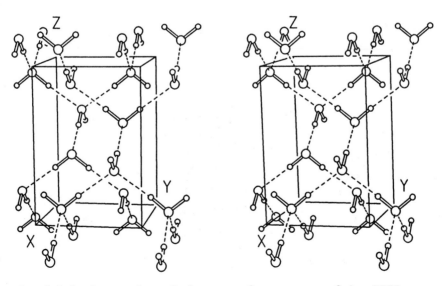

Fig. 3.1-8 Stereoview of the crystal structure of ice VIII.

The bond lengths and bond angles in ice VIII show a slight distortion from perfect tetrahedral coordination. The bonded O...O distance is 2.892Å; O...O...O angles of the type O...D-O-D...O and O-D...O...D-O are 107.96°, while angles of the type O...D-O...D-O are 110.23°. One nonbonded O...O distance is shorter (2.740Å), and the other is longer (3.319Å), than the bonded distance. The D_2O molecule has an O-D bond length of 0.973Å and a D-O-D angle of 104°. This is in excellent agreement with the measured dimensions of the D_2O molecules in other high-pressure ice structures. The O-D...O bond is very slightly bent, with a bond angle of 177° at the deuterium atom.

An interesting feature common to ice polymorphs VI, VII and VIII is the interpenetration of hydrogen-bonded frameworks. In the case of ice VII two interpenetrating, but unconnected, diamond-type frameworks co-exist.

8. Summary of structural features

The principal structural characteristics of the high-pressure polymorphs of ice are listed in Table 3.1-7. The corresponding data for ice Ih and Ic are also included for comparison.

Table 3.1-7 Structural features of the polymorphs of ice.[10]

Poly-morph	Hydrogen positions	Density* ($g\ cm^{-3}$)	Number of nearest neighbours under 3Å	Hydrogen bonded O...O distance*	Nearest non-bonded O...O distance*	O...O...O Bond angle*
Ih	disordered	0.93	4	2.74	4.5	109
Ic	disordered	0.93	4	2.75(-130°C)	4.5	109.5
II	ordered	1.18	4	2.75-2.84	3.24	80-128
III	disordered	1.16	4	2.76-2.80	3.43	87-141
IV	disordered	1.27	4	2.79-2.92	3.14	
V	disordered	1.23	4	2.76-2.87	3.28	84-128
VI	disordered	1.31	4	2.80-2.82	3.51	76-128
VII	disordered	1.49	8	2.95	2.95	109.5
VIII	ordered	1.49	6	2.80-2.96	2.80	109
IX	ordered	1.16	4	2.76-2.80	3.51	87-140

* Reduced to values at 1 atm and -163°C unless otherwise noted.

The structures of all the phases are variants on two simple themes: four-coordination of water molecules and the possibility of order or disorder in their orientations. Both themes have their origin in a property that is unique to water molecules, namely their capacity to form four hydrogen bonds in a tetrahedral arrangement. The molecule donates its two hydrogen atoms to form two hydrogen bonds to two of its neighbours, and two other neighbours each donate one hydrogen atom to an hydrogen bond to the central molecule. It is clear that each molecule can form its hydrogen bonds in six different ways, which is the number of arrangements of the two near and two distant hydrogen atoms around a given oxygen atom. Consequently, orientational disorder is possible. Each water molecule in the crystal cannot occupy all orientations because each hydrogen bond can contain only one hydrogen atom. If disorder is present, it has profound effects on some of the properties of the phase. All polymorphs of ice except ice II are disordered at high temperatures and become partly or completely ordered at low temperature.

References

[1] B. Kamb, W.C. Hamilton, S.J. LaPlaca and A. Prakash, *J. Chem. Phys.* **55**, 1934 (1971); B. Kamb, *Acta Crystallogr.* **17**, 1437 (1964).

[2] B. Kamb and A. Prakash, *Acta Crystallogr., Sect. B* **24**, 1317 (1968).

[3] H. Engelhardt and B. Kamb, *J Chem. Phys.* **75**, 5887 (1981).

[4] B. Kamb, A. Prakash and C. Knobler, *Acta Crystallogr.* **22**, 706 (1967).

[5] B. Kamb, *Science (Washington)* **150**, 205 (1965).

[6] E. Weir, S. Block and G. Piermasini, *J. Res. Natl. Bur. Stand.* **69c**, 275 (1965).

[7] J.D. Jorgensen, R.A. Beyerlein, N. Watanabe and T.G. Worlton, *J. Chem. Phys.* **81**, 3211 (1984).

[8] S.J. LaPlaca, W.C. Hamilton, B. Kamb and A. Prakash, *J. Chem. Phys.* **58**, 567 (1973).

[9] P.W. Bridgman, *J. Chem. Phys.* **5**, 946 (1937).

[10] P.V. Hobbs, *Ice Physics*, Clarendon Press, Oxford, 1974.

[11] W.C. Hamilton, B. Kamb, S.J. LaPlaca and A. Prakash, *Physics of Ice*, Plenum Press, New York, 1969, pp. 44-58.

[12] A.F. Wells, *Structural Inorganic Chemistry*, 5th ed. Clarendon Press, Oxford, 1984, pp. 653-666.

[13] D. Eisenberg and W. Kauzmann, *The Structure and Properties of water*, Oxford University press, New York, 1969.

[14] E. Whalley in P. Schuster, G. Zundel and C. Sandorfy (eds.), *The Hydrogen Bond*, Vol. 3, North-Holland, Amsterdam, 1976, chap. 29.

3.2 Potassium Hydrogen Fluoride (Potassium Bifluoride)
KHF₂

Crystal Data

Tetragonal, space group *I4/mcm* (No. 140)

$a = 5.67$, $c = 6.81$Å, $Z = 4$

Atom	Position	x	y	z
K	4(a)	0	0	1/4
F	8(h)	.1420	.6420	0
H	4(d)	0	1/2	0

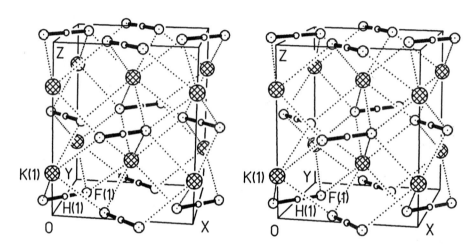

Fig. 3.2-1 Stereoview of the crystal structure of KHF₂. The square-antiprismatic coordination environment of the potassium ion is indicated by dotted lines to its nearest neighbouring fluorine atoms.

Crystal Structure

Potassium bifluoride has been subjected to intensive study because of its importance in the theory of hydrogen bonding, especially in the context of a *symmetrical* hydrogen bond. As early as 1923, Bozorth[1] determined its crystal structure and pointed out that the structure is composed of potassium atoms and F_2 "dumb-bells" (in which the two fluorine atoms are separated by 2.25Å) lying in planes perpendicular to the tetragonal axis (Fig. 3.2-1). For the hydrogen atoms there are two possible positions, one of which is in the middle of the dumb-bell, forming an HF_2^- ion.

In 1952, a single crystal neutron diffraction study on KHF₂[2] showed that the hydrogen atom is symmetrically located within the linear [F-H-F]⁻ system. It was difficult, however, to decide whether a short hydrogen bond is symmetric or asymmetric using diffraction data only. The neutron diffraction

measurements could be equally well explained by two models: either (i) the
proton vibrates with a large amplitude about a single potential minimum or
(ii) it oscillates with a smaller amplitude, but is also statistically
disordered between two close sites. Additional experimental evidence from
another source was therefore required to settle this point.

In 1964, Ibers[3] refined the structure of KHF_2 using new neutron
diffraction data. The agreement between the differences in mean square
amplitudes of vibration of hydrogen and fluorine along the bond as obtained
from an analysis of the thermal parameters and as calculated from the
spectroscopic data lent support to the linear symmetric model of the
bifluoride ion. The F-H-F bond distance was updated to 2.277(6)Å.

Remarks

1. The strongest hydrogen bonds

The strongest hydrogen bonds are observed in the bifluoride anions of
compounds of the types MHF_2 (M = Li, Na, K, Rb, NH_4, Cs) and $MF(HF_2)$ (M = Sr,
Ba). The F...F distances from X-ray analyses are approximately the same for
all compounds, though in the alkaline-earth compounds a slightly asymmetrical
hydrogen bridge is likely, whereas in the alkali-metal compounds symmetrical
hydrogen bonds are found. Table 3.2-1 lists some F...F bond distances
observed in bifluoride compounds.

Table 3.2-1 The F...F distances in some bifluorides.

Compound	F...F	(F...F)$_{corr.}$*	Ref.
$NaHF_2$	2.264	2.288	[4]
KHF_2	2.277	2.293	[4]
NH_4HF_2	2.269	2.291	[4]
	2.275	2.297	[4]
$BaF(HF_2)$	2.269	2.281	[5]
$SrF(HF_2)$	2.266	2.269	[5]

* Corrected for thermal motion.

2. Comparison between the structures of NH_4HF_2 and KHF_2

Potassium and ammonium salts are often isostructural, the K^+ and NH_4^+
ions having the same charge and very similar sizes.[6] In KHF_2 and NH_4HF_2,[7]
however, the HF_2^- ions are oriented differently in the two structures (Fig.
3.2-1 and Fig. 3.2-2). In KHF_2 each K^+ ion is surrounded by eight equidistant
F⁻ neighbours at 2.77Å; but in NH_4HF_2 the NH_4^+ ion has only four nearest F⁻
neighbours at 2.797 and 2.822Å, and the next set of four F⁻ neighbours occur
at distances of 3.02 and 3.40Å. In NH_4HF_2 there are two independent HF_2^- ions

with F...F distances of 2.273 and 2.268Å. The breakdown of the coordination group of eight F⁻ neighbours into two sets of four is to be attributed to the formation of N-H...F bonds.

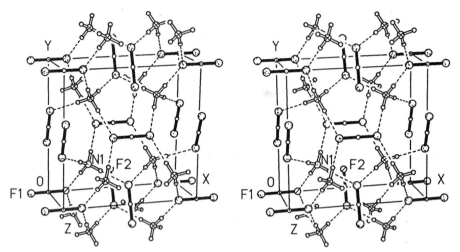

Fig. 3.2-2 Stereoview of the crystal structure of NH_4HF_2. All H atoms and N-H...F hydrogen bonds are shown.

3. The asymmetric strong hydrogen bond

Williams and Schneemeyer[8] determined the crystal structure of *p*-toluidinium bifluoride (4-$MeC_6H_4NH_3^+HF_2^-$) by neutron diffraction and found that the linear bifluoride ion has the following bond lengths:

$$F \xrightarrow{\ 1.025\text{Å}\ } H \xrightarrow{\ 1.235\text{Å}\ } F$$

This is the first known example of a HF_2^- ion possessing an asymmetric hydrogen bond. Other strong asymmetric hydrogen bonds are observed in solid HF (F...F = 2.49Å),[9] while oxonium bifluoride $(H_3O)(HF_2)$ shows an intermediate behavior (F...F = 2.367Å)[10] similar to that found in KH_2F_3,[11] which forms two bent [F-H...F-H...F]⁻ ions with F...F distances of (2.35, 2.34) and (2.29, 2.34)Å.

4. Molecular orbital treatment of the HF_2^- ion

A useful bonding picture in F-H-F may be obtained from a molecular orbital approach. If the bond axis is taken along the z direction, the $1s$ orbital of the H atom and the two $2p_z$ orbitals of the F atoms overlap with each other, as shown in Fig. 3.2-3(a), to form three molecular orbitals:

$$\psi_1 = N_1[2p_z(A) + 2P_z(B) + C1s],$$
$$\psi_2 = N_2[2p_z(A) - 2p_z(B)], \text{ and}$$
$$\psi_3 = N_3[2p_z(A) + 2p_z(B) - C1s],$$

where C is a weighting coefficient and N_1, N_2, N_3 are normalizing constants.

The ordering of the molecular orbitals are shown qualitatively in Fig. 3.2-3(b). Since there are four valence electrons, the bonding (ψ_1) and nonbonding (ψ_2) molecular orbitals are both occupied to yield a 3-center, 4-electron bond. The bond order of each F-H link in HF_2^- is 0.5, which can be compared with that in the HF molecule as follows:

Molecule	Bond order	d/Å	k(m.dyn./Å)
HF	1	0.93	8.9
HF_2^-	0.5	1.13	2.3

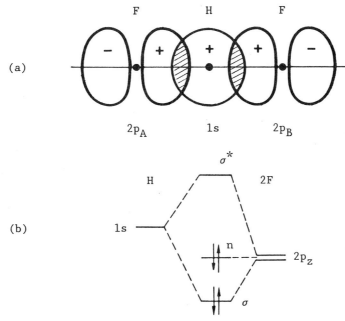

Fig. 3.2-3 Bonding in HF_2^-: (a) orbital overlap; (b) qualitative molecular orbital correlation diagram.

5. Structures of the $H_nF_{n+1}^-$ anions and related compounds

A series of homologous $H_nF_{n+1}^-$ anions have been prepared and characterized as their K^+, NH_4^+, $(CH_3)_4N^+$, and $C_5H_5NH^+$ salts. Table 3.2-2 lists the structural data of some $H_nF_{n+1}^-$ anions and Fig. 3.2-4 shows the structures.

The averaged hydrogen bonding F-H...F distances (in Å) in the known intermediary compounds of the system KF-HF steadily increase with increasing HF content, i.e., with increasing average size of the hydrogen-bonded structural unit. The upper limit of the series is obviously given by the F...F distance in the infinite chain of the pure HF solid, which is 2.500(1)Å at -146°C. Similar trends, though with less data for other counter cations, can be seen in the complex acid fluoride structures listed in Table 3.2-2.

Table 3.2-2   Structural data of $H_nF_{n+1}^-$ anions.

Anion	Compound	Hydrogen bond length (Å)	Figure	Ref.
HF_2^-	$K(HF_2)$	2.277	3.2-1	[3]
	$NH_4(HF_2)$	2.291, 2.297	3.2-2	[4]
	$(C_5H_5NH)(HF_2)$	2.326		[12]
$H_2F_3^-$	$(CH_3)_4N(H_2F_3)$	2.302, 2.316	3.2-4	[13]
	$K_2(H_2F_3)(H_3F_4)$	2.323, 2.352		[14]
	$K(H_2F_3)$	2.331 (av.)		[11]
	$(C_5H_5NH)(H_2F_3)$	2.343, 2.345		[12]
$H_3F_4^-$	$K(H_3F_4)$	2.401, 2x2.40	3.2-4	[14]
	$K_2(H_2F_3)(H_3F_4)$	2.281, 2.402, 2.441		[14]
	$(CH_3)_4N(H_3F_4)$			[13]
	$(NH_4)(H_3F_4)$			[15]
	$(NO)(H_3F_4)$			[16]
	$(C_5H_5NH)(H_3F_4)$	2.361, 2x2.389		[12]
$H_4F_5^-$	KH_4F_5	2.453 (av.)		[17]
$H_5F_6^-$	$(CH_3)_4N(H_5F_6)$	2.266, 4x2.484	3.2-4	[13]

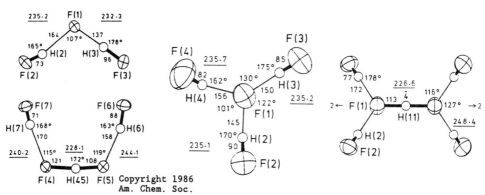

Copyright 1986 Am. Chem. Soc.

Fig. 3.2-4 Structures of $H_nF_{n+1}^-$ anions with interatomic distances (pm) and angles, underlined for F...F. (After refs. 13 and 14).

The crystal structures of four complexes of the system pyridine-hydrogen fluoride ($C_5H_5N.nHF$, $n = 1,2,3,4$) have been determined, as shown in Fig. 3.2-5. In $C_5H_5N.HF$ the acidic H atom lies distinctly closer to the fluorine than to the nitrogen atom, i.e., the hydrogen bond is of the type F-H...N rather than N-H...F [Fig. 3.2-5(a)]. It is actually the first example of this type and the shortest between a nitrogen and a fluorine atom established so far by crystal-structure analysis. The complex $C_5H_5N.2HF$, according to its crystal structure, can be reformulated as $C_5H_5NH^+(HF_2)^-$ [Fig. 3.2-5(b)]. The hydrogen bond within the HF$_2^-$ ion is almost as strong as that in the respective alkali metal salts, but it contains the H atom in a definitely off-center position close to the outer F atom. The other hydrogen bond, between the N atom and the inner F atom, is of the type N-H...F and still very short.

Ionic formulae are also appropriate for the 1:3 and 1:4 compounds of pyridine with hydrogen fluoride, i.e., C$_5$H$_5$NH$^+$(H$_2$F$_3$)$^-$ and C$_5$H$_5$NH$^+$(H$_3$F$_4$)$^-$, respectively. But only in the 1:3 compound is the unit of structure again an internally neutralized, tightly bound zwitterionic molecular complex [Fig. 3.2-5(c)]. In the crystal structure of the 1:4 compound, there are infinite ribbons, along the *b* direction, formed by trifurcated hydrogen N-H(...F)$_3$ bonds between the cation and three anions, and *vice versa* [Fig. 3.2-5(d)].

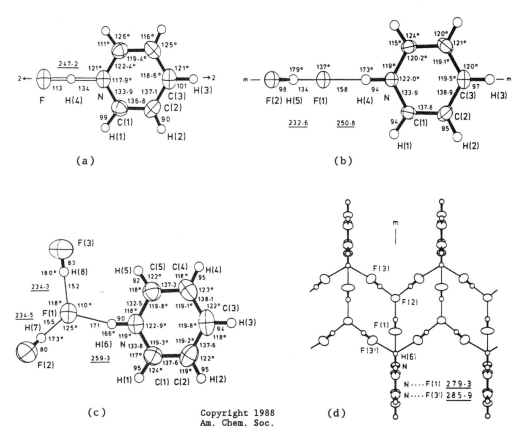

(a)   (b)

(c)   Copyright 1988   (d)
             Am. Chem. Soc.

Fig. 3.2-5 Structures of (a) C$_5$H$_5$N.HF, (b) C$_5$H$_5$N.2HF, (c) C$_5$H$_5$N.3HF and (d) C$_5$H$_5$N.4HF. (After ref. 12).

The crystal structures of (H$_3$F$_2$)$^+$(Sb$_2$F$_{11}$)$^-$ and (H$_2$F)$^+$(Sb$_2$F$_{11}$)$^-$ have been determined.[18] The (H$_3$F$_2$)$^+$ ion of (H$_3$F$_2$)(Sb$_2$F$_{11}$) [Fig. 3.2-6(a)] is located on a crystallographic symmetry center and thus has a *trans* conformation. With a F...F distance of 2.30(1)Å, its inner hydrogen bond is only slightly weaker than the strongest known hydrogen bonds (F...F down to 2.26Å) in the series of hydrogen difluorides and poly(hydrogen fluorides). By longer hydrogen bonds of 2.41(1)Å the cations are linked with the anions in infinite chains. The dioctahedral $Sb_2F_{11}^-$ ion is also centrosymmetric with a linear central fluorine bridge.

In the structure of $(H_2F)^+(Sb_2F_{11})^-$ the ions occupy general positions. With much longer and less gradated F...F distances to the anions (six distances between 2.64 and 2.78Å) the H_2F^+ ion exhibits orientational disorder. The Sb-F-Sb bridge angle of the anion amounts to 144.2(3)°.

In the crystal structure of hypofluorous acid, HOF, the molecules are linked in zigzag chains by O-H...O type hydrogen bonds and not by O-H...F hydrogen bonds, as shown in Fig. 3.2-6(b).[19]

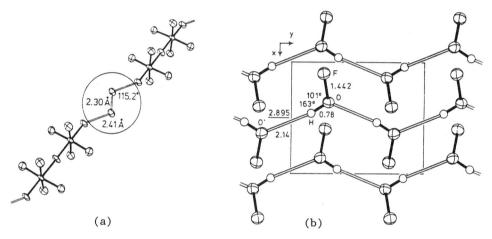

(a)   (b)

Fig. 3.2-6 (a) Structure of $(H_3F_2)^+(Sb_2F_{11})^-$. The fluoronium ion $(H_3F_2)^+$ with its hydrogen bonds is shown in the circle. (b) Crystal structure of HOF showing interatomic distances (Å) and angles (°). (After refs. 18 and 19).

6. Structure of the low-melting poly(hydrogen chloride) $Me_3N.5HCl$

Crystal structure analysis of a 1:5 adduct (mp -54°C) of trimethylamine with hydrogen chloride has revealed the ionic pair $(Me_3NH)^+[Cl(HCl)_4]^-$ as its basic unit.[20] The protonated cation and complex anion are linked by a N-H...Cl(1) bridge that deviates considerably from linearity, and distorted trigonal bipyramidal coordination at the central Cl⁻ ion is completed by uncommon Cl-H...Cl(1) hydrogen bonds to the remaining four HCl molecules (Fig. 3.2-7). A much weaker branched interaction of the Me_3NH^+ ion with atom Cl(5) of an adjacent ion pair involves N...Cl and H...Cl distances of 3.547 and 2.96Å, respectively, and an angle of 128° at the H atom.

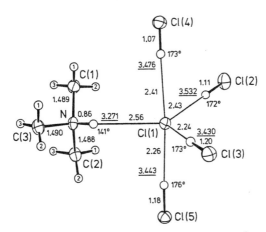

Fig. 3.2-7 Geometry and selected bond distances (Å) and angles (°) of the $(Me_3NH)^+[Cl(HCl)_4]^-$ ion pair. Underlined values correspond to overall Cl...Cl and N...Cl distances in the hydrogen bonds. (After ref. 20).

References

[1]   R.M. Bozorth, *J. Am. Chem. Soc.* **45**, 2128 (1923).

[2]   S.W. Peterson and H.A. Levy, *J. Chem. Phys.* **20**, 704 (1952).

[3]   J.A. Ibers, *J. Chem. Phys.* **40**, 402 (1964).

[4]   H.C. Carrell and J. Donohue, *Isr. J. Chem.* **10**, 195 (1972).

[5]   W. Massa and E. Herdtweck, *Acta Crystallogr., Sect. C* **39**, 509 (1983).

[6]   A.F. Wells, *Structural Inorganic Chemistry*, 5th ed., Clarendon press, Oxford, 1984.

[7]   T.R.R. McDonald, *Acta Crystallogr.* **13**, 113 (1960).

[8]   J.M. Williams and L.F. Schneemeyer, *J. Am. Chem. Soc.* **95**, 5780 (1973).

[9]   M. Atoji and W.N. Lipscomb, *Acta Crystallogr.* **7**, 173 (1954).

[10] D. Mootz, U. Ohms and W. Poll, *Z. Anorg. Allg. Chem.* **479**, 75 (1981).

[11] J.D. Forrester, M.E. Senko, A. Zalkin and D.H. Templeton, *Acta Crystallogr.* **16**, 58 (1963).

[12] D. Boenigk and D. Mootz, *J. Am. Chem. Soc.* **110**, 2135 (1988).

[13] D. Mootz and D. Boenigk, *Z. Anorg. Allg. Chem.* **544**, 159 (1987).

[14] D. Mootz and D. Boenigk, *J. Am. Chem. Soc.* **108**, 6634 (1986).

[15] D. Mootz and W. Poll, *Z. Naturforsch.* **39b**, 290 (1984).

[16] D. Mootz and W. Poll, *Z. Naturforsch.* **39b**, 1300 (1984).

[17] B.A. Coyle, L.W. Schroeder and J.A. Ibers, *J. Solid State Chem.* **1**, 386 (1970).

[18] D. Mootz and K. Bartmann, *Angew. Chem. Int. Ed. Engl.* **27**, 391 (1988).

[19] W. Poll, G. Pawelke, D. Mootz and E.H. Appelman, *Angew. Chem. Int. Ed. Engl.* **27**, 392 (1988).

[20] D. Mootz and J. Hocken, *Angew. Chem. Int. Ed. Engl.* **28**, 1697 (1989).

3.3   Hydrochloric Acid Dihydrate
$HCl \cdot 2H_2O$ ($H_5O_2^+Cl^-$)

Crystal Data

Monoclinic, space group $P2_1/c$ (No. 14)

$a = 3.991(5)$, $b = 12.055(1)$, $c = 6.698(1)$Å, $\beta = 100.58(2)°$, $Z = 4$

Atom	x	y	z
Cl	.0182(4)	.3277(1)	.1482(2)
O(1)	.5533(13)	.1279(4)	.0244(7)
O(2)	.2949(18)	.0572(4)	.2974(8)
H(1)	.430	.093	.160
H(2)	.380	.160	-.075
H(3)	.730	.190	.075
H(4)	.140	-.010	.280
H(5)	.220	.110	.410

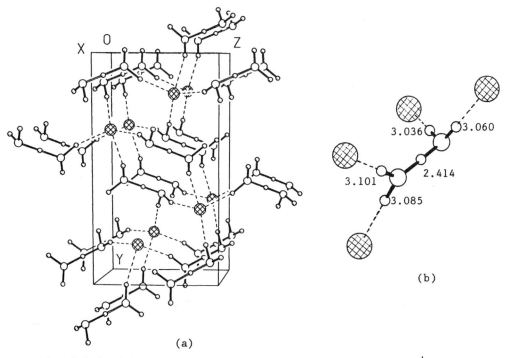

(a)

(b)

Fig. 3.3-1 (a) Crystal structure of H_5O_2Cl. (b) The $H_5O_2^+$ ion in hydrochloric acid dihydrate; the e.s.d.'s of hydrogen bond lengths are in the range 0.005-0.007Å. (Adapted from ref. 15).

Crystal Structure

The phase diagram of the hydrochloric acid-water system indicates the existence of three intermediate compounds; their melting points and structural formulae are as follows:

$HCl \cdot H_2O$: mp -15.4°C, $(H_3O)^+Cl^-$

$HCl \cdot 2H_2O$: mp -17.4°C, $(H_5O_2)^+Cl^-$

$HCl \cdot 3H_2O$: mp -24.9°C, $(H_5O_2)^+Cl^- \cdot H_2O$

The crystal structure of hydrochloric acid dihydrate contains puckered layers of Cl^- ions connected by hydrogen bonds from the water molecules. The water molecules are bonded to each other in pairs by a very short bond (2.41Å), forming $H_5O_2^+$ ions. These $H_5O_2^+$ ions connect the chlorine layers: one end [O(1)] is bonded to two chlorine ions belonging to the same layer (bond lengths are 3.04 and 3.06Å); the other end [O(2)] is connected to two chlorine ions in different layers (bond lengths are 3.09 and 3.10Å). Each oxygen atom is accordingly surrounded by three neighbours, which are pyramidally arranged.[1]

In $H_5O_2^+$, hydrogen atom H(1) seems to be situated very near the center of the O(1)...O(2) bond. From the X-ray data it cannot of course be decided whether the bond is symmetrical or asymmetrical. However, as the bond is extremely short the conditions for a symmetric single maximum distribution might be fulfilled.

Remarks
1. The O-H...O hydrogen bond

The subject of strong hydrogen bonds has been extensively reviewed in recent years.[2] The strength of a hydrogen bond A-H-B is most conveniently defined by the distance between the atoms A and B. If this distance is shorter than the sum of their van der Waals radii by more than 0.5Å, the bond is "very strong". A difference between 0.3Å and 0.5Å defines a "strong" hydrogen bond, while <0.3Å characterizes a "weak" hydrogen bond. For the homonuclear O-H-O bond, the corresponding categories are: shorter than 2.5Å for a "very strong" bond, 2.5 to 2.75Å for a "strong" bond, and over 2.75Å for a "weak" bond.

A decrease in the separation between the oxygen atoms is accompanied by a lowering of the potential well in which the hydrogen resides. The corresponding hydrogen bond energies are: >100 kJ/mole for a "very strong" bond, >50 kJ/mole for a "strong" bond, and <50 kJ/mole (often <30 kJ/mole) for a "weak" bond.

The O...O distance also affects the shape of the potential energy function of the proton. It is generally agreed that the shortest and strongest H-bonds have a single minimum halfway, or approximately halfway, between the oxygen atoms. This minimum becomes shallower as the separation increases and eventually splits into two minima separated by an energy barrier that becomes higher as the separation further increases.

2. The $H_5O_2^+$ ion

As discussed in Section 3.2, the strongest symmetrical hydrogen bond in HF_2^- has a bond order of 1/2. The $H_5O_2^+$ ion, which may well be expected to have a strong hydrogen bond, is a well-documented chemical entity as found in over 20 crystal structures. Some relevant data are presented in Table 3.3-1.

Table 3.3-1   Some compounds containing the $H_5O_2^+$ ion.

Compound	Diffraction method	O...O/Å	Geometry of H-bond	Ref.
$(H_5O_2)Cl.H_2O$	X-ray	2.434		[3]
$(H_5O_2)ClO_4$	X-ray	2.424		[4]
$(H_5O_2)Br$	neutron	2.40	O —1.17— H —1.22— O, 174.7°	[5]
$(H_5O_2)Br.H_2O$	X-ray	2.47		[6]
$(H_5O_2)_2SO_4.2H_2O$	X-ray	2.43		[7]
$(H_5O_2)[trans-Co(en)_2Cl_2]Cl_2$	neutron	2.431	centered	[8]
$(H_5O_2)[C_6H_2(NO_2)_3SO_3].2H_2O$	neutron	2.436	O —1.128— H —1.310— O, 175°	[9]
$(H_5O_2)[C_6H_4(COOH)SO_3].H_2O$	X-ray, neutron	2.414	O —1.201— H —1.219— O	[10]
$(H_5O_2)[PW_{12}O_{40}]$	X-ray, neutron	2.414	centered	[11]
$(H_5O_2)[(C_2H_5)_4N]_3[Mo_2Cl_8H]^- [MoCl_4O(H_2O)]$	X-ray	2.336	-	[12]
$[H_5O_2][Mn(H_2O)_2(SO_4)_2]$	X-ray	2.426	centered	[13]

The common characteristic of these compounds is a very short O...O distance (2.40-2.45Å), which clearly justifies the formulation of hydrogen chloride dihydrate as $(H_5O_2)Cl$.

3. The hydronium ion H_3O^+

The H_3O^+ ion has been referred to as proton hydrate, oxonium ion, and hydronium ion. Various experimental studies not only established the pyramidal H_3O^+ ion as a stable entity but also yielded other structural information. The O-H bonds have the same length as in ice, and the apical angles generally fall between 110° and 115°. Diffraction methods have provided direct confirmation of H_3O^+ ions in dilute aqueous acids, and have enabled measurement of the O-H or O-D bond length (1.02Å) of the ion in concentrated hydrochloric acid. The H_3O^+ ion (Fig. 3.3-2) is coordinated to four water molecules, three of which are linked by strong H bonds, and the fourth by much weaker charge-dipole forces.[14]

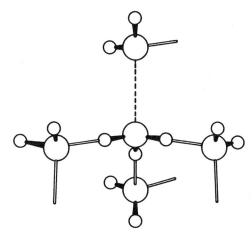

Fig. 3.3-2 Model illustrating the average coordination environment of the H_3O^+ ion in dilute aqueous acids.   [After ref. 14].

Present knowledge about the H_3O^+ ion may be summarized as follows. As a chemical entity H_3O^+ is just as real as its counterpart, the hydroxide ion OH^-. In water and aqueous solutions both ions are equally short-lived owing to rapid proton transfer. Their average lifetimes at 25°C are respectively 2.2 and 4.0 psec, as measured by nmr. The existence of discrete H_3O^+ ions has been confirmed in crystals and in acid solutions. In aqueous acids the H_3O^+ ion is strongly hydrogen-bonded to three H_2O molecules, with O...O distances (2.52Å) much shorter than in pure water (2.83Å). Proton transfer between H_3O^+ and water (or water and OH^-) is a concerted process accompanied by a rearrangement of the H-bond network. It takes less than 1 percent of the average lifetime of the H_3O^+ ion.

Since protons tend to associate with water to form H_3O^+ ions, the conventional structural formulae for some hydrates should be re-written as follows:

$HCl.H_2O$	$(H_3O)Cl$
$HBr.H_2O$	$(H_3O)Br$
$HNO_3.H_2O$	$(H_3O)NO_3$
$H_2SO_4.H_2O$	$(H_3O)(HSO_4)$
$H_2SeO_4.H_2O$	$(H_3O)(HSeO_4)$
$H_2SO_4.2H_2O$	$(H_3O)_2SO_4$
$CF_3SO_3H.H_2O$	$(H_3O)(CF_3SO_3)$
$HClO_4.2\frac{1}{2}H_2O$	$(H_3O)(ClO_4).1\frac{1}{2}H_2O$

4. Proton-water complexes in hydrates[15,16]

Besides H_3O^+ and $H_5O_2^+$, many other hydronium complexes have been proposed, for example $H_7O_3^+$, $H_9O_4^+$, $H_{13}O_6^+$, and $H_{14}O_6^{2+}$. If 2.45Å is taken as the upper limit for the O-H...O distance within a hydronium ion, longer O-H...O distances represent hydrogen bonds between a hydronium ion and neighbouring water molecules. According to this criterion, reasonable and unambiguous structural formulae can be written for many acid hydrates of known structures (Table 3.3-2).

Table 3.3-2 Structural formulae for some acid hydrates.

Acid hydrate	Empirical formula	Geometry	Structural formula
$HNO_3 \cdot 3H_2O$	$(H_7O_3)NO_3$	2.482 O 2.576	$(H_3O)NO_3 \cdot 2H_2O$
$HClO_4 \cdot 3H_2O$	$(H_7O_3)ClO_4$	2.49 O 2.54	$(H_3O)ClO_4 \cdot 2H_2O$
$HCl \cdot 3H_2O$	$(H_7O_3)Cl$	2.43 O 2.65	$(H_5O_2)Cl \cdot H_2O$
$HSbCl_6 \cdot 3H_2O$	$(H_7O_3)SbCl_6$	2.41 O 2.66	$(H_5O_2)SbCl_6 \cdot H_2O$
$HCl \cdot 6H_2O$	$(H_9O_4)Cl \cdot 2H_2O$	2.540 2.514 2.514	$(H_3O)Cl \cdot 5H_2O$
$CF_3SO_3H \cdot 4H_2O$	$(H_9O_4)CF_3SO_3$	2.502 2.532 2.572	$(H_3O)(CF_3SO_3) \cdot 3H_2O$
$HBr \cdot 4H_2O$	$\frac{1}{2}(H_9O_4)(H_7O_3)(Br)_2 \cdot H_2O$	2.47 2.50 / 2.50 2.59 2.59	$(H_3O)Br \cdot 3H_2O$
$[(C_9H_{18})_3(NH)_2Cl]$ $Cl \cdot HCl \cdot 6H_2O$	$[(C_9H_{18})_3(NH)_2Cl]$ $(H_{13}O_6)Cl_2$	2.52 2.29 2.52 2.52 2.52	$[(C_9H_{18})_2(NH)_2Cl]$ $(H_5O_2)Cl_2 \cdot 4H_2O$
$HSbCl_6 \cdot 3H_2O$	$(H_{14}O_6)\frac{1}{2}(SbCl_6)$	2.66 2.69 2.41 2.41 2.69 2.66	$(H_5O_2)(SbCl_6)H_2O$

References

[1] J.O. Lundgren and I. Olovsson, *Acta Crystallogr.* 23, 966 (1967).

[2] J. Emsley, *Chem. Soc. Rev.* 9, 91 (1980).

[3] J.O. Lundgren and I. Olovsson, *Acta Crystallogr.* 23, 971 (1967).

[4] I. Olovsson, *J. Chem. Phys.* **49**, 1063 (1968).

[5] R. Attig and J.M. Williams, *Angew. Chem. Int. Ed. Engl.* **15**, 491 (1976).

[6] J.O. Lundgren, *Acta Crystallogr., Sect. B* **26**, 1893 (1970).

[7] T. Kjällman and I. Olovsson, *Acta Crystallogr., Sect. B* **28**, 1692 (1972).

[8] J. Roziere and J.M. Williams, *Inorg. Chem.* **15**, 1174 (1976).

[9] J.O. Lundgren and R. Tellgren, *Acta Crystallogr., Sect. B* **30**, 1937 (1974).

[10] R. Attig and J.M. Williams, *Inorg. Chem.* **15**, 3057 (1976).

[11] G.M. Brown, M.R. Noe-Spirlet, W.R. Busing and H.A. Levy, *Acta Crystallogr., Sect. B* **33**, 1038 (1977).

[12] A. Bino and F.A. Cotton, *J. Am. Chem. Soc.* **101**, 4150 (1979).

[13] F.M. Chang, M. Jansen and D. Schmitz, *Acta Crystallogr., Sect. C* **39**, 1497 (1983).

[14] P.A. Giguère, *J. Chem. Educ.* **56**, 571 (1979).

[15] P. Schuster, G. Zundel and C. Sandorfy (eds.), *The Hydrogen Bond*, Vol. 2, North-Holland, Amsterdam, 1976, chap. 10.

[16] A.F. Wells, *Structural Inorganic Chemistry*, 5th ed., Clarendon Press, Oxford, 1984.

3.4 Complex Containing Trinuclear Molybdenum Clusters Bridged by the Hydrogen Oxide Ligand $H_3O_2^-$

$$\{[Mo_3O_2(C_2H_5COO)_6(H_2O)_2]_2(H_3O_2)\}Br_3 \cdot 6H_2O$$

Crystal Data

Triclinic, space group $P\bar{1}$ (No. 2)

$a = 14.358(4)$, $b = 12.114(3)$, $c = 11.402(2)$Å, $\alpha = 111.08(2)$,

$\beta = 105.69(3)$, $\gamma = 66.74(2)°$, $Z = 1$

Atom*	x	y	z
Mo(1)	.35307(6)	.13578(8)	.17044(7)
Mo(2)	.20617(6)	.24760(8)	.32832(7)
Mo(3)	.17784(6)	.06851(8)	.10030(7)
O(1)	.2883(4)	.0703(6)	.2521(5)
O(2)	.2037(4)	.2301(6)	.1468(5)
Ow(1)	.4963(5)	.1151(7)	.1324(7)
Ow(2)	.1547(5)	.3764(7)	.4969(6)
Ow(3)	.0897(5)	-.0310(7)	-.0217(6)
H(1)	0	0	0
H(2)	.109(9)	-.06(1)	-.08(1)

* Only atoms of structural interest are listed.

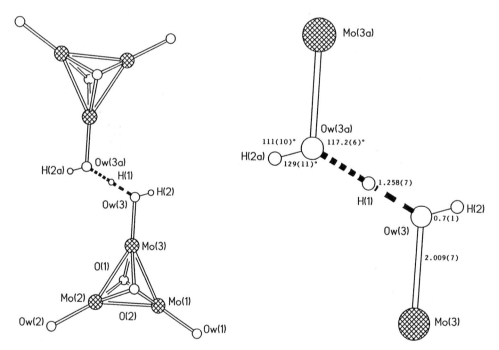

Fig. 3.4-1 (a) Skeletal structure of $\{[Mo_3O_2(C_2H_5COO)_6(H_2O)_2]_2(H_3O_2)\}^{3+}$; the propionato groups have been omitted for clarity. (b) The structure and dimensions of the $H_3O_2^-$ bridging ligand. (After ref. 2).

Crystal Structure

The existence of a hydrogen oxide bridging ligand, $H_3O_2^-$, between metal atoms was first reported in compounds of trinuclear, triangular clusters of molybdenum and tungsten.[1,2] It has since been shown to play an important and quite general role in classical coordination chemistry. This ligand is formed by means of a short and very strong hydrogen bond between a hydroxo-ligand of one metal atom and an aqua ligand of another metal atom, and the resulting $H_3O_2^-$ group constitutes a bridge as shown in Fig. 3.4-1.

In the title compound, each Mo atom has a coordination number of 9, counting the positions of the two neighbouring metal atoms and the coordinated atoms of the various ligands. The trinuclear $[Mo_3O_2(C_2H_5COO)_6(H_2O)_3]^{2+}$ cluster cation contains molybdenum atoms with a formal oxidation state of +4 and two triply-bridging capping oxygen atoms. Each pair of Mo atoms is bridged by two carboxylate ligands and each metal atom is coordinated to a water ligand.

The most prominent feature of these trinuclear species is their unusual kinetic stability in aqueous solution. All attempts to substitute ligands in the primary coordination sphere of the metal cluster have failed. This stability is attributed to the high coordination number of the metal atoms in the cluster.

Remarks

1. The $H_3O_2^-$ bridging ligand

The existence of the $H_3O_2^-$ ligand was demonstrated by single crystal X-ray studies of cluster ions of Mo(IV) and W(IV), and of classical hydroxoaqua ions of Cr(III) and Co(III).[2] It was shown to exist also in a Ru(III) hydroxoaqua complex whose structure had been determined earlier. It is now known that this ligand exists in most hydroxoaqua metal ions. These ions, in the crystalline state, are not mononuclear, but binuclear or polynuclear. Furthermore, a distinction between a hydroxo-ligand and an aqua-ligand is impossible, if the proton forming the hydrogen bond of $H_3O_2^-$ is equidistant from both oxygen atoms. The $H_3O_2^-$ bridge was also shown to exist in compounds which were previously mistaken for double-salts such as *trans*-$[Co(en)_2N_3(H_2O)]\cdot[Co(en)_2N_3(OH)](ClO_4)_3$. These salts are, in reality, not made up of aqua-ions and hydroxo-ions at a 1:1 ratio but of symmetrical binuclear ions having a $H_3O_2^-$ ligand between the metal atoms.

The important structural features of some compounds in this class are presented in Table 3.4-1. The last two entries are dinuclear cations in which the two metal atoms are doubly bridged by two *cis* $H_3O_2^-$ ligands.

Table 3.4-1   Structural data for $H_3O_2^-$ bridging ligands in some metal
complexes[2].

Compound	O...O Å	M-O Å	M...M Å	M-O-O degree
$\{[Mo_3O_2(C_2H_5COO)_6(H_2O)_2]_2(H_3O_2)\}Br_3 \cdot 6H_2O$ dicluster	2.52(1)	2.009(7)	5.63	117.1
$\{[W_3O_2(C_2H_5COO)_6(H_2O)_2]_2(H_3O_2)\}(NCS)_3 \cdot H_2O$ dicluster	2.46(1)	2.04(1)	5.73	120.3
$[W_3O_2(CH_3COO)_6(H_2O)(H_3O_2)]NCS$ polycluster	2.44(1)	2.02(1) 2.07(1)	5.95	132.9 130.0
$trans$-$[Co(en)_2(H_3O_2)](ClO_4)_2$	2.441(2)	1.916(1)	5.72	130.4(1)
$trans$-$[Ru(bpy)_2(H_3O_2)](ClO_4)_2$	2.538(6)	2.007(3)	5.79	127.3(1)
cis-$[Cr(bpy)_2(H_3O_2)]_2I_4 \cdot 2H_2O$ binuclear, bridged by two $H_3O_2^-$	2.446(5)	1.925(3) 1.928(3)	5.03	127.1(2) 126.2(2)
cis-$[Cr(bpy)_2(H_3O_2)]_2(NO_3)_4$ binuclear, bridged by two $H_3O_2^-$	2.442(4)	1.934(3) 1.913(4)	4.96	128.7(2) 125.6(2)

2. Structure of so-called "double salts"

    Crystal structure analysis[3] of $trans$-$\{[Co(en)_2(NO_2)]_2(H_3O_2)\}$-
$(ClO_4)_3 \cdot 2H_2O$ (Fig. 3.4-2) and $trans$-$\{[Co(en)_2(NCS)]_2(H_3O_2)\}(CF_3SO_3)_3 \cdot H_2O$ have
shown that they are definitely not double salts, but salts of a single,
binuclear cation $[(en)_2XCo(H_3O_2)CoX(en)_2]^{3+}$ (X = NO_2, NCS). The two Co atoms
in this cation have the same coordination environment.

    The binuclear $H_3O_2^-$-bridged structure found in these compounds may be
expected in other so-called double salts because the strong hydrogen bond in
the $H_3O_2^-$ bridge (~ 100 kJ/mol^{-1}) makes this structure more stable than a
hypothetical structure with distinct mononuclear hydroxo and aqua ions.
However, before the general role of $H_3O_2^-$ bridging in coordination chemistry
was outlined, a double salt formulation seemed to be more justified than a
symmetric formulation such as $[Co(en)_2(NO_2)(OH)]_2(ClO_4)_2 \cdot HClO_4$, which would
imply an acid salt of a hydroxo-complex.

    Ardon and Bino[2] have noted that a 1907 paper by Alfred Werner[5]
actually proposed an essentially correct formulation for this class of
compounds, which unfortunately were later mistaken for double salts. This
paper, bearing the title "Über anomale anorganische Oxonium Salze, eine neue
Klasse basischer Salze" (on anomalous oxonium salts, a new class of basic
salts) reported the composition of basic salts of $trans$-$[Co(NH_3)_4(NO_2)H_2O]^{2+}$.
Werner's idea that a hydrogen atom may be coordinated with two (or more)
oxygen atoms was proposed in an even earlier paper (1902),[6] in which he

$$Cl \begin{bmatrix} O_2N \\ (H_3N)_4 \end{bmatrix} \begin{matrix} COOH \\ \\ \end{matrix}$$
$$Cl \begin{bmatrix} O_2N \\ (H_3N)_4 \end{bmatrix} \begin{matrix} COOH \\ \\ \end{matrix} HCl,$$

first introduced the concept of a symmetric hydrogen bond (without using this term explicitly). It is noteworthy that Werner applied the idea of hydrogen bonding to structural problems as early as 1902, long before its advocation by Latimer and Rodebush as an explanation of the physical properties of water and related liquids.[7]

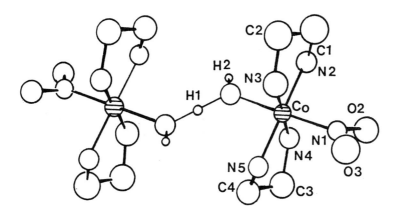

Fig. 3.4-2   The structure of *trans*-{[Co(en)$_2$(NO$_2$)]$_2$(H$_3$O$_2$)}$^{3+}$.   (After ref. 3).

References

[1] A. Bino and D. Gibson, *J. Am. Chem. Soc.* **104**, 4383 (1982).

[2] M. Ardon and A. Bino, *Structure and Bonding* **65**, 1 (1987).

[3] M. Ardon, A. Bino and W.G. Jackson, *Polyhedron* **6**, 181 (1987).

[4] W.G. Jackson and A.M. Sargeson, *Inorg. Chem.* **17**, 1348 (1978).

[5] A. Werner, *Ber.* **40**, 4122 (1907).

[6] A. Werner, *Liebig. Ann. Chem.* **322**, 296 (1902).

[7] W.M. Latimer and W.H. Rodebush, *J. Am. Chem. Soc.* **42**, 1419 (1920).

3.5   Lithium Nitride

Li₃N

Crystal Data

α-Li₃N: Hexagonal, space group *P6/mmm* (No. 191)

   a = 3.648(1), *c* = 3.875(1)Å, *Z* = 1

β-Li₃N: Hexagonal, space group *P6₃/mmc* (No. 194)

   a = 3.552(1), *c* = 6.311(3)Å, *Z* = 2

| | α-Li₃N | | | β-Li₃N | | |
Atom	Position	x	y	z	Position	x	y	z
N	1(a)	0	0	0	2(c)	1/3	2/3	1/4
Li(1)	1(b)	0	0	1/2	2(b)	0	0	1/4
Li(2)	2(c)	1/3	2/3	0	4(f)	1/3	2/3	.583

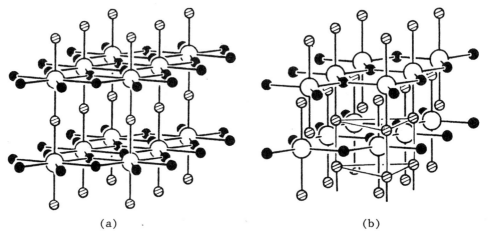

(a)   (b)

Fig. 3.5-1 The crystal structure of Li₃N: (a) α-form, (b) β-form. Large circles: nitrogen; small circles (black or shaded): lithium. The trigonal prismatic coordination in the second sphere of the β-form is outlined for a nitrogen atom. (After ref. 1).

Crystal Structure

Among binary compounds A₃B of the alkali metals (A = Li, Na, K, Rb, Cs) with elements of the nitrogen group (B = N, P, As, Sb, Bi), lithium nitride, Li₃N, is unique in that it is so far the only known nitride of the alkali metals. At normal atmospheric pressure it occurs in the α form, which is made up of planar Li₂N-layers and pure Li layers [Fig. 3.5-1(a)]. In each Li₂N-layer, the Li(2) atoms form a simple hexagonal arrangement (just like a carbon layer in graphite) with a N atom at the centre of each ring, and midway between these layers are layers containing only Li(1) atoms. The resulting coordination environments are:

> Li(1):   2 N collinear at 1.94Å;
>
> Li(2):   3 N trigonal planar at 2.13Å; and
>
> N: 8 Li hexagonal bipyramidal.

Under high pressure (> 0.6 GPa), α-Li$_3$N undergoes a phase transformation to β-Li$_3$N. The crystal structure of β-Li$_3$N [Fig. 3.5-1(b)] consists of planar layers of hexagonal symmetry with a composition LiN (Li-N distance in the layers 2.055Å). The layers are stacked in the sequence ABAB... in the c direction. Each nitrogen atom binds two further lithium atoms at a distance of 2.095Å above and below the plane, thus completing its first coordination sphere to give a trigonal bipyramid. As a result of the interpenetration of the Li$_3$N layer packets, each nitrogen atom is additionally coordinated trigonal prismatically by six lithium atoms further away at 2.313Å in a second coordination sphere. Thus, altogether each nitrogen atom is surrounded by a fully capped trigonal prismatic array of lithium atoms.

Remarks

1. Relationship between the structures of α-Li$_3$N and β-Li$_3$N

Li$_3$N is an ionic compound, in which the N^{3-} ion has a much greater size than the Li$^+$ ion. In β-Li$_3$N the N^{3-} anions are arranged in hexagonal closest packing, with Li$^+$ cations inserted into all tetrahedral interstices [Li(2)] and half of the triangular interstices [Li(1)].

Fig. 3.5-2 shows the structural relationship between α-Li$_3$N and β-Li$_3$N. For the transformation of α-Li$_3$N into β-Li$_3$N the Li-N bonds marked with $\approx$ have to be opened and the layer packets shifted towards each other such that the packing of the nitrogen atoms changes from a simple hexagonal arrangement (α-Li$_3$N) a hexagonal closest packed arrangement (β-Li$_3$N). The lithium atoms shown in black then have to be merely shifted in the direction of the nitrogen atoms of the neighbouring layer.

Since the phase transformation takes place at the relatively low pressure of 0.6 GPa, it can hardly be rule out that at least a partial transformation of α-Li$_3$N into β-Li$_3$N takes place on preparing powdered samples by grinding in a mortar, so that X-ray powder photographs of α-Li$_3$N invariably show small amounts of β-Li$_3$N.

When Li$_3$N is subjected to pressures of more than 10 GPa, a F-centered cubic phase is formed. It is assumed that this highest pressure phase of Li$_3$N crystallizes in the Li$_3$Bi-type structure, in which the nitride ions form a ccp lattice with Li$^+$ ions inserted into all the octahedral and tetrahedral interstices.

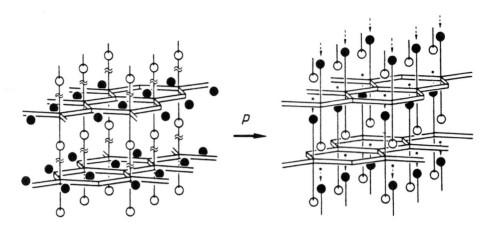

Fig. 3.5-2 Structural relationship between α-Li$_3$N and β-Li$_3$N. The (LiN)$_n$ structural unit is represented schematically and simply as a hexagonal network. The small dots indicate the positions from which the lithium atoms have to be shifted in the direction of the arrow. (After ref. 1).

2. Li$_3$N as a solid ionic conductor

The crystal structure of α-Li$_3$N is very loosely packed, and conductivity arises from lithium vacancies within the Li$_2$N layers. The difference electron density map (Fig. 3.5-3) shows that Li$_3$N is an ionic compound, and that both the cation and anion are almost fully ionized. The interaction between the carrier ions (Li$^+$) and the "fixed" ions (N^{3-}) is relatively weak,

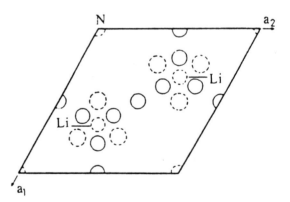

Fig. 3.5-3 Difference electron density map, through the Li$_2$N-layer, for α-Li$_3$N at 20°C calculated under the assumption of the presence of Li$^+$ and N^{3-} ions. Full and broken lines correspond to positive and negative densities, respectively. The lines are drawn at 0.05 eÅ^{-3}. (After ref. 3).

so that ion migration can be quite effective. Experimental data have been gathered in support of α-Li_3N as a solid ionic conductor.[4]

3. Ionic nitrides and azides

Li_3N is a typical example of the "salt-like" nitrides, which include compounds of Group IIA (M_3N_2 with M = Be, Mg, Ca, Sr, Ba), Group IB (M_3N with M = Cu, Ag) and Group IIB (M_3N_2 with M = Zn, Cd, Hg) metals. Although the species N^{3-} (ionic radius 1.46Å) can be identified in these compounds, the charge separation is usually incomplete. The azides NaN_3, KN_3, $Sr(N_3)_2$, $Ba(N_3)_2$, etc. are colorless crystalline salts which contain the symmetrical linear N_3^- group with N-N bonds close to 1.18Å (mean value). The B subgroup azides such as AgN_3, $Cu(N_3)_2$, and $Pb(N_3)_2$ are far less ionic and have complex polymeric crystal structures; they are also shock-sensitive and detonate readily.

4. Group III and IV nitrides

The covalent nitrides MN (M = B, Al, Ga, In, Tl) belong to the class of "III-V semiconductors" whose structures are summarized in Table 3.5-1.

Table 3.5-1 Structures of III-V compounds MX.*

X	M =	B	Al	Ga	In
N		L,S	W	W	W
P		S	S	S	S
As		S	S	S	S
Sb		-	S	S	S

*   L = BN layer lattice, see Section 2.5
     S = sphalerite (zinc blende, cubic ZnS), see Section 2.9
     W = wurzite (hexagonal ZnS), see Section 2.10

Silicon nitride, Si_3N_4, is an important ceramic material with extremely desirable properties: high strength and wear durability, high decomposition temperature and oxidation passivity, resistance to thermal shock and corrosive environments, low coefficient of friction, etc. Si_3N_4 exists in two forms, α and β, which stand in the same relationship as cristobalite and tridymite. The more common α-form crystallizes in space group $P31c$ with a = 7.818, c = 5.591Å and Z = 4.[5] The β-form, like Ge_3N_4, has a structure [Fig. 3.5-4(a)] related to that of phenacite (Be_2SiO_4), with a = 7.595, c = 2.9023Å and Z = 2 in space group $P6_3$.[6] Each N atom is common to three SiN_4 tetrahedra, with Si-N = 1.704-1.767(5)Å, N-Si-N = 106.6-113.5, and Si-N-Si = 113.5-125.1°.

A hypothetical crystal structure of β-C_3N_4 has been proposed, which consists of a network of corner-linked CN_4 tetrahedra with a hexagonal unit cell, as shown in Fig. 3.5-4(b).[8] If β-C_3N_4 could be synthesized, its bulk modulus and hardness would be comparable to or even exceed that of diamond, because this hypothetical structure has no obvious planes of weakness.

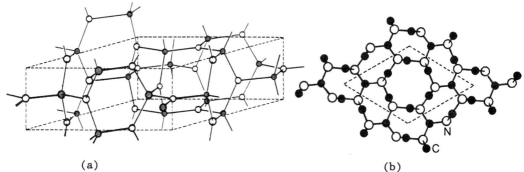

(a)   (b)

Fig. 3.5-4 (a) Crystal structure of β-Si_3N_4, the open and shaded circles representing the N and Si atoms, respectively. (After ref. 7). (b) Hypothetical crystal structure of β-C_3N_4. Mean planes illustrated are located at $z = \pm 1/4$. The structure consists of these buckled planes stacked in AAA...sequence. The unit cell is outlined. (Adapted from ref. 8).

5. Nitrides of early transition metals

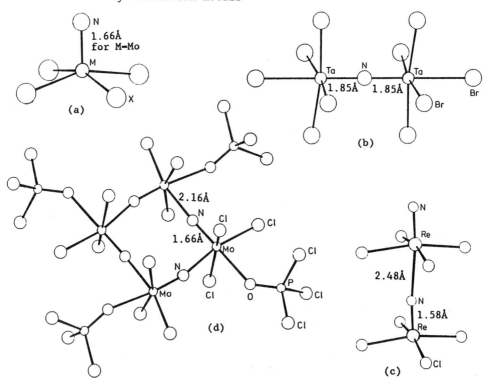

Fig. 3.5-5 Structures of some nitrido complexes: (a) anions $[MNX_4]^-$ (M = Mo, Re, Ru, Os; X = F, Cl, Br, I), (b) symmetrically bridged $[Ta_2NBr_{10}]^{3-}$, (c) two asymmetrically bridged $ReNCl_4$ molecules as part of a column, and (d) tetrameric $[MoNCl_3.OPCl_3]_4$ with asymmetric nitrido bridges. (After ref. 9)

The early transition metals are known to form mononitrides, dinuclear mononitrides, trinuclear dinitrides, tetramers, and linear polymers.[9] The structures of some representative compounds are illustrated in Fig. 3.5-5.

Theoretical studies have led to the conclusion that close analogies exist among transition-metal nitrides, organic polyenes and phosphazenes, and that benzene-like cyclic trimers should be particularly stable.[10] The first triazatrimetallabenzene, $[Cp^*TaN(Cl)]_3$ ($Cp^* = \eta^5\text{-}C_5Me_5$), has been synthesized by the following reaction and characterized by crystal structure analysis.[11]

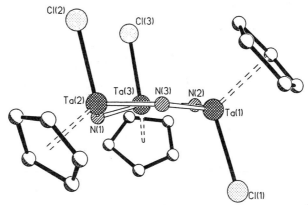

$$3\ [Cp^*TaCl_4] + 3\ N(SnMe_3)_3 \xrightarrow[-\ 9\ Me_3SnCl]{}$$

As shown in Fig. 3.5-6, the six-membered $(TaN)_3$ ring exists in a boat conformation with Ta(1) and N(1) deviating by 0.13 and 0.39Å, respectively, from the plane of the other four atoms (within 3×10^{-4}Å). The molecule has a non-crystallographic mirror plane through Cl(1), Ta(1) and N(1).

Fig. 3.5-6 Lateral view of the molecular structure of $[Cp^*TaN(Cl)]_3$.
Selected bond distances(Å) and angles(°): Ta(1)-N(2) 1.90(3), N(2)-Ta(3) 1.86(3), Ta(3)-N(1) 1.88(2), N(1)-Ta(2) 1.90(2), Ta(2)-N(3) 1.91(2), N(3)-Ta(1) 1.84(2), Ta(1)-Cl(1) 2.38(1), Ta(2)-Cl(2) 2.35(1), Ta(3)-Cl(3) 2.35(1); average N-Ta-N 109.8(10), Ta-N-Ta 127.9(12), N-Ta-Cl 103.6(9). (After ref. 11).

6. Structure of $[H_2\overline{C(CH_2)_5N}Li]_6$[12] and $Li(ND_3)_4$[13]

The crystal structure of $[H_2\overline{C(CH_2)_5N}Li]_6$ reveals a hexamer, which can be viewed as a cyclized ladder composed of six NLi rungs, or as a stack of two six-membered, trimeric $(NLi)_3$ rings, as shown in Fig. 3.5-7.

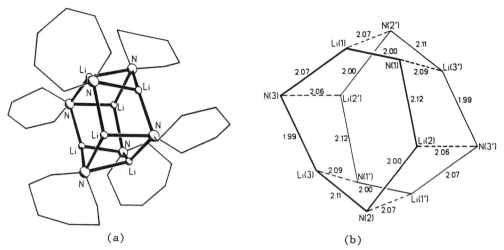

(a)   (b)

Fig. 3.5-7   (a) Molecular structure of $[H_2\overline{C(CH_2)_5}NLi]_6$; the molecule has crystallographic inversion symmetry.   (b) Li-N bond lengths; hexamethylene units have been omitted for clarity.   (After ref. 12).

The N-Li distances fall clearly into two sets: (i) short, 1.99(1)-2.00(2)Å, and (ii) long, 2.06(1)-2.12(2)Å.   The structure is one of alternating short and long bonds within the six-membered rings, which are held together by medium-length bonds.

The crystal structure of ^{7}Li(ND$_3$)$_4$ (phase II) above the antiferromagnetic transition temperature (25K) has been determined using neutron powder diffraction data.[13]   The space group was uniquely determined to be $I\overline{4}3d$, with a = 14.813(3)Å.   Fig. 3.5-8(a) shows the molecular structure of Li(ND$_3$)$_4$, in which one N atom bonds to the Li atom at a much longer

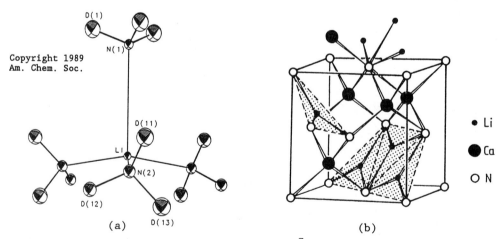

(a)   (b)

Fig. 3.5-8   (a) Molecular structure of ^{7}Li(ND$_3$)$_4$.   (After ref. 13).   (b) Section from the crystal structure of LiCaN.   (After ref. 14).

distance [2.488(16)Å] than the other three bonds [1.984(4)Å], so that the Li(ND₃)₄ complex is perhaps better described by the formula [7]Li(ND₃)₃.ND₃. The more weakly bound ND₃ can provide some insight into the nature of the phase I - phase II transition that occurs in Li(NH₃)₄ at 82K. Upon warming, the weakly bound NH₃ molecules may reorient or dissociate from the tetraamine unit, presumably providing an impetus to an order-disorder phase transition.

7. Ternary nitrides[14]

Ternary nitrides satisfying $r(Li^+)/r(M^{n+}) \geq 1$ preferentially crystallize in the fluorite structure. The cations occupy the tetrahedral holes of the approximately ccp matrix of nitride ions either statistically (LiMgN) or in an ordered manner (LiZnN; Li₃MN₂, M = Al, Ga, Fe; Li₅MN₃, M = Ge, Ti; Li₆MN₄, M = Cr, Mo, W; Li₇MN₄, M = V, Mn; Li₁₅Cr₂N₉). LiSrN crystallizes in the YCoC structure,[15] which is energetically more favorable than the hypothetical fluorite type for $r(Li^+)/r(Sr^{2+}) < 1$.

The crystal structure of orthorhombic LiCaN [Fig. 3.5-8(b)] is related to the fluorite structure.[14] The Ca atoms occupy half of the tetrahedral holes, whereas the Li atoms are displaced from their ideal tetrahedral sites to within 0.135Å of the layers of nitrogen atoms. Each Li atom is thus coordinated by three nitrogen atoms at Li-N = 2.071(4)-2.108(8)Å, the fourth nitrogen being 3.570(8)Å away. The resultant infinite $^1_\infty[LiN_{3/3}]$ bands, running in the [010]$_{ortho}$ direction, are one-dimensional sections of the Li₂N-layer of the Li₃N structure, which through N atoms shared with with CaN₄ tetrahedra, become interconnected into a three-dimensional framework. Each N atom is in a distorted pentagonal bipyramidal coordination environment.

The series of ternary nitrides Ca₃EN where E = P, As, Sb, Bi, Ge, Sn, or Pb have the cubic anti-perovskite structure [Fig. 3.5-9(a)]; the P and As compounds are distorted variants.[16] The stoichiometry and interatomic distances show that the post-transition element (Bi, Sb, or Pb) behaves as an anion of oxidation state -3. Electrical measurements show that the P compound is an insulator, the Bi compound is a degenerate semiconductor at room temperature, and the "electron-deficient" Pb compound has a much higher electrical conductivity.

The nitride CaNiN, in which the Ni cation has the uncommon formal valence of +1, behaves as a good metal down to 4.2K. It adopts a simple layer structure [Fig. 3.5-9(b)],[16] which is isomorphous with that of RCoC, where R is Y or a heavy rare-earth element.[17] The closest non-bonded Ni...Ni distance of 3.50Å occurs between adjacent layers.

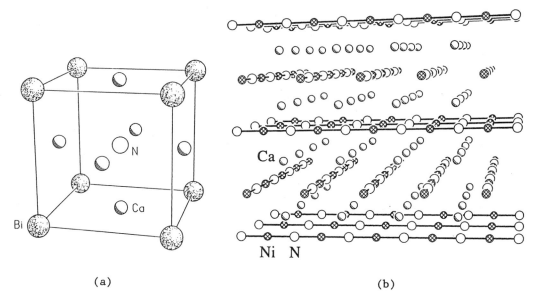

(a)   (b)

Fig. 3.5-9 (a) Unit cell of Ca_3BiN, with $a = 4.89$Å. (b) Crystal structure of tetragonal CaNiN. The Ni and N atoms are each bonded to two neighbours of the other kind to form a linear chain. The chains are packed in layers, which are separated by the Ca atoms. The chain directions in alternate layers are orthogonal to each other. (Adapted from ref. 16).

References

[1] H.J. Beister, S. Haag, R. Kniep, K. Strössner and K. Syassen, *Angew. Chem. Int. Ed. Engl.* **27**, 1101 (1988).

[2] A. Rabenau and H. Schulz, *J. Less-Common Met.* **50**, 155 (1976).

[3] H. Schulz and K. Schwarz, *Acta Crystallogr., Sect. A* **34**, 999 (1978).

[4] A. Rabenau, *Solid State Ionics* **6**, 24 (1982).

[5] K. Kato, Z. Inoue, K. Kijima, I. Kawada, H. Tanaka and T. Yamane, *J. Am. Ceram. Soc.* **58**, 90 (1975).

[6] R. Grün, *Acta Crystallogr., Sect. B* **35**, 800 (1979).

[7] A.F. Wells, *Structural Inorganic Chemistry*, 5th ed., Clarendon Press, Oxford, 1984.

[8] A.Y. Liu and M.L. Cohen, *Science (Washington)* **245**, 841 (1989).

[9] K. Dehnicke and J. Strähle, *Angew. Chem. Int. Ed. Engl.* **20**, 413 (1981).

[10] R.A. Wheeler, R. Hoffmann and J. Strähle, *J. Am. Chem. Soc.* **108**, 5381 (1986).

[11] H. Plenio, H.W. Roesky, M. Noltemeyer and G.M. Sheldrick, *Angew. Chem. Int. Ed. Engl.* **27**, 1330 (1988).

[12] D. Barr, W. Clegg, S.M. Hodgson, G.R. Lamming, R.E. Mulvey, A.J. Scott, R. Snaith and D.S. Wright, *Angew. Chem. Int. Ed. Engl.* **28**, 1241 (1989).

[13] V.G. Young, Jr., W.S. Glaunsinger and R.B. von Dreele, *J. Am. Chem. Soc.* **111**, 9260 (1989).

[14] G. Cordier, A. Gudat, R. Kniep and A. Rabenau, *Angew. Chem. Int. Ed. Engl.* **28**, 1702 (1989).

[15] G. Cordier, A. Gudat, R. Kniep and A. Rabenau, *Angew. Chem. Int. Ed. Engl.* **28**, 201 (1989).

[16] F.J. DiSalvo, *Science (Washington)* **247**, 649 (1990).

[17] M.H. Gerss and W. Jeitschko, *Z. Naturforsch., Teil B* **41**, 946 (1986).

Note The structure of $Li_3[FeN_2]$ is described in A. Gudat, R. Kniep, A. Rabenau, W. Bronger and U. Ruschewitz, *J. Less-Common Met.* **161**, 31 (1990); $Li_4[FeN_2]$, a defect variant of the Li_3N type, in A. Gudat, R. Kniep and R. Rabenau, *Angew. Chem. Int. Ed. Engl.* **30**, 199 (1991).

A comprehensive review of the crystallographic data for coordination compounds of lithium is given in U. Olsher, R.M. Izatt, J.S. Bradshaw and N.K. Dalley, *Chem. Rev.* **91**, 137 (1991).

The lithium phosphorus(V) nitride $Li_{10}P_4N_{10}$ contains the complex anion $P_4N_{10}^{10-}$ that is analogous to molecular P_4O_{10}. See W. Schnick and U. Berger, *Angew. Chem. Int. Ed. Engl.* **30**, 830 (1991).

Si_3N_4 as a component of advanced ceramic materials (with full citation of literature references on the α and β phases) is reviewed in H. Lange, G. Wötting and G. Winter, *Angew. Chem. Int. Ed. Engl.* **30**, 1579 (1991).

3.6 Complex Containing a Cryptated Sodium Cation and a Natride Anion

$$Na^+(C_{18}H_{36}N_2O_6)\cdot Na^-$$

Crystal Data

Rhombohedral, space group $R32$ (No. 155)

$a = 8.83(1)$, $c = 29.26(2)$Å, $Z = 3$ (hexagonal setting)

Atom	x	y	z
Na$^+$	0	0	0
Na$^-$	1/3	2/3	1/6
N	0	0	.0928
O	.308	.176	.0347
C(1)	.184	.090	.1076
C(2)	.321	.182	.0829
C(3)	.391	.350	.0197

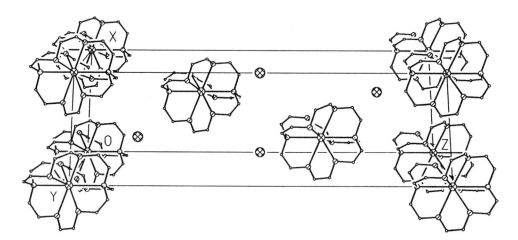

Fig. 3.6-1 Crystal structure of $Na^+(C_{18}H_{36}N_2O_6)\cdot Na^-$. Large circles represent Na$^-$ (natride) anions.

Crystal Structure

Elemental sodium dissolves only very slightly ($\sim 10^{-6}$ M) in ethylamine, but when cryptand[2,2,2] (molecular formula $N[CH_2CH_2OCH_2CH_2OCH_2CH_2]_3N$, also abbreviated as C222) is added, the solubility increases dramatically up to 0.2M according to

$$2Na(s) + cryptand[2,2,2] \longrightarrow Na^+(C222) + Na^-.$$

When cooled to -15°C or below, the solution deposited shiny, gold-coloured thin hexagonal plates which, after washing with ether and drying in an inert atmosphere, were characterized as $C_{18}H_{36}N_2O_6Na_2$ by CHN elemental analysis.

Crystal structure analysis of this complex provided the most convincing evidence that the Na− (natride, previously called sodide) anion is indeed present.[1] Fig. 3.6-1 shows that the two sodium atoms are in very different environments in the crystal. One is located inside the cryptand at distances from the nitrogen and oxygen atoms which are characteristic of a trapped Na+ cation; the resulting encapsulated cation occupies a site of symmetry 32. The other sodium species, a Na− ion, is at a large distance away from all other atoms. The disposition of the ether oxygen atoms in the cryptand chains indicates a repulsive interaction with the Na− ion as expected.

The close analogy of the natride ion with an iodide ion is brought out clearly by comparing the present structure with that of sodium iodide cryptate, [Na+C(222)].I−, which crystallizes in space group *P*31*c* (No. 159).[2] Though the space groups of the two crystals are different, corresponding ion-atom distances match remarkably well:

	[Na+C(222)].Na−		[Na+C(222)].I−
Na+ − N	2.72(1)Å	Na+ − N	2.75(2)Å
Na+ − O	2.57(1)	Na+ − O	2.58(2)
Na− ...Na+	7.06(1)	I− ...Na+	7.40(1)
Na− ...N	5.54(2)	I− ...N	5.39(2)
Na− ...O	5.72(2)	I− ...O	5.22(2)

Remarks

1. Alkali metal anions ("alkalides")

Compounds containing alkali metals in the -1 oxidation state are limited neither to salts of Na− nor to complexes with cryptand[2,2,2]. In 1987, Dye *et al.*[3] reported that they have synthesized over 30 "alkalides" and eight electrides using various complexants (mainly 18C6, C222 and HMHCY) since the first natride, Na+(C222).Na−, was prepared in 1974,[4,5] and determined the structures of two cesides, namely Cs+(18C6)2.Cs− (1) and Cs+(C222).Cs− (2) .

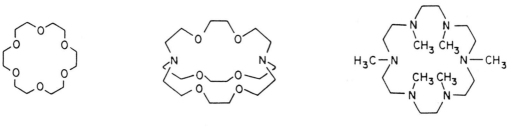

18-crown-6	cryptand[2,2,2]	hexamethyl hexacyclen
(18C6)	(C222)	(HMHCY)

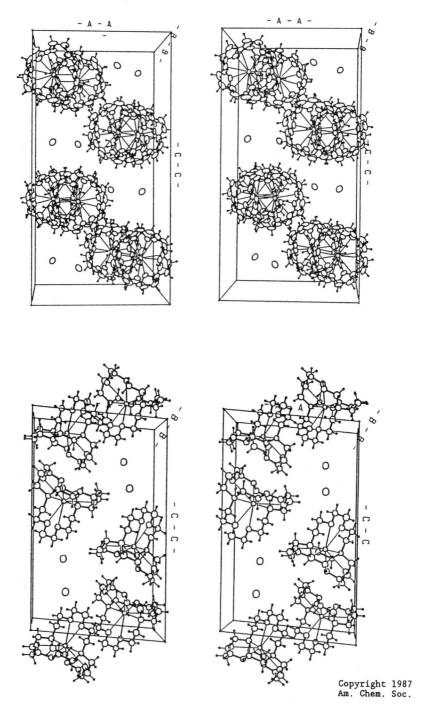

Fig. 3.6-2 ORTEP stereo packing diagrams of $Cs^+(18C6)_2 \cdot Cs^-$ (top) and $Cs^+(C222) \cdot Cs^-$ (bottom). The Cs^+ ions are represented by isolated circles. (After ref. 3).

In both crystal structures (Fig. 3.6-2), the cesium anions are in pockets lined with H atoms from the crown ether or cryptand. The closest Cs$^-$...H distance is 4.29Å in 1 and 4.37Å in 2; the average distances to the 15 closest atoms (all are H atoms) are 4.66Å (1) and 4.70Å (2). If the van der Waals radius of H is taken as 1.2Å, a minimum radius of 3.1Å can be assigned to Cs$^-$; as Cs$^-$ is expected to be rather polarizable, the H atoms might penetrate somewhat into its outer electronic shell. The average distance to neighbouring H atoms yields an effective radius of 3.5Å, showing that Cs$^-$ occupies more than four times the volume of I$^-$.

2. Sizes of alkali-metal anions (alkalides)

The sizes of alkali-metal anions can be estimated from the crystal structures of alkalides by assuming that they are in contact with their nearest neighbours, which are hydrogen atoms having a van der Waals radius of 1.2Å. Another set of figures can be derived by subtracting the radius of an alkali-metal cation from the interatomic distance in the metal. Table 3.6-1

Table 3.6-1 Estimated diameters (Å) of alkali-metal anions from structures of alkalides and alkali metals.[*]

Compound	r_{M^-} (min)	r_{M^-} (av)	d_{atom}	r_{M^+}	r_{M^-}
Na metal			3.72	.95	2.77
K$^+$(C222).Na$^-$	2.55	2.73(14)			
Cs$^+$(18C6)$_2$.Na$^-$	2.34	2.64(16)			
Rb$^+$(15C5)$_2$.Na$^-$	2.60	2.89(16)			
K$^+$(HMHCY).Na$^-$	2.48	2.77(10)			
Cs$^+$(HMHCY).Na$^-$	2.35	2.79(8)			
K metal			4.63	1.33	3.30
K$^+$(C222).K$^-$	2.94	3.12(10)			
Cs$^+$(15C5)$_2$.K$^-$	2.77	3.14(16)			
Rb metal			4.85	1.48	3.37
Rb$^+$(C222).Rb$^-$	3.00	3.21(14)			
Rb$^+$(18C6).Rb$^-$	2.99	3.23(9)			
Rb$^+$(15C5)$_2$.Rb$^-$	2.64	3.06(16)			
Cs metal			5.27	1.67	3.60
Cs$^+$(C222).Cs$^-$	3.17	3.50(15)			
Cs$^+$(18C6)$_2$.Cs$^-$	3.09	3.46(15)			

* HMHCY stands for hexamethyl hexacyclen, which is the common name of 1,4,7,10,13,16- hexaaza- 1,4,7,10,13,16-hexamethyl cyclooctadecane. r_{M^-} (min) is equal to the distance between an anion and its nearest hydrogen minus the van der Waals radius of hydrogen, 1.2Å. r_{M^-} (av) is the average radius over the nearest hydrogen atoms. The numbers in brackets are the numbers of hydrogen atoms for averaging. d_{atom} is the interatomic distance in the metal. r_{M^-} is equal to d_{atom} minus r_{M^+}.

shows that the calculated radii of alkali-metal anions from the two methods are in good general agreement.[6]

3. Alkali-metal-anion chains and dimers in alkalide structures

In the structure of $Cs^+(C222).Cs^-$ (Fig. 3.6-2), the Cs^- anions form zigzag chains along the b direction with $Cs^-...Cs^-$ contacts of 6.38Å, ~0.6Å shorter than the "effective" contact distance of 7.0Å for a pair of Cs^- ions (Table 3.6-1). The same chain arrangement also occurs in $Rb^+(18C6).Rb^-$ with $Rb^-...Rb^-$ distances of only 5.13Å, nearly 0.9Å shorter than twice the minimum radius of Rb and 1.3Å shorter than twice its effective radius.[6] The alkalides $K^+(C222).K^-$ and $Rb^+(C222).Rb^-$ are isostructural, and the most striking feature is that the anions occur in pairs with unusually short interionic distances of 4.90 and 5.13Å, respectively.[6]

Since none of the known natride structures exhibits the above type of chain and dimeric association of alkali-metal anions, it has been suggested that the d-orbital character of the heavier alkali metals may be involved in bonding or substantial distortion of the K^-, Rb^- and Cs^- anions from spherical shape.

4. Electrides

When Dye and co-workers synthesized the first salt of an alkali-metal anion in 1974,[4] they also obtained "blue, strongly paramagnetic solids" by evaporation of ethylamine solutions that contained cryptated cations and solvated electrons. This discovery permitted observation of the optical spectrum of a solvent-free "electride" and prompted intensive efforts to isolate crystalline electrides. Rapid autocatalytic decomposition of solutions frustrated these attempts until 1983, when shiny black plates of $Cs^+(18\text{-crown-}6)_2.e^-$ were successfully isolated.[7]

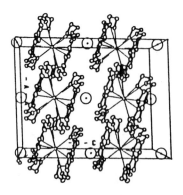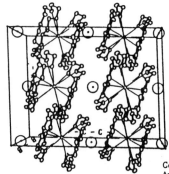

Fig. 3.6-3 ORTEP stereo packing diagram of $Cs^+(18C6)_2.e^-$. The anionic holes are indicated by circles with centers. (After ref. 8).

The electride $Cs^+(18C6)_2.e^-$ crystallizes in the monoclinic space group $C2/c$ with $a = 13.075(5)$, $b = 15.840(7)$, $c = 17.359(8)$Å, $\beta = 92.30(3)°$ and $Z = 4$.[8]   The crystal structure (Fig. 3.6-3), determined at -57°C, consists of only complexed cations and noise-level electron density (0.07 eÅ^{-3}) at the anionic sites, supporting the picture of an electride as a salt with localized electrons as the component anions.

The elongated electron cavity is completely enclosed by eight complexed cations. The cavities are stacked along c at intervals of 8.68Å, each having a nearly square cross-section ~5Å on a side midway between the planes of complexed cesium cations [Fig. 3.6-4(a)]. Constriction regions of minimum diameter ~2Å connect adjacent cavities to form channels in the c direction [Fig. 3.6-4(b)]. Connecting channels of smaller diameter and longer inter-cavity separation (10.2Å) occur along the [110] and [1$\bar{1}$0] directions.[9]

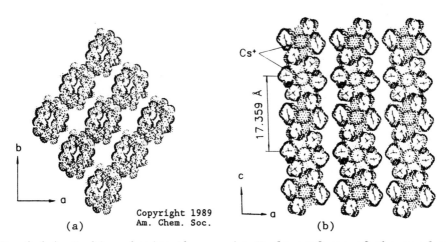

(a)   (b)

Fig. 3.6-4 Packing showing the van der Waals surfaces of the complexed cations in $Cs^+(18C6)_2.e^-$. (a) Cross-sectional view showing the maximum cross section of a cavity and the channels that connect the cavities in the *ab* plane. Similar channels at right angles are hidden in this view by projections of the crown ether molecules. (b) Cross-sectional view showing the principal channels along the c axis. (After ref. 9).

NMR evidence and the electrical, magnetic, and optical properties of $Cs^+(18C6)_2.e^-$, as well as its isomorphism with $Cs^+(18C6)_2.Na^-$, strongly favor a "stoichiometric F-center" model of the electride in which the electrons are trapped at anionic sites and interact only weakly with one another and with the complexed cations.

More recently, the structure of potassium cryptand[2,2,2] electride, $K^+(C222).e^-$, has been shown to have a crystal structure[10] that is very similar to those of the corresponding kalide (previously potasside),

$K^+(C222).K^-$ and rubidide, $Rb^+(C222).Rb^-$, in which anion pairs in close contact are accommodated in elongated dumbbell-shaped cavities. The electride contains interlocking sheets of complexed cations with open channels that connect very large cavities ($\sim$12x4x6Å), each presumably capable of trapping a pair of electrons. A view down the most open channel, in the *c* direction and perpendicular to the sheets, is shown in Fig. 3.6-5(a). Each channel is bounded by hydrogen van der Waals surfaces and has a minimum diameter of about 4Å. Adjacent anionic sites between a pair of sheets are also connected by somewhat more constricted channels in one direction but nearly completely blocked in the third direction. A schematic diagram showing the system of roughly two-dimensional channels and cavities is shown in Fig. 3.6-5(b).

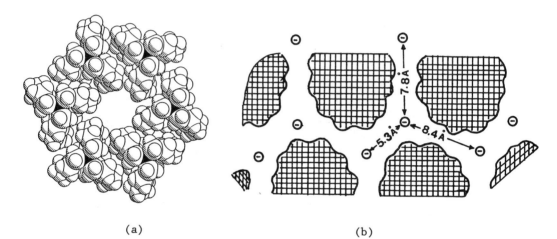

(a)   (b)

Fig. 3.6-5 (a) Space-filling representation of the packing in $K^+(C222).e^-$ showing a central hole (surrounded by six complexed cations), which connects elongated cavities that contain electron pairs. (b) Schematic diagram of the open channels in $K^+(C222).e^-$ showing probable locations of the trapped electrons and the inter-site separations. In the corresponding kalide $K^+(C222).K^-$, which has a similar packing in a different space group, the closest K^- to K^- distances are 4.90, 7.95 and 9.26Å. (Adapted form ref. 11).

The crystal structure of triclinic $Cs^+(15C5)_2.e^-$ exhibits a large cavity (minimum radius 2.0Å, average radius 2.35Å) around the center of the unit cell, which is the most likely site for the trapped electron.[12]

5. Mixed-metal alkalides

Improved synthetic techniques have yielded a range of new alkalides and electrides,[13,14] including mixed-metal alkalides such as $Rb^+(15C5)_2.Na^-$ and $Cs^+(18C6)_2.Na^-$. In these two structures the large complexed ions form closed packed arrays with well-separated anionic sites, each being surrounded by eight complexed cations and six other anionic sites farther away.[9] The natride $Cs^+(18C6)_2 \cdot Na^-$ [space group $C2/c$, $a = 13.581(3)$ $b = 15.684(2)$, $c = 17.429(4)$, $\beta = 93.16(2)°$ and $Z = 4$] is isostructural with the corresponding electride. On the other hand, the structure of $K^+(C222) \cdot Na^-$[15] is completely different from that of $K^+(C222) \cdot e^-$, which shows evidence of paired electrons. Both $Cs^+(15C5)_2 \cdot K^-$ and $Rb^+(15C5)_2 \cdot Rb^-$[16] are isostructural with $Rb^+(15C5)_2 \cdot Na^-$.

References

[1] F.J. Tehan, B.L. Barnett and J.L. Dye, *J. Am. Chem. Soc.* **96**, 7203 (1974).

[2] D. Moras and R. Weiss, *Acta Crystallogr., Sect. B* **29**, 396 (1973).

[3] R.H. Huang, D.L. Ward, M.E. Kuchenmeister and J.L. Dye, *J. Am. Chem. Soc.* **109**, 5561 (1987).

[4] J.L. Dye, J.M. Ceraso, M.T. Lok, B.L. Barnett and F.J. Tehan, *J. Am. Chem. Soc.* **96**, 608 (1974).

[5] J.L. Dye, *J. Chem. Educ.* **54**, 332 (1977).

[6] R.H. Huang, D.L. Ward and J.L. Dye, *J. Am. Chem. Soc.* **111**, 5707 (1989).

[7] A. Ellaboudy, J.L. Dye and P.B. Smith, *J. Am. Chem. Soc.* **105**, 6490 (1983).

[8] S.B. Dawes, D.L. Ward, R.H. Huang and J.L. Dye, *J. Am. Chem. Soc.* **108**, 3534 (1986).

[9] S.B. Dawes, D.L. Ward, O. Fussa-Rydel, R.H. Huang and J.L. Dye, *Inorg. Chem.* **28**, 2132 (1989).

[10] D.L. Ward, R.H. Huang and J.L. Dye, *Acta Crystallogr., Sect. C* **44**, 1374 (1988).

[11] J.L. Dye, *Science (Washington)* **247**, 663 (1990).

[12] D.L. Ward, R.H. Huang, M.E. Kuchenmeister and J.L. Dye, *Acta Crystallogr., Sect C* **46**, 1831 (1990).

[13] J.L. Dye, *J. Phys. Chem.* **88**, 3842 (1984).

[14] J.L. Dye, *Prog. Inorg. Chem.* **32**, 327 (1984).

[15] D.L. Ward, R.H. Huang and J.L. Dye, *Acta Crystallogr., Sect. C* **46**, 1833 (1990).

[16] D.L. Ward, R.H. Huang and J.L. Dye, *Acta Crystallogr., Sect. C* **46**, 1838 (1990).

Note Lattice parameters for several polycrystalline alkalides and electrides containing 15-crown-5 are reported in S. Docuff, K.-L. Tsai and J.L. Dye, *Inorg. Chem.* **30**, 767 (1991).

3.7   Cesium Suboxide

Cs_7O

Crystal Data

Hexagonal, space group $P\bar{6}m2$ (No. 187)

$a = 16.244(2)$, $c = 9.145(5)$Å; $Z = 3$ (at -170°C)

Atom	x	y	z
Cs(1)	.8136	-1 + 2x	.2199
Cs(2)	2/3	1/3	.2032
Cs(3)	.5509	-1 + 2x	0
Cs(4)	.4493	2x	1/2
Cs(5)	.2158	2x	0
Cs(6)	.1123	2x	1/2
Cs(7)	0	0	0
O	.7481	-1 + 2x	0

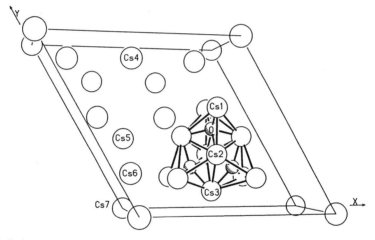

Fig. 3.7-1 Crystal structure of Cs_7O viewed in a direction nearly parallel to the c axis.

Crystal Structure

The unit cell of Cs_7O[1] contains ten Cs atoms and a $Cs_{11}O_3$ cluster of D_{3h} symmetry, which comprises three octahedral Cs_6O groups each sharing two adjacent faces with the others (Fig. 3.7-1). These clusters form chains along (001) and are also surrounded by the other Cs atoms.

Distances within the cluster are Cs-Cs = 3.72-4.27Å, Cs-O = 2.73-3.96Å, and these do not change much with temperature. The Cs-Cs distances between clusters is 5.27Å, which is also the shortest distance between any particular Cs atom in a cluster and the other ten Cs atoms, and is similar to the interatomic distance in Cs metal.

Remarks

1. The $Cs_{11}O_3$ and Rb_9O_2 clusters[2]

A small region of stability of alkali metal clusters in the solid state

exists in the suboxides of the heavy alkali metals, Rb and Cs, although only
at relatively low temperatures. Similar compounds of the lighter metals and
the higher homologs of O do not exist because of the ease of formation of the
"normal" oxides M_2O.

The Rb and Cs suboxides contain Rb_9O_2 and $Cs_{11}O_3$ clusters, which are
composed of two and three face-sharing octahedral M_6O units, respectively, as
shown in Fig. 3.7-2.

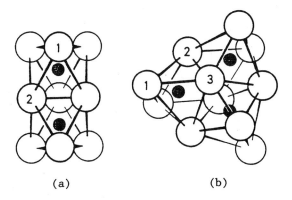

(a)   (b)

Fig. 3.7-2 Rb_9O_2 (a) and $Cs_{11}O_3$ (b) clusters (open circles: Rb, Cs;
black circles: O). (After ref. 2).

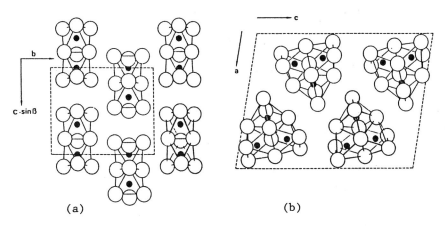

(a)   (b)

Fig. 3.7-3 Crystal structures of Rb_9O_2 (a) and $Cs_{11}O_3$ (b).
(After ref. 2).

"Molecular crystals" consisting of Rb_9O_2 and $Cs_{11}O_3$ clusters only are
shown in Fig. 3.7-3. These clusters also occur together with a stoichiometric
amount of Rb and Cs in a variety of phases with higher M:O ratios. Fig. 3.7-4
shows the phase diagrams of $Cs_{11}O_3$-Cs and $Cs_{11}O_3$-Rb. Rb_9O_2 and $Cs_{11}O_3$
clusters do not occur in the gas phase, as has been demonstrated by mass
spectroscopic investigation of the partially oxidized metal vapors.

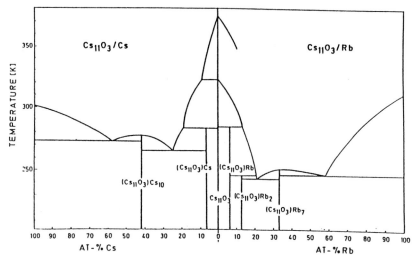

Fig. 3.7-4 Phase diagrams of the $Cs_{11}O_3$-Cs and $Cs_{11}O_3$-Rb systems. (After ref. 2).

2. Crystal structures of alkali metal suboxides[2]

The crystal structures of all binary suboxides (with the exception of "Cs_3O") as well as the structures of $Cs_{11}O_3Rb$, $Cs_{11}O_3Rb_2$, and $Cs_{11}O_3Rb_7$ have

Table 3.7-1   Structural data of alkali metal suboxides.

Compound	Space group	Structural feature
$Cs_{11}O_3$	$P2_1/c$	$Cs_{11}O_3$ cluster
Cs_4O	$Pna2$	$(Cs_{11}O_3)Cs$; one Cs atom per $Cs_{11}O_3$ cluster
Cs_7O	$P6m2$	$(Cs_{11}O_3)Cs_{10}$; ten Cs atoms per $Cs_{11}O_3$ cluster
$Cs_{11}O_3Rb$	$Pmn2_1$	one Rb atom per $Cs_{11}O_3$ cluster
$Cs_{11}O_3Rb_2$	$P2_1$	two Rb atoms per $Cs_{11}O_3$ cluster
$Cs_{11}O_3Rb_7$	$P2_12_12_1$	seven Rb atoms per $Cs_{11}O_3$ cluster
Rb_9O_2	$P2_1/m$	Rb_9O_2 cluster
Rb_6O	$P6_3/m$	$(Rb_9O_2)Rb_3$; three Rb atoms per Rb_9O_2 cluster

been solved by single crystal X-ray methods (Fig. 3.7-5). The available data are summarized in Table 3.7-1. All structures conform to the following rules:

a) The O atoms are surrounded by octahedra of Rb or Cs atoms.

b) Face-sharing of two such octahedra results in the cluster Rb_9O_2; three equivalent octahedra form the cluster $Cs_{11}O_3$.

c) The O-M distances are near the values expected of M^+ and O^{2-} ions. The ionic character of the M atoms is expressed in short intra-cluster M-M distances.

d) The inter-cluster M-M distances are comparable to the distances in metallic Rb and Cs.

e) The clusters and additional alkali metal atoms form compounds of new stoichiometries.

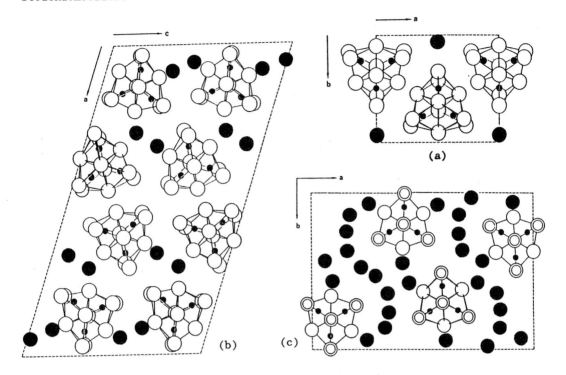

Fig. 3.7-5 Crystal structures of $(Cs_{11}O_3)Rb$ (a), $(Cs_{11}O_3)Rb_2$ (b) and $(Cs_{11}O_3)Rb_7$ (c). (After ref. 2).

3. Chemical bonding in the alkali metal suboxides[2,3]

A first insight into chemical bonding within the alkali metal suboxides is gained by comparing the interatomic distances in the compounds Rb_9O_2 and $Cs_{11}O_3$ with those in the "normal" oxides and in metallic Rb and Cs. The M-O distances nearly correspond to the sum of ionic radii. The large inter-cluster M-M distances correspond to the distances in elemental M (Rb or Cs). Therefore, the formulations $(Rb^+)_9(O^{2-})_2(e^-)_5$ and $(Cs^+)_{11}(O^{2-})_3(e^-)_5$ represent a rather realistic description of the bonding in these clusters. Their stability is due to strong O-M bonds and additional weaker M-M bonds. For a stable configuration, M-M bonds are necessary, which in the electrostatic model are taken into account as a partial shielding of the positive charges. $Cs_{11}O_3$, Rb_9O_2 and all other alkali metal suboxides have metallic luster and are good metallic conductors. Thus the electrons in M-M bonding states do not stay localized within the clusters but delocalize throughout the crystal due to the close contacts between the clusters.

The metal clusters Rb_9O_2 and $Cs_{11}O_3$ form "intermetallic compounds" with Rb and Cs, e.g. $(Rb_9O_2)Rb_3$, $(Cs_{11}O_3)Rb$, $(Cs_{11}O_3)Rb_2$, $(Cs_{11}O_3)Rb_7$, and $(Cs_{11}O_3)Cs_{10}$ (Table 3.7-1). The molar volumes of these compounds closely correspond to the sum of the molar volumes of Rb_9O_2 and $Cs_{11}O_3$ and the atomic volumes of Rb and Cs, respectively. For example, the molar volumes of $(Cs_{11}O_3)Cs$ and $(Cs_{11}O_3)Cs_{10}$ exceed the volume of $Cs_{11}O_3$ by 69.9 and 696.5 cm^3 mol^{-1}, respectively, which may be compared with the value 69.4 cm^3 mol^{-1} for elemental Cs at -50°C. Clearly, the additional Cs atoms in these suboxides form purely metallic bonds to the $Cs_{11}O_3$ cluster. In support of this view, the photoelectron spectrum of $(Cs_{11}O_3)Cs_{10}$ looks like a superposition of the spectra of $Cs_{11}O_3$ and Cs.

References

[1] A. Simon, *Z. Anorg. Allg. Chem.* **422**, 208 (1976).

[2] A. Simon, *Structure and Bonding* **36**, 81 (1979).

[3] A. Simon, *Angew. Chem. Int. Ed. Engl.* **27**, 159 (1988).

3.8 Basic Beryllium Acetate

$Be_4O(CH_3COO)_6$

Crystal Data

Cubic, space group *Fd3* (No. 203)

$a = 15.74(1)$Å, $Z = 8$

Atom	Position	x	y	z	Occupancy
Be	32(e)	-.0611	.0611	.0611	1
O(1)	8(a)	0	0	0	1
O(2)	96(g)	-.0386	.1617	.0576	1
C(1)	48(f)	0	.1997	0	1
C(2)	48(f)	0	.2950	0	1
H(1)	96(g)	-.038	.052	.317	1/2
H(2)	96(g)	.025	.061	.317	1/2
H(3)	96(g)	.065	.008	.317	1/2

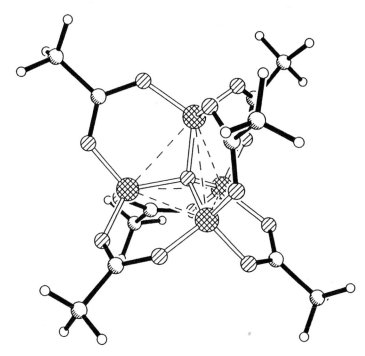

Fig. 3.8-1 Molecular structure of $Be_4O(CH_3COO)_6$.

Crystal Structure

X-Ray analysis has shown that the basic acetate consists of discrete molecules $Be_4O(CH_3COO)_6$, which has the structure illustrated in Fig. 3.8-1.

In the crystal, the symmetry of the molecule is 23. The central oxygen atom is tetrahedrally surrounded by four beryllium atoms, and each beryllium atom is tetrahedrally surrounded by four oxygen atoms. The six acetate groups are attached symmetrically to the six edges of the tetrahedron. Hydrogen atoms of statistical weight 1/2 are distributed over two equivalent sets of positions, corresponding to two equivalent orientations of the methyl groups. Therefore, the methyl hydrogen atoms lie approximately between *cis* and *trans* configurations with respect to the carboxyl oxygen atoms.

The bond lengths and bond angles are quite normal:

Be-O(1)	1.666(4)Å	C(2)-H	1.1Å
Be-O(2)	1.624(10)Å	O(2)-C(1)-O(2)	123.4(3)°
C(1)-C(2)	1.500(6)Å	O(1)-Be-O(2)	115.2(2)°
C(1)-O(2)	1.264(8)Å	O(2)-Be-O(2)	102.7(3)°

The Be-O distances can be compared with those found in hexagonal beryllium oxide, which also contains beryllium-oxygen tetrahedra[5] and possesses two different Be-O distances (1.655 and 1.647Å). The C(1)-O(2) distance is in qualitative agreement with a plethora of other determinations of the C-O distance in a carboxylate group. The C(1)-C(2) distance seems to be short, only 1.500Å. However, since it is adjacent to a double bond, the methyl H atoms can enter into hyperconjugation. The net effect of this would be to shorten the C-C distance.

Remarks

1. The structural chemistry of beryllium

Beryllium, because of its small size and simple set of valence orbitals, almost invariably exhibits tetrahedral 4-coordination in its compounds. Thus EDTA, which coordinates strongly to Mg, Ca and Al, does not chelate Be appreciably. BeO has the wurtzite (hexagonal ZnS) structure, whereas other Be chalcogenides adopt the zinc blende (cubic ZnS) modification. BeF_2 adopts the cristobalite (a modification of SiO_2) structure and has only a very low electrical conductivity when fused. Be_2C and Be_2B have the antifluorite structure with 4-coordinate Be and 8-coordinate C or B. In phenacite, Be_2SiO_4, both Be and Si are tetrahedrally coordinated, and Li_2BeF_4 has the same structure (see Fig. 3.5-4). $[Be(H_2O)_4]SO_4$ features a tetrahedral aquo-ion (Be-O = 1.61Å) which is hydrogen bonded to the surrounding sulfate groups with O-H...O distances of 2.62 and 2.68Å.

BeH_2, $BeCl_2$ and $Be(CH_3)_2$ appear to be polymerized through BeHBe, BeClBe and $Be(CH_3)Be$ 3-center bonds and achieve tetrahedral coordination about Be. Other configurations involving linear 2-coordination (e.g. $BeBu^t_2$) or trigonal 3-coordination [e.g. cyclic $(MeBeNMe_2)_2$] are rare.

2. M_4OX_6 and $M_4OX_6Y_4$ type compounds

Beryllium is unique in forming a series of stable, volatile, molecular oxide-carboxylates of the general formula $Be_4O(RCOO)_6$, where R = H, Me, Et, Ph, etc. These are all white crystalline compounds which are soluble in

organic solvents, even alkanes, but insoluble in water and lower alcohols.
They are inert to water but are hydrolyzed by dilute acids. In solution they
are un-ionized and monomeric.

Beryllium also forms a basic beryllium nitrate, $Be_4O(NO_3)_6$, which is
believed to have a structure involving a tetrahedral array of beryllium atoms
edge-bridged by nitrate groups, and having a quadridentate oxygen atom
coordinating to each beryllium atom from the center of the tetrahedron [Fig.
3.8-2(a)].[6]

Other M_4OX_6 compounds, for example $Zn_4O(CH_3COO)_6$ and $Zn_4O(S_2CNPr_2)_6$,
have been prepared. The structure of the central core in these compounds is
the same as in $Be_4O(CH_3COO)_6$.

Contaminating a bromo-Grignard reagent with oxygen leads to the
isolation, in small yields, of a compound formulated as $Mg_4OBr_6(Et_2O)_4$.[8] In
the molecular structure [Fig. 3.8-2(b)], the magnesium atom is 5-coordinate,
not 4-coordinate as in the beryllium compounds. Thus each Mg atom is bonded
to not only the central oxygen atom and three bidentate bridging bromide ions
but to an external ether ligand as well.

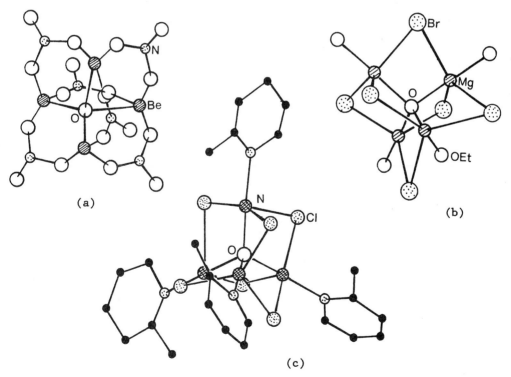

Fig. 3.8-2 Molecular structures of (a) $Be_4O(NO_3)_6$, (b) $Mg_4OBr_6(Et_2O)_4$,
and (c) $Cu_4OCl_6(2\text{-Mepy})_4$. (Adapted from refs. 6 and 7).

The structure of $Cu_4OCl_6(2\text{-Mepy})_4$ is shown in Fig. 3.8-2(c).[7] The central unit Cu_4OCl_6 is of the same type as that in $Mg_4OBr_6(Et_2O)_4$,[8] with a trigonal bipyramidal arrangement of bonds about Cu(II). The analogous $Cu_4OCl_6(OPPh_3)_4$ molecule is notable as one of the comparatively rare examples of a ligand O atom forming two collinear bonds.

References

[1] A. Tulinsky and C.R. Worthington, *Acta Crystallogr.* **10**, 748 (1957).

[2] A. Tulinsky, C.R. Worthington and E. Pignataro, *Acta Crystallogr.* **12**, 623 (1959).

[3] A. Tulinsky and C.R. Worthington, *Acta Crystallogr.* **12**, 626 (1959).

[4] A. Tulinsky, *Acta Crystallogr.* **12**, 634, (1959).

[5] G.A. Jeffrey, G.S. Parry and R.L. Mozzi, *J. Chem. Phys.* **25**, 1024 (1956).

[6] D.A. Armitage, *Inorganic Rings and Cages*, Edward Arnold, London, 1972, chap. 2.

[7] N.S. Gill and M. Sterns, *Inorg. Chem.* **9**, 1619 (1970).

[8] J.A. Bertrand and J.A. Kelly, *Inorg. Chem.* **8**, 1982 (1969).

3.9 α-Rhombohedral Boron (α-Boron, R12-Boron)

B

Crystal Data

Rhombohedral, space group $R\bar{3}m$ (No. 166)

Hexagonal cell: $a = 4.908(3)$, $c = 12.567(7)$Å, $Z = 36$

Rhombohedral cell: $a = 5.057(3)$, $\alpha = 58.07(5)°$, $Z = 12$

Atom	x	y	z
B(1)	.1177	.2354	-.1073
B(2)	.1961	.3922	.0245

Referred to a hexagonal cell; all atoms are in position 18(h).

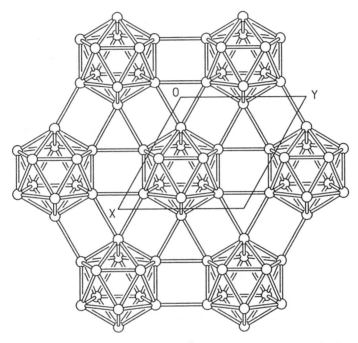

Fig. 3.9-1 A close-packed layer of interlinked icosahedra in the crystal structure of α-rhombohedral boron. (Adapted from ref. 2).

Crystal Structure

Elemental boron exists in many allotropes, three of which have been studied by X-ray diffraction, namely α-R12, β-R105 and α-T50; here R and T indicate the rhombohedral and tetragonal systems respectively, and the numerals give the number of atoms in the primitive unit cell.[2,3] These structures are all characterized by icosahedral B_{12} units that are linked together in a three-dimensional framework. The icosahedron is a regular polyhedron with 12 vertices and 20 equilateral triangular faces. The B atoms are located at the vertices of the icosahedron, as shown in Fig. 3.9-3.

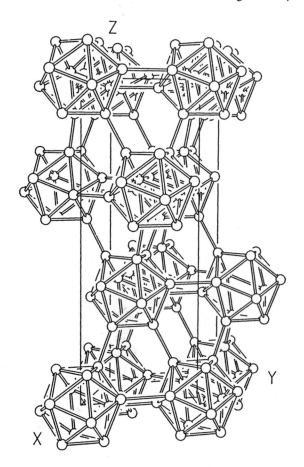

Fig. 3.9-2 Perspective view of the crystal crysture of α-R12 boron
showing the bonds between layers of icosahedra at $z = 0$, $z = 1/3$,
$z = 2/3$ and $z = 1$.

The simplest structure of elemental boron is that of the α-rhombohedral
form (α-R12 boron), the unit cell of which contains one B_{12} icosahedron.[1]
The icosahedra are arranged in approximately cubic closest packing. A layer
of interlinked icosahedra perpendicular to the three-fold symmetry axis is
illustrated in Fig. 3.9-1, and Fig. 3.9-2 shows a perspective view of the
crystal structure approximately along the a axis. Each B atom has five
neighbours (at 1.77Å) located at the corners of a pentagon, so that it
constitutues the apex of a pentagonal pyramid. Half of the atoms in every B_{12}
icosahedron each has also a boron atom of an adjacent icosahedron 1.71Å away
on the local C_5 axis passing through it; the remaining six atoms of the
icosahedron are 2.03Å distant from each of two atoms belonging to different
icosahedra. Therefore, of the atoms in any one icosahedron, half have one

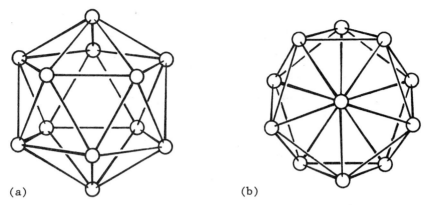

Fig. 3.9-3 The icosahedral B_{12} unit viewed along one of its symmetry axes: (a) three-fold; (b) five-fold.

neighbour at 1.71Å and five at 1.77Å, and the other half have five neighbours at 1.77Å and two at 2.03Å.

Remarks

1. Crystal structure of α-tetragonal boron

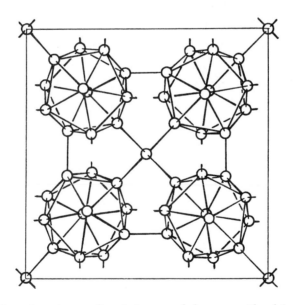

Fig. 3.9-4 The structure of α-tetragonal boron. The bisphenoidal array of four icosahedra in one unit cell is projected down the tetragonal *c* axis. The distortion of some of the inter-icosahedral bonds from the preferred pentagonal pyramidal coordination geometry is evident in this projection. (Adapted from ref. 5).

The structure of α-tetragonal boron (α-T50 boron) is comparatively simple.[4] The unit cell has dimensions *a* = 8.75, *c* = 5.06Å, and the space group is $P4_2/nnm$ (No. 134). There are 50 atoms in the unit cell, of which 48 form four nearly regular icosahedral groups, which are linked together both directly and through the remaining B atoms, as shown in Fig. 3.9-4. The latter B atoms form tetrahedral bonds, but the atoms of the icosahedra form six bonds, five to their neighbours in the B_{12} group and a sixth to an atom in another B_{12} or to a 4-coordinate B atom. Although all the icosahedral B atoms form pentagonal pyramidal bonds, for 8 of the 12 B atoms there is considerable deviation (20°) of the external bond from the five-fold axis of the icosahedron. The bond lengths in this structure are 1.60Å for the 4-coordinated B, 1.68Å for bonds between B_{12} groups, and 1.81Å within the B_{12} groups.

2. Crystal structure of β-rhombohedral boron[2,6,7]

The hexagonal unit cell of β-rhombohedral boron (β-R105 boron) contains 3x105 = 305 boron atoms with a = 10.944 and c = 23.81Å in space group $R\bar{3}m$ (No. 166).[6] The crystal structure can be described in terms of three types of B_{12} icosahedra: type **A** (site symmetry $\bar{3}m$), type **B** (2/m) and type **C**; the icosahedra of the latter type are fused together to form trimeric groups C_3 (3m) each composed of 28 atoms. Each type **A** icosahedron is bonded radially to six type **B** icosahedra, three above and three below, and to six C_3 groups in the equatoral plane [Fig. 3.9-5(b)]. The resulting geometrical figure enclosing each type **A** icosahedron is thus a truncated icosahedron which, like buckminsterfullerene (C_{60}, as illustrated in Fig. 2.5-8), has sixty equivalent vertices, ninety edges, twelve pentagonal faces and twenty hexagonal faces. This B_{60} unit is linked to the central type **A** icosahedron through an additional set of twelve boron atoms lying along the five-fold axes, giving rise to the "B_{84} unit" shown in Fig. 3.9-5(a). Alternatively, the B_{84} unit may also be regarded as a B_{12} isocahedron linked radially to twelve B_6 "half-icosahedra", each attached like an inverted umbrella to an icosahrdral vertex. The complicated framework of boron atoms is generated by completing the half-icosahedra of each B_{84} unit and continuing to add those groups to which these are bonded [Fig. 3.9-5(b)]. The unit cell contains three icosahedra of type **A**, nine of type **B**, six trimeric C_3 groups, and three boron atoms ($\bar{3}m$) each connecting a pair of C_3 groups, for a total of 3(12+36+56+1) = 315 atoms.

The interstices between the icosahedra readily accommodate atoms of transition metals and atoms with partially-filled p orbitals; for example, $NiB_{48.5}$ is a solid solution of Ni in β-rhombohedral boron, and similar boron-rich phases such as $SiB_{\sim36}$, $GeB_{\sim90}$, $CrB_{\sim41}$, $FeB_{\sim49}$, $VB_{\sim65}$ and $VB_{\sim165}$ have been characterized.[8]

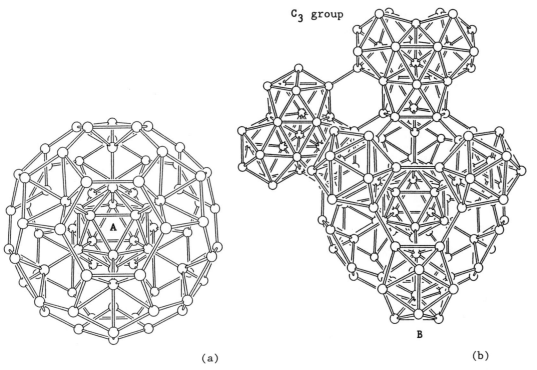

Fig. 3.9-5 Structure of β-rhombohedral boron viewed in the c direction:
(a) the B_{84} unit; (b) three type B icosahedra (above) and two C_3 groups
(equatorially) fitting into the half-icosahedral depressions of a B_{84} unit.

Since each boron atom is linked to more than four neighbours, it is
obvious that one cannot identify an electron pair bond with each edge of the
icosahedron. The bonding in these species is multi-centered and best
understood in terms of the delocalized molecular orbital theory.[9,10]

3. Icosahedral borides

The structural and bonding characteristics of boron, as revealed in the
studies of its allotropes, are retained to a large extent in many of its
compounds. For example, there exist a number of boron-rich borides built of
icosahedral B_{12} groups and additional atoms, as shown in Table 3.9-1.[11] The
majority of boranes and carboranes are based on complete or partial
icosahedral molecular skeletons as discussed in Sections 3.10 and 3.11.

4. Crystal structure of boron carbides and related compounds[12]

The potential usefulness of refractory p-type thermoelectric materials
has prompted intensive studies of the electronic transport properties of boron
carbides, $B_{1-x}C_x$ with $0.1 \leq x \leq 0.2$. The carbides of composition $B_{13}C_2$ and

Table 3.9-1   Icosahedral borides.

Crystal	Structural units
B_4C	$B_{11}C$, CBC chain
$B_{13}C_2$	B_{12}, CBC chain
$B_{12}P_2$	B_{12}, P-P link
$B_{12}As_2$	B_{12}, As-As link
BeB_{12}	B_{12}, Be
AlC_4B_{40}	$3B_{12}$, chains CBC, CBB CAlB
NaB_{15}	B_{12}, 3B, Na
$MgAlB_{14}$	B_{12}, 2B, Mg, Al
NiB_{25}	$2B_{12}$, B, Ni

B_4C, as well as the related compounds $B_{12}P_2$ and $B_{12}As_2$, crystallize in space group *R3m* having a rhombohedral unit cell of dimensions $a_R \simeq 5.2$Å and $\alpha_R \simeq 66°$ (or alternatively a hexagonal cell with $a_H \simeq 5.6$Å and $c_H = 12.1$Å). Structurally these materials are derived from α-boron with the addition of a two- or three-atom link directed along the rhombohedral [111] direction. The crystal structure of an idealized boron carbide is shown in Fig. 3.9-6.

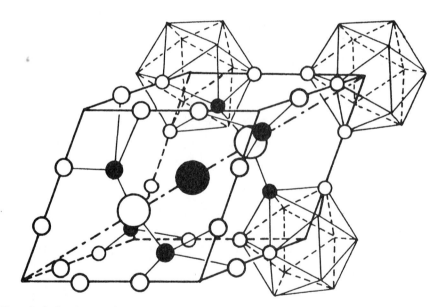

Fig. 3.9-6 Crystal structure of boron carbide, showing rhombohedral unit cell, [111] chain, and icosahedral structures. Small dark circles represent $6(h)_1$ sites; and small open circles, $6(h)_2$ sites. The two large open circles represent carbon atoms in $2(c)$ sites at the ends of the chain, and the large dark circle represents the $1(b)$ site at the center of the chain. (After ref. 12).

Detailed single-crystal structure determination, abetted by ^{10}B, ^{11}B and ^{13}C NMR studies, has led to the view that $B_{13}C_2$ [or more correctly, $B_{12}(CBC)$] is composed of B_{12} icosahedra linked by C-B-C chains. The structure of the carbon-rich crystalline phase B_4C has been shown to be $B_{11}C(CBC)$ rather than $B_{12}(CCC)$, with statistical distribution of a C atom over the vertices of the $B_{11}C$ icosahedron with a bias towards the $6(h)_2$ sites (Fig. 3.9-6). A structural basis for understanding the boron carbides is as follows: if $B_{12}(CBC)$ is taken as the standard composition, a decrease in carbon content may result in some C-B-C chains replaced by either linear B-B-C or rhombic B-B_2-B groups, whereas for carbon-rich materials a statistical carbon substitution in the icosahedra tends toward the $B_{11}C(CBC)$ configuration.

For boron phosphide and boron arsenide, two-atom P-P or As-As links directly substitute for the three-atom chains in the carbides, with a marked expansion of a_H (~ 6.1Å) and a slight contraction of c_H (~ 11.9Å). Fig. 3.9-7 compares the measured bond lengths in the series α-boron, B_4C, $B_{12}P_2$ and $B_{12}As_2$. The icosahedral B-B bonds in the polar triangle become longer in progressing along the series while the equatorial icosahedral B-B bonds become slightly shorter.

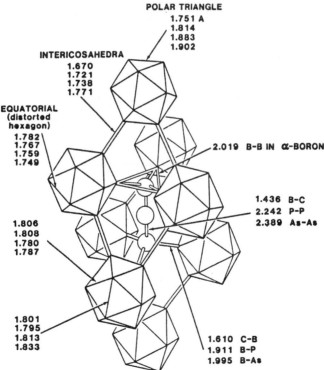

POLAR TRIANGLE
1.751 A
1.814
1.883
1.902

INTERICOSAHEDRA
1.670
1.721
1.738
1.771

**EQUATORIAL
(distorted
hexagon)**
1.782
1.767
1.759
1.749

2.019 B-B IN α-BORON

1.436 B-C
2.242 P-P
2.389 As-As

1.806
1.808
1.780
1.787

1.801
1.795
1.813
1.833

1.610 C-B
1.911 B-P
1.995 B-As

Fig. 3.9-7 A comparison of bond lengths: values in Å are listed in the order α-boron, boron carbide B_4C, boron phosphide and boron arsenide. (After ref. 12).

References

[1] B.F. Decker and J.S. Kasper, *Acta Crystallogr.* **12**, 503 (1959).

[2] J. Donohue, *The Structures of the Elements*, Wiley-Interscience, New York, 1974.

[3] N.N. Greenwood, *The Chemistry of Boron*, Pergamon Press, Oxford, 1975.

[4] J.L. Hoard, R.E. Hughes and D.E. Sands, *J. Am. Chem. Soc.* **80**, 4507 (1958).

[5] E.L. Mutterties, *The Chemistry of Boron and Its Compounds*, Wiley, New York, 1967.

[6] J.L. Hoard, D.B. Sullenger, C.H.L. Kennard and R.E. Hughes, *J. Solid State Chem.* **1**, 268 (1970); B. Callmer, *Acta Crystallogr., Sect. B* **33**, 1951 (1977).

[7] R. Naslain in V.I. Matkovich (ed.), *Boron and Refractory Borides*, Springer-Verlag, Berlin, 1977, p. 139.

[8] T. Lundström and L.-E. Torgenius, *Z. Kristallogr.* **167**, 235 (1984); M.F. Garbauskas, J.S. Kasper and G.A. Slack, *J. Solid State Chem.* **63**, 424 (1986).

[9] W.N. Lipscomb, *Boron Hydrides*, Benjamin, New York, 1963.

[10] E.L. Muetterties (ed.), *Boron Hydride Chemistry*, Academic Press, New York, 1975.

[11] A.F. Wells, *Structural Inorganic Chemistry*, 5th ed., Clarendon Press, Oxford, 1984.

[12] D. Emin, T. Aselage, C.L. Beckel, I.A. Howard and C. Wood (eds.), *Boron-Rich Solids*, American Institute of Physics, New York, 1986.

3.10 Decaborane(14)

$B_{10}H_{14}$

Crystal Data

Monoclinic, space group $C2/a$ (No. 13, unconventional non-primitive setting with two-fold axis in the c direction)

$a = 14.32(2)$, $b = 20.36(2)$, $c = 5.62(1)$Å, $\gamma = 90.06(2)°$, $Z = 8$[1]

Atom	x	y	z	Atom	x	y	z
B(1)	.0334	.3315	-.0001	B(1')	.2843	.4186	.4955
B(2)	.0993	.2739	.1726	B(2')	.3496	.4768	.3233
B(3)	.1187	.2784	-.1309	B(3')	.3690	.4123	.6273
B(4)	.0980	.2001	.0006	B(4')	.3472	.5505	.4959
B(5)	.0202	.2087	.2388	B(5')	.2695	.5413	.2568
H(1)	.0443	.3892	-.0066	H(1')	.2959	.3611	.4994
H(2)	.1593	.2883	.3085	H(2')	.4095	.4631	.1867
H(3)	.1858	.2945	-.2370	H(3')	.4362	.4565	.7355
H(4)	.1575	.1594	-.0053	H(4')	.4062	.5917	.5016
H(5)	.0365	.1748	.4064	H(5')	.2858	.5755	.0896
H(6)	.0444	.3108	-.2223	H(6')	.2960	.4385	.7172
H(7)	.0897	.2208	-.2217	H(7')	.3380	.5294	.7184

Origin at 1 on glide plane a. Equipoints at $(0,0,0; 1/2,1/2,0) +$
$x,y,z; -x,-y,-z; 1/2+x,y,-z; 1/2-x,-y,z$.

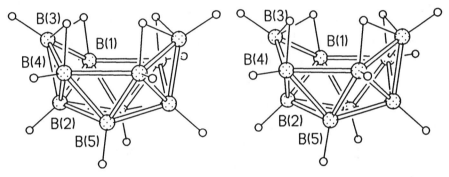

Fig. 3.10-1 Stereoview of the molecular structure of $B_{10}H_{14}$.

Crystal Structure

Crystalline $B_{10}H_{14}$ prepared by sublimation at or above room temperature shows a high degree of polysynthetic twinning. In the untwinned condition (ordered structure), the crystal is monoclinic but pseudo-orthorhombic.

The solution of the crystal structure by Kasper, Lucht and Harker[2] resulted from the first successful application of phase inequalities,[3] a prototype of modern "direct methods". These authors determined the disordered structure in the orthorhombic space group Pnnm ($a = 7.225$, $b = 10.44$, $c = 5.68$Å and $Z = 2$), placing all atoms in general position $8(h)$ with half occupancy. Tippe and Hamilton determined the ordered structure with neutron diffraction data collected at $-160°C$.[4] To achieve higher accuracy the hydrogen atoms were exchanged for deuterium, and ^{10}B was substituted for ^{11}B in view of the high neutron cross-section of the lighter isotope. An improved structure was derived by Brill, Dietrich and Dierks using accurate low-temperature X-ray and neutron diffraction data.[1] The asymmetric unit consists of two independent half-melecules. To a remarkably good approximation, the atomic coordinates of molecules I and II are related by the non-crystallographic symmetry transformation ($x' = 1/4 + x$, $y' = 3/4 - y$, $z' = 1/2 - z$).

The molecule of $B_{10}H_{14}$ has site symmetry 2 and idealized symmetry *mm2*. The boron atoms are at ten of the vertices of a somewhat distorted regular icosahedron, two neighbouring vertices of which are unoccupied. The resulting arrangement comprises two regular pentagonal pyramids with a common base edge. The angle between the basal planes of these pyramids is 76°. Each of ten terminal hydrogen atoms is attached to a single boron atom in the direction of a five-fold axis of the icosahedron. Each of the remaining four hydrogens each bridges two boron atoms, as shown in Fig. 3.10-1.

Crystal Structure analysis yielded the following interatomic distances: B-B bonds in the range 1.731-1.793Å, except that the pair joining the two icosahedra is longer at 1.987Å; bridged B-H bonds are in the range 1.321-1.354Å, and terminal B-H bonds fall within 1.177-1.191Å.[1]

Remarks

1. Structural studies of boranes

The structure determination of $B_{10}H_{14}$ provides an excellent example of the use of X-ray methods in opening up a new area, namely the chemistry of electron-deficient compounds.

The element boron forms a large number of compounds with hydrogen called "boron hydrides" or "boranes". The simplest is diborane, B_2H_6, which has an empirical formula which parallels that of ethane, C_2H_6, but there are not sufficient electrons to form an ethane-like configuration with two electrons in each bond. The correct structure of diborane as predicted by Dilthey in 1921[5] was established by Price in 1947[6] from its infrared spectrum.

In 1948 Kasper and co-workers[2] published the crystal structure of $B_{10}H_{14}$, which was completely different from any structure predicted at that time. From 1951 onwards, Lipscomb[7,8] carried out structure determinations on many other boranes and their derivatives. Fig. 3.10-2 shows the molecular structures of some boranes.[9,10]

During the past four decades the chemistry of boranes and the related carboranes has been one of the major fields of research in inorganic chemistry.[11-13] The importance of boranes stems from three factors: firstly, the completely novel structural principles involved; secondly, the

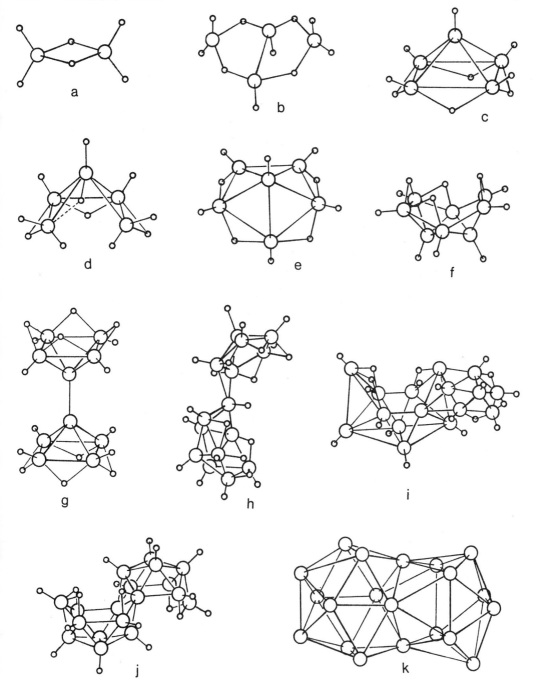

Fig. 3.10-2 Molecular structures of some boranes: (a) B_2H_6 (D_{2h}),
(b) B_4H_{10} (C_{2v}), (c) B_5H_9 (C_{4v}), (d) B_5H_{11} (idealized C_s) [in the solid the
apical *endo*-terminal H atom binds weakly to B(2), but in solution it
equilibrates rapidly between B(2) and B(5)], (e) B_6H_{10} (C_s), (f) B_8H_{12} (C_s),
(g) $B_{10}H_{16}$ (D_{4h}), (h) $B_{15}H_{23}$ (C_1), (i) $B_{16}H_{20}$ (C_1), (j) *anti*-$B_{18}H_{22}$ (C_i) and
(k) $B_{20}H_{16}$ (D_{2d}).   (After refs. 9 and 10).

need to extend MO bonding theory considerably to cope with unusual stoichiometries; and finally, the emergence of a versatile and extremely extensive reaction chemistry which parallels but is quite distinct from that of organic and organometallic chemistry. This efflorescence of activity is highlighted by the award of the 1976 Nobel Prize in Chemistry to Lipscomb for his studies of boranes which illuminated the nature of chemical bonding in electron-deficient compounds.[14] A recent article by Fehlner and Housecroft[12] gives an excellent overview of the cluster structure, bonding, energetics, and reactivity of boranes and heteroboranes, with particular emphasis on their central role as "pattern setters" for other molecular cluster systems. Beaudet has given a comprehensive and updated review of the results of molecular structure determinations of boranes and carboranes.[13]

Recently the mixed hydride gallaborane, $GaBH_6$, has been synthesized by metathesis involving $[H_2GaCl]_2$ and $LiBH_4$ and characterized as a diborane-like molecule by electron diffraction.[15] The structural parameters of the related molecules $Me_2Ga(\mu-H)_2BH_2$ and $HGa\{(\mu-H)_2BH_2\}_2$[16] have also been determined.[17]

2. Bonding in boranes

In B_2H_6, each boron atom is considered to be sp^3 hybridized. The two terminal B-H bonds on each B atom are σ bonds having a pair of electrons each. This accounts for eight of the total of twelve electrons available for bonding. Each of the B-H-B three-center, two-electron bridging bond [Fig. 3.10-3(a)] results from the overlap of two boron sp^3 hybrid orbitals and the 1s orbital of the H atom.

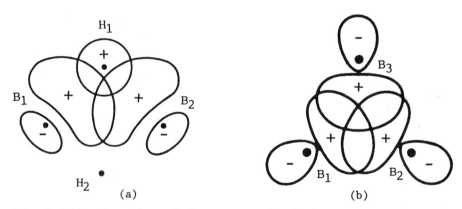

Fig. 3.10-3 Formation of three-centee bonds from atomic orbitals: (a) bridging bond B-H-B; (b) central BBB bond.

Lipscomb has developed a topological treatment of higher boranes of the general formula $B_pH_{p+q}^{z-}$ in which each B atom has a terminal H atom bonded to it.[5] Let the following two-electron bond types be defined:

s = number of three-center B-H-B bonds;

t = number of three-center central BBB bonds [Fig. 3.10-3(b)];

y = number of normal B-B bonds; and

x = number of *additional* terminal B-H bonds.

The equations of orbital balance, electron balance, and hydrogen balance
impose the following restrictions on the "*styx*" numbers:

$$x = q - s$$
$$t = p - s - z$$
$$2y = s - x + 3z$$

For example, the (*styx*) codes for B_2H_6, B_5H_{11} and B_6H_{10} are (2002), (3203) and
(4220), respectively; for B_4H_{10} both (3103) and (4012) are acceptable
solutions, but only the latter corresponds to reality.

The structure of $B_{12}H_{12}^{2-}$ is a regular icosahedron with twenty
equilateral triangles forming the faces. All of the hydrogen atoms are
external to the boron icosahedron and are attached by terminal B-H bonds. The
$B_{12}H_{12}^{2-}$ anion has 50 electrons for bonding. In accordance with the *styx*
formula (0,10,3,0), twelve terminal B-H bonds and three B-B bonds account for
30 electrons, and the remaining 20 electrons for 10 three-center BBB bonds.
The icosahedron itself involves a resonance hybrid of several canonical forms
of the type shown in Fig. 3.10-4.

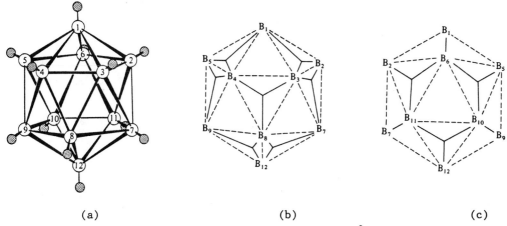

(a)   (b)   (c)

Fig. 3.10-4 (a) Molecular structure of the $B_{12}H_{12}^{2-}$ anion; (b) front
and (c) back sides of the $B_{12}H_{12}^{2-}$ framework showing one of many
canonical forms with localized B-B and BBB bonds. (After refs. 11 and 25).

3. Direct methods

In a single-crystal X-ray diffraction experiment the intensities of a
large number of reflections are recorded, but the phases which relate them to

the atomic arrangement in the structure were previously considered to be lost in the experiment. Determining a crystal structure thus requires solving the "phase problem". The pioneers of the direct methods, Harker, Kasper, Hauptman, Karle, Zachariasen, Sayre, Cochran, and others proved that the phase information is contained in the diffracted intensities and developed the theory and mathematical procedures to retrieve the phase data.[18,19] Many structures may now be solved almost automatically, using powerful computer program packages such as MULTAN[20], SHELX[21] and XTAL[22].

The development of direct methods in the past four decades has transformed the whole field of crystallography and opened a new era of research in structural chemistry, culminating in the award of the 1985 Nobel Prize in Chemistry to Herbert Hauptman[23] and Jerome Karle[24] for their pioneering efforts.

References

[1] R. Brill, H. Dietrich and H. Dierks, *Acta Crystallogr., Sect B* **27**, 2003 (1971).

[2] J.S. Kasper, C.M. Lucht and D. Harker, *J. Am. Chem. Soc.* **70**, 881 (1948); *Acta Crystallogr.* **3**, 436 (1950).

[3] D. Harker and J.S. Kasper, *Acta Crystallogr.* **1**, 70 (1948).

[4] A. Tippe and W.C. Hamilton, *Inorg. Chem.* **8**, 464 (1969).

[5] W. Dilthey, *Z. Angew. Chem.* **34**, 596 (1921).

[6] W.C. Price, *J. Chem. Phys.* **16**, 894 (1948).

[7] W.N. Lipscomb, *J. Chem. Phys.* **22**, 985 (1954).

[8] W.N. Lipscomb, *Boron Hydrides*, Benjamin, New York, 1963.

[9] N.N. Greenwood in J.C. Bailar, H.J. Eméleus, R. Nyholm and A.F. Trotman-Dickenson (eds.), *Comprehensive Inorganic Chemistry*, Vol. **1**, Pergamon Press, Oxford, 1973, chap. 11.

[10] J.D. Kennedy, *Prog. Inorg. Chem.* **32**, 519 (1984); **34**, 211 (1986).

[11] E.L. Muetterties, *The Chemistry of Boron and Its Compounds*, Wiley, New York, 1967.

[12] T.P. Fehlner and C.E. Housecroft, in J.F. Liebman and A. Greenberg (eds.), *Molecular Structure and Energetics*, Vol. *1. Chemical Bonding Models*, VCH Publishers, Weinheim, 1986, chap. 6.

[13] R.A. Beaudet in J.F. Liebman, A. Greenberg and R.E. Williams (eds.), *Molecular Structure and Energetics*, Vol. *5. Advances in Boron and the Boranes*, VCH Publishers, Weinheim, 1988, chap. 20.

[14] W.N. Lipscomb, *Science (Washington)* **196**, 1047 (1977). [Nobel Lecture].

[15] C.R. Pulham, P.T. Brain, A.J. Downs, D.W.H. Rankin and H.E. Robertson, *J. Chem. Soc., Chem. Commun.*, 177 (1990).

[16] M.T. Barlow, A.J. Downs, P.D.P. Thomas and D.W.H. Rankin, *J. Chem. Soc., Dalton Trans.*, 1793 (1979).

[17] A.J. Downs and P.D.P. Thomas, *J. Chem. Soc., Dalton Trans.*, 809 (1978).

[18] H.A. Hauptman, *Crystal Structure Determination: The Role of the Cosine Invariants*, Plenum Press, New York, 1972.

[19] M.F.C. Ladd and R.A. Palmer, *Theory and Practice of Direct Methods in Crystallography*, Plenum Press, New York, 1977.

[20] P. Main in G.M. Sheldrick, C. Krüger and R. Goddard (eds.), *Crystallographic Computing 3: Data Collection, Structure Determination, Proteins, and Databases*, Clarendon Press, Oxford, 1985, p. 206, and references cited therein.

[21] G.M. Sheldrick in D. Sayre (ed.), *Computational Crystallography*, Clarendon Press, Oxford, 1982, p. 506; G.M. Sheldrick in G.M. Sheldrick, C. Krüger and R. Goddard (eds.), *Crystallographic Computing 3: Data Collection, Structure Determination, Proteins, and Databases*, Clarendon Press, Oxford, 1985, p. 175.

[22] J.M. Stewart and S.R. Hall (eds.), *XTAL User's Manual*, Technical Report TR1364, Computer Science Center, Univ. Maryland, 1983.

[23] H.A. Hauptman, *Angew. Chem. Int. Ed. Engl.* **25**, 603 (1986). [Nobel Lecture].

[24] J. Karle, *Angew. Chem. Int. Ed. Engl.* **25**, 614 (1986). [Nobel Lecture].

[25] W.L. Jolly, *The Chemistry of the Non-Metals*, Prentice-Hall, Englewood Cliffs, N.J., 1966, chap. 12.

Note Recent advances in borane and carborane chemistry are covered in G.A. Olah, K. Wade and R.E. Williams (eds.), *Electron Deficient Boron and Carbon Clusters*, Wiley, New York, 1991.
Analogues of $B_{12}H_{12}{}^{2-}$ reported in the recent literature include *closo*-(MeP)$B_{11}H_{11}$ in T.D. Getman, H.-B. Deng, L.-Y. Hsu and S.G. Shore, *Inorg. Chem.* **28**, 3612 (1989); *closo*-NB$_{11}H_{12}$ in J. Müller, J. Runsink and P. Paetzold, *Angew. Chem. Int. Ed. Engl.* **30**, 175 (1991); and $K_2[Al_{12}iBu_{12}]$ in W. Hiller, K.-W. Klinkhammer, W. Uhl and J. Wagner, *ibid.*, p. 179.
Polyhedral boron halides and diboron tetrahalides are reviewed in J.A. Morison, *Chem. Rev.* **91**, 35 (1991).
Dimeric adducts of the elusive alane that are stabilized by tertiary amines, $[LH_2Al(\mu\text{-}H)]_2$ with L = NMe_3, NMe_2CH_2Ph and $\overline{NMeCH_2CH_2CH{=}CH(CH_2)}$, are reported in J.T. Atwood, F.R. Bennett, F.M. Elms, C. Jones, C.L. Raston and K.D. Robinson, *J. Am. Chem. Soc.* **113**, 8183 (1991).

3.11 π-Cyclopentadienyl-π-(1)-2,3-dicarbollyliron(III), a Metallocarborane

$C_5H_5FeB_9C_2H_{11}$

Crystal Data

Monoclinic, space group $P2_1/c$ (No. 14)

$a = 11.470$, $b = 6.629$, $c = 16.808\text{Å}$, $\beta = 99.86°$, $Z = 4$

Atom	x	y	z	Atom	x	y	z
Fe	.2244	-.0331	.1897	C(3)	.2008	-.1537	.2978
$C_5H_5^-$ group				B(4)	.1354	.0749	.2794
C(1R)	.0917	-.0331	.0909	B(5)	.2502	.2337	.2580
C(2R)	.1817	.0862	.0767	B(6)	.3764	.0795	.2621
C(3R)	.2779	-.0361	.0780	B(7)	.3035	-.1630	.3822
C(4R)	.2433	-.2285	.0942	B(8)	.1786	-.0172	.3790
C(5R)	.1278	-.2228	.1026	B(9)	.2127	.2301	.3555
$B_9C_2H_{11}^{2-}$ group				B(10)	.3607	.2328	.3447
C(2)	.3345	-.1476	.2871	B(11)	.4159	-.0112	.3609
				B(12)	.3173	.0825	.4194

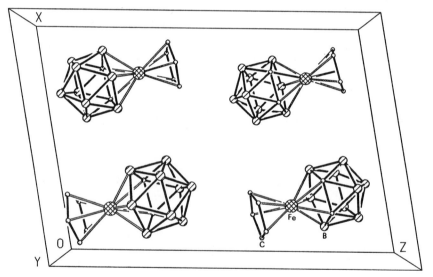

Fig. 3.11-1   Crystal structure of $(C_5H_5)FeB_9C_2H_{11}$.

Crystal Structure

The determination of the crystal structure of the metallocarborane (carbametallaborane) $(C_5H_5)FeB_9C_2H_{11}$ provided the first confirmation[1] of the structure of the $nido$-$B_9C_2H_{11}^{2-}$ ion. The geometry of this Fe complex is of the "sandwich-type" as predicted by Hawthorne and Pilling.[2]

The skeleton of the molecule is shown in Fig. 3.11-1. The atoms of the carborane part occupy eleven of the twelve corners of an icosahedron, which is approximately regular. The iron atom is at the twelfth corner, but somewhat farther outward. Within the experimental accuracy of about 0.2Å, each hydrogen atom is on the line passing through its parent B or C atom and the opposite atom of the icosahedron. The iron atom is almost equidistant (all at approximately 2.07Å) from the carbon atoms in the planar cyclopentadienyl ring and the five neighbours in the decapped icosahedron. The configuration is an almost exactly eclipsed rather than a staggered one.

Remarks

1. Carboranes

   Carboranes[3,4] and metallocarboranes occupy a strategic position in the chemistry of the elements since they overlap and give coherence to several other large areas including the chemistry of polyhedral boranes, transition-metal complexes, metal-cluster compounds, and organometallic chemistry.

   Carboranes have the general formula $[(CH)_a(BH)_mH_b]^{c-}$ with a CH units and m BH units at the polyhedral vertices, plus b "extra" H atoms. The number of electrons available for skeletal bonding is three per CH unit, two per BH unit, one per H atom, and c from the anionic charge. Hence the total number of skeletal bonding electron pairs is:

   $$(3a + 2m + b + c)/2 = n + (a + b + c)/2$$

where $n(= a + m)$ is the number of occupied vertices of the polyhedron.

   Closo-carboranes have $(n+1)$ pairs of skeletal bonding electrons (i.e. $a + b + c = 2$).

   Nido-carboranes have $(n+2)$ pairs of skeletal bonding electrons (i.e. $a + b + c = 4$).

   Arachno-carboranes have $(n+3)$ pairs of skeletal bonding electrons (i.e. $a + b + c = 6$).

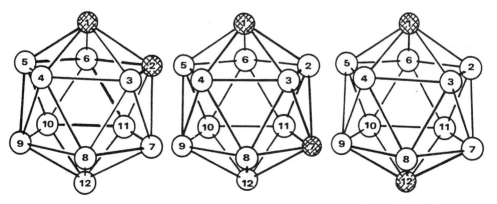

closo-1,2-$C_2B_{10}H_{12}$(o-carborane)   closo-1,7-$C_2B_{10}H_{12}$(m-carborane)   closo-1,12-$C_2B_{10}H_{12}$(p-carborane)

Fig. 3.11-2 The three isomeric icosahedral carboranes. (After ref. 4).

Closo-carboranes are the most numerous and the most thermodynamically stable of the carboranes. Most *closo*-carboranes are stable to at least 400°C, though they may undergo rearrangement to more stable isomers in which the distance between the C atoms is increased. Fig. 3.11-2 shows the three isomeric icosahedral carboranes ($C_2B_{10}H_{12}$, dicarba*closo*dodecaboranes), which are unusual both in their ease of preparation and their remarkable stability in air. *Ortho*-carborane ($1,2$-$C_2B_{10}H_{12}$) transforms to *meta*-carborane ($1,7$-$C_2B_{10}H_{12}$) at 450°C, and when the temperature is raised to 620°C, the latter transforms to the *para* isomer ($1,12$-$C_2B_{10}H_{12}$). An extensive derivative chemistry of the icosahedral carboranes has been developed.

2. Metalloboranes and metallocarboranes[5,6]

Strong bases attack the icosahedral carborane $1,2$-$C_2B_{10}H_{12}$, extruding a boron atom to give the *nido*-$C_2B_9H_{11}^{2-}$ (commonly referred to as dicarbollide) anion which has the probable structure shown in Fig. 3.11-3(a). Each of the three boron atoms and the two carbon atoms on the open face of the cage directs an orbital towards the apical position vacated by the removal of a boron atom. These orbitals, which contain a total of six electrons, thus bear a striking resemblance to the p orbitals in the π system of the cyclopentadienyl anion, $C_5H_5^-$.

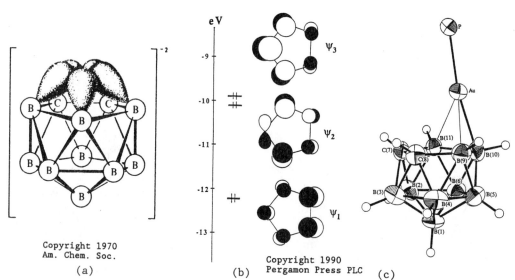

Copyright 1970
Am. Chem. Soc.

(a)

Copyright 1990
Pergamon Press PLC

(b)

(c)

Fig. 3.11-3   (a) Probable structure of the *nido*-$C_2B_9H_{11}^{2-}$ ion. (b) The occupied π-frontier molecular orbitals of *nido*-$C_2B_9H_{11}^{2-}$. (c) Molecular structure of [10-*endo*-(PPh$_3$Au)-7,8-*nido*-$C_2B_9H_{11}$]$^-$. (After refs. 12 and 10).

The structure of $(C_5H_5)Fe^{III}C_2B_9H_{11}$ implies that a BH unit of the *closo*-icosahedral carborane $C_2B_{10}H_{12}$ is replaceable by a Fe(C_5H_5) fragment. The analogy of $C_2B_9H_{11}{}^{2-}$ to $C_5H_5{}^-$ is further substantiated by the existence of $[Fe^{II}(C_2B_9H_{11})]^{2-}$ that consists of two iscosahedra sharing a common vertex. A wide range of metal-ligand combinations can formally substitute for a BH or a CH unit in a polyhedral borane or carborane skeleton. In the resulting metalloboranes and metallocarboranes, all ligands have in common a planar or nearly planar arrangement of the ligating atoms, with a significant degree of electron delocalization and hence aromatic character. Metal complexes of the boron ligands are frequently more stable than their metallocene relatives, and can exhibit structural isomerism and thermal migration of metal atoms in the molecular framework, as well as other kinds of novel stereochemistry. Many metalloboranes and metallocarboranes have had their structures determined by X-ray and neutron diffraction. The formulae of some representative examples of the *closo* type are given in Table 3.11-1, and the molecular structures are shown in Fig. 3.11-4. Examples with *nido* structures are given in Table 3.11-2 and illustrated in Fig. 3.11-5.

Table 3.11-1   Some metalloboranes and metallocarboranes with *closo* structures.

Number of skeletal atoms	Shape	Examples	Fig. 3.11-4
5	Trigonal bipyramid	$BH_2Fe_4(\mu\text{-}H)(CO)_{12}$	(a)
6	Octahedron	$B_5H_3(CO)_2Fe(CO)_3$	
		$B_3H_3Co_3(\eta^5\text{-}C_5H_5)_3$	(b)
7	Pentagonal bipyramid	$C_2B_4H_6Fe(CO)_3$	
		$B_2H_2PPhCo_4(\eta^5\text{-}C_5H_5)_4$	(c)
8	Dodecahedron	$C_2B_4Me_2H_4SnCo(\eta^5\text{-}C_5H_5)$	
		$B_4H_4Ni_4(\eta^5\text{-}C_5H_5)_4$	(d)
9	Tricapped trigonal prism	$C_2B_6H_8Pt(PMe_3)_2$	
		$[CB_7H_8Co(\eta^5\text{-}C_5H_5)]^-$	
10	Bicapped square antiprism	$C_2B_7H_9Co(\eta^5\text{-}C_5H_5)$	
		$B_9H_7Cl_2Ni(PMe_2Ph)_3$	(e)
11	Octadecahedron	$C_2B_8H_{10}Co(\eta^5\text{-}C_5H_5)$	
		$B_{10}H_9(OMe)Rh(\eta^5\text{-}C_5Me_5)$	(f)
12	Icosahedron	$B_{10}H_9(PMe_2Ph)Pt_2Cl(PMe_2Ph)_3$	(g)
		$[(C_2B_9H_{11})Co(C_2B_8H_{10})Co(C_2B_9H_{11})]^{2-}$	(h)
		$[(C_2B_9H_{11})_2M]^{n-}$ [M = Fe(II), Co(III), Ni(IV)]	
		$[(C_2B_9H_{11})_2M]^{n-}$ [M = Cu(II), Au(III), Pd(II)]	(i)
		(slipp-distorted structure)	
		$(C_2B_9H_{10})_2(HCS_2)Co$	(j)
13	1,5,6,1 Polyhedron	$C_2B_{10}H_{12}Co(\eta^5\text{-}C_5H_5)$	
		$[(C_2B_{10}Me_2H_{10})_2Ti]^{2-}$	(k)
14	Bicapped hexagonal prism	$C_2B_{10}H_{12}Co_2(\eta^5\text{-}C_5H_5)_2$	
		$C_4B_8Me_4H_8Fe_2(\eta^5\text{-}C_5H_5)_2$	

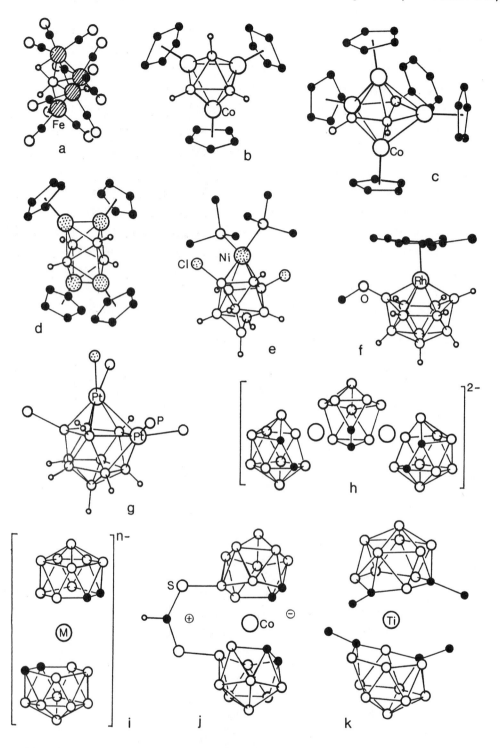

Fig. 3.11-4 Molecular structures of some *closo* metalloboranes and metallocarboranes. (After refs. 5-7).

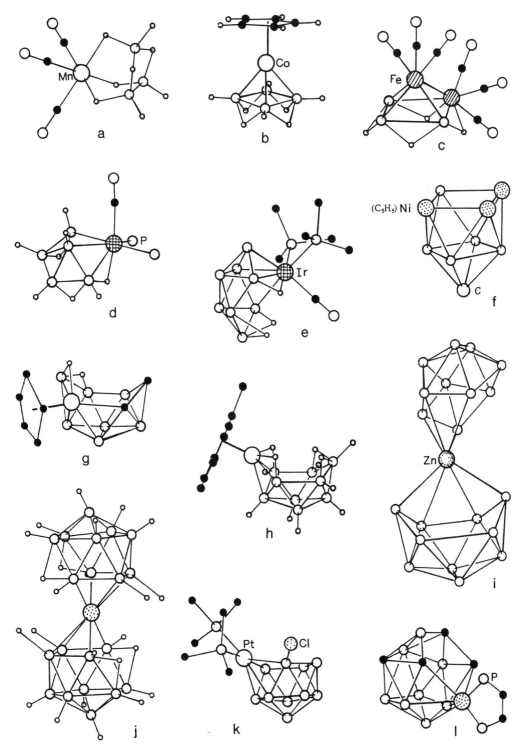

Fig. 3.11-5 Molecular structures of some *nido* metallobaranes and metallocarboranes. (After refs. 5 and 6).

Table 3.11-2   Some metalloboranes and metallocarboranes with *nido* structures.

Number of skeletal atoms	Shape	Examples	Fig. 3.11-5
4	Butterfly	$B_3H_8Mn(CO)_3$	(a)
5	Square pyramid	$B_4H_8Co(\eta^5\text{-}C_5H_5)$	(b)
		$B_3H_7Fe_2(CO)_6$	(c)
6	Pentagonal pyramid	$B_5H_8Ir(CO)(PPh_3)_2$	(d)
		$C_2B_3H_7Fe(CO)_3$	
7	Hexagonal prism	$C_5BH_5PhMn(CO)_3$	
		$C_2B_2N_2Et_2Me_4Cr(CO)_3$	
8	Bicapped square antiprism missing an equatorial vertex	$B_8H_{11}Ir(CO)(PMe_3)_3$	(e)
9	Monocapped square antiprism	$CB_5H_6Ni_3(\eta^5\text{-}C_5H_5)_3$	(f)
		$C_2B_6R_2H_6Pt(PR_3)_2$	
10	$B_{10}H_{14}$-type	$C_2B_7H_{11}Co(\eta^5\text{-}C_5H_5)$	(g)
		$B_9H_{13}Co(\eta^5\text{-}C_5Me_5)$	(h)
11	Icosahedral fragment	$[Zn(\eta^4\text{-}B_{10}H_{12})_2]^{2-}$	(i)
		$[Ni(\eta^4\text{-}B_{10}H_{12})_2]^{2-}$	(j)
		$B_{10}H_{11}ClPt(PMe_2Ph)_2$	(k)
13	14-Vertex closo polyhedron missing one vertex	$C_4B_8Me_4H_8Ni(PPh_2CH_2)_2$	(l)

3. Isolobal analogy

Hoffmann has employed frontier orbital similarities to correlate various metal-ligand fragments with the building blocks of organic chemistry (CH_3, CH_2, CH and C), thereby accounting for the structural formulae of a wide variety of organometallic complexes.[8]   For instance, d^7-$Mn(CO)_5$ resembles the methyl radical in having a singly occupied hybrid orbital pointing away from the fragment, and the analogy accords with the existence of the dimer $Mn_2(CO)_{10}$ and of $(CH_3)Mn(CO)_5$. Two fragments are said to be "isolobal" if the number, symmetry properties, approximate energy and shape of their frontier orbitals and the number of electrons in them are similar. Thus $Mn(CO)_5$ is isolobal with CH_3, d^{10}-$Pt(PMe_3)_2$ with CH_2, and d^9-$Ni(\eta^5\text{-}C_5H_5)$ with CH; each isolobal connection is symbolized by a "two-headed arrow on top of an orbital lobe" (Fig. 3.11-6).

The general rules applicable to the construction of conceptual bridges between inorganic and organic chemistry are elegantly summarized by Hoffmann in Table 3.11-3.[8]

Stone has compiled a list of metal-ligand combinations that are useful for recognizing structural relationships between organometallic and organic molecules and devising logical synthetic schemes (Table 3.11-4).[9]

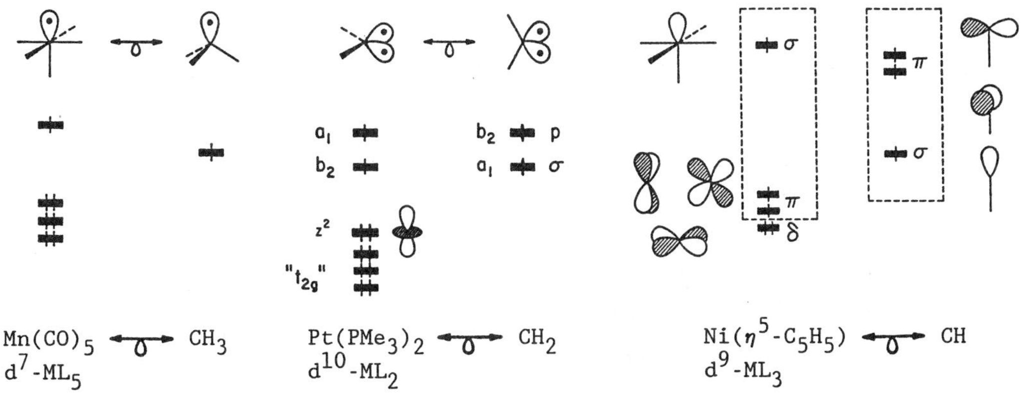

$Mn(CO)_5 \xleftrightarrow{\sigma} CH_3$
$d^7\text{-}ML_5$

$Pt(PMe_3)_2 \xleftrightarrow{\sigma} CH_2$
$d^{10}\text{-}ML_2$

$Ni(\eta^5\text{-}C_5H_5) \xleftrightarrow{\sigma} CH$
$d^9\text{-}ML_3$

Fig. 3.11-6 Isolobal analogies between metal-ligand and hydrocarbon fragments. (Adapted from ref. 8).

Table 3.11-3 Isolobal analogies.

Organic fragment	Coordination number of the transition metal on which the analogy is based				
	9	8	7	6	5
CH_3	$d^1\text{-}ML_8$	$d^3\text{-}ML_7$	$d^5\text{-}ML_6$	$d^7\text{-}ML_5$	$d^9\text{-}ML_4$
CH_2	$d^2\text{-}ML_7$	$d^4\text{-}ML_6$	$d^6\text{-}ML_5$	$d^8\text{-}ML_4$	$d^{10}\text{-}ML_3$
CH	$d^3\text{-}ML_6$	$d^5\text{-}ML_5$	$d^7\text{-}ML_4$	$d^9\text{-}ML_3$	

L = neutral two electron ligand. A bidentate ligand is equivalent to L_2.

Table 3.11-4 Isolobal relationships between hydrocarbon and metal-ligand fragments.

CH_3	CH_2	CH
$Mn(CO)_5$	$Fe(CO)_4$	$Co(CO)_3$
$Fe(CO)_2(\eta^5\text{-}C_5H_5)$	$Rh(CO)(\eta^5\text{-}C_5Me_5)$	$Ni(\eta^5\text{-}C_5H_5)$
$Mo(CO)_3(\eta^5\text{-}C_5H_5)$	$Re(CO)_2(\eta^5\text{-}C_5H_5)$	$W(CO)_2(\eta^5\text{-}C_5H_5)$
	$Cr(CO)(NO)(\eta^5\text{-}C_5H_5)$	$[Mn(CO)_2(\eta^5\text{-}C_5H_5)]^+$
$Co(CO)_4$	$Cr(CO)_5$	$Re(CO)_4$
$PtH(PPh_3)_2$	$Pt(PMe_3)_2$	$Sn(C_6H_5)$
$Zn(\eta^5\text{-}C_5H_5)$	$Cu(\eta^5\text{-}C_5Me_5)$	$Rh(\eta^6\text{-}C_6H_6)$
$Au(PPh_3)$	$IrCl(CO)_2$	$Cu(PPh_3)$
$Rh(PPh_3)_2(\eta^5\text{-}C_2B_9H_{11})$	$Fe(CO)_2(\eta^5\text{-}C_2B_9H_{11})$	$Mn(CO)_2(\eta^5\text{-}C_2B_9H_{11})$
	$TaMe(\eta^5\text{-}C_5H_5)_2$	$TaCl(PMe_3)_2(\eta^5\text{-}C_5H_5)$
CH_3^+	CH_2^+	CH^+
$Cr(CO)_5$	$Mn(CO)_4$	$Fe(CO)_3$
$Mn(CO)_2(\eta^5\text{-}C_5H_5)$	$Fe(CO)(\eta^5\text{-}C_5H_5)$	$Rh(\eta^5\text{-}C_5H_5)$
BH_3	BH_2	BH

Note that $C_5H_5^-$ is electronically equivalent to a tridentate ligand. The ubiquitous $Pt(PR_3)_2$ fragment has two high-flying orbitals that are half filled, the remaining eight electrons of Pt^0 filling the nonbonding d_{z^2} orbital and the t_{2g} set.

The occupied π-frontier molecular orbitals of $nido$-$C_2B_9H_{11}^{2-}$ are illustrated in Fig. 3.11-3(b).[10] The unoccupied frontier molecular orbitals of R_3PAu^I and R_3PCu^I comprise a valence (s - p_z) hybrid orbital and a degenerate pair of metal (p_x, p_y) atomic orbitals. For copper the latter are also valence orbitals, but for gold they are high-lying and hence responsible for the linear coordination of gold(I) *versus* the tetrahedral coordination of copper(I). In [(PPh$_3$)CuC$_2$B$_9$H$_{11}$]$^-$ the PPh$_3$Cu fragment is bonded fairly symmetrically to the five facial atoms of the carborane ligand.[11] On the other hand, in the gold(I) analogue the metal atom is essentially *endo-σ*-bonded to B(10), slipping towards it by 0.87Å from the center of the slightly nonplanar open C_2B_3 face [Fig. 3.11-3(c)].[10] The structural features of [(PPh$_3$)AuC$_2$B$_9$H$_{11}$]$^-$ reflect the strong interaction between the acceptor gold (s - p_z) orbital and the occupied ψ_3 orbital of $nido$-$C_2B_9H_{11}^{2-}$, whereas the equilibrium geometry of [(PPh$_3$)CuC$_2$B$_9$H$_{11}$]$^-$ results from maximum overlap of all three occupied carborane ligand π-orbitals with the appropriate copper acceptor orbitals.

References

[1] A. Zalkin, D.H. Templeton and T.E. Hopkins, *J. Am. Chem. Soc.* **87**, 3988 (1965).

[2] M.F. Hawthorne and R.L. Pilling, *J. Am. Chem. Soc.* **87**, 3987 (1965).

[3] K.P. Callahan and M.F. Hawthorne, *Adv. Organomet. Chem.* **14**, 45 (1976).

[4] N.N. Greenwood in J.C. Bailar, H.J. Emeléus, R. Nyholm and A.F. Trotman-Dickenson (eds.), *Comprehensive Inorganic Chemistry*, Vol. 1, Pergamon Press, Oxford, 1973, chap. 11.

[5] R.N. Grimes (ed.), *Metal Interactions with Boron Clusters*, Plenum Press, New York, 1982.

[6] J.D. Kennedy, *Prog. Inorg. Chem.* **32**, 519 (1984); **34**, 211 (1986).

[7] F. Jiang, T.P. Fehlner and A.L. Rheingold, *J. Chem. Soc., Chem. Commun.*, 1395 (1987).

[8] R. Hoffmann, *Angew. Chem. Int. Ed. Engl.* **21**, 711 (1982). [Nobel Lecture.]

[9] F.G.A. Stone, *Angew. Chem. Int. Ed. Engl.* **23**, 89 (1984).

[10] E.J.M. Hamilton and A.J. Welch, *Polyhedron* **9**, 2407 (1990).

[11] Y. Do, H.C. Kang, C.B. Knobler and M.F. Hawthorne, *Inorg. Chem.* **26**, 2348 (1987).

[12] R.G. Adler and M.F. Hawthorne, *J. Am. Chem. Soc.* **92**, 6174 (1970).

Note Recent advances in polyhedral borane chemistry, including aza- and oxa-metallaboranes, are presented in N.N. Greenwood and J.D. Kennedy, *Pure Appl. Chem.* **63**, 317 (1991).
     Main group heterocarboranes published since 1982 are reviewed in N.S. Hosmane and J.A. Maguire, *Adv. Organomet. Chem.* **30**, 99 (1990).
     A comprehensive review of recent work on boron and carbon clusters is covered in G.A. Olah, K. Wade and R.E. Williams, *Electron Deficient Boron and Carbon Clusters*, Wiley, New York, 1991.

3.12 Borax
Na₂B₄O₅(OH)₄·8H₂O

Crystal Data

Monoclinic, space group $C2/c$ (No. 15)

$a = 11.885(1)$, $b = 10.654(1)$, $c = 12.206(1)$Å, $\beta = 106.623(5)°$

$Z = 4$

Atom*	x	y	z	Atom	x	y	z
Na(1)	0	0	0	O(5)	.1622	.5163	.4895
Na(2)	0	.8469	1/4	H(4)	.7616	.2597	.2807
				H(5)	.1127	.4589	.0379
B₄O₅(OH)₄²⁻					Water molecules		
B(1)	.0852	.3452	.2151	O(6)	.1240	.8463	.4493
B(2)	.0978	.4566	.3918	O(7)	.1233	.0009	.1956
O(1)	0	.2672	1/4	O(8)	.1197	.1647	.4615
O(2)	.1544	.4194	.3146	O(9)	.1171	.7049	.1718
O(3)	.0194	.4346	.1243				
O(4)	.1614	.2712	.1679				

* The water H atoms have been omitted.

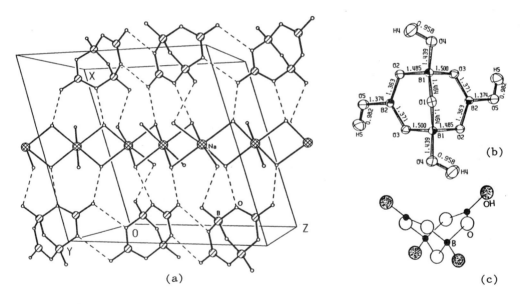

Fig. 3.12-1 (a) Crystal structure of borax. (b) Bond lengths and (c) perspective view of the B₄O₅(OH)₄²⁻ ion. (Adapted from ref. 2).

Crystal Structure

The crystal structure of borax was determined by X-ray diffraction in 1956[1] and by neutron diffraction in 1978.[2] It consists of B₄O₅(OH)₄²⁻ and Na⁺ ions and water molecules.

The isolated boron-oxygen polyanion is composed of two boron-oxygen tetrahedra and two boron-oxygen triangles linked through corners. Four oxygen atoms, each linked to only one boron, belong to the hydroxyl groups. The resulting $B_4O_5(OH)_4^{2-}$ ion (Fig. 3.12-1) has diad rotational symmetry in the crystal; bond lengths as determined by neutron diffraction are shown in Fig. 3.12-1(b).

Adjacent $B_4O_5(OH)_4^{2-}$ anions, related by centers of symmetry, are linked by two equivalent hydrogen bonds O(5)-H...O(3) to form infinite chains approximately parallel to the *c* axis. The sodium ions are octahedrally coordinated by water molecules (the range of Na-O distances is 2.40-2.46Å), and the octahedra share edges to form another set of continuous chains approximately parallel to the *c* axis (Fig. 3.12-2). The links between the two kinds of chains can only be weak hydrogen bonds between the OH groups of $B_4O_5(OH)_4^{2-}$ anions and the ligand water molecules, and this explains the degree of softness of borax, 2 to 2.5 in Mohs' scale. The perfect cleavage {100} and the good cleavage {110} are parallel to these chains.

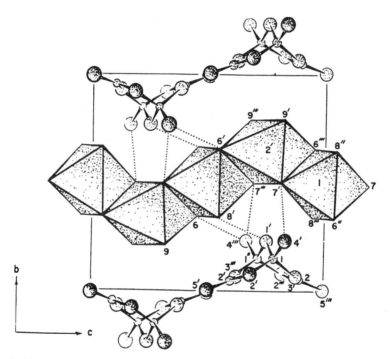

Fig. 3.12-2 Projection of the borax structure on (100). Small open circles, boron; large open circles, O^{2-} or OH^-; Na^+ enclosed within an octahedron of H_2O ligands. Dotted lines indicate some of the hydrogen bonds. (After ref. 1).

The present structure is satisfactory on both crystallographic and chemical grounds, so that there is no justification for retaining the familiar chemical formula, $Na_2B_4O_7 \cdot 10H_2O$, for borax.

Remarks

1. Structural units in borates

The many structural units in borates (Fig. 3.12-3) may be divided into
three kinds:

a) Units containing B in planar BO_3 coordination only

Monomeric triangular $[BO_3]^{3-}$ units are found in the rare-earth
orthoborates $LnBO_3$ and $Mg_3(BO_3)_2$. Binuclear trigonal planar units occur in
the pyroborates $Mg_2B_2O_5$, CoB_2O_5, $Fe_2B_2O_5$. Trinuclear cyclic units occur in
the metaborates $NaBO_2$ and KBO_2, which should therefore be written as $M_3B_3O_6$.
Polynuclear linkage of BO_3 units into infinite chains of stoichiometry BO_2
occur in $Ca(BO_2)_2$, and three-dimensional linkage of planar BO_3 units occurs in
the borosilicate mineral tourmaline and in glassy B_2O_3.

b) Units containing B in tetrahedral BO_4 coordination only

Monomeric tetrahedral BO_4 units are found in the zircon-type compound
$TaBO_4$ and in the minerals $(Ta,Nb)BO_4$ and $Ca_2H_4BAsO_8$. The related tetrahedral
unit $[B(OH)_4]^-$ occurs in $Na_2[B(OH)_4]Cl$ and $Cu[B(OH)_4]Cl$. Binuclear

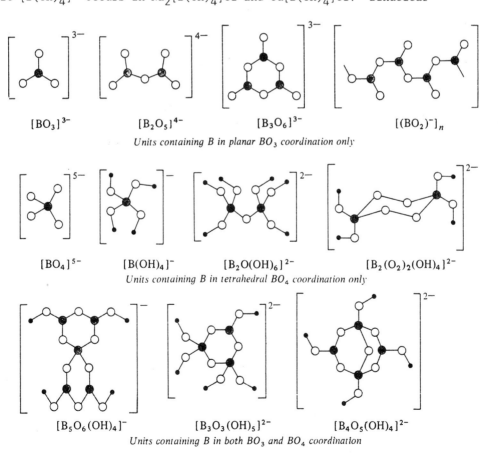

$[BO_3]^{3-}$ $[B_2O_5]^{4-}$ $[B_3O_6]^{3-}$ $[(BO_2)^-]_n$

Units containing B in planar BO_3 coordination only

$[BO_4]^{5-}$ $[B(OH)_4]^-$ $[B_2O(OH)_6]^{2-}$ $[B_2(O_2)_2(OH)_4]^{2-}$

Units containing B in tetrahedral BO_4 coordination only

$[B_5O_6(OH)_4]^-$ $[B_3O_3(OH)_5]^{2-}$ $[B_4O_5(OH)_4]^{2-}$

Units containing B in both BO_3 and BO_4 coordination

Fig. 3.12-3 Some structural units in borates. (After ref. 3).

tetrahedral units have been found in $Mg[B_2O(OH)_6]$, and a cyclic binuclear tetrahedral structure characterizes the peroxoanion $[B_2(O_2)_2(OH)_4]^{2-}$ in $Na_2[B_2(O_2)_2(OH)_4] \cdot 6H_2O$. A more complex polynuclear structure comprising sheets of tetrahedrally coordinated $BO_3(OH)$ units occurs in the borosilicate mineral $CaB(OH)SiO_4$, and a fully three-dimensional polynuclear structure is found in BPO_4, $BAsO_4$, and the minerals $NaBSi_3O_8$ and $Zn_4B_6O_{13}$.

 c) Units containing B in both BO_3 and BO_4 coordination

 The structure of monoclinic HBO_2 contains both planar BO_3 and tetrahedral BO_4 units joined by shared common O atoms. A structure in which the ring has but one BO_4 unit is the spiroanion $[B_5O_6(OH)_4]^-$, which occurs in hydrated potassium pentaborate $KB_5O_8 \cdot 4H_2O$, i.e. $K[B_5O_6(OH)_4] \cdot 2H_2O$. The anhydrous pentaborate KB_5O_8 has the same $[B_5O_6(OH)_4]^-$ structural unit, but dehydration of the OH groups links the spiroanions sideways into ribbon-like helical chains. The mineral $CaB_3O_3(OH)_5 \cdot H_2O$ has two BO_4 units in the six-membered heterocycle $[B_3O_3(OH)_5]^{2-}$, and related chain anions $[B_3O_4(OH)_3^{2-}]_n$ linked by a common O atom are found in the important mineral colemanite $Ca_2B_6O_{11} \cdot 5H_2O$, i.e. $[CaB_3O_4(OH)_3] \cdot H_2O$. It is clear from these examples that, without structural data, the stoichiometry of a borate mineral gives little indication of its constitution.

2. Structural principles

 Several principles have been proposed in connection with the structural chemistry of borates:[4-8]

 a) In borates, the boron combines with oxygen either in trigonal planar three-fold or tetrahedral four-fold coordination. In addition to the mononuclear anions BO_3^{3-} and BO_4^{5-} and their partially fully protonated equivalents, there exists an extensive series of polynuclear anions formed by corner-sharing of triangular and/or tetrahedral boron-oxygen units. The polyanions may form separate boron-oxygen complexes, rings, chains, layers, and frameworks. It is not yet clearly understood which factors determine the coordination number of boron in a given boron-oxygen compound.

 b) In boron-oxygen polynuclear anions, most of the oxygen atoms are linked to one or two boron atoms, and a few can bridge three boron atoms.

 c) In borates, the hydrogen atoms do not attach to boron atoms directly, but always to oxygens which are not shared by two borons, thereby generating hydroxyl groups.

 d) Most of the boron-oxygen rings are six-membered comprising three B atoms and three oxygen atoms.

 e) In a borate crystal there may be two or more different boron-oxygen units. For example, boric acid, $B(OH)_3$,[9] may exist in the isolated form in the presence of more complex polyanions, or such insular groups may polymerize and attach themselves to side chains of more complex polyanions.

3. Compounds of boron with coordination number 2

In the majority of its compounds boron is tricoordinate or tetracoordinate, but can also attain higher coordination numbers as in metal borides, polyboranes, carboranes and metallaboranes. In recent years the chemistry of dicoordinate boron compounds has undergone rapid development, and the classes of compounds known so far are listed in Table 3.12-1.

Table 3.12-1 Some well-characterized two-coordinated boron compounds.

Formula	Compound class	Example*	Fig.	Ref.
X–B=O	boroxane**	Cl–B=O		[10]
R–B≡N–R'	iminoborane	$(Me_3Si)_3Si$–B≡N–tBu	3.12-4(a)	[13]
R_2N=B=CR$_2$'	alkylidene aminoborane (aminomethyleneborane)	iPr_2N=B=C$(SiMe_3)_2$	3.12-4(b)	[16]
$[R_2$N=B=NR$_2$']$^+$	bis(dialkylamino)borinium ion	$[(tmp)$=B=NMe$_2]^+$	3.12-4(c)	[17]
R_2N=B=NR'	amino iminoborane	(tmp)=B=NAr		[14]
R_2N=B=PR'	amino boranylidenephosphane (stablized as metal carbonyl Lewis adduct)	(tmp)=B=PCMe$_3$.Cr(CO)$_5$	3.12-4(d)	[18]

* tmp = 2,2,6,6-tetramethylpiperidino; Ar = 2,4,6-tri-tert-butylphenyl.

** The formula is often written as X–B≡O; borthianes X–B=S and borselenanes X–B=Se are also known.

The boroxanes X–B=O (X = H, F, Cl, Me) and their S and Se analogues are linear, highly reactive molecules which exist only in the gas phase. An infrared diode laser and microwave spectroscopic study of eight isotopic species of Cl–B=O yielded r_s(Cl–B) = 1.68274(19) and r_s(B–O) = 1.20622(21)Å.[10] Comparison of the latter value with the interatomic distance of 1.2045Å in diatomic BO[11] indicates that the molecular structure is more faithfully represented by Cl–B≡O.

The iminoboranes R–B≡N–R'[12] are isoelectronic and isostructural with alkynes R–C≡C–R'; other matching pairs are aminoboranes R_2B=NR$_2$' with alkenes, and amine-boranes X_3B–NR$_3$ with alkanes. For purposes of comparison "typical" single, double and triple BN bond lengths are taken as 1.58, 1.41 and 1.26Å, respectively.[12] In the crystal structure of $(Me_3Si)_3$Si–B≡N–tBu determined at -60°C, the C-N-B-Si backbone is nearly linear (Si-B-N 176.6, B-N-C 177.4°), and the B-N bond distance is 1.221Å [Fig. 3.12-4(a)].[13]

Amino iminoboranes can be described with the use of resonance structures A-C, the boron atom in A and B bearing a formal negative charge. In the literature formula B or D is used to emphasize the linearity and π-electron delocalization of the N-B-N skeleton.[14]

$$R_2\ddot{N}-B\equiv N-R' \longleftrightarrow R_2N=B=\ddot{N}-R' \longleftrightarrow R_2\ddot{N}-B=\ddot{N}-R' \qquad R_2N\dot{=}B\equiv N-R'$$

A   B   C   D

I   II   III

In the crystal structure of (tmp)=B=N-Ar (tmp = 2,2,6,6-tetramethyl-piperidino, Ar = 2,4,6-tri-*tert*-butylphenyl), I, the average BN distance for the two independent molecules is 1.38Å to the tmp group and 1.25Å to the Ar group.[14] The metal carbonyl stabilized adduct II has BN bonds of 1.368Å and

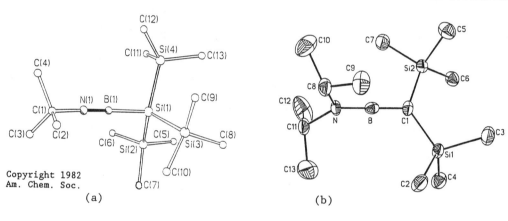

Copyright 1982
Am. Chem. Soc.

(a)

(b)

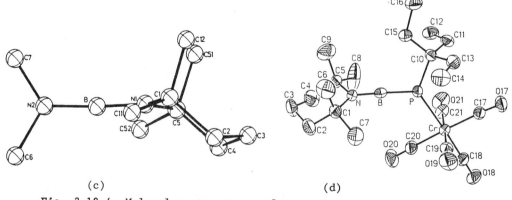

(c)

(d)

Fig. 3.12-4 Molecular structures of some dicoordinate boron compounds: (a) $(Me_3Si)_3Si-B\equiv N-tBu$, (b) $iPr_2N=B=C(SiMe_3)_2$, (c) $[(tmp)=B=NMe_2]^+$ and (d) $(tmp)=B=PCMe_3 \cdot Cr(CO)_5$. (After refs. 13 and 16-18).

1.295Å to the tmp and NC(Cr) groups, respectively.[15] A Lewis-acidic metal halide such as $PdCl_2$ adds two amino iminoborane molucules to form III, in which the tmp units are almost perpendicular to the $PdCl_2N_2$ plane; the two BN. bond lengths are 1.35 and 1.32Å to the tmp and NC(Pd) groups, respectively.[14]

   In the allene-like alkylidene aminoborane shown in Fig. 3.12-4(b), the N=B and B=C bond distances are 1.363 and 1.391Å, respectively.[16] In the analogous allene-like structure of the bis(dialkylamino)boron(1+) ion shown in Fig. 3.12-4(c), the BN bond distances to the tmp and NMe_2 groups are 1.30 and 1.42Å, respectively.[17] Recent work has led to the isolation of an amino boranylidenephosphane, complexed at P, as depicted in Fig. 3.12-4(d).[18] The N=B and B=P bond distances in this adduct are 1.339 and 1.743Å, respectively.

References

[1] N. Morimoto, *Miner. J. (Japan)* **2**, 1 (1956); through *Structure Reports* **20**, 376 (for 1956).

[2] H.A. Levy and G.C. Lisensky, *Acta Crystallogr., Sect. B* **34**, 3502 (1978).

[3] N.N. Greenwood and A. Earnshaw, *Chemistry of the Elements*, Pergamon Press, Oxford, 1986, pp. 231-234.

[4] Zhou Gongdu, *Huaxue Tongbao*, 113 (1960). [In Chinese].

[5] C.L. Christ, *Am. Miner.* **45**, 334 (1960).

[6] J.R. Clark and C.L. Christ, *Am. Miner.* **56**, 1934 (1971).

[7] C.L. Christ and J.R. Clark, *Phys. Chem. Miner.* **2**, 59 (1977).

[8] G. Heller, *Top. Curr. Chem.* **131**, 39 (1986).

[9] M. Gajhede, S. Larsen and S. Rettrup, *Acta Crystallogr., Sect. B* **42**, 545 (1986).

[10] K. Kawaguchi, Y. Endo and E. Hirota, *J. Mol. Spectrosc.* **93**, 38 (1982).

[11] K.P. Huber and G. Herzberg, *Molecular Spectra and Molecular Structure*, Vol. **IV**, Van Nostrand Reinhold, New York, 1979.

[12] P. Paetzold, *Adv. Inorg. Chem.* **31**, 123 (1987); *Pure Appl. Chem.* **63**, 345 (1991).

[13] M. Haase, U. Klingebiel, R. Boese and M. Polk, *Chem. Ber.* **119**, 1117 (1986).

[14] H. Nöth, *Angew. Chem. Int. Ed. Engl.* **27**, 1603 (1988).

[15] H. Nöth, W. Rattay and V. Wietelmann, *Chem. Ber.* **120**, 859 (1987).

[16] R. Boese, P. Paetzold and A. Tapper, *Chem. Ber.* **120**, 1069 (1987).

[17] H. Nöth, R. Staudigl and H.-U. Wagner, *Inorg. Chem.* **21**, 706 (1982).

[18] G. Linti, H. Nöth, K. Polborn and R.T. Paine, *Angew. Chem. Int. Ed. Engl.* **29**, 682 (1990).

3.13   Sodium β–Alumina
$Na_2O \cdot 11Al_2O_3$

Crystal Data

Hexagonal, space group $P6_3/mmc$ (No. 194)

$a = 5.594$, $c = 22.53$Å, $Z = 1$

Atom	Position	x	y	z	Occupancy
Al(1)	12(K)	-.16775	-x	.10630	~ 1
Al(2)	4(f)	1/3	2/3	.02477	~ 1
Al(3)	4(f)	1/3	2/3	.17555	~ 1
Al(4)	2(a)	0	0	0	~ 1
O(1)	12(K)	.15711	-x	.05011	~ 1
O(2)	12(K)	.50318	-x	.14678	~ 1
O(3)	4(f)	2/3	1/3	.05552	~ 1
O(4)	4(e)	0	0	.14253	~ 1
O(5)	2(c)	1/3	2/3	1/4	~ 1
Na(1)	6(h)	-.2938	-x	1/4	0.250
Na(2)	6(h)	-.1269	-x	1/4	0.174

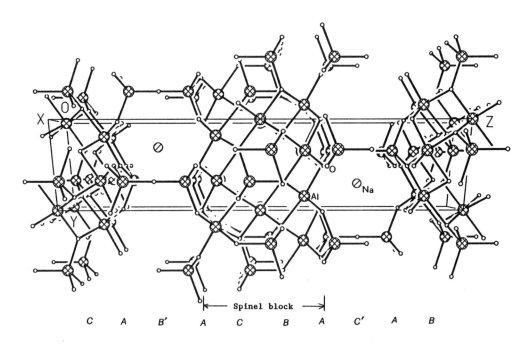

	C	A	B'	A	C	B	A	C'	A	B

Fig. 3.13-1   Crystal structure of β-alumina, $Na_2O \cdot 11Al_2O_3$.

Crystal Structure

The main features of the β-alumina structure were deduced by Bragg, Gottfried and West[1] in 1931; these authors were troubled by the small percentage of sodium found by chemical analysis, which was inconsistent with the space group symmetry, and accordingly suggested a degree of randomness in the structure. In 1937, Beevers and Ross[2] redetermined the structure and arrived at the formula $Na_2O \cdot 11Al_2O_3$ with precisely described ordered positions for sodium and all the other atoms. This formula seemed to agree reasonably well with the analytical data then available and is now referred to as the "ideal formula". In 1971, Peters *et al.*[3] again redetermined the structure of β-alumina, the parameters of which are given in the table above.

The idealized β-alumina structure is shown in Fig. 3.13-1. It consists of spinel-like slices or "spinel blocks", each comprising four closest packed O layers, which are joined through O layers in which the atoms are also in the positions of cubic closest packing but occupy only one quarter of those positions. A unit cell contains two "spinel-blocks", related to each other by the two basal mirror planes at $z = 1/4$ and $z = 3/4$. Each block consists of four *ABCA* oxygen layers; three aluminum atoms are sandwiched between each pair of layers, giving the typical Al_3O_4 alternation of spinel. The aluminum atoms are in both octahedral and tetrahedral interstices of the oxygen atoms. They assume the same set of positions as the combined sets of positions of magnesium and aluminum in $MgAl_2O_4$ spinel. Across the basal mirror planes the two spinel blocks are joined by an Al-O-Al column. This linkage can alternatively be described as two AlO_4 tetrahedra with a common oxygen vertex in the mirror plane, the two sets of three basal oxygen atoms being part of each spinel block. The Al-O-Al "spacer column" keeps the two spinel blocks apart by about 4.8Å.

In β-alumina, the sequence of O layers perpendicular to the hexagonal axis is

   ---B'(ACBA)C'(ABCA)B'---,

where each pair of parentheses encloses the four closest packed layers of one spinel block and the primed symbols refer to the one-quarter filled layers.

The sodium distribution in β-alumina is considerably more complex. The

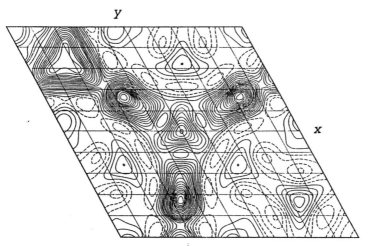

Fig. 3.13-2 Distribution of Na$^+$ ions in the mirror plane for a β-alumina sample. (After ref. 4).

crystal, which Peters *et al.* used for X-ray analysis, was apparently representative in that both neutron activation analysis and X-ray intensity analysis found 29% excess soda relative to the classical formula $Na_2O.11Al_2O_3$, corresponding to a unit cell content of $Na_{2.58}Al_{21.87}O_{34}$. In the averaged unit cell, the sodium ions are smeared out in a complex pattern in the basal mirror plane. Fig. 3.13-2 shows the distribution of Na^+ ions in the mirror plane for a β-alumina sample.

Remarks

1. β-Alumina as a superionic conductor[4]

Superionic conductors are also called solid electrolytes. As a rule, these are defect structures. Some cations in them are very weakly bound to the other atoms; their thermal-vibration amplitudes are so large that they are comparable with the distances between the possible crystallographic positions which these ions may occupy. As a result, some cations can easily migrate through the crystal, this property being reflected in the term "solid electrolyte". Their ionic conductivity σ is comparable with, and for some compounds even higher than that of liquid electrolytes. β-Alumina is a good superionic conductor, in which the carrier Na^+ ions can travel through the crystal along the mirror planes, which lie between the spinel blocks. The migration paths of the Na^+ ions are mapped by the distribution of the Na^+ ions, as shown in Fig. 3.13-2. β-Alumina has been used for the purification of metallic sodium and in the manufacture of sodium-sulfur storage cells.

2. β''-Alumina[5]

β''-Alumina has the approximate stoichiometric formula $Na_2O.5Al_2O_3$ and is usually stabilized by the addition of some Mg and/or Li. It is worth mentioning that a compound designated as β'-alumina later turned out to be β-alumina that is rich in Na_2O.

Structurally β''-alumina (Fig. 3.13-3) is closely related to β-alumina. It is rhombohedral, space group $R3m$, and the triply-primitive hexagonal cell has $a = 5.614$ and $c = 33.85$Å; thus the a axis is about the same as that in β-alumina, but the c axis is 1.5 times as long. Whereas β-alumina has two ~11Å spinel blocks related by a two-fold screw axis parallel to c, β''-alumina has three such blocks related by a three-fold screw axis. The sequence of O layers perpendicular to the trigonal axis is

$$---B'(CABC)A'(BCAB)C'(ABCA)B'---.$$

As in β-alumina, the blocks are spaced
well apart by Al-O-Al columns. The alkali
ions reside in the incompletely filled
basal plane containing the column oxygen
and can diffuse quite freely in this plane.

The actual structures of the blocks of
β- and β"-alumina are somewhat distorted
from this idealized scheme, as the oxygen
layers are not quite planar. In the calcium
β"-alumina of idealized composition
$Ca(Mg,Al)_{11}O_{17}$ the Ca atoms are distributed
along edge-linked hexagonal pathways.[6] In
barium β"-alumina, $Ba_{0.82}Mg_{0.63}Al_{10.37}O_{17}$,
the Ba atoms are non-centrosymmetric at
short-range and occupy predominantly the $6(c)$
sites [Ba(1), $z = 0.16952$, 54.0(4)%; Ba(1'),
$z = -0.17157$, 15.3(5)%], but significant Ba
density [Ba(2) at (1/3, 1/6, 1/6), 4.1(5)%]
also lies on the conduction pathways
between adjacent $6(c)$ sites.[7]

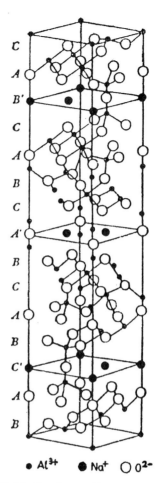

• Al^{3+} ● Na^+ ○ O^{2-}

Fig. 3.13-3 Crystal structure of
β"-alumina, $Na_2O.5Al_2O_3$. (After
ref. 21).

The mobile ion distributions in β"-alumina samples with 23%, 40%, 75%,
and 100% of the Na^+ substituted by Eu^{2+} have been studied.[8] Single-crystal
X-ray structure refinements of the resulting isomorphous materials have shown
that there are two well-defined sites for the mobile cations in the conduction
planes: the tetrahedral site (as Na^+ ion positions shown in Fig. 3.13-3) and
the eight-coordinate mid-oxygen site (in the mid-point between the two
neighbouring oxygen atoms). In Na β"-alumina, the Na^+ ions occupy five-sixths
of the available tetrahedral sites, and conduction occurs by a vacancy
mechanism. The Na^+ distribution is highly disordered at room temperature.[9]
At low Eu^{2+} concentrations, the structure is dominated by Na^+-Eu^{2+} repulsions,
there is short-range ordering among the mobile cations, and the Eu^{2+} ions
occupy poorly defined sites between the eight-coordinate and tetrahedral

sites. As the Eu^{2+} concentration increases, mobile ion-framework and Eu^{2+}-Eu^{2+} interactions become more important, the Eu^{2+} ions occupy well-defined eight-coordination sites, and a long-range ordered arrangement of mobile cations is favoured at low temperature.

3. β'''-Alumina and β''''-alumina

β'''-Alumina, idealized formula $Na_2O.4MgO.15Al_2O_3$, belongs to space group $P6_3/mmc$ with $a = 5.62$ and $c = 31.8$Å.[10] The basal mirror planes are 15.9Å apart, very loosely packed, and contain the sodium ions. Between these planes are spinel-like blocks, consisting of six layers of ccp oxygens with Mg and Al in their interstices. The sequence of O layers is

$$---A'(BCABCA)B'(ACBACB)A'---.$$

The spinel blocks are spaced apart by linear Al-O-Al columns; these spacers, together with the loose packing, provide for an extraordinarily high two-dimensional sodium mobility.

Rhombohedral β''''-alumina, the structure of which consists of three spinel blocks of six oxygen layers each, bears the same relation to β'''-alumina as β''-alumina does to β-alumina.[11]

There exist many analogous and similar structures based on Ga_2O_3 or Fe_2O_3 in place of Al_2O_3 in the β-aluminas.[12]

4. Structure of alums

The alums are a series of double salts of the general formula

$$M^IM^{III}(XO_4)_2.12H_2O$$

in which M^I is a monovalent metal, such as Na^+, K^+, Rb^+, Cs^+, NH_4^+ or Tl^+; M^{III} is a trivalent metal, such as Al^{3+}, Cr^{3+}, Fe^{3+}, Rh^{3+}, In^{3+} or Ga^{3+}; and X is S or Se.[13]

The alums crystallize in the space group $Pa3$ but occur in three distinct types, α, β and γ.[14] There are only a few known γ alums with sodium as the monovalent M^I ion. Most cesium sulfate alums adopt the β structure, but exceptions occur when M^{III} is cobalt, rhodium or iridium, for which the cesium sulfate alums have the α structure. The best known alum, $KAl(SO_4)_2.12H_2O$, belongs to the α type.

The first alum crystal structure was determined by Beevers and Lipson in one of the earliest applications of isomorphous replacement.[15] They measured the structure amplitudes of $KAl(SO_4)_2.12H_2O$, $KCr(SO_4)_2.12H_2O$ and $KAl(SeO_4)_2.12H_2O$, and from their differences were able to derive the phases for all reflections, calculate an electron-density map, and hence locate all the atoms. Since then the structures of a large number of α[16,17], β[18,19] and γ alums[20] have been refined in X-ray and neutron diffraction studies.

In γ alums the sulfate ion is oriented with a single oxygen coordinated to the monovalent ion [Fig. 3.13-4(a)]. The sodium ion is eight-coordinate, being surrounded by six water molecules and two sulfate oxygen atoms. In α and β alums the orientation of the sulfate group is reversed, although sometimes with disorder in α alums, so that the monovalent ion is twelve-coordinate, having as nearest neighbours six water molecules and six sulfate oxygen atoms [Fig. 3.13-4(b)]. In the β alums, the $M^I(H_2O)_6$ unit is planar and oriented normal to the threefold axis of the unit cell, so that all $O-M^I-O$ angles within it are exactly 60°. In the α alums, the six water molecules are arranged in a crown configuration about M^I with $O-M^I-O$ angles increased to 64.9-66.6°. In the cesium α alums the water molecules in the crown configuration approach the Cs^+ ion about 0.12-0.13Å closer than in the planar geometry of the β alums. Conversely the distance from Cs^+ to a sulfate oxygen atom is about 0.3Å longer in the α type as compared to the β type. There are no significant differences in the $M^{III}-O$ bond lengths between the sulfate and selenate alums. This indicates that the difference between the α and β alum structures does not perturb the M^{III} coordination sphere. Table 3.13-1 gives some examples and summarizes the structural features of the three types of alums.

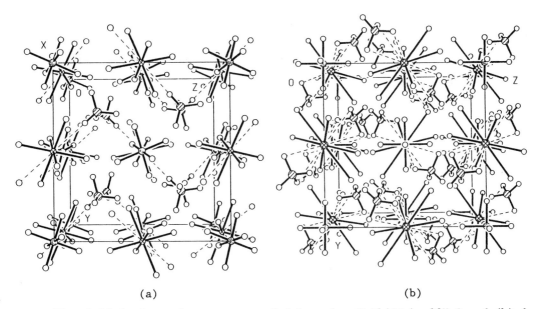

(a)   (b)

Fig. 3.13-4 Crystal structures of (a) γ alum NaAl(SO₄)₂.12H₂O and (b) β alum CsAl(SO₄)₂.12H₂O. For clarity the bonds between the M^I atoms and the sulfate groups are represented by broken lines.

Table 3.13-1 Structural types of alums.

Type	Example	Structural features
α	$KAl(SO_4)_2 \cdot 12H_2O$ $CsFe(SeO_4)_2 \cdot 12H_2O$ $NaCr(SO_4)_2 \cdot 12H_2O$	The SO_4^{2-} group lies on a threefold axis of the cubic unit cell and is oriented with three oxygen atoms coordinated to the M^I atom. M^I is 12-coordinate and surrounded by six sulfate oxygen atoms and six H_2O; the water molecules occur in a crown configuration with all $O-M^I-O$ angles > 60°.
β	$CsAl(SO_4)_2 \cdot 12H_2O$ $CsFe(SO_4)_2 \cdot 12H_2O$ $CsIn(SO_4)_2 \cdot 12H_2O$	The SO_4^{2-} group lies on a threefold axis of the cubic unit cell and is oriented with three oxygen atoms coordinated to the M^I atom. M^I is 12-coordinate and surrounded by six sulfate oxygen atoms and six H_2O; the water molecules are coplanar with M^I so that all $O-M^I-O$ angles are exactly 60°.
γ	$NaAl(SO_4)_2 \cdot 12H_2O$	The SO_4^{2-} group lies on a threefold axis of the cubic unit cell and is oriented with a single oxygen coordinated to the M^I atom. M^I is 8-coordinate and surrounded by two sulfate oxygen atoms and six H_2O.

References

[1] W.L. Bragg, C. Gottfried and J. West, *Z. Kristallogr.* **77**, 255 (1931).

[2] C.A. Beevers and M.A. Ross, *Z. Kristallogr.* **97**, 59 (1937).

[3] C.R. Peters, M. Bettman, J.W. Moore and M.D. Glick, *Acta Crystallogr., Sect. B* **27**, 1826 (1971).

[4] W.L. Roth, F. Reidinger and S. LaPlaca, *Superionic Conductors*, Plenum Press, New York, 1976.

[5] M. Bettman and C.R. Peters, *J. Phys. Chem.* **73**, 1774 (1969).

[6] M. Aldén, J.O. Thomas and G.C. Farrington, *Acta Crystallogr., Sect. C* **40**, 1763 (1984).

[7] J.O. Thomas, M. Aldén, G.J. McIntyre and G.C. Farrington, *Acta Crystallogr., Sect. B* **40**, 208 (1984).

[8] M.A. Saltzberg, J.O. Thomas and G.C. Farrington, *Chem. Mater.* **1**, 19 (1989).

[9] K.G. Frase, J.O. Thomas and G.C. Farrington, *Solid State Ionics*, **9**, 307 (1983).

[10] M. Bettman and L.L. Terner, *Inorg. Chem.* **10**, 1442 (1971).

[11] N. Weber and A. Venero, Paper 1-JV-70. Proc. 72nd Annu. Meet. Am. Ceram. Soc., Philadelphia, 1970.

[12] Y. Matsui, Y. Bando, Y. Kitami and R.S. Roth, *Acta Crystallogr., Sect. B* **41**, 27 (1985), and references cited therein.

[13] S. Haussühl, *Z. Kristallogr.* **116**, 371 (1961).

[14] H. Lipson, *Proc. Roy. Soc. London* **A151**, 347 (1935).

[15] C.A. Beevers and H. Lipson, *Nature (London)* **134**, 327 (1934).

[16] A.C. Larson and D.T. Cromer, *Acta Crystallogr.* **22**, 793 (1967).

[17] R.S. Armstrong, J.K. Beattie, S.P. Best, G.P. Braithwaite, P.D. Favero, B.W. Skelton and A.H. White, *Aust. J. Chem.* **43**, 393 (1990).

[18] D.T. Cromer, M.I. Kay and A.C. Larson, *Acta Crystallogr.* **21**, 383 (1966).

[19] J.K. Beattie, S.P. Best, B.W. Skelton and A.H. White, *J. Chem. Soc. Dalton Trans.*, 2105 (1981).

[20] D.T. Cromer, M.I. Kay and A.C. Larson, *Acta Crystallogr.* **22**, 182 (1967).

[21] R. Collongues, J. Thery and J.P. Boilot in P. Hagenmüller and W. van Gool (eds.), *Solid Electrolytes*, Academic Press, New York, 1978.

Note The latest refinement of the structure of sodium β-alumina and subsequent Fourier calculations are reported in K. Edström, J.O. Thomas and G.C. Farrington, *Acta Crystallogr., Sect. B* **47**, 210 (1991).

The Na$^+$ ions in sodium β-alumina can be wholly or partially exchanged for other monovalent and a few divalent cations. The structure of Ag$^+$ β-alumina is reported in J.P. Boilot, P. Colomban, R. Collongues, G. Collin and R. Comès, *J. Phy. Chem. Solids* **41**, 253 (1980). Cd^{2+} is the only divalent ion that can replace Na$^+$ to produce a range of crystals with variable vacancy densities corresponding to the approximate formula Na$_{1.22-y}$Cd$_{y/2}$Al$_{11}$O$_{17.11}$, as reported in P.H. Sutter, L. Cratty, M. Salzberg and D.C. Farrington, *Solid State Ionics* **9-10**, 295 (1983).

Structural aspects of the ion-exchange process Na$^+ \longrightarrow$ Cd^{2+} in Na$^+$ β-alumina and Cd$^{2+} \longrightarrow$ Ag$^+$ in Ag$^+$ β-alumina are described in K. Edström, J.O. Thomas and G.C. Farrington, *Acta Crystallogr., Sect. B* **47**, 635 (1991); *ibid.*, p. 643.

3.14 Gallium(I) Tetrachlorogallate(III) ("Gallium Dichloride")

$$Ga(GaCl_4)$$

Crystal Data

Orthorhombic, space group *Pnna* (No. 52)

$a = 7.24(2)$, $b = 9.72(2)$, $c = 9.50(2)$Å, $Z = 4$

Atom	x	y	z
Ga(I)	.681	1/4	1/4
Ga(III)	1/4	0	.183
Cl(1)	.339	.174	.054
Cl(2)	.010	.048	.315

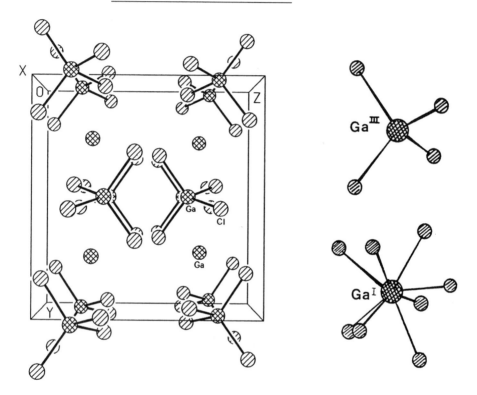

Fig. 3.14-1 Crystal structure of $Ga(GaCl_4)$ and coordination geometries about the Ga^{III} and Ga^I ions in crystalline $Ga(GaCl_4)$. (Adapted from ref. 3).

Crystal Structure

The crystal structure of gallium dichloride consists of tetrahedral $(GaCl_4)^-$ ions and Ga^I cations.[1] The length of the Ga-Cl bond in the $(GaCl_4)^-$ tetrahedron is 2.19Å, which may be compared with the length of 2.22Å for the Ga-Cl bonds in the vapour of Ga_2Cl_6 as determined by electron diffraction.[2]

In the crystal structure a kinked row of Ga^{III} atoms ringed by Cl atoms runs parallel to the *b* axis. "Chains" of this kind are centered on

(1/4,0,1/4) and (3/4,0,3/4). The Ga^I cations lie between these chains in similar kinked rows centered on (1/4,0,3/4) and (3/4,0,1/4). Each Ga^I ion is surrounded by eight Cl atoms, from six different tetrahedra, at the corners of an irregular dodecahedron. Four Cl atoms are at 3.19Å, two at 3.15Å, and the other two at 3.27Å from the Ga^I ion. The shortest Cl...Cl distance in the coordination polyhedron is 3.8Å. Fig. 3.14-1 shows the coordination polyhedra of the Ga^{III} and Ga^I ions. The diamagnetism of "gallium dichloride" is consistent with the present structural formulation.

Remarks

1. Mixed-valence Ga, In and Tl halides[3]

Gallium, indium and thallium can all occur in univalent and trivalent oxidation states in their compounds. Although this simple generalization at first appears to be violated by the existence of several halides of the composition MX_2, the possibility that they are M^I,M^{III} mixed-valence compounds rather than compounds of M^{II} has been experimentally substantiated.

The halides $GaBr_2$, GaI_2; InF_2, $InCl_2$, $InBr_2$; $TlCl_2$, $TlBr_2$ have all been prepared and shown to be diamagnetic. By analogy to $Ga^I(Ga^{III}Cl_4)$, these halides can be formulated as $M^I(M^{III}X_4)$. In the type of $M'^I(M^{III}X_4)$ compounds, $In^I(Al^{III}Cl_4)$, $In^I(Ga^{III}Cl_4)$ and $Tl^I(In^{III}Cl_4)$ are known.

M_2X_3 type compounds, such as Ga_2I_3, In_2Cl_3, Tl_2Cl_3, and Tl_2Br_3 are all mixed-valence materials with the general formula $M^I_3(M^{III}X_6)$: namely $Ga^I_3(Ga^{III}I_6)$, $In^I_3(In^{III}Cl_6)$, $Tl^I_3(Tl^{III}Cl_6)$ and $Tl^I_3(Tl^{III}Br_6)$, respectively. For Tl_3I_4, the formula $Tl^I_5(Tl^{III}I_8)$ seems more appropriate than $Tl^I_2(Tl^{III}I_4)$. The intensely colored, diamagnetic compound Ga_3I_5 is a mixed-valence complex with the formula $Ga^I_2(Ga^{III}I_5)$. In_4Br_6, Tl_4Cl_6 and In_4Cl_7 are all I,III mixed-valence compounds.

Some related complexes such as $(Ga^IL_4)(Ga^{III}Cl_4)$ with a wide range of N,As,O,S, and Se donors have been prepared.

2. Crystal structures of In_4Se_3 and Tl_4S_3[4]

The chalcogenides of Ga, In and Tl are numerous and have been extensively studied not only because of their intriguing stoichiometries, but also because many of them are semiconductors, semi-metals, photoconductors, or light emitters. These compounds, as expected from the positions of their component atoms in the Periodic Table, are far from ionic, but the formal oxidation states remain a useful device for electron counting and for checking the overall charge balance. Many of these are mixed-valence compounds. The structures of In_4Se_3[5] and Tl_4S_3[6] have been determined.

The crystal structure of In_4Se_3 can be regarded to a first approximation

as $In^I(In_3)^V(Se^{-II})_3$, but the compound does not really comprise discrete ions.
The triatomic unit $(In^{III}-In^{III}-In^{III})$, having a formal charge of +5, is bent,
the angle at the central atom being 158° and the In-In distances 2.79Å (*cf.*
3.24-3.26Å in metallic In). However, it is also possible to discern non-
planar five-membered heterocycles in the structure formed by joining two In
atoms from one (In_3) unit to the terminal In atom of an adjacent (In_3) unit
via two bridging Se atoms, so that the structure can be represented
schematically as in Fig. 3.14-2. The In^{III}-Se distances average 2.69Å as
compared with the closest In^I-Se contact of 2.97Å.

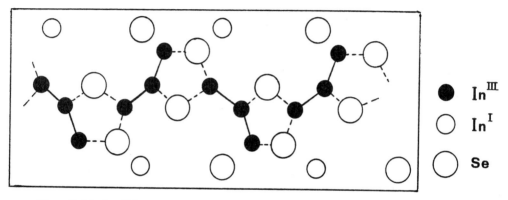

Fig. 3.14-2 Schematic structure of In_4Se_3. (Adapted from ref. 4).

The compound Tl_4S_3, which has the same stoichiometry as In_4Se_3, exhibits
a different structure (Fig. 3.14-3). In the crystal the chains of corner-
shared $(Tl^{III}S_4)$ tetrahedra of overall stoichiometry (TlS_3) are bound together
by Tl^I ions; within the chain the Tl^{III}-S distance is 2.54Å, whereas the Tl^I-S
distances vary between 2.90-3.36Å.

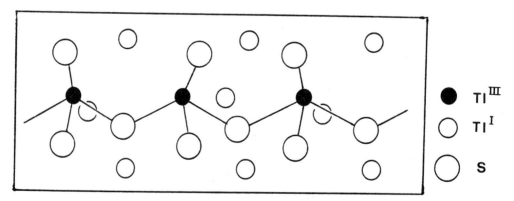

Fig. 3.14-3 Schematic structure of Tl_4S_3. (Adapted from ref. 4).

3. Compounds of the type $(L_nMEL'_n)_x$ (M = Ga, In; E = P, As, Sb)

These compounds feature strong σ-bonding between the group III (M) and V (E) elements, together with substituents that are capable of facile thermal elimination. They are potentially useful as single-source presursors to the production of gallium arsenide and related semiconductor materials.[7]

The complexes of 1:1 stoichiometry are trimeric in cases where the ligands L and L' are of moderate size, while sterically demanding ligands results in dimer formation. The available structural data for these compounds have been summarized by Cowley and Jones, as shown in Table 3.14-1.[7] Representative examples of the cyclic dimeric and trimeric structures are illustrated in Fig. 3.14-4.

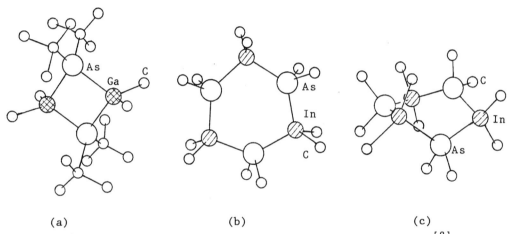

(a)   (b)   (c)

Fig. 3.14-4 Molecular structures of (a) [Me₂GaAs(*t*-Bu)₂]₂,[8] (b) approximately planar and (c) puckered conformations of [Me₂InAsMe₂]₃.[9] (After refs. 8 and 9).

The cyclic III/V compounds $(L_nMEL_n')_x$ exhibit several structural features (Table 3.41-1):

a) Both the M and E atoms adopt approximately tetrahedral coordination geometries.

b) The M_2E_2 rings are planar or nearly so, and the M_3E_3 rings take several conformations that are close in energy.

c) In all dimeric structures the endocyclic bond angle at E is larger than that at M, while in the M_3E_3 rings there is a wide variation in the endocyclic angles as expected from their conformational flexibility.

d) The average endocyclic M-E bond lengths in the trimers are shorter than those in the corresponding dimers, which in turn are shorter than those found in the corresponding binary compounds GaP (2.360Å), GaAs (2.448Å), InP (2.541Å), InAs (2.614Å) and InSb (2.805Å).[10] Exocyclic M-E bonds, if present, are always shorter than those within the cyclic rings.

Table 3.14-1 Structural data for some $(L_nMEL_n')_x$ (x = 2, 3) compounds.*

Compounds	Conformation	M-E	M-E-M	E-M-E	L-M-L	L'-E-L'
$[Me_2GaP(t-Bu)_2]_2$	planar	2.474(5)	93.5(1)	86.5(1)	106.3(5)	111.1(5)
$[(n-Bu)_2GaP(t-Bu)_2]_2$	planar	2.476(5)	93.3(2)	86.7(2)	107.0(6)	109.5(7)
$[Et_2InP(t-Bu)_2]_2$	planar	2.635(2)	94.4(1)	85.6(1)	-	112.7(6)
$[Ga\{As(CH_2SiMe_3)_2\}_3]_2$**	puckered	2.555(4)	95.29(4)	83.81(4)	118.02(5)	103.8(4)
$[Me_2GaAs(t-Bu)_2]_2$	planar	2.549(1)	95.69(2)	84.31(2)	109.3(3)	110.3(2)
$[(n-Bu)_2GaAs(t-Bu)_2]_2$	planar	2.552(3)	95.07(8)	84.92(8)	110.9(7)	109.0(7)
$[(t-Bu)_2GaPh_2]_3$	planar	2.439(3)	138.5(4)	101.5(3)	122.0(7)	99.0(1)
$[Me_2GaAs(i-Pr)_2]_3$§	boat	2.517(2)	123.30(6)	105.71(6)	115.0(6)	100.6(5)
$[Br_2GaAs(CH_2SiMe_3)_2]_3$	twist-boat	2.450(2)	106.67(7)	117.40(8)	112.29(9)	108.4(5)
$[MeGaSb(t-Bu)_2]_3$	twist-boat	2.720(2)	121.74(5)	106.37(5)	115.8(7)	104.3(5)
$[Cl_2GaSb(t-Bu)_2]_3$	irregular boat	2.661(9)	112.11(9)	118.6(1)	108.0(4)	110.0(1)
$[Me_2InAsMe_2]_3$†	planar	2.679(2)	134.26(8)	105.18(7)	123.0(1)	99.0(1)
	puckered	2.669(2)	124.69(9)	96.86(8)	124.0(1)	100.0(1)
$[Me_2InSb(t-Bu)_2]_3$	twist-boat	2.855(2)	121.64(6)	106.95(5)	117.0(1)	107.1(8)

*The average values of the bond lengths (Å) and bond angles (°) are given. **Weakly-bonded dimer.
§Anti-isomer. †Planar and puckered forms are present in same asymmetric unit.

References

[1] G. Garton and H.M. Powell, *J. Inorg. Nucl. Chem.* **4**, 84 (1957).

[2] D.P. Stevenson and V. Schomaker, *J. Am. Chem. Soc.* **64**, 2514 (1942).

[3] M.B. Robin and P. Day, *Adv. Inorg. Chem. Radiochem.* **10**, 248 (1967).

[4] N.N. Greenwood and A. Earnshaw, *Chemistry of the Elements*, Pergamon press, Oxford, 1986, chap. 7.

[5] J.H.C. Hogg, H.H. Sutherland and D.J. Williams, *J. Chem. Soc., Chem. Commun.*, 1568 (1971).

[6] B. Leclerc and M. Bailly, *Acta Crystallogr., Sect. B* **29**, 2334 (1973).

[7] A.H. Cowley and R.A. Jones, *Angew. Chem. Int. Ed. Engl.* **28**, 1208 (1989).

[8] A.M. Arif, B.L. Benac, A.H. Cowley, R.L. Geerts, R.A. Jones, K.B. Kidd, J.M. Power and S.T. Schwab, *J. Chem. Soc., Chem. Commun.*, 1543 (1986).

[9] A.H. Cowley, R.A. Jones, K.B. Kidd, C.M. Nunn and D.L. Westmoreland, *J. Organomet. Chem.* **341**, C1 (1988).

[10] R.W.G. Wyckoff, *Crystal Structures*, 2nd ed., Vol. 1, Wiley-Interscience, New York, 1963, p. 110.

Note The compounds In^ICl, In_2Cl_3 ($In_3^IIn^{III}Cl_6$), In_5Cl_9 ($In_3^IIn_2^{III}Cl_9$), $InCl_3$ and In_7Cl_9 have been unambiguously charaterized in the indium-chlorine system. See H.P. Beck and D. Wilhelm, *Angew. Chem. Int. Ed. Engl.* **30**, 824 (1991).

3.15 Calcium Carbide
CaC$_2$

Crystal Data

Tetragonal, space group $I4/mmm$ (No. 139)

$a = 3.89(1)$, $c = 6.38(1)$Å, $Z = 2$

Atom	x	y	z
Ca	0	0	0
C	0	0	.406

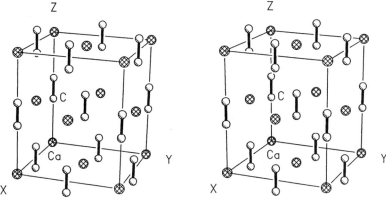

Fig. 3.15-1 Stereoview of the crystal structure of CaC$_2$. The non-primitive face-centred tetragonal unit cell with $a' = 5.50$, $c = 6.38$Å, and $Z = 4$ shows its relationship to the NaCl structure.

Crystal Structure[1]

Calcium carbide has been in the past the major source of ethylene for the chemical industry and for oxy-acetylene welding. It has at least four modifications whose occurrence depends on both the temperature of formation and the impurities that are present. CaC$_2$ I, a tetragonal crystalline phase which is stable between 298K and 720K, is the common form occurring in commercial calcium carbide. The crystal consists of a close packing of Ca^{2+} and C$_2{}^{2-}$ ions (Fig. 3.15-1). The C$_2{}^{2-}$ group lies on a C_4 axis, the C-C bond distance being 1.20Å, and the significant Ca...C contacts are 2.59 and 2.82Å. The structure may also be described as a distorted NaCl-type arrangement of Ca^{2+} and C$_2{}^{2-}$ ions, the symmetry being lowered from cubic to tetragonal owing to the parallel alignment of the "dumbbell" anions. The coordination number of the Ca^{2+} ion is ten.

Calcium carbide has many of the properties associated with ionic compounds. It forms colourless, transparent crystals which at ordinary temperatures do not conduct electricity. It is decomposed by water or dilute acids, and since the anion then becomes unstable, acetylene is evolved.

Remarks

1. The modifications of CaC_2

CaC_2 I is stable at room temperature; below 273K CaC_2 II is said to be the stable modification. CaC_2 IV is stable above 720K, and can transform to CaC_2 III, when it is kept pure and below 720K. The structures of these forms have been determined, the relevant crystal data being summarized in Table 3.15-1.

Table 3.15-1   Crystal data of CaC_2 II-IV.[2]

Modification	II	III	IV
system	triclinic	monoclinic	cubic
cell parameters	a = 8.42(2)Å b = 11.84(3) c = 3.94(1) α = 93.4(5)° β = 92.5(5) γ = 89.9(5)	a = 8.36(2)Å b = 4.20(1) c = 11.25(3) β = 96.3(5)°	a = 5.889(5)Å
space group*	C1	B2$_1$/c	Fm3
z	8	8	4
C-C distance/Å	1.14	1.24	1.20
C.N. of Ca^{2+}	7	8	6

* For CaC_2 II space group C1 has been selected to show the resemblance to the cubic superstructure with a $\overset{\sim}{=}$ 5.9Å. CaC_2 IV could be isomorphous with pyrite (FeS_2).

2. Compounds isomorphous with CaC_2 I.

There are many compounds isomorphous with tetragonal CaC_2. Acetylides of the alkaline and rare earths, silicides of molybdenum and tungsten, and peroxides of both alkalis and alkaline earths have structures of this type. Table 3.15-2 lists some crystals with the tetragonal CaC_2 arrangement.

Table 3.15-2   Crystals with the tetragonal CaC_2 structure.[3]

BaC_2	LaC_2	SrC_2	BaO_2	$MoSi_2$
CeC_2	LuC_2	TbC_2	CaO_2	$(Na,Al)Si_2$
DyC_2	MgC_2	TmC_2	CsO_2	WSi_2
ErC_2	NdC_2	UC_2	KO_2	
GdC_2	PrC_2	YC_2	RbO_2	
HoC_2	SmC_2	YbC_2	SnO_2	

In O_2^{2-}, the O-O distance is 1.49Å, similar to that determined in H_2O_2 and considerably greater than the value of 1.28Å in O_2^-.

3. Structure of ThC_2

The crystal structure of ThC_2 has been determined by neutron diffraction.[6] The space group is $C2/c$, and the unit parameters are: a = 6.53, b = 4.24, c = 6.56Å, β = 104°, and Z = 4.

The structure of ThC_2 is very similar to the CaC_2 I structure but of lower symmetry. The similarity to the NaCl structure is not readily seen from the conventional diagram showing a unit cell of the structure, but can be recognized from Fig. 3.15-2, where the relation to the monoclinic axes is indicated.

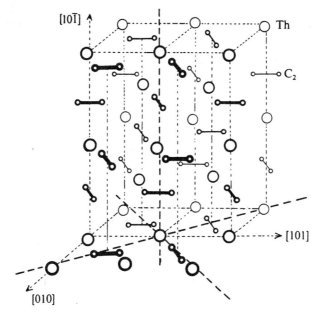

Fig. 3.15-2 Crystal structure of ThC_2. The heavy broken lines are edges of an orthogonal NaCl-type cell with dimensions: 11.7, 11.7 and 10.3Å. (After ref. 3).

4. The C-C bond distance and carbide-metal interactions in MC_2 and $M_4(C_2)_3$[3]

An interesting feature of the MC_2 and $M_4(C_2)_3$ compounds is the variation in the C-C distance as measured by neutron diffraction. Typical values (in Å) are:

CaC_2	1.192	$La_4(C_2)_3$	1.236
YC_2	1.275	$Ce_4(C_2)_3$	1.276
CeC_2	1.283	$U_4(C_2)_3$	1.295
LaC_2	1.303		
UC_2	1.350		

The C-C distance in CaC_2 is close to that in acetylene (1.205Å), and it has been suggested that the observed increase in the lanthanide and actinide carbides results from a partial localization of the supernumerary electron in the antibonding orbital of the $[C{\equiv}C]^{2-}$ ion. The effect is more noticeable in the sesquicarbides than in the dicarbides.

5. Heavy metal acetylides

Acetylene reacts with certain heavy metal ions, chiefly Ag^+ and Cu^+, to form insoluble acetylides, Cu_2C_2 and Ag_2C_2. Both are explosive when dry; owing to the lack of success in preparing single crystals, their structures remain unknown. However, crystalline $Ag_2C_2.6AgNO_3$ has been prepared and subjected to crystal structure analysis.[4] $Ag_2C_2.6AgNO_3$ is rhombohedral, space group $R\bar{3}$, with $a = 7.99$Å, $\alpha = 106.2°$, and $Z = 1$. Photoelectron spectroscopy has confirmed the presence of two kinds of silver atoms: Ag(1) in a general position and Ag(2) lying on a $\bar{3}$ axis. In the resulting cationic cage the C atom occupies a general position close to the inversion center, so that the C≡C group is disordered over three orientations (Fig. 3.15-3). The crystal structure consists of a packing of $[Ag_8C_2]^{6+}$ and NO_3^- ions.

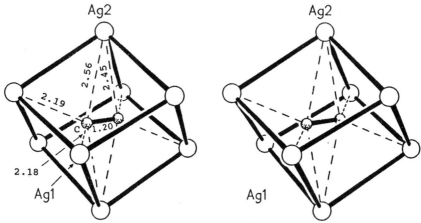

Fig. 3.15-3 Stereoview of the $[Ag_8C_2]^{6+}$ ion showing the C≡C group in one possible oriention. The shorter Ag...C contacts are in Å.

In $Ag_2C_2.6AgNO_3$ and other compounds with acetylene side-on coordinated to a transition metal atom, the orbital interactions may be described qualitatively as shown in Fig. 3.15-4.[5] Subscripts || and ⊥ indicate that the orbitals concerned are parallel and perpendicular, respectively, to the $M-C_2$ plane. The extent of overlap decreases in the order (a) > (b) > (c) > (d). Interaction (a) is bonding, (b) is normally bonding since most transition metals have d electrons which may occupy $d\pi_{||}$ orbitals. In early transition metals with vacant $d\pi_\perp$ orbitals (c) is bonding, but in the later transition metal complexes, especially those with d^{10} configuration, the interaction should be repulsive and antibonding. Interaction (d) has a negligible effect because of poor overlap.

More recently C_2 units were found in many ternary carbides of the rare-earths with transition metals.[8]

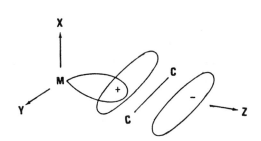

(a) Vacant metal σ (combination of s, p$_z$, and d$_z$2) with filled ligand $\pi_{||}$

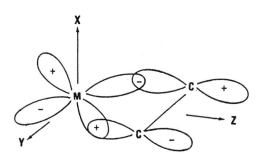

(b) Filled metal π (combination of d$_{yz}$, and p$_y$) with vacant ligand $\pi_{||}^{*}$

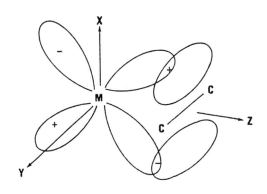

(c) Vacant or filled metal π (combination of d$_{xz}$, and p$_x$) with filled ligand $\pi_{\perp}$

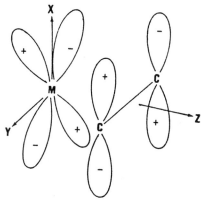

(d) Filled or vacant metal δ (d$_{xy}$) with vacant ligand $\pi_{\perp}^{*}$

Fig. 3.15-4 The MO overlap of the side-on coordination of C$_2^{2-}$ to a transition metal atom M. (After ref. 7).

6. Crystal structure of Sc$_3$C$_4$

The carbide Sc$_3$C$_4$ crystallizes in space group *P4/mnc* with a = 7.4873(5), c = 15.026(2)Å and Z = 10.[9] The unit cell contains eight C$_3$ units, two C$_2$ pairs and twelve isolated C atoms [Fig. 3.15-5(a)]; the C-C bond lengths in the nearly linear (176°) C$_3$ units (2 x 1.34Å) and in the C$_2$ pairs (1.25Å) indicate double and triple bond characters, respectively. The atomic layers **ABC** and **DEF** are of the distorted NaCl type and adjacent layers are connected by the C$_3$ units; the lattice parameters of the Sc$_3$C$_4$ and NaCl structures are related by $a(Sc_3C_4) \sim (5/2)^{\frac{1}{2}}a(NaCl)$ and $c(Sc_3C_4) \sim 3a(NaCl)$ [Fig. 3.15-5(b)].

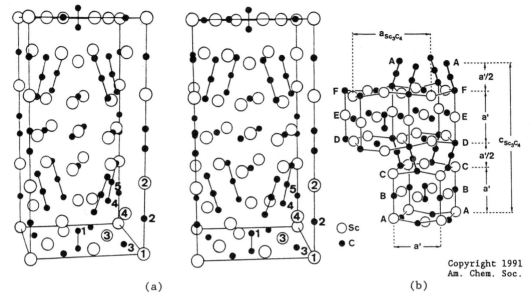

(a)   (b)

Fig. 3.15-5 Crystal structure of Sc_3C_4: (a) stereoview and (b) cutout emphasizing NaCl-like layers (lattice parameter a') connected by C_3 units. (After ref. 9).

References

[1] U. Dehlinger and R. Glocker, *Z. Kristallogr.* **64**, 296 (1926).

[2] N.G. Vannerberg, *Acta Chem. Scand.* **15**, 769 (1961); **16**, 1212 (1962).

[3] A.F. Wells, *Structural Inorganic Chemistry*, 5th ed., Clarendon Press, Oxford, 1984.

[4] Jin Xianglin, Zhou Gongdu, Wu Nianzu, Tang Youqi and Huang Haochun, *Acta Chimica Sinica* **48**, 232 (1990). [In Chinese.]

[5] P.M. Maitlis, *Organic Chemistry of Palladium*, Vol. **1**, Academic Press, New York, 1974, pp. 123-130.

[6] E.B. Hunt and R.E. Rundle, *J. Am. Chem. Soc.* **73**, 4777 (1951).

[7] T.C.W. Mak, K.Y. Hui, O.W. Lau and W.-K. Li, *Problems in Inorganic and Structural Chemistry*, The Chinese University Press, Hong Kong, 1982.

[8] R.-D. Hoffmann, W. Jeitschko and L. Boonk, *Chem. Mater.* **1**, 580 (1989).

[9] R. Pöttgen and W. Jeitschko, *Inorg. Chem.* **30**, 427 (1991).

3.16   α–Oxalic Acid Dihydrate

$$H_2C_2O_4 \cdot 2H_2O$$

Crystal Data

Monoclinic, space group $P2_1/n$ (No. 14)

$a = 6.0968(7)$, $b = 3.4975(4)$, $c = 11.9462(15)$Å, $\beta = 105.78(1)°$

(100K, X-ray diffraction)[1] $Z = 2$

Atom	x	y	z
C(1)	-.04507	.05869	.05196
O(1)	.08539	-.05636	.15013
O(2)	-.22146	.24248	.03627
O(3)	-.45139	.63112	.17877
H(1)	.0332	.0299	.2137
H(2)	-.5659	.6806	.1159
H(3)	-.3714	.4731	.1563

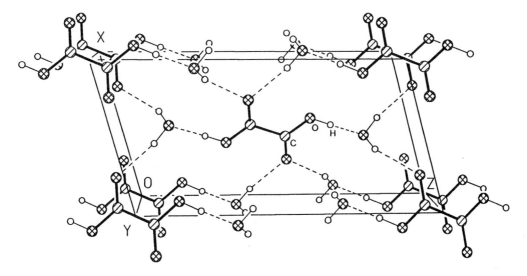

Fig. 3.16-1   Crystal structure of α-oxalic acid dihydrate.

Crystal Structure

Oxalic acid is the simplest dicarboxylic acid and an important reagent. The structure of α-oxalic acid dihydrate has been extensively studied by X-ray and neutron diffraction methods from the 1930's to the 1980's.[1-7]   The crystal consists of oxalic acid molecules and water molecules; bond lengths (in Å) and angles are given in Fig. 3.16-2.   The oxalic acid molecule is located at a crystallographic center of symmetry and all atoms, including the two hydrogens, are found to be coplanar.

In α-oxalic acid dihydrate, the oxygen atom of the water molecule is an acceptor for a strong hydrogen bond [O(1)-H(1)...O(3") = 2.487Å] with the hydroxyl oxygen of the oxalic acid molecule.   The hydrogen atom is completely

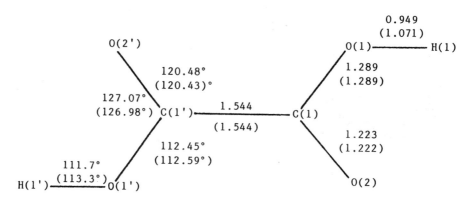

Fig. 3.16-2 Measured dimensions of the oxalic acid molecule. The 100K
neutron diffraction data are in parentheses. Symmetry transformations:
' - x, - y, - z; " -1/2 - x, -1/2 + y, 1/2 - z; ''' -1 - x, 1 - y, - z.[1]

associated with the hydroxyl oxygen atom as no evidence was found for any
disordering of the hydrogen position or the existence of a H_3O^+ ion. The
O(1)-H(1)...O(3") angle of 174.6° indicates that the hydrogen atom lies very
close to the line joining O(1) and O(3"). The weaker hydrogen bonds [O(3)-
H(2)...O(2"') = 2.826Å and O(3)-H(3)...O(2) = 2.830Å] are far from linear
(169.9° and 153.8° at H, respectively) and form closed quadrilaterals which
connect adjacent oxalic acid molecules. Infinite helical chains of hydrogen
bonds extend parallel to the b axis. The sequence of the chain is ...H(3")-
O(3")...H(1)-O(1)-C(1)-O(2)...H(3)-O(3)...H(1")-O(1")-C(1")-O(2")..., where
O(2") is hydrogen-bonded to a hydrogen atom displaced one unit cell up the b
axis from the starting point, H(3").

Remarks

1. The deformation density in oxalic acid

At the IUCr Conference in 1975 an international project was proposed,
under which a number of laboratories would study the charge density in one
selected crystalline material in order to assess the reproducibility of
experimental electron density maps. Oxalic acid dihydrate was the unanimous
choice since it is a relatively hard, stable substance of which good crystal
can be easily grown.

Four X-ray and five neutron data sets were analyzed by comparison of
thermal parameters, positional parameters and X-N electron density maps.[7]
Three sets of theoretical calculations were also included in the comparison.
Several chemically significant features were reproduced in all the
experimental density maps, and positional parameters were reproducible to a
precision of 0.001Å or better.

Fig. 3.16-3 shows the experimental deformation density in the plane of the oxalic acid molecule. Fig. 3.16-4 shows a section of deformation density perpendicular to the C-C bond axis through its midpoint. The most interesting feature of the electron density distribution in the oxalic acid molecule is the large peak at the center of the C-C bond. The elongation of the peak, as shown in Fig. 3.16-4, means that the C-C bond contains significant π character.

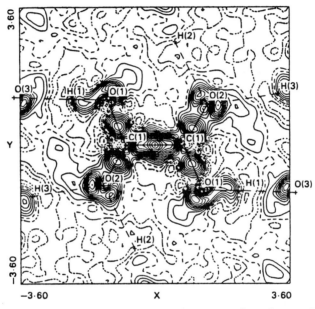

Fig. 3.16-3 Experimental deformation density in the plane of the oxalic acid molecule (100K). Contours are at 0.05 $e\text{Å}^3$ intervals with the zero and negative contours dashed. (After ref. 1).

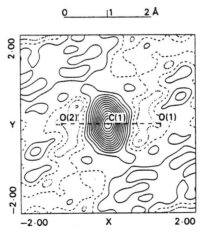

Fig. 3.16-4 Section of the deformation density perpendicular to the C-C axis through the midpoint of the bond. Contours as in Fig. 3.16-3. (After ref. 1).

2. Abnormally long C-C bonds in $H_2C_2O_4$, $HC_2O_4^-$ and $C_2O_4^{2-}$

The available structural data show that the C-C bonds in $H_2C_2O_4$ (1.538-1.554Å with a mean of 1.548Å; 1.548Å from electron diffraction[8]), $HC_2O_4^-$ (1.544-1.553Å in ten compounds with a mean of 1.550Å[9]), and $C_2O_4^{2-}$ (1.559-1.574Å in seven compounds with a mean of 1.567Å[10]) are all significantly longer than the normal C-C single bond in aliphatic compounds. On the other hand, the simple valence bond structural formula of the planar oxalate skeleton satisfies the condition for conjugation, and the existence of delocalized π bonding has been substantiated by deformation density studies.

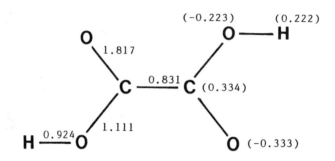

Fig. 3.16-5 The bond orders and net atomic charges (shown in parentheses) in oxalic acid molecule.

A MNDO computation[11] has yielded the bond orders (according to the Wiberg definition) and net atomic charges in the oxalic acid molecule (Fig. 3.16-5). Of the 17 fully occupied molecular orbitals, four are π molecular orbitals which are alternately bonding and antibonding in between the two carbon atoms. These π molecular orbitals contribute little to bonding in the $H_2C_2O_4$ molecule; the Wiberg bond order of the C-C bond is made up of a σ component of 0.816 and a π component of only 0.015. The σ bond order is less than unity mainly because each of the two carbon atoms is bonded to two highly electronegative oxygen atoms, leading to a reduction of charge density on the carbon atoms.

The good reducing properties of oxalic acid are associated with its relatively weak C-C bond which breaks during a reaction, and the resulting fragments are then oxidized to carbon dioxide.

A neutron diffraction study of $K_2C_2O_4 \cdot H_2O_2$ at 123K showed a centrosymmetric, planar oxalate ion with C-C = 1.5675(5), C-O(1) = 1.2653(3) and C-O(2) = 1.2460(3)Å, O(1) being the hydrogen-bond acceptor to the hydrogen peroxide molecule, which has dimensions O-O = 1.4578(4),

O-H = 1.0117(5)Å, O-O-H = 100.69(3) and H-O-O-H = 101.6°. The potassium ion
has eight nearest neighbours in a distorted dodecahedral arrangement with
normal K-O distances in the range 2.708(1)-2.997(1)Å.[12]

3. Deformation density maps of hydrogen bonds

    In crystalline oxalic acid dihydrate are found two types of hydrogen
bonds:

    Short: O(1)-H(1)...O(3")   2.487Å

    Long: O(3)-H(2)...O(2")   2.826Å,   O(3)-H(3)...O(2)   2.830Å

The density distributions in these hydrogen bonds are shown in Fig. 3.16-6.

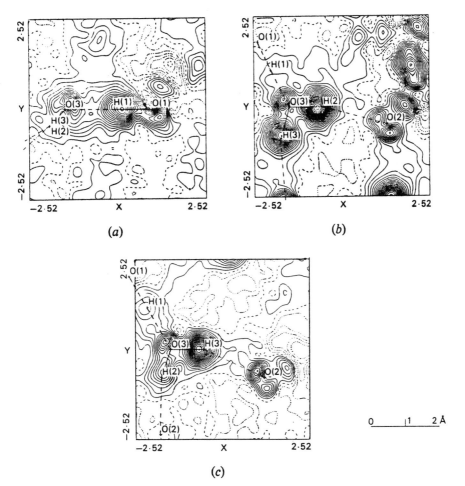

(a)   (b)

(c)

Fig. 3.16-6 Deformation density calculated without subtracting the
hydrogen atom density: (a) section perpendicular to the plane of the
water molecule and containing the O(3")...H(1)-O(1) short hydrogen bond,
(b) plane defined by the positions of O(3), H(2) and O(2"), (c) plane
defined by the positions of O(3), H(3) and O(2). (After ref. 1).

A difference between the short and long hydrogen bonds is apparent in deformation density maps for which the density of the hydrogen atom has not been substracted (Fig. 3.16-6). For the long hydrogen bonds, a clear break is present in the H(2)...O(2) region [Fig. 3.16-6(b)] and only a slight positive peak in the H(3)...O(2) region [Fig. 3.16-6(c)]. On the other hand, the short hydrogen bond shows a continuous ridge of density with a saddle point of 0.24(3) $e\mathring{A}^3$ between the oxalic acid proton and the water O atom. This result has been interpreted as giving evidence for a covalent contribution to the short hydrogen bond. The covalent contributions are not limited to the shortest symmetrical hydrogen bond but decrease gradually as the hydrogen bond length increases.

References

[1]   E.D. Stevens and P. Coppens, *Acta Crystallogr., Sect. B* **36**, 1864 (1980).

[2]   W.H. Zachariasen, *Z. Kristallogr.* **89**, 442 (1934).

[3]   R.G. Delaplane and J.A. Ibers, *Acta Crystallogr., Sect. B* **25**, 2423 (1969).

[4]   T.M. Sabine, G.W. Cox and B.M. Craven, *Acta Crystallogr., Sect. B* **25**, 2437 (1969).

[5]   P. Coppens and T.M. Sabine, *Acta Crystallogr., Sect. B* **25**, 2442 (1969).

[6]   E.D. Stevens, *Acta Crystallogr., Sect. B* **36**, 1876 (1980).

[7]   Project Reporter: P. Coppens (USA). Co-authors: J. Dam, S. Harkema, D. Feil (The Netherlands); R. Feld, M.S. Lehmann (France); R. Goddard, C. Krüger (FRG); E. Hellner (FRG); H. Johansen (Denmark); F. K. Larsen (Denmark); T.F. Koetzle, R.K. McMullan (USA); E.N. Maslen (Australia); E.D. Stevens, P. Coppens (USA), *Acta Crystallogr, Sect. A* **40**, 184 (1984).

[8]   Z. Nahlovska, B. Nahlovsky and T.G. Strand, *Acta Chem. Scand.* **24**, 2617 (1970).

[9]   J.O. Thomas and N. Renne, *Acta Crystallogr., Sect. B* **31**, 2161 (1975).

[10] H. Küppers, *Acta Crystallogr., Sect. B* **29**, 318 (1973).

[11] Zhou Gongdu, Li Qi and Liu Ruozhuang, *Huaxue Tongbao*, 43 (1988). [In Chinese].

[12] B.F. Pedersen and Å. Kvick, *Acta Crystallogr., Sect. C* **46**, 21 (1990).

Note The more acidic hydrogen atoms in organic molecules (alkyne > quinone > alkene > aromatic > aliphatic) tend to form C-H...O hydrogen bonds in crystals and, as far as possible, these bonds are accommodated within the framework of stronger interactions such as O-H...O and N-H...O hydrogen bonds. For recent reviews see R. Taylor and O. Kennard, *J. Am. Chem. Soc.* **104**, 5063 (1982); G.D. Desiraju, *Acc. Chem. Res.* **24**, 290 (1991).

3.17   Forsterite (Olivine)
Mg₂SiO₄

Crystal Data

Orthorhombic, space group *Pbnm* (No. 62)

$a = 4.744$, $b = 10.22$, $c = 6.004$Å, $Z = 4$

Atom	x	y	z
Si	0.074	0.405	1/4
Mg(1)	0	0	0
Mg(2)	0.010	0.278	3/4
O(1)	0.230	0.092	3/4
O(2)	0.282	0.051	1/4
O(3)	0.223	0.337	0.460

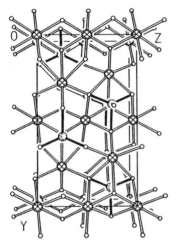 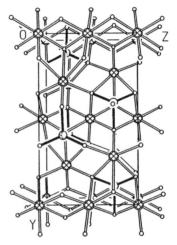

Fig. 3.17-1 Stereoview of the crystal structure of forsterite. Si-O bonds are shown in solid black.

Crystal Structure

Forsterite, Mg_2SiO_4, is a mineral in the olivine group which contains separate [SiO₄] tetrahedra. Bragg and Brown[1] first determined the structure of a magnesium-rich olivine of composition $9Mg_2SiO_4 \cdot Fe_2SiO_4$. Later, Belov *et al.*[2] and Hanke and Zemann[3] refined the forsterite structure.

The structure of forsterite is shown in a stereoview in Fig. 3.17-1. The main features of the structure are as follows: i) the oxygen atom positions are close to those of hcp; ii) each silicon is surrounded by a tetrahedral group of four oxygen atoms, forming a single [SiO₄] tetrahedron with a mean Si-O bond length of 1.62Å; and iii) every magnesium atom is surrounded octahedrally by six oxygen atoms, with Mg-O distances in the range 2.07-2.17Å with a mean of 2.10Å. The Mg atoms occupy two kinds of Wyckoff positions; the [SiO₄] tetrahedra point alternately in opposite directions along *a* and *b* in accordance with the holohedral symmetry.

Remarks

1. Classification of silicates

Silicates constitute the largest part of the earth's crust and mantle. They play an important role as raw materials as well as products in technical processes, such as natural building materials, cements, glasses, ceramics, refractories, etc.

Before the fundamental structures of the silicates were determined, attempts were made to calssify them in terms of silicic acids by analogy with other inorganic salts made up of metal ions and acid radicals, with much resultant confusion. After the methods of X-ray crystallography were applied to silicates, it was found that the overwhelming majority of them contain Si in tetrahedral coordination, and the $[SiO_4]$ tetrahedra are either isolated or share corners with other tetrahedra, giving rise to an enormous variety of silicate structures. In many silicates, silicon may be replaced to a certain extent by aluminium, so the structure of silicates are further extended to cover cases where such partial substitution occurs.

Machatschki[4] was the first to realize that the kind and the degree of linkage of $[SiO_4]$ tetrahedra constitute a basis for the classification of silicates. Since in his time only a few silicate structures had been determined, this was an extremely far-sighted proposal. Very soon afterwards, Bragg[5] and Naray-Szabó,[6] having access to the structures of the most important rock-forming silicate minerals, followed Machatschki's suggestion and created what is now usually refered to as "Bragg's classification of silicates":[7]

a)	Monosilicates	$[SiO_4]$
b)	Group silicates	
	disilicates	$[Si_2O_7]$
	cyclosilicates	$[Si_3O_9]$, $[Si_6O_{18}]$
c)	Chain silicates	
	single chain silicates	$[SiO_3]_n$
	double chain silicates	$[Si_4O_{11}]_n$
d)	Layer silicates	$[Si_4O_{10}]_n$
e)	Framework silicates	$[SiO_2]_n$

Extensions of Bragg's classification of silicates have been suggested by Zoltai[8] and Liebau.[9]

In this book, we choose forsterite (monosilicate), beryl (cyclosilicate; Section 3.18), diopside (chain silicate; Section 2.22), muscovite (layer silicate; Section 3.19), feldspar (Section 3.20) and zeolite A (framework silicate; Section 3.21) as examples to illustrate the relations between the chemical composition, structure, and properties of silicates.

2. Replacement of cations in silicates

In silicates, there is extensive substitution of cations. The name olivine is applied to members of the forsterite-fayalite series, $(Mg,Fe)_2SiO_4$, in which iron may freely replace magnesium since the two kinds of atoms have closely similar ionic radii. In monticellite, $CaMgSiO_4$, one-half of the magnesium atoms of forsterite are substituted by calcium atoms. Replacement of some of the cations in a given crystal structure by cations of another kind may result in retention of complete ordering, statistically random distribution, or the creation of vacant sites. Hence many structurally related minerals are found with very different physical properties.

3. $[SiO_4]$ and $[SiO_6]$ polyhedra

In general, metal ions are larger and have a lower valence than silicon so that the latter forms bonds to oxygen ions more strongly than do the metal ions, yielding $[SiO_4]$ tetrahedra with a mean bond length Si-O = 1.62Å.

On the other hand, if M is a metalloid the strength of the M-O bond is roughly comparable to that of the Si-O bond. The small metalloid atoms which have a high formal charge can then compete successfully with the Si atoms for the O atoms. As a result, the Si-O bond distance in such compounds is lengthened to about 1.77Å, leaving enough space for six O atoms to gather around each Si atom. Fig. 3.17-2 shows the $[SiO_4]$ tetrahedron and $[SiO_6]$ octahedron and their average dimensions.

X-Ray crystal structure determinations of silicates containing octahedrally coordinated silicon are not common. Until 1984, 13 such structures had been published with measured dimensions for 19

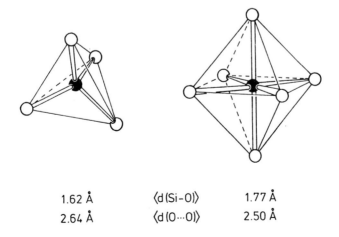

| 1.62 Å | $\langle d(Si-O)\rangle$ | 1.77 Å |
| 2.64 Å | $\langle d(O\cdots O)\rangle$ | 2.50 Å |

Fig. 3.17-2 The $[SiO_4]$ tetrahedron and $[SiO_6]$ octahedron and their average dimensions. (After ref. 9).

crystallographically independent [SiO$_6$] octahedra.[9] The data clearly
demonstrate that in phases containing [SiO$_6$] octahedra the Si-O bonds are
weaker than the ones existing in phases containing [SiO$_4$] tetrahedra.
Therefore, under ordinary conditions for a given composition the phase
containing silicon in octahedral coordination should be energetically less
favorable than the one with tetrahedrally coordinated silicon. This may also
be deduced from the shorter O...O contacts of 2.50Å (and hence, higher
repulsive forces between the oxygen ions) in a [SiO$_6$] octahedron relative to
the corresponding distance of 2.64Å in the [SiO$_4$] tetrahedron.

No silicate has yet been found in which the [SiO$_6$] octahedron has ideal
symmetry *m*3*m*; instead, the octahedra invariably show angular and/or bond
length distortions, and individual Si-O distances vary between 1.70 and 1.84Å.

In [SiO$_6$], the Si-O bond is generally considered to be partly covalent
and partly ionic, but less covalent than the Si-O bond in [SiO$_4$]. In
tetrahedral silicon, in addition to the four σ bonds of the sp^3 hybrid, there
is also some π overlap of the remaining 2p orbitals of oxygen with the 3d0
orbitals of silicon. In octahedral silicon, the d$_{x^2-y^2}$ and d$_{z^2}$ orbitals
participate simultaneously in σ-bonding (sp^{3}d^2 hybridization), and the d$_{xy}$,
d$_{yz}$ and d$_{xz}$ orbitals can be used for π-bonding with the appropriate orbitals
of the oxygen atoms.

The atomic charges of Si and O atoms in a silicate crystal may be used
to describe the ionicity of the Si-O bond. For example, in crystalline
K$_2$Si^{VI}Si$^{IV}_3$O$_9$,[10] the atomic charges of Si^{VI}, Si^{IV} and O are +3.3, +2.5 and   -
1.4, respectively, which suggest that an increase in the coordination number
of a Si atom from 4 to 6 significantly increases its ionic character.

References

[1] W.L. Bragg and G.B. Brown, *Z. Kristallogr.* **63**, 538 (1926).
[2] N.V. Belov, E.N. Belova, N.H. Andrianova and P.F. Smirnova, *Dokl. Akad.
     Nauk SSSR* **81**, 399 (1951). [In Russian].
[3] K. Hanke and J. Zemann, *Naturwissenschalten* **50**, 91 (1963).
[4] F. Machatschki, *Zentralbl. Miner.* **A3**, 97 (1928).
[5] W.L. Bragg, *Z. Kristallogr.* **74**, 237 (1930).
[6] St. Náray-Szabó, *Z. Phys. Chem.* **B9**, 356 (1930).
[7] L. Bragg and G.F. Claringbull, *Crystal Structures of Minerals*, Bell,
     London, 1965.
[8] T. Zoltai, *Am. Miner.* **45**, 960 (1960).
[9] F. Liebau, *Structural Chemistry of Silicates*, Springer-Verlag, Berlin,
     1985, chap. 3.
[10] D.K. Swanson and C.T. Prewitt, *Am. Miner.* **68**, 581 (1983).

Note A comprehensive review of the crystal chemistry of six-coordinated
     silicon, especially silicates synthesized at high pressures and
     temperatures, is given in L.W. Finger and R.M. Hazen, *Acta Crystallogr.,
     Sect. B* **47**, 561 (1991).

3.18 Beryl

Be₃Al₂[Si₆O₁₈]

Crystal Data

Hexagonal, space group *P6/mcc* (No. 192)

$a = 9.2088(5)$, $c = 9.1896(7)$Å, $Z = 2$ (from ref. 3)

Atom	x	y	z	Atom	x	y	z
Si	0.3875	0.1158	0	O(1)	0.3100	0.2366	0
Be	1/2	0	1/4	O(2)	0.4988	0.1455	0.1453
Al	2/3	1/3	1/4	O(3)	0	0	1/4

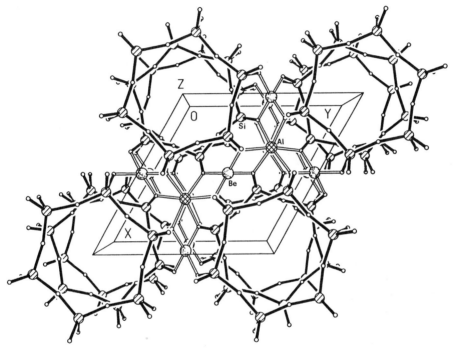

Fig. 3.18-1 Crystal structure of beryl. (Adapted from ref. 6).

Crystal Structure

The structure of beryl was determined by Bragg and West,[1] and refined by Gibbs *et al.*,[2] Morosin,[3] and Hazen *et al.* at several pressures up to 57 kbar.[4]

The structure of beryl consists of [SiO₄] tetrahedra, which are connected by sharing oxygen atoms of the type O(1) so as to form a six-fold ring. The resulting $(Si_6O_{18})^{2-}$ cyclic anions are arranged in parallel layers about the z = 0 and 1/4 planes. They are further linked together by bonding, through oxygen atoms of the type O(2), to Be[II] and Al[III] ions, to give a three-dimensional framework (Fig. 3.18-1). The bond distances are Si-O = 1.592 to 1.620Å in the [SiO₄] tetrahedron, 1.653Å in the [BeO₄] tetrahedron, and 1.904Å in the [AlO$_6$] octahedron.

Remarks

1. The structure of BaTi(Si$_3$O$_9$)

The mineral benitoite, BaTi(Si$_3$O$_9$), consists of three-membered single-ring anions (Si$_3$O$_9$)$^{6-}$ and Ba^{2+} and Ti^{4+} cations. Its structure is similar to beryl; the cyclic ions (Si$_3$O$_9$)$^{6-}$ are also arranged in layers with their planes parallel, as shown in Fig. 3.18-2. The metal ions Ba^{2+} and Ti^{4+} are both six-coordinated, lying between layers of anions to bind them together at different levels.

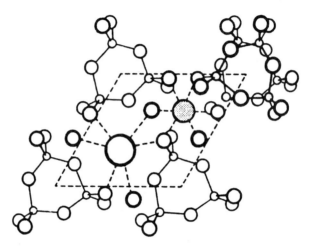

Fig. 3.18-2 Crystal structure of benitoite: largest circle: Ba^{2+}; shaded circle: Ti^{4+}. Many Si and O atoms have been omitted for the sake of clarity. (After ref. 6).

2. Varieties of rings in silicates

To date, single rings composed of 3, 4, 6, 8, 9 and 12 [SiO$_4$] tetrahedra have been found in crystalline silicates. They may be referred to as a three-membered single ring, a four-membered single ring, etc. Condensation of two single rings *via* corners of the tetrahedra results in the formation of a double ring. Depending on the number of [SiO$_4$] tetrahedra in the single rings that form the double ring, the latter may be called a three-membered double ring, a four-membered double ring, etc. Fig. 3.18-3 shows the single rings (a to f) and double rings (g to i) found in crystalline silicates:

(a) [Si$_3$O$_9$] in benitoite, BaTi[Si$_3$O$_9$];

(b) [Si$_4$O$_{12}$] in tatamellite, Ba$_4$(Fe,Ti)$_4$B$_2$[Si$_4$O$_{12}$]O$_5$Cl$_x$;

(c) [Si$_6$O$_{18}$] in beryl; Be$_3$Al$_2$[Si$_6$O$_{18}$];

(d) [Si$_8$O$_{24}$] in muirite, Ba$_{10}$(Ca,Mn,Ti)$_4$[Si$_8$O$_{24}$](Cl,O,OH)$_{12}$·4H$_2$O;

(e) $[Si_9(O,OH)_{27}]$ in eudialyte, $Na_{12}(Ca,RE)_6(Fe,Mn,Mg)_3Zr_3$ and
  $(Zr,Nb)_x[Si_3O_9]_2[Si_9(O,OH)_{27}]_2Cl_y$;

(f) $[Si_{12}O_{36}]$ in synthetic $K_{16}Sr_4[Si_{12}O_{36}]$;

(g) $[Si_6O_{15}]$ in synthetic $[Ni(en)_3]_3[Si_6O_{15}].26H_2O$;

(h) $[Si_8O_{20}]$ in steacyite, $K_{1-x}(Na,Ca)_{2-y}Th_{1-z}[Si_8O_{20}]$; and

(i) $[Si_{12}O_{30}]$ in milarite, $KCa_2(Be_2Al)[Si_{12}O_{30}].3/4H_2O$.

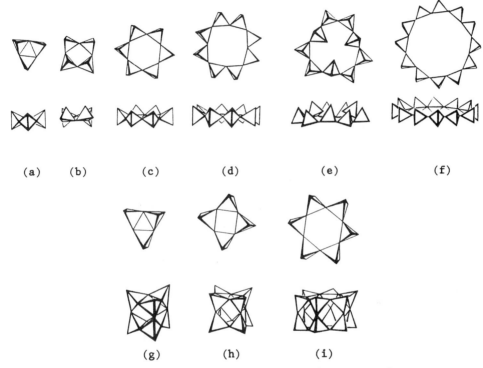

Fig. 3.18-3 Single rings and double rings found in silicates. Two views of each ring are illustrated. (After ref. 5).

References

[1] W.L. Bragg and J. West, *Proc. Roy. Soc. London* **111A**, 691 (1926).

[2] G.V. Gibbs, D.W. Breck and E.P. Meagher, *Lithos* **1**, 275 (1968).

[3] B. Morosin, *Acta Crystallogr.*, *Sect. B* **28**, 1899 (1972).

[4] R.M. Hazen, A.Y. Au and L.W. Finger, *Am. Miner.* **71**, 977 (1986).

[5] F. Liebau, *Structural Chemistry of Silicates*, Springer-Verlag, Berlin, 1985.

[6] A.F. Wells, *Structural Inorganic Chemistry*, 5th ed., Clarendon Press, Oxford, 1984.

3.19 Muscovite (Mica)

$KAl_2[AlSi_3O_{10}](OH)_2$

Crystal Data

Monoclinic, space group $C2/c$ (No. 15)

$a = 5.189(10)$, $b = 8.995(20)$, $c = 20.097(5)$Å, $\beta = 95.18(8)°$, $Z = 4$

Atom	x	y	z	Atom	x	y	z
K	.0000	.1016	.2500	O(2)	.2450	-.1980	.1620
Al	.2484	.0871	.0016	O(3)	.7629	-.1287	.1674
(Si,Al)	.4625	-.0758	.1372	O(4)	.4250	.2600	.0542
Si(2)	.4593	.2550	.1365	O(5)	.4650	-.0550	.0527
O(1)	.4080	.0960	.1680	OH	-.0470	.0580	.0520

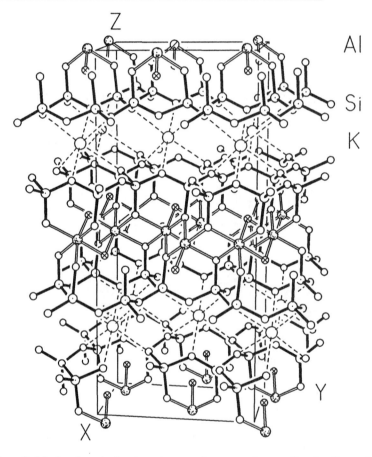

Fig. 3.19-1 Crystal structure of muscovite. Bonds from O to Si, Al and K atoms are indicated by solid, open and broken lines, respectively; OH groups are represented by crossed circles.

Crystal Structure

The structure of muscovite (Fig. 3.19-1) was firstly determined by Jackson and West,[1] and was later refined by Radoslovich[2] and Rothbauer.[3]

Muscovite is a layer (or sheet) silicate of the mica group. The layer

has the composition $[AlSi_3O_{10}]_n$ in which the $[SiO_4]$ and $[AlO_4]$ tetrahedra, by
linking three corners of each tetrahedron to neighbours, form a hexagonal
layer with hydroxyl ions located in the plane of their vertices and at the
centers of all six-fold rings, as shown in Fig. 3.19-2(a). Two of these
layers are placed together with the vertices of their tetrahedra pointing
inwards. These vertices are cross-linked by Al atoms, which are six-
coordinated. Hydroxyl ions are incorporated, linked to Al alone, and form a
firmly bound double layer, with the bases of the tetrahedra on each outer
side. The double layer is shown in detail in Fig. 3.19-2(b).

The bases of the tetrahedra are symmetrically opposed like mirror images
across a central plane. Two opposite hexagonal rings thus outline a large
cavity which accommodates a K atom in twelve-coordination. The K-O
electrostatic bonds are weak and easily broken, and micas accordingly possess

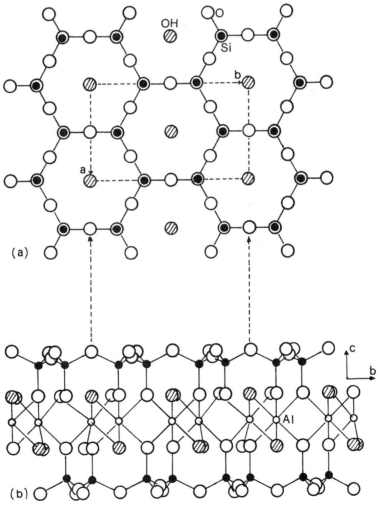

Fig. 3.19-2 Portions of the layer structure of muscovite. (After ref. 6).

very perfect cleavage parallel to the *ab* plane.

The single layer network of $[AlSi_3O_{10}]_n$ with the OH groups has hexagonal symmetry, which however will be lost when two single layers are staggered to form a double layer linked by Al. This lowers the lattice symmetry to monoclinic, and the basal triads of oxygen atoms are rotated 13° from the ideal hexagonal positions.

Remarks

1. Layer silicates

Muscovite is an end member of the mica group. The micas, hydrous micas, talc, pyrophyllite, chlorites, vermiculites, montmorillonites, and kaolinites, and the clay minerals all have a layer structure of pseudo-hexagonal symmetry and they cleave easily, owing to their being based on the hexagonal network of

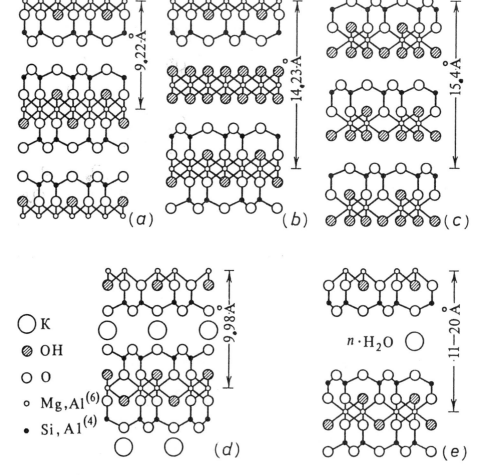

Fig. 3.19-3 Structure of some layer silicates: (a) talc, (b) chlorite, (c) kaolin, (d) mica, (e) montmorillonite. (After ref. 9).

linked tetrahedra. These layer-type minerals are susceptible to substitution of the metal ions, so the actual compositions can be very complex. The structural schemes of some layer silicate minerals are shown in Fig. 3.19-3.

The clay minerals are characteristically composed of very small particles, generally less than one micron across, which are commonly hexagonal flakes. There are many and complex variations of structure based on one or more of the layer types. The vertices of the tetrahedra can form part of a hydroxide-like layer. Their bases can either be directly opposed, or can be fitted to hydroxyl or water layers. When layers of different types succeed each other with more or less regularity, the changes in stacking these layers together occur in an endless variety of ways. This irregularity of structure is no doubt related to their capacity to take up or lose water and mediate ionic exchange. In many cases the structures swell or contract in different states of hydration, and contain replaceable ions. On account of these properties clay minerals are important as soil constituents.

2. Chrysotile (major component of commerical asbestos)

Chrysotile, $Mg_3[Si_2O_5](OH)_4$, occurs as bundles of parallel fibers. Diffraction patterns show that the fibrous bundles are neither single crystals nor individuals in a more or less parallel orientation. Its structure is that of a composite sheet with a silicon-oxygen layer on one side and a layer

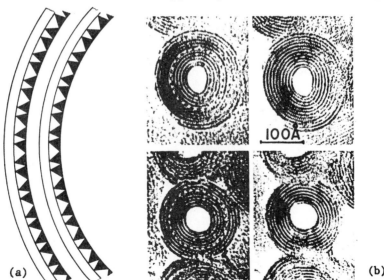

(a)   (b)

Fig. 3.19-4 (a) The schema of the sheet curl between octahedral layers (white bars) and tetrahedral layers (black triangles) in chrysotile; (b) high resolution electron micrographs of concentric structure in the cross-section of a tube of chrysotile. (Adapted from ref. 8).

resembling magnesium hydroxide (brucite) on the other. This structure would
have a strong tendency to curl, because the dimensions of the latter layer
structure are greater than those of the former. There is therefore a strong
expansive force on the hydroxyl side which tends to make each sheet curl into
a cylinder, with the silicon-oxygen sheets on the inner side. It appears that
the relatively weak forces holding the sheets together are insufficient to
resist this tendency and that the structure comprises a series of such curved
sheets stacked on each other. The fibers of chrysotile asbestos are thus
composed of long curved laths or tubes, which have been observed not only by
X-ray diffraction,[4] but also by high resolution electron microscopy.[5]
Fig. 3.19-4 shows (a) the schema of the sheet curl between octahedral and
tetrahedral layers in chrysotile; and (b) high resolution electron micrographs
of concentric structure in the cross-section of a tube of chrysotile.

References

[1] W.W. Jackson and J. West, *Z. Kristallogr.* **76**, 211 (1931); **85**, 160 (1933).

[2] E.W. Radoslovich, *Acta Crystallogr.* **13**, 919 (1960).

[3] R. Rothbauer, *Neues Jahrb. Miner. Mh.*, 143 (1971).

[4] E.J.W. Whittaker, *Acta Crystallogr.* **9**, 855 (1956).

[5] K. Yada, *Acta Crystallogr., Sect. A* **27**, 659 (1971).

[6] L. Bragg and G.F. Claringbull, *Crystal Structures of Minerals*, Bell,
    London, 1965.

[7] A.S. Povarennykh, *Crystal Chemical Classification of Minerals*, Vol. **1**,
    Plenum Press, New York, 1972, p. 442.

[8] F. Liebau, *Structural Chemistry of Silicates*, Springer-Verlag, Berlin,
    1985, chap. 10.

[9] B.K. Vainshtein, V.M. Fridkin and V.L. Indenbom, *Modern Crystallography
    II, Structure of Crystals*, Springer-Verlag, Berlin, 1982.

Note Apophyllite, an "optically anomolous" silicate of empirical formula
     $KCa_4Si_8O_{20}F \cdot 8H_2O$ with the sheet structure, was discovered by Brewster in
     1815 to exhibit birefringence when observed through crossed polarizers
     (45°). A recent review of optically anomomalous crystals, whose optical
     symmetry is lower than would be expected from their external form and
     X-ray diffraction pattern, is given in B. Kahr and J.M. McBride, *Angew.
     Chem. Int. Ed. Engl.* **31**, 1 (1992).

3.20   Low-Albite (Feldspar)
Na[AlSi$_3$O$_8$]

Crystal Data

Triclinic, space group $C\bar{1}$ (No. 2)[*]

a = 8.1151(8), b = 12.7621(25), c = 7.1576(6)Å, α = 94.218(12),

β = 116.803(8), γ = 87.707(13)° (13K, from ref. 4), Z = 4

[*] The unconventional space group $C\bar{1}$ is chosen to correspond to the unit cell used originally by Taylor.[5]

Atom	x	y	z	Atom	x	y	z
Na	.26470(6)	.98952(1)	.14441(7)	O(3)	.31264(3)	.60794(2)	.19019(4)
Al	.50896(5)	.66774(3)	.20773(6)	O(4)	.81931(3)	.85141(2)	.25924(4)
Si(1)	.00359(4)	.82131(3)	.23701(5)	O(5)	.51020(3)	.80192(2)	.26814(4)
Si(2)	.19041(4)	.60996(3)	.31323(5)	O(6)	.02294(3)	.69437(2)	.22592(1)
Si(3)	.67890(4)	.88190(3)	.35960(5)	O(7)	.78987(3)	.89099(2)	.61058(4)
O(1)	-.00565(3)	.12907(2)	.96644(4)	O(8)	.18503(3)	.86821(2)	.43708(4)
O(2)	.58864(3)	.99755(2)	.27889(4)				

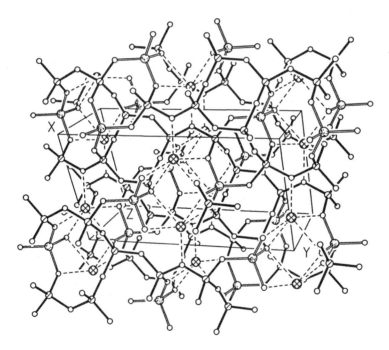

Fig. 3.20-1 Crystal structure of albite, Na[AlSi$_3$O$_8$]. Bonds from O to Al and Na atoms are represented by open and broken lines, respectively.

Crystal Structure

The structure of the low-temperature form of albite, Na[AlSi$_3$O$_8$], was first determined by Taylor in 1934.[1] It was refined in 1969,[2] and more recently with three-dimensional neutron and X-ray diffraction intensity data sets in 1980[3] and low-temperature (13K) neutron diffraction data in 1986.[4]

Albite is the one of the end members of the feldspar group. The feldspars are the most important rock-forming minerals, comprising some two-thirds of igneous rocks; for example, granite is composed of feldspars, quartz, and micas.

The feldspars are all closely related in framework structure, form, and physical properties. They may be divided into two main groups: orthoclase and plagioclase. Feldspars included in the first group are the mineral orthoclase itself, sanidine ($KAlSi_3O_8$), and celsian ($BaAl_2Si_2O_8$). The angle between the prominent cleavages (001) and (010) is 90°. The plagioclases include the minerals albite ($NaAlSi_3O_8$) and anorthite ($CaAl_2Si_2O_8$), long regarded as the end members of a continuous isomorphous series in which (NaSi) is replaced by (CaAl); the angle between cleavages is inclined at about 86°.

The structural scheme common to all feldspars is a three-dimensional framework of linked tetrahedral $[SiO_4]$ and $[AlO_4]$ groups, with cations occupying cavities within the framework. With a knowledge of the structural scheme of sanidine (first X-ray analysis of a feldspar),[5] it is possible to account for many features of the feldspar group as a whole. A study of albite provides a structural interpretation of the division of the feldspar family into orthoclase and plagioclase groups. The difference between the symmetries of the two groups is associated with the difference in size of the positive ions, for the framework is essentially the same, but contracts slightly around the smaller sodium and calcium ions in the plagioclase group.

The framework of feldspars may be regarded as built from layers of tetrahedra placed at the points of a planar net composed of squares and octagons, the fourth vertex of each tetrahedron pointing either upwards or

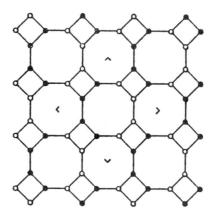

Fig. 3.20-2 Simplified representation of a layer of the albite framework. Tetrahedra pointing up and down are represented by solid and open circles, respectively. (After ref. 6)

downwards. These layers are then joined through the projecting vertices so
that adjacent layers are related by planes of symmetry. In each layer of the
albite structure half of the tetrahedra point upwards, and the other half
downwards, as shown in Fig. 3.20-2. Each Na$^+$ ion in low-albite is irregularly
surrounded by seven oxygen atoms with Na-O distances of 2.3606(5)-2.6140(7)Å.

Remarks

1. Low-albite and high-albite

Samples of a feldspar of given composition, obtained from different
sources, may differ in their physical properties; in particular, the optical
properties may depend to a very marked degree on the thermal history of the
material. This has led to the recognition of the existence of a range of
"transitional" states between the extremes of so-called "low-temperature" and
"high-temperature" materials. The term "low-temperature" may mean either
formation at a low temperature or very prolonged annealing (under geological
conditions) at low temperature; the term "high-temperature" means formation at
high temperature for a suitable period; some materials may then be preserved
in the state characteristic of the high-temperature form on quenching at a
lower temperature.[5]

The difference between the low- and high-temperature forms of albite is
the ordering of Si and Al in the tetrahedral sites. In low-albite there is an
ordered arrangement of Al in one-quarter of the tetrahedra and Si in the
remaining sites, with mean Si-O = 1.614Å and Al-O = 1.743Å, but there is a
random arrangement in high-albite, with (Si,Al)-O = 1.64Å.

2. Feldspars and geological evolution of the earth

The fact that the aluminosilicate framework is highly adaptable to minor
structural modification and stoichiometric variation accounts for the dominant
occurrence of the feldspar family of minerals in the earth's crust. The
chemical composition and structural characteristics of any feldspar species
are a faithfully preserved record of its primordial formation and subsequent
changes in the course of geological evolution. Thus crystallographic studies
on the feldspars, as well as other silicate minerals, provide a wealth of
valuable information towards our understanding of the geological history of
the earth.

3. General rules governing the structures of natural silicates

The structures of natural silicates obey the following general rules:[6]

a) With some exceptions, such as stishovite, the Si atoms form [SiO₄]
tetrahedra with little deviation of bond lengths and bond angles from the mean
values:

$$\langle Si-O\rangle = 1.62\text{Å}$$
$$\langle O-Si-O\rangle = 109.47°$$
$$\langle Si-O-Si\rangle = 140°$$

b) The most important and widespread substitution is that of Al for Si in tetrahedral coordination, so that most of the "silicates" that occur in nature are in fact aluminosilicates. This substitution must be accompanied by the incorporation of cations to balance the charge on the silicon-oxygen anion. The mean Al-O of bond length in $[AlO_4]$ is 1.76Å. In almost all silicates which occur as minerals the anions are tetrahedral $[(Si,Al)O_4]$ groups. The Al atoms also occupy positions of octahedral coordination in aluminosilicates.

c) The $[(Si,Al)O_4]$ tetrahedra are linked to one another by corners rather than edges or faces. According to Loewenstein's aluminum avoidance rule,[7] two Si-O-Al groups have a lower energy content than one Al-O-Al plus one Si-O-Si group. The rule essentially states that AlO_4 tetrahedra do not share corners in tetrahedral framework structures if this can be avoided. The few known violations of Loewenstein's rule are connected with severe distortions of the tetrahedral groups.[8]

d) One oxygen atom can belong to no more than two $[SiO_4]$ tetrahedra.

e) If s is the number of oxygen atoms of a $[SiO_4]$ tetrahedron shared with other $[SiO_4]$ tetrahedra, then for a given silicate anion the difference between the s values of all $[SiO_4]$ tetrahedra tends to be small.

References

[1] W.H. Taylor, J.A. Darbyshire and H. Strunz, *Z. Kristallogr.* **87**, 464 (1934).

[2] P.H. Ribbe, H.D. Megaw, W.H. Taylor, R.B. Ferguson and R.J. Traill, *Acta Crystallor., Sect. B* **25**, 1503 (1969).

[3] G.E. Harlow and G.E. Brown, *Am. Miner.* **65**, 986 (1980).

[4] J.V. Smith, G. Artioli and Å. Kvick, *Am. Miner.* **71**, 727 (1986).

[5] W.H. Taylor in L. Bragg and G.F. Claringbull, *Crystal Structures of Minerals*, Bell, London, 1965, chap. 14.

[6] F. Liebau, *Structural Chemistry of Silicates*, Springer-Verlag, Berlin, 1985, chaps. 8 and 10.

[7] W. Loewenstein, *Am. Miner.* **39**, 92 (1954).

[8] W. Depmeier, *Acta Crystallogr., Sect. B* **40**, 185 (1984).

3.21 Dehydrated Zeolite 4A (Linde Molecular Sieve Type 4A)
$Na_{12}[Al_{12}Si_{12}O_{48}]$

Crystal Data

Cubic, space group $Pm3m$ (No. 221)

$a = 12.292(2)$Å, $Z = 1$

Atom	Position	x	y	z	Occupancy factor
(Si,Al)	24(k)	0	.1836	.3718	1
O(1)	12(h)	0	.2277	1/2	1
O(2)	12(i)	0	.2917	.2917	1
O(3)	24(m)	.1123	.1123	.3418	1
Na(1)	8(g)	.2014	.2014	.2014	1
Na(2)	12(i)	0	.4302	.4302	1/4
Na(3)	12(j)	.2307	.2307	1/2	1/12

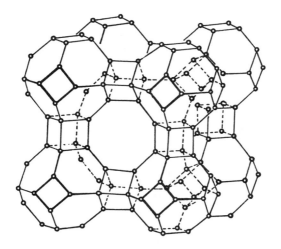

Fig. 3.21-1 The framework structure of zeolite A; the small circles represent alternate Si and Al atoms at the corners of the space-filling polyhedra.

Crystal Structure

Zeolites are a family of crystalline aluminosilicates of the general formula $M_r^I M_s^{II}[Al_p Si_q O_{2(p+q)}] \cdot mH_2O$ (with $r + 2s = p$), whose structure is characterized by a system of polyhedral cavities interconnected by large tunnels such that the enclosed ions or molecules can readily diffuse through the crystal lattice. Zeolites generally contain loosely held water, which can be driven off by heating the crystals and subsequently regained on exposure to a moist atmosphere. When occluded water is removed, it can be replaced by other small molecules. There is a close relationship between the size of molecules that can diffuse through the framework and the dimensions of the aperture or "bottle-neck" connecting one cavity to the next. This property of the zeolites to admit molecules below a certain limiting size, while refusing passage to larger molecules, has led to their being termed "molecular sieves".

In addition to the natural zeolites, many new zeolites have been synthesized and found to possess desirable properties: absorption and ion-exchange capability, behaviour as molecular sieves, and catalytic activity.

Zeolite A is a synthetic zeolite that has not been found in nature. Its structure was determined by polycrystalline X-ray diffraction in 1956[2] and later refined in 1971.[3] The crystal structure of dehydrated zeolite 4A was determined by single crystal X-ray analysis in 1977.[1]

The framework structure of dehydrated zeolite 4A, $Na_{12}[Al_{12}Si_{12}O_{48}]$, is shown in Fig. 3.21-1. The $[(Si,Al)O_4]$ tetrahedra are linked to form a cubo-octahedron (truncated octahedron, β-cage) whose vertices correspond to the (Si,Al) atoms (Fig. 3.21-2).

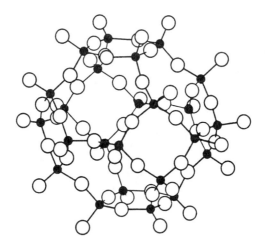

Fig. 3.21-2 The arrangement of $[AlO_4]$ and $[SiO_4]$ tetrahedra which constitutes the cubo-octahedral cage in some zeolites. The solid circles represent Si or Al. (After ref. 19).

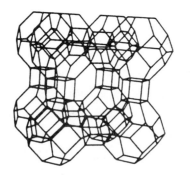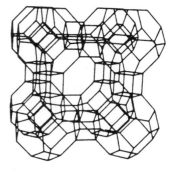

Fig. 3.21-3 Stereoview of the framework of zeolite A. (After ref. 4).

In zeolite A, cubo-octahedra of this type lie at the corners of the unit cell, generating small cubes at the centers of the cell edges and eight-rings at the face centers. Access through these eight-rings to a very large cavity (truncated cubo-octahedron, α-cage) at the center of the unit cell is possible. Fig. 3.21-3 shows a stereoview of the framework of zeolite A.

In dehydrated zeolite 4A, the Na^+ ions are distributed over three crystallographically-different sites: eight lie on three-fold axes near the centers of six-rings, three are on eight-ring planes, and the twelfth is loosely held in the large cavity on a two-fold axis opposite a four-ring.

The K^+ ions in dehydrated K_{12}-A (K^+-exchanged zeolite A with 12 K^+ ions per unit cell) can all be reduced by 0.1 torr of cesium vapor at 350°C to give dehydrated fully Cs^+-exchanged zeolite A containing extra cesium atoms occurring primarily as (Cs_4^{3+}) clusters. The crystal structure of Cs_{12}-A.xCs [x *ca.*3/4; space group *Pm3m*, $a = 12.281(1)$Å] may be viewed as a homogeneous mixture of Cs_{13}-A and Cs_{12}-A whose populations are about 3/4 and 1/4, respectively.[5] In each of these, per unit cell, three Cs^+ ions are located at the centers of the 8-rings. Within the β-cage, two Cs^+ are found 4.05(2)Å apart on a unique threefold axis. Each β-cage Cs^+ in Cs_{13}-A associates further with an α-cage cation at 3.85(1)Å to give the linear (Cs_4)$^{3+}$ cluster. All eight 6-ring sites in the α-cage are occupied by Cs^+ in Cs_{13}-A, but only six are so occupied in Cs_{12}-A. Each Cs^+ ion is coordinated to three framework oxygens at 2.96Å and extends 1.84Å into the α-cage from the [111] plane at O(3). A stereoview of this is shown in Fig. 3.21-4.

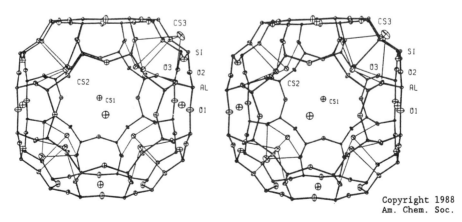

Fig. 3.21-4 Stereoview of an α-cage in Cs_{13}-A. The zeolite A framework is drawn with heavy bonds between tetrahedrally coordinated (Si,Al) and oxygen atoms. Cs^+ ion coordination by framework oxygens is indicated by fine line. (After ref. 5).

Remarks

1. Framework structures and polyhedral cages in zeolites[4-7]

    The structure of zeolites can be represented by the topology of the anionic framework and the size, charge and locations of the exchange ions within the framework. Table 3.21-1 lists some important zeolites, whose structures are shown in Fig. 3.21-5 and Fig. 3.21-6.

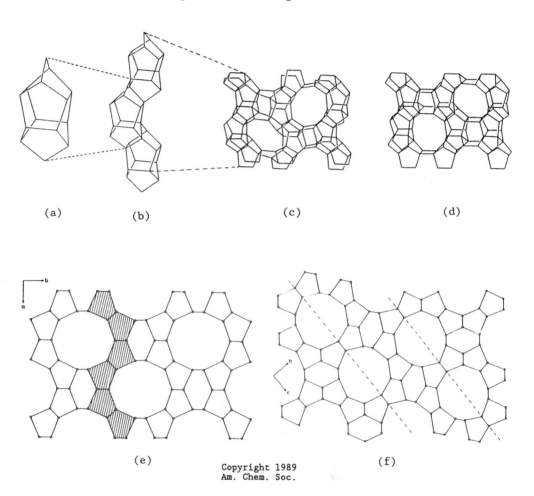

(a)   (b)   (c)   (d)

(e)   (f)

Copyright 1989
Am. Chem. Soc.

Fig. 3.21-5 Schematic representations of (a) the pentasil building unit, (b) this unit joined into chains, (c) a projection of the layers of such chains joined to form the zeolite ZSM-5 lattice (*i*-stacking), (d) a similar projection of the layers joined to form the zeolite ZSM-11 lattice (*σ*-stacking). (e) Projection of the zeolite ZSM-22 framework along (001). One chain of edge-sharing five-membered rings is emphasized by hatching. (f) Projection of zeolite ZSM-23 along (100). (After refs. 13 and 14).

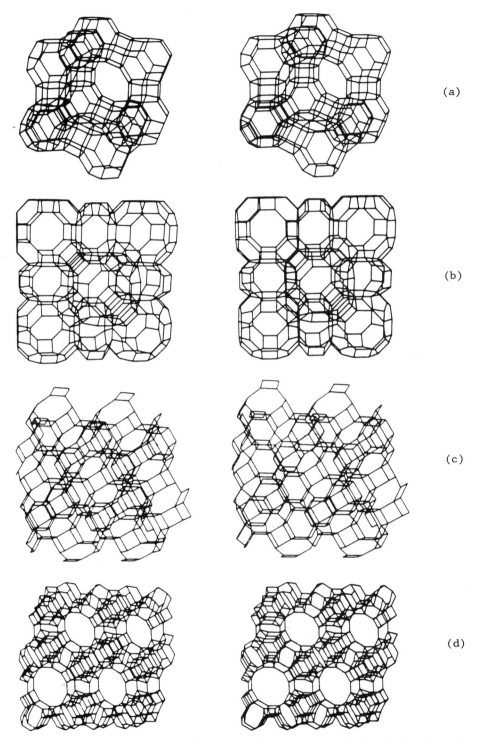

Fig. 3.21-6 Stereoview of framework structure of (a) faujasite viewed along [111], (b) zeolite ZK-5 viwed along [100], (c) chabasite viewed along [001] and (d) zeolite L viwed along [001]. (After ref. 4).

Table 3.21-1 Some important zeolite structures.

Name	Formula, unit cell content	Crystallographic data (Å)	Ref.
Zeolite A, Na_{12}-A	$Na_{12}[Al_{12}Si_{12}O_{48}]$	Pm3m, a = 12.292	[1]
Cs_{12}-A.3/4Cs	$Cs_{12}[Al_{12}Si_{12}O_{48}].3/4Cs$	Pm3m, a = 12.281(1)	[5]
Faujasite	$Na_{13}Ca_{12}Mg_{11}[Al_{59}Si_{133}O_{384}].260H_2O$	Fd3m, a = 24.67	[8]
(Zeolite X and Y)			
Zeolite ZK-5	$Na_{30}[Al_{30}Si_{66}O_{192}].98H_2O$	Im3m, a = 18.7	[9]
Chabazite	$Ca_2[Al_4Si_8O_{24}].13H_2O$	R$\bar{3}$m, a = 13.78, c = 15.06	[10]
Zeolite L	$K_6Na_3[Al_9Si_{27}O_{72}].21H_2O$	P6/mmm, a = 18.4, c = 7.5	[11]
Mordenite	$Na_8[Al_8Si_{40}O_{96}].24H_2O$	Cmcm, a = 18.13, b = 20.49, c = 7.52	[12]
Zeolite ZSM-11	$96SiO_2.4[N(C_4H_9)_4OH]$	I$\bar{4}$m2, a = 20.067(1), c = 13.411(1)	[13]
Zeolite ZSM-22	$24SiO_2.HN(C_2H_5)_2$	Cmc2$_1$, a = 13.859(3), b = 17.420(4)	[14]
		c = 5.038(2)	

Some polyhedral cages occuring in the zeolites are shown in Fig. 3.21-7.[15] The designation n^m means m faces, each of which is a n-membered ring. In other words, skeletons of zeolites are formed by fusion of these polyhedral cages. The silicates with polyhedral cages having windows too small to let through the encaged molecules are termed "clathrasils". Geometric shapes and approximate free diameters of the cages which occur in the silica framework of clathrasils are described in Sections **2.15** and **7.1**.

2. Structure-directing role of cations in zeolite synthesis

The structure-directing role of cations in zeolite synthesis has been reviewed by Barrer.[16] According to this work:

a) Na^+ in the reaction mixture favors the formation of zeolite A, faujasites, gismondine types, sodalite and cancrinite hydrates, and gmelinites;

b) K^+ favors the formation of chabazite and zeolite L;

c) solutions containing Ca^{2+} favor the formation of thomsonite and epistilbite;

d) Sr^{2+} favors zeolites of the heudandite and ferrierite types;

e) Ba^{2+} and Li^+ ions favor different kinds of synthetic zeolite framework.

The structure-directing influence of the cations is exerted in at least two ways:

a) by changing the kinetics of formation of the crystalline zeolite from the low-molecular-weight species in aqueous solution by a condensation process, and

b) by stabilizing the crystalline zeolite framework.

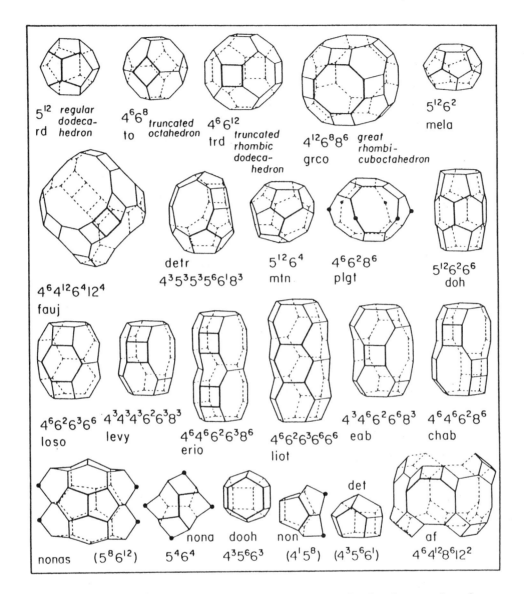

Fig. 3.21-7 Selected cages and building units in the 3D frameworks of zeolites and related materials. Some small distortions from clinographic projection were made to reduce overlap at the boundaries. Only some of the species are planar convex polyhedra. Vertices at the intersection of only two edges are marked by small filled circles. Each species is labeled with the face symbol, and some are labeled with a specific name, or mnemonic code, or both. The following codes are from the name of zeolites: chabazite, eab, erionite, faujasite, levynite, liottite, losod, paulingite, mtn (ZSM-thirtynine). Others are from clathrasils. (After ref. 15).

3. Framework topology of ZSM-18

ZSM-18 is an aluminosilicate which crystallizes with product ratios of SiO_2 to Al_2O_3 ranging from 10 to 30. Its synthesis is facilitated by the presence of the triquat cation as an effective template which, upon calcination, leaves a large pore in the zeolite product (Fig. 3.21-8).

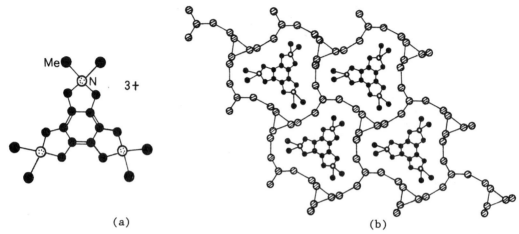

(a)   (b)

Fig. 3.21-8 (a) Structural formula of the triquat cation. (b)
Illustration of the influence of the triquat cation as a template on the
formation of ZSM-18, showing a view of the (Si,Al) framework between
$z = 0$ and $z = 1/2$ parallel to the c axis. (Adapted from ref. 17).

The framework topology of ZSM-18 has been determined by model building
and subsequent constrained distance and angle least-squares refinement of X-
ray powder diffraction data.[17] It is the first example of an
aluminosilicate zeolite containing 3-rings of tetrahedral (Si,Al)-O species.
The structure is characterized by a linear 12-ring channel with a pore opening

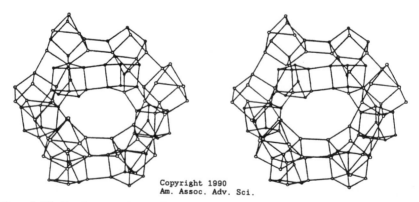

Fig. 3.21-9 Stereoview of the ZSM-18 framework in the direction of an
open channel through the 12-ring apertures. (After ref. 17).

of ~7.0Å (Fig. 3.21-9). In addition, the channels are lined with pockets that are capped by 7-rings of dimensions 2.8 by 3.5Å.

4. A double ring silicate with a zeolite-like structure

Cubic $[(n\text{-}C_4H_9)_4N]H_7[Si_8O_{20}]\cdot(128/24)H_2O$ crystallizes in space group *Fm3c* with $a = 28.61(1)$Å and $Z = 24$.[18] The crystal structure (Fig. 3.21-10) resembles that of zeolite A with Si-O...H...O-Si linkages replacing the Si-O-Si bonds of the zeolite. The H^+ ions form short hydrogen bonds of 2.6Å between terminal oxygen atoms of the $[Si_8O_{20}]$ double rings. The silicate "framework" contains two types of cages which resemble 4^66^8 truncated octahedra and $4^{12}6^88^6$ truncated cubo-octahedra.

The N atoms of the $(n\text{-}C_4H_9)_4N^+$ ions are located at the centers of the "octagonal" faces common to two large cages such that the alkyl legs extend pairwise into these adjacent cages. Each large cage thus contain a total of twelve *n*-butyl groups.

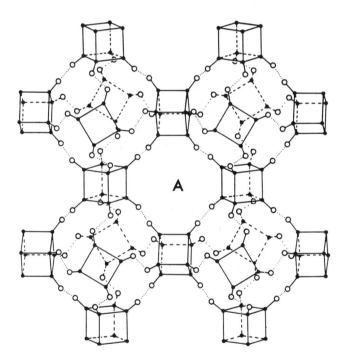

Fig. 3.21-10 Simplified diagram of the crystal structure of $[(n\text{-}C_4H_9)_4N]H_7[Si_8O_{20}]\cdot(128/24)H_2O$. One half of the unit cell $0 \leq z \leq 1/2$ is shown, and the Si and O atoms are represented by small dots and open circles, respectively. Letter **A** indicates the center of one face common to two large cages and is the site occupied by the nitrogen atom of the organic cation. (After ref. 18).

Within each truncated octahedron is a cubic arrangement of hydrogen-bonded water molecules (O...O $ca.$ 2.8Å) with their oxygen atoms lying on C_3 axes. Eight other water oxygen atoms are located near the centers of the "hexagonal" faces of the truncated octahedron on the C_3 axes. They are hydrogen-bonded to three of the six terminal oxygen atoms which belong to three different [Si_8O_{20}] groups, and to the inner water cube (all O...O distances are about 2.8Å).

References

[1]   V. Subramanian and K. Seff, *J. Phy. Chem.* **81**, 2249 (1977).

[2]   T.B. Reed and D.W. Breck, *J. Am. Chem. Soc.* **78**, 5972 (1956).

[3]   V. Gramlich and W.M. Meier, *Z. Kristallogr.* **133**, 134 (1971).

[4]   R.M. Barrer, *Zeolites and Clay Minerals as Sorbents and Molecular Sieves*, Academic Press, London, 1978, chap. 2.

[5]   N.H. Heo and K. Seff in W.H. Flank and T.E. Whyte, Jr. (eds.), *Perspectives in Molecular Sieve Science*, American Chemical Society, Washington, DC, 1988, chap. 11.

[6]   F. Liebau, *Structural Chemistry of Silicates*, Springer-Verlag, Berlin, 1985.

[7]   D.W. Breck, *Zeolite Molecular Sieves*, Wiley, New York, 1974.

[8]   W.H. Bauer, *Am. Miner.* **49**, 697 (1964).

[9]   W.M. Meier and G.T. Kokotailo, *Z. Kristallogr.* **121**, 211 (1965).

[10] J.V. Smith, F. Rinaldi and L.S. Dent Glasser, *Acta Crystallogr.* **16**, 45 (1963).

[11] R.M. Barrer and H. Villiger, *Z. Kristallogr.* **128**, 352 (1969).

[12] W.M. Meier, *Z. Kristallogr.* **115**, 439 (1961).

[13] C.A. Fyfe, H. Gies, G.T. Kokotailo, C. Pasztor, H. Strobl and D.E. Cox, *J. Am. Chem. Soc.* **111**, 2470 (1989).

[14] B. Marler, *Zeolites* **7**, 393 (1987).

[15] J.V. Smith, *Chem. Rev.* **88**, 149 (1988).

[16] R.M. Barrer, *Zeolites* **1**, 130 (1981).

[17] S.L. Lawton and W.J. Rohrbaugh, *Science (Washington)* **247**, 1319 (1990).

[18] G. Bissert and F. Liebau, *Acta Crystallogr., Sect. A* **40**, Supplement, C232 (1984).

[19] A.F. Wells, *Structural Inorganic Chemistry*, 5th ed., Clarendon Press, Oxford, 1984.

Note Modern techniques for the structural elucidation of aluminosilicate catalysts are discussed in J.M. Thomas and C.R.A. Catlow, *Prog. Inorg. Chem.* **35**, 1 (1987).

3.22 Phenyl-(2,2',2"-nitrilotriethoxy)silane (Silatrane)
$(C_6H_5)\overset{\frown}{Si}(OCH_2CH_2)_3\overset{\shortmid}{N}$

Crystal Data

Orthorhombic, space group *Pbca* (No. 61)

$a = 13.220(15)$, $b = 18.524(8)$, $c = 10.050(1)$Å, $Z = 8$

Atom	x	y	z	Atom	x	y	z
Si	.21311	.37401	.43623	C(3)'	.39108	.44225	.32193
O(1)	.27803	.30277	.48072	C(4)	.36212	.27569	.40673
O(2)	.13437	.36616	.30803	C(5)	.16666	.37277	.16748
O(3)	.25701	.45562	.47388	C(6)	.33645	.48928	.40485
N	.32834	.38048	.27990	C(7)	.11310	.37203	.56923
C(1)	.40744	.33624	.33700	C(8)	.02323	.40776	.55562
C(1)'	.37927	.30887	.26743	C(9)	-.04922	.41079	.65849
C(2)	.27862	.35125	.16345	C(10)	-.03493	.37601	.77527
C(2)'	.27023	.40468	.15896	C(11)	.05451	.33674	.79370
C(3)	.35192	.45739	.26356	C(12)	.12740	.33423	.69270

Atoms C(1), C(1)'; C(2), C(2)'; C(3), C(3)' represent three sets of two-fold disordered methylene groups.

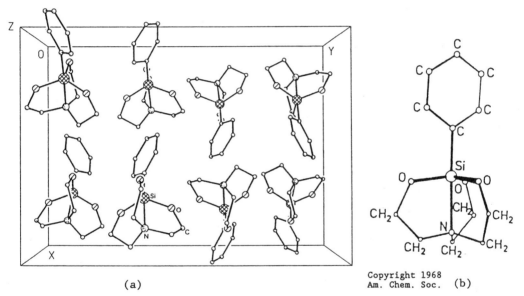

Fig. 3.22-1 Crystal structure (a) and molecular structure (b) of $(C_6H_5)\overset{\frown}{Si}(OCH_2CH_2)_3\overset{\shortmid}{N}$. (After ref. 1).

Crystal Structure

The crystal structure of silatrane, $(C_6H_5)\overset{\frown}{Si(OCH_2CH_2)_3}\overset{\shortmid}{N}$, as determined by Turley and Boer,[1] confirmed the predicated penta-coordination of silicon. The distorted trigonal-bipyramidal coordination geometry of the silicon atom has the three equatorial oxygens bent away from the phenyl group toward the nitrogen, as shown by the average C-Si-O angle of 97.1° and the average O-Si-O angle of 118.5°. Bond distances to silicon are Si-O(1) = 1.638Å, Si-O(2) = 1.664Å, Si-O(3) = 1.665Å, <Si-O> = 1.656Å, Si-C = 1.882Å and Si-N = 2.193Å.

The Si-N distance is much shorter than the sum of the respective van der Waals radii, about 3.5Å, thereby confirming the existence of a transannular Si←N bond (Fig. 3.22-1).

Remarks

1. Structure of silatranes, $XSi(OCH_2CH_2)_3N$

   One of the important recent developments in the structural chemistry of silicon involves non-tetra-coordinate derivatives. In the last two decades, there was an ever-increasing interest in penta- and hexa-coordinate silicon compounds. This progress was stimulated by the development of new experimental techniques and theoretical approaches as well as the specific biological activity findings for some silicon compounds with an expanded coordination sphere.

   Among the stable compounds of penta-coordinate silicon, the most studied are the silatranes, $XSi(OCH_2CH_2)_3N$.[2,3] The axial nitrogen atom is linked through three $(CH_2)_2$ links to oxygen atoms at the triangular base. The tetradentate tripodal ligand occupies four coordination positions around the central Si atom, but for penta-coordinate species the steric requirements of the ligands are such that the trigonal-bipyramidal structure results. In these compounds there is an intramolecular donor-acceptor interaction between the nitrogen and silicon atoms.

   The Si←N distance in about 50 silatranes containing three condensed rings varies within the range 2.0-2.4Å. The longer the Si←N bond, the more planar the configuration of the NC_3 grouping and the structure of the $XSiO_3$

Table 3.22-1 Structure data of some silatranes.[3]

Compound	Si←N(Å)	Torsion angle (°) (O-Si-N-C)
$ClSi(OCH_2CH_2)_3N$	2.036	12.7, 15.2 13.5
$Cl_2CHSi(OCH_2CH_2)_3N$	2.062	18.6, 19.3, 18.5
$ClCH_2Si(OCH_2CH_2)_2(OCOCH_2)N$	2.085	19.0, 17.7, 16.3
$ClCH_2Si(OCH_2CH_2)_3N$	2.120	
$C_6H_5NHCH_2Si(OCH_2CH_2)_3N$	2.143	13.1, 11.2, 11.7
$CH_2=CHSi(OCH_2CH_2)_3N$	2.150	
$Cl(CH_2)_3Si(OCH_2CH_2)_2(OCOCH_2)N$	2.153	20.9, 19.3, 17.3
$NC(CH_2)_3Si(OCH_2CH_2)_3N$	2.164	
$p-(CH_3)C_6H_4Si(OCH_2CH_2)_3N$	2.181	16.1, 17.0, 13.2
$ClCH_2Si[OCH(CH_3)CH_2]_2(OC_6H_4)N$	2.181	9.4, 12.0, 11.6
$C_6H_5Si[OCH(CH_3)CH_2]_2(OC_6H_4)N$	2.193	1.8, 2.8, 3.4
$NCS(CH_2)_3Si(OCH_2CH_2)_3N$	2.209	13.6, 13.2, 14.0

fragment approaches tetrahedral. Table 3.22-1 gives the structural data of some silatranes.[3]

2. Factors governing the stability of penta-coordinate silicon compounds

   Theoretical and experimental studies on penta-coordinate silicon derivatives have demonstrated that their existence is determined by a combination of factors: electronegativity of the substituents and steric interactions between substituents. One should also emphasize the role and significance of the chelate effect, the size, number and constitution of the chelate rings invloving the silicon atom. For example, the different Si←N bond lengths in $ClCH_2\overset{\longleftarrow}{Si}(OCH_2CH_2)_3\overset{\longrightarrow}{N}$ and $ClCH_2\overset{\longleftarrow}{Si}(OCH_2CH_2)_2(OCOCH_2)N$ (Table 3.22-1) can be rationalized by assuming that π conjugation between oxygen and its carbonyl neighbour in the keto compound mitigates π donation by oxygen to silicon, thus making the latter a stronger acceptor in its dative bond with nitrogen. The N_{1s} XPS values of 402.7 eV in the keto compound and 401.9 eV in the other support this view.

   There are two types of favored structures, the trigonal bipyramid and the square pyramid, for penta-coordinate silicon. The former is more stable than the later. Lehn and co-workers[4] have described the increased stability of complexes by several orders of magnitude over monocyclic analogues upon the addition of another connecting bridge onto the macrocyclic ring to form a macrobicyclic ligand or cryptand. It is thus reasonable to suggest that the stability of the complexes increases for the four bonding configurations in the following sequence:

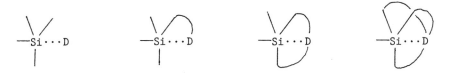

| Intermolecular | Intramolecular | Transannular | Intra-bridgehead |

3. Model for the pathway of nucleophilic substitution at silicon

   X-Ray analysis of a series of penta-coordinated organosilicon compounds (I)-(VIII) nicely maps the entire reaction pathway of nucleophilic substitution at the Si atom (Fig. 3.22-2).[5] In the iodo (I) and bromo (II) derivatives the secondary Si...X bonds are very weak and predominantly ionic (both compounds are good electrical conductors in polar solvents). Accordingly, the Pauling-Bürgi bond numbers, n, correspond to the weak Si...X and strong Si-O interactions. The $OSiC_3$ tetrahedron is somewhat deformed, the exocyclic O-Si-C angles being 106° and 103°. In contrast, in the chloro

derivatives (III),[6] (IV), (V) and (VI) the "secondary" Si...Cl interaction
becomes comparable in strength with the "primary" Si-O bond, since their bond
numbers are now practically equal. In these four compounds the Si atom adopts
an almost distortion-free trigonal bipyramidal configuration (Cl-Si-C 92-94°,
exocyclic O-Si-C 91-90°, and "axial" O-Si-C 162-173°). In the fluoro

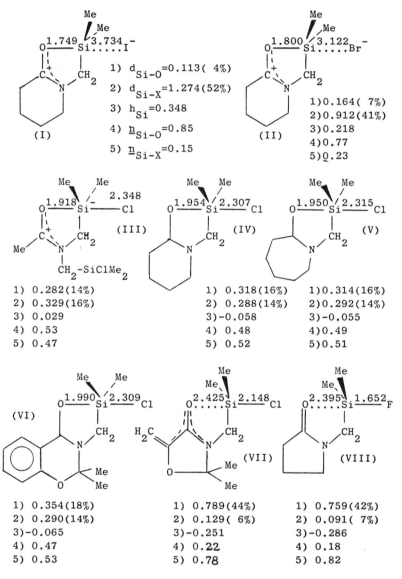

Fig. 3.22-2 Structural parameters of a series of penta-coordinated
silicon lactam derivatives. The d values given in items 1) and 2) are
the actual and percentage lengthening of the bond relative to a "normal"
single bond; h_{Si} is the deviation of the Si atom from the plane of the
"equatorial" carbon atoms. (After ref. 5).

derivative (**VIII**) the stereochemistry at Si is inverted as the Si-F bond becomes primary covalent and the Si...O bond becomes secondary, the F-Si-C and O...Si-C angles being 101-102° and 85-84°, respectively. Modification of the lactam ring in (**VII**) causes this chloro derivative to approach the limiting case of (**VIII**), possibly owing to conjugation of the exocyclic double bond with the peptide system.

4. Azasilatranes

Azasilatranes of the type (**IX**) with R = H were first reported in 1977.[7] A systematic multinuclear NMR spectroscopy study of hydro- and hydrocarbon-substituted azasilatranes appeared a decade later,[8] and a recent X-ray analysis of the phenyl derivative (**IXa**; R = H, X = Ph) provided the first set of structural parameters (Si←N_{ax} 2.132, Si-N_{eq} 1.739Å, N_{ax}-Si-N_{eq} 82.1°) for an azasilatrane.[9]

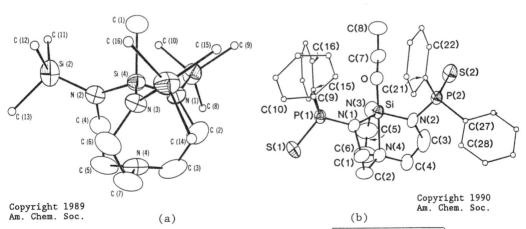

(IX)　　　　　(X)　　　　　(XI)　　　　　(XII; M = Mo, W)

(a)
(b)

Fig. 3.22-3 Molecular structures of (a) MeSi(Me₃SiNCH₂CH₂)₃N̄ (**IXc**) and (b) EtOSi(S=PPh₂NCH₂CH₂)₂(HNCH₂CH₂)N̄ (**X**). (After refs. 11 and 12).

The introduction of substituents at the equatorial NH functionalities has greatly expanded the scope of azasilatrane chemistry. The crystal structures of trisubstituted azasilatrane (**IXb**; R = Me, X = OEt)[10] and (**IXc**; R = SiMe$_3$, X = Me) [Fig. 3.22-3(a)][11] have been determined. The geometry of the hypercoordination at silicon in (**IXb**) (Si←N$_{ax}$ 2.135, Si-N$_{eq}$ 1.755Å, N$_{ax}$-Si-N$_{eq}$ 83.0°) is very similar to that in (**IXa**), whereas (**IXc**) was found to have the longest ever recorded Si←N$_{ax}$ distance of 2.775(7)Å.

In the disubstituted azasilatrane (**X**) [Fig. 3.22-3(b)],[12] the transannular Si←N$_{ax}$ distance of 2.214(3)Å is lengthened relative to that in (**IXb**), whereas the Si-O distance of 1.658(3)Å is shortened as compared to the corresponding value of 1.699(2) in (**IXb**). These data indicate a decrease in the Si←N$_{ax}$ dative bond strength in (**X**) with a concomitant transfer of electron density into the collinear Si-O bond.

The azasilatrane EtOSi(Ph$_2$PNCH$_2$CH$_2$)$_2$(HNCH$_2$CH$_2$)N, the desulfurized derivative of (**X**), is capable of acting as a bidentate (P,P') or a tridentate (P,P',O) ligand. The bidentate diphosphine chelating mode has been found in a crystal structure analysis of (**XI**), whereas IR (carbonyl bands indicating a *fac*-M(CO)$_3$ group) and NMR data strongly support the rather unusual coordination of the silyl ether to a transition metal in (**XII**).[12]

5. Other atrane compounds

"Atranes" having an intramolecular transannular dative bond of the types N→B (boratranes), N→Al (alumatranes), and N→Ti (titatranes) have been synthesized and investigated using NMR (^{1}H, ^{15}N, ^{11}B, ^{27}Al) and XPS techniques.[3] Some typical examples are:

M(OCH$_2$CH$_2$)$_3$N   (M = B, Al)   M(OCH$_2$CH$_2$)(OCOCH$_2$)$_2$N   (M = B, Al)

M(OCH$_2$CH$_2$)$_2$(OCH$_2$CH$_2$CH$_2$)N   (M = B)   CH$_3$COOTi(OCH$_2$CH$_2$)$_3$N

M(OCH$_2$CH$_2$)$_2$(OCOCH$_2$)N   (M = B, Al)   p-RC$_6$H$_4$COOTi(OCH$_2$CH$_2$)$_3$N   (R = CH$_3$, CH$_3$O, F, Cl, Br)

X-Ray analysis of B(OCH$_2$CH$_2$)$_2$(OCOCH$_2$CH$_2$)N and B(OCH$_2$CH$_2$)$_2$(OCOCH$_2$)N have shown that the N→B bond distance (1.637Å) in the former boratrane, which contains one six-membered ring, is shorter than that (1.663Å) in the latter.[13,14]

References

[1]   J.W. Turley and F.P. Boer, *J. Am. Chem. Soc.* **90**, 4026 (1968).

[2]   M.G. Voronkov, V.M. Dyakov and S.V. Kirpichenko, *J. Organomet. Chem.* **233**, 1 (1982); St.N. Tandura, N.V. Alekseev and M.G. Voronkov, *Top. Curr. Chem.* **131**, 99 (1986).

[3] Guanli Wu, Kaijuan Lu and Yexin Wu in J.F. Liebman and A. Greenberg (eds.), *Modern Models of Bonding and Delocalization*, VCH, New York, 1988, p. 195.

[4] J.-M. Lehn and J.-P. Sauvage, *J. Am. Chem. Soc.* **97**, 6700 (1975).

[5] Yu.T. Struchkov in J.J. Stezowski, J.-L. Huang and M.-C. Shao (eds.), *Molecular Structure: Chemical Reactivity and Biological Activity*, Oxford University Press, New York, 1988, p. 443.

[6] K.D. Onan, A.T. McPhail, C.H. Yoder and R.W. Hillard, Jr., *J. Chem. Soc., Chem. Commun.*, 209 (1978).

[7] E. Lukevics, G.I. Zelčans, I.I. Solomennikova, E.E. Liepiņš, I.S. Yankovska and I.B. Mažheika, *Zh. Obsch. Khim.* **47**, 109 (1977); *J. Gen. Chem. USSR (Engl. Transl.)* **47**, 98 (1977).

[8] Ē. Kupče, E. Liepiņš, A. Lapsiņa, G.I. Zelčans and E. Lukevics, *J. Organomet. Chem.* **333**, 1 (1987).

[9] A.A. Macharashvili, V.E. Shklover, Yu.T. Struchkov, A. Lapsina, G. Zelčans and E. Lukevics, *J. Organomet. Chem.* **349**, 23 (1988).

[10] D. Gudat and J.G. Verkade, *Organometallics* **8**, 2772 (1989).

[11] D. Gudat, L.M. Daniels and J.G. Verkade, *J. Am. Chem. Soc.* **111**, 8520 (1989).

[12] D. Gudat, L.M. Daniels and J.G. Verkade, *Organometallics* **9**, 1464 (1990).

[13] Dou Shi-qi, Wu Ye-xin and Wu Guan-li, *Jiegou Huaxue (J. Struct. Chem.)* **2**, 273 (1983). [In Chinese.]

[14] Kuang Bao, Dai Jin-bi, Wu Ye-xin and Wu Guan-li, *Jiegou Huaxue (J. Struct. Chem.)* **2**, 277 (1983). [In Chinese.]

Note An updated compendium of structural data for silatranes and correlations between them are presented in A. Greenberg and G. Wu, *Struc. Chem.* **1**, 79 (1990).

3.23   Complex Containing Nonanuclear Germanium Anions

$$[K(crypt)^+]_6 Ge_9^{2-} Ge_9^{4-} \cdot 2\tfrac{1}{2} en$$

$$[C_{18}H_{36}O_6N_2K^+]_6 Ge_9^{2-} Ge_9^{4-} \cdot 2\tfrac{1}{2}(C_2H_8N_2)$$

Crystal Data

Triclinic, space group $P\bar{1}$ (No. 2)

$a = 20.037(2)$, $b = 28.944(2)$, $c = 14.546(2)$Å

$\alpha = 99.356(8)$, $\beta = 94.077(8)$, $\gamma = 87.60(1)°$, $Z = 2$

Atom*	x	y	z	Atom	x	y	z
Ge_{11}	.7707	.0642	.4146	Ge_{21}	.1334	.3666	.0364
Ge_{12}	.7941	.1692	.5109	Ge_{22}	.2787	.3690	.1948
Ge_{13}	.8530	.1220	.3755	Ge_{23}	.2137	.3039	.0906
Ge_{14}	.7019	.1148	.5375	Ge_{24}	.2041	.4312	.1327
Ge_{15}	.6696	.1861	.4557	Ge_{25}	.3246	.4185	.0755
Ge_{16}	.7765	.1905	.3430	Ge_{26}	.3321	.3240	.0526
Ge_{17}	.6487	.0934	.3680	Ge_{27}	.2102	.4170	.9531
Ge_{18}	.7533	.0959	.2573	Ge_{28}	.2119	.3202	.9212
Ge_{19}	.6622	.1608	.2781	Ge_{29}	.3163	.3670	.9132

* Only the Ge positional parameters are listed.

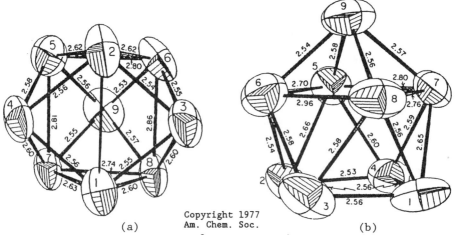

Copyright 1977
Am. Chem. Soc.

(a)   (b)

Fig. 3.23-1 Structures of the Ge_9^{2-} (a) and Ge_9^{4-} (b) clusters. (After ref. 1).

Crystal Structure

The crystal structure[1] consists of Ge_9^{2-} and Ge_9^{4-} anions, $[K(crypt)^+]$ cations, and neutral ethylenediamine molecules. The most remarkable feature is the presence of two different anionic clusters containing nine Ge atoms each.

The Ge_9^{2-} ion has the shape of a tricapped trigonal prism [Fig. 3.23-1(a)] which exhibits essentially C_{2v} symmetry but is clearly derived from the D_{3h} limit, with 2.81, 2.86 and 3.17Å for the parallel edges of the

trigonal prism, 171° for the angle between opposed faces therein, and capping
atoms separated by 4.00, 4.16 and 4.22Å.

The monocapped square-antiprismatic configuration of the Ge_9^{4-} ion [Fig.
3.23-1(b)] in quite close to ideal C_{4v} symmetry, with a nearly square base,
which has 3.58 and 3.64 diagonal distances and a vicinal dihedral angle 5.3°.
The angles for the characteristic opposed faces parallel to the four-fold axis
are 162° and 156°.

The structures found for all the independent $[K(crypt)^+]$ cations are
comparable to those determined for similar symmetry-unconstrained units in
$[K(crypt)]_2Te_3$.[2] The K-O (2.76 to 2.92Å) and K-N (2.95 to 3.04Å) distances,
as well as those within the ligand, exhibit a wider range than those reported
in $[K(crypt)]I$. Fig. 3.23-2 shows the structure of one of the $[K(crypt)]^+$
ions.

The closest approach of the nitrogen atoms in ethylenediamine to any
atom in the cluster, 3.67Å, suggests that hydrogen bonding between the two is
not important.

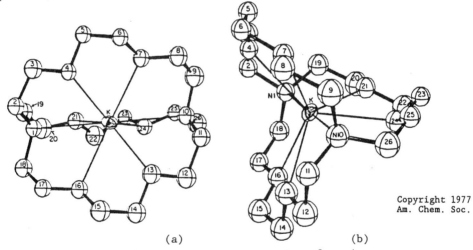

Copyright 1977
Am. Chem. Soc.

(a)   (b)

Fig. 3.23-2 The $[K(crypt)]^+$ cation in $[K(crypt)]_6^+ Ge_9^{2-}Ge_9^{4-}\cdot2.5en$: (a) N-K-N
axis horizontal; (b) tilted for a nearer end-on view from N10. (After ref. 1).

Remarks

1. Polyatomic anions of Ge, Sn and Pb[3]

It has been known since the early 1930s that reduction of Ge, Sn and Pb
by sodium in liquid ammonia gives polyatomic metal anions, and crystalline
compounds can be isolated by using ethylenediamine, e.g. $Na_4(en)_5Ge_9$ and
$Na_4(en)_7Sn_9$. A dramatic advance was achieved by means of the polydentate
crypt ligand. Thus the reaction of crypt in ethylenediamine with the alloys
$NaSn_{1-1.7}$ and $NaPb_{1.7-2}$ gave red crystalline salts $[Na(crypt)]_2Sn_5$ and

[Na(crypt)]$_2$Pb$_5$ containing the D_{3h} cluster anions Sn$_5{}^{2-}$ and Pb$_5{}^{2-}$, respectively (Fig. 3.23-3).[4] Similarly, the alloys NaSn$_{\sim 2.25}$ and KGe, respectively, react with crypt in ethylenediamine to give dark-red crystals of [Na(crypt)]$_4$Sn$_9$[5] and the deep-red compound [K(crypt)]$_6$Ge$_9$.Ge$_9$.2½en.[1] Some examples of polyatomic anions of Ge, Sn and Pb are listed in Table 3.23-1.

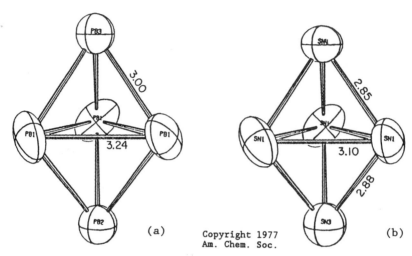

(a)   Copyright 1977   (b)
             Am. Chem. Soc.

Fig. 3.23-3 Configurations of and interatomic distances within the (a) Pb$_5{}^{2-}$ and (b) Sn$_5{}^{2-}$ ions. (After ref. 4).

Table 3.23-1 Polyatomic anions of Ge, Sn and Pb.

Anion	Shape	M-M/Å	Crystal	Ref.
Ge$_4{}^{2-}$	tetrahedron	2.77-2.79	[K(crypt)]$_2$Ge$_4$	[6]
Sn$_4{}^{2-}$	tetrahedron	2.93-2.97	[K(crypt)]$_2$Sn$_4$	[6]
Sn$_5{}^{2-}$	trigonal bipyramid	2.89(x6), 3.10(x3)	[Na(crypt)]$_2$Sn$_5$	[4]
Pb$_5{}^{2-}$	trigonal bipyramid	3.00(x6), 3.23(x3)	[Na(crypt)]$_2$Pb$_5$	[4]
Ge$_9{}^{2-}$	tricapped trigonal prism	2.53-2.86	[K(crypt)]$_6$Ge$_9$.Ge$_9$.2½en	[1]
Ge$_9{}^{4-}$	monocapped antiprism	2.53-2.96		
Sn$_9{}^{4-}$	monocapped antiprism	2.93-3.31	[Na(crypt)]$_4$Sn$_9$	[5]
Sn$_8$Tl^{3-}	tricapped trigonal prism	2.88-3.16	[K(crypt)]$_3$(Sn$_8$Tl,Sn$_9$Tl)$_{0.5}$.en	[7]
Sn$_9$Tl^{3-}	bicapped square antiprism	2.86-3.35		
[KSn$_9$]$^{3-}{}_\infty$	monocapped square-antisprismatic	2.93-3.22	[K(crypt)]$_3$(KSn$_9$)	[8]
	Sn$_9{}^{4-}$ clusters bridged by K$^+$			
	ions to form an infinite chain			

2. Heteropolyatomic anions of tin and lead

Tricapped trigonal-prismatic Sn$_8$Tl^{3-}, bicapped square-antiprismatic Sn$_9$Tl^{3-},[7] and the two series Sn$_{9-x}$Ge$_x{}^{4-}$ (x = 0-9) and Sn$_{9-x}$Pb$_x{}^{4-}$ (x = 0-9)[9] are known. In crystalline [K(crypt)]$_3$(Sn$_9$Tl^{3-},Sn$_8$Tl^{3-})$_{0.5}$.en (crypt = 4,7,13,16,21,24-hexaoxa-1,10-diazatricyclo[8.8.8]hexacosane), the two clusters Sn_9Tl^{3-} and Sn_8Tl^{3-} exhibit an unusual 50:50 occupational disorder in a single anionic site, with seven atoms coincident in both species. The structures of Sn_8Tl^{3-} and Sn_9Tl^{3-} are shown in Fig. 3.23-4; the latter is isostructural and isoelectronic with $B_{10}H_{10}^{2-}$.[10]

Dihedral angles for nine-atom clusters provide a better delineation of the degree of distortion from an ideal model, D_{3h} or C_{4v}. The most important of the dihedral angles is that between the opposite faces of the trigonal prism, which is 180° for an ideal D_{3h} geometry such as Bi_9^{5+} (Section 3.31), dropping progressively to 177° in Sn_6Tl^{3-} (C_{2v}), 171° in Ge_9^{2-} (~ C_{2v}), 162° in Ge_9^{4-} (~ C_{4v}), and 158° in Sn_9^{4-} (C_{4v}). The value of 177° for $TlSn_8^{3-}$ indicates only a very minor distortion.

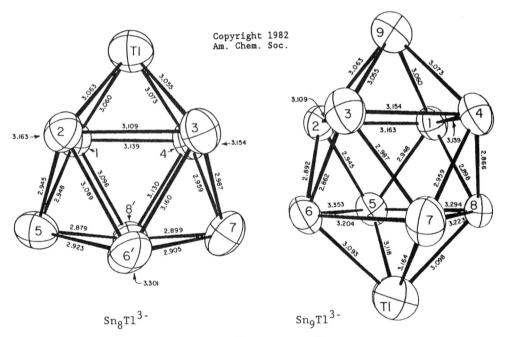

Sn_8Tl^{3-} Sn_9Tl^{3-}

Fig. 3.23-4 Structures of the Sn_8Tl^{3-} and Sn_9Tl^{3-} clusters. (After ref. 7).

The first members of a new class of mixed d-block/p-block polyhedral clusters, $[K(crypt)]_4[M_9Cr(CO)_3]$ (M = Sn,[11] Pb[12]), have been synthesized and characterized only recently. Both $[Sn_9Cr(CO)_3]^{4-}$ and $[Pb_9Cr(CO)_3]^{4-}$ are isostructural and isoelectronic with Sn_9Tl^{3-}, all obeying Wade's rules[13] for a *closo* cluster with 22 electrons for ten vertices.

References

[1] C.H.E. Belin, J.D. Corbett and A. Cisar, *J. Am. Chem. Soc.* **99**, 7163 (1977).

[2] A. Cisar and J.D. Corbett, *Inorg. Chem.* **16**, 632 (1977).

[3] J.D. Corbett, *Prog. Inorg. Chem.* **21**, 129 (1976).

[4] P.A. Edwards and J.D. Corbett, *Inorg. Chem.* **16**, 903 (1977).

[5] J.D. Corbett and P.A. Edwards, *J. Am. Chem. Soc.* **99**, 3313 (1977).

[6] S.C. Critchlow and J.D. Corbett, *J. Chem. Soc., Chem. Commun.*, 236 (1981).

[7] R.C. Burns and J.D. Corbett, *J. Am. Chem. Soc.* **104**, 2804 (1982).

[8] R.C. Burns and J.D. Corbett, *Inorg. Chem.* **24**, 1489 (1985).

[9] R.W. Rudolph and W.L. Wilson, *J. Am. Chem. Soc.* **103**, 2480 (1981).

[10] C.H. Schwalbe and W.N. Lipscomb, *Inorg. Chem.* **10**, 160 (1971).

[11] B.W. Eichhorn, R.C. Haushalter and W.T. Pennington, *J. Am. Chem. Soc.* **110**, 8704 (1988).

[12] B.W. Eichhorn and R.C. Haushalter, *J. Chem. Soc., Chem. Commun.*, 937 (1990).

[13] K. Wade, *J. Chem. Soc., Chem. Commun.*, 792 (1971); *Adv. Inorg. Chem. Radiochem.* **18**, 1 (1976).

3.24 Tin(II) Chloride Dihydrate
$[SnCl_2(H_2O)] \cdot H_2O$

Crystal Data

Monoclinic, space group $P2_1/c$ (No. 14)

$a = 9.313$, $b = 7.250$, $c = 8.970$Å, $\beta = 114.91°$, $Z = 4$

Atom	x	y	z
Sn	.37441	.26178	.53584
Cl(1)	.2864	.4923	.6875
Cl(2)	.3061	.5000	.3050
O(1)	.1094	.1756	.4128
O(2)	-.0674	.2066	.5984

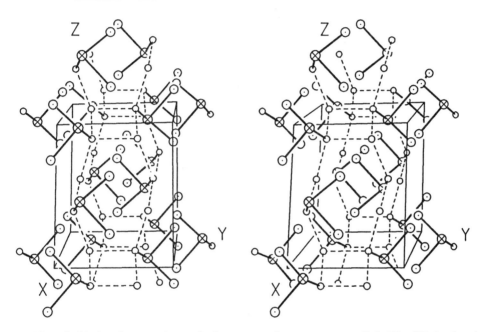

Fig. 3.24-1 Stereoview of the crystal structure of $SnCl_2 \cdot 2H_2O$ showing the pyramidal $[SnCl_2(H_2O)]$ group and the hydrogen-bonded layers.

Crystal Structure

Crystalline tin(II) chloride dihydrate has a predominant layer structure parallel to the (100) plane (Fig. 3.24-1).[1,2] It is built up from double layers of the dichloroaquotin(II) complex, $[SnCl_2(H_2O)]$, and the intervening space is filled by layers of the second water molecule. These layers are mainly held together by hydrogen bonds involving both types of water molecules (O...O = 2.74, 2.79, and 2.80Å), so that the crystal structure is more appropriately represented by the formula $[SnCl_2(H_2O)] \cdot H_2O$. In the pyramidal $[SnCl_2(H_2O)]$ complex Sn-Cl = 2.500 and 2.562Å, and Sn-O(aquo ligand) = 2.325Å; the observed bond angles (87.9°, 86.9° and 85.0°) are considerably

less than the tetrahedral value required by the sp^3 hybrid configuration
observed in many tin(II) compounds, reflecting the strong repulsion exerted by
the lone pair of electrons on the tin atom. Four additional Cl atoms at
distances of 3.209, 3.336, 3.416 and 3.654Å complete the coordination
environment of the tin atom.

Remarks

1. The structure and uses of tin(II) chloride[3,4]

    In the vapor phase the SnCl$_2$ molecule is bent with a bond angle of 95°
and a bond length of 2.42Å, [Fig. 3.24-2(a)].[5] Crystalline anhydrous SnCl$_2$
has a layer structure with chains of corner-shared trigonal pyramidal [SnCl$_3$]
groups [Fig. 3.24-2(b)]. The dihydrate also has a three-coordinated structure
with only one of the H$_2$O molecules directly bonded to the tin atom [Fig. 3.24-
2(c)]. In CsSnCl$_3$ and K$_2$[SnCl$_3$]Cl.H$_2$O, the crystals contain pyramidal SnCl$_3^-$
with Sn-Cl = 2.52 and 2.59Å, respectively [Fig. 3.24-2(d)].

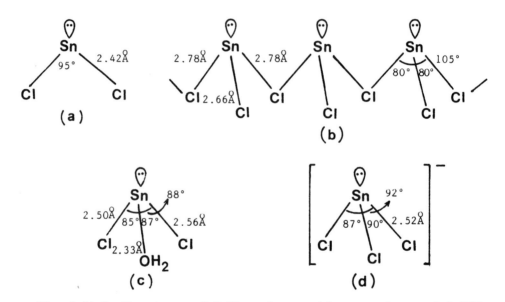

Fig. 3.24-2 Structures of SnCl$_2$ and some chloro complexes of Sn(II):
(a) SnCl$_2$(vapour), (b) SnCl$_2$(crystalline), (c) the neutral aquo complex
in [SnCl$_2$(H$_2$O)].H$_2$O, and (d) the (SnCl$_3$)$^-$ ion in CsSnCl$_3$.

Apart from its structural interest, SnCl$_2$ is important as a widely used
mild reducing agent in acid solution. The dihydrate is commercially available
for use in electrolytic tin-plating baths, as a sensitizer in silvering
mirrors and in the plating of plastics, and as a perfume stabilizer in toilet
soaps.

2. Structures of SnF_2 and some related compounds[3,4]

Tin(II) fluoride is apparently trimorphic, having in addition to the normal (α) form two metastable forms. The α form consists of tetrameric molecules [Fig. 3.24-3(a)], in which the mean Sn-F bond length in the $[Sn_4F_4]$ ring is 2.18Å and the mean exo-cyclic bond length is 2.05Å. There are Sn(II) atoms with two quite different arrangements of ligands as shown in Fig. 3.24-3(b) and (c), where broken lines represent intermolecular bonds, but the three shortest bonds from each are those within the Sn_4F_8 molecule and are pyramidal with a mean bond angle of 84°. However, having regard to all the ligands the bond arrangements may be described as tetrahedral SnF_3E and octahedral SnF_5E, where E stands for the lone pair occupying an orbital.

In the 1:1 compound $SnF_2 \cdot AsF_5$ there are cyclic cations $Sn_3F_3{}^{3+}$ [Fig. 3.24-3(d)] and the structural formula is therefore $(Sn_3F_3)(AsF_6)_3$. Here the Sn atom is characterized by "one-sided" 4-coordination [Fig. 3.24-3(e)], variously described as square pyramidal or as derived from a trigonal bipyramid, SnF_4E, with Sn-F(axial) = 2.59Å and Sn-F(equatorial) = 2.10Å. There are two further sets of neighbours: two F at 2.85Å and two F at 3.05Å.

There are many halides of Sn(II) containing different halogens. The structure of SnClF can be described as a chain structure [Fig. 3.24-3(f)]; the four close neighbours are one F at 2.18Å, two F at 2.39Å, and one Cl at 2.52Å.

3. Structural features in tin chemistry

The structural chemistry of the Group IVB elements affords abundant illustrations of the trend to be expected from increasing atomic size, increasing electropositivity, and increasing tendency to form M(II) compounds.

The ability of Sn to form polyatomic cluster anions of very low formal oxidation state, $Sn_5{}^{2-}$, $Sn_9{}^{4-}$,...., reflects the now well-established tendency of the heavier B subgroup elements to form chain, ring or cluster homopolyatomic ions.

Sn(IV) compounds, besides tetrahedral coordination, often occur in five- and six-coordination. SnF_4 is polymeric with octahedral coordination about Sn, the $[SnF_6]$ units being joined into planar layers by edge-sharing of four equatorial F atoms. Complexes with a wide range of organic and inorganic ligands are known, particularly the six-coordinate *cis*- and *trans*-L_2SnX_4 and occasionally the 1:1 complex $LSnX_4$. Five-coordinate trigonal bipyramidal complexes have been established for $SnCl_5{}^-$ and $Me_2SnCl_3{}^-$. A novel rectangular pyramidal geometry for Sn(IV) has been revealed in $[(MeC_6H_3S_2)_2SnCl]^-$.[6]

In Sn(II) compounds, Sn(II) rarely adopts structures typical of spherically symmetrical ions because the nonbonding pair of electrons, which is $5s^2$ in the free gaseous ion, can be easily distorted in the condensed phase. Thus the nonbonding pair of Sn readily serves in a donor capacity with respect to a neighbouring atom bearing vacant orbitals, and the "vacant" Sn 5p and 5d orbitals can act as acceptors in forming additional covalent bonds.

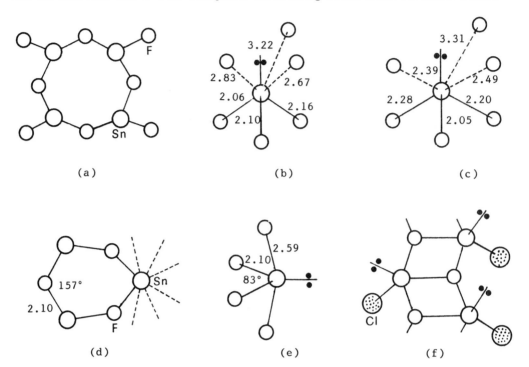

Fig. 3.24-3 Structures of SnF_2 and some of its derivatives. (After ref. 7).

References

[1] B. Kamenar and D. Grdenić, *J. Chem. Soc.*, 3954 (1961).

[2] H. Kiriyama, K. Kitahama, O. Nakamura and R. Kiriyama, *Bull. Chem. Soc. Jpn.* **46**, 1389 (1973).

[3] J.A. Zubieta and J.J. Zuckerman, *Prog. Inorg. Chem.* **24**, 251 (1978).

[4] N.N. Greenwood and A. Earnshaw, *Chemistry of the Elements*, Pergamon Press,
     Oxford, 1986, p. 439.

[5] M.W. Lister and L.E. Sutton, *Trans. Faraday Soc.* **37**, 406 (1941).

[6] A.C. Sau, R.O. Day and R.R. Holmes, *J. Am. Chem. Soc.* **103**, 1264 (1981).

[7] A.F. Wells, *Structural Inorganic Chemistry*, 5th ed., Clarendon Press,
     Oxford, 1984.

Note The inorganic chemistry of tin, including tin-metal bonded compounds, is summarized in P.G. Harrison (ed.), *Chemistry of Tin*, Blackie, Glasgow and London, 1989.

3.25 Hexanuclear Basic Lead(II) Perchlorate Hydrate
[Pb$_6$O(OH)$_6$](ClO$_4$)$_4$·H$_2$O

Crystal Data

Orthorhombic, space group *Pbca* (No. 61)

a = 10.814(5), b = 16.706(6), c = 26.273(8)Å, Z = 8

Atom*	x	y	z	Atom	x	y	z
Pb(1)	.3391	−.0421	.2050	Cl(4)	.107	.335	.084
Pb(2)	.2347	.1505	.1530	O(11)	.820	.359	.208
Pb(3)	.5672	.1527	.1990	O(12)	.884	.222	.207
Pb(4)	.5078	.0543	.0866	O(13)	.833	.275	.276
Pb(5)	.4734	.2794	.0863	O(14)	.028	.315	.235
Pb(6)	.7958	.2123	.0935	O(21)	.853	−.033	.174
O(1)	.440	.166	.133	O(22)	.056	−.065	.136
O(2)	.301	.025	.129	O(23)	−.022	.067	.134
O(3)	.309	.096	.222	O(24)	.044	.016	.207
O(4)	.630	.267	.141	O(31)	.315	.028	−.013
O(5)	.605	.192	.054	O(32)	.135	.074	.026
O(6)	.689	.111	.136	O(33)	.197	.138	−.044
O(7)	.519	.012	.176	O(34)	.317	.152	.024
O(8)	.388	.116	.482	O(41)	.141	.398	.045
Cl(1)	.890	.296	.233	O(42)	.090	.258	.064
Cl(2)	.978	−.002	.162	O(43)	0	.365	.109
Cl(3)	.241	.099	−.002	O(44)	.194	.335	.120

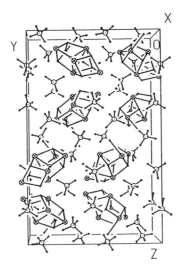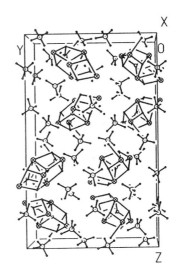

Fig. 3.25-1 Stereoview of the molecular packing in
[Pb$_6$O(OH)$_6$](ClO$_4$)$_4$·H$_2$O. Hydrogen bonds involving the hydroxyl and aqua
ligands have been omitted for clarity.

Crystal Structure

To a good approximation, the crystal comprises a packing of discrete
[Pb$_6$O(OH)$_6$(H$_2$O)]$^{4+}$ clusters and ClO$_4^-$ ions (Fig. 3.25-1).[1] The cluster
cation consists of three face-sharing distorted tetrahedra of Pb atoms, one of
which bears a terminal aqua ligand. Four of the lead atoms, Pb(2) to Pb(5),
are located at the corners of a central tetrahedron, while the remaining two

cap two of the tetrahedral faces, thereby generating two exterior tetrahedra, as shown in Fig. 3.25-2. The adjacent Pb...Pb distances average 3.81Å and range from 3.44 to 4.09Å.

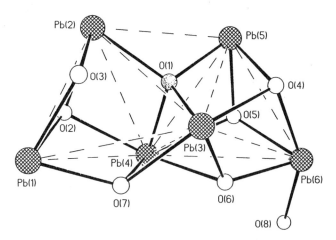

Fig. 3.25-2 Structure of the $[Pb_6O(OH)_6(H_2O)]^{4+}$ cluster. The three face-sharing tetrahedra of Pb atoms are indicated by broken lines. The central O atom is distinguished by shading, and O(8) represents the terminal aqua ligand.

Apart from the perchlorate ions, the asymmetric unit contains eight oxygen atoms of three distinct types: an oxide ion, O(1), six hydroxide ions, O(2) to O(7), and a water molecule, O(8). The unique oxygen atom O(1) is located at the center of the central lead tetrahedron, and the six OH⁻ ions over the six external faces of the two exterior lead tetrahedra. The Pb-O distances in the $[Pb_6O(OH)_6(H_2O)]^{4+}$ cluster are listed in Table 3.25-1.

Table 3.25-1 The Pb-O distances (Å) in the $[Pb_6O(OH)_6(H_2O)]^{4+}$ cluster.

	Pb(1)	Pb(2)	Pb(3)	Pb(4)	Pb(5)	Pb(6)
O(1)	–	2.30	2.22	2.35	2.29	–
O(2)	2.33	2.30	–	2.55	–	–
O(3)	2.37	2.18	3.01	–	–	–
O(4)	–	–	2.53	–	2.23	2.37
O(5)	–	–	–	2.67	2.21	2.33
O(6)	–	–	2.23	2.53	–	2.33
O(7)	2.28	–	2.48	2.46	–	–
O(8)	–	–	–	–	–	2.74

Remarks

1. Relationship between the $[Pb_6O(OH)_6(H_2O)]^{4+}$ and PbO structures

   In the $[Pb_6O(OH)_6(H_2O)]^{4+}$ cluster, the oxide ion is near the centroid of the central tetrahedron at an average distance of 2.29Å from its lead neighbours. A tetrahedral environment for the oxide ion (Pb-O = 2.28Å) is also found in the colourless adamantane-like complex, $Pb_4O(OSiPh_3)_6$,[2] obtained as a 1:1 benzene solvate by the reaction of Ph_3SiOH with $Pb(C_5H_5)_2$.

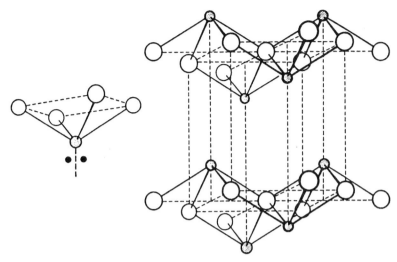

Fig. 3.25-3 The crystal structure of tetragonal PbO (and SnO). The small shaded circles represent metal atoms. The arrangement of bonds from a metal atom is shown at the right, where the two dots represent an "inert pair" of electrons. (After ref. 4).

Tetragonal PbO, stable at room temperature, is the most widely used inorganic compound of lead. It has the layer structure of Fig. 3.25-3 in which the lead atom is bonded to four oxygen atoms arranged in a square to one side of it, with the lone pair of electrons presumably occupying the apex of the tetragonal pyramid.[3] Each oxygen atom is surrounded tetrahedrally by four lead neighbours at 2.30Å in a manner similar to that in the $[Pb_6O(OH)_6]^{4+}$ cluster, so that the latter may be considered as a fragment of the PbO lattice.

2. The oxides of lead

Structural data are available for several oxides of lead and their polymorphs. The results from X-ray and neutron diffraction studies are summarized in Table 3.25-2.[3] The crystal structure of red lead, Pb_3O_4, is illustrated in Fig. 3.25-4; this structure is isomorphous with that of $ZnSb_2O_4$.

3. Stereochemically inert lone pairs in subvalent Group IV compounds

The influence of unshared (lone) pairs of electrons on the geometry of main-group compounds, as popularized by the VSEPR model, is amply verified by a wealth of structural data. However, authenticated molecular structures which exhibit stereochemically inactive lone-pair electrons have become increasingly numerous and better known in recent years.[5,6] A striking

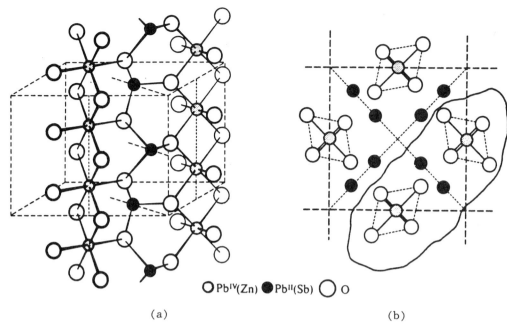

$\bigcirc$ Pb^{IV}(Zn) $\bullet$ Pb^{II}(Sb) $\bigcirc$ O

(a)   (b)

Fig. 3.25-4 Crystal structure of Pb_3O_4 (and $ZnSb_2O_4$). (a) Portion of the structure outlined in the projection (b) showing the chains of $Pb^{IV}O_6$ octahedra (mean Pb-O 2.14Å) joined by pyramidally coordinated Pb^{II} atoms (one 2.13, two at 2.18Å, mean O-Pb-O 76°). [After ref. 4].

Table 3.25-2   The oxides of lead.

Form	System	Structural features
PbO, red (litharge)	tetragonal	see remark 1 and Fig. 3.25-3
PbO, yellow (massicot)	orthorhombic	related to red form; zigzag chain (Pb-O 2.20Å, Pb-O-Pb 148°) connected into layers by longer bonds (2.49Å)
Pb_3O_4, red (red lead)	tetragonal	see Fig. 3.25-4
Pb_2O_3, black	monoclinic	Pb^{II} atoms (very irregular 6-coordination; 2.31, 2.43, 2.44, 2.64, 2.91, 3.00Å) situated between layers of distorted $Pb^{IV}O_6$ octahedra (mean 2.18Å)
PbO_2, maroon	tetragonal	rutile-type (mean 2.18Å)
PbO_2, black	orthorhombic	a-PbO_2 structure derived from hcp layers with half of octahedral sites filled (see p. 171 of ref. 4)

example is decaphenylstannocene, $[\eta^5\text{-}(C_6H_5)_5C_5]_2Sn$, which belongs to point group S_{10} [Fig. 3.25-5].[7] The tin atom is located at an inversion center between a pair of perfectly planar and staggered cyclopentadienyl rings

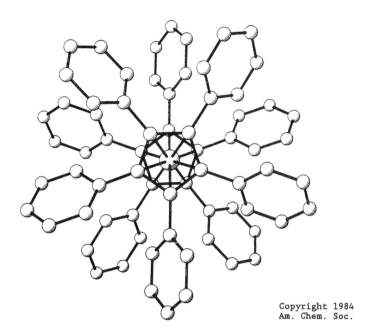

Fig. 3.25-5 Decaphenylstannocene viewed along its principal symmetry
axis. (After ref. 7).

(2.401Å to the ring centers; average Sn-C = 2.692Å). The attached phenyl
groups are inclined to each cyclopentadienyl ring oppositely in an opposed
paddle wheel fashion, and there are no short contacts in the molecular
packing.

   An example of a divalent lead complex in which the lone pair is
stereochemically inert is $Pb[SC(NH_2)_2]_6(ClO_4)_2$.[8] The metal atom lies at the
origin of the triclinic unit cell (space group $P\bar{1}$), being equidistant from six
S atoms and at the center of a slightly distorted octahedron (Fig. 3.25-6).

   Based on a structural survey of compounds in which subvalent Group IV
atoms occupy perfectly symmetrical sites (Table 3.25-3), Ng and Zuckerman[6]
have called attention to the following patterns:

   a) Certain subvalent, fourth-group compounds undergo single or a series
of first-order phase transformations at increasing temperatures from their
expected distorted phases to give successively more symmetrical structures.

   b) If the material does not melt first, this may include a phase in
which the subvalent group-four atom occupies a site of perfect cubic symmetry.

   c) The materials which contain perfectly symmetrical sites for the
subvalent fourth-group atom have regular octahedral (six-) or cubic (eight-)
or dodecahedral (12-coordinated) environments consisting of equidistant

Table 3.25-3 Symmetrical subvalent compounds of heavier Group IV elements.

Molecular

Ge: $[\eta^5\text{-}C_5H_5Mn(CO)_2]_2Ge$

Sn: $[\eta^5\text{-}(C_6H_5)_5C_5]_2Sn$

Extended Lattice

Ge: $MGeCl_3$ (M = Cs^+, Me_4N^+)

GeSe at >651°C

GeTe at >300°C

$GeBi_2Te_4$ film on NaCl

$Ge_3Bi_2Te_6$ film on NaCl

$K_2Ge_2O_3$

Sn: $CsSnCl_3$ at >117°C

$MSnBr_3$ (M = Cs^+, $MeNH_3^+$)

$MSnI_3$ (M = Cs^+ at >152°C, $MeNH_3^+$, Et_4N^+)

Cs_4SnBr_6

SnTe

SnI_2

$[Rh(NH_3)_6]_3^{3+}[Rh(SnCl_3)_4SnCl_4]^{5-}[SnCl_6]^{4-}.4H_2O$

Pb: PbF_2 at >315°C

PbI_2

$CsPbF_3$ at >615°C

$MPbCl_3$ (M = Cs^+ at >46.9°C, $MeNH_3^+$)

$MPbBr_3$ (M = Cs^+ at >130°C, $MeNH_3^+$)

$MPbI_3$ (M = Cs^+ at >305°C, $MeNH_3^+$)

PbS

PbSe

PbTe

$M_2PbCo(NO_2)_6$ (M = Rb^+, Cs^+)

$K_2PbNi(NO_2)_6$

$M_2PbCu(NO_2)_6$ (M = K^+, Rb^+, Cs^+, Tl^+)

halogen (F, Cl, Br or I) or chalcogen (O, S, Se or Te) nearest neighbours. In
the regular cubic examples only oxygen has been found thus far to be present;
all the regular cubic geometries discovered have lead(II) atoms surrounded by
12 oxygen atoms. Examples with asymmetric counter ions exist, as do examples
arising from epitaxial effects on materials deposited on rock-salt surfaces.

d) Evidence for the cubic symmetry can come from the observation of
single lines in Mössbauer (for tin), pure nuclear quadrupole resonance (NQR),
or electron-spin resonance (ESR) spectra, or from X-ray or neutron diffraction
studies.

e) The transformations to the cubic phase are accompanied by isotropic

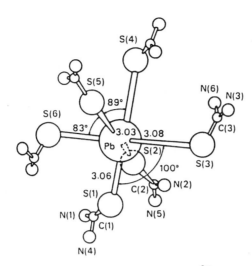

Fig. 3.25-6 Geometry of the Pb[SC(NH$_2$)$_2$]$_6$$^{2+}$ cation in the perchlorate salt. (After ref. 8).

expansion of the lattices and large thermal parameters for the fourth-group atom.

f) The resulting cubic phases are often intensely colored and electrically conducting.

There seems at the present time no general theory available that would allow one to predict which compositions would exhibit these unusual behaviors (*cf.* ref. 9).

References

[1] T.G. Spiro, D.H. Templeton and A. Zalkin, *Inorg. Chem.* **8**, 856 (1969).

[2] C. Gaffney, P.G. Harrison and T.J. King, *J. Chem. Soc., Chem. Commun.*, 1251 (1980).

[3] J. Leciejewicz, *Acta Crystallogr.* **14**, 1304 (1961).

[4] A.F. Wells, *Structural Inorganic Chemistry*, 5th ed., Clarendon Press, Oxford, 1984, pp. 556-559.

[5] J.K. Burdett, *Molecular Shapes*, Wiley, New York, 1980.

[6] S.-W. Ng and J.J. Zuckerman, *Adv. Inorg. Chem. Radiochem.* **29**, 297 (1985).

[7] M.J. Heeg, C. Janiak and J.J. Zuckerman, *J. Am. Chem. Soc.* **106**, 4259 (1984).

[8] I. Goldberg and F.H. Herbstein, *Acta Crystallogr., Sect. B* **28**, 400 (1972).

[9] D.L. Cullen and E.C. Lingafelter, *Inorg. Chem.* **10**, 1264 (1971).

3.26   Nitrogen Triiodide-Ammonia (1/1)

$NI_3 \cdot NH_3$

Crystal Data

Monoclinic, space group $P2_1/m$ (No. 11)

$a = 7.132(3)$, $b = 7.506(3)$, $c = 6.324(4)$Å, $\beta = 98.6(1)°$, $Z = 2$

Atom	x	y	z
I(1)	1/2	0	1/2
I(2)	.0629(2)	1/4	.3173(2)
I(3)	.2439(2)	1/4	.8568(2)
N(1)	.322(2)	1/4	.540(2)
N(2)	-.227(2)	1/4	.036(3)
H(1)	-.18	1/4	-.12
H(2)	-.31	.13	.05

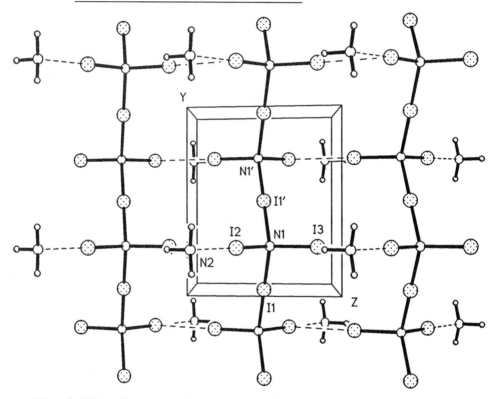

Fig. 3.26-1 The crystal structure of nitrogen triiodide-ammonia viewed parallel to the *a* axis.

Crystal Structure

Pure NI_3 has not been isolated, but the structure of its 1:1 adduct with NH_3 has been elucidated.[1] The adduct is an explosive material, obtained as brown-black lustrous needles, and sensitive to shock, light and heat. Crystalline $NI_3 \cdot NH_3$ has a polymeric structure in which tetrahedral $[NI_4]$ units are corner-linked into infinite -N-I-N-I- chains (with bond lengths 2.15 and

2.30Å) as shown in Fig. 3.26-1. The chain has 2_1 symmetry extending in the [010] direction. The other I atoms of the tetrahedra strongly interact with those in adjacent chains, I...I = 3.36Å, being linked into sheets in the *c* direction. In addition, one I of each [NI₄] unit is also weakly attached to an ammonia molecule (N...I 2.53Å) that projects into the space between the sheets of tetrahedra.

A further interesting feature is the presence of linear symmetrical N(1)-I(1')-N(1') and almost linear N(1)-I(2)-N(2) (176.2°) and N(1)-I(3)-I(2, neighbouring chain) (172.3°) groups, which suggest the presence of 3-center, 4-electron bonds characteristic of polyhalides.

Remarks

1. Adducts of NI₃

The principal structural characteristic of NI₃.NH₃ is that it does not contain discrete NI₃ molecules but instead consists of chains of [NI₄] tetrahedra sharing two vertices, with one NH₃ attached by a weaker bond to alternate unshared I atoms along the chain, as represented schematically in Fig. 3.26-2(a).

The ammonia moiety in NI₃.NH₃ is relatively easily substituted by other nitrogen bases without change in the polymeric N-I framework. The resulting adducts are usually formed by direct reaction of preformed NI₃.NH₃ with an excess amount of base (pyridine, quinuclidine, hexamethylenetetramine, etc.), with or without additional solvent (usually water), and ammonia is set free.

The X-ray structural investigation of the pyridine adduct NI₃.py[3] has established that it is structurally very similar to NI₃.NH₃. Chains of tetrahedra with infinite -N-I-N-I- linkages showing a characteristic translational period of 7.5Å parallel to the *b* axis are again found [Fig. 3.26-2(b)]. The adduct pyridine molecules are bonded through nitrogen N(2) to I(2). Table 3.26-1 summarizing the corresponding distances and angles emphasizes the structural analogy and also shows that in the pyridine adduct the cohesion of the separate chains of tetrahedra to the sheets of tetrahedron in the *bc* plane is not as strong as in the ammonia adduct. The angle N(1)-I(3)-I(2, neighbouring chain) is distorted to 153° by the bulky pyridine ring, and formation of the normally linear 3-center, 4-electron bond between the three atoms is made more difficult. The I(3)-I(2, neighbouring chain) distance between the chains of tetrahedra is, as a result, increased from 3.36 to 3.93Å and can be regarded as only a very weak I...I contact.

In the (CH₃)₂NI crystal,[4] the chains of [(CH₃)₂NI₂] tetrahedra with infinite -N-I-N-I- linkages [Fig. 3.26-2(c)] are similar to those in NI₃.NH₃.

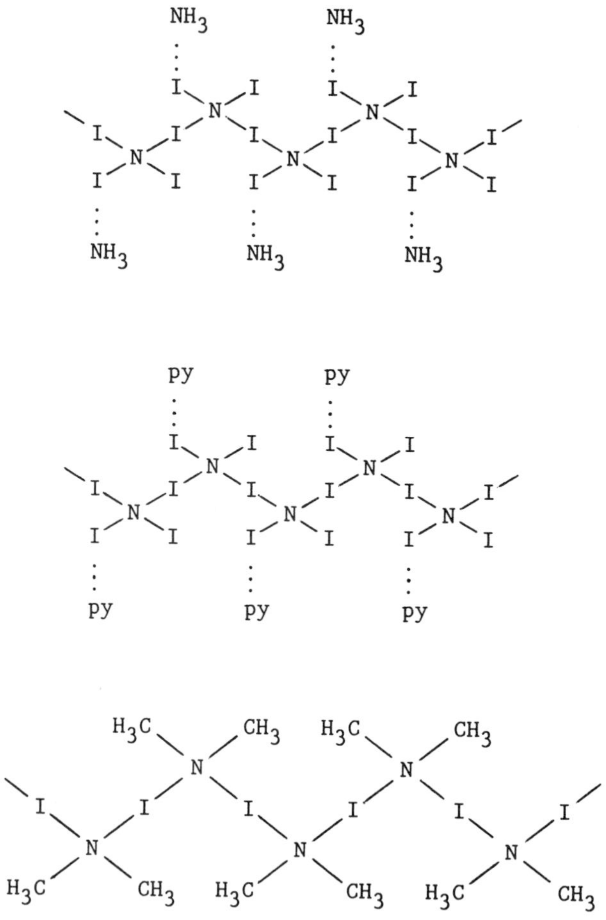

(a)

(b)

(c)

Fig. 3.26-2 Schematic representation of the structure of (a) $NI_3 \cdot NH_3$, (b) $NI_3 \cdot py$, and (c) $(CH_3)_2NI$.

Table 3.26-1 Bond distances (Å) and angles (deg) in $NI_3 \cdot NH_3$ and $NI_3 \cdot py$.

Distance or angle measured	$NI_3 \cdot py$	$NI_3 \cdot NH_3$
I(3)-I(2, neighbouring chain)	3.93	3.36
N(1)-I(1)	2.36	2.30
N(1)-I(2)	2.10	2.15
I(2)-N(2)	2.59	2.53
N(1)-I(1')-N(1')	180	180
N(1)-I(2)-N(2)	174	176
N(1)-I(3)-I(2, neighbouring chain)	153	172

Brick red $NI_3 \cdot I_2 \cdot (CH_2)_6N_4$ is the most stable NI_3 complex so far investigated. The crystal structure (Fig. 3.26-3) shows the nitrogen triiodide and hexamethylenetetramine molecules each to be approximately tetrahedrally surrounded by two molecules of one type and two of another. Iodine molecules make up one of the four links between NI_3 and $(CH_2)_6N_4$. The geometrical arrangement and measured distances indicate that there is marked intermolecular bonding between the iodine atoms of NI_3 and the nitrogen atoms in $(CH_2)_6N_4$. In addition, the nitrogen atom of NI_3 is involved in intermolecular bonding to the iodine molecule which in turn is similarly bonded to hexamethylenetetramine.

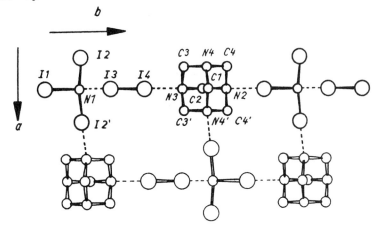

Fig. 3.26-3 Projection of the structure of $NI_3.I_2.(CH_2)_6N_4$ on the (001) plane. Intermolecular bonds are shown by dashed lines. Principal distances are $I-N(NI_3)$ 2.140 and 2.144Å, $I(I_2)...N(NI_3)$ 2.474Å, $I(I_2)...N(C_6H_{12}N_4)$ 3.232Å, $I-I$ 2.808Å, $I(NI_3)...N(C_6H_{12}N_4)$ 2.567 and 2.582Å. The $I-N-I$ angles at the NI_3 moiety are 104.9-110.9°. (After ref. 5).

2. The halides of nitrogen

Two halides of nitrogen are well characterized, NF_3 and NCl_3. The former is quite stable when pure and has been studied as a vapor, whereas the explosive NCl_3 has to be examined in the crystalline state at -125°C.[6] Partially halogenated compounds which have been prepared include NH_2F, NHF_2, NH_2Cl and $NHCl_2$. Pure NBr_3 at low temperature is a deep red, very temperature-sensitive, volatile solid.

X-Ray structural analysis of solid NCl_3 has revealed a structure made up of NCl_3 pyramids (idealized C_{3v} molecular symmetry is lowered to *m* in the crystal), which are stacked in layers parallel to the *ac* plane [Fig. 3.26-4(a)]. Pyramids within a layer are arranged in the same orientation, whereas those in neighbouring layers have an anti-parallel orientation [Fig. 3.26-4(b)]. The pyramids form undulating layers parallel to the *bc* plane. Their projection on the *bc* plane shows that the chlorine atoms are approximately hexagonally close-packed. The N-Cl bond distance and the Cl-N-Cl bond angle within a NCl_3 molecule vary appreciably (from 1.71 to 1.78Å with a weighted mean of 1.75Å, and from 105.1° to 109.6°).

In the NCl_3 crystal, the distances between molecules lie in the range

3.19-3.68Å; the fact that the linear N-Cl-N bond is the shortest implies the presence of a 3-center, 4-electron bond analogous to that in polyhalides.

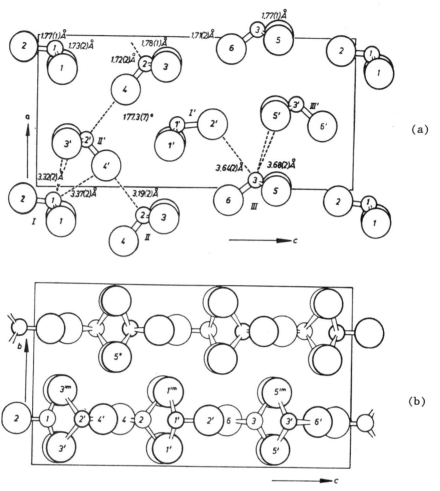

(a)

(b)

Fig. 3.26-4 Projections of the crystal structure of NCl_3: (a) a layer on the (010) plane; (b) on the (100) plane. (After ref. 6).

References

[1] H. Hartl, H. Bärnighausen and J. Jander, *Z. Anorg. Allg. Chem.* **357**, 225 (1968).

[2] J. Jander, *Adv. Inorg. Chem. Radiochem.* **19**, 1 (1976).

[3] H. Hartl and D. Ullrich, *Z. Anorg. Allg. Chem.* **409**, 228 (1974).

[4] R. Hagedorn, H. Pritzkow and J. Jander, *Acta Crystallogr., Sect. B* **33**, 3209 (1977).

[5] H. Pritzkow, *Z. Anorg. Allg. Chem.* **409**, 237 (1974).

[6] H. Hartl, J. Schöner, J. Jander and H. Schulz, *Z. Anorg. Allg. Chem.* **413**, 61 (1975).

3.27   Orthorhombic Black Phosphorus
P

Crystal Data

Orthorhombic, space group *Cmca* (No. 64)

$a = 3.3136(5)$, $b = 10.478(1)$, $c = 4.3763(5)$Å, $Z = 8$

Atom	Position	x	y	z
P	8(f)	0	.10168	.08056

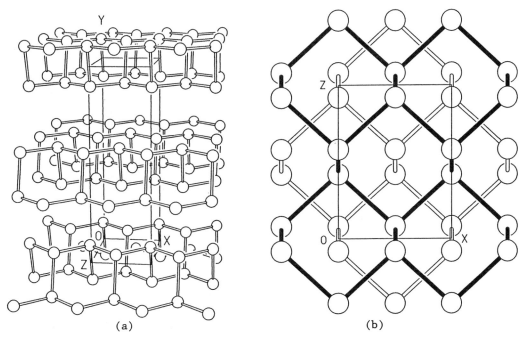

(a)   (b)

Fig. 3.27-1   (a) Perspective view of the crystal structure of black phosphorus.   (b) Projection showing two of the puckered layers (parallel to *ac* plane).

Crystal Structure

Orthorhombic black phosphorus[1] has a continuous double-layer structure [Fig. 3.27-1(a)], which is a heavily puckered hexagonal net [Fig. 3.27-1(b)], in which each P atom is covalently bound to three others within the same layer with the following bond distances and angles:

P-P   2.244Å (one),   2.224Å (two);

P-P-P   96.34° (one),   102.09° (two).

In addition, each P atom has eight neighbours in the same layer at distances of 3.314(x4), 3.334(x2), and 3.475(x2)Å, which may be contrasted with the shortest interlayer contact of 3.801Å.

The structure relates well to the fact that orthorhombic black phosphorus is a semiconductor and exhibits flakiness.

Remarks

1. Allotropes of phosphorus

Phosphorus exists in many allotropic modifications. At least five crystalline polymorphs are known and there are also several amorphous or vitreous forms. All forms, however, melt to give the same liquid which consists of symmetrical P_4 tetrahedral molecules. The same molecular entity exists in the gas phase, the P-P bond length being 2.21Å, but at high temperatures (above ~800°C) and low pressures P_4 is in equilibrium with diatomic P_2 (P-P 1.895Å) species. Fig. 3.27-2 shows the interconversion of the various forms of elemental phosphorus.

The commonest form of phosphorus, and the one which is usually formed by condensation from the gaseous or liquid states, is the waxy, cubic white α-P_4 form (or α-white phosphorus) which is stable from -77°C to its melting point (44.1°C) at atmospheric pressure. The crystal data of α-P_4 are a = 18.51(3)Å, space group $I\bar{4}3m$, $I432$ or $Im3m$, with $Z = 56(P_4)$,[2] but its crystal structure is still unknown.

2. Rhombohedral black phosphorus and cubic black phosphorus

Under very high pressures, orthorhombic black phosphorus undergoes further reversible transitions successively to denser rhombohedral and primitive cubic forms.

Rhombohedral black phosphorus, built from a simple hexagonal network (space group $R\bar{3}m$, a = 3.377, c = 8.806Å, and Z = 6) of phosphorus atoms, is considerably less puckered than the orthorhombic form, as shown in Fig. 3.27-3(a).[3] Deductions from powder diffraction data gave P-P = 2.13Å and P-P-P = 95°.

In the cubic form, each phosphorus atom is octahedrally coordinated by other phosphorus atoms at a = 2.38Å [Fig. 3.27-3(b)].[3] If this form of phosphorus is considered as "metallic", the calculated radius for 12-coordination[4] is 1.04 x a/2 = 1.24Å, which can be compared with the value of 1.18Å deduced from metal-rich phosphide structures.

The structures of the three crystalline black forms of phosphorus are related simply to each other as indicated in Fig. 3.27-1 and Fig. 3.27-3. The cubic form is derived geometrically from the rhombohedral form by reducing the inter-bond angle to 90° and compressing successive hexagonal layers of the latter. In this way all the points of a simple cubic lattice can be occupied by phosphorus atoms. In a similar way, the simple cubic lattice can be

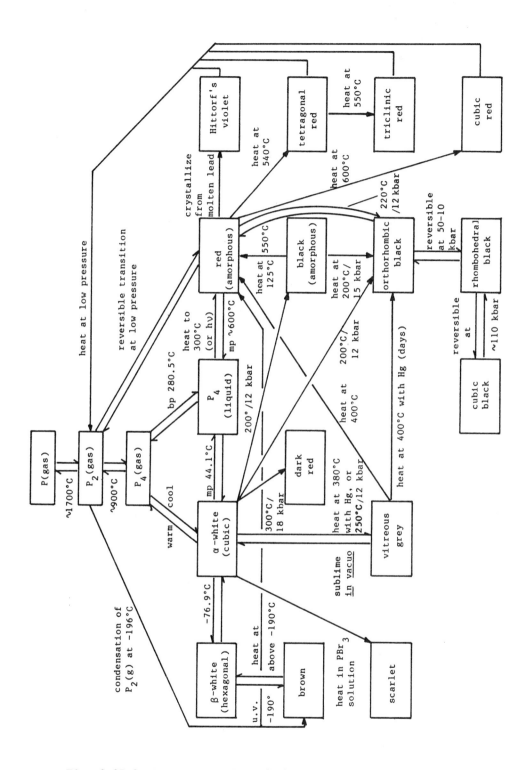

Fig. 3.27-2 Interconversion of the various forms of elemental phosphorus. (After ref. 8).

derived from the orthorhombic form by reducing the inter-bond angles to 90°
and compressing successive double layers.

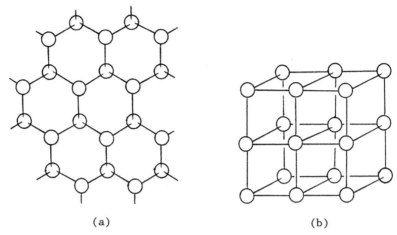

(a)   (b)

Fig. 3.27-3 The structures of (a) rhombohedral black phosphorus,
showing a portion of one hexagonal layer; (b) cubic black phosphorus,
showing four unit cells.   (After ref. 6).

3. Hittorf's violet phosphorus (monoclinic violet phosphorus)

This form of phosphorus is named after Hittorf who first prepared it in
1865. The crystal system is monoclinic, space group $P2/c$, with $a = 9.21$, $b =$
9.15, $c = 22.60$Å, $\beta = 106.1°$, and $Z = 84(P)$.[5]

The structure consists of cage-like P_8 and P_9 groups, which are linked
alternately by pairs of phosphorus atoms to form tubes of pentagonal cross-
section. Parallel "pentagonal tubes" form "double layers", each consisting of
two systems of tubes which snugly fit together with no chemical bonds between
them. The tube systems in neighbouring double layers are approximately
perpendicular to each other, with intermittent cross-linkages between them. A
schematic general view of the structure is shown in Fig. 3.27-4. The
geometrical parameters of a pentagonal tube are given in Fig. 3.27-5. The
bond distances fall into three groups. In the eclipsed system - one where the
central bond is P(2)-P(10), P(4)-P(12), P(6)-P(14), P(8)-P(16) or P(19)-P(20),
the P-P bond distances [average being 2.278(14)Å] are significantly longer
than the others. In the staggered systems, those where the central bond is
P(5)-P(19), P(7)-P(20), P(13)-P(19) or P(15)-P(20), the distances are
intermediate, at 2.238(2)Å. The remaining systems are all *trans*, in part,
with the group of shortest P-P bond distances averaging 2.203(8)Å.

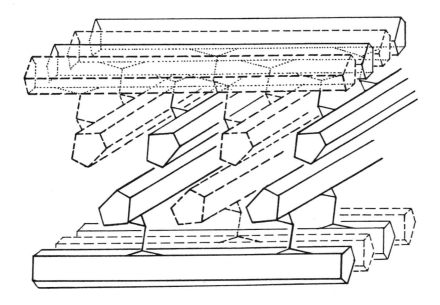

Fig. 3.27-4 A schematic representation of the structure of monoclinic (Hittorf's) phosphorus. (Adapted from ref. 7).

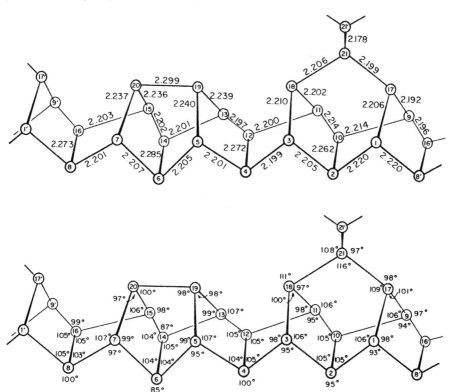

Fig. 3.27-5 Bond distances (in Å) and bond angles in monoclinic (Hittorf's) phosphorus. (After ref. 7).

4. Allotropes of arsenic[7]

Elemental arsenic is known in six allotropic forms: α (grey, metallic, rhombohedral, ordinary), β, γ, δ, ϵ (orthorhombic, arsenolamprite), and yellow (cubic). The β, γ and δ varieties are amorphous.

α-Arsenic crystallizes in space group $R\bar{3}m$ with a hexagonal unit cell of parameters a_H = 3.7598(1), c_H = 10.5475Å (at 300K) and Z = 6; the As atom occupies a 6(c) site with coordinates (0, 0, .22764).[9]

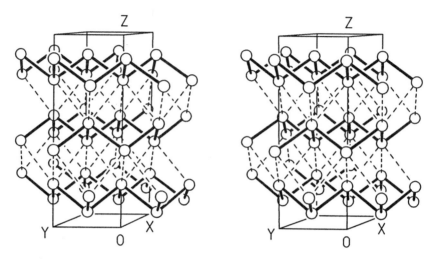

Fig. 3.27-6 Stereoview of the crystal structure of α-arsenic; intralayer covalent bonds are shown by solid lines and interlayer contacts as dashed lines.

The structure of α-arsenic consists of puckered sheets of covalently bonded arsenic atoms stacked in layers perpendicular to the hexagonal c axis (Fig. 3.27-6). Each arsenic atom is bonded to three others (As-As-As bond angle = 96.7°) in the same layer at 2.517Å and has three next nearest neighbours in an adjacent layer at 3.120Å.

ϵ-Arsenic is isostructural with black phosphorus, with lattice parameters a = 3.63, b = 4.45 and c = 10.96Å.[10]

Yellow arsenic, formed as a sublimation product, is cubic and presumably consists of As_4 molecules, but structural data have not been obtained because the crystal readily decomposes in the X-ray beam.

5. Compounds containing P_x and As_x fragments

Homonuclear aggregates of P and As exist in many compounds and complexes. For example, the following phosphides are formed between the alkali metals and phosporus: M_3P, MP_x (x < 1), MP, M_4P_6, M_3P_7, M_3P_{11}, MP_5, MP_7, $MP_{10.3}$, MP_{11}, and MP_{15}.[11] These substances are coloured black up to the composition M_4P_6, and they react spontaneously with protic solvents.

The phosphides M_3P_7 and M_3P_{11} are yellow or orange, but just as reactive.   The remaining compounds are red or brown and, with increasing P content, more and more inert to mineral acids.   Fig. 3.27-7 shows the structures of some homonuclear E_x (E = P, As, Sb) groups and chain phosphide ions in which the two-connected P atoms are associated with formal negative charges.

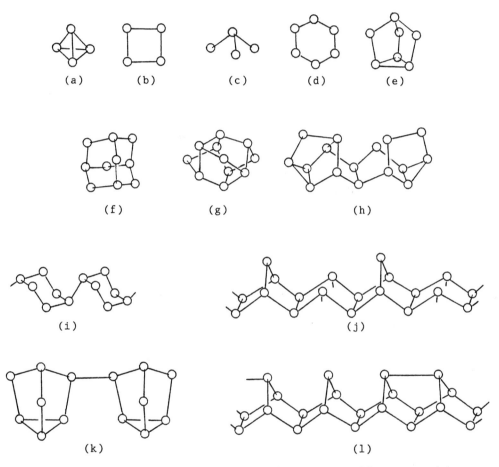

Fig. 3.27-7 Structures of homonuclear E_x (E = P, As, Sb) groups: (a) P_4, As_4; (b) P_4^{4-}, As_4^{4-}; (c) P_4^{6-}; (d) P_6^{2-}; (e) P_7^{3-}, As_7^{3-}, Sb_7^{3-}; (f) P_{10}^{6-}; (g) P_{11}^{3-}, As_{11}^{3-}; and (h) P_{16}^{2-}.   Chain phosphide ionic structures in (i) BaP_3, (j) TlP_5, (k) RbP_7 and (l) KP_{15}.   (Adapted from refs. 11 and 12).

The largest known homoatomic polyarsenide cluster exists in red, crystalline $[Rb(crypt)]_4As_{22}$.4DMF.[13]   The centrosymmetric anion consists of two As_{11} cages linked by a single As-As bond of length 2.432(6)Å.   Each As_{11} subunit in As_{22}^{4-} possesses the trishomocubane structure found in As_{11}^{3-} [14] and P_{11}^{3-};[15] three edges of the "reference cube" are bridged by As(1), As(4) and As(7), with an approximate D_3 molecular axis passing through As_a and As_b [Fig. 3.27-8(a)].

The bonding in the E_x groups can be summarized thus: i) electron clouds tend to repel each other, ii) isoelectronic units behave practically in the same way, iii) the eight-electron and eighteen-electron rules hold, iv) formal ions can be used with the bonding of isoelectronic elements or groups, v) for the same number of bonds various structural configurations are possible, and vi) homonuclear bonds are supplemented by heteronuclear bonds in the sense of formal donor-acceptor interactions.

Several heteroatomic polyarsenide clusters have been synthesized and characterized by X-ray analysis. The $As_{11}Te^{3-}$ anion in $[K(crypt)]_3As_{11}Te.en$ has an exocyclic tellurium atom attached to a trishomocubane-like As_{11} cluster of approximate D_3 symmetry [threefold axis through As(1) and As(5) in Fig. 3.27-8(b)].[16] The $As_{10}Te_3^{2-}$ cluster anion in its Ph_4P^+ salt is related to the As_{11}^{3-} type but has one of the three two-connected As atoms replaced by a bridging Te(1) atom, and the two remaining tellurium atoms are linked in a terminal fashion to As(7) and As(10) [Fig. 3.27-8(c)]; the cluster has

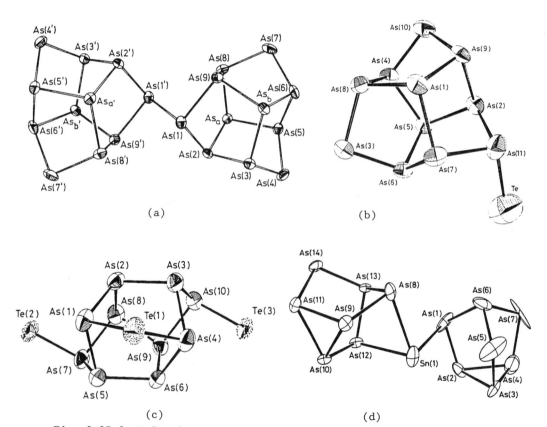

(a)   (b)

(c)   (d)

Fig. 3.27-8 Molecular structures of some polyarsenide cluster anions: (a) As_{22}^{4-}, (b) $As_{11}Te^{3-}$, (c) $As_{10}Te_3^{2-}$ and (d) $As_{14}Sn^{4-}$. (After refs. 13, 16 and 17).

approximate C_2 symmetry, and the average As-Te(bridging) and As-Te(terminal) bond lengths are 2.608(4) and 2.542(5)Å, respectively.[17] The unusual geometry of the $As_{14}Sn^{4-}$ anion in $[K(crypt)]_4As_{14}Sn$ is shown in Fig. 3.27-8(d).[13] Two As_7 cages of the type found in $Rb_3As_7 \cdot 3en$[18] are asymmetrically bridged by the Sn(1) atom, which forms bonds of 2.84(1)Å to As(1), 2.98(1)Å to As(8), and 2.90(1)Å to As(12).

The crystal structure of $[P_4\{2,6-(MeO)_2C_6H_3\}_6](Me_3SnF_2)_2$ consists of dications $[P_4\{2,6-(MeO)_2C_6H_3\}_6]^{2+}$ and anions $(Me_3SnF_2)^-$.[19] The centrosymmetric cation has a planar four-membered ring, in which λ^3P and λ^4P atoms alternate and the substituents on the λ^3P atoms are *trans* to each other [Fig. 3.27-9(a)]. The P-P bond lengths (2.231 and 2.232Å) and the P-C bond lengths (1.783-1.809Å) lie in the usual ranges. The $(Me_3SnF_2)^-$ anion is the first example of the $(R_3SnF_2)^-$ class and displays a slightly distorted trigonal-bipyramidal geometry with axial F atoms. The Sn-F (2.596 and 2.607Å) and Sn-C (2.138-2.155Å) distances are somewhat longer than those in similar pentacoordinated tin compounds.

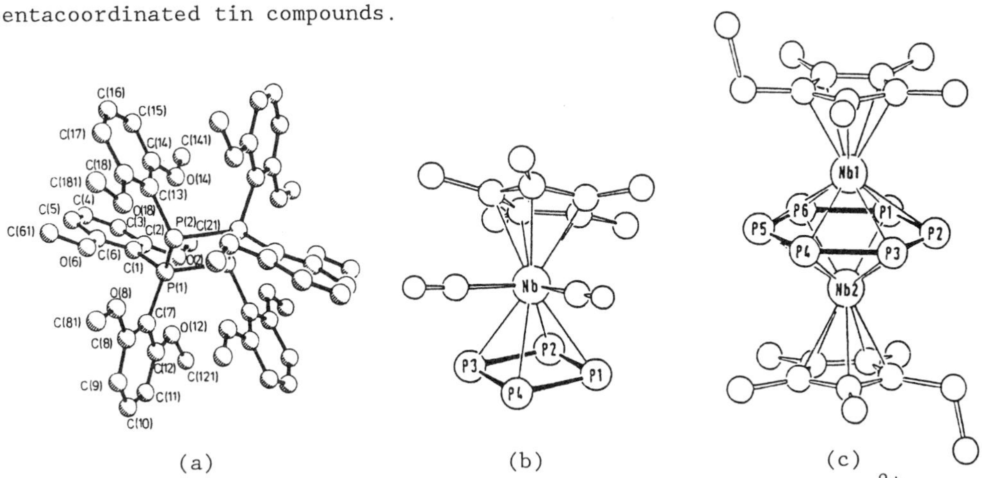

(a)   (b)   (c)

Fig. 3.27-9 Molecular structure of (a) $[P_4\{2,6-(MeO)_2C_6H_3\}_6]^{2+}$, (b) $Cp^*(CO)_2Nb(\eta^4-P_4)$ and (c) $(C_5Me_4Et)_2Nb_2P_6$. (After refs. 19 and 21).

6. Metal complexes containing planar P_x and As_x fragments

The structures of some metal complexes containing planar P_x and As_x fragments are shown in Fig. 3.27-9(b,c) and Fig. 3.27-10.

The sandwich complex $(\eta^5-C_5Me_5)Ni(\eta^3-P_3)$ crystallizes in space group $P2_1/m$ (incorrectly given as $P2_1/c$ in the source paper) with a crystallographic mirror plane passing through the Ni and P(1) atoms and a C-Me fragment.[20] The P_3 fragment is nearly equilateral [P(1)-P(2) 2.091(3), P(2)-(2') 2.119(2)Å], and the metal-ligand bond distances are: Ni-P(1) 2.222(3), Ni-P(2) 2.219(2), Ni-P_3(centroid) 1.860, Ni-C_5Me_5(centroid) 1.683Å. The angle C_5Me_5(centroid)-Ni-P_3(centroid) is 179.3°.

The crystal structure analysis of $Cp^*(CO)_2Nb(\eta^4\text{-}P_4)$ shows that the molecule exhibits nearly C_s symmetry [Fig. 3.27-9(b)].[21] The four P atoms are coplanar and form a slightly distorted (kite-shaped) square [P(3)-P(4) 2.181(2), P(4)-P(1) 2.178(2), P(1)-P(2) 2.136(2), P(2)-P(3) 2.141(2)Å]; the P(2)-Nb bond [2.607(1)Å] is slightly shorter than the other three P-Nb bonds (average 2.635Å).

In the triple-decker complex $(C_5Me_4Et)_2Nb_2P_6$ the middle deck can be regarded formally as consisting of two allyl-like P_3^- units joined to form the $cyclo\text{-}P_6^{2-}$ ligand [Fig. 3.27-9(c)].[21] There are two long [P(1)-P(2)

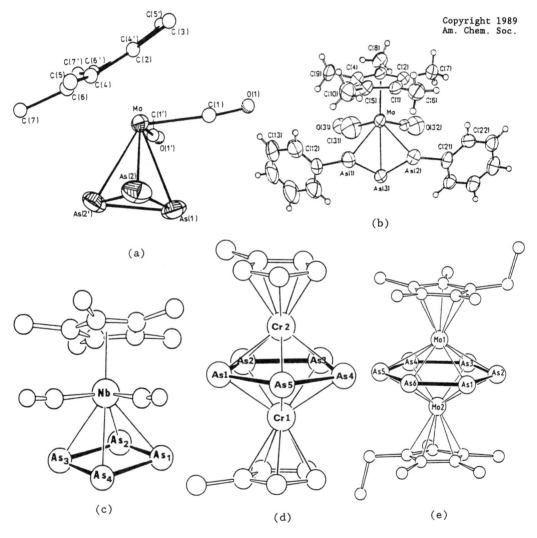

(a)

(b)

(c)

(d)

(e)

Fig. 3.27-10 Structures of (a) $(C_5Me_5)(CO)_2Mo(\eta^3\text{-}As_3)$, (b) $Cp^*Mo(CO)_2(\eta^3\text{-}PhAsAsAsPh)$, (c) $(\eta^5\text{-}C_5Me_5)(CO)_2Nb(\eta^4\text{-}As_4)$, (d) $(\eta^5\text{-}C_5H_4Me)Cr(\mu,\eta^5\text{-}As_5)Cr(\eta^5\text{-}C_5H_4Me)$, and (e) $(\eta^5\text{-}C_5Me_4Et)_2Mo_2(\mu,\eta^6\text{-}As_6)$. (After refs. 22-26).

2.241(10), P(4)-P(5) 2.243(9)Å] and four short [2.140(9)-2.181(9), average
2.157Å] P-P distances.

The crystal structure of $(C_5Me_5)(CO)_2Mo(\eta^3-As_3)$ indicates that the Mo
atom forms the apex of a tetrahedron with a practically equilateral As_3
triangle as its base [Fig. 3.27-9(a)], with a crystallographic mirror plane
through the atoms As(1), Mo, C(2) and C(3).[22] The bond lengths in the As_3
ring are As(1)-As(2) = 2.372(1) and As(2)-As(2') = 2.377(2)Å.

The crystal structure of $Cp^*Mo(CO)_2(\eta^3-PhAsAsAsPh)$ [Fig. 3.27-10(b)]
shows that the η^3-PhAsAsAsPh complex is an isoelectronic and isolobal analogue
of π-allyl complexes.[23] In the structure, the Mo atom is nearly equidistant
to all three As atoms [2.657(2), 2.665(2) and 2.685(2)Å], and the As_3 fragment
has dimensions: As(1)-As(3) 2.363(2), As(2)-As(3) 2.360(2)Å, and
As(1)-As(3)-As(2) 83.2(1)°.

There are two molecules of the same structure [Fig. 3.27-10(c)] in the
asymmetric unit of $(\eta^5-C_5Me_5)(CO)_2Nb(\eta^4-As_4)$.[24] A planar cyclo-$As_4$ ligand
coordinates to the Nb atom with As-As 2.345(4)-2.409(4)Å for molecule (1), and
As-As 2.348(5)-2.400(4)Å for molecule (2).

In the triple-decker sandwich complex $(\eta^5-C_5H_4Me)Cr(\mu,\eta^5-As_5)Cr(\eta^5-$
$C_5H_4Me)$ the (μ,η^5-As_5) group has a planar configuration with bond lengths
As(1)-As(2) 2.425(5), As(2)-As(3) 2.427(6), As(3)-As(4) 2.405(6), As(4)-As(5)
2.395(6), and As(1)-As(5) 2.436(5)Å [Fig. 3.27-10(d)].[25]

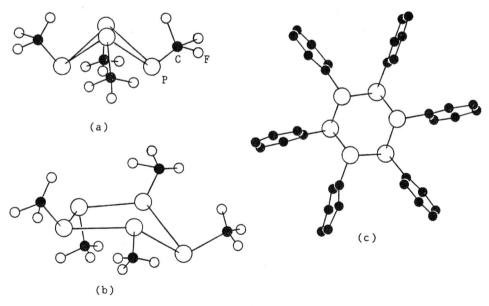

(a)

(b)

(c)

Fig. 3.27-11 Molecular structures of (a) $(PCF_3)_4$, (b) $(PCF_3)_5$ and (c) $(PPh)_6$.
(After ref. 12).

The crystal structure analysis of $(\eta^5\text{-}C_5Me_4Et)_2Mo_2(\mu,\eta^6\text{-}As_6)$ shows that the As_6-ring and both five-membered rings are planar and parallel to one another [Fig. 3.27-10(e)].[26] The mean As-As distance (2.35Å) lies almost exactly between the values for an As-As single bond (e.g. As_4 2.44Å) and an As=As double bond (*ca.* 2.24Å).

7. Organocyclophosphanes

Organocyclophosphanes $(PR)_n$ with $n = 3$ to 6 are well known.[27] Small alkyl substituents as well as the phenyl group favour the formation of large rings, whereas bulky groups favour $(PR)_4$ and even $(PR)_3$. In $(PR)_3$ compounds two of the R groups lie on one side of the ring. The larger rings are invariably puckered as illustrated in Fig. 3.27-11.

References

[1] A. Brown and S. Rundqvist, *Acta Crystallogr.* 19 684 (1965).

[2] D.E.C. Corbridge and E.J. Lowe, *Nature (London)* **170**, 629 (1952).

[3] J.C. Jamieson, *Science (Washington)* **139**, 1291 (1963).

[4] L. Pauling, *The Nature of the Chemical Bond*, 3rd ed., Cornell University Press, Ithaca, 1960, p. 537.

[5] H. Thurn and H. Krebs, *Acta Crystallogr., Sect. B* **25**, 125 (1969).

[6] D.E.C. Corbridge, *The Structural Chemistry of Phosphorus*, Elsevier, Amsterdam, 1974, chap. 2.

[7] J. Donohue, *The Structures of the Elements*, Wiley, New York, 1974.

[8] N.N. Greenwood and A. Earnshaw, *Chemistry of the Elements*, Pergamon Press, Oxford, 1986, chap. 12.

[9] D. Schiferl and C.S. Barrett, *J. Appl. Cryst.* 2, 30 (1969).

[10] Z. Johan, *Chem. Erde* 20, 71 (1959).

[11] H.G. von Schnering, *Angew. Chem. Int. Ed. Engl.* **20**, 33 (1981).

[12] D.E.C. Corbridge, *Phosphorus, An Outline of its Chemistry, Biochemistry and Technology*, 3rd ed., Elsevier, Amsterdam, 1985.

[13] R.C. Haushalter, B.W. Eichhorn, A.L. Rheingold and S.J. Geib, *J. Chem. Soc., Chem. Commun.*, 1027 (1988).

[14] C.H.E. Belin, *J. Am. Chem. Soc.* **102**, 6036 (1980).

[15] W. Wichelhaus and H.-G. von Schnering, *Naturwissenschaften* **60**, 141 (1986).

[16] C. Belin and H. Mercier, *J. Chem. Soc., Chem. Commun.*, 190 (1987).

[17] R.C. Haushalter, *J. Chem. Soc., Chem. Commun.*, 196 (1987).

[18] H.-G. von Schnering in A.H. Cowley (ed.), *Rings, Clusters, and Polymers of the Main Group Elements*, American Chemical Society, Washington, DC, 1983, p. 69.

[19] L. Heuer, L. Ernst, R. Schmutzler and D. Schomburg, *Angew. Chem. Int. Ed. Engl.* **28**, 1507 (1989).

[20] O.J. Scherer, J. Braun and G. Wolmershäuser, *Chem. Ber.* **123**, 471 (1990).

[21] O.J. Scherer, J. Vondung and G. Wolmershäuser, *Angew. Chem. Int. Ed. Engl.* **28**, 1355 (1989).

[22] I. Bernal, H. Brunner, W. Meier, H. Pfisterer, J. Wachter and M.L. Ziegler, *Angew. Chem. Int. Ed. Engl.* **23**, 438 (1984).

[23] J.R. Harper, M.E. Fountain and A.L. Rheingold, *Organometallics* **8**, 2316 (1989).

[24] O.J. Scherer, J. Vondung and G. Wolmershäuser, *J. Organomet. Chem.* **376**, C35 (1989).

[25] O.J. Scherer, W. Wiedemann and G. Wolmershäuser, *J. Organomet. Chem.* **361**, C11 (1989).

[26] O.J. Scherer, H. Sitzmann and G. Wolmershäuser, *Angew. Chem. Int. Ed. Engl.* **28**, 212 (1989).

[27] M. Baudler, *Pure Appl. Chem.* **52**, 755 (1980); *Angew. Chem. Int. Ed. Engl.* **21**, 492 (1982); *Phosphorus Sulfur* **18**, 57 (1983).

Note Complexes with substituent-free acyclic and cyclic E_n ligands (E = P, As, Sb, Bi) are reviewed in O.J. Scherer, *Angew. Chem. Int. Ed. Engl.* **29**, 1104 (1990).

Norbornadiene-like As_7^- and crown-shaped *cyclo*-As_8^{8-} units exist in the ionic compounds $[As_7\{Cr(CO)_3\}]^{3-}$ and $\frac{1}{\infty}[Rb\{NbAs_8\}]^{2-}$, respectively; as a ligand in the related $[(Cp''Nb)_2(\mu,\eta^{4:4}\text{-}As_8)]$ and $[(Cp''Nb)_2\{As_8Cr(CO)_5\}]$ ($Cp'' = \eta^5\text{-}1,3\text{-}tBu_2C_5H_3$) complexes, *cyclo*-$As_8$ takes a tub shape like cyclooctatetraene but the ring is strongly distorted and puckered. See O.J. Scherer, R. Winter, G. Heckmann and G. Wolmerschäuser, *Angew. Chem. Int. Ed. Engl.* **30**, 850 (1991).

3.28   Hexachlorocyclophosphazene
(PNCl$_2$)$_3$

Crystal Data

Orthorhombic, space group *Pnma* (No. 62)

$a = 14.15(2)$, $b = 12.99(2)$, $c = 6.19(1)$Å, $Z = 4$

Atom	x	y	z
P(1)	.442(0)	1/4	.403(1)
P(2)	.595(8)	.144(2)	.563(7)
N(1)	.650(9)	1/4	.604(1)
N(2)	.495(6)	.143(3)	.456(5)
Cl(1)	.318(7)	1/4	.552(9)
Cl(2)	.403(0)	1/4	.097(8)
Cl(3)	.586(9)	.069(7)	.842(9)
Cl(4)	.681(3)	.054(7)	.396(5)

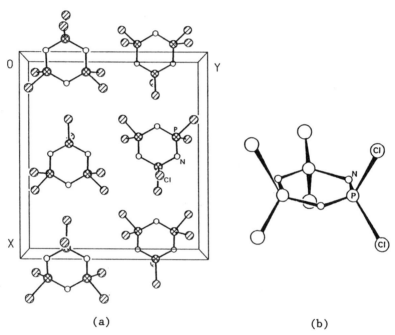

(a)   (b)

Fig. 3.28-1 (a) Crystal structure and (b) molecular structure of (PNCl$_2$)$_3$.

Crystal Structure

The crystal is composed of discrete (PNCl$_2$)$_3$ molecules which are held together by van der Waals forces. Structure analysis of the cyclic trimer has demonstrated the six-membered ring to be planar within experimental error. The bonds within the ring are of the same length (1.58Å), and the interior angles are all close to 120° (N-P-N = 118.4°, P-N-P = 121.4°). The Cl-P-Cl planes are perpendicular to the plane of the central ring, and Cl-P-Cl = 102° (Fig. 3.28-1).[1]

The most notable features of the structure are the equality and shortness of the P-N bonds, the close approximation to planarity of the ring, and the near equality of the angles within it. The equality of the P-N bonds excludes the possibility of alternate single and double bonds; the length is very much shorter than the value 1.8Å expected for a single bond between phosphorus and nitrogen, implying that the bond order is greater than unity. Since the average angle at the N atom deviates little from 120°, the P-N bond cannot be strengthened appreciably by charge transfer in the sense P^+N^-. All these considerations point to aromatic character of the heterocyclic ring.

Remarks

1. Cyclic phosphonitrilic halides

   Detailed X-ray investigations have demonstrated the presence of an alternating ring of nitrogen and phosphorus atoms in the trimers (PNX₂)₃ and tetramers $(PNX_2)_4$ (X = F, Cl, Br), and in the pentamer (PNCl₂)₅. Compounds in the series $(PNF_2)_n$ with n up to 17 are known, and crystallographic data have been obtained for $(PNCl_2)_6$ and $(PNCl_2)_8$.[2]

   The cyclic trimer $(PNF_2)_3$ has an exactly planar six-membered ring in which all six P-N distances are equal (molecular symmetry D_{3h}). Most other trimers are also more or less planar with equal P-N distances. The cyclic tetramer $(PNF_2)_4$ is also a planar heterocycle with even shorter P-N bonds,

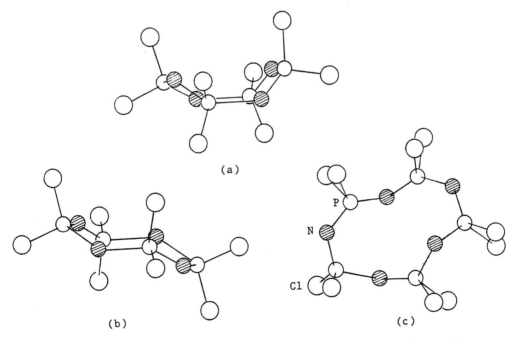

Fig. 3.28-2 Molecular structures of (a) (PNCl₂)₄ boat form, (b) (PNCl₂)₄ chair form, and (c) (PNCl₂)₅. (After ref. 2).

1.51Å, and with ring angles of 122.7° and 147.4° at P and N, respectively. The compound $(PNCl_2)_4$ exists in both the metastable boat form and the stable chair form (Fig. 3.28-2). The remarkable diversity of molecular conformations observed for the eight-membered ring suggests that the particular conformation adopted in each case results from a delicate balance of intra- and inter- molecular forces including the details of skeletal bonding, the orientation of substituents and their polar and steric nature, crystal-packing effects, etc.

The pentamer $(PNCl_2)_5$ contains a ten-membered ring which is nearly planar but considerably distorted form D_{5h} symmetry as two nitrogen atoms are puckered towards the center of the ring (Fig. 3.28-2). The N-P-N angle remains close to 120° in all the halide ring structures, but the P-N distance is shorter and the P-N-P angle greater in $(PNF_2)_4$ than in the other rings.

2. Bonding in phosphazenes[3,4]

All phosphazenes, whether cyclic or chain, contain the formally unsaturated group -N=P- with two-coordinate N and four-coordinate P atoms. The following experimental facts have been gathered in regard to their properties and configurations:

a) the rings and chains are very stable;

b) the skeletal interatomic distances are equal within the ring or along the chain, unless there is differing substitution at the various P atoms;

c) the P-N distances are shorter than expected for a covalent single bond (~1.77Å) and are usually in the range 1.58±2Å;

d) the N-P-N angles are usually in the range 120±2°; but the P-N-P angles in various compounds span the range 120-148.6°;

e) skeletal N atoms are weakly basic and can coordinate to metals or be protonated, especially when there are electron-releasing groups on P;

f) unlike many aromatic systems the phosphazene skeleton is difficult to reduce electrochemically; and

g) spectral effects associated with organic π-systems are not found.

Briefly stated, the bonding in phosphazenes is not adequately represented by a sequence of alternating double and single bonds -N=P-N=P-, yet it differs from aromatic σ-π systems in which there is extensive electron delocalization through p_π-p_π bonding. The possibility of d_π-p_π bonding in P-N systems has been considered. In such systems, the σ bonds formed from the phosphorus sp^3 orbitals overlapping with the nitrogen sp^2 orbitals are supplemented by π bonding between the nitrogen p orbitals and the phosphorus d orbitals. Delocalized π bonding thus occurs over the entire ring.

In the cyclic phosphonitrilic compounds, any given ring may, in the first instance, adopt a configuration most favourable to π bonding involving the nitrogen p_z orbital (z axis perpendicular to the ring plane) and the d_{xz} and d_{yz} orbitals of phosphorus. Secondly, "in-plane" electron delocalization probably arises from overlap of the lone pair orbitals on nitrogen with the d_{xy} orbitals on phosphorus, forming additional π' bonds in the plane of the ring. Thirdly, direct electrical effects between exocyclic groups themselves may influence ring conformation and the presence of bulky substituents may rule out certain configuration or sterically prevent the reversion of one form into another. In Fig. 3.28-3 the possible schemes of orbital overlap for π bonding in a trimeric phosphazene are indicated.

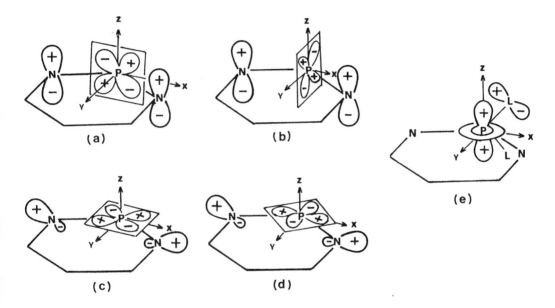

Fig. 3.28-3 Orbital overlap schemes in (PNX)₃: (a) and (b) represent ring π bonding, (c) and (d) ring π' bonding, and (e) exocyclic π bonding with ligand L. The d_{xz} and d_{xy} orbitals are also involved in exocyclic π/π' bonding. (Adapted from ref. 5).

The trimeric halides (PNX₂)₃ and their fully substituted derivatives all appear to adopt near planar configurations which favour a higher degree of π bonding but relatively weak π' bonding. Crystal packing effects may influence the puckering in some of these compounds.

3. Anticancer agents[6]

The geometrical structure of Rosenberg's active platinum drugs (Section 4.20) has led to a suggestion that their anticancer activity arises from the prevention of the replication of vicious DNA through a process of intercalation between plates (A...T and G...C) of bases. Thus, if a given molecule contains both (i) pairs of Cl atoms in a "square-planar like 3.4Å situation" and (ii) a planar ring with highly basic endocyclic N or O atoms, one may reasonably speculate that it stands a chance of showing antitumor activity.

This has proved to be the case with hexachlorocyclophosphazene and related compounds. For $(PNCl_2)_3$, the distance between two Cl atoms bonded to the same phosphorus atom and located perpendicular to the plane of ring is nearly 3.4Å, and the planar ring does contain N atoms. The cyclophosphazenes have been found to exhibit significant antitumor activity in some preliminary screening trials.

4. Polyphosphazenes[7-9]

In recent years many polyphosphazenes, $[PNR_2]_n$, with a variety of substituents at phosphorus have been prepared. They often exhibit useful properties including low-temperature flexibility, resistance to chemical attack, flame retardancy, stability to UV radiation, and reasonably high thermal stability. For example, fibers of poly[bis(trifluoroethoxy)-phosphazene] repel water, are resistant to hydrolysis or strong sunlight, and do not burn; polyphosphazene elastomers are now being manufactured for use in

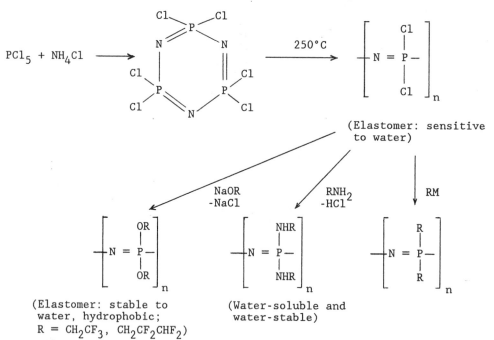

Fig. 3.28-4 Scheme illustrating the synthesis of poly(dichlorophosphazene) and poly(organophosphazene). (After ref. 7).

fuel lines, gaskets, O-rings, shock absorbers, and carburettor components.

The most commonly used synthetic route to polyphosphazenes is the ring opening/substitution method. This procedure (Fig. 3.28-4) involves the initial preparation of polydichlorophosphazene, $(PNCl_2)_n$, by the ring-opening polymerization of the cyclic trimer and subsequent nucleophilic displacement of the chlorine atom along the chain. Thus the cyclic trimers $(PNR_2)_3$, as useful starting materials for polymerization, have been subjected to intensive studies in academic and industrial institutions.

5. Phosphazanes[10]

Phosphazanes are compounds containing *formal* phosphorus-nitrogen single bonds. The eight-membered ring methoxyphosphazene $[NP(OMe)_2]_4$ (I) rearranges to the oxophosphazane $[MeNP(O)OMe]_4$ (II) which exists in two isomeric forms. Compound I has a saddle ring structure; isomer IIa has a 2-*trans*-4-*cis*-6-

trans-8 structure with equatorial phosphoryl groups and a slightly distorted boat ring, while the other isomer IIb has a 2-*cis*-4-*trans*-6-*trans*-8 structure in the chair conformation.

The tetrameric acid $[HNP(O)OH]_4 \cdot 2H_2O$ (III) and its metal salts $K_4[HNP(O)O]_4 \cdot 4H_2O$ (IV) and $Cs_4[HNP(O)O]_4 \cdot 6H_2O$ (V) have cyclotetraphosphazane rings in the boat, chair, and saddle conformations, respectively. Two short and distinct P-O bond lengths were found in the acid, leading to its proper formulation as a hydronium salt, $(H_3O)_2[N_4P_4O_6(OH)_2]$.

The structural data of compounds I-V are summarized in Table 3.28-1.

Table 3.28-1 Structural data of some eight-membered cyclophosphazanes.

	I	IIa	IIb	III	IV	V
Mean bond lengths (Å)						
P-N	1.57	1.673	1.670	1.661	1.673	1.676
P-O		1.469	1.467	1.517	1.500	1.498
P-OMe	1.58	1.579	1.572			
Mean bond angles (°)						
N-P-N	121.0	106.9	108.2	107.3	108.1	106.9
P-N-P	132.2	122.5	123.0	125.6	131.9	128.9
O-P-O	105.5	115.8	116.0	116.1	117.8	117.7
Ring conformation	saddle	slightly distorted boat	chair	boat	chair	saddle

6. Iminophosphanes (phosphazenes)

According to *ab initio* calculations the parent system HP=NH has a planar bent structure with the (*E*)-conformation slightly favoured over the (*Z*). Crystallographic studies on about 40 R-P=N-R' compounds amply support this view.[10] The P(III) monophosphazene $(Me_3Si)_2NP=NSiMe_3$ has a planar Si_2NPNSi skeleton, the P(V) atom of $(Me_3Si)_2NP(=NSiMe_3)_2$ is trigonal planar, and the phosphazane dimer $[Me_3SiN-P-N(SiMe_3)_2]_2$ has a planar four-membered ring (Fig. 3.28-5).[11] The phosphazane and phosphazene linkages in these

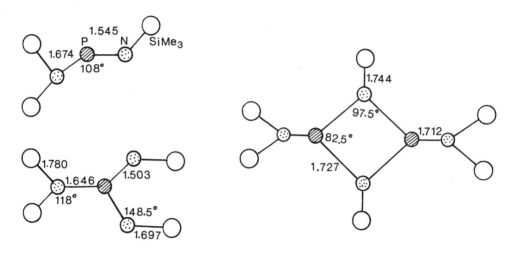

Fig. 3.28-5 Structures of some related compounds containing phosphazane and phosphazene P-N bonds. (Adapted from ref. 11).

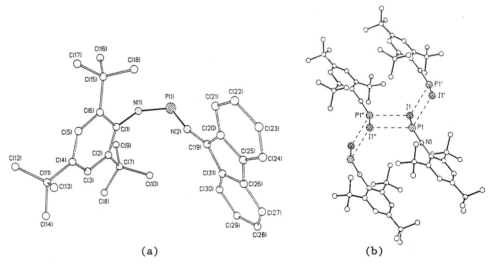

(a)   (b)

Fig. 3.28-6 Iminophosphanes R-P=N-R' with (a) the (*Z*)-configuration and (b) a twisted *cis* arrangement of the R and R' groups about the P=N bond. (After refs. 13 and 10).

compounds are clearly distinguishable, and the P-N bonds in the dimer are markedly longer than the two types of bonds present in the monomer. The phosphazene (and to a lesser extent, phosphazane) P-N bond involving P(III) is invariably longer than that involving P(V) in a pair of related compounds.[12]

A (Z)-configuration has been found in (fluorenyl)=N-P=N-Ar (Ar = 2,4,6-$tBu_3C_6H_2$), where steric congestion accounts for orthogonality of the aryl ligand to the plane of the P=N double bond [Fig. 3.28-6(a)].[13] The phosphorus-iodine bond in I-P=N-Ar is considerably longer than that in PI_3 (2.895 vs. 2.52Å) owing to n(N) → σ^*(PI) charge transfer accompanied by a short intermolecular P...I contact of 3.605Å; the I-P-N-C torsion angle has the unusual value of -140° [Fig. 3.26-6(b)].[10]

References

[1] A. Wilson and D.F. Carroll, *J. Chem. Soc.*, 2548 (1960).

[2] D.E.C. Corbridge, *The Structural Chemistry of Phosphorus*, Elsevier, Amsterdam, 1974, chap. 12.

[3] N.N. Greenwood and A. Earnshaw, *Chemistry of the Elements*, Pergamon Press, Oxford, 1986.

[4] D.P. Craig and N.L. Paddock, *J. Chem. Soc.*, 4118 (1962).

[5] D.E.C. Corbridge, *Phosphorus, An Outline of its Chemistry, Biochemistry and Technology*, 3rd ed., Elsevier, Amsterdam, 1985.

[6] J.-F. Labarre, *Top. Curr. Chem.* **102**, 1 (1982).

[7] M. Zeldin, K.J. Wynne and H.R. Allcock (eds.), *Inorganic and Organometallic Polymers*, American Chemical Society, Washington, D.C., 1988, chap. 19-25.

[8] R.E. Singler, G.L. Hagnauer and R.W. Sicka in J.C. Arthur, Jr. (ed.), *Polymers for Fibres and Elastomers*, American Chemical Society, Washington, D.C., 1984, p. 143.

[9] H.R. Allcock, *Chem. & Eng. News*, March 18, 22 (1985).

[10] E. Niecke and Gudat, *Angew. Chem. Int. Ed. Engl.* **30**, 217 (1991).

[11] R.A. Shaw, *Phosphorus and Sulfur* **4**, 101 (1978).

[12] R.A. Shaw, *Pure Appl. Chem.* **44**, 317 (1975).

[13] E. Niecke, M. Nieger, C. Gärtner-Winkhaus and B. Kramer, *Chem. Ber.* **123**, 477 (1990).

Note The lithium phosphorus(V) nitride $Li_{10}P_4N_{10}$ contains the complex anion $P_4N_{10}^{10-}$ that is analogous to molecular P_4O_{10}. See W. Schnick and U. Berger, *Angew. Chem. Int. Ed. Engl.* **30**, 830 (1991).

3.29 Tetraphosphorus Trisulfide

P_4S_3

Crystal Data

Orthorhombic, space group *Pmnb* (No. 62)

$a = 9.660(5)$, $b = 10.597(5)$, $c = 13.671(5)$Å, $Z = 8$

Atom	x	y	z
P(1)	.6341	.2324	.8376
P(2)	3/4	.0801	.7580
P(3)	3/4	.0219	.9934
S(1)	.5853	.1474	.9711
S(2)	3/4	-.0699	.8580
P(4)	.1345	.4456	.7536
P(5)	1/4	.2873	.8223
P(6)	1/4	.5301	.9683
S(3)	.0851	.5613	.8728
S(4)	1/4	.3347	.9709

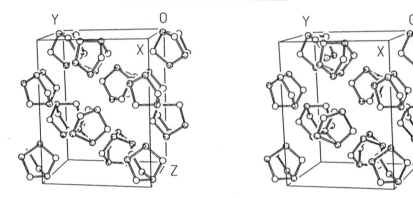

Fig. 3.29-1 Stereoview of the crystal structure of P_4S_3. The dotted circles represent P atoms.

Crystal Structure

The crystal structure consists of molecules of P_4S_3 with shortest intermolecular distances of approximately 3.6Å (Fig. 3.29-1).[1] Both independent molecules in the unit cell lie in mirror planes. The intramolecular atomic distances and bond angles have the following average values:

P-P = 2.235Å	P-S = 2.090Å
S-P-S = 99.4°	P-S-P = 103.0°
S-P-P = 103.1°	P-P-P = 60.0°

Remarks

1. Compounds isostructural with P_4S_3

As_4S_3, As_4Se_3 and P_4Se_3 are all isostructural with P_4S_3. As_4S_3 occurs in both the α- and β-form of the orange-yellow mineral dimorphite, the two forms differing only in the arrangement of the molecular unit.[2] The β-form of As_4S_3 is the stable modification at room temperature, and the α-form above 130°C.

The cations $As_3S_4^+$ and $As_3Se_4^+$ in $(As_3S_4)(SbCl_6)$ and $(As_3Se_4)(SbCl_6)$, respectively, are isoelectronic clusters of the same structure as P_4S_3 (Fig. 3.29-2).[3]

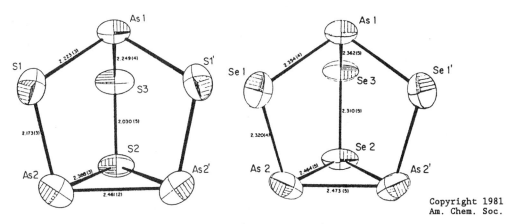

Fig. 3.29-2 Structures of $As_3S_4^+$ and $As_3Se_4^+$. (After ref. 3).

The homopolyatomic anions P_7^{3-} in $Ba_3(P_7)_2$, As_7^{3-} [Fig. 3.29-3(a)] in $Ba_3(As_7)_2$, and Sb_7^{3-} in $[Na(crypt)]_3(Sb_7)$ have the same molecular skeleton as P_4S_3 . The compound $As_7(SiMe_3)_3$, whose skeleton of $[As_7]$ [Fig. 3.29-3(b)] has the same shape as As_7^{3-}, is stable to air and moisture for several hours.[4]

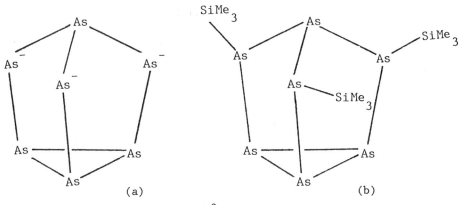

Fig. 3.29-3 Structure of As_7^{3-} and $As_7(SiMe_3)_3$.

2. Phosphorus sulphides

A series of phosphorus sulphides are known: P_4S_3, P_4S_4, P_4S_5, P_4S_7, P_4S_9, and P_4S_{10}. Structural studies have shown that the molecules are all based on the P_4 tetrahedron but there is no tetrahedral P_4 unit common to them. The P_4 molecule is held together by six equivalent covalent bonds, P-P = 2.21Å, but in the P_4S_n molecules some P-P bonds are replaced by sulfur bridges. Fig. 3.29-4 shows the molecular structures of the P_4S_n series.[5,6]

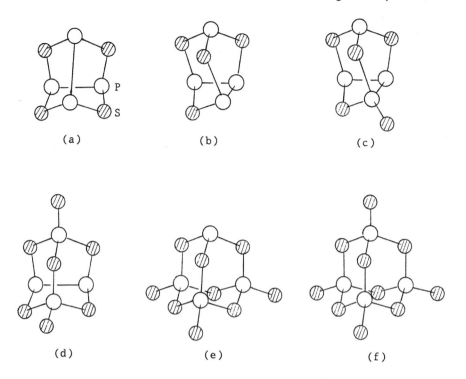

Fig. 3.29-4 Structures of phosphorus sulfides (schematic).
(a) α-P_4S_4 (D_{2d}), (b) β-P_4S_4 (C_s), (c) P_4S_5 (C_i), (d) P_4S_7 (C_{2v}),
(e) P_4S_9 (C_{3v}), (f) P_4S_{10} (T_d). (After ref. 7).

The pnictide chalcogenides shown in Table 3.29-1 are known to form a
series of isostructural molecules.[8]

Table 3.29-1   Isostructural pnictide chalcogenide molecules.

No. of atoms	8	8	9	11	14
P_4S_3	α-P_4S_4	β-P_4S_4	P_4S_5	P_4S_7	P_4S_{10}
P_4Se_3	P_4Se_4		P_4Se_5	P_4Se_7	P_4Se_{10}
P_7^{3-}	N_4S_4		$S_4N_5^-$		
As_7^{3-}	N_4Se_4				
$As_3S_4^+$	As_4S_4	As_4S_4	As_4S_5		As_4S_{10}
$As_3Se_4^+$	As_4Se_4				$P_4S_9N^-$
As_4Se_3					

Tetraphosphorus decasulfide is used in organic chemistry to convert OH,
C=O, COOH or $CONH_2$ groups into their sulfur analogues. This sulfide is an
important industrial chemical for the manufacture of insecticides and zinc
dialkyl or diaryl dithiophosphates as oil additives. Some of the reactions of
P_4S_{10} are summarized in Fig. 3.29-5.

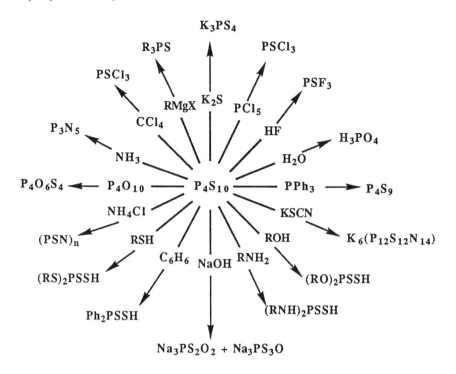

Fig. 3.29-5 Some reactions of P_4S_{10}. (Adapted from ref. 8).

3. Phosphorus oxides and oxosulfides

The discrete molecules P_4O_6, P_4O_7, P_4O_8, P_4O_9, P_4O_{10} and $P_4O_4S_6$, $P_4O_6S_4$ are all based on the P_4 tetrahedron as a common core. Their structures are shown in Fig. 3.29-6.

4. Realgar

The crystal structure of the mineral realgar, As_4S_4, was first determined by Ito and co-workers in 1952 by means of Harker-Kasper inequalities,[9] and refinement was carried out by Mullen and Nowacki in 1972.[10] The crystal consists of a packing of cradle-like, covalently-bonded As_4S_4 molecules held together by van der Waals forces.

The As_4S_4 molecule in realgar has eight independent As-S bond lengths ranging from 2.228(2)Å to 2.247(2)Å, and two As-As bond distances of 2.566(1)Å and 2.571(1)Å. The S-As-S angles range from 94.5° to 95.1°, the As-As-S angles from 98.7° to 100.0°, and the As-S-As angles from 100.8° to 101.3° [Fig. 3.29-7(a)].

The shape of the As_4S_4 molecule (ideal symmetry D_{2d}) in realgar is similar to its gaseous configuration as described by Lu and Donohue.[11] Note that the four S atoms are almost coplanar, and the As_4S_4 structure may be

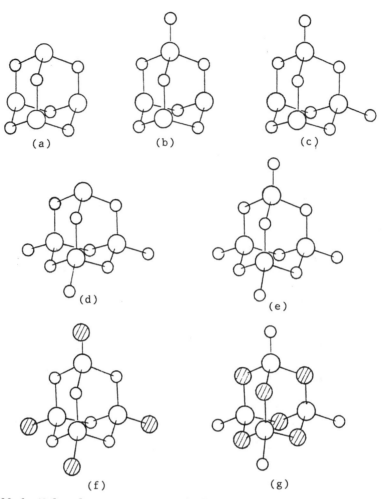

Fig. 3.29-6 Molecular structures of phosphorus oxides and oxosulfides, (a) P_4O_6 (T_d), (b) P_4O_7 (C_{3v}), (c) P_4O_8 (C_s), (d) P_4O_9 (C_{3v}), (e) P_4O_{10} (T_d), (f) $P_4O_6S_4$ (T_d), (g) $P_4O_4S_6$ (T_d). (After ref. 7).

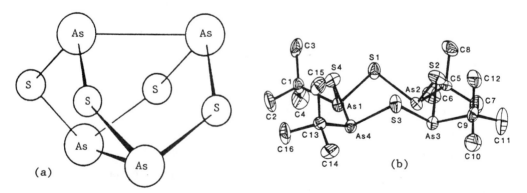

Fig. 3.29-7 Molecular structures of (a) As_4S_4 in realgar and (b) (*tert*-butyl)$_4As_4S_4$. (After ref. 12).

matched with that adopted by N_4S_4 (Section 3.34) in which the S atoms form a tetrahedron and the N atoms a square. A second form of As_4S_4 consists of geometrical isomers having a structure analogous to β-P_4S_4.

In the crystal structure of (*tert*-butyl)$_4As_4S_4$,[12] the discrete molecule is composed of an eight-membered As_4S_4 ring with one exocyclic *tert*-butyl group bonded to each As atom [Fig. 3.29-7(b)]. The central ring approximates a crown (S_8) conformation and the *exo* groups are equatorial. The As-S distances vary from 2.232(3) to 2.249(3)Å; the intra-ring angles at As average 101.5(7) and those at S average 95.8(15)°. The ring is flexed such that the angles between the planes of the As-S-As units and the plane of the four As atoms are 133.3(1), 110.0(1), 133.5(1) and 108.9(1)°, making the S...S cross-ring distances 5.374(5) and 4.327(5)Å. The analogous $(C_6H_5)_4As_4S_4$ molecule also has the crown conformation, but its symmetry is constrained to C_4 by the lattice and all phenyl groups are equatorial.[13] The As...As, As-S, and S...S distances are 3.37, 2.25 and 2.26, and 3.51Å, respectively.

References

[1] Yuen Chu Leung, J. Waser, S. van Houten, A. Vos, G.A. Wiegers and E.H. Wiebenga, *Acta Crystallogr.* **10**, 574 (1957).

[2] H.J. Whitfield, *J. Chem. Soc. (A)*, 1800 (1970); *J. Chem. Soc., Dalton Trans.*, 1737 (1973).

[3] B.H. Christian, R.J. Gillespie and J.F. Sawyer, *Inorg. Chem.* **20**, 3410 (1981).

[4] H.G. von Schnering, D. Fenske, W. Hönle, M. Binnewies and K. Peters, *Angew. Chem. Int. Ed. Engl.* **18**, 679 (1979).

[5] A. Vos, R. Olthof, F. van Bolhuis and R. Botterweg, *Acta Crystallogr.* **19**, 864 (1965).

[6] A.M. Griffin, P.C. Minshall and G.M. Sheldrick, *J. Chem. Soc., Chem. Commun.* 809 (1976).

[7] D.E.C. Corbridge, *The Structural Chemistry of Phosphorus*, Elsevier, Amsterdam, 1974, chap. 4.

[8] D.E.C. Corbridge, *Phosphorus: An Outline of its Chemistry, Biochemistry and Technology*, 4th ed., Elsevier, Amsterdam, 1990.

[9] T. Ito, N. Morimoto and R. Sadanaga, *Acta Crystallogr.* **5**, 775 (1952).

[10] D.J.E. Mullen and W. Nowacki, *Z. Kristallogr.* **136**, 48 (1972).

[11] C.-S. Lu and J. Donohue, *J. Am. Chem. Soc.* **66**, 818 (1944).

[12] J.T. Shore, W.T. Pennington and A.W. Cordes, *Acta Crystallogr.*, Sect. C **44**, 1831 (1988).

[13] G. Bergerhoff and H. Namgung, *Z. Kristallogr.* **150**, 209 (1979).

Note The crystal structure of P_4O_6S is reported in F. Frick, M. Jansen, P.J. Bruna and S.D. Peyerimhoff, *Chem. Ber.* **124**, 1711 (1991). Refinement of the crystal structure of P_4O_9 is reported in B. Lüer and M. Jansen, *Z. Kristallogr.* **197**, 247 (1991).

3.30   Potassium Antimony Tartrate Trihydrate (Tartar Emetic)
$K_2[Sb_2(d-C_4H_2O_6)_2] \cdot 3H_2O$

Crystal Data

Orthorhombic, space group $C222_1$ (No. 20)

$a = 11.192(2)$, $b = 11.696(3)$, $c = 25.932(5)$Å, $Z = 8$

Atom	x	y	z	Atom	x	y	z
Sb(1)	.3788	.3369	.0487	O(14)	.0794	.2232	.1244
Sb(2)	.1272	.1790	.1955	*O(15)	.1855	.0375	.1491
K(1)	.3664	0	0	O(16)	.1689	-.0407	.0714
K(2)	0	.5784	.2500	C(1)	.5011	.2816	.1452
K(3)	.3464	.0634	.3938	C(2)	.4103	.3768	.1583
O(1)	.4989	.2496	.0988	C(3)	.3312	.3283	.2039
O(2)	.3396	.4083	.1165	C(4)	.2275	.4090	.2095
O(3)	.5679	.2420	.1791	C(11)	.1237	.3148	.0286
O(4)	.3024	.2156	.1943	C(12)	.1624	.1924	.0396
O(5)	.1221	.3585	.2059	C(13)	.0912	.1433	.0853
O(6)	.2393	.5086	.2180	C(14)	.1561	.0349	.1031
O(11)	.1969	.3938	.0339	O(W1)	.2948	-.2563	.0747
O(12)	.2879	.1926	.0528	O(W2)	.3291	-.2421	.1854
O(13)	.0134	.3277	.0162	O(W3)	.4441	-.0191	.1393

* The y value for O(15) (ref. 2) has been changed from 0.3075 to 0.0375.

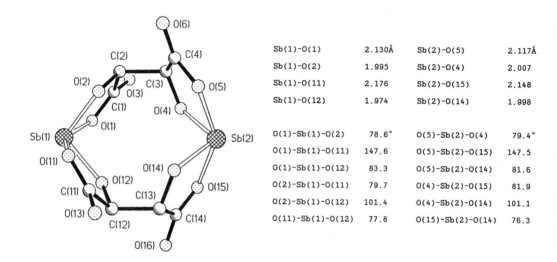

Sb(1)-O(1)	2.130Å		Sb(2)-O(5)	2.117Å	
Sb(1)-O(2)	1.995		Sb(2)-O(4)	2.007	
Sb(1)-O(11)	2.176		Sb(2)-O(15)	2.148	
Sb(1)-O(12)	1.974		Sb(2)-O(14)	1.998	
O(1)-Sb(1)-O(2)	78.6°		O(5)-Sb(2)-O(4)	79.4°	
O(1)-Sb(1)-O(11)	147.6		O(5)-Sb(2)-O(15)	147.5	
O(1)-Sb(1)-O(12)	83.3		O(5)-Sb(2)-O(14)	81.6	
O(2)-Sb(1)-O(11)	79.7		O(4)-Sb(2)-O(15)	81.9	
O(2)-Sb(1)-O(12)	101.4		O(4)-Sb(2)-O(14)	101.1	
O(11)-Sb(1)-O(12)	77.8		O(15)-Sb(2)-O(14)	76.3	

Fig. 3.30-1 Molecular structure of $[Sb_2(d-C_4H_2O_6)_2]^{2-}$ and coordination geometry about the two antimony atoms.

Crystal Structure

Optically active $K_2[Sb_2(d-C_4H_2O_6)_2] \cdot 3H_2O$ is widely used as "tartar emitic" in medicine for treating various parasitic diseases. Its crystal structure was first elucidated in 1966[1] and further refined in 1974.[2]

In the unit cell there are eight antimony-tartrate dimers, which are tartrato-(4)-bridged binuclear complexes in which the two antimony atoms are each coordinated to a carboxyl oxygen atom and an α-hydroxyl atom from two tartrate groups (Fig. 3.30-1). It can be seen that the complex possesses approximate D_2 point symmetry. The quadridentate tartrate ligands form four nearly planar five-membered rings with the two antimony atoms, with dihedral angles of 99° between tartrate ligands at the antimony atoms.

   The coordination of the two d-tartaric acid groups to the antimony atoms can be described in terms of either a distortion of a trigonal-bipyramidal or a square-planar bonding configuration. The axial bonds are antimony-carboxyl oxygen bonds (bond distances are 2.13, 2.18, 2.12 and 2.21Å) and the equatorial bonds are antimony-hydroxyl oxygen bonds (1.97, 2.00, 1.99 and 2.01Å). The axial bond angles (for two symmetry independent Sb atoms) O(1)-Sb(1)-O(11) and O(5)-Sb(2)-O(15) are 148° and 149°, respectively, and the equatorial angles O(2)-Sb(1)-O(12) and O(4)-Sb(2)-O(14) are both 101°.

   The three water molecules are hydrogen bonded to one another as O(W1)...O(W2)...O(W3) chains, and are hydrogen bonded to tartrate oxygen atoms in different antimony tartrate dimers related by the C-centering operation. These hydrogen bonds connect the antimony tartrate dimers into infinite sheets parallel to the ab plane. The potassium atoms are situated between these layers and stabilize the solid by electrostatic interactions with tartrate and water oxygen atoms.

   The crystal structure of $(NH_4)_2[Sb_2(d-C_4H_2O_6)_2].3H_2O$ has also been determined.[3]

Remarks

1. Coordination geometry about antimony(III)[4]

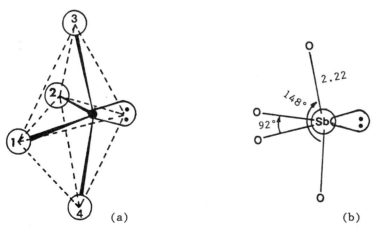

Fig. 3.30-2 Coordination about antimony(III) (a) and geometry of the $[SbO_4]$ polyhedron (b). (After ref. 4).

Antimony(III) forms a variety of compounds with oxoacids. In these compounds, the geometry about antimony is a distorted trigonal bipyramid owing to the presence of a lone pair (E) (Fig. 3.30-2). From the available structural data of some of these compounds, the mean equatorial distance (Sb-1 and Sb-2) is 2.01Å, and the mean axial distance (Sb-3 and Sb-4) 2.22Å, with mean angles of 92.0° (1-Sb-2) and 148.0° (3-Sb-4).

2. Crystal structure of $K_2[Sb_2(d,\ell\text{-}C_4H_2O_6)_2]\cdot3H_2O$

The racemic salt $K_2[Sb_2(d,\ell\text{-}C_4H_2O_6)_2]\cdot3H_2O$ has been investigated.[5] The compound crystallizes in space group $Pca2_1$ with $a = 8.79(2)$, $b = 16.32(2)$, $c = 12.19(2)$Å, and $Z = 4$. The crystal is a racemic mixture of potassium di-μ-(+)-tartratodiantimonate(III) and di-μ-(-)-tartratodiantimonate(III). Two antimony atoms are bridged by two tartrate groups which act as double bidentate ligands through the deprotonated carboxyl group and α-hydroxyl oxygen atoms with mean Sb-O distances of 2.20 and 2.01Å, respectively (Fig. 3.30-3). The chelate rings are nearly planar and the oxygen atoms are arranged about each antimony atom in the same way as in the d-form (Fig. 3.30-2).

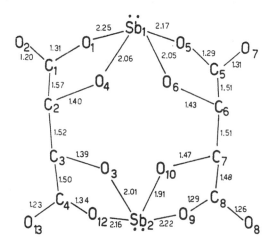

Fig. 3.30-3 Bond lengths in the $[Sb_2(d,\ell\text{-}C_4H_2O_6)_2]^{2-}$ ion. (After ref. 5).

3. Mixed-valence compounds of antimony

Most compounds of antimony are colorless, but salts containing octahedral SbX_6^- and SbX_6^{3-} (X = halogen) are either blue or red, depending on the nature of the cation. The crystal structure of $(NH_4)_2SbBr_6$ is shown in Fig. 3.30-4.[6] This compound is diamagnetic, and the difference in bond length between the two kinds of $SbBr_6$ octahedra, δ, is 0.24Å. In the analogous chloride salt $(C_3H_7NH_3)_4SbCl_6(Cl)_2$ δ has the value 0.28Å.[7] Both the far infrared spectra[8] and ^{121}Sb Mössbauer spectra[9] of these two compounds exhibit separate peaks, which substantiate the presence of Sb(III) and Sb(V).

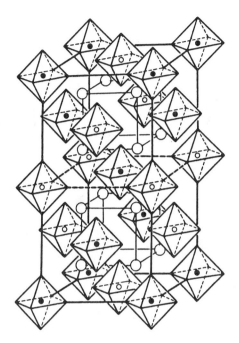

Copyright 1966
Am. Chem. Soc.

Fig. 3.30-4 Crystal structure of $(NH_4)_2SbBr_6$. (After ref. 6).

References

[1] Hsiang-chi Mu, *Kexue Tongbao* **17**, 502 (1966). [In Chinese].

[2] M.E. Gress and R.A. Jacobson, *Inorg. Chim. Acta* **8**, 209 (1974).

[3] G.A. Kiosse, N.I. Golovastikov, A.V. Ablov and N.V. Belov, *Doklady Akad. Nauk. SSSR* **177**, 329 (1967). [*Sov. Phys.-Dokl.* **12**, 990 (1968)].

[4] C.A. McAuliffe in G. Wilkinson, R.D. Gillard and J.A. McCleverty (eds.), *Comprehensive Coordination Chemistry*, Vol. 3, Pergamon Press, Oxford, 1987, chap. 28.

[5] B. Kamenar, D. Grdenic and C.K. Prout, *Acta Crystallogr., Sect. B* **26**, 181 (1970).

[6] S.L. Lawton and R.A. Jacobson, *Inorg. Chem.* **5**, 743 (1966).

[7] G. Birke, H.P. Latscha and H. Pritzkow, *Z. Naturforsch., Teil B* **31**, 1285 (1976).

[8] T. Barrowcliffe, I.R. Beattie, P. Day and K. Livingston, *J. Chem. Soc. (A)*, 1810 (1967).

[9] A.Y. Aleksandrov, S.P. Ionov, A.M. Pritchard and V.I. Goldanskii, *J.E.T.P. Lett.* **13**, 8 (1971).

3.31 Bismuth Subchloride

$$Bi_{24}Cl_{28}, \text{ or } (Bi_9^{5+})_2(BiCl_5^{2-})_4(Bi_2Cl_8^{2-})$$

Crystal Data

Orthorhombic, space group *Pnnm* (No. 58)

$a = 23.057(2)$, $b = 15.040(7)$, $c = 8.761(3)$Å, $Z = 2$

Atom	x	y	z	Atom	x	y	z
Bi(1)	.0459	.2229	.1876	Cl(2)	.052	.418	1/2
Bi(2)	.2002	.1544	.1815	Cl(3)	.109	.126	1/2
Bi(3)	.4101	.4550	0	Cl(4)	.264	.291	1/2
Bi(4)	.0984	.0692	0	Cl(5)	.290	.044	1/2
Bi(5)	.0723	.3982	0	Cl(6)	.315	.010	0
Bi(6)	.4139	.1009	0	Cl(7)	.338	.216	.194
Bi(7)	.3578	.1930	1/2	Cl(8)	.332	.449	.216
Bi(8)	.2390	.3296	0	Cl(9)	.433	.337	1/2
Bi(9)	.1525	.3425	.2669	Cl(10)	.445	.086	.302
Cl(1)	0	0	.288	Cl(11)	.466	.274	0

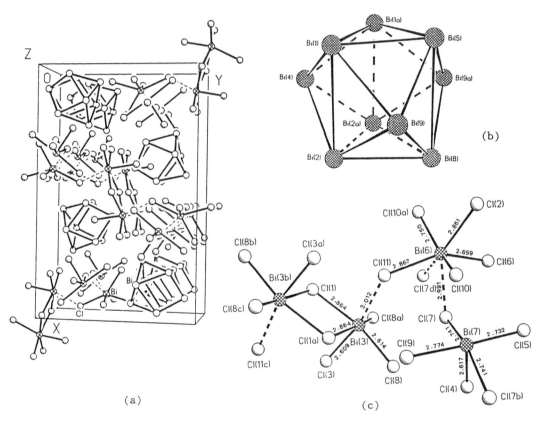

(a)

(b)

(c)

Fig. 3.31-1 Crystal structure of $Bi_{24}Cl_{28}$ (a) and structures of the Bi_9^{5+} (b), $BiCl_5^{2-}$ and $Bi_2Cl_8^{2-}$ groups (c).

Crystal Structure

The unit cell of $Bi_{24}Cl_{28}$ contains four Bi_9^{5+} polyhedra, eight $BiCl_5^{2-}$ groups, and two $Bi_2Cl_8^{2-}$ groups (assuming that the Bi-Cl distances >2.9Å are non-bonded). The molecular packing is illustrated in Fig. 3.31-1(a). The $Bi_9{}^{5+}$ polyhedron has the form of a slightly distorted trigonal prism with three Bi atoms capped at its faces, as shown in Fig. 3.31-1(b). The anion $BiCl_5{}^{2-}$ has square-pyramidal coordination, as shown in Fig. 3.31-1(c); in $Bi_2Cl_8{}^{2-}$ two such pyramids share a basal edge in the *trans* manner, as shown in Fig. 3.31-1(c). Weak Bi...Cl interactions form the only links between these groups. Thus the black, diamagnetic crystalline lower chloride of composition Bi_6Cl_7 or $Bi_{24}Cl_{28}$ should be formulated $(Bi_9{}^{5+})_2(BiCl_5{}^{2-})_4(Bi_2Cl_8{}^{2-})$.[1]

The bond distances in the $Bi_9{}^{5+}$ cation are to be compared with those in $(Bi^+)(Bi_9{}^{5+})(HfCl_6{}^{2-})_3$,[3] as shown in Fig. 3.31-2.

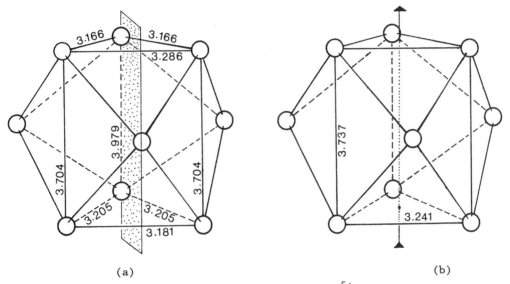

(a)	(b)

Fig. 3.31-2 Bond distances found in the $Bi_9{}^{5+}$ cation (a) in $Bi_{24}Cl_{28}$ and (b) in $Bi_{10}(HfCl_6)_3$. (After ref. 2).

Remarks

1. Cationic bismuth clusters

Bismuth is known to form several cationic clusters, some examples of which are listed in Table 3.31-1.

Table 3.31-1 Cationic bismuth clusters.

Cation	Crystal	Cluster structure	Point group symmetry
Bi^+	$Bi_{10}(HfCl_6)_3$	–	–
$Bi_5{}^{3+}$	$Bi_5(AlCl_4)_3$	trigonal bipyramid	D_{3h}
$Bi_8{}^{2+}$	$Bi_8(AlCl_4)_2$	square antiprism	D_{4h}
$Bi_9{}^{5+}$	$Bi_{24}Cl_{28}$	tricapped trigonal prism	$C_{3h}(\sim D_{3h})$

The crystal structure determination of black needles of $Bi_{10}(HfCl_6)_3$, reveals the ionic formulation $(Bi^+)(Bi_9^{5+})(HfCl_6^{2-})_3$. This is the first unambiguous identification of Bi^+ in the solid state.[3]

Bi_5^{3+} and Bi_8^{2+} have been established in crystalline $Bi_5(AlCl_4)_3$[4] and $Bi_8(AlCl_4)_2$,[5] respectively. The Bi_8^{2+} cation in the latter lies on a mirror plane and deviates only slightly from idealized D_{4d} symmetry (Fig. 3.31-3). The average of the observed Bi-Bi distances is 3.100Å, which corresponds approximately to the value in metallic bismuth. The Bi-Bi-Bi bond angles in the square and triangular planes of the Bi_8^{2+} polyhedron are between 88.1 and 91.8°, and between 59.3 and 60.7°, respectively.

The structural relationship of Bi_8^{2+} to the Bi_9^{5+} cluster, which is richer by one Bi^{3+} unit, is of special interest as it may give some indication of the possible paths of interconversion between the two species. The square-pyramidal Bi_8^{2+} can be transformed by very slight shifts of the atoms plus addition of one Bi^{3+} into the trigonal-prismatic tricapped Bi_9^{5+}.

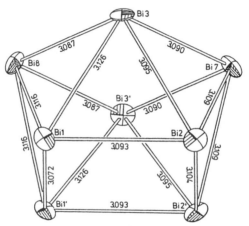

Fig. 3.31-3 Structure of the Bi_8^{2+} cluster. (After ref. 5).

2. Bonding in bismuth clusters

There are many factors in favour of a MO treatment of the bonding in bismuth clusters:[2]

a) The clusters contain only one type of atom, a factor that greatly simplifies or improves the necessary approximations, especially relative to estimations of the resonance integrals necessary for useful heteroatomic results.

b) To a good approximation, the good bonding orbitals are only the 6p set. For the gaseous Bi^+ ion ($6s^26p^2$), the lowest state involving promotion to sp^3 is 8.5 eV above ground and the lowest to $s^2p^1d^1$, 10eV. Thus $6s^2$ is indeed a good "inert pair" so that the basis set is substantially iso-orbital as well, making the problem still easier to handle and the results more credible. At this point inclusion of all overlap becomes relatively simple.

c) The relatively high symmetry of Bi$_9^{5+}$(D_{3h}), the first example studied, provides a great assistance.[6]

d) The simplest results are "one-electron" solutions, but the high delocalization in the ion should reduce the importance of interelectronic effects.

The original application to Bi$_9^{5+}$ gave a rationalization of the observed stoichiometry and symmetry to the extent that (a) the calculated bonding orbitals accommodated only the needed 22 electrons, with an appreciable gap to the lowest antibonding orbtial, (b) neither conclusion was affected by rather appreciable variations in overlap scaling parameters, and (c) the bond orders and therefore charges were relatively uniformly distributed over the nine bismuth atoms.

Application of the calculation methods to Bi$_5^{3+}$ and Bi$_8^{2+}$ led to unambiguous prediction of the correct configuration in both cases, there being only one for which the indicated 12 and 22 bonding electrons will first fill bonding MOs and give a diamagnetic result.

3. Stereochemistry of group V pentaphenyls

The pentaphenyl compounds E(C$_6$H$_5$)$_5$ (E = P, As, Sb) are colourless, stable solids at room temperature whereas Bi(C$_6$H$_5$)$_5$ is violet coloured and highly sensitive to heat, light, and moisture. P(C$_6$H$_5$)$_5$ and As(C$_6$H$_5$)$_5$ have very similar monoclinic unit cells, and in the crystalline state their phenyl groups adopt a trigonal-bipyramidal (TBP) arrangement with longer axial E-C bonds.[7] Triclinic Sb(C$_6$H$_5$)$_5$[8] and Bi(C$_6$H$_5$)$_5$[9] likewise constitute an isomorphous pair but the molecular geometry is square pyramidal with particularly short E-C(axial) and longer E-C(basal) bonds [Fig. 3.31-4(a)]. With well-formed crystals of Bi(C$_6$H$_5$)$_5$ a violet/colourless dichroism was observed. Maximum absorption was found in the direction parallel to the Bi-C(axial) bond and maximum transmission in the plane normal to it. However, light transmitted in this plane is strongly polarized: when the electric vector vibrates *parallel* to the plane, the crystal appears violet, and when the electric vector is in the *perpendicular* orientation, the crystal becomes colourless. Fig. 3.31-4(b) shows that a beam of light is polarized after passing parallel to the basal plane of the square pyramid.

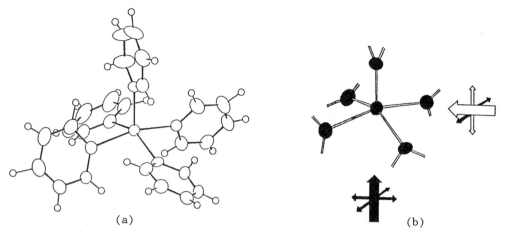

Fig. 3.31-4 (a) Molecular structure of $Bi(C_6H_5)_5$. (b) Optical behavior of a $Bi(C_6H_5)_5$ crystal relative to the molecular orientation. The large arrow indicates the direction of light, the small arrow the vibration plane of the electric vector, and filled and open arrows represent absorption and transmission, respectively. (After ref. 9).

Table 3.31-2   Stereochemistry and average bond lengths of group VA
                        pentaphenyls and substituted derivatives.

Compound	Molecular geometry	Average bond lengths (Å) about central atom E*		
		E–C(axial)	E–C(equatorial)	E–C(basal)
$P(C_6H_5)_5$	TBP	1.987(6)	1.850(6)	–
$As(C_6H_5)_5$**	TBP	–	–	–
$Sb(C_6H_5)_5$	SP	2.115(5)	–	2.216(7)
$Sb(4\text{-}MeC_6H_4)_5$	TBP	2.25(1)	2.16(1)	–
$Bi(C_6H_5)_5$	SP	2.221(9)	–	2.326(9)
$Bi(C_6H_5)_3(2\text{-}FC_6H_4)_2$	SP	2.216(5)	–	2.285(7), 2.355(5)F
$Bi(C_6H_5)_3(2,6\text{-}F_2C_6H_3)_2$***	SP	2.26(3)	–	2.26(3), 2.37(3)F
		2.23(2)	–	2.23(3), 2.37(3)F
$Bi(4\text{-}FC_6H_4)_3(C_6F_5)_2$	SP	2.22(2)	–	2.24(2)F, 2.23(2)F$_5$
$Bi(4\text{-}MeC_6H_4)_3(C_6F_5)_2$	SP	2.18(2)	–	2.24(1), 2.44(2)F$_5$
$Bi(4\text{-}MeC_6H_4)_3(2\text{-}FC_6H_4)_2$	TBP	2.391(2)F	2.184(1)	–

* The number of F substituents in the phenyl ring is indicated after the listed value.

** Crystal is isomorphous with the P analogue but a complete structure analysis is not available.

*** There are two independent molecules of similar structure in the unit cell.

The available structural data for group VA pentaphenyls are summarized in Table 3.31-2. Note in particular that $Sb(C_6H_5)_5$ and $Sb(4\text{-}MeC_6H_4)_5$[10] have different molecular structures. In Bi complexes where some of the ligands bear fluoro or methyl substituents, a less bulky phenyl group preferentially

occupies an equatorial position of the TBP structure and the axial position of
the SP structure,[11,12] as illustrated in Fig. 3.31-5.

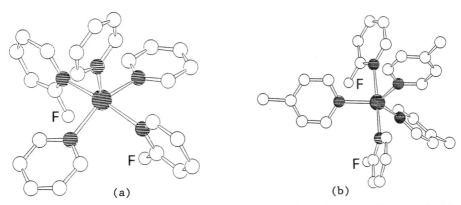

(a)   (b)

Fig. 3.31-5 Molecular structures of (a) $Bi(C_6H_5)_3(2\text{-}FC_6H_4)_2$ and (b)
$Bi(4\text{-}MeC_6H_4)_3(2\text{-}FC_6H_4)_2$. (After ref. 12).

The substituted pentaphenylbismuth $Bi(C_6H_5)_3(2\text{-}FC_6H_4)_2$ is, like
$Bi(C_6H_5)_5$, violet and dichroic in the solid state, but exhibits a reddish
colour in solution. The methylfluoro derivative $Bi(4\text{-}MeC_6H_4)_3(2\text{-}FC_6H_4)_2$ has
the same colour in solution, but forms orange crystals. Obviously the Berry
pseudorotation mechanism is operative in solution, and the solid state
structure is determined by lattice forces.[12]

References

[1] A. Hershaft and J.D. Corbett, *Inorg. Chem.* **2**, 979 (1963).

[2] J.D. Corbett, *Prog. Inorg. Chem.* **21**, 129 (1976).

[3] R.M. Friedman and J.D. Corbett, *J. Chem. Soc., Chem. Commun.*, 422 (1971).

[4] R.C. Burns, R.J. Gillespie and W.-C. Luk, *Inorg. Chem.* **17**, 3596 (1978).

[5] B. Krebs, M. Hucke and C.J. Brendel, *Angew. Chem. Int. Ed. Engl.* **21**, 445
(1982).

[6] J.D. Corbett and R.E. Rundle, *Inorg. Chem.* **3**, 1408 (1964).

[7] P.J. Wheatley and G. Wittig, *Proc. Chem. Soc. London*, 251 (1962); P.J.
Wheatley, *J. Chem. Soc.*, 2206 (1964).

[8] A.L. Beauchamp, M.J. Bennett and F.A. Cotton, *J. Am. Chem. Soc.* **90**, 6675
(1968).

[9] A. Schmuck, J. Buschmann, J. Fuchs and K. Seppelt, *Angew. Chem. Int. Ed.
Engl.* **26**, 1180 (1987).

[10] C. Brabant, J. Hubert and A.L. Beauchamp, *Can. J. Chem.* **51**, 2952 (1973).

[11] A. Schmuck and K. Seppelt, *Chem. Ber.* **122**, 803 (1989).

[12] A. Schmuck, P. Pyykkö and K. Seppelt, *Angew. Chem. Int. Ed. Engl.* **29**, 213
(1990).

3.32   Hydrogen Peroxide

H_2O_2

Crystal Data

Tetragonal, space group, $P4_12_12$ (No. 92)

$a = 4.004(8)$, $c = 7.832(13)$Å (110K), $Z = 4$

Atom	x	y	z
O	.0762(3)	.1670(3)	.2204(1)
H	-.0483(7)	.2834(6)	.1296(3)

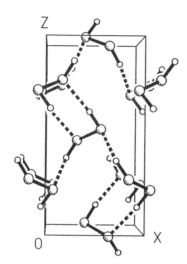 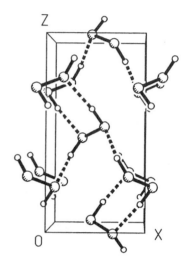

Fig. 3.32-1   Stereoview of the crystal structure of hydrogen peroxide.

Crystal Structure

Thenard in 1818 first synthesized liquid hydrogen peroxide, and a great number of experimental and theoretical studies of this compound have since been published.   In 1951 Abrahams *et al.*[1] determined the crystal structure of hydrogen peroxide by X-ray diffraction, and in 1965 Busing and Levy[2] redetermined the structure by neutron diffraction.   The two analyses are in good agreement (Fig. 3.32-1).   In 1980, Savariault and Lehmann[3] used new X-ray and neutron diffraction measurements at 110K to determine the deformation electron density of hydrogen peroxide.

In the crystal the H_2O_2 molecule adopts a skew configuration with O-O = 1.458(4)Å, O-H = 0.988(5)Å, O-O-H = 101.9(1)°, and dihedral angle H-O-O-H = 90.2(4)°.   The length of the hydrogen bond O-H...O between H_2O_2 molecules is 2.761(5)Å, quite similar to the average distance of 2.758Å in heavy ice.[4] In the gas phase, the dihedral angle is 111.5°.[5]   Fig. 3.32-2 shows the molecular structure of H_2O_2.

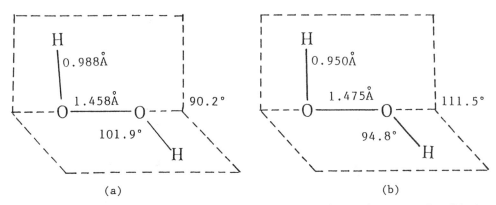

Fig. 3.32-2 Molecular structure of H_2O_2: (a) in the crystal, (b) in the gas phase.

The principal binding forces between the H_2O_2 molecules are the hydrogen bonds which extend in infinite helices around the four-fold screw axes. Each oxygen atom has the same environment as every other one, and there are two such close approaches to each atom. One of these involves sharing its hydrogen with an oxygen atom of an adjacent molecule, and the other is formed by the oxygen atom donating an electron pair to another molecule.

Remarks

1. The oxygen lone pairs and the hydrogen bond in hydrogen peroxide[3]

Fig. 3.32-3 shows a section of $\Delta\rho_{(X-N)}$ through the oxygen lone pair region orthogonal to the oxygen-oxygen axis. Three maxima are found in this region. One of these corresponds to the intramolecular oxygen-hydrogen bond;

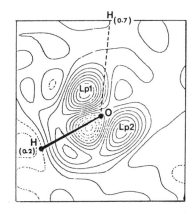

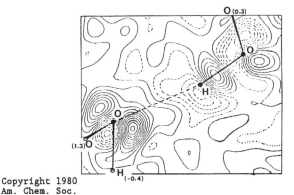

Fig. 3.32-3 Section of $\Delta\rho_{(X-N)}$ orthogonal to the oxygen-oxygen bond through the oxygen atom. Interval between contours 0.1 e$Å^3$. (After ref. 3).

Fig. 3.32-4 Section of $\Delta\rho_{(X-N)}$ through the atoms O...H-O involved in the hydrogen bond. Interval between contours 0.1 e$Å^3$. (After ref. 3).

the other two are taken to indicate the location of the lone-pair distributions, LP_1 and LP_2.

The distances of the lone-pair LP_1 and LP_2 maxima to the oxygen atom are
$O-LP_1 = 0.36\text{Å}$ and $O-LP_2 = 0.37\text{Å}$; the angle LP_1-O-LP_2 is $150.7°$.

Fig. 3.32-4 shows a section of $\Delta\rho_{(X-N)}$ through the atoms $O...H-O$. The
features are typical for a long hydrogen bond, where the main interaction is
of electrostatic nature, and as often observed the oxygen-hydrogen bond points
towards a region of high electron density, i.e. the lone pair, which in this
case in LP_1. Normally discussion of this type of directionality can only
indicate trends, as many interactions and constraints are operating when
molecules pack together, and it is unlikely that the oxygen-hydrogen bond can
orient itself freely.

In a hydrogen bond system $O-LP...H-O$, the lone pair may point towards
the hydrogen atom ("lone pair directionality"), and hydrogen bond interactions
will then dominate over intra-atomic interactions, or alternatively the
oxygen-hydrogen bond may point toward the lone pair ("hydrogen
directionality") and dominate intermolecular interactions. In H_2O_2 the first
case would correspond to LP_1 moving into the line $O...H$, creating a nearly
perfect tetrahedron as well as reducing the lone pair-hydrogen distance. As
this is contrary to the observations of a small dihedral angle $LP_1-O-O-H$
($6.7°$), and an angle $O-H...LP_1$ of $176.6°$, one may conclude that H_2O_2 satisfies
both the concept of intramolecular domination of the lone pair and "hydrogen
directionality". In the simple picture of hydrogen bonding, electrons are
transferred from the donor oxygen to the acceptor hydrogen, so in the present
case one would expect LP_1 to be smaller than LP_2, but this is contrary to
observation.

2. The dihedral angle in H_2O_2

H_2O_2 is the smallest molecule known to show hindered rotation about a
single bond, the rotational barriers being 4.62 and 29.45 kJ mol^{-1} for the
trans and *cis* conformations, respectively. The dihedral angle ranges from 90°
in crystalline H_2O_2 to 180° in $Na_2C_2O_4.H_2O_2$, where the H_2O_2 molecule (O-O-H =
97°) is found to have a planar *trans* configuration. This large range of
values for the dihedral angle indicates that the configuration of H_2O_2 is very
sensitive to its surroundings. The value of the dihedral angle can easily be
altered by hydrogen bonding, and also in $Na_2C_2O_4.H_2O_2$ by the proximity of two
Na^+ ions which complete a tetrahedron around the O atom. Values of the
dihedral angle found in organic peroxides also cover a wide range. Table
3.32-1 lists the values of the dihedral angle for some related compounds.

Table 3.32-1 Dihedral angles R-O-O-R'[6]

Compound	Dihedral angle	Compound	Dihedral angle
H_2O_2 (crystal) (gas)	90.2° 111.5°	$Li_2C_2O_4 \cdot H_2O_2$	180°
$H_2O_2 \cdot 2H_2O$	129°	$Na_2C_2O_4 \cdot H_2O_2$	180°
$K_2C_2O_4 \cdot H_2O_2$	101.6°	$NH_4F \cdot H_2O_2$	180°
$Rb_2C_2O_4 \cdot H_2O_2$	103.4°	O_2R_2 (organic)	81°~146°
O_2F_2	87.5°	theoretical	90°~120°

3. Chemical properties of H_2O_2[7]

Hydrogen peroxide, which can be converted to H_2O, O_2, OH^-, H_2OOH^+, OOH^-, and O_2^- under different conditions, has a rich and varied chemistry.

a) It can act either as an oxidizing or a reducing agent in both acid and alkaline solution. Typical examples are:

In acid solution, as an oxidizing agent:

$$2Fe^{2+} + H_2O_2 + 2H^+ \longrightarrow 2Fe^{3+} + 2H_2O$$

as a reducing agent:

$$2Ce^{4+} + H_2O_2 \longrightarrow 2Ce^{3+} + 2H^+ + O_2$$

In alkaline solution, as an oxidizing agent:

$$Mn^{2+} + H_2O_2 \longrightarrow Mn^{4+} + 2OH^-$$

as a reducing agent:

$$2Fe^{3+} + H_2O_2 + 2OH^- \longrightarrow 2Fe^{2+} + 2H_2O + O_2$$

b) It undergoes proton acid/base reactions to form peroxonium salts H_2OOH^+, hydroperoxides OOH^-, and peroxides.

c) It reacts to give peroxometal complexes and peroxoacid anions. Treatment of alkaline aqueous solutions of chromate(IV) with H_2O_2 yields the stable red paramagnetic tetraperoxochromate(V) compound $[Cr^V(O_2)_4]^{3-}$, whereas treatment of chromate(VI) with H_2O_2 in acid solution followed by extraction with ether and coordination with pyridine yields the neutral peroxochromate(VI) complex $[CrO(O_2)_2py]$.

d) It can form addition compounds with other molecules through hydrogen bonding, such as $Na_2C_2O_4 \cdot H_2O_2$, $NH_4F \cdot H_2O_2$, $H_2O_2 \cdot 2H_2O$, etc.

4. Parameters of the O-O bond

Many molecules and ions contain an O-O bond in their structures. Table 3.32-2 lists the bond length, bond energy and bond order of the O-O bonds in some related molecules.

Table 3.32-2   Parameters of O-O bonds in some related molecules.[6]

Molecule or ion	O-O/Å	Bond energy/ kJ mol^{-1}	Bond order
O_2^+ [in $O_2(AsF_6)$]	1.123	625.1	2.5
O_2 ($^3\Sigma_g$)	1.207	490.4	2
O_2 ($^1\Delta_g$)	1.2107	396.2	2
O_3	1.278	-	-
O_2^- (in α-KO_2)	1.28	-	1.5
O_2^{2-} (in Na_2O_2)	1.49	204.2	1
H_2O_2 (gas phase) (crystal)	1.475 1.467	- 213	1 1

5. Ionic ozonides

Ionic ozonides are red, crystalline solids containing the bent O_3^- molecular ion. The alkali-metal ozonides MO_3 (M = K, Rb, Cs)[8,9] are extremely sensitive to moisture and also thermally labile, but R_4NO_3 (R = CH_3, C_2H_5)[10,11] are more stable. Low-temperature X-ray structure analyses have yielded the results summarized in Table 3.32-3.

Table 3.32-3   Crystal data and structural parameters of ionic ozonides.

Ozonide	Space group	a/Å	b/Å	c/Å	Z	O-O/Å	O-O-O/°
KO_3	I4/mcm	8.648(1)		7.164(1)	8	1.346(2)	113.5(1)
RbO_3	P2$_1$/c	6.441(2)	6.030(4)	8.746(4)	4	1.343(6)	113.7(5)
$(CH_3)_4NO_3$	Pmmn	8.510(3)	6.900(3)	5.444(2)	2	1.288(3)	119.6(4)
$(C_2H_5)_4NO_3$	P3$_1$21	7.066(2)		17.526(4)	3	1.302(4)	118.4(4)

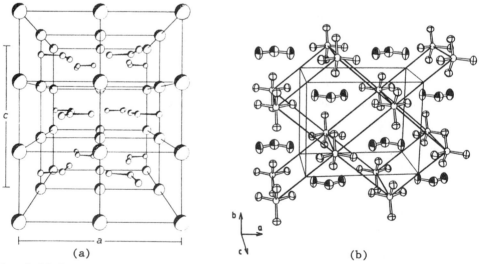

(a)   (b)

Fig. 3.32-5 Perspective diagram of the crystal structure of (a) KO_3 and (b) $(CH_3)_4NO_3$; reduced unit cells showing the structural relationship to CsCl are outlined. (After refs. 9 and 10).

The first three crystal structures shown in Table 3.23-3, as well as that of CsO_3 (with $O-O = 1.332(8)$Å and $O-O-O = 114.6(6)°$),[12] are all related to the CsCl type despite the difference in space groups and unit-cell parameters. In the alkali-metal ozonides the intermolecular O...O contacts are short (3.005Å and 2.992Å in KO_3 and RbO_3, respectively) and the terminal (negatively charged) O atoms of each O_3^- ion are directed toward the bridging (positively polarized) O atoms of neighbouring O_3^- groups [Fig. 3.32-5(a)]. The surprisingly high stability of $(CH_3)_4NO_3$ can be attributed to the "antiferroelectric" arrangement of O_3^- dipoles with much longer intermolecular O...O distances of 4.388Å [Fig. 3.32-5(b)].

$(C_2H_5)_4NO_3$ is the only known ozonide that exhibits an ionic packing reminiscent of the NaCl type.[11] This structure difference can be attributed to the increasing size of the cation in the sequence $K-Rb-Cs-(CH_3)_4N-(C_2H_5)_4N$. The cation/anion radius ratio r_+/r_- of the first four ozonides lies in the range 0.747-1.483, for which the lattice energy attains a minimum at a coordination number of 8 (CsCl type). On the other hand, the considerably larger r_+/r_- value of 2.185 in $(C_2H_5)_4NO_3$ favours a reduction in coordination number to 6 to give a structure of the NaCl type.

The crystal structures of both $(CH_3)_4NO_3$ and $(C_2H_5)_4NO_3$ are stabilized by strong C-H...O hydrogen bonds, the presence of which is supported by IR spectral data.[11]

References

[1] S.C. Abrahams, R.L. Collin and W.N. Lipscomb, *Acta Crystallogr.* 4, 15 (1951).

[2] W.R. Busing and H.A. Levy, *J. Chem. Phys.* 42, 3054 (1965).

[3] J.-M. Savariault and M.S. Lehmann, *J. Am. Chem. Soc.* 102, 1298 (1980).

[4] S.W. Peterson and H.A. Levy, *Acta Crystallogr.* 10, 70 (1957).

[5] R.L. Redington, W.B. Olson and P.C. Cross, *J. Chem. Phys.* 36, 1311 (1962).

[6] A.F. Wells, *Structural Inorganic Chemistry*, 5th ed., Clarendon Press, Oxford, 1984.

[7] N.N. Greenwood and A. Earnshaw, *Chemistry of the Elements*, Pergamon Press, Oxford, 1986.

[8] W. Schnick and M. Jansen, *Rev. Chim. Miner.* 24, 446 (1987).

[9] W. Schnick and M. Jansen, *Angew. Chem. Int. Ed. Engl.* 24, 54 (1985).

[10] W. Hesse and M. Jansen, *Angew. Chem. Int. Ed. Engl.* 27, 1341 (1988).

[11] W. Hesse and M. Jansen, *Inorg. Chem.* 30, 4380 (1991).

[12] M. Jansen and W. Assenmacher, *Z. Kristallogr.*, in press.

3.33   Orthorhombic α-Sulfur

S_8

Crystal Data

Orthorhombic, space group *Fddd* (No. 70)

$a = 10.3849(15)$, $b = 12.7549(23)$, $c = 24.4098(15)$Å (100K), $Z = 16$

Atom	x	y	z
S(1)	.85576(3)	.95367(2)	-.04937(1)
S(2)	.78417(3)	1.03215(2)	.07599(1)
S(3)	.70603(3)	.98106(2)	.00361(1)
S(4)	.78528(3)	.90851(2)	.12969(1)

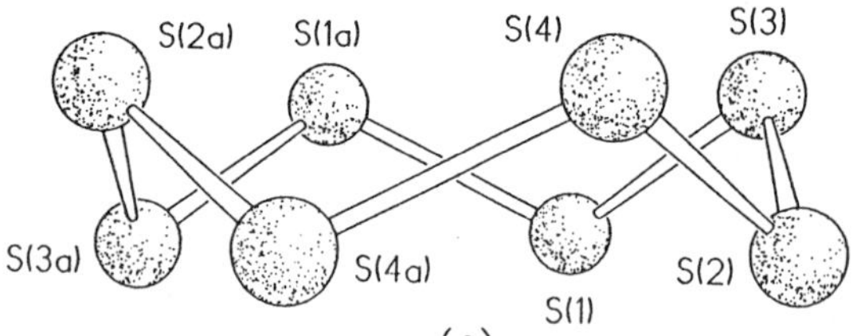

(a)

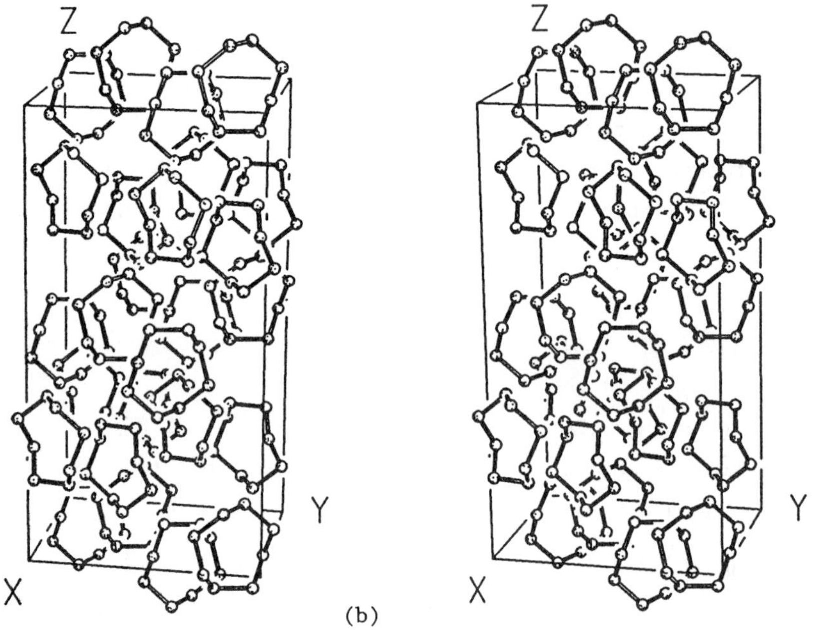

(b)

Fig. 3.33-1   (a) Conformation of the S_8 molecule with atom labelling.
(b) Stereoview of the orthorhombic α-sulfur crystal structure.

Crystal Structure

All modifications of crystalline sulfur contain either sulfur rings, which may have 6, 8, 10, 12, 18 or 20 sulfur atoms, or chains of sulfur atoms. Cyclo-S_8 is the most common form and occurs in three main allotropes: orthorhombic α-sulfur, monoclinic β-sulfur and monoclinic γ-sulfur.

The crystal structure of orthorhombic α-sulfur was first determined in 1935 by Warren and Burwell,[3] who assumed a staggered ring conformation for the S_8 molecule. Accurate structure refinements were carried out by Abrahams (1955),[4] Cooper (1962),[5] Caron and Donohue (1965),[6] Pawley and Rinaldi (1972),[7] and Rettig and Trotter (1987).[8] P. Coppens *et al.* used X-ray and neutron diffraction measurements to examine the charge density in α-sulfur.[1]

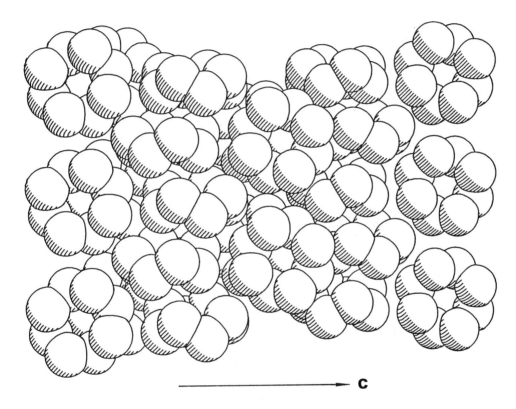

C

Fig. 3.33-2 Packing of S_8 molecules represented as space-filling models.

The molecular packing in the crystal is very complex (Fig. 3.33-1). A portion of the structure, viewed perpendicular to the mean plane of half of the molecules, a direction which is not rational with respect to the crystal axes, is shown in Fig. 3.33-2. From these two figures it is apparent why the molecular packing in this crystal is sometimes referred to as the "crankshaft" arrangement. As may be seen, these crankshafts extend in two different directions, a result leading to the great complexity of the complete crystal structure. The usual depiction of the orthorhombic α-sulfur structure as one in which the rings stack *directly* over one another is clearly incorrect, as is easily seen from Figs. 3.33-1 and 3.33-2.[2]

There are 128 sulfur atoms that account for 16 S_8 molecules in the unit

cell. Although the crystal symmetry imposes an exact two-fold axis on the
molecule, the higher symmetry of $D_{4d}(\bar{8}2m)$ is closely approximated. Distances
and angles within a single molecule are presented in Fig. 3.33-3. The average
values are: S-S bond length = 2.060(3)Å, S-S-S bond angle = 108.0(7)°, and
S-S-S-S torsion angle = 98(2)°.

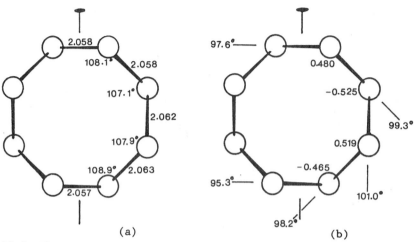

(a)   (b)

Fig. 3.33-3 The S_8 molecule in α-sulfur: (a) bond distances and angles; (b)
torsion angles and distances from the mean molecular plane. (After ref. 2).

Remarks

1. Polymorphs of cyclo-S_8

     All three polymorphs, orthorhombic α-, monoclinic β- and monoclinic γ-
sulfur, are composed of cyclo-S_8 molecules. α-Sulfur is the stable form at
room temperature and atmospheric pressure. In the phase equilibrium diagram
of sulfur (Fig. 3.33-4), only α- and β-sulfur are shown since γ-sulfur is a
metastable phase. The area BCF gives the conditions for stable monoclinic β-
sulfur relative to α (BF), liquid (CF) or sulfur vapor (BC). If α-sulfur is
slowly heated the transition into β-sulfur is observed at 95.5°C (B). This
melts at 119.25°C (C). Rapid heating causes α-sulfur to melt at 114.5° (E)
before its transition into β-sulfur, and the curve EF shows the effect of
pressure on the melting point of α-sulfur. Cooling sulfur slowly at
atmospheric pressure gives large crystals of β-sulfur which revert to small
crystals of α-sulfur below 95.5°C. So the large crystals of α-sulfur found
naturally are believed to have been formed directly from liquid sulfur by
cooling at pressures above those at the triple point F, i.e., about 1300 atm.

     At room temperature, monoclinic β-sulfur has the cell parameters[9] a =
10.926(2), b = 10.855(2), c = 10.790(3)Å, β = 95.92(2)°, Z = 6[S_8] in space
group $P2_1/c$. Since each S_8 molecule has the noncentrosymmetric crown

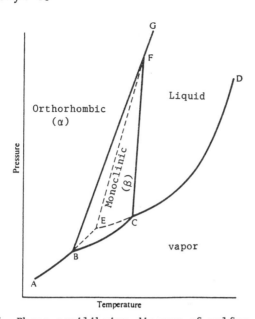

Fig. 3.33-4   Phase equilibrium diagram of sulfur (schematic).

configuration, two molecules in the unit cell are required to be two-fold
disordered.

The transition temperature T_c is 198K (λ transition); below the λ point,
the inversion center should vanish and the disorder should be gradually
reduced, the space group turning to $P2_1$. At 113K, the cell parameters become
$a = 10.799(2)$, $b = 10.684(2)$, $c = 10.663(2)$Å, $\beta = 95.71(1)°$, and $Z = 6[S_8]$.

Crystalline monoclinic γ-sulfur was obtained by evaporation of a
pyridine solution of cuprous ethylxanthate, $CuSSCOC_2H_5$. γ-Sulfur has the cell
parameters[10] $a = 8.44(3)$, $b = 13.025(10)$, $c = 9.36(5)$Å, $\beta = 125.0(3)°$, $Z = 4[S_8]$ in space group $P2/c$. In the crystal, the molecules of S_8 form a pseudo-
hexagonal close-packed structure with the c axis corresponding to the unique
axis of the hcp structure.

2. Homocyclic sulfur allotropes[11]

Many homocyclic sulfur allotropes have been characterized by X-ray
diffraction using single crystals, and the data obtained are summarized in
Table 3.33-1. All neutral sulfur rings form puckered structures.

The allotrope of cyclohexasulfur has a structure consisting of staggered
S_6 molecules. The crystal data of this structure are: space group $R\bar{3}$ (No.
148), $a_H = 10.818(2)$, $c_H = 4.280(1)$Å, and $Z = 3[S_6]$. The molecular packing
is very efficient. Each atom has two neighbours at 3.50Å and one at 3.53Å;
there are thus 18 short contacts per molecule. Fig. 3.33-5 shows the
structure as viewed perpendicular to the c axis. The molecular structures of
cyclo-S_7, S_{10}, S_{12} and S_{18} are shown in Fig. 3.33-6.

Table 3.33-1   Structural data of homocyclic sulfur allotropes.

Molecule	Space group	Site symmetry	S-S Bond lengths	Bond angles(°)	Torsional angles	Ref.
S_6	$R\bar{3}$	D_{3d}	2.068	102.6	73.8	[12]
γ-S_7	$P2_1/c$	C_1	1.998-2.175	101.9-107.4	0.4-108.8	[13]
δ-S_7	$P2_1/n$	C_1	1.995-2.182	101.5-107.5	0.3-108.0	[13]
α-S_8	Fddd	C_2	2.046-2.052	107.3-109.0	98.5	[1]
β-S_8	$P2_1/c$	C_1	2.047-2.057	105.8-108.3	96.4-101.3	[9]
γ-S_8	P2/c	C_2	2.023-2.060	106.8-108.5	97.9-100.1	[10]
S_{10}	C2/c	C_2	2.033-2.078	103.3-110.2	75.4-123.7	[14]
S_{11}	$Pca2_1$	C_1	2.032-2.110	103.3-108.6	69.3-140.5	[18]
S_{12}	Pnnm	C_{2h}	2.048-2.057	105.4-107.4	86.0-89.4	[15]
S_{13}	$P2_1/c$	C_1	1.978-2.113	102.8-111.1	29.5-116.3	[18]
α-S_{18}	$P2_12_12_1$	C_{2h}	2.044-2.067	103.8-108.3	79.5-89.0	[16]
β-S_{18}	$P2_1/n$	C_i	2.053-2.103	104.2-109.3	66.5-87.8	[17]
S_{20}	Pbcn	C_2	2.023-2.104	104.6-107.7	66.3-89.9	[16]

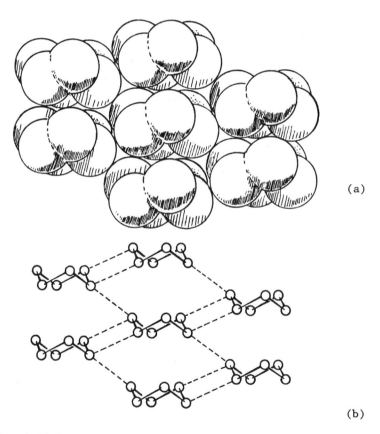

(a)

(b)

Fig. 3.33-5 Projection of the S_6 structure perpendicular to the c axis: (a) packing drawing of the view, and (b) the intermolecular contacts shown as dashed lines. (Adapted from ref. 2).

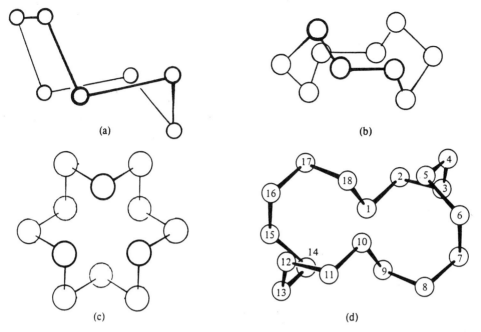

Fig. 3.33-6 The molecular structures of cyclo-S_7 (a), cyclo-S_{10} (b), cyclo-S_{12} (c) and cyclo-S_{18} molecule (d).

3. Prospects for research on elemental sulfur

Sulfur is one of the most common elements. The crystallization of sulfur, especially the allotrope which consists of S_8 molecules, used to be one of the important production industries. Despite several decades of extensive research activities in the field of elemental sulfur, there remain open questions concerning its structure and properties. Basic data concerning the thermodynamic properties of most homocyclic sulfur molecules are lacking, and even the homolytic dissociation energy of the S_8 ring is not accurately known. Fundamental reactions of sulfur rings like interconversions are only poorly understood, and further examples of the formation of homocyclic derivatives directly from the corresponding S_n parent molecules are likely to be discovered. Only a few molecular orbital calculations regarding the structure, conformational change, and bonding of S_n molecules have been published. Some observed data still await clarification, such as why the difference electron density at the midpoint of the S-S bond in α-S_8 is so low, being only about 0.05 eÅ^3.[1] Much more work needs to be done before the chemistry of elemental sulfur can be regarded as being "well known".[11]

4. Cyclic selenium sulfides

Two classes of compounds with general formulae of Se_nS_{8-n} and Se_nS_{12-n}

are known.[19]   In addition to their various stoichiometric compositions, there are many possibilities for positional and optical isomerism.

Eight-membered ring selenium sulfides constitute two isomorphic series: the sulfur-rich phases ($Se_{1.1}S_{6.9}$ to $Se_{3.7}S_{4.3}$) crystallize in the monoclinic γ-S_8 lattice and the selenium-rich species ($Se_{4.7}S_{3.3}$ to Se_5S_3) in the monoclinic α-Se_8 lattice (Fig. 3.33-7). X-Ray analysis have shown that in all cases the sulfur and selenium atoms are distributed statistically over the atomic sites.

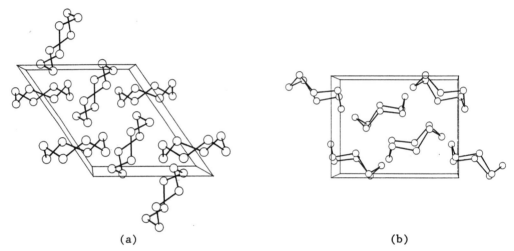

(a)   (b)

Fig. 3.33-7 Molecular packing in the unit cell of (a) γ-S_8, space group $P2/c$ and (b) α-Se_8, space group $P2_1/n$. (After ref. 19).

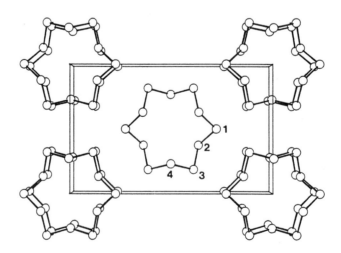

Fig. 3.33-8 Crystal structure of Se_nS_{12-n} with labeling of the four independent atomic sites. (After ref. 20).

The twelve-membered selenium sulfide Se_nS_{12-n} (Fig. 3.33-8) crystallizes from benzene to give rhombic crystals (space group *Pnmm*, $a = 4.774$, $b = 9.193$, and $c = 14.680$Å) and is isostructural with S_{12}.[20] Of the four independent sites two (2 and 4) are occupied by sulfur only while the other two are each 25% occupied by selenium. Crystallization of Se_nS_{12-n} in carbon disulfide yielded a 1:1 adduct (space group $R\bar{3}m$) having the same structure as $S_{12} \cdot CS_2$, but again with a statistical distribution of selenium over some of the sulfur sites.

References

[1]   P. Coppens, Y.W. Yang, R.H. Blessing, W.F. Cooper and F.K. Larsen, *J. Am. Chem. Soc.* **99**, 760 (1977).

[2]   J. Donohue, *The Structures of the Elements*, Wiley, New York, 1974.

[3]   B.E. Warren and J.T. Burwell, *J. Chem. Phys.* **3**, 6 (1935).

[4]   S.C. Abrahams, *Acta Crystallogr.* **8**, 661 (1955).

[5]   A.S. Cooper, *Acta Crystallogr.* **15**, 578 (1962).

[6]   A. Caron and J. Donohue, *Acta Crystallogr.* **18**, 562 (1965).

[7]   G.S. Pawley and R.P. Rinaldi, *Acta Crystallogr., Sect. B* **28**, 3605 (1972).

[8]   S.J. Rettig and J. Trotter, *Acta Crystallogr., Sect. C* **43**, 2260 (1987).

[9]   L.M. Goldsmith and C.W. Strouse, *J. Am. Chem. Soc.* **99**, 7580 (1977).

[10] Y. Watanabe, *Acta Crystallogr., Sect. B* **30**, 1396 (1974).

[11] R. Steudel, *Top. Curr. Chem.* **102**, 149 (1982).

[12] J. Steidel, J. Pickardt and R. Steudel, *Z. Naturforsch., Teil B* **33**, 1554 (1978).

[13] R. Steudel, J. Steidel, J. Pickardt, F. Schuster and R. Reinhardt, *Z. Naturforsch., Teil B* **35**, 1378 (1980).

[14] R. Reinhardt, R. Steudel and F. Schuster, *Angew. Chem. Int. Ed. Engl.* **17**, 57 (1978).

[15] J. Steidel, R. Steudel and A. Kutoglu, *Z. Anorg. Allg. Chem.* **476**, 171 (1981).

[16] M. Schmidt, E. Wilhelm, T. Debaerdemaeker, E. Hellner and A. Kutoglu, *Z. Anorg. Allg. Chem.* **405**, 153 (1974).

[17] T. Debaerdemaeker and A. Kutoglu, *Cryst. Struct. Commun.* **3**, 611 (1974).

[18] R. Steudel, J. Steidel and T. Sandow, *Z. Naturforsch., Teil B* **41**, 958 (1986).

[19] R. Steudel and R. Laitinen, *Top. Curr. Chem.* **102**, 177 (1982).

[20] J. Weiss and W. Bachtler, *Z. Naturforsch., Teil B* **28**, 523 (1973).

3.34   Tetrasulfur Tetranitride

$$S_4N_4$$

Crystal Data

Monoclinic, space group $P2_1/n$ (No. 14)

$a = 8.752(2)$, $b = 7.084(7)$, $c = 8.629(2)$Å, $\beta = 93.68(5)°$, $Z = 4$

Atom	x	y	z
S(1)	.00514	.92167	.30433
S(2)	.15149	.70853	.09900
S(3)	-.14966	.79887	.05795
S(4)	-.04076	.54600	.27272
N(1)	.0100	.7641	-.0245
N(2)	-.0272	.7230	.3915
N(3)	-.1818	.6035	.1481
N(4)	.1641	.8843	.2193

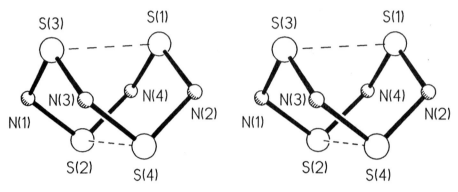

Fig. 3.34-1 Stereoview of the S_4N_4 molecule.

Crystal Structure

The crystal structure of S_4N_4 was determined at room temperature by Clark[2] and refined by Sharma and Donohue,[3] and at low temperature by Delucia and Coppens.[1] The molecule in the crystal conforms closely to D_{2d} symmetry and may be considered as an eight-membered ring, folded back on itself so as to form two cross-ring S...S linkages of 2.576 and 2.586Å at room temperature, and 2.595 and 2.590Å at 120K (Fig. 3.34-1).

In this cage molecule, the S-N bond lengths vary from 1.619 to 1.633Å, with a mean value of 1.626Å. The bond angles are: N-S-N = 104.18° to 105.09°, mean value 104.50°; S-N-S = 112.50° to 112.82°, mean value 112.66°. Though the S...S distances are more than 1Å shorter than twice the van der Waals radius of sulfur, they are still considerably longer than a single S-S bond, which in cyclo-S_8 is 2.047Å.

The low-temperature results show that even though S_4N_4 is thermochromic and loses its orange color on cooling, no structural transition occurs between room temperature and 120K. While intermolecular distances generally contract 1-2% in this temperature range, the S...S intramolecular distance remains remarkably constant, suggesting a relatively strong interaction with a well-defined minimum in the potential energy versus distance curve.

Refinements of the occupancy of a spherical valence shell with a variable radial function for each type of atom consistently show a charge transfer of 0.2-0.3 e from S to N, in qualitative agreement with their relative electronegativities and with theoretical results.[4]

In S_4N_4, an electron density peak of about 0.20 eÅ^{-3} is observed at the center of the molecule between the two perpendicular S...S bonds, which are at a distance of only 2Å from each other. The peak in the center of the molecule connects through two saddle point with each of the S...S bond peaks. The nature of this interaction presents an interesting problem.

The structural determination of S_4N_4 by electron diffraction has been attempted several times. All these reports agree that S_4N_4 exists as a molecular species of D_{2d} symmetry. In the first study,[5] it was suggested that the molecule consists of a square plane of four S atoms linked to a tetrahedron of four N atoms, a structure analogous to that found in β-realgar, As_4S_4 (see Section 3.29). In the second study,[6] the molecule was found to have a plane of four N atoms linked to an approximate tetrahedron of four S atoms, an arrangement as shown in Fig. 3.34-1. However, in adducts formed with BF_3[7], $SbCl_5$[8] or SO_3[9], the S_4N_4 unit adopts the β-realgar form. Recently, an electron diffraction study of gaseous S_4N_4[10] shows that the molecule has the same configuration as that in the crystalline state (Fig. 3.34-1). The bonded S-N distance is 1.623(4)Å, while the short unbridged S...S distance is 2.666(14)Å. The longer nitrogen-bridged S...S distance is 2.725(10)Å. The bond angles are N-S-N 105.3(7)°, S-N-S 114.2(6)°, and S...S-N 88.4(9)°. The angle between the plane of the four N atoms and the N-S-N plane is 92.5(3)°.

There are many theoretical discussions at both the semi-empirical and *ab initio* levels about S_4N_4. The unusual structure of S_4N_4, in which the sulfur atoms are three-coordinate and the nitrogen atoms two-coordinate, is the reverse of that found for the isoelectronic cages α-P_4S_4 and As_4S_4. There is thus a preference for the more electronegative atoms to occupy the lowest coordination sites in these molecules.

Remarks

1. Sulfur-nitrogen heterocycles and cages

The variety of chemical transformations and structures found for sulfur-nitrogen heterocycles have attracted much experimental and theoretical interest. The wide variety of sulfur-nitrogen compounds can be attributed to the range of oxidation states (0→6) and coordination numbers (2→4) available to sulfur in combination with nitrogen or other sulfur atoms. In addition nitrogen may be two-coordinate (unsaturated) or three-coordinate (saturated).

For example, the molecule S_4N_4 exhibits very versatile chemical behavior and provides a source of many S-N heterocycles. The syntheses of other S-N rings from S_4N_4 can be classified as follows: reactions of S_4N_4 with i) halogens or other oxidizing agents, ii) nucleophiles or reducing agents, iii) metal halides or organometallic reagents, and iv) acetylenes.[12]

Some molecules and ions of sulfur-nitrogen heterocycles and cages are shown in Fig. 3.34-2.[11,12]

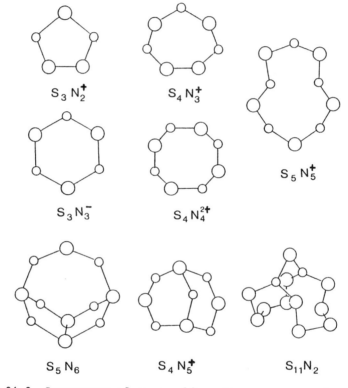

$S_3N_2^+$ $S_4N_3^+$

$S_3N_3^-$ $S_4N_4^{2+}$ $S_5N_5^+$

S_5N_6 $S_4N_5^+$ $S_{11}N_2$

Fig. 3.34-2 Structures of some sulfur-nitrogen compounds.

Many aspects of the molecular structure, chemical behavior, and spectroscopic properties of electron-rich S-N heterocycles derive from the fact that the HOMO and LUMO are both normally π^* or $n\pi$ orbitals. The result of the partial occupation of π^* orbitals is a weakened π framework which is readily deformed or even disrupted under the influence of heat or upon adduct formation. The intense colors of many S-N rings are also a manifestation of their electron richness and can be attributed to $\pi^* \rightarrow \pi^*$ or $n\pi \rightarrow \pi^*$ transitions.

The crystal structures of $(\overline{SeNSeNSe})(AsF_6)_2$ and $(\overline{SeNSeNSe})_2(AsF_6)_2$ containing the first stable binary selenium-nitrogen species have been determined.[13] In $(\overline{SeNSeNSe})(AsF_6)_2$ the 6-πe $(\overline{SeNSeNSe})^{2+}$ cation has a planar configuration with Se(1)-Se(2) 2.334(3), Se(1)-N(2) 1.73(2), Se(3)-N(2) 1.70(2), Se(3)-N(1) 1.69(2) and Se(2)-N(1) 1.74(2)Å, as shown in Fig. 3.34-3(a). In $(\overline{SeNSeNSe})_2(AsF_6)_2$, the two five-membered rings are weakly linked into a centrosymmetric dimer [Fig. 3.34-3(b)] through overlap of the two singly occupied molecular orbitals of the selenium portion of each ring. For both crystallographically different dimers in the crystal, the long Se...Se distances are 2x3.123(3) and 2x3.149(3)Å. The bond lengths in the dimers are Se(1)-Se(2) 2.398(3), 2.395(3); Se(1)-N(1) 1.76(2), 1.74(1); Se(3)-N(1) 1.69(2), 1.72(1); Se(3)-N(2) 1.69(1), 1.70(1) and Se(2)-N(2) 1.76(1), 1.77(1)Å. These are all longer than the corresponding ones in the monomer.

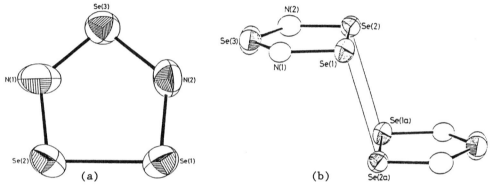

Fig. 3.34-3 Structures of (a) $Se_3N_2^{2+}$ and (b) $(Se_3N_2)_2^{2+}$. (After ref. 13).

2. Adducts of S_4N_4 with transition metal halides

There have been numerous claims of the preparation of adducts of S_4N_4 with transition-metal halides, e.g. WCl_4, $MoCl_5$ and $TiCl_4$, based on IR spectra and chemical analysis. When they are supported by X-ray structural characterization, these S_4N_4 adducts may be divided into three kinds:[14]

a) Treatment of S_4N_4 with transition-metal halids can lead to fragmentation of the S_4N_4 ring to give S_2N_2-bridged complexes, such as $(\mu\text{-}S_2N_2)(AlCl_3)_2$, $(\mu\text{-}S_2N_2)(SbCl_5)_2$, $[Ph_4P][(\mu\text{-}S_2N_2)(VCl_5)_2]$, and $(\mu\text{-}S_2N_2)[MoCl_4(NSCl)]_2$.

b) In the crystals of $S_4N_4 \cdot TaCl_5$, $S_4N_4 \cdot FeCl_3$, $S_4N_4 \cdot SbCl_5$, $S_4N_4 \cdot BF_3$ and $S_4N_4 \cdot AsF_5$, the S_4N_4 ligand is coordinated through one of the N atoms. The structural consequence of this coordination is the disruption of the cross-ring S-S bonds of S_4N_4 to give a boat-shaped eight-membered ring with approximately coplanar S atoms (1), and a remarkable regularity in the S-N bond distances.

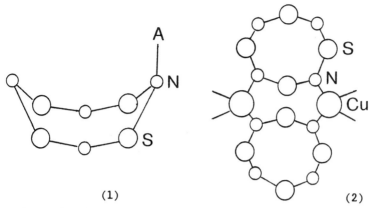

(1)   (2)

c) In the three complexes $S_4N_4 \cdot CuCl$, $S_4N_4 \cdot CuBr$ (2), and $S_4N_4 \cdot CuCl_2$, the Cu atoms of neighbouring $[Cu-X]_x$ zig-zag chains are each bridged by two S_4N_4 rings. The conformation and structural parameters for these bidentate S_4N_4 ligands are not significantly different from those of uncoordinated S_4N_4. In $S_4N_4 \cdot CuCl_2$, the Cu atom is octahedrally surrounded by Cl and S_4N_4.

3. S_4N_4 as a tridentate ligand

In most of the adducts of S_4N_4 and transition metal halides, S_4N_4 is coordinated to the metal atom through an electron pair on a nitrogen atom. But the Vaska complex, $IrCl(CO)(PPh_3)_2$, reacts with S_4N_4 with the elimination of triphenylphosphine to give $IrCl(CO)(PPh_3)(S_4N_4)$.[15] The distinguishing feature of the structure of this adduct is that the $[IrCl(CO)PPh_3]$ fragment has inserted into a sulfur-nitrogen bond, $S(4)-N(1)$ of S_4N_4, while $S(2)$ is coordinated to the iridium atom, thus resulting in the formation of a bicyclo[4.3.0]-skeleton, as shown in Fig. 3.34-4. Of the two sulfur atoms coordinating the Ir atom, $S(4)$ has a coordination number of two, and $S(2)$ a coordination number of three. The S atom with the lower coordination number

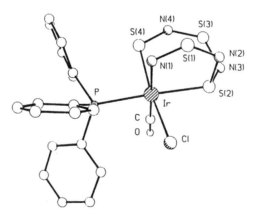

Fig. 3.34-4 Molecular structure of $IrCl(CO)(PPh_3)(S_4N_4)$. (After ref. 15).

should have the higher electron density and consequently the shorter Ir-S distance; this is indeed found to be the case: Ir-S(4) = 2.335 and Ir-S(2) = 2.391Å.

4. $S_4(NH)_4$ and $N_4(SF)_4$

The NH group is "isoelectronic" with S and can therefore subrogate it in *cyclo*-S_8. The reduction of S_4N_4 with dithionite or with $SnCl_2$ in boiling ethanol/benzene yields tetrasulfur tetraimide, $S_4(NH)_4$. An experimental charge density study[16] has shown that the idealized *4mm* symmetry of the isolated molecule is reduced to *m* due to extensive hydrogen bonding in the solid state (Fig. 3.34-5). Bent S-N bonds of order >1.0 with endocyclic maxima are found.

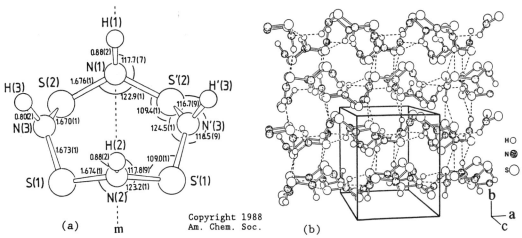

Copyright 1988
Am. Chem. Soc.

Fig. 3.34-5 Structure of $S_4(NH)_4$: (a) molecular geometry and dimensions; (b) packing and hydrogen bonding. (After ref. 16).

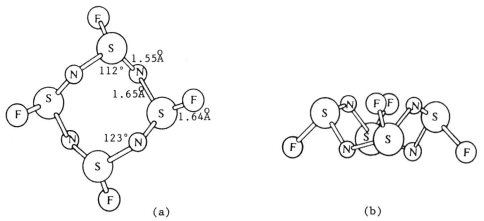

Fig. 3.34-6 Molecular structure of $N_4(SF)_4$: (a) top view and (b) side view. (After ref. 17).

Cyclo-$N_4(SF)_4$ can be made by fluorinating S_4N_4 with a hot slurry of AgF_2/CCl_4. The conformation of the N_4S_4 ring of this molecule (Fig. 3.34-6) is very different from that in S_4N_4 (Fig. 3.34-1) or $S_4(NH)_4$ (Fig. 3.34-5). The $N_4(SF)_4$ molecule shows a very pronounced alternation in S-N bond lengths and the S-F bonds point alternately up and down around the puckered ring, conforming to an idealized molecular symmetry of S_4.[17]

References

[1]   M.L. DeLucia and P. Coppens, *Inorg. Chem.* **17**, 2336 (1978).

[2]   D. Clark, *J. Chem. Soc.*, 1615 (1952).

[3]   B.D. Sharma and J. Donohue, *Acta Crystallogr.* **16**, 891 (1963).

[4]   D.R. Salahub and R.P. Messmer, *J. Chem. Phys.* **64**, 2039 (1976).

[5]   O. Hassel and H. Viervoll, *Tids. Kjemi Bergvesen Met.* **3**, 7 (1943).

[6]   C.-S. Lu and J. Donohue, *J. Am. Chem. Soc.* **66**, 818 (1944).

[7]   M.G.B. Drew, D.H. Templeton and A. Zalkin, *Inorg. Chem.* **6**, 1906 (1967).

[8]   D. Neubauer and J. Weiss, *Z. Anorg. Allg. Chem.* **303**, 28 (1960).

[9]   A. Gieren and B. Dederer, *Z. Anorg. Allg. Chem.* **440**, 119 (1978).

[10] M.J. Almond, G.A. Forsyth, D.A. Rice, A.J. Downs, T.L. Jeffery and
      K. Hagen, *Polyhedron* **8**, 2631 (1989).

[11] T. Chivers and R.T. Oakley, *Top. Curr. Chem.* **102**, 117 (1982).

[12] T. Chivers, *Chem. Rev.* **85**, 341 (1985).

[13] E.G. Awere, J. Passmore, P.S. White and T. Klapötke, *J. Chem. Soc., Chem.
      Commun.*, 1415 (1989).

[14] T. Chivers and F. Edelmann, *Polyhedron* **5**, 1661 (1986).

[15] F. Edelmann, H.W. Roesky, C. Spang, M. Noltemeyer and G.M. Sheldrick,
      Angew. Chem. Int. Ed. Engl. **25**, 931 (1986).

[16] D. Gregson, G. Klebe and H. Fuess, *J. Am. Chem. Soc.* **110**, 8488 (1988).

[17] G.A. Wiegers and A. Vos, *Acta Crystallogr.* **14**, 562 (1961); **16**, 152
      (1963).

Note the cobaltocenium salt $[CoCp_2][S_3N_3]$ contains continuous stacks of alternating cations and anions, the Cp and S_3N_3 ring planes being orthogonal to each other with multiple C-H...N...H-C (p_π) interactions. See P.N. Jagg, P.F. Kelly, H.S. Rzepa, D.J. Williams, J.D. Woolins and W. Wylie, *J. Chem. Soc., Chem. Commun.*, 942 (1991).

3.35 Polythiazyl (Polymeric Sulfur Nitride)

(SN)$_x$

Crystal Data

Monoclinic, space group $P2_1/c$ (No. 14)

$a = 4.153(6)$, $b = 4.439(5)$, $c = 7.637(12)$Å, $\beta = 109.7(1)°$, $Z = 4$

Atom	x	y	z
S	.1790(8)	.7873(6)	.3443(4)
N	.141(3)	.431(2)	.322(2)

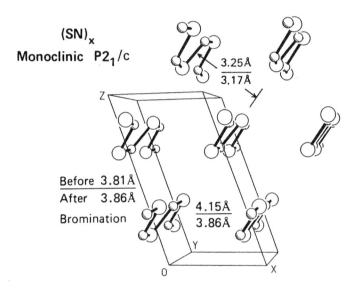

Fig. 3.35-1 Projection of the crystal structure of (SN)$_x$ on (010), showing the inter-chain separations before and after bromination. (Adapted from ref. 6).

Crystal Structure

Polythiazyl was first prepared by Burt[1] in 1910, using a method that is still practised today, namely the solid-state polymerization of crystalline S_2N_2 at room temperature.

Structure determination of the lustrous golden crystals of (SN)$_x$ has been carried out by X-ray analysis.[2] The monoclinic unit cell contains two translationally inequivalent, centrosymmetrically related chains. The chains all lie in the (102) plane where the two types of inequivalent chains alternate (Fig. 3.35-1). The (SN)$_x$ chains were found to be planar at -145°C and nearly planar at room temperature, the maximum deviation of an atom from the mean plane being 0.01Å.

A single (SN)$_x$ chain consists of roughly equal S-N bonds with lengths of

1.593 and 1.628Å at room temperature, and N-S-N and S-N-S bond angles of 106.2° and 119.9°, respectively. Fig. 3.35-2 shows the packing of the (SN)$_x$ chains in one layer of the crystal.

A redetermination of the crystal structure of (SN)$_x$ using neutron diffraction[2] yielded results which are consistent with the X-ray study.

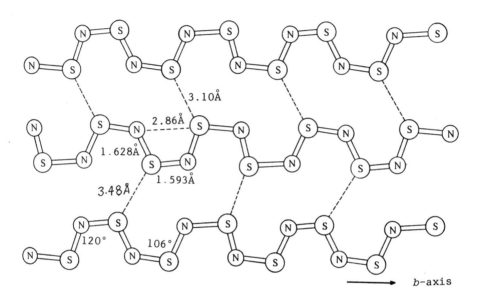

Fig. 3.35-2 Polymeric chains in one layer of (SN)$_x$ and the important structural parameters.

Remarks

1. Solid-state polymerization of S_2N_2 to (SN)$_x$

Crystalline S_2N_2 consists of square-planar molecules packed in a monoclinic lattice with space group $P2_1/c$, $a = 4.485(2)$, $b = 3.767(1)$, $c = 8.452(3)$Å, $\beta = 106.43(3)°$, and $Z = 2$ at -130°C. When colorless S_2N_2 crystals are allowed to stand at room temperature for a period, golden (SN)$_x$ crystals will be gradually formed. The (SN)$_x$ chain can conceivably be generated from adjacent square-planar S_2N_2 molecules, and a free radical mechanism has been suggested. Since polymerization can take place with only minor movements of the atoms, the starting material and product are pseudomorphs and the crystallinity of the former is maintained. Fig. 3.35-3 shows the stepwise polymerization of S_2N_2 molecules to form (SN)$_x$ chains.[3]

2. Electrical conductivity of (SN)$_x$

(SN)$_x$ has some unusual properties. For example, it has a bronze colour and metallic luster, and its electrical conductivity is about that of mercury metal. Values of the conductivity σ of (SN)$_x$ depend on the purity and

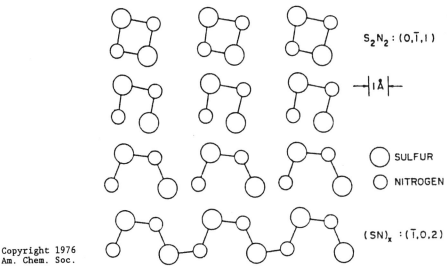

$S_2N_2 : (0,\bar{1},1)$

SULFUR

NITROGEN

$(SN)_x : (\bar{1},0,2)$

Fig. 3.35-3 The polymerization of S_2N_2 to $(SN)_x$. The top view is a projection of the S_2N_2 structure onto the $(0\bar{1}1)$ plane with the *a* axis horizontal. The bottom view is a projection of the $(SN)_x$ structure onto the $(\bar{1}02)$ plane with the *b* axis horizontal. The middle views schematically show the polymerization process. (After ref. 3).

crystallinity of the polymer and on the direction of measurement, being much greater along the fibers (*b* axis) than across them. At room temperature typical values of $\sigma||$ are 1000-4000 ohm^{-1}cm^{-1}, and this increases by as much as 1000-fold on cooling to 4.2K. Typical values of the anisotropy ratio $\sigma||/\sigma\perp$ are ~50 at room temperature and ~1000 at 40K.

In 1975, Greene[4] found that very pure $(SN)_x$ becomes superconducting below 0.26K. This is the first example of a non-metallic polymeric superconductor.

Why is it that the $(SN)_x$ chains exhibit metallic luster and electrical conductivity? A conjugated single-bond/double-bond system can be formulated, in which every S-N unit has one antibonding, π^*, electron. The half-filled, overlapping π^* orbitals will combine to form a half-filled conduction band in much the same way as the half-filled ns orbitals of alkali metal atoms form a conduction band. However, in $(SN)_x$, the conduction band lies only along the direction of the $(SN)_x$ fibers, so the polymer behaves as a "one-dimensional metal".

3. Halogenated polysulfur nitride

There is intense current interest in one-dimensional molecular metals[5,6] and several related, partially-halogenated derivatives of $(SN)_x$ have also been made, some of which exhibit even higher electrical conductivity. For example, blue-black single crystals of $(SNBr_{0.4})_x$ have a room temperature conductivity of 2×10^4 $ohm^{-1}cm^{-1}$, i.e., an order of magnitude greater than that of the parent $(SN)_x$ polymer.

Exposure of $(SN)_x$ to bromine vapor at room temperature leads to a material of composition $(SNBr_{0.55})_x$ which when left in vacuum (10^{-5} Torr) for one hour gives $(SNBr_{0.4})_x$. At this final composition the crystal has expanded by ~50% in volume perpendicular to the chain axis accompanied by some exfoliation and microscopic cracking, but the external habit and fibrous nature remain intact. Combined use of electron diffraction, X-ray powder diffraction, X-ray precession photography, EXAFS, and Raman techniques has revealed that the incorporated bromine exists as Br_2 and Br_3^-, both of which are aligned with their axis parallel to the $(SN)_x$ chain axis. A possible packing of the Br_3^- ion between $(SN)_x$ chains, with maximization of S...Br interactions, is shown in Fig. 3.35-4.

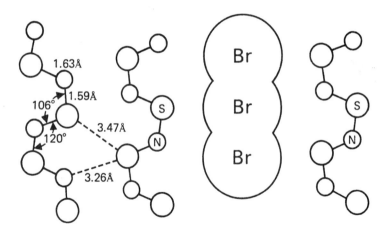

Fig. 3.35-4 Structure of brominated $(SN)_x$ showing a possible packing scheme for Br_3^-. (After ref. 6).

Another mode of attachment of bromine to $(SN)_x$ is in the form of a monolayer on the outside surface of the fibers. For ~30Å diameter fibers of $(SNBr_{0.4})_x$ all the bromine can be accommodated in this way (Fig. 3.35-5). It has been estimated that 10% of the bromine enters the $(SN)_x$ lattice while some 90% resides in the interfibrillar regions.

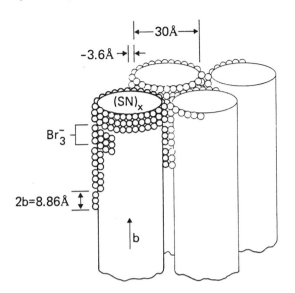

Fig. 3.35-5 Bromine cladding the outer surface of $(SN)_x$ fibers.
(After ref. 6).

References

[1] F.P. Burt, *J. Chem. Soc.*, 1171 (1910).

[2] G. Heger, S. Klein, L. Pintschovious and H. Kahlert, *J. Solid State Chem.*
     23, 341 (1978).

[3] M.J. Cohen, A.F. Garito, A.J. Heeger, A.G. MacDiarmid, C.M. Mikulski,
     M.S. Saran and J. Kleppinger, *J. Am. Chem. Soc.* **98**, 3844 (1976).

[4] R.L. Greene, G.B. Street and L.J. Suter, *Phys. Rev. Lett.* **34**, 577 (1975).

[5] M.M. Labes, P. Love and L.F. Nichols, *Chem. Rev.* **79**, 1 (1979).

[6] G.B. Street and W.D. Gill in W.E. Hatfield (ed.), *Molecular Metals*,
     Plenum Press, New York, 1979, p. 301.

Note Recent literature on polythiazl, $(SN)_x$, and its metastable open-chain
oligomeric nitric oxide analogues, $(NO)_n$ ($n = 2 - 12$), is covered in
W.H. Jones, *J. Phys. Chem.* **95**, 2588 (1991).

3.36   Decaselenium Hexafluoroantimonate

$$Se_{10}(SbF_6)_2$$

Crystal Data

Monoclinic, space group $P2_1/c$ (No. 14)

$a = 19.869(5)$, $b = 16.200(4)$, $c = 13.282(3)$Å, $b = 109.24(2)°$

$Z = 8$

Atom	x	y	z	Atom	x	y	z
Se_{10}^{2+} cation 1				Se_{10}^{2+} cation 2			
Se(11)	.5175	.7596	.0397	Se(21)	1.0210	.7565	.2244
Se(12)	.4660	.8890	.0659	Se(22)	.9675	.8858	.1429
Se(13)	.4041	.9305	-.1020	Se(23)	.9028	.9267	.2451
Se(14)	.2873	.8920	-.1177	Se(24)	.7875	.8794	.1481
Se(15)	.2625	.7871	-.2373	Se(25)	.7663	.7797	.2525
Se(16)	.2885	.6609	-.1326	Se(26)	.7984	.6513	.1829
Se(17)	.3388	.7185	.0448	Se(27)	.9092	.6149	.3140
Se(18)	.4486	.6620	.6620	Se(28)	.9618	.7396	.3562
Se(19)	.4547	.7394	-.1517	Se(29)	.9545	.6573	.0961
Se(110)	.3992	.6182	-.1544	Se(210)	.8436	.7073	.0480

Atoms in the SbF_6^- anions are omitted.

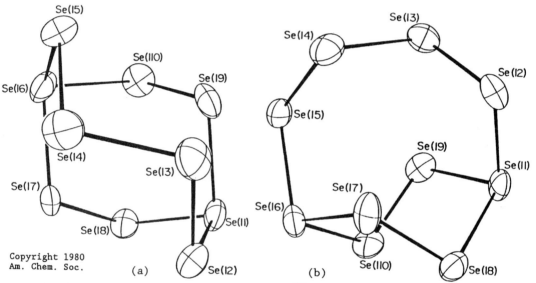

(a)   (b)

Fig. 3.36-1 Molecular structure of Se_{10}^{2+} cation 1: (a) viewed
approximately down the b axis; (b) viewed approximately down the c axis.
(After ref. 1)

Crystal Structure

The crystal of $Se_{10}(SbF_6)_2$ consists of Se_{10}^{2+} cations and SbF_6^-
anions.[1] There are two Se_{10}^{2+} cations in the asymmetric unit, each with
approximate C_2 symmetry. The equivalent distances and angles in both cations
match one another within 2-3 standard deviations. It should be noted that the
Se_{10}^{2+} structure is chiral and that these two cations represent the two

enantiomorphs. Each enantiomorph of Se_{10}^{2+} has a bicyclo[4.2.2]decane skeleton and can alternatively be viewed as either a six-membered boat-shaped ring linked across the middle by a chain of four selenium atoms or as two fused eight-membered rings, as shown in Fig. 3.36-1.

The bond lengths in Se_{10}^{2+} lie in the range 2.24-2.44Å, and they alternate in length on moving away from the two three-coordinate selenium atoms along any of the chains. The bond lengths fall into three rather distinct groups: (1) the six bonds adjacent to the three-coordinate selenium atoms, which carry a formal positive charge, are the longest, having lengths of 2.40-2.46Å; (2) the bonds adjacent to these long bonds are shorter with lengths of 2.25-2.27Å; and (3) the unique central bond in the four-atom backbone has an intermediate length of 2.35-2.36Å (Fig. 3.36-2).

The bond angles in the two Se_{10}^{2+} cations are very similar and range from 97° to 106°. The angles at the three-coordinate positions are slightly smaller, 97°-103°, than the angles at the two-coordinate atoms, 102°-106°. The rather small range of the bond angles suggests that the Se_{10}^{2+} cation is relatively free of angle strain (Fig. 3.36-2).

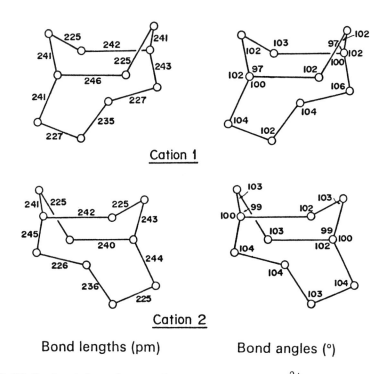

Cation 1

Cation 2

Bond lengths (pm)   Bond angles (°)

Fig. 3.36-2 Bond lengths and bond angles in Se_{10}^{2+}. Values are given for the two crystallographically non-equivalent ions in the structure. (After ref. 3).

There are three notable intramolecular contacts in Se_{10}^{2+} in the range 3.30-3.49Å. Although not as short as the transannular bond in Se_8^{2+} (2.84Å), they are significantly shorter than twice the van der Waals radus of selenium, ~4.0Å. One contact [Se(18)...Se(110) = 3.30Å; Se(27)...Se(29) = 3.37Å] runs across the middle of the six-membered ring in the base of each Se_{10}^{2+}, whereas the other two contacts [Se(13)...Se(19) = 3.39Å, Se(14)...Se(17) = 3.49Å; Se(23)...S(28) = 3.41Å, Se(24)...Se(210) = 3.43Å] bridge each eight-membered ring to form four- and six-membered rings if they are considered as weak "bonds". The bicyclic Se_{10}^{2+} ion has also been found in the complex salt $Se_{10}(SO_3F)_2$.[2]

The SbF_6^- anions are octahedral and the average Sb-F bond length in the four SbF_6^- anions is 1.87Å.

Remarks

1. Polyatomic cations of sulfur, selenium and tellurium[3,4]

Sulfur, selenium and tellurium may be oxidized by several different oxidizing agents to yield a variety of polyatomic cations. The presently known species which have been characterized by X-ray crystallography are summarized in Table 3.36-1.

Table 3.36-1 Polyatomic cations of S, Se and Te.

E_4^{2+}	S_4^{2+}	Se_4^{2+}	Te_4^{2+}	$Te_2Se_2^{2+}$	
E_6^{2+}			Te_6^{2+}	$Te_3S_3^{2+}$	$Te_2Se_4^{2+}$
E_6^{4+}			Te_6^{4+}		
E_8^{2+}	S_8^{2+}	Se_8^{2+}	Te_8^{2-}	$Te_2Se_6^{2+}$	
E_{10}^{2+}		Se_{10}^{2+}		$Te_2S_8^{2+}$	$Te_2Se_8^{2+}$
E_{19}^{2+}	S_{19}^{2+}				

The ion $Te_{3.0}Se_{1.0}^{2+}$ corresponds to a disordered mixture of Te_3Se^{2+}, Te_4^{2+} and $Te_2Se_2^{2+}$.[5]

Compounds which contain these cations invariably appear deeply colored. For example, dark-green solutions of selenium contain the Se_8^{2+} cation, and red solutions of tellurium the Te_4^{2+} ion.

2. The $Te_2Se_8^{2+}$ ion

The reaction between Te and $Se_8(AsF_6)_2$ leads to the formation of the black crystalline compound $Te_2Se_8(AsF_6)_2 \cdot SO_2$, which is the first example of a ten-atom polyatomic cation.[6] The cation $Te_2Se_8^{2+}$ has the same basic structure of Se_{10}^{2+} which consists of a six-membered boat-shaped ring bridged by a chain of four atoms to form an eight-membered ring. Fig. 3.36-3 shows the bond lengths and bond angles in $Te_2Se_8^{2+}$. Slightly different dimensions for this cation have been observed in the unsolvated complex $Te_2Se_8(AsF_6)_2$ and in $(Te_2Se_6)(Te_2Se_8)(AsF_6)_4 \cdot 2SO_2$.[4]

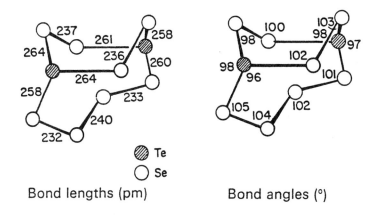

Bond lengths (pm)   Bond angles (°)

Fig. 3.36-3   Bond lengths and bond angles in $Te_2Se_8^{2+}$.   (After ref. 3).

3. Square-planar S_4^{2+}, Se_4^{2+}, Te_4^{2+} and $Te_2Se_2^{2+}$ systems

X-Ray crystallographic studies on the salts $S_4(S_7I)_4(AsF_6)_6$, $Se_4(HS_2O_7)_2$, $Se_4(Sb_8F_{33})(SbF_6)$, $Te_4(AlCl_4)_2$, $Te_4(Al_2Cl_7)_2$, and $Te_2Se_2(Sb_3F_{14})(SbF_6)$ [5] have shown that the cations S_4^{2+}, Se_4^{2+}, Te_4^{2+} and $Te_2Se_2^{2+}$ all have square-planar structures.   The bond lengths in the four ions are slightly shorter than the normally accepted values for the corresponding single bonds, and are listed in Table 3.36-2.

Table 3.36-2   Bond lengths in some square-planar ions related to S_4^{2+}.[3]

Ions	Bond length/Å	Single bond length/Å	Crystal
S_4^{2+}	1.98	2.04	$S_4(S_7I)_4(AsF_6)_6$
Se_4^{2+}	2.28	2.34	$Se_4(HS_2O_7)_2$
			$Se_4(Sb_8F_{33})(SbF_6)$
Te_4^{2+}	2.66	2.74	$Te_4(AlCl_4)_2$, $Te_4(Al_2Cl_7)_2$
trans-$Te_2Se_2^{2+}$	2.47	2.54	$Te_2Se_2(Sb_3F_{14})(SbF_6)$

The trend is consistent with a valence-bond description of these ions in terms of four equivalent resonance structures such as (I), or alternatively with a simple molecular orbital description in which three of the four p molecular orbitals, the two almost non-bonding (e_g) orbitals and the (a_{1g})

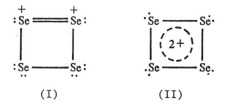

        (I)   (II)

bonding orbitals are filled and the antibonding (b_{2u}) orbital is empty.   Thus these molecules may be described as 6p-electron "aromatic" systems (II).

4. The S_8^{2+} and Se_8^{2+} ions

X-Ray crystallographic studies of $S_8(AsF_6)_2$ and $Se_8(AlCl_4)_2$ have shown that the structures of S_8^{2+} and Se_8^{2+} are very similar.   Fig. 3.36-4 shows the structures of the S_8^{2+} (or Se_8^{2+}) cation, and Fig. 3.36-5 gives the bond

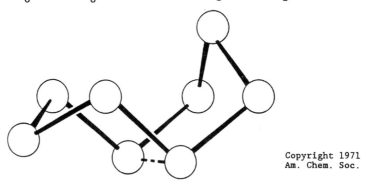

Fig. 3.36-4 Molecular structure of S_8^{2+} (or Se_8^{2+}).   (After ref. 7).

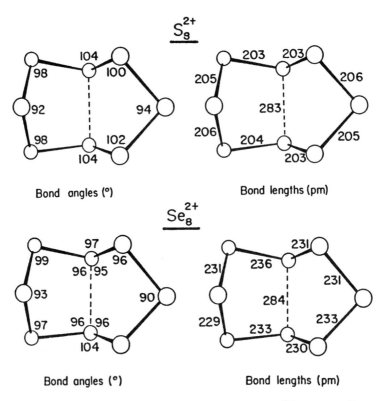

Bond angles (°)   Bond lengths (pm)

Bond angles (°)   Bond lengths (pm)

Fig. 3.36-5 Bond lengths and bond angles in S_8^{2+} and Se_8^{2+}.   (After ref. 3).

lengths and bond angles for S_8^{2+} and Se_8^{2+}.

   The eight-membered ring has an exo-endo conformation and a rather long transannular bond which is relatively shorter and stronger in the case of Se_8^{2+}. This structure may be regarded as being half-way between those of the cage molecule S_4N_4 and the crown-shaped ring of S_8 and Se_8. Indeed, since S_4N_4 has two electrons less than S_8^{2+} and is isoelectronic with the unknown S_8^{4+} (which appears to be unstable with respect to S_4^{2+}), it is conceivable that as two electrons are removed from S_8 one end folds up to generate a transannular bond. Then, with the removal of two more electrons, the other end also folds up and another bond is formed to give the S_4N_4 structure (Fig. 3.36-6).

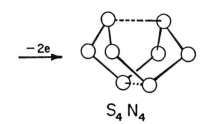

Fig. 3.36-6 Relationship between the structures of S_8, S_8^{2+} and S_4N_4. (After ref. 3).

   The Se_8^{2+} (and likewise S_8^{2+}) and Se_{10}^{2+} structures are related in a simple way.[4] As illustrated in Fig. 3.36-7, the removal of atoms Se(9) and Se(10) in Se_{10}^{2+} and the formation of a long transannular bond between Se(1) and Se(5) result in the exo-endo eight-membered ring of Se_8^{2+}.

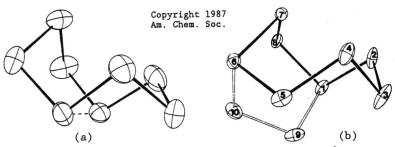

(a)   (b)

Fig. 3.36-7 Relationship between the cations Se_8^{2+} (a) and Se_{10}^{2+} (b) showing the former (filled-in bonds) as a fragment of the latter. (After ref. 4).

5. The $Te_2Se_6{}^{2+}$ cation

The $Te_2Se_6{}^{2+}$ ion, as found in $(Te_2Se_6)(Te_2Se_8)(AsF_6)_4 \cdot 2SO_2$, is not isostructural with $Se_8{}^{2+}$ but instead possesses a bicyclo(2.2.2)octane skeleton with tellurium atoms occupying the bridgehead positions.[4] Fig. 3.36-8 shows two views of the $Te_2Se_6{}^{2+}$ cationic framework which approximates to a distorted cube with three long (unbonded) edges. The structure of $Te_2Se_6{}^{2+}$ can be considered as being derived from that of $Te_2Se_8{}^{2+}$ by removal of two atoms from the four-atom selenium chain.

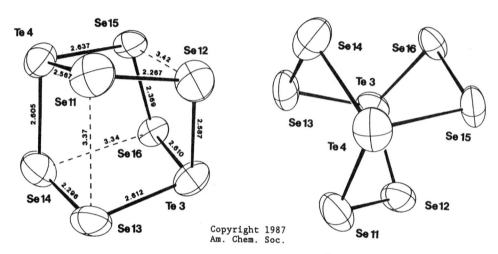

Copyright 1987
Am. Chem. Soc.

Fig. 3.36-8 Two views of the $Te_2Se_6{}^{2+}$ ion in
$(Te_2Se_6)(Te_2Se_8)(AsF_6)_4 \cdot 2SO_2$. (After ref. 4).

6. The $Te_8{}^{2+}$ cation

The tellurium polycation $Te_8{}^{2+}$ occurring in $Te_8[WCl_6]_2$ has a bicyclic structure composed of two five-membered ring envelopes sharing a common edge [Fig. 3.36-9(a)].[8] The molecule has a crystallographically imposed C_2 axis, and the transannular Te1-Te1^I bond is longer than a single bond (e.g. 2.712Å in PhTeTePh and 2.740Å in $MgTe_2$) but still markedly shorter than the edges (average 3.121Å) of the $Te_6{}^{4+}$ prism (Section 3.37). Its structure is thus clearly different from those of its $S_8{}^{2+}$ and $Se_8{}^{2+}$ homologues which exhibit approximate C_s symmetry and only weak transannular interactions.

The $Te_8{}^{2+}$ ions are associated to form band-shaped chains [Fig. 3.36-9(b)] in the crystal. The edges Te3-Te4 amd Te3^I-Te4^I of each polycation form planar, rectangular Te_4 rings with the corresponding edges of neighbouring ions. The Te-Te distance between neighbouring polycations (3.424Å) lies in the range of weak bonding, being shorter than the inter-chain Te...Te

separation (3.491Å) in elemental tellurium[9] and the sum of the van der Waals
radii (4.40Å).

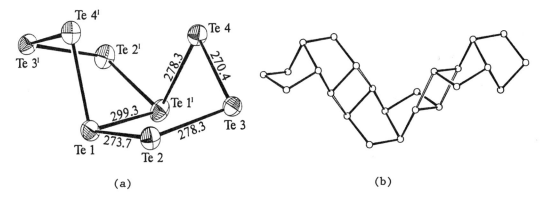

(a)   (b)

Fig. 3.36-9 (a) Molecular structure of Te_8^{2+}. (b) Linkage (open bonds)
between adjacent polycations to form an infinite chain. (After ref. 8).

7. The S_{19}^{2+} cation

X-Ray crystallographic study on a product of the oxidation of sulfur in
oleum (or with SbF_5 in HF or SO_2) has shown that the compound has the totally
unexpected formulation $(S_{19}^{2+})(AsF_6^-)_2$.[10] The novel S_{19}^{2+} cation consists
of two seven-membered rings joined by a five-atom chain. As shown in Fig.
3.36-10, one of the rings has a boat conformation while the other is
disordered, existing as a 4:1 mixture of chair and boat conformations. The S-
S distances vary greatly from 1.87 to 2.39Å, and S-S-S angles from 91.9° to
127.6°.

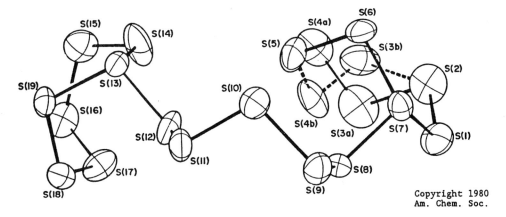

Fig. 3.36-10 Molecular structure of the disordered cation S_{19}^{2+}. The broken
lines indicate the less favored boat conformation. (After ref. 10).

References

[1] R.C. Burns, W.-L. Chan, R.J. Gillespie, W.-C. Luk, J.F. Sawyer and D.R. Slim, *Inorg. Chem.* **19**, 1432 (1980).

[2] M.J. Collins, R.J. Gillespie, J.F. Sawyer and G.J. Schrobilgen, *Acta Crystallogr., Sect. C* **42**, 13 (1986).

[3] R.J. Gillespie, *Chem. Soc. Rev.* **8**, 315 (1979).

[4] M.J. Collins, R.J. Gillespie and J.F. Sawyer, *Inorg. Chem.* **26**, 1476 (1987).

[5] P. Boldrini, I.D. Brown, M.J. Collins, R.J. Gillespie, E. Maharajh, D.R. Slim and J.F. Sawyer, *Inorg. Chem.* **24**, 4302 (1985).

[6] P. Boldrini, I.D. Brown, R.J. Gillespie, P.R. Ireland, W.-C. Luk, D.R. Slim and J.E. Vekris, *Inorg. Chem.* **15**, 765 (1976).

[7] C.G. Davies, R.J. Gillespie, J.J. Park and J. Passmore, *Inorg. Chem.* **10**, 2781 (1971).

[8] J. Beck, *Angew. Chem. Int. Ed. Engl.* **29**, 293 (1990).

[9] C. Adenis, V. Langer and O. Lindqvist, *Acta Crystallogr., Sect. C* **45**, 941 (1989).

[10] R.C. Burns, R.J. Gillespie and J.F. Sawyer, *Inorg. Chem.* **19**, 1423 (1980).

Note A polymeric Te_7^{2+} cation with Te-Te distances in the range 2.747-2.976Å exists in Te_7WOBr_5, as reported in J. Beck, *Angew. Chem. Int. Ed. Engl.* **30**, 1128 (1991).

3.37 Hexatellurium Tetrakis(hexafluoroarsenate)-
Arsenic Trifluoride (1/2)

$$Te_6(AsF_6)_4 \cdot 2AsF_3$$

Crystal Data

Monoclinic, space group $C2/c$ (No. 15)

$a = 14.832(9)$, $b = 12.242(8)$, $c = 15.301(9)$Å, $\beta = 96.59(7)°$

$Z = 4$

Atom	x	y	z	Atom	x	y	z
Te(1)	.0891	.0673	.3114	F(6)	.419	.149	.300
Te(2)	.1370	.2539	.2388	F(7)	.489	-.017	.357
Te(3)	.0490	.2591	.3809	F(8)	.354	-.023	.274
As(1)	.3875	.2235	.1090	F(9)	.372	-.061	.429
As(2)	.3915	.0474	.3674	F(10)	.114	.156	.549
As(3)	.1682	.0433	.5894	F(11)	.162	.088	.693
F(1)	.402	.270	.006	F(12)	.271	.098	.592
F(2)	.288	.286	.110	F(13)	.171	-.005	.487
F(3)	.443	.334	.155	F(14)	.065	-.018	.595
F(4)	.293	.109	.378	F(15)	.215	-.073	.633
F(5)	.443	.116	.452				

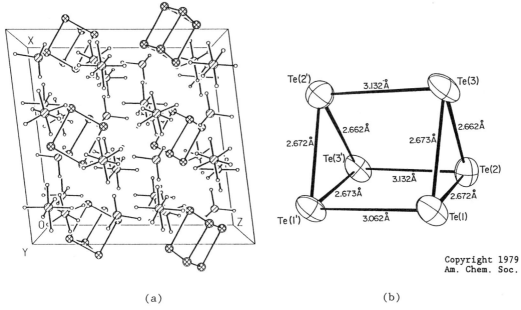

(a)   (b)

Fig. 3.37-1 (a) Crystal structure of $Te_6(AsF_6)_4 \cdot 2AsF_3$ and (b) structure
of the Te_6^{4+} cation. (Adapted from ref. 1).

Crystal Structure

The crystal consists of Te_6^{4+} and AsF_6^- ions plus AsF_3 molecules.[1]
The Te_6^{4+} cation represents the first reported example of an isolated
hexaatomic trigonal-prismatic species.

The bond lengths of Te_6^{4+} found for $Te_6(AsF_6) \cdot 2AsF_3$ are shown in Fig.
3.37-1. The end triangular faces are nearly parallel, while the atoms in each

of the rectangular faces are near coplanar. The structure of Te_6^{4+} in $Te_6(AsF_6)_4 \cdot 2SO_2$ has very similar dimensions.

From the two crystal structures, it is seen that the Te-Te bond distances in the end faces of the Te_6^{4+} cations range from 2.662 to 2.694Å, with an average of 2.675Å, which is slightly shorter than twice the covalent radius of tellurium (2.74Å). The long bonds between the end faces range from 3.062 to 3.148Å, with an average of 3.121Å, which is longer than twice the covalent radius for tellurium, but still shorter than the long-range interactions observed in elemental tellurium.

Remarks

1. $Te_3S_3^{2+}$ and $Te_2Se_4^{2+}$

The mixed cations $Te_3S_3^{2+}$ and $Te_2Se_4^{2+}$, with two electrons more than Te_6^{4+}, do not have a trigonal-prismatic configuration. The homopolyatomic species S_6^{2+} and Se_6^{2+} are as yet unknown, but the compounds $Te_3S_3(AsF_6)_2$, $Te_2Se_4(SbF_6)_2$ and $Te_2Se_4(AsF_6)_2$ have been isolated and characterized by crystal structure analysis.[3] Both cations, $Te_3S_3^{2+}$ and $Te_2Se_4^{2+}$, are based on the same skeleton consisting of a three-membered ring to which is attached a three-atom chain to give a five-membered ring, as shown in Fig. 3.37-2. Alternatively the structure may be described as a six-membered ring with a

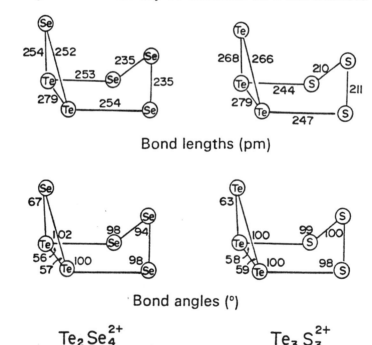

Bond lengths (pm)

Bond angles (°)

$Te_2Se_4^{2+}$ $Te_3S_3^{2+}$

Fig. 3.37-2 Bond lengths and bond angles in $Te_2Se_4^{2+}$ and $Te_3S_3^{2+}$. (After ref. 2).

pronounced boat conformation and one cross-ring bond.

The neutral S_6 ring has the expected chair conformation, but the ions $Te_3S_3^{2+}$ and $Te_2Se_4^{2+}$ have boat conformations. Furthermore Te_6^{4+} has a trigonal prismatic structure. Presumably the hypothetical neutral Te_3S_3 and Te_2Se_4 rings would have the chair conformation. The removal of two electrons thus leads to the folding up of the molecule into a pronounced boat conformation and the formation of a cross-ring bond which, in this case, is rather strong. Thus $Te_3S_3^{2+}$ and $Te_2Se_4^{2+}$ have structures which are intermediate between the chair conformation of S_6 and the trigonal prism of Te_6^{4+} (Fig. 3.37-3). As electrons are removed from the neutral six-membered rings there is a strong tendency for them to fold up into more compact cluster structures.

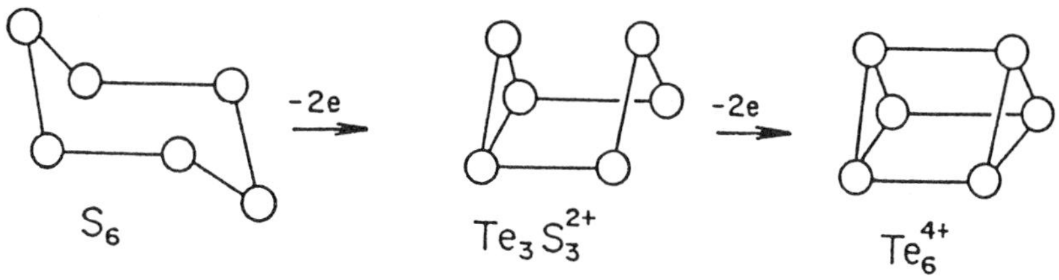

Fig. 3.37-3 Relationship between the structures of S_6, $Te_3S_3^{2+}$ and Te_6^{4+}. (After ref. 2).

2. Derivation of the cage structures from the basic cluster structures

Gillespie[2] has pointed out that the structures of cages may be derived from the basic cluster structures according to the following rules:

a) One or more edges of a cluster A_n may be replaced by a bridging atom B.

b) One or more edges of a cluster may be broken by the addition of an electron pair, so that a bonding electron pair is replaced by two non-bonding pairs.

c) One or more atoms may be removed from a cluster leaving the bonding electron pairs in place as non-bonding pairs.

d) The removal of a two-coordinate bridging atom B from a cage is tantamount to the addition of an electron pair to an edge of the corresponding cluster structure (b).

These rules are summarized in equation (1)-(4). It is an important feature of these rules that in processes (2), (3) and (4) in which bonding electron pairs are replaced by non-bonding pairs the atoms nevertheless remain clearly, if weakly, bonded to each other, and although the interatomic

$$\text{(1)}$$

$$\text{(2)}$$

$$\text{(3)}$$

$$\text{(4)}$$

distance necessarily increases the overall structure still retains the same
general shape. Since main group element cages contain predominantly P, S and
heavier elements it is reasonable to suppose that the attractive force between
formally non-bonded atoms arises from the presence of unfilled orbitals, e.g.
d orbitals on these heavy atoms. These vacant orbitals permit some
delocalization of electron density from the formally non-bonding pairs on
neighbouring atoms, which then become slightly bonding. Process (2) can be
described in molecular orbital terms as the filling of the antibonding orbital
associated with the A-A bond. However, rather than the resultant situation
being somewhat antibonding, as is usually assumed, it is actually slightly
bonding. Since two pairs of electrons are involved in this interaction it can
be described as a weak double bond.

3. Polychalcogenide anions

The majority of polysulfide and polyselenide anions have acyclic
structures, as exemplied by the S_n^{2-} ions in BaS_3, BaS_4, K_2S_5 and Cs_2S_6, the
Se_n^{2-} ions in $[Ba(en)_4]Se_4$ and $(NBu_4)_2Se_6$, as well as the Te_n^{2-} ions in K_2Te_3,
$[Na(2,2,2\text{-crypt})]_2Te_4$ and $(NBu_4)_2Te_5$.[4] For an unbranched polychalcogenide
E_n^{2-} may exist in two enantiomeric *cis* and two enantiomeric *trans* forms;
generally the two *cis* form are idential (*meso* forms).[6] The Se_5^{2-} species is
trans in Rb_2Se_5 (as a right-handed helix)[5] and Cs_2Se_5 but *cis* in [Cs(18-
crown-6)]$_2Se_5$.[7] Rb_2Te_5 and Cs_2Te_5 are unusual in that they contain infinite
one-dimensional chains of Te atoms.[8]

Cyclic polychalcogenide anions, in contrast to the well-established
cationic species, are rare and some examples became known only recently. The
compound Re_2Te_5 , formulated as $Re_6Te_8^{2+}Te_7^{2-}$, features a butterfly-like anion

[Fig. 3.37-4(a)].[9] The bicyclic decaselenide ion in $[Ph_3PNPPh_3]_2Se_{10}.DMF$ has been found to consist of two six-membered chair-shaped rings sharing a common edge [Fig. 3.37-4(b)]; the measured bond distances are consistent with a central $^+Se-Se^+$ fragment linking two Se_4^{2-} ions.[10] The undecaselenide ion in $(PPh_4)_2Se_{11}$ has a spirocyclic structure which may be regarded as a central Se^{2+} cation coordinated in a square-planar manner by two Se_5^{2-} chelates [Fig. 3.37-4(c)].[11] The Se_{16}^{4-} fragment in the hexadecaselenide Cs_4Se_{16} consists of a Se_6 ring and two Se_5^{2-} chains [Fig. 3.37-4(d)].[12]

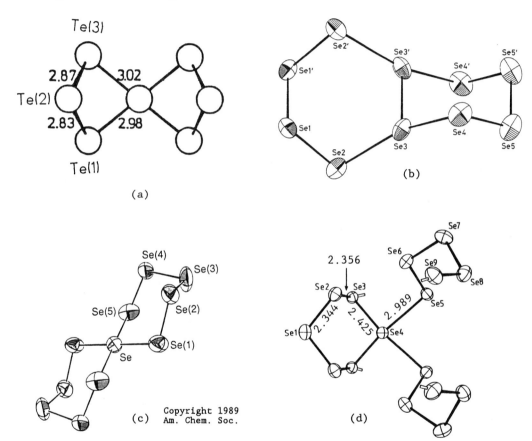

Fig. 3.37-4 Molecular structures of some cyclic polychalcogenide anions: (a) Te_7^{2-}, (b) Se_{10}^{2-} (molecular symmetry C_2), (c) Se_{11}^{2-}, and (d) Se_{16}^{4-} (molecular symmetry C_2). (After refs. 9-12).

4. Polyselenide and polytelluride ligands

Metal complexes containing polyselenides and polytellurides as ligands have been reviewed.[4] Some examples are shown in Fig. 3.37-5. In $(\eta^5-C_5H_4Me)_2V_2Se_5$ [Fig. 3.37-5(a)] a metal-metal bond, 2.779(4)Å in length, is bridged by dianionic ligands of the types μ-Se, μ-η-Se_2 [Se-Se = 2.295(2)Å]

and $syn\text{-}\mu\text{-}Se_2$ [Se-Se = 2.290(2)Å].[13]   The structure of $[(\eta^5\text{-}C_5H_4Me)_2Ti]_2Se_4$
features a Ti_2Se_4 core in the chair conformation, the Se-Se bond distance
being 2.343(1)Å [Fig. 3.37-5(b)].[14]   The compound $[NBu_4]_4[Hg_4Te_{12}]$ contains
a centrosymmetric cluster anion in which four coplanar Hg atoms are bridged by
pairs of $\mu\text{-}Te^{2-}$, $\mu\text{-}Te_2^{2-}$, and $\eta\text{-}Te_3^{2-}$ ligands [Fig. 3.37-5(c)].[15]   The anion
in $[PPh_4]_4[Pd(Te_4)_2]$ consists of two half-chair Pd-Te_4 rings sharing the same
metal atom [Fig. 3.37-5(d)], with normal average Pd-Te and Te-Te bond
distances of 2.59(2) and 2.74(4)Å, respectively.[16]

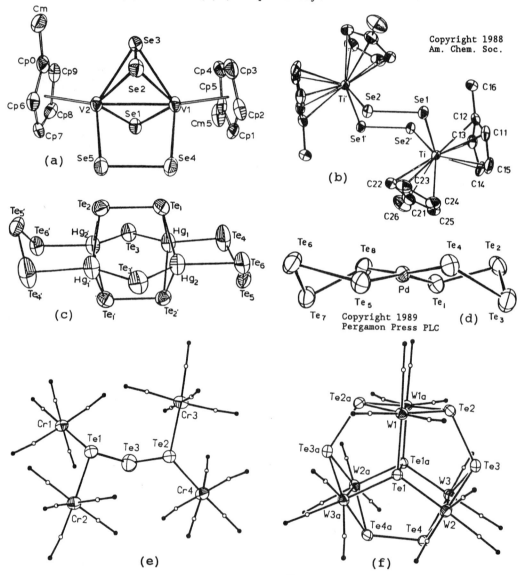

Fig. 3.37-5 Molecular structures of (a) $(\eta^5\text{-}C_5H_4Me)_2V_2Se_5$, (b) $[(\eta^5\text{-}$
$C_5H_4Me)_2Ti]_2Se_4$, (c) $[Hg_4Te_{12}]^{4-}$, (d) $[Pd(Te_4)_2]^{4-}$, (e) $[Cr_4(CO)_{20}(Te_3)]^{2-}$
and (f) $[\{W(CO)_3\}_6(Te_2)_4]^{2-}$.   (After refs. 13-17).

The reaction of $Cr(CO)_6$ with polytellurides such as K_2Te_3, K_2Te_4, $(Ph_4P)_2Te_4$ or $[(t\text{-}Bu)_4N]_2Te_5$ leads to formation of the extremely stable $[Cr_4(CO)_{20}(Te_3)]^{2-}$ anion shown in Fig. 3.37-5(e).[17] $W(CO)_6$ reacts with $(Ph_4P)_2Te_4$ to generate the cluster $[\{W(CO)_3\}_6(Te_2)_4]^{2-}$ in which all tungsten atoms are in a formal +1 oxidation state. This molecule resembles a pinwheel with a Te_2^{2-} unit as the axis and three W-W bonds acting as paddles, the rim being completed by three bridging Te_2^{2-} groups [Fig. 3.37-5(f)].[17]

References

[1] R.C. Burns, R.J. Gillespie, W.-C. Luk and D.R. Slim, *Inorg. Chem.* **18**, 3086 (1979).

[2] R.J. Gillespie, *Chem. Soc. Rev.* **8**, 315 (1979).

[3] R.J. Gillespie, W.-C. Luk, E. Maharajh and D.R. Slim, *Inorg. Chem.* **16**, 892 (1977).

[4] M.A. Ansari and J.A. Ibers, *Coord. Chem. Rev.* **100**, 223 (1990).

[5] P. Böttcher, *Z. Kristallogr.* **150**, 65 (1979).

[6] P. Laur in A. Senning (ed.), *Sulfur in Organic and Inorganic Chemistry*, Vol. 3, Marcel Dekker, New York, 1971.

[7] N.E. Brese, C.R. Randall and J.A. Ibers, *Inorg. Chem.* **27**, 940 (1988).

[8] P. Böttcher and U. Kretschmann, *J. Less-Common Met.* **95**, 81 (1983); *Z. Anorg. Allg. Chem.* **491**, 39 (1982).

[9] F. Klaiber, W. Petter and F. Hulliger, *J. Solid State Chem.* **46**, 112 (1983).

[10] D. Fenske, G. Kräuter and K. Dehnicke, *Angew. Chem. Int. Ed. Engl.* **29**, 390 (1990).

[11] M.G. Kanatzidis and S.-P. Huang, *Inorg. Chem.* **28**, 4667 (1989).

[12] W.S. Sheldrick and H.G. Braunbach, *Z. Naturforsch., Teil B* **44**, 1397 (1989).

[13] A.L. Rheingold, C.M. Bolinger and T.B. Rauchfuss, *Acta Crystallogr., Sect. C* **42**, 1878 (1986).

[14] D.M. Giolando, M. Papavassiliou, J. Pickardt and T.B. Rauchfuss, *Inorg. Chem.* **27**, 2596 (1988).

[15] R.C. Haushalter, *Angew. Chem. Int. Ed. Engl.* **24**, 433 (1985).

[16] R.D. Adams, T.A. Wolfe, B.W. Eichhorn and R.C. Haushalter, *Polyhedron* **8**, 701 (1989).

[17] J.W. Kolis, *Coord. Chem. Rev.* **105**, 195 (1990).

Note The silver(I) polyselenide complexes $[(Ph_4P)Ag(Se_4)]_n$, $[(Me_4N)Ag(Se_5)]_n$, $[(Et_4N)Ag(Se_4)]_4$, and $(Pr_4N)_2[Ag_4(Se_4)_3]$ are described in S.-P. Huang and M.G. Kanatzidis, *Inorg. Chem.* **30**, 1455 (1991).

The first transition-metal complex with a terminal tellurido ligand, *trans*-$W(PMe_3)_4(Te)_2$, is reported in D. Rabinovich and G. Parkin, *J. Am. Chem. Soc.* **113**, 9421 (1991).

3.38   Xenon Difluoride
XeF$_2$

Crystal Data

Tetragonal, space group *I4/mmm* (No. 139)

a = 4.315(3), c = 6.990(4)Å, Z = 2

Atom	x	y	z
Xe	0	0	0
F	0	0	.2837(3)*

* Neutron diffraction data.[2]

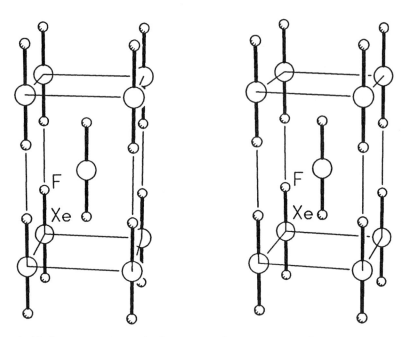

Fig. 3.38-1   Stereview of the crystal structure of XeF$_2$.

Crystal Structure

The crystal structure of xenon difluoride has been determined by
X-ray[1] and neutron diffraction.[2]   The symmetric linear XeF$_2$ molecules are
aligned on the tetrad axes with a Xe-F distance of 1.983(2)Å (Fig. 3.38-1).
The anisotropic thermal parameters suggest that the molecules undergo thermal
displacements, of which the root mean square is described by a cone of
precession about the symmetry axis with a half angle of about 7°.   This
implies that the observed Xe-F distance is foreshortened; a corrected mean
separation, based on the assumption that F atoms "ride" on Xe, is 2.00(1)Å.

Each F atom has one F neighbour at 3.02Å, and four at 3.08Å.   Eight non-
bonded F neighbours coordinate each XeF$_2$ molecule at the vertices of a square
prism at 3.42Å from Xe.

Remarks

1. The discovery of rare-gas compounds

　　Pauling was the first to suggest that "inert" gases could form covalent bonds in 1933,[3] but his hypothesis was not verified for 30 years.

　　In 1962, while investigating the chemistry of PtF_6, Bartlett and Lohmann noticed that its accidental exposure to air produced a change in colour, a phenomenon which they established to be associated with the formation of the new compound $O_2^+[PtF_6]^-$.[4] Recognizing that PtF_6 must therefore be an oxidizing agent of unprecedented power, Bartlett reasoned that Xe should similarly be oxidizable by this reagent since the first ionization energy of Xe is comparable to that of molecular oxygen (1170 kJ mol^{-1} for Xe → Xe$^+$ + e$^-$; 1175 kJ mol^{-1} for O_2 → O_2^+ + e$^-$). He proceeded to show that deep-red PtF_6 vapour spontaneously oxidized Xe to produce an orange-yellow solid (now known to have the composition $Xe(PtF_6)_x$ with $1 < x < 2$) and communicated the result in a brief note.[5]

　　Within a few months the synthesis of XeF_2 and XeF_4 were accomplished. A two-day conference held at Argonne National Laboratory in April, 1963 led to the publication of an impressive monograph[6] of over 400 pages in the same year. A historical account[7] of the conception and attempted synthesis of noble-gas compounds prior to Bartlett's discovery makes extremely interesting reading.

2. Structure of XeF_4 and $XeF_2 \cdot XeF_4$

　　XeF_4 is monoclinic, space group $P2_1/n$, with $a = 5.050(3)$, $b = 5.922(3)$, $c = 5.771(3)$Å, $\beta = 99.6(1)°$ and $Z = 2$. Three-dimensional neutron-diffraction

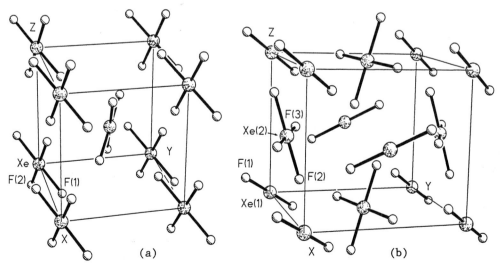

Fig. 3.38-2 Molecular packing in (a) XeF_4 and (b) $XeF_2 \cdot XeF_4$.

analysis[8] showed that the XeF_4 molecule conforms to idealized D_{4h} symmetry to high precision with an average Xe-F bond length, corrected for thermal motion, of 1.95(1)Å. The molecular packing [Fig. 3.38-2(a)] is very efficient, each fluorine making eight contacts with fluorines of other molecules at distances of 2.98 to 3.26Å.

The adduct $XeF_2.XeF_4$ crystallizes in space group $P2_1/c$ with $a = 6.64(1)$, $b = 7.33(1)$, $c = 6.40(1)$Å, $\beta = 92.67(8)°$ and $Z = 2$.[9] The structure contains a packing of discrete XeF_2 [Xe-F = 2.010(6)Å] and XeF_4 [Xe-F = 1.945(7), 1.972(7)Å] molecules [Fig. 3.38-2(b)] that have essentially the same dimensions as in neat XeF_2 and XeF_4.

3. Compounds of Xenon

The chemistry of Xe is by far the most extensive among Group 0 elements.[10] Xe can react directly with fluorine to form fluorides, from which other compounds of Xe are prepared, by reactions which fall mostly into four classes:

a) with F^- acceptors, yielding fluorocations of xenon, as in $(XeF_5)(AsF_6)$, $(XeF_5)(PtF_6)$;

b) with F^- donors, yielding fluoroanions of xenon, as in $(NO)_2(XeF_8)$, $Cs(XeF_7)$;

Table 3.38-1   Some compounds of xenon with fluorine and oxygen.

Compound	Configuration	Xe-F/Å	Xe-O/Å
XeF_2	linear, $D_{\infty h}$	2.00	
XeF_4	square planar, D_{4h}	1.93	
XeF_6	distorted octahedral, C_{3v}	1.89	
XeF_3^+ in $(XeF_3)(SbF_5)$	T-shaped, C_{2v}	1.84-1.91	
XeF_5^+ in $(XeF_5)(PtF_6)$	square pyramidal, C_{4v}	1.79-1.85	
XeF_5^- in $(NMe_4)(XeF_5)$*	pentagonal planar, D_{5h}	1.979-2.034	
XeF_8^{2-} in $(NO)_2(XeF_8)$	square antiprismatic, D_{4d}	1.96-2.08	
$Xe_2F_3^+$	V-shaped, C_{2v}	1.90-2.14	
$XeOF_4$	square pyramidal, C_{4v}	1.90	1.70
XeO_2F_2	trigonal pyramidal, C_{2v}	1.90	1.71
XeO_3F_2	trigonal bipyramidal, D_{3h}	-	-
XeO_3F^-	square pyramidal (chain)	2.36-2.48	1.77
XeO_3	pyramidal, C_{3v}		1.76
XeO_4	tetrahedral, T_d		1.74
XeO_6^{4-} in $K_4XeO_6.9H_2O$	octahedral, O_h		1.86
$Xe(OSeF_5)_2$	linear about Xe		2.12
$XeF(OSO_2F)$	linear about Xe	1.94	2.16

* K.O. Christe, E.C. Curtis, D.A. Dixon, H.P. Mercier, J.C.P. Sanders and G.J. Schrobilgen, J. Am. Chem. Soc. 113, 3351 (1991).

c) F/H metathesis between XeF$_2$ and an anhydrous acid, such as

XeF$_2$ + HOClO$_3$ → F-Xe-OClO$_3$ + HF; and

d) hydrolysis, yielding oxofluorides, oxides and xenates, such as

XeF$_6$ + H$_2$O → XeOF$_4$ + 2HF.

Some compounds of xenon with fluorine and oxygen are listed in Table 3.38-1. The structures of some compounds of xenon are shown in Fig. 3.38-3.

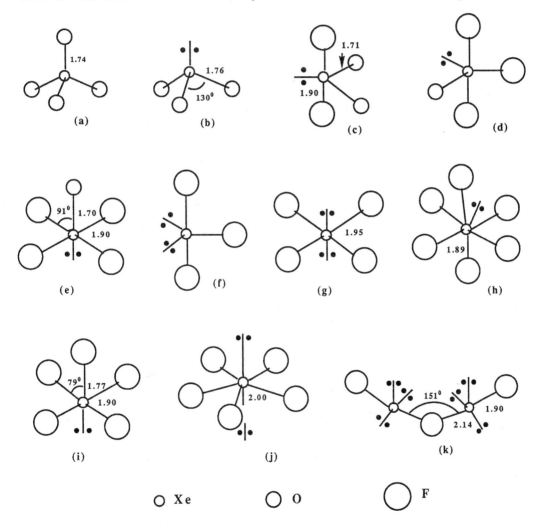

Fig. 3.38-3 Structures of some compounds of xenon: (a) XeO$_4$, (b) XeO$_3$, (c) XeO$_2$F$_2$, (d) XeOF$_3^+$, (e) XeOF$_4$, (f) XeF$_3^+$, (g) XeF$_4$, (h) XeF$_6$, (i) XeF$_5^+$, (j) XeF$_5^-$, and (k) Xe$_2$F$_3^+$.

4. Bonding in XeF$_2$ and other rare-gas compounds

XeF$_2$ is a typical rare-gas compound. In a simple molecular orbital model the valence orbitals are taken to be the 5p$_z$ of xenon and the 2p$_z$ of each fluorine atom. Construction of the bonding (ψ_b), nonbonding (ψ_n) and antibonding (ψ_a) molecular orbitals are shown in Fig. 3.38-4. Filling in the four valence electrons results in a 3-center, 4-electron σ bond extending over the entire F-Xe-F system. Hence the formal bond order of the Xe-F bond can be taken as 1/2.

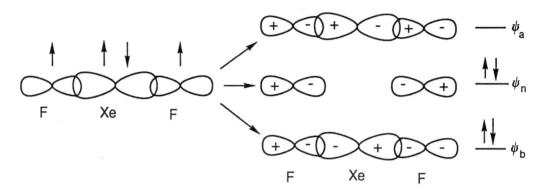

Fig. 3.38-4 Molecular orbitals of XeF_2.

According to the valence-shell electron-pair repulsion (VSEPR) theory, the Xe core in XeF_2 is surrounded by ten electrons (eight from Xe and one from each F atom) distributed in five pairs: two bonding and three nonbonding. The five pairs are directed to the corners of a trigonal bipyramid and, because of their greater mutual repulsive interaction, the three nonbonding pairs are situated in the equatorial plane at 120° to each other, leaving the two bonding pairs perpendicular to the plane and so protruding a linear F-Xe-F molecule.

In the same way XeF_4, with six electron pairs, is considered as pseudo-octahedral with its two nonbonding pairs *trans* to each other, leaving the four F atoms in a plane around Xe. More distinctively, the seven electron-pairs of XeF_6 suggest the possibility of a distorted structure based on either a monocapped octahedral or a pentagonal pyramidal arrangement of electron pairs, with the Xe-F bonds bending away from the protruding nonbonding pair.

5. The structure of xenon hexafluoride[11]

Gaseous XeF_6 is monomeric with no detectable dipole moment. Its lowest energy conformation has the C_{3v} structure shown in Fig. 3.38-5, the geometrical parameters listed being derived from electron diffraction.[12]

If XeF_6 were exactly octahedral, $\phi(F_1)$ and $\phi(F_2)$ would be 54.8° and 125.2°, respectively. The observed ground-state geometry can be readily explained by the repulsion of the sterically active lone pair. X-Ray powder diffraction has shown that the chemical analogue $Xe(OTeF_5)_6$ exhibits this XeF_6 gas phase structure in the crystalline state.[13]

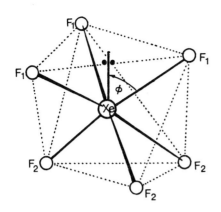

$$\text{Xe-F}_1 = 1.941\text{Å}$$
$$\text{Xe-F}_2 = 1.850\text{Å}$$
$$\phi(\text{F}_1) = 67.5°$$
$$\phi(\text{F}_2) = 127.7°$$
$$\text{F}_1...\text{F}_2 = 3.106\text{Å}$$
$$\text{F}_2...\text{F}_2 = 2.535\text{Å}$$
$$\text{F}_1...\text{F}_2 = 2.498\text{Å}$$

Fig. 3.38-5 Geometry of XeF_6 in the ground state. The lone pair orbital points in the positive direction of the C_{3v} symmetry axis, and ϕ is the angle between this axis and the Xe-F bond. (After ref. 11).

All known experimental data are consistent with the following nonrigid model proposed by Bartell.[14] Starting from the C_{3v} structure, the electron pair can pass between two fluorine atoms to an equivalent position surrounded by three fluorine atoms. This continuous molecular rearrangement, designated the $C_{3v} \to C_{2v} \to C_{3v}$ transformation, involves only modest changes in bond angles and virtually no change in bond lengths.

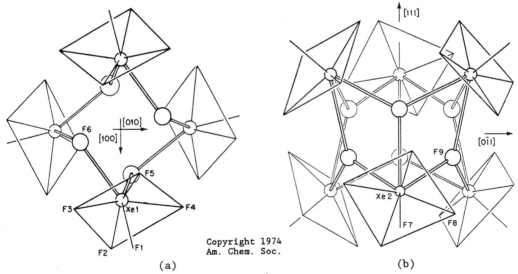

(a)   (b)

Fig. 3.38-6 (a) Tetramer of $XeF_5^{+}F^{-}$. Xenon atoms are represented by small circles and bridging fluoride ions by large circles; the XeF_5^{+} ions are shown in skeletal form to preserve clarity. (b) Hexamer of $XeF_5^{+}F^{-}$. (After ref. 15).

In condensed phases XeF_6 has a pronounced tendency to auto-associate. Solid phases I to IV have been characterized by X-ray crystallography, but all structural studies suffer from insufficient data, absorption problems, and especially disorder. As an example, the cubic modification IV crystallizes in space group *Fm3c* with a = 25.06(5)Å (193K) and Z = 144.[15] The structure is based on the association of XeF_5^+ and F^- ions into roughly spherical cyclic tetrameric (symmetry $\bar{4}$) and hexameric (symmetry 32) units (Fig. 3.38-6). Fluorine atoms in positions 192(j) and 48(f), which are well resolved, form tetragonal pyramids with the Xe atoms of the octahedra. The remaining bridging F atoms are disordered with two possible orientations for the tetramer and four for the hexamer.

6. Structure of the XeF_5^+ ion

Salts of the type $XeF_5^+MF_6^-$ (M = Au, Pt, Ru, Nb) and $XeF_5^+MF_4^-$ (M = Ag, Au) are known. The crystal structure of $XeF_5^+AuF_4^-$ contains double layers of square-pyramidal (C_{4v}) cations partitioned by layers of essentially square-planar anions [Fig. 3.38-7(a)].[16] Each XeF_5^+ ion interacts with four adjacent AuF_4^- ions to generate a capped Archimedian antiprism about the Xe atom; in the cation the axial Xe-F bond is longer than the equatorial bond [Fig. 3.38-7(b)]. In contrast to this, although the same coordination geometry occurs in the $XeF_5^+MF_6^-$ salts, the axial Xe-F bond distance (about 1.80Å) is invariably shorter than the equatorial by 0.03 to 0.06Å.[17]

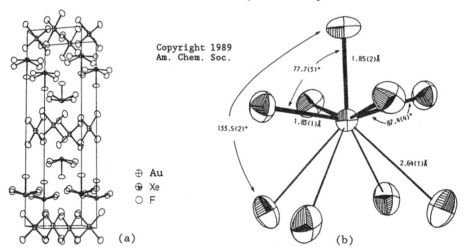

Copyright 1989
Am. Chem. Soc.

77.7(3)°
1.85(2)Å
155.5(2)°
1.83(1)Å
87.9(4)°
2.64(1)Å

⊕ Au
⊕ Xe
○ F

(a)   (b)

Fig. 3.38-7 (a) Crystal structure of $XeF_5^+AuF_4^-$. (b) Molecular structure of XeF_5^+ and its interactions with neighbouring F atoms. (After ref. 16).

7. Structure of the $[(XeOF_4)_3F]^-$ and $[(IF_5)_3F]^-$ ions

The reaction of excess $XeOF_4$ with CsF yielded the salt $Cs[(XeOF_4)_3F]$ which crystallizes in space group *Pa3* with a = 13.933(7)Å and Z = 8.[18] The anion possesses C_3 symmetry (Fig. 3.38-8) and consists of distorted $XeOF_4E$ octahedra [E = lone electron pair; mean Xe-F = 1.90(3), Xe=O = 1.70(5)Å] linked through the central bridging fluorine atom [Xe...F = 2.62(1)Å, Xe...F...Xe = 116.5(8)°] which lies 0.49Å above the plane of the three Xe atoms. Atoms F(4)-F(7) and Xe lie within 0.03Å of their least-squares plane.

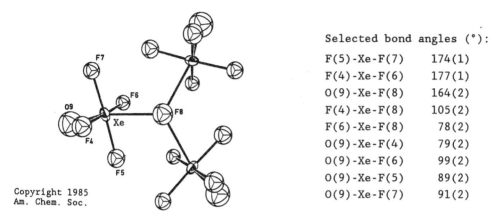

Selected bond angles (°):

F(5)-Xe-F(7)	174(1)
F(4)-Xe-F(6)	177(1)
O(9)-Xe-F(8)	164(2)
F(4)-Xe-F(8)	105(2)
F(6)-Xe-F(8)	78(2)
O(9)-Xe-F(4)	79(2)
O(9)-Xe-F(6)	99(2)
O(9)-Xe-F(5)	89(2)
O(9)-Xe-F(7)	91(2)

Fig. 3.38-8 Structure of the $[(XeOF_4)_3F]^-$ anion. (After ref. 18).

The anion $[(IF_5)_3F]^-$ is composed of three square-pyramidal IF_5 fragments joined by a central bridging fluoride ion,[19] bearing a striking resemblance to the structure of $[(XeOF_4)_3F]^-$. However, the space groups of $K[(IF_5)_3F]$, $Cs[(IF_5)_3F]$, and $Cs[(XeOF_4)_3F]$ are all different, being $P2_1/c$ [a = 13.535, b = 13.638(2), c = 13.421(4)Å, β = 90.49(3)°, Z = 8] and $P\bar{4}3n$ [a = 13.790(11)Å, Z = 8] for the first two compounds, respectively.

8. Structure of the $(Xe_2F_{11})^+$ ion

    $(Xe_2F_{11}^+)_2NiF_6^{2-}$ has been prepared by the reaction between nickel difluoride, krypton difluoride, and xenon hexafluoride in anhydrous hydrogen fluoride.[20] The essentially octahedral NiF_6^{2-} group resides at an inversion center, and the $(Xe_2F_{11})^+$ ion consists of two square-pyramidal XeF_5 moieties bridged by an additional fluorine atom F_b (Fig. 3.38-9). All four Xe atoms and the Ni atom necessarily lie in the same plane. The $Xe(1)-F_b$ and $Xe(2)-F_b$ distances are sufficiently short [2.35(1) and 2.21(1)Å, respectively] to justify the formulation of a $(Xe_2F_{11})^+$ species. The $Xe(1)-F_b-Xe(2)$ angle is 140.3(6)°. In the two XeF_5 square pyramids, the $Xe-F_{ax}$ bond length [1.81(2) and 1.82(1)Å, respectively] is shorter than the average $Xe-F_{eq}$ bond length of 1.86Å, and the $F_{ax}-Xe-F_{eq}$ angles are close to 80°.

    The existence of the cation $(Xe_2F_{11})^+$ was crystallographically verified first in the compound $(Xe_2F_{11})^+(AuF_6)^-$.[21] The crystal structure of $(Xe_2F_{11}^+)_2NiF_6^{2-}$ represents the first example of two $Xe_2F_{11}^+$ cations linked to the same anion. As in $(Xe_2F_{11})^+(AuF_6)^-$ and $(XeF_2XeF_5)^+(AsF_6)^-$,[22] each XeF_5 unit in $(Xe_2F_{11}^+)_2(NiF_6)^{2-}$ interacts strongly with two F ligands.

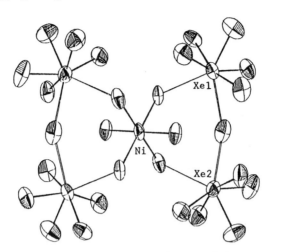

Fig. 3.38-9 Structure of $(Xe_2F_{11}^+)_2NiF_6^{2-}$. (After ref. 20).

9. Structure of $[MeCN-Xe-C_6F_5]^+[(C_6F_5)_2BF_2]^-\cdot MeCN$

This compound has no significant fluorine bridge between cation and anion, as the shortest intermolecular Xe...F distance is 3.135(5)Å. The $[C_6F_5Xe]^+$ cation, which contains the first structural characterization of a Xe-C bond, is coordinated by the nitrogen atom of a MeCN molecule as shown in Fig. 3.38-10.[23] An additional MeCN molecule per formula unit is disordered and displays no contact to the cation. The C-Xe-N bond angle is 174.5(3)°.

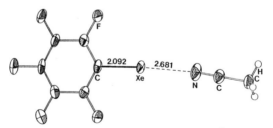

Fig. 3.38-10 Structure of the $[MeCN-Xe-C_6F_5]^+$ ion. (After ref. 23).

References

[1] S. Siegel and E. Gebert, *J. Am. Chem. Soc.* **85**, 240 (1963).

[2] H.A. Levy and P.A. Agron, *J. Am. Chem. Soc.* **85**, 241 (1963).

[3] L. Pauling, *J. Chem. Phys.* **1**, 56 (1933).

[4] N. Bartlett and D.H. Lohmann, *Proc. Chem. Soc. (London)*, 115 (1962).

[5] N. Bartlett, *Proc. Chem. Soc. (London)*, 218 (1962).

[6] H.H. Hyman (ed.), *Noble-Gas Compounds*, The University of Chicago Press, Chicago, 1963.

[7] P. Laszlo and G.J. Schrobilgen, *Angew. Chem. Int. Ed. Engl.* **27**, 479 (1988).

[8] D.H. Templeton, A. Zalkin, J.D. Forrester and S.M. Williamson, *J. Am. Chem. Soc.* **85**, 242 (1963); J.H. Burns, P.A. Agron and H.A. Levy in ref. 6, p. 211.

[9] J.H. Burns, R.D. Ellison and H.A. Levy, *Acta Crystallogr.* **18**, 11 (1965).

[10] N. Bartlett and F.O. Sladky in G. Wilkinson (ed.), *Comprehensive Inorganic Chemistry*, Vol. 1, Pergamon Press, Oxford, 1973, chap. 6.

[11] K. Seppelt and D. Lentz, *Prog. Inorg. Chem.* **29**, 167 (1982).

[12] H. Rupp and K. Seppelt, *Angew. Chem. Int. Ed. Engl.* **13**, 612 (1974).

[13] E. Jacob, D. Lentz, K. Seppelt and A. Simon, *Z. Anorg. Allg. Chem.* **472**, 7 (1981).

[14] R.M. Gavin, Jr. and L.S. Bartell, *J. Chem. Phys.* **48**, 2460 (1968).

[15] R.D. Burbank and G.R. Jones, *J. Am. Chem. Soc.* **96**, 43 (1974).

[16] K. Lutar, A. Jesih, I. Leban, B. Žemva and N. Bartlett, *Inorg. Chem.* **28**, 3467 (1989).

[17] B. Žemva, *Croat. Chim. Acta* **61**, 163 (1988).

[18] J.H. Holloway, V. Kaucic, D. Martin-Rovet, D.R. Russell, G.J. Schrobilgen and H. Selig, *Inorg. Chem.* **24**, 678 (1985).

[19] A.R. Mahjoub, A. Hoser, J. Fuchs and K. Seppelt, *Angew. Chem. Int. Ed. Engl.* **28**, 1526 (1989).

[20] A. Jesih, K. Lutar, I. Leban and B. Žemva, *Inorg. Chem.* **28**, 2911 (1989).

[21] K. Leary, A. Zalkin and N. Bartlett, *Inorg. Chem.* **13**, 775 (1974).

[22] B. Žemva, A. Jesih, D.H. Templeton, A. Zalkin, A.K. Cheetham and N. Bartlett, *J. Am. Chem. Soc.* **109**, 7420 (1987).

[23] H.J. Frohn, S. Jakobs and G. Henkel, *Angew. Chem. Int. Ed. Engl.* **28**, 1506 (1989).

Note In $Me_4N^+IF_6^-$ the anion is a C_{3v} distorted octahedron with a sterically active electron pair (unlike regular octahedral BrF_6^-), whereas $NO^+IF_6^-$ exists as a tetramer resembling the $(XeF_5^+F^-)_4$ unit in crystalline XeF_6, as reported in A.-R. Mahjoub and K. Seppelt, *Angew. Chem. Int. Ed. Engl.* **30**, 323 (1991); IF_8^-, like XeF_8^-, has a square-antiprismatic structure with a stereochemically inactive lone pair. See A.-R. Mahjoub and K. Seppelt, *Angew. Chem. Int. Ed. Engl.* **30**, 876 (1991).

Besides KrF_2 the cationic species KrF^+ and $Kr_2F_3^+$ are found in salts such as $KrF^+SbF_6^-$ and $Kr_2F_3^+AsF_6^-$. The cation $[HC\equiv N-Kr-F]^+$ is reported in G.J. Schrobilgen, *J. Chem. Soc., Chem. Commun.*, 863 (1988).

Experimental and theoretical studies in the chemistry of He, Ar and Ne are reviewed in G. Frenking and D. Cremer, *Structure and Bonding* **73**, 17 (1990).

3.39   Gamma Brass

Cu$_5$Zn$_8$

Crystal Data

Cubic, space group $I\bar{4}3m$ (No. 217)

$a = 8.869(2)$Å, $Z = 4$

Atom	Position	x	y	z
Zn(1)	8(c)	.110	.110	.110
Cu(1)	8(c)	-.172	-.172	-.172
Cu(2)	12(e)	.355	0	0
Zn(2)	24(g)	.313	.313	.036

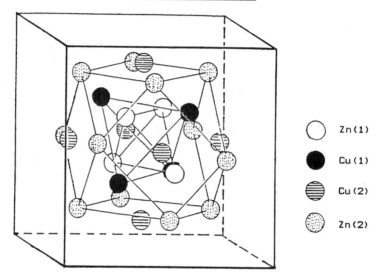

○	Zn(1)
●	Cu(1)
⊖	Cu(2)
⊙	Zn(2)

Fig. 3.39-1 Structure of Cu$_5$Zn$_8$ (γ-brass). The unit cell consists of two clusters: **A** and **B** centered at (0,0,0) and (1/2,1/2,1/2), respectively. This figure only shows the atomic sites in cluster **A**. (After ref. 1).

Crystal Structure

Cu$_5$Zn$_8$ is the prototype of the γ-brass structure and was characterized as such by Westgren and Phragmén in 1925.[2] The crystal structure was subsequently investigated by Bradley and Thewlis,[3] and refined by Bradley and Gregory,[4] who determined the positional parameters of the atoms, but could only infer the ordering scheme by analogy to the ordering in Au$_5$Zn$_8$ since, with X-rays, it was not possible to tell the difference between Cu and Zn. More recently in 1968, Heidenstam and co-workers[1] refined the structure using neutron powder diffraction data.

The structure of Cu$_5$Zn$_8$ may be described in terms of two clusters each comprising 26 atoms. One cluster, **A**, has its center at the origin (0,0,0); the other, **B**, is centered on (1/2,1/2,1/2). In space group $I\bar{4}3m$ the clusters are necessarily idential.

Each cluster is built of atoms constituting the vertices of an inner tetrahedron [Zn(1)], an outer tetrahedron [Cu(1)], an octahedron [Cu(2)], and a somewhat distorted cubo-octahedron [Zn(2)], as shown in Fig. 3.39-1. The measured metal-metal bond distances lie in the range 2.53-2.83Å.

Remarks

1. Structures of Cu_5Cd_8 and Cu_9Al_4

The structures of the alloys Cu_5Cd_8 and Cu_9Al_4 have been refined with single-crystal diffractometer data.[1] Cu_5Cd_8 is isostructural with Cu_5Zn_8. Cu_9Al_4 has nearly the same structure but crystallizes in space group $P\bar{4}3m$. Consequently the atomic distributions in clusters **A** and **B** are different, and the correspondence between the positional parameters of the two alloys is only approximate (Table 3.39-1).

Table 3.39-1   Structural parameters of Cu_5Cd_8 and Cu_9Al_4.

	Cu_5Cd_8	Cu_9Al_4	
		Cluster A	Cluster B
space group	$I\bar{4}3m$	$P\bar{4}3m$	
a/Å	9.5888(3)	8.7023(5)	
Z	4	4	
inner tetrahedral position	Cu	Al	Cu
x	.0939	.1144	.6046
outer tetrahedral position	Cu	Cu	Cu
x	-.1617	-.1690	.3248
octahedral position	(Cd,Cu)	Cu	Cu
x	.3506	.3565	.8554
cubo-octahedral position	(Cd,Cu)	Cu	Al
x	.2980	.3142	.8108
z	.0577	.0337	.5367

2. Electron compounds and the Cu-Zn alloy system

Copper is notable in forming an extensive series of alloys with many other metals, many of which have played an important part in the development of technology through the ages. In most cases the alloys can be thought of as non-stoichiometric intermetallic compounds of definite structural types, and despite the apparently bizarre formulae that emerge from the succession of phases, they can conveniently be classified by a set of rules first outlined by Hume-Rothery in 1926.[5] The determining factor is the ratio of the total number of valence electrons to the total number of atoms, i.e. the "electron concentration", and because of this the phases are sometimes referred to as "electron compounds".

The phase diagram of the Cu-Zn system is shown in Fig. 3.39-2. Copper has one valence electron per atom ($3d^{10}4s^1$), and the pure metal has the ccp

structure (Section 2.1). Admixture with zinc increases the electron
concentration in the primary alloy (α-phase) which can be described as a ccp
solid solution of Zn in Cu. This continues until the electron concentration
approaches $21/14 = 3/2$, when the ccp structure then becomes less stable than
a bcc arrangement that crystallizes as β-brass, CuZn. Further increase in
electron concentration results in the formation of γ-brass, Cu_5Zn_8, at an
electron concentration of $21/13$. This γ-phase can itself take up more Zn
until a third critical concentration is reached near $21/12 = 7/4$, giving rise
to the ϵ-phase of $CuZn_3$ with the hcp structure. Further addition of Zn yields
the η-phase which is a solid solution of Cu in Zn. Table 3.39-2 lists the
structural data of Cu-Zn alloys.[6]

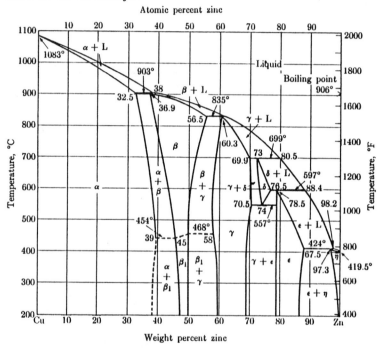

Cu-Zn diagram. (*ASM Metals Handbook.* Metals Park, O.: American Society for Metals.)

Fig. 3.39-2 Phase diagram of the CuZn system.

Table 3.39-2 Structural data of Cu-Zn alloys.

Phase	Formula	System	Space group	Unit cell parameters/Å
α	Cu(Zn)	cubic	Fm3m	a = 3.6074 (100% Cu)
β	CuZn	cubic	Im3m	a = 2.9907 (at 879°C)
β'	CuZn	cubic	Pm3m	a = 2.9539 (47.66 at % Zn)
γ	Cu_5Zn_8	cubic	I$\bar{4}$3m	a = 8.69 (39.7 at % Cu)
δ	$CuZn_3$	cubic	Pm3m	a = 2.995 (600°C, 25.5 at % Cu)
ϵ	$CuZn_3$	hexagonal	P6$_3$/mmc	a = 2.735, c = 4.286 (21.0 at % Cu)
η	Zn(Cu)	hexagonal	P6$_3$/mmc	a = 2.6595, c = 4.9368 (100% Zn)

In calculating the electron concentration the valence electrons for different types of atoms are counted as follows:

1e:	Cu, Ag, Au;
2e:	Be, Mg, Zn, Cd, Hg;
3e:	Al, Ga, In;
4e:	Ge, Sn, Pb;
5e:	Sb, Bi;
0e:	Fe, Co, Ni, Ru, Rh, Pd, Pt, Ir, Os.

A selection of electron compounds with the β, γ and ϵ structures is given in Table 3.39-3.

Table 3.39-3 Examples of electron compounds.

Electron concentration	Phase (structure)	Examples8
~ 21/14	β-phase (bcc and β-Mn type)	CuBe, CuZn; Cu_3Al, Cu_3Ga, Cu_3In; Cu_5Si, Cu_5Sn
		AgMg, AgZn, AgCd; Ag_3Al, Ag_3In
		AuMg, AuZn, AuCd; Au_3Al
	β-phase (hcp)	Ag_3Ga, Au_3In; Cu_5Ge, Ag_5Sn, Au_5Sn; Ag_7Sb
~ 21/13	γ-phase (complex cubic with 52 atoms per unit cell)	Cu_5Zn_8, Cu_5Cd_8, Cu_5Hg_8; Cu_9Al_4, Cu_9Ga_4, Cu_9In_4
		$Cu_{31}Si_8$, $Cu_{31}Sn_8$
		Ag_5Zn_8, Ag_5Cd_8, Ag_5Hg_8; Ag_9In_4
		Au_5Zn_8, Au_5Cd_8; Au_9In_4
~ 21/12	ϵ-phase (hcp)	$CuBe_3$, $CuZn_3$, $CuCd_3$; Cu_3Si, Cu_3Ge, Cu_3Sn; $Cu_{13}Sb_3$
		$AgZn_3$, $AgCd_3$; Ag_5Al_3; Ag_3Sn; $Ag_{13}Sb_3$
		$AuZn_3$, $AuCd_3$; Au_5Al_3; Au_3Sn

3. **Structure of the intermetallic cubic phase** $(Al,Zn)_{49}Mg_{32}$

$(Al,Zn)_{49}Mg_{32}$ is one of the most complex alloys known.[7] It crystallizes in space group *Im*3 with 162 atoms in a unit cell of edge a = 14.25(3)Å. Fig. 3.39-3 shows the multi-shell atomic arrangement around each lattice point. (a) Atom **A** at (1/2,1/2,1/2) is surrounded by an icosahedron of 12**B** atoms. (b) A pentagonal dodecahedron is formed by placing 8**D** and 12**E** atoms above the faces of the icosahedron. (c) Capping the faces of the latter with 12**C** atoms generates a rhombic triacontahedron, whose 32 vertices include the 8**D** and 12**E** atoms. (d) The next shell is an irregular truncated icosahedron consisting of 48**F** and 12**G** atoms, each located above the center of a triangle that constitutes one-half of one of the 30 rhombs bounding the triacontahedron. (e) The **H** atoms constitute the vertices of a distorted cubo-octahedron with a pair of **G** atoms located inside each "square face" and a ring of six **F** atoms inside each puckered "hexagonal face". Such cubo-octahedra can be packed to fill all space by sharing the 72 atoms (48F + 12G + 24H/2) associated with each polyhedron. Since within each cubo-octahedron there is a nucleus of 45 atoms (A + 12B + 8D + 12E + 12C), the total number of atoms in the body-centered unit cell is 2(45 + 72/2) = 162.

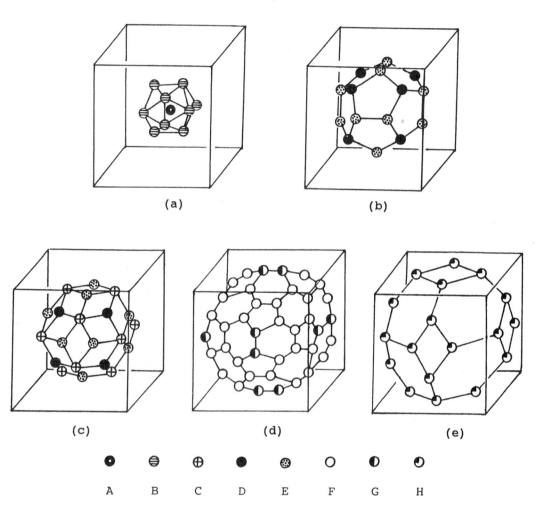

⊙	⊖	⊕	●	⊛	○	◑	◕
A	B	C	D	E	F	G	H

Fig. 3.39-3 Atomic arrangement in the cubic alloy $(Al,Zn)_{49}Mg_{32}$. For clarity atoms in the back side of each polyhedron are not shown.

Cubic Al_5CuLi_3 (known as the R phase)[8] has been shown to be isostructural with $(Al,Zn)_{49}Mg_{32}$; the atomic parameters for these two alloys are listed in Table 3.39-4.

Audier and co-workers have pointed out that the H atoms actually define 24 vertices of a *larger* triacontahedron whose remaining 8 vertices are unoccupied. The $(Al,Zn)_{49}Mg_{32}$ cubic structure type can then be described elegantly in terms of a close-packed arrangement of multi-shell "137-atom

Table 3.39-4   Atomic parameters for cubic $(Al,Zn)_{49}Mg_{32}$ and Al_5CuLi_3.

Atom type	Wyckoff notation	x	y	z	CN	Atoms in alloy system Al-Zn-Mg[*]	Al-Cu-Li[**]
A	2(a)	0	0	0	12	Al	Al,Cu
B	24(g)	0	.0908	.1501	12	Al,Zn(81)	Al,Cu
C	24(g)	0	.1748	.3007	12	Al,Zn(57)	Al,Cu
D	16(f)	.1836	.1836	.1836	16	Mg	Li
E	24(g)	0	.2942	.1194	16	Mg	Li
F	48(h)	.1680	.1860	.4031	12	Al,Zn(64)	Al,Cu
G	12(e)	.4002	0	1/2	14	Mg	Li
H	12(e)	.1797	0	1/2	15	Mg	Al

[*] Site occupancy of position 2(a) is 0.80; % occupancy of Zn atoms enclosed in parentheses.   [**] Stoichiometry of the isostructural alloy is $Al_{56.5}Cu_{11.3}Li_{32.1}$.

double triacontahedra" [45 + 60 + 24 + 8(virtual atoms)] which share rhomb faces along <100> and partially overlap along <111> (Fig. 3.39-4).[9]

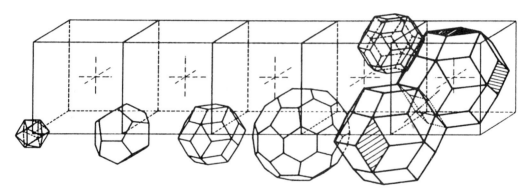

Fig. 3.39-4   Schematic representation of the *Im*3 $(Al,Zn)_{49}Mg_{32}$ phase showing the successive shells of atoms and two types of connection between the large triacontahedra.   (After ref. 9).

4.   Pentagonal Frank-Kaspar phases

Pentagonal Frank-Kaspar (FK) phases are an important family of alloy phases characterized by tetrahedral coordination of all atoms, which lie on planar or somewhat rumpled layers consisting of pentagons and triangles.[9] The pentagons from different layers stack on top of each other in a staggered manner and the CN12 icosahedra thus formed, together with CN14, CN15 and/or CN16 polyhedra, constitute the building blocks of a close-packed crystal structure.   Cubic $(Al,Zn)_{49}Mg_{32}$ is a pentagonal FK phase, and the available crystallographic data for various Frank-Kaspar and related phases are shown in Table 3.39-5.[10,11]

Table 3.39-5   Crystal data of pentagonal Frank-Kasper and related phases.

Type	Example	Space group	a(Å)	b(Å)	c(Å)	β	N	12	24	15	16	Existence of quasicrystal
μ	Mo_6Co_7	R3m	4.762		25.615		39	54	15	15	15	
pσ	$W_6(Fe,Si)_7$	Pbam	9.283	7.817	4.755		26	54	15	15	15	
M	$Nb_{48}Ni_{39}Al_{13}$	Pnam	9.303	16.266	4.933		52	54	15	15	15	
I	$V_{41}Ni_{36}Si_{23}$	Cc	13.462	23.381	8.940	100.3°	228	58	10.5	10.5	21	Yes
C	$V_2(Co,Si)_3$	C2/m	17.17	4.66	7.55	99.2°	50	60	7.5	7.5	25	
T	$(Al,Zn)_{49}Mg_{32}$	Im3	14.16				162	61	7	7	25	Yes
T	Mg_4CuAl_6	Im3					162	61	7	7	25	Yes
T	Al_5CuLi_3	Im3					162	61	7	7	25	Yes
X	$Mn_{45}Ci_{40}Si_{15}$	Pnnm	15.42	12.39	4.74		74	63	5	5	27	
-	Mg_4Zn_7	C2/m	25.96	5.24	14.28	102.5°	110	64	3.5	3.5	29	
C14 $MgZn_2$	Mn-Ni-Si	$P6_3/mmc$	4.76		7.50		12	67			33	Yes
C15 $MgCu_2$	$MgCu_2$	Fd3m	7.08				24	67			33	
G	$Ti_6Ni_{16}Si_7$	Fd3m	11.15					48				
η	$NiTi_2$	Fd3m	11.2				96	50				Yes
Cu_4Cd_3	Cu_4Cd_3	F$\bar{4}$3m	25.87				1124	51				Yes
$β-Mg_2Al_3$	Mg_2Al_3	Fd3m	28.24				1168	58				

N = number of atoms in the unit cell; CN = coordination number.

5. Quasicrystalline materials

A "quasicrystal" (shortened form of quasiperiodic crystal) is a solid phase whose structure has a long-range *quasiperiodic translational order* and long-range *orientational order* with *disallowed* crystallographic symmetry. The long-range translational order differentiates quasicrystals from glasses, and noncrystallographic orientational symmetry not only distinguishes quasicrystals from ordinary (periodic) and incommensurate crystals, but also forces quasiperiodicity by imposing geometric constrains on the incommensurate length ratios.[12,13]

The science of quasicrystalline materials began in 1984 with the discovery of the metastable "icosahedral phase" $Al_{86}Mn_{14}$, which produces sharp diffraction spots exhibiting noncrystallographic point-group symmetry $m\bar{3}\bar{5}$ that cannot be indexed in any Bravais lattice.[14] This phase forms in rapidly solidified Al-Mn alloys in the composition range 10-20 at % Mn. At a higher composition range of 18-22 at % Mn, the icosahedral phase is replaced by a "decagonal phase" (or *T* phase) whose diffraction pattern exhibits one-dimensional translational symmetry and 10/*m* (or 10/*mmm*) symmetry.[15] In many alloy systems the icosahedral phase and other quasiperiodic crystals can be easily made by a variety of methods, mainly rapid solidification from the melt and solid-state transformation in a thin deposited film. Quasicrystalline and ordinary crystalline phases are often obtained in the same preparation; for example, in the Al-Cu-Li system both cubic R-Al$_5$CuLi$_3$ and the icosahedral phase T$_2$-Al$_6$CuLi$_3$ coexist after conventional ingot casting. Icosahedral quasicrystals often form from Cu, Zn, Mn, Al, Cr, Ni, V, Si... alloys that contain icosahedral units in their equilibrium phases.

The detailed arrangement of atoms in quasicrystals is a central problem that remains incompletely solved. The icosahedron, the pentagonal dodecahedron, and in particular the rhombic triacontahedron all figure prominently in many attempts to construct structural models by "decorating" two or more coexisting unit-cell types in a quasilattice with atoms and atomic clusters.[9,13,16,17].

References

[1]   O. von Heidenstam, A. Johansson and S. Westman, *Acta Chem. Scand.* **22**, 653 (1968).

[2]   A. Westgren and G. Phragmén, *Phil. Mag.* **50**, 311 (1925).

[3]   A.J. Bradley and J. Thewlis, *Proc. Roy. Soc. London* **A112**, 678 (1926).

[4]   A.J. Bradley and C.H. Gregory, *Phil. Mag.* **62**, 143 (1931).

[5]   W. Hume-Rothery, *J. Inst. Metals* **35**, 295, 307 (1926).

[6]   W.B. Pearson, *Handbook of Lattice Spacings and Structures of Metals and Alloys*, Vol. 1, 1958, Vol. 2, 1967, Pergamon Press, New York.

[7]   G. Bergman, J.L.T. Waugh and L. Pauling, *Acta Crystallogr.* **10**, 254 (1957).

[8]   E.E. Cherkashin, P.I. Kripyakevich and G.I. Oleksiv, *Soviet Phys. Cryst.* **8**, 681 (1964).

[9]   M. Audier, P. Sainfort and B. Dubost, *Phil. Mag. B* **54**, L105 (1986).

[10] C.B. Shoemaker and D.P. Shoemaker, *Acta Crystallogr., Sect. B* **37**, 1 (1981); *ibid.* **42**, 3 (1986).

[11] K.H. Kuo, D.S. Zhou and D.X. Li, *Phil. Mag. B* **55**, 33 (1987).

[12] P.W. Stephens and A.I. Goldman, *Scientific American* **264**, No. 4, April 1991, p. 24.

[13] J.P. Steinhardt and S. Ostlund, *The Physics of Quasicrystals*, World Scientific, Singapore, 1987; Ch. Janot and J.M. Dubois (eds.), *Quasicrystalline Materials*, World Scientific, Singapore, 1988.

[14] D. Shechtman, I. Blech, D. Gratias and J.W. Cahn, *Phys. Rev. Lett.* **53**, 1951 (1984).

[15] L. Bendersky, *Phys. Rev. Lett.* **55**, 1461 (1985).

[16] Q.B. Yang, *Phil. Mag. B* **61**, 155 (1990).

[17] W. Steurer, *Z. Kristallogr.* **190**, 179 (1990).

3.40 α-Silver Iodide
AgI

Crystal Data

Cubic, space group $Im3m$ (No. 229)

$a = 5.03(9)$Å (150°C), 5.11(1)Å (478°C), $Z = 2$

Atom	Position	x	y	z	Occupancy
I	2(a)	0	0	0	1
Ag	6(b)	0	1/2	1/2	
Ag	12(d)	1/4	0	1/2	statistically
Ag	24(h)	0	1/8	1/8	distributed

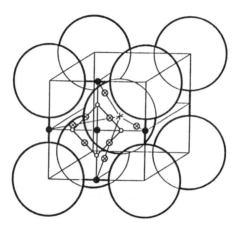

Fig. 3.40-1 Unit cell of α-AgI. Large circles represent I⁻; Ag⁺ ions
statistically distributed in 6(b) ●, 12(d) o, and 24(h) ⊗ positions.

Crystal Structure

In crystalline α-AgI, the I⁻ ions constitute a comparatively rigid body-centered-cubic lattice, in which the Ag⁺ ions distribute themselves statistically among the 6(b), 12(d) and 24(h) equipoints of space group $Im3m$ (Fig. 3.40-1),[1,2] and also partially populate the passageways between these positions. This model is supported by the experimental finding that the "molten" state of Ag⁺ ions takes on the prominent ionic transport properties inherent in α-AgI. Every unit cell of this structure provides 42 possible positions for two Ag⁺; the Ag⁺...I⁻ distances are as follows:

6(b) positions having 2 I⁻ neighbours at 2.52Å,

12(d) positions having 3 I⁻ neighbours at 2.67Å, and

24(h) positions having 4 I⁻ neighbours at 2.86Å.

Remarks

1. Polymorphs of silver and copper(I) halides

At room temperature the stable form of silver iodide is γ-AgI, which has

the zinc blende type of structure (Section 2.9). β-AgI, which has the
wurtzite structure (Section 2.10), is the stable form between 136° and 146°C.
This structure is closely related to that of common ice Ih (Section 2.16) and
AgI has been found to be particularly effective in inducing the precipitation
of rain. At 146°C β-AgI undergoes a phase change to cubic α-AgI.

When β-AgI transforms to α-AgI, there is a dramatic effect on the ionic
electrical conductivity of the solid, which leaps from 3.4 x 10^{-4} to 1.3 ohm^{-1}
cm^{-1}, a factor of nearly 4000. This is because in α-AgI the silver ions move
freely between positions of 2-, 3-, and 4-coordination between the easily
deformed iodide ions (Fig. 3.40-1).

In β- and γ-AgI, the Ag atoms have tetrahedral coordination, in which
covalent bonding has dominance, but in α- and high-pressure modifications
ionic bonding becomes important. Thus different structures correspond to
different bond types and properties.

The phase transitions and structures of silver and copper(I) halides are
listed in Table 3.40-1.

Table 3.40-1 Polymorphs of silver and copper(I) halides.[3]

Substance	Modification	Structure type	Region of stability
AgCl		NaCl	below mp
AgBr		NaCl	below mp
AgI	α	bcc (aver.)	146°C-mp
	β	wurtzite	136°-146°C
	γ	zinc blende	<136°C
	high-pressure	NaCl	>3020 atm
CuCl	β	wurtzite (aver.)	435°C-mp
	γ	zinc blende	<435°C
CuBr	α	bcc (aver.)	485°C-mp
	β	wurtzite (aver.)	405°-485°C
	γ	zinc blende	<405°C
CuI	α	zinc blende (aver.)	430°C-mp
	β	wurtzite (aver.)	390°-430°C
	γ	zinc blende	<390°C

2. Structures of $RbAg_4I_5$ and $(C_5H_5NH)Ag_5I_6$

The reaction of AgI with RbI forms $RbAg_4I_5$, which has been found to have
the highest specific electrolytic conductivity, 0.27 $ohm^{-1}cm^{-1}$ (at room
temperature), of any solid measured and remains so at low temperature.

$RbAg_4I_5$ is cubic, space group $P4_132$ (or $P4_332$), with a = 11.24(2)Å, and
Z = 4. The positional parameters as determined by Geller[4] are listed in
Table 3.40-2.

Table 3.40-2   Positional parameters of $RbAg_4I_5$.

Atom	Wyckoff position	x	y	z	Occupancy factor
Rb	4(a)	3/8	3/8	3/8	1
I(1)	8(c)	.0306	x	x	1
I(2)	12(d)	3/8	-.1780	(3/4)-y	1
Ag(1)	8(c)	.1739	x	x	.111
Ag(2)	24(e)	.5299	.2713	.7980	.391
Ag(3)	24(e)	.9964	.8506	.2154	.229

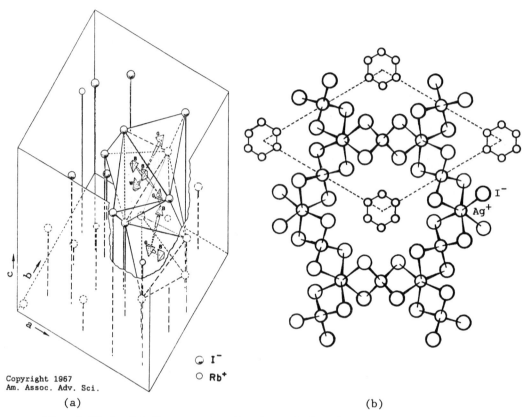

(a)   (b)

Fig. 3.40-2   (a) Perspective drawing of the $RbAg_4I_5$ structure.   Eight
face-sharing iodide tetrahedra are sketched.   The small tetrahedra
labeled W, B and G correspond to silver ions Ag(1), Ag(2) and Ag(3) of
Table 3.40-2, and the diffusion paths are indicated.   (b) Plan view of
the structure of $(C_5H_5NH)Ag_5I_6$ at -30°C.   The *m* sites are in the empty
tetrahedra.   (After refs. 4 and 6).

In $RbAg_4I_5$ the arrangement of the iodide ions is that of the Mn atoms in
β-Mn,[5] constituting 56 face-sharing tetrahedra per unit cell.   Each Rb^+ ion
is surrounded by a distorted octahedron of iodide ions.   The sixteen Ag^+ ions
in the unit cell are disordered over three sets of nonequivalent sites.
Through faces shared by iodide tetrahedra [Fig. 3.40-2(a)], a silver ion in an
Ag(1) site may diffuse to any one of three adjacent Ag(2) sites; a silver ion

in an Ag(2) site has access to an Ag(1) site, to another Ag(2) site, and to one of two Ag(3) sites; a silver ion in an Ag(3) site may move only to two Ag(2) sites. The silver "inter-site" distances are in the range 1.68-1.91Å, which are much shorter than the Ag-I distances in the range 2.73-2.94Å. Alternating Ag(2) and Ag(3) sites form channels parallel to the three axial directions, and the Ag(1) sites serve to connect these principal diffusion paths [Fig. 3.40-2(a)].

There are two low-temperature phase transformations in $RbAg_4I_5$; the space groups and equipoint relationships of the three phases are shown in Table 3.40-3.[6] The transition at 209K is second order and results mainly in a redistribution of the Ag^+ ions. In the phase that exists below 122K it is possible for the Ag^+ ions to be ordered.

Table 3.40-3 Space groups and equipoint transformations of $RbAg_4I_5$ phases.

T > 209K $P4_132$	209K > T > 122K R32	T < 122K P321
Rb in 4a (32)	⎰ 1a(32) ⎱ 3e(2)	⎰ 1a(32) / 2d(3) ⎱ 3f(2) / 6g(1)
Ag(1), I(1) in 8c (3)	⎰ 2c(3) ⎱ 6f(1)	⎰ 2c(3) / 2 sets of 2d(3) ⎱ 3 sets of 6g(1)
I(2) in 12d (2)	⎰ 3d(2) / 3e(2) ⎱ 6f(1)	⎰ 3e(2) / 6g(1) ⎱ 3f(2) / 6g(1) / 3 sets of 6g(1)
Ag(2), Ag(3) in 24e (1)	⎰ 4 sets of 6f each; ⎱ 8 sets of 6f, total	⎰ 3 sets of 6g each; ⎱ 24 sets of 6g, total

Numbers in parentheses designate point symmetries. At 130K, a = 11.17(1)Å, α = 90.10(5)° and Z = 4. At 90K, a = 15.776(5), C = 19.320(5)Å and Z = 12.

Another interesting silver iodide based solid electrolyte is $(C_5H_5NH)Ag_5I_6$. At -30°C, the structure belongs to space group *P6/mcc* with *a* = 11.97(2), *c* = 7.41(1)Å and *Z* = 2. In the structure [Fig. 3.40-2(b)] the ten Ag^+ ions occupy sites 4*c* and 6*f*; the former are octahedrally coordinated, and the latter tetrahedrally coordinated. There are 24 tetrahedral sites (24*m*) that are empty. Every I^- ion is at the corner of one of these "*m*-type" tetrahedra.

The octahedra share faces that are parallel to the hexagonal (001) plane, thus generating straight channels along the c axis. Short inter-site distances prevent the occupancy of a f site and an adjacent m site, or two neighbouring m sites, but both m and c sites may be occupied simultanously. As the temperature is increased from -30°C, Ag^+ ions in the c and f sites are excited into the m sites. However, the process must begin with Ag^+ ions moving from f to m sites. When a Ag^+ ion moves from a f site to a m site, one other m-type tetrahedron is free for occupance by Ag^+ ions from the octahedra.

3. Structure of Ag_2HgI_4

Ag_2HgI_4 exists in two forms: the low temperature form (β) is a yellow tetragonal crystal, whereas the high temperature form (α) is an orange-red cubic crystal (Fig. 3.40-3), the transition temperature being 50.7°C. When

$$\beta\text{-}Ag_2HgI_4 \xrightarrow{\ 50.7°C\ } \alpha\text{-}Ag_2HgI_4$$
$$\text{yellow} \qquad\qquad\qquad \text{orange-red}$$

the β-form transforms to the α-form, the Ag and Hg atoms become almost randomly distributed on the possible sites, thus accounting for the fact that α-Ag_2HgI_4 is a good superionic conductor.

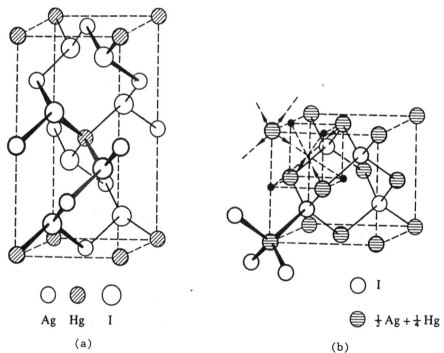

Ag   Hg   I   I

                                                          ½ Ag + ¼ Hg

(a)   (b)

Fig. 3.40-3 (a) Low-temperature and (b) high-temperature forms of Ag_2HgI_4. (After ref. 7).

4. Properties of AgI-type superionic solids[8-11]

All of the AgI-type superionic solids have the following properties to a
lesser or greater extent:

a) The cations are structurally disordered and the cation sublattice is
"liquid-like". The number of cations is less than the number of sites (voids)
available for them and these sites are occupied at random.

b) The anions are arranged in such a way that the local potentials are
rather flat along certain lines which interconnect neighbouring sites. Along
these lines the difference in potential energy is of the order of the thermal
energy. The common structural motif in the conducting compounds is the
existence of passageways of face-shared iodide ion polyhedra.

c) As a consequence of a) and b) a large fraction or all of the cations
can move from one site to another with a very low activation energy of the
order of thermal energy and thus, in general, participate in the cation
diffusion process. The diffusion path is channel-like and not exactly liquid-
like.

d) The disordering process may follow first-order or second-order
kinetics. At the first-order transition temperature the conductivity changes
abruptly and generally involves a change in lattice symmetry and latent heat.
In a second-order phase transition, there is no abrupt change in conductivity
with only a small change in the slope. Generally, there is little or no
change in the lattice symmetry, only a small distortion, and it is accompanied
by a power-law divergence in the specific heat.

e) Ion-ion correlation is an important factor in bringing about
disorder.

f) The transport of ions is possibly a jump-diffusion process on which
an additional local cation motion is superimposed.

5. Relation of conductivity to structure

Several generalizations of the relation between the conductivity of
solid electrolytes and their structures have been made:[6]

a) Almost all solid electrolytes of any kind have networks of
passageways formed from the face-sharing of anion polyhedra.

b) In structures in which the sites available to the current carriers
are not crystallographically equivalent, the distribution of carriers over the
different sites is markedly non-uniform.

c) The conductivity is associated with the nature of the passageways:
the simpler they are, the higher will be the conductivity. Three-dimensional
networks exhibit higher average conductivities than two-dimensional networks.

A larger number of available sites and/or a larger volume of crystal space occupied by the conduction passageways tend to give higher conductivities.

  d) The stability of the Ag^+ and Cu^+ ions in both four- and three-coordination and their monovalency are responsible for their being the mobile ions in most good solid electrolytes.

References

[1]   S. Hoshino, *J. Phys. Soc. Jpn.* **12**, 315, 838 (1957).

[2]   L.W. Strock, *Z. Phys. Chem.* **B31**, 132 (1935).

[3]   B.R. Lawn, *Acta Crystallogr.* **17**, 1341 (1964).

[4]   S. Geller, *Science (Washington)* **157**, 310 (1967); J.N. Bradley and P.D. Greene, *Trans. Faraday Soc.* **63**, 2516 (1967); L. Bonpunt and F. Leroy, *Acta Crystallogr., Sect. B* **44**, 553 (1988).

[5]   G.D. Preston, *Phil. Mag.* **5**, 1207 (1928); P.I. Kripiakevich, *Kristallografiya* **5**, 273 (1960). [*Sov. Phys.-Crystallogr. (Engl. Transl.)*].

[6]   S. Geller, *Science (Washington)* **176**, 1016 (1972); *Acc. Chem. Res.* **11**, 87 (1978).

[7]   A.F. Wells, *Structural Inorganic Chemistry*, Clarendon Press, Oxford, 3rd ed., 1962, 5th ed., 1984.

[8]   K. Funke, *Prog. Solid State Chem.* **11**, 345 (1976).

[9]   S. Chandra, *Superionic Solids*, North-Holland, Amsterdam, 1981.

[10] J.C. Philips, *J. Electrochem. Soc.* **123**, 934 (1976).

[11] P. Hagenmuller and W. van Gool (eds.), *Solid Electrolytes*, Academic Press, New York, 1978.

Note The structural chemistry of copper(I) halide complexes with amines is reviewed in K.G. Caulton, G. Davis and E.M. Holt, *Polyhedron* **9**, 2319 (1990).

3.41 Silver Nitrate Oxide
$Ag(Ag_6O_8)NO_3$

Crystal Data

Cubic, space group $Fm3m$ (No. 225)

$a = 9.890(2)$Å, $Z = 4$

Atom	Position	x	y	z	Occupancy
Ag(1)	4(a)	0	0	0	1
Ag(2)	24(d)	0	1/4	1/4	1
O(1)	32(f)	.147	.147	.147	1
O(2)	96(k)	.546	.546	.592	1/8
N	4(b)	1/2	1/2	1/2	1

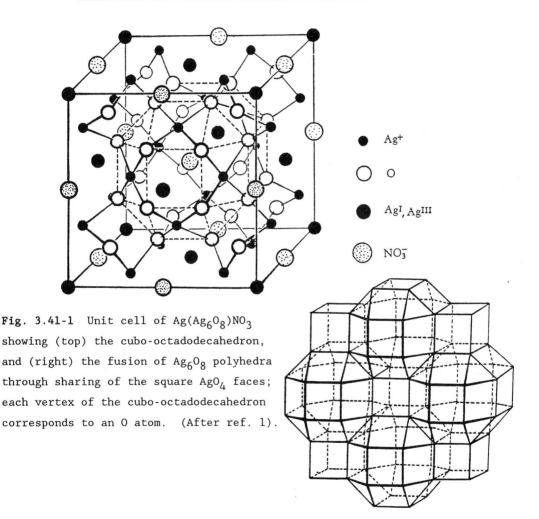

●	Ag^+
○	O
●	Ag^I, Ag^{III}
◉	NO_3^-

Fig. 3.41-1 Unit cell of $Ag(Ag_6O_8)NO_3$ showing (top) the cubo-octadodecahedron, and (right) the fusion of Ag_6O_8 polyhedra through sharing of the square AgO_4 faces; each vertex of the cubo-octadodecahedron corresponds to an O atom. (After ref. 1).

Crystal Structure

Electrolysis of an acidic solution of silver nitrate leads to the formation of a black, crystalline substance at the anode having the composition Ag_7NO_{11}. Although this material has been the subject of a great many studies, only its crystal structure has been determined and its electronic structure is still in some doubt. Following an earlier crystallographic study, the detailed structure has been established for $Ag(Ag_6O_8)NO_3$ by both X-ray and neutron diffraction.[1,2]

The structure of $Ag(Ag_6O_8)NO_3$ is built of a polyhedral framework, as shown in Fig. 3.41-1. Each cubo-octadodecahedron is formed from twelve square-planar AgO_4 units, which result from the bonding of Ag atoms (using dsp^2 hybrid orbitals) with O atoms at Ag-O = 2.05Å. The polyhedra themselves are joined by a sharing of the AgO_4 faces, generating smaller cubes to fill the intervening space. Thus in this neutral Ag_6O_8 polyhedral framework there are two types of cavities: the larger type is a cubo-octadodecahedron with a NO_3^- ion within it, and the smaller cavity is a cube containing a Ag^+ ion with Ag...O = 2.50Å. The NO_3^- groups lie perpendicular to the three-fold axis; the O atoms are located statistically in the 96(k) positions, each being close to two Ag atoms at Ag...O = 2.50Å.

Remarks

1. Properties of $Ag(Ag_6O_8)NO_3$ [3,5]

Anodic oxidation of AgX in acid solution gives crystalline $Ag(Ag_6O_8)X$, the most stable examples being those with X = NO_3^- or HF_2^-, although samples with X = ClO_4^-, BF_4^- or HSO_4^- have also been prepared. Crystalline $Ag(Ag_6O_8)HF_2$, $Ag(Ag_6O_8)ClO_4$ and $Ag(Ag_6O_8)BF_4$ have unit-cell dimensions matching almost exactly those of the nitrate, so the NO_3^- anion can be replaced while keeping the (Ag_6O_8) framework intact. Hence the crystals can be considered as clathrate salts, with ions as guests within the very rigid Ag_6O_8 cages.

Single crystals of $Ag(Ag_6O_8)NO_3$ exhibit high electrical conductivity at room temperature, *ca.* 2.1×10^2 ohm^{-1} cm^{-1}. The molar susceptibility has been reported by various authors as 398×10^{-6}, 688×10^{-6} and $6379 \pm 1750 \times 10^{-6}$ cgs, suggesting that the substance may contain a variable amount of paramagnetic impurity and that the pure substance is diamagnetic. Since the diamagnetism argues against the presence of any Ag^{II}, the (Ag_6O_8) framework may be formulated as $(Ag^IAg_5^{III}O_8)$, and the compound described as $Ag^+(Ag^IAg_5^{III}O_8)(NO_3)^-$. Mixed-valent compounds are common, but compounds which contain two covalent oxidation states and one ionic oxidation state are quite rare.

The $Ag(Ag_6O_8)X$ salts decompose on heating in a complex manner, evolve oxygen to form AgO and AgX in boiling water, and have oxidizing properties similar to those of AgO.

2. Mixed-valent compounds

Mixed-valent compounds play an important role in many scientific fields.[4,5] Although the major impact of the subject has been in chemistry, its importance has become increasingly clear in solid state physics, geology, and biology. Extensive interest and effort in the field of molecular metals have demonstrated that mixed valence is a prerequisite for high electrical conductivity. The intense colours of many minerals have been shown to be due to mixed valence, and the electron-transfer properties of certain mixed-valent metalloproteins are important in molecular biology.

Structural data provide valuable information as to whether the static coordination geometries around the metal ions of a mixed-valent compound are equivalent or not. Moreover, crystal and molecular structures of a few representative mixed-valent compounds serve to illustrate the wide range of chemical classes covered by these substances. A coarse distinction is made between two groups, one comprising small molecular units and the other consisting of polynuclear crystalline solids.

The mixed-valent compounds of main group elements are always found in the heavier post-transition region with electron configuration ns^2 as well as ns^0. Some representative examples are listed in Table 3.41-1.[4]

Table 3.41-1 Mixed-valent compounds of main group elements.

Element	Valence	Examples
Cu	0, 1	KCu_4S_3
	1, 2	$Cu_4Cu_2(S_2)_2S_2$
	1, 2	$Cu_2(NH_3)_3(NCS)_3$, $Cu_3(SO_3)_2 \cdot 2H_2O$
Ag	0, 1	Ag_2F
	1, 3	AgO, $Ag(Ag_6O_8)NO_3$
Au	1, 3	$CsAuCl_3$, $Au(dmg)_2(AuCl_2)$
Ga	1, 3	$GaCl_2$, $(GaL_4)(GaCl_4)$
In	1, 3	$InCl_2$, In_4Cl_7, In_4Br_6
Tl	1, 3	TlO, TlS, $TlCl_2$, Tl_4Cl_6
Sn	2, 4	$Sn_2Cl_6(H_2O)_2$ (aq. solution)
Pb	2, 4	Pb_3O_4, $[Co(NH_3)_6]PbCl_6$
Sb	3, 5	Sb_2O_4, Cs_2SbCl_6, $(NH_4)_2SbBr_6$
Bi	3, 5	$BaBiO_3$, $Bi_{12}Cl_{14}$
I	0, -1	I_n^-
Xe	6, 8	$K_4Xe_3O_{12}$

The crystal structures of $GaCl_2$, Pb_3O_4, $(NH_4)_2SbBr_6$, $Bi_{12}Cl_{14}$ and I_n^- are discussed in Sections 3.14, 3.25, 3.30, 3.31 and 2.6, respectively.

3. Mixed-valent compounds of copper[6]

Robin and Day[3] have discussed the general behaviour of mixed-valent compounds and classified them into three classes, I-III, as exemplied[7] in Table 3.41-2 for the element copper. The stereochemistries of some class I mixed-valent copper compounds are shown in Table 3.41-3, and some representaive structures are illustrated in Fig. 3.41-2.

Table 3.41-2 Classification of mixed-valent Cu^I/Cu^{II} compounds[3]

Class	I	II	III
Stereochemistry	Localized stereochemistry different for different valence	Same stereochemistry for different valence – distinguishable	Same stereochemistry for different valence – indistinguishable (A) Clusters; (B) infinite arrays
Electronic properties	Localized Cu^I and Cu^{II}	Delocalized electron	Delocalized
Magnetic properties	Cu^I-diamagnetic Cu^{II}-paramagnetic	–	Metallic ferromagnetic
Conductivity	Insulator	Semiconductor	Conductor
Electronic spectra	Cu^I-colourless Cu^{II}-normal d-d	Cu^I-colourless Cu^{II}-near normal d-d Characteristic(I)-(II) charge transfer spectra	No spectra of constituent Cu^I and Cu^{II} ions Characteristic charge transfer spectra Metallic reflectivity
Examples	See Fig. 3.41-2	$[Cu^I(2,5\text{-}DTH)_2]$- $[Cu^{II}(2,5\text{-}DTH)_2](ClO_4)_{8/3}$ $[Cu_2Cl_3(MeNN)_2]$	CuS, Cu_2S, $K[Cu_4S_3]$

Table 3.41-3 Some mixed-valent Cu^I/Cu^{II} compounds of class I.[6]

Compound	Cu^{II}	Stereochemistry*	Cu^I	Stereochemistry*	ESR(g)
$[Cu(NH_3)_4][CuCl_2].H_2O$	CuN_4Cl_2	ERO	$CuCl_2$	L	-
$[Cu(NH_3)_4][CuBr_2]$	CuN_4Br_2	ERO	$CuBr_2$	L	-
$[Cu(NH_3)_4][CuI_2]$	CuN_4I_2	ERO	CuI_2	L	-
$[Cu(bipy)_2Cl][CuCl_2]_2$	CuN_4Cl	TB	$CuCl_2$	L	2.091
			$CuCl_4$	Td	-
$[Cu(C_{44}N_4H_{60})Cl][CuCl_2]$	CuN_4Cl	SBP	$CuCl_2$	L	2.066, 2.088
$[Cu(Ph_3AsO)_4][CuCl_2]$	CuO_4	SP	$CuCl_2$	-	2.06
$[Cu_2(SO_3)Cu(SO_3)].2H_2O$	CuO_6	ERO	$CuSO_3$	Td	-
$[Cu(NH_3)_3(CN)_4]$	CuN_6	ERO	CuC_3	Tr	-

* ERO = elongated rhombic octahedral;   TB = trigonal bipyramidal;

  SBP = square-based pyramidal;   L = linear;

   Td = tetrahedral;   Tr = trigonal.

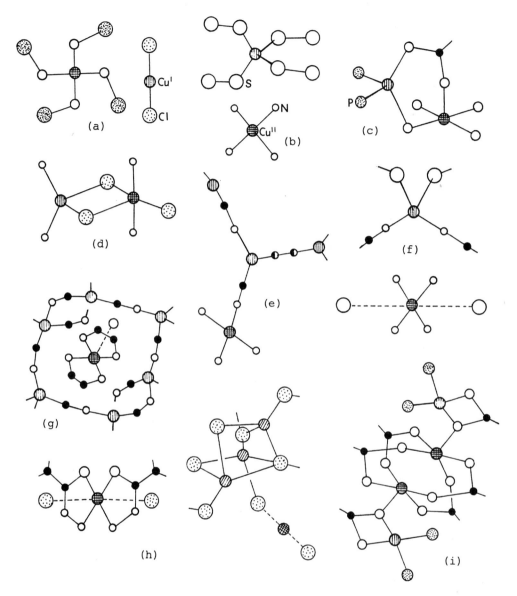

Fig. 3.41-2 Structures of some mixed-valent Cu^I/Cu^{II} compounds:

(a) $[Cu^{II}(Ph_3AsO)_4](Cu^ICl_2)$, (b) $Na_4[Cu^{II}(NH_3)_4][Cu^I(S_2O_3)_2]\cdot NH_3$,

(b) $[Cu_2(acacP)_2(3-MeOC_6H_4CO_2)]$, (d) $[Cu^ICu^{II}(4-metz)_4Cl_3]$,

(e) $[Cu_4^I(CN)_6\{Cu^{II}(NH_3)_2\}]_n$, (f) $Cu_2(NCS)_3(NH_3)_3]$, (g) $[Cu_3(en)_2(CN)_4\cdot H_2O]$

(h) $[Cu^{II}(N\text{-benzoylhydrazine})][Cu_3^ICl_5]$, and (i) $[Cu_4(O_2CMe)_6(Pph_3)_4]$.

The crystal structure of the class II complex $[Cu_2Cl_3(MeNN)_2]$, (where MeNN = 4-methyl-1,8-naphythyridine), has been determined.[8] Both CuN_2Cl_2 chromophores have distorted tetrahedral stereochemistry; they are nearly equivalent crystallographically and bridged by a single Cl^- ligand, resulting

in a short Cu-Cu separation of 2.89Å, as shown in Fig. 3.41-3(a). The spin-
only magnetic moment of 1.99 BM per Cu_2 unit indicates a $Cu^{I/II}$ species.

   In the extreme class III behaviour, two types of structures are
envisaged: clusters and infinite arrays. The double-layer structure of
$K[Cu_4S_3]$, which exhibits the electrical conductivity and the reflectivity
typical of a metal, belongs to class IIIB [Fig. 3.41-3(b)].[9] The class IIIA
behaviour has been sought in the polynuclear $Cu_{4-8}X_n$ clusters of Cu^I,
especially where X = sulfur.

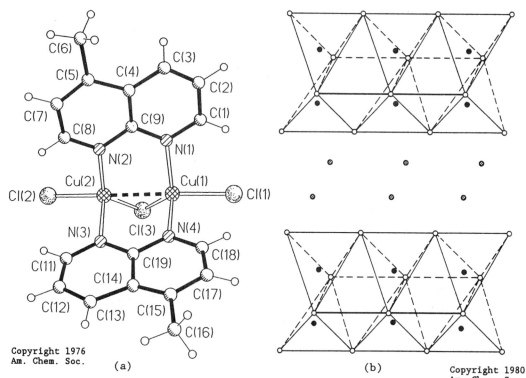

            (a)   (b)

Fig. 3.41-3   (a) Structure of $[Cu_2Cl_3(MeNN)_2]$. (b) Schematic
representation of a portion of adjacent double layers showing the
octahedral holes of $K[Cu_4S_3]$. The shaded circles represent Cu positions,
the open circles S, and the crossed circles K. (After refs. 8 and 9).

4. Mixed-valent compounds of gold[10]

   Many complexes that appear to contain Au^{II} are really mixed-valent Au^I-
Au^{III} compounds.[10] Some early examples were the demonstration that $CsAuCl_3$
is $Cs_2[Au^ICl_2][Au^{III}Cl_4]$, $AuCl_2S(CH_2Ph)_2$ is $[Au^IClS(CH_2Ph)_2][Au^{III}Cl_3S(CH_2Ph)_2]$
and $AuCl(DMG)$ is $[Au^{III}(DMG)_2][Au^ICl_2]$ (DMG = dimethylglyoxime). The
crystal structure of $CsAuCl_3$ is closely related to that of perovskite (section
2.19), the structure being distorted so that instead of six octahedral

neighbours the two kinds of gold atoms form linear $Au^ICl_2^-$ and square-planar $Au^{III}Cl_4^-$ ions, as can be seen in Fig. 3.41-4(a).[11] An X-ray study on $AuCl_2$ has shown that it actually comprises $Au_2^IAu_2^{III}Cl_8$ molecules [Fig. 3.41-4(b)].[12] A number of other well-characterized examples are given in Table 3.41-4.

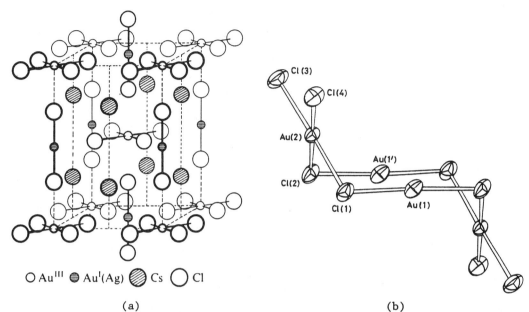

$\bigcirc$ Au^{III} $\ominus$ Au^I(Ag) $\oslash$ Cs $\bigcirc$ Cl

           (a)   (b)

Fig. 3.41-4 (a) Crystal structure of $CsAuCl_3$; $Cs_2Ag^IAu^{III}Cl_6$ also has this structure. (b) Molecular structure of Au_4Cl_8. (After refs. 11 and 12).

Table 3.41-4 Some mixed-valent Au^I-Au^{III} compounds.[10]

Formula	Best formulation	Formula	Best formulation
AuO	$Au^IAu^{III}O_2$	$Rb_2AgAu_3I_8$	$Rb_2Ag[Au^II_2]_2[Au^{III}I_4]$
AuS	$Au^IAu^{III}S_2$	$K_5Au_5(CN)_{10}I_2$	$K_5[Au^I(CN)_2]_4[Au^{III}(CN)_2I_2]$
AuSe	$Au^IAu^{III}Se_2$	$Au_2Cl_4(CO)$	$[Cl_3Au^{III}(\mu\text{-}Cl)Au^I(CO)]$
$RbAuBr_3$	$Rb_2[Au^IBr_2][Au^{III}Br_4]$	$Au_2Cl_4(MeCCMe)_2$	$[Au^I(MeCCMe)_2][Au^{III}Cl_4]$
$KAuI_3$	$K_2[Au^II_2][Au^{III}I_4]$	$Au(mut^*)PPh_3$	$[Au^I(PPh_3)_2][Au^{III}(mut)_2]$
$Rb_3Au_3Cl_8$	$Rb_3[Au^ICl_2]_2[Au^{III}Cl_4]$	$Au(dtc^{**})Br$	$[Au^{III}(dtc)_2][Au^IBr_2]$

* mut = maleonitriledithiolate,   ** dtc = N,N-dibutyldithiocarbamate

The compound $[\{Au^I(CH_2)_2PPh_2\}_2Au^{III}(C_6F_5)_3]$ has been shown to contain a gold(I)-gold(III) bond of 2.572(1)Å unsupported by covalent bridges; the $Au^I\cdots Au^I$ contact is 2.769(1)Å [Fig. 3.41-5(a)].[13] A linear $[Au^{II}]_2$-Au^I-$[Au^{II}]_2$ chain occurs in the centrosymmetric cation [Fig. 3.41-5(b)] of the $[C_6F_5\{Au(CH_2)_2PPh_2\}_2Au(C_6F_5)_2\{Au(CH_2)_2PPh_2\}_2C_6F_5]^+[Au^{III}(C_6F_5)_4]^-$ complex salt; the bond lengths are Au1-Au2 = 2.755(1) and Au2-Au3 = 2.64091)Å.[14]

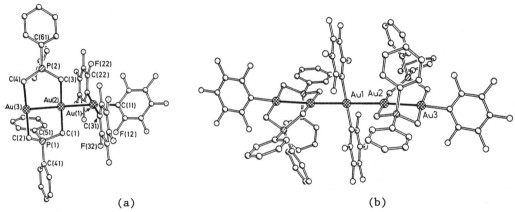

(a)   (b)

Fig. 3.41-5 Structures of two mixed-valent oligonuclear complexes containing linear Au$_x$ chains and unsupported gold-gold bonds. (After refs. 13 and 14).

References

[1]   Chou Kung-du (Zhou Gongdu), *Sci. Sinica (Peking)* **12**, 139 (1963).

[2]   I. Náray-Szabó, G. Argay and P. Szabó, *Acta Crystallogr.* **19**, 180 (1965).

[3]   M.B. Robin and P. Day, *Adv. Inorg. Chem. Radiochem.* **10**, 247 (1967).

[4]   P. Day in D.B. Brown (ed.), *Mixed-Valence Compounds*, D. Reidel, Dordrecht, 1980, pp. 3-24.

[5]   W. Levason and M.D. Spicer, *Coord. Chem. Rev.* **76**, 45 (1987).

[6]   B.J. Hathaway in G. Wilkinson, R.D. Gillard and J.A. McCleverty (eds.), *Comprehensive Coordination Chemistry*, Pergamon Press, Oxford, 1987, Vol. **5**, chap. 53.

[7]   P. Day, *Endeavour*, **20**, 45 (1970).

[8]   D. Gatteschi, C. Mealli and L. Sacconi, *Inorg. Chem.* **15**, 2774 (1976).

[9]   D.B. Brown, J.A. Zubieta, P.A. Vella, J.T. Wrobleski, T. Watt, W.E. Hatfield and P. Day, *Inorg. Chem.* **19**, 1945 (1980).

[10] R.J. Puddephatt in G. Wilkinson, R.D. Gillard and J.A. McCleverty (eds.), *Comprehensive Coordination Chemistry*, Pergamon Press, Oxford, 1987, Vol. **5**, chap. 55.

[11] A.F. Wells, *Structural Inorganic Chemistry*, 5th ed., Clarendon Press, Oxford, 1984, p. 467.

[12] D.B. Dell'Amico, F. Calderazzo, F. Marchetti and S. Merlino, *J. Chem. Soc., Dalton Trans.*, 2257 (1982).

[13] R. Usón, A. Laguna, M. Laguna, M.T. Tartón and P.G. Jones, *J. Chem. Soc., Chem. Commun.*, 740 (1988).

[14] R. Usón, A. Laguna, M. Laguna, J. Jiménez and P.G. Jones, *Angew. Chem. Int. Ed. Engl.* **30**, 198 (1991).

3.42   Tri-iodoheptakis(tri-p-fluorophenyl-phosphine)undecagold
$$Au_{11}I_3[P(p\text{-}FC_6H_4)_3]_7$$

Crystal Data

Trigonal, space group $R3$ (No. 146)

$a = 15.96(2)$Å, $\alpha = 108.4(1)°$; $Z = 1$

Atom*	x	y	z
Au(1)	0	0	0
Au(2)	-.1548	-.1548	-.1548
Au(3)	-.1319	.0482	-.0979
Au(4)	.0513	.0865	-.1099
Au(5)	.0786	.1982	.0856
I	.1052	.1560	-.2215
P(2)	-.2865	-.2865	-.2865
P(3)	-.2473	.0942	-.1661
P(5)	.1311	.3664	.1486

* The parameters of C and F atoms are omitted.

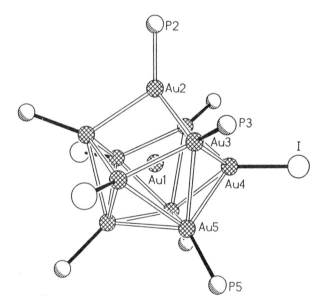

Fig. 3.42-1   Structure of the $Au_{11}I_3P_7$ moiety.   (Adapt from ref. 1).

Crystal Structure

The crystal consists of discrete molecules $Au_{11}I_3[P(p\text{-}FC_6H_4)_3]_7$ each lying on a crystallographic three-fold rotation axis.[1] An inspection of all non-bonding interactions within each molecule and between different molecules does not reveal unusually short contacts.

The metal atom cluster is an incomplete icosahedron in which one triangular face has been substituted by a single gold atom, Au(2), with retention of C_3 symmetry (Fig. 3.42-1). The eleven Au atoms fall into five non-equivalent sets exhibiting different coordination patterns as well as different coordination numbers: the central atom Au(1), which lies on the three-fold axis, is ten-coordinate; the apical atom Au(2), also on the three-fold axis, is five-coordinate; three Au(3) and three Au(4) atoms, which form an equatorial chair-like hexagon, are six-coordinate; and finally, the three basal Au(5) atoms are seven-coordinate. The coordination polyhedron of the inner atom is bordered by ten triangular faces, and by three irregularly shaped squares sharing an apex in Au(2). Au(1) is the first example of a metal atom whose coordination sphere is occupied exclusively by other metal atoms in a discrete molecular compound.

The Au-Au interactions are: mean center-to-periphery 2.68Å, peripheral distances 2.836(4)-3.187(3)Å, mean 2.98Å. There are three independent sets of Au-P bonds in the range 2.21(1)-2.29(1)Å. The Au-I distances are 2.600(5)Å.

Remarks

1. Gold clusters[2,3]

Many gold clusters have had their structures determined, for example:

Au_4 : $Au_4(PPh_3)_4I_2$;

Au_5 : $[Au_5(dppm)_3(dppm-H)]^{2+}$;

Au_6 : $[Au_6(PPh_3)_6]^{2+}$, $[Au_6(dppp)_4]^{2+}$, $[Au_6(Ptol_3)_6]^{2+}$;

Au_7 : $[Au_7(PPh_3)_7]^+$;

Au_8 : $[Au_8(PPh_3)_7]^{2+}$, $[Au_8(PPh_3)_6I]^+$, $[Au_8(PPh_3)_8]^{2+}$;

Au_9 : $[Au_9(PPh_3)_8]^+$, $[Au_9(PAr_3)_8]^{3+}$;

Au_{10}: $[Au_{10}Cl_3(PCy_2Ph)_6]^+$;

Au_{11}: $Au_{11}(PAr_3)_7I_3$, $[Au_{11}(PPh_3)_8X_2]^+$, $Au_{11}(PPh_3)_7(SCN)_3$;

Au_{13}: $[Au_{13}(PPhMe_2)_{10}Cl_2]^{3+}$, $[Au_{13}(dppm)_6]^{5+}$; and

Au_{55}: $Au_{55}(PPh_3)_{12}Cl_6$.

The structures of some low-nuclearity gold clusters with the general formula $[Au_mL_m]^{x+}$ are shown in Fig. 3.42-2. Their geometries are critically dependent on the electron count.

High-nuclearity gold clusters having the general formula $[Au(AuL)_m]^{x+}$ may be divided into two sub-classes, according to whether the outer gold atoms adopt a spherical or toroidal topology. Mingos *et al.* have concluded that "spherical" clusters [Fig. 3.42-3(a)] are characterized by (12m+18) cluster valence electrons. Thus all 9 valence orbitals of the central gold atom are involved in bonding with the radial sp hybrids of the outer AuL groups, each of which contributes two metal-ligand bonding and ten non-bonding (d) electrons to the total count. "Toroidal" clusters [Fig. 3.42-3(b)] are characterized by (12m+16) cluster valence electrons, since there is no low-lying cluster orbital which can be symmetry-matched to one of the p orbitals of the central gold atom.

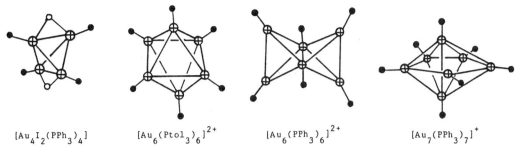

$[Au_4I_2(PPh_3)_4]$ $[Au_6(Ptol_3)_6]^{2+}$ $[Au_6(PPh_3)_6]^{2+}$ $[Au_7(PPh_3)_7]^+$

Fig. 3.42-2 Structure of some low-nuclearity gold clusters. (After ref. 2).

(a) Spherical clusters: (12m + 18) valence electrons

$[Au_9(PR_3)_8]^+$ $[Au_{11}I_3(PR_3)_7]$ $[Au_{13}Cl_2(PR_3)_{10}]^{3+}$

(b) Toroidal clusters: (12m + 16) valence electrons

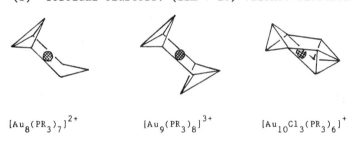

$[Au_8(PR_3)_7]^{2+}$ $[Au_9(PR_3)_8]^{3+}$ $[Au_{10}Cl_3(PR_3)_6]^+$

Fig. 3.42-3 Structures of some high-nuclearity gold clusters $[Au(AuPR_3)_m]^{x+}$. (After ref. 2).

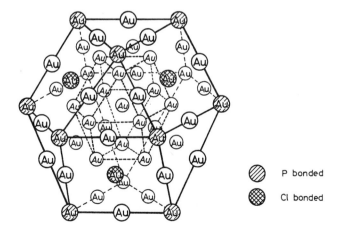

P bonded

Cl bonded

Fig. 3.42-4 Probable structure of the Au₅₅ cluster. (After ref. 23).

Schmid has synthesized the highest-nuclearity gold cluster known to date and formulated it as $Au_{55}(PPh_3)_{12}Cl_6$.[4]   The proposed model (Fig. 3.42-4) has a Au_{13} core, which is a centered icosahedron or cuboctahedron, enclosed within an Au_{42} cuboctahedral shell that may be fluxional through an M_{42} icosahedron. The phosphine ligands are postulated to terminate at the twelve vertices of the outer polyhedron, and the Cl ligands to radiate from the metal atoms central to the six square faces of the cuboctahedron.

2. Supraclusters based on vertex-sharing icosahedra

The crystal structure of $[Au_{13}Ag_{12}(PPh_3)_{12}Cl_6]Cl_m \cdot n$EtOH has been determined.[5]   The idealized structure of the 25-atom cluster $[Au_{13}Ag_{12}(PPh_3)_{12}Cl_6]^{m+}$ is shown in Fig. 3.42-5(a). The idealized metal framework consists of a pair of icosahedra, each centered by an Au atom, with a shared vertex (the central Au atom) and a common five-fold axis. About this axis the metal atom layer sequences are Ag-Au_5-Au-Ag_5-Au-Ag_5-Au-Au_5-Ag with successive M_5 pentagons mutually staggered, so that the configuration of the metal framework may be designated as **S-S-S** (**S** stands for staggered). The twelve phosphine ligands are terminal to the pentagonal-pyramidal Au_5Ag caps at each end of the molecule, while the Cl ligands doubly bridge six of the ten

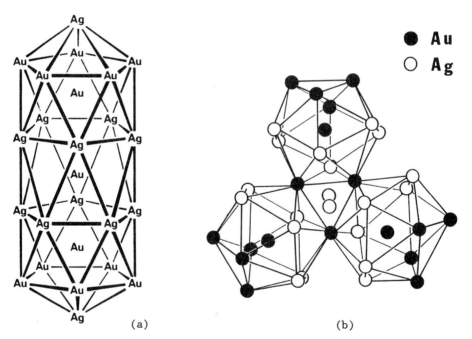

(a)   (b)

Fig. 3.42-5 (a) Idealized S-S-S framework structure of the $Au_{13}Ag_{12}$ cluster in $[Au_{13}Ag_{12}(PPh_3)_{12}Cl_6]^{m+}$. (b) The metal framework of the 38-atom cluster $Au_{18}Ag_{20}(Pp\text{-}tol_3)_{12}(\mu\text{-}Cl)_6(\mu_3\text{-}Cl)_6Cl_2$; some metal-metal bonds have been omitted for clarity. (After ref. 3 and 9).

edges of the Ag_{10} pentagonal antiprism around the equator of the cluster. The M-M distances are in the range 2.5-3.3Å, except for two of the Cl-bridged edges of 3.5Å. The cluster cation in $[Au_{13}Ag_{12}(Pp\text{-}tol_3)_{10}(\mu\text{-}X)_6X_2]PF_6$ (X = Br, Cl) has the same kind of S-S-S metal framework; here each gold atom in a pentagon is coordinated by one phosphine ligand, two silver-atom pentagons are bridged by halides, and the top and bottom silver atoms are each bonded by one halide.[6] On the other hand, the framework of the 25-atom cluster cations in $[Au_{13}Ag_{12}(Pp\text{-}tol_3)_{10}(\mu\text{-}Cl_5)Cl_2]SbF_6$[6] and $[Au_{13}Ag_{12}(PPh_3)_{10}(\mu\text{-}Cl)_6Cl_2]Cl$. $nEtOH$[7] take the S-E-S configuration, where E refers to the mutually eclipsed arrangement of the two Ag_5 pentagons.

Similar 37-atom, 38-atom, and 46-atom bimetallic frameworks have been found in the Au-Ag cluster compounds $[Au_{18}Ag_{19}(Pp\text{-}tol_3)_{12}(\mu\text{-}Br)_6(\mu_3\text{-}Br)_4Br](AsF_6)_2$,[6,8] $[Au_{18}Ag_{20}(Pp\text{-}tol_3)_{12}(\mu\text{-}Cl)_6(\mu_3\text{-}Cl)_6Cl_2]$,[6,9] and $[Au_{22}Ag_{24}(PPh_3)_{12}Cl_{10}]$,[10] respectively. The most interesting structural feature of this series of 25-, 37-, 38- and 46-atom clusters is that they can be built from 13-atom, centered icosahedral units (Fig. 3.42-6).[11] Thus the 25-atom (2x13 - 1 = 25) cluster is composed of *two* icosahedra sharing a vertex (designated as s_2); the 37- and 38-atom clusters are constructed from *three* icosahedra sharing three vertices in a cyclic manner (3x13 - 3 = 36, s_3) plus one and two capping atoms, respectively [Fig. 3.42-5(b)]. The 46-

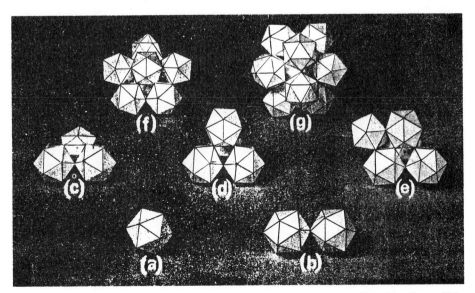

Fig. 3.42-6 Supraclusters s_n formed by *n*-centered icosahedra (nuclearity given in parentheses): (a) $s_1(13)$, (b) $s_2(25)$, (c) $s_3(36)$, (d) $s_4(46)$, (e) $s_5(56)$, (f) $s_7(76)$, (g) $s_{12}(127)$. (After ref. 11).

atom cluster s_4 comprises four Au_7Ag_6 gold-centered icosahedra sharing six gold vertices; the twelve Ph_3P ligands are coordinated to the twelve peripheral gold atoms, and the ten chloride ligands to the twenty-four silver atoms.

The structural characteristics of this series of bimetallic Au-Ag clusters led naturally to the concept of "cluster of clusters", which may open up new pathways to novel high-nuclearity "supraclusters" using vertex-, edge-, and face-sharing and/or close-packing of smaller cluster units as building blocks.[11] It is predicted that trigonal-bipyramidal (s_5) and pentagonal-bipyramidal (s_7) arrays of vertex-sharing icosahedra will give rise to 56- and 76-atom clusters (nuclearity = $10n + 6$), respectively, as depicted in Fig. 3.42-6(e) and (f). In the 127-atom cluster s_{12} [Fig. 3.42-6(g)], twelve centered icosahedra share thirty corners and the central hole thus created is filled by one additional atom.

The $Au_{55}(PPh_3)_{12}Cl_{16}$ cluster of Schmid has been re-formulated as $Au_{67}(PPh_3)_{14}Cl_8$ in a ^{252}Cf plasma desorption mass spectral study.[12] The postulated structure consists of two tetrahedral s_4(46) units sharing *two* icosahedra to form a centrosymmetric s_6(67) cluster with eleven shared vertices (Fig. 3.42-7). The observed high mass zones are explicable in terms of the fragmentation of $Au_{67}(PPh_3)_{14}Cl_8$ (s_6; *ca.* 16600 *m/z*), $Au_{46}(PPh_3)_{12}Cl_6$(S_4, *ca.* 12800 *m/z*) and $Au_{25}(PPh_3)_{12}Cl_6$ (s_2, *ca.* 8590 *m/z*).

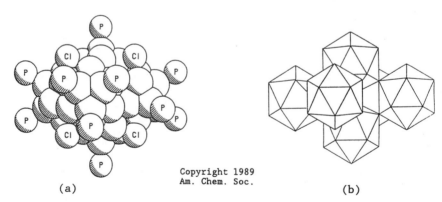

(a)   (b)

Fig. 3.42-7 Space-filling representation (a) and vertex-sharing icosahedral framework (b) of the proposed s_6 structure for $Au_{67}(PPh_3)_{14}Cl_8$. (After ref. 12).

3. Gold clusters based on polyauriomethanes

Recent work on polyauriomethanes has revealed that highly aurated compounds containing two or more gold atoms on *one* carbon atom readily bind further gold(I) units LAu^+ (L = R_3P), a property termed "aurophilicity".[12] The coordination number of the polyaurated C atom is thereby increased beyond four, and close Au...Au contacts are formed. Novel clusters synthesized according to this strategy include $(Me_3P)C(AuPPh_3)_3$,[13] the cationic gold clusters in the (triauriomethyl)oxazoline complex **1**,[14] $[(Ph_3PAu)_5C]^+BF_4^-$ (**2**),[15] and $[\{C_6H_4(CH_2CH_2PPh_2Au)_2\}_3C](BF_4)_2$ (**3**).[16]

The cation in **1** possesses a crystallographic C_2 axis through Au4 and Au5 (Fig. 3.42-8). The pentacoordinated C1 and C1' atoms are each in a tetragonal-pyramidal environment (Au1...Au2 2.839, Au2...Au3 2.839, Au3...Au5 2.799, Au5...Au1 2.893; average Au-C 2.16Å, and Au5 is the common corner of the two pyramids. The Au4...Au5 distance is 2.811Å, so that all eight Au atoms are components of a single cluster. The novel counterion $[(CH_3)_3SiF_2]^-$ nearly attains D_{3h} molecular geometry.

The cation in **2** has crystallographic C_3 symmetry with the five gold atoms at the corners of a slightly distorted, C-centered pyramid. The average Au-C bond distance is 2.084Å. The $Au_{ax}...Au_{eq}$ and $Au_{eq}...Au_{eq}$ contacts are 2.900 and 3.600Å, respectively. The analogous tris-chelated cation in **3** features a distorted carbon-centered octahedron exhibiting C_2 symmetry [Fig. 3.42-9(a)]. The average Au-C bond distance is 2.13Å and Au...Au distances lie between 2.942 and 3.132Å.

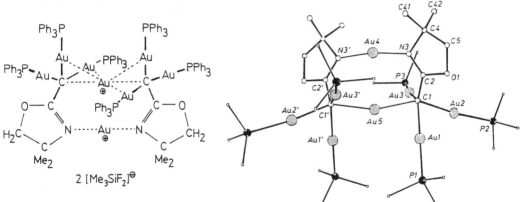

Fig. 3.42-8 Structural formula of compound **1** and molecular geometry of its dicationic gold cluster. Each pair of primed and unprimed atoms are related by two-fold rotational symmetry. (After ref. 14).

Recent research has demonstrated that nitrogen, like carbon, can attain penta- or hexa-coordination in polyaurated species with the gold(I)-phosphine units clustering around it. The extremely stable dication $[(Ph_3PAu)_5N]^{2+}$ obtained in the reaction:

$$[(Ph_3PAu)_4N]^+BF_4^- + [Ph_3PAu]^+BF_4^- \xrightarrow[\text{THF}]{(Me_2N)_3PO} [(Ph_3PAu)_5N]^{2+} + 2BF_4^-$$

has a nitrogen-centered trigonal-bipyramidal structure [Fig. 3.42-9(b)] in the crystalline adduct $[(Ph_3PAu)_5N](BF_4)_2 \cdot 2THF$.[17]   The trication $[(Ph_3PAu)_6N]^{3+}$ has been prepared *via* a different route.

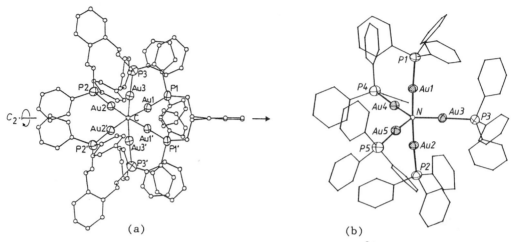

(a)   (b)

Fig. 3.42-9   (a) Structure of $[\{C_6H_4(CH_2CH_2PPh_2Au)_2\}_3C]^{2+}$.   (b) Structure of $[(Ph_3PAu)_5N]^{2+}$.   Selected bond distances and angles: N-Au1 2.116(6), N-Au2 2.112(6), N-Au3 2.051(7), N-Au4 2.081(6), N-Au5 2.066(7)Å; Au1-N-Au2 174.9(4), Au3-N-Au4 129.0(3), Au4-N-Au5 109.6(3), Au3-N-Au5 121.4(3)°.   (After refs. 16 and 17).

4. Relativistic effects in gold clusters

    The importance of relativistic effects in chemistry has gained attention recently.[18-20]   Among transition metals, the magnitude of relativistic stabilization of outer-shell s electrons reaches a maximum for gold as indicated by, for example, the stability of the auride ion and the yellow colour of metallic gold.   Due to the relativistic expansion of the 5d and contraction of the 6s orbitals, there is considerable interaction between formally nonbonded gold atoms separated by more than the 2.88Å interatomic distance found in the metal.   A small positive Au...Au overlap population of 0.0435 has been calculated for an Au_2^{2+} unit at an Au...Au separation of 3.04Å.[21]   Also, several aggregates of LAu^IX neutral molecules with short (3.07-3.72Å) Au...Au intermolecular contacts have been observed.

    In the dimeric Au^{III} complex *trans,trans*-$[Au(CH_2)_2PPh_2]_2Br_4$, the eight-membered organometallic ring is in a chair conformation (C_{2h} symmetry), with the P atoms at the tips of the chair and unexceptional bond lengths and angles (Fig. 3.42-10).[22]   The Au_2Br_4 unit is planar within 0.016Å, with an Au...Au distance of 3.069(2)Å, and long Au-Br bonds of 2.535(4) and 2.576(4)Å. The "*trans*" Br-Au-Br angle has the unusual value of 149.5(2)°, being nearly

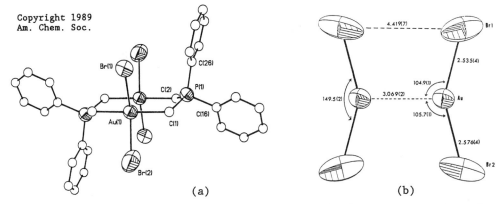

(a)   (b)

Fig. 3.42-10 (a) Molecular structure and (b) the Au_2Br_4 portion of *trans,trans*-[Au(CH₂)₂PPh₂]₂Br₄. (After ref. 22).

equally bisected by the Au...Au axis. This geometry is halfway between an ideal 4-coordinate, square-planar Au^{III} center (Br-Au-Br = 180°) and a 5-coordinate, trigonal-bipyramidal structure (Br-Au-Br = 120°). The latter would require the existence of an Au-Au bond in the equatorial plane occupying the fifth coordination site. The observed deviation from square-planar geometry is too large to be accounted for by packing forces influencing the Br atom positions. The shortening of the Au...Au distance is consistent with the square-planar dsp^2 to trigonal-bipyramidal d^2sp^2 transformation.

References

[1] P. Bellon, M. Manassero and M. Sansoni, *J. Chem. Soc., Dalton Trans.*, 1481 (1972).

[2] D.M.P. Mingos and R.L. Johnston, *Structure and Bonding*, **68**, 31 (1987); D.M.P. Mingos and A.S. May in D.F. Shriver, H.D. Kaesz and R.D. Adams (eds.), *The Chemistry of Metal Cluster Complexes*, VCH Publishers, New York, 1990, chap. 2; K.P. Hall and D.M.P. Mingos, *Prog. Inorg. Chem.* **32**, 237 (1984).

[3] I.G. Dance in G. Wilkinson, R.D. Gillard and J.A. McCleverty (eds.), *Comprehensive Coordination Chemistry*, Vol 1, Pergamon Press, Oxford, 1987, chap. 4.

[4] G. Schmid, N. Klein, L. Korste, U. Kreibig and D. Schönauer, *Polyhedron*, **7**, 605 (1988); G. Schmid, R. Pfeil, R. Boese, F. Bandermann, S. Meyer, G.H.M. Calis and J.W.A. van der Velden, *Chem. Ber.* **114**, 3634 (1981).

[5] B.K. Teo and K. Keating, *J. Am. Chem. Soc.* **106**, 2224 (1984).

[6] M.-C. Hong, Z.-Y. Huang, H.-Q. Liu, B.K. Teo and D.-B. Huang, *Jiegou Huaxue (J. Struct. Chem.)* **9**, 94 (1990).

[7] M. Hong, Z. Huang, B. Kang, X. Lei and H. Liu, *Inorg. Chim. Acta* **168**, 163
     (1990); erratum, **173**, 128 (1990).

[8] B.K. Teo, M.C. Hong, H. Zhang and D.B. Huang, *Angew. Chem. Int. Ed.
     Engl.* **26**, 897 (1987).

[9] B.K. Teo, M.C. Hong, H. Zhang, and D.B. Huang and X. Shi, *J. Chem. Soc.,
     Chem. Commun.*, 204 (1988).

[10] B.K. Teo, X. Shi and H. Zhang, *Chem. Eng. News* **67**(2), 6 (1989).

[11] B.K. Teo, *Polyhedron* **7**, 2317 (1988).

[12] J.P. Fackler, Jr., C.J. McNeal, R.E.P. Winpenny and L.H. Pignolet, *J. Am.
     Chem. Soc.* **111**, 6434 (1989).

[13] H. Schmidbaur, F. Scherbaum, B. Huber and G. Müller, *Angew. Chem. Int.
     Ed. Engl.* **27**, 419 (1988).

[14] F. Scherbaum, B. Huber, G. Müller and H. Schmidbaur, *Angew. Chem. Int.
     Ed. Engl.* **27**, 1542 (1988).

[15] F. Scherbaum, A. Grohmann, G. Müller and H. Schmidbaur, *Angew. Chem. Int.
     Ed. Engl.* **28**, 463 (1989).

[16] O. Steigelmann, P. Bissinger and H. Schmidbaur, *Angew. Chem. Int. Ed.
     Engl.* **29**, 1399 (1990).

[17] A. Grohmann, J. Riede and H. Schmidbaur, *Nature (London)* **345**, 140 (1990).

[18] P. Pyykkö and J.-P. Desclaux, *Acc. Chem. Res.* **12**, 276 (1979).

[19] P. Pyykkö, *Chem. Rev.* **88**, 563 (1988).

[20] K.S. Pitzer, *Acc. Chem. Res.* **12**, 271 (1979).

[21] Y. Jiang, S. Alvarez and R. Hoffmann, *Inorg. Chem.* **24**, 749 (1985).

[22] R.G. Raptis, J.P. Fackler, Jr., H.H. Murray and L.C. Porter, *Inorg. Chem.*
     28, 4057 (1989).

[23] G. Schmid, *Structure and Bonding* **62**, 51 (1985).

Note The historical development of theoretical models in metal cluster
     chemistry is described in D.M.P. Mingos, *Pure Appl. Chem.* **63**, 807 (1991).
     For a balanced account of the entire field, including non-metallic
     clusters, see D.M.P. Mingos and D.J. Wales, *Introduction to Cluster
     Chemistry*, Prentice-Hall, Englewood Cliffs, NJ, 1991. Recent advances
     are covered in D.F. Shriver, H.D. Kaesz and R.D. Adams (eds.), *The
     Chemistry of Metal Cluster Complexes*, VCH Publishers, Weinheim, 1990.

     A theoretical discussion of the shell structure of clusters is given in
     T.P. Martin, T. Bergmann, H. Göhlich and T. Lange, *J. Phys. Chem.* **95**,
     6421 (1991).

     The 25-atom cluster in $[Au_{13}Ag_{12}(PPh_3)_{10}(\mu\text{-}Br_6)Br_2]SbF_6 \cdot 20EtOH$ has an
     exact **S-E-S** configuration; see B.K. Teo, X. Shi and H. Zhang, *J. Am.
     Chem. Soc.* **113**, 4329 (1991). A similar C_2 cluster having a nearly **S-E-S**
     configuration occurs in $[Au_{13}Ag_{12}(Pp\text{-}tol_3)_{10}(\mu\text{-}Cl_5)Cl_2](SbF_6)_2 \cdot nEtOH$, as
     reported in B.K. Teo and H. Zhang, *Inorg. Chem.* **30**, 3116 (1991).

3.43 Calomel (Mercury(I) Chloride, Merchlite)
Hg_2Cl_2

Crystal Data

Tetragonal, space group $I4/mmm$ (No. 139)

$a = 4.482(2)$, $c = 10.910(3)$Å, $Z = 2$

Atom	x	y	z
Hg	0	0	.1158(3)
Cl	0	0	.3380(4)

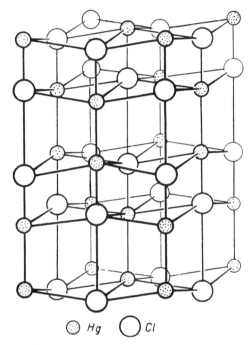

$\odot$ Hg $\bigcirc$ Cl

Fig. 3.43-1 Crystal structure of Hg_2Cl_2. (After ref. 3).

Crystal Structure

The crystal structure of calomel consists of linear Cl-Hg-Hg-Cl groups each lying on a four-fold axis. The Cl-Hg-Hg-Cl molecules are link-coupled by their Hg-Cl ends to form a framework with maximum Cl...Cl separations. Thus, parallel to (001) there are double corrugated layers (Fig. 3.43-1), but the strength of the Hg-Hg bonds between these is almost equal to the strength of the bonds in the perpendicular (100) direction.

Every Hg atom is surrounded by five Cl atoms plus one other Hg atom. The interatomic distances are: Hg-Hg = 2.526(6)Å, Hg-Cl = 2.43(4)Å (one) and 3.209(6)Å (four), and Cl...Cl = 3.52Å.[1]

The structure of Hg_2Cl_2 was recently refined using the Rietveld profile-fitting technique on neutron powder data.[2] The parameters obtained [a =

4.4795(5), c = 10.9054(9)Å; z(Hg) = .1190(1), z(Cl) = .3356(8); Hg-Hg = 2.5955(2), Hg-Cl = 2.3622(2), Hg...Cl = 3.2059(3)Å] are in good agreement with the single-crystal X-ray results.

Remarks

1. Experimental evidence for the existence of the dinuclear Hg_2^{2+} ion

Most of the mercury(I) compounds are insoluble or only sparingly soluble in water. In all cases the dinuclear Hg_2^{2+} ion is present rather than mononuclear Hg^+. The evidence for this is overwhelming and includes the following:[4]

a) In crystalline mercury(I) compounds, instead of the sequence of alternating M^+ and X^- expected for MX compounds, Hg-Hg pairs are found in which the separation, though not constant, lies in the range 2.50-2.70Å,[4] which is shorter than the Hg-Hg separation of 3.00Å found in the metal itself.

b) The Raman spectra of mercury(I) compounds in the solid state and in solution always exhibit the characteristic Hg-Hg stretching vibration.

c) Mercury(I) compounds are diamagnetic, whereas the monatomic Hg^+ ion has a $d^{10}s^1$ configuration and would therefore be paramagnetic.

d) The measured emfs of concentration cells of mercury(I) salts are only explicable on the assumption that a two-electron transfer is involved.

e) It is found that "equilibrium constants" are in fact only constant if the concentration $[Hg_2^{2+}]$ is employed rather than $[Hg^+]^2$.

2. Structures of Hg_2^{2+} salts

Mercury(I) halides, Hg_2X_2, are most common and readily obtained. Hg_2F_2 is decomposed by water, whereas the other halides are insoluble in water and are usually obtained by precipitation from an aqueous mercury(I) nitrate solution with the addition of an alkali halide. Although F, Cl and Br have different electronegativities, the Hg-Hg distances in the halides are not significantly different.

A number of stable oxysalts of mercury(I), for example nitrate, chlorate, bromate, iodate, sulfate, selenate, perchlorate, and carbonate are readily prepared. The structures of those mercury(I) salts that have been determined generally show the presence of an essentially linear O-Hg-Hg-O grouping, the terminal oxygens being provided either by water molecules or by the oxyanions. When the oxysalts react with a nitrogen ligand, such as 4-cyano-pyridine, the N-Hg-Hg-N unit is obtained.

The structural data of some mercury(I) compounds with Hg_2 pairs in the structure are listed in Table 3.43-1.

Table 3.43-1   Structural data of some compounds with Hg_2 pairs in the structure.[5,6]

Compound	Hg-Hg/Å	Structural feature
Hg_2F_2	2.507	F-Hg-Hg-F, linear, Hg-F 2.14Å
Hg_2Cl_2	2.526	Cl-Hg-Hg-Cl, linear, Hg-Cl 2.43Å
Hg_2Br_2	2.49	Br-Hg-Hg-Br, linear, Hg-Br 2.71Å
$Hg_2(NO_2)_2$	2.516	O-Hg-Hg-O, Hg-Hg-O 174.2°, Hg-O 2.244
$Hg_2(H_2O)_2(NO_3)_2$	2.54	H_2O-Hg-Hg-OH_2, linear, Hg-O 2.15Å
$Hg_2(BrO_3)_2$	2.507	O-Hg-Hg-O, linear, Hg-O 2.16Å
Hg_2SO_4	2.50	O-Hg-Hg-O, Hg-Hg-O 164.9°, Hg-O 2.24Å
Hg_2SeO_4	2.51	O-Hg-Hg-O, Hg-Hg-O 160°, Hg-O 2.21Å
$Hg_2(ClO_4)_2 \cdot 4H_2O$	2.50	O-Hg-Hg-O, linear, Hg-O 2.1Å
$Hg_2SiF_6 \cdot 2H_2O$	2.495	H_2O-Hg-Hg-OH_2, Hg-Hg-O 170.9°, Hg-O 2.20Å
$Hg_2(4\text{-}CN\text{-}py)_2(ClO_4)_2$	2.498	N-Hg-Hg-N, Hg-Hg-N 176°, Hg-N 2.21Å
$Hg_2(3\text{-}Cl\text{-}py)_2(ClO_4)_2$	2.487	N-Hg-Hg-N, Hg-Hg-N 167.4°, Hg-N 2.21Å
$[(Hg_2)_2O(NO_3)]NO_3 \cdot HNO_3$	2.573(av.)	Hg-Hg 2.513(3), 2.502(2), 2.507(3)Å
$[(Hg_2)_5(OH)_4(NO_3)_2](NO_3)_4$	2.502(av.)	Hg-Hg 2.500(2), 2.495(1), 2.511(1)Å
$[Hg_2(OHg)_2](NO_3)_2$	2.510(2)	

The crystal structures of a series of basic mercury(I) nitrates $[(Hg_2)_2O(NO_3)]NO_3 \cdot HNO_3$ (1), $[(Hg_2)_5(OH)_4(NO_3)_2](NO_3)_4$ (2), and $[Hg_2(OHg)_2](NO_3)_2$ (3), have been determined.[7] In the structure of (1) the principal feature is an infinite chain with the solvated HNO_3 molecule hydrogen-bonded to one nitrate ion, in the structure (2) a finite four-oxonium-link chain, and in the structure of (3) an infinite folded layer containing both Hg(I) and Hg(II) ions (Fig 3.43-2). The Hg-Hg distances in these compounds are listed in Table 3.43-1.

3. Polymercury cations

Table 3.43-2   Structural data of some polymercury clusters.[7,8]

Compound	Geometry of Hg_n	Hg-Hg (Å)
$Hg_3(AsF_6)_2$	Hg_3, linear	2.552(4)
$Hg_3(AlCl_4)_2$	Hg_3, linear	2.56
$Hg_4(AsF_6)_2$	Hg_4, linear	2.70(inner), 2.57(outer)
$Hg_{2.85}AsF_6$	Hg_∞, linear	2.64
$[HgOs_3(CO)_{11}]_3$	Hg_3, triangle	3.122(3), 3.082(3), 3.097(3)
$Hg_3(\mu\text{-}dppm)_3(SO_4)_2$	Hg_3, triangle	2.764(1), 2.764(1), 2.802(1)
$[Ru_{18}(C)_2(CO)_{42}Hg_3]^{2-}$	Hg_3, triangle	2.919(5), 2.922(5), 2.934(6)
$[HgMn(CO)_2C_5H_4Me]_4$	Hg_4, square	2.888(2)
$Hg_6[Rh(PMe_3)_3]_4$	Hg_6, octahedron	3.131(3)-3.149(3)
$Hg_9[Co(CO)_3]_6$	Hg_9, tricapped trigonal prism	3.094(av. trig.-trig.)
		3.151(av. trig.-sq.)

The structures of many polymercury clusters have been determined.   Their structural data are listed in Table 3.43-2.

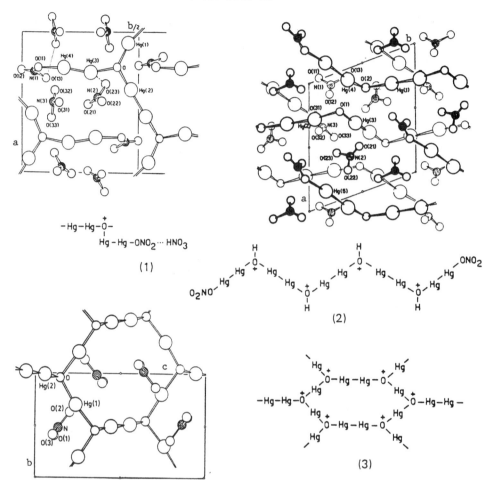

Fig. 3.43-2 Structures of $[(Hg_2)_2O(NO_3)]NO_3 \cdot HNO_3$ (1), $[(Hg_2)_5(OH)_4-(NO_3)_2](NO_3)_4$ (2) and $[Hg_2(OHg)_2](NO_3)_2$ (3). (After ref. 7).

The reduction of Hg^{2+} in a $HgCl_2:Hg:AlCl_3$ mixture of 1:2:2 molar ratio at 240°C after six days yielded $Hg_3(AlCl_4)_2$, which contains the Hg_3^{2+} ion with the following structure:[9]

$$\left[Hg \overset{2.551Å}{\underset{174.4°}{\rule{2cm}{0.4pt}}} Hg \overset{2.562Å}{\rule{2cm}{0.4pt}} Hg \right]^{2+}$$

$Hg_3(AsF_6)_2$ formed from mercury and arsenic pentafluoride contains the linear and symmetrical Hg_3^{2+}:[10]

$$\left[Hg \overset{2.552Å}{\rule{2cm}{0.4pt}} Hg \overset{2.552Å}{\rule{2cm}{0.4pt}} Hg \right]^{2+}$$

The reaction of mercury with AsF_5 in liquid sulfur dioxide yields dark red $Hg_4(AsF_6)_2$, which contains the centrosymmetric Hg_4^{2+} ion.[11]

$$Hg \cdots Hg \underset{124°}{---} Hg \overset{2.58Å}{\underset{177°}{------}} Hg \overset{2.62}{------} Hg \overset{2.98Å}{\cdots} Hg$$

The molecular structure of $[M_{18}Hg_3(C)_2(CO)_{42}]^{2-}$ (M = Ru, Os) has virtual C_3 symmetry and the mean twist angle of 30° between the central Hg_3 triangle and the two adjacent Ru_3 triangles (which together form a "spiral" linking unit) results in the two Ru_9 faces being exactly staggered with a total twist angle of 60° between them [Fig. 3.43-3(a)].[12] The metal framework of the cluster $Hg_9[Co(CO)_3]_6$ is best described as a trigonal prism

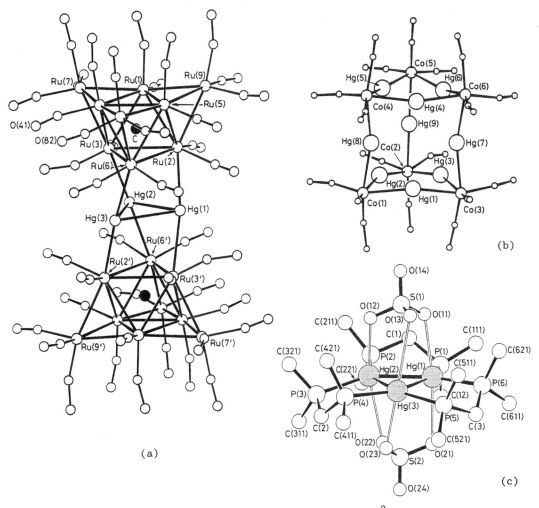

Fig. 3.43-3 Structures of (a) $[Ru_{18}Hg_3(C)_2(CO)_{42}]^{2-}$, (b) $Hg_9[Co(CO)_3]_6$ and (c) $Hg_3(\mu\text{-dppm})_3(SO_4)_2$. (After refs. 12-14).

with the Co atoms at each corner and a Hg atom at the mid-point of each edge
[Fig. 3.43-3(b) and Fig. 3.43-4(d)][13]. Mercury atoms Hg(1)-Hg(6) on the
edges of the triangular faces are distorted from linearity by being bent away
from the center of the triangle and inside the two Co_3 planes. Mercury atoms
Hg(7)-Hg(9) on the edges of the square faces are bent towards the center of
the cluster. The structure of $Hg_3(\mu\text{-dppm})_3(SO_4)_2$ has a triangular Hg_3^{4+}
cluster with all the edges spanned by bridging dppm ligands [Fig. 3.43-
3(c)].[14] The two SO_4^{2-} counter anions are weakly coordinated to the three
Hg atoms. An overall C_{3v} symmetry of the cluster is only violated by the
orientation of the phenyl rings and the conformation of the five-membered Hg-
Hg-P-CH_2-P rings.

   $[HgOs_3(CO)_{11}]_3$[15] and $[HgMn(CO)_2C_5H_4Me]_4$[16] contain triangular and
square clusters of Hg atoms, respectively, as shown in Fig. 3.43-4 (a) and
(b). In $Hg_6[Rh(PMe_3)_3]_4$[17] and $Hg_9[Co(CO)_3]_6$[13] the Hg_6 cluster takes the
form of an octahedron and a tricapped trigonal prism, respectively, as shown
in Fig. 3.43-4 (c) and (d).

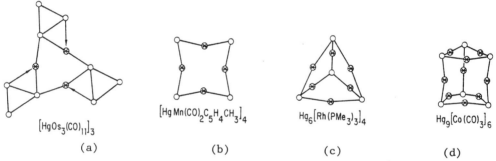

Fig. 3.43-4   Structure of (a) $[HgOs_3(CO)_{11}]_3$, (b) $[HgMn(CO)_2C_5H_4Me]_4$,
   (c) $Hg_6[Rh(PMe_3)_3]_4$ and (d) $Hg_9[Co(CO)_3]_6$. (After ref. 8).

4. $Hg_{3-\delta}AsF_6$, an anisotropic low-temperature superconductor

   An excess of mercury reacts with AsF_5 in liquid sulfur dioxide at room
temperature to yield an insoluble golden crystalline tetragonal compound,
originally formulated as "Hg_3AsF_6". X-Ray crystallography has shown it to be
$Hg_{2.86}AsF_6$, which contains infinite non-intersecting chains of mercury atoms
running through the ordered AsF_6^- sublattice (space group $I4_1/amd$) in mutually
perpendicular directions (Fig. 3.43-5).[18] Each Hg position is only
partially occupied, and the average Hg-Hg distance is 2.64Å. The compound has
a conductivity approaching that of a metal, and each mercury chain is
essentially a "one-dimensional" wire.[19] The structure, one-dimensional
fluctuation, chain-ordering transformation, lattice dynamics, temperature
dependent stoichiometry, optical properties, NMR spectra, electrical

transport, and anisotropic superconductivity of this "incommensurate linear-chain metal" have been reviewed.[20]

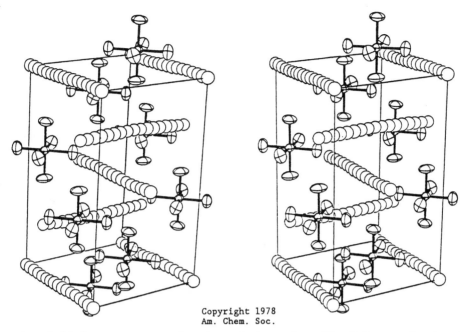

Fig. 3.43-5 A stereoscopic view of the unit cell of $Hg_{3-\delta}AsF_6$ showing the disordered model for the Hg chains. The mercury chains are not required crystallographically to be linear, so that a slight buckling of the chains is observed. (After ref. 18).

The compounds $Hg_{3-\delta}MF_6$ (M = Sb, Nb, Ta)[21,22] are isostructural with the $Hg_{3-\delta}AsF_6$ prototype. The golden crystals of tetragonal $Hg_{3-\delta}MF_6$ (M = Nb, Ta) slowly transform to silver crystals of trigonal [space group $P31m$, a = 5.02, c = 7.68Å (for Nb), Z = 1] Hg_3MF_6, the structure of which consists of hexagonal sheets of Hg atoms that are separated by sheets of MF_6^- ions.[23]

5. Alkali-metal amalgams[8]

Alkali-metal amalgams of known structure include the sodium amalgams $NaHg_2$, $NaHg$ and Na_3Hg_2,[24,25] the potassium amalgams KHg_2 and KHg,[26] the cesium amalgam $CsHg$,[27] and the rubidium amalgam $Rb_{15}Hg_{16}$.[28] The most highly reduced of these species, namely Na_3Hg_2, contains discrete 2.99Å by 2.99Å square-planar Hg_4^{6-} clusters which are not isoelectronic with other square-planar post-transition element clusters such as Bi_4^{2-}, Se_4^{2-} and Te_4^{2-}. In gold coloured NaHg slightly distorted Hg_4 rectangles (3.05 by 3.22Å) are fused into a zigzag ribbon, whereas in the likewise gold coloured KHg slightly distorted (93.6 rather than 90°) Hg_4 squares of 3.03Å edges are linked by 3.36Å Hg-Hg bonds. In KHg_2 the Hg_4 rectangles (sides 3.00 and 3.08Å) share

common edges to form a zigzag double chain. Thus the prototypal building
blocks for alkali-metal amalgams are derived from the Hg_4^{6-} squares found in
Na_3Hg_2. Fig. 3.43-6 shows the structures of Na_3Hg_2, KHg and KHg_2.

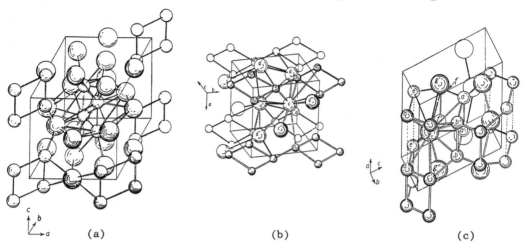

(a)   (b)   (c)

Fig. 3.43-6 Structures of (a) Na_3Hg_2, (b) KHg and (c) KHg_2.
(After refs. 24 and 26).

The amalgam $Rb_{15}Hg_{16}$[28] belongs to space group $I4_1/a$ with $a =$
16.65(3), $c = 18.13(4)$Å and $Z = 4$, and is a defect variant of the CsCl type.
Its structure and lattice constants can be interpreted in terms of an ensemble
embodying 4 x 4 x 4 = 64 CsCl unit cells (each with $a \sim 4.10$Å). This large
cubic cell of idealized content $Rb_{64}Hg_{64}$ is somewhat elongated in one
direction, such that it corresponds to the real c axis of the tetragonal unit
cell of $Rb_{15}Hg_{16}$. The 64 Hg atoms of the idealized unit cell, arranged in a
primitive cubic array, "relax" to a real structure comprising eight Hg_4
squares and four Hg_8 cubes with Hg-Hg distances of ca. 3.00Å, as shown in Fig.
3.43-7. The Rb atoms occupying the centers of the cubes in the ideal
structure are, as a result of the reduced Hg-Hg distance, no longer present in
the real structure, so that the composition becomes $Rb_{60}Hg_{64} = Rb_{15}Hg_{16} =$
$RbHg_{1.067}$.

According to King,[8] the mercury atoms in Hg_4^{6-} can be considered to
have seven-orbital spd^5 cylindrical bonding manifolds. Six of these seven
orbitals, namely the s orbital and the five d orbitals, are external orbitals,
whereas the single p orbital is an internal orbital. A neutral Hg atom with
six external orbitals is a zero skeletal electron donor, so that the Hg_4^{6-}
anion has six skeletal electrons. Since the overlap of the external p
orbitals in Hg_4^{6-} has the geometry of the square, the anion is a 4m+2 aromatic
system with a π-electron network analogous to that of benzene. However, Hg_4^{6-}

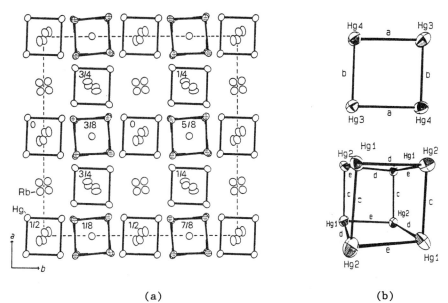

(a)   (b)

Fig. 3.43-7 Structure of $Rb_{15}Hg_{16}$. (a) Projection of the crystal structure of $Rb_{15}Hg_{16}$ along [001]: dotted Hg-atoms form Hg_8 cubes, undotted Hg_4 squares; the numbers indicate the approximate coordinate of the center of the respective square or cube in z. The Rb atoms are shown uncoupled. (b) Hg_4 square and Hg_8 cube in $Rb_{15}Hg_{16}$; distance [Å]: $a = 2.956(3)$, $b = 3.042(3)$, $c = 2.955(3)$, $d = 2.979(3)$, $e = 2.939(3)$; interatomic angles [°] at Hg1: 85.66(7), 86.38(7), 86.16(7); Hg2: 94.26(7), 93.44(8), 93.69(7); Hg3: 90.32(7); Hg4: 89.68(8). (After ref. 28).

has neither the electrons nor the orbitals for any Hg-Hg σ-bonding. Thus the eight-electron difference in the skeletal electron counts of the square clusters Bi_4^{2-} and Hg_4^{6-} relates to the respective presence and absence of σ-bonding.

References

[1]   E. Dorm, *J. Chem. Soc., Chem. Commun.*, 466 (1971).

[2]   N.J. Calos, C.H.K. Kennard and R.L. Davis, *Z. Kristallogr.* **187**, 305 (1989).

[3]   A.S. Povarennykh, *Crystal Chemical Classification of Minerals*, Vol. 2, Plenum Press, New York, 1972.

[4]   N.N. Greenwood and A. Earnshaw, *Chemistry of the Elements*, Pergamon Press, Oxford, 1986.

[5]   C.A. McAuliffe (ed.), *The Chemistry of Mercury*, Macmillan, London, 1977.

[6]   D. Grdenić in G. Dodgson, J.P. Glusker and D. Sayre (eds.), *Structural Studies on Molecules of Biological Interest*, Clarendon Press, Oxford, 1981, pp. 207-221.

[7] B. Kamenar, D. Matkovic-Calogovic and A. Nagl, *Acta Crystallogr., Sect. C* **42**, 385 (1986).

[8] R.B. King, *Polyhedron* **7**, 1813 (1988).

[9] R.D. Ellison, H.A. Levy and K.W. Fung, *Inorg. Chem.* **11**, 833 (1972).

[10] B.D. Cutforth, C.G. Davies, P.A.W. Dean, R.J. Gillespie, P.R. Ireland and P.K. Ummat, *Inorg. Chem.* **12**, 1343 (1973).

[11] B.D. Cutforth, R.J. Gillespie and P.R. Ireland, J.F. Sawyer and P.K. Ummat, *Inorg. Chem.* **22**, 1344 (1983).

[12] P.J. Bailey, B.F.G. Johnson, J. Lewis, M. McPartlin and H.R. Powell, *J. Chem. Soc., Chem. Commun.*, 1513 (1989); L.H. Gade, B.F.G. Johnson, J. Lewis, M. McPartlin and H.R. Powell, *ibid.*, 110 (1990).

[13] J.M. Ragosta and J.M. Burlitch, *J. Chem. Soc., Chem. Commun.*, 1187 (1985).

[14] B. Hämmerle, E.P. Müller, D.L. Wilkinson, G. Müller and P. Peringer, *J. Chem. Soc., Chem. Commun.*, 1527 (1989).

[15] M. Fajardo, H.D. Holden, B.F.G. Johnson, J. Lewis and P.R. Raithby, *J. Chem. Soc., Chem. Commun.*, 24 (1984).

[16] W. Gäde and E. Weiss, *Angew. Chem. Int. Ed. Engl.* **20**, 803 (1981).

[17] R.A. Jones, F.M. Real, G. Wilkinson, A.M.R. Galas and M.B. Hursthouse, *J. Chem. Soc., Dalton Trans.*, 126 (1981).

[18] A.J. Schultz, J.M. Willams, N.D. Miro, A.G. MacDiarmid and A.J. Heeger, *Inorg. Chem.* **17**, 646 (1978).

[19] I.D. Brown, B.D. Cutforth, C.G. Davies, R.J. Gillespie, P.R. Ireland and J.E. Vekris, *Can. J. Chem.* **52**, 791 (1974).

[20] W.E. Hatfield (ed.), *Molecular Metals*, Plenum Press, New York, 1979, pp. 419-469.

[21] Z. Tun and I.D. Brown, *Acta Crystallogr., Sect. B* **38**, 2321 (1982).

[22] Z. Tun and I.D. Brown, *Acta Crystallogr., Sect. B* **42**, 209 (1986).

[23] I.D. Brown, R.J. Gillespie, K.R. Morgan, Z. Tun and P.K. Ummat, *Inorg. Chem.* **23**, 4506 (1984).

[24] J.W. Nielsen and N.C. Baenziger, *Acta Crystallogr.* **7**, 277 (1954).

[25] J.D. Corbett, *Inorg. Nucl. Chem. Lett.* **5**, 81 (1969).

[26] E.J. Duwell and N.C. Baenziger, *Acta Crystallogr.* **8**, 705 (1955).

[27] H.-J. Deiseroth and A. Strunck, *Angew. Chem. Int. Ed. Engl.* **26**, 687 (1987).

[28] H.-J. Deiseroth and A. Strunck, *Angew. Chem. Int. Ed. Engl.* **28**, 1251 (1989).

Chapter 4

Inorganic Compounds (Transition Elements)

4.1.	*Vanadyl Bisacetylacetonate* $VO(C_5H_7O_2)_2$	484
4.2.	*Potassium Octacyanomolybdate(IV) Dihydrate* $K_4Mo(CN)_8 \cdot 2H_2O$	492
4.3.	*Cesium Iron Fluoride* $Cs_3Fe_2F_9$	501
4.4.	*Sodium Tris(ethylenediamine)cobalt Chloride Hexahydrate* $(+)_{589}[Co(en)_3]_2Cl_6 \cdot NaCl \cdot 6H_2O$	509
4.5.	*(1,3,6,8,10,13,16,19-Octaazabicyclo[6.6.6]eicosane)cobalt(II) Dithionate Monohydrate (Cobalt Sepulchrate Complex)* $(+)_{490}[Co(C_{12}H_{30}N_8)]S_2O_6 \cdot H_2O, [Co(sep)]S_2O_6 \cdot H_2O$	517
4.6.	*Chlorotris(triphenylphosphine)rhodium(I) (Wilkinson's Catalyst)* $RhCl[P(C_6H_5)_3]_3$	534
4.7.	*Potassium Rhenium Hydride* K_2ReH_9	543
4.8.	*Trinuclear Molybdenum–Sulfur Cluster Complex* $(NH_4)_4[Mo_3S_2(NO)_3(S_2)_4(S_3NO)] \cdot 3H_2O$	559
4.9.	*Pentakis(triphenylphosphine)hexakis(4-chlorobenzenethiolato) hexasilver(I) Toluene (1/2) Solvate* $Ag_6(SC_6H_4Cl)_6(PPh_3)_5 \cdot 2C_6H_5CH_3$	569
4.10.	*Iron(II) Phthalocyanine (FePc)* $C_{32}H_{16}N_8Fe$	578
4.11.	*Copper–Oxygen Based High-Temperature Superconductor* $Tl_2Ba_2Ca_2Cu_3O_{10}$	589
4.12.	*Ternary Molybdenum Oxide System* $K_2Mo_8O_{16}$	600
4.13.	*Potassium Tetracyanoplatinate (1.75/1) Sesquihydrate* $K_{1.75}[Pt(CN)_4] \cdot 1.5 H_2O$	610
4.14.	*Dodecatungstophosphoric Acid Hexahydrate (Keggin Structure)* $(H_5O_2)_3[PW_{12}O_{40}]$	618
4.15.	*Nitrogenpentammineruthenium(II) Dichloride* $[Ru(NH_3)_5(N_2)]Cl_2$	633
4.16.	*Iron–Sulfur Tetranuclear Cluster* $[(CH_3)_4N]_2[Fe_4S_4(SC_6H_5)_4]$	644
4.17.	*Dioxygen Adduct of Chlorocarbonylbis(triphenylphosphine)iridium (Vaska's Compound)* $IrO_2Cl(CO)[P(C_6H_5)_3]_2$	659
4.18.	*(meso-Tetraphenylporphinato)iron(II) [Fe(TPP)]* $C_{44}H_{28}N_4Fe$	669
4.19.	*Dioxygen Adduct of (2-Methylimidazole)-meso-tetra (α,α,α,α–o-pivalamidophenyl)porphyrinatoiron(II)-Ethanol* $C_{70}H_{76}N_{10}O_7Fe$	686
4.20.	*cis-Dichlorodiammineplatinum(II) (cisplatin, cis-DDP)* $PtCl_2(NH_3)_2$	694
4.21.	*Cobalt Inosine 5′-Phosphate Heptahydrate* $C_{10}H_{11}O_8N_4PCo \cdot 7H_2O$	702
4.22.	*Dipotassium Octachlorodirhenate(III) Dihydrate* $K_2[Re_2Cl_8] \cdot 2H_2O$	718
4.23.	*Benzyltriphenylphosphonium Dodeca-μ-carbonyl-dodecacarbonyl-dihydrido-polyhedro-tridecarhodate* $[P(CH_2Ph)Ph_3]_3[Rh_{13}H_2(CO)_{24}]$	730
	Addendum to Chapter 4	732

4.1   Vanadyl Bisacetylacetonate

VO(C₅H₇O₂)₂

$$VO(C_5H_7O_2)_2$$

Crystal Data

Triclinic, space group $P\bar{1}$ (No. 2)

$a = 7.53(2)$, $b = 8.23(3)$, $c = 11.24(4)$Å, $\alpha = 73.0°$, $\beta = 71.3°$,

$\gamma = 66.6°$, $Z = 2$[1]

Atom[2]	x	y	z	Atom	x	y	z
V	.1439	.2900	.2230	C(9)	-.2166	.1965	.4665
O(2)	-.0600	.4143	.3589	C(10)	-.1326	.0918	.3706
O(3)	-.0028	.1222	.2689	C(11)	-.2090	-.0593	.3814
O(4)	.1482	.5362	.1361	C(12)	.2228	.7832	-.0116
O(5)	.2201	.2300	.0530	C(13)	.2216	.5920	.0196
O(6)	.3333	.1867	.2799	C(14)	.2971	.4901	-.0753
C(7)	-.2756	.4635	.5600	C(15)	.2926	.3136	-.0562
C(8)	-.1770	.3521	.4559	C(16)	.3700	.2140	-.1656

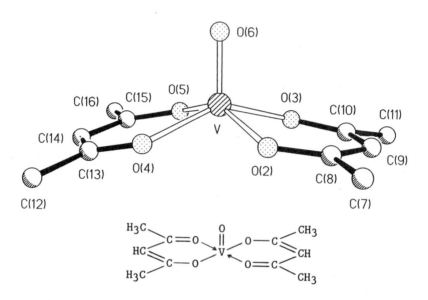

Fig. 4.1-1   Structure of VO(acac)₂.

Crystal Structure

The crystal structure of vanadyl bisacetylacetonate, VO(acac)₂, has been determined to establish the geometry of the bonds at the five-coordinate vanadium(IV) atom (Fig. 4.1-1).[1,2] The measured bond distances and angles (Table 4.1-1) show that the molecule approximates to C_{2v} symmetry.

The five ligand O atoms occupy the corners of a rectangular (nearly square) pyramid, with the vanadium atom lying 0.55Å above the mean plane of the basal atoms O(2)-O(5). The acetylacetone skeletons, each accurately planar as expected, make an angle of 158.3°.

Table 4.1-1 Bond lengths and angles in VO(acac)$_2$.

Bond lengths/Å		Bond angles/deg	
V-O$_6$	1.571	O$_2$-V-O$_3$	86.9
V-O$_2$	1.975	O$_4$-V-O$_5$	87.8
V-O$_3$	1.955	O$_2$-V-O$_4$	83.9
V-O$_4$	1.984	O$_3$-V-O$_5$	83.5
V-O$_5$	1.962	O$_2$-V-O$_5$	149.6
O$_2$-C$_8$	1.282	O$_3$-V-O$_4$	145.4
O$_3$-C$_{10}$	1.278	O$_2$-V-O$_6$	104.7
O$_4$-C$_{13}$	1.272	O$_3$-V-O$_6$	106.2
O$_5$-C$_{15}$	1.286	O$_4$-V-O$_6$	108.4
		O$_5$-V-O$_6$	105.6

Remarks

1. Structures of VE(Q)$_4$ (E = O,S) compounds

Vanadyl (VO^{2+}) complexes are frequently five-coordinate, and the stereochemistry is almost invariably square-pyramidal as exemplified by VO(acac)$_2$. The V=O bond length lies in the range 1.56-1.63Å, which is about 0.4Å shorter than the four equatorial V-O bonds. Thiovanadyl complexes containing the VS^{2+} unit have analogous structures, and Table 4.1-2 lists the structural data of some related compounds.[3-6]

Table 4.1-2 Structural data of some VE(Q)$_4$ (E = O,S) compounds.

Compound*	VO(tsalen)	VO(salen)	[VO(edt)$_2$]$^{2-}$	[VS(edt)$_2$]$^{2-}$	[VS(SPh)$_4$]$^{2-}$	VS(acen)
V=E	1.598	1.590	1.625	2.078	2.078	2.061
V-O (av.)	-	1.921	-	-	-	1.958
V-S (av.)	2.346	-	2.378	2.364	2.391	-
V-N (av.)	2.080	2.050	-	-	-	2.028
V-plane	0.608	0.609	0.668	0.784	0.807	0.607

* The ligands tsalen^{2-}, salen^{2-}, edt^{2-} and acen^{2-} are N,N'-ethylene-bis(thiosalicylideneaminate), N,N'-ethylenebis(salicylideneaminate), ethane-1,2-dithiolate and N,N'-ethylenebis(acetylacetonylideniminato), respectively. V-plane is the displacement of the vanadium atom above the mean plane of the four equatorial ligand atoms.

2. Structures of VO(Q)$_4$L compounds and their MO description

VO(acac)$_2$ and other VO(Q)$_4$ compounds have the ability to accommodate further a sixth ligand *trans* to the V=O bond to produce a distorted octahedral structure. The structural data for some VO(Q)$_4$L complexes are listed in Table 4.1-3. The molecular structures of VO(acac)$_2$(py) and [(VO)$_6$(CO$_3$)$_4$(OH)$_9$]$^{5-}$ are shown in Fig. 4.1-2. The crown-shaped hexanuclear anion is consolidated by bridging hydroxo and carbonato groups, the latter functioning in the common $\mu_2(2L,2M_{syn,syn})$ mode as well as in an unprecedented $\mu_6(3L,3M)$ mode.[13]

Table 4.1-3   Structural data of VO(Q)$_4$L compounds (Å).

Compound*	V=Oa	<V-Oe>	V...L	<Oa-V-Oe>	V-plane(Oe)	Ref.
VO(acac)$_2$(py)	1.57	2.01	2.48(N)	99°	0.32	[7]
VO(acac)$_2$(γ-pic)	1.58	1.99	2.45(N)	100°	0.34	[7]
[VO(H$_2$O)$_4$.H$_2$O]$^{2+}$	1.584	2.01	2.181(O)	98°	-	[8]
[VO(NCS)$_4$.H$_2$O]$^{2-}$	1.62	2.04(N)	2.22(O)	97°	-	[9]
[VOF$_4$(H$_2$O)]$^{2-}$	1.62	1.92(N)	2.27(O)	99°	0.29	[10]
[VO(C$_2$O$_4$).H$_2$O]$^{2-}$	1.594	2.01	2.184(O)	99°	0.30	[11]
VO{(O$_2$C)$_2$C$_5$H$_3$N}.2H$_2$O	1.591	2.02	2.184(N)	102°	-	[12]
[(VO)$_6$(CO$_3$)$_4$(OH)$_9$]$^{5-}$	1.616	2.00	2.287	99°	0.31	[13]

* Oa: axial oxygen, Oe: equatorial oxygen, γ-pic stands for γ-picoline.

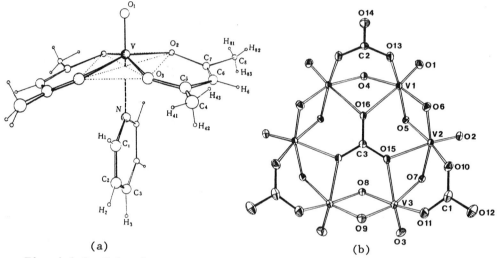

(a)   (b)

Fig. 4.1-2   Molecular geometry of (a) VO(acac)$_2$(py) and
(b) [(VO)$_6$(CO$_3$)$_4$(OH)$_9$]$^{5-}$.   (After refs. 7 and 13).

Table 4.1-4   Nature of chemical bonding in VO(Q)$_4$L.

AO of V	AO or MO of ligands	Chemical bond
4s + 3d$_{z^2}$	σ_{sp}	
3d$_{xz}$	2p$_x$ (O atom)	V=O bond
3d$_{yz}$	2p$_y$	
4s - 3d$_{z^2}$	$(\sigma_1 + \sigma_2 + \sigma_3 + \sigma_4)/2$	
4p$_x$	$(\sigma_1 - \sigma_3)/\sqrt{2}$	four equatorial
4p$_y$	$(\sigma_2 - \sigma_4)/\sqrt{2}$	V-Q bonds
3d$_{x^2-y^2}$	$(\sigma_1 - \sigma_2 + \sigma_3 - \sigma_4)/2$	
4p$_z$	σ_{sp} or σ_{sp}^2 (L atom)	V....L bond
3d$_{xy}$	-	AO with one nonbonding electron

The electronic structure of this class of $VO(Q)_4L$ compounds is well described in terms of molecular orbital theory (Table 4.1-4), as suggested by Ballhausen and Gray.[14] The V=O bond possesses rather higher multiple bond character: accompanying the σ bonding along the z axis, two d-p type π interactions arise with participation of the d_{xz} and d_{yz} orbitals of vanadium. In the equatorial positions, the metal atom forms four single bonds with the symmetry-adapted group orbitals of the Q ligands. The $3d_{xy}$ atomic orbital of vanadium is occupied by one non-bonding electron, while the remaining $4p_z$ orbital is available for forming the sixth weak bond.

3. Mononuclear and dinuclear vanadium complexes

Vanadium compounds with formal oxidation states from -3 to +5 are known except for -2. The most stable oxidation states are +4 and +5, and those lower than +2 generally occur in organometallic compounds. Table 4.1-5 lists selected examples of mono- and dinuclear vanadium compounds in various oxidation states, and their structures are illustrated in Fig. 4.1-3 and Fig. 4.1-4, respectively.

Table 4.1-5   Oxidation states and stereochemistry of some mono- and dinuclear vanadium complexes.

Oxidation state	Coord. No.	Coordination polyhedron	Examples	Fig.	Ref.
-3	7	monocapped trigonal prism	$[(Ph_3Sn)_2V(CO)_5]^-$	4.1-3(a)	[15]
-1	6	octahedron	$[V(CO)_6]^-$	-	[16]
0	6	octahedron	$trans$-$V(CO)_2(dmpe)_2$	-	[17]
1	7	pseudo octahedron	$[V(CO)_2(dmpe)_2(MeCN)]^+$	4.1-3(b)	[17]
1	7	monocapped octahedron	$[V(CO)_3(PMe_3)_4]^+$	4.1-3(c)	[16]
1	7	sandwich structure	$[V(CO)_4(C_6H_2Me_4-1,2,4,5)]^+$	4.1-3(d)	[18]
2	6	octahedron	$[V(H_2O)_6]^{2+}$	-	[19]
2	6	octahedron	$[\mu-N_2]\{[(o-Me_2NCH_2)C_6H_4]_2V(py)\}_2$	4.1-4(a)	[20]
3	6	octahedron	$V_2OCl_4(py)_6$	4.1-4(d)	[21]
3	7	pentagonal bipyramid	$[V_2L(H_2O)_4]^{4-}$	4.1-4(c)*	[22]
4	5	trigonal bipyramid; py ligands axial	$V(O-2,6-C_6H_3Me_2)_3(py)_2$	-	[23]
4	6	octahedron	$[V(acac)_3]^+$	-	[24]
4	8	trigonal prism capped on two rectangular faces	$V_2(\mu-\eta^2-S_2)_2(S_2CMe)_4$	4.1-4(b)	[25]
5	6	octahedron	$[cis-VO_2(C_2O_4)_2]^{3-}$	4.1-3(e)	[26]
5	7	pentagonal bipyramid	$VO(O_2CBu^t)_3$	4.1-3(f)	[27]

* H_2L = 1,7,14,20-tetramethyl-2,6,15,19-tetraaza[7,7](2,6)-pyridinophane-4,7-diol.

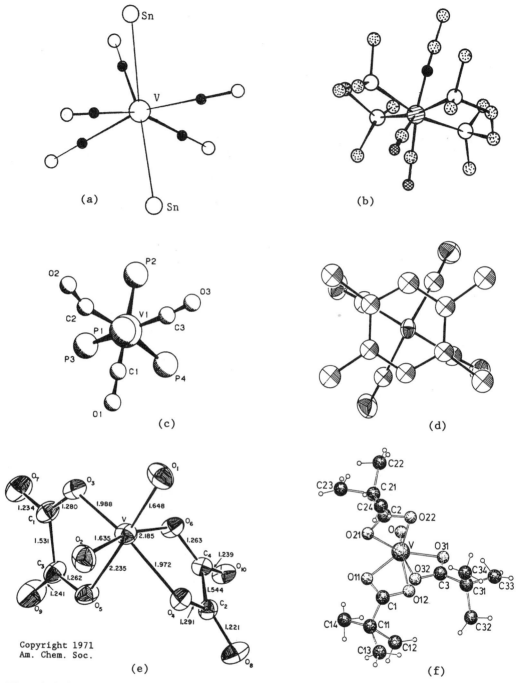

Copyright 1971
Am. Chem. Soc.

Fig. 4.1-3 Structures of some mononuclear vanadium complexes:
(a) $[(Ph_3Sn)_2V(CO)_5]^-$ (phenyl groups omitted), (b) $[V(CO)_2(dmpe)_2(MeCN)]^+$,
(c) $[V(CO)_3(PMe_3)_4]^+$ (methyl groups omitted), (d) $[V(CO)_4(C_6H_2Me_4\text{-}1,2,4,5)]^+$,
(e) $[cis\text{-}VO_2(C_2O_4)_2]^{3-}$ and (f) $VO(O_2CBu^t)_3$. (After refs. 15, 16, 17, 18, 26
and 27).

The coordination geometry of [(Ph$_3$Sn)$_2$V(CO)$_5$]$^-$ can be described as a trigonal prism with Sn(1) capping the rectangular face defined by C(2), C(3), C(4) and C(5) [Fig. 4.1-3(a)].[15] Both [V(CO)$_2$(dmpe)$_2$(CH$_3$CN)]$^+$ [Fig. 4.1-3(b)] and [V(CO)$_2$(dmpe)$_2$(O$_2$CEt)] have an analogous seven-coordinate "pseudo octahedral" structure with "equatorial" *bis*-dmpe chelating ligands and *one* axial site accommodating *two* carbonyl groups.[17] In [V(CO)$_3$(PMe$_3$)$_4$]$^+$ the coordination polyhedron is a monocapped octahedron of nearly C_{3v} symmetry [Fig. 4.1-3(c)].[16] In [V(CO)$_4$(C$_6$H$_2$Me$_4$-1,2,4,5)]$^+$[V(CO)$_6$]$^-$ the anion is a nearly regular octahedron, and the cation has the expected half-sandwich structure with the V(CO)$_4$ fragment eclipsing two of the methyl groups [Fig. 4.1-3(d)].[18] The crystal structure of VSO$_4$.6H$_2$O consists of a packing of two kinds of independent, centrosymmetric [V(H$_2$O)$_6$]$^{2+}$ octahedra and SO$_4^{2-}$ tetrahedra.[19] The structure and molecular dimensions of [*cis*-VO$_2$(C$_2$O$_4$)$_2$]$^{3-}$ are shown in Fig. 4.1-3(e), the O=V=O bond angle being 103.8°.[26] The complex VO(O$_2$CBu^t)$_3$ has a pentagonal-bipyramidal structure with one of the three bidentate pivalato groups coordinated to the metal in an axial/equatorial fashion [Fig. 4.1-3(f)].[27]

In crystalline (μ-N$_2$){[(*o*-Me$_2$NCH$_2$)C$_6$H$_4$]$_2$V(py)}$_2$.2THF the dinuclear complex consists of two almost identical fragments linked by a bridging dinitrogen ligand [Fig. 4.1-4(a)].[20] The V-(μ-N$_2$)-V unit is approximately linear [V1-N1-N2 171.6(3)°, V2-N2-N1 171.3(3)°], and the V-N (average) and N-N bond distances are 1.833(3) and 1.228(4)Å, respectively. The V$_2$(μ-η^2-S$_2$)$_2$(S$_2$CMe)$_4$ molecule contains a metal-metal bond of length 2.800(2)Å. The coordination polyhedron around each vanadium center is a trigonal prism capped on two rectangular faces, and the pair of μ-η^2-S$_2$ ligands constitute the common rectangular face of the two polyhedra [Fig. 4.1-4(b)]. The centrosymmetric anion [V$_2$L(H$_2$O)$_4$]$^{4-}$ contains a planar di-μ-alkoxy core with pentagonal-bipyramidal coordination geometry around each metal center [Fig. 4.1-4(c)].[22] In crystalline V$_2$OCl$_4$(py)$_6$.CH$_3$CN there are two independent dinuclear molecules each having imposed two-fold symmetry; the C_2 axis passes through the N-V-O-V-N unit in one oxo-bridged dimer but only through the O atom in the other [Fig. 4.1-4(d)].[21]

4. Mixed-valent complexes of vanadium

The synthesis and characterization of mixed-valent vanadium complexes have attracted great interest in recent years. In [{VO(salen)}$_2$][I$_5$].MeCN the cations are stacked parallel to the *b* axis; the distinct V-O distances along the column [O(61')···V(2) 2.41(4), V(2)=O(62) 1.67(5), O(62)-V(1) 2.06(5), V(1)=O(61) 1.58(3)Å] are consistent with the oxo-bridged mixed-valent

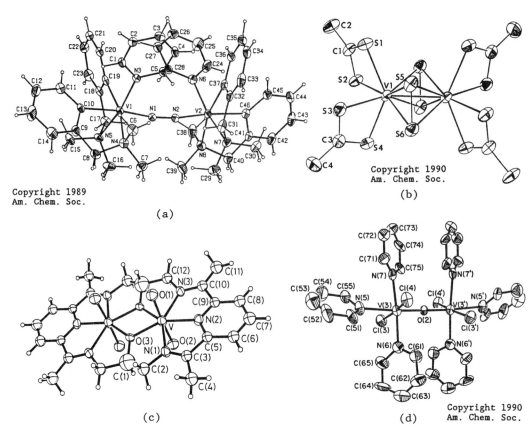

Copyright 1989
Am. Chem. Soc.

(a)

Copyright 1990
Am. Chem. Soc.

(b)

(c)

Copyright 1990
Am. Chem. Soc.

(d)

Fig. 4.1-4 Structure of some dinuclear vanadium complexes: (a) $(\mu\text{-}N_2)\{[(o\text{-}Me_2NCH_2)C_6H_4]_2V(py)\}_2$, (b) $V_2(\mu\text{-}\eta^2\text{-}S_2)_2(S_2CMe)_4$, (c) $[V_2L(H_2O)_4]^{4-}$ and (d) $V_2OCl_4(py)_6$ (one of the two independent molecules is shown). (After refs. 20, 25, 22 and 21).

formulation $[V^{IV}(salen)=O \longrightarrow V^V(salen)=O]^+$ for the cationic backbone [Fig. 4.1-5(a)].[28] The hexanuclear macrocycle $[V_6O_6(\mu\text{-}O)_4(\mu\text{-}O_2CPh)_9]$ consists of one binuclear fragment containing octahedral vanadium atoms (V5,V6) and one tetranuclear fragment in which the metal coordination corresponds to a tetragonal pyramid (V2,V3) distorted towards a trigonal bipyramid (V1,V4) [Fig. 4.1-5(b)].[27] The mixed-valent vanadium pyrophosphate $V_2(VO)(P_2O_7)_2$ consists of linear trimers containing two V^{III} octahedra joined through edges to a central V^{IV} square pyramid [Fig. 4.1-5(c)]; these trimers are linked in three dimensions through the $P_2O_7^{4-}$ groups.[29] The anion $[V_3(mp)_6]^-$ (H_2mp = 2-mercaptophenol) possesses pseudo-S_6 symmetry with the central V(III) atom located on an inversion center. The metal cluster is consolidated by μ_2-aryloxy bridges between V(III) and the other two V(IV) atoms, and the thiolato groups function only as terminal ligands [Fig. 4.1-5(d)].[30]

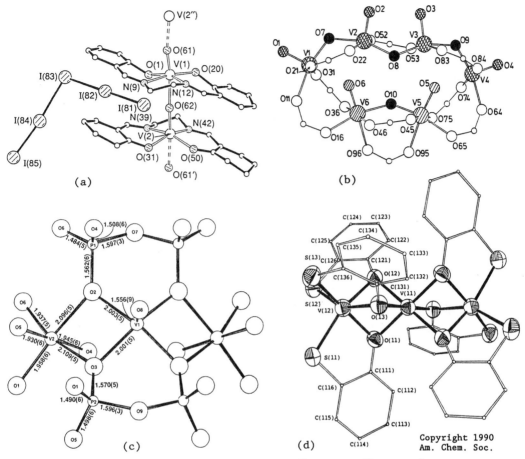

Fig. 4.1-5 Structures of some mixed-valent vanadium complexes:
(a) [(salen)VOVO(salen)]⁺ (I₅⁻ also shown), (b) [V₆O₆(μ-O)₄(μ-O₂CPh)₉],
(c) V₂(VO)(P₂O₇)₂ and (d) [V₃(mp)₆]⁻. (After refs. 28, 27, 29 and 30).

References

[1] R.P. Dodge, D.H. Templeton and A. Zalkin, *J. Chem. Phys.* **35**, 55 (1961).

[2] P.-K. Hon, R.L. Belford and C.E. Pfluger, *J. Chem. Phys.* **43**, 3111 (1965).

[3] J.C. Dutton, G.D. Fallon and K.S. Murray, *Inorg. Chem.* **27**, 34 (1988).

[4] J.R. Nicholson, J.C. Huffman, D.M. Ho and G. Christou, *Inorg. Chem.* **26**, 3030 (1987).

[5] J.K. Money, J.C. Huffman and G. Christou, *Inorg. Chem.* **24**, 3297 (1985).

[6] M. Sato, K.M. Miller, J.H. Enemark, C.E. Strouse and K.P. Callahan, *Inorrg. Chem.* **20**, 3571 (1981).

[7] Shao Meicheng, Wang Lifeng, Zhang Zeying and Tang Youchi, *J. Mol. Sci.* **2**, 17 (1984).

[8] P.M. Tachez, F. Theobald, K.J. Watson and R. Mercier, *Acta Crystallogr.*, *Sect. B* **35**, 1545 (1979).

[9] A.C. Hazell, *J. Chem. Soc.*, 5745 (1963).

[10] K. Waltersson, *J. Solid State Chem.* **29**, 195 (1979).

[11] R.E. Oughtred, E.S. Raper and H.M.M. Shearer, *Acta Crystallogr., Sect. B* **32**, 82 (1976).

[12] B.H. Bersted, R.L. Belford and I.C. Paul. *Inorg. Chem.* **7**, 1557 (1968).

[13] T.C.W. Mak, P. Li, C. Zheng and K. Huang, *J. Chem. Soc., Chem. Commun.*, 1597 (1986).

[14] C.J. Ballhausen and H.B. Gray, *Inorg. Chem.* **1**, 111 (1962).

[15] J.E. Ellis, T.G. Hayes and R.E. Stevens, *J. Organomet. Chem.* **216**, 191 (1981).

[16] J.-P. Charland, E.J. Gabe, J.E. McCall, J.R. Morton and K.F. Preston, *Acta Crystallogr., Sect. C* **43**, 48 (1987).

[17] F.J. Wells, G. Wilkinson, M. Motevalli and M.B. Hursthouse, *Polyhedron* **6**, 1351 (1987).

[18] F. Calderazzo, G. Pampaloni, D. Vitali and P.F. Zanazzi, *J. Chem. Soc., Dalton Trans*, 1993 (1982).

[19] F.A. Cotton, L.R. Falvello, R. Llusar, E. Libby, C.A. Murillo and W. Schwotzer, *Inorg. Chem.* **25**, 3423 (1986).

[20] J.J.H. Edema, A. Meetsma and S. Gambarotta, *J. Am. Chem. Soc.* **111**, 6878 (1989).

[21] Y. Zhang and R.H. Holm, *Inorg. Chem.* **29**, 911 (1990).

[22] J.C. Dutton, G.D. Fallon and K.S. Murray, *J. Chem. Soc., Chem. Commun.*, 64 (1990).

[23] S. Gambarotta, F. van Bolhuis and M.Y. Chiang, *Inorg. Chem.* **26**, 4301 (1987).

[24] T.W. Hambley, C.J. Hawkins and T.A. Kabanos, *Inorg. Chem.* **26**, 3740 (1987).

[25] S.A. Duraj, M.T. Andras and P.A. Kibala, *Inorg. Chem.* **29**, 1232 (1990).

[26] W.R. Scheidt, C.-C. Tsai and J.L. Hoard, *J. Am. Chem. Soc.* **93**, 3867 (1971).

[27] D. Rehder, W. Priebsch and M. von Oeynhausen, *Angew. Chem. Int. Ed. Engl.* **28**, 1221 (1989).

[28] A. Hills, D.L. Hughes, G.J. Leigh and J.R. Sanders, *J. Chem. Soc., Dalton Trans.*, 61 (1991).

[29] J.W. Johnson, D.C. Johnston, H.E. King, Jr., T.R. Halbert, J.F. Brody and D.P. Goshorn, *Inorg. Chem.* **27**, 1646 (1988).

[30] B. Kang, L. Weng, H. Liu, D. Wu, L. Huang, C. Lu, J. Cai, X. Chen and J. Lu, *Inorg. Chem.* **29**, 4873 (1990).

4.2 Potassium Octacyanomolybdate(IV) Dihydrate
$K_4Mo(CN)_8 \cdot 2H_2O$

Crystal Data

Orthorhombic, space group *Pnma* (No. 62)

$a = 16.64(1)$, $b = 11.660(7)$, $c = 8.710(5)$Å, $Z = 4$

Atom	x	y	z	Atom	x	y	z
Mo	.13648	1/4	.09680	N(6)	.1381	1/4	.4761
K(1)	.1480	.0477	.6903	C(1)	.0381	.1396	.1654
K(2)	.8625	1/4	.9741	C(2)	.1724	.0729	.0697
K(3)	.9603	1/4	.4791	C(3)	.0608	1/4	-.1060
N(1)	-.0136	.0833	.2095	C(4)	.2168	1/4	-.0997
N(2)	.1921	-.0209	.0499	C(5)	.2589	1/4	.1776
N(3)	.0192	1/4	-.2118	C(6)	.1403	1/4	.3440
N(4)	.2591	1/4	-.2019	O	.8890	.0567	.5913
N(5)	.3263	1/4	.2146				

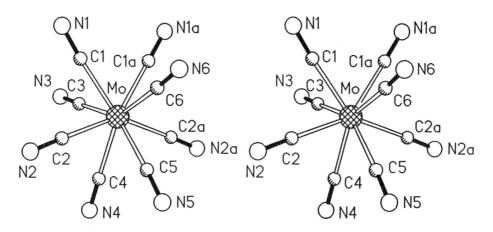

Fig. 4.2-1 Stereoview of the molecular structure of Mo(CN)$_8^{4-}$ with atom labelling. (Adapted from ref. 1).

Crystal Structure

The crystal structure of $K_4Mo(CN)_8 \cdot 2H_2O$ was firstly determined a half-century ago by Hoard and Nordsieck in a two-dimensional X-ray analysis.[2] Refinement using three-dimensional data demonstrated that the coordination polyhedron of Mo, of required C_s symmetry, closely approaches the shape of the ideal D_{2d} dodecahedron (Fig. 4.2-1).[1] The averaged dimensions are: Mo-C, 2.163(5)Å; C≡N, 1.152(6)Å; Mo-N, 3.314(5)Å; and Mo-C-N, 177.5(9)°.

The distortion of the coordination polyhedron from D_{2d} symmetry and the departure of the Mo-C≡N chains from strict linearity are both attributable to the complicated packing relations that directly involve the potassium and the nitrogen atoms. The coordination groups around each of the three structurally distinct K$^+$ ions are not describable as simply distorted versions of familiar polyhedra. The water molecules serve to complete the coordination spheres of the two classes of K^+ ions.

Remarks

1. Configuration of $M(CN)_8^{n-}$ (M = Mo or W; n = 3 or 4) complexes

The ions $M(CN)_8^{n-}$ occur in at least three definite stereochemical configurations, as found by crystal structure determinations.

a) The more frequently occurring ones are the D_{2d} dodecahedron [Fig. 4.2-2(a)] in $K_4Mo(CN)_8 \cdot 2H_2O$, $[n\text{-}Bu_4N]_3Mo(CN)_8$[3] and $[C_6H_6NO_2]_4Mo(CN)_8$.[4]

b) The D_{4d} square antiprism coordination geometry [Fig. 4.2-2(b)] is adopted in $H_4W(CN)_8 \cdot 6H_2O$[5] and $H_4W(CN)_8 \cdot 4HCl \cdot 12H_2O$.[6]

c) The geometries of the complex anions in $Na_3W(CN)_8 \cdot 4H_2O$[7] and $[HN(C_2H_5)_3]_2[H_3O]_2Mo(CN)_8$[8] are about halfway between those of a square antiprism or dodecahedron and a C_{2v} 4,4-bicapped trigonal prism, respectively.

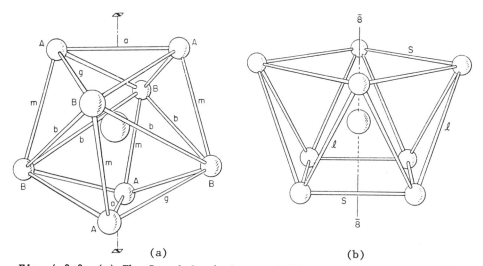

(a)   (b)

Fig. 4.2-2 (a) The D_{2d} dodecahedron and (b) D_{4d} square antiprism, showing the usual labeling scheme. (After ref. 9).

Interest in the stereochemistry of eight-coordinate complexes such as $Mo(CN)_8^{4-}$ was stimulated by a postulate of Orgel.[10] Using simple symmetry arguments, he pointed out that d^2 MX_4Y_4 molecules might adopt dodecahedral rather than square-antiprismatic geometry if the ligands X and Y had substantially different π acceptor properties.

2. Conditions leading to the formation of high-coordinate complexes[11]

The complexes of higher coordination number: seven, eight, nine, and more are extremely interesting because numerous geometries are possible, and, as the energies of the various polytopes are often quite similar, inter-conversion is feasible and frequently observed. Several factors lead to the formation of high-coordinate complexes.

a) Balance of charge and ion size

With a given ligand, the ability of a metal ion to achieve a high coordination number appears to depend in part on a subtle balance of charge and ion size. For example, eight-coordinate complexes are found predominantly for 5th and 6th period transition metal ions in Groups III to VI in oxidation states +3 to +5. These ions are large enough to accommodate more ions or molecules in a coordination sphere without severe ligand-ligand repulsions. These relatively highly charged ions exercise an attraction on the ligands that is sufficient to overcome the force of ligand-ligand repulsions.

b) d Orbital population

The 3+, 4+ and 5+ ions of Groups III to VIB have electron configurations of d^0, d^1 or d^2. This implies that d orbital occupation has some influence on the formation of high-coordinate complexes. Low d orbital populations will minimize the σ anti-bonding role of the d electrons. In any complex of coordination number greater than six, there will be at least one d-like MO that is not strongly anti-bonding; therefore, population by one or two electrons will not seriously affect the σ bonding in the complex.

c) Nature of the ligand

If the polarizability of an ionic ligand is high, a large amount of charge density is easily transferred to the metal ion, and fewer ligands will be necessary to arrive at electroneutrality. As a consequence, complexes of higher coordination number are formed readily by ligands that are small and/or of low polarizability, e.g. H^-, F^-, OH^-. Chelating ligands with small donor atoms, such as C, N, and O, have low steric demands and tend to form high-coordinate complexes.

3. The stereochemistry of high coordination[12,13]

The most commonly occurring coordination numbers for transition metals are 4 and 6, and a few examples are higher than 8.

a) Coordination number 9

The most favorable arrangement of nine atoms around a central atom is in the form of a tricapped trigonal prism of D_{3h} symmetry [Fig. 4.2-3(a)]; ReH_9^{2-} and $[M(H_2O)_9]^{3+}$ (M = Sm, Pr, Yb, Y, Ho) have this coordination geometry. The monocapped square antiprism of C_{4v} symmetry is another possible mode of nine-coordination [Fig. 4.2-3(b)], which has been observed in the complex $[Th(CF_3COCHCOCH_3)_4(H_2O)]$.

b) Coordination number 10

Three related symmetrical structures of nearly equal energy are the

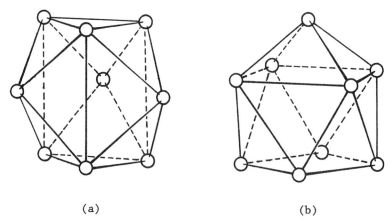

(a)   (b)

Fig. 4.2-3 The tricapped trigonal prism (a) and the monocapped square
antiprism (b).

bicapped square antiprism [Fig. 4.2-4(a)], the tetracapped trigonal prism
[Fig. 4.2-4(b)] and the tetradecahedron [Fig. 4.2-4(c)], all of which are
substantially more stable than the pentagonal antiprism [Fig. 4.2-4(d)], the
pentagonal prism [Fig. 4.2-4(e)] and the bicapped square prism [Fig. 4.2-
4(f)]. The coordination polyhedron of $K_4[Th(O_2CCO_2)_4(H_2O)_2].2H_2O$ is a
bicapped D_{4d} square antiprism.

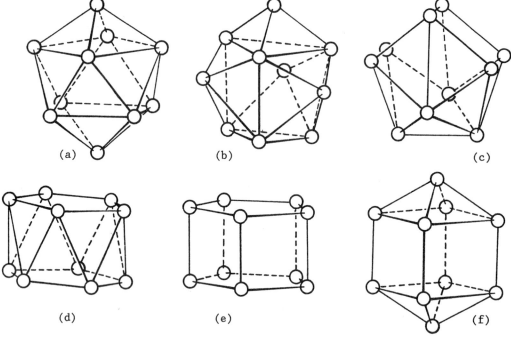

(a)   (b)   (c)

(d)   (e)   (f)

Fig. 4.2-4 Coordination polyhedra for ten-coordination. (After ref. 12).

c) Coordination number 12

The symmetrical coordination polyhedra for twelve-coordination are the icosahedron and the cubooctahedron which have all edge lengths identical, as shown in Fig. 4.2-5. The complex ion $[Ce(NO_3)_6]^{3-}$ has been found to possess distorted icosahedral stereochemistry.

Higher coordination number, up to 16, are known, particularly among organometallic compounds and metal borohydrides.

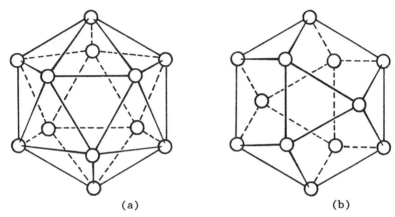

(a)   (b)

Fig. 4.2-5 The icosahedron (a) and the cubooctahedron (b). (After ref. 12).

4. Structure of Prussian blue

Addition of $K_4[Fe^{II}(CN)_6]$ to an aqueous Fe^{III} solution produces the intensely blue fine precipitate, Prussian blue. The X-ray powder pattern and Mössbauer spectrum of this compound are the same as those of Turnbull's blue, which is produced by the converse addition of $K_3[Fe^{III}(CN)_6]$ to aqueous Fe^{II}. By varying the conditions and proportions of the reactants, a whole range of these blue materials of varying composition can be produced with some, which are actually colloidal, described as soluble Prussian blue. They have found application as pigments in the manufacture of inks and paints.

Prussian blue can be easily dissolved in concentrated HCl and reprecipitated by dilution with water. By allowing only a very slow diffusion of water vapour into the HCl solution of Prussian blue it became possible to grow single crystals suitable for X-ray work. The structural data of three small single crystals are listed in Table 4.2-1. Fig.4.2-6 shows the unit cell in space group $Pm3m$.[14]

According to the reliably determined analytical and density data, a number of positions in this unit cell are only partially occupied, the vacant sites being randomly distributed. The X-ray analysis yielded average bond distances of Fe(II)-C = 1.92Å, C-N = 1.13Å, and Fe(III)-N = 2.03Å.

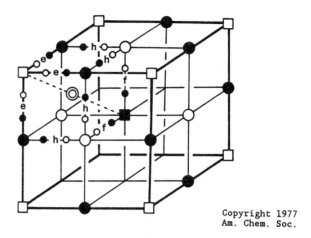

Fig. 4.2-6 The unit cell of Prussian blue in space group $Pm3m$. The atom types and crystallographic positions are partially indicated by the following symbols: (□) Fe(III), 1(a); (◯) Fe(III), 3(c); (■) Fe(II), 1(b); (●) Fe(II), 3(d); (◎) O, 8(g); (•) C, (○) N or O. (After ref. 14).

Table 4.2-1 Structural data of Prussian blue, $Fe_4[Fe(CN)_6]_3 \cdot nH_2O$, ($n = 14$-16), cubic, space group $Pm3m$ (No. 221), $a = 10.166(3)$Å, $Z = 1$.

Atom	Position	Crystal I		Crystal II		Crystal III	
		x	occupancy	x	occupany	x	occupancy
Fe(III)	1(a) 0,0,0		1		1		1
Fe(III)	3(c) 0,1/2,1/2		1		1		1
Fe(II)	1(b) 1/2,1/2,1/2		.267		.665		.824
Fe(II)	3(d) 1/2,0,0		.911		.778		.725
C	6(e) x,0,0	.3108	.911	.3125	.778	.3087	.725
C	6(f) x,1/2,1/2	.3108	.267	.3125	.665	.3087	.824
C	12(h) x,1/2,0	.1887	.911	.1875	.778	.1913	.725
N	6(e) x,0,0	.2005	.911	.1994	.778	.1979	.725
N	6(f) x,1/2,1/2	.2005	.267	.1994	.665	.1979	.725
N	12(h) x,1/2,0	.2995	.911	.3006	.778	.3021	.725
O	6(e) x,0,0	.2100	.089	.2138	.222	.2074	.725
O	6(f) x,1/2,1/2	.2100	.733	.2138	.335	.2074	.176
O	12(h) x,1/2,0	.2900	.089	.2862	.222	.2926	.275
O	8(g) x,x,x	.2608	1	.2517	1	.2577	1

A neutron powder diffraction study of $Fe_4[Fe(CN)_6]_3 \cdot 14D_2O$[15] has revealed two distinguishable kinds of water molecules: six are coordinated to Fe(III) at empty nitrogen sites, and approximately eight others either occupy the centers of the unit-cell octants or are connected by hydrogen bonds to the coordinated ones at O-D...O distances of 2.87Å. The disordered overall structure can be described as a superposition of various ordered substructures. Magnetic contributions to the neutron intensities below the Curie temperature of 5.6K indicate ferromagnetism, which **originates from the** high-spin Fe(III) ions.

The available structural information, as well as the properties of Prussian blue and Turnbull's blue, led to the proposal of a prototype skeleton

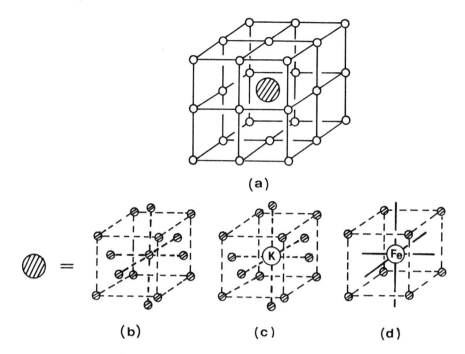

Fig. 4.2-7 (a) Structure of Prussian blue skeleton. Small circles represent Fe atoms, and large shaded circle represents the accommodated group. The lines between two Fe atoms represent -C≡N-. (b), (c) and (d) Atomic arrangement of $(H_2O)_{15}$, $[K(H_2O)_{14}]^+$ and $[Fe(CN)_6(H_2O)_8]^{3-/4-}$, respectively.

structure [Fig. 4.2-7(a)] with the following features:

a) In the skeleton, there are two coordination types for iron atoms: octahedron and planar square. The Fe(II) ions preferentially occupy the square-planar sites, regardless of the combination of the starting complexes.

b) In the skeleton, when the Fe(II) and Fe(III) are arranged alternatively, the charge-transfer process readily occurs:

$$Fe(II)-C≡N-Fe(III) \rightleftharpoons Fe(III)-C≡N-Fe(II)$$

The intense blue colour is due to this process.

c) In the skeleton, there is a hole per cubic unit cell (a = 10.2Å).

With a radius of about 9Å, this hole can accommodate a large group, such as $[H_2O]_{15}$, $[K(H_2O)_{14}]^+$, $[Cl(H_2O)_{14}]^-$, $[Fe(CN)_6(H_2O)_8]^{3-/4-}$, ..., which is determined by the composition and condition of solutions. For the ideal formula $Fe_4^{III}[Fe^{II}(CN)_6]_3.15H_2O$, the $[H_2O]_{15}$ group comprises a six-capped cube with a centered H_2O molecule [Fig. 4.2-7(b)].

d) Prussian blue exhibits a very strong tendency to incorporate small

amounts of K^+ into its lattice. Solid samples treated with saturated KCl
solution for a long time produce a product containing only 2.8% K^+, i.e. one
hole accommodates a K^+ ion to form $[K(H_2O)_{14}]^+$ as shown in Fig. 4.2-7(c). Its
ideal composition is $Fe_4^{II}[Fe^{III}(CN)_6]_3[K(H_2O)_{14}]$, in which the Fe^{II} atoms are
located at the origin and face-centers, the Fe^{III} atoms at the mid-points of
unit cell edges, and the C atoms bind to the Fe^{III} atoms. This is perhaps the
prototype structure of these compounds.

e) The holes always accommodate some $[Fe(CN)_6(H_2O)_8]^{3-/4-}$ groups [Fig.
4.2-7(d)], the quantity of which is determined by the composition of the
solution and treatment method for the precipitates. In the solid these groups
are orientated randomly to conform to overall $Pm3m$ symmetry.

References

[1]   J.L. Hoard, T.A. Hamor and M.D. Glick, *J. Am. Chem. Soc.* **90**, 3177 (1968).

[2]   J.L. Hoard and H.H. Nordsieck, *J. Am. Chem. Soc.* **61**, 2853 (1939).

[3]   B.J. Corden, J.A. Cunningham and R. Eisenberg, *Inorg. Chem.* **9**, 356
      (1970).

[4]   S.S. Basson, J.G. Leipoldt and A.J. van Wyk, *Acta Crystallogr., Sect. B*
      36, 2025 (1980).

[5]   S.S. Basson, L.D.C. Bok and J.G. Leipoldt, *Acta Crystallogr., Sect. B* **26**,
      1209 (1970).

[6]   L.D.C. Bok, J.F. Leipoldt and S.S. Basson, *Z. Anorg. Allg. Chem.* **392**, 303
      (1972).

[7]   L.D.C. Bok, J.G. Leipoldt and S.S. Basson, *Acta Crystallogr., Sect. B* **26**,
      684 (1970).

[8]   D.L. Kepert, *Prog. Inorg. Chem.* **24**, 179 (1978).

[9]   S.J. Lippard, *Prog. Inorg. Chem.* **21**, 91 (1976).

[10] L.E. Orgel, *J. Inorg. Nucl. Chem.* **14**, 136 (1960).

[11] K.F. Purcell and J.C. Kotz, *Inorganic Chemistry*, Saunders, Philadelphia,
      1977.

[12] M.C. Favas and D.L. Kepert, *Prog. Inorg. Chem.* **28**, 309 (1981).

[13] D.L. Kepert, *Inorganic Stereochemistry*, Springer-Verlag, Berlin, 1982;
      D.L. Kepert in G. Wilkinson, R.D. Gillard and J.A. McCleverty (eds.),
      Comprehensive Coordination Chemistry, Vol. 1, Pergamon Press, Oxford,
      1987, p. 31.

[14] H.J. Buser, D. Schwarzenbach, W. Petter and A. Ludi, *Inorg. Chem.* **16**,
      2704 (1977).

[15] F. Herren, P. Fischer, A. Ludi and W. Hälg, *Inorg. Chem.* **19**, 956 (1980).

4.3   Cesium Iron Fluoride

$$Cs_3Fe_2F_9$$

Crystal Data

Hexagonal, space group $P6_3/mmc$ (No. 194)

$a = 6.347(1)$, $c = 14.805(3)$Å, $Z = 2$

Atom	Position	x	y	z
Cs(1)	2d	1/3	2/3	3/4
Cs(2)	4f	1/3	2/3	.43271
Fe	4e	0	0	.15153
F(1)	6h	.1312	.2624	1/4
F(2)	12k	.1494	.2988	.5940

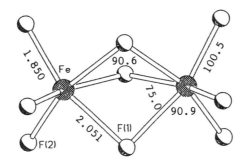

Fig. 4.3-1   The binuclear $Fe_2F_9^{3-}$ unit in $Cs_3Fe_2F_9$.

Crystal Structure

Cs₃Fe₂F₉ is an exceptional example for an iron(III) compound in which predominant ferromagnetic interactions occur between some Fe^{3+} ions, an unexpected feature for the d^5 electron configuration. The first determination of the crystal structure was based on film intensities.[2] A redetermination using diffractometer data yielded a more precise crystal structure,[1] which comprises a packing of binuclear entities of face-sharing octahedra containing iron(III). The bond lengths and angles are shown in Fig. 4.3-1. The non-bonded Fe...Fe distance in the binuclear unit is 2.916Å.

Remarks

1.   Structure of transition-metal fluoro compounds with single bridges[2]

In most oxidation states of d transition-metal ions M, the radius ratio $r_M:r_F$ falls within the range 0.41-0.73 for stable ionic contact in octahedral coordination. This coordination mode in fact occurs in the majority of cases and it is quite unaffected by the F:M stoichiometry of the compound.

The most important fluoride structure types with only corner-sharing are listed in Table 4.3-1. Most chain, layer, and framework structures belong to this group.

Table 4.3-1　Fluoride structure types displaying single bridges.

Chain structures

single chain			double chain		triple chain	
type	cis-bridged	trans-bridged	type	example	type	example
MF_5	VF_5					
$A_2^I MF_5$	K_2FeF_5	Tl_2AlF_5	$A^I A^{II} M_2^{III} F_9$	$KPbCr_2F_9$	$A^I MF_4$	$KCrF_4$
	Rb_2CrF_5			$NaBaFe_2F_9$		$CsCrF_4$
$A^{II}MF_5$	$SrFeF_5$	$CaCrF_5$	$A_2^{II} M^{II} M^{III} F_9$	Ba_2CoFeF_9		
	$Na_2Ba_3Cr_4F_{20}$	$CaFeF_5$				
		$BaFeF_5$				

Layer structures

single layer			double layer		triple layer	
type	trans-terminal F	cis-terminal F	type	example	type	example
MF_4		NbF_4				
$A^I MF_4$	$NaTiF_4$	$NaCrF_4$	$A_3^I M_2^{II} F_7$	$K_3Zn_2F_7$	$A_4^I M^I M_4^{III} F_{18}$	$Cs_4CoCr_4F_{18}$
	$KFeF_4$					
	$AMnF_4$					
$A_2^I MF_4$	K_2NiF_4					
$A^I A_2^{II} M_2 F_9$		$CsBa_2Ni_2F_9$				

Framework structures

type		example
AMF_3:	perovskites	$NaNiF_3$, $K_4Mn_3F_{12}$
$A_x MF_3$:	bronzes	$K_{0.6}FeF_3$ TTB, $Cs_{0.2}Zn_{0.2}Fe_{0.8}F_3$ HTB
		$FeF_3.1/3H_2O$ HTB
$A^I M^{II} M^{III} F_6$:	pyrochlores	$RbNiCrF_6$, $CsAgFeF_6$, NH_4MnFeF_6, $NH_4Fe_2F_6$
		$CsNi_2F_6$
$A_2^I M^{II} M^{III} F_7$:	weberites	Na_2MnFeF_7, $Fe_2F_5.2H_2O$

TTB and HTB refer to the tetragonal and hexagonal tungsten bronze structures, respectively.

a) Isotropically linked $[MF_6]$ groups

Only three-dimensional corner-sharing between identical $[MF_6]$ groups, resulting in the composition F:M = 3, leads to chemical and geometrical isotropy. Apart from the trifluoride structures, this is found in the cubic AMF_3 compounds of the perovskite type and of the modified pyrochlore type, $RbNiCrF_6$. Related to both but no longer strictly isotropic are the tetragonal (TTB) and hexagonal (HTB) tungsten bronze structures. Fig. 4.3-2 shows the smallest units of N = 4, 6 and 8 polyhedra in three-dimensional corner-sharing of the pyrochlore, bronze, and perovskite structures.

(b) Anisotropically linked $[MF_6]$ groups

NbF_5 and TaF_5 are isolated tetramers with ring structures. Fig. 4.3-3

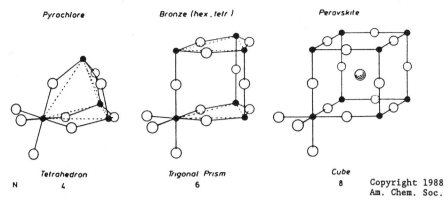

Fig. 4.3-2 Smallest units of N = 4,6, and 8 polyhedra in three-dimensional corner-sharing of the pyrochlore, bronze, and perovskite structures. (After ref. 2).

to Fig. 4.3-6 show different connections of corner-sharing octahedra. Fig. 4.3-3 shows the order of octahedral vacancies in the framework structure of the cation-deficient perovskites $K_4Mn_3F_{12}$ and $Cs_2Ba_2Cu_3F_{12}$ and the chiolite-like arrangement of Jahn-Teller elongated octahedra in its basal plane.

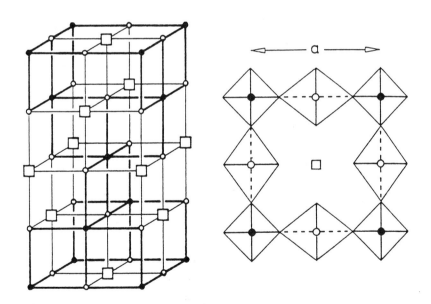

Fig. 4.3-3 Structure of $K_4Mn_3F_{12}$. (o and ● represent Mn atoms, □ represents vacancy, the K^+ ions are not shown). (After ref. 3).

Fig. 4.3-4 shows (a) the puckered layer structure of $CsBa_2Ni_2F_9$, a cation-deficient hexagonal perovskite, and (b) the triple-layer structure of $Cs_4CoCr_4F_{18}$ derived from the pyrochlore framework. In both layer types (a) and (b), each shown in different projections, the octahedra have three face-terminal ligands at the surface of the sheet.

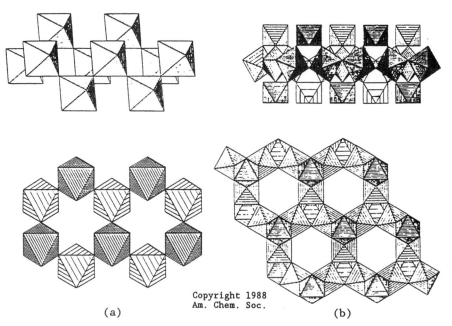

(a)   (b)

Fig. 4.3-4 Layer structure derived from (a) CsBa$_2$Ni$_2$F$_9$ and (b) Cs$_4$CoCr$_4$F$_{18}$. (After ref. 2).

Fig. 4.3-5 shows the triple-chain structure of CsCrF$_4$. If short sections of AMF$_4$ layer structures are curled up, which is made possible by the all-*cis* connection of three parallel strands, even a chain structure may be

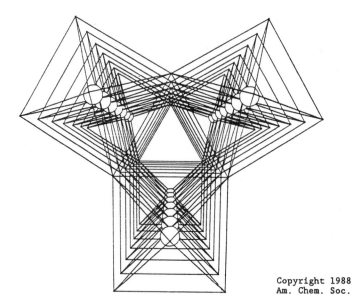

Fig. 4.3-5 Triple-chain structure of CsCrF$_4$ as seen along the threefold channel. (After ref. 2)

achieved for this composition, F:M = 4. Such a triple-chain structure has been observed in a tilted and an untilted version in the chromium compounds $KCrF_4$ and $CsCrF_4$, respectively.

Several variants of a single *trans*-connected chain of octahedra with composition F:M = 5 are known (Fig. 4.3-6). An important distinction may be made according to the packing of chains, which in Mn(III) compounds was found to be either tetragonal or (pseudo)hexagonal. Another kind of chain structure is realized by continued *cis* bridging, resulting in zigzag chains of a sometimes helical array, as found in the $SrFeF_5$ structure.

Joining two parallel *cis*-bridged chains leads to double-chain structures of composition F:M = 4.5, intermediate between the chain structure of $BaGaF_5$ and the layer structure of $BaZnF_4$, as shown schematically in Fig. 4.3-6 on the left.

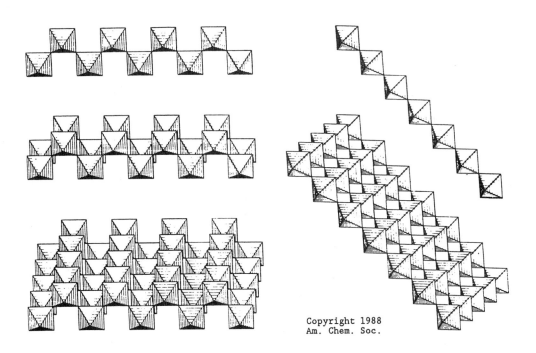

Fig. 4.3-6 Idealized *cis*- and *trans*-connected chains, double chains, and layers of octahedra in ternary fluorides of composition A_mMF_5, $A_mM_2F_9$, and A_mMF_4, respectively. (After ref. 2).

2. Structure of transition-metal fluoro compounds with multiple bridges[2]

Transition-metal fluorides which contain groups of n octahedra joined by multiple bridges are listed in Table 4.3-2.

Table 4.3-2 Fluoride crystal structures containing groups of n
octahedra joined by multiple bridges.

Double bridges (edge sharing)			Triple bridges (face sharing)		
F:M	n	Example	F:M	n	Example
		Isolated pair			
-	-	-	4.5	2	$Cs_3Fe_2F_9$
		Chain structures			
4	∞	Na_2CuF_4	3	∞	$CsNiF_3$
		Layer structures			
4	2	$Ba_2CuV_2F_{12}$	3.75	2	$Cs_7Ni_4F_{15}$
3.5	3	$BaMnGaF_7$	3.33	3	$Cs_4Ni_3F_{10}$
			3.2	5	$Cs_6Ni_5F_{16}$
		Framework structures			
3.71	∞	$Ba_6Zn_7F_{26}$	3	2	$CsMnF_3$
3.5	2	$BaMnFeF_7$	3	3	$CsCoF_3$
3.33	∞	$Ba_2Ni_3F_{10}$	3	3	$CsMnNiF_6$
3.09	3	$Ba_6Cu_{11}F_{34}$	3	4	$Cs_5CdNi_4F_{15}$
3	∞	$NaMnCrF_6$			
3	2	$(NH_4)MnFeF_6$			
2.8	2+∞	α-$Ba_2Cu_5F_{14}$			
2.57	3+4	$Ba_2Zn_7F_{18}$			
2.5	∞	Cr_2F_5			

a) Pure forms of either edge- or face-sharing of octahedra

These forms are known only from the structure types of Na_2CuF_4 and
$CsNiF_3$, respectively. Both are chain structures, as shown in Fig. 4.3-7.

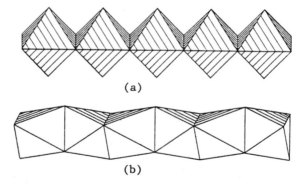

(a)

(b)

Fig. 4.3-7 Structures of (a) Na_2CuF_4 (edge-sharing) and (b) $CsNiF_3$
(face-sharing). (After ref. 2).

b) Edge-sharing in combination with corner-sharing

The rutile structure (see Section 2.14) of difluorides are well known.
Cr_2F_5 and $AlMnF_5$ and some of the ternary BaF_2-MF_2 fluorides contain the
infinite and linear chains of *trans* edge-sharing octahedra of the parent

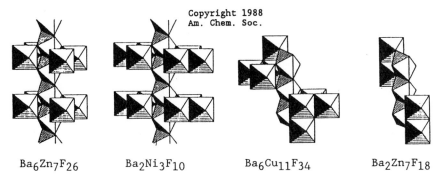

| Ba$_6$Zn$_7$F$_{26}$ | Ba$_2$Ni$_3$F$_{10}$ | Ba$_6$Cu$_{11}$F$_{34}$ | Ba$_2$Zn$_7$F$_{18}$ |

Fig. 4.3-8 Edge-sharing octahedral units in some BaF$_2$-MF$_2$ systems. (After ref. 2).

rutile structure. For example, in α-Ba$_2$Cu$_5$F$_{14}$ the infinite chain is zigzag by consecutive *cis-trans*-sharing of edges as known for instance from the ZrCl$_4$ structure. In addition, or alternatively, bi-, tri-, and tetra-nuclear edge-sharing units are found in other interesting barium compounds, some of which are shown in Fig. 4.3-8. In some structures, edge-sharing binuclear groups are connected via the remaining corners to form a layer or framework structure, as shown in Fig. 4.3-9.

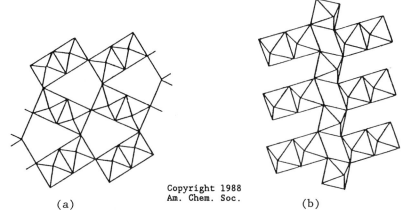

| (a) | (b) |

Fig. 4.3-9 Binuclear edge-sharing groups in (a) the framework structure of (NH$_4$)MnFeF$_6$ and (b) the layer structure of Ba$_2$CuV$_2$F$_{12}$. (After ref. 2).

c) Face-sharing structures

In addition to the one-dimensionally widened framework structures formed by AMF$_3$ compounds, three related layer structure types have been found in the compounds Cs$_7$Ni$_4$F$_{15}$, Cs$_4$Ni$_3$F$_{10}$, and Cs$_6$Ni$_5$F$_{16}$. They contain the chain fragments of 2, 3 and 5 face-sharing octahedra shown in Fig. 4.3-10. These units are only two-dimensionally linked by corners and since F:M > 3, some ligands are terminal. In the units of face-sharing octahedra, the values of

M-M distances are listed in Table 4.3-3. Metal-metal bonding in these compounds may be present to a significant degree.

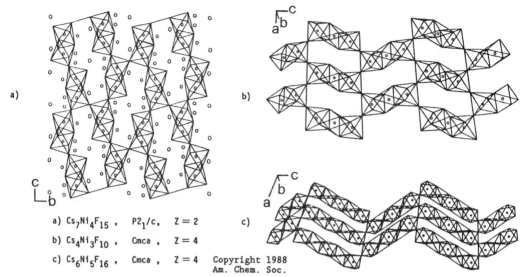

a) $Cs_7Ni_4F_{15}$, $P2_1/c$, $Z = 2$

b) $Cs_4Ni_3F_{10}$,   Cmca , $Z = 4$

c) $Cs_6Ni_5F_{16}$,   Cmca , $Z = 4$ Copyright 1988
                                          Am. Chem. Soc.

Fig. 4.3-10 Bi-, tri-, and penta-nuclear groups of face-sharing octahedra in some layer structures of CsF-NiF$_2$ phases: (a) $Cs_7Ni_4F_{15}$, (b) $Cs_4Ni_3F_{10}$, and (c) $Cs_6Ni_5F_{16}$.[4] (After ref. 2).

Table 4.3-3 M-M and M-F distances (Å) in some N face-sharing octahedra in transition metal fluorides.

Compound	M-M	M-F	M-M/M-F	Compound	M-M	M-F	M-M/M-F
	N = 2				N = 3		
$CsCoF_6$	2.817	2.061	1.37	$CsNiF_3$(HP)	2.707	2.034	1.33
$CsFeF_3$	2.923	2.102	1.39	$CsCoF_3$	2.753	2.061	1.34
$CsMnF_3$	2.995	2.140	1.40	$Cs_4Ni_3F_{10}$	2.727	2.028	1.34
$Cs_7Ni_4F_{15}$	2.782	2.037	1.37	$Cs_4Zn_3F_{10}$	2.794	2.051	1.36
$Cs_7Co_4F_{15}$	2.835	2.065	1.37	$Cs_4Co_3F_{10}$	2.779	2.057	1.35
	N = 5				N = ∞		
$Cs_6Ni_5F_{16}$	2.672	2.019	1.32	$CsNiF_3$(NP)	2.621	2.013	1.30

HP = high-pressure form; NP = normal-pressure form.

References

[1] J.M. Dance, J. Mur, J. Darriet, P. Hagenmuller, W. Massa, S. Kummer and D. Babel, *J. Solid State Chem.* **63**, 446 (1986).

[2] W. Massa and D. Babel, *Chem. Rev.* **88**, 275 (1988).

[3] G. Frenzen, S. Kummer, W. Massa and D. Babel, *Z. Anorg. Allg. Chem.* **553**, 75 (1987).

[4] R.E. Schmidt and D. Babel, *Z. Anorg. Allg. Chem.* **516**, 187 (1984).

4.4   Sodium Tris(ethylenediamine)cobalt Chloride Hexahydrate
$(+)_{589}[Co(en)_3]_2Cl_6 \cdot NaCl \cdot 6H_2O$

Crystal Data

Trigonal, space group *P3* (No. 143)

$a = 11.47(3)$, $c = 8.06(2)$Å, $Z = 1$[1]

Atom	x	y	z	Atom	x	y	z
Co	2/3	1/3	0	Co'	1/3	2/3	1/2
N(1)	.783	.295	.854	N(1')	.217	.705	.354
N(2)	.647	.183	.146	N(2')	.353	.817	.646
C(1)	.773	.167	.907	C(1')	.227	.833	.407
C(2)	.742	.140	.093	C(2')	.258	.860	.593
Cl(1)	.617	.100	.528	Cl(1')	.383	.900	.028
O(1)	.897	.098	.560	Na	0	0	.770
O(2)	.098	.201	.980	Cl(2)	0	0	.270

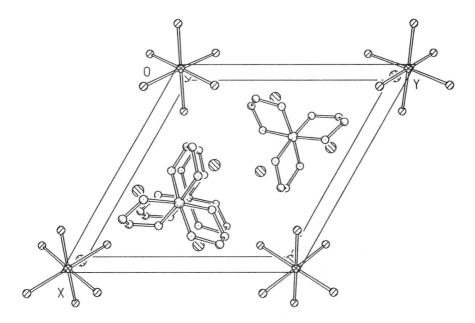

Fig. 4.4-1 Crystal structure of $(+)[Co(en)_3]_2Cl_6 \cdot NaCl \cdot 6H_2O$ projected along the *c* axis.

Crystal Structure

The crystal structure consists of a packing of $[Co(en)_3]^{3+}$, $[Na(H_2O)_6]^+$ and Cl⁻ ions (Fig. 4.4-1). The symmetry of the structure approaches closely to space group $P6_3$. In the $\Lambda(+)$-$[Co(en)_3]^{3+}$ ion, the ethylenediamine molecules take "gauche" forms and six nitrogen atoms form a slightly distorted octahedron around the central cobalt atom (Fig. 4.4-2). This is the first metal chelate complex whose absolute configuration was determined by means of X-ray crystallography.[2,3]

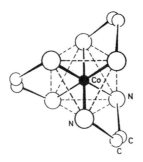

Fig. 4.4-2   Structure of $\Lambda(+)$-$[Co(en)_3]^{3+}$.   (After ref. 4).

The $\Lambda(+)$-$[Co(en)_3]^{3+}$ complex has D_3 symmetry within the limits of experimental error. The shape and size of the cobalt-ethylenediamine chelate ring are described by the following parameters:

          Co-N = 1.978(4)Å   N-Co-N = 85.4(3)°

          N-C = 1.497(10)Å   Co-N-C = 108.4(5)°

          C-C = 1.510(10)Å   N-C-C = 105.8(7)°

          torsion angle N-C-C-N = 55.0°

Remarks

1. Designation of absolute configuration[4]

     Optical isomerism occurs when a complex and its mirror image are not superimposable. The most general statement of the criterion for the appearance of optical isomerism is that the complex must not possess an improper rotation axis. The resolution of a metal chelate into enantiomeric forms merely means that one form is the mirror image of the other. There is no direct correlation between optical rotation and chirality, as the direction of rotation of the plane of polarization of the solution of a particular isomer may change its sign with the wavelength of the light, with the solvent, or even with the concentration of the solution being examined.

     In order to determine the absolute configuration of mirror-image-related isomers, it is necessary to resort to X-ray crystal analysis. Using a technique (Section 5.4) developed by Bijvoet and co-workers,[5] Saito and co-workers[1] determined the absolute configuration of the $(+)_{589}[Co(en)_3]^{3+}$ ion, which is dextrorotatory at the sodium D-line (λ = 589 nm).

     Absolute configuration concerning six-coordinate complexes based on the octahedron is designated according to the IUPAC proposal:[6]

     Two skew line AA' and BB' define a right-handed helix, namely AA' determines an axis of a helix and BB' constitutes a tangent to the helix and defines its pitch [Fig. 4.4-3 (a) and (b)]. As far as a qualitative measure of helicity is concerned, the steepness of a helix (pitch) is in general of no consequence. Alternately, if BB' is chosen an as axis of the helix and AA' as its pitch, a helix of the same handedness is obtained. In the same way Fig. 4.4-3(d) defines a left-handed helix. The Greek letter delta

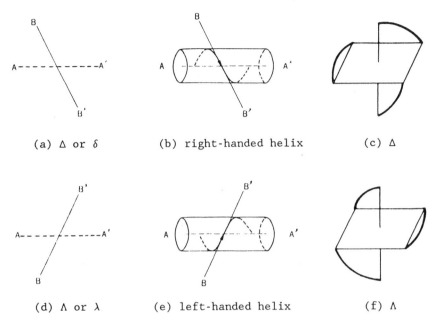

(a) Δ or δ   (b) right-handed helix   (c) Δ

(d) Λ or λ   (e) left-handed helix   (f) Λ

Fig. 4.4-3 Pairs of non-orthogonal skew lines are shown in projection along the common normal. The full line BB' is above the plane of the paper, the dotted line AA' below the plane. (Adapted from ref. 4).

(Δ referring to configuration, δ to conformation) is associated with the two skew lines shown in Fig. 4.4-3(a), and the Greek letter lambda (Λ for configuration, λ for conformation) is associated with the two skew lines shown in Fig. 4.4-3(d). The absolute configurations of the two optical isomers of a *tris*-bidentate complex with symmetrical ligands are shown in Fig. 4.4-3(c) and (f), where any pair of two edges of an octahedron on which a chelate ring is spanned defines two skew lines Δ or Λ.

2. Structure of some optically-active complexes with five-membered chelate
   rings

     When three bidentate ligands, such as ethylenediamine (en), propylenediamine (pn), *trans*-1,2-diaminocyclohexane (chxn) and *trans*-1,2-diaminocyclopentane (cptn), are coordinated octahedrally to a central metal atom, two optical isomers (Λ and Δ) can occur. Several crystal structures of en chelate complexes have been fully investigated: $(+)_{589}[Co(en)_3]Br_3 \cdot H_2O$, $(+)_{589}[Co(en)_3]Cl_3 \cdot H_2O$, $(+)_{589}[Co(en)_3](NO_3)_3$, and $[Co(en)_3]_2(HPO_4)_3 \cdot 9H_2O$.[4] In these crystals, the geometry of the complex ion $[Co(en)_3]^{3+}$ is similar to that in $(+)_{589}[Co(en)_3]_2Cl_6 \cdot NaCl \cdot 6H_2O$.

A neutron and X-ray diffraction study of $(-)_{589}[Co(pn)_3]Br_3$[4] gave a C-C bond length of 1.518Å. The octahedron formed by the six nitrogen atoms is trigonally twisted and slightly compressed along the three-fold axis: the upper triangle formed by the three nitrogen atoms is rotated counterclockwise by about 5° with respect to the lower triangle formed by the remaining three nitrogen atoms from the position expected for a regular octahedron. The Co-N bond makes an angle of 55.9° with respect to the three-fold axis of the complex ion, as compared to the value of 54.75° for a regular octahedron. A packing diagram of $(-)_{589}[Co(pn)_3]Br_3$ is shown in Fig. 4.4-4.

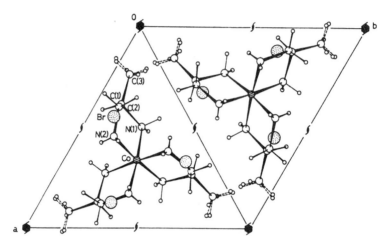

Fig. 4.4-4 Crystal structure of $(-)_{589}[Co(pn)_3]Br_3$. (After ref. 4).

When the dianion of *S*-aspartic acid, $HO_2C-CH(NH_2)-CH_2-CO_2H$, acts as a terdentate ligand (through two amino N atoms and four carboxylic O atoms) to form an octahedral complex, there exist three possible isomers as follows:

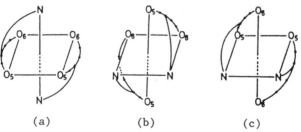

(a)   (b)   (c)

Five-membered, six-membered and seven-membered chelate rings are formed. In the structures found, the N atoms are in *cis*-positions, as shown in Fig. 4.4-5.[7] The coordination octahedra are slightly distorted, probably owing to the non-bonded hydrogen interactions as well as to the formation of strained chelate rings.

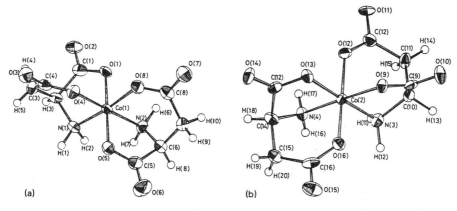

Fig. 4.4-5 Structure of (a) *cis*(N)-*trans*(O_5)- and (b) *cis*(N)-*trans*(O_6)-[Co(*S*-asp)₂]⁻; O_5 and O_6 refer to those oxygen atoms that form five- and six-membered chelate rings with amino N atoms, respectively.   (After ref. 7).

3. Structure of $K_{2(1-x)}Na_{1+2x}[Co(NO_2)_6]$, $x = 0.18(1)$[8]

This compound is non-stoichiometric with the idealized empirical formula $K_2Na[Co(NO_2)_6]$. The crystal is cubic, with space group $Fm3$, $a = 10.245(1)$Å, and $Z = 4$. The Na⁺ and $[Co(NO_2)_6]^{3-}$ ions are arranged like the ions in sodium chloride and each potassium ion is located at the center of a cube whose corners are alternately occupied by four Na⁺ and four $[Co(NO_2)_6]^{3-}$ ions. The specimen used in the X-ray crystallographic study has the composition $K_{1.64}Na_{1.36}[Co(NO_2)_6]$, the excess Na⁺ being found to occupy the K⁺-site randomly. Six nitro groups coordinate to Co(III) octahedrally at a Co-N distance of 1.952Å.

Fig. 4.4-6(a) shows a section of the final difference synthesis through the cobalt atom and the two three-fold axes of rotation. A peak due to bonding electrons is located on the Co-N bond at 1.43Å from the cobalt atom. Fig. 4.4-6(b) is a section through the plane perpendicular to the Co-N bond and at 0.246Å from the cobalt nucleus. This is a section through the cube face formed by the residual electron density peaks. Eight peaks with heights of $1.7(1)$ eÅ⁻³ are arranged at the eight corners of a cube at 0.43Å ($0.246\sqrt{3}$Å) from the cobalt nucleus. This feature is exactly the same as that predicted by ligand field theory for 3d electrons in non-bonding orbitals.

This example nicely demonstrates that X-ray diffraction can be used to determine accurate charge-densities in transition metal complexes that have moderate complexity and are of sufficient chemical interest.

Firstly, the central metal atom is largely neutralized by donation of electrons from the ligating atoms. Accordingly, Pauling's electroneutrality rule has indeed been verified for transition-metal complexes. Secondly, the

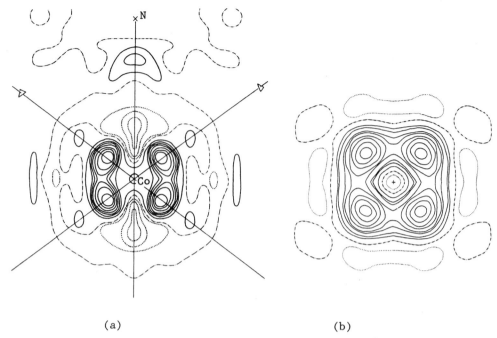

(a)   (b)

Fig. 4.4-6 Difference electron-density in $[Co(NO_2)_6]^{3-}$: (a) section
through Co and the two three-fold axes; (b) section through the plane
perpendicular to the Co-N bond at 0.246Å from the cobalt nucleus.
Contours are drawn at intervals of $0.2e\text{Å}^{-3}$. (After ref. 8).

charge-density of d electrons in non-bonding orbitals of a transition-metal
atom in a complex is aspherical owing to the ligand field. The distribution
of non-bonding electrons has a tendency to avoid regions of high field arising
from the ligands. The distribution can be reasonably accounted for by ligand
field theory. Thirdly, the distribution of bonding electrons and the location
of lone-pair electrons can be observed, and the effective charge on each atom
estimated with reasonable certainty. In some chelate complexes, the "lone-
pair" orbitals of the ligating atoms are not directed toward the central metal
atom. If the electron-density distribution and geometrical arrangement of the
atomic nuclei are all known, it is possible to predict the physical and
chemical properties of the complex on the basis of quantum mechanical
calculations.

4. Structure of $[Cr(en)_3][Ni(CN)_5] \cdot 1.5H_2O$[9]

The crystal consists of discrete $[Cr(en)_3]^{3+}$ and $[Ni(CN)_5]^{3-}$ ions linked
by intermolecular hydrogen bonds involving the water molecules of
crystallization. There are two crystallographically independent $[Ni(CN)_5]^{3-}$

ions in the compound: one is a regular square pyramid and the other a
distorted trigonal bipyramid [Fig. 4.4-7(a) and (b)]. The square pyramid has
axial and average equatorial Ni-C bond lengths of 2.168(14) and 1.862(6)Å,
respectively. The average C-Ni-C angle between the opposing basal carbon
atoms is 159.5(4)°, so that the nickel atom is 0.34Å above the basal plane of
carbon atoms. The distorted trigonal-bipyramidal $[Ni(CN)_5]^{3-}$ group has Ni-C
axial bonds that are significantly shorter than the Ni-C equatorial bonds.
The average axial Ni-C bond length is 1.837(9)Å. The C-Ni-C angle between the
axial carbon atoms is 172.8(5)°. There are two equivalent equatorial Ni-C
bonds of average length 1.907(9)Å and one longer equatorial bond of
1.992(14)Å. This longer bond and the larger C-Ni-C angle, 141.2(5)°, between
the other two equatorial carbon atoms are the primary deviation from regular
trigonal-bipyramidal geometry.

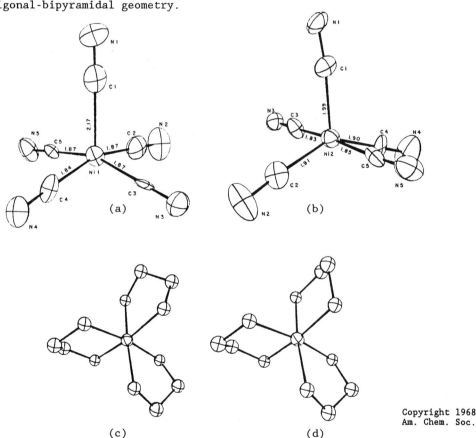

Fig. 4.4-7 Structures of (a) square-pyramidal $[Ni(CN)_5]^{3-}$, (b)
trigonal-bipyramidal $[Ni(CN)_5]^{3-}$, (c) $[Cr(en)_3]^{3+}$, $\Lambda\delta\lambda\lambda$, and (d)
$[Cr(en)_3]^{3+}$, $\Lambda\delta\delta\lambda$. (After ref. 9).

Both independent $[Cr(en)_3]^{3+}$ ions have distorted octahedral geometry. Figs. 4.4-7(c) and (d) show the coordination environment of each of the Cr atoms; the configurations around the metals are arbitrarily chosen as Λ, since the centrosymmetric space group $(P2_1/c)$ requires that both the Λ and the Δ configurations of each cation be present in equal numbers. The conformation of the Cr(1) cation [Fig. 4.4-7(c)] is ΛδλλλΛ; the conformation of the Cr(2) cation [Fig. 4.4-7(d)] is Λδδλ. These are the first examples of conformations for tris(ethylenediamine) metal complexes that differ from Λδδδ.

References

[1] K. Nakatsu, M. Shiro, Y. Saito and H. Kuroya, *Bull. Chem. Soc. Jpn.* **30**, 158 (1957).

[2] Y. Saito, K. Nakatsu, M. Shiro and H. Kuroya, *Bull. Chem. Soc. Jpn.* **30**, 795 (1957).

[3] Y. Saito, K. Nakatsu, M. Shiro and H. Kuroya, *Acta Crystallogr.* **7**, 636 (1954); **8**, 729 (1955).

[4] Y. Saito, *Inorganic Molecular Dissymmetry*, Springer-Verlag, Berlin, 1979.

[5] J.M. Bijvoet, A.F. Peerdeman and A.J. van Bommel, *Nature (London)* **168**, 271 (1951).

[6] U. Thewalt, K.A. Jensen and C.E. Schäffer, *Inorg. Chem.* **11**, 2129 (1972).

[7] I. Oonishi, M. Shibata, F. Marumo and Y. Saito, *Acta Crystallogr., Sect. B* **29**, 2448 (1973); I. Oonishi, S. Sato and Y. Saito, *Acta Crystallogr., Sect. B* **31**, 1318 (1975).

[8] S. Ohba, K. Toriumi, S. Sato and Y. Saito, *Acta Crystallogr., Sect. B* **34**, 3535 (1978).

[9] K.N. Raymond, P.W.R. Corfield and J.A. Ibers, *Inorg. Chem.* **7**, 1362 (1968).

4.5 (1,3,6,8,10,13,16,19-Octaazabicyclo[6.6.6]eicosane)-cobalt(II) Dithionate Monohydrate (Cobalt Sepulchrate Complex)

$(+)_{490}[Co(C_{12}H_{30}N_8)]S_2O_6 \cdot H_2O$, $[Co(sep)]S_2O_6 \cdot H_2O$

Crystal Data

Orthorhombic, space group $P2_12_12_1$ (No. 19)

$a = 15.67(1)$, $b = 15.411(6)$, $c = 8.757(4)$Å, $Z = 4$

Atom	x	y	z	Atom	x	y	z
Co	.14990	.25343	.24567	C(8)	-.0351	.2293	.2615
N(1)	.2001	.1633	.0805	C(9)	.2108	.3696	.4905
N(2)	.2594	.2302	.3912	C(10)	.0023	.3709	.1416
N(3)	.0862	.1478	.3611	C(11)	.1410	.4248	.0633
N(4)	.1998	.3472	.0816	C(12)	.0961	.4394	.3307
N(5)	.1297	.3579	.4044	S(1)	.43128	.37899	.21562
N(6)	.0226	.2764	.1571	S(2)	.55249	.44173	.21001
N(7)	.2263	.0780	.3160	O(1)	.4432	.3027	.1224
N(8)	.0721	.4278	.1731	O(2)	.4196	.3568	.3736
C(1)	.2433	.0876	.1577	O(3)	.3695	.4402	.1588
C(2)	.2736	.1330	.4216	O(4)	.6098	.3767	.2740
C(3)	.1372	.0650	.3527	O(5)	.5421	.5188	.3045
C(4)	.2561	.2124	-.0268	O(6)	.5671	.4599	.0519
C(5)	.2445	.2801	.5342	O(W)	.5155	.1709	.3380
C(6)	-.0001	.1378	.2873	H(N2)	.3076	.2523	.3440
C(7)	.2140	.2979	-.0615	H(N4)	.2517	.3669	.1149

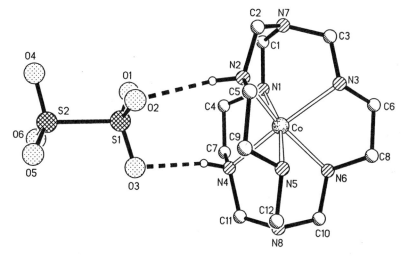

Fig. 4.5-1 Bonding interaction between cation and anion in [Co(sep)]S$_2$O$_6$·H$_2$O. (After ref. 1).

Crystal Structure

The crystal structure of the title compound consists of a packing of [Co(sep)]$^{2+}$ cations, S$_2$O$_6$$^{2-}$ anions, and water molecules. Fig. 4.5-1 shows the [Co(sep)]$^{2+}$ complex hydrogen bonded to the dithionate group. In the crystal lattice, the cations and anions are linked through a two-dimensional hydrogen-bonding network with N(3), N(5), and O(5) excluded from the scheme. The water molecule is hydrogen bonded to oxygens O(1) and O(2) at one end of the dithionate ion, with a longer contact to O(4) at the other end. The dithionate ion also forms pairs of contacts with the protons of nitrogens N(2), N(4) (Fig. 4.5-1) and N(1), N(6). These O...N distances lie within the range 3.004 to 3.182Å and satisfy the empirical geometric criteria for significant hydrogen bonding (Table 4.5-1).

Table 4.5-1 lists selected interatomic distances. The Co-N bond lengths average 2.164Å, and individual values differ within 0.033Å. This value indicates that the Co-N bonds may be slightly lengthened by the ligand cage, as compared with Co-N = 2.11Å in $[Co(NH_3)_6]Cl_2$. The C-N bonds alternate in length along the chain from the cap nitrogen, N(cap), to the first carbon of the ethylene bridge, C(en). The distances averaged over D_3 equivalents are given in Table 4.5-1 and vary sequentially from 1.440(6) to 1.513(6) to 1.490(6)Å in the chain N(cap)-C(cap)-N(ligator)-C(en). A similar averaged variation of 1.443(8) to 1.523(7) to 1.493(9)Å was found in the $(-)_{490}$-$[Co(sep)]Cl_3.H_2O$ structure.[2] Each of the individual chains in both of these structures shows an equivalent C-N bond-length variance. The average C(cap)-N(cap)-C(cap) angle in $[Co(sep)]S_2O_6.H_2O$ is 116.1° as compared with the value 113.9° found in $[Co(sep)]Cl_3.H_2O$.

Table 4.5-1   Interatomic distances (Å) for $[Co(sep)]S_2O_6.H_2O$.

Co-N (av.)	2.164	O(2)...N(2)	3.182
C-N(cap)(av.)	1.440	O(3)...N(4)	3.095
C(cap)-N(lig)(av.)	1.513	O(1)...O(w)	2.996
C(en)-N(lig)(av.)	1.490	O(2)...O(w)	3.250
C-C (av.)	1.520	O(1)...N(6')	3.004
S-O (av.)	1.444	O(6)...N(1')	3.048
S(1)-S(2)	2.132		

Remarks

1. Clathrochelating ligands sepulchrate (sep) and sarcophagine (sar)

Sepulchrate and sarcophagine are sexidentate ligands of the macropolycyclic type involving N atom donors, as shown in Fig. 4.5-2.[3] These molecular cages have the capacity to encapsulate other molecules or ions. The structures of the Co(III) and Co(II) sepulchrate ions are illustrated in Fig. 4.5-3.

Numerous variations are possible for the introduction of different substituents in the apical position of sarcophagine. They include $-NO_2$, - NHOH, -NO, $-NH_2$, $-NH_3^+$, $-N(CH_3)_3^+$, -OH, -Cl, -Br, -I, -COOH, -COOR, $-NHCOCH_3$, -CN, $-CONH_2$, and $-CH_2OH$. These substituents impart a wide range of properties to the molecules and ions. In general, the synthetic strategy involves the use of suitable capping reagents; for example, nitromethane (instend of ammonia) and formaldehyde have been used to construct a nitromethyl cage with C apices. The nitro group an be reduced to the amine group with zinc dust, nitrosation of the amine in aqueous solution leads to the hydroxy and chloro

(in the presence of Cl⁻) derivatives, and reductive elimination replaces Cl by
H to give the parent cage ion.

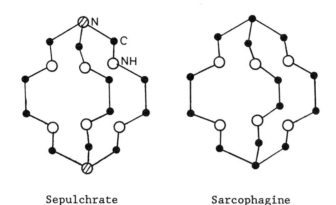

Sepulchrate   Sarcophagine

Fig. 4.5-2 Stable hexaazamacrobicyclic cage ligands.

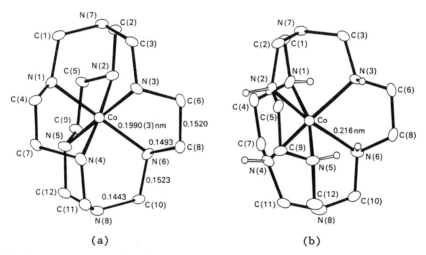

(a)   (b)

Fig. 4.5-3 Structures of the caged metal ions in (a) [Co(III)sepul-
chrate]Cl₃[2] and (b) [Co(II)sepulchrate]ZnCl₄ salts. (After ref. 4).

2. Structures of [Rh(MENOsar)]ZnCl₄.Cl.3H₂O, [Rh(MENOtar)]ZnCl₄.Cl.3H₂O and
   [Cr(diamsar)]Cl₃.H₂O

     The structures of the two closely related Rh^{3+} metal-cage species,
[Rh(MENOsar)]ZnCl₄.Cl.3H₂O (**a**), and [Rh(MENOtar)]ZnCl₄.Cl.3H₂O (**b**), have been
determined.[5] The trimethylene linkages in (**b**) give rise to a larger cavity
at the center of the ligand and force the Rh-N distances [2.102(3) to
2.117(3)Å] to be significantly longer than in (**a**) [2.056(3) to 2.077(3)Å].
The twist angle about the approximate three-fold axis changes from 52° for (**a**)
to 66° for (**b**).

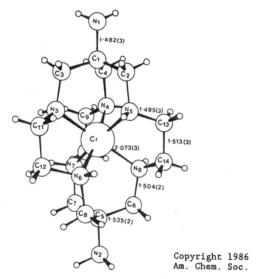

A similar structure for [Cr(diamsar)]Cl₃.H₂O has been reported,[6] as shown in Fig. 4.5-4. The average Cr-N distance in the encapsulated ion is 2.070Å, which is similar to that in the $[Cr(NH_3)_6]^{3+}$ ion (2.064Å). The twist angle is ~50° compared to 60° for an idealized octahedral complex.

Fig. 4.5-4 Structure of the Cr(diamsar)³⁺ cation. (After ref. 6).

3. Phospha- and arsa-capped cage molecules[7,8]

Based-induced condensation of [Co(sen)]³⁺, where sen = 4,4',4"-ethylidynetris(3-azabutan-1-amine), with paraformaldehyde and phosphine or arsine in triethylamine has led to the formation of encapsulated cobalt(III) ions with apical P or As atoms. X-Ray analysis of [Co(Mearsasar)](PF₆)₃.3H₂O revealed an unusual oblique conformation of the arsena-capped encapsulating ligand, as shown in Fig. 4.5-5(c).[8] The enantiomeric conformational stereoisomers found in the crystal (space group $C2/c$) may be designated Δ-$ob_3(R,R)$ and Λ-$ob_3(S,S)$, where ob_3 indicates that the en C-C bonds are oblique to the pseudo C_3 axis of the ion, and (R,R) and (S,S) describe the right- or left-handed helical relationship between the N-C bonds of the caps and the C_3 axis.[9]

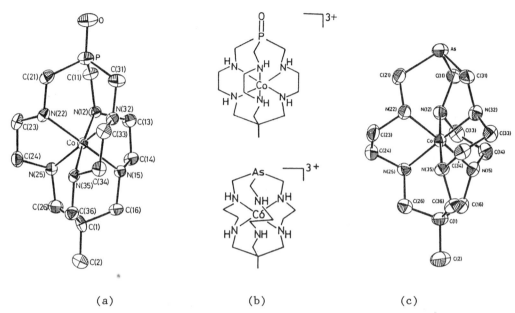

(a)   (b)   (c)

Fig. 4.5-5 (a) Molecular structure of [Co(Me,Ophosphasar)]$^{3+}$ in the Cl[ZnCl$_4$] salt; Co-N($n2$) 1.982(5)-1.993(5), Co-N($n5$) 1.969(4)-1.976(5)Å. (b) Structural formulae of the cage molecular ions. (c) Molecular structure of the cation in [Co(Mearsasar)](PF$_6$)$_3$.3H$_2$O; Co-N($n2$) 1.998(2)-2.008(2), Co-N($n5$) 1.997(2)-1.994(2)Å. (After refs. 7 and 8).

The phosphine oxide capped cage cation in [Co(Me,Ophosphasar)]Cl[ZnCl$_4$] has been shown by X-ray analysis to have the lel_3 conformation [Fig. 4.5-5(a)], so that the enantiomeric forms Δ-$lel_3(R,R)$ and Λ-$lel_3(S,S)$ coexist in the crystal (space group $P2_1/n$).[7] The structure also shows that the larger phosphorus cap causes an expansion of the adjacent Co-N($n2$) bonds compared with those of the aza-, arsa- and carbon-capped analogues. The contrasting spectral and electrochemical features of the aqueous [Co(Me,Ophosphasar)]$^{3+}$ and [Co(Mephospharsar)]$^{3+}$ complexes closely resemble those found for carbon-capped cage analogues with constrained Δ-$lel_3(R,R)$ and Δ-$ob_3(S,S)$ comformations, respectively. This implies that [Co(Mephospharsar)]$^{3+}$ has the same Δ-$ob_3(S,S)$ conformation as its arsa analogue.

4. Properties of caged metal ions

The unique properties of encapsulated metal ion complexes have been
discussed by Sargeson and co-workers:

a) Metal-centered redox properties[10]

The redox properties of the macrobicyclic hexaamine cage complexes
parallel those of the simple hexaamine complexes, such as $[Co(NH_3)_6]^{3+}$ or
$[Co(en)_3]^{3+}$, since reduction proceeds by a one-electron step to Co(II), which
is followed by an irreversible reduction of Co(II). From the studies of the
electrochemistry of cobalt amine donor complexes, it appears that the
stabilities of the various oxidation states are influenced by factors such as
(i) the nature of the substituents on the ligands, (ii) the preferred cavity
size of the ligand, (iii) the ability to delocalize electron density from the
cobalt to the ligand, (iv) the ability or inability to undergo stereochemical
change in order to attain a preferred geometry, and (v) outer-sphere effects
such as solvation and ion pairing.

b) Substituents in the apical position and redox potentials[3]

Numerous variations are available for substituents in the apical
position and these have been fully explored.[10,11] The cage substituents
lead to a wide range of redox potentials for the caged ions. They span at
least 0.6V in redox range, i.e., ten orders of magnitude in equilibrium
constant or five orders of magnitude in redox reaction rate for the
Co(II)/(III) systems.[10] If the other metal ion couples are included the
total redox span is ~2V (~ +1V → -1V) at intervals of -0.1V using the various
substituents, as listed in Table 4.5-2. This gives an impressive array of
redox reagents, although they are not all equally accessible. The magnitude
of the variation is somewhat surprising in view of the distance of the
substituent from the metal ion and it implies relatively close communication
between the apical substituents and the metal centers. This communication is
also reflected in the NMR chemical shifts of the apical C atoms and the large
coupling with the metal in the cage, but it is not reflected much in the
ligand field spectra. It has been argued therefore that these effects arise
from the repulsion of electronic states of the same symmetry transmitted
through the bonds.

c) Strain effect and LFSE[3]

For the Co(III)/(II) sepulchrate 3+/2+ ions, despite the "insulating
organic coat", the electron self-exchange rate is ~10^5 fold greater than that
for the parent $Co(en)_3^{2+/3+}$ ions. In both sets of ions the ligand field
spectroscopy, magnetism and Co-N bond lengths (±.01Å) match closely, i.e. the
electronic conditions of the pairs match. The only explanation for the

Table 4.5-2   Redox potentials at 25°C, $\mu = 0.2$ versus NHE.

				E/V	$[Msar]^{2+/3+}$	E/V
O_2N-	Co	$-NO_2$	2+/3+	+0.06	M =	
H_3N-	Co	$-NH_3$	4+/5+	+0.04	Cr	-1.0
H_3C-	Co	$-NH_3$	3+/4+	-0.33	Mn	+0.56
H_2N-	Co	$-NH_2$	2+/3+	-0.32	Fe	+0.10
H-	Co	-H	2+/3+	-0.40	Co	-0.40
Cl-	Co	-Cl	2+/3+	-0.13	Ni	+0.90
HO-	Co	-OH	2+/3+	-0.20	Cu	+1.2
$Co(Me,ClCH)_2absar$			2+/3+	-0.50		
Cosep			2+/3+	-0.30		

increased self-exchange range is the strain generated in the ligand by the encapsulation of the ion.   This shows up in observed distortions of bond angles and torsion angles and in molecular mechanics calculations.[12]   The cage "hole" appears to be a little too large for Co(III) and a little too small for Co(II).   The strain generated in the ligand in both oxidation states aids the stretching of the Co(III)-N bonds and compression of the Co(II)-N bonds.   In this way the $Co(sep)^{3+}$-$Co(sep)^{2+}$ assembly is helped towards the transition state even though the strain effect does not appear in the ground-state metal-ligand bond lengths.   One of the most interesting aspects of this strain effect is the stereochemistry it engenders about the metal for the different ions.   When each cage complex ion is viewed along the C_3 axis and at the angle which the trigonal faces related by the C_3 axis bear to each other, interesting facts emerge from the data as shown in Fig. 4.5-6.

It is evident that for all those ions where the ligand field stabilization energies (LFSE) are small their structures are not much different from that of the metal-free diprotonated ligand.   Also the trend is more towards the trigonal prism than towards the octahedron.   In those ions where the ligand field effect is larger the structures progressively approach that of an octahedron Ni(II) < Cr(III) < Fe(III) < Co(III).   Clearly, the ligand dominates the former cases and the metal the latter.   An apparent exception is the $[Co(sep)]^{2+}$ structure.   Here, however, there are substantial distortions at the angles around the N cap of the ligand and in this manner the distortion about the metal center is relieved.   By comparison, the $[Co(II)di(NH_3)sar]^{4+}$ structure shows relatively regular geometry in the vicinity of the C-cap and the expected trigonal distortion at the metal center.   These distortions are also reflected in the spectroscopy of the complexes in solution and in the solid state.   The triad of Zn(II), Cd(II),

and Hg(II) complexes does not differ greatly in structure despite the range of
M-N bond lengths (2.19, 2.30, 2.35Å) and the three members tend towards the
trigonal prism as the central ion becomes larger.

Not only does the Co-N bond length change substantially during the
electron transfer process (0.17Å) but the stereochemistry of the cage also
alters sharply [twist angle ϕ 56° Co(III) to ϕ 28° Co(II) for Co(sar)$^{2+/3+}$],
and there is a regularity in the sar and di(NH$_2$)sar chemistry. It might be
expected therefore that all the Co(II)/(III) substituted sar and di(NH$_2$)sar
type of cages would have fairly similar electron self exchange rates.

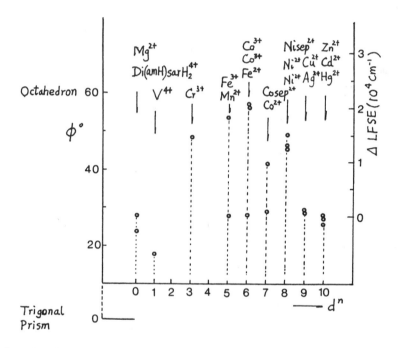

Fig. 4.5-6 The proximity of the metal ion cages to an octahedral or
trigonal prismatic structure. (Adapted from ref. 3).

5. Metal ion template synthesis of molecular knots

The design of synthetic receptors for substrate binding has been
actively pursued in an effort to understand molecular recognition in
biological systems.[13,14] In the synthesis of metal ion receptors, the
preorganization of the reactants within the coordination sphere of a metal
directs the course of the reaction. The enchanced yields resulting from this
"template effect" can be considered as an expression of molecular
recognition.[15]

The elegant synthesis by Sargeson and co-workers of the sepulchrate ion
from [Co(en)$_3$]Cl$_3$, formaldehyde, and ammonia nicely illustrates the principles

of the thermodynamic coordination template effect (Fig. 4.5-7).[2]   The
kinetically inert Co(en)$_3^{3+}$ complex has a rigid C_3 skeleton that is well suited
to the formation of the two aza-caps. The reversible nature of the carbon-
nitrogen bond forming reactions allows the thermally stable products to
accumulate, hence leading to extraordinarily high yields exceeding 90%.

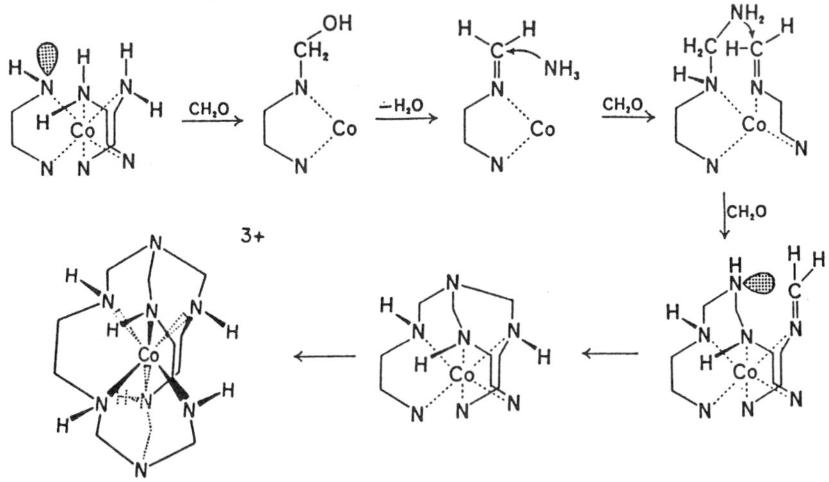

Fig. 4.5-7 Proposed synthesis of the [Co(sep)]$^{3+}$ ion. (After ref. 4).

Some of the most aesthetically appealing molecules assembled with the
template methodology are the "catenands", a novel class of interlocked
macrocycles (catenanes) bearing coordinating sites.[16] The synthetic
strategies based on a three-dimensional templte effect around a transition
metal are outlined in Fig. 4.5-8.[17]

The functionalized ligand 2,9-bis(p-hydroxyphenyl)-1,10-phenanthroline
(I) is an important precursor of catenane synthesis. In the presence of
[Cu(CH$_3$CN)$_4$]BF$_4$, two ligands I fit together to form the very stable copper(I)
complex II, which yields the catenate III by cyclization with the diiodo
derivative of pentaethylene glycol. Demetallation of III with potassium
cyanide affords the free ligand IV with the catenane structure
(Fig. 4.5-9).[18]

The shapes of III and IV, as established by crystallographic studies,
are quite different despite the fact that both have the same bond connectivity
and molecular topology (Fig. 4.5-10).[19]

The synthesis of an intriguing "molecular trefoil knot" has been
realized using two transition metal centers connected by two bis-chelated
molecular links (Fig. 4.5-8, bottom).[17] Starting with the precursor V, the
mononuclear complex VI is formed along with a smaller fraction of the double-

helical complex **VII**. Ring-closure reactions with $ICH_2(CH_2OCH_2)_5CH_2I$, carried
out on the reaction mixture, yield **VIII** and a small amount of the knotted
system **IX**. X-Ray analysis has shown that **VIII** has pseudo-D_2 symmetry, the two
metal atoms being located 6.3Å apart inside a double helix.[20]

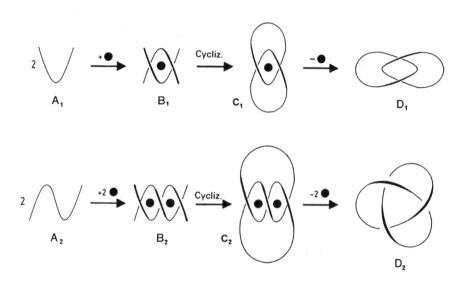

Fig. 4.5-8 Template synthesis of interlocked macrocyclic ligands, the
catenands. Top: the function of the transition metal center is to
gather two linear fragments A_1 and to entwine them, leading to B_1.
After cyclization of B_1 using the proper connections, C_1 is formed. Its
demetalation affords the catenand D_1. Bottom: strategy towards a
trefoil knot using two transition metal centers as templating species.
Two coordinating molecular threads A_2 are interlaced on two transition
metal centers, forming a double helix B_2. After cyclization to C_2 and
demetalation, a knotted system D_2 is obtained. (After ref. 17).

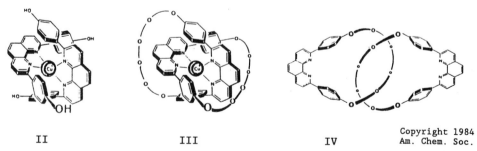

Fig. 4.5-9 One-pot template synthesis of catenate **III** and its
decomplexation, leading to the catenand **IV**. (After ref. 18).

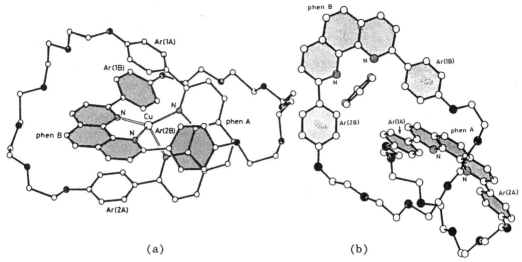

Fig. 4.5-10 Molecular structures of (a) the copper(I) catenate **III** and (b) the corresponding catenand **IV** (an adduct benzene molecule is also shown). (After ref. 19).

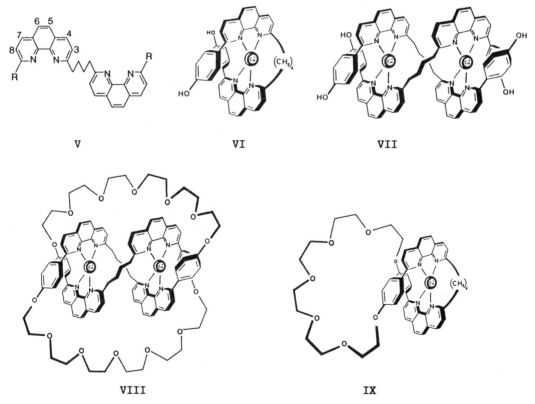

From a tetrafunctional double helicoidal precursor, the most probable 2+2 connections are depicted in Fig. 4.5-11.

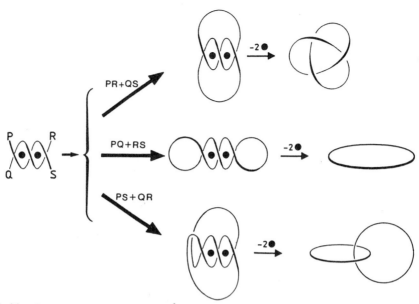

Fig. 4.5-11 Some possible cyclization pathways for a tetra-functionalized dinuclear helicoidal complex. With 1,10-phenanthroline-copper(I) based systems, the most probable connections correspond to reaction PR + QS, leading to the desired knot **IX**. (After ref. 17).

Quantitative demetallation of the copper(I) complexes **VIII** and **IX** leads to the 43-membered ring **X** and the free trefoil knot **XI**, respectively. A tiny amount of **XII**, the unraveled topological iscmer of **XI**, was also obtained by demetallating impure fractions from the chromatographic purification of **IX**.

6. C-H Activation in a coordinated catenand[21]

The insertion of Ni(II), Cu(I), Zn(II), Co(II), and Li into the catenand ligand cat30 results in a distorted tetrahedral stereochemistry at the metal center.[16] Reaction of Pd(OAc)$_2$ with cat30 in refluxing CH$_2$Cl$_2$ yielded the orange compound [Pd(cat30-H)]PF$_6$. An X-ray crystallographic study revealed

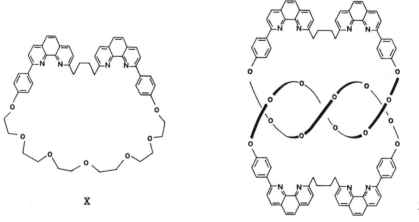

X   XI

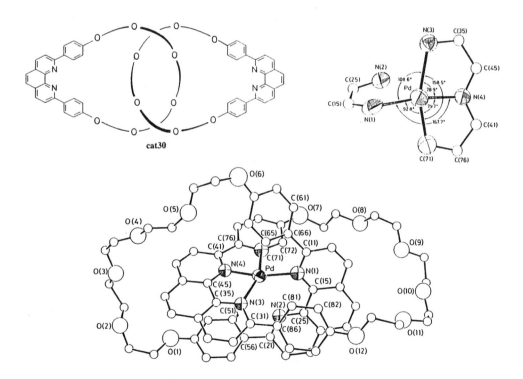

XII

that the monocation was formed via *ortho*-metallation of one of the ligand
phenyl rings, resulting in distorted square-planar coordination of Pd(II) with
Pd-N(1) = 2.139(11), Pd-N(3) = 2.286(10), Pd-N(4) = 1.961(10) and Pd-C(71) =
2.017(7)Å (Fig. 4.5-12).[21] The N(2) atom is 2.559(11)Å from the metal
center, so that one macrocyclic component of the catenand serves as a
tridentate ligand while the other acts in a monodentate fashion.

Fig. 4.5-12. Structural formula of cat30, molecular structure of the
[Pd(cat30-H)]⁺ ion, and coordination geometry about the Pd(II) center.
(After ref. 21).

7. Inorganic double helices

    Double-stranded helical polymetallic complexes have been synthesized by the template strategy outlined in Fig. 4.5-8 (species B_2). Suitable ligands for generating the flexible but non-labile double-helical polynucleating environments are the oligopyridines[22] and, in particular, oligobipyridines containing n 2,2'-bipyridine subunits separated by $(n-1)$ -CH_2OCH_2- bridges (n = 2 to 5).[23,24] In the double-helical binuclear cation in [Pd(quinquepy)]$(PF_6)_2$·2MeCN, each metal atom is in an irregular five-coordinate environment with normal bonds (1.941-2.085Å) to a "terpyridyl" fragment (rings A, B and C) of one ligand and a terminal pyridine (E') from the other ligand, and a long bond of ~2.6Å to the remaining pyridine (D') of the second ligand (Fig. 4.5-13).[22]

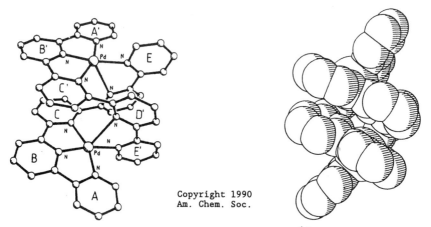

Fig. 4.5-13 Two views of the [Pd$_2$(quinquepy)$_2$]$^{4+}$ ion. (Afte ref. 22).

    The most fruitful approach to the spontaneous assembly of "double-stranded helicates" employs Cu(I) ions and oligobipyridine ligands such as BP_2 and BP_3. The -CH_2OCH_2- "spacers" facilitate strain-free pseudo-tetrahedral coordination at successive metal centers by the bipyridine subunits from two ligand chains [Fig. 4.5-14(a) and (b)].

BP$_2$

BP$_3$

    In the crystal structure of [Cu$_3$(BP$_3$)$_2$]$(CF_3COO)_3$·5H$_2$O, the trinuclear cation does not possess full D_2 molecular symmetry as represented in Fig. 4.5-14(a) and (b).[23] Whereas three bipyridine units are parallel to one

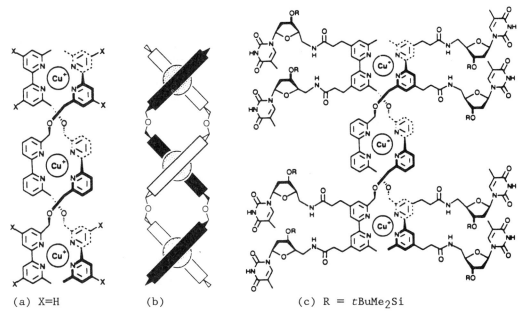

(a) X=H   (b)   (c) R = tBuMe$_2$Si

Fig. 4.5-14 Structual formula (a) and schematical representation (b) of the double-stranded trihelicate complex cation in [Cu$_3$(BP$_3$)$_2$]$^{3+}$(CF$_3$COO$^-$)$_3$.   (c) Deoxyribonucleohelicate structure.   (After refs. 23 and 26).

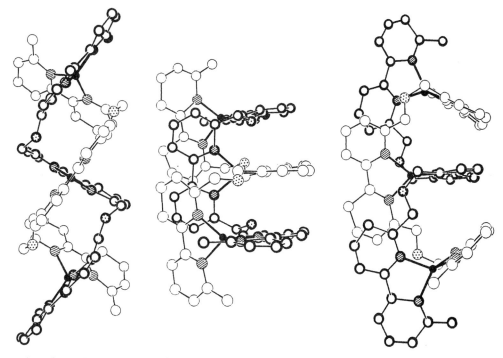

Fig. 4.5-15 Three views of the crystal structure of the double-stranded helicate complex [Cu$_3$(BP$_3$)$_2$]$^{3+}$; to facilitate perception of the structure one strand has been drawn in heavy lines; solid circles, Cu; hatched circles, N; stippled circles, O.   (After ref. 23).

another, the other three are not, and consequently only one molecular C_2 axis is retained (Fig. 4.5-15). In this inorganic double helix both strands are of the same chirality [Fig. 4.5-15(a)], and each has a pitch of ~12Å and a radius of ~6Å for the circumscribed cylinder. The parallel bipyridines [Fig. 4.5-15(b)] have an inter-stack separation of 3.6Å, and the shortest distance between the nonparallel bipyridines [Fig. 4.5-15(c)] is ~3.4Å. Whereas the nonparallel bipyridines overlap by more than one pyridine ring [bases at the left side of Fig. 4.5-15(b)], the overlap of the parallel bipyridines is reduced to about 1/3 of a pyridine ring [Fig. 4.5-15(c)]. It should be noted that a dynamic helix bending-twisting process must be present in solution, rapidly interchanging the roles of the two sets of bipyridines to confer average D_2 molecular symmetry to the $[Cu_3(BP_3)_2]^{3+}$ complex.

The complexation of the Ag(I) ion by oligobipyridine ligands in helix self-assembly has been achieved in the synthesis of trinuclear, tetranuclear, and pentanuclear helicates.[25] X-Ray analysis of $[Ag_3(BP_3)_2](CF_3SO_3)_3$ has revealed a double helical structure for the trinuclear cation, which is analogous to that of the corresponding Cu(I) helicate.

The use of suitably functionalized oligobipyridine compounds, such as the ligand shown in Fig. 4.5-14(a) with $X = CH_2CH_2CO_2Bu^t$, allows the introduction of deoxynucleoside substituents onto the periphery of the double-stranded helicate backbone.[26] The structure of a deoxyribonucleohelicate containing eight 6'-amino-3'-deoxythymidine (as the soluble silylated derivative) residues is illustrated in Fig. 4.5-14(c). Several other oligonucleosidic trihelicates and pentahelicates having aminodeoxy-thymidine and aminodeoxy-adenosine substituents have been synthesized.[26] These interesting deoxyribonucleohelicates are "inside-out analogues of DNA" carrying an overall positive charge within the strands. Their novel structures offer vast opportunities for the investigation of an entirely new aspect of DNA-molecule interactions.

References

[1] I.I. Creaser, R.J. Geue, J. MacB. Harrowfield, A.J. Herlt, A.M. Sargeson, M.R. Snow and J. Springborg, *J. Am. Chem. Soc.* **104**, 6016 (1982).

[2] I.I. Creaser, J. MacB. Harrowfield, A.J. Herlt, A.M. Sargeson, J. Springborg, R.J. Geue and M.R. Snow, *J. Am. Chem. Soc.* **99**, 3181 (1977).

[3] A.M. Sargeson, *Pure Appl. Chem.* **56**, 1603 (1984).

[4] A.M. Sargeson, *Chem. Brit.* **15**, 23 (1979).

[5] R.J. Geue, M.B. McDonnell, A.M. Sargeson and A.C. Willis, *Acta Crystallogr., Sect. A* **43** Supplement, C-181 (1987).

[6] P. Comba, I.I. Creaser, L.R. Gahan, J.M. Harrowfield, G.A. Lawrance, L.L. Martin, A.W.H. Mau, A.M. Sargeson, W.H.F. Sasse and M.R. Snow, *Inorg. Chem.* **25**, 384 (1986).

[7] A. Höhn, R.J. Geue, A.M. Sargeson and A.C. Willis, *J. Chem. Soc., Chem. Commun.*, 1644 (1989).

[8] A. Höhn, R.J. Geue, A.M. Sargeson and A.C. Willis, *J. Chem. Soc., Chem. Commun.*, 1648 (1989).

[9] *Inorg. Chem.* **9**, 1 (1970). [IUPAC nomenclature.]

[10] A.M. Bond, G.A. Lawrance, P.A. Lay, and A.M. Sargeson, *Inorg. Chem.* **22**, 2010 (1983).

[11] A.M. Sargeson, *Pure Appl. Chem.* **50**, 905 (1978).

[12] D. Geselowitz, *Inorg. Chem.* **20**, 4457 (1981).

[13] J.-M. Lehn, *Science (Washington)* **227**, 849 (1985).

[14] T.J. Meade and D.H. Busch, *Prog. Inorg. Chem.* **33**, 59 (1985).

[15] T.J. McMurry, K.N. Raymond and P.H. Smith, *Science (Washington)* **244**, 938 (1989); D.H. Busch and N.A. Stephenson, *Coord. Chem. Rev.* **100**, 119 (1990).

[16] C.O. Dietrich-Buchecker and J.-P. Sauvage, *Chem. Rev.* **87**, 795 (1987); C.O. Dietrich-Buchecker, C. Hemmert, A.-K. Khémiss and J.-P. Sauvage, *J. Am. Chem. Soc.* **112**, 8002 (1990).

[17] C.O. Dietrich-Buchecker and J.-P. Sauvage, *Angew. Chem. Int. Ed. Engl.* **28**, 189 (1989).

[18] C.O. Dietrich-Buchecker, J.-P. Sauvage and J.-M. Kern, *J. Am. Chem. Soc.* **106**, 3043 (1984).

[19] M. Cesario, C.O. Dietrich-Buchecker, J. Guilhem, C. Pascard and J.-P. Sauvage, *J. Chem. Soc., Chem. Commun.*, 244 (1985).

[20] C.O. Dietrich-Buchecker, J. Guilhem, C. Pascard and J.-P. Sauvage, *Angew. Chem. Int. Ed. Engl.* **29**, 1154 (1990).

[21] A.J. Blake, C.O. Dietrich-Buchecker, T.I. Hyde, J.-P. Sauvage and M. Schröder, *J. Chem. Soc., Chem. Commun.*, 1663 (1989).

[22] E.C. Constable, S.M. Elder, J. Healy and M.D. Ward, *J. Am. Chem. Soc.* **112**, 4590 (1990).

[23] J.-M. Lehn, A. Rigault, J. Siegel, J. Harrowfield, B. Chevrier and D. Moras, *Proc. Natl. Acad. Sci. USA* **84**, 2565 (1987).

[24] J.-M. Lehn and A. Rigault, *Angew. Chem. Int. Ed. Engl.* **27**, 1095 (1988).

[25] T.M. Garrett, U. Koert, J.-M. Lehn, A. Rigault, D. Meyer and J. Fischer, *J. Chem. Soc., Chem. Commun.*, 557 (1990).

[26] U. Koert, M.M. Harding and J.-M. Lehn, *Nature (London)* **346**, 339 (1990).

Note Electro-reductive crystallization of the macrocyclic sodium tris(bipyridine) cryptate, [Na$^+$⊂tris(bpy)]Br$^-$, in DMF yields deep-violet *sodio-cryptatium*, a neutral species formulated as [Na$^+$ ⊂(bpy$^-$)(bpy)$_2$]$^\bullet$ in which the encaged Na$^+$ ion lies closer to one of the bpy units on which the unpaired electron is located. See L. Echegoyen, A. DeCian, J. Fischer and J.-M. Lehn, *Angew. Chem. Int. Ed. Engl.* **30**, 838 (1991).

4.6   Chlorotris(triphenylphosphine)rhodium(I)
(Wilkinson's Catalyst)
$$RhCl[P(C_6H_5)_3]_3$$

Crystal Data

Orange form

Orthorhombic, space group *Pna2₁* (No. 33)

$a = 19.470(3)$, $b = 12.689(2)$, $c = 18.202(3)$Å, $Z = 4$

Red form

Orthorhombic, space group *Pna2₁* (No. 33)

$a = 32.96(1)$, $b = 12.271(2)$, $c = 11.007(2)$Å, $Z = 4$

Atom*	Orange form			Red form		
	x	y	z	x	y	z
Rh	.06304	.02892	1/4	.11298	.12204	-1/4
Cl	-.0508	.0726	.2088	.0770	.0895	-.4336
P(1)	.0800	.2085	.2555	.12326	-.0662	-.2471
P(2)	.1723	-.0199	.2607	.1655	.1689	-.1363
P(3)	.0154	-.1362	.2757	.0729	.2767	-.2305

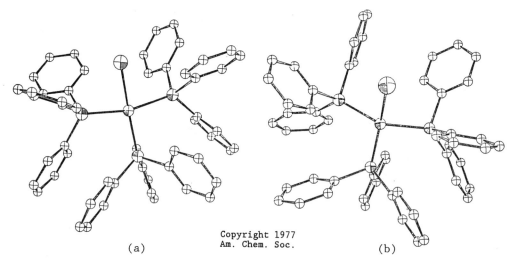

(a)   (b)

Fig. 4.6-1 Molecular structure of RhCl(PPh₃)₃: (a) orange form and (b) red form. (After ref. 1).

Crystal Structure

The important catalytic reagent "Wilkinson's catalyst", RhCl(PPh₃)₃, crystallizes in orange and red forms depending upon the conditions of synthesis. The molecular structure in both allotropes can be described to a first approximation in terms of square-planar coordination. There is however a distortion toward tetrahedral geometry which is more marked in the case of the red allotrope. Fig. 4.6-1 shows the structures of the two forms projected onto the mean molecular plane. Selected bond lengths and intramolecular contacts are listed in Table 4.6-1. The pattern of bond lengths is consistent with the relative π acidity of the ligands, the *trans* effect of phosphorus being greater than that of chlorine.[2] So the Rh-P distance *trans* to a triphenylphosphine ligand is 0.1Å longer than that *trans* to the chloro ligand.

Table 4.6-1 Selected bond lengths and intramolecular contacts.

Atoms	Distance/Å	
	Orange form	Red form
Rh-Cl	2.404	2.376
Rh-P(1)	2.304	2.334
Rh-P(2)	2.225	2.214
Rh-P(3)	2.338	2.322
Rh...H(222)	2.84	-
Rh...H(236)	2.94	-
Rh...H(112)	-	2.77
Rh...H(216)	-	2.94
Cl-Rh-P(1)	85.3°	85.3°
Cl-Rh-P(3)	84.5°	86.1°
P(2)-Rh-P(3)	96.4°	100.4°
P(1)-Rh-P(2)	97.7°	97.9°

Remarks

1. Wilkinson's catalyst

The prototype Wilkinson's catalyst, $RhCl(PPh_3)_3$, was discovered in 1965.[3] It undergoes a variety of reactions, most of which involve either replacement of a phosphine ligand (e.g. with CO, CS, C_2H_4, O_2 giving *trans* products) or oxidative addition (e.g. with H_2, CH_3I) to form Rh(III), but its importance arises from its effectiveness as a homogeneous catalyst for highly selective hydrogenation of complicated organic molecules at ambient temperatures and pressures.[4-6] Wilkinson's catalysts may be either neutral molecules or cationic ions. There are three main types of compounds:[7]

a) Those without a metal-hydride bond such as $RhCl(PPh_3)_3$ and $[Rh(S)_2(PR_3)_3]^+$ (S = solvent), which react, usually reversibly, with molecular hydrogen.

b) Those with an M-H bond such as $RhH(CO)(PPh_3)_3$ or $RuHCl(PPh_3)_3$, which do not usually react with H_2.

c) f-Block element hydrides such as $(Cp_2^*LuH)_2$, whose catalytic cycles do not involve oxidative-addition reactions.

The precise mechanism is complicated and has been the subject of much speculation and controversy; Fig. 4.6-2 shows a simplified but reasonable scheme. The complex $RhCl(PPh_3)_3$ dissociates one of its PPh_3 ligands (L) in solution to give complex [A], which is coordinated by a solvent molecule S. Oxidative addition of H_2 to [A] gives a dihydride complex [B]. The same dihydride can also be produced, though much less rapidly, by direct oxidative addition of H_2 to undissociated $RhCl(PPh_3)_3$, giving [E] and subsequent dissociation of triphenylphosphine. The octahedral complex [C] is formed by displacement of S from [B] by an olefin. The olefin inserts into the Rh-H bond to give the hydride-alkyl species [D], and the alkane is reductively eliminated to regenerate [A]. The coordinatively unsaturated complex [A] further reacts with H_2 to continue the catalytic cycle.

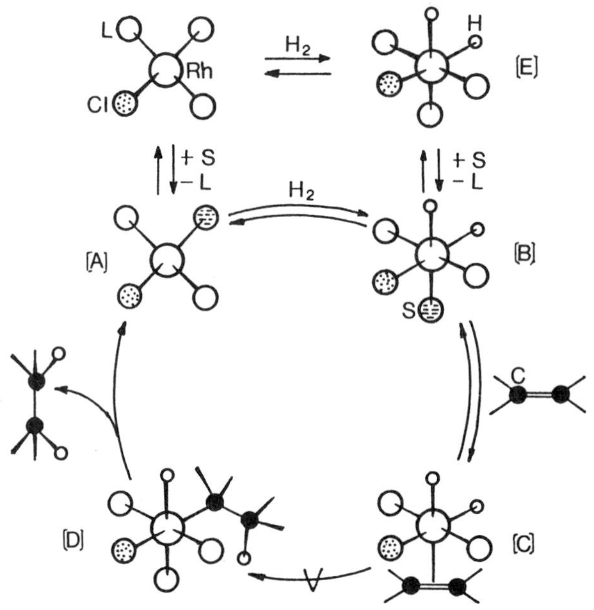

Fig. 4.6-2 Machanism of olefin hydrogenation by Wilkinson's complex. (After ref. 8).

2. Rhodium-triphenylphosphine catalyzed hydroformylation of olefins.[9]

The discovery of the catalytic properties of $RhCl(PPh_3)_3$ naturally brought about a widespread search for other rhodium phosphine complexes with catalytic activity. $[Rh(CO)_2H(PPh_3)_2]$, which selectively catalyses the hydrogenation of 1-alkenes rather than 2-alkenes, has been used in the hydroformylation of alkenes. This is a process of enormous industrial importance, it being used to convert 1-alkenes into aldehydes which can then be transformed to alcohols for the production of polyvinylchloride and polyalkenes and, in the case of the long-chain alcohols, for the production of detergents:

$$RCH=CH_2 + H_2 + CO \xrightarrow{\text{catalyst}} RCH_2CH_2CHO$$

Wilkinson has proposed both dissociative and associative mechanisms, as outlined in Fig. 4.6-3.[10] The complex $Rh(CO)_2H(PPh_3)_2$ was selected as the key intermediate, even though an equilibrium between several species may exist in solution. This selection was substantiated by the observation that, if Rh(CO)H(PPh₃)₂ and Rh(CO)₂H(PPh₃)₂ are present together in solution, only the latter reacts with ethylene at 25°C and 1 atm, as shown by NMR measurements.

Inspection of Fig. 4.6-3 shows that the associative pathway (b) affords more steric hindrance to the coordinating olefin and would be expected to provide preferential formation of the linear alkyl rhodium intermediate. The associative mechanism is preferred at high concentrations of catalyst and triphenylphosphine.

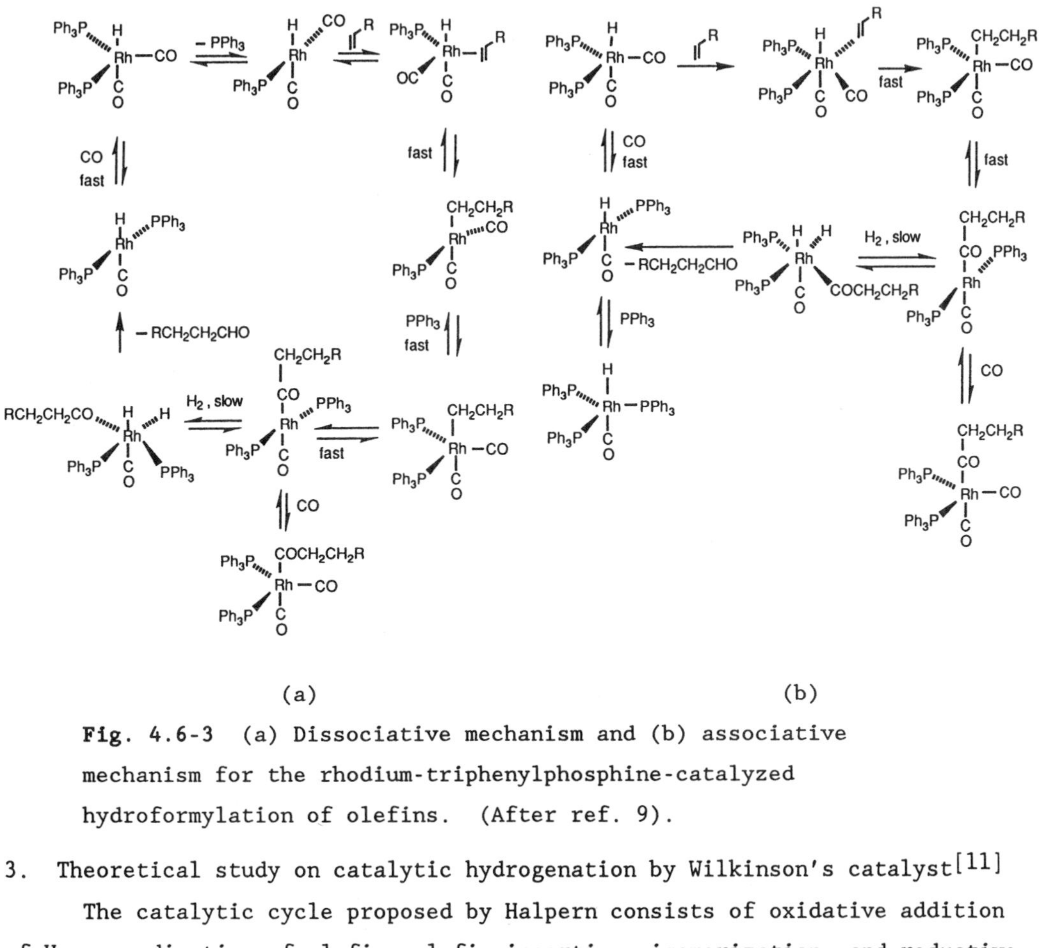

(a)   (b)

Fig. 4.6-3 (a) Dissociative mechanism and (b) associative mechanism for the rhodium-triphenylphosphine-catalyzed hydroformylation of olefins. (After ref. 9).

3. Theoretical study on catalytic hydrogenation by Wilkinson's catalyst[11]

The catalytic cycle proposed by Halpern consists of oxidative addition of H_2, coordination of olefin, olefin insertion, isomerization, and reductive elimination of alkane [Fig. 4.6-4(a)].[12] The rate-determining step is believed to be the olefin-insertion step. In a recent theoretical study, PH_3 was used instead of PR_3 and C_2H_2 as a model of olefin in the simplified catalytic cycle shown in Fig. 4.6-4(b). All the equilibrium and transition

state structures were optimized by the RHF energy gradient method. The basis
functions used were: 3-21G for ethylene and hydrides, STO-2G for "spectator
ligands" PH_3 and Cl, and valence double zeta for Rh with effective core
potential replacing the core electrons (up to 4p). The MP2 calculations at
selected, RHF-optimized structures employed a larger basis set consisting of
uncontracted (3s,3p,4d) functions, the valence double zeta set for Rh, 4-31G
for the ethyl group, (10s,7p)/[3s2p] for P and Cl, and (4s)/[3s] for the

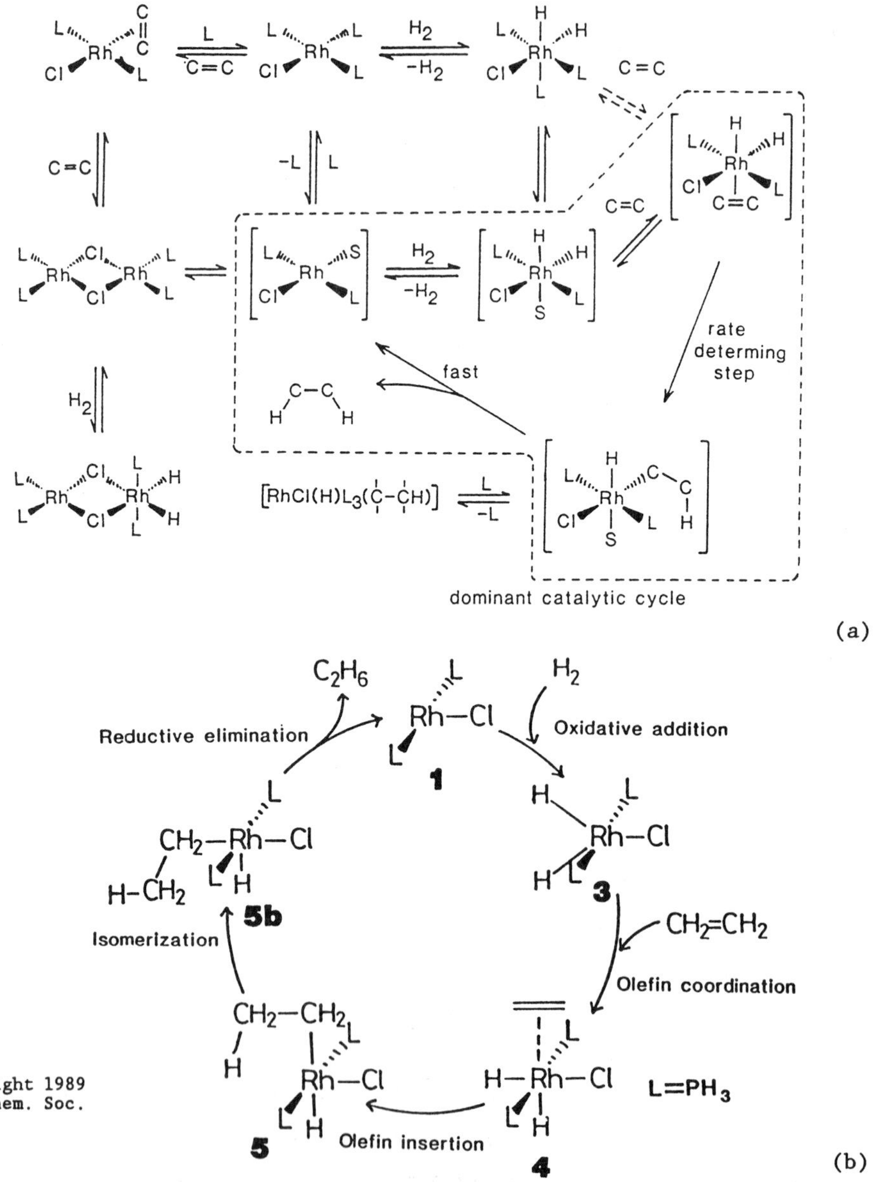

(a)

(b)

Fig. 4.6-4 (a) The catalytic cycle of hydrogenation by Wilkinson's catalyst
as proposed by Halpern. (b) The model catalytic cycle. (After ref. 11).

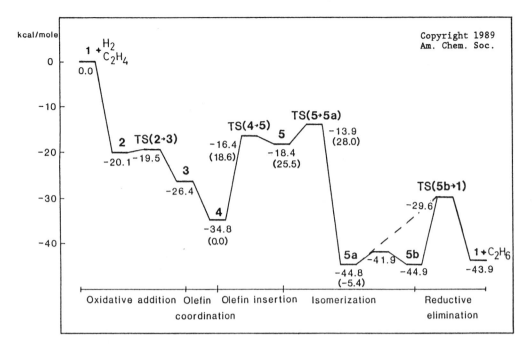

Fig. 4.6-5 Potential energy profile of the entire catalytic cycle in the Halpern mechanism for olefin hydrogenation, in kcal mol^{-1} at the RHF level, relative to $1 + C_2H_4 + H_2$. Numbers in parentheses are the MP2 energy at the RHF optimized geometries, relative to 4. (After ref. 11).

hydrides. The potential energy profile shown in Fig. 4.6-5 is constructed from the energetics of the elementary reactions involved in the Halpern mechanism. The optimized structures are shown in Fig. 4.6-6.

The first step of the H_2 oxidative addition is exothermic and leads to the dihydride complex 3. During this step, there may be an H_2 complex 2 from which oxidative addition takes place with almost no activation barrier. The ethylene coordination that follows requires no activation energy. The resulting ethylene dihydride complex 4 is in the valley of the potential energy surface of the catalytic cycle. Ethylene insertion requires a much higher activation energy of 18 kcal/mol and is endothermic by 16 kcal/mol at the RHF level. The *trans* ethyl hydride complex 5, the direct product of ethylene insertion, is unstable owing to the *cis* effect of Cl and the *trans* effect of H and C_2H_5. Therefore isomerization takes place to give more stable ethyl hydride complexes, which have ethyl and hydride ligands *cis* to each fsother. The final reductive elimination step requires a substantial energy barrier of 15 kcal mol^{-1}.

The potential energy profile is smooth without excessive barriers and too stable intermediates which would disrupt the sequence of steps (Fig. 4.6-5). The rate-determining step is found to be olefin insertion followed by isomerization, supporting the Halpern mechanism. The activation barrier of reductive elimination is smaller than that of the reverse of the rate-determining step, which is an important requirement of a good olefin hydrogenation catalyst.

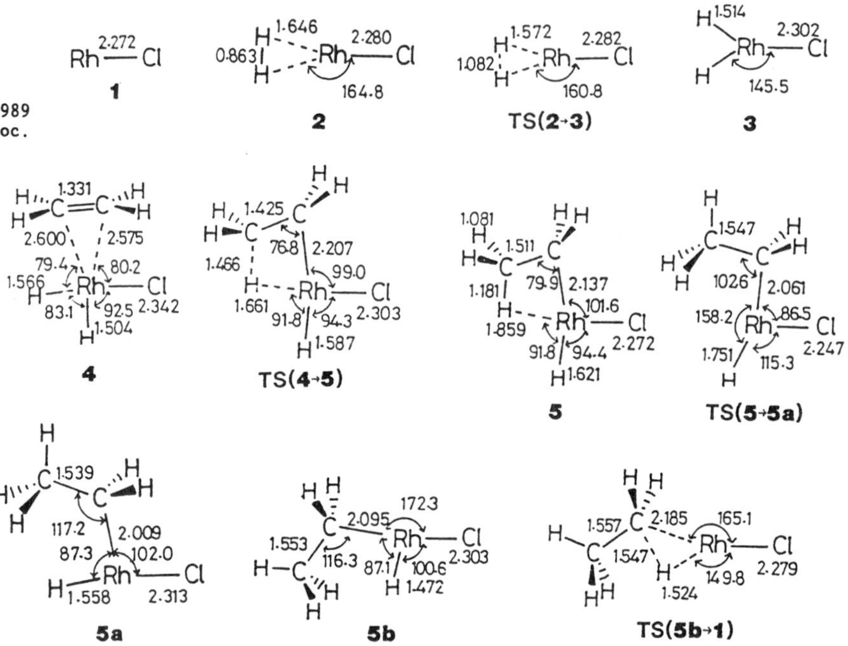

Fig. 4.6-6 Optimized structures (in Å and deg) of some important species. TS(2→3), for instance, denotes the transition state connecting 2 and 3. Though practically all the geometrical parameters were optimized, only essential values are shown. Two PH_3 groups, one above and one below the plane of paper, are omitted for clarity. (After ref. 11).

4. Structure of $RhHCl(SiR_3)(PPh_3)_2$ and hydrosilylation[13]

The Si-H bond of hydrosilanes is easily cleaved by oxidative addition to $RhCl(PPh_3)_3$:

$$RhCl(PPh_3)_3 + SiHCl_3 \longrightarrow RhHCl(SiCl_3)(PPh_3)_2 + PPh_3$$

The crystal structure of a hydridorhodium(III) silyl complex has been determined.[14] The complex is trigonal bipyramidal with the H and Cl ligands located at the axial positions. The five-coordinate species are potent catalytic intermediates towards substrates containing multiple interatomic linkages or an active hydrogen atom, as shown in Fig. 4.6-7.

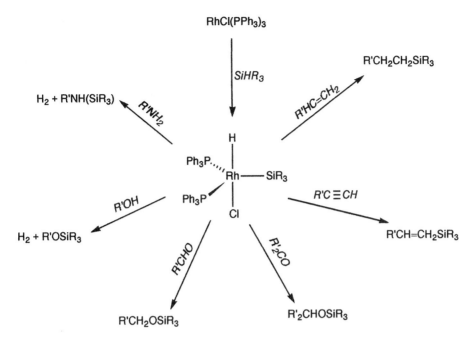

Fig. 4.6-7 Reactions of hydridorhodium(III) silyl complexes. (After ref. 13).

References

[1] M.J. Bennett and P.B. Donaldson, *Inorg. Chem.* **16**, 655 (1977).

[2] R. Mason and A.D.C. Towl, *J. Chem. Soc. (A)*, 1601 (1970).

[3] J.F. Young, J.A. Osborne, F.H. Jardine and G. Wilkinson, *J. Chem. Soc., Chem. Commun.*, 131 (1965).

[4] G.W. Parshall, *Homogeneous Catalysis*, Wiley-Interscience, New York, 1980.

[5] D.P. Arnold, M.A. Bennett, G.T. Crisp and J.C. Jeffrey in E.C. Alyea and D.W. Meek (eds.) *Catalytic Aspects of Metal Phosphine Complexes*, American Chemical Society, Washington, D.C., 1982, chap. 11.

[6] L.H. Pignolet (ed.), *Homogeneous Catalysis by Metal Phosphine Complexes*, Plenum, New York, 1983.

[7] F.A. Cotton and G. Wilkinson, *Advanced Inorganic Chemistry*, 5th ed., Wiley, New York, 1988, p. 1245.

[8] A. Yamamoto, *Organotransition Metal Chemistry*, Wiley, New York, 1986, pp. 362-364.

[9] R.L. Pruett, *Adv. Organomet. Chem.* **17**, 1 (1979).

[10] C.K. Brown and G. Wilkinson, *J. Chem. Soc. (A)*, 2753 (1970).

[11] N. Koga and K. Morokuma in D.R. Salahub and M.C. Zerner (eds.), *The Challenge of d and f Electrons, Theory and Computation,* American Chemical Society, Washington, DC, 1989, chap. 6.

[12] J. Halpern and C.S. Wong, *J. Chem. Soc., Chem. Commun.,* 629, (1973); J. Halpern in Y. Ishii and M. Tsutsui (eds.), *Organotransition Metal Chemistry,* Plenum, New York, 1975, p. 109; J. Halpern, T. Okamoto and A. Zakhariev, *J. Mol. Catal.* 2, 65 (1976).

[13] F.H. Jardine, *Prog. Inorg. Chem.* **28**, 63 (1981).

[14] K.W. Muir and J.A. Ibers, *Inorg. Chem.* **9**, 440 (1970).

<h1 style="text-align:center">4.7 Potassium Rhenium Hydride</h1>

<h2 style="text-align:center">K₂ReH₉</h2>

Crystal Data

Hexagonal, space group $P\bar{6}2m$ (No. 189)

$a = 9.607(5)$, $c = 5.508(5)$Å, $Z = 3$

Atom	x	y	z
Re(1)	0	0	0
Re(2)	1/3	2/3	1/2
K(1)	.5881	0	0
K(2)	.2610	0	1/2
H(1)	.1789	0	0
H(2)	.8789	0	.2171
H(3)	.1483	.6250	1/2
H(4)	.2240	.5254	.7125

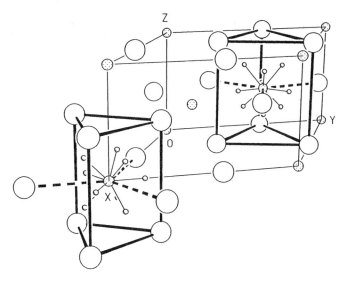

Fig. 4.7-1 Unit cell of K₂ReH₉ showing the coordination environments of the two non-equivalent Re atoms.

Crystal Structrue

The crystal structure of the classical transition-metal hydride complex, K₂ReH₉, has been determined by X-ray[2] and neutron diffraction.[1] In view of the large distances between the rhenium atoms [5.51Å from Re(1) to Re(1) along the c axis and 5.55Å from Re(2) to Re(2), with the Re(1)...Re(2) distance even larger], direct and bridge bonding between rhenium atoms are precluded. Although the Re atoms are crystallographically non-equivalent, their coordination environments turn out to be essentially the same. Each Re is surrounded by nine K⁺ arranged in a trigonal prism with three ions above the centers of the rectangular faces, as shown in Fig. 4.7-1. The averages of the nine Re...K distances in the two different prisms are 3.78 and 3.88Å, respectively. Bond lengths and bond angles in the discrete ReH_9^{2-} ion are given in Table 4.7-1.

Table 4.7-1   Bond lengths(Å) and angles(°) in ReH_9^{2-}.

Re(1)-H(1)	1.72	H(2)-Re(1)-H(2)	91.6°
Re(1)-H(2)	1.67	H(4)-Re(2)-H(4)	95.6°
Re(2)-H(3)	1.61	H(3)-Re(2)-H(4), projected along trigonal	
Re(2)-H(4)	1.70	axis onto (001), 59.1°	
Re-H (av)	1.68		

Remarks

1. Structure types of transition-metal hydride complexes[3]

Structure determinations of several key transition-metal hydride complexes laid the foundation toward an understanding of the geometry of M-H bonds. According to the bonding modes, the complexes may be classified as follows.

a) Terminal M-H bonded compounds

The first hydride complexes to be investigated by the neutron diffraction technique were compounds containing the terminal M-H linkage. The classic investigations of K_2ReH_9[1] and $HMn(CO)_5$[4] established unequivocally that the length of a M-H single bond is in the range 1.6-1.7Å, consistent with what is expected for a normal covalent linkage.

b) M-H-M bridged compounds

The M-H-M bridge bond is of particular interest because it is a member of a select family of bonds: electron-deficient three-center-two-electron bonds. The classic M-H-M bridged system is the $[HCr_2(CO)_{10}]^-$ anion, for which neutron diffraction found that the bridging H atom is in an off-axis position, situated about 0.3Å from the center of the Cr-Cr bond, with Cr-H-Cr = 158.9°, and Cr-H = 1.737 and 1.707Å.[5]

c) Triply-bridged (μ_3-H)M_3 compounds

An H atom covalently bound simultaneously to three other atoms has been found in some metal cluster complexes. The structures of two tetrahedral clusters, $HFeCo_3(CO)_9[P(OMe)_3]_3$ and $H_3Ni_4Cp_4$, have been studied.[6,7]

d) Interstitial hydride compounds

Two kinds of hydride compounds have H atoms inside cavities surrounded by metal atoms. In the first, hydrogen combines with a transition-metal element to form binary hydrides MH_n or M_mH_n. All lanthanides and actinides and V, Nb, Ta, Cr, Cu, Zn, Pd form hydride phases of variable compositions. Some alloys, such as $LaNi_5$, can absorb a large amount of hydrogen to form $LaNi_5H_6$. In the second type of hydride, an interstitial H atom is located within a transition-metal cluster. The $[HCo_6(CO)_{15}]^-$, $[HNi_{12}(CO)_{21}]^{3-}$, and $[H_2Ni_{12}(CO)_{21}]^{2-}$ anions have definitively shown the existence of hydrides inside the octahedral holes of their individual clusters.

e) Molecular hydrogen complexes

Since the discovery of the first stable dihydrogen complex, $W(CO)_3(PPr_3)_2(\eta^2\text{-}H_2)$, many molecular hydrogen complexes have been synthesized and investigated.[8]

2. Structure of $HMn(CO)_5$ and other metal carbonyl hydrides

The structure of $HMn(CO)_5$ as determined by neutron diffraction has C_{4v} symmetry, with the hydrogen directly bonded to the metal: Mn-H = 1.60Å, Mn-C(ax) = 1.822(12)Å, and Mn-C(eq) = 1.853(12)Å.[4]

The molecular structure of the series $HMn(CO)_5$, $H_2Fe(CO)_4$ and $HCo(CO)_4$ has been investigated by gas-phase electron diffraction.[9] The results for $HMn(CO)_5$ are in good agreement with those from the neutron diffraction study, as shown in Fig. 4.7-2(a). In $H_2Fe(CO)_4$ the $Fe(CO)_4$ group has C_{2v} symmetry with a configuration intermediate between octahedral and tetrahedral geometries. The hydrogen atoms are *cis*, with H-Fe-H = 100(10)°. The C-Fe-C angles for the axial and equatorial sets are 148.5(1.5) and 96.0(6)°, respectively. The Fe-H distance is 1.56(2)Å, as shown in Fig. 4.7-2(b). The $HCo(CO)_4$ structure has C_{3v} symmetry, with the hydrogen directly bonded to the cobalt atom at 1.56(2)Å, as shown in Fig. 4.7-2(c).

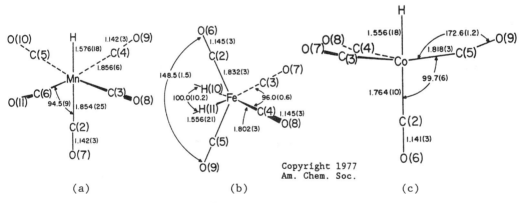

Copyright 1977
Am. Chem. Soc.

(a)   (b)   (c)

Fig. 4.7-2 Molecular configurations of (a) $HMn(CO)_5$, (b) $H_2Fe(CO)_4$, (c) $HCo(CO)_4$. (After ref. 9).

3. Structure of M-H-M bridged compounds

Many transition-metal complexes with M-H-M bonds have had their structures determined by X-ray and neutron diffraction. All contain bent M-H-M bonds, and Fig. 4.7-3 shows the structures of $[HCr_2(CO)_{10}]^-$ and $[HW_2(CO)_{10}]^-$.

The molecular structure of $Ta_2Cl_4(PMe_3)_4H_4$ is shown in Fig. 4.7-4.[10] The pseudo-square planes of chloro and PMe_3 ligands are in the eclipsed

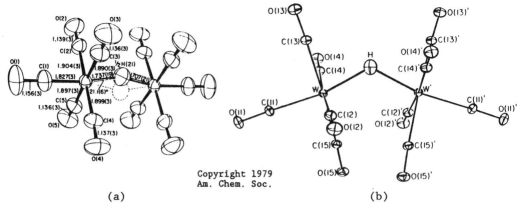

(a)   (b)

Fig. 4.7-3 Structures of (a) $[HCr_2(CO)_{10}]^-$ and (b) $[HW_2(CO)_{10}]^-$.
(After ref. 3).

conformation, the phosphine ligands are staggered among themselves, and the
four bridging H atoms are staggered by 45° with respect to the end groups.
The molecular symmetry, D_{2d}, is imposed by the tetragonal space group.

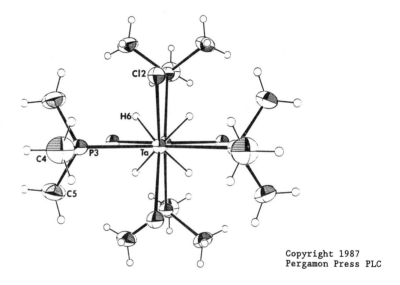

Fig. 4.7-4 Molecular structure of $Ta_2Cl_4(PMe_3)_4H_4$. (After ref. 10).

The bond lengths are Ta-Ta = 2.511(2)Å, Ta-Cl = 2.461(5)Å, and Ta-P =
2.604(5)Å. In the bridge region, the Ta-H distance is 1.81(21)Å, and the
Ta-H-Ta angle 88(4)°. The bond lengths and bond angles of the trimethyl-
phosphine group are normal. Another structurally characterized $(\mu\text{-H})_4$ dimer
is $Re_2H_8(PEt_2Ph)_4$, in which Re-Re = 2.538(4)Å, Re-H(bridge) = 1.878(7)Å, Re-
H-Re = 85.0(3)°, and the two $Re(PEt_2Ph)_2$ groups are in the eclipsed
conformation, as shown in Fig. 4.7-5(a) (neutron diffraction study).[11]

The nature of the bond between bridging hydride ligands and binuclear

tantalum atoms is an interesting problem. It has been found that the bridging
hydrides in $Ta_2Cl_4(PMe_3)_4H_4$ do not rotate, whereas the bridging hydrides in
$Ta_2Cl_4(PMe_3)_4H_2$ rotate rapidly about the Ta-Ta bond axis. To elucidate
this phenomenon Scioly and co-workers[10] calculated the ground state
electronic structures of the model compounds $Ta_2Cl_4(PH_3)_4H_4$ and $Ta_2Cl_4(PH_3)_4H_2$
by the multiple-scattering Xα method.

In $Ta_2Cl_4(PH_3)_4H_4$, the HOMO is a Ta-Ta σ-bonding level, below which lie
four groups of levels: (i) the Ta-P σ-bonding orbitals; (ii) the Cl lone pair
orbitals; (iii) the Ta-Cl σ-bonding orbitals; (iv) the P-H σ-bonding orbitals.
The four remaining occupied levels are those that describe the σ-, π-, π-, and
δ-type interactions between the Ta and four bridging hydrides.

The $Ta_2Cl_4(PH_3)_4H_2$ system is similar to $Ta_2Cl_4(PH_3)_4H_4$. The most
striking change is the splitting of the π-orbitals of $Ta_2Cl_4(PH_3)_4H_4$, one of
which forms a Ta-Ta π-bonding orbital of $Ta_2Cl_4(PH_3)_4H_2$, and the δ-orbital of
Ta-H bonding of $Ta_2Cl_4(PH_3)_4H_4$ changes to a LUMO of $Ta_2Cl_4(PH_3)_4H_2$ and loses
all of its hydridic character. Accordingly, in $Ta_2Cl_4(PMe_3)_4H_4$, the presence
of a δ-interaction effectively "locks" the hydride ligands into the staggered
conformation, and a similar conclusion has been reached in the case of
$Re_2H_8(PH_3)_4$,[12] a model for $Re_2H_8(PEt_2Ph)_4$. In $Ta_2Cl_4(PMe_3)_4H_2$, the loss of
δ-interaction lowers the rotation barrier about the Ta-Ta axis, so the
bridging hydrides rotate rapidly.

The structure of the complex $Re_2H_8(PPh_3)_4$ in the crystalline state
depends on the solvents used to grow crystals.[13] With acetone as solvent,
the crystal composition is $Re_2H_8(PPh_3)_4 \cdot (CH_3)_2CO$, in which the two $Re(PPh_3)_2$
groups are in a staggered arrangement, as shown in Fig. 4.7-5(b). If crystals

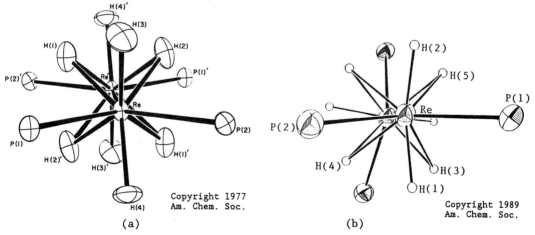

Copyright 1977
Am. Chem. Soc.

(a)

Copyright 1989
Am. Chem. Soc.

(b)

Fig. 4.7-5 Comparison of the $Re_2H_8P_4$ cores of (a) $Re_2H_8(PEt_2Ph)_4$ and of
(b) $Re_2H_8(PPh_3)_4 \cdot (CH_3)_2CO$. (After refs. 11 and 13).

are obtained from a THF solution, the composition is $Re_2H_8(PPh_3)_4 \cdot 2C_4H_8O$, in which the two $Re(PPh_3)_2$ groups have a planar eclipsed conformation as found in $Re_2H_8(PEt_2Ph)_4$ [Fig. 4.7-5(a)].

Some examples of M-H-M bridged compounds of iridium, which have been characterized by X-ray and neutron diffraction, are shown in Fig. 4.7-6.[14]

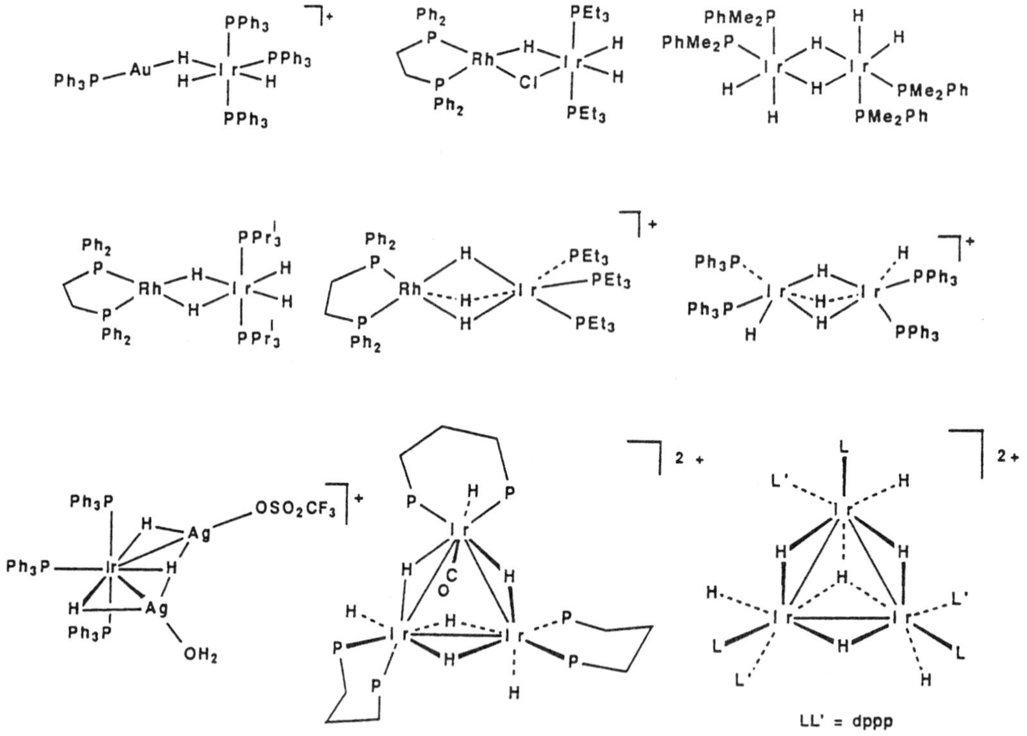

Fig. 4.7-6 Structure of some M-H-M bridged compounds.   (Adapted from ref. 14).

4. Metal hydrido complexes containing tetrahedral or trigonal-bipyramidal cluster frameworks

In $HFeCo_3(CO)_9[P(OMe)_3]_3$, the $FeCo_3$ core forms a tetrahedral cluster, with the H atom displaced 0.978(3)Å from the Co_3 face [Fig. 4.7-7(a)].[6]   The average Co-H and Co-H-Co values are 1.734(4)Å and 91.8(2)°, respectively.   In $H_3Ni_4Cp_4$, the H_3Ni_4 core resembles a cube with a missing corner [Fig. 4.7-7(b)].[7]   The average Ni-H distance is 1.691(8)Å, and the H atoms are displaced outward by 0.907(6)Å from the faces of the tetrahedron.   These two tetrahedral clusters were structurally characterized by neutron diffraction.

The structure of $[AuIr_3(\mu_2\text{-}H)_3H_3(NO)_3(dppe)_3]BF_4$ [Fig. 4.7-8(a)] revealed the $AuIr_3$ unit to be approximately tetrahedral with Ir-Ir =

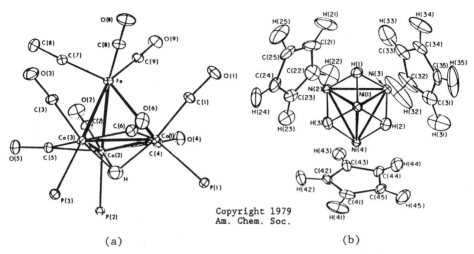

Copyright 1979
Am. Chem. Soc.

(a)   (b)

Fig. 4.7-7 (a) Molecular structure of HFeCo₃(CO)₉[P(OMe)₃]₃, with methoxy groups removed for clarity. (b) Molecular structure of H₃Ni₄Cp₄, with one of the cyclopentadienyl rings removed for clarity. (After ref. 3).

2.805(1)-2.861(1)Å and Ir-Au = 2.696(1)-2.718(1)Å.[15] The X-ray crystal structure of [IrAu₄(PPh₃)₆(H)₂]BF₄ [Fig. 4.7-8(b)] shows that the IrAu₄ core consists of an approximately trigonal-bipyramidal cluster, with the Ir(PPh₃)₂ unit occupying an equatorial position, with Ir-Au = 2.637(1)-2.753(1)Å and Au-Au = 2.794(1)-3.142(1)Å.[16] The positions of the two hydride ligands could not be determined from the X-ray analysis. In the suggested assignment, they bridge the Ir-Au bonds of the two apical gold atoms. The crystal structure of [Ir₂Cu₃H₆(MeCN)₃(PMe₂Ph)₆][PF₆]₃ [Fig. 4.7-8(c)] shows that the metal atoms are in a trigonal-bipyramidal arrangement with the two Ir atoms in the axial positions, with Cu-Ir = 2.793(9) and Cu-Cu = 2.570(22)Å.[17]

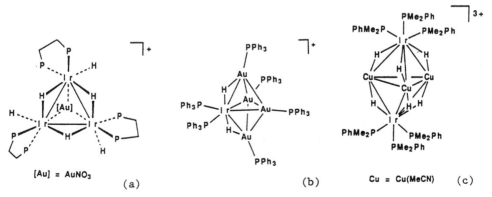

Fig. 4.7-8 Molecular structures of (a) [AuIr₃(μ₂-H)₃H₃(NO)₃(dppe)₃]⁺, (b) [IrAu₄(PPh₃)₆(H)₂]⁺ and (c) [Ir₂Cu₃H₆(MeCN)₃(PMe₂Ph)₆]³⁺. (Adapted from ref. 14).

5. Structures of interstitial hydride complexes

The structure determination of $[HCo_6(CO)_{15}]^-$ [Fig. 4.7-9(a)] is significant in that, for the first time, an "interstitial" H atom in a hexanuclear metal cluster has been unequivocally characterized by single-crystal neutron diffraction.[18] The H-Co distances are nearly equivalent (average = 1.82Å), and the H atom is displaced by less than one standard deviation from the geometric center of the octahedral Co_6 cluster. The Co-Co distances range from 2.50 to 2.67Å, with an average of 2.579Å, resulting in an apparent radius for the interstitial H atom of 1.82-(2.58/2) = 0.53Å. The insertion of a H atom into the cluster causes a small but appreciable swelling of the octahedron: the average Co-Co distance increases from 2.509(8)Å in $[Co_6(CO)_{15}]^{2-}$ to 2.579(15)Å in $[HCo_6(CO)_{15}]^-$. In $[HNi_{12}(CO)_{21}]^{3-}$ and $[H_2Ni_{12}(CO)_{21}]^{2-}$ [Fig. 4.7-9(b)] the individual Ni-Ni distances for octahedral holes containing H are 0.02-0.07Å longer than those for empty holes.[19]

6. Structure of molecular hydrogen complexes

The activation of hydrogen by metal centers is an important chemical reaction from both the commercial and scientific standpoints. The H-H bond is strong (103 kcal/mol), and H_2 addition to unsaturated organic and other compounds must be mediated by metal centers whose roles constitute the basis of catalytic hydrogenation. In catalytic mechanisms, hydride complexes formed by the cleavage of H_2 are regarded as key intermediates.[8]

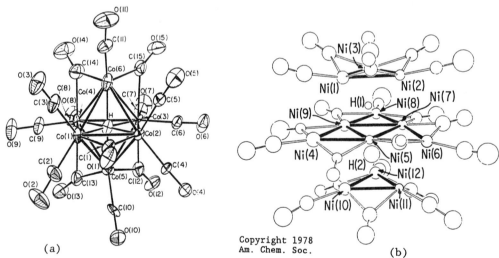

(a)   (b)

Fig. 4.7-9 Molecular structures of (a) $[HCo_6(CO)_{15}]^-$ (approximate C_{2v} symmetry with ten terminal, one symmetrically bridging, and four unsymmetrically bridging carbonyl groups) and (b) $[H_2Ni_{12}(CO)_{21}]^{2-}$ (D_{3h} symmetry; in the isostructural trianion the H atom is localized in one of the two octahedral interstices. (After refs. 18 and 19).

The first isolable transition-metal complex containing a coordinated dihydrogen molecule is W(CO)$_3$[P(C$_3$H$_7$)$_3$]$_2$(H$_2$). X-Ray and neutron diffraction studies and a variety of spectroscopic methods have shown that it possesses η^2-bonded H$_2$,[20] as illustrated in Fig. 4.7-10. The resulting geometry about

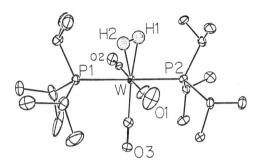

Fig. 4.7-10 Molecular structure of W(CO)$_3$[P(C$_3$H$_7$)$_3$]$_2$(H$_2$), with H atoms of the P(C$_3$H$_7$)$_3$ groups omitted for clarity. (After ref. 20).

the W atom is that of a regular octahedron with *cis* inter-ligand angles about tungsten ranging from 88.0 to 92.0°. The dihydrogen ligand is symmetrically coordinated in an η^2 mode with average W-H distances of 1.95Å (X-ray) and 1.75Å (neutron, ΔF) at -100°C. The H-H distance is 0.75Å (X-ray) and 0.84Å (neutron, ΔF), slightly larger than that of the free H$_2$ molecule (0.74Å).

The bonding of H$_2$ to the metal appears to involve either the transfer of σ-electrons of H$_2$ to a vacant metal d orbital, or the transfer of electrons from an occupied metal d orbital to the antibonding orbital of H$_2$, as shown in Fig. 4.7-11. Dihydrogen binding and eventual H-H cleavage would then be aided by back-donation of electrons from the metal to σ^* of H$_2$.

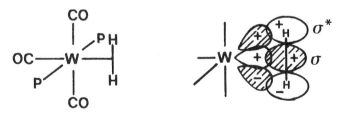

Fig. 4.7-11 Bonding in W-η^2-H$_2$.

The structure of *trans*-[Fe(η^2-H$_2$)(H)(PPh$_2$CH$_2$CH$_2$PPh$_2$)$_2$]BPh$_4$ has been determined by neutron diffraction at 20K, as shown in Fig. 4.7-12.[21] The H-H bond distance of 0.816(16)Å is the same as that found in the complex W(η^2-H$_2$)(CO)$_3$[P(C$_3$H$_7$)$_3$]$_2$ despite the fact that the tungsten complex has a more labile H$_2$ ligand. The terminal hydride-iron distance of 1.535(12)Å is shorter than the distances to the dihydrogen ligand, H-Fe = 1.616(10)Å. This is the first experimental demonstration of the expected difference in metal bonding to hydride and H_2.

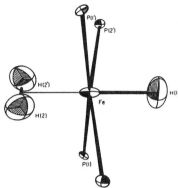

Fig. 4.7-12 Octahedral coordination environment about the metal atom in *trans*-$[Fe(\eta_2\text{-}H_2)(H)(PPh_2CH_2CH_2PPh_2)_2]BPh_4$. (After ref. 21).

7. Structures of some polyhydride complexes

   a) Mg_2FeH_6 [22]

   The ternary metal hydride Mg_2FeH_6 and its deuteride have been prepared and characterized. X-Ray and neutron powder diffraction analysis showed cubic symmetry [hydride $a = 6.443(1)$Å; deuteride $a = 6.430(1)$Å] and the K_2PtCl_6 structure type. The main structural unit of Mg_2FeD_6 is represented in Fig. 4.7-13(a). The octahedral FeD_6 group is surrounded by eight Mg atoms in a cubic configuration. These groups are arranged on a fcc lattice such that each Mg atom is tetrahedrally surrounded by four FeD_6 groups and has twelve D atoms as nearest neighbors. The metal-deuterium bond distances are Fe-D = $1.556(5)$Å and Mg-D = $2.2739(3)$Å, and the shortest separations between the deuterium atoms within (intra) and between (inter) the FeD_6 groups are $2.201(6)$ and $2.346(6)$Å, respectively. Mössbauer, Raman, and IR measurements show the presence of octahedral low-spin $[FeH_6]^{4-}$ ions.

   b) $FeH_6Mg_4Br_{3.5}Cl_{0.5}(THF)_8$ [23]

   The crystal structure of this compound reveals an octahedral $[FeH_6]^{4-}$ core with $[MgBr(THF)_2]^+$ units situated over four of the eight faces of the octahedron. The distances are Fe-H = $1.609(2)$Å and Mg-H = $2.045(18)$Å. These data support the idea that the Fe-H bond is largely "covalent" in nature while the Mg-H interaction is largely "ionic". The $[FeH_6]^{4-}$ core structure of $FeH_6Mg_4X_4(THF)_8$ (neutron diffraction) is shown in Fig. 4.7-13(b).

   c) $ReH_7[Ph_2P(CH_2)_2PPh_2]$ [24]

   The structure of $ReH_7[Ph_2P(CH_2)_2PPh_2]\cdot2THF$ has been determined by neutron diffraction. The polyhydride complex $ReH_7[Ph_2P(CH_2)_2PPh_2]$ has C_{2v} coordination geometry with seven discrete "classical" hydride ligands, as

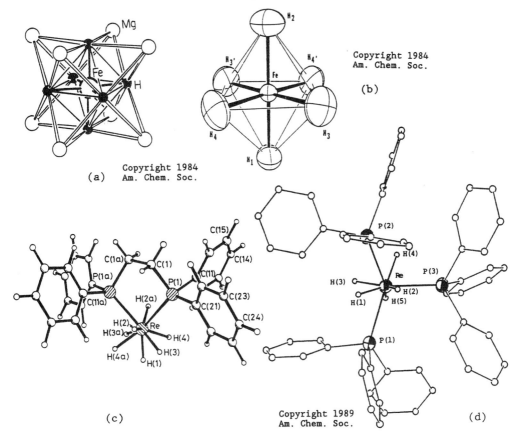

Fig. 4.7-13 Structures of some polyhydride complexes: (a) structural
unit of Mg_8FeH_6, (b) $[FeH_6]^{4-}$ core in $FeH_6Mg_4X_4(THF)_8$, (c)
$ReH_7[Ph_2P(CH_2)_2PPh_2]$, and (d) $ReH_5(PPh_3)_3$. (After refs. 22-25).

shown in Fig. 4.7-13(c). The rhenium-hydrogen bond lengths are: Re-H(1) =
1.660(9), Re-H(2) = 1.689(6), Re-H(3) = 1.669(7), and Re-H(4) = 1.671(6)Å.
There are no intramolecular H...H separations less than 1.77Å.

    d) $ReH_5(PPh_3)_3$ [25]

    The X-ray structure determination of this complex gave an average Re-H
distance of 1.54(5)Å [Fig. 4.7-13(d)]. The closest H...H distance of
1.60(13)Å is too great to be associated with a molecular hydrogen ligand.

8. Metal complexes containing agostic bonds

    The term "agostic" refers to the various manifestations of covalent
interactions of the type M←H-C in organometallic compounds, where the "half
arrow" indicates formal donation of two electrons from the H_α atom to the
metal center.[26] Adoption of this half-arrow convention facilitates electron
counting in a molecule in the context of the 18-electron rule.[27] As in all

three-centered, two-electron bridging systems involving only *three* valence orbitals, the M-H-C fragment is bent. The agostic C-H distance is in the range 1.13-1.19Å, about 5-10% longer than a non-bridging C-H bond; the M-H distance is also elongated by 10-20% relative to a normal terminal M-H bond. NMR spectroscopy can be conveniently used for the diagnosis of *static* agostic systems which exhibit low $J(C-H_\alpha)$ values owing to the reduced $C-H_\alpha$ bond order. Typical values of $J(C-H_\alpha)$ are in the range 60-90 Hz, which are significantly lower than those (120-130 Hz) expected for $C(sp^3)$-H bonds. The principal types of compounds with agostic bonds are shown in Fig. 4.7-14.[28]

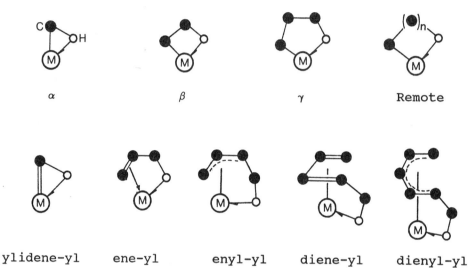

Fig. 4.7-14 Structural types of agostic alkyl and unsaturated-hydrocarbon-yl complexes. (After ref. 28).

Fig. 4.7-15 shows the molecular structures of some representative transition-metal complexes in each of which an agostic interaction has been characterized by either X-ray or neutron diffraction; relevant geometrical and NMR data are given in Table 4.7-2.

Table 4.7-2 Structural data for some compounds containing agostic bonds.

Compound*	M-C(Å)	M-H(Å)	C-H(Å)	$J(C-H_\alpha)$(Hz)	Ref.
(a)	2.122(2)	2.447(3)	1.095(3)	129.5	[29]
(b)	2.038(2)	1.874(3)	1.164(3)		[30]
(c)	1.898(2)	2.119(4)	1.131(3)	101	[31]
(d)	1.946(3)	2.042(5)	1.135(5)	74	[31]
(e)	1.927(2)	1.753(4)	1.191(4)	103	[32]
	1.920(2)	1.747(4)	1.176(5)		
(f)	2.276(10)	2.19(2)		145	[33]

* As shown in Fig. 4.7-15.

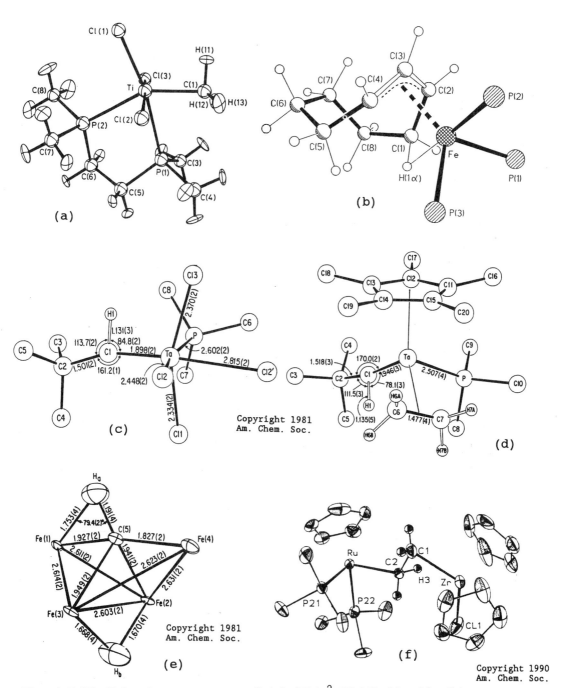

Fig. 4.7-15 Molecular structures of (a) $[Ti(\eta^2\text{-}CH_3)Cl_3(dmpe)]$, (b) $[Fe\{P(OMe)_3\}_3(\eta^3\text{-}C_8H_{13})]^+$ in its BF_4^- salt, (c) $[Ta(CHCMe_3)(PMe_3)Cl_3]_2$ showing only one-half of the dimeric unit, (d) $[Ta(\eta^5\text{-}C_5Me_5)(CHCMe_3)(\eta^2\text{-}C_2H_4)(PMe_3)]$, (e) $[HFe_4(\eta^2\text{-}CH)(CO)_{12}]$ with the carbonyl groups omitted for clarity, and (f) $[(\eta^5\text{-}C_5H_5)(PMe_3)_2RuCH_2CH_2Zr(\eta^5\text{-}C_5H_5)_2Cl]$. (After refs. 29-33).

A neutron diffraction study of [Ti(η^2-CH$_3$)Cl$_3$(dmpe)] has shown that one Ti-C-H angle (93.5°) is much smaller than the other two (112.9° and 118.4°), but surprisingly all three C-H bonds do not differ significantly from 1.10Å [Fig. 4.7-15(a)].[29] In [Fe{P(OMe)$_3$}$_3$(η^3-C$_8$H$_{13}$)]$^+$ [Fig. 4.7-15(b)] the occurrence of an agostic M-H-C bond was demonstrated for the first time on the basis of NMR and neutron diffraction studies.[30] For the two related tantalum-neopentylidene complexes shown in Fig. 4.7-15(c,d),[31] the Ta=C distances as determined from low-temperature neutron diffraction are much shorter than the "normal" Ta=C double bond length of 2.03Å in 18-electron complexes such as [Ta(η^5-C$_5$H$_5$)$_2$(CH$_2$)(CH$_3$)].[34] The marked lengthening of the C$_\alpha$-H$_\alpha$ bonds, short Ta-H$_\alpha$ distances, and small Ta-H$_\alpha$-C$_\alpha$ angles are all indicative of agostic M-H-C bonding. [HFe$_4$(η^2-CH)(CO)$_{12}$] was the first cluster to have an agostic bond completely characterized by neutron diffraction; one of the two independent molecules is shown in Fig. 4.7-15(e).[32] The complex shown in Fig. 4.7-15(f) contains a C$_2$ bridge between an electron-rich metal and an electron-deficient metal, and the agostic interaction occurs between one H atom of the RuCH$_2$ fragment and the Zr center.[33]

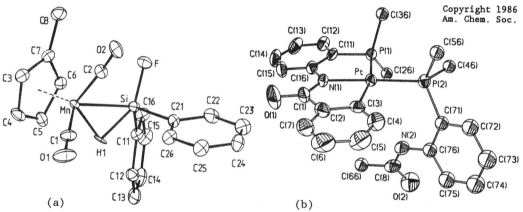

(a)   (b)

Fig. 4.7-16 Molecular structures of (a) [Mn(HSiFPh$_2$)(η^5-C$_5$H$_4$Me)(CO)$_2$] and (b) *cis*-Pt(*o*-Ph$_2$PC$_6$H$_4$NC(O)C$_6$H$_4$)(*o*-Ph$_2$PC$_6$H$_4$NHC(O)Ph). Only the pivotal atom of each terminal phenyl group is shown. (After refs. 35 and 36).

A neutron-diffraction analysis has revealed a M←H-Si bond in the compound [Mn(HSiFPh$_2$)(η^5-C$_5$H$_4$Me)(CO)$_2$] [Fig. 4.7-16(a)].[35] The structural parameters of the Mn←H-Si fragment in a series of related compounds in the class [Mn(HSiR$_3$)(η^5-Cp')(CO)(L)] are compared in Table 4.7-3.

The cyclometalated platinum(II) complex shown in Fig. 4.7-16(b) has been found to possess a M←H-N interaction.[36] The agostic H atom was located at a distance of 2.318(22)Å from Pt, and the ν(N-H) band at 3200 cm^{-1} is shifted by 150 cm^{-1} to lower wave numbers relative to its expected value.

Table 4.7-3 Structural data for [Mn(HSiR$_3$)(η^5-Cp')(CO)(L)].[35]

SiR$_3$	Cp'	L	d(Mn-Si)(Å)	d(Mn-H)(Å)	d(Si-H)(Å)
SiCl$_3$	C$_5$H$_4$Me	CO	2.254(1)	1.47(3)	1.79(4)
SiCl$_2$Ph	C$_5$H$_5$	CO	2.310(2)	1.49(6)	1.79(6)
SiFPh$_2$	C$_5$H$_4$Me	CO	2.352(4)	1.569(4)	1.802(5)
SiHPh$_2$	C$_5$H$_4$Me	CO	2.364(2)		
SiPh$_3$	C$_5$H$_4$Me	CO	2.424(2)	1.55(4)	1.76(4)
SiMePh(CH$_2$CMe$_3$)	C$_5$H$_4$Me	CO	2.461(7)		
SiHPh$_2$	C$_5$H$_4$Me	PMe$_3$	2.327(1)	1.49(4)	1.78(4)
SiHPh$_2$	C$_5$Me$_5$	CO	2.395(1)	1.52(3)	1.77(3)

References

[1] S.C. Abrahams, A.P. Ginsberg and K. Knox, *Inorg. Chem.* **3**, 558 (1964).

[2] K. Knox and A.P. Ginsberg, *Inorg. Chem.* **3**, 555 (1964).

[3] R. Bau, R.G. Teller, S.W. Kirtley and T.F. Koetzle, *Acc. Chem. Res.* **12**, 176 (1979).

[4] S.J. LaPlaca, W.C. Hamilton, J.A. Ibers and A. Davison, *Inorg. Chem.* **8**, 1928 (1969).

[5] J. Roziere, J.M. Williams, R.P. Stewart, Jr. J.L. Petersen and L.F. Dahl, *J. Am. Chem. Soc.* **99**, 4497 (1977).

[6] R.G. Teller, R.D. Wilson, R.K. McMullan, T.F. Koetzle and R. Bau, *J. Am. Chem. Soc.* **100**, 3071 (1978).

[7] T.F. Koetzle, R.K. McMullan, R. Bau, D.W. Hart, R.G. Teller, D.L. Tipton and R.D. Wilson in R. Bau (ed.), *Transition Metal Hydrides*, American Chemical Society, Washington, D.C., 1978, p. 61.

[8] G.J. Kubas, *Acc. Chem. Res.* **21**, 120 (1988).

[9] E.A. McNeill, and F.R. Scholer, *J. Am. Chem. Soc.* **99**, 6243 (1977).

[10] A.J. Scioly, M.L. Luetkens Jr., R.B. Wilson Jr., J.C. Huffman and A.P. Sattelberger, *Polyhedron* **6**, 741 (1987).

[11] R. Bau, W.E. Carroll, R.G. Teller and T.F. Koetzle, *J. Am. Chem. Soc.* **99**, 3872 (1977).

[12] A. Dedieu, T.A. Albright and R. Hoffmann, *J. Am. Chem. Soc.* **101**, 3141 (1979).

[13] F.A. Cotton and R.L. Luck, *Inorg. Chem.* **28**, 4522 (1989).

[14] T.M.G. Carneiro, D. Matt and P. Braunstein, *Coord. Chem. Rev.* **96**, 49 (1989), and references cited therein.

[15] A.L. Casalnuovo, Pignolet, J.W.A. van der Velden, J.J. Bour and J.J. Steggerda *J. Am. Chem. Soc.* **105**, 5957 (1983).

[16] A.L. Casalnuovo, J.A. Casalnuovo, P.V. Nilsson and L.H. Pignolet, *Inorg. Chem.* **24**, 2554 (1985).

[17] L.F. Rhodes, J.C. Huffman and K.G. Caulton, *J. Am. Chem. Soc.* **107**, 1759 (1985).

[18] D.W. Hart, R.G. Teller, C.-Y. Wei, R. Bau, G. Longoni, S. Campanella, P. Chini and T.F. Koetzle, *Angew. Chem. Int. Ed. Engl.* **18**, 80 (1979).

[19] R.W. Broach, L.F. Dahl, G. Longoni, P. Chini, A.J. Schultz and J.M. Williams in R. Bau (ed.), *Transition Metal Hydrides*, American Chemical Society, Washington, D.C., 1978, p. 93.

[20] G.J. Kubas, R.R. Ryan, B.I. Swanson, P.J. Vergamini and H.J. Wasserman, *J. Am. Chem. Soc.* **106**, 451 (1984).

[21] J.S. Ricci, T.F. Koetzle, M.T. Bautista, T.M. Hofstede, R.H. Morris and J.F. Sawyer, *J. Am. Chem. Soc.* **111**, 8823 (1989).

[22] J.-J. Didisheim, P. Zolliker, K. Yvon, P. Fisher, J. Schefer, M. Gubelmann and A.F. Williams, *Inorg. Chem.* **23**, 1953 (1984).

[23] R. Bau, M.Y. Chiang, D.M. Ho, S.G. Gibbins, T.J. Emge and T.F. Koetzle, *Inorg. Chem.* **23**, 2823 (1984).

[24] J.A.K. Howard, S.A. Mason, O. Johnson, I.C. Diamond, S. Crennell, P.A. Keller and J.L. Spencer, *J. Chem. Soc., Chem. Commun.*, 1502 (1988).

[25] F.A. Cotton and R.L. Luck, *J. Am. Chem. Soc.* **111**, 5757 (1989).

[26] M. Brookhart and M.L.H. Green, *J. Organomet. Chem.* **250**, 395 (1983).

[27] M. Berry, N.J. Cooper, M.L.H. Green and S.J. Simpson, *J. Chem. Soc., Dalton Trans.*, 29 (1980).

[28] M. Brookhart, M.L.H. Green and L.-L. Wong, *Prog. Inorg. Chem.* **36**, 1 (1988).

[29] Z. Dawoodi, M.L.H. Green, V.S.B. Mtetwa, K. Prout, A.J. Schultz, J.M. Williams and T.F. Koetzle, *J. Chem. Soc., Dalton Trans.*, 1629 (1986).

[30] R.K. Brown, J.M. Williams, A.J. Schultz, G.D. Stucky, S.D. Ittel and R.L. Harlow, *J. Am. Chem. Soc.* **102**, 981 (1980).

[31] A.J. Schultz, R.K. Brown, J.M. Williams and R.R. Schrock, *J. Am. Chem. Soc.* **103**, 169 (1981).

[32] M.A. Beno, J.M. Williams, M. Tachikawa and E.L. Muetterties, *J. Am. Chem. Soc.* **103**, 1485 (1981).

[33] R.M. Bullock, F.R. Lemke and D.J. Szalda, *J. Am. Chem. Soc.* **112**, 3244 (1990).

[34] L.W. Messerle, P. Jennische, R.R. Schrock and G. Stucky, *J. Am. Chem. Soc.* **102**, 6744 (1980).

[35] U. Schubert, G. Scholz, J. Müller, K. Ackermann, B. Wörle and R.F.D. Stansfield, *J. Organomet. Chem.* **306**, 303 (1986).

[36] D. Hedden, D.M. Roundhill, W.C. Fultz and A.L. Rheingold, *Organometallics* **5**, 336 (1986).

Note Complex hydrides with the composition $A_xM_yH_z$ (A = alkali or alkaline-earth metal; M = transition metal) are reviewed in W. Bronger, *Angew. Chem. Int. Ed. Engl.* **30**, 759 (1991).

η^2-Coordination of the Si-H σ bond to transition metals, in a manner analogous to the better known agostic η^2-CH bond, is reviewed in U. Schubert, *Adv. Organomet. Chem.* **30**, 151 (1991).

4.8 Trinuclear Molybdenum–Sulfur Cluster Complex $(NH_4)_4[Mo_3S_2(NO)_3(S_2)_4(S_3NO)]\cdot3H_2O$

Crystal Data

Triclinic, space group $P\bar{1}$ (No. 2)

$a = 9.535(2)$, $b = 9.740(2)$, $c = 15.795(4)$Å, $\alpha = 89.85(2)°$,
$\beta = 91.80(2)°$, $\gamma = 99.37(1)°$, $Z = 2$

Atom	x	y	z	Atom	x	y	z
Mo(1)	.2786	.7896	.1327	N(L1)	.4429	-.1617	.0874
Mo(2)	.0962	.0046	.1313	N(L2)	-.0833	.0346	.2145
Mo(3)	.2296	.7709	.3511	N(L3)	.3597	-.1963	.4331
S(T1)	.0533	.7468	.2227	N(B1)	.7006	.4605	.2520
S(T2)	.3511	.9511	.2535	N(A1)	.5706	.1729	.1181
S(S1)	.1292	.7007	.0068	N(A2)	.2103	.2624	.9218
S(S2)	.2162	.5531	.0733	N(A3)	.6390	-.0031	.3772
S(S3)	.2222	.1759	.1339	N(A4)	.9843	.3947	.2011
S(S4)	.1730	.1600	.3528	O(L1)	.5547	-.1242	.0537
S(S5)	.0322	.6594	.4368	O(L2)	-.2065	.0563	.2017
S(S6)	.1465	.5252	.3818	O(L3)	.4501	-.1758	.4940
S(E1)	.1507	.9943	.0792	O(B1)	.6560	.4689	.1711
S(E2)	.0833	-.0398	.3869	O(W1)	.8345	.2396	.3772
S(C1)	.3904	.6713	.2529	O(W2)	.2677	.2614	.6595
S(C2)	.3546	.4529	.2405	O(W3)	.3829	.3921	.5046
S(C3)	.5280	.3974	.2905				

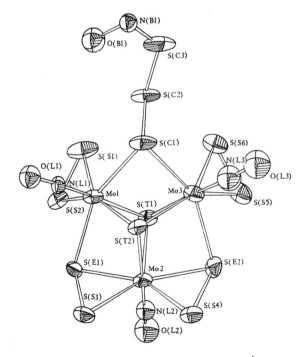

Fig. 4.8-1 Structure of $[Mo_3S_2(NO)_3(S_2)_4(S_3NO)]^{4-}$. (After ref. 1).

Crystal Structure

The crystal structure shows that the bonding modes of molybdenum and sulfur are diversified. In the $[Mo_3S_2(NO)_3(S_2)_4(S_3NO)]^{4-}$ anion, the Mo atoms seem to be in a divalent state. The NO ligand coordinated to Mo forms a linear Mo-N-O group and may be considered as NO^+. The nitric oxide ligand bonded to the trisulfur chain, in a bent configuration, exist as NO^- and gives rise to a negative monovalent $(S_3NO)^-$ group. If all S_n ligands are taken to be divalent anions, the formal oxidation state of Mo in the trinuclear complex is +2. After pairing its d electrons, Mo^{2+} should have three empty 4d orbitals available for seven-coordinate d^3sp^3 hybridization.

In the $[Mo_3S_2(NO)_3(S_2)_4(S_3NO)]^{4-}$ complex, the average Mo-Mo distance is 3.394Å. Three of NO groups are coordinated to Mo atoms: the average bond lengths of Mo-N and N-O are 1.76Å and 1.24Å, respectively, and the average Mo-N-O bond angle is 178°. The fourth NO group is bonded to the end of trisulfur chain, with S-N = 1.79(3)Å, N-O = 1.34(3)Å, and S-N-O = 95(2)°. There are four ligand types of S_n: (i) $S(T_1)$ and $S(T_2)$ form μ_3-bridges to three Mo atoms, (ii) two μ_2-S_2 coordinate side-on to one Mo atom, with S-S = 2.05Å and Mo-S = 2.46Å, (iii) two S_2 groups are side-on *and* end-on coordinated to Mo atoms, as shown in Fig. 4.8-1, and (iv) the S_3NO group bridging two Mo atoms, with Mo-S = 2.52Å.

Remarks

1. Types of coordination compounds with Mo-S and W-S bonds

During the last dozen years a large number of crystal structures have been studied for transition metal complexes and clusters containing S_n^{2-} ligands, especially by Müller and co-workers. A systematic treatment of the coordination chemistry of Mo-S and W-S compounds has been reviewed.[2] The types of these coordination compounds are listed in Table 4.8-1.

2. Some examples of mononuclear complexes with S_n^{2-} ligands[3]

Fig. 4.8-2(a) shows the structure of $Ti(\eta\text{-}C_5Me_5)_2(S_3)$ that has a dihedral angle of 49° between the TiS(1)S(3) and S(1)S(2)S(3) fragments.[4] The Mo(IV) complex $[(C_5Me_5)Mo(S_4)(S_2CO)]^-$ contains both S_4^{2-} and the novel dithiocarbonate ligand [Fig. 4.8-2(b)].[5] The anion in $(Et_4N)_2[Ni(S_4)_2]$ has been shown to contain two NiS_4 rings in the half-chair conformation [Fig. 4.8-2(c)].[6] A twist-boat ThS_5 ring exists in $(\eta^5\text{-}C_5Me_5)_2ThS_5$ (C_2 symmetry) [Fig. 4.8-2(d)]; the Th-S(1) and Th-S(2) bonds are best described as ionic and dative, respectively.[7] The anion in $(NH_4)_2[Pt(S_5)_3 \cdot 2H_2O]$[8] has three chair-shaped PtS_5 rings arranged like a propeller [Fig. 4.8-2(e)]; it has been resolved by forming diastereoisomers with optically-active $[Ru(LL)_3]^{2+}$ (LL = bpy or phen), thus constituting a rare example of purely inorganic chirality and the first such compound to involve a third-row transition element.[9] The related anion in $(Ph_4P)_2[Pt(S_6)_3]$ has a C_2 axis passing through Pt and the

Table 4.8-1 Types of coordination compounds with Mo-S and W-S bonds.[2]

Types		Compound	M-M/Å	M-S/Å
$\{Mo_2S_2\}^{n+}$	n=6, Mo(V)	$[Mo_2O_2S_2(S_2)_2]^{2-}$	2.829	2.304-2.338
	n=4, Mo(IV)	$[Mo_2S_2(CN)_8]^{4-}$	2.758	2.293-2.298
	n=2, Mo(III)	$[Mo_2S_2(CN)_8]^{6-}$	2.644	2.36-2.37
	(n=2), $\{Mo_2S_2(NO)_2\}$	$[Mo_2S_2(NO)_2(CN)_6]^{6-}$		
$\{Mo_3S_4\}^{n+}$	n=4, Mo(IV)	$[Mo_3S_4(CN)_9]^{5-}$	2.769	2.312-2.368
	n=3,2Mo(IV),1Mo(III)	$[Mo_3S_4(Cp)_3]$		
$\{Mo_4S_4\}^{n+}$	n=4, Mo(III)	$[Mo_4S_4(CN)_{12}]^{8-}$	2.853, 2.855	2.381, 2.382
	n=5, 3Mo(III),1Mo(IV)	$[Mo_4S_4(C_5H_4i\text{-}Pr)_4]^+$	2.860-2.923	2.343(av.)
	n=6, 2Mo(III)/(IV)	$[Mo_4S_4(C_5H_4i\text{-}Pr)_4]^{2+}$	2.790-2.902	2.343(av.)
	(n=4), $\{Mo_4S_4(NO)_4\}$	$[Mo_4S_4(NO)_4(CN)_8]^{8-}$	2.99, 3.67	2.35, 2.76
$\{M=S_t\}^{n+}$	n=4, Mo(VI)	MoS_4^{2-}		2.171-2.186
	n=3, M(V)	$[M_2S_4(S_4)_2]^{2-}$, M = Mo	2.846	2.114-2.123
		M = W	2.834	2.089-2.117
	n=2, M(IV)	$[MS(MS_4)_2]^{2-}$, M = Mo	2.802-3.082	2.086
		M = W	2.882-3.047	2.070
$\{M=S=M\}^{n+}$	n=6, Mo(IV)	$[(CN)_6MoSMo(CN)_6]^{6-}$		2.172
$\{M(S_2)_t\}^{n+}$	n=4, Mo(VI)	$[(Cl_2Fe)S_2MoO(S_2)]^{2-}$	[2.752]	2.396-2.421
	n=3, Mo(V)	$[Mo_2O_2S_2(S_2)_2]^{2-}$	2.829	2.384-2.426
	n=2, Mo(IV)	$[Mo_3S(S_2)_6]^{2-}$	2.719, 2.725	2.403-2.459
$\{M(S_4)_t\}^{n+}$	n=3, W(V)	$[W_2S_4(S_4)_2]^{2-}$	2.834	2.389-2.435
	n=2, Mo(IV)	$[MoS(S_4)_2]^{2-}$		2.331-2.387
	n=2, Mo(IV)	$[MoO(S_4)_2]^{2-}$		2.346-2.386
$\{M(S_2)M\}^{n+}$	n=8, Mo(V)	$[Mo_2(S_2)_6]^{2-}$	2.826, 2.828	2.382-2.507
	n=6, Mo(IV)	$[Mo_3S(S_2)_6]^{2-}$	2.719, 2.725	2.413-2.490
	(n=6), $\{Mo(NO)\}^{3+}$	$[Mo_4(NO)_4(S_2)_5S_3]^{4-}$	3.342-3.458	2.441-2.547
	(n=6), $\{MoNO\}^{3+}$	$[Mo_4(NO)_4(S_2)_6O]^{2-}$	3.309	2.444-2.524
	(n=6), Mo(IV)	$[Mo_2(S_2)(SO_2)(CN)_8]^{4-}$	2.684	2.422-2.437
	(n=5, Mo(III), Mo(IV))	$[Mo_2(S_2)(SO_2)(CN)_8]^{5-}$	2.790	2.436-2.448
$\{M(S_x)M\}^{n+}$	(n=6), $\{MoNO\}^{3+}$	$[Mo_2(NO)_2(S_2)_3(S_5)(OH)]^{3-}$	3.413	2.475, 2.485
$\{M(SH)\}^{n+}$	n=5, W(VI)	$[WS_3(SH)]^-$		
	n=3, W(IV)	$[W(CN)_7(SH)]^{4-}$		2.621
$\{M_2S_2(SH)_2\}^{n+}$	(n=6), Mo(IV)	$[Mo_2S_2(SH)_2(Cp\text{-}Me_n)_2]$		

midpoint of the S(9)-S(9^i) bond [Fig. 4.8-2(f)].[10] The anion in $(Et_4N)_2[Hg(S_6)_2]$ has two bidentate S_6^{2-} ligands bonded tetrahedrally to the metal center [Fig. 4.8-2(g)].[11] Examples of the S_9^{2-} ion as a ligand are found in $[Ag(S_9)]^{-}$[12] and $[Au(S_9)]^{-}$,[13] as shown in Figs. 4.8-2(h) and (i), respectively.

3. Some examples of dinuclear complexes with S_n^{2-} ligands

The anion in $(Ph_4P)_2[W_2S_4(S_2)(S_4)]\cdot0.5DMF$ contains four types of S_n^{2-} ligands with the WS_4 ring in an envelope conformation [Fig. 4.8-3(a)].[2] The

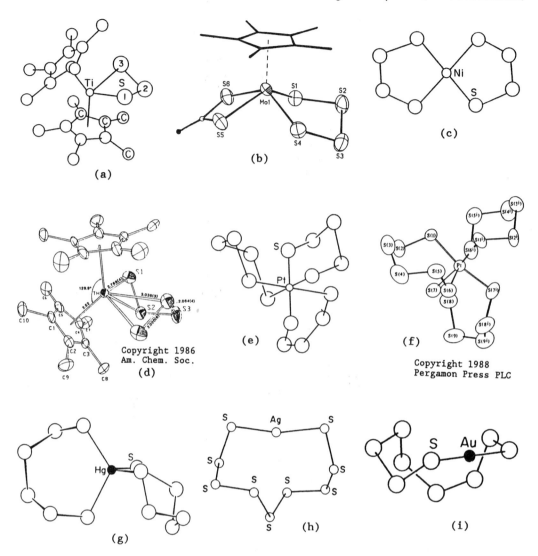

Fig. 4.8-2 Some mononuclear complexes containing S_n^{2-} ligands. (After refs. 3, 5, 7 and 10).

$[(S_5)FeS_2MoS_2]^{2-}$ anion in its Ph_4P^+ salt has a planar $Fe(\mu-S)_2Mo$ unit and a chair-shape FeS_5 ring [Fig. 4.8-3(b)].[14] The $[Mo_2(\mu-S)_2(S_2)_4]^{2-}$ ion shown in Fig. 4.8-3(c) occurs in both $(NH_4)_2[Mo_2(S_2)_6]\cdot 2H_2O$[15] and $(NH_4)_2[Mo_2(S_2)_6]\cdot 8/3H_2O$.[16] In $(\eta-C_5Me_5)_2Rh_2(S_4)_2$ the two chelated RhS_4 rings couple together to give a puckered Rh_2S_3 core [Fig. 4.8-3(d)].[17] The $[Re_2(CO)_6(S_4)_2]^{2-}$ dianion has a centrosymmetric structure [Fig. 4.8-3(e)] in its $n-Bu_4N^+$ salt.[18] The $[(S_6)Cu(S_8)Cu(S_6)]^{4-}$ anion in its Ph_4P^+ salt consists of two CuS_6 rings (approximately C_2 and C_s) linked by a S_8^{2-} chain [Fig. 4.8-3(f)].[19]

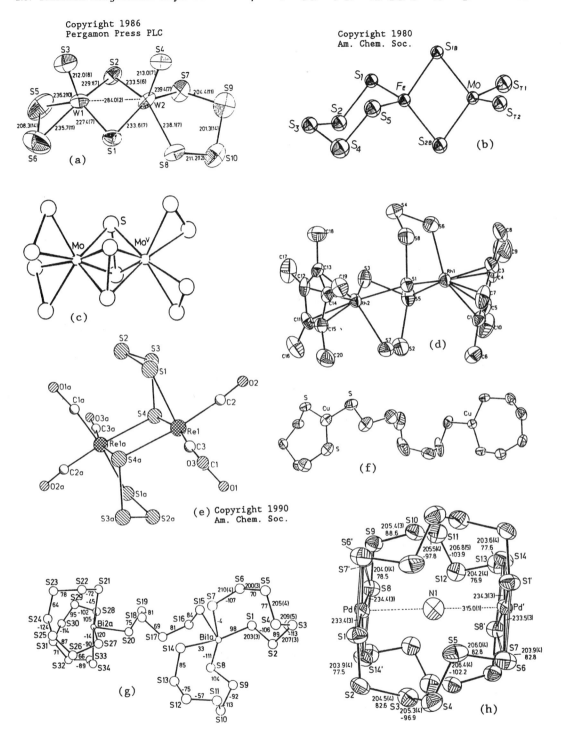

Fig. 4.8-3 Some dinuclear complexes containing S_n^{2-} ligands. (After refs. 2,3,14 and 17-21).

The anion in $(Ph_4As)_4[Bi_2S_{34}]$ is hitherto the sulfur-richest metal complex.[20] The two Bi(III) atoms, linked by a S_6^{2-} chain, are each chelated by two S_7^{2-} ligands. As shown in Fig. 4.8-3(g), the coordination polyhedron of each Bi atom is a strongly distorted square pyramid with a terminal S atom of the S_6^{2-} chain at its apex. The anion of $(PPh_4)_2(NH_2Me_2)[Pd_2(S_7)_4(NH_4)]$, which exhibits approximate D_4 symmetry, has a cage structure formed by linking two staggered PdS_4 units by four S_7^{2-} chains; the central cavity is occupied by the NH_4^+ ion [Fig. 4.8-3(h)].[21]

4. Some examples of polynuclear clusters with S_n^{2-} ligands

Among the Mo clusters that are presently known, the trinuclear ones exhibit rather wide variations in both structure and chemistry. They are either monocapped or bicapped species.[22] The bond lengths of the monocapped trinuclear molybdenum complex $[Mo_3S(S_2)_6]^{2-}$ are given in Fig. 4.8-4(a).[2] In $[Mo_3S(S_2)_3Cl_7]^{3-}$, the three Mo atoms form an exact equilateral triangle with sides 2.750Å [Fig. 4.8-4(b)].[23] The triangular plane is capped by μ_3-S and μ_3-Cl atoms. Each pair of Mo atoms is also connected by a bridging S_2 ligand (S-S = 2.029Å) such that one of the S atoms is situated in the same plane, and the other one is directed towards the side opposite to the μ_3-S atom. In addition, two terminal Cl ligands are attached to each Mo atom so that the coordination geometry may be described as distorted pentagonal bipyramidal.

In $(NH_4)_4[Mo_4(NO)_4(S_2)_5(S)_3].2H_2O$, the orange-red anion comprises two equilateral triangular arrays of Mo atoms [Mo-Mo 3.375(2)Å] joined by a common edge (dihedral angle 127.59°), with two μ_3-S ligands located normal to each Mo_3-triangle [Mo-S 2.501(5)Å] and one μ_4-S ligand [Mo-S 2.616(5)Å] on the other side of the triangular arrays [Fig. 4.8-4(c)].[24] Four of the total of five S_2^{2-} ligands each functions as an unsymmetrical bridge between two Mo atoms, one S atom being bound to only one Mo center [Mo-S 2.465(5)Å], and the other to both Mo centers [Mo-S 2.492(5)Å]. The interatomic distances in the S_2^{2-} ligands are almost equal [S-S 2.048(7)Å]. In the Mo-N-O group, the Mo-N and N-O distances are 1.742(16) and 1.219(22)Å, respectively.

In the isostructural compounds $(NH_4)_2[Mo_4(NO)_4(S_2)_6O].2H_2O$ and $(NH_4)K[Mo_4(NO)_4(S_2)_6O].2H_2O$, the metal atoms of the complex anion form a tetragonal disphenoid, over whose edges two handle-shaped and four roof-shaped S_2^{2-} ligands are situated [Fig. 4.8-4(d)].[25] An O atom is located at the center. Since an NO ligand is, in addition, coordinated to each Mo atom, every metal center is in a pentagonal-bipyramidal environment.

The anion in $(Ph_4P)_2(NH_4)[Cu_3(S_4)_3].2CH_3OH$ has almost C_3 symmetry, being composed of three CuS_4 half-chair rings and a central Cu_3S_3 chair ring with

alternating short and long Cu-S bond lengths [Fig. 4.8-4(e)].[19]   The
hexanuclear $[Cu_6(S_4)_3(S_5)]^{2-}$ anion in its Ph_4P^+ salt has the C_2 structure
shown in Fig. 4.8-4(f).[26]   The complex contains ten μ_3-S atoms and the
aggregate of six Cu(I) atoms can be approximately described as two Cu_4
tetrahedra sharing one edge.   The cluster anion $[Re_4S_4(S_3)_6]^{4-}$, located at a
site of $\bar{4}3m$ symmetry, has a highly symmetrical cubane skeleton with S_3^{2-}
ligands bridging pairs of Re(IV) atoms on the faces of the cube.[27]   As shown
in Fig. 4.8-4(g), the Re_2S_3 rings are envelope-shaped but in the crystal S3 is
disordered over two sites at 0.691Å above and below the S2-Re-Rec-S2c plane.

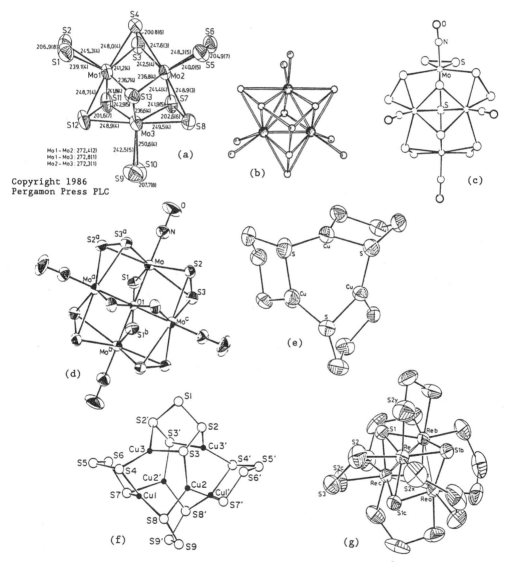

Fig. 4.8-4 Some polynuclear complexes containing S_n^{2-} ligands.   (After
refs. 2, 3, 19, 23, 25 and 27).

5. S_n^{2-} as ligands in metal complexes

The S_n^{2-} ligand is attracting increasing attention since no other simple ligand is as versatile in regard to the variety of coordination modes.[3,28]

a) S^{2-} It can act either as a terminal or a bridging ligand. Ligands in which S acts as a donor atom are usually classified as soft Lewis bases, in contrast to oxygen donor-atom ligands which tend to be hard Lewis bases. The larger size of the S atom and the consequent greater deformability of its electron cloud give a qualitative rationalization of this difference, and the participation of metal $d\pi$ orbitals in bonding to sulfur is well established. Examples of S^{2-} as a terminal ligand are given in Fig. 4.8-3 and Fig. 4.8-4. In the μ_2 bridging mode S is usually regarded as a 2-electron donor. In the μ_3 triply bridging mode S can be regarded as a 4-electron donor, using both its unpaired electrons and one lone-pair. The psuedo-cubane structure adopted by some of the μ_3-S compounds is assuming added significance as a crucial structural unit in many biologically important systems, e.g. the $[(RS)MS]_4$ (M = Mo, Fe) units which cross-link the polypeptide chains in nitrogenase and ferredoxins. In μ_4-bridging 4 or 6 electrons are involved, depending on the geometry of the mode.[29] No molecular compounds are known in which S bridges 6 or 8 metal atoms has yet been synthesized, but encapsulated (interstitial) sulfur is well known.

b) S_2^{2-} At least 8 modes of coordination of the disulfide ligands are known, as listed in Table 4.8-2.[30]

Table 4.8-2 Types of metal-disulfide complexes.[15]

Type	Example	d_{S-S}/Å
	$[Mo_2O_2S_2(S_2)_2]^{2-}$	2.08
	$[Mo_4(NO)_4S_{13}]^{4-}$	2.048
	$Mn_4(CO)_{15}(S_2)_2$	2.07
	$Mn_4(CO)_{15}(S_2)_2$	2.09
	$[Ru_2(NH_3)_{10}S_2]^{4+}$	2.014
	$Co_4(C_5H_5)_4S_2(S_2)_2$	2.01
	$\{SCo_3(CO)_7\}_2(S_2)$	2.042
	$[Mo_2(S_2)_6]^{2-}$	2.043

c) S_n^{2-} (n = 2-7)   These anions generally have a chain structure, whereby the average S-S bond length is smaller in S_n^{2-} ions (n > 2) than in S_2^{2-}, and the length of the S-S terminal bond decreases from S_3^{2-} (2.15Å) to S_7^{2-} (1.992Å), as shown in Table 4.8-3.[31] These facts indicate that the negative charge (filling a π^* antibonding MO) is delocalized over the whole chain, but the delocalization along the chain is less in higher polysulfides. These considerations are important in comparing the S-S bond lengths of the free ions with those of their metal complexes.

Table 4.8-3   Comparison of selected dimensions in some polysulfide ions.

Polysulfide	S_t-S (Å)	S_t-S-S (°)
S_2^{2-}	2.13	-
S_3^{2-}	2.15	103
S_4^{2-}	2.074(1)*	109.76(2)*
S_5^{2-}	2.043(4)*	109.2(1)*
S_6^{2-}	2.03(2)*	109.6(20)*
S_7^{2-}	1.992(2)*	111.3(1)*

* Average values involving the terminal sulfur atoms.

References

[1]   Tang Kaluo, Jin Xianglin and Tang Youqi, *Scientia Sinica* **B17**, 657 (1984).

[2]   A. Müller, *Polyhedron* **5**, 323 (1986).

[3]   A. Müller and E. Diemann in G. Wilkinson, R.D. Gillard and J.A. McCleverty (eds.), *Comprehensive Coordination Chemistry*, Vol. 2, Pergamon Press, Oxford, 1987, chap. 16.1.

[4]   P.H. Bird, J.M. McCall, A. Shaver and U. Siriwardane, *Angew. Chem. Int. Ed. Engl.* **21**, 384 (1982).

[5]   J.W. Kolis, *Coord. Chem. Rev.* **105**, 195 (1990).

[6]   A. Müller, E. Krickemeyer, H. Bögge, W. Clegg and G.M. Sheldrick, *Angew. Chem. Int. Ed. Engl.* **22**, 1006 (1983).

[7]   D.A. Wrobleski, D.T. Cromer, J.V. Ortiz, T.B. Rauchfuss, R.R. Ryan and A.P. Satteberger, *J. Am. Chem. Soc.* **108**, 174 (1986).

[8]   P.E. Jones and L. Katz, *Acta Crystallogr.*, *Sect. B* **25**, 745 (1969).

[9]   R.D. Gillard and F.L. Wimmer, *J. Chem. Soc., Chem. Commun.*, 936 (2978).

[10]   R. Sillanpaa, P.S. Cartwright, R.D. Gillard and J. Valkonen, *Polyhedron* **7**, 1801 (1988).

[11]   A. Müller, J. Schimanski and U. Schimanski, *Angew. Chem. Int. Ed. Engl.* **23**, 159 (1984).

[12]   A. Müller, J. Schimanski, M. Römer, H. Bögge, F.W. Baumann, W. Eltzner, E. Krickemeyr and U. Billerbeck, *Chimia* **39**, 25 (1985).

[13]   G. Marbach and J. Strähle, *Angew. Chem. Int. Ed. Engl.* **23**, 246 (1984).

[14] D. Coucouvanis, N.C. Baenziger, E.D. Simhon, P. Stremple, D. Swenson, A. Kostikas, A. Simopoulos, V. Petrouleas and V. Papaefthymiou, *J. Am. Chem. Soc.* **102**, 1730 (1980).

[15] A. Müller, W.O. Nolte, and B. Krebs, *Inorg. Chem.* **19**, 2835 (1980).

[16] Zhou Gongdu, Tang Kaluo and Xu Xiaojie, *Acta Chim. Sinica* **41**, 385 (1983). [In Chinese].

[17] H. Brunner, N. Janietz, W. Meier, B. Nuber, J. Wachter and M.L. Ziegler, *Angew. Chem. Int. Ed. Engl.* **27**, 708 (1988).

[18] T.S.A. Hor, B. Wagner and W. Beck, *Organometallics* **9**, 2183 (1990).

[19] A. Müller, F.-W. Baumann, H. Bögge, M. Römer, E. Krickemeyer and K. Schmitz, *Angew. Chem. Int. Ed. Engl.* **23**, 632 (1984).

[20] A. Müller, M. Zimmermann and H. Bögge, *Angew. Chem. Int. Ed. Engl.* **25**, 273 (1986).

[21] A. Müller, K. Schmitz, E. Krickemeyer, M. Penk and H. Bögge, *Angew. Chem. Int. Ed. Engl.* **25**, 453 (1986).

[22] J.L. Huang, J.Q. Huang, M.Y. Shang, H.H. Zhuang, S.F. Lu, X.T. Lin, Y.H. Lin, M.D. Huang and J.X. Lu in J.J. Stezowski, Jin-Ling Huang and Mei-Cheng Shao (eds.), *Molecular Structure: Chemical Reactivity and Biological Activity*, Oxford University Press, 1988, pp. 462-480.

[23] Lu Shaofang, Shang Maoyu and Huang Jinling, *Jiegou Huaxue (J. Struct. Chem.)* **3**, 9 (1984).

[24] A. Müller, W. Eltzner, and N. Mohan, *Angew. Chem. Int. Ed. Engl.* **18** 168 (1979).

[25] A. Müller, W. Eltzner, H. Bögge and S. Sarkar, *Angew. Chem. Int. Ed. Engl.* **21**, 535 (1982).

[26] A. Müller, M. Römer, H. Bögge, E. Krickemeyer and D. Bergmann, *J. Chem. Soc., Chem. Commun.*, 348 (1984); *Z. Anorg. Allg. Chem.* **511**, 84 (1984).

[27] A. Müller, E. Krickemeyer and H. Bögge, *Angew. Chem. Int. Ed. Engl.* **25**, 272 (1986).

[28] H. Vahrenkamp, *Angew. Chem. Int. Ed. Engl.* **14**, 322 (1975).

[29] R.D. Adams and T.S.A. Hor, *Organometallics* **3**, 1915 (1984).

[30] N.N. Greenwood and A. Earnshaw, *Chemistry of the Elements*, Pergamon Press, Oxford, 1986, p. 790.

[31] M.G. Kanatzidis, N.C. Baenziger and D. Coucouvanis, *Inorg. Chem.* **22**, 290 (1983).

Note The mixed-metal cluster anion $[Cu_{12}Mo_8S_{32}]^{4-}$ containing μ_1-, μ_2-, μ_3-, and μ_4-S ligands is described in J. Li, X. Xin, Z. Zhou and K. Yu, *J. Chem. Soc., Chem. Commun.*, 249 (1991).

Mercury polysulfide complexes $[Hg(S_x)(S_y)]^{2-}$ ($x = y = 4$; $x = 4$, $y = 5$; $x = y = 5$) are reported in T.D. Bailey, R.M.H. Banda, D.C. Craig, I.G. Dance and I.N.L. Ma, *Inorg. Chem.* **30**, 187 (1991).

4.9 Pentakis(triphenylphosphine)hexakis(4-chlorobenzene-thiolato)hexasilver(I) Toluene (1/2) Solvate

$$Ag_6(SC_6H_4Cl)_6(PPh_3)_5 \cdot 2C_6H_5CH_3$$

Crystal Data

Triclinic, space group $P\bar{1}$ (No. 2)

$a = 15.024(5)$, $b = 18.488(5)$, $c = 26.506(10)$Å, $\alpha = 82.14(2)°$,

$\beta = 86.19(2)°$, $\gamma = 66.45(3)°$, $Z = 2$

Atom*	x	y	z	Atom	x	y	z
Ag(1)	.88495	.32121	.17821	S(4)	.77382	.2400	.1845
Ag(2)	.69727	.19732	.26769	S(5)	1.03937	.2312	.2388
Ag(3)	1.00249	.15370	.35486	S(6)	.85402	.1184	.3360
Ag(4)	.90058	.19178	.26048	P(1)	.9440	.3767	.0999
Ag(5)	.79827	.34905	.30467	P(2)	.5943	.1215	.2788
Ag(6)	.58831	.48530	.28092	P(3)	1.1530	.0622	.3931
S(1)	.90140	.2902	.3776	P(4)	.5647	.5732	.3477
S(2)	.62789	.3332	.3057	P(5)	.4611	.5455	.2124
S(3)	.75246	.4388	.2248				

* C, H, Cl atoms have been omitted.

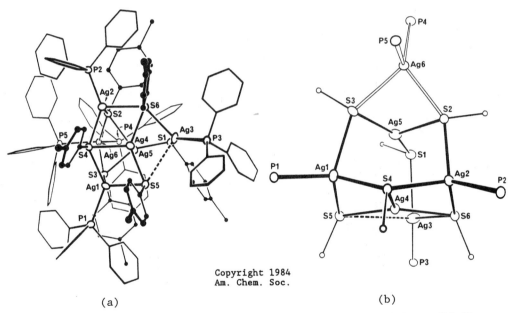

Copyright 1984
Am. Chem. Soc.

(a)   (b)

Fig. 4.9-1 (a) Molecular structure of $Ag_6(SC_6H_4Cl)_6(PPh_3)_5$. (b) The $Ag_6S_6P_5$ core, including the thiolate S-C bonds. (After ref. 1).

Crystal Structures

The crystal structure contains a discrete $Ag_6(SC_6H_4Cl)_6(PPh_3)_5$ molecule together with two molecules of toluene per asymmetric unit. The 21 aryl substituents of the hexanuclear molecule almost completely envelope it, as shown in Fig. 4.9-1(a).

At first examination the molecular core [Fig. 4.9-1(b)] does not appear to possess the high symmetry often present in metal thiolate cages. Two silver atoms [Ag(4),Ag(5)] have trigonal AgS_3 coordination; two [Ag(1),Ag(2)] have tetrahedral AgS_3P coordination; Ag(6) has tetrahedral AgS_2P_2 coordination, while Ag(3) has distorted AgS_3P coordination with one very long Ag-S bond, marked as a broken line in Fig. 4.9-1. Five of the thiolate ligands are triply bridging, while that at S(1) is doubly bridging. The Ag-S distances range from 2.40 to 2.84Å, except for Ag(3)-Ag(5), which is 3.32Å. There is an approximate mirror plane applicable to the $Ag_6S_6P_5$ core, passing through atoms P(4), P(5), Ag(6), Ag(5), S(1), Ag(3), P(3), Ag(4) and S(4).

The $Ag_5(SR)_6$ cage, when symmetrized, contains a basal unit comprising an irregular hexagon of alternating silver and sulfur atoms, centered by a silver atom with trigonal-planar $(S)_3$ coordination. A trigonal-planar podal unit, (AgS_3), is parallel to the basal plane and is attached to it by three S-Ag bonds. Tetrahedral coordination of each Ag(basal) atom is completed by a terminal phosphine ligand. The additional $Ag(PPh_3)_2$ moiety bridges to S(podal) atoms.

Remarks

1. Organothiolate anion as a ligand

The organothiolate anion (RS^-) is a fundamental ligand. Together with the homologous organoselenolate (RSe^-) and organotellurolate (RTe^-), and alkoxide (RO^-) and aryloxide (ArO^-) anions, the thiolate anion may be classified as a pseudohalide, comparable as a ligand to Cl^-, Br^- and I^-. In metal complexes, RS^- can be compared with HS^- and S^{2-}. The substituent R in RS^- can be manipulated and modified to effect steric and electronic control of ligation ability.

The categorization of RS^- as a pseudohalide rests in part on the one-electron oxidation:

$$RS^- \rightleftharpoons \tfrac{1}{2}RSSR + e^-,$$

quantification of which is complicated by chemical irreversibility, and in part on the basicity:

$$RS^- + H^+ \rightleftharpoons RSH.$$

Thiolates are stronger bases than halides, with pK_a values in water ranging from 11.1 (t-BuSH) through 9.4 ($PhCH_2SH$) to 6.6 (PhSH).

The propensity of the thiolate ligand to occur as a pyramidal double bridge M-$(\mu$-SR)-M and thereby to form aggregated metal complexes is amply demonstrated. Thiolate groups also function readily as terminal ligands, less frequently as triple bridges, and rarely in the quadruply-bridging mode.

Metal 1,1-dithiolates have been discussed in detail by Coucouvanis,[4] and will not be dealt with in the present section.

2. Molecular structure types of metal thiolate complexes

A rich variety of structure types have been established by diffraction methods. The examples are all selected from a Polyhedron Report by Dance[2] and supplemented by other sources.[3-6] The structures of non-molecular metal thiolate complexes are not discussed here.

The diagrams of idealized structure types in Figs. 4.9-2 to 4.9-9 show M atoms as small open circles, while the SR ligands are usually presented as larger dotted circles at the S atom positions only, since the substituent configurations and conformations are variable. Where substituents are included the small filled circles represent α-carbon atoms. Stipled circles denote heteroatoms.

a) Monometallic structures

Some monometallic structure types are shown in Fig. 4.9-2. Some selected examples are: **1a** $Hg(SEt)_2$, **1b** $Sn(SAr)_2$ (Ar = 2,4,6-tri-t-butylphenyl), **1c** $[Cu(SPh)_3]^{2-}$, **1d** $[Pb(SPh)_3]^-$, **1e** $[Ni(SPh)_4]^{2-}$, **1f** $[Sn(SPh)_5]^-$, and **1g** $[OMo(SPh)_4]^-$.

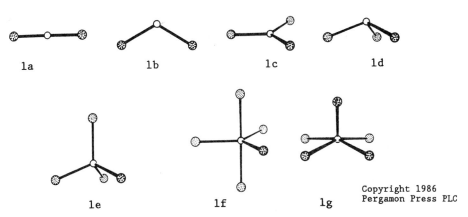

Copyright 1986
Pergamon Press PLC

Fig. 4.9-2 Monometallic structure types. (Adapted from ref. 2).

b) Bimetallic structures

Some bimetallic structure types are shown in Fig. 4.9-3. The examples are: **2a** $[Fe_2(SEt)_6]^{2-}$, **2b** $[Mn_2(SCH_2CH_2S)_4]^{2-}$, **2c** $[(CO)_3Fe(\mu\text{-}SMe)_3Fe(CO)_3]^+$, **2d** $(\mu\text{-}SCH_2Ph)_3\{OMo(SCH_2Ph)_2\}_2$ and **2e** $[Co_2\{o\text{-}(SCH_2)_2C_6H_4\}_3]^{2-}$.

c) Trimetallic structures

Some trimetallic structure types are shown in Fig. 4.9-4. The selected examples are: **3a** $(RS)Pb(\mu\text{-}SR)_2Pb(\mu\text{-}SR)_2Pb(SR)$ (R = 2,6-diisopropylphenyl), **3b** $[(PhS)_2Fe(\mu\text{-}S)_2Fe(\mu\text{-}S)_2Fe(SPh)_2]^{3-}$, **3c** $[Ni_3(S_2\text{-}o\text{-}xyl)_4]^{2-}$,

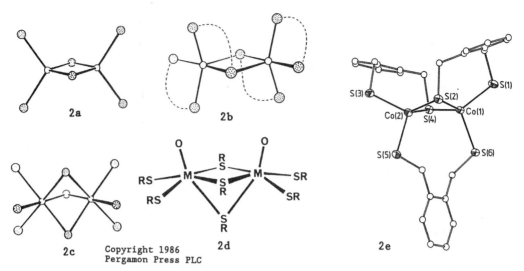

Fig. 4.9-3 Bimetallic structure types. (Adapted from refs. 2 and 3).

3d $(\mu\text{-SPh})_3(CuPPh_3)_2Cu(PPh_3)_2$, 3e $(\mu_3\text{-S})(\mu\text{-SR})_3(MSR)_3$ (M = Fe, Co, Ni),
3f $[(\mu\text{-SPh})_3(FeCl_2)_3]^{3-}$, 3g $(\mu\text{-SEt})_3Pd_3(S_2CSEt)_3$, 3h $[Mo_3S_4(SCH_2CH_2S)_3]^{2-}$, and
3i $(\mu\text{-SEt})_5(\mu\text{-CO})(CoCO)_3$.

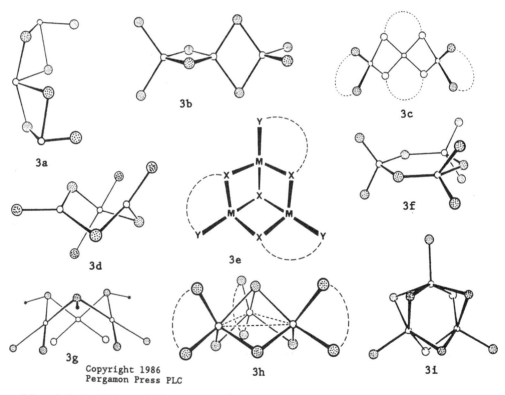

Fig. 4.9-4 Trimetallic structure types. (Adapted from ref. 2).

d) Tetrametallic structures

Some tetrametallic structure types are shown in Fig. 4.9-5. The examples are: 4a $(Bu^tSCu)_4(PPh_3)_2$, 4b $(Et_2MeCSAg)_8(PPh_3)_2$, 4c $(\mu\text{-SEt})_8(CoCO)_4$, 4d $[Cu_4(SMe)_6]^{2-}$, 4e $[Co_4(SPh)_{10}]^{2-}$, 4f $Cu_4(SPh)_4(PPh_3)_4$, and 4g $[(RS)_3Fe_3(\mu_3\text{-}S)_4Mo(cat)(RS)]^{2-}$.

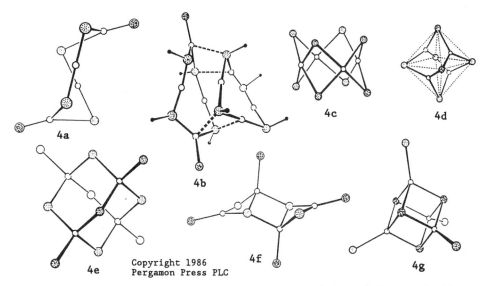

Copyright 1986
Pergamon Press PLC

Fig. 4.9-5 Tetrametallic structure types. (Adapted from ref. 2).

e) Pentametallic structures

Some pentametallic structure types are shown in Fig. 4.9-6. The examples are: 5a $[Cu_5(SPh)_7]^{2-}$, 5b $[Ag_5(SBu^t)_6]^-$, 5c $(MeZnSBu^t)_5$, and idealization of this structure as 5d.

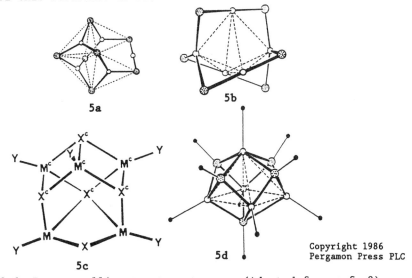

Copyright 1986
Pergamon Press PLC

Fig. 4.9-6 Pentametallic structure types. (Adapted from ref. 2).

f) Hexametallic structures

Some hexametallic structure types are shown in Fig. 4.9-7. The examples are: **6a** $[Ag_6(SPh)_8]^{2-}$, **6b** $Ni_6(SEt)_{12}$, **6c** $Ag_6(SC_6H_4Cl)_6(PPh_3)_5$, **6d** $[(\mu_4-S)(\mu_3-S)_2(\mu-S)_6Fe_6(SEt)_2]^{4-}$, **6e** $[Fe_6S_6(SC_6H_4Me)_6]^{3-}$, and **6f** $[Na_2\{Fe_6S_9(SMe)\}_2]^{6-}$.

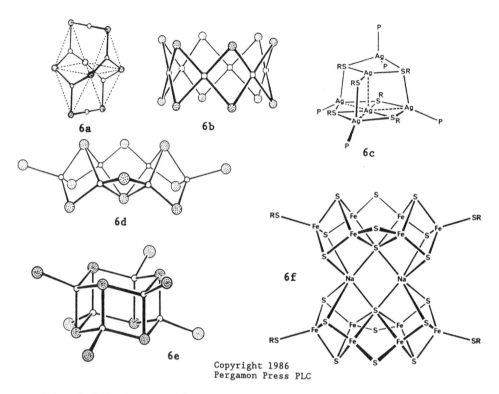

Fig. 4.9-7   Hexametallic structure types.   (Adapted from ref. 2).

g) Octametallic structures

Some octametallic structure types are shown in Fig. 4.9-8. The examples are: **8a** $[ICd_8(SCH_2CH_2OH)_{12}]^{3+}$, **8b** $[ClZn_8(SPh)_{16}]^-$, **8c** $Ni_8(SCH_2COOEt)_{16}$, **8d** $(MeZnSPr^i)_8$, **8e** $[S_6(CoSPh)_8]^{5-}$, **8f** $[(\mu-OMe)_3\{Mo(\mu_3-S)_4(FeSPh)_3\}_2]^{3-}$, **8g** $Cu_8(SC_5H_{11})_4(S_2CSC_5H_{11})_4$, **8h** $[Cu(SC_6H_2Pr^i_3)]_8$, and **8i** $[Ni_8S(SBu^t)_9]^-$.

h) High-(> 8)metallic structures

Some high-metallic structure types are shown in Fig. 4.9-9. The examples are: **9a** $[W_2Fe_7S_8(SEt)_{12}]^{4-}$, **10a** $[S_4Cd_{10}(SPh)_{16}]^{4-}$, **10b** $[Cd_{10}(SCH_2CH_2OH)_{16}]^{4+}$, **12a** $Ag_{12}(SC_6H_{11})_{12}$, **12b** $[Ag_6(SPh)_8]_2^{2-}$, and **14a** $(\mu-SBu^t)_{14}Ag_{14}(PPh_3)_4$.

3. Structural principles for metal thiolate compounds

Dance has formulated the following set of principles governing the structures of metal thiolate compounds:[2]

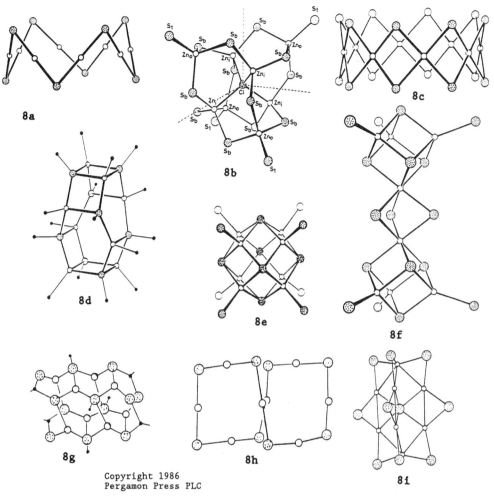

Fig. 4.9-8 Octametallic structure types. (Adapted from refs. 2,3,5-7).

(1) Metal coordination numbers are at the low ends of normal ranges.
This is attributed to the strongly electron-releasing characteristic and
polarizability of thiolate ligands. M(I) is usually two-coordinate, M(II) is
usually four-coordinate, and M(III) and M(IV) are usually five- and six-
coordinate.

(2) Thiolate ligands do not have any prominent influence on metal
coordination stereochemistry for the coordination numbers (4-6) where dual
stereochemistry can occur.

(3) Thiolate ligands are usually doubly-bridging or terminal. Triple
bridging can be achieved by conventional thiolate ligands, sometimes
with a dissymmetry that is similar to a double primary bridging plus a
secondary bridging connection. Thiolate bridging can be prevented by
incorporation of a sterically bulky R group.

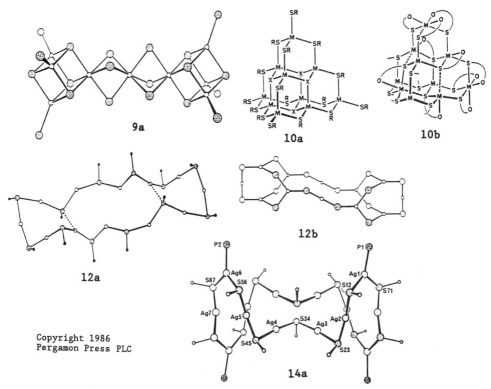

Fig. 4.9-9 High-metallic structure types.   (Adapted from ref. 2).

(4) Polymetallic cages are maintained by thiolate bridges, not metal-
metal bonds, and are normally comprised of fused $M_2(SR)_2$, $M_3(SR)_3$ or $M_4(SR)_4$
cyclic moieties.

(5) Polymetallic cages constructed of two-connected thiolate and two-,
three- or four-connected M atoms have insufficient bonds for framework
rigidity and are subject to angular distortions from idealized high-symmetry
structures. Non-rigidity and distortions increase with cage size and with
size (n) of the $M_n(SR)_n$ cycles which comprise the cage.

(6) Polymetal thiolate aggregates can frequently be idealized in terms
of highly symmetric cyclic or globular structures containing regular polygons
or polyhedra (deltahedra) of M atoms. These idealized structures are related
in series by polyhedral expansion or polyhedron-capping operations.

(7) Low connectivities at M and thiolate S atoms, and the consequent
cage non-rigidity, allow additional secondary (elongated) M-S connections that
increase rigidity: secondary bridging may be intra- or intermolecular.

(8) Cage expansion is limited by the reduction in rigidity which it
incurs. A mechanism for mechanical stabilization of globular cages with more
than 10-14 M and S atoms involves incorporation of an internal (spherical)
heteroligand.

(9) Compounds with two-coordinate M atoms and doubly-bridging thiolate ligands are comprised of zigzag strands of {-(μ-SR)-M-(μ-SR)-M-} segments linear at M and bent at S. Although coordinatively saturated, these strands can associate in opposed pairs, presumably in order to maximize the coverage by substituents of the remaining molecular surface and to maximize crystal density. A skein-type folding of the strand occurs in cyclic molecules, while two strands are intertwined to form the chain of one-dimensionally non-molecular structures. Recognizable structural functions that: (i) extend a strand (zigzag segments), (ii) intertwine double strands (crossover segments), and (iii) terminate a chain (bend and end segments) may be combined in the construction of molecules.

(10) Doubly-bridging thiolate normally adopts pyramidal stereochemistry at the S atom, but inversion is fast in solution, and in the crystalline phase the inversion barrier (*ca.* 40-70 kJ mol^{-1}) can be overcome by steric forces.

(11) In (μ-SR)$_3$M$_3$ cycles in a chair conformation the three R substituents are usually mixed equatorial and axial in configuration, but three axial substituents are possible, although with some expansion of S-M-S angles as a consequence.

(12) Substituent conflict can occur in folded (μ-SR)$_4$M$_4$ cycles, limiting the configurations available to the cycles and the structures of cages that contain fused cycles of this type.

References

[1] I.G. Dance, L.J. Fitzpatrick and M.L. Scudder, *Inorg. Chem.* **23**, 2276 (1984).

[2] I.G. Dance, *Polyhedron* **5**, 1037 (1986).

[3] W. Tremel, B. Krebs and G. Henkel, *Angew. Chem. Int. Ed. Engl.* **23**, 634 (1984).

[4] D. Coucouvanis, *Prog. Inorg. Chem.* **11**, 233 (1970); **26**, 301 (1979).

[5] Q.-C. Yang, K.-L. Tang, H. Liao, Y.-Z. Han, Z.-G. Chen and Y.-Q. Tang, *J. Chem. Soc., Chem. Commun.*, 1076 (1987).

[6] T. Krüger, B. Krebs and G. Henkel, *Angew. Chem. Int. Ed. Engl.* **28**, 61 (1989).

[7] R. Chadha, R. Kumar and D.G. Tuck, *J. Chem. Soc., Chem. Commun.*, 188 (1986).

Note Homoleptic and heteroleptic polynuclear transition-metal complexes with thiolates or mixed sulfide-thiolate ligands are reviewed in B. Krebs and G. Henkel, *Angew. Chem. Int. Ed. Engl.* **30**, 769 (1991).

New copper(I) and silver(I) cluster compounds with sulfur-containing ligands are described by K.-L. Tang and Y.-Q. Tang in E. Brock (ed.), *Heteroatom Chemistry*, VCH Publishers, Weinheim, 1990.

4.10   Iron(II) Phthalocyanine (FePc)

$$C_{32}H_{16}N_8Fe$$

Crystal Data

Monoclinic, space group $P2_1/a$ (No. 14)

$a = 19.246(7)$, $b = 4.776(1)$, $c = 14.538(5)$Å, $\beta = 120.86(3)°$, $Z = 2$
(110K)[1]

Atom	x	y	z	Atom	x	y	z
Fe	0	0	0	C(11)	.29720	.4542	.29347
N(1)	-.07890	-.5221	.07187	C(12)	.34512	.6588	.28558
N(2)	.03302	-.1974	.25335	C(13)	.31995	.8046	.18988
N(3)	.16138	.0282	.25335	C(14)	.24577	.7477	.09881
N(4)	.19740	.2207	.07336	C(15)	.19792	.5416	.10663
C(1)	-.00929	-.4091	.14486	C(16)	.11962	.4288	.02839
C(2)	.03275	-.5001	.25579	H(3)	-.0328	-.812	.2736
C(3)	.01345	-.7016	.30839	H(4)	.0566	-.869	.4584
C(4)	.06696	-.7337	.41769	H(5)	.1747	-.595	.5478
C(5)	.13709	-.5702	.47239	H(6)	.2042	-.259	.4564
C(6)	.15640	-.3705	.41915	H(11)	.3133	.362	.3557
C(7)	.10257	-.3380	.30959	H(12)	.3970	.701	.3484
C(8)	.10240	-.1515	.23043	H(13)	.3531	.938	.1871
C(9)	.15879	.1985	.17967	H(14)	.2304	.852	.0378
C(10)	.22281	.3963	.20199				

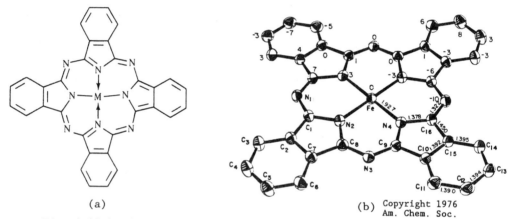

(a)

(b) Copyright 1976
Am. Chem. Soc.

Fig. 4.10-1   (a) Structural formula of a metal phthalocyanine complex.
(b) Structure of the centrosymmetric FePc molecule. On the lower half
of the diagram the atom numbering and averaged bond lengths are shown;
on the upper half the symbol for each atom is replaced by its
perpendicular displacement, in units of 0.01Å, from the mean plane of
the entire molecule. (After ref. 2).

Crystal Structure

The basic stereochemistry of four-coordinate metal complexes of
phthalocyanine (hereafter abbreviated as Pc) [Fig. 4.10-1(a)] was established
by J. Monteath Robertson in a series of classic studies.[3,4] The crystal
structures of Fe(II)Pc [Fig. 4.10-1(b)] and Mn(II)Pc were later determined by
X-ray diffraction at room temperature.[2] The electron density study of

Fe(II)Pc at 110K was reported in 1984, and the measured bond lengths are listed in Table 4.10-1.[1] The molecule is planar, the deviation of any atom from exact planarity being quite small. Individual units of the macrocycle, i.e., the pyrrole ring and benzene rings, are each planar to within 0.01Å. The dihedral angles between an individual pyrrole ring and its associated benzene ring in the four isoindole moieties are all less than 2°.

Table 4.10-1   Bond lengths of FePc.

Bond	100K[1]	Room temp.[2]	Bond	100K	Room temp.
Fe-N(2)	1.928	1.927	C(3)-C(4)	1.389	1.387
Fe-N(4)	1.926	1.926	C(4)-C(5)	1.401	1.394
N(1)-C(1)	1.324	1.320	C(5)-C(6)	1.394	1.396
N(1)-C(16)	1.329	1.324	C(6)-C(7)	1.394	1.394
N(2)-C(1)	1.371	1.374	C(7)-C(8)	1.454	1.454
N(2)-C(8)	1.387	1.382	C(9)-C(10)	1.451	1.452
N(3)-C(8)	1.322	1.322	C(10)-C(15)	1.397	1.393
N(3)-C(9)	1.325	1.322	C(10)-C(11)	1.393	1.397
N(4)-C(9)	1.387	1.382	C(11)-C(12)	1.388	1.386
N(4)-C(16)	1.373	1.375	C(12)-C(13)	1.401	1.395
C(1)-C(2)	1.451	1.449	C(13)-C(14)	1.388	1.389
C(2)-C(7)	1.392	1.390	C(14)-C(15)	1.392	1.393
C(2)-C(3)	1.393	1.395	C(15)-C(16)	1.448	1.446

The pattern of intermolecular contacts in FePc is essentially the same as that reported for CuPc.[5] Of particular interest is the contact of 3.24Å between the iron atom and the azamethine nitrogen atom N(1) of an adjacent molecule. The N(1)-Fe-N(4) and N(1)-Fe-N(2) angles are 85.5° and 89.5°, respectively. The deviation of N(1) from the mean plane is toward the Fe atom of the adjacent molecule. It has been suggested that this Fe...N(1) "interaction" found in the β polymorphs stabilizes the crystal structure.

The size of the "central hole" of phthalocyanine complexes is found to be somewhat smaller than those in the analogous porphyrin complexes. The experimental results suggest a greater ligand field strength for the phthalocyanato ligand compared to porphinato ligands.

Remarks

1. First application of the isomorphous replacement and heavy atom methods

Phthalocyanine, a tetrabenzotetraazaporphyrin of formula $C_{32}H_{18}N_8$, and numerous metal phthalocyanines $C_{32}H_{16}N_8M$ occur as a beautifully crystalline series of macrocyclic organic pigments of extraordinary stability.[6] The copper derivative, for example, sublimes without decomposition at 580 °C. Their close structural relationship to the natural porphyrins makes it desirable to conduct a careful study of the geometry of these molecules as well as the stereochemistry of the metals with which they exist in combination. They are also of historical interest in connection with the application of the isomorphous replacement and heavy atom methods for the detemination of crystal structures.

   Metal-free and nickel phthalocyanines were selected for the first
detailed X-ray studies in this series.[3-4] These two compounds are very
closely isomorphous, and the structure analysis proceeded straightforwardly
without any structural assumption or appeal to chemical theory. When viewed
along the short monoclinic b axis, the symmetry centers lie on an effectively
primitive lattice. Assuming complete isomorphism and comparing the absolute
values of the structure factors of nickel and metal-free phthalocyanine, the
phases of the $h0\ell$ reflections may be determined by the equation:

$$F(\mathrm{NiPc}) - F(\mathrm{Pc}) = F(\mathrm{Ni}).$$

In this equation the first two structure factors, which are measured
quantities, may be either positive or negative, so that four possible
combinations of these quantities are possible. However, only two need be
considered since the two Ni atoms at the symmetry centers always make a
positive contribution. The alternatives are plotted in Fig. 4.10-2 for a
large number of reflections.

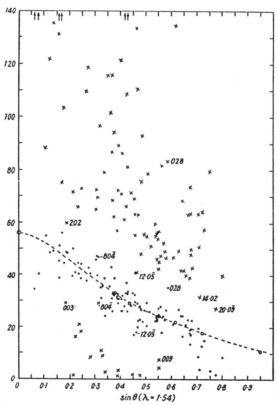

Fig. 4.10-2 Values of $F_{(\mathrm{NiPc})}$-$F_{(\mathrm{Pc})}$. The resultant curve for $F_{(\mathrm{Ni})}$ has
an upper limit at 56, twice the atomic number of nickel since there are
two molecules in the unit cell. (After ref. 3).

It is found that one choice of signs always gives a value for $F_{(Ni)}$ that
lies near the expected atomic scattering curve indicated by the broken line,
and this value for each reflection is indicated by a dot in the diagram, the
other possibility being shown by a cross. It is clear that there is no
ambiguity in the choice of signs required to satisfy the equation.

After the phases for all the $h0\ell$ reflections in metal-free
and nickel phthalocyanines are determined in this manner, the corresponding
electron density projection maps can be calculated. Fig. 4.10-3 shows the map
for metal-free phthalocyanine.

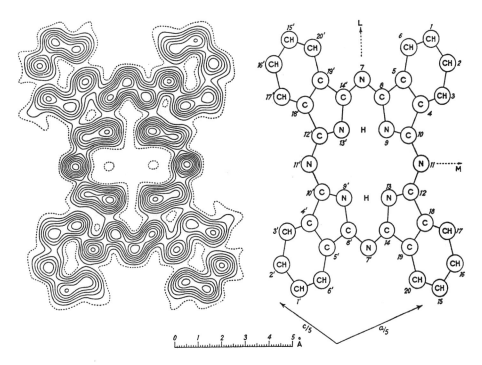

Fig. 4.10-3 Electron density projection map for metal-free
phthalocyanine and the corresponding atomic positions. (After ref. 3).

2. Electron distribution around the Fe atom of the ground state of FePc[1]

In order to analyze the asymmetry in the atomic charge distribution, X-X
deformation density maps were calculated. They are defined as

$$\rho_{X-X}(r) = \frac{2}{V} \sum_{0}^{H} (F_{obs} - F_{calc,high\ order}) \exp 2\pi iH.r$$

where the calculated amplitudes are based on parameters from the high-order
refinement and spherical atom scattering factors, and F_{obs} is assigned the
sign of F_{calc}. The function ρ_{X-X} has an advantage over the conventional

difference map as it is based on high order parameters which are much less biased by the spherical atom approximation used in the refinement model. Aspherical features in the atomic electron distribution are therefore in general more pronounced in the X-X maps.

The description of the Fe valence electron distribution in terms of the multipole parameters is equivalent to the expression of the charge distribution in terms of the Fe atomic orbitals and their occupancies, as listed in Table 4.10-2. The figures given in column 6 represent a quantitative estimate of the features observed in the deformation density maps. For example, the d_{xy} orbital is found to be more heavily populated than might be expected for the spherical atom (column 7), while the d_{z^2} orbital is underpopulated. A significant population is also found for the $d_{x^2-y^2}$ orbital, which is largely attributed to significant covalent character of the Fe-N interaction. The ratio of the occupancies of the d_{xy}, $d_{xz,yz}$ and d_{z^2} orbitals is equal to 2:2.6(2):1.1(1), as compared with 2:3:1, 2:2:2, 1:4:1, and 1:3:2 for the ^3E_gA, ^3A_g, $^3B_{2g}$, and ^3E_gB states, respectively. (The term symbol nomenclature is described in ref. 7). Within the error limits the observation is only compatible with the ^3E_gA state which is therefore the leading contributor to the ground state of the iron atom.

It should be noted that the formalism used to derive d-orbital occupancies presumes that the 4p orbitals are either not populated, or populated evenly to give an approximately spherically p-density distribution. The d-orbital occupancies from refinement excluding 4s electrons are identical within the limits of experimental error.

Table 4.10-2   Electron occupancies in square-planar $3d^6$ electron configurations and comparison with experimental results[*].

Term symbol	^3E_gA	^3A_g	$^3B_{2g}$	^3E_gB	Experimental		Spherical atom
$d_{x^2-y^2}$	-	-	-	-	0.70(7)	(12.9%)	1.2(20%)
d_{z^2}	1(17%)	2(33%)	1(17%)	2(33%)	0.93(6)	(17.1%)	1.2(20%)
$d_{xz,yz}$	3(49%)	2(33%)	4(67%)	3(49%)	2.12(7)	(39.1%)	2.4(40%)
d_{xy}	2(33%)	2(33%)	1(17%)	1(17%)	1.68(10)	(30.9%)	1.2(20%)

[*] The z axis is the four-fold symmetry axis perpendicular to the molecular plane. The x and y axes are in the plane along the Fe-N bonds.

3. Crystal structure of β-CuPc

Copper phthalocyanine (CuPc) is a compound of historical significance and great commercial value; β-CuPc is the normal stable form at room temperature and pressure.

The structure of β-CuPc[5] is isomorphous with those of the metal-free and nickel phthalocyanines. The essential difference between the various polymorphic forms of CuPc is that the copper atom coordinates to different nitrogen atoms of neighbouring rings, resulting in different molecular packing arrangements and hence different crystal structures. In order to produce any given polymorphic form, it becomes necessary to block in some way the N atoms which are not required for coordination. For example, the α-polymorph is readily formed by treating the β-polymorph with sulphuric acid and thus blocking the N(1) and N(3) atoms with sulphate residues; the copper atom then coordinates to N(2) or N(4).

In contrast to β-CuPc, where the Cu atom appears to adopt distorted octahedral coordination with four Cu-N distances of 1.93Å and two Cu-N of 3.28Å, the Pt atom in the α- and γ-forms of PtPc seems to be satisfied with four Pt-N bonds in a square-planar environment at a mean distance of 1.98Å. The closest approaches to the Pt atom perpendicular to the plane of the Pc molecule are 3.57Å to N(4), 3.83 to N(2), 3.84 to N(1), 3.73 to C(16), and 3.90 to C(1) in the α-form, and 3.70, 3.80, 3.76, 3.81 and 3.77Å to the corresponding atoms in the γ-form. These distances are comparable to those between the carbon and nitrogen atoms of neighbouring rings and are long enough to preclude the likelihood of electron interaction. It appears that the preference of Pt for square-planar coordination accounts for the α- and γ-forms being the stable polymorphs of PtPc at room temperature and pressure, whereas for copper and many other metal phthalocyanines the β-phase is the normal stable form. Certainly this is consistent with the fact that a β-type polymorph of PtPc has not yet been prepared.

4. Structure of phthalocyanine sandwich compounds

The structure of α-SnPc$_2$ can be represented by two phthalocyanine ring systems between which the tin(IV) atom is sandwiched. The ring systems are rotated 42° with respect to one another to form a square antiprism of N atoms around the central Sn atom [Fig. 4.10-4(a)].[8] The two N$_4$ squares are nearly parallel to each other (1.1° between normals) and 2.70Å apart. The average Sn-N distance is 2.347(7)Å. The two Pc ring systems are buckled and drawn in towards the metal atom like a pair of back-to-back saucers. The pyrrole rings bend outward at angles of 1-12° relative to the plane defined by the square of four N atoms bound to the Sn atom. The phenyl rings make angles of 2-14° with respect to the same plane.

Several other metallo-phthalocyanine complexes like ZrPc$_2$[9,10] and LuPc(Pc·)[11] (where Pc· is the phthalocyaninato radical monoanion) also

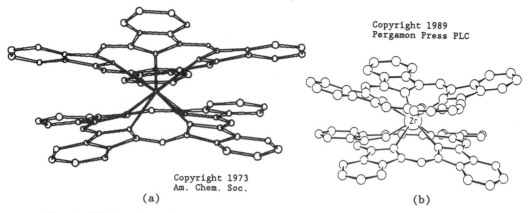

(a)   (b)

Fig. 4.10-4 Molecular structure of (a) α-SnPc$_2$ and (b) ZrPc$_2$. (After
refs. 8 and 9).

exhibit the same type of distorted square-antiprismatic coodination geometry
about the metal center. Table 4.10-3 gives a comparison of the ring
distortions and structural parameters for some $M^{IV}Pc_2$ and $M^{III}Pc(Pc\cdot)$
complexes.[10] The wok-like distortion in the Pc rings is most severe in
ZrPc$_2$, which is isostructural with β-SnPc$_2$ but not with α-SnPc$_2$.

Table 4.10-3 Comparison of structural features of some $M^{IV}Pc_2$ and
$M^{III}Pc(Pc\cdot)$ complexes.

	Mean distance between N$_4$ planes	Staggering angle of rings	Average M-N bond distance	Range of disphacements of outermost C atoms
ThPc$_2$	2.96Å	37°	2.48Å	0.27 - 1.30Å
NdPc(Pc·)	2.94	38	2.47	
UPc$_2$	2.81	37	2.43	0.49 - 1.10
LuPc(Pc·)	2.69	45	2.38	0.25 - 1.10
α-SnPc$_2$	2.70	42	2.35	0.18 - 1.01
ZrPc$_2$	2.20	42	2.30	0.49 - 1.47

5. Phthalocyanine derivatives as materials for optical data storage[12]

Phthalocyanines are attractive as optical recording media: (i) they
absorb well in the He/Ne diode laser region (633 nm) with $\epsilon > 10^5$, (ii) they
are thermally stable, unaffected by atomspheric effects and are non-toxic, and
(iii) they may be applied to the substrate by sublimation or, with appropriate
substitution, by spin-coating. If two protons of Pc are replaced by various
trivalent or higher-valent metal atoms, the resulting MPc ligand can take up
one or two axial ligand(s) leading to a significant improvement in the
solubility of the virtually insoluble original substance [Fig. 4.10-5(a)].

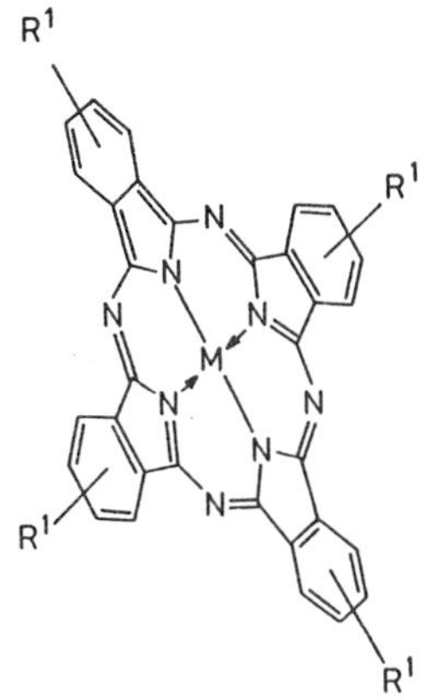

λ_{max} = 600-800 nm

M   = $Cu^{2\oplus}, VO^{2\oplus}, Al(OAlkyl)^{2\oplus}, Si(OAlkyl)_2^{2\oplus}, Ge(OAlkyl)_2^{2\oplus}$

R^1 = H, 4F, OAlkyl, CH_2O-Alkyl

(a)

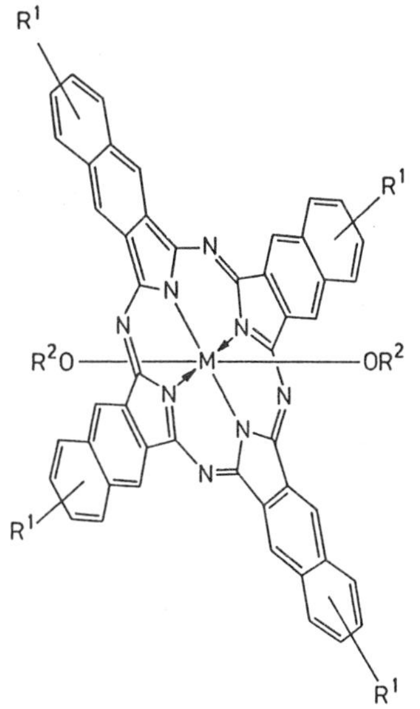

λ_{max} = 720-820 nm

M   = Si, Ge

R^1 = H, t-C_4H_9, $(OCH_2-CH_2)_n-O$Alkyl, COOAlkyl

R^2 = $Si(Alkyl)_3, Si(Alkyl)_2-O-$Alkyl

(b)

Fig. 4.10-5 (a) Peripherally substituted phthalocyanine derivatives and (b) peripherally and axially substituted naphthalocyanine derivatives for WORM media. (After ref. 12).

In 1981 Kivits and co-workers proposed the use of vanadylphthalocyanine, VOPc, as a viable absorbing dye in storage disks of the WORM (*Write Once Read Many*) type.[13] Thermal treatment of this material brings about a bathochromic shift from its absorption maximum at 730 nm as well as a broadening of the long-wavelength absorption band beyond 800 nm, permitting the use of a Kr laser (799 nm). Recent industrial efforts have concentrated on structural modification of MPc systems by peripheral and/or axial substitution to achieve improved properties in regard to absorption, solubility, and thermal stability. Enlargement of the macrocyclic π system by benzannelation to give the naphthalocyanines (Nc) [Fig. 4.10-5(b)] usually results in a substantial bathochromic shift, although these derivatives are thermally less stable and much more difficult to synthesize. As with the MPc compounds, the MNc derivatives may be rendered soluble by the use of bulky peripheral substituents such as the *tert*-butyl group and by long-chain alkyl axial substituents bearing hydrophilic functions (carboxylic acid, alcohol, or ether). For example, the biaxial derivatives of silicon-naphthalocyanine [$SiNc^{2+}$ with R^1 = H and R^2 = $Si(CH_3)_2O(CH_2)_6NHCO(CH_2)_{14}CH_3$] exhibit high absorption in the range 780-830 nm, a reflectivity exceeding 30%, and a signal-to-noise ratio of > 50dB. The most recent patent applications involve axially and peripherally substituted aluminum-, germanium- and silicon-naphthalocyanines, and prototypes of WORM disks based on these materials have been constructed.

6. "Face-to-face" $[M(Pc)O]_n$ and halogen-doped polymers

     The class of cofacially joined phthalocyaninatopolymetalloxanes $[M(Pc)O]_n$ (M = Si, Ge, Sn) has attracted considerable interest because of their high conductivity after doping with iodine or other electron acceptors. These "face-to-face" polymers and their halogen-doped derivatives $\{M(Pc)OX_y\}_n$ (X = Br , I; $y \leq 1.1$) have been investigated using X-ray powder diffractometry, and the crystallographic data are summarized in Table 4.10-4.[14,15] The halogen-doped metallomacrocyclic polymers are essentially isomorphous with Ni(Pc)I, which consists of stacks of staggered NiPc units and parallel chains of I_3^- counter anions.[16] The crystal structures of $[Si(Pc)O]_n$ and Ni(Pc)I are shown in Fig. 4.10-6(a) and (b), respectively. Microcrystalline $[Ge(TBP)O]_n$ (TBPH$_2$ = tetrabenzoporphyrin) has been shown to be isostructural with $[Ge(Pc)O]_n$ with an interstack distance of 3.46Å.[17]

Table 4.10-4 Crystallographic data for $[M(Pc)O]_n$ and $\{[M(Pc)X_y\}_n$ materials.

Polymer	Space group	Z	Unit cell parameters, Å	Density, g cm^{-3} calcd	found	Interplanar spacing, Å	Staggering angle($\emptyset$), deg
$[Si(Pc)O]_n$	Ibam	4	a = 13.80(5),b = 27.59(5), c = 6.66(4)	1.458(21)	1.432(10)	3.33(2)	39(3)
$[Ge(Pc)O]_n$	P4/m	1	a = 13.27(5),c = 3.53(2)	1.609(28)	1.512(2)	3.53(2)	0(5)
	I4/m	2	a = 18.76(5),c = 3.57(2)	1.589(26)		3.57(2)	0(5)
$[Sn(Pc)O]_n$	P4/m	1	a = 12.81(5),c = 3.82(2)	1.715(25)	1.719(10)	3.82(2)	probably eclipsed*
Ni(Pc)I	P4/mcc	2	a = 13.936(6),c = 6.488(3)	1.84	1.78(4)	3.244(2)	39.5
$\{[Si(pc)O]I_{1.12}\}_n$	P4/mcc	2	a = 13.97(5),c = 6.60(4)	1.802(24)	1.744(10)	3.30(2)	39(3)
$\{[Si(Pc)O]Br_{1.12}\}_n$	P4/mcc	2	a = 13.97(5),c = 6.60(4)			3.30(2)	39(3)
$\{[Ge(Pc)O]I_{1.07}\}_n$	P4/mcc	2	a = 13.96(5),c = 6.96(4)	1.805(23)	1.774(10)	3.48(2)	40(4)

*Formally this space group requires a staggering angle of 0°.

7. Uranyl "superphthalocyanine"

     The reaction of o-dicyanobenzene with anhydrous uranyl chloride yields a cyclic five-subunit "superphthalocyanine" complex [Fig. 4.10-7(a)], the

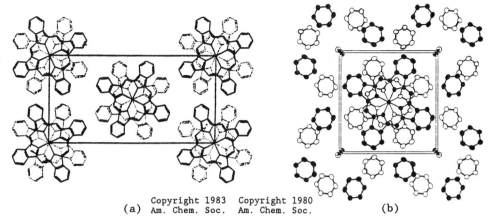

(a)   (b)

Fig. 4.10-6 Crystal structures of (a) $[Si(Pc)O]_n$ and (b) Ni(Pc)I viewed parallel to the stacking direction. (After refs. 14 and 16).

formation of which is attributed to the large uranium atom as a template in cyclization and condensation.[18] In the structure determined from X-ray analysis, the uranium atom has the expected compressed pentagonal bipyramidal coordination geometry with average U=O and U-N bond lengths of 1.744(8) and 2.524(9)Å, respectively. The cyclopentakis(2-iminoisoindoline) ligand is severely and irregularly distorted from planarity, presumably as a consequence of appreciable steric strain within the macrocycle [Fig. 4.10-7(b)].

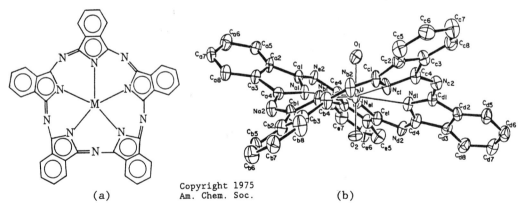

(a)
(b)

Fig. 4.10-7 (a) Structural formula of dioxocyclopentakis(2-iminoisoindoline)uranium(VI), where M represents the UO_2 group. (b) Perspective view of the $UO_2(N_2C_8H_4)_5$ molecule showing the non-planar conformation of the macrocyclic ligand. (After ref. 18).

References

[1] P. Coppens and L. Li, *J. Chem. Phys.* **81**, 1983 (1984).

[2] J.F. Kirner, W. Dow and W.R. Scheidt, *Inorg. Chem.* **15**, 1685 (1976).

[3] J.M. Robertson, *Organic Crystals and Molecules*, Cornell University Press, Ithaca, N.Y. 1953, p. 263.

[4] J.M. Robertson and I. Woodward, *J. Chem. Soc.*, 219 (1937); R.P. Linstead and J.M. Robertson, *J. Chem. Soc.*, 1736 (1936); J.M. Robertson, *J. Chem. Soc.*, 615 (1935).

[5] C.J. Brown, *J. Chem. Soc. (A)*, 2488, 2494, (1968).

[6] F.H. Moser and A.L. Thomas, *The Phthalocyanines*, Vols. I and II, CRC Press, Boca Rota, Florida, 1983.

[7] S. Obara and H. Kashiwagi, *J. Chem. Phys.* **77**, 3155 (1982).

[8] W.E. Bennett, D.E. Broberg and N.C. Baenziger, *Inorg. Chem.* **12**, 930 (1973).

[9] J. Silver, P.J. Lukes, P.K. Hey and J.M. O'Connor, *Polyhedron* **8**, 1631 (1989).

[10] J. Silver, P. Lukes, S.D. Howe and B. Howlin, *J. Mat. Chem.* **1**, 29 (1991).

[11] A. de Cian, M. Moussavi, J. Fischer and R. Weiss, *Inorg. Chem.* **24**, 3126 (1985).

[12] M. Emmelius, G. Pawlowski and H.W. Vollmann, *Ang. Chem. Int. Ed. Engl.* **28**, 1445 (1989), and references cited therein.

[13] P. Kivits, R. de Bont and J. Van der Veen, *Appl. Phys. A* **26**, 101 (1981).

[14] C.W. Kirk, T. Inabe, K.F. Schoch and T.J. Marks, *J. Am. Chem. Soc.* **105**, 1539 (1983).

[15] B.N. Diel, T. Inabe, J.W. Lyding, K.F. Schoch, C.R. Kannewurf and T.J. Marks, *J. Am. Chem. Soc.* **105**, 1551 (1983).

[16] C.S. Schramm, R.P. Scaringe, D.R. Stojakovic, B.M. Hoffman, J.A. Ibers and T.J. Marks, *J. Am. Chem. Soc.* **102**, 6702 (1980).

[17] M. Hanack and T. Zipplies, *J. Am. Chem. Soc.* **107**, 6127 (1985).

[18] V.W. Day, T.J. Marks and W.A. Wachter, *J. Am. Chem. Soc.* **97**, 4519 (1975).

Note Metal complexes of phthalocyanines and related compounds are reviewed in H. Schultz, H. Lehmann, M. Rein and M. Hanack, *Structure and Bonding*, **74**, 41 (1990).

For the most recent diffraction and theoretical results on metal(II) phthalocyanines, see B.N. Figgis, E.S. Kucharski and P.A. Reynolds, *J. Am. Chem. Soc.* **111**, 1683 (1989); P.A. Reynolds, B.N. Figgis, E.S. Kucharski and A.S. Mason, *Acta Crystallogr.*, Sect. *B* **47**, 899 (1991); P.A. Reynolds amd B.N. Figgis, *Inorg. Chem.* **30**, 2294 (1991), and referenes cited therein.

The structure and stereochemistry of linear oligopyrroles are covered in H. Falk, *The Chemistry of Linear Oligopyrroles and Bile Pigments*, Springer-Verlag, Berlin, 1989.

4.11 Copper-Oxygen Based High-Temperature Superconductor
$Tl_2Ba_2Ca_2Cu_3O_{10}$

Crystal Data

Tetragonal, space group $I4/mmm$ (No. 139)

$a = 3.8503(6)$, $c = 35.88(3)$Å, $Z = 2$

Atom	Position	x	y	z
Tl	4(e)	1/2	1/2	.2201
Ba	4(e)	0	0	.1448
Cu(1)	2(b)	1/2	1/2	0
Cu(2)	4(e)	1/2	1/2	.0896
Ca	2(a)	0	0	.0463
O(1)	4(c)	1/2	0	0
O(2)	8(g)	1/2	0	.0875
O(3)	4(e)	1/2	1/2	.1588
O(4)*	4(e)	1/2	1/2	.2719

* Refining this atom at a statistically-distributed 16(n) site gives x = 0.60(1),

y = 1/2, and z = 0.2724(9).

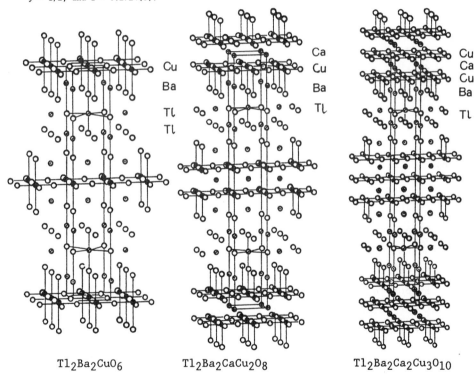

$Tl_2Ba_2CuO_6$ $Tl_2Ba_2CaCu_2O_8$ $Tl_2Ba_2Ca_2Cu_3O_{10}$

Fig. 4.11-1 A comparison of the structures of $Tl_2Ba_2Ca_{n-1}Cu_nO_{4+2n}$ for n = 1, 2, and 3. Metal atoms are shaded and Cu-O bonds are shown. (Adapted from ref. 1).

Crystal Structure

$Tl_2Ba_2Ca_2Cu_3O_{10}$ has the highest transition temperature (~125K) of any presently known bulk superconductor. No superstructure is observed, and the material is essentially twin-free.

The crystal structures of the $Tl_2Ba_2Ca_{n-1}Cu_nO_{4+2n}$ series are shown in Fig. 4.11-1. The structures differ from one another by the number of consecutive Cu-O sheets. In $Tl_2Ba_2Ca_2Cu_3O_{10}$, triple sheets of corner-sharing square-planar CuO_4 groups are oriented parallel to the (001) plane. Additional oxygen atoms are located above and below the triple Cu-O sheets and are positioned at a distance of 2.5Å from the copper atoms. There are no oxygen atoms between the triple Cu-O sheets. Interatomic distances and angles are listed in Table 4.11-1.

Table 4.11-1   Interatomic distances and angles in $Tl_2Ba_2Ca_2Cu_3O_{10}$.

Distances (Å)		Angles (degrees)		
Cu(1)-O(1)	1.9252(3) (x 4)	O(1)-Cu(1)-O(1)	180.0	(x 2)
Cu(2)-O(2)	1.927(1) (x 4)	O(1)-Cu(1)-O(1)	90.0	(x 4)
Cu(2)-O(3)	2.48(5) (x 1)	Cu(1)-O(1)-Cu(1)	180.0	(x 1)
Cu(1)-Cu(2)	3.214(6) (intersheet)	O(2)-Cu(2)-O(2)	175(1)	(x 2)
Tl(1)-O(3)	2.20(5) (x 1)	O(2)-Cu(2)-O(2)	89.91(6)	(x 4)
Tl(1)-O(4)*	1.92(3) (x 1)	O(2)-Cu(2)-O(3)	92.3(7)	(x 4)
Tl(1)-O(4)*	2.48(3) (x 2)	Cu(2)-O(2)-Cu(2)	175(1)	(x 1)
Tl(1)-O(4)*	3.02(4) (x 2)			
Ba(1)-O(2)	2.82(2) (x 4)			
Ba(1)-O(3)	2.768(9) (x 4)			
Ba(1)-O(4)*	2.99(3) (x 1)			
Ca(1)-O(1)	2.542(5) (x 4)			
Ca(1)-O(2)	2.427(14) (x 4)			

The Cu-O bond distances in the middle sheet are 1.925Å whereas those in the two outer sheets are slightly longer at 1.927Å. The individual Cu-O sheets are separated by 3.2Å (Cu-Cu distance) with calcium (and some thallium) ions coordinated to eight O atoms with *4mm* site symmetry and an average Ca-O distance of 2.48Å.

Barium ions reside just above and below the Cu-O triple sheets in nine-coordination with oxygen. The Ba-Cu-Ca-Cu-Ca-Cu-Ba slabs in $Tl_2Ba_2Ca_2Cu_3O_{10}$ alternate with a double thallium-oxygen layer, giving a layer repeat sequence of ...Tl-Tl-Ba-Cu-Ca-Cu-Ca-Cu-Ba... along the *z* direction. Thallium forms bonds to six oxygen atoms in a distorted octahedral arrangement where the octahedra share edges within the *ab* plane. A subtle disorder is suggested by the relatively larger basal thermal parameters for Tl and O(4). This is evidence for the existence of a locally distorted thallium environment where the Tl and O atoms within the sheets are statistically positioned around the 4(*e*) sites in order to form shorter Tl-O bonds. The average Tl-O bond distance within the sheet, with Tl and O(4) at the 4(*e*) positions, is 2.72Å and is significantly longer than the sum of ionic radii, 2.28Å, suggesting that Tl and O actually reside off this special position. With O(4) shifted to

the 16(n) positions, two short and two long Tl-O bonds are formed, 2.48 and
3.02Å. This is identical to the situation seen in the thallium-oxygen sheets
of the $Tl_2Ba_2CuO_6$ and $Tl_2Ba_2CaCu_2O_8$ compounds, whose structures are shown in
Fig. 4.11-1.

Fig. 4.11-2(a-c) show the coordination polyhedra around Cu, Tl, and Ca.
In Fig. 4.11-2(a) the tetragonal-pyramidal coordination of the copper ion is
indicated by hatching. As in all the superconductors with CuO_5 polyhedra
described up to now, the bases of the pyramids are oriented parallel to the
a/b plane and are linked at all four basal corners. The apices of the
pyramids are stretched to an unusual extent: 39% relative to the distances in
the base of the pyramid. Fig. 4.11-2(a) shows that there are CuO polygons at
z = 0 and 1/2, and these are joined to form infinite two-dimensional grids.

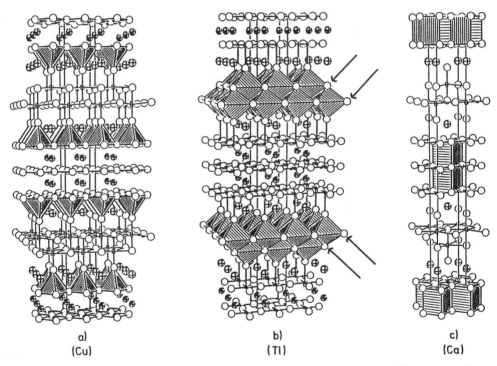

a)
(Cu)

b)
(Tl)

c)
(Ca)

Fig. 4.11-2 Coordination polyhedra in $Tl_2Ba_2Ca_2Cu_3O_{10}$, ⊕ = Ba^{2+}, ● = Ca^{2+}, O
= O^{2-}. (a) CuO_5 pyramids around copper (hatched) and planar polygons at z =
0 and 1/2 in the unit cell. (b) Double layers of TlO_6 octahedra, shown
hatched. The oxygen sites indicated by the arrows are O(4), only 25%
occupied. (c) Coordination polyhedra around Ca/Tl. (After ref. 2).

Within the crystal structure there are regions where copper ions are crowded
together. Fig. 4.11-2(b) shows the polyhedra around the thallium. These
compressed octahedra are joined together to form a double layer as in the

other thallium superconductors. Along the [001] direction the CuO_5 pyramids and the TlO_6 octahedra are joined at their corners by shared oxygen ions, thus forming the spaces which contain the Ca^{2+} ions and the Ba^{2+} ions. The alkaline-earth ions are packed into the crystal lattice in an ordered pattern. Ba^{2+} has the 8+1 coordination (which is typical of the K^+ ion in the K_2NiF_4 structure) and Ca^{2+} has cubic coordination, as shown in Fig. 4.11-2(c).

Remarks

1. Development of oxide superconductors[3]

Superconductivity was first discovered in mercury metal in 1911; the transition temperature (T_c) at which mercury becomes superconducting is 4.1K. Other materials were discovered to be superconducting, and by 1975 the highest T_c had risen slowly to 23K for the intermetallic compound Nb_3Sn.[4] Subsequently, a period of some pessimism set in, reinforced by theoretical predictions that T_c would never rise above 30K.

The field of oxide superconductors emerged in the early 1960s; its historical development is sketched in Table 4.11-1. The first superconducting oxides, NbO and TiO, may be viewed merely as metals containing some dissolved

Table 4.11-1 History of discovery of oxide superconductors.

Compound	T_c	Date discovered
TiO, NbO	1K	1964
$SrTiO_{3-x}$	0.7K	1964
Bronzes		
A_xWO_3	6K	1965
A_xMoO_3	4K	1969
A_xReO_3	4K	1969
Ag_7O_8X	1K	1966
$LiTi_2O_4$	13K	1974
$Ba(Pb,Bi)O_3$	13K	1975
$(La,Ba)_2CuO_4$	35K	1986
$YBa_2Cu_3O_7$	95K	1987
Bi/Sr/Cu/O	22K	1987
Bi/Sr/Ca/Cu/O	90K	1987
Tl/Ba/Ca/Cu/O	125K	1988
$(Ln,Ce)_2CuO_4$	24K	1989

oxygen. They have NaCl-related structures, but the direct metal-metal interactions are sufficiently strong to produce metallic properties. For the other oxide superconductors, the direct metal-metal interaction is weak, and the conduction band arises from strong metal-oxygen covalent bonding. Initially, all such oxide superconductors were based on transition metal cations having, on the average, a fraction of one d electron per cation. Later it became evident that superconductivity can also occur in oxides with a fraction of one s electron per cation.

The perovskite type superconductors $YBa_2Cu_3O_{7-x}$ and $La_{2-x}Sr_xCuO_4$ are discussed in Section 2.19.

2. Structure of $(Tl,Pb)Sr_2Ca_2Cu_3O_9$[5]

High-temperature superconductors based on oxides of thallium and copper but not containing barium, $(Tl,Pb)Sr_2Ca_{n-1}Cu_nO_x$ $(n = 2,3)$, have been prepared. A T_c of about 85K is found for $(Tl_{0.5}Pb_{0.5})Sr_2CaCu_2O_7$, whereas $(Tl_{0.5}Pb_{0.5})Sr_2Ca_2Cu_3O_9$ has a Tc of about 120K. Both materials possess tetragonal symmetry with $a = 3.80$, $c = 12.05$Å for $(Tl_{0.5}Pb_{0.5})Sr_2CaCu_2O_7$, and $a = 3.81$, $c = 15.23$Å for $(Tl_{0.5}Pb_{0.5})Sr_2Ca_2Cu_3O_9$. The structure of the latter phase (Fig. 4.11-3), based on single-crystal X-ray diffraction data, is essentially the same as that of $TlBa_2Ca_2Cu_3O_9$,[6] and is closely related to that of $Tl_2Ba_2Ca_2Cu_3O_{10}$.[1] Triple sheets of corner-sharing CuO_4 groups are oriented parallel to the (001) plane. The middle sheet is exactly planar and the CuO_4 units have $4/mmm$ site symmetry. The Cu-O bond length in this sheet is 1.904Å whereas that in the outer CuO_2 sheets is 1.906Å because of a slight puckering of these sheets. Additional oxygen atoms are bonded to the copper atoms of the two outer sheets at a distance of 2.37Å, giving these copper atoms a square-pyramidal environment with $4mm$ symmetry.

The triple Cu-O sheets of $(Tl,Pb)Sr_2Ca_2Cu_3O_9$ are separated by calcium ions that bond to eight oxygen atoms at an average distance of 2.48Å as found in the structures of $Tl_2Ba_2Ca_{n-1}Cu_nO_{2n+4}$ $(n = 2,3)$.[1,7]

3. Structures of $Bi_2Sr_{3-x}Ca_xCu_2O_{8+y}$[8] and related bismuth superconductors[9]

The structure of the superconductor $Bi_2Sr_{3-x}Ca_xCu_2O_{8+y}$, with x ranging from about 0.4 to 0.9, has been determined by single crystal X-ray diffraction. The crystals are pseudo-tetragonal, with an A-centered orthorhombic subcell of dimensions $a = 5.399$, $b = 5.414$ and $c = 30.904$Å. The structure is made up of alternating double copper-oxygen sheets and double bismuth-oxygen sheets, as shown in Fig. 4.11-4. The Cu-Cu separation between the sheets in the double layer is 3.25Å. Calcium and strontium ions are located between the Cu-O sheets with Ca(Sr)-O bond lengths ranging between 2.45 and 2.55Å. Strontium cations reside just above and below the Cu-O double sheets. These Sr-Cu-Ca(Sr)-Cu-Sr slabs alternate with a double bismuth layer, resulting in a layer repeat sequence of ...Bi-Bi-Sr-Cu-Ca(Sr)-Cu-Sr... . Electrical measurements on a single crystal of $Bi_2Sr_{3-x}Ca_xCu_2O_{8+y}$ show a resistivity drop at about 116K and apparent zero resistivity at 91K.

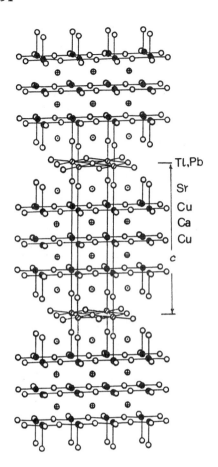

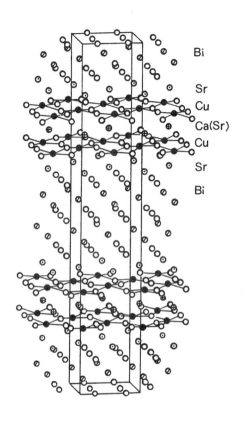

Fig. 4.11-3 The structure of $(Tl,Pb)Sr_2Ca_2Cu_3O_9$. Atoms in the Tl,PbO layers are shown in their ideal positions. Metal atoms are shaded and Cu-O bonds are drawn. (Adapted from ref. 5).

Fig. 4.11-4 Structure of $Bi_2Sr_{3-x}Ca_xCu_2O_{8+y}$ showing the CuO_2 layers. Metal atoms are shaded and only Cu-O bonds are shown. Oxygen atoms for the Bi layers are idealized. (Adapted from ref. 8).

The structures of $Bi_2(Sr,Ca)_2CuO_{8-x}$ and $Bi_2(Sr,Ca)_3Cu_2O_{10-x}$ have been determined by single-crystal X-ray and neutron diffraction.[9] Fig. 4.11-5 shows a perspective view of the different coordination polyhedra surrounding Cu, Bi, and Sr/Ca in $Bi_2(Sr,Ca)_2CuO_{8-x}$. In Fig. 4.11-5(a) the CuO_6 octahedra, stretched along the [001] direction, are indicated by hatching. The Cu^{2+} and Cu^{3+} ions have the same coordination. The bismuth ion occupies a tetragonal-pyramidal environment, which is shown in an idealized representation in Fig. 4.11-5(b). Fig. 4.11-5(c) shows the typical corner linkage of the BiO_5-CuO-BiO_5 polyhedra along the [001] direction. The alkaline-earth ions are arranged within the Cu/Bi/O crystal packing [Fig. 4.11-5(d)].

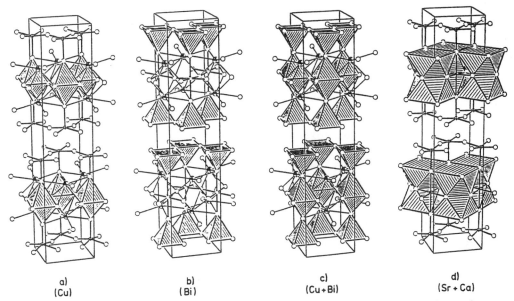

a)
(Cu)   b)
(Bi)   c)
(Cu+Bi)   d)
(Sr+Ca)

Fig. 4.11-5 Coordination polyhedra of $Bi_2(Sr,Ca)_2CuO_{8-x}$. (a) Octahedral coordination around the copper ions indicated by hatching. (b) Idealized BiO_5 polyhedra with ball-and-stick model of the Cu/O coordination. (c) Linking between the CuO_6 octahedra and the idealized BiO_5 pyramids. (d) The rectangular antiprism around the alkaline-earth ions. (After ref. 2).

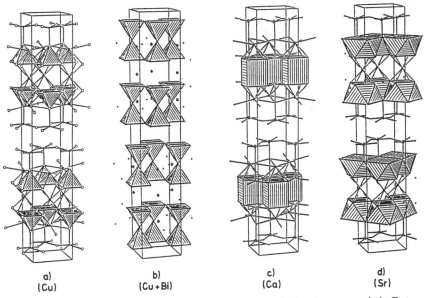

a)
(Cu)   b)
(Cu+Bi)   c)
(Ca)   d)
(Sr)

Fig. 4.11-6 Coordination polyhedra in $Bi_2(Sr,Ca)_3Cu_2O_{10-x}$. (a) Tetragonal-pyramidal coordination of Cu^{2+}/Cu^{3+} by O^{2-}. The BiO_5 environment is shown in a ball-and-stick model. (b) Linking of the CuO_5 and BiO_5 polyhedra; the BiO_5 pyramids are shown in an idealized representation. (c) Coordination of Ca^{2+} by O^{2-}. (d) Coordination of Sr^{2+} by O^{2-}. (After ref. 2).

The coordination polyhedra of $Bi_2(Sr,Ca)_3Cu_2O_{10-x}$ are shown in Fig. 4.11-6. In Fig. 4.11-6(a) the tetragonal pyramids of O^{2-} ions around Cu^{2+} and Cu^{3+} are indicated by hatching. Fig. 4.11-6(b) shows in addition the tetragonal pyramids around Bi^{3+} and Bi^{5+}, which are linked to those of Fig. 4.11-6(a). The Ca^{2+} ions intercalate between the layers of CuO_5 pyramids; as can be seen in Fig. 4.11-6(c), these have a stretched cubic coordination. The bulky Sr^{2+} ions occupy rectangular antiprisms [Fgi. 4.11-6(d)].

4. Structure of $(B^{III})_1(A^{II})_2Ca_{n-1}Cu_nO_{2n+2.5+\delta}$ [10]

The superconducting cuprates containing thallium or bismuth are layered compounds with the general formula $(B^{III})_1(A^{II})_2Ca_{n-1}Cu_nO_{2n+2.5+\delta}$, where B^{III} = Tl or Bi, A^{II} = Ba or Sr. Their structural data and superconducting properties are listed in Table 4.11-2, and their primitive tetragonal unit cells (space group $P4/mmm$) with ideal atomic positions are illustrated in Fig. 4.11-7.

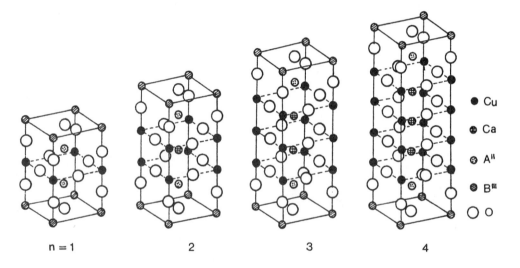

					● Cu
					⊕ Ca
n = 1	2	3	4		⊗ A^{II}
					⊕ B^{III}
					○ O

Fig. 4.11-7 Unit cells (with idealized atomic positions) of the first four members of the homologous series $Tl_1(A^{II})_2Ca_{n-1}Cu_nO_{2n+2.5+\delta}$. (After ref. 10).

Table 4.11-2 Structural data and superconducting properties of $(B^{III})_1(A^{II})_2Ca_{n-1}Cu_nO_{2n+2.5+\delta}$.

Components	n	a(Å)	c(Å)	$T_{c(onset)}$(K)	$T_{c(R=0)}$(K)
B^{III} = Tl	2	3.847	12.74	98	90
A^{II} = (Ba,Ca)	3	3.849	15.87	114	110
	4	3.850	19.01	125	122
B^{III} = (Tl,Bi)	1	3.745	9.00	50	25
A^{II} = (Sr,Ca)	2	3.800	12.07	90	75

5. Coordination of copper in high-temperature ceramic superconductors[11]

A general feature in ceramic superconductors is the occurrence of a small number of Cu-O coordination polyhedra. Copper is located in one of the three configurations: (a) at the center of a square array of coplanar oxygen atoms (square-planar); (b) at the center of the square base of a pyramid with oxygen at the vertices (pyramidal 4+1); and (c) at the center of an axially elongated octahedron with oxygen at the vertices (distorted octahedral 4+2) (Fig. 4.11-8). The distance X is about 1.95Å and Y about 2.3Å.

In the crystal structures, the coordination polyhedra are arranged so that the square-planar configurations are perpendicular to the *c* axis of the unit cell and the long axis of the pyramids and octahedra are parallel to it. Thus all structures can be viewed as having parallel planes of copper-oxygen sheets with typical Cu-O separations of 1.95Å.

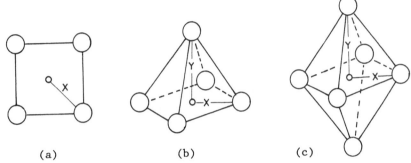

(a)   (b)   (c)

Fig. 4.11-8 Coordination polyhedra of copper and oxygen atoms in high Tc superconductors: (a) square-planar, (b) pyramidal 4+1 and (c) distorted octahedral 4+2. (After ref. 11).

6. Factors influencing the T_c of oxide superconductors

Knowledge of the crystal structures of oxide superconductors has led to a consideration of some factors which could affect the T_c of these materials.

a) The principal characteristic of oxide superconductors is that they all contain mixed-valent cations. It would even seem that three oxidation states need to be accessible; for example, Cu^I, Cu^{II} and Cu^{III} or Bi^{III}, Bi^{IV} and Bi^V. This is of course a requirement of the disproportionation mechanisms for superconductivity.[2]

b) High-temperature superconductivity only occurs in oxides where there is very high covalency, which is necessary to develop bands capable of supporting metallic behavior. Conversely, a very broad conduction band might be expected to produce only low T_c superconductors.[2]

c) The CuO_2 networks with 180° Cu-O-Cu connections seem to be important for superconducting behaviour. A relationship between Cu-O distance and T_c

has been found for $La_{2-x}A_xCuO_4$ phases. The corresponding distances in the $Bi_2Sr_2Ca_{n-1}Cu_nO_{4+2n}$ systems are shorter, yet T_c's are generally lower.[1]

d) For the $(B^{III})_2A^{II}_2Ca_{n-1}Cu_nO_{2+xn}$ phases where B^{III} is Bi or Tl, A^{II} is Ba or Sr, and n is the number of Cu-O sheets stacked consecutively, there is a general trend toward higher T_c as n increases. If the relation is linear, room temperature superconductivity might be achieved for $n = 10$.[1]

e) Basic or electropositive cations such as Ba^{II} and Y^{III} or La^{III} seem to help in the formation of superconducting oxides. In part, this may be simply that they favour networks with 180° Cu-O-Cu connections. However, these electropositive cations have other effects that seem to be important. They increase the covalency of the Cu-O bond[12] and also stabilize the high oxidation states of cations such as Cu^{III} and Bi^V.

7. The local charge picture of copper oxide superconductors

According to the "local charge" picture, all the layered copper oxide superconductors are built of electronically active CuO_x superconducting layers sandwiched between other structural layers which serve as spacers and electronic charge reservoirs.[13] Table 4.11-3 lists some representative examples of the superconducting compounds with two CuO_2 layers. The stacking sequences which are responsible for the formation of the double CuO_5 pyramidal layers are the same for all the compounds: $-MO-CuO_2-M'-CuO_2-MO-$, the charge reservoir layers being enclosed in square brackets to emphasize the close structural similarity among the different families of high-temperature superconductors.

Table 4.11-3 Superconductor stacking sequences based on two CuO_2 layers.[13]

Formula	Stacking sequence of one elementary structural unit			
$TlBa_2CaCu_2O_7$	$-Ca-CuO_2-BaO-[$		TlO	$]-BaO-CuO_2-$
$Tl_2Ba_2CaCu_2O_8$	$-Ca-CuO_2-BaO-[$	TlO-	-TlO	$]-BaO-CuO_2-$
$Ba_2YCu_3O_{6+d}$	$-Y-CuO_2-BaO-[$		$-CuO_d-$	$]-BaO-CuO_2-$
$Ba_2YCu_4O_8$	$-Y-CuO_2-BaO-[$	CuO-	-CuO	$]-BaO-CuO_2-$
$Pb_2Sr_2(Y,Ca)Cu_3O_{8+d}$	$-(Y,Ca)-CuO_2-SrO-[$	PbO $-CuO_d-$ PbO		$]-SrO-CuO_2-$
$La_2CaCu_2O_6$	$-Ca-CuO_2-LaO-[$			$]-LaO-CuO_2-$

The last compound is related to the others in terms of the active pyramidal copper-oxygen layers but has no charge reservoir layer.

References

[1] C.C. Torardi, M.A. Subramanian, J.C. Calabrese, J. Gopalakrishnan, K.J. Morrissey, T.R. Askew, R.B. Flippen, U. Chowdhry and A.W. Sleight, *Science (Washington)* 240, 631 (1988).

[2] H. Müller-Buschbaum, *Angew. Chem. Int. Ed. Engl.* 28, 1472 (1989).

[3] A.W. Sleight, *Science (Washington)* **242**, 1519 (1988); C.P. Poole, Jr., T. Datta and H.A. Farach, *Copper Oxide Superconductors*, Wiley, New York, 1988; D.M. Ginsberg (ed.), *Physical Properties of High Temperature Superconductors*, World Scientific, Singapore, Vol. I, 1989, Vol. II, 1990.

[4] T.H. Geballe and J.K. Hulm, *Science (Washington)* **239**, 367 (1988).

[5] M.A. Subramanian, C.C. Torardi, J. Gopalakrishnan, P.L. Gai, J.C. Calabrese, T.R. Askew, R.B. Flippen and A.W. Sleight, *Science (Washington)* **242**, 249 (1988).

[6] S.S.P. Parkin, V.Y. Lee, A.I. Nazzal, R. Savoy, R.Beyers and S.J. LaPlaca, *Phys. Rev. Lett.* **61**, 750 (1988).

[7] M.A. Subramanian, J.C. Calabrese, C.C. Torardi, J. Gopalakrishnan, T.R. Askew, R.B. Flippen, K.J. Morrissey, U. Chowdhry and A.W. Sleight, *Nature (London)* **332**, 420 (1988).

[8] M.A. Subramanian, C.C. Torardi, J.C. Calabrese, J. Gopalakrishnan, K.J. Morrissey, T.R. Askew, R.B. Flippen, U. Chowdhry and A.W. Sleight, *Science (Washington)* **239**, 1015 (1988).

[9] H.G. von Schnering, L. Walz, M. Schwarz, W. Becker, M. Hartweg, T. Popp, B. Hettich, P. Müller and G. Kämpf, *Angew. Chem. Int. Ed. Engl.* **27**, 574 (1988).

[10] P. Haldar, K. Chen, B. Maheswaran, A. Roig-Janicki, N.K. Jaggi, R.S. Markiewicz and B.C. Giessen, *Science (Washington)* **241**, 1198 (1988).

[11] L.E. Toth, M. Osofsky, S.A. Wolf, E.F. Skelton, S.B. Qadri, W.W. Fuller, D.U. Gubser, J. Wallace, C.S. Pande, A.K. Singh, S. Lawrence, W.T. Elam, B. Bender and J.R. Spann, in D.L. Nelson, M.S. Whittingham and T.F. George (eds.), *Chemistry of High-temperature Superconductors*, American Chemical Society, Washington, DC, 1987, chap. 22.

[12] A.W. Sleight in W.E. Hatfield and J.H. Miller (eds.) *High-temperature Superconducting Materials*, Marcel Dekker, New York, 1988, p. 1.

[13] R.J. Cava, *Science (Washington)* **247**, 656 (1990).

Note Updated crystallographic data on 19 distinct structural types of layered copper-oxide superconductors are summarized by R.M. Hazen in D.M. Ginsberg (ed.), *Physical Properties of High Temperature Superconductors II*, World Scientific, Singapore, 1990, chap. 3.

The discovery of superconductivity at 40K in the new infinite-layer compound $Sr_{1-y}Nd_yCuO_2$ is reported in M.G. Smith, A. Manthiram, J. Zhou, J.B. Goodenough and J.T. Markert, *Nature (London)* **351**, 549 (1991).

The crystal chemistry of copper oxometallates is reviewed and illustrated with computer colour graphics in H. Müller-Buschbaum, *Angew. Chem. Int. Ed. Engl.* **30**, 723 (1991).

4.12   Ternary Molybdenum Oxide System

$K_2Mo_8O_{16}$

Crystal Data

Monoclinic, space group $P2/n$ (No. 13)

$a = 10.232(3)$, $b = 10.286(4)$, $c = 5.758(1)$Å, $\gamma = 90.14(3)°$, $Z = 2$

Atom	x	y	z	Atom	x	y	z
Mo(1)	.1021	.9207	.8455	O(2)	.0941	.0535	.1170
Mo(2)	.0825	.9273	.4015	O(3)	.0818	.7798	.6246
Mo(3)	.5767	.0994	.8351	O(4)	.5817	.2896	.8794
Mo(4)	.4301	.9196	.6144	O(5)	.4192	.1183	.6133
K(1)	3/4	3/4	.6437	O(6)	.5572	.9036	.8839
K(2)	3/4	3/4	.1290	O(7)	.7189	.0827	.6157
O(1)	.1165	.0795	.6217	O(8)	.2878	.9178	.8845

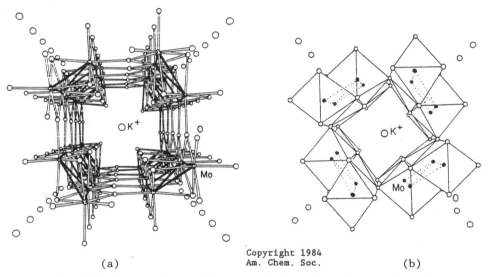

(a)   (b)

Fig. 4.12-1   (a) Three-dimensional view down the c axis of $K_2Mo_8O_{16}$.
(b) Corner sharing of the double strings creating channels where the
potassium ions are located.   (After ref. 1).

Crystal Structure

The structure of $K_2Mo_8O_{16}$ consists of double strings of edge-sharing
MoO_6 octahedra (each sharing four edges with four others), which are
interconnected via corner sharing of oxygen atoms (Fig. 4.12-2).[1]
Molybdenum atoms within the strings are shifted from their octahedral centers
so as to form planar tetranuclear clusters.

An alternative description views the structure as consisting of
molybdenum oxide cluster chains extended parallel to the c axis.   The chains
are built from Mo_4O_{16} clusters that share oxygen atoms on opposite edges of
the planar tetrameric molybdenum atom cluster.   The K^+ ions occupy sites along

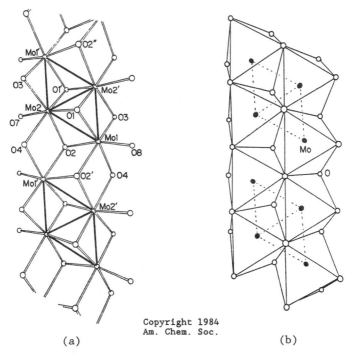

(a)   (b)

Fig. 4.12-2 (a) Section of one type of molybdenum oxide cluster chain in $K_2Mo_8O_{16}$. The other type differs mainly in the Mo(1)-Mo(2) and Mo(1')-Mo(2') bond lengths. Mo-Mo bonds are represented as solid black lines. (b) Double strings of edge-sharing MoO_6 octahedra in $K_2Mo_8O_{16}$. Tetrameric molybdenum atom clusters as shown. (After ref. 1).

the c axis in channels formed by four metal oxide cluster chains crosslinked by molybdenum-oxygen bonds, as shown in Fig. 4.12-1. Potassium is coordinated to eight oxygen atoms in a distorted square-prismatic arrangement. The spacings between K^+ along a channel are 2.79 and 2.96Å.

The Mo atom clusters and connectivity within and between chains via the sharing of oxygen atoms are shown in Fig. 4.12-2. The tetranuclear units contain two types of Mo atoms: (i) the apical atoms Mo(1) each bonded to two other Mo atoms and six O atoms, and (ii) the atoms Mo(2) each bonded to three Mo atoms and six O atoms. The cluster chains contain four types of oxygen atoms, each being bonded to three Mo atoms. Intrachain atom types O(1) and O(2) are triply bridging to Mo atoms within and between Mo_4O_{16} clusters, respectively, and are also coordinated to K^+ ions. These oxygen atoms may be considered to be sp^3 hybridized. Interchain atom types, O(3) and O(7), and O(4) and O(8), connect the individual cluster chains and are in

trigonal-planar-like coordination with molybdenum. Atoms O(3) and O(7) are
each shared between two separate Mo_4O_{16} clusters, while atoms O(4) and O(8)
are each shared by three cluster units. These oxygen atoms may be considered
to be sp^2 hybridized since the sum of the Mo-O-Mo bond angles around each atom
is 359.8-360.0°. The inter- and intrachain linking of the cluster units may
be represented by the formula $Mo_4O_2{}^{intra}O_{4/2}{}^{intra}O_{4/2}{}^{inter}O_{6/3}{}^{inter}$.

The structure displays two slightly different types of cluster chains.
Both types possess inversion symmetry within and between the planar Mo_4 units,
but the five Mo-Mo bonds in the clusters of one chain are closer to being
equivalent than in the clusters of the other chain.

Remarks

1. Structure of $Ba_{1.14}Mo_8O_{16}$[2]

Triclinic $Ba_{1.14}Mo_8O_{16}$ is structurally similar to $K_2Mo_8O_{16}$, but the two
Mo_4 clusters are more regular and more distorted, respectively, than in
$K_2Mo_8O_{16}$. The 2.697Å sides of the more regular cluster units in the potassium
compound [Mo(1)-Mo(2)] are elongated to 2.837Å in the more distorted units,
whereas the dimensions of the other bonds of the two cluster types are listed
in Table 4.12-1. The valence of each crystallographically different Mo atom
can be estimated by a summation of the Mo-O bond strengths for that atom.
Bond strengths are calculated from

$$s = (d/1.882)^{-6.0}$$

where s = bond strength of an Mo-O bond, d = crystallographic Mo-O bond
length, and the values 1.882 (Mo-O bond of unit strength) and -6.0 are fitted
parameters.[3] A comparison of bond valence for $K_2Mo_8O_{16}$ and $Ba_{1.14}Mo_8O_{16}$ is
given in Table 4.12-1.

As the Mo(1)-Mo(2) and Mo(1')-Mo(2') edges become longer, the remaining
three Mo-Mo bonds become shorter. For this reason, the calculated valence for
Mo(2), which is the atom that forms the unique central Mo(2)-Mo(2') bond,
changes only slightly as the number of cluster electron is varied. The apical
atoms Mo(1) are most affected by changes in the number of cluster electrons.

2. Structure of $NaMo_4O_6$ and $Ba_{0.62}Mo_4O_6$

These two compounds adopt structures with infinite metal-metal bonded
chains consisting of octahedral cluster units fused on opposite edges.
Because of the difference in effective ionic radii of the cations concerned,
their lattice types are different. $NaMo_4O_6$ is tetragonal with a = 9.559 , c
= 2.860Å, and Z = 2.[4] $Ba_{0.62}Mo_4O_6$ is orthorhombic with a = 9.509, b =
9.825, c = 2.853Å, and Z = 2.[2] The infinite chains running parallel to c
are interlinked in such a way that channels accommodating the cations are

Table 4.12-1 Comparison of bond lengths (Å) and calculated bond valence
values (given below the broken line) for Mo atoms within the
cluster units of $K_2Mo_8O_{16}$ and $Ba_{1.14}Mo_8O_{16}$.

Atom	$K_2Mo_8O_{16}$ regular clusters	distorted clusters	$Ba_{1.14}Mo_8O_{16}$ regular clusters	distorted clusters
Mo(1)-Mo(2)	2.697	2.837	2.616	2.847
Mo(1)-Mo(2')	2.596	2.565	2.578	2.546
Mo(2)-Mo(2')	2.551	2.527	2.578	2.560
Mo(1)-O(1)	2.066	2.086	2.079	2.082
Mo(1)-O(2)	2.043	2.037	2.034	2.046
Mo(1)-O(2')	2.121	2.078	2.095	2.104
Mo(1)-O(3)	1.935	1.939	1.936	1.931
Mo(1)-O(4)	2.135	2.090	2.143	2.079
Mo(1)-O(8)	1.974	1.914	2.022	1.894
Mo(2)-O(1)	2.047	2.042	2.053	2.051
Mo(2)-O(1')	2.062	2.044	2.053	2.038
Mo(2)-O(2)	2.032	2.093	2.055	2.062
Mo(2)-O(3)	2.020	1.988	2.023	2.003
Mo(2)-O(4)	2.130	2.150	2.128	2.119
Mo(2)-O(7)	2.055	2.037	2.043	2.030
Mo(1)	3.74	3.99	3.66	4.03
Mo(2)	3.53	3.54	3.51	3.61
Cluster av.	3.64	3.77	3.59	3.82
No. of cluster electrons	9.4	8.9	9.6	8.7

provided. Fig. 4.12-3 shows a segment of one $(Mo_4O_6^-)_\infty$ chain in $NaMo_4O_6$ which
shows intrachain Mo-Mo and Mo-O bonding. Within the Mo_4O_6 repeat unit of each
chain there are 13 Mo-Mo bonds of average bond length 2.796Å for $NaMo_4O_6$,

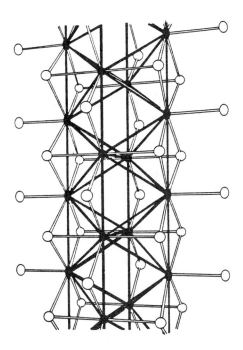

Fig. 4.12-3 Segment of one $(Mo_4O_6^-)_\infty$ chain in $NaMo_4O_6$ showing
intrachain Mo-Mo and Mo-O bonding. The Mo and O atoms are represented
by filled and open circles, respectively. (After ref. 5).

2.778Å for $Ba_{0.62}Mo_4O_6$, and Mo-O 2.038Å for $NaMo_4O_6$, 2.048Å for $Ba_{0.62}Mo_4O_6$.

Fig. 4.12-4 shows that the infinite chains bind each other through Mo-O-Mo bridges to form three-dimensional frameworks of $NaMo_4O_6$ (a) and $Ba_{0.62}Mo_4O_6$ (b).

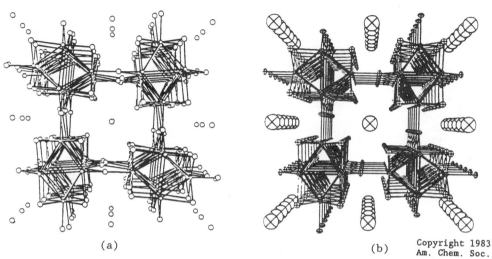

(a)   (b)

Fig. 4.12-4 Three-dimensional perspective view of the structures of (a) $NaMo_4O_6$ and (b) $Ba_{0.62}Mo_4O_6$ down the unique axis of chain growth and tunnel formation. (After ref. 5).

3. Structure of ternary molybdenum chalcogenides, $A_xMo_{3n+3}X_{3n+5}$

Compounds of the general composition $A_xMo_{3n+3}X_{3n+5}$ can contain either one kind of cluster exclusively or different clusters together in an ordered array, as listed in Table 4.12-2.[6] The structures containing only one kind of cluster exhibit (with the exception of the Ag compounds) a particularly close relationship with the structure of the first member of the series. All compounds crystallize in space group $R\bar{3}$ as do the Chevrel phases $A_xMo_6X_8$. The Mo_6X_8 cluster is stretched along its threefold axis, and inserted in the channels between the elongated clusters are $n-1$ planar Mo_3X_3 fragments together with the corresponding number of A atoms. Fig. 4.12-5 shows the structure of the clusters of $Mo_{3n+3}X_{3n+5}$. This genesis of the oligomeric clusters allows for a suitable fragmentation procedure to estimate the optimal electron counts. All intra- and intercluster contacts of the Mo_3X_4 fragments are the same as in the Chevrel phases. Both fragments together can have 20 to 24 d electrons for M-M bonding. The number of electrons in M-M bonding states per Mo_3X_3 fragment in the infinite chain is 13. Therefore the number of electrons (Z) in M-M bonding states of an oligomeric cluster with n condensed Mo_6 octahedra should be within the limits $13n + 7 \leq Z \leq 13n + 11$, where the upper limit corresponds to the closed configuration. This estimate is in agreement with the results of MO calculations for the ions $[Mo_9S_{11}]^{4-}$ and $[Mo_{12}S_{14}]^{6-}$.[7] With $Z = 13n + 11$ and $n = 2$ and 3 one gets $Z = 37$ and $Z = 50$, respectively. The MO calculation gives 50 electrons in M-M bonding states for $n = 3$. In the case where $n = 2$ (and all clusters with even n) there is some ambiguity in the electron count due to a non-bonding state in the HOMO-LUMO gap. The cluster contains 38 electrons when this level is occupied and 36 if unoccupied.

Table 4.12-2 Ternary molybdenum chalcogenides $A_xMo_{n+3}X_{3n+5}$ characterized by single-crystal structural investigations.

n	Compound	Clusters	Charge	Z/Mo Experimental	Estimated
I. Clusters of one kind					
1	$A_xMo_6X_8$ (X = S,Se,Te)	Mo_6X_8	0 to 4-	4.00-3.33	-
2	$Ag_{2.3}CsMo_9Se_{11}$	Mo_9Se_{11}	3.3-	3.92	4.11-3.67
2	$Ag_{4.4}ClMo_9Se_{11}$	Mo_9Se_{11}	3.4-	3.93	4.11-3.67
2	$Ag_{3.6}Mo_9Se_{11}$	Mo_9Se_{11}	3.6-	3.96	4.11-3.67
3	$Cs_2Mo_{12}Se_{14}$	$Mo_{12}Se_{14}$	2-	3.83	4.17-3.83
5	$Rb_4Mo_{18}Se_{20}$	$Mo_{18}Se_{20}$	4-	4.00	4.22-4.00
7	$Cs_6Mo_{24}Se_{26}$	$Mo_{24}Se_{26}$	6-	4.08	4.25-4.08
9	$Cs_8Mo_{30}Se_{32}$	$Mo_{30}Se_{32}$	8-	4.13	4.27-4.13
0	$TlMo_3Se_3$	$Mo_{6/2}Se_{6/2}$	1-	4.33	-
II. Clusters of different kinds					
1	$In_{\sim 3}Mo_{15}Se_{19}$	Mo_6Se_8			
2		Mo_9Se_{11}			
1	$In_2Mo_{15}Se_{19}$	Mo_6Se_8			
2		Mo_9Se_{11}			
1	$Tl_4Mo_{18}S_{22}$	Mo_6S_8			
3		$Mo_{12}S_{14}$			

In Table 4.12-2, those in the first group contain one individual kind of clusters with n face-condensed Mo_6 octahedra; those in the second group contain different kinds of clusters.[8] In the first group the number of electrons per Mo atom, Z/Mo, in M-M bonding states as derived from the composition is compared to the optimum range estimated through a fragmentation of the clusters. All the examples in this table have less d electrons than expected for the closed configuration. There is no ambiguity in the electron counts for the large cluster compounds with alkali metal cations; in each case four electrons are missing for the closed configuration.

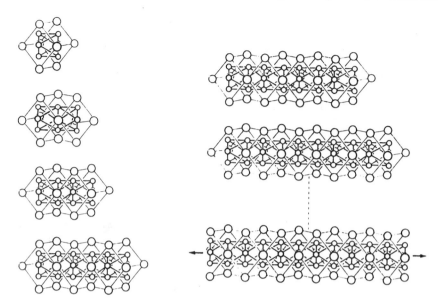

Fig. 4.12-5 Monomeric, oligomeric and polymeric clusters $Mo_{3n+3}X_{3n+5}$ (Mo small circles) containing Mo_6 octahedra condensed via faces.[8] The compositions of the clusters are Mo_6X_8, Mo_9X_{11}, $Mo_{12}X_{14}$, $Mo_{18}X_{20}$, $Mo_{24}X_{26}$, $Mo_{30}X_{32}$ and for the polymeric chain Mo_3X_3. (After ref. 6).

The infinite chains in the compound $LiMo_3Se_3$, when dissolved in organic solvents, remain fairly rigid owing to the strong metal-metal bonding between the molybdenum atoms.[9] Thus unlike most organic polymers which usually fold into a loose spherical shape in solution, this "inorganic polymer" behaves as ultra-thin "metallic wires" of only 6Å in diameter.

4. Alkoxide clusters of Mo and W

A variety of cluster alkoxides of Mo and W has been synthesized and characterized, in which alkoxide bridges and metal-metal bonds are present. A new structural relationship between metal oxides and metal alkoxides has emerged.[10] The molecular alkoxide-supported clusters of Mo and W contain M_xO_y cores that are analogues of the cluster subunits found in the solid-state structures of reduced metal oxides. For example, the $[W_4(OEt)_{16}]$ cluster contains the same $[M_4(\mu_3\text{-}O)_2(\mu\text{-}O)_4O_{10}]$ unit [structure (f) in Fig. 4.12-6] as that in the $[Ba_{1.14}Mo_8O_{16}]$ structure [Fig. 4.12-2(a)].

The structures of some alkoxides clusters are shown in Fig. 4.12-6: (a) ethane-like structure $M_2(OR)_6$ (M = Mo,W); (b) bridged dimer structure $M_2(OR)_6$; (c) $W_2(OR)_5(\mu\text{-}OR)py(\mu\text{-}C_2H_2)$; (d) $W_3(OR)_6(\mu_2\text{-}OR)_3(\mu_3\text{-}CMe)$; (e) $W_3(OR)_6(\mu\text{-}OR)_3(\mu_3\text{-}OR)(\mu_3\text{-}O)$; (f) $[W_4(OEt)_{16}]$; (g) $W_4(OR)_7(\mu_2\text{-}OR)_4(\mu_3\text{-}OR)$, with a butterfly arrangement of W atoms; and (h) [Mo$_6$(μ_3-OMe)$_8$(OMe)$_6$]$^{2-}$ with a M$_6$ octahedral cluster.

The alkoxide cluster compounds of Mo and W offer many attractive features for the development of a rich area of organometallic chemistry:[11]

a) By a suitable choice of alkoxide ligand, steric access to the metal may be made substrate-selective.

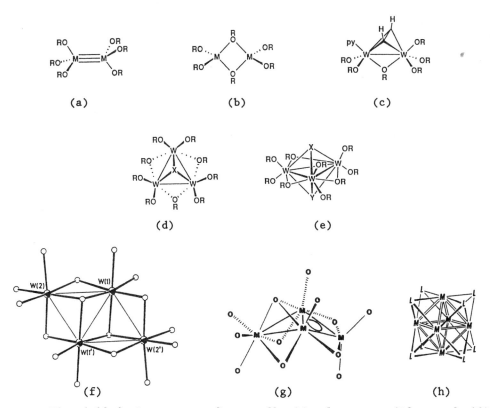

Fig. 4.12-6 Structures of some alkoxide clusters. (After ref. 10).

b) The metal atoms behave as Lewis acidic centers and the presence of M-M bonds provides reservoirs of electrons for π-acid substrates.

c) Alkoxide ligands are electronically flexible and readily move between terminal and bridging modes. Their electron-releasing property may be fine-tuned by manipulating R. Response to substrate uptake or release may involve a change in M-O-C angle or terminal-bridge site exchange.

d) Synthetic procedures have been established to incorporate a variety of functional groups of interest into the alkoxide clusters.

e) Unlike most organometallic systems, it is not necessary to generate a reactive species by chemical or physicochemical means.

Chisholm and co-workers have explored the similarities and differences between alkoxide and carbonyl cluster chemistry and summarized the results as follows:[10]

'Alkoxide and carbonyl ligands complement each other because they both behave as "π buffers" to transition metals. Alkoxides, which are π donors, stabilize early transition metals in high oxidation states by donating electrons into vacant d_π orbitals, whereas carbonyls, which are π acceptors, stabilize later transition elements in their lower oxidation states by accepting electrons from filled d_π orbitals. Both ligands readily form bridges that span M-M bonds. In solution fluxional processes that involve bridge-terminal ligand exchange are common to both alkoxide and carbonyl ligands. The fragments $[W(OR)_3]$, $[CpW(CO)_2]$, $[Co(CO)_3]$, and CH are related by the isolobal analogy. Thus the compounds $[(RO)_3W\equiv W(OR)_3]$, $[Cp(CO)_2W\equiv W(CO)_2Cp]$, hypothetical $[(CO)_3Co\equiv Co(CO)_3]$, and HC$\equiv$CH are isolobal. Alkoxide and carbonyl cluster compounds often exhibit striking similarities with respect to substrate binding —— e.g., $[W_3(\mu_3\text{-}CR))(OR')_9]$ versus $[Co_3(\mu_3\text{-}CR)(CO)_9]$ and $[W_4(C)(NMe)(OPr^i)_{12}]$ versus $[Fe_4(C)(CO)_{13}]$ —— but differ with respect to M-M bonding. The carbonyl clusters use e_g-type orbitals for M-M bonding whereas the alkoxide clusters employ t_{2g}-type orbitals. Another point of difference involves electronic saturation. In general, each metal atom in a metal carbonyl cluster has an 18-electron count; thus, activation of the cluster often requires thermal or photochemical CO expulsion or M-M bond homolysis. Alkoxide clusters, on the other hand, behave as electronically unsaturated species because the π electrons are ligand-centered and the LUMO metal-centered. Also, access to the metal centers may be sterically controlled in metal alkoxide clusters by choice of alkoxide groups whereas ancillary ligands such as tertiary phosphanes or cyclopentadienes must be introduced if steric factors are to be modified in carbonyl clusters. When alkynes and ethylene react with dinuclear carbonyl compounds, CO ligand loss is a prerequisite for substrate uptake and subsequent activation. For $[M_2(OR)_6]$ compounds (M = Mo and W) the nature of substrate uptake and activation is dependent upon the choice of M and R, leading to a more diverse chemistry.'

References

[1]　C.C. Torardi and J.C. Calabrese, *Inorg. Chem.* **23**, 3281 (1984).

[2]　C.C. Torardi and R.E. McCarley, *J. Solid State Chem.* **37**, 393 (1981).

[3]　J.C.J. Bart and V. Ragain in H.F. Barry and P.C.H. Mitchell (eds.), *Proceedings of the Climax Third International Conference on the Chemistry and Uses of Molybdenum*, Climax Molybdenum Co., Ann Arbor, 1979, p. 19.

[4] C.C. Torardi and R.E. McCarley, *J. Am. Chem. Soc.* **101**, 3963 (1979).

[5] R.E. McCarley in M.H. Chisholm (ed.), *Inorganic Chemistry: Towards the 21st Century*, American Chemical Society, Washington, D.C., 1983, chap. 18.

[6] A. Simon, *Angew. Chem. Int. Ed. Engl.* **27**, 159 (1988).

[7] T. Hughbanks and R. Hoffmann, *J. Am. Chem. Soc.* **105**, 1150 (1983).

[8] R. Chevrel, P. Gougeon, M. Potel and M. Sergent, *J. Solid State Chem.* **57**, 25 (1985).

[9] F.J. DiSalvo, *Science (Washington)* **247**, 649 (1990).

[10] M.H. Chisholm, D.L. Clark, M.J. Hampden-Smith and D.H. Hoffman, *Angew. Chem. Int. Ed. Engl.* **28**, 432 (1989).

[11] W.E. Buhro and M.H. Chisholm, *Adv. Organomet. Chem.* **27**, 311 (1987).

4.13   Potassium Tetracyanoplatinate (1.75/1) Sesquihydrate
$$K_{1.75}[Pt(CN)_4] \cdot 1.5H_2O$$

Crystal Data

Triclinic, space group $P\bar{1}$ (No. 2)

$a = 10.36(2)$, $b = 11.83(2)$, $c = 9.30(2)$Å, $\alpha = 102.4(1)°$, $\beta = 106.4(1)°$,

$\gamma = 114.7(1)°$, $Z = 4$

Atom	x	y	z	Atom	x	y	z
Pt(1)	0	0	0	N(31)	-.3658	.1172	.3478
Pt(2)	.0093	-.0211	.2568	N(32)	-.0267	.3570	.4106
Pt(3)	0	0	1/2	K(1)	.3999	.3952	.3293
C(11)	.2308	-.0621	.0930	K(2)	-.2717	-.3853	.3918
C(12)	.0513	-.2368	.0523	K(3)	-.1121	.5011	.1294
C(21)	-.1202	-.1477	.2317	K(4)	.5079	.0265	.0139
C(22)	.1991	-.2337	.3658	O(11)	.3512	.2525	.0969
C(23)	.1388	.1064	.2771	O(12)	.4031	.2475	.0737
C(24)	-.1856	.1836	.1514	O(2)	.5474	.5189	.1997
C(31)	-.2315	.0754	.4032	O(3)	.4476	.1694	.5506
C(32)	-.0201	.2276	.4432	H(11,1)	.4524	.2084	.1706
N(11)	.3652	-.1030	.1471	H(11,2)	.3020	.1954	.1272
N(12)	.0879	-.3753	.0844	H(12,2)	.4786	.2682	.0776
N(21)	-.1958	-.2205	.2154	H(21)	.4659	.5945	.1140
N(22)	.3018	-.3610	.4301	H(22)	.5997	.4351	.1795
N(23)	-.2077	.1861	.2820	H(31)	.5391	.1121	.6342
N(24)	-.2996	.3022	.0924	H(32)	.4044	.0988	.5296

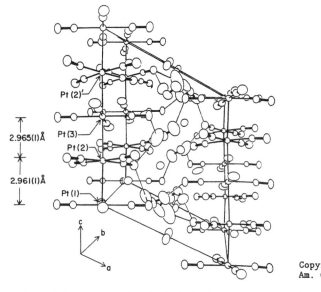

2.965(1)Å

2.961(1)Å

Pt (2)′

Pt (3)

Pt (2)

Pt (1)

Fig. 4.13-1 Unit cell of $K_{1.75}[Pt(CN)_4] \cdot 1.5H_2O$ showing the nonlinear
Pt(1)-Pt(2)-Pt(3) chain extending along c. Inversion centers occur at
Pt(1) and Pt(3) only. The Pt(2) atom is displaced 0.170Å perpendicular
to the c axis. Hydrogen bonds from H_2O to cyanide nitrogen atoms
(N...H-O < 2.6Å) are indicated by faint lines, and K^+ ions are shown
without bonding interactions. (After ref. 1).

Crystal Structure

The structure of this "one-dimensional electrical conductor" comprises an unusual "zigzag" metal atom chain containing three crystallographically independent Pt atoms with a Pt(1)-Pt(2)-Pt(3) bond angle of 173.25(3)°. Inversion centers occur at Pt(1) and Pt(3). The most surprising finding is that the two independent metal atom separations are equal [2.961(1) and 2.965(1)Å], just as in the case of $K_2[Pt(CN)_4]Br_{0.3} \cdot 3H_2O$,[3] where they are 2.888(6) and 2.892(6)Å. The short Pt-Pt separations and almost totally non-eclipsed configuration of adjacent $[Pt(CN)_4]^{-1.75}$ groups (torsion angles between adjacent platinocyanide groups ranging from 38.46 to 51.82°) are indicative of considerable $5d_{z^2}(Pt)$ orbital overlap, resulting in strong metal-metal bond formation, as well as repulsive π-π cyanide interactions. While Pt(1) and Pt(3) reside on the *c* axis, Pt(2) is displaced 0.170(1)Å normal to *c*. The deformation of the Pt-atom chain is the result of an unsymmetrical electrostatic environment about Pt(2) involving K^+...$^-N\equiv C$ interactions. The water molecules play an important role in the structure of the crystal, i.e., in addition to forming single and bifurcated hydrogen bonds that bind adjacent $[Pt(CN)_4]^{-1.75}$ groups in a single strand of a Pt-Pt chain, they also serve to cross-link $[Pt(CN)_4]^{-1.75}$ groups of different stacks, as shown in Fig. 4.13-2. In addition to normal H-bonding interactions, all lone-pair orbitals appear to be directed toward K^+ ions, increasing the binding of

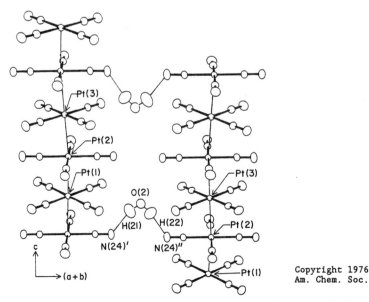

Fig. 4.13-2 View of the {110} section showing both the interchain platinocyanide group linked by $H_2O(2)$ and the Pt chain distortion. Hydrogen bonds are drawn as faint lines. (After ref. 1).

the water molecules. Also involved in the overall hydrogen bonding scheme is the disorder of one K^+ ion and one H_2O molecule.

Remarks

1. Partially oxidized tetracyanoplatinate (POTCP) complexes

A summary of the physical property data for a wide range of well-characterized POTCP complexes is given in Table 4.13-1. There exist two basic types of POCTP complex:

I. Anion-deficient complexes. The general formula for these materials is $M_i[Pt(CN)_4]X_n \cdot jH_2O$, [where M is a monovalent cation, i an integer (1-3), n nonintegral and < 1.0, and j has values of 0-3]. The molecular structures of salts of this type vary significantly, as do the electrical properties, between the hydrated and anhydrous derivatives.

II. Cation-deficient complexes. The general formula for these materials is $M_i[Pt(CN)_4] \cdot jH_2O$, [where i is nonintegral (1 < i < 2) and j appears to be always > 1].

Table 4.13-1 Crystal structure and conductivity data of POTCP complexes.[2]

Complex	Space group	d_{Pt-Pt}/Å (T=298K)	Conductivity ($\Omega^{-1}cm^{-1}$)	Color
Pt metal (for comparison)	Fm3m	2.775	9.4×10^4	metallic
$K_{1.75}[Pt(CN)_4] \cdot 1.5H_2O$	$P\bar{1}$	2.965, 2.961	115-125	bronze
$Rb_{1.75}[Pt(CN)_4] \cdot xH_2O$		2.94	1	bronze
$Cs_{1.75}[Pt(CN)_4] \cdot xH_2O$		2.88	~25	bronze
$K_2[Pt(CN)_4]Br_{0.30} \cdot 3H_2O$	P4mm	2.89	4-1050	bronze
$K_2[Pt(CN)_4]Cl_{0.30} \cdot 3H_2O$	P4mm	2.87	~200	bronze
$Rb_2[Pt(CN)_4]Cl_{0.30} \cdot 3H_2O$	P4mm	2.877, 2.924	~10	bronze
		2.885, 2.862 (T=110K)		
$Cs_2[Pt(CN)_4]Cl_{0.30}$	I4/mcm	2.859	~200	bronze
$Cs_2[Pt(CN)_4](N_3)_{0.25} \cdot 0.5H_2O$	$P\bar{4}b2$	2.877	40-270	reddish-copper
$K_2[Pt(CN)_4](FHF)_{0.30} \cdot 3H_2O$	P4mm	2.918, 2.928	-	reddish-bronze
$Rb_2[Pt(CN)_4](FHF)_{0.40}$	I4/mcm	2.798	1600-2300	gold
$Rb_2[Pt(CN)_4](FHF)_{0.26} \cdot 1.7H_2O$	C2/c	2.89	-	greenish-bronze
$Cs_2[Pt(CN)_4](FHF)_{0.23}$	I4/mcm	2.872	250-350	reddish-bronze
$Cs_2[Pt(CN)_4]F_{0.19}$	Immm	2.866	-	reddish-gold

The structural and electronic properties of one-dimensional POTCP complexes are highly dependent on the nature of the species that constitute the crystal lattice. In general, as the degree of partial oxidation increases

and d_{Pt-Pt} decreases the conductivity of a POTCP salt increases. Some
examples are discussed below:

   a) $K_2[Pt(CN)_4]Br_{0.3}.3H_2O$

   The crystal structure of $K_2[Pt(CN)_4]Br_{0.3}.3H_2O$ was first reported by
Krogmann and Hausen,[4] and the structural description was revised in a
neutron diffraction study by Williams and co-workers.[5]

   In the crystal, the degree of partial oxidation of Pt is 0.3, and the
$[Pt(CN)_4]^{1.7-}$ units stack to form linear, Pt-atom chains that have equal Pt-Pt
spacings, 2.89Å. The K^+ ions are situated in one-half of the unit cell and
the H_2O molecules, which are located in the other half, form a network of
hydrogen bonds between the CN^- ligands and either the Br^- anion at the unit-
cell center or the CN^- ligand from a $[Pt(CN)_4]$ group on an adjacent chain.
Adjacent $[Pt(CN)_4]_\infty$ chains are separated by 9.89Å, which accounts in part for
the pronounced anisotropy in the physical properties of this crystal. The
cations and water molecules do not occupy the same molecular plane as the
$[Pt(CN)_4]$ groups. Fig. 4.13-3 shows the structure of $K_2[Pt(CN)_4]Br_{0.3}.3H_2O$.
The central portion of the unit cell is alternatively occupied by either a Br^-
ion (at 1/2,1/2,1/2, labeled Br^*, 60% occupancy) or an H_2O molecule (at
1/2,1/2,0.67, labeled H_2O^*, 40% occupancy). Accompanying the very short
intrachain Pt-Pt separation of 2.89Å is an intrachain torsion angle of 45°
between adjacent $[Pt(CN)_4]^{1.7-}$ groups, which apparently minimizes ligand
steric hindrance.

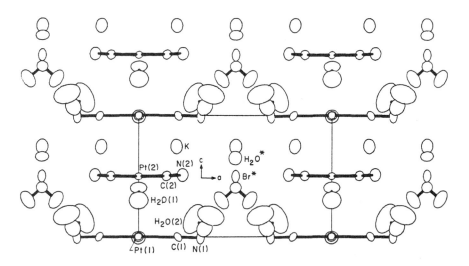

Fig. 4.13-3 Structure of $K_2[Pt(CN)_4]Br_{0.3}.3H_2O$, shown as a half-cell
projection along the *b* axis. (After ref. 5).

b) $Rb_2[Pt(CN)_4]Cl_{0.3} \cdot 3H_2O$

This crystal has been characterized by single-crystal neutron-diffraction at 110K,[6] and appears to be isostructural with $K_2[Pt(CN)_4]Br_{0.3} \cdot 3H_2O$ with only one halide site per unit cell. Another significant structural feature is that, though not required by the space group symmetry, all intrachain Pt-Pt separations in $K_2[Pt(CN)_4]Br_{0.3} \cdot 3H_2O$ are equal. However, in $Rb_2[Pt(CN)_4]Cl_{0.3} \cdot 3H_2O$ there is a definite dimerization of the linear Pt-atom chain, as shown in Fig. 4.13-4. In comparison with $K_2[Pt(CN)_4]Br_{0.3} \cdot 3H_2O$, it appears that the replacement of the K^+ ion by the larger Rb^+ ion results in lattice expansion along c, which in turn produces larger Pt-Pt spacings in $Rb_2[Pt(CN)_4]Cl_{0.3} \cdot 3H_2O$. It is therefore reasonable to expect that the electrical conductivity of the Rb salt might be less than that of the K salt due to the longer, dimerized Pt chain. A summary of the structural results derived from numerous diffraction studies, in which monovalent cations of differing radius (r^+) have been substituted for K^+, is as follows.[7] (i) For a given d_{Pt-Pt}, as r^+ increases the tendency for dimerization also increases. (ii) For a given r^+, as d_{Pt-Pt} decreases the probability that dimerization will occur decreases. (iii) For a given r^+, d_{Pt-Pt} tends to depend on the degree of partial oxidation of Pt, and therefore the degree of chain dimerization may depend on the degree of partial oxidation.

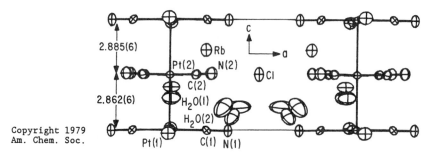

Fig. 4.13-4 Unit cell of $Rb_2[Pt(CN)_4]Cl_{0.3} \cdot 3H_2O$. (After ref. 6).

c) $Rb_2[Pt(CN)_4](FHF)_{0.4}$

This crystal is unique in that it contains the shortest Pt-Pt spacing (2.798Å) and highest room-temperature electrical conductivity of any known POTCP salt.[8-9] Fig. 4.13-5 shows the crystal structure.

2. One-dimensional band theory

POTCP complexes exhibit metallic properties. In this case, the structural requirements for a one-dimensional metal are satisfied by square-planar complexes which can stack above one another to form a columnar

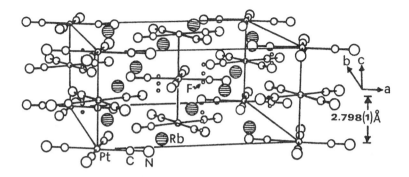

Fig. 4.13-5 Perspective view of the unit cell of $Rb_2[Pt(CN)_4](FHF)_{0.4}$. (After ref. 8).

stack, as shown in Fig. 4.13-6. The ligands surrounding the metal atoms ensure that the inter-chain separation is large, ~9Å, compared with the average intra-chain metal-metal distance which is usually less than 3Å. This very large difference between the intra- and interchain metal-metal separation results in much higher degrees of anisotropy in this class of one-dimensional metal.[10]

Krogmann has discussed the metallic properties of this class of complexes in terms of a partially filled one-dimensional band in the metal-atom chain direction.[11] In the columnar stacked structure, orbital overlap can occur between d_{z^2}, p_z and to a lesser extent the d_{xz}, d_{yz} and $d_{x^2-y^2}$ orbitals of the metal atoms. If the metal atoms are close enough this will result in the formation of a one-dimensional band throughout the length of the crystal, and if the band is only partially filled, one-dimensional metallic properties will be found in the chain direction. Since an ion with a d^8 configuration possesses an even number of electrons, a partially filled band can arise only if: (i) overlap occurs between the highest energy filled and the lowest energy empty bands [Fig. 4.13-7(a)], or (ii) electrons are removed from the top part of the filled band of highest energy, i.e. partial oxidation of the d^8 ion [Fig. 4.13-7(b)]. The latter is the case when partial oxidation of a d^8 ion occurs.

A very comprehensive molecular orbital treatment of POTCP complexes has been given by Whangbo and Hoffmann.[12]

3. Generalizations of POTCP complexes[2]

A few generalizations for understanding the chemistry and physics of POTCP complexes are listed below.

a) Partially oxidized tetracyanoplatinate (POTCP) complexes that are

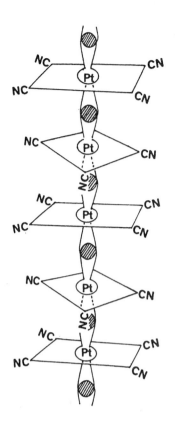

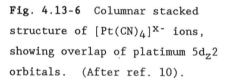

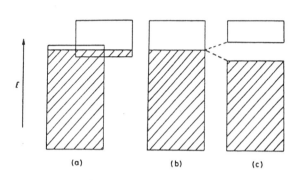

Fig. 4.13-6 Columnar stacked
structure of $[Pt(CN)_4]^{X-}$ ions,
showing overlap of platimum $5d_{z^2}$
orbitals. (After ref. 10).

Fig. 4.13-7 Simple band structure
of a one-dimensional metallic
complex: (a) overlap of full and
empty bands; (b) partially filled
band; (c) splitting of partially
filled band by the Peierls
distortion. (After ref. 10).

hydrated form primitive tetragonal lattice (*P4mm*), with the M^+ cations located
in one-half of the unit cell and water molecules in the other half. More
importantly, the cations are always located between the planes of the
$[Pt(CN)_4]$ groups.

b) POTCP complexes that are anhydrous form body-centered tetragonal
lattices (*I4/mcm*) in which the cations occupy sites in the same plane as the
$[Pt(CN)_4]$ groups. Some of these salts, e.g. $Rb_2[Pt(CN)_4](FHF)_{0.4}$, have the
shortest intra- and interchain Pt-Pt spacings and highest electrical
conductivities of all known POTCP salts.

c) Electrical conductivities parallel to the Pt-atom chain, $\sigma_{||}$, in
POTCP metals vary inversely with the intrachain Pt-Pt spacing d_{Pt-Pt}. The
d_{Pt-Pt} spacings are highly dependent on two factors. (i) The degree of

partial oxidation and d_{Pt-Pt} are inversely related, that is, the Pt-atom chain in a POTCP complex behaves in an accordion-like fashion, depending on the degree of partial oxidation. (ii) In a series of isostructural compounds those containing the smaller alkali-metal cations have the shrotest d_{Pt-Pt}. In addition, POTCP salts containing smaller cations have a higher tendency to be hydrated which tends to increase d_{Pt-Pt}. POTCP materials have metallic behavior near room-temperature, with $\sigma_{||}$ as high as approximately $2000\ \Omega^{-1}cm^{-1}$, but they become semiconductors at low temperature.

  d) POTCP complexes have intrachain Pt-Pt spacings of 2.8-2.96Å, that is, as short as in Pt metal itself (2.775Å). The d_{Pt-Pt} spacing may be calculated from the degree of partial oxidation (DPO):[13]

$$d_{Pt-Pt}(\text{Å}) \;=\; 2.59 - 0.60\log(DPO).$$

  e) The body-centered tetragonal anhydrous salts, with the shortest d_{Pt-Pt}, have the lowest three-dimensional ordering temperatures (T_{3D}). Therefore, they have the highest electrical conductivities down to the lowest temperatures. These properties are due in part to the absence of hydrogen-bonding interactions.

  f) There is as yet no known Na^+ or anhydrous Li^+ POTCP complex, which may be expected to have Pt-Pt separations less than that in Pt metal.

References

[1] J.M. Williams, K.D. Keefer, D.M. Washecheck and N.P. Enright, *Inorg. Chem.* **15**, 2446 (1976).

[2] J.M. Williams, *Adv. Inorg. Chem. Radiochem.* **26**, 235 (1983).

[3] J.M. Williams, M. Iwata, F.K. Ross, J.L. Petersen and S.W. Peterson, *Mater. Res. Bull.* **10**, 411 (1975).

[4] K. Krogmann and H.D. Hausen, *Z. Anorg. Allg. Chem.* **358**, 67 (1968).

[5] J.M. Williams, J.L. Petersen, H.M. Gerdes and S.W. Peterson, *Phys. Rev. Lett.* **33**, 1079 (1974).

[6] R.K. Brown and J.M. Williams, *Inorg. Chem.* **18**, 1922 (1979).

[7] R.K. Brown and J.M. Williams, *Inorg. Chem.* **17**, 2607 (1978).

[8] A.J. Schultz, C.C. Coffey, G.C. Lee and J.M. Williams, *Inorg. Chem.* **16**, 2129 (1977).

[9] D.J. Wood, A.E. Underhill, A.J. Schultz and J.M. Williams, *Solid State Commun.* **30**, 501 (1979).

[10] A.E. Underhill and D.M. Watkins, *Chem. Soc. Rev.* **9**, 429 (1980).

[11] K. Krogmann, *Angew. Chem. Int. Ed. Engl.* **8**, 35 (1969).

[12] M.-H. Whangbo and R.J. Hoffmann, *J. Am. Chem. Soc.* **100**, 6093 (1978).

[13] J.M. Williams, *Inorg. Nucl. Chem. Lett.* **12**, 651 (1976).

4.14   Dodecatungstophosphoric Acid Hexahydrate
(Keggin Structure)
$(H_5O_2)_3[PW_{12}O_{40}]$

Crystal Data

Cubic, space group $Pn3m$ (No. 224)

$a = 12.1506(5)$Å, $Z = 2$

Atom	X-ray(Mo Kα)			Neutron (λ = 1.246Å)		
	x	y	z	x	y	z
P	3/4	3/4	3/4	3/4	3/4	3/4
W	.75821	.95680	.95680	.75822	.95688	.95688
O(1)	.82252	.82252	.82252	.82273	.82273	.82273
O(2)	.65617	.84383	.99326	.65683	.84317	.99316
O(3)	.87223	.87223	1.02398	.87155	.87155	1.02447
O(4)	.73311	1.05452	1.05452	.73276	1.05441	1.05441
O(W)	3/4	1.15472	1/4	3/4	1.15248	1/4
H(W)	.74503	1.10148	1.19100	.74525	1.11259	1.18288
H(A)	3/4	1/4	1/4	3/4	1/4	1/4

Crystal Structure

Knowledge of the atomic configuration of $[PW_{12}O_{40}]^{3-}$ is of central importance in relation to the structures of heteropoly acids. The anion and its analogues with other atoms replacing either P or W or both are known to occur in at least 80 crystalline compounds.[3] In 1934 Keggin determined the structure of the title compound using X-ray powder diffraction data.[2] Reinvestigation by X-ray and neutron single-crystal diffraction methods shows that Keggin's "pentahydrate" is actually a hexahydrate with the correct formula $(H_5O_2)_3[PW_{12}O_{40}]$.[1]

The structure of the "Keggin molecule" $[PW_{12}O_{40}]^{3-}$ ($\bar{4}3m$ symmetry) is shown in Fig. 4.14-1. It comprises four trigonal groups of edge-shared WO_6

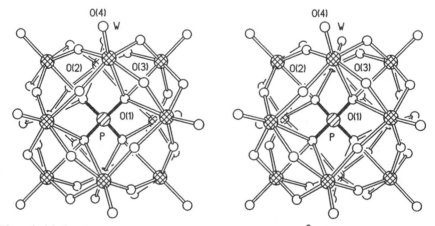

Fig. 4.14-1 Stereoscopic view of the $[PW_{12}O_{40}]^{3-}$ anion along -c; a is along the horizontal from left to right. The atoms of the asymmetric unit that generate the cluster anion are labeled.

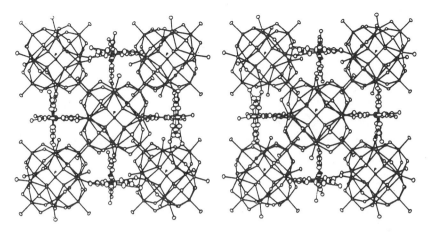

Fig. 4.14-2 Stereoscopic drawing showing the packing of anions and diaquahydrogen ions. Orientation is the same as in Fig. 4.14-1. (After ref. 1).

octahedra, each sharing corners with its neighbours and with the encapsulated central PO_4 tetrahedron. The packing of anions and diaquahydrogen ions is shown in Fig. 4.14-2. Interatomic distances and angles within the anion as determined in the neutron refinement are given in Table 4.14-1.

Table 4.14-1 Distances and angles within the $[PW_{12}O_{40}]^{3-}$ anion.

P-O(1)	1.5305Å	O(1)-P-O(1')	109.47°
W-O(1)	2.4349	P-O(1)-W	125.96°
W-O(2)	1.9029	P-O(1)-O(2)	93.46°
W-O(3)	1.9094	P-O(1)-O(3)	144.16°
W-O(4)	1.7041	O(1)-W-O(2)	82.96
O(1)...O(1')	2.499	O(1)-W-O(3)	72.08
O(1)...O(2)	2.901	O(2)-W-O(3)	88.45
O(1)...O(3)	2.591	O(2)-W-O(2')	85.45
O(2)...O(2')	2.577	O(2)-W-O(3')	154.82
O(2)...O(3)	2.659	O(2)-W-O(4)	103.08
O(2)...O(4)	2.827	O(3)-W-O(3')	86.96
O(3)...O(3')	2.628	O(3)-W-O(4)	102.09
O(3)...O(4)	2.813	W-O(3)-W"	126.75
		W-O(2)-W'	152.43

The six water molecules are paired to form nearly planar diaquahydrogen ions $H_5O_2^+$ (symmetry 222), which have linear centered O...H...O bonds only 2.370(5)Å long (2.414Å corrected for thermal motion), each such ion being in twofold disorder on a $\overline{4}2m$ site. The H_2O bond angle is 118.7(5)°. An alternative interpretation consistent with the thermal parameters is that each $H_2O...H$ fragment of the $H_5O_2^+$ ion has the geometry of a rather flat pyramid and that additional disorder of the O atoms about the twofold axis on which they appear to lie produces apparent 222 symmetry for the $H_5O_2^+$ ion (Fig. 4.14-3).

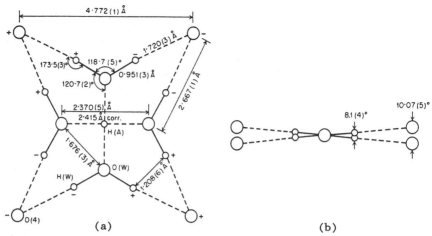

(a)   (b)

Fig. 4.14-3 (a) Structure of the diaquahydrogen ion $H_5O_2^+$ and its hydrogen-bonding to O(4) atoms of neighbouring anions. Atom H(A) is on a $\bar{4}2m$ site, and the $H_5O_2^+$ ion (symmetry 222) is in twofold disorder about this site. Atoms marked with + signs are above the plane through the four O(W) sites, and those marked with - signs are below the plane. (b) View along the O(W)...H(A)...O(W) axis of the $H_5O_2^+$ ion oriented vertically in (a), showing the twist from coplanarity. (After ref. 1).

Remarks

1. Isopolyanions of V, Nb and Ta

The structures of several crystalline isopolyanions of vanadates, niobates, and tantalates have been determined. Fig. 4.14-4 shows three typical structures: $[Nb_6O_{19}]^{8-}$, $[V_{10}O_{28}]^{6-}$ and $[V_{18}O_{42}]^{12-}$.

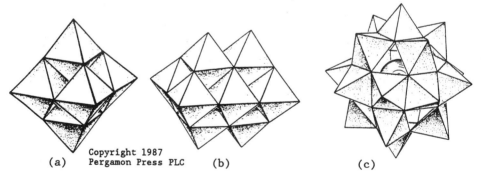

(a)   (b)   (c)

Fig. 4.14-4 Structures of (a) $[Nb_6O_{19}]^{8-}$, (b) $[V_{10}O_{28}]^{6-}$, and (c) $[V_{18}O_{42}]^{12-}$. (After ref. 4).

The structure of the anion in $Na_7HNb_6O_{19}.15H_2O$[5] is shown in Fig. 4.14-4(a). The Nb atoms are displaced towards the unshared oxygen atoms. The Nb-O bond lengths are 1.75-1.78Å (terminal), 1.970-2.056Å (bridging), and 2.371-2.386Å (central). The isopolytantalate, $[Ta_6O_{19}]^{8-}$, is isostructural with the corresponding niobate. The bond lengths in $K_4Na_2H_2Ta_6O_{19}.2H_2O$ are Ta-O (terminal), 1.786-1.817; Ta-O (bridging), 1.976-2.012; and Ta-O (central), 2.356-2.426Å.[4]

The structure of the decavanadate anion as established by several X-ray investigations[6,7] is shown in Fig. 4.14-4(b). The anion consists of an arrangement of ten edge-shared VO_6 octahedra with D_{2d} symmetry. The octahedra are distorted in order to maintain approximate valence balance at terminal and bridging oxygens, and bond lengths range from 1.60Å for V-O (terminal) to 2.32Å for V-O (central). The $[Nb_{10}O_{28}]^{6-}$ anion is exactly analogous to $[V_{10}O_{28}]^{6-}$.

The structure of the anion $[V_{18}O_{42}]^{12-}$, shown in Fig. 4.14-4(c), consists of an almost spherical arrangement of edge- and corner-shared VO_5 square pyramids with axial V-O bonds directed outwards. The central cavity has a radius of 4.5Å and is fractionally occupied by K^+ or H_2O in the crystal.[8]

The structure of $(NMe_4)_6[V_{15}O_{36}]Cl.4H_2O$ has been determined.[9] The $[V_{15}O_{36}]^{5-}$ anion is composed of 15 tetragonal VO_5 pyramids linked to form a spheroidal D_{3h} species. The V atoms reside on the surface of a sphere of radius 3.45±0.1Å centered around the encapsulated Cl^- ion (Fig. 4.14-5).

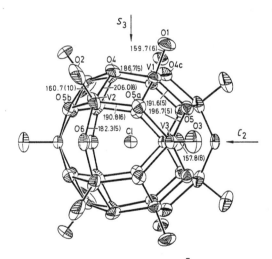

Fig. 4.14-5 Structure of the $[V_{15}O_{36}]^{5-}$ anion with entrapped Cl^- ion in $(NMe_4)_6[V_{15}O_{36}]Cl.4H_2O$ (bond lengths in pm). (After ref. 9).

The compound $(NH_4)_8[V_{19}O_{41}(OH)_9].11H_2O$ has been prepared and characterized by physical, spectroscopic and crystallographic methods.[10] The anion $[V_{19}O_{41}(OH)_9]^{8-}$ has the shape of an ellipsoid of approximate C_3 (or, without the central V atom, D_3) symmetry and is generated by joining 12 VO_6

octahedra (V^{IV}) and 6 VO_4 tetrahedra (V^V), as well as a central V^VO_4 unit, as shown in Fig. 4.14-6. Within the anion are a VO_4 tetrahedron and a μ_3-O atom disordered over two lattice sites in a ratio of 1:1.

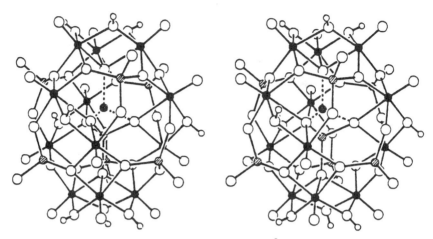

Fig. 4.14-6 Stereoview of the $[V_{19}O_{41}(OH)_9]^{8-}$ ion viewed perpendicular to the pseudo C_3 axis; V^{IV} = black, V^V = hatched, central disordered V hatched in one alternate site and cross-hatched with V-O bonds drawn with broken lines in the other site; O and H = large and small open circles, respectively. (After ref. 10).

2. Isopolyanions of Mo and W

Crystals of about a dozen isopolymolybdates and isopolytungstates have been isolated. Some representative structures are shown in Fig. 4.14-7.

The structure of $[Mo_7O_{24}]^{6-}$ can be viewed as 7/10 of the decavanadate structure and involves an arrangement of octahedra with *cis*-MoO_2 stereo-chemistry for each metal atom.[12] The structures of α- and β-$[Mo_6O_{26}]^{4-}$, $[W_{10}O_{32}]^{4-}$, and $[H_2W_{12}O_{42}]^{10-}$ all consist of MO_6 octahedra. The anion in $K_8[Mo_{36}O_{112}(H_2O)_{16}]\cdot xH_2O$ (x ~ 38-40) consists of two Mo_{18} units related by inversion symmetry.[13] Each unit contains a Mo_7O_{24} moiety surrounded by edge- and corner-sharing MoO_6 octahedra linked by terminal and bridging water molecules (indicated by open circles) as shown in Fig. 4.14-7(f). Two of the Mo atoms in the Mo_7 units become seven-coordinate (quasi pentagonal bipyramidal).

3. Coordination types of heteroatom in heteropolyanions

By the present count, about 70 elements (other than Mo and W) have been observed as heteroatoms in polyanions.[4] Since each element may form more than one heteropolyanion and combinations of several heteroatoms occur frequently, the total number of known heteropolyanions is extremely large.

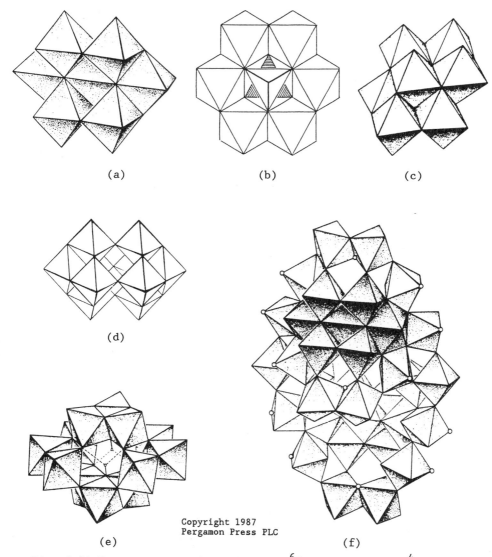

Fig. 4.14-7 Structures of (a) $[Mo_7O_{24}]^{6-}$, (b) α-$[Mo_8O_{26}]^{4-}$, (c) β-$[Mo_8O_{26}]^{4-}$, (d) $[W_{10}O_{32}]^{4-}$, (e) $[H_2W_{12}O_{42}]^{10-}$, and (f) $[Mo_{36}O_{112}(H_2O)_{16}]^{8-}$. (After refs. 4 and 11).

According to the coordination types of heteroatom, the structures of heteropolyanions can be classified as shown in Table 4.14-2.

4. Structures of some heteropolyanions with tetrahedral coordination of the heteroatom

There are probably more heteropolyanions of phosphorus than of any other heteroatom, and most of these have structures derived from the Keggin anion $[PW_{12}O_{40}]^{3-}$.

Table 4.14-2   Coordination types of heteroatom in heteropolyanions.

Coordination of heteroatom X		Examples
Pyramidal	AsO_3	$[As_6V_{15}O_{42}(H_2O)]^{6-}$
Tetrahedral	PO_4	$\alpha-[PW_{12}O_{40}]^{3-}$ (Keggin structure)
		$\alpha-[P_2W_{18}O_{62}]^{6-}$ (Dawson structure)
		$[P_5W_{30}O_{110}]^{15-}$
		$[H_7P_8W_{48}O_{184}]^{33-}$
		$[P_2Mo_5O_{23}]^{6-}$
	AsO_4	$[As_2Mo_6O_{26}]^{6-}$
	BO_4	$[B(Co,W_{11})O_{39}(H_2O)]^{6-}$
Octahedral	TeO_6	$[TeMo_6O_{24}]^{6-}$ (Anderson structure)
	MnO_6	$[MnV_{13}O_{38}]^{7-}$
	AlO_6	$[AlV_{14}O_{40}]^{9-}$
Antiprismatic	CeO_8	$[CeW_{10}O_{36}]^{8-}$
Icosahedral	CeO_{12}	$[CeMo_{12}O_{42}]^{8-}$

a) Reduced Keggin-type heteropolyanions

The β isomer of the $[XM_{12}O_{40}]^{n-}$ anions results from a 60° rotation about the C_3 axis of one trigonal edge-sharing M_3O_{13} unit of the α Keggin structure. Other possible skeletal isomers (γ, δ and ϵ) are derived fom the α structure by 60° rotations of two, three and four M_3O_{13} units about their respective C_3 axes.[14]   The ϵ structure is adopted by $[Al_{13}O_4(OH)_{24}(H_2O)_{12}]^{7+}$.[14]

Reduction of the Keggin anion $\alpha-[PMo_{12}O_{40}]^{3-}$ leads, in aqueous acidic solutions, to stable "heteropoly blues" derived from an isomeric β anion.   The structure of the four-electron reduced $\beta-[PMo_{12}O_{40}]^{7-}$ ion conforms closely to idealized C_{3v} symmetry.[15]   The metatungstate $[H_2W_{12}O_{40}]^{6-}$ and heteropolyanions $[XW_{12}O_{40}]^{n-}$ with the Keggin structure can be reduced to "brown forms", in which the number of electrons introduced by reduction is always a multiple of 6.   The six-electron reduced anion in $Rb_4H_8[H_2W_{12}O_{40}].\sim18H_2O$ exhibits the α-Keggin structure with the metal atom disordered over two positions in the WO_6 octahedron.[16]

The anions of two blue salts, $(HNMe_3)_3[H_4AsMo_{12}O_{40}]$ (1) and $(HNMe_3)_2[H_2AsMo_{12}O_{40}]$ (2), have been structurally characterized by X-ray analysis: $[H_4As^VMo^V_4Mo^{VI}_8O_{40}]^{3-}$ is a four-electron reduced β-Keggin anion, whereas two-electron reduced $[H_2As^VMo^V_2Mo^{VI}_{10}O_{40}]^{3-}$ has the α-Keggin structure (Fig. 4.14-8).[17]

The novel centrosymmetric $[H_6As^{III}_{10}Mo^V_8Mo^{VI}_{16}O_{90}]^{8-}$ anion of $(HNMe_3)_8[H_6As_{10}Mo_{24}O_{90}].9H_2O$ (3), the precursor of 1 and 2, is composed of two $As_4Mo_{12}O_{42}$ spherical fragments of approximate C_{3v} symmetry with a central AsO_3

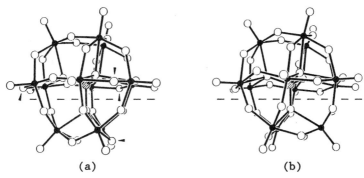

(a)   (b)

Fig. 4.14-8 Structures of (a) the β-Keggin anion $[H_4AsMo_{12}O_{40}]^{3-}$ (protonated O atoms marked with arrows) and (b) the α-Keggin anion $[H_2AsMo_{12}O_{40}]^{3-}$ (the central AsO_4 tetrahedron is shown in one of the disordered orientations). The common $AsMo_9$ spherical fragments lie above the broken line. (After ref. 17).

group (Fig. 4.14-9).[17] Each half contains an $AsMo_9$ Keggin fragment (*cf.* structural unit above the broken line in Fig. 4.14-8) capped by a ring built of three AsO_3 and three $Mo^{VI}O_6$ polyhedra, and the two units are linked by four bridges. Mild oxidation of the blue compound **3** affords **1** and **2**. It is interesting that the framework of **3** degrades to the Keggin-type structure of **1** and **2** with retention of the $AsMo_9$ spherical fragment, and merely As^{III} is oxidized to As^V in going from **3** to **1**.

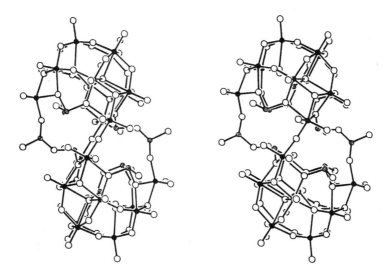

Fig. 4.14-9 Stereoview of the structure of $[H_6As_{10}Mo_{24}O_{90}]^{8-}$ (Mo black, As hatched, O unfilled). (After ref. 17).

b) Structure of α-$[P_2W_{18}O_{62}]^{6-}$

The structure of α-$[P_2W_{18}O_{62}]^{6-}$, the "Dawson anion", is shown in Fig. 4.14-10(a)[18]. It is seen to be a fused dimer of $[PW_9O_{34}]$ units.

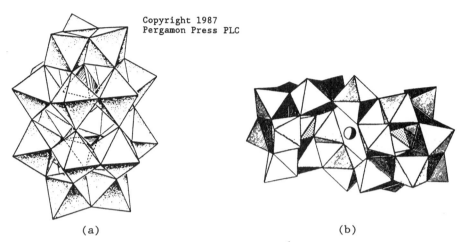

(a)   (b)

Fig. 4.14-10 Structure of (a) α-$[P_2W_{18}O_{62}]^{6-}$, the "Dawson anion" and (b) the $[KP_2W_{20}O_{72}]^{13-}$ anion. (After refs. 4 and 19).

In $K_{13}[KP_2W_{20}O_{72}]\cdot xH_2O$ the polyanion contains two PW_9O_{34} units bridged by two WO_6 octahedra, with K in the central cavity coordinated to eight oxygen atoms in a hexagonal bipyramidal environment [Fig. 4.14-10(b)].[19]

c) Structure of $[NaP_5W_{30}O_{110}]^{14-}$

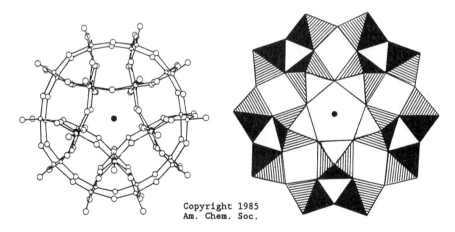

Fig. 4.14-11 Structure of $[NaP_5W_{30}O_{110}]^{14-}$ viewed approximately along the virtual C_5 symmetry axis: (left) O and W atoms are represented as large and small open circles, respectively, and the Na^+ by the central filled circle. The P atoms are shown as filled circles that are almost entirely obscured by the inner pentagon of five W atoms; (right) shows the WO_6 octahedra in the upper half of the anion. (After ref. 20).

The anion in $(NH_4)_{14}[NaP_5W_{30}O_{110}] \cdot 31H_2O$ has approximate D_{5h} symmetry and consists of a cyclic assembly of five $[PW_6O_{22}]$ units, each derived from the Keggin anion, $[PW_{12}O_{40}]^{3-}$, by removal of two sets of three corner-shared WO_6 octahedra [Fig. 4.14-11].[20] A sodium ion is located within the polyanion on the fivefold axis and 1.25Å above the pseudo mirror plane that contains the five P atoms.

d) Structure of $[H_7P_8W_{48}O_{184}]^{33-}$

The anion in $K_{28}Li_5H_7[P_8W_{48}O_{184}] \cdot 92H_2O$ has a crown shape formed by the linkage of four $P_2W_{12}O_{48}$ subunits, as shown in Fig. 4.14-12.[21] These subunits are derived from the Dawson structure of $[P_2O_{18}O_{62}]^{6-}$ by loss of six adjacent WO_6 octahedra, two from the cap and four from the belt polyhedra. The polyanion has approximately point symmetry D_{4h}.

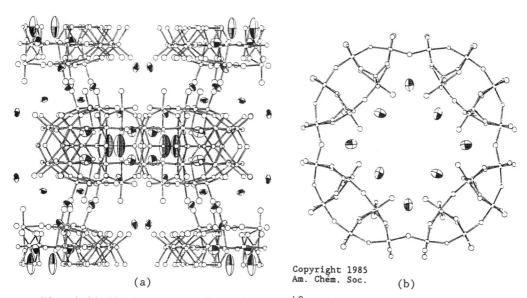

Copyright 1985
Am. Chem. Soc.

(a)   (b)

Fig. 4.14-12 Structure of $[P_8W_{48}O_{184}]^{40-}$: (a) unit cell projected along the *b* axis, and (b) along the *c* axis (with K$^+$ cations). (After ref. 21).

e) Structures of $[P_2Mo_5O_{23}]^{6-}$ and $[As_2Mo_6O_{26}]^{6-}$

The $[P_2Mo_5O_{23}]^{6-}$ ion has a C_2 structure that leaves a vertex of each PO_4 tetrahedron unshared, as shown in Fig. 4.14-13(a). The $[As_2Mo_6O_{26}]^{6-}$ ion has a D_{3h} structure that leaves a vertex of each AsO_4 tetrahedron unshared, as shown in Fig. 4.14-13(b).

f) Structure of $Ba_3[BCoW_{11}O_{39}(H_2O)] \cdot 26H_2O$

In $Ba_3[BCoW_{11}O_{39}(H_2O)] \cdot 26H_2O$ the polyanion has the structure shown in Fig. 4.14-14(a).[22] The heteroatom B is in tetrahedral coordination. The Co and W atoms are disordered and distributed randomly as $[Co_{1/12}, W_{11/12}]$.

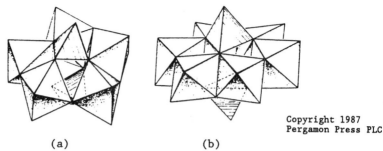

(a)   (b)

Fig. 4.14-13 Structure of (a) $[P_2Mo_5O_{23}]^{6-}$ and (b) $[As_2Mo_6O_{26}]^{6-}$ anions. (After ref. 4).

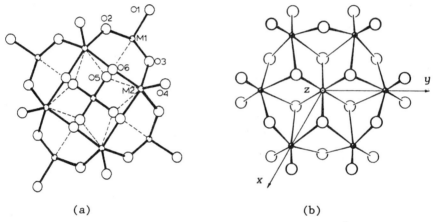

(a)   (b)

Fig. 4.14-14 (a) Structure of the $[BCoW_{11}O_{39}(H_2O)]^{6-}$ heteropolyanion. (b) Structure of the "Anderson anion" in $(NH_4)_6[TeMo_6O_{24}].Te(OH)_6.7H_2O$. (After refs. 22 and 23).

5. Some heteropolyanions with octahedral coordination of the heteroatom

   a) Structure of $[TeMo_6O_{24}]^{6-}$

   In 1936 Anderson proposed that several 1:6 heteropolyanions had the planar $[D_{3d}]$ structure shown in Fig. 4.14-14(b). The structure of the "Anderson" ion $[TeMo_6O_{24}]^{6-}$ was determined by Evans in 1974.[23]

   b) Structures of $[MnV_{13}O_{38}]^{7-}$ and $[AlV_{14}O_{40}]^{9-}$ [4]

   The Anderson structure discussed above can be viewed as a close-packed layer of edge-sharing MO_6 octahedra. Addition of three octahedra above and three octahedra below this layer produces a $XM_{12}O_{38}$ cluster of octahedral symmetry, in which the central XO_6 octahedron shares each of its edges with an MO_6 octahedron. The mineral sherwoodite contains the only naturally occurring heteropoly anion, $[AlV_2^{IV}V_{12}^{V}O_{40}]^{9-}$, and has a structure consisting of the $XM_{12}O_{38}$ cluster with two additional VO_6 octahedra placed in *trans* positions on a four-fold axis, as shown in Fig. 4.14-15(b). The synthetic heteropolyvanadates $[XV_{13}O_{38}]^{7-}$ [X = Mn(IV), Ni(IV)] have the sherwoodite structure less one of the equatorial VO_6 octahedra [Fig. 4.14-15(a)].

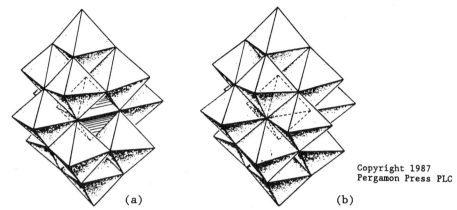

(a)   (b)

Fig. 4.14-15 Structures of the (a) $[MnV_{13}O_{38}]^{7-}$ and (b) $[AlV_{14}O_{40}]^{9-}$ anions. (After ref. 4).

6. Structures of $[CeW_{10}O_{36}]^{8-}$ and $[CeMo_{12}O_{42}]^{8-}$

The structure of $[CeW_{10}O_{36}]^{8-}$ can be regarded as the attachment of two quadridentate ligands, derived from W_6O_{19} through the loss of one WO_6 octahedron, to the heteroatom (Ce). The Ce atom has antiprismatic coordination and approximate site symmetry D_{4d} [Fig. 4.14-16(a)].[23] Salts containing $[CeW_{10}O_{35}]^{6-}$ and $[LnW_{10}O_{35}]^{7-}$ are known.[24] The $[CeMo_{12}O_{42}]^{8-}$ ion features an icosahedrally coordinated heteroatom and face-shared pairs of MoO_6 octahedra, as shown in Fig. 4.14-16(b).[25]

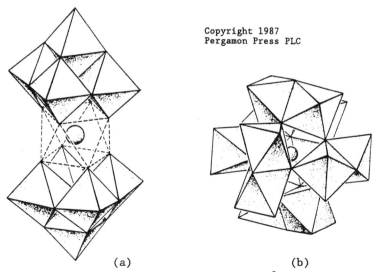

(a)   (b)

Fig. 4.14-16 Structures of the (a) $[CeW_{10}O_{36}]^{8-}$ and (b) $[CeMo_{12}O_{42}]^{8-}$ anions. (After ref. 4).

7. Structure of $[As_6V_{15}O_{42}(H_2O)]^{6-}$ [26]

The anion $[As_6V_{15}O_{42}(H_2O)]^{6-}$ in $K_6[As_6V_{15}O_{42}(H_2O)]\cdot 8H_2O$ has been shown to have crystallographic D_3 symmetry (Fig. 4.14-17). It consists of 15 distorted tetragonal VO_5 pyramids and 6 trigonal AsO_3 pyramids, with a statistically disordered H_2O molecule entrapped at its center. The VO_5 pyramids are linked with one another through vertices. Two AsO_3 groups are joined to each other via an oxygen bridge, forming a handle-like As_2O_5 moiety.

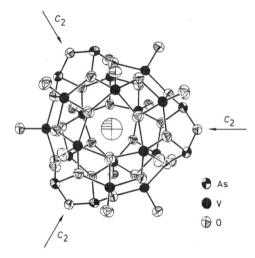

Fig. 4.14-17 Structure of $[As_6V_{15}O_{42}(H_2O)]^{6-}$. (After ref. 26).

The VO_5 pyramids can be divided into three differently linked groups. The first group includes six polyhedra, each of which is joined through two edges and three vertices with its neighbouring pyramids and through one vertex with an AsO_3 groups. The second group contains six VO_5 pyramids, which are joined through three edges with other VO_5 pyramids and through two vertices with two AsO_3 groups. Finally, the third group consists of three VO_5 pyramids, which are mutually joined through two vertices and two edges with four VO_5 pyramids and through two vertices with two As_2O_5 moieties.

8. Heteropolyanions $[M_4(H_2O)_2(XW_9O_{34})_2]^{10-}$ and $[WM_3(H_2O)_2(XW_9O_{34})_2]^{12-}$

Polytungstometallates $K_{10}[M_4(H_2O)_2(XW_9O_{34})_2]\cdot nH_2O$ (M = Co^{II}, X = P, $n = 22$; M = Zn, X = As, $n = 23$),[27] D,L-$Na_{12}[WM_3(H_2O)_2(XW_9O_{34})_2]\cdot nH_2O$ (1, M = X = Zn, $n = 46$; 2, M = Cu, X = Zn, $n = 48$)[28] and D,L-$K_{12}[WZnV_2O_2(ZnW_9O_{34})_2]\cdot 30H_2O$ (3)[28] have been synthesized and characterized by X-ray crystallography. The $[M_4(H_2O)_2(XW_9O_{34})_2]^{-10}$ anion is centrosymmetric and nearly attains idealized symmetry 2/m; it consists of two α-B-$XW_9O_{34}^{9-}$ units (each derived from the α-Keggin anion by the removal of a triad of edge-sharing octahedra)[4] sharing two M(1) atoms, through which the pseudo twofold axis passes (Fig. 4.14-18). The remaining two M(2) atoms are each surrounded by five skeletal O atoms and a terminal aqua ligand.

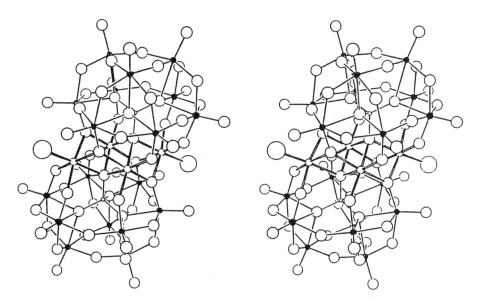

Fig. 4.14-18 Stereoview of the $[Co_4(H_2O)_2(PW_9O_{34})_2]^{10-}$ anion. (After ref. 27).

The polyanions in 1, 2 and 3 have analogous structures except that the M(1) sites are occupied by a W atom and another one: Zn in 1 or 3 and Cu in 2. the M(2) sites remain equivalent and accommodate Zn in 1, Cu in 2, or V in 3; for these atoms, the aqua ligand in 1 or 2 is replaced by an oxo-ligand in 3.

References

[1] G.M. Brown, M.-R. Noe-Spirlet, W.R. Busing and H.A. Levy, *Acta Crystallogr. Sect. B* **33**, 1038 (1977).

[2] J.F. Keggin, *Proc. Roy. Soc. London* **A144**, 75 (1934).

[3] H.T. Evans, Jr. in J.D. Dunitz and J.A. Ibers (eds.), *Perspectives in Structural Chemistry*, Vol. 4, Wiley, New York, 1971, p. 1.

[4] M.T. Pope in G. Wilkinson, R.D. Gillard and J.A. McCleverty (eds.), *Comprehensive Coordination Chemistry*, Vol. 3, Pergammon Press, 1987, chap. 38; M.T. Pope, *Heteropoly and Isopoly Oxometalates*, Springer-Verlag, Berlin, 1983.

[5] A. Goiffon, E. Philippot and M. Maurin, *Rev. Chim. Miner.* **17**, 466 (1980).

[6] H.T. Evans, Jr., *Inorg. Chem.* **5**, 967 (1966).

[7] Shao Meicheng, Wang Lifeng, Zhang Zeying and Tang Youqi, *Scientia Sinica* **B17**, 137 (1984).

[8] G.K. Johnson and E.O. Schlemper, *J. Am. Chem. Soc.* **100**, 3645 (1978).

[9] A. Müller, E. Krickemeyer, M. Penk, H.-J. Walberg and H. Bögge, *Angew. Chem. Int. Ed. Engl.* **26**, 1045 (1987).

[10] A. Müller, M. Penk, E. Krickemeyer, H. Bögge and H.-J. Walberg, *Angew Chem. Int. Ed. Engl.* **27**, 1719 (1988).

[11] A.F. Wells, *Structural Inorganic Chemistry*, 5th ed., Clarendon Press, Oxford, 1984.

[12] H.T. Evans, Jr., B.M. Gatehouse and P. Leverett, *J. Chem. Soc., Dalton Trans.*, 505 (1975).

[13] B. Krebs, and I. Paulat-Boeschen, *Acta Crystallogr. Sect. B* **38**, 1710 (1982).

[14] G. Johansson, *Acta Chem. Scand.* **14**, 771 (1960).

[15] J.N. Barrows, G.B. Jameson and M.T. Pope, *J. Am. Chem. Soc.* **107**, 1771 (1985).

[16] Y. Jeannin, J.P. Launay and M.A. Seid Sedjadi, *Inorg. Chem.* **19**, 2933 (1980).

[17] A. Müller, E. Krickemeyer, M. Penk, V. Wittneben and J. Döring, *Angew. Chem. Int. Ed. Engl.* **29**, 88 (1990).

[18] R. Massart, R. Contant, J.-M. Fruchart, J.-P. Ciabrini and M. Fournier, *Inorg. Chem.* **16**, 2916 (1977).

[19] J. Fuchs and R. Palm, *Z. Naturforsch., Teil B* **39**, 757 (1984).

[20] M.H. Alizadeh, S.P. Harmalker, Y. Jeannin, J. Martin-Frére and M.T. Pope, *J. Am. Chem. Soc.* **107**, 2662 (1985).

[21] R. Contant and A. Tézé, *Inorg. Chem.* **24**, 4610 (1985).

[22] T.J.R. Weakley, *Acta Crystallogr. Sect. C* **40**, 16 (1984).

[23] H.T. Evans, Jr., *Acta Crystallogr. Sect. B* **30**, 2095 (1974).

[24] R.D. Peacock and T.J.R. Weakley, *J. Chem. Soc. (A)*, 1836 (1971).

[25] D.D. Dexter and J.V. Silverton, *J. Am. Chem. Soc.* **90**, 3589 (1968).

[26] A. Müller and J. Döring, *Angew. Chem. Int. Ed. Engl.* **27**, 1721 (1988).

[27] H.T. Evans, Jr., C.M. Tourné, G.F. Tourné and T.J.R. Weakley, *J. Chem. Soc., Dalton Trans.*, 2699 (1986).

[28] C.M. Tourné, G.F. Tourné and F. Zonnevijlle, *J. Chem. Soc., Dalton Trans.*, 143 (1991).

Note Some recent advances in polyoxometallate chemistry are covered in M.T. Pope and A. Müller, *Angew. Chem. Int. Ed. Engl.* **30**, 34 (1991).

New developments in polyvanadate chemistry from cages and clusters to baskets, belts, bowls, and barrels are reviewed in W.G. Klemperer, T.A. Marquart and O.M. Yaghi, *Angew. Chem. Int. Ed. Engl.* **31**, 49 (1992).

Novel polyoxometallate cluster shells that are structurally complementary to encapsulated "anionic guests" can be assembled by template-controlled design. Recent examples include $[HV_{10}O_{44}(NO_3)]^{10-}$, $[HV_{22}O_{54}(ClO_4)]^{6-}$ and $[H_2V_{18}O_{44}(N_3)]^{5-}$ in which the central anionic groups unexpectedly do not exhibit disorder. See A. Müller, *Nature* **352**, 115 (1991); A. Müller, E. Krickemeyer, M. Penk, R. Rohlfing, A. Armatage and H. Bögge, *Angew. Chem. Int. Ed. Engl.* **30**, 1674 (1991).

4.15 Nitrogenpentammineruthenium(II) Dichloride
$$[\mathrm{Ru(NH_3)_5(N_2)]Cl_2}$$

Crystal Data

Cubic, space group $Fm3m$ (No. 225), or $F432$ (No. 211), or $F43m$ (No. 217)

$a = 10.141(1)$Å, $Z = 4$

Atom	x	y	z
Ru	0	0	0
Cl	1/4	1/4	1/4
N(1)	.2071(9)	0	0
N(2)	.3172(55)	0	0

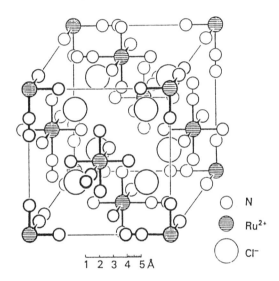

1 2 3 4 5Å

○ N

◉ Ru^{2+}

○ Cl^-

Fig. 4.15-1 Unit cell of $[Ru(NH_3)_5(N_2)]Cl_2$. Orientation of the N_2 molecule is arbitrary. (After ref. 1).

Crystal Structure

Transition metals, in the form of complexes or in the metallic form, are known to be necessary for the fixation of molecular nitrogen. The first complex containing molecular nitrogen as a ligand, $[Ru(NH_3)_5(N_2)]Cl_2$, was synthesized by Allen and Senoff in 1965.[2]

In the crystal structure determination of $[Ru(NH_3)_5(N_2)]Cl_2$, the space group could not be unequivocally established because all the non-hydrogen atoms are in special positions and the nitrogen molecule randomly occupies one of the six octahedral positions around the ruthenium ion.

Within the limits of accuracy imposed by the disorder, the Ru-N-N unit is linear with Ru-N(1) = 2.10 and N(1)-N(2) = 1.12Å. However, as the thermal vibrations of N(1) and N(2) are significantly greater perpendicular to the bond axis than along it, the Ru-N-N unit may not be precisely linear.

Other salts $[Ru(NH_3)_5(N_2)]X_2$ (X = Br, I, BF_4, and PF_6) are isomorphous with the dichloride, with the unit-cell parameter a = 10.41, 10.94, 11.166, and 11.79Å, respectively.

Remarks

1. Coordination modes of dinitrogen

Complexes of dinitrogen are of considerable interest as models for biological nitrogen fixation and as intermediates in synthetic applications. The known coordination modes of dinitrogen are shown in Table 4.15-1.

Table 4.15-1 Coordination modes of dinitrogen.

Coordination mode	Example	$d_{N-N}/$Å	Ref.
(a) η^1-N_2, M-N≡N	$[Ru(NH_3)_5(N_2)]^{2+}$	1.12	[1]
(b) μ(bis-η^1)-N_2, M-N≡N-M	$[(C_5Me_5)_2Ti]_2(N_2)$	1.16 (av.)	Fig. 6.11-1
(c) μ_3-N_2, M-N-N-M \ / M	$(\mu_3$-$N_2)[(C_{10}H_8)(C_5H_5)_2Ti_2]$- $[(C_5H_4)(C_5H_5)_3Ti_2]$	1.301	[3]
(d) μ_3-N_2, M-N-N$\overset{M'}{\underset{M}{\diamond}}$N-N-M	$[WCl(py)(PMePh)_3(\mu_3$-$N_2)]_2(AlCl_2)_2$		[4]
(e) $(\mu$-η^2:$\eta^2)$-N_2, M$\overset{N}{\underset{N}{\diamond}}$M (planar)	$[\{i$-$Pr_2PCH_2SiMe_2)_2N\}ZrCl]_2(N_2)$	1.548	[8]
(f) μ(bis-η^2)-N_2, (non-planar)	$[(PhLi)_6Ni_2(N_2)(Et_2O)]_2$	1.35	[6]

Most examples of stable dinitrogen complexes have been found to belong to the η^1-N_2 category. A typical feature of this class of compounds is an only slightly elongated N-N bond length as compared to that in gaseous dinitrogen (1.0976Å). Some structural data are listed in Table 4.15-2.

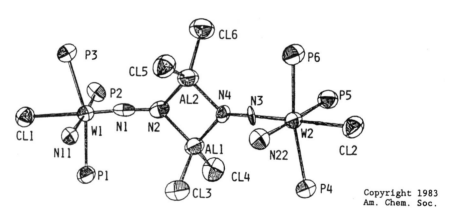

Fig. 4.15-2 Molecular configuration of $[WCl(py)(PMePh)_3(\mu_3$-$N_2)]_2(AlCl_2)_2$ in its 1:1 crystalline benzene solvate. (After ref. 4).

In the mixed-metal complex shown in Fig. 4.15-2, both W-N-N linkages are essentially linear, and the four metal atoms and two μ_3-N_2 ligands almost lie in the same plane.[4] The N1-N2 and N3-N4 distances are 1.46(4) and 1.25(3)Å, respectively, but their reliability is affected by the abnormal thermal motion of the N1 and N3 atoms.

Table 4.15-2 X-Ray structural data for some dinitrogen complexes.[7]

Complex	N-N/Å	M-N/Å	M-N-N/deg	Comments
trans-[Mo(dppe)$_2$(N$_2$)$_2$]	1.10(2)	2.01(1)	171.8(11)	-
trans-[ReCl(Me$_2$PhP)$_4$N$_2$]	1.06(3)	1.97(2)	177(1)	Partial disorder of Cl, N$_2$
[Ru(NH$_3$)$_5$N$_2$]Cl$_2$	~1.12	~2.10	180	N$_2$, NH$_3$ disordered
[RuN$_3$(en)$_2$N$_2$]PF$_6$	1.106(11)	1.894(9)	179.3(9)	
[(Ru(NH$_3$)$_5$)$_2$N$_2$](BF$_4$)$_4$·2H$_2$O	1.124(15)	1.928(6)	178.3(5)	trans-Ru-NH$_3$ = 2.146(6),
				cis-Ru-NH$_3$(av.) = 2.12Å
[Os(NH$_3$)$_5$N$_2$]Cl$_2$	1.12(2)	1.842(13)	178.3(13)	
[CoH(Ph$_3$P)$_3$N$_2$]	1.101(12)	1.784(13)	178(2)	Two independent molecules per unit
	1.123(13)	1.829(12)	178(1)	cell
[(((C$_6$H$_{11}$)$_3$P)$_2$Ni)$_2$N$_2$]	1.12	1.77	178.2	
		1.79	178.3	
[((η^5-C$_5$Me$_5$)$_2$ZrN$_2$)$_2$N$_2$]	1.116(8)	-	~180°	Terminal N$_2$ distances
	1.114(7)			
	1.182(5)			Bridging N$_2$ distance
[{Mo(Me$_2$P(CH$_2$)$_2$PMe$_2$)-				
(η^6-1,3,5(CH$_3$)$_3$C$_6$H$_3$)}$_2$N$_2$]	1.145(7)	2.042(4)	175.6(4)	
[((PhLi)$_3$Ni)$_2$N$_2$(Et)$_2$O)$_2$]$_2$	1.35	1.91-1.94	~90°	
[(Me$_2$PhP)$_4$ClReN$_2$MoCl$_4$(OMe)]		1.90(1)(Mo-N)	178.7(9)	
	1.18(3)	1.81(1)(Re-N)	179.6	

The simple side-on coordination mode of dinitrogen to a metal center (M⟨N‖N) has not yet been found. In the [(η^5:η^5-C$_{10}$H$_8$)(η^5-C$_5$H$_5$)$_2$Ti$_2$][(η^1:η^5-C$_5$H$_4$)(η^5-C$_5$H$_5$)$_3$Ti$_2$](μ_3-N$_2$) complex the dinitrogen ligand is coordinated simultaneously to three titanium atoms (Fig. 4.15-3). Bond lengths are: Ti(3)-N(1) 2.181(10), Ti(3)-N(2) 2.097(11), Ti(2)-N(1) 1.953(11), Ti(4)-N(2) 1.857(11), and N(1)-N(2) 1.301(12)Å.[3] Other related dinitrogen bridged dinuclear Ti and Zr complexes are discussed in Section 6.11.

[(C$_5$Me$_5$)$_2$Sm]$_2$(μ-η^2:η^2-N$_2$) is the first example of planar, side-on bonding between two metals and a dinitrogen ligand [Fig. 4.15-4(a)].[5] The molecule has C_2 symmetry, and the Sm(1)-N(1) and Sm(2)-N(1) distances are 2.347(6) and 2.368(6)Å, respectively. The short N(1)-N(1') distance of 1.088(12)Å is not consistent with that of the free N$_2$ molecule. Since the NMR spectrum of this complex is similar to that of [(C$_5$Me$_5$)$_2$Sm]$_2$(μ-η^1:η^1-N$_2$Ph$_2$), the true length may be longer than 1.088Å.

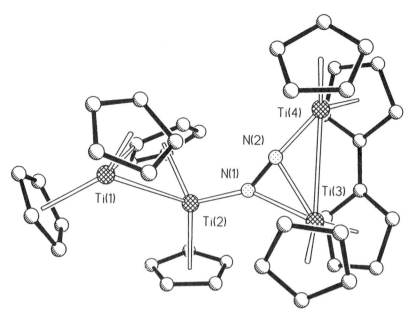

Fig. 4.15-3 Structure of the $[(C_{10}H_8)(C_5H_5)_2Ti_2][(C_5H_4)(C_5H_5)_3Ti_2](\mu_3-N_2)$ complex with Ti and N atom labels. (Adapted from ref. 3).

The centrosymmetric $[\{(i\text{-}Pr_2PCH_2SiMe_2)_2N\}ZrCl]_2(\mu\text{-}\eta^2:\eta^2\text{-}N_2)$ complex also features a side-on-bridging N_2 ligand [Fig. 4.15-4(b)].[8] The N-N bond length of 1.548(7)Å is the longest ever reported for a transition-metal dinitrogen complex, and the planar Zr_2N_2 core is symmetrical with essentially identical, relatively short Zr-N bond distances [Zr-N2, 2.024(4); Zr-N2', 2.027(4)Å].

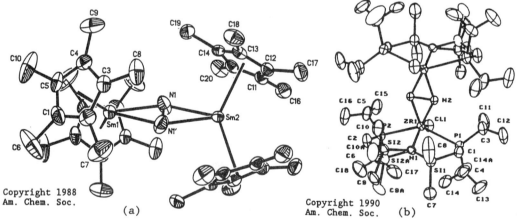

(a)   (b)

Fig. 4.15-4 Molecular Structures of (a) $[(C_5Me_5)_2Sm]_2(N_2)$ and (b) $[\{i\text{-}Pr_2PCH_2SiMe_2)_2N\}ZrCl]_2(N_2)$. (After refs. 5 and 8).

The geometry of the centrosymmetric dimeric $[(PhLi)_6Ni_2N_2(Et_2O)_2]_2$ molecule is shown in Fig. 4.15-5(a).[6] Fig. 4.15-5(b) depicts the internal skeleton of the strongly polar complex (Ni, N, and Li atoms). A considerably elongated Ni-Ni linkage (2.687 Å) is bridged by a nitrogen molecule almost at right angles with respect to the N-N bond. The two independent Ni atoms, the Li(1) atom, and N_2 display a strongly distorted trigonal-pyramidal geometry. The trigonal planes contain two Ni atoms and the N_2 molecule, and the apices are occupied by Li atoms.

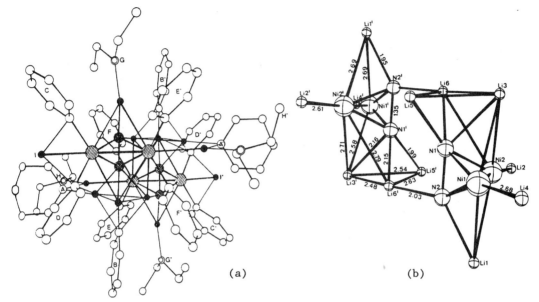

(a)   (b)

Fig. 4.15-5 (a) Molecular structure of $[(PhLi)_6Ni_2N_2(Et_2O)_2]_2$ and (b) internal skeleton of the complex. (After ref. 6).

2. Bonding nature of transition metal complexes of molecular N_2

Bonding of dinitrogen to transition metals may be postulated to proceed via several stereochemical possibilities, which may be divided into the "end-on" and "side-on" categories. In the end-on arrangement, the fixation of the N_2 ligand is accomplished by a σ bond between the $3\sigma_g$ orbital of the nitrogen and hybrid orbitals of the metal and by back bonding from the metal to the vacant $1\pi_g{}^*$ orbital of the nitrogen, as shown in Fig. 4.15-6(a). In the side-on arrangement, the bonding is considered to arise from two interdependent components as illustrated schematically in Fig. 4.15-6(b).

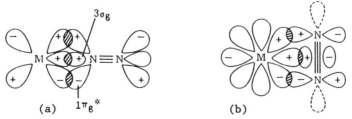

(a) $1\pi_g{}^*$ (b)

Fig. 4.15-6 Orbital interactions of N_2 and M in transition metal complexes of dinitrogen. (After ref. 6).

In the first part, σ overlap between the filled π orbital of N_2 and a suitably directed vacant hybrid metal orbital forms the donor bond. In the second part, π overlap between a filled metal d orbital with the vacant antibonding orbital of N_2. This π back bonding reinforces the σ bond synergically as in the Dewar-Chatt-Duncanson model for metal-olefin π complexes (Section **6.4**).

The transition metal dinitrogen complexes have been investigated by MO calculations, and the results, in the light of experimental data reported to date, suggest the following generalizations.[9]

a) Both σ donation and π back-donation are related to the formation of the metal-nitrogen bond, the former interaction being more important. The π back-donation contributes to the weakening of the N-N bond.

b) The N-N bond of the side-on complex is appreciably weakened because of the electron donation from the bonding π and σ orbitals of the N_2 ligand to the unoccupied MO's of the metal. Moreover, there is electron donation from the occupied metal MO's to the antibonding π^* orbital of the ligand. Therefore, the end-on coordination mode is considered to be more favourable than the side-on one. The weak N-N bond in the side-on complex indicates that the N_2 ligand in this type of compound is fairly activated. The reduction of the coordinated nitrogen molecule may proceed through this activated form.

The formation of metal-nitrogen bonds is likely to be a key feature of many reactions involving ligating dinitrogen. A new and well-defined area of organic synthesis is based on the direct formation of carbon-nitrogen bonds by reaction of coordinated dinitrogen, or a protonated derivative, with a range of organic compounds including alkyl and acyl halides, carboxylic acid anhydrides, aldehydes, ketones, and activated aryl or vinyl halides.[10]

A significant development has been the demonstration that a cyclic system can be produced for the conversion of N_2 to NH_3 using the following sequence of reactions.[11,12]

$$[W(N_2)_2(dppe)_2] \xrightarrow{\quad CF_3C_6H_4SO_3H \quad} [(CF_3C_6H_4SO_3)W(dppe)_2(NNH_2)]^+$$

$$\uparrow N_2 \downarrow \qquad\qquad\qquad\qquad\qquad \downarrow \text{electrochemical reduction}$$

$$NH_3 \xleftarrow{\quad \text{reduction, } H^+ \quad} [W(dppe)_2(NNH_2)]$$

3. Structures of polymeric $(RO)_3M{\equiv}N$ (M = W, Mo; R = *t*-Bu, *i*-Pr)[13,14]

In the crystalline state, $(RO)_3M{\equiv}N$ forms a linear polymer involving ...M${\equiv}$N...M${\equiv}$N... with alternating short M${\equiv}$N and long M...N bonds, corresponding formally to triple and weak dative bonds, respectively. ORTEP plots of the monomeric $(t\text{-}BuO)Mo{\equiv}N$ and $(i\text{-}PrO)_3Mo{\equiv}N$ units are shown in Fig. 4.15-7.

Table 4.15-3 Structural parameters for the (RO)₃M≡N compounds .

Parameter	M = W R = t-Bu	M = Mo		
		R = t-Bu*	R = t-Bu**	R = i-Pr
space group	P6₃cm	P6₃cm	P6₃	P6₃cm
M≡N /Å	1.740	1.661	1.673	1.597
M...N /Å	2.661	2.883	2.844	2.515
M-O /Å	1.872	1.882	1.888	1.894
N≡N-O/deg.	101.6	103.27	103.45	101.1
M-O-C/deg.	136.6	135.1	134.33	128.6
O-M-O/deg.	116.1	114.9	114.76	116.4

* at -90°C; ** at -160°C.

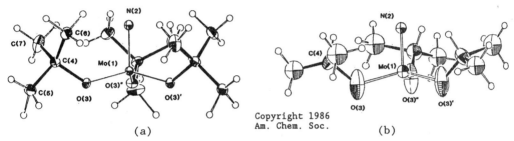

Copyright 1986
Am. Chem. Soc.

(a)   (b)

Fig. 4.15-7 ORTEP drawing of (a) (*t*-BuO)₃Mo≡N unit and (b) (*i*-PrO)₃Mo≡N unit. (After ref. 13).

A comparison of pertinent structural parameters for the compounds (RO₃)M≡N, where M = Mo and W, is given in Table 4.15-3. The small but significant differences in the M≡N and M...N distances of the Mo and W complexes underscore the physicochemical properties of the compounds and the different reactivities of the M≡N bond. The following rationalization is based on a consideration of electronic factors.

It is well recognized that π-acceptor ligands, such as NO and CO, bind more strongly to W(OR)₃ centers than to Mo(OR)₃ centers in either mononuclear or dinuclear compounds. This reflects the greater π back-bonding capabilities of W relative to Mo in its medium and higher oxidation states. In oxidation state +6, W is a weaker oxidizing agent than Mo and will interact less strongly with the N^{3-} ligand, leaving some net negative charge on the nitride ligand. This supposition is supported by molecular orbital calculations. Mo, being more electronegative than W, matches the energy of the nitride ligand better than W; thus there is less charge separation and more covalent bonding in the Mo case.

The reactivity of the M≡N unit is dependent on both the metal and the ligand. Addition of alcohol (> 6 equiv.) to a hydrocarbon solution of (*t*-BuO)₃W≡N gives W(OR)₆ and NH₃, while under analogous conditions the related reactions employing (*t*-BuO)₃Mo≡N give (RO)₃Mo≡N:

$$(t\text{-BuO})_3W\equiv N + 6ROH \longrightarrow W(OR)_6 + 3t\text{-BuOH} + NH_3$$

$$(t\text{-BuO})_3Mo\equiv N + ROH(\text{excess}) \longrightarrow (RO)_3Mo\equiv N + 3t\text{-BuOH}$$

$$R = Et, \ i\text{-Pr}, \ neo\text{-Pentyl}$$

This difference in reactivity may in part arise from the greater negative charge located on the nitride for $(t\text{-BuO})_3W\equiv N$, making it easier to be protonated.

4. Complexes containing N_2R, N_2R_2 and $N_2R_2^{2-}$ (R = H or an organic group)[15]

The diazenido (N_2R), diazene (N_2R_2), and hydrazido(2-) ($N_2R_2^{2-}$) ligands can adopt the following possible configurations in mononuclear transition metal complexes, many of which have been structurally characterized by X-ray crystallography.

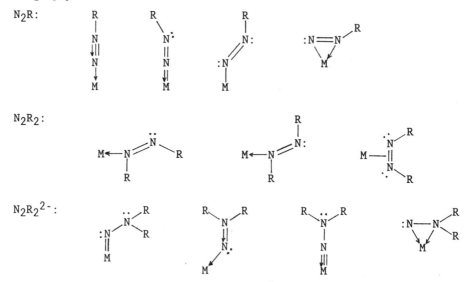

As a bridging ligand N_2R, N_2R_2 and $N_2R_2^{2-}$ can adopt the following coordination modes:

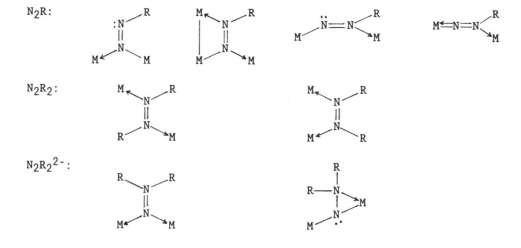

Fig. 4.15-8 shows the structure of a compound of the M-N=N-R type, $[Rh(PhP((CH_2)_3PPh_2)_2)Cl(N_2Ph)]^+$, [16] and the structure of a compound of the M=N-NR₂ type, $[MoO(NNMe_2)(C_9H_6NO)_2]$. [17]

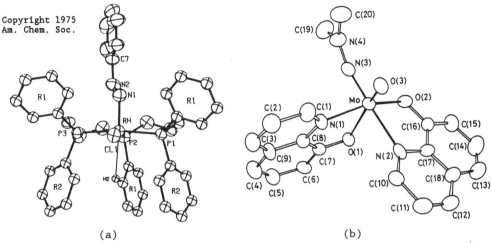

(a)   (b)

Fig. 4.15-8 Structures of (a) the $[Rh(PhP((CH_2)_3PPh_2)_2)Cl(N_2Ph)]^+$ cation, and (b) $[MoO(NNMe_2)(C_9H_6NO)_2]$. (After refs. 16 and 17).

5. Nitrogenase model compounds containing N_2H_x ligands

The recent characterization of a nitrogenase type containing iron only has cast doubt to the prevalent presumption that the heterometal center is the

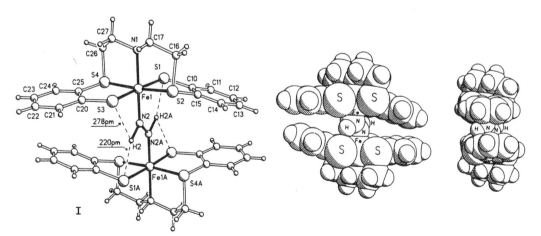

Fig. 4.15-9 Molecular structure and space-filling frontal and side views of nitrogenase model compound I. Important bond distances (Å) and angles (°): N2-N2A 1.300(7), N2-H2 1.16(5), Fe1-N2 1.867(4), Fe1-N1 2.037(4), Fe1-S1 2.318(2), Fe1-S2 2.234(1), Fe1-S3 2.288(2), Fe1-S4 2.251(1); Fe1-N2-N2A 132.2(5), N2A-N2-H2 103.1, Fe1-N2-H2 124.7, N1-Fe-N2 179.5(2), N2-Fe-S1 91.5(1), N2-Fe-S2 93.1(1), N2-Fe-S3 87.7(1), N1-Fe-S4 93.4(1). (After ref. 19).

N_2 binding site in Mo/Fe or V/Fe nitrogenases.[18] An X-ray analysis of
$[\mu\text{-}N_2H_2\{Fe("N_HS_4")\}_2]$, I, revealed that (i) it is the first example of
coordination of diazene, HN=NH, to iron, (ii) being ligated by $"N_HS_4"^{2-}$, the
iron center is exclusively coordinated by biologically occurring donor atoms,
and (iii) the diazene is stabilized not only by coordination to iron but also
by strong tricentric (forked) N-H···S hydrogen bridges (Fig. 4.15-9).[19]
Compound I is centrosymmetric, and the *trans*-N_2H_2 protons were located by a
difference Fourier synthesis.

If in nitrogenase centers the same stabilization of the dizaene ligand
by N-H...S hydrogen bridges does occur, the $N_2 \rightarrow N_2H_2$ conversion, the most
difficult step of nitrogen fixation, should be facilitated and may even become
exoenergetic.

In the salt $[(tripod)Co(\eta^2\text{-}N_2H_3)](BPh_4)\cdot2THF$ (tripod = $CH_3C(CH_2PPh_2)_3$),
II, the geometry of the complex cation is best described as a tetrahedron with
the N_2H_3 unit occupying one ligand site; the magnetic moment of $1.8\mu_B$
corresponds to a low-spin d^7 configuration at Co^{II}. The related hydrazine
complex $[(tripod)Co(\eta^2\text{-}N_2H_4)](BF_4)(BPh_4)\cdot THF$, III, has a high-spin d^7 Co^{II}
atom (magnetic moment $3.84\mu_B$) in a square-pyramidal coordination environment
with the Co-P(apical) bond marginally longer than the Co-P(basal) bonds (Fig.
4.15-10).[20]

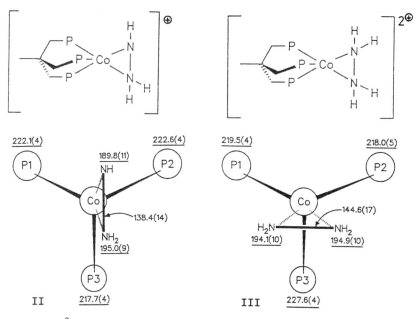

Fig. 4.15-10 η^2-Bonding (top) and coordination geometry (bottom) of the
complex cations in **II** and **III**. The underlined numerical values indicate
cobalt-ligand distances in pm. (After ref. 20).

References

[1] F. Bottomley and S.C. Nyburg, *Acta Crystallogr. Sect. B* **24**, 1289 (1968).

[2] A.D. Allen and C.V. Senoff, *J. Chem. Soc., Chem. Commun.*, 621 (1965); C.V. Senoff, *J. Chem. Educ.* **67**, 368 (1990).

[3] G.P. Pez, P. Apgar and R.K. Crissey, *J. Am. Chem. Soc.* **104**, 482 (1982).

[4] T. Takahashi, T. Kodama, A. Watakabe, Y. Uchida and M. Hidai, *J. Am. Chem. Soc.* **105**, 1680 (1983).

[5] W.J. Evans, T.A. Ulibarri and J.W. Ziller, *J. Am. Chem. Soc.* **110**, 6877 (1988).

[6] C. Krüger and Y.-H. Tsay, *Angew. Chem. Int. Ed. Engl.* **12**, 998 (1973).

[7] F. Bottomley in R.W.F. Hardy, F. Bottomley and R.C. Burns (eds.), *A Treatise on Dinitrogen Fixation*, Wiley, New York, 1979, chap. 3.

[8] M.D. Fryzuk, T.S. Haddad and S.J. Rettig, *J. Am. Chem. Soc.* **112**, 8185 (1990).

[9] T. Yamabe, K. Hori, T. Minato and K. Fukui, *Inorg. Chem.* **19**, 2154 (1980).

[10] H.M. Colquhoun, *Acc. Chem. Res.* **17**, 23 (1984).

[11] G.J. Leigh, *Transition Met. Chem.* **11**, 118 (1986).

[12] C.J. Pickett and J. Talarmin, *Nature (London)* **317**, 652 (1985).

[13] D.M.-T. Chan, M.H. Chisholm, K. Folting, J.C. Huffman and N.S. Marchant, *Inorg. Chem.* **25**, 4170 (1986).

[14] M.H. Chisholm, J.C. Huffman and D.M. Hoffman, *Inorg. Chem.* **22**, 2903 (1983).

[15] R.A. Henderson, G.J. Leigh and C.J. Pickett, *Adv. Inorg. Chem. Radiochem.* **27**, 197 (1983).

[16] A.P. Gaughan Jr. and J.A. Ibers, *Inorg. Chem.* **14**, 352 (1975).

[17] J. Chatt, B.A.L. Crichton, J.R. Dilworth, P. Dahlstrom and J.A. Zubieta, *J. Chem. Soc., Dalton Trans.*, 1041 (1982).

[18] J.R. Chisnell, R. Premakumar and P.E. Bishop, *J. Bacteriol.* **170**, 27 (1988).

[19] D. Sellmann, W. Soglowek, F. Knoch and M. Moll, *Angew. Chem. Int. Ed. Engl.* **28**, 1271 (1989).

[20] S. Vogel, A. Barth, G. Huttner, T. Klein, L. Zsolnai and R. Kremer, *Angew. Chem. Int. Ed. Engl.* **30**, 303 (1991).

Note Side-on and end-on coordination modes of dinitrogen occur in the Ti^{II} amido complex {[(Me₃Si)₂N]TiCl(TMEDA)}₂(μ-N₂) and the mixed-valent Ti^{I}/Ti^{II} complex [{[(Me₃Si)₂N]₂Ti}₂(μ-η^2:η^2-N₂)₂]⁻, respectively. See R. Duchateau, S. Gambarotta, N. Beydoun and C. Bensimon, *J. Am. Chem. Soc.* **113**, 8986 (1991).

Ab initio MO studies of the stereochemistry and coordination modes of transition-metal complexes of N₂, CO, and CO₂ are reviewed by S. Sakaki in I. Bernal (ed.), *Stereochemistry of Organometallic and Inorganic Compounds, Vol. 4: Stereochemical Control, Bonding and Steric Rearrangements*, Elsevier, Amsterdam, 1990.

The remarkably stable, biphenylene-bridged μ_2-N₂ complex of a ruthenium cofacial metallodiporphyrin complex and its putative reduction intermediates (namely the μ_2-diazene, μ_2-hydrazine, and bis-diammine complexes) are reported in J.P. Collman, J.E. Hutchison, M.A. Lopez, R. Guilard, and R.A. Reed, *J. Am. Chem. Soc.* **113**, 2794 (1991).

4.16   Iron-Sulfur Tetranuclear Cluster

$[(CH_3)_4N]_2[Fe_4S_4(SC_6H_5)_4]$

Crystal Data

Orthorhombic, space group $P2_12_12_1$ (No. 19)

$a = 11.704(11)$, $b = 23.944(16)$, $c = 14.876(10)$Å, $Z = 4$

Atom*	x	y	z	Atom	x	y	z
Fe(1)	.18143	.32134	.38918	R(2)C(1)	-.0158	.4310	.0779
Fe(2)	.09834	.35208	.22479	R(2)C(2)	-.0635	.4323	-.0078
Fe(3)	.32662	.36383	.26278	R(2)C(3)	-.0257	.4713	-.0705
Fe(4)	.23532	.26010	.23906	R(2)C(4)	.0598	.5091	-.0474
S(1)	.2474	.3303	.1344	R(2)C(5)	.1075	.5079	.0384
S(2)	.3559	.2873	.3530	R(2)C(6)	.0697	.4689	.1010
S(3)	.0592	.2703	.2993	R(3)C(1)	.4986	.4276	.1207
S(4)	.1795	.4099	.3307	R(3)C(2)	.4034	.4534	.0828
S(5)	.1117	.3078	.5289	R(3)C(3)	.4084	.4736	-.0048
S(6)	-.0638	.3796	.1547	R(3)C(4)	.5086	.4681	-.0546
S(7)	.4978	.4049	.2340	R(3)C(5)	.6037	.4423	-.0167
S(8)	.2810	.1858	.1527	R(3)C(6)	.5987	.4220	.0710
R(1)C(1)	.1950	.3496	.6026	R(4)C(1)	.2615	.1268	.2223
R(1)C(2)	.3047	.3323	.6260	R(4)C(2)	.1525	.1047	.2345
R(1)C(3)	.3653	.3609	.6920	R(4)C(3)	.1382	.0553	.2825
R(1)C(4)	.3163	.4068	.7346	R(4)C(4)	.2329	.0280	.3184
R(1)C(5)	.2067	.4241	.7112	R(4)C(5)	.3418	.0501	.3062
R(1)C(6)	.1460	.3955	.6452	R(4)C(6)	.3561	.0995	.2582

*The atomic coordiantes of the $[(CH_3)_4N]^+$ cation have been omitted.

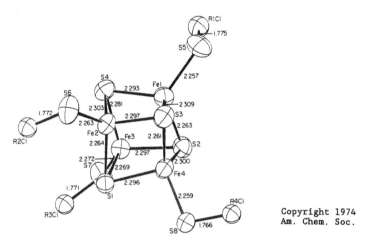

Copyright 1974
Am. Chem. Soc.

Fig. 4.16-1 Structure of a portion of the $[Fe_4S_4(SPh)_4]^{2-}$ core, showing principal bond lengths and the atom lebeling scheme.   (After ref. 1).

Crystal Structure

The structure consists of a packing of discrete cations and anions.   The principal dimensions of the $Fe_4S^*_4S_4$ portion of the anion (asterisk indicates bridging S atom) together with the atom lebeling scheme are shown in Fig. 4.16-1.

The tetranuclear cluster anion $[Fe_4S_4(SPh)_4]^{2-}$, like its benzyl analogue $[Fe_4S_4(SCH_2Ph)_4]^{2-}$, possesses the cubane-type of stereochemistry.   The $Fe_4S^*_4$ core has effective D_{2d} symmetry with an average Fe...Fe distance of

2.736Å, but deviation from T_d symmetry is relative small. The bonded Fe-S*
distances occur as sets of four 2.267(5)Å and eight 2.296(4)Å, giving an
average of 2.286Å. The average terminal Fe-S distance is 2.263Å.

Remarks

1. Structure of $[Fe_4X_4(SPh)_4]^n$ (X = chalcogenide; n = -2 and -3) and related
    iron-sulfur proteins

    The structures of the cubane-type clusters $[Fe_4X_4(SPh)_4]^n$ (X = S, Se or
Te; n = -2 or -3) are schematically compared in Fig. 4.16-2 with emphasis on
core stereochemistry.[2,3] The eletronic ground state (S = 0) of the
dianionic species exhibit four short and eight long Fe-X bonds, corresponding
to *compressed* tetragonal distortion with respect to the *idealized* 4 axis of
each Fe_4X_4 core of D_{2d} symmetry. The cluster trianions have virtually
identical structures except that the cores are *elongated* along the idealied 4
axis, and the ground states become S = 3/2 or spin-admixed (S = 1/2 + 3/2).
The $[Fe_4X_4(SPh)_4]^{3-}$ structures also show a greater deviation from idealized
elongated D_{2d} symmetry as the chalcogen atomic number is increased.

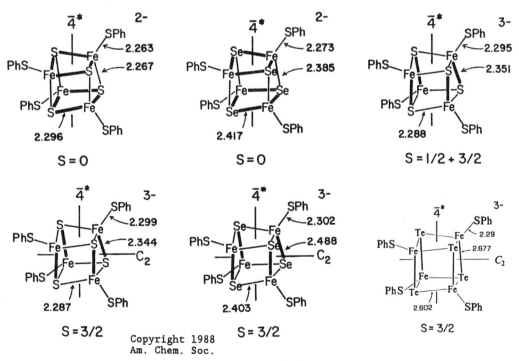

Fig. 4.16-2 Schematic representation of the structures of clusters in
(Me₄N)₂[Fe₄X₄(SPh)₄] (X = S, Se) and (Et₃MeN)₃[Fe₄S₄(SPh)₄] (top row)
and (Me₄N)₃[Fe₄X₄(SPPh)₄].2MeCN (X = S, Se) and (Et₄N)₃[Fe₄Te₄(SPh)₄]
(bottom row). Mean values of long (bold line) and short Fe-X core bonds
and Fe-S terminal bonds, crystallographic and idealized (*) symmetry
axes, and ground spin states are indicated. (After refs. 2 and 3).

The existence of three physiologically significant total oxidation
levels of the $[Fe_4S_4(S-Cys)_4]$ clusters present in non-heme iron-sulfur redox
proteins, ferredoxin (Fd) and high potential iron proteins (HiPIP), is firmly
established.   Proteins containing iron have two major functions: (i) oxygen
transport and storage, and (ii) electron transfer.   The latter function
usually involves one-electron redox reactions (ox, red and s-red stand for
oxidized, reduced and super-reduced forms, respectively):

$$[Fe_4S_4)(SR)_4]^{3-} \rightleftharpoons [Fe_4S_4(SR)_4]^{2-} + e$$

$$Fd_{red}, HiPIP_{s-red} \qquad\qquad Fd_{ox}, HiPIP_{red}$$

On the basis that an elongated tetragonal geometry is the intrinsically
stable core structure of the reduced clusters $[Fe_4S_4(SR)_4]^{3-}$, it is speculated
that the native form of "high-potential", but not Fd, proteins may have
evolved to resist the *ca.* 0.08Å axial core expansion found in passing from
analogous dianion to trianion.   The former proteins are reducible only in the
unfolded state.   Fig. 4.16-3 shows the structure of the 8Fe-ferredoxin from
Peptococcus aerogenes.[4]   This ferredoxin contains two separate $[Fe_4S_4]$ cores
with a cubane-like configuration, the centers of which are some 12Å apart, and
is enveloped by a "linear" polypeptide of 54 amino acids.   The $[Fe_4S_4]$ centers
in both of these proteins are surrounded principally by non-polar amino acid
side chains.

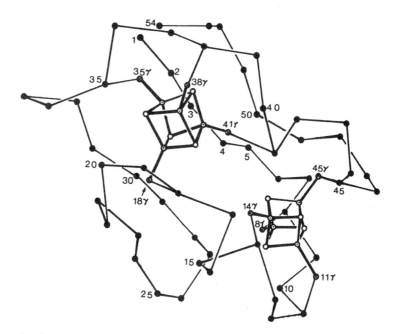

Fig. 4.16-3 Structure of the 8Fe-ferredoxin from *Peptococcus
aerogenes.* (After ref. 5).

Several proteins containing $[Fe_4(S\text{-}Cys)_4(\mu_3\text{-}S)_4]$ centers have been characterized, and Table 4.16-1 lists the average dimensions of these centers.[5]

Table 4.16-1 Average dimensions of $[Fe_4(S\text{-}Cys)_4(\mu_3\text{-}S)_4]$ centers in some proteins.

	HiPIP$_{ox}$	HiPIP$_{red}$	Fd$_{ox}$I	Fd$_{ox}$II
Distance (Å)				
Fe...Fe	2.72(4)	2.81(4)	2.73(6)	2.67(7)
S....S	3.55(6)	3.65(11)	3.52(7)	3.49(12)
Fe...S	2.26(8)	2.32(9)	2.23(12)	2.22(14)
Fe...S(Cys)	2.20(2)	2.22(3)	2.22(16)	2.25(17)
Angles (deg)				
Fe-S-Fe	74(1)	76(2)	75(2)	75(2)
S-Fe-S	104(2)	104(3)	103	104
S-Fe-S(Cys)	115(5)	116(5)	115(3)	115(3)

2. Structure of $[Fe_4S_4(SH)_4]^{2-}$ and $[Fe_4S_4(\eta^5\text{-}Cp)_4]^n$ systems ($n = -1$ to $+3$)

The cubane-type cluster anion in $(PPh_4)_2[Fe_4S_4(SH)_4]$ is the simplest synthetic analogue for a ferredoxin, and has crystallographic symmetry C_2.[6] The $[Fe_4S_4(SH)_4]^{2-}$ dianion also shows tetragonal distortion from idealized symmetry ($T_d \rightarrow D_{2d}$). The measured Fe-Fe, Fe-S and Fe-S(H) bond distances [Fig. 4.16-4(a)] compare well with those in other related clusters like $[Fe_4S_4(SPh)_4]^{2-}$.

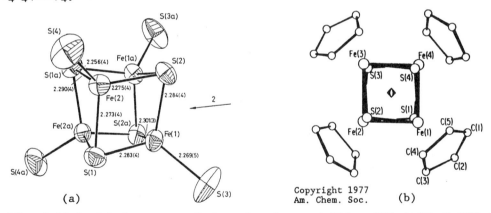

Fig. 4.16-4 (a) Structure of the anion in crystalline $(PPh_4)_2[Fe_4S_4(SH)_4]$. (b) Cubane-like $[Fe_4(\mu_3\text{-}S)_4(\eta^5\text{-}Cp)_4]^{2+}$, which ideally conforms to D_{2d} symmetry, viewed along its crystallographic 4 axis. (After refs. 6 and 7).

Crystalline $[Fe_4(\mu_3\text{-}S)_4(\eta^5\text{-}Cp)_4][PF_6]_2$ contains discrete $[Fe_4S_4Cp_4]^{2+}$ and PF_6^- ions. Fig. 4.16-4(b) shows the configuration of the dication whose center is located on a crystallographic 4 axis. Under D_{2d} symmetry, the six Fe-Fe distances break down into sets of (2.834Å x 4) and (3.254Å x 2), and the twelve Fe-S bond lengths into sets of (2.156Å x 4), (2.204Å x 4) and (2.212Å x 4).

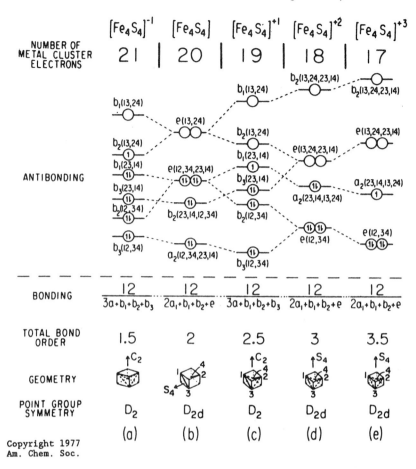

Fig. 4.16-5 Qualitative energy-level diagram for the tetrametal cluster orbitals in the $[Fe_4S_4Cp_4]^n$ system ($n = -1, 0, +1, +2, +3$). (After ref. 7).

A qualitative metal cluster MO description of $[Fe_4S_4Cp_4]^n$ accounts for the observed structural variation in the Fe_4S_4 core of the dication from those of the neutral molecule and monocation and also provides a prediction of the probable structures for $n = -1$ to $+3$ (Fig. 4.16-5).

3. Structures of $[Fe_6S_6]$, $[Fe_7S_6]$ and $[Fe_6S_6Mo_2]$ cores and related compounds

The crystal structure of $(PPN)_3[Fe_6S_6Cl_6]$, PPN = $[(Ph_3P)_2N]^+$, and $Fe_6S_6(PBu^n_3)_4Cl_2$ have been determined.[8] Fig. 4.16-6 shows the structures of the $[Fe_6S_6Cl_6]^{3-}$ anion and the $Fe_6S_6P_4Cl_2$ portion of $Fe_6S_6(PBu_3^n)_4Cl_2$.

The $[Fe_6S_6Cl_6]^{3-}$ anion has imposed centrosymmetry and takes the form of a slightly distorted hexagonal prism. The $[Fe_6(\mu_2\text{-}S)(\mu_3\text{-}S)_4(\mu_4\text{-}S)]^{2+}$ core of $Fe_6S_6(PBu^n_3)_4Cl_2$, which formally contains 4Fe(II) + 2Fe(III), is built through the fusion of six nonplanar Fe_2S_2 rhombs to form an open basket with the bridging group $Fe-(\mu_2-S)-Fe$ (bond angle 75.5°) as its handle.

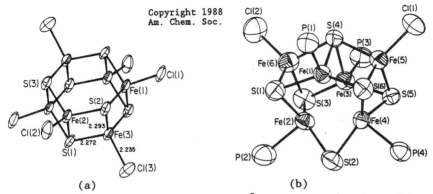

Fig. **4.16-6** (a) Structure of $[Fe_6S_6Cl_6]^{3-}$ in its PPN^+ salt. (b) Structure of the $Fe_6S_6P_4Cl_2$ portion of $Fe_6S_6(PBu_3{}^n)_4Cl_2$. (After ref. 8).

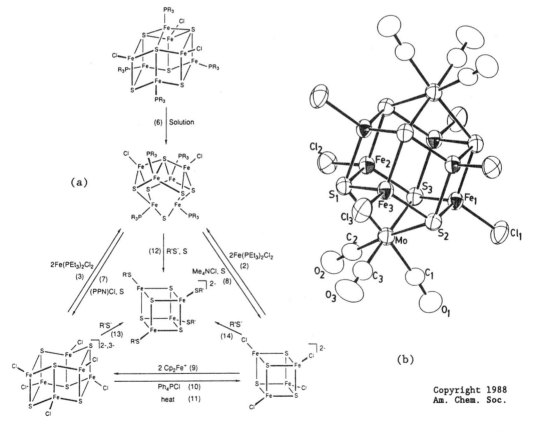

Fig. **4.16-7** (a) Schematic depiction of core conversion and related reactions of Fe_4S_4, Fe_6S_6 and Fe_7S_6 clusters (R = Et, R' = p-tol). (b) Structure and labeling of the $[Fe_6S_6X_6(Mo(CO)_3)_2]^{n-}$ anions in I and II (n = 4, X = Cl), III (n = 3, X = Cl), and IV (n = 3, X = Br). (After refs. 8 and 10).

Based on the structural characteristics of $Fe_7S_6(PEt_3)_4Cl_2$,[9] $[Fe_6S_6Cl_6]^{3-}$, $Fe_6S_6(PBu^n{}_3)_4Cl_2$, and $Fe_4S_4X_4$, the core conversion and related reactions are depicted schematically in Fig. 4.16-7(a).

The crystal structures of $(Et_4N)_4Fe_6S_6Cl_6[Mo(CO)_3]_2 \cdot 2CH_3CN$ (I), $(Ph_4P)_4Fe_6S_6Cl_6[Mo(CO)_3]_2 \cdot 2CH_3CN$ (II), $(Et_4N)_3Fe_6S_6Cl_6[Mo(CO)_3]_2$ (III), and $(Et_4N)_3Fe_6S_6Br_6[Mo(CO)_3]_2$ (IV) have been determined.[10] Complexes I and II contain the $[Mo_2Fe_6S_6]^{2+}$ core, and complexes III and IV contain the $[Mo_2Fe_6S_6]^{3+}$ core. Coupling of the $Mo(CO)_3$ units to the Fe_6S_6 central cage results in an elongation of the latter along the 3 axis in III and idealized 3 axis in I, II, and IV. As a result of this elongation, the Fe-S bonds parallel to 3 axes in I-IV, in the range 2.31-2.33Å, are significantly longer than the corresponding bonds in the parent prismanes $[Fe_6S_6X_6]^{n-}$ ($n = 3$, X = Cl, Br; $n = 2$, X = Cl; 2.27-2.28Å). Fig. 4.16-7(b) shows the structure of the $[Fe_6S_6X_6\{Mo(CO)_3\}_2]^{n-}$ anions.

4. Structure of $[Na_2Fe_{18}S_{30}]^{8-}$ [11]

$[Na_2Fe_{18}S_{30}]^{8-}$, a high-nuclearity cyclic cluster generated solely by iron-sulfur bridge bonding, is depicted in two views in Fig. 4.16-8. The $Fe_{18}S_{30}$ portion is a cyclic cluster of toroidal shape and is the largest Fe-S cluster yet prepared. The cluster has an imposed inversion center but closely approaches C_{2h} symmetry. The eighteen Fe atoms are nearly coplanar, the largest deviation from the mean plane being ±0.154Å by Fe(4); the majority are under 0.11Å. The maximum thickness of the toroid is *ca.* 3.3Å.

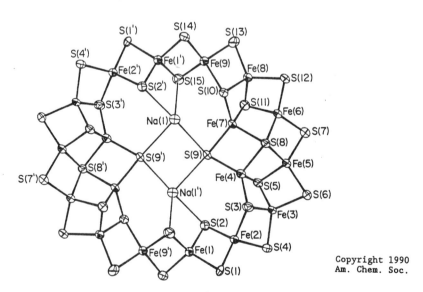

Fig. 4.16-8 Structure of $[Na_2Fe_{18}S_{30}]^{8-}$; primed and unprimed atoms are related by the inversion operation. (After ref. 11).

$[Na_2Fe_{18}S_{30}]^{8-}$ is constructed by the fusion of nonplanar Fe_2S_2 rhombs. Here 24 rhombs are connected by a combination of edge- and corner-sharing such that there are twenty μ_2-S, eight μ_3-S, and two μ_4-S atoms, excluding interactions with Na^+. Every FeS_4 unit is tetrahedral. The structure is recognizable as the conjoint of $2Fe_3S_4 + 2Fe_6S_9$ cores, as found in $[Fe_3S_4(SR)_4]^{3-}$ and $[Fe_6S_9(SR)_2]^{4-}$,[12] respectively, supported by four additional μ_2-S bridges. The mean Fe-Fe distance is 2.727Å. The trend of Fe-S mean bond lengths is Fe-(μ_4-S) 2.35Å > Fe-(μ_3-S) 2.29Å > Fe-(μ_2-S) 2.25Å. Although the cluster is mixed-valence [14Fe(III) + 4Fe(II)], there is no structural evidence for localized valence sites.

Encapsulated within the toroidal cavity, in a manner similar to a crown ether, are two sodium ions, Na(1) and Na(1'). These are monosolvated and make four bonding contacts to S atoms over the range 2.871-2.910Å. Two additional sodium atoms, Na(2) and Na(2'), are situated above and below the cluster at Na-S 3.007-3.027Å and are trisolvated.

5. Some cubane-type clusters of structural interest

    a) Bridged double cubanes

    In the singly-bridged double-cubane core of the anion in $(n\text{-}Bu_4N)_4(Ph_4P)_2[(Fe_4S_4Cl_3)_2S]$, a crystallographic C_2 axis passes through the μ_2-S ligand, and inter-cluster S...S repulsions are relieved by a twist of one cubane subunit relative to the other [Fig. 4.16-9(a)].[13]

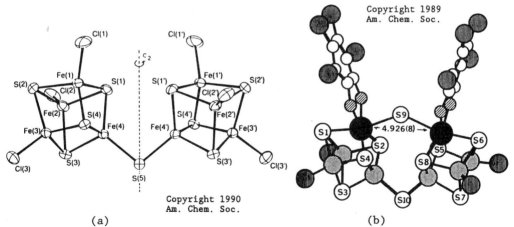

(a)   (b)

Fig. 4.16-9 Strutures of double cubanes (a) $[(Fe_4S_4Cl_3)_2S]^{4-}$ and (b) $[\{MoFe_3S_4Cl_2(Cl_4cat)\}_2(\mu_2\text{-}S)_2]^{6-}$. (After refs. 13 and 15).

Doubly-bridged double cubanes $[\{MoF_3S_4Cl_2(Cl_4cat)\}_2(\mu_2\text{-}S)(\mu_2\text{-}X)]^{n-}$ (X = S, n = 6; X = OH or CN, n = 5; X = N₂H₄, n = 4) have been synthesized and characterized by X-ray crystallography.[14] Fig. 4.16-9(b) shows the hexaanion as a doubly-bridged double cubane with two homometallic M-(μ_2-S)-M bridges.[15] The other anions have analogous structures with the μ_2-X group bridging the two Mo atoms. All anions have short Fe-S bonds in the Fe-S-Fe bridge, and the Mo-X-Mo bond angles and Mo...Mo separations are: X = S, 137.2(7)°, 4.926(8)Å; X = OH, 158(2)°, 4.248(9)Å; X = CN, Mo-C-N 161°, 5.221(8)Å; X = N_2H_4, Mo-N-N 162°, 5.22(1)Å.

In another series of doubly-bridged double cubanes of the type [{MFe$_3$S$_4$(SR)$_2$(R'$_2$cat)}$_2$(μ-SR)$_2$]$^{4-}$ (M = Mo, W), the cubane cores are linked by two heterometallic M-(μ-SR)-Fe bridges in a centrosymmetric arrangement.[16]

Triply-bridged double cubanes of the types [Mo$_2$Fe$_7$S$_8$(μ-SR)$_6$L$_6$]$^{n-}$ and [Mo$_2$Fe$_6$S$_8$(μ-SR)$_3$L$_6$]$^{n-}$ are known for over a decade.[17] The reaction of [Mo$_2$Fe$_7$S$_8$(μ-SR)$_6$(SR)$_6$]$^{4-}$ (R = Ph, m-tolyl) with acetyl chloride results in extrusion of the Fe^{II}(SR)$_3$ bridge unit and ligand substitation at the Fe centers; the structures of the anions are shown in Fig. 4.16-10.[18]

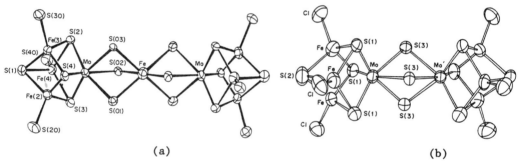

(a)   (b)

Fig. 4.16-10 Structures of [Me$_2$Fe$_7$S$_8$(μ-SR)$_6$(SR)$_6$]$^{4-}$ (R = Ph or m-totyl; molecular symmetry C_i) and [Mo$_2$Fe$_6$S$_8$(μ-SR)$_3$Cl$_6$]$^{3-}$ (R = m-tolyl; molecular symmetry D_{3h}). (After ref. 18).

b) [(n-BuSn(S)(O$_2$PPh$_2$))$_3$O]$_2$Sn

In the crystal structure of [(n-BuSn(S)(O$_2$PPh$_2$))$_3$O]$_2$Sn.3C$_6$H$_6$[19] the tin-sulfur cluster forms a double cube. Fig. 4.16-11(a) shows the chelating phosphinate ligands arranged around the outside of one of the symmetry-related Sn$_4$S$_3$O cubes. The core of the molecule is shown in Fig. 4.16-11(b) as two cubes connected at a common corner occupied by Sn(1). Distortions from cubic geometry are caused primarily by the geometric requirements of the oxygen atom O(3). The four-membered Sn$_2$S$_2$ rings are nearly planar with angles that deviate only slightly from 90°. The angles at O(3) are nearly tetrahedral, and the four-membered rings containing O(3) are not planar.

An interesting analogy exists between the topology of tin-oxygen or tin-sulfur clusters and iron-sulfur clusters. Recent work has demonstrated core conversions among Fe$_4$S$_4$, Fe$_6$S$_6$, and Fe$_7$S$_6$ clusters that resemble the core

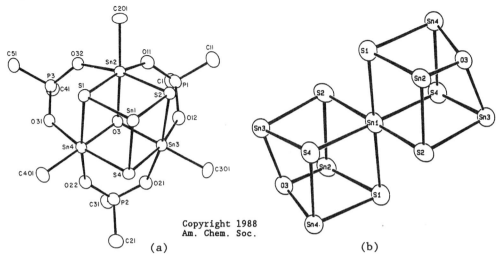

Copyright 1988
Am. Chem. Soc.

(a)   (b)

Fig. 4.16-11 (a) The asymmetric unit of the [(n-BuSn(S)(O₂PPh₂))₃O]₂Sn molecule. Atom Sn(1) occupies a crystallographic inversion center. Pendant carbon atoms of the six phenyl groups and of the three n-butyl groups are omitted for clarity. (b) The tin-sulfur-oxygen core of the molecule. (After ref. 19).

compositions of the Sn_4O_4 cubes, Sn_6O_6 drums, and Sn_7S_6 bicube. There is also a structural analogy among the iron-sulfur and tin-oxygen cubes and drums. Although the Fe_7S_6 and Sn_7S_6 cores refer to compounds having different geometries (the former resembles an iron-capped drum, while the latter is a double cube), insight into the existence of additional interesting structural forms may be gained by these comparisons.

c) $[(H_2O)_9Mo_3S_4SnS_4Mo_3(H_2O)_9]^{8+}$

Crystal structure analysis of $[(H_2O)_9Mo_3S_4SnS_4Mo_3(H_2O)_9](CH_3C_6H_4SO_3)_8 \cdot$ $26H_2O$ revealed the existence of a heterometal double-cubane $Mo_3S_4SnS_4Mo_3$ core, as shown in Fig. 4.16-12.[20] The Mo-Mo distances (2.680, 2.691 and 2.694Å) are among the shortest in complexes with the cubane-type Mo_3MS_4 or incomplete cubane-type Mo_3S_4 core. The Mo-Sn distances (3.670, 3.729 and 3.739Å) are

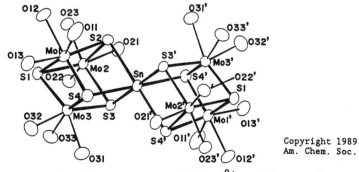

Copyright 1989
Am. Chem. Soc.

Fig. 4.16-12. Structure of $[(H_2O)_9Mo_3S_4SnS_4Mo_3(H_2O)_9]^{8+}$. (After ref. 20).

longer than in other related clusters.

   d) $[Cl_7Cu_5(ReS_4)]^{3-}$

   The trianion in $(PPh_4)_2(NEt_4)[Cl_7Cu_5(ReS_4)]$ possesses a "face-sharing double-cubane" structure, as shown in Fig. 4.16-13.[21] The structural features in this cluster are: the six metal atoms form approximately a basal-plane-centered rectangular pyramid; very short Re-Cu distances of 2.637(2)-2.668(4)Å; the zigzag chain arrangement of alternating Cu and Cl atoms (a CuCl molecule and two Cu_2Cl_3 fragments); and finally, the coordination number four for a sulfur ligand, which is relatively rare in inorganic complexes and allows for face sharing in the double cubane.

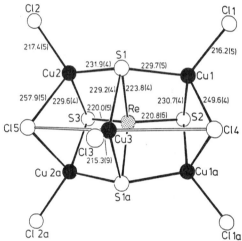

Fig. 4.16-13. Structure of $[Cl_7Cu_5(ReS_4)]^{3-}$. (After ref. 21).

6. Iron-sulfur cluster core formation based on the fusion of Fe_2S_2 rhombs

Table 4.16-2 Stereochemistry of iron-sulfur clusters.

Formula*	Core	Structure
$[Fe_2S_2L_4]^{2-}$ †	$[Fe_2(\mu_2\text{-}S)_2]^{2+}$	planar rhomb, D_{2h} (1)
$[Fe_3S_4(SR)_4]^{3-}$	$[Fe_3(\mu_2\text{-}S)_4]^{+}$	vertex-shared bis-rhomb, D_{2d} (3, 14)
$[Fe_3S_4(SR)_3]^{2-}$ †	$[Fe_3(\mu_3\text{-}S)(\mu_2\text{-}S)_2]^{+}$	Fe-voided cubane, C_{3v} (5)
$[Fe_4S_4L_4]^{1-,2-,3-}$ †	$[Fe_4(\mu_3\text{-}S)_4]^{3+,2+,1+}$	cubane, T_d(6)
$[Fe_4S_3(NO)_7]^{-}$	$[Fe_4(\mu_3\text{-}S)_3]^{-}$	S-voided cubane, C_{3v}
$[Fe_6S_6L_6]^{2-,3-}$	$[Fe_6(\mu_3\text{-}S)_6]^{3+,2+}$	prismane, D_{3h}
$[Fe_6S_6(PR_3)_4L_2]$	$[Fe_6(\mu_4\text{-}S)(\mu_3\text{-}S)_4(\mu_2\text{-}S)]^{2+}$	basket, C_{2v}
$[Fe_6S_8(PR_3)_6]^{2+,1+}$	$[Fe_6(\mu_3\text{-}S)_8]^{2+,1+}$	stellated octahedron,§ O_h
$[Fe_6S_9L_2]^{4-}$	$[Fe_6(\mu_4\text{-}S)(\mu_3\text{-}S)_2(\mu_2\text{-}S)_6]^{2-}$	C_{2v} (13, 15)
$[Fe_7S_6(PR_3)_4Cl_3]$	$[Fe_7(\mu_3\text{-}S)_6]^{3+}$	monocapped prismane, C_{3v}
$[Fe_8S_6I_8]^{3-}$	$[Fe_8(\mu_4\text{-}S)_6]^{5+}$	stellated octahedron,§ O_h
$[Na_2Fe_{18}S_{30}]^{8-}$	$[Fe_{18}(\mu_4\text{-}S)_2(\mu_3\text{-}S)_8(\mu_2\text{-}S)_{20}]^{10-}$	toroid, C_{2h}

* L = RS-, RO-, halide. † Biological cluster. § Alternatively, bicapped prismane.

The conceptual assembly of iron-sulfur clusters from Fe_2S_2 rhombs is illustrated in Fig. 4.16-14. Its application to the core structures of various types of clusters is shown in Table 4.16-2.[11]

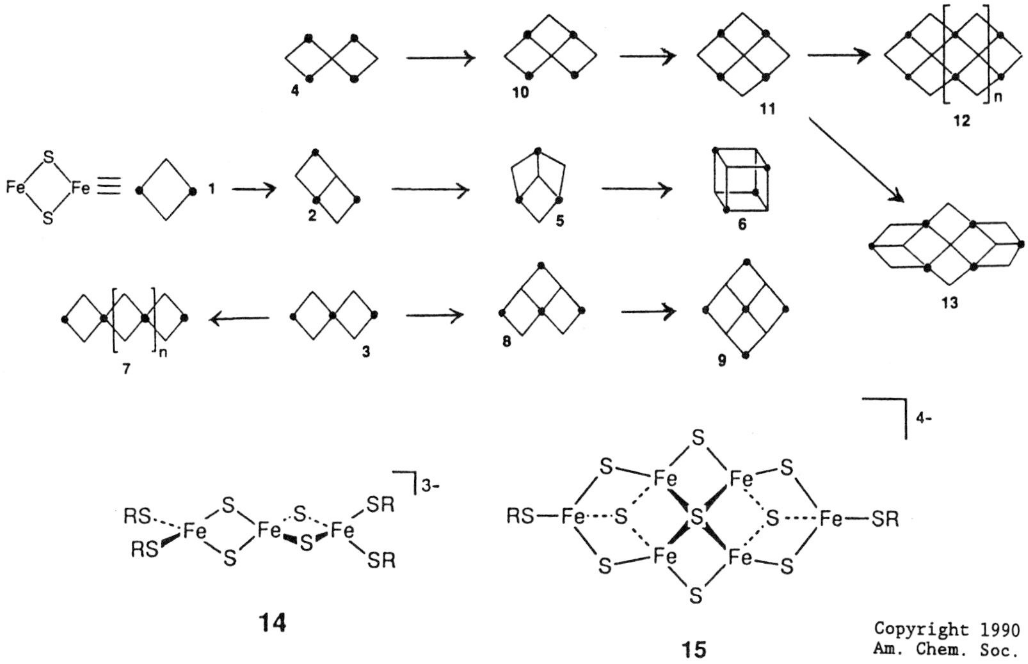

Fig. 4.16-14 Buildup of core units of clusters by connection of Fe_2S_2 rhombs 1 by edge-sharing (2), Fe vertex-sharing (3), and S vertex-sharing (4). Structures 5-13 are derived by additional connectivity of these modes. Two representative molecules are labelled as 14 and 15. (After ref. 11).

7. Molybdoferredoxin and nitrogen fixation

Nitrogen fixation takes place in a wide variety of bacteria, the best known being rhizobium which is found in nodules on the roots of leguminous plants such as peas, beans and soya. It has been known since 1930 that traces of molybdenum are necessary for the growth of these bacteria. It is now recognized that essential constituents of all nitrogen-fixing bacteria are:

    (i)   adenosine triphosphate (ATP), which is a highly active energy-transfer agent;

    (ii)   ferredoxin, $Fe_4S_4(SR)_4$, which is an efficient electron-transfer agent; and

    (iii) a metallo-enzyme.

The best known metallo-enzyme is nitrogenase, which consists of two distinct proteins.[22-24] One of these conains both Fe and Mo and is therefore known as the "FeMo protein" or "molybdoferredoxin". It is brown, air-sensitive, and has a molecular weight of about 240,000; it involves 2 atoms of Mo, 24-36 atoms of Fe, and about the same number of S atoms. The other protein, yellow and extremely air-sensitive, contains Fe but no Mo and is known as "Fe protein" or "azoferredoxin". It has a molecular weight of 65,000 and the structure involves a butterfly-shaped dimer of two identical subunits held together by a Fe_4S_4 ferredoxin core at its head.

Extensive spectroscopic investigations have established that the FeMo protein has an $\alpha^2\beta^2$ subunit structure and contains two types of metal-sulfur cluster in a 2:1 ratio: (i) four "P-clusters" each having a highly distorted Fe_4S_4 core in which three Fe atoms are distinguished from the fourth, and (ii) two "FeMo-cofactors" each comprising 1 Mo, 6-7 Fe, and 8-9 S. It is generally

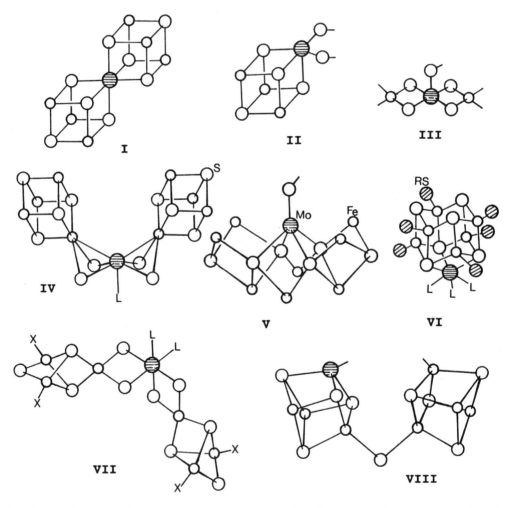

Fig. 4.16-15 Core structures of some proposed models for the FeMo-cofactor in the FeMo protein of nitrogenase. (Adapted from refs. 15, 25 and 27).

accepted that the FeMo-cofactor is a novel Mo-Fe-S cluster with 3-4 Fe atoms linked to Mo via bridging sulfur atoms, and ligation by oxygen or nitrogen donors to the metal atoms is likely.

The "Dominant Hypothesis" designates the FeMo-cofactor as the substrate binding and reducing site in the biological fixation of N_2.[23] A plethora of structural models (Fig. 4.16-15) have been proposed for the nitrogenase active site and extensive efforts directed towards the synthesis of satisfactory analogues.[14,25-27]

Presently it appears that proposed structures **VI**, **VII** and **VIII** best accommodate most of the available data on the FeMo-cofactor in a chemically reasonable fashion. Structure **VI** is based on the M_8S_6 core found in the mineral pentlandite, $[Fe_8S_6I_8]^{3-}$, $[Co_8S_6(SPh)_8]^{4-,5-}$, and the $Mo(CO)_3$-capped $[Fe_6S_6L_6]^{n-}$ prismanes [see Fig. 4.16-7(b)].[27] Model **VIII** is a composite cluster with Fe_4S_4 and $MoFe_3S_4$ units linked by a single $Fe-(\mu_2-S)-Fe$ bridge; it is based on the stepwise synthesis of the doubly-bridged double cubanes by the coupling of $[MoFe_3S_4]$ clusters, and the structure is shown in Fig. 4.16-9(b).[15]

Two research groups have undertaken the difficult task of solving the crystal structure of the FeMo protein,[28,29] and the race between synthetic work using speculative models and protein crystallography in elucidating the structure of the FeMo-cofactor continues to intensify ongoing research in bioinorganic chemistry.

References

[1] L. Que, M.A. Bobrik, J.A. Ibers and R.H. Holm, *J. Am. Chem. Soc.* **96**, 4168 (1974).

[2] M.J. Carney, G.C. Papaefthymiou, M.A. Whitener, K. Spartalian, R.B. Frankel and R.H. Holm, *Inorg. Chem.* **27**, 346 (1988).

[3] P. Barbaro, A. Bencini, I. Bertini, F. Briganti and S. Midollini, *J. Am. Chem. Soc.* **112**, 7238 (1990).

[4] E.T. Adman, L.C. Sieker and L.H. Jensen, *J. Biol. Chem.* **25**, 3801 (1976).

[5] C.D. Garner in B.F.G. Johnson (ed.), *Transition Metal Clusters*, Wiley, Chichester, 1980, chaps. 3 and 4.

[6] A. Müller, N.H. Schladerbeck and H. Bögge, *J. Chem. Soc., Chem. Commun.*, 35 (1987).

[7] Trinh-Toan, B.K. Teo, J.A. Ferguson, T.J. Meyer and L.F. Dahl. *J. Am. Chem. Soc.* **99**, 408 (1977).

[8] B.S. Snyder and R.H. Holm, *Inorg. Chem.* **27**, 2339 (1988).

[9] I. Noda, B.S. Snyder and R.H. Holm, *Inorg. Chem.* **25**, 3851 (1986).

[10] D. Coucouvanis, A. Salifoglou, M.G. Kanatzidis, W.R. Dunham, A. Simopoulous and A. Kostikas, *Inorg. Chem.* **27**, 4066 (1988).

[11] J.-F. You, B.S. Snyder, G.C. Papaefthymiou and R.H. Holm, *J. Am. Chem. Soc.* **112**, 1067 (1990).

[12] K.S. Hagen, A.D. Watson and R.H. Holm, *J. Am. Chem. Soc.* **105**, 3905 (1983).

[13] P.R. Challen, S.-M. Koo, W.R. Dunham and D. Coucouvanis, *J. Am. Chem. Soc.* **112**, 2455 (1990).

[14] D. Coucouvanis, *Acc. Chem. Res.* **24**, 1 (1991).

[15] D. Coucouvanis, P.R. Challen, S.-M. Koo, W.M. Davis, W. Butler and W.R. Dunham, *Inorg. Chem.* **28**, 4181 (1989).

[16] W.H. Armstrong, P.K. Mascharak and R.H. Holm, *J. Am. Chem. Soc.* **104**, 4373 (1982).

[17] R.H. Holm, *Chem. Soc. Rev.* **10**, 455 (1981).

[18] B. Kang, H. Liu, J. Cai, L. Huang, Q. Liu, D. Wu, L. Weng and J. Lu, *Transition Met. Chem.* **14**, 427 (1989).

[19] K.C.K. Swamy, R.O. Day and R.R. Holmes, *J. Am. Chem. Soc.* **110**, 7543 (1988).

[20] H. Akashi and T. Shibahara, *Inorg. Chem.* **28**, 2906 (1989).

[21] A. Müller, E. Krickemeyer and H. Bögge, *Angew. Chem. Int. Ed. Engl.* **25**, 990 (1986).

[22] E.I. Stiefel, H. Thomann, H. Jin, R.E. Bare, T.V. Morgan, S.J.N. Burgmayer and C.L. Coyle in L. Que, Jr. (ed.), *Metal Clusters in Proteins*, American Chemical Society, Washington, DC, 1988, p. 373, and references cited therein.

[23] P.J. Stephens in T.G. Spiro (ed.), *Molybdenum Enzymes*, Wiley, New York, 1985, p. 117.

[24] B.K. Burgess, *Chem. Rev.* **90**, 1377 (1990).

[25] B.A. Averill, *Structure and Bonding* **53**, 61 (1983).

[26] B.A. Averill in ref. 22, p. 259.

[27] D. Coucouvanis in ref. 22, p. 390; S.A. Al-Ahmad, A. Salifoglou, M.G. Kanatzidis, W.R. Dunham and D. Coucouvanis, *Inorg. Chem.* **29**, 927 (1990).

[28] T. Yamane, M.S. Weininger, L.E. Mortenson and M.G. Rossmann, *J. Biol. Chem.* **257**, 1221 (1982).

[29] N.I. Sosfenov, V.I. Andrianov, A.A. Vagin, B.V. Strokopytov, B.K. Vainstein, A.E. Shilov, R.I. Gvozdev, G.I. Likhtenshtein and I.Z. Blazhchuk, *Dokl. Akad. Nauk SSSR* **291**, 1123 (1986).

Note The synthesis and structure of the heterometallic trinuclear incomplete cubane-like clusters $[(CH_3CH_2)_4N][(M_2CuS_4)(edt)_2(PPh_3)]$ (M = Mo, W) are described in N. Zhu, Y. Zheng and X. Wu, *Polyhedron*, 10, 2743 (1991).

The cluster $[(Fe(\mu-'S_4'))_2(\mu-S_2)]$, where the tetradentate chelating ligand $'S_4'^{2-}$ = 2,2'-(ethylenedithio)dibenzenethiolate, is reported in D. Sellmann, G. Mahr and F. Knoch, *Angew. Chem. Int. Ed. Engl.* 30, 1477 (1991).

Recent studies of the crystal structures of the nitrogenase Fe and MoFe proteins are reported in P.M. Gresshoff, L.E. Roth, G. Stacey and W.E. Newton (eds.), *Nitrogen Fixation: Achievements and Objectives*, Proc. 8th Int. Congr., Chapman and Hall, New York, 1990, pp. 111-116, 117-124.

4.17   Dioxygen Adduct of Chlorocarbonylbis(triphenyl-phosphine)iridium (Vaska's Compound)

$IrO_2Cl(CO)[P(C_6H_5)_3]_2$

Crystal Data

Triclinic, space group $P\bar{1}$ (No. 2)

$a = 19.02(3)$, $b = 9.83(2)$, $c = 9.93(2)$Å, $\alpha = 94.0(1)°$, $\beta = 64.9(1)°$,

$\gamma = 93.2(1)°$, $Z = 2$

Atom	x	y	z	Atom	x	y	z
Ir	.2342	.2100	.0068	C(17)	.251	.670	.258
X(1)	.2873	.1153	.1618	C(18)	.230	.534	.245
X(2)	.1467	.0293	.0019	P(2)	.3430	.1266	-.2024
O(1)	.224	.366	-.117	C(19)	.361	.217	-.363
O(2)	.279	.397	-.073	C(20)	.435	.271	-.448
P(1)	.1335	.3207	.2136	C(21)	.449	.334	-.579
C(1)	.039	.317	.203	C(22)	.390	.342	-.625
C(2)	.036	.325	.066	C(23)	.316	.288	-.540
C(3)	-.036	.321	.058	C(24)	.302	.225	-.409
C(4)	-.104	.309	.188	C(25)	.431	.138	-.178
C(5)	-.101	.302	.324	C(26)	.444	.257	-.107
C(6)	-.030	.306	.332	C(27)	.509	.271	-.075
C(7)	.111	.257	.394	C(28)	.561	.166	-.114
C(8)	.098	.118	.410	C(29)	.548	.046	-.185
C(9)	.085	.063	.545	C(30)	.483	.033	-.216
C(10)	.085	.148	.663	C(31)	.333	-.052	-.250
C(11)	.098	.288	.646	C(32)	.352	-.102	-.396
C(12)	.111	.342	.512	C(33)	.349	-.241	-.427
C(13)	.160	.499	.238	C(34)	.326	-.331	-.313
C(14)	.111	.601	.243	C(35)	.307	-.282	-.167
C(15)	.132	.737	.256	C(36)	.310	-.142	-.136
C(16)	.202	.772	.263				

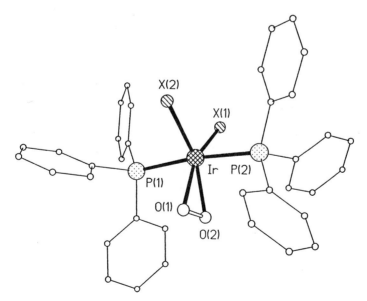

Fig. 4.17-1 Molecular structure of $IrO_2Cl(CO)[P(C_6H_5)_3]_2$.

Crystal Structure

The molecular structure of $IrO_2Cl(CO)[P(C_6H_5)_3]_2$ in the crystal is

essentially that deduced by Vaska[2] from spectroscopic and magnetic data
(Fig. 4.17-1). The carbonyl and chlorine ligands are disordered and
indistinguishable, and were approximated by chlorine atoms labelled as X(1)
and X(2); the phenyl rings were treated as rigid groups in structure
refinement. The Ir, O(1), O(2), X(1) and X(2) atoms are co-planar. The
difference electron density in this plane illustrates rather well the near
equivalence of the X(1) and X(2) peaks and the fact that while the O(1)-O(2)
peaks are readily resolved, there is no indication of CO resolution (Fig.
4.17-2).

Fig. 4.17-2 Electron density at the X(1), X(2), O(1) and O(2) positions
in the best least-squares plane containing these positions and the Ir
atom, which is marked with a cross. The contour interval is 0.26 $e\text{Å}^{-3}$
and contours from 1.31 to 5.24 $e\text{Å}^{-3}$ are shown. (After ref. 1).

The total disorder of the Cl and CO positions may be rationalized in the
following way. The compound results from the attack above or below the
square-planar $IrCl(CO)[P(C_6H_5)_3]_2$ molecule by O_2 in benzene solution. Since
the triphenylphosphine ligands are equivalent in solution, only one isomer is
formed, and if the relative positions of Cl and CO do not affect the
crystallization process, then total disorder would be expected. This is not
unreasonable, for the molecular packing is determined almost entirely by the
triphenylphosphine groups: the volume per triphenylphosphine in this structure
is only 15% greater than in triphenylphosphine itself.

Principal intramolecular distances and angles are listed in Table
4.17-1. The two oxygen atoms are almost equidistant from Ir and the O-O
distance of 1.30(3)Å, while longer than that in molecular oxygen (1.207Å), is
significantly less than that in a typical peroxide (1.49Å). This equivalence
of the oxygen atoms is consistent with the model of π-bonding of molecular

Table 4.17-1 Bond lengths and angles in $IrO_2Cl(CO)[P(C_6H_5)_3]_2$.

Bond lengths (Å)		Bond angles (°)	
Ir-P(1)	2.38(1)	P(1)-Ir-P(2)	172.8(5)
Ir-P(2)	2.36(1)	P(1)-Ir-X(1)	93.0(5)
Ir-X(1)	2.42(2)	P(1)-Ir-X(2)	90.9(5)
Ir-X(2)	2.38(2)	X(1)-Ir-X(2)	100.1(6)
Ir-O(1)	2.09(3)	X(1)-Ir-O(1)	152.4(8)
Ir-O(2)	2.04(3)	X(1)-Ir-O(2)	115.9(10)
O(1)-O(2)	1.30(3)	O(1)-Ir-O(2)	36.7(9)

Remarks

1. Molecular oxygen carriers and Vaska's compounds[4]

The compound *trans*-IrCl(CO)(PPh₃)₂ was discovered in 1961 by Vaska and Di Luzio.[3] This planar 16-electron complex can act as a reversible oxygen carrier by means of the equilibrium:

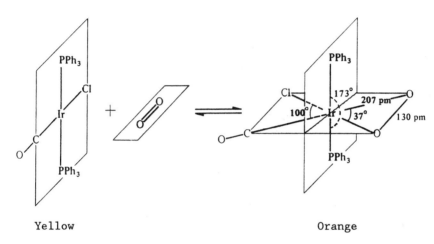

Yellow   Orange

The facile absorption of O₂ by a solution of Vaska's compound is accompanied by a change in colour from yellow to orange which may be reversed by flushing with N₂. This is one of the most widely studied synthetic oxygen-carrying systems. The O-O distance of 1.30Å in the oxygenated product is rather close to the 1.28Å value of the superoxide ion, O_2^-, but this would imply Ir(II) which is paramagnetic whereas the compound is actually diamagnetic. The oxygenation is instead normally treated as an oxidative addition with O₂ acting as a bidentate peroxide ion, O_2^{2-}, to give a hexacoordinate Ir(III) product. However, in view of the small "bite" of this ligand the alternative formulation in which the O₂ acts as a neutral unidentate ligand giving a five-coordinate Ir(I) product has also been proposed.

Oxygen-carrying properties are evidently critically dependent on the precise charge distribution and steric factors within the molecule. Replacement of the Cl ligand in Vaska's compound with I causes loss of oxygen-carrying ability, the oxygenation being irreversible. This can be rationalized by noting that the lower electronegativity of the iodide would enhance the electron density on the metal, thus facilitating M→O_2 π donation: this increases the strength of the M-O_2 bond, and by placing charges in the antibonding orbitals of the O_2 ligand, causes an increase in the O-O distance from 1.30Å to 1.51Å.

2. Ligand types of dioxygen-metal complexes

    In 1976 Vaska[5] published an important paper classifying dioxygen complexes according to their molecular geometry. Vaska's structural types constitute the first four entries in Table 4.17-2: the super compounds (types Ia and Ib) in which the O-O distance is roughly constant (~1.3Å) and close to the value reported for the superoxide anion, and the peroxo compounds (types IIa and IIb) in which the O-O distance is close to the values reported for

Table 4.17-2 Structural classification of dioxygen complexes.[6]

Structure type	Structural designation	Vaska classification	Example
η^1 dioxygen (O-O-M)	η^1 dioxygen	Type Ia (superoxo)	$[Co(CN)_5O_2]^{3-}$
η^2 dioxygen (M bonded to both O)	η^2 dioxygen	Type IIa (peroxo)	$(Ph_3P)_2PtO_2$
M—O, O—M	$\eta^1{:}\eta^1$ dioxygen	Type Ib (superoxo)	$[(H_3N)_5CoO_2Co(NH_3)_5]^{5+}$
M—O, O—M	$\eta^1{:}\eta^1$ dioxygen	Type IIb (peroxo)	$[(H_3N)_5CoO_2Co(NH_3)_5]^{4+}$
M, M bridged by O-O	$\eta^2{:}\eta^2$ dioxygen	-	$[(UO_2Cl_3)_2O_2]^{4-}$
M, O—M	$\eta^1{:}\eta^2$ dioxygen	-	$[(Ph_3P)_2ClRhO_2]_2$

H_2O_2 and O_2^{2-} (~1.48Å). The a or b classification distinguishes complexes in which the dioxygen is bound to one metal atom (type a) or bridges two metal atoms (type b). Gubelmann and co-workers has introduced a "hapto" nomenclature in which the structures are classified by the number of atoms of dioxygen bound to the metal ion, thus avoiding the assignment of a possibly misleading oxidation state to the dioxygen.[6]

    The stretching frequencies attributed to the O-O vibration are closely related to the structural type.[5] Type I complexes show O-O stretching vibrations around 1125 cm^{-1} and type II around 860 cm^{-1}. This sharp difference enables the O-O stretching frequency as measured by infrared or Raman spectroscopy to be used for structure type classification.

3. Examples of different structural types of dioxygen-metal complexes

a) Co(3-*t*-Busalen)(py)(O$_2$), (type Ia)

Fig. 4.17-3 shows the structure of the oxygenated complex. The cobalt-oxygen bond is nearly perpendicular to the Co-salen plane. The measured dimensions are: Co-O-O 116.4(5)°, O-O 1.350(11)Å, Co-O 1.870(6)Å, O(3)-Co-N(3) 175.6(3)°, and the Co-O-O plane approximately bisects an in-plane O-Co-N angle. The cobalt-dioxygen geometry is qualitatively the same as other oxygenated complexes, as shown in Table 4.17-3. The reasons for the variation in the O-O distances in these complexes are still unknown.

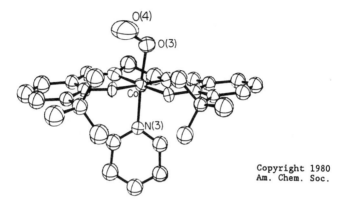

Fig. 4.17-3 Structure of Co(3-*t*-Busalen)(O$_2$)(py). (After ref. 7).

Table 4.17-3 Selected dimensions of oxygenated cobalt complexes

Compound	Co-O/Å	O-O/Å	Co-base/Å	Co-O-O(deg)
Co(3-t-Busalen)(O$_2$)(pyridine)	1.870(6)	1.350(11)	2.018(6)	116.4(5)
Co(3-t-Busaltmen)(O$_2$)(N-benzylimidazole)	1.877(7)	1.273(10)	1.974(8)	117.5(5)
Co(3-F-saltmen)(O$_2$)(N-methylimidazole)	1.881(2)	1.320(3)	2.004(3)	117.4(2)
Co(Saltmen)(O$_2$)(N-benzylimidazole)	1.889(2)	1.277(3)	2.011(2)	120.0(2)
*Co(3-MeO-Saltmen)(O$_2$)(aqua)	1.877(15)	1.25(2)	2.048(5)	117(1)
Average	1.879(7)	1.294(40)	2.011(27)	117.7(1.4)

* Average values.

b) [(NH$_3$)$_5$CoO$_2$Co(NH$_3$)$_5$](SO$_4$)(HSO$_4$)$_3$, (type Ib)[8]

The coordinating ligands about the Co atoms form nearly regular octahedra with Co-N and Co-O distances of 1.95 and 1.89Å. The Co-O-O-Co group is planar, with the bridging peroxide group skewed towards the Co-Co axis; the O-O distance is 1.31Å and the Co-O-O angles are 118°. By sharing electron pairs with six ligands, the Co atoms attain a rare gas configuration. In addition, this description explains satisfactorily the following: (i) the Co-N and Co-O distances, 1.95 (average) and 1.89Å; the sums of the single-bond

radii are 1.95 and 1.91Å; (ii) the O-O distance, 1.31Å, which is in
satisfactory agreement with the distance predicted for a single bond plus a
three-electron bond; (iii) the planarity of the Co-O-O-Co group; this is
required by the use of two orbitals of each oxygen atom in forming the O-O
bond, and in contrast, the O-O bond in H_2O_2 uses but one orbital of each
oxygen atom and the molecule is non-planar, and (iv) the octahedral
coordination about the cobalt atoms; this results from $3d^24s4p^3$ hybridization.
The Co-O-O angles, 118°, are slightly smaller than the predicted value of
125°. A valence bond representation of the cation is shown in Fig. 4.17-4.

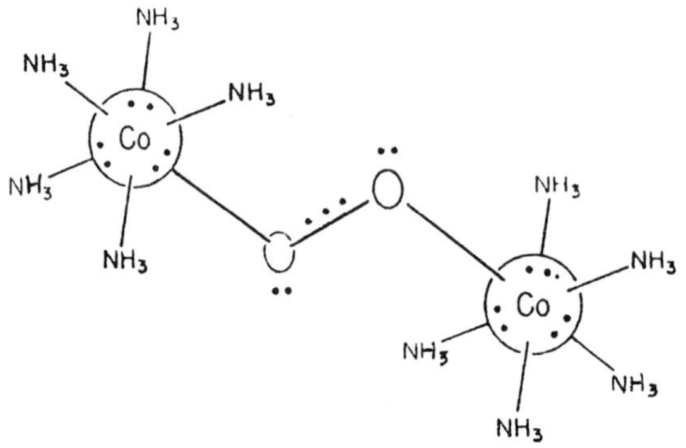

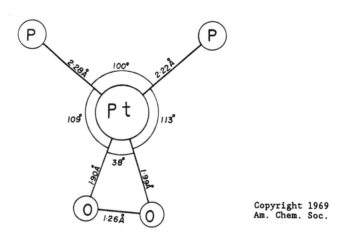

Fig. 4.17-4 A valence-bond representation of $[(NH_3)_5CoO_2Co(NH_3)_5]^{5+}$.
(After ref. 8).

c) $(Ph_3P)_2PtO_2$ (type IIa)[9]

Fig. 4.17-5 shows the approximate configuration of this complex in the
neighbourhood of the Pt atom. All the atoms in this diagram are near-coplanar
as evidenced by the sum of the bond angles at the Pt atom.

Fig. 4.17-5 Structure of the complex $[P(C_6H_5)_3]_2Pt(O_2)$. The phenyl
groups have been omitted. (After ref. 9).

Table 4.17-4 η^2 Dioxygen (Vaska type IIa) complexes.[6]

Complex*	O-O/Å	Complex	O-O/Å
Ti(OEP)(O₂)	1.458,1.445	[Co(2=phos)₂(O₂)]⁺	1.42
[{Ti(NTA)(O₂)}₂O]⁴⁻	1.469,1.481	L₃(NC)₂Co(μ-CN)Co(CN)L₂(O₂), L = PPhMe₂	1.44
[Nb(O₂)F₅]²⁻	1.17	[Rh(PPhMe₂)₄(O₂)]⁺	1.43
Nb(η⁵-C₅H₅)(O₂)Cl	1.47	[Ta(O₂)F₅]²⁻	1.39
[RhCl(O₂)(PPh₃)₂]₂	1.44	RhCl(O₂)(PPh₃)₂	1.413
[{Ta(O₂)F₄}₂O]⁴⁻	1.64	MoO(O₂)(pycc)	1.447
[Ir(2=phos)₂(O₂)]⁺	1.38	[Ir(PPhMe₂)₄(O₂)]⁺	1.49
MoO(O₂)[PhCON(Ph)O]₂	1.21	IrCl(PPh₂Et)₂(CO)(O₂)	1.47
MoO(O₂)₂[(S)-MeCH(OH)CONMe₂]	1.459,1.451	Pd[PPh(t-Bu)₂]₂(O₂)	1.37
[WO(O₂)F₄]²⁻	1.20	Pt[PPh(t-Bu)₂]₂(O₂)	1.43

* OEP = octaethylporphinate, NTA = nitrilotriacetate, pycc = dipicolinate, 2=phos = cis-1,2- bis-
(diphenylphosphino)ethene

Of the four types of Vaska's complexes, the type IIa peroxo complexes are by far the most widespread amongst the transition metals, though many are not reversible oxygen carriers and some are formed by deprotonation of H_2O_2 rather than coordination of molecular O_2. Some of these complexes are listed in Table 4.17-4.

d) $[Co_2(C_{16}H_{23}N_5)_2(O_2)]I_4$ (type IIb)[10]

The structure consists of binuclear cations [LCo-O-O-CoL]⁴⁺ of symmetry C_2 and iodide counter-ions, where L is

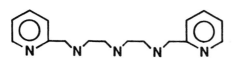

pydien

The binuclear cation consists of Co(III) chelates linked together by a peroxide bridge (Fig. 4.17-6). The peroxo nature of the dioxygen bridge is confirmed by the O-O distance, 1.489(8)Å. This value is very close to the bond distance reported for H_2O_2 (O-O 1.48Å).

e) $[(UO_2Cl_3)_2O_2]^{4-}$ ($\eta^2:\eta^2$ dioxygen)[11]

The crystal structure of $[N(PhCH_2)Me_3]_4[Cl_3O_2U(\mu_2-O_2)UO_2Cl_3]$ reveals that the binuclear anion consists of two identical, distorted pentagonal bipyramids sharing the peroxo-group as a common edge. This arrangement of side-on bridging by a peroxo-group [Fig. 4.17-7] was not discussed by Vaska. Each uranium atom lies in the plane through its five equatorial ligand atoms. The maximum deviation from the least-squares plane through the uranium, the three chlorine atoms, and the peroxo-group is 0.14Å. The O-O distance (1.49Å) in the peroxo-group is within the expected range.

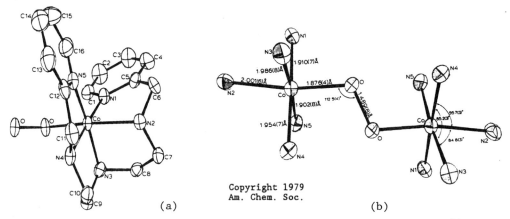

(a)   (b)

Fig. 4.17-6 Structure of the complex cation $[Co(pydien)_2(O_2)]^{4+}$: (a) geometric arrangement of the ligands around the Co atom, and (b) molecular dimensions. (After ref. 10).

Fig. 4.17-7 Structure of the anion $[Cl_3O_2U(\mu_2\text{-}O_2)UO_2Cl_3]^{4-}$. (After ref. 11).

f) $[(Ph_3P)_2ClRhO_2]_2$ (η^1:η^2 dioxygen)[12]

The structure of the centrosymmetric complex is shown in Fig. 4.17-8. The coordination of the rhodium atom can be described in terms of a trigonal bipyramid. The equatorial sites are occupied by a triphenylphosphine P(2), a chlorine atom Cl(1), and a dioxygen molecule composed of O(1) and O(2). The axial sites are occupied by the other triphenylphosphine P(1) and an oxygen atom O(2') of the inversion related subunit. Thus the two dioxygen molecules of the dimer act as unusual bridges and exhibit a geometry that is a combination of the more common sideways π-bonded dioxygen complexes and the *trans* peroxo-bridged species. The bond lengths are O-O = 1.44(1)Å, Rh-O(1) = 1.980(7), and Rh-O(2) = 2.198(7)Å.

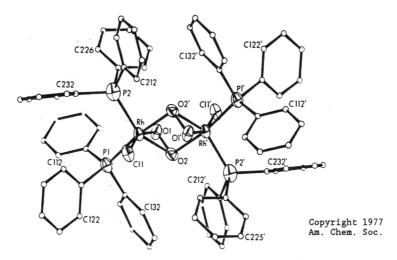

Fig. 4.17-8 Structure of [(Ph₃P)₂ClRhO₂]₂. (After ref. 12).

4. Structure of IrCl(CO)(PPh₃)₂·Ag(B₁₁CH₁₂)[13]

The metathesis of silver salts is a widely used method of halide ion abstraction from labile coordination compounds, and the formation of the perchlorato analogue of Vaska's compound is a typical example:

$$\text{IrCl(CO)(PPh}_3)_2 + \text{AgClO}_4 \xrightarrow{\text{benzene}} \text{Ir(OClO}_3)(\text{CO})(\text{PPh}_3)_2 + \text{AgCl(s)}$$

In contrast, when a toluene solution of IrCl(CO)(PPh₃)₂ is treated with Ag(B₁₁CH₁₂), there is no precipitate of AgCl, but a 1:1 adduct forms. Fig. 4.17-9 shows the nature of the adduct. The iridium to silver metal-metal bond

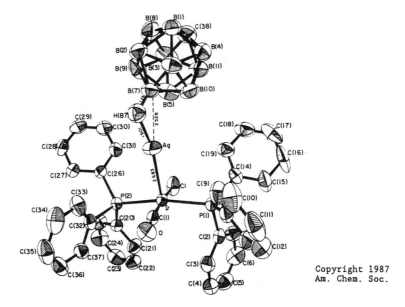

Fig. 4.17-9 Structure of IrCl(CO)(PPh₃)₂·Ag(B₁₁CH₁₂). (After ref. 13).

is one of the clearest illustrations of pure donor-acceptor metal-metal bonding. The preferential bonding of the electron-deficient silver atom to iridium over chloride in adduct formation indicates that the iridium atom is the most basic site in Vaska's compound. The unconsummated metathesis achieved with $Ag(B_{11}CH_{12})$ and $IrCl(CO)(PPh_3)_2$ suggests that the carborane anion has extremely low nucleophilicity.

References

[1] S.J. LaPlaca and J.A. Ibers, *J. Am. Chem. Soc.* **87**, 2581 (1965).

[2] L. Vaska, *Science (Washington)* **140**, 809 (1963).

[3] L. Vaska and J.W. Di Luzio, *J. Am. Chem. Soc.* **83**, 2784 (1961).

[4] N.N. Greenwood and A. Earnshaw, *Chemistry of the Elements*, Pergamon Press, Oxford, 1986.

[5] L. Vaska, *Acc. Chem. Res.* **9**, 175 (1976).

[6] M.H. Gubelmann and A.F. Williams, *Structure and Bonding* **55**, 1 (1983).

[7] W.P. Schaefer, B.T. Huie, M.G. Kurilla and S.E. Ealick, *Inorg. Chem.* **19**, 340 (1980).

[8] W.P. Schaefer and R.E. Marsh, *Acta Crystallogr.* **21**, 735 (1966).

[9] C.D. Cook, P.-T. Cheng and S.C. Nyburg, *J. Am. Chem. Soc.* **91**, 2123 (1969).

[10] J.H. Timmons, R.H. Niswander, A. Clearfield and A.E. Martell, *Inorg. Chem.* **18**, 2977 (1979).

[11] J.C.A. Boeyens and R. Haegele, *J. Chem. Soc., Dalton Trans.*, 648 (1977).

[12] M.J. Bennett and P.B. Donaldson, *Inorg. Chem.* **16**, 1585 (1977).

[13] D.J. Liston, C.A. Reed, C.W. Eigenbrot and W.R. Scheidt, *Inorg. Chem.* **26**, 2739 (1987).

Note Vaska's compound and its 1:2 dichloromethane solvate both exhibit disorder of the Cl^- and CO ligands, which could be resolved in the structure analyses. See M.R. Churchill, J.C. Fettinger, L.A. Buttrey, M.D. Barkan and J.S. Thompson, *J. Organomet. Chem.* **340**, 257 (1988); A.J. Blake, E.A.V. Ebsworth, H.M. Murdoch and L.J. Yellowlees, *Acta Crystallogr., Sect. C* **47**, 657 (1991).

Under a nitrogen atmosphere, $Ir(CO)Cl(PPh_3)_2$ reacts with C_n in benzene to form $(\eta^2-C_n)Ir(CO)Cl(PPh_3)_2 \cdot mC_6H_6$ ($n = 60$, $m = 5$; $n = 70$, $m = 2.5$), in which the Ir atom is attached to the C_{60} moiety through a 6-6 ring fusion. See A.L. Balch, V.J. Catalano and J.W. Lee, *Inorg. Chem.* **30**, 3980 (1991); A.L. Balch, V.J. Catalano, J.W. Lee, M.M. Olmstead and S.R. Parkin, *J. Am. Chem. Soc.* **113**, 8953 (1991).

4.18   (meso-Tetraphenylporphinato)iron(II) [Fe(TPP)]
$C_{44}H_{28}N_4Fe$

Crystal Data

Tetragonal, space group $I\bar{4}2d$ (No. 122)

$a = 14.992(2)$, $c = 13.778(2)$Å, $Z = 4$

Atom	x	y	z	Atom	x	y	z
Fe	0	0	0	C(9)	.1330	.4852	-.1058
N	.1134	.0658	-.0025	C(10)	.1105	.4652	-.0107
C(1)	.1973	.0329	.0166	C(11)	.0850	.3786	.0145
C(2)	.2623	.1035	.0161	H(2)	.330	.094	.029
C(3)	.2175	.1798	-.0072	H(3)	.241	.244	-.016
C(4)	.1249	.1561	-.0168	H(7)	.103	.282	-.203
C(5)	.0569	.2183	-.0300	H(8)	.145	.434	-.248
C(6)	.0822	.3116	-.0556	H(9)	.152	.550	-.124
C(7)	.1047	.3324	-.1516	H(10)	.112	.515	.043
C(8)	.1300	.4189	-.1766	H(11)	.069	.362	.087

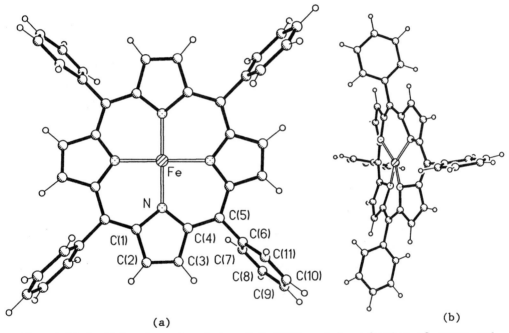

(a)   (b)

Fig. 4.18-1 Molecular structure of Fe(TPP): (a) numbering of atoms and
(b) oblique view showing the deviations from planarity of the porphyrin
skeleton.

Crystal Structure

The electron density distribution in Fe(TPP) has been analyzed using
accurate X-ray diffraction data collected at 120K.[1] The structural results
are in agreement with those of the room temperature study.[2] The principal
bond lengths and angles are listed in Table 4.18-1. The Fe-N bond distance is
within the range of those in low- and intermediate-spin iron porphyrins.[3]

Fe(TPP) has a ruffled structure as shown in Fig. 4.18-1(b). The dihedral angles between the porphyrin and the pyrrole planes, and between the porphyrin and the benzene planes, are 12.8° and 78.9°, respectively.

Table 4.18-1 Bond lengths (Å) and bond angles in Fe(TPP) (deg).

Fe-N	1.966	N-Fe-Na	90.0
N-C(1)	1.376	Fe-N-C(1)	127.4
N-C(4)	1.379	Fe-N-C(4)	127.1
C(1)-C(2)	1.439	C(1)-N-C(4)	105.4
C(2)-C(3)	1.365	N-C(1)-C(2)	110.8
C(3)-C(4)	1.438	N-C(1)-C(5a)	125.3
C(4)-C(5)	1.394	C(2)-C(1)-C(5)	123.9
C(5)-C(6)	1.491	C(1)-C(2)-C(3)	106.5
C(1)-C(5a)	1.395	C(4)-(5)-C(1a)	122.5

The deformation density map through the porphyrin mean plane shows lone-pair peaks at the nitrogen atoms, a larger double bond character for the peripheral $C(2)-C(3)$ bond than the other bonds in the pyrrole ring, as well as preferential ocupancy of the Fe d_{xy} orbital [Fig. 4.18-2(a)]. A significant population of electron density resides in the $d_{x^2-y^2}$ orbital, which is mainly attributed to σ donation from the porphyrin ligand. The d_{z^2} population shown in Fig. 4.18-2(b) suggests that the principal contributor to the ground state of Fe(TPP) is $^3A_{2g}$, as opposed to ^3E_g for FePc (see Setion 4.10).

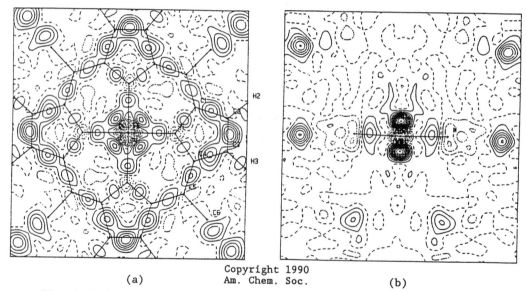

(a)   (b)

Fig. 4.18-2 Deformation density section (a) through the porphyrin plane and (b) perpendicular to the porphyrin plane and containing a linear N-Fe-N fragment; contours are drawn at 0.1 eÅ^{-3} with the zero and negative contours broken. (After ref. 1).

In Table 4.18-2 the d-orbital occupancies derived from the multipole refinement parameters are compared with theoretical values from SCF-CI calculations[4] and the populations for the spherical Fe(II) atom.

Table 4.18-2   Comparison of experimental and theoretical d-orbital occupancies in Fe(TPP)

	$d_{x^2-y^2}$	d_{z^2}	d_{xz}, d_{yz}	d_{xy}	4p	total d
Experimental	0.38	2.31	2.81	1.79		7.29
$^3A_{2g}$ Anharmonic Fe	0.37	2.45	2.40	1.62		6.84
$^3A_{2g}$ (SCF-CI)	0.18	1.94	1.98	2.02	0.13	6.12
^3E_g (SCF-CI)	0.18	0.99	2.96	2.03	0.16	6.16
Spherical	1.2	1.2	2.4	1.2		6.00

Remarks

1. Porphinato-iron stereochemistry[3]

The stereochemistry of porphyrins and related tetrapyrrole macrocycles is of interest because of the common occurrence of this type of macrocycle in biological systems. A number of reviews have appeared.[5]

a) Porphinato-iron(III) derivatives

The ferric ion in porphinato-iron(III) complexes is invariably found to be bound to one or two axial ligands. Under tetragonal symmetry (or nearly so) the five d electrons of the ferric ion can be formally arranged into four possible spin states as follows:

i)   low-spin $S = 1/2$ state, $(d_{xy})^2(d_{xz}, d_{yz})^3(d_{z^2})^0(d_{x^2-y^2})^0$;

ii)   high-spin $S = 5/2$ state, $(d_{xy})^1(d_{xz}, d_{yz})^2(d_{z^2})^1(d_{x^2-y^2})^1$;

iii)   intermediate-spin $S = 3/2$ state, $(d_{xy})^2(d_{xz}, d_{yz})^2(d_{z^2})^1(d_{x^2-y^2})^0$;

and iv)   admixed intermediate-spin $S = 3/2$, $5/2$ state, which is quite distinct from a spin equilibrium and conceptually difficult to visualize because the one-electron model of orbital population breaks down. However, partial occupation of the $d_{x^2-y^2}$ orbital can be envisaged as portrayed in Table 4.18-3.

The first important generalization is that the spin state and stereochemistry of the iron(III) center is controlled almost entirely by the nature and number of axial ligands. The coordination of strong field ligands leads to low-spin six-coordinate hemes, e.g., bis(imidazole)iron(III) derivatives. Weaker-field ligands, typically anionic ones such as Cl^-, N_3^-, etc., lead to five-coordinate high-spin derivatives. The effective axial ligand field of the complexes listed in Table 4.18-3 decreases from left to right. Consequently the energy of the d_{z^2} orbital also decreases, giving sequentially low, high, and intermediate states. The stereochemical consequences of the various orbital occupations can be rationalized with remarkable simplicity by considering whether or not the antibonding d orbitals, $d_{x^2-y^2}$ and d_{z^2}, are occupied. Occupation of $d_{x^2-y^2}$ in a high-spin derivative either gives rise to an expanded porphinato core as in the six-coordinate complexes, or extrudes the iron atom out of the porphinato plane as in the five-coordinate complexes. In both cases, the Fe-N_p distance increases markedly from that of the low-spin complexes (Fe-N_p = 1.99Å). It is important to note that the larger size of the high-spin iron(III) atom does not preclude its insertion into the porphinato plane. When high- and low-spin six-coordinate complexes are compared occupation of d_{z^2} causes extension of the axial iron-ligand bonds by $\geq$ 0.1Å.

Table 4.18-3   Spin-state/stereochemical relationships for porphinatoiron(III) complexes.*

Spin	Low (S=1/2)	High (S=5/2)			Admixed intermed. (S=3/2,5/2)	Intermed. (S=3/2)
Coord. No.	6	5	or	6	5	6
Examples	$[Fe(TPP)(HIm)_2]^+$	Fe(TPP)Cl		$[Fe(TPP)((CH_2)_4SO)_2]^+$	Fe(TPP)(ClO$_4$)	$[Fe(TPP)C(CN)_3]_n$
	$[Fe(TPP)(PMS)_2]^+$	$[Fe(TPP)]_2O$		$[Fe(TPP)(H_2O)_2]^+$	Fe(OEP)(ClO$_4$)	
	$[Fe(TPP)(CN)_2]^-$	Fe(TPP)(NCS)		$[Fe(OEP)(3\text{-}Cl.py)_2]^+$		
Fe-N_p	[1.990]	[2.069]		2.045	1.994-2.001	1.995
Fe-Ct_p	0.0-0.11	[0.51]		0.0	0.26-0.30	0.0
Fe-Ct_N	0.0-0.09	[0.47]		0.0	0.26-0.28	0.0
Ct_N-N_p	[1.99]	[2.015]		2.045	1.981	1.995
Fe-N_{ax}	1.957-2.013	-		2.08-2.316	2.029	2.317

* All lengths are in Å; the values given in square brackets are the average values for all members of the class given in ref. 3. Ct_p stands for the center of the porphyrin ring.

b) Porphinato-iron(II) stereochemistry

The d^6 iron(II) ion can exhibit three spin states as follows:

i)   low-spin S=0 state, $(d_{xy})^2(d_{zx},d_{yz})^4(d_{z^2})^0(d_{x^2-y^2})^0$;

ii)   high-spin S=2 state, $(d_{xy})^2(d_{xz},d_{yz})^2(d_{z^2})^1(d_{x^2-y^2})^1$; and

iii) intermediate-spin S=1 state, $(d_{xy})^2(d_{z^2})^2)(d_{xz},d_{yz})^2(d_{x^2-y^2})^0$.

The principles that govern iron(II) spin states and stereochemistry are quite similar to those for iron(III). As a consequence of decreased charge there is a small increase in radii for iron(II) compared to iron(III). This leads to increased bond lengths in the coordination group, as listed in Table 4.18-4.

The population of the $d_{x^2-y^2}$ orbital in high-spin iron(II) complexes is manifested structurally in long Fe-N_p bonds. When five-coordinate, a high-spin iron(II) atom has significant out-of-plane displacement (Fe...Ct_p ~ 0.5Å). Comparison can be made with low-spin five-coordinate Fe(TPP)(NO) having Fe...Ct_p of only 0.21Å and the numerous low-spin six-coordinate complexes where Fe...Ct_p is always less than 0.11Å. Both low- and intermediate-spin complexes have $d_{x^2-y^2}$ empty; the shorter Fe-N_p distances in the intermediate-spin Fe(TPP) (1.97Å) compared with low-spin FeL_2(TPP) (~ 2.00Å) is readily understood by considering the proportionately greater charge attraction of the ligands to the iron when the coordination number is four. This causes a severe S_4 ruffling of the porphyrin core in Fe(TPP). As far as Fe-axial ligand distances are concerned, the population of the d_z2 orbital is manifested in longer axial bond in high-spin complexes than in low-spin complexes.

Table 4.18-4 Spin-state/stereochemical relationships for porphinato-iron(II) derivatives.*

Spin	L (S=0)		H (S=2)		Intermed. (S=1)
Coord. No.	5 or	6	5 or	6	4
Examples	Fe(TPP)(NO)	Fe(TPP)(Pip)	Fe(TPP)(2-MeIm)	Fe(TPP)(THF)$_2$	Fe(TPP)
	Fe(OEP)(CS)	Fe(TPP)(CO)(py)	[Fe(TPP)(SR)]$^-$		
		Fe(TPP)(1-MeIm)	Fe(PF)(2-MeIm)		
Fe-N$_p$	1.981-2.001	[2.002]	[2.085]	2.057	1.967
Fe...Ct$_p$	0.21-0.23	0.0-0.10	[0.53]	0.0	0.0
Fe...Ct$_N$	0.21-0.23	0.0-0.11	[0.45]	0.0	0.0
Ct$_N$...N$_p$	1.970-1.990	[2.000]	[2.036]	2.057	1.967
Fe-N$_{ax}$	1.717	2.014	2.095-2.161	2.351	-

* All lengths are in Å; the values given in square brackets are the average values for all members of the given class in ref. 3.

2. Trends in metalloporphyrin stereochemistry[6]

The porphinato ligand, which loses the two pyrrolic protons on complexation of the metal ion, is a dianion. With very few exceptions, the porphinato dianion acts as a tetradentate ligand with metal ions. Thus the usual minimum coordination number of the metal ion in a metalloporphyrin is four. The extensive electronic delocalization which occurs in the porphinato ligand leads to a substantial planarity of the macrocycle and an essentially square-planar environment for the metal ion in the four-coordinate complexes. Coordination numbers greater than four result from the addition of other ligands, either neutral or anionic, and metalloporphyrins in which the metal ion has a coordination number of five, six, seven or eight have been characterized. The five-coordinate complexes have a square-pyramidal geometry with the single axial ligand occupying the apex of the square pyramid. The two axial ligands of the six-coordinate metalloporphyrins are found on opposite sides of the porphinato plane, yielding complexes with tetragonal geometries. The seven- and eight-coordinate derivatives have three or four donor groups bonded to the metal ion on the same side of the porphinato plane.

Table 4.18-5 lists the structural parameters for four- and six-coordinate metalloporphyrins. The four-coordinate d^4 Cr porphyrin, known to be high-spin with $3d_{x^2-y^2}$ empty, has substantially shorter $M-N_p$ bonds than the Mn(II) derivative. The six-coordinate metalloporphyrins of Co(II) and Co(III) are all low-spin complexes. Thus for the Co(III) complexes both $d_{x^2-y^2}$ and d_{z^2} are unoccupied, while d_{z^2} is singly occupied in the d^7 Co(II) derivatives, reasulting in significant elongation of the axial bonds.

Table 4.18-5   Structural parameters for four- and six-coordinate
metalloporphyrins.

Four-coordinate			Six-coordinate		
Metal ion	$M-N_p$/Å	Spin-state, S	Metal ion	$M-N_p$/Å	Spin-state, S
d^4Cr	2.033	2	d^4Cr	2.03	1
d^5Mn	>2.082	5/2	d^4Mn	2.01-2.03	2
d^6Fe	1.972	1	d^5Fe	1.99	1/2
d^7Co	1.949	1/2	d^6Fe	2.00	0
d^8Ni	1.928	0	d^6Co	1.95-1.98	0
d^9Cu	1.981	1/2	d^7Co	1.99	1/2
d^{10}Zn	2.036	0	d^8Ni	2.04	1

The structural parameters of importance for a series of square-pyramidal metalloporphyrins are listed in Table 4.18-6. The displacement of the metal atom out of the porphinato plane, $M...Ct_p$, allows relatively large $M-N_p$ bond distances without requiring large radial expansion of the porphinato core. The magnitude of the metal atom displacements is correlated with the "size" of the metal atom with respect to M-N bond formation.

Table 4.18-6   Structural parameters of five-coordinate metalloporphyrins.[6,7]

Metalloporphyrin	Distances/Å				Spin-state
	$M-N_p$	$Ct...N_p$	$M...Ct_p$	M-L(ax)	(S)
Mg(TPP)(OH$_2$)	2.072	2.054	0.27	2.099	0
Zn(TPP)(py)	2.073	2.047	0.33	2.143	0
Ti(OEP)Me$_2$(O)	2.110	2.031	0.58	1.619	0
Mn(TPP)(Cl)	2.008	1.990	0.27	2.373	2
Mn(TPP)(N$_3$)	2.005	1.992	0.23	2.045	2
Fe(Porp)(X)	2.067	2.015	0.45	-	5/2
Mn(TPP)(1-MeIm)	2.128	2.065	0.56	2.192	5/2
Fe(TPP)(2-MeIm)	2.086	2.044	0.42	2.161	2
Co(TPP)(-CH$_2$COCH$_3$)	1.948	1.945	0.11	2.028	0
Co(TPP)(1-MeIm)	1.977	1.973	0.13	2.157	1/2
Co(TPP)(1,2-DiMeIm)	1.985	1.979	0.15	2.216	1/2
Tl(OEP)(Cl)	2.212	2.10	0.69	2.452	0
[Fe(OEP)(3-Clpy)]$^+$	1.979	1.967	0.22	2.126	1
[Fe(OEP)(NO)]$^+$	1.994	1.973	0.29	1.644	0

3. Out-of-plane metalloporphyrin complexes[8]

Porphyrins, like other π-macrocycles, have a central "core" that can be altered by puckering. This consequently limits the core size to a narrow range of variation, which has been observed to be 2.098-1.929Å. In certain complexes the metal ion is unable to fit into this hole and therefore lies out of the pyrrole N atom plane. An example of this phenomenon is the ferrous-ferric ion system. If the difference between their empirical ionic radii is taken into account, the prediction that the ferric ion lies in plane and the ferrous iron out-of-plane appears valid. In hemoglobin, the iron atom in the oxygenated form lies roughly in-plane, while the metal lies out-of-plane in the deoxy form.

$Zr^{IV}(OEP)(OAc)_2$ and its isostructural Hf analogue each contains two bidentate acetate ligands; the metal atom exhibits square-antiprismatic coordination and lies out of the porphyrin plane [Fig. 4.18-3(a)].[9] In (H-TPP)Re(CO)$_3$ the Re atom lies off the S_2 axis normal to the porphyrin plane [Fig. 4.18-3(b)].[10]

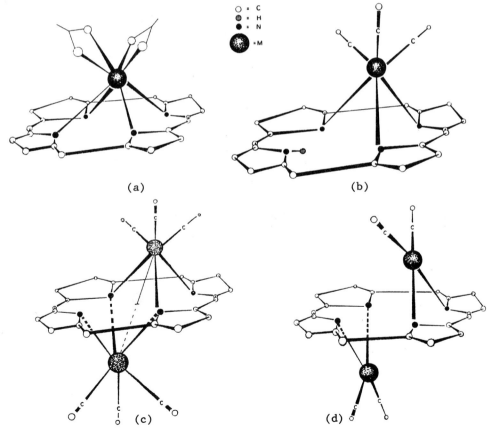

Fig. 4.18-3 Structures of (a) $Zr^{IV}(OEP)(OAc)_2$, (b) (H-TPP)Re(CO)$_3$, (c) [Tc(CO)$_3$](Por)[Re(CO)$_3$] and (d) Rh(CO)$_2$(OEP). (After ref. 8).

In [Tc(CO)$_3$](Por)[Re(CO)$_3$] the disordered metal atoms lie above and below the porphyrin plane but off the S_2 axis normal to it [Fig. 4.18-3(c)]. The distance between the metal atoms is ~3.1Å, too long for a formal bond but short enough to allow some metal interaction.[11] In Rh(CO)$_2$(OEP) the metal atoms likewise lie above and below the plane of the macrocycle and off the S_2 axis normal to it [4.18-3(d)].[12]

4. Nonplanar porphyrins

In Fe(TPP)(CB$_{11}$H$_{12}$).C$_7$H$_8$ the [B$_{11}$CH$_{12}$]$^-$ anion of rather diffuse charge coordinates to the Fe(III) atom of [Fe(TPP)]$^+$ through an open Fe-H-B bond (Fig. 4.18-4).[13] The measured dimensions are Fe-H = 1.82(4)Å, B-H = 1.25(4)Å, and Fe-H-B = 151(3)°. The long Fe-H distance, as compared with those in the range 1.56-1.64Å found in ferraboranes with Fe-H-B interactions, is in keeping with the high oxidation state of the central iron atom in the present complex.

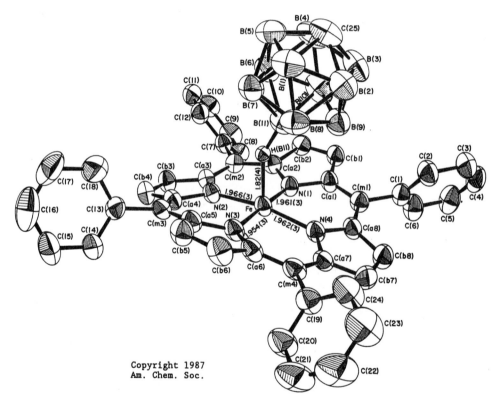

Fig. 4.18-4 Molecular structure of the [Fe(TPP)(CB$_{11}$H$_{12}$)]. (After ref. 13).

The species exhibits a near-D_{2d} ruffling of the core with substantial displacements of individual atoms from the mean plane of the 24 atoms (Fig.

4.18-4). The short Fe-N_p bond distances (average 1.961Å), are consistent with values expected for admixed intermediate-spin state with predominant S = 3/2 character and not for a high-spin state. The observed saddle-shaped core conformation appears to be best suited to achieve the small dihedral angles that allow closer approach of the two cores in molecular packing (Fig. 4.18-5).

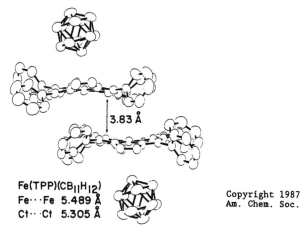

Fe(TPP)(CB$_{11}$H$_{12}$)
Fe···Fe 5.489 Å
Ct···Ct 5.305 Å

Fig. 4.18-5 Edge-on view of a dimer of the Fe(TPP)(CB$_{11}$H$_{12}$).C$_7$H$_8$ system. The Fe-H-B bond and C$_7$H$_8$ molecule are not shown. (After ref. 13).

Crystal structure analysis of the peripherally crowded porphyrins Zn(OMTPP).py.2CHCl$_3$ and Zn(OETPP).MeOH.2MeOH have shown that both macrocycles are markedly nonplanar and saddle-shaped.[14] In the Zn(OETPP).MeOH molecule each pyrrole ring with its β-alkyl substituents is displaced alternately above and below the mean plane defined by the four nitrogen atoms (C$_\beta$ atoms of

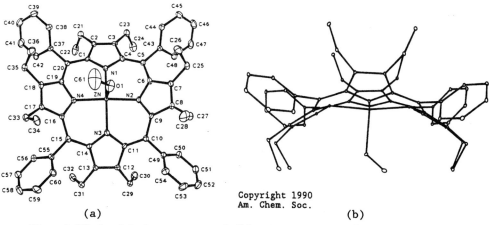

(a)   (b)

Fig. 4.18-6 (a) Structure and (b) edge-on view of the Zn(OETPP).MeOH molecule. (After ref. 14).

adjacent pyrrole by more than 1Å), and the phenyl groups are tilted to minimize steric repulsion (Fig. 4.18-6). The octamethyl analogue Zn(OMTPP).py has a very similar structure with pyridine replacing methanol as the axial ligand.

5. Structure of [Ru(OEP)]$_2$ and [Rh(OEP)-In(OEP)]

Fig. 4.18-7 shows the molecular structures of [Ru(OEP)]$_2$.2C$_5$H$_{12}$[15] and [Rh(OEP)-In(OEP)][16] [OEP = 2,3,7,8,12,13,17,18-octaethyl-porphyrinato].

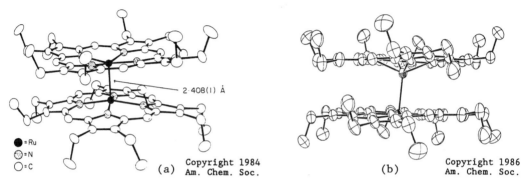

2·408(1) Å

● = Ru
◉ = N
○ = C

(a)   Copyright 1984
        Am. Chem. Soc.

(b)   Copyright 1986
        Am. Chem. Soc.

Fig. 4.18-7 Molecular structure of (a) [Ru(OEP)]$_2$ and (b) [Rh(OEP)-In(OEP)]. (After refs. 15 and 16).

In [Ru(OEP)]$_2$, two cofacial ruthenium octaethylporphyrin groups are bound together by a metal-metal bond of 2.408(1)Å, which is formally a Ru=Ru bond. The two halves of the dimer differ from each other in regard to ethyl group orientations; for one porphyrin macrocycle two of the peripheral ethyl groups are oriented towards the central plane of the molecule, while for the other porphyrin only one ethyl group is oriented in this manner. Each porphyrin skeleton has a dome-shaped conformation, with displacement of the ruthenium atom from the plane of the four coordinating N atoms (0.30Å for both Ru atoms) in the direction of the other Ru atom. The two porphyrin skeletons are essentially parallel (dihedral angle between the two 24-atom cores is 0.2°) with an interplanar separation of 3.26Å, which indicates that significant orbital overlap between the porphyrin aromatic systems is possible.

The molecular structure of [Rh(OEP)-In(OEP)] consists of two subunits joined facially by a Rh-In bond of length 2.584(1)Å, which is shorter than the value estimated for a Rh-In single bond (2.62Å). The two porphyrin skeletons are virtually parallel with a dihedral angle of 1.4° between the two 24-atom cores. The interplanar separation of 3.41Å is relatively long compared to that of [Ru(OEP)]$_2$ (3.26Å), indicating little interaction between the two porphyrin ligand units. The In(OEP) fragment is twisted 21.8° relative to the Rh(OEP) unit. Each porphyrin group has seven ethyl groups oriented away from the center of the molecule, with no evidence for side-chain disorder. The two metal atoms are displaced from the plane of the four coordinated pyrrole N atoms towards each other (0.01Å for Rh, 0.83Å for In).

6. Structure of hydroporphinoid Ni(II) complexes

Fig. 4.18-8 shows the structure of Ni(II) *ccccc*-octaethyl-pyrrocorphinate, which is representative of all available crystal structures of tetra- and hexahydroporphinoid nickel(II) complexes.[17] The ligand system of the tetracoordinated, diamagnetic Ni(II) complex is not planar but instead exhibits a saddle-shaped deformation ("ruffling") with approximate S_4 symmetry. Each of the four *meso* carbon atoms lies, on the average, 0.74Å alternatingly above and below an equatorial plane roughly defined by the four nitrogen centers and the metal coordination center. The three hydropyrrolic five-membered rings are in half-chair conformations, resulting in quasi-equatorial and quasi-axial positions for the peripheral substituents of these rings, as revealed by the cylindrical projection of the structure. The three half-chairs are conformationally coupled with each other in the sense that their puckering alternates in such a way as to mimic the wave form of the saddle-shaped ligand periphery.

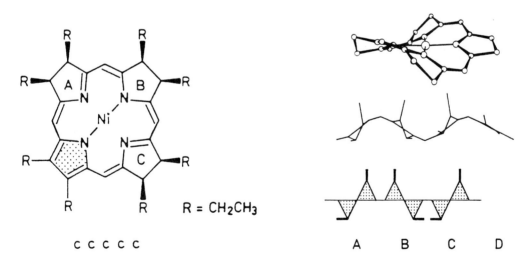

R = CH₂CH₃

c c c c c

A   B   C   D

Fig. 4.18-8 X-ray structure of Ni(II) *ccccc*-octaethyl-pyrrocorphinate. (After ref. 18).

The ligand ruffling process is schematically depicted by the geometric model in Fig. 4.18-9. Coplanar, concentric circles of constant circumference undergo deformation, forming saddle-shaped surfaces of increasing steepness and, at the same time, resulting in the increasing convergence of four equidistant points (representing the four nitrogen centers) towards the center of the original circles, the points remaining in a common plane.

Fig. 4.18-9 Geometric model of the ruffling of the ligand periphery of hydroporphinoid Ni(II) complexes by Ni(II)-induced contraction of the inner coordination hole. (After ref. 18).

7. Oxo-bridged binuclear iron(III) complexes

Iron complexes of the type LFe-O-FeL (H_2L = porphyrins, Schiff bases and similar macrocyclic and chelating ligands) are of interest as model compounds for some biological systems (such as haemerythrins), and also because of their antiferromagnetic properties resulting from a strong spin-spin exchange operating through the Fe-O-Fe bridge. The structural data for some oxo-bridged iron(III) complexes are given in Table 4.18-7.[19]

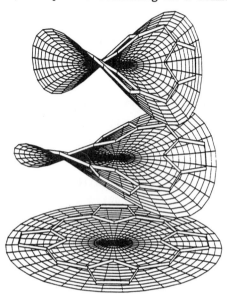

1   2   3

The tetradentate macrocyclic ligand TAAB (1) undergoes a two-electron reduction to give the anion 2 in forming complexes with Cu^{2+}, Ni^{2+}, Co^{2+}, Zn^{2+}, Fe^{2+}, Co^{3+}, and Fe^{3+} ions.[20] Oxidation of the Fe(II)-TAAB complex yields a stable iron(III) oxo-bridged dimeric [Fe$_2$(TAAB)$_2$O]$^{4+}$ cation, which readily reacts with the methoxide ion to form the neutral binuclear complex [Fe(TAAB)(OCH$_3$)$_2$]$_2$O (3).[19] The Fe-O-Fe linkage is very strong and can only be cleaved with hydrofluoric acid.

Table 4.18-7   Structural data for some iron(III) complexes containing a
Fe-O-Fe bridging system.

Complex	Fe-Ob (Å)	Fe-L (Å)	CN of Fe	d(Å)	Fe-O-Fe angle(°)	Fe...Fe separation (Å)
[Fe(TAAB)(OCH$_3$)$_2$]$_2$O	1.777	2.067(N)	5	0.50	176.0	3.55
[Fe$_3$(TPP)]$_2$O	1.763	2.087(N)	5	0.50	174.5	3.53
[Fe(salen)]$_2$O.2py	1.80	2.087(N) 1.918(O)	5	0.56	139.1	3.36
[Fe(salen)]$_2$O	1.78	2.12(N) 1.92(O)	5	0.58	144.6	3.39
[Fe(2-Mequin)$_2$]$_2$O.CHCl$_3$	1.780	2.190(N) 1.918(O)	5	0.64	151.6	3.45
[Fe(salen)]$_2$O.CH$_2$Cl$_2$	1.794	2.105(N) 1.923(O)	5	0.56	142.2	3.40
[Fe(C$_{22}$H$_{22}$N$_4$)]$_2$O.CH$_3$CN	1.792	2.054(N)	5	0.698	142.8	3.40
[Fe(OOC)$_2$pyCl(H$_2$O)$_2$]$_2$O.4H$_2$O	1.773	2.094(O) 2.056(OH$_2$) 2.105(N)	6	0.17(Fe 1) 0 (Fe 2)	180	3.55

(OOC)$_2$pyCl = 4-chloro-2,6-pyridinedicarboxylate; TAAB = tetrabenzo[b,f,j,n] [1,5,9,13]tetraaza-
cyclohexadecine; TPP = tetraphenylporphyrin; sal = salicyladehydate; 2-Mequin = 8-hydroxy-2-
methylquinolinate; salen = N,N'-ethylenebis(salicylideneiminate); C$_{22}$H$_{22}$N$_4$ = [7,16-dihydro-6,8,15,17]-
tetramethyldibenzo[b,i][1,4,8,11]tetraazacyclotetradecinate. d = displacement of the iron atom out of the
plane defined by the equatorial donor atoms.

The molecular structure of 3 as determined from X-ray diffraction is illustrated in Fig. 4.18-10. Both TAAB ligands are saddle-shaped with methoxy groups *cis* to each other. Each iron atom is displaced from the plane of the four coordinating nitrogen atoms by about 0.50Å towards the bridging oxygen atom, a feature common to all complexes shown in Table 4.18-7.

The factors affecting the Fe-O-Fe bridge angle, which ranges from 139.1° to 180° (Table 4.18-7), are still not well understood. The Fe-O (bridging) bond length is invariably shorter than that usually taken as the Fe-O single bond length (~ 1.92Å), implying a significant degree of π-bonding character. A larger Fe-O-Fe angle may thus be associated with increased multiple bonding involving the bridging oxygen atom. On the other hand, since the Fe-O bond

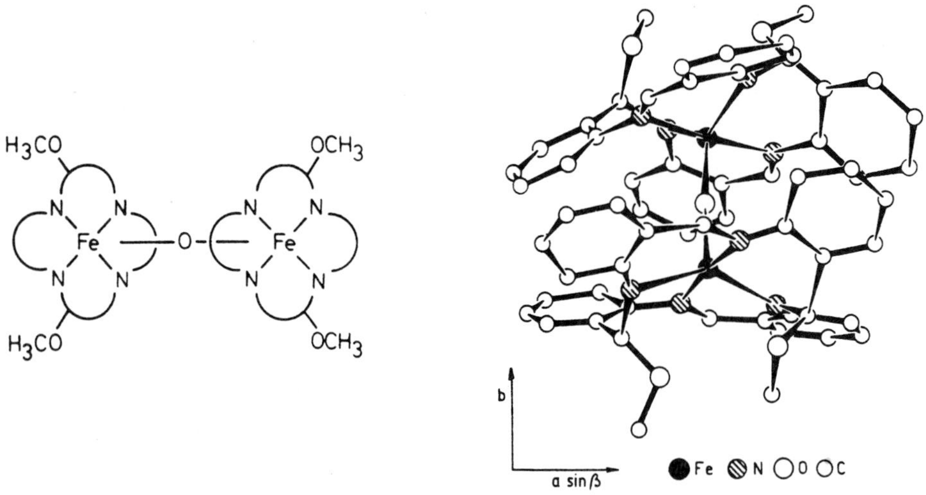

Fig. 4.18-10 Structural formula and molecular structure of [Fe(TAAB)(OCH₃)₂]₂O (3). (After ref. 19).

lengths are more or less constant for all known LFe-O-FeL structures, it seems likely that steric bulk of the L ligands as well as crystal packing play a dominant role in determining the bridge angle.

8. Sapphyrin and its metal complexes

Sapphyrin 1 [Fig. 4.18-11(a)] is a 22-π-electron pentapyrrolic "expanded porphyrin" in the form of a dark blue solid (hence its name), which was first reported by Woodward in 1966.[21] The crystal structure of a mixed hexafluorophosphate fluoride salt of the decaalkyl sapphyrin 2 (H₃sap), [H₅sap²⁺·F⁻][PF₆], has been determined.[22] In the cation [Fig. 4.18-11(b)], the five N atoms of the diprotonated macrocycle and the central fluoride ion are coplanar to within 0.03Å, and no H atom deviates by more than 0.14Å from this plane. The hydrogen-bonded N...F distances range from 2.697(3) to 2.788(3)Å, which are appropriate for a central core size of ~5.5Å in diameter.

Although the parent sapphyrin 1 and its substituted derivatives 2-4 [Fig. 4.18-11(a)] are potentially pentadentate ligands, they give rise to only out-of-plane metal complexes. In both [Ir(CO)₂(H₃sap)]⁺ and [{Rh(CO)₂}₂(Hsap)] each metal center bridges an imino and an amino nitrogen atom, and the approximately square coordination plane makes an angle of about 47° with the mean plane of the considerably ruffled N₅C₂₄ sapphyrin skeleton (Fig. 4.18-12).[23] The pyrrole rings are distorted in such a way that a pair of adjacent ligand N atoms are bent towards their bonded metal center; the dirhodium complex nearly attains C_2 symmetry. The Rh-N and Ir-N bond distances are in the ranges 2.061(4)-2.081(4) and 2.064(5)-2.091(6)Å, respectively.

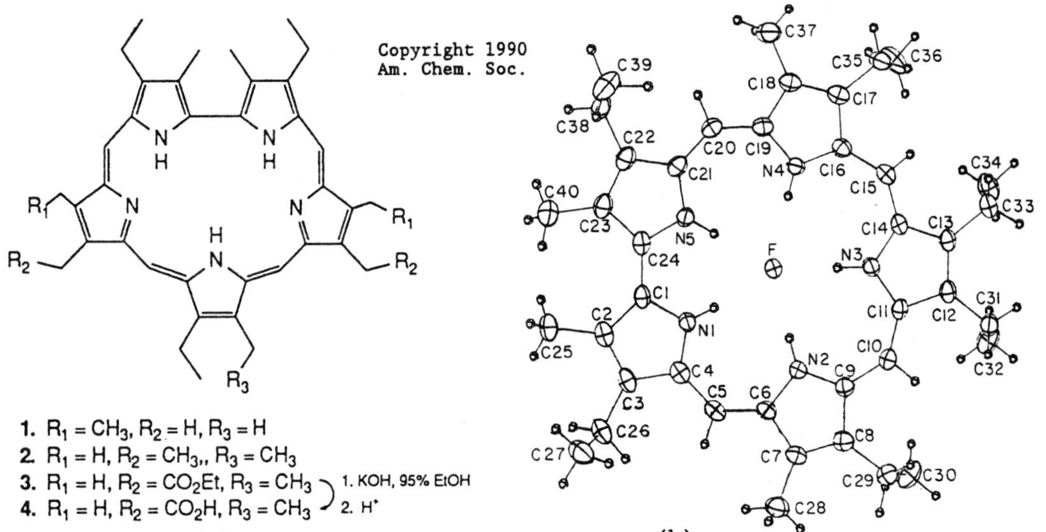

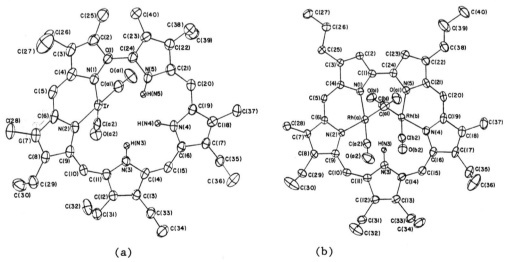

Copyright 1990
Am. Chem. Soc.

1. $R_1 = CH_3$, $R_2 = H$, $R_3 = H$
2. $R_1 = H$, $R_2 = CH_3$, $R_3 = CH_3$
3. $R_1 = H$, $R_2 = CO_2Et$, $R_3 = CH_3$ ⎤ 1. KOH, 95% EtOH
4. $R_1 = H$, $R_2 = CO_2H$, $R_3 = CH_3$ ⎦ 2. H⁺

(a)   (b)

Fig. 4.18-11 (a) Structural formulae of sapphyrin and some of its derivatives. (b) Structure of the $[(H_5sap)F]^+$ ion. (After ref. 22).

(a)   (b)

Fig. 4.18-12 Structures of (a) the cation of $[Ir(CO)_2(H_3sap)]Cl$ and (b) $[\{Rh(CO)_2\}_2(Hsap)]$. (After ref. 23).

9. Molecular pentad that mimics photosynthesis

A "molecular pentad" (Fig. 4.18-13) has been designed and synthesized to reproduce the essential features of photosynthetic solar energy coversion.[24] The molecule consists of two covalently-linked porphyrin moieties, one metallated with zinc (P_{Zn}) and the other present as the free base (P), to which are attached a carotenoid polyene (C) and a diquinone (Q_A-Q_B), respectively. Excitation of P in a chloroform solution of the pentad yields an initial charge-separated state, $C\text{-}P_{Zn}\text{-}P^{\cdot+}\text{-}Q_A^{\cdot-}\text{-}Q_B$, with a quantum yield of 0.85. Subsequent electron transfer steps lead to an essentially quantitative yield of the final charge-separated state, $C^{\cdot+}\text{-}P_{Zn}\text{-}P\text{-}Q_A\text{-}Q_B^{\cdot-}$, which has a lifetime of 55 microseconds. The synthetic five-part molecular device preserves a significant fraction (~1.0 eV) of the initial excitation energy (1.9 eV) in the long-lived, charge-separated state.

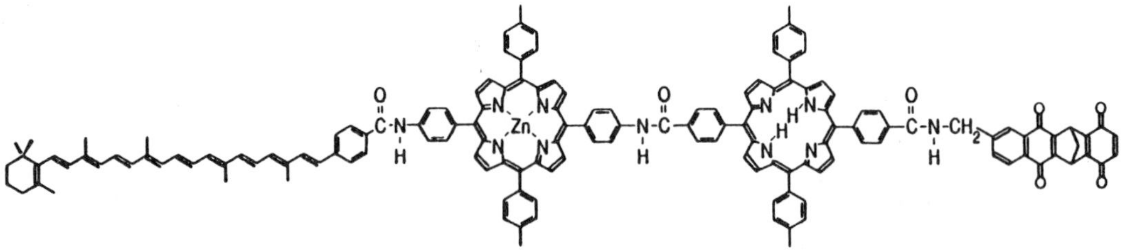

Fig. 4.18-13 Molecular structure of the pentad $C\text{-}P_{Zn}\text{-}P\text{-}Q_A\text{-}Q_B$.

References

[1]   N. Li, Z. Su, P. Coppens and J. Landrum, *J. Am. Chem. Soc.* **112**, 7294 (1990).

[2]   J.P. Collman, J.L. Hoard, N. Kim, G. Lang and C.A. Reed, *J. Am. Chem. Soc.* **97**, 2676 (1975).

[3]   W.R. Scheidt and C.A. Reed, *Chem. Rev.* **81**, 543 (1981).

[4]   M.-M. Rohmer, *Chem. Phys. Lett.* **116**, 44 (1985).

[5]   W.R. Scheidt and Y.J. Lee, *Structure and Bonding*, **64**, 1 (1987), and references cited therein.

[6]   W.R. Scheidt *Acc. Chem. Res.* **10**, 339 (1977).

[7]   W.R. Scheidt, D.K. Geiger, Y.J. Lee, C.A. Reed and G. Lang, *Inorg. Chem.* **26**, 1039 (1987); W.R. Scheidt, Y.J. Lee and K. Hatano, *J. Am. Chem. Soc.* **106**, 3191 (1984).

[8]   G.A. Taylor and M. Tsutsui, *J. Chem. Educ.* **52**, 715 (1975).

[9]   W.R. Scheidt in D. Dolphin (ed.), *The Porphyrins*, Vol. 3, Academic Press, New York, 1978, chap. 10.

[10] D. Ostfeld, M. Tsutsui, C.P. Hrung and D.C. Conway, *J. Coord. Chem.* **2**, 101 (1972).

[11] D. Cullen, E. Meyer, T.S. Srivastava and M. Tsutsui, *J. Am. Chem. Soc.* **94**, 7603 (1972).

[12] A. Takenaka, Y. Sasada, T. Omura, H. Ogoshi and Z. Yoshida, *J. Chem. Soc., Chem. Commun.* 792 (1973).

[13] G.P. Gupta, G. Lang, Y.J. Lee, W.R. Scheidt, K. Shelly and C.A. Reed, *Inorg. Chem.* **26**, 3022 (1987).

[14] K.M. Barkigia, M.D. Berber, J. Fajer, C.J. Medforth, M.W. Renner and K.M. Smith, *J. Am. Chem. Soc.* **112**, 8851 (1990).

[15] J.P. Collman, C.E. Barnes, P.N. Swepston and J.A. Ibers, *J. Am. Chem. Soc.* **106**, 3500 (1984).

[16] N.L. Jones, P.J. Carroll and B.B. Wayland, *Organometallics* **5**, 33 (1986).

[17] C. Kratky, R. Waditschatka, C. Angst, J.E. Johansen, J.C. Plaquevent, J. Schreiber and A. Eschenmoser, *Helv. Chim. Acta* **68**, 1312 (1985).

[18] A. Eschenmoser, *Angew. Chem. Int. Ed. Engl.* **27**, 5 (1988).

[19] B. Kamenar and B. Kaitner in G. Dodgson, J.P. Glusker and D. Sayre (eds.), *Structural Studies on Molecules of Biological Interest*, Clarendon Press, Oxford, 1981, chap. 15.

[20] V. Katovic, S.C. Vergez and D.H. Busch, *Inorg. Chem.* **16**, 1716 (1977).

[21] R.B. Woodward, *Aromatic Conference*, Sheffield, England, 1966; V.J. Bauer, D.L.J. Clive, D. Dolphin, J.B. Paine III, F.L. Harris, M.M. King, J. Loder, S.-W.C. Wang and R.B. Woodward, *J. Am. Chem. Soc.* **105**, 6429 (1983).

[22] J.L. Sessler, M.J. Cyr and V. Lynch, *J. Am. Chem. Soc.* **112**, 2810 (1990).

[23] A.K. Burrell, J.L. Sessler, M.J. Cyr, E. McGhee and J.A. Ibers, *Angew. Chem. Int. Ed. Engl.* **30**, 91 (1991).

[24] D. Gust, T.A. Moore, A.L. Moore, S.-J. Lee, E. Bittersmann, D.K. Luttrull, A.A. Rehms, J.M. DeGraziano, X.C. Ma, F. Gao, R.E. Belford and T.T. Trier, *Science (Washington)* **248**, 199 (1990).

Note The strutures of a family of TPP-based lattice clathrates is described in M.P. Byrn, C.J. Curtis, I. Goldberg, Y. Hsiou, S.I. Khan, P.A. Sawin, S.K. Tendick and C.E. Strouse, *J. Am. Chem. Soc.* **113**, 6549 (1991).

The furan, thiophene, and selenophene analogues of porphyrin, which exist as dications, are covered in E. Vogel, *Pure Appl. Chem.* **62**, 557 (1990); also reviewed are the porphyrin isomer polyphycene and its derivatives. The *meso*-octamethyl porphyrinogen tetraanion forms high-valent square-planar Fe(III) and square-pyramidal oxo Mo(V) complexes, as reported in D. Jacoby, C. Floriani, A. Chiesi-Villa and C. Rizzoli, *J. Chem. Soc., Chem. Commun.*, 220 (1991).

The structure of uranylpentaphyrin is reported in A.K. Burrell, G. Hemmi, V. Lynch and J.L. Sessler, *J. Am. Chem. Soc.* **113**, 4690 (1991). An alkyl-substituted derivative of rubyrin, a new hexapyrrolic expanded porphyrin, is reported in J.L. Sessler, T. Morishima and V. Lynch, *Angew. Chem. Int. Ed. Engl.* **30**, 977 (1991).

The remarkably stable, biphenylene-bridged μ_2-N$_2$ complex of a ruthenium cofacial metallodiporphyrin complex and its putative reduction intermediates (namely the μ_2-diazene, μ_2-hydrazine, and bis-diammine complexes) are reported in J.P. Collman, J.E. Hutchison, M.A. Lopez, R. Guilard, and R.A. Reed, *J. Am. Chem. Soc.* **113**, 2794 (1991).

Additional references are listed in p. 742.

4.19   Dioxygen Adduct of (2-Methylimidazole)-meso-tetra($\alpha,\alpha,\alpha,\alpha$-o-pivalamidophenyl)porphyrinatoiron(II)-Ethanol

$Fe(O_2)(T_{piv}PP)(2\text{-MeIm}).EtOH$

$C_{70}H_{76}N_{10}O_7Fe$

Crystal Data

Monoclinic, space group $C2/c$ (No. 15)

$a = 18.864(5)$, $b = 19.451(5)$, $c = 18.287(5)$Å, $\beta = 91.45(2)°$, $Z = 4$

Atom*	x	y	z	Atom	x	y	z
Fe	0	.13407	1/4	C(A10)	.1977	.4125	.2805
O(1)	0	.2317	1/4	C(A11)	.2158	.4404	.4011
O(2A)	.0163	.2707	.2021	N(B1)	-.1095	.2509	.5023
O(2B)	.0470	.2711	.2718	C(B7)	-.1118	.3123	.5230
N(1)	.0964	.1400	.2078	O(B1)	-.1221	.3269	.5901
C(1)	.1132	.1405	.1343	C(B8)	-.1085	.3716	.4726
C(2)	.1895	.1458	.1270	C(B9)	-.1793	.3972	.4556
C(3)	.2176	.1481	.1949	C(B10)	-.0728	.3527	.4057
C(4)	.1601	.1441	.2454	C(B11)	-.0660	.4292	.5023
C(5)	.1695	.1427	.3207	C(A1)	.2435	.1453	.3514
N(2)	.0426	.1370	.3510	C(A2)	.2841	.0865	.3566
C(6)	.1144	.1394	.3703	C(A3)	.3540	.0893	.3815
C(7)	.1230	.1379	.4479	C(A4)	.3820	.1505	.4039
C(8)	.0596	.1354	.4761	C(A5)	.3432	.2096	.4015
C(9)	.0083	.1347	.4156	C(A6)	.2741	.2073	.3744
C(10)	-.0650	.1363	.4236	C(B1)	-.0934	.1294	.4999
N(A1)	.2325	.2674	.3690	C(B2)	-.0969	.0652	.5323
C(A7)	.2531	.3312	.3596	C(B3)	-.1216	.0593	.6034
O(A1)	.3144	.3458	.3629	C(B4)	-.1427	.1149	.6402
C(A8)	.1972	.3859	.3502	C(B5)	-.1390	.1799	.6087
C(A9)	.1266	.3690	.3682	C(B6)	-.1143	.1869	.5382

* Hydrogen atoms, and atoms comprising the 2-methylimidazole group and the ethanol molecule have been omitted.

Crystal Structure

Exposure of $Fe(T_{piv}PP)(2\text{-MeIm}).EtOH$ [space group $C2/c$, $a = 18.871(12)$, $b = 19.425(13)$, $c = 18.434(11)$Å, $\beta = 91.48(3)°$, $Z = 4$] to dioxygen yields a 1:1 dioxygen adduct, $Fe(O_2)(T_{piv}PP)(2\text{-MeIm}).EtOH$, with only minor changes in unit-cell dimensions. The direct, precise determination of the stereochemical changes accompanying oxygenation of this iron(II)(porphyrinato)(base) complex has been elucidated.[1]

In the presence of ethanol the deoxy complex binds dioxygen reversibly, non-cooperatively, and with lower affinity than when the sample is desolvated; in the latter case dioxygen uptake has been found to be cooperative.

Some selected parameters for the coordination spheres (Fig. 4.19-1), with those in square brackets pertaining to the deoxy complex, are: Fe-N_p = 1.997(4), 1.995(4)Å [2.068(5), 2.075(5)Å]; Fe-N_{Im} = 2.107(4)Å [2.095(6)Å]. The iron atom is displaced 0.086Å [0.399Å] from the least-squares plane of the porphyrinato nitrogen atoms toward the imidazole ligand. The dioxygen ligand exhibits disorder with an average O-O separation of 1.22(2)Å. The Fe-O bond length is 1.898(7)Å, and the Fe-O-O angle is 129(1)°.

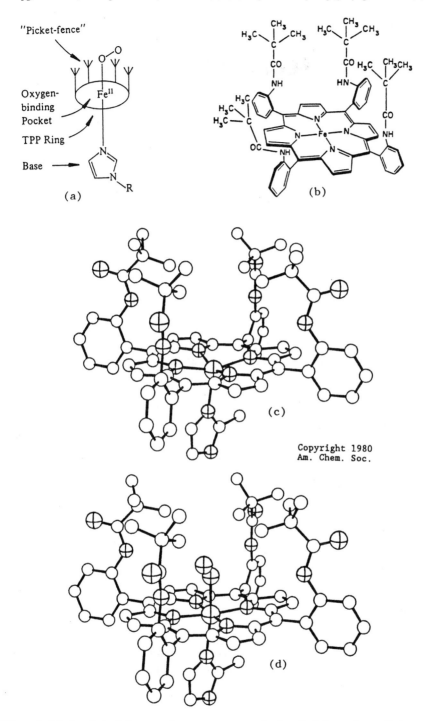

Copyright 1980
Am. Chem. Soc.

Fig. 4.19-1 (a) Scheme of components of the "picket fence" iron porphyrin dioxygen complex, (b) structural formula of $Fe(T_{piv}PP)$, (c) perspective view of $Fe(T_{piv}PP)(2\text{-MeIm})$ and (d) perspective view of $Fe(O_2)(T_{piv}PP)(2\text{-MeIm})$. (After refs. 1 and 3).

Table 4.19-1 lists the structural data in comparison with hemoproteins and their model analogues. The measured bond lengths indicate that the Fe(II) atom is typically low-spin in the oxy complex and high-spin in the deoxy complex.

Table 4.19-1   Stereochemistry of hemoproteins and their model analogues (interatomic distances in Å).[1]

Complex	M...Ct	M-N$_p$	M-N$_{Im}$	M-O	O-O	M-O-O(°)
Fe(TPP)(2-MeIm).EtOH	.42	2.086	2.161			
Fe(T$_{piv}$PP)(2-MeIm).EtOH	.399	2.072	2.095			
Myoglobin (Mb)	.42	2.06	2.1			
Erythrocruorin	.23	2.02	2.2			
Fe(O$_2$)(T$_{piv}$PP)(1-MeIm)[2]	-.02	1.98	2.07	1.75	1.16	131
Mb(O$_2$)	.24	2.07	2.03	2.02	1.21	111
Oxycobaltomyoglobin	.25	1.99	1.97	1.89	1.26	131
Oxyerythrocruorin	.38	2.04	2.1	1.8	1.25	170
Fe(O$_2$)(T$_{piv}$PP)(2-MeIm).EtOH	.086	1.996	2.107	1.898	1.22	129

Remarks

1. Control of oxidation and coordination in model porphyrin systems[3]

The heme proteins are responsible for oxygen transport and storage, electron transport, oxygen reduction, hydrogen peroxide utilization and decomposition, and hydrocarbon oxidation. The active site in each case contains an iron porphyrin (the prosthetic group), the nitrogens of the porphyrin ring occupying four essentially planar coordination sites of the metal. Their diversity of function must therefore be dictated by the number and nature of the axial ligands, the spin and oxidation state of the metal, and the nature of the polypeptide chain.

Oxygen binding heme proteins are five-coordinate high-spin iron(II) species, which upon oxygenation become six-coordinate low-spin. The difficulty in reproducing this behaviour is dominated by two problems: (a) the irreversible oxidation of iron(II) porphyrin on exposure to oxygen, and (b) the difficulty in obtaining well-defined five-coordinate iron porphyrins.

Various approaches have been used to control oxidation and coordination in model porphyrin systems:

a) *Excess ligand.* The presence of excess base (imidazole, pyridine) will minimize the concentration of five-coordinate heme and reduce μ-peroxo complex formation.

b) *Low temperatures*. Iron(II)-O$_2$ porphyrin complexes are stable at low temperatures (~ -60°C), where the irreversible oxidation reactions are slowed down.

c) *Kinetic measurements*. Fast spectroscopic methods may be used to observe reversible oxygen binding even under conditions where irreversible oxidation will occurr.

d) *Metal substitution*. Replacement of iron with cobalt or ruthenium leads to metalloporphyrins that are more inert to oxidation and possess different coordination properties.

e) *Immobilization*. This approach attempts to prevent irreversible oxidation by anchoring the porphyrin to a solid support.

f) *Steric encumbrance*. By sterically blocking one or both faces of the porphyrin with functional groups covalently bound to the ring, μ-oxo bridge formation may be prevented, and the steric hindrance is designed (i) to direct base binding to the open face, ensuring five-coordination, and (ii) to allow O$_2$ to bind on the hindered face, steric hindrance preventing μ-oxo bridge formation.

g) *Chelated heme*. Covalent attachment of a potential ligand to the porphyrin periphery allows for control of the extent of coordination.

2. "Picket-fence" porphyrins

In "picket-fence" porphyrins, steric encumbrance at the metal site of these substituted TPP molecules depends on two factors:

a) owing to steric repulsion, the TPP will adopt a conformation in which the four *meso*-phenyl rings are essentially perpendicular to the porphyrin ring; substituents at the *ortho*-positions of the phenyl rings will lie above and below the porphyrin plane, and

b) for TPP molecules containing mono-*ortho*-substituted phenyl rings, separation and, depending on the bulk of the substituent, interconversion of the four possible atropisomers may be achieved.

Two typical five-coordinate "picket fence" porphyrins are shown in Fig. 4.19-2.

3. "Capped" porphyrins

In "capped" porphyrin molecules a benzene or naphthalene ring is covalently attached to all four *ortho*-positions of the *meso*-phenyl rings, enclosing a volume of space above one face of the porphyrin ring [Fig. 4.19-3(a)]. If the cap is sufficiently tight the binding of bases should be prevented on the enclosed face; binding on the open face would result in a five-coordinate species such as Fe^{III}(Cl)(C$_2$-Cap) [Fig. 4.19-3(b)]. On the

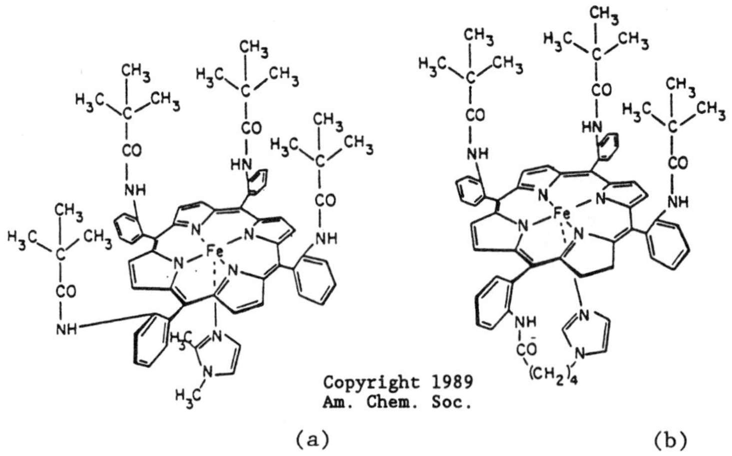

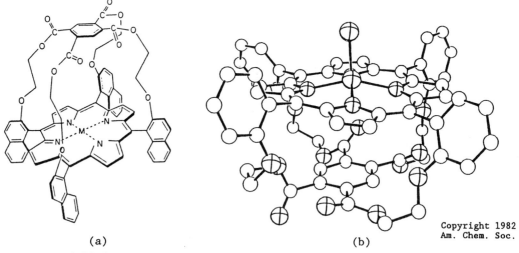

Copyright 1989
Am. Chem. Soc.

(a)   (b)

Fig. 4.19-2 Five-coordinate "picket fence" porphyrins: (a) FeT$_{piv}$PP-(1,2-Me$_2$Im), (b) FePiv$_3$(5CImP). (After ref. 4).

Copyright 1982
Am. Chem. Soc.

(a)   (b)

Fig. 4.19-3 Structures of (a) "capped" iron porphyrin and (b) Fe^{III}(Cl)(C$_2$-Cap). (After refs. 3 and 5).

other hand, the smaller dioxygen molecule would be able to fit under the cap, which should provide a physical barrier to μ-oxo complex formation.

In the modified "capped" porphyrin shown in Fig. 4.19-4, a pyridine is covalently bound to two opposite *meso*-aromatic rings of the parent "capped" porphyrin to forms some kind of "strap" over the porphyrin face opposite to the cap. The resulting iron(II) "capped strapped" porphyrin exhibits reversible oxygen binding in toluene solution at room temperature and shows good stability to autoxidation.[6] Furthermore, decomposition does not result in a μ-oxo complex, both strap and cap contriving to prevent dimer formation.

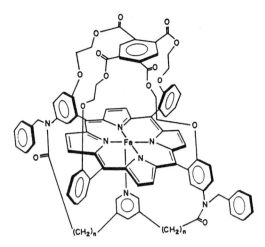

Fig. 4.19-4 A "capped strapped" iron(II) porphyrin. (After ref. 3).

4. "Strapped" porphyrins

The "strapped" porphyrin class of heme protein models embraces all those compounds in which some group is covalently linked to two corners (usually diagonally opposite) of a porphyrin macrocycle. These porphyrins may be singly- or doubly-strapped and may be classified according to the nature of the chain:

a) Simple non-functionalized alkyl chains whose role is to span one face of the porphyrin, discouraging μ-oxo bridging and providing a more hydrophobic environment.

b) Straps incorporating some bulky group which will provide more steric encumbrance than a simple alkyl chain.

c) Functionalized straps which incorporate some group capable of binding to or interacting with the metal at the porphyrin core. These may be used to maintain five-coordination or to form six-coordinate mixed ligand systems L-M-L', where one ligand binds poorly to the metal.

Porphyrins strapped with simple alkyl chains are poor models for oxygen binding heme proteins. In most cases the strap is too "floppy" and can be pushed to one side allowing μ-oxo bridge formation. In addition, the base is not prevented from binding under the strap, leading to oxygen binding on the open face, and irreversible oxidation. The logical extension is to incorporate some bulky group into the strap to increase the steric encumbrance about one face. Fig. 4.19-5 shows an example of this type of porphyrin containing a bulky blocking group in the strap.[7]

The incorporation of potential ligands into the porphyrin strap has three advantages. (i) A stoichiometric amount of ligand is built into the

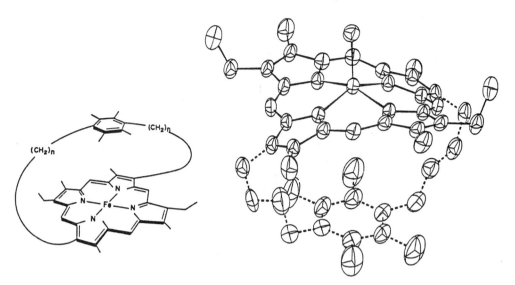

Fig. 4.19-5 Structure of "strapped" porphyrin containing a bulky blocking group. Hydrogen atoms have been omitted; the dashed bonds are used to distinguish between the "strap" and the porphyrin skeleton. (After ref. 3).

system, ensuring five-coordination without the addition of external ligand. In the case of nitrogen bases, mixtures of six- and four-coordinate complexes are avoided. (ii) For ligands which bind poorly to iron(II), coordination would be favoured by constraining the ligand into a position suitable for binding to the metal. (iii) Because the strap is fixed to the porphyrin and the ligand in two positions, complications due to ligand replacement or dissociation will be minimized. The strapped porphyrin approach is also useful for orientating other interactive groups (e.g. metal binding sites, electron donor/acceptor groups) into specific geometries with respect to the porphyrin.

5. Structure of $Fe(\beta\text{-PocPivP})(1,2\text{-Me}_2\text{Im})(CO)$[4]

The structures of a five-coordinate "pocket" iron(II) porphyrin and its CO complex are shown in Fig. 4.19-6.

The important bond lengths and angles in the CO complex are: Fe-C = 1.768(8)Å, C-O = 1.148(7)Å, Fe-C-O = 172.5(6)°, Fe-N$_{Im}$ = 2.079(5)Å, Fe-N = 1.973(8)Å, and C-Fe-N$_{Im}$ = 176.3(3)°. The Fe atom is 0.001Å out of the 24-atom least-squares plane toward the CO ligand. The modest distortion of the carbonyl subunit is accompanied by considerable ruffling of the porphyrin periphery and significant shifting of the benzene "cap" away from the bound carbonyl ligand. The lower CO affinities observed for the encumbered "pocket"

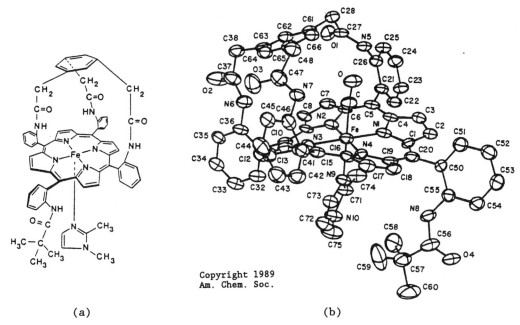

(a)   (b)

Fig. 4.19-6 Structures of (a) five-coordinate "pocket" porphyrin Fe(β-PocPivP)(1,2-Me$_2$Im) and (b) the CO complex of this iron(II) porphyrin. (After ref. 4).

porphyrins, as compared with unhindered "picket fence" control systems, reflect a combination of structural changes involving both localized ligand distortions and overall porphyrin skeletal deformations. These results could have significant implications for the interpretation of the protein structures of carbonylated heme systems.

References

[1] G.B. Jameson, F.S. Molinaro, J.A. Ibers, J.P. Collman, J.I. Brauman, E. Rose and K.S. Suslick, *J. Am. Chem. Soc.* **102**, 3224 (1980).

[2] G.B. Jameson, G.A. Rodley, W.T. Robinson, R.R. Gagne, C.A. Reed and J.P. Collman, *Inorg. Chem.* **17**, 850 (1978).

[3] B. Morgan and D. Dolphin, *Structure and Bonding* **64**, 115 (1987).

[4] K. Kim, J. Fettinger, J.L. Sessler, M. Cyr, J. Hugdahl, J.P. Collman and J.A. Ibers, *J. Am. Chem. Soc.* **111**, 403 (1989).

[5] M. Sabat and J.A. Ibers, *J. Am. Chem. Soc.* **104**, 3715 (1982).

[6] J.E. Baldwin, J.H. Cameron, M.J. Crossley, I.J. Dagley, S.R. Hall and T. Klose, *J. Chem. Soc., Dalton Trans.*, 1739 (1984).

[7] S. David, D. Dolphin, B.R. James, J.B. Paine III, T.P. Wijesekera, F.W.B. Einstein and T. Jones, *Can. J. Chem.* **64**, 208 (1986).

4.20 cis-Dichlorodiammineplatinum(II) (cisplatin, cis-DDP)
$$PtCl_2(NH_3)_2$$

Crystal Data

Triclinic, space group $P\bar{1}$ (No. 2)

$a = 6.75(2)$, $b = 6.55(2)$, $c = 6.23(2)$Å, $\alpha = 92.2(3)°$, $\beta = 84.6(3)°$,

$\gamma = 110.7(3)°$, $Z = 2$ (120K)

Atom	x	y	z
Pt	.248	.005	.026
Cl(1)	.183	-.318	.206
Cl(2)	.301	.208	.342
N(1)	.304	.274	-.130
N(2)	.205	-.154	-.265

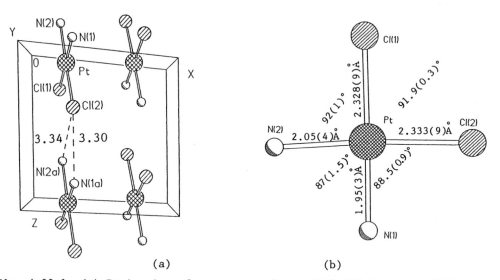

(a)   (b)

Fig. 4.20-1   (a) Projection of structure of *cis*-PtCl$_2$(NH$_3$)$_2$ down [010].
(b) Bond lengths and angles in *cis*-PtCl$_2$(NH$_3$)$_2$. (Adapted from ref. 1).

Crystal Structure

The crystal and molecular structure of the *cis*-isomer of dichloro-
diammineplatinum(II), *cis*-DDP, is shown in Fig. 4.20-1. The planar molecules
are parallel to one another and stacked along the *a* axis. The Pt...Pt
distances along the stack are alternately 3.372 and 3.409Å, which may be short
enough to indicate some significant interaction. The crystals are not
pleochroic. The molecular parameters are Pt-Cl = 2.33(1)Å (av.), Pt-N =
2.01(4)Å (av.), Cl-Pt-Cl = 91.9(4)°. There are possible intermolecular
hydrogen bonds of the type N(2)...Cl(1) 3.41Å, N(2)...Cl(2) 3.34Å, and
N(1)...Cl(2) 3.30Å; the relevant bond angles are Pt-N(2)...Cl(1) 169°, Pt-
N(2)...Cl(2) 110°, and Pt-N(1)...Cl(2) 114°.

cis-[PtCl$_2$(NH$_3$)$_2$] is a drug that is remarkably successful for treating some forms of cancer. The current consensus is that it acts by binding to the guanine units of DNA, being stacked between base pairs in the double helix, so that the nonbonding distance of Cl(1)...Cl(2) = 3.4Å is highly relevant to its antitumor function.[2-5] The affinity order is often stated to be: G > A > C > T. Intercalative binding of this type of anticancer platinum drugs to DNA is illustrated schematically in Fig. 4.20-2.

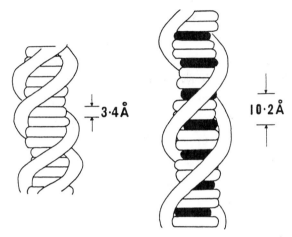

Fig. 4.20-2 Schematic representation of a double-stranded DNA without (left) and with (right) a bound intercalator (darkened). (After ref. 9).

Remarks

1. Structure of *trans*-DDP and the "*trans*-effect"

   trans-DDP is monoclinic, space group $P2_1/a$, with a = 7.99(1), b = 6.00(1), c = 5.45(1)Å, β = 95.2(2)°, and Z = 2 (120K). The crystal structure is shown in Fig. 4.20-3(a), and the bond lengths and angles in Fig. 4.20-3(b).

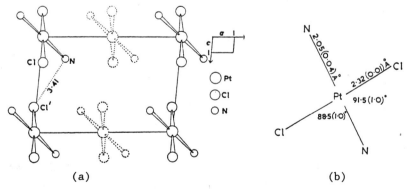

(a)   (b)

Fig. 4.20-3 (a) Projection of the structure of *trans*-DDP down [010]. The molecules shown by full lines lie at (0,0,0), and those shown by broken lines at (1/2,1/2,0). (b) Bond lengths and angles in *trans*-DDP. (After ref. 1).

Since the early investigations revealed that the ammine-platinum compounds with antitumor properties all have the *cis*-configuration, a consideration of the "*trans*-effect" is of relevance in the present context. In the 1920's Chernyaev noticed that when there are alternative positions at which an incoming ligand might effect a substitution, the choice depends primarily on the nature of the ligand *trans* to it. This became known as the *trans*-effect and has had a considerable influence on the synthetic coordination chemistry of Pt(II). Although incoming and leaving ligands modify this somewhat, the general order of *tran*-directing power, namely the relative ability of a ligand to labilize the ligand *trans* to it, is

$$CN^- \sim C_2H_4 \sim CO \sim NO > PR_3 \sim H\text{-} \sim SC(NH_2)_2 > CH_3^- > C_6H_5^- > SCN^- >$$
$$NO_2^- > I^- > Br^- > Cl^- > py \sim NH_3 > OH^- > H_2O.$$

By choosing the right sequence of ligand substitutions, one can exercise kinetic control of the reaction products to obtain a desired isomer of a square-planar complex. For example, the *cis*- and *trans*-isomers of $[PtCl_2(NO_2)(NH_3)]^-$ can be prepared from $PtCl_4^{2-}$ using the following synthetic scheme:

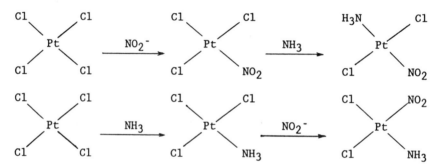

2. Structure of the intercalated $[Pt(terpy)(HET)]_2 \cdot [d(CpG)]_2$ complex

The first structurally characterized example of a metallo-intercalator bound to a short oligodeoxynucleotide fragment was found in a 2:2 complex of $[Pt(terpy)(HET)]^+$ with d(CpG)$^-$ [Fig. 4.20-4(a)].[6] In this complex, one cation intercalates symmetrically within a Watson-Crick base-paired $[d(CpG)]_2^{2-}$ duplex with the HET chain projecting into the minor groove of the miniature DNA helix, while the other stacks between $[d(CpG)]_2^{2-}$ units to form an infinite column [Fig. 4.20-4(b)]. Intercalation lengthens the usual G-C base pair separation of 3.4Å to 6.8Å, and the right-handed dinucleotide helix is also underwound by ~20° relative to the usual winding angle of 34° per base pair in an idealized B-DNA structure. An addition feature in this intercalation complex is a change in sugar pucker of the 5′ dC base from the normal C2′-*endo* to a C3′-*endo* conformation.

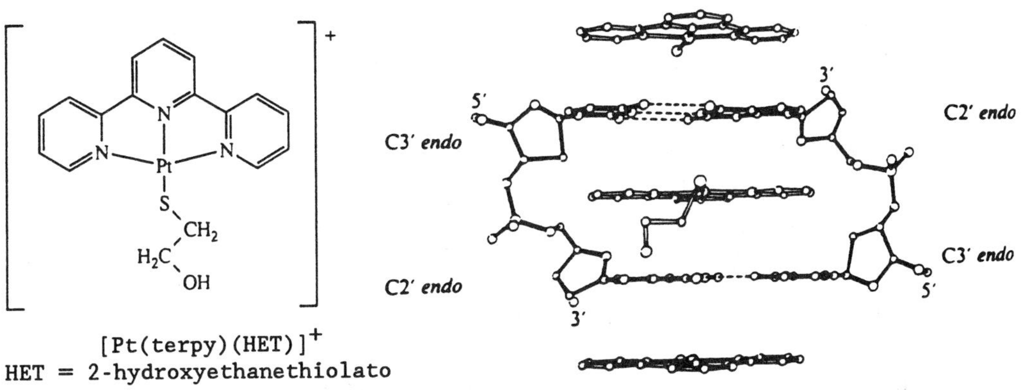

C3′ endo
C2′ endo
C2′ endo
C3′ endo
5′
3′
3′
5′
[Pt(terpy)(HET)]+
HET = 2-hydroxyethanethiolato
(a)
(b)

Fig. 4.20-4 (a) Structural formula of $[Pt(terpy)(HET)]^+$. (b) Crystal structure of the intercalated $[Pt(terpy)(HET)]_2.[d(CpG)]_2$ complex. (After ref. 6).

3. Regioselectivity and structures of *cis*- and *trans*-DDP adducts of DNA

In the adduct *cis*-$Pt(NH_3)_2$(d(pGpG)) one phosphate group is involved in a hydrogen bond with a NH_3 ligand at platinum [Fig. 4.20-5(a)].[7] This could well be an important explanation for the observation that platinum antitumor drugs need an acid N-H group to donate a hydrogen bond. The $O_3PO...H-NH_2$ interaction could stabilize and induce the distortion of the DNA, thereby hampering or blocking the replication processes.

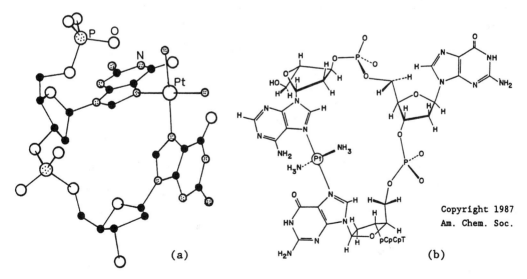

P
N
Pt
(a)
pCpCpT
Copyright 1987
Am. Chem. Soc.
(b)

Fig. 4.20-5 (a) Structure of *cis*-$Pt(NH_3)_2$(d(pGpG)). (b) Structure of the adduct formed by *trans*-DDP with the oligonucleotide d(ApGpGpCpCpT). (After refs. 7 and 11).

From this structure and other related studies,[8,9] e.g. an X-ray analysis of *cis*-Pt(NH$_3$)$_2$(d(CpGpG)),[10] the following general conclusions can be drawn:

a) When adjacent guanines are present in an oligonucleotide, cisplatin chelates to these at the N(7) positions.

b) When GBG (B = any nucleobase) sequences are present in an oligonucleotide, cisplatin chelates also to both guanines; when the central base is adenine then also an AG (A = adenine) chelate is formed.

c) Double-stranded oligonucleotides may be formed with the above-mentioned platinated GG- and GBG- containing oligonucleotides (at least eight units seem to be required).

d) The distortion of the double helix is rather small in case of GG-chelation by cisplatin.

trans-DDP forms a variety of DNA adducts including 1,3-intrastrand crosslinks. Fig. 4.20-5(b) shows the structure of one such example, in which two purine N7 atoms are linked by platinum to close a 23-membered ring.[8]

4. Linked Pt-intercalator complexes

Recent development involves the design and synthesis of complexes in which an intercalator and a chemically reactive platinum functionality are covalently linked. Two examples are shown in Fig. 4.20-6.[9] In the cation [Pt{AO(CH$_2$)$_6$en}(ox)]$^+$ (AO = acridine orange, ox = oxalato) the {Pt(en)}$^{2+}$ moiety binds covalently to DNA while the AO ring intercalates one or two base pairs away [Fig. 4.20-6(a)]. This multifunctional mode of binding unwinds DNA to an extent greater than that observed for either component alone. In *cis*-

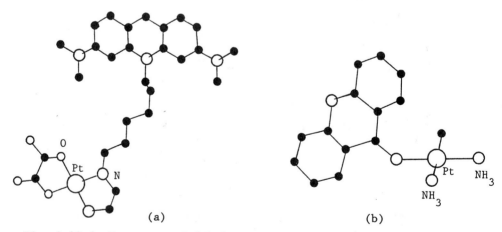

(a)   (b)

Fig. 4.20-6 Structure of (a) [Pt{AO(CH$_2$)$_6$en}(ox)]$^+$ and (b) *cis*-[Pt(NH$_3$)$_2$(*N9*-9-AA)Cl]$^+$. (After ref. 9).

$[Pt(NH_3)_2(N9-9-AA)Cl]^+$ the metal is bonded to the N9 position of the imino tautomer of 9-aminoacridine (9-AA). This complex is of structural interest because the rigid geometry of coordinated 9-AA forces the apical H1 hydrogen to lie within 2.4Å of the platinum center [Fig. 4.20-6(b)].

5. Relation between molecular structure and anticancer activity

The study of platinum complexes as anticancer drugs points to the existence of a link between structure and activity. The principal findings are summarized in a set of "rules of thumb" by Rosenberg:[2,3]

a) The complexes exchange only some of their ligands quickly in reactions with biological molecules.

b) The complexes should be electrically neutral, although the active form may be charged after undergoing ligand exchanges in the body.

c) The geometry of the complexes is either square planar or octahedral.

d) Two *cis* monodentate or one bidentate leaving group (exchangeable ligands) are required: the corresponding *trans* isomers of the monodentate leaving groups are generally inactive.

e) The rates of exchange of these groups should fall into a restricted region, since too high a reactivity will mean that the chemical reacts immediately with blood constituents and never gets to the tumor cell, while too low a reactivity would allow it to get to the cells, but it would lie dormant once there.

f) The leaving groups should be approximately 3.4Å apart on the molecule, which is the same as the spacing between the steps of the Watson-Crick DNA ladder.

g) The groups across the molecule from the leaving groups should be strongly bonded, relatively inert amine-type systems.

Reedijk and co-workers.[12] have proposed that platinum compounds with antitumor activity must fulfill all of the following structural requirements:

a) The two amine ligands in the Pt-compound should be in a *cis*-orientation. In a bidentate chelating ligand, this geometric requirement is automatically fulfilled. The general formulae should be $cis\text{-}PtX_2(Am)_2$ and $cis\text{-}PtY_2X_2(Am)_2$, (Am = amine ligand). The number of variations studied in the case of Pt(IV) is however still rather limited.

b) The ligands X, usually anions, should consist of groups that have intermediate binding strength to Pt(II), or that are (for other reasons) easily leaving (i.e. by enzymatic activation). Common examples are: Cl^-, SO_4^{2-}, citrate(3-), oxalate(2-) and other carboxylic acid residues. For the

Pt(IV) compounds, the Y group is often OH⁻ (with the two Y ligands in *trans*-orientation); only a few anti-tumor active compounds of this class have been reported.

  c) The amine ligands, either monodentate or bidentate, should have at least one N-H group as required for the hydrogen-bond donor function. All compounds with both amine ligands lacking such a H-bond donor property have been found to be inactive. The role of this N-H group in the biological activity could be either kinetic (i.e. play a role in the approach of the DNA), or thermodynamic (e.g. give an additional (de)stabilization after binding to the biological target DNA). However, steric effects and a role in transport through the cell wall cannot be excluded.

  Some *cis*-Pt-compounds have superior properties in the treatment of tumors in patients who do not respond well to cisplatin. The most promising compounds currently available are shown in Fig. 4.20-7.

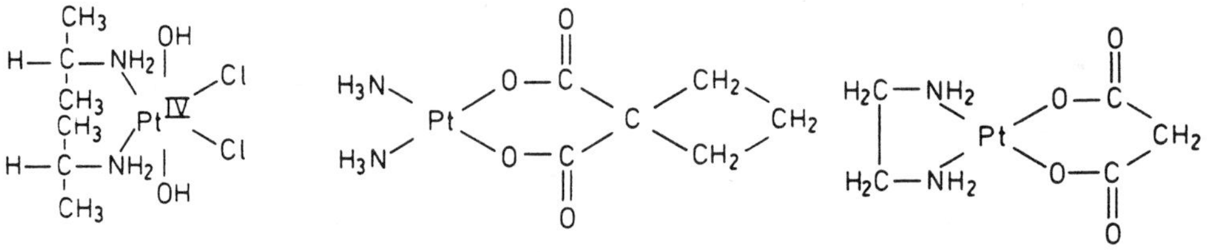

CHIP   CBDCA

Fig. 4.20-7 Active analogues of cisplatin being used on a large scale: two compounds are known by their abbreviated names, the third is (ethylenediamine)malonatoplatinum(II). (After ref. 12).

References

[1] G.H.W. Milburn and M.R. Truter, *J. Chem. Soc.* (A), 1609 (1966).

[2] B. Rosenberg, L. VanCamp, J.E. Trosko and V.H. Mansour, *Nature (London)* **222**, 385 (1969).

[3] B. Rosenberg in T.G. Spiro (ed.), *Nucleic Acid-Metal Ion Interactions*, Wiley, New York, 1980.

[4] M. Green and M. Green, *Transition Met. Chem.* **10**, 196 (1985).

[5] J.K. Barton and S.J. Lippard, *Met. Ions Biol.* **1**, 31 (1980).

[6] A.H.-J. Wang, J. Nathans, G.A. van der Marel, J.H. van Boom and A. Rich, *Nature (London)* **276**, 471 (1978).

[7] S.E. Sherman, D. Gibson, A.H.-J. Wang and S.J. Lippard, *Science (Washington)* **230**, 412 (1985); *J. Am. Chem. Soc.* **110**, 7368 (1988).

[8] S.E. Sherman and S.J. Lippard, *Chem. Rev.* **87**, 1153 (1987).

[9] W.I. Sundquist and S.J. Lippard, *Coord. Chem. Rev.* **100**, 293 (1990).

[10] G. Admiraal, J.L. van der Veer, R.A.G. de Graaff, J.H.J. den Hartog and J. Reedijk, *J. Am. Chem. Soc.* **109**, 592 (1987).

[11] C.A. Lepre, K.G. Strothkamp and S.J. Lippard, *Biochemistry* **26**, 5651 (1987).

[12] J. Reedijk, A.M.J. Fichtinger-Schepman, A.T. van Oosterom and P. van de Putte, *Structure and Bonding* **67**, 53 (1987).

Note Further elaboration of the *trans*-effect in the isomerization of Pt(II) and Pd(II) complexes is given in Yu.N. Kukushkin, *Platinum Metals Rev.* **35**, 28 (1991); *Koord. Khim.* **5**, 1856 (1979); *ibid.* **8**, 201 (1982); *Chemistry of Coordination Compounds*, Vysshaya Shkola, Moscow, 1985, p. 455. [In Russian.]

The chemistry of Pt electrophiles with nucleobases and related model systems is reviewed in B. Lippert, *Prog. Inorg. Chem.* **37**, 1 (1989).

4.21   Cobalt Inosine 5'-Phosphate Heptahydrate
$C_{10}H_{11}O_8N_4PCo \cdot 7H_2O$

Crystal Data

Orthorhombic, space group $P2_12_12_1$ (No. 19)

$a = 10.859(5)$, $b = 25.987(5)$, $c = 6.845(5)$Å, $Z = 4$

Atom*	x	y	z	Atom	x	y	z
Co	.0783	.3547	.2756	C(1')	.5530	.4364	.2480
P	.4204	.2436	.2287	C(2')	.6249	.4326	.4465
O(5')	.5000	.2921	.3027	C(3')	.6391	.3741	.4631
O(7)	.5081	.1974	.2238	C(4')	.6570	.3559	.2502
O(8)	.3685	.2574	.0296	C(5')	.6193	.3015	.2097
O(9)	.3225	.2377	.3872	O(1')	.5800	.3921	.1372
N(1)	.1837	.5530	.2995	O(2')	.7458	.4530	.4099
C(2)	.3060	.5645	.2913	O(3')	.7425	.3628	.5819
N(3)	.3930	.5299	.2885	O(W1)	-.0207	.3841	.0331
C(4)	.3479	.4815	.2911	O(W2)	-.0236	.3966	.4766
C(5)	.2274	.4652	.2891	O(W3)	.1729	.3155	.0568
C(6)	.1332	.5036	.3027	O(W4)	.1788	.3175	.4993
O(6)	.0212	.4982	.3154	O(W5)	-.0529	.2966	.2850
N(7)	.2230	.4118	.2859	O(W6)	.8993	.1557	.2383
C(8)	.3371	.3971	.2726	O(W7)	.7265	.5578	.3873
N(9)	.4159	.4374	.2752				

* The parameters of hydrogen atoms have been omitted.

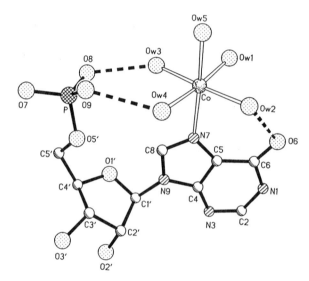

Fig. 4.21-1 Molecular structure of the 1:1 complex of cobalt with inosine 5'-phosphate. Broken lines represent hydrogen bonds. (Adapted from ref. 1).

Crystal Structure

    The structure consists of four moieties: hypoxanthine ring, ribose ring, phosphate group, and cobalt complex ion (Fig. 4.21-1).

    The bond lengths and bond angles in the hypoxanthine ring are in agreement with those found in inosine, inosine dihydrate, and monosodium inosine 5'-phosphate.[2] A common and characteristic feature of the

hypoxanthine moieties concerns the angles around the C(6)-O(6) bond of length
1.227Å: the C(5)-C(6)-O(6) angle (129.3°) is appreciably larger than N(1)-
C(6)-O(6) (119.7°), due to the steric interaction between O(6) and N(7). The
atoms composing the purine ring show a significant degree of puckering, which
often occurs in purine and pyrimidine derivatives.

The bond lengths and bond angles in the ribose ring have their usual
values.[2] An important stereochemical parameter in nucleosides and
nucleotides is the glycosyl torsion angle, C(8)-N(9)-C(1')-O(1'), which
describes the relative orientation of the base with respect to the sugar.
This angle is 25.4°, so that the conformation about the glycosyl bond is *anti*
in the complex. The ribose carbon atom C(1') is 0.134Å out of the base plane,
and the glycosidic C(1')-N(9) bond makes an angle of about 5.0° with it. The
ribose ring is puckered with C(3') displaced from the best four-atom plane of
the furanose ring by 0.560Å on the same side as C(5'), and the conformation of
the ribose ring is C(3')-*endo*. The conformation around the C(4')-C(5') bond,
described by the angles O(1')-C(4')-C(5')-O(5') and C(3')-C(4')-C(5')-O(5'),
is *gauche-gauche*. The dihedral angle between the base and the ribose plane is
67.3°. The ester linkage, C(5')-O(5'), shows the elongated conformation as
usual, the torsion angle C(4')-C(5')-O(5')-P being 161.6°.

The coordination geometry about the cobalt atom is approximately
octahedral with the coordination sites occupied by five water oxygen atoms and
atom N(7) of the hypoxanthine moiety, a site which is accessible and
presumably favorable in the biopolymer, and is similar to that found in the
cobalt-theophylline complex.[3] Although in the vitamin B_{12} derivatives, the
analogous N(7) site of the dimethylbenzimidazole base is coordinated to the
corrin-bound cobalt atom, the present complex is the first nucleotide showing
the transition metal directly bonded to the base in an analogous manner. The
Co-O distances range from 2.075 to 2.119Å with an average value of 2.094Å.
The Co-N distance is 2.162Å. It is also notable that the phosphate oxygen
atoms do not partake in metal coordination.

Remarks
1. Nucleic acid constituents and chemical structure of DNA and RNA[2,4]

Nucleic acids are composed of organic nitrogen bases (e.g. adenine),
sugars (e.g. ribose or 2'-deoxyribose), and phosphates. The meanings of the
following terms are:

nucleoside: a combination of an organic nitrogen base and a sugar; a
typical nucleoside is adenosine.

nucleotide: a nucleoside phosphate ester.

nucleic acid: a nucleotide polymer, i.e., a polymeric diester of orthophosphoric acid (e.g. DNA, RNA).

polypeptide: an oligomer formed from organic aminoacids, $RCH(NH_2)CO_2H$, by means of peptide bonding -CO-NH-.

a) Geometry of bases

The common constituent bases, their numbering schemes, and geometric data of their planar skeletons are displayed in Fig. 4.21-2. The pyrimidine bases have a six-membered heterocyclic ring, while the purine bases have a

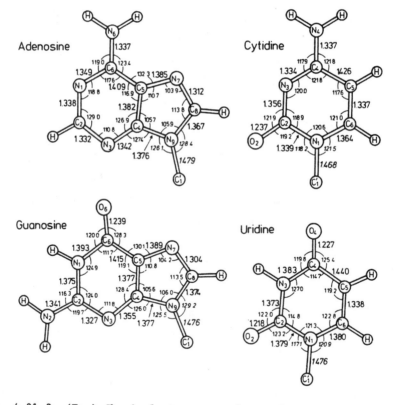

Fig. 4.21-2 (Top) Chemical structure of some bases. (Bottom) Average bond angles (°) and distances (Å) in 9-substituted adenine and guanine and in 1-substituted uracil and cytosine. (After ref. 2).

bicyclic five- and six-membered ring system. Pyrimidine and purine bases have been investigated as individual units and as moieties substituted at the glycosyl nitrogen, N(1) in pyrimidines and N(9) in purines.

b) Geometry of ribose and deoxyribose in nucleosides

Rotation of the base relative to the sugar is sterically restricted and two conformational states are preferred, *syn* and *anti*. The *anti* conformation predominates and is associated with C₃'-*endo* for riboses, while pyrimidine deoxyribonucleosides prefer C₂'-*endo* to C₃'-*endo* puckering. Fig. 4.21-3 shows the definition of sugar puckering modes and the geometrical data for ribose and deoxyribose in nucleosides.

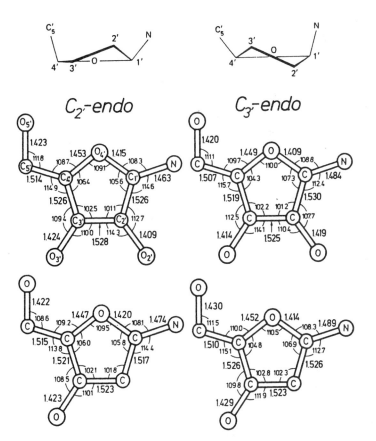

Fig. 4.21-3 Definition of sugar puckering modes and geometrical data for ribose and deoxyribose in nucleosides with C₂'-*endo* and C₃'-*endo* puckering. Data are averages obtained from well-refined crystal structures (R < 0.08). (After ref. 2).

c) Chemical structure of RNA and DNA

There are two classes of nucleotides depending on whether the 2'-position on the furanose ring bears a hydroxyl group (ribonucleotides) or not (deoxyribonucleotides). RNAs are polyribonucleotides and DNAs are polydeoxyribonucleotides. Fig. 4.21-4 shows the chemical structure of RNA and DNA. For oligonucleotides the prefixes d and r are used to indicate deoxyribose and ribose, respectively.

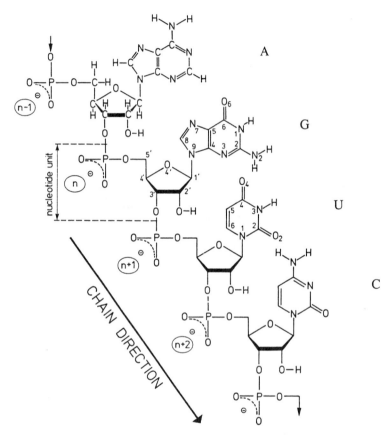

Fig. 4.21-4 Fragment of ribonucleic acid (RNA) with sequence adenosine (A), guanosine (G), uridine (U), cytidine (C) linked by 3',5'-phosphodiester bonds. Chain direction is from 5'- to 3'-end as shown by the arrow. Atom numbering scheme is indicated in one framed nucleotide unit, 5'-GMP. All hydrogen atoms are drawn for A and only functional hydrogens in other nucleotides. In short notation, this fragment would be pApGpUpCp or pAGUCp. In deoxyribonucleic acid (DNA), the hydroxyl attached to C2' is replaced by hydrogen and uracil, by thymine. (After ref. 2).

d) DNA double helix and base-pairs

In 1953, Watson and Crick[5] proposed as a model for DNA conformation a complementary double helix, in which hydrogen-bonded base-pairs of adenine with thymine (A-T) and guanine with cytosine (G-C) are stacked like stairs

3.4Å apart, and right-handed rotation of about 36° between adjacent base-pairs produces a double helix with 10 base-pairs per turn, as shown in Fig. 4.21-5. A model of the helix was constructed with bases located along the helix axis and sugar-phosphate backbones winding in antiparallel orientation along the periphery. This discovery led to an explosion of new developments in many areas of biochemistry and genetics, and Watson and Crick together with Wilkins were honoured in 1962 with the Nobel Prize in Medicine-Physiology.

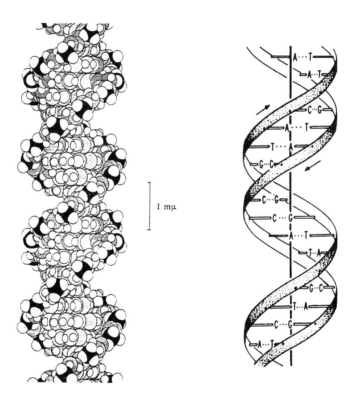

Fig. 4.21-5 Schematic drawing of the DNA double helix. The sugar phosphate backbones run at the periphery of the helix in antiparallel orientation. Base pairs (A-T and G-C) drawn symbolically as bars between chains are stacked along the center of the helix. (After ref. 2).

     Hydrogen-bonds play a key role in the stabilization of nucleic acid and protein secondary structure. Most important are hydrogen-bonded base pairs between homo- and hetero-bases which can be arranged in 28 different ways. Among the heteropairs, only those of the Watson-Crick type lead to regular double-helical structures. Fig. 4.21-6 shows the geometry of Watson-Crick type base-pairs.

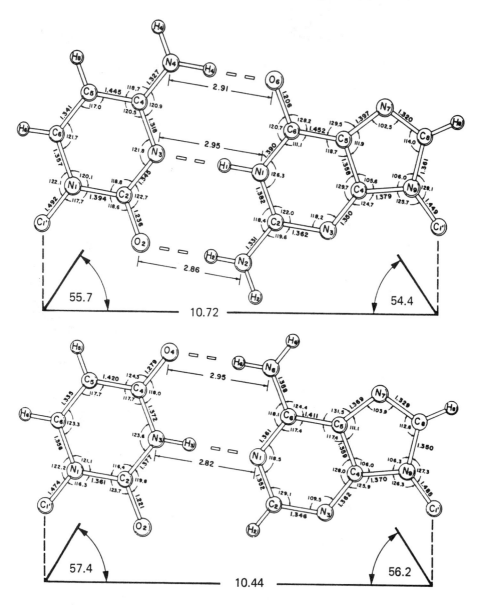

Fig. 4.21-6 Watson-Crick base pairs observed in crystal structures of
GpC (top) and ApU (bottom). Hydrogen atoms were not located
experimentally but are calculated from the positions of the other atoms.
(After ref. 2).

2. Conformation of nucleic acid[6]

From the structures of nucleic acid fragments determined by single-
crystal X-ray diffraction, detailed information on a wide variety of DNA
conformations has been gained. Two different diffraction patterns can be
produced from DNA fibers depending upon the relative humidity of the local

environments: A-type and B-type. The familiar Watson and Crick double helical
DNA model was derived using the B-type diffraction pattern.

A-DNA and B-DNA are different in several ways. a) In B-DNA the bases
occupy the center of the molecule and are almost perpendicular to the axis.
In A-DNA the base pairs are removed from the central axis of the molecule and
have a considerable tilt in the opposite direction. b) In B-DNA the rise per
residue along the axis is 3.34Å, which is the thickness of the unsaturated
purine-pyrimidine base pair. Due to the tilting of the bases in A-DNA the
rise per residue is 2.56Å along the axis. The diameter of the A-DNA molecule
is slightly larger. c) In A-DNA the minor groove is flat and wide while the
major groove is deep and extends through the central axis of the molecule.
The tilt of the base pairs brings the phosphate groups closer together on
opposite strands across the major groove of A-DNA than they are in B-DNA.
d) The vertical distance between the terminal phosphates of the two chains is
much smaller in A-DNA than in B-DNA.

The interesting octamer d(GpGpCpCpGpGpCpC)[7] has a central segment
containing the two different families of sugar pucker so that the molecular
structure has mixed A-DNA and B-DNA characteristics. Fig. 4.21-7 illustrates
some of the features of this structure in comparison with A- and B-DNA.

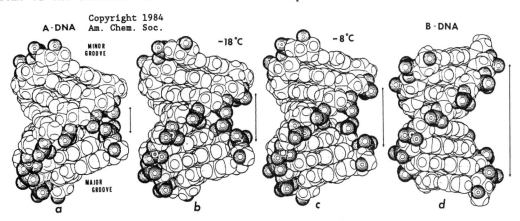

Fig. 4.21-7 Van der Waals drawing of the molecular structure of
d(GpGpCpCpGpGpCpC). The central diagrams b and c show the form of the octamer
observed experimentally at -18°C and at -8°C. Models of pure A-DNA (a) and B-
DNA (d) are shown on either side for comparison. The helix axis is vertical
and a horizontal two-fold axis is located in the plane of the paper. The
arrows to the right of each structure represent the vertical distance between
the terminal phosphates of the two chains. The phosphorus and oxygen atoms
have a heavier shading than the other atoms. The tilt to the bases from the
horizontal is 19° in A-DNA and -7° in B-DNA, while base pairs in b and c are
tilted approximately 14° and 12° respectively. (After ref. 6).

B-DNA to A-DNA conformational transition is a small change in comparison
to the molecular rearrangement which occurs when DNA adopts the left-handed

Z-DNA conformation. This left-handed structure was discovered in a
hexanucleoside pentaphosphate with the sequence $(dC-dG)_3$[8] and subsequently
with $(dC-dG)_2$.[9] Fig. 4.21-8 shows a van der Waals drawing of Z-DNA in
comparison to the more familiar right-handed B-DNA. The left-handed double
helix has the phosphate groups arranged in a zigzag array, hence the name
Z-DNA. The structure is favored by sequences which have an alternation of
purines and pyrimidines.

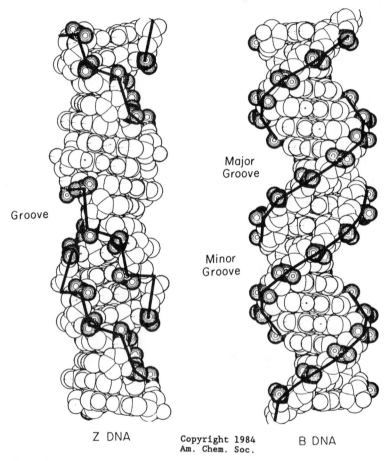

Groove

Major
Groove

Minor
Groove

Z DNA   Copyright 1984   B DNA
                Am. Chem. Soc.

Fig. 4.21-8 Van der Waals side views of Z-DNA and B-DNA. The zigzag
path of the sugar-phosphate backbone is shown by the heavy lines. The
groove in Z-DNA is deep, extending to the axis of the double helix.
This is in contrast to B-DNA which has two grooves. (After ref. 6).

The Z-DNA helix has 12 base pairs per turn of the helix with a pitch of
44.6Å, whereas in right-handed DNA it has 10 base pairs per turn occupying a
distance of 34Å. The diameter of the Z-DNA helix is only 18Å compared to the
somewhat thicker 20Å found in B-DNA. There is only one groove in Z-DNA

compared to two in B-DNA. The Watson-Crick base pairs that form the outer convex surface of the molecule in Z-DNA correspond to the major groove of the B-DNA helix.

DNA and RNA can form DNA-RNA hybrid molecules self-complementarily. The oligonucleotide r(GpCpG)d(TpApTpApCpGpC) contains three ribonucleotides at the 5' end connected covalently to seven deoxynucleotides.[10] This molecule is self-complementary and forms two DNA-RNA hybrid segments with three base pairs surrounding a central region of four base pairs of double helical DNA. All three parts of the molecule adopt a conformation which is close to that in A-DNA or in the 11-fold RNA double helix. Fig. 4.21-9 shows the structure in three different orientations 90° apart from each other about the dashed vertical axis. The left is a view into the broad, flat minor groove of the molecule. Phosphate groups on opposite strands are 8.6Å apart across the minor groove. The heavy black lines are drawn from phosphate to phosphate to show the flow of the polynucleotide chains. The ribonucleotides are shaded and the ribose 2' oxygen atoms are black. The center view shows both the minor groove on the upper part of the molecule and the major groove at the lower right. The shaded ribonucleotide backbone segments are close to each other on either side of the major groove. The bases are tilted 19° from the vertical axis. In this center view the tilt reverses itself in going from the upper part to the lower part of the molecule. The view on the right looks

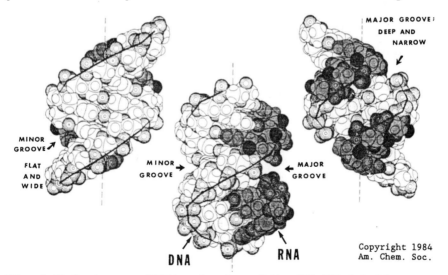

Fig. 4.21-9 A space-filling drawing of the DNA-RNA hybrid r(GpCpG)d(TpApTpApCpGpC). Oxygen atoms are solid circles; nitrogen atoms are stippled; phosphorus atoms have spiked circles. (After ref. 6).

down into the deep and narrow major groove of the molecule. The phosphate groups are 4.7Å away from each other across this groove.

3. Modes and sites of metal binding to nucleic acid components[11]

The significance of metal binding to nucleic acids and to enzymes involved in processes related to DNA replication, transcription, and messenger RNA translation has long been recognized. Nucleotides offer four different binding sites for metal ions: phosphate oxygens, ribose hydroxyls, base ring nitrogens, and keto substituents. Base amino groups do not participate in metal coordination. While transition metals bind to phosphate oxygens and base nitrogens, alkaline earth cations prefer phosphate oxygens and sugar hydroxyls over base nitrogens. Alkaline metal ions complex equally well to all three kinds of ligands. The affinity of metal ions for nucleotide-binding sites is rather specific, complexation to purine N(7) accommodating six-coordinate metals whereas pyrimidine N(3) allows only lower coordination numbers. As a rule, in purine bases the imidazole nitrogens are better ligands than pyrimidine nitrogens.

Current knowledge of this important topic is summarized in Table 4.21-1 with a few comments:

a) Binding sites N(1) and N(6)H_2 of A; N(3)H and O(4) of T; O(6), N(1)H, and N(2)H_2 of G; and O(2), N(3) and N(4)H_2 of C are employed in Watson-Crick interbase hydrogen-bonding modes. Metal binding at any of these sites must therefore have some effect on the ability of the bases to form hydrogen bonds of the Watson-Crick type. Some sites [N(1)H of G, N(1) of A, and N(3)H of T or N(3) of C] necessarily preclude the simultaneous presence of metal binding and Watson-Crick base pairing; other sites are less restrictive and may only weaken the base pairing. Metal bonding at N(7) of purines may, on balance, strengthen base pairing.

b) Many of the sites noted in Table 4.21-1 require deprotonation of the base.

c) Several "bonding modes" other than those given in Table 4.21-1 have been observed or suggested. For example, many dimeric compounds consisting of two nucleosides or nucleotides and two metal centers have been proposed. "Macrochelates" involving several phosphate oxygens and some ring nitrogens simultaneously bound to a metal have also been postulated. Evidence also exists that stacked complexes involving the interaction of the base portion of an uncomplexed nucleotide with the base portion of a complexed nucleotide can be formed. A coordinated water molecule may hydrogen bond to a ring nitrogen or a phosphate oxygen of a nucleotide that is directly bound to the metal via

Table 4.21-1 Possible metal binding modes and sites in nucleosides and
nucleotides.[11]

A. Monodentate binding modes	
I. **Base sites**	
(1) A	N(7), N(1), -N(6)H$_2$ (high pH)
(2) G	N(7), N(1) (high pH), -CO(6)
(3) C	N(3), -C=(2), -N(4)H$_2$ (high pH)
(4) U and T	-C=O(2), -C=O(4), N(3) (high pH)
II. **Sugar hydroxyls**	Weak and probably not important
III. **Phosphate oxygens**	Highly probable, especially with hard metals
B. Bidentate binding modes	
I. **Base site plus exocyclic functional group**	
(1) A	N(7), -N(6)H$_2$ chelation - a weakly possible mode at high pH, definitive evidence is not available.
(2) G	N(7), -C=O(6) or N(1) (deprotonated), -C=O(6) chelation - highly controversial, frequently speculated about in solution studies, only weakly in evidence from solid state work
(3) C	N(3), -C=O(2) chelation weak, but well established; N(3), -N(4)H$_2$ chelation - a weak possible mode at high pH, definitive evidence is not available
(4) U and T	N(3) (deprotonated) and -C=O(2) or -C=O(4) chelation - weak, but examples exist for Hg(II)
II. **N(ring), O(sugar)**	No true chelate of this type is known
III. **N(ring), O(phosphate)**	Likely, but definitive evidence is lacking
IV. **O, O(phosphate)**	Well established, especially for hard metals; tridentate chelate also has strong support
V. **O, O(sugar)**	An established bonding mode which is most commonly found at high pH

a phosphate oxygen (or a ring nitrogen). Such "indirect" chelates are also
possible when the coordinated hydrogen-bond donor is an amine. Furthermore,
exocyclic groups are likely to participate in such "indirect" chelates,
particularly O(6) and N(6)H$_2$ of N(7) bound purines.

4. Structural information on metal-nucleic acid interactions[11,12]

A large number of crystal structure determinations of metal-base
complexes and metal-nucleotide complexes have been performed. Some examples
are discussed below:

    a) Bis(acetylacetonato)(nitro)(deoxyadenosine)cobalt(III)[13]

The molecular structure of the deoxyadenosine complex is presented in
Fig. 4.21-10(a). The plane of the deoxyadenosine ligand approximately bisects
the O-Co-O bonds of two of the acetylacetonato oxygen atoms; in the adopted

molecular conformation the exocyclic amino group forms a bifurcated inter-
ligand hydrogen bond system.

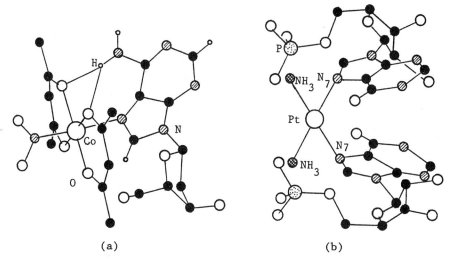

(a)   (b)

Fig. 4.21-10 Molecular structures of (a) bis(acetylacetonato)(nitro)-
(deoxyadenosine)cobalt(III) and (b) bis(inosine-5'-monophosphate)-
(diammine)platinum(II). (After refs. 13 and 14).

b) Bis(inosine-5'-monophosphate)(diammine)platinum(II)[14]

In this complex, Pt(II) binds through equatorial sites to two inosine-
5'-monophosphate ligands, which are related by a crystallographic C_2 axis
[Fig. 4.21-10(b)]. An important aspect of the complex is its nonstoichiometry, with only about 56% of the Pt sites occupied in the crystal. In the
isostructural diethyleneamine complex only about 36% of the Pt sites are
occupied.

c) Bis(1-methylthyminato)mercury(II)[15]

Deprotonation at N(3) of thymidine followed by metal attack at the
deprotonated site yields strong metal-N(3) bonding. In the Hg(II) complex
(Fig. 4.21-11) two strong linear Hg(II)-N(3) bonds as well as a number of weak
intramolecular and intermolecular interactions with O(2) and O(4) of the
deprotonated base are found.

d) (Nitrato)(1-methylcytosine)silver(I)[16]

This structure provides some insight into the binding of Ag(I) to
polynucleotides (Fig. 4.21-12). The most pronounced feature is the formation
of centrosymmetrical dimers in which the 1-methyl-cytosine ligands bridge two
symmetry-related Ag^+ ions. Within each dimers there are two strong metal-
ligand bonds: one to N(3) [2.225(2)Å] and one to O(2) [2.367(2)Å].
Interestingly, the dimeric units are packed into columns and connected by a

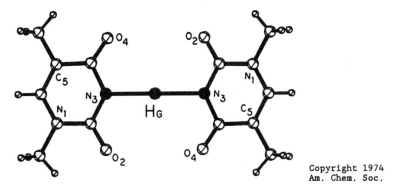

Fig. 4.21-11 Molecular structure of bis(1-methylthyminato)mercury(II). (After ref. 15).

second Ag-O(2) bond at 2.564(3)Å. Within each column there is significant base-base overlap (mean distance 3.34Å). The Ag...Ag repeat length (3.64Å) in the polymeric columnar stacks is reminiscent of the base-base stacking distance of about 3.5Å in duplex DNA. The nearly commensurate Ag...Ag and base-base distances suggest that cooperative propagation of base-Ag-base polymers parallel to the helical axis of a duplex DNA could be induced. In the light of the versatility of O(2) of cytosine residues, such polymeric fragments could be readily accommodated in regions of high G-C content.

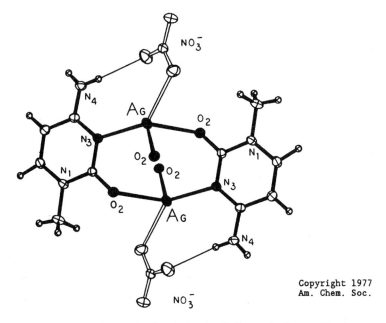

Fig. 4.21-12 Basic dimeric unit in (nitrato)(1-methylcytosine)-silver(I). (After ref. 16).

e) $[Co_2(H_2O)_4(5'-UMP)_2]_n$ (UMP, uridinemonophosphate)[17]

In this complex the metal binds exclusively to the phosphate group, and a polymeric chain is propagated through alternating metal-(phosphate oxygen)-metal binding, as shown in Fig. 4.21-13.

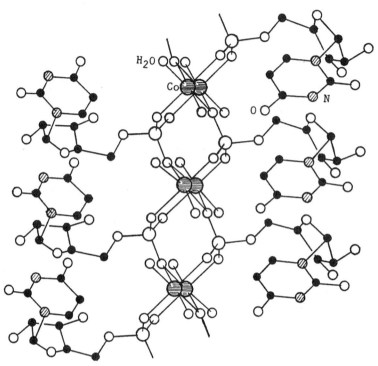

Fig. 4.21-13 Structure of a polymeric chain in $[Co_2(H_2O)_4(5'-UMP)_2]_n$, which is propagated by phosphate groups bridging neighbouring metal atoms. (After ref. 17).

f) Complex of terpyridineethanethiolatoplatinum(II) with d(CpG)[6,18]

The complex of $[Pt(terpy)(HET)]^+$ with the dinucleoside monophosphate d(CpG) exemplifies the stabilization generated by the stacking interaction mode [Fig. 4.20-4(b)]. The ring nitrogen atoms of the tightly chelating terpyridine ligand and the sulfur atom of the ethanethiolato anion comprise the primary coordination sphere of Pt(II). Guanine-cytosine base pairs are stacked above and below the platinum coordination plane at a distance of 3.4Å. The resulting stereochemistry resembles that of an intercalated fragment of a double-helical DNA polymer. The sugar rings adopt different puckering angles on either side of the intercalator, having a C(3') *endo* conformation at the 5'-end of the dinucleoside phosphate and the normal C(2') *endo* conformation at the 3'-end. The metal-coordinated terpyridine ligand thus serves to stabilize a specific binding mode. This complex is also discussed in Section 4.20.

References

[1] K. Aoki, *Bull. Chem. Soc. Jpn.* **48**, 1260 (1975).

[2] W. Saenger, *Principles of Nucleic Acid Structure*, Springer-Verlag, New York, 1984.

[3] L.G. Marzilli, T.J. Kistenmacher and C.-H. Chang, *J. Am. Chem. Soc.* **95**, 7507 (1973).

[4] Y. Mizumo, *The Organic Chemistry of Nucleic Acids*, Elsevier and Kodansha, Amsterdam, 1986.

[5] J.D. Watson and F.H.C. Crick, *Nature (London)* **171**, 737 (1953).

[6] A.H.-J. Wang in J.A. Vida and M. Gordon (eds.), *Conformationally Directed Drug Design*, American Chemical Society, Washington, DC, 1984, chap. 5.

[7] A.H.-J. Wang, S. Fujii, J.H. van Boom and A. Rich, *Proc. Natl. Acad. Sci. USA* **79**, 3968 (1982).

[8] A.H.-J. Wang, G.J. Quigley, F.J. Kolpak, G. van der Marel, J.H. van Boom and A. Rich, *Science (Washington)* **211**, 171 (1981).

[9] H. Drew, T. Takano, S. Tanaka, K. Itakura and R.E. Dickerson, *Nature (London)* **286**, 567 (1980).

[10] A.H.-J. Wang, S. Fujii, J.H. van Boom, G.A. van der Marel, S.A.A. van Boeckel and A. Rich, *Nature (London)* **299**, 601 (1982).

[11] D.J. Hodgson, *Prog. Inorg. Chem.* **23**, 211 (1977); T.G. Spiro (ed.), *Nucleic Acid-Metal Ion Interactions*, Wiley-Interscience, New York, 1980.

[12] S.E. Sherman and S.J. Lippard, *Chem. Rev.* **87**, 1153 (1987); W.I. Sundquist and S.J. Lippard, *Coord. Chem. Rev.* **100**, 293 (1990).

[13] T. Sorrell, L.A. Epps, T.J. Kistenmacher and L.G. Marzilli, *J. Am. Chem. Soc.* **99**, 2173 (1977).

[14] D.M.L. Goodgame, I. Jeeves, F.L. Phillips and A.C. Skapski, *Biochim. Biophys. Acta* **378**, 153 (1975).

[15] L.D. Kosturko, C. Folzer and R.F. Stewart, *Biochemistry* **13**, 3949 (1974).

[16] L.G. Marzilli, T.J. Kistenmacher and M. Rossi, *J. Am. Chem. Soc.* **99**, 2797 (1977).

[17] B.A. Cartwright, D.M.L. Goodgame, I. Jeeves and A.C. Skapski, *Biochim. Biophys. Acta* **477**, 195 (1977).

[18] A.H.-J. Wang, J. Nathans, G. van der Marel, J.H. van Boom and A. Rich, *Nature (London)* **276**, 471 (1978).

Note A comprehensive survey of the crystallographic structural information on oligonucleotides and oligonucleotide-drug complexes is presented in O. Kennard and W.H. Hunter, *Angew. Chem. Int. Ed. Engl.* **30**, 1254 (1991).

4.22   Dipotassium Octachlorodirhenate(III) Dihydrate
$K_2[Re_2Cl_8]\cdot 2H_2O$

Crystal Data

Triclinic, space group $P\bar{1}$ (No. 2)

$a = 6.752$, $b = 7.855$, $c = 7.610$Å, $\alpha = 102.0°$, $\beta = 108.9°$, $\gamma = 104.8°$,

$Z = 1$

Atom	x	y	z
Re	.8193	.9854	.9504
Cl(1)	.754	.046	.226
Cl(2)	.826	.279	.957
Cl(3)	.709	.909	.615
Cl(4)	.639	.675	.892
K	.878	.659	.326
O	.78	.46	.56

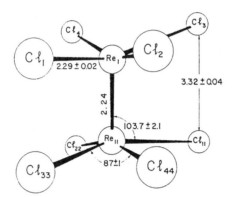

Copyright 1965
Am. Chem. Soc.

Fig. 4.22-1 Structure of the $Re_2Cl_8^{2-}$ ion in $K_2[Re_2Cl_8]\cdot 2H_2O$. (After ref. 1).

Crystal Structure

The crystal structure of $K_2[Re_2Cl_8]\cdot 2H_2O$ consists of K^+ and $[Re_2Cl_8]^{2-}$ ions and H_2O molecules. The $[Re_2Cl_8]^{2-}$ ion is located at a crystallographic center of symmetry, so that its two halves are in the eclipsed orientation relative to each other (Fig. 4.22-1). The principal dimensions indicate that the symmetry of the ion is D_{4h} within experimental error (Table 4.22-1).

Remarks

1. Metal-metal quadruple bonds

The first shortest rhenium-rhenium bond distance, 2.22Å, was discovered in a compound erroneously formulated as dimeric "(pyH)HReCl$_4$".[2] Since 1965, when Cotton and co-workers elucidated the nature of the quadruple bond in $K_2[Re_2Cl_8]\cdot 2H_2O$,[1] the subject of multiple bonds between metal atoms has evolved into a major research discipline.[3-6] Structural data for some compounds containing M≡M quadruple bonds are listed in Table 4.22-2.

Table 4.22-1   Bond lengths (Å) and angles (deg.) in $Re_2Cl_8^{2-}$.

Re_I-R_{II}	2.241(7)	Cl_1-Re_I-Cl_2	85.4(8)
Re_I-Cl_1	2.26(2)	Cl_2-Re_I-Cl_3	88(1)
Re_I-Cl_2	2.29(2)	Cl_3-Re_I-Cl_4	86.0(9)
Re_I-Cl_3	2.31(2)	Cl_4-Re_I-Cl_1	87.9(8)
Re_I-Cl_4	2.31(2)	Re_{II}-Re_I-Cl_1	105.8(6)
Cl_2-Cl_{44}	3.31(5)	Re_{II}-Re_I-Cl_2	103.4(7)
Cl_3-Cl_{11}	3.33(4)	Re_{II}-Re_I-Cl_3	101.7(6)
		Re_{II}-Re_I-Cl_4	103.6(6)

Table 4.22-2   Metal-metal quadruple bonds.

Bond	Compound	Bond length/Å
$Cr \equiv Cr$	$Cr_2(mhp)_4$	1.879(1)
$Mo \equiv Mo$	$Mo_2Cl_4(PMe_3)_4$	2.130(1)
$W \equiv W$	$W_2(O_2CPh)_4(THF)_2$	2.196(1)
$Tc \equiv Tc$	$(Bu_4N)_2Tc_2Cl_8$	2.147(4)
$Re \equiv Re$	$K_2[Re_2Cl_8] \cdot 2H_2O$	2.241(7)

mhp = 2-hydroxy-6-methylpyridine

In the bonding description of $Re_2Cl_8^{2-}$, the z coordinate axis is taken as the metal-metal bond axis, and the $d_{x^2-y^2}$ orbitals are assumed to become heavily involved in metal-ligand σ bonding. The d-orbital overlap scheme for the formation of a quadruple bond between two MX_4 units is shown in Fig. 4.22-2. The $[Re_2Cl_8]^{2-}$ ion has eight electrons to be placed in the bonding molecular orbitals, leading to a $\sigma^2\pi^4\delta^2$ configuration and hence a bond order of four.

For $Cr_2(mhp)_4$ the square shape of the experimental deformation electron density section through the midpoint of the metal-metal bond and perpendicular to the bond axis indicates a δ-contribution in addition to σ- and π-bonding between the chromium atoms (Fig. 4.22-3).[8]

The quadruple bond can undergo a variety of interesting reactions as outlined in Fig. 4.22-4.[7] (1) There is a rich chemistry in which ligands are exchanged and virtually every type of ligand can be used except for strongly π-accepting ones. (2) Addition of a mononuclear species to a $M \equiv M$ bond can yield a trinuclear cluster, e.g., $MoO_4^- + Mo_2(O_2CCH_3)_4 \longrightarrow [Mo_3O_2(O_2CCH_3)_6(H_2O)_3]^{2+}$. (3) Two quadruple bonds can combine to form a metallacyclobutadiyne ring. (4) Oxidative addition of acids to generate $M \equiv M$ bonds (particularly $W \equiv W$ bonds) is a key part of their chemistry. (5) Phosphines can act as reducing agents as well as ligands to give products with triple bonds of the $\sigma^2\pi^4\delta^2\delta^{*2}$ type. (6) Photoexcitation by the $\delta \rightarrow \delta^*$ transition can lead to reactive species which are potentially useful in

photosensitizing various reactions, including water splitting.   (7)
Electrochemical oxidation or reduction reduces the bond order and generates
reactive intermediates.   (8) With π-acceptor ligands the M≡M bonds are
usually cleaved to give mononuclear products, which are sometimes inaccessible
by any other synthetic route.

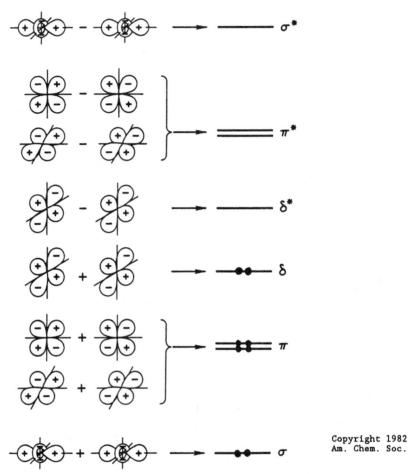

Fig. 4.22-2 Scheme of d-orbital overlap for the formation of a
quadruple bond between two metal atoms.   (After ref. 7).

2. Metal-metal triple bonds

When only six electrons are available for metal-metal bonding, the bond
order is three for the resulting $\sigma^2\pi^4$ configuration.   Addition of electrons
beyond the eight needed for a quadruple bond also gives rise to a triple bond
with the $\sigma^2\pi^4\delta^2\delta^{*2}$ configuration.   The first M≡M triple bond was discovered in
$Re_2Cl_6(dth)_2$ (dth = dithiahexane) in 1966.   The structure of $Mo_2(CH_2SiMe_3)_6$,
which contains the first reported Mo≡Mo triple bond, was determined in
1973.[9]   Table 4.22-3 lists some dinuclear compounds which contain metal-
metal triple bonds.

Fig. 4.22-3 Deformation density section through a plane perpendicular to the metal-metal bond through the bond midpoint in the $Cr_2(mhp)_4$ molecule; contour intervals at 0.05 $e\text{Å}^{-3}$. (After ref. 8).

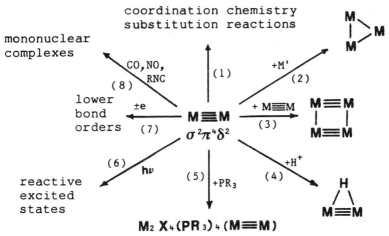

Fig. 4.22-4 Some reaction types of dimetal compounds containing quadruple bonds. (After ref. 7).

Table 4.22-3 Metal-metal triple bonds.[3,4,6]

Bond	Compound	Bond length/Å
V≡V	$V_2(C_8H_9O_2)_4$	2.200(2)
Nb≡Nb	$(NEt_4)_2[Nb_2Cl_6(\mu\text{-tetrathiophene})_3]$	2.632
Ta≡Ta	$Na_2[Ta_2Cl_6(\mu\text{-tetrathiophene})_3]$	2.626
Cr≡Cr	$Cr_2(CO)_4(\eta^5\text{-}C_5Me_5)_2$	2.280(2)
Mo≡Mo	$Mo_2(NMe_2)_6$	2.211(2), 2.217(2)
W≡W	$W_2(CH_2SiMe_3)_6$	2.255(2)
Re≡Re	$Re_2Cl_4(PEt_3)_4$	2.232(6)
Fe≡Fe	$Fe_2(\mu_2\text{-}CO)_3[C_4(Ph)_2(Bu)_2]_2$	2.177(3)
Ru≡Ru	$Ru_2(O_2CC_3H_7)_4Cl_2$	2.281(4)
Os≡Os	$Os_2(O_2CCH_3)_4Cl_2$	2.200(2)

The M≡M triple bond provides reactive functionalities and reaction types illustrating the chemistry of the metal-metal triple bond are shown in Fig. 4.22-5.[7] (1) Coordination chemistry through ligand substitution reactions

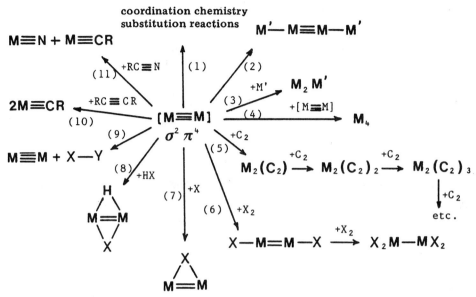

Fig. 4.22-5 Some reaction types of Mo≡Mo (or W≡W) compounds. (After ref. 7).

(both S_N1 and S_N2 mechanisms) is extensive, and substitution at one metal atom may influence the course of substitution at the other. (2) Polynuclear chain compounds incorporating the M≡M bond can be synthesized. (3) Homo- or heteronuclear trimetal clusters can be obtained by reacting a d^0 complex such as $MoO(OPr^i)_4$ with a triply-bonded molecule, e.g., $Mo_2(OPr^i)_6$; in this particular case one obtains the cluster $Mo_3(\mu_3\text{-}O)(\mu_3\text{-}OR)(\mu_2\text{-}OR)_3(OR)_6$, R = $CHMe_2$. (4) Replacement of alkoxy groups in $Mo_2(OR)_6$ compounds by halides, X, has provided a general route to compounds of formula $Mo_4X_4(OR)_8$ that may be viewed as dimers of $M_2X_2(OR)_4$ (M≡M). (5) The central Mo_2^{6+} unit provides a template for the oligomerization of alkynes and the co-oligomerization of alkynes and nitriles. (6-8) Oxidative addition reactions of various kinds are well established. (9) Reductive elimination reactions are also known. (10-11) Good examples of reactions that have no analogies in mononuclear metal chemistry are the triple bond metatheses whereby M≡CR and M≡N bonds may be formed.

3. Metal-metal double bond

    The first M=M bond was found in $Cs_3Re_3Cl_{12}$ which contains a triangular $[Re_3Cl_{12}]^{3-}$ anion, as shown in Fig. 4.22-6.[10] The mean Re-Re distance is

2.48Å, which is notably shorter than those (2.74 and 2.76Å) in rhenium metal[11] and indicates the extent of direct intermetal bonding in the Re₃ cluster. Some compounds containing M≡M bonds are listed in Table 4.22-4.

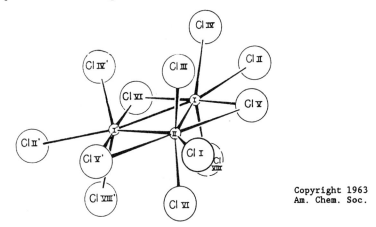

Fig. 4.22-6 Structure of the [Re₃Cl₁₂]³⁻ ion. (After ref. 10).

The chemistry of the metal-metal double bond is less well developed than that of the higher bond orders, but the reaction types shown in Fig. 4.22-7 suggest that a rich chemistry may be expected: (1) simple oxidative addition; (2) coupling of ketones concommitant with oxidative addition; (3) for Nb and Ta compounds, addition of H₂, leading to an intermediate that can be reduced and then subjected to further oxidative additions; (4) with Nb and Ta compounds, the cyclotrimerization or polymerization of internal acetylenes is catalyzed; and (5) again with Nb and Ta compounds, nitriles are coupled to form C=C bonds.

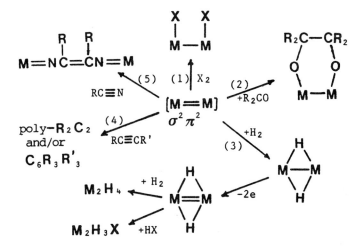

Fig. 4.22-7 Some reaction types of dinuclear compounds containing metal-metal double bonds. (After ref. 7).

Table 4.22-4   Metal-metal double bonds.[3,4]

Bond	Compound	Bond length/Å
V=V	$V_2(CO)_5(\eta_5-C_5H_5)_2$	2.462(2)
Nb=Nb	$Nb_2Br_2(\mu-THT)_3$	2.728(5)
Ta=Ta	$Ta_2Cl_2(\mu-THT)_3$	2.681(1)
Mo=Mo	$Mo_2(NMe_2)_4$	2.523(1)
W=W	$W_2Cl_4(OCH_3)_4(CH_3OH)_2$	2.479(1)
Re=Re	$Cs_3Re_3Cl_{12}$	2.477(3)
Fe=Fe	$Fe_2(CO)_6(DTBA)$	2.316(1)
Ru=Ru	$Ru_2(C_{22}H_{22}N_4)_2(BF_4)_2$	2.379(1)
Os=Os	$Os_3(\mu_2-H)_2(CO)_{10}$	2.683(1)
Co=Co	$[As_3Co(\mu-H)_3CoAs_3]^+$	2.377(8)

4. Metal-metal single bonds and bonds of non-integral order

Hg_2X_2 (Section 3.43) and $M_2(CO)_{10}$ (M = Mn, Tc, Re) (Section 6.7) are common examples of dinuclear compounds containing metal-metal single bonds. In the context of the present discussion, relevant examples of transition-metal compounds known to possess M-M single bonds are listed in Table 4.22-5.

Table 4.22-5   Metal-metal single bonds.[3,4,6]

Bond	Compound	Bond length/Å
Zr-Zr	$[ZrCl_2(PEt_3)_2](\mu-Cl)_2$	3.169
V-V	$(Et_4N)_2[\{V(S_2C_2H_4)\}_2(\mu,\eta^2-S_2C_2H_4)_2]$	2.626
Nb-Nb	$Nb_2Cl_8(PPhMe_2)_4$	2.838
Ta-Ta	$[TaH_2\{OSi(Bu^t)_3\}_2]_2$	2.720
Mo-Mo	$Mo_2(OPr^i)_6Cl_4$	2.731
W-W	$W_2S_2(S_2CNEt_2)_2(OMe)_4$	2.791
Fe-Fe	$[PPN^+]_2[Fe(CO)_8]^{2-}$	2.79
Os-Os	$Os_2(\mu-O_2CMe)_2(CO)_6$	2.731
Co-Co	$Co_2(triaz)_4$ quadruply bridged	2.265
Rh-Rh	$Rh_2(O_2CMe)_4(H_2O)_2$	2.39
Ir-Ir	$Ir_2(form)_4$ quadruply bridged	2.524
Ni-Ni	$[Ni_2(CN)_6]^{4-}$	2.32
Pd-Pd	$[Pd_2(CNMe)_6]^{2+}$	2.531
Pt-Pt	$Pt_2\{(Bu^t)_2PCH_2CH_2CH_2P(Bu^t)_2\}_2$	2.765

The electronic configuration giving rise to most of these bonds is presumably $\sigma^2\pi^4\delta^2\delta^{*2}\pi^{*4}$. When a species is oxidized to give a positive ion, such as $[Rh_2(O_2CMe)_4(H_2O)_2]^+$, the bond order becomes 1.5 with electronic configuration $\sigma^2\pi^4\delta^2\delta^{*2}\pi^{*3}$.

A large number of diruthenium compounds of the type $Ru_2(O_2CR)_4X$ is known, in which the bond order of 2.5 is based on electronic configuration $\sigma^2\pi^4\delta^2\delta^{*2}\pi^*$ (or $\sigma^2\pi^4\delta^2\pi^{*2}\delta^*$). In 1965, the bond order 3.5 based on $\sigma^2\pi^4\delta^2\delta^*$

was first recognized in $[Tc_2Cl_8]^{3-}$, and in 1973 $[Mo_2(SO_4)_4]^{3-}$ was found to have a metal-metal bond of order 3.5 based on $\sigma^2\pi^4\delta$.

In dealing with multiple bonds between atoms, especially the δ components of such bonds, it has been found that the bond length and bond order are not necessarily inversely related, as shown by the data for the following two series of complexes.[4]

Species	Configuration	Bond order	Bond length/Å
$[Tc_2Cl_8]^{3-}$	$\sigma^2\pi^4\delta^2\delta^*$	3.5	2.110(5)
$[Tc_2Cl_8]^{2-}$	$\sigma^2\pi^4\delta^2$	4.0	2.150(5)
$[Re_2Cl_4(PMe_2Ph)_4]^0$	$\sigma^2\pi^4\delta^2\delta^{*2}$	3.0	2.241(1)
$[Re_2Cl_4(PMe_2Ph)_4]^+$	$\sigma^2\pi^4\delta^2\delta^*$	3.5	2.218(1)
$[Re_2Cl_4(PMe_2Ph)_4]^{2+}$	$\sigma^2\pi^4\delta^2$	4.0	2.215(1)

As the mean oxidation number of the metal atom increases, the metal d orbitals contract, thus weakening the M-M bonding, including the dominant σ and π components.   There are therefore two countervailing trends: (i) because of the increasing oxidation number of the metal atom, the bonds tend to get longer; and (ii) in view of the increasing bond order, the bonds tend to become shorter.

5. Tetra-bridged diplatinum complexes

Many tetra-bridged diplatinum complexes have been prepared and characterized.   Table 4.22-6 are listed some examples of the sulfato-bridged,

Table 4.22-6   Metal-metal bond lengths in diplatinum compounds.

Compound	Pt-Pt Bond lengths/Å
diplatinum(III,III) complexes; $(\sigma)^2$	
$K_2[Pt_2(SO_4)_4(H_2O)_2]$	2.461(1)
$K_2[Pt_2(SO_4)_4(OSMe_2)_2] \cdot 4H_2O$	2.471(1)
$Na_2[Pt_2(HPO_4)_4(H_2O)_2]$	2.486(1)
$(3,4\text{-}Me_2pyH)[Pt_2(HPO_4)_4(3,4\text{-}Me_2py)_2]$	2.494(1)
$(Bu_4N)_2[Pt_2(P_2O_5H_2)_4(CH_3CN)_2]$	2.676(1)
$K_4[Pt_2(P_2O_5H_2)_4Cl_2]$	2.695(1)
$K_4[Pt_2(P_2O_5H_2)_4(Im)_2] \cdot 7H_2O$	2.745(1)
$K_4[Pt_2(P_2O_5H_2)_4(NO_2)_2] \cdot 2KNO_2 \cdot 2H_2O$	2.754(1)
$K_4[Pt_2(P_2O_5H_2)_4(SCN)_2] \cdot 2H_2O$	2.760(1)
$K_4[Pt_2(P_2O_5H_2)_4(CH_3)I] \cdot 2H_2O$	2.782(1)
diplatinum(II,III) complex; $(\sigma)^2(\sigma^*)^1$	
$K_4[Pt_2(P_2O_5H_2)_4Br] \cdot 2H_2O$	2.793(1)
diplatinum(II,II) complex; $(\sigma)^2(\sigma^*)^2$	
$K_4[Pt_2(P_2O_5H_2)_4] \cdot 2H_2O$	2.925(1)

ImH = imidazole

orthophosphato-bridged and diphosphito-bridged diplatinum compounds and their Pt-Pt bond lengths.[4,12,13] In the $[Pt_2(P_2O_5H_2)_4XY]^{4-}$ species the Pt-Pt bond length decreases in the order XY = CH_3I > $(SCN)_2$ > $(NO_2)_2$ > $(Im)_2$ > Cl_2, which parallels the corresponding decrease in axial-ligand donor strength in a manner similar to the established structural *trans*-influence series in mononuclear platinum(II) complexes.[14] The higher *trans* influence of SCN^- over the NO_2^- and Im^- ligands indicates that σ-electronic delocalization between the Pt-Pt and Pt-X bonds plays a crucial role in the weakening of the metal-metal bond.

Fig. 4.22-8 shows the structures of a sulfato-bridged diplatinum compound and an orthophosphato-bridged diplatinum compound. Fig. 4.22-9 shows two diphosphito-bridged diplatinum complexes.

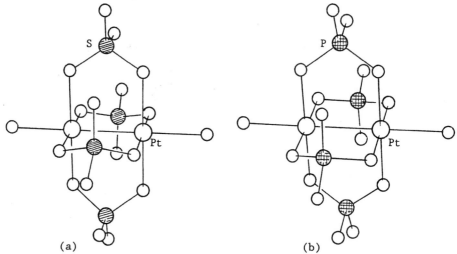

(a)   (b)

Fig. 4.22-8 Structure of (a) the $[Pt_2(SO_4)_4(H_2O)_2]^{2-}$ anion in $K_2[Pt_2(SO_4)_4(H_2O)_2]$ and (b) the $[Pt_2(HPO_4)_4(H_2O)_2]^{2-}$ anion in $Na_2[Pt_2(HPO_4)_4(H_2O)_2]$. (After ref. 4).

6.   Some bis(diphenylphosphino)methane-bridged dinuclear complexes

Diphosphine ligands of the type $R_2P(CH_2)_nPR_2$ can form either chelated or bridged complexes.[15] One of the most widely used bifunctional bridging phosphine ligand is bis(diphenylphosphino)methane, (abbreviated as dppm),[16] which gives rise to three classes of dinuclear compounds:

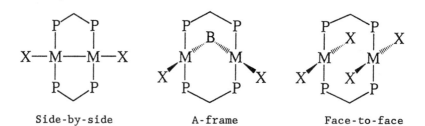

Side-by-side   A-frame   Face-to-face

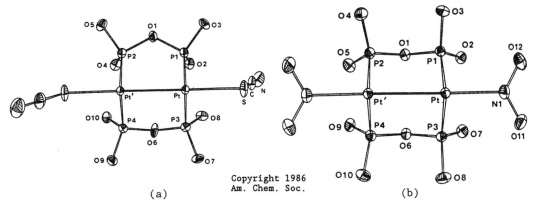

Copyright 1986
Am. Chem. Soc.

(a)   (b)

Fig. 4.22-9 Structure of the centrosymmetric dinuclear anion in
(a) $K_4[Pt_2(P_2O_5H_2)_4(SCN)_2] \cdot 2H_2O$ and (b) $K_4[Pt_2(P_2O_5H_2)_4(NO_2)_2] \cdot 2H_2O$.
For clarity, two of the four bridging diphosphite ligands have been
omitted. (After ref. 12).

Owing to the proximity of the metal centers, these complexes exhibit
interesting chemistry involving (i) oxidation of one or both metals with
formation of a M-M bond, (ii) addition of neutral molecules with M-M bond
formation, or (iii) formation of bridging groups such as μ-CO, μ-CH$_2$, η^2-CO
or η^2-RNC.

Fig. 4.22-10 shows two dinuclear complexes bridged by dppm ligands with
little, if any, bonding interaction between the metal atoms. The doubly-

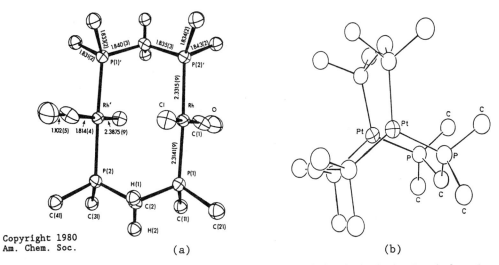

Copyright 1980
Am. Chem. Soc.   (a)   (b)

Fig. 4.22-10 Molecular structures of (a) *trans*-[Rh$_2$Cl$_2$(CO)$_2$(μ-dppm)$_2$] and
(b) Pt$_2$(μ-dppm)$_3$, with all but *ipso*-carbon atoms of the phenyl groups omitted
for clarity. (After refs. 17 and 18).

bridged, face-to-face *trans*-[$Rh_2Cl_2(CO)_2(\mu$-dppm)$_2$] dimer is centrosymmetric
with a non-bonded Rh...Rh separation of 3.2386(5)Å.[17] The $Pt_2(\mu$-dppm)$_3$
molecule adopts a manxane-type (bicyclo[3.3.3]undecane-type, idealized
symmetry C_{3h}) structure with trigonal-planar coordination of the Pt(0) atoms,
the Pt...Pt separation being 3.023(1)Å.[18] $Pt_2(\mu$-dppm)$_3$ can be easily
protonated to yield [$Pt_2H(\eta^1$-dppm)(μ-dppm)$_2$]$^+$, and this oxidative-addition
reaction occurs with the cleavage of a Pt-P bond in the parent compound and
the formation of a Pt(I)-Pt(I) bond [2.769(1)Å] in the product.

Fig. 4.22-11(a) shows the structure of an A-frame triplatinium complex
bridged by a pair of μ-dppm ligands.[19] The average Pt-Pt bond length is
2.593Å, and the Pt(1)−Pt(2)−Pt(3) bond angle is 80°. The coordination
geometry about each Pt atom is planar.

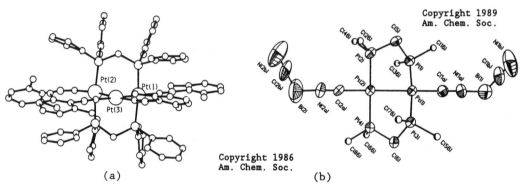

              (a)   (b)

Fig. 4.22-11 Molecular structures of (a) the trinuclear dication in
[$Pt_3(2,6$-$Me_2C_6H_3NC)_4(\mu$-dppm)$_2$](PF_6)$_2 \cdot CH_2Cl_2$ and (b) [$Pt(\mu$-dppm)($CNBH_2CN$)]$_2$;
only *ipso*-carbon atoms of phenyl rings are shown. (After refs. 19 and 20).

Many dipalladium and diplatinum complexes doubly bridged by μ-dppm
ligands have been prepared; the metal atoms have formal oxidation state +1.
In the following series of [$Pt_2(\mu$-dppm)$_2LL'$]$^{n+}$ complexes, abbreviated as
[L...L']$^{n+}$, the Pt-Pt bonds lie in the range 2.620-2.769Å.[20,21] A
representative example is shown in Fig. 4.22-11(b).[20]

[Cl....CO]$^+$	2.620	[NCBH$_2$NC....CNBH$_2$CN]	2.665
[OC....CO]$^{2+}$	2.642	[H$_3$BNC....CNBH$_3$]	2.667
[Cl....Cl]	2.651	[NC....CN]	2.704
[Cl....PPh$_3$]$^+$	2.665	[H....(η^1-dppm)]$^+$	2.769

References

[1] F.A. Cotton and C.B. Harris, *Inorg. Chem.* 4, 330, (1965); F.A. Cotton,
     ibid., p. 334.

[2] V.G. Kuznetsov and P.A. Kozmin, *Zh. Strukt. Khim.* 4, 55 (1963).

[3] F.A. Cotton and R.A. Walton, *Multiple Bonds between Metal Atoms*, Wiley,
     New York, 1982.

[4] F.A. Cotton and R.A. Walton, *Structure and Bonding* **62**, 1 (1985).

[5] J. Templeton, *Prog. Inorg. Chem.* **26**, 211 (1979).

[6] L. Messerle, *Chem. Rev.* **88**, 1229 (1988); F.A. Cotton and R. Poli, *Inorg. Chem.* **26**, 3652 (1987); *Polyhedron* **6**, 1625 (1987).

[7] F.A. Cotton, *J. Chem. Educ.* **60**, 713 (1983).

[8] A. Mitschler, B. Rees, R. Wiest and M. Benard, *J. Am. Chem. Soc.* **104**, 7501 (1982).

[9] W. Mowat and G. Wilkinson, *J. Chem. Soc., Dalton Trans.*, 1120 (1973).

[10] J.A. Bertrand, F.A. Cotton and W.A. Dollase, *Inorg. Chem.* **2**, 1166 (1963).

[11] C.T. Sims, C.M. Craighead and R.I. Jaffee, *J. Metal.* **7**, 168 (1955).

[12] C.-M. Che, W.-M. Lee, T.C.W. Mak and H.B. Gray, *J. Am. Chem. Soc.* **108**, 4446 (1986).

[13] C.-M. Che, T.C.W. Mak, V.M. Miskowski and H.B. Gray, *J. Am. Chem. Soc.* **108**, 7840 (1986); and references cited therein.

[14] T.G. Appleton, H.C. Clark and L.E. Manzer, *Coord. Chem. Rev.* **10**, 335 (1973).

[15] F.A. Cotton and G. Wilkinson, *Advanced Inorganic Chemistry*, 5th ed., Wiley-Interscience, New York, 1988, pp. 434, 900.

[16] R.J. Puddephatt, *Chem. Soc. Rev.* **12**, 99 (1983).

[17] M. Cowie and S.K. Dwight, *Inorg. Chem.* **19**, 2500 (1980).

[18] L. Manojlovic-Muir and K.W. Muir, *J. Chem. Soc., Chem. Commun.*, 1155 (1982).

[19] Y. Yamamoto, K. Takahashi and H. Yamazaki, *J. Am. Chem. Soc.* **108**, 2458 (1986).

[20] M.N.I. Khan, C. King, J.-C. Wang, S. Wang and J.P. Fackler, Jr. *Inorg. Chem.* **28**, 4656 (1989).

[21] R.J. Blau and J.H. Espenson, *Inorg. Chem.* **25**, 878 (1986).

Note The Co=Co bond length in the new "dimetal sandwich" $[(\eta^5\text{-}C_5Me_5)_2Co_2]$ is 2.253(1)Å, as reported in J.J. Schneider, R. Goddard, S. Werner and C. Krüger, *Angew. Chem. Int. Ed. Engl.* **30**, 1124 (1991).

Recent advances in the field of multiple metal-metal bonding are reviewed in J.P. Fackler, Jr. (ed.), *Metal-Metal Bonds and Clusters in Chemistry and Catalysis*, Plenum Press, New York, 1990.

4.23   Benzyltriphenylphosphonium Dodeca-μ-carbonyl-dodecacarbonyldihydrido-*polyhedro*-tridecarhodate

$$[P(CH_2Ph)Ph_3]_3[Rh_{13}H_2(CO)_{24}]$$

Crystal Data

Monoclinic, space group $P2_1/n$ (No. 14)

$a = 14.334(2)$, $b = 29.231(4)$, $c = 24.626(3)$Å, $\beta = 94.61(3)°$, $Z = 4$

Atom*	x	y	z	Atom	x	y	z
Rh(1)	.2938	.1312	.3071	Rh(8)	.0151	.2234	.2903
Rh(2)	.2030	.2130	.3257	Rh(9)	.0464	.1979	.1835
Rh(3)	.2338	.1907	.2192	Rh(10)	-.0464	.1152	.2023
Rh(4)	.1333	.1146	.1677	Rh(11)	.0107	.0612	.2901
Rh(5)	.1894	.0572	.2547	Rh(12)	-.0816	.1424	.3073
Rh(6)	.1643	.0831	.3611	Rh(13)	.1051	.1413	.2745
Rh(7)	.0717	.1642	.3794				

* The coordinates of the other atoms are omitted.

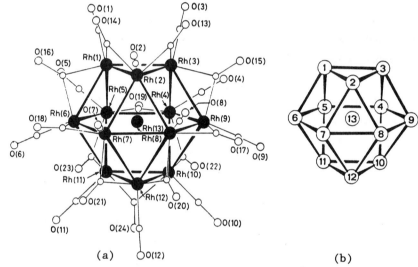

Fig. 4.23-1   (a) Structure of the $[Rh_{13}H_2(CO)_{24}]^{3-}$ anion.   (b) A view of the Rh_{13} Cluster.   (After ref. 1).

Crystal Structure

The stereochemistry of the non-hydrogen atoms of the $[Rh_{13}H_2(CO)_{24}]^{3-}$ cluster is illustrated in Fig. 4.23-1(a); it bears a very close resemblance to that of $[Rh_{13}H_3(CO)_{24}]^{2-}$.[2] The metal atom cluster consists of a closed polyhedron, of D_{3h} idealized symmetry, representing a fragment of hexagonal close packing, as shown in Fig. 4.23-1(b). The 13 atoms are disposed in three layers: two outer triangles with the same orientation and a central centered hexagon. The unique central atom exhibits true metallic 12-coordination, while the surface atoms each has five metal-metal connections. The polyhedron contains both triangular (eight) and square (six) faces. The Rh-Rh distances are in the range 2.746-2.887Å; these values are not very scattered and can be considered as normal.

Half of the carbonyl ligands are terminally bound, one per surface Rh atom, and the remaining twelve symmetrically bridge half of the polyhedron edges in such a way as to give three Rh-C connections per metal atom. In this way, bridged edges in the upper part of the molecule correspond to unbridged ones in the lower part and *vice versa*. The average Rh-C and C-O distances for the terminal and bridging groups are 1.81, 1.17 and 2.00, 1.20Å, respectively.

In order to locate the hydridic hydrogen atoms, the polyhedron described by the carbon atoms of the ligands is illustrated in Fig. 4.23-2. The largest holes on the cluster surface have been found on three square faces defined by atoms Rh(1,3,4,5), Rh(6,7,11,12), and Rh(8,9,10,12). In fact, the bridging ligands located around these faces are one, two and two, respectively, while the remaining square faces bear three or four bridging groups. The first face is bisected by the idealized symmetry plane; the other two are related by the plane. On the basis of steric arguments, therefore, the hydrogen atoms can be located in the vicinity of these three faces, either as outer ligands or in semi-interstitial positions.

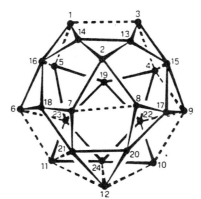

Fig. 4.23-2 Polyhedron described by the carbon atoms on the cluster surface. Full lines indicate distances under 3.10Å, broken lines those in the range 3.12-3.16Å. The largest holes in the vacinity of the faces C(1,3,4,19,5), C(9,10,12,20,17) and C(6,11,12,21,18) correspond to the cluster cavities occupied by hydrogen atoms. (After ref. 1).

Remarks

1. Structure of high nuclearity transition metal clusters

A large number of high nuclearity clusters, with structures resembling fragments of bulk metal lattices, i.e. body centered cubic (bcc), cubic close packed (ccp) and hexagonal close packed (hcp), have been characterized.[3]

Fig. 4.23-3 and Fig. 4.23-4 show some examples of high nuclearity transition metal clusters.[4]

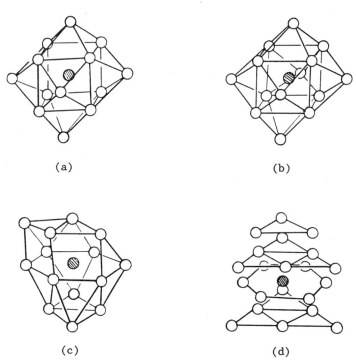

(a)   (b)

(c)   (d)

Fig. 4.23-3 Structures of some high nuclearity rhodium clusters: (a) $[Rh_{14}(CO)_{26}]^{2-}$ (bcc), (b) $[Rh_{15}(CO)_{30}]^{3-}$ (bcc), (c) $[Rh_{15}(CO)_{27}]^{3-}$ (bcc/hcp), (d) $[Rh_{22}(CO)_{37}]^{4-}$ (hcp/ccp). (After ref. 4).

2.   Principle of polyhedral inclusion[4]

Mingos has developed an electron counting procedure (which may be described as "polyhedral inclusion") that is based on the formal division of the high nuclearity cluster into an internal (encapsulated) polyhedron and an external (surface) polyhedron.[5] Three sub-classes of "close-packed" cluster have been identified:

a)   Clusters where radial metal-metal bonding predominates

In this case, the surface atoms act as ligands towards the central atom or cluster, to which they are bonded by radial (σ) bonds. The surface tangential bonding is not strong in these clusters. Cluster valence electron counts (Ne) for such clusters are given by:

$$Ne = 12n_s + \Delta_i$$

where n_s is the number of surface atoms and Δ_i is the electron count characteristic of the central atom or cluster, as listed in Table 4.23-1.

Table 4.23-1 Characteristic electron counts (Δ_i) for interstitial
moieties in high nuclearity clusters.

No. of interstitial atoms	Δ_i	Molecular analogue
1	18	$Mo(CO)_6$
2 (dimer)	34	$Mn_2(CO)_{10}$
3 (triangle)	48 (and 50*)	$Os_3(CO)_{12}$
3 (linear fragment)	50	$OsRe_2(CO)_{14}$
4 (tetrahedron)	60	$Ir_4(CO)_{12}$
6 (octahedron)	86	$Rh_6(CO)_{16}$

* Although isolated triangular clusters are charaterized by 48 electrons, the
presence of bridging metal atoms can lead to the stabilization of an a_2' MO and
a valence electron count of 50 for the central triangle.[6]

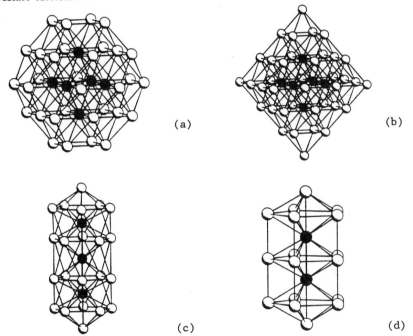

(a) (b)

(c) (d)

Fig. 4.23-4 Structures of high nuclearity clusters: (a) $[Pt_{38}(CO)_{44}H_m]^{2-}$
(ccp), (b) $[Ni_{38}Pt_6(CO)_{48}H_{6-n}]^{n-}$ (ccp), (c) $[Au_{13}Ag_{12}Cl_6(PPh_3)_{12}]^{m+}$
(icp)*, (d) $[Pt_{19}(CO)_{22}]^{4-}$ (icp)*. *"icp" = "icosahedral close
packing", a term used to denote high nuclearity clusters with 5-fold
symmetry.[5] (After ref. 4).

Examples of this class of cluster are listed in Table 4.23-2. An
important feature of these clusters is that their structures are independent
of the electron count of the surface polyhedron. Since radial metal-metal
bonding predominates, there is a soft potential energy surface for cluster
rearrangement. This has been experimentally observed for gold clusters.

Table 4.23-2   Examples of high-nuclearity cluster compounds where radial metal-metal bonding interactions predominate.

Cluster	n_i	n_s	Structure	Electron count
$[Au_9(PPh_3)_8]^+$	1	8	bcc	114
$[Au_{11}I_3(PPH_3)_7]$	1	10	bcc/icp	$138 \; (= 12n_s + 18)$
$[Au_{13}Cl_2(PMePh_2O_{10}]^{3+}$	1	12	icp	162
$[Pt_{19}(CO)_{22}]^{4-}$	2	17	icp	238
$[Rh_{22}(CO)_{35}H_{4-q}]^{4-}$	2	20	ccp/bcc	$273 + 1 \; (= 12n_s + 34)$
$[Au_{13}Ag_{12}Cl_6(PPh_3)_{12}]^{3+}$	3	22	icp	$314 \; (= 12n_s + 50)$
$[Pt_{26}(CO)_{32}H_x]$	3	23	hcp	$324 + x \; (= 12n_s + 48 + x)$
$[Ni_{38}Pt_6(CO)_{48}H_{6-n}]^{n-}$	6	38	ccp	$542 \; (= 12n_s + 86)$
$[Pt_{38}(CO)_{44}]^{2-}$	6	32	ccp	$470 \; (= 12n_s + 86)$

b)   Clusters with a partical bonding contribution from the surface tangential orbitals

All of the clusters of this type (Table 4.23-3) possess a single interstitial atom and are characterized by $(12n_s + 18 + x)$ cluster valence electrons. These clusters generally have 3 extra occupied skeletal MO's (i.e. $x = 6$), which are exclusively involved in tangential bonding in the external cluster polyhedron.

c)   Clusters with a complete bonding contribution from the surface tangential orbitals

The electron count in these clusters, such as $[Rh_{13}(CO)_{24}H_5]$ (Fig. 4.23-1) and $[Rh_{15}(CO)_{30}]^{3-}$ [Fig. 4.23-3(b)], as for main group clusters, is governed by the arrangement of the skeletal (surface) atoms. For *closo*-type structures the electron count is $(14n_s + 2)$.[7]

Table 4.23-3   Examples of high-nuclearity cluster compounds where tangential interactions make a partial contribution to metal-metal bonding.

Cluster	n_i	n_s	Structure	Electron count
$[Rh_{14}(CO)_{25}]^{4-}$	1	13	bcc	180
$[Rh_{14}(CO)_{26}]^{2-}$	1	13	bcc	180
$[Rh_{15}(CO)_{27}]^{3-}$	1	14	bcc/hcp	$192 \; (= 12n_s + 24)$
$[Rh_{17}(CO)_{30}]^{3-}$	1	16	hcp	216
$[Rh_{22}(CO)_{37}]^{4-}$	1	21	ccp/hcp	276
$[Pt_{24}(CO)_{30}]^{2-}$	1	23	ccp	$302 \; (= 12n_s + 26)$

3.   M$_{13}$ clusters as building blocks of "superclusters"

If the M_{13} cluster is regarded as spherically shaped, a number of them would tend to adopt a cubic closest packed structure, which in the simplest case would be made up of thirteen M_{13} clusters. Conceivably these $(M_{13})_{13}$

clusters can also aggregate in an organized way to form a second superstructure $[(M_{13})_{13}]_n$ consisting of closest packed $(M_{13})_{13}$ clusters.

Schmid and co-workers[8] have confirmed this concept from Debye-Scherrer diffraction patterns of the powder samples. The metal atoms originating from the outer shell of the cluster upon degradation form "normal" metal (M_∞) which gives rise to the correspondingly well-known X-ray diffraction pattern. In addition, reflections are observed that are in very good agreement with the most important lattice planes to be expected for the superstructures. Thus, for all metals, one finds three new lattice planes characterizing the superstructures that are integral multiples of the interplanar spacings (*d*) of the metals. Rhodium, platinum and gold all crystallize in the cubic system, as do also the superstructures. Ruthenium, on the other hand, has a hexagonal structure; nevertheless, the three additional *d* spacings are identical to those found with (the equally large) rhodium, i.e. ruthenium also forms $(Ru_{13})_{13}$ superclusters with cubic structure upon cluster degradation, as shown in Table 4.23-4.

Table 4.23-4 Powder diffraction data of $(M_{13})_{13}$, $[(M_{13})_{13}]_n$, and (for comparison) M_∞.

M	M_∞ d/Å	hkl	$(M_{13})_{13}$ d/Å	$[(M_{13})_{13}]_n$ d/Å	Occupation	$2\theta(°)$
Rh	1.902	200		15.3	8 x d(M_∞) = 15.216	5.8
	0.8725	331	6.11(d^2)		7 x d(M_∞) = 6.1075	14.5
	1.0979	222	3.24(d^1)		3 x d(M_∞) = 3.2937	7.5
Ru*				15.3		5.8
			6.11(d^2)			14.5
			3.24(d^1)			27.5
Au	2.093	200		16.8	8 x d(M_∞) = 16.312	5.3
	0.9358	331	6.8(d^2)		7 x d(M_∞) = 6.5551	13.0
	1.1774	222	3.5(d^1)		3 x d(M_∞) = 3.5322	14.1

* Ru_∞ crystallizes hexagonally; consequently, its structural data are not to be compared with those for $(Ru_{13})_{13}$.

In Table 4.23-4 the new 2θ-values and lattice planes are collected together with the corresponding values for the normal metals M_∞, where M = Rh, Ru and Au. Fig. 4.23-5 shows the connection between the observed X-ray reflections and the architecture of the superclusters. Fig. 4.23-5(a) demonstrates that, looking along the [111] direction of the marked M_{13} ccp cluster, the lattice planes are also common to the $(M_{13})_{13}$ supercluster and the $[(M_{13})_{13}]_n$ superstructure. Lattice planes corresponding to new periodicities for two neighbouring M_{13} clusters are illustrated in Fig.

4.23-5(b) and (c). In Fig. 4.23-5(d) is shown the crucial 8 x d_{200}
periodicity, which is a measure of the diameter of a $(M_{13})_{13}$ supercluster.
That $[(M_{13})_{13}]_n$ is composed of closest packed $(M_{13})_{13}$ clusters is thus
substantiated by the occurrence of lattice planes with d spacings of 15.3 and
16.8Å in Table 4.23-4. The magnitude of n is much larger than 13 since
microcrystals of gold are known to have diameters of a few nanometers.

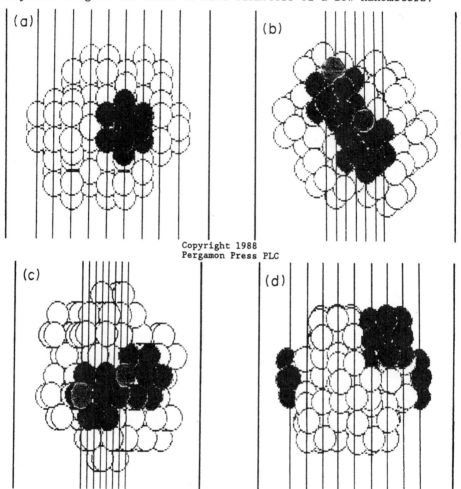

Fig. 4.23-5 Different views of a $(M_{13})_{13}$ particle. (a) Viewed along the
[111] direction of the ccp M_{13} clusters. One M_{13} cluster is drawn in black to
emphasize the structure principle. (b) Viewed along the [222] direction of
the M_{13} clusters. The two grey shaded atoms show a new periodicity (3 x d_{222})
of the superstructure. (c) Viewed along the [331] direction of the M_{13}
clusters. Two M_{13} clusters are drawn in black. The grey shaded atoms show a
new periodicity (7 x d_{331}) of the superstructure. (d) Viewed along the [200]
direction of the M_{13} clusters. A new periodicity (8 x d_{200}) is determined by
the distance from the first (black) row to the last (black) row. (After
ref. 9).

4. Ligand-stabilized large transition-metal clusters

Many medium to large transition-metal clusters possess central cores which resemble chunks of metal lattices,[9] the outer atoms being protected by suitable ligands to prevent intermolecular interactions and coagulation processes like those observed for naked clusters.[8] According to the shell model, a central metal atom is surrounded by 12 others in ccp or hcp packing, yielding a "magic number" of 13 for a one-shell cluster. In general, the n-th shell consists of $(10n^2 + 2)$ atoms, so that transition-metal full-shell clusters containing 13, 55, 147, 309, 561, ..., $[1 + \Sigma(10n^2 + 2)]$ atoms are expected to be especially stable.[10] Fig. 4.23-6 shows the first five Mackay icosahedra.

Table 4.23-5 shows some ligand-stabilized, full-shell transition-metal clusters which have been synthesized. The M_{55} and larger clusters have been investigated extensively by a range of physical techniques such as high-resolution transmission electron microscopy, X-ray powder diffraction, electron diffraction, EXAFS, NMR spectroscopy, ultracentrifuge measurements, etc. For the giant M_{309} and M_{561} clusters the exact compositions cannot be ascertained, although electron microscopy clearly reveals the number of [110] and [100] atom planes (shells) which justifies assignment of the idealized number of atoms.

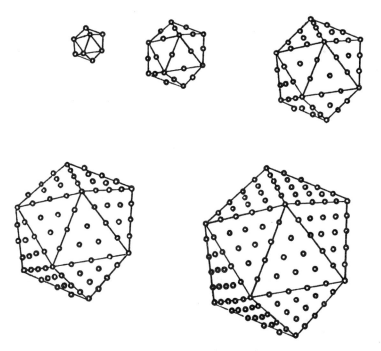

Fig. 4.23-6 The first five Mackay icosahedra, corresponding to n = 13, 55, 147, 309 and 561, respectively. (Adapted from ref. 11).

Table 4.23-5 Ligand-stabilized large transition-metal clusters.

Type	No. of metal atoms	Examples	Ref.
one-shell	13	$[Rh_{13}H_2(CO)_{24}]^{3-}$	[1]
		$[Rh_{13}H_3(CO)_{24}]^{2-}$	[2]
		$[Au_{13}(PMePh_2)_{10}Cl_2]^{3+}$	[12]
two-shell	55	$Au_{55}(PPh_3)_{12}Cl_6$	[13]
		$Au_{55}(Ph_2PC_6H_4SO_3)_{12}]^{12-}$	[14]
		$Ru_{55}(PBu^t_3)_{12}Cl_{20}$	[15]
		$Rh_{55}(PBu^t_3)_{12}Cl_{20}$	[16]
		$Pt_{55}(AsBu^t_3)_{12}Cl_{20}$	[16]
three-shell	147	-	-
four-shell	309	$Pt_{309}(phen^*)_{36}O_{30\pm10}$	[17]
five-shell	561	$[Pd_9(phen)(O)_3(OAc)_3]_{63.5\pm3.5}$	[18]
		$Pd_{561}(phen)_{36}O_{190-200}$	[19]

phen = phenanthroline, $phen^* = 4,7-(C_6H_4SO_3^-Na^+)_2phen.2H_2O$

5. Large metal clusters containing selenide and phosphine ligands

Fenske and co-workers have synthesized several novel clusters by reacting transition metal halides with $Se(SiMe_3)_2$ in the presence of tertiary phosphines.[20]

The centrosymmetric cluster $Ni_{34}Se_{22}(PPh_3)_{10}$ possesses a central Ni_{14} unit consisting of two staggered planar Ni_5 rings with a Ni_4 plane [Ni(9), Ni(16), Ni(9'), Ni(16')] between them [Fig. 4.23-7(a)].[21] Atom Ni(9) is disordered. The Ni_{34} polyhedron is then formally built up by adding five Ni_4 "butterfly" fragments to the pentagonal antiprismatic Ni_{14} unit. The faces of the Ni_{34} unit are finally capped by two μ_5-Se ligands over the Ni_5 rings and twenty μ_4-Se atoms (Ni-μ_4Se 2.26-2.46Å, Ni-μ_5Se 2.36-2.46Å). Additionally, one PPh_3 group is attached to each of the ten peripheral metal atoms [Ni(5), Ni(7), Ni(8), Ni(10), Ni(13) and their symmetry equivalents]. All the Ni atoms are effectively covered by the Se atoms and PPh_3 ligands [Fig. 4.23-7(b)].

New copper clusters include $Cu_{20}Se_{13}(PEt_3)_{12}$ (I), $Cu_{29}Se_{15}(PPr^i_3)_{12}$(II), $Cu_{30}Se_{15}(PPr^i_3)_{12}$ (III), $Cu_{36}Se_{18}(PBu^t_3)_{12}$ (IV) and $Cu_{70}Se_{35}(PEt_3)_{22}$ (V). The metal framework in I consists of six trigonal prisms linked through common edges, with a μ_8-Se1 atom at the center of the resulting Cu_{28} cube [Fig. 4.23-8(a)].[22] Each peripheral Cu1 atom is coordinated by a PEt_3 ligand, and μ_5-Se2 atoms cap all faces of the polyhedron. Cluster II has five layers of Cu atoms bridged by μ_4 to μ_6-Se ligands to give a central core of approximate D_{3h} symmetry [Fig. 4.23-8(b)].[23] Cluster III has the same structure as II with

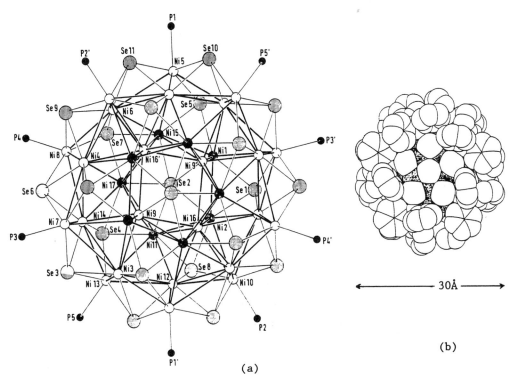

(a)

(b)

Fig. 4.23-7 (a) Molecular structure (phenyl groups have been omitted), and (b) space-filling model (H atoms have been omitted; the free Ni cluster surface is the dark area) of $Ni_{34}Se_{22}(PPh_3)_{10}$. (After ref. 20).

an additional Cu atom in its center. The Cu_{36}[23] and Cu_{70}[22] clusters have rather more complex core structures.

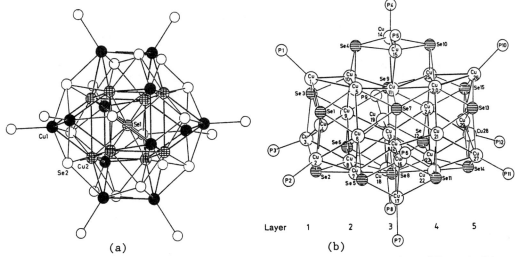

(a)

(b)

Fig. 4.23-8 Framework structures of (a) $Cu_{20}Se_{13}(PEt_3)_{12}$ (I) and (b) $Cu_{29}Se_{15}(PPr^i_3)_{12}$ (II). (After refs. 22 and 23).

6. High-nuclearity clusters with ccp metal cores

Clusters of the general formula $[Ni_{38}Pt_6(CO)_{48}H_{6-n}]^{n-}$ ($n = 3,4,5,6$) have been synthesized.[24] The anionic clusters in $(AsPh_4)_4[Ni_{38}Pt_6(CO)_{48}H_2]$ and $(AsPh_4)_2(NBu_4)_3[Ni_{38}Pt_6(CO)_{48}H]$ have the same metal framework consisting of an inner Pt_6 octahedron fully encapsulated in a ν_3 octahedron of 38 Ni atoms [Fig. 4.23-9(a)]. The bimetallic core corresponds to an octahedral fragment of a cubic-close-packed (ccp) array of metal atoms. The carbonyl ligands are of three types (18 terminal, 12 edge-bridging and 18 face-bridging) such that two opposite ν_3 triangular faces of the metal core differ from the other six, resulting in idealized D_{3d} symmetry for the anionic cluster.

The cluster anion $[Os_{20}(CO)_{40}]^{2-}$ in its PBu_4^+ salt possesses a metal core that consitutes a tetrahedral fragment of a ccp lattice.[25] The ligand shell comprises 40 terminal carbonyl groups in three different sets, matching the overall idealized T_d symmetry of the cluster [Fig. 4.23-9(b)]. The four unique Os atoms at the face-centers of the tetrahedral framework have the highest number of nearest metal neighbours and an average metal-metal bond distance of 2.678(8)Å that approaches the value of 2.6754Å for the bulk metal.

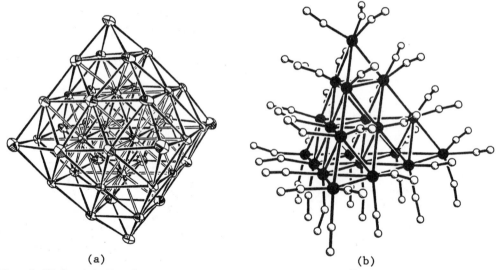

(a)   (b)

Fig. 4.23-9   (a) Metal core of the $[Ni_{38}Pt_6(CO)_{48}H_{6-n}]^{n-}$ ($n = 4,5$) clusters. (b) Structure of the $[Os_{20}(CO)_{40}]^{2-}$ dianion.   (After refs. 24 and 25).

References

[1] V.G. Albano, G. Ciani, S. Martinengo and A. Sironi, *J. Chem. Soc., Dalton Trans.*, 978 (1979).

[2] V.G. Albano, A. Ceriotti, P. Chini, G. Ciani, S. Martinengo and W.M. Anker, *J. Chem. Soc., Chem. Commun.*, 859 (1975).

[3]   P. Chini, *J. Organomet. Chem.* **200**, 37(1980).

[4]   D.M.P. Mingos and R.L. Johnston, *Structure and Bonding* **68**, 31 (1987).

[5]   D.M.P. Mingos, *Chem. Soc. Rev.* **15**, 31 (1986).

[6]   D.G. Evans and D.M.P. Mingos, *Organometallics* **2**, 435 (1983).

[7]   P. Chini, G. Longoni and V.G. Albano, *Adv. Organomet. Chem.* **14**, 285 (1976); J.A. Ferguson and T.J. Meyer, *J. Am. Chem. Soc.* **94**, 3409 (1972).

[8]   G. Schmid and N. Klein, *Angew. Chem. Int. Ed. Engl.* **25**, 922 (1986).

[9]   G. Schmid, *Polyhedron* **7**, 2321 (1988); *Endeavour* **14**, 172 (1990).

[10] A.L. Mackay, *Acta Crystallogr.* **15**, 916 (1962).

[11] M.R. Hoare, *Adv. Chem. Phys.* **40**, 49 (1979).

[12] C.E. Briant, B.R.C. Theobald, J.W. White, L.K. Bell, D.M.P. Mingos and A.J. Welch, *J. Chem. Soc., Chem. Commun.*, 201 (1981).

[13] G. Schmid, R. Pfiel, R. Boese, F. Bandermann, S. Meyer, G.H.M. Calis and J.W.A. van der Velden, *Chem. Ber.* **114**, 3634 (1981).

[14] G. Schmid, N. Klein, L. Korste, U. Kreibig and D. Schönauer, *Polyhedron* **7**, 605 (1988).

[15] G. Schmid and W. Huster, *Z. Naturforsch., Teil B* **41**, 1028 (1986).

[16] G. Schmid, U. Giebel, W. Huster and A. Schwenk, *Inorg. Chim. Acta* **85**, 97 (1984).

[17] G. Schmid, B. Morun and J.-O. Malm, *Angew. Chem. Int. Ed. Engl.* **28**, 778 (1989).

[18] M.N. Vargaftik, V.P. Zagorodnikov, I.P. Stolyarov, I.I. Moiseev, V.A. Likholobov, D.I. Kochubey, A.L. Chuvilin, V.I. Zaikovsky, K.I. Zamaraev and G.I. Timofeeva, *J. Chem. Soc., Chem. Commun.*, 937 (1985); P. Chini, *Gazz. Chim. Ital.* **109**, 225 (1979).

[19] G. Schmid, *Nachr. Chem. Tech. Lab.* **34**, 249 (1987).

[20] D. Fenske, J. Ohmer, J. Hachgenei and K. Merzweiler, *Angew. Chem. Int. Ed. Engl.* **27**, 1277 (1988).

[21] D. Fenske, J. Ohmer and J. Hachgenei, *Angew. Chem. Int. Ed. Engl.* **24**, 993 (1985).

[22] D. Fenske and H. Krautscheid, *Angew. Chem. Int. Ed. Engl.* **29**, 1452 (1990).

[23] D. Fenske, H. Krautscheid and S. Balter, *Angew. Chem. Int. Ed. Engl.* **29**, 796 (1990).

[24] A. Ceriotti, F. Demartin, G. Longoni, M. Manassero, M. Marchionna, G. Piva and M. Sansoni, *Angew. Chem. Int. Ed. Engl.* **24**, 697 (1985).

[25] A.J. Amoroso, L.H. Gade, B.F.G. Johnson, J. Lewis, P.R. Raithby and W.-T. Wong, *Angew. Chem. Int. Ed. Engl.* **30**, 107 (1991).

Addendum to Chapter 4

Section 4.15

New azoalkane- and nitrene-bridged carbonyl metal clusters such as $Ru_4(EtN=NEt)(CO)_{12}$ and $Fe_4(\mu_4\text{-}NEt)_2(CO)_{11}$ are reported in B. Hansert, A.K. Powell and H. Vahrenkamp, *Chem. Ber.* **124**, 2697 (1991).

Section 4.18

A steroid-capped porphyrin bearing convergent hydroxy groups has a reactive site that can recognise and bind to a variety of simple and functionalized amines. See R.P. Bonnar-Law and J.K.M. Sanders, *J. Chem. Soc., Chem. Commun.*, 574 (1991).

Porphyrins have been used in the design of efficient catalysts capable of mimicking enzymes. Template synthesis has yielded a cyclic trimeric supermolecule of the formula $[(C_{32}H_{34}N_4Zn)(C_{28}H_{38}P_2Pt)]_3.(C_{10}H_{10}NO_2)_3Al$, in which the porphyrins linked *via* alkyne-Pt-alkyne units wrap themselves around a central tris(pyridinylacetatoacetate)aluminum moiety, as reported in L.G. Mackay, H.L. Anderson and J.K.M. Sanders, *J. Chem. Soc., Chem. Commun.*, 43 (1992).

The synthesis and structure of the first "furochlorophin" system (a fully conjugated 18-π tetraazamacrocycle containing three pyrrolic units and a condensed furan ring) are reported in C.K. Chang, W. Wu, S.-S. Chern and S.-M. Peng, *Angew. Chem. Int. Ed. Engl.* **31**, 63 (1992).

Chapter 5

Organic Compounds

5.1. *Prostaglandin PGF$_{1\beta}$ (Tri-p-bromobenzoate Methyl Ester)* $C_{42}H_{47}O_8Br_3$ **744**

5.2. *α-Glycine* $^+H_3N\text{–}CH_2\text{–}CO_2^-$ **757**

5.3. *Perdeuterio-α-Glycylglycine* $^+D_3N\text{–}CD_2\text{–}CO\text{–}ND\text{–}CD_2\text{–}CO_2^-$ **767**

5.4. *Potassium Dihydrogen Isocitrate* $K(C_6H_7O_7)$ **786**

5.5. *Methyl p-Bromocinnamate* $BrC_6H_4CH{=}CHCOOCH_3$ **794**

5.6. *Cyclooctatetraene (COT)* C_8H_8 .... **799**

5.7. *Polyethylene Adipate (PEA)* $[\text{–}CO(CH_2)_4CO\cdot O(CH_2)_2O\text{–}]_n$ **811**

5.8. *8,8-Dichlorotricyclo[3.2.1.0^{1,5}]octane* $C_8H_{10}Cl_2$ **821**

5.9. *Cubane* C_8H_8 .... **830**

5.10. *1′,8′:3,5-Naphtho[5.2.2]propella-3,8,10-triene* $C_{18}H_{14}$ **839**

5.11. *[2.2]Paracyclophane* $C_{16}H_{16}$ **849**

5.12. *1,1′-Binaphthyl* $C_{20}H_{14}$ **862**

5.13. *Kekulene (Cyclo[d.e.d.e.d.e.d.e.d.e.]dodecakisbenzene)* $C_{48}H_{24}$ **873**

5.14. *s-Triazine* $C_3H_3N_3$ **881**

5.15. *Di(2,3,6,7-tetramethyl-1,4,5,8-tetraselenafulvalenium) Hexafluorophosphate* $(C_{10}H_{12}Se_4)_2PF_6$ **890**

5.16. *Silenes (Silaethenes)* I $Me_2Si{=}C(SiMe_3)(SiBu^t{}_2Me)\cdot C_4H_8O$
II $Me_2Si{=}C(SiMe_3)(SiBu^t{}_2Me)$ **902**

5.17. *Sucrose* $C_{12}H_{22}O_{11}$ **911**

5.18. *(+)-8-Bromocamphor* $C_{10}H_{15}OBr$ **922**

5.19. *Cholesterol Hemiethanolate* $C_{27}H_{46}O\cdot\frac{1}{2}C_2H_5OH$ **932**

5.20. *Codeine Hydrobromide Dihydrate* $C_{18}H_{21}O_3N\cdot HBr\cdot 2H_2O$ **947**

5.21. *Sodium Benzylpenicillin (Sodium Penicillin G)* $C_{16}H_{17}N_2O_4SNa$ **956**

5.22. *Vitamin B$_{12}$ Coenzyme* $C_{72}H_{100}CoN_{18}O_{17}P\cdot 17H_2O$ **968**

5.23. *Valinomycin* $C_{54}H_{90}N_6O_{18}\cdot 3(CH_3)_2SO$ **984**

5.1   Prostaglandin PGF$_{1\beta}$ (Tri-p-bromobenzoate Methyl Ester)
C$_{42}$H$_{47}$O$_8$Br$_3$

Crystal Data

Orthorhombic, space group $P2_12_12_1$ (No. 19)

$a = 26.14$, $b = 33.93$, $c = 4.76$Å, $Z = 4$

Atom	x	y	z	Atom	x	y	z
Br(1)	.63174	.47212	.01758	C(17)	.59564	.30890	-.10939
Br(2)	.63476	.35299	-.21371	C(18)	.60688	.27266	-.24344
Br(3)	.29836	.19886	.16054	C(19)	.57848	.23938	-.15468
O(1)	.34242	.12472	-.51624	C(20)	.47724	.03968	-.28187
O(2)	.37917	.08772	-.17795	C(21)	.49046	.00396	-.08600
O(3)	.51648	.17763	-.03904	C(22)	.49111	.47002	-.23930
O(4)	.47548	.21145	.27676	C(23)	.48345	.43217	-.40866
O(5)	.39422	.33494	-.06161	C(24)	.45817	.39924	-.22800
O(6)	.42681	.29515	-.37595	C(25)	.45393	.36086	-.41734
O(7)	.64002	.16036	.40039	C(26)	.42221	.33257	-.23383
O(8)	.68815	.12388	.11804	C(27)	.39942	.26075	-.25072
C(1)	.19407	.05198	-.11009	C(28)	.55798	.09409	.09676
C(2)	.19771	.08252	-.28724	C(29)	.60572	.09796	-.01240
C(3)	.24290	.09588	-.37575	C(30)	.65203	.09038	.15999
C(4)	.28700	.08330	-.24648	C(31)	.67528	.15801	.24266
C(5)	.28642	.05488	-.04880	C(32)	.70630	.19000	.13500
C(6)	.23859	.03885	.02603	C(33)	.69792	.22707	.24057
C(7)	.33866	.10228	-.32869	C(34)	.72300	.26250	.14650
C(8)	.42725	.10450	-.23528	C(35)	.76164	.25635	-.03000
C(9)	.43282	.14464	-.10760	C(36)	.77360	.21976	-.14132
C(10)	.48675	.14279	.04589	C(37)	.74627	.18892	-.03900
C(11)	.51282	.10674	-.10328	C(38)	.68358	.05368	.07943
C(12)	.46710	.07703	-.12347	C(39)	.65506	.01565	.14440
C(13)	.50569	.20957	.10708	C(40)	.31319	.47755	-.56601
C(14)	.54020	.24402	.02812	C(41)	.33840	.44075	-.64926
C(15)	.52861	.28079	.12427	C(42)	.30974	.40358	-.57723
C(16)	.55678	.31405	.10016				

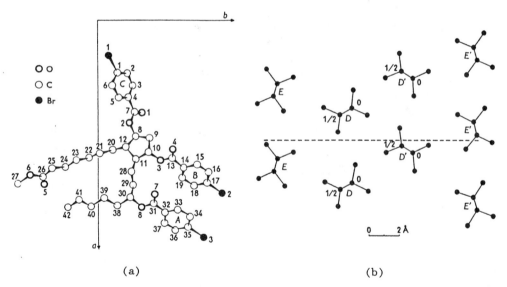

(a)   (b)

Fig. 5.1-1   (a) Molecular structure of prostaglandin PGF$_{1\beta}$ tri-p-bromo-benzoate methyl ester (correct absolute configuration) with the atom numbering scheme and designation of bromobenzoate groups.   (b) Schematic representation of the hydrocarbon chain packing as seen along the chain axes.   The numbers 0 and 1/2 are fractions of the repeat distance (2.55Å) along a chain.   Dotted line indicates the plane $z = 0$.   (After ref. 1).

Crystal Structure

Prostaglandin $PGF_{1\beta}$ is a trihydroxy acid with two hydroxyl groups attached directly to a cyclopentane ring and the third to a carbon side chain containing a *trans* double bond between C(28) and C(29) [Fig. 5.1-1]. The carboxyl group is at the end of a second side chain. X-Ray analysis of the tri-*p*-bromobenzoate methyl ester of $PGF_{1\beta}$ established the stereochemistry of the molecule (Fig. 5.1-1).[1] The two side chains and the bromobenzoate groups *A* and *B* are in close contact whereas other molecules can approach the five-membered ring on both sides of group *C*. Similar parts of the molecules are in contact along the short *c* axis, but this is not possible in other directions and carbon chains have to pack with benzene rings. Nevertheless, the carbon chains stick together as far as possible and alternate with regions of bromobenzoate groups.

All bromobenzoate groups are each planar within 0.08Å. Groups *A* and *B* are approximately parallel; their planes make an angle of 42° with the (001) plane and their twofold axes are tilted 15° to the same plane. The third benzoate group *C* shows very similar angles but has its diad axis roughly perpendicular to those of the other two when the three groups are in packing contact.

The hydrocarbon chains are surprisingly regular, considering their small length and their packing contacts with benzene rings. Atoms C(21) to C(25) are planar within 0.05Å whereas C(20) is 0.16Å off the plane. Atoms C(39) to C(42) lie within 0.002Å from a plane with C(38) off by 0.12Å. The axes of the two chains are roughly parallel to the (001) plane but not mutually parallel. The carbon-chain arrangement is illustrated schematically in Fig. 5.1-1(b). The plane of chain *D* is inclined 28° to the (001) plane and that of chain *E* 76° in the same sense.

Remarks

1. Prostaglandins

The prostaglandins are a family of biologically-potent C_{20} hydroxylated fatty acids first discovered in seminal fluids. Their physiological activities include contraction of uterine smooth muscle, menstruation, vasodilation, bronchial dilation, lowering of blood pressure, and platelet aggregation.[2] In 1983 S. Bergström, B. Samuelsson and J.R. Vane shared the Nobel Prize in Physiology or Medicine for their contributions in this field.[3-5]

The classical prostaglandins are formally derivatives of the hypothetical parent prostanoic acid, and are biosynthesized from polysaturated

straight-chain C_{20} carboxylic acids (such as arachidonic acid, the most
prevalent substrate in primates) in a metabolic chain beginning with the
action of the enzyme cyclooxygenase.

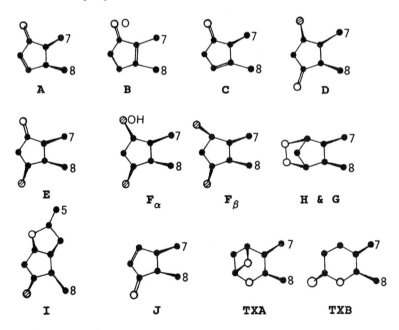

nat-Prostanoic acid   Arachidonic Acid   9α,11α,15(S)-Trihydroxy-5-cis-13-trans-
                                                        prostadienoic acid = $PGF_{2α}$

Prostaglandins (PG) are classified into types according to the
functionalities present in the cyclopentane ring (Fig. 5.1-2). A subscript is
used to indicate the number of double bonds in the aliphatic side chains and
the configuration of the hydroxyl group at C(9). The prostacyclins have an
additional oxolane ring, an ether linkage from C(9) to C(6), and can be named
as 6,9α-oxoprostanoic acid derivatives. For thromboxanes, the TXA designation
is used for those compounds having both the oxane and oxetane rings, whereas
solvolysis products retaining only the oxane ring are TXB derivatives.

The first crystal structure determination of a prostaglandin, on a tri-
p-bromobenzoate-substituted derivative of $PGF_{1β}$, by Abrahamsson in 1963
revealed a "hairpin" motif characterized by the approximate parallel
arrangment of the alkyl portions of the α and ω side chains.[1] Subtle

Fig. 5.1-2 Designation scheme for prostaglandins and thromboxanes based
on their ring moieties. The α and ω side chains are denoted by C_7 and
C_8, respectively. (Adapted from ref. 6).

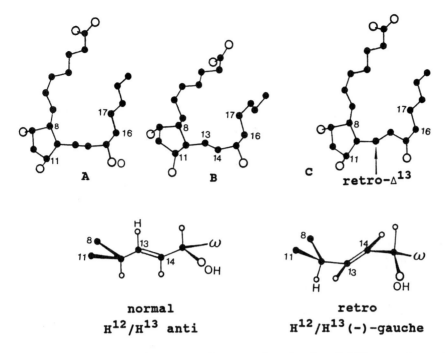

Fig. 5.1-3 Three different conformers (A, B, C) of $PGF_{2\alpha}$ projected on the planes defined by atoms C(12), C(15) and O(15). (Adapted from ref. 6).

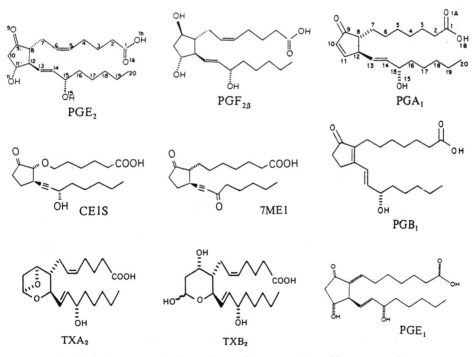

Fig. 5.1-4 Structural formulae of some naturally occurring prostaglandins, thromboxanes, and synthetic analogues.

structural variations such as Δ^{13}-double bond geometry and ring conformations are illustrated by the observed structures of $PGF_{2\alpha}$ (Fig. 5.1-3).[6,7]

   A variety of naturally occurring eicosanoids and synthetic analogues has been subjected to X-ray analysis.[8,9]   Some examples are shown in Fig. 5.1-4. Most prostaglandins adopt the hairpin conformation, and the atypical "L-shape" conformation has been observed only for PGB_1[10] and the synthetic compound 13,14-didehydro-11-deoxy-15-ketoprostaglandin E, (referred to as 7ME1)[11] (Fig. 5.1-5).

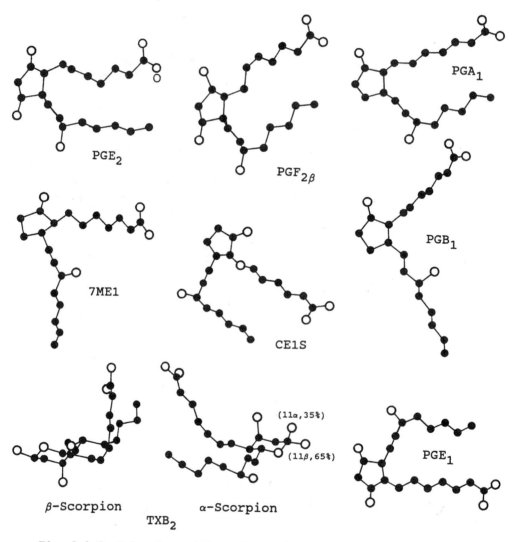

Fig. 5.1-5 Molecular conformations of some prostanoids and thromboxane TXB_2. CE1S stands for 11-deoxy-7-oxa-13,14-didehydroprostaglandin E_1. For the α-tail scorpion conformer of TXB_2, the equilibrium percentages for the anomeric disorder of the O(11) hydroxyl group are given. (Adated from refs. 9-14).

The elusive thromboxane hormone TXA_2 has a half-life of about 30 seconds and it breaks down to the inactive TXB_2, which was found to crystallize with two conformationally distinct molecules in the unit cell.[9] In the "β-tail scorpion" conformer, the carboxylic acid chain turns up through a right angle in the direction of the β-face of the molecule; in the other "α-tail scorpion" conformer, this side chain turns down through a right angle in the direction of the α-face (see bottom part of Fig. 5.1-5).

The conformational distinction between prostaglandin structures is illustrated by the data given in Table 5.1-1,[14] which reveal a broad range of ring/α-chain geometries. The ring/α-chain junction geometries of $PGF_{2\alpha}$, $PGF_{2\beta}$, PGF_{2I} and CE1S are similar; likewise PGE_1, PGA_1, and 7ME1 constitute another group. The ring/ω-chain junction geometries are similar for all molecules in Table 5.1-1 except for PGB_1, CEIS, and 7MEI; it is noted that PGB_1 takes the "L-shape" conformation, and presence of the alkyne function in CE1S and 7ME1 precludes a quantitative comparison with the other prostaglandin structures.

Table 5.1-1   Comparison of conformations of some prostaglandin structures.*   (Adapted from ref. 14).

Compound	Ring/chain junction geometry(°)				Endocyclic trosion angles(°)					Ring con-formation	Ref.
	α chain		ω chain								
	T1	TX1	T2	TX2	TR1	TR2	TR3	TR4	TR5		
PGE_1 [a]	-69	168	-137	106	5	-28	40	-37	20	C(12) envelope	[12]
	-62	179	-142	98	5	-25	37	-34	19	C(12) envelope	[12]
PGE_2	62	-61	-126	113	9	-28	35	-29	14	C(12) envelope	[13]
PGA_1 [b]	-74	165	-135	107	-10	-2	13	-17	17	relatively flat	[10]
PGA_1 [c]	-65	175	-132	113	-7	3	2	-6	8	nearly flat	[16]
PGB_1	88	-89	-172	8	2	-2	2	-1	-1	nearly flat	[10]
$PGF_{1\beta}$ [d]	98	-145	133	-116	12	19	-41	47	-37	half chair	[1]
$PGF_{2\beta}$	172	54	-129	115	1	-25	38	-37	22	C(12) envelope	[13]
PGF_{2I} [a,e]	-176	72	-118	124	-42	28	-3	-20	38	C(9) envelope	[15]
	169	56	-116	127	-51	33	-1	-31	48	C(9) envelope	[15]
CE1S	-167	76	-77	165	1	-25	41	-40	25	C(12) envelope	[14]
7ME1	-72	168	-136	102	7	-28	39	-33	16	C(12) envelope	[11]

*The α chain consists of atoms C(1) through C(7); the ω chain contains atoms C(13) through C(20). Torsion angle definitions: T1 = Ω(6-7-8-12); TX1 = Ω(6-7-8-9); T2 = Ω(8-12-13-14); TX2 = Ω(11-12-13-14); TR1 = Ω(8-9-10-11); TR2 = Ω(9-10-11-12); TR3 = Ω(10-11-12-8); TR4 = Ω(11-12-8-9); TR5 = Ω(12-8-9-10). [a] Two independent molecules. [b] Monoclinic form. [c] Orthorhombic form. [d] Tri-p-bromobenzoate methyl ester. [e] p-Iodophenacyl ester of 15(S)-methyl $PGF_{2\alpha}$.

The carbocycle conformations of some postaglandins are presented in Table 5.1-1. The endocyclic torsion angles *TR1-TR5* show that the five-atom rings of PGE_1, PGE_2, $PGF_{2\beta}$, CE1S and 7ME1 adopt C(12) envelope conformations. The two independent molecules of $PGF_{2\alpha}$, studied as the tris(hydroxymethyl)-methylamine salt, adopt different C(8) and C(9) envelope conformations.[7]

2. Crystal structures of alkanes[17]

   n-Alkanes, C_nH_{2n+2}, commonly exhibit polymorphic forms. If sufficiently pure, the straight chain alkanes with *n* even (6 < *n* < 26) crystallize in the triclinic system, and in the hexagonal form when near the melting point, whereas those with *n* odd (11 < *n* < 43) are orthorhombic. In the crystal structures of *n*-alkanes, the axes of all molecules always run parallel to each other, regardless of the crystalline modification.

   In 1925 Müller, an early pioneer of the study of *n*-alkanes, grew a single crystal of $n\text{-}C_{29}H_{60}$ and determined its structure.[18] The unit cell is orthorhombic with *a* = 7.45, *b* = 4.97, and *c* = 77.2Å. Fig. 5.1-6 shows the *ab* face of the unit cell perpendicular to the chain axes; it was called by Müller a basic (main) plane. The transverse arrangement of the molecular chains is highly characteristic of most *n*-alkanes and many other long chain compounds, including a number of high-polymers. The most widely encountered type of subcell is the orthorhombic polyethylene unit cell with *a* = 7.45, *b* = 4.97, and *c* = 2.54Å (the aliphatic zigzag repeat distance).

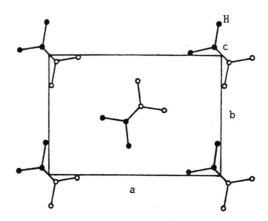

Fig. 5.1-6 Face *ab* of the unit cell of orthorhombic *n*-alkanes. (After ref. 17).

   For $n\text{-}C_{18}H_{38}$, a triclinic cell with *a* = 4.28, *b* = 4.82, *c* = 23.07Å, α = 91.1, β = 92.1, γ = 107.3°, and *Z* = 1 has been observed. In the case of *n*-$C_{36}H_{74}$ the true cell is monoclinic with *a* = 5.57, *b* = 7.42, *c* = 48.35Å, β = 119.1°, space group $P2_1/a$, and *Z* = 2. The orthorhombic subcell and the coordinates of the carbon atoms in the subcell are in close agreement with those found for polyethylene.

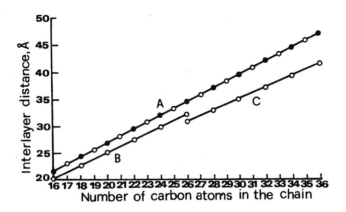

Fig. 5.1-7 Dependence of interlayer distance on n, the number of carbon atoms in the chain. (After ref. 17).

The distance between the planes of the crystal lattice passing through the end groups of the chain molecules is of the order of the chain length of one molecule. Fig. 5.1-7 shows the relation of interlayer distance and the number of carbon atoms in the chain. The points for all the odd n-alkanes lie on the straight line A. The points of triclinic crystals of even numbers of the homologous series up to C_{24} lie on line B. The points for even n-alkanes from C_{28} and above having orthorhombic subcells lie on line C.

The molecular packing of n-alkanes can be discussed in two aspects: the packing of the molecules in a layer and the packing of layers.

If the chains are azimuthally chaotic (prone to rotation around their axes), their average cross sections are circular, giving rise to a hexagonal packing mode as shown in Fig. 5.1-8(a). For molecules of an arbitrary cross section, two types of close-packed layers are possible: one with an oblique and the other with a rectangular cell, as shown in Fig. 5.1-8(b) and (c).

The crystal structures of n-pentane,[19] n-heptane,[20] and n-octadecane[21] have been determined precisely. The C-C bonds are somewhat shorter and the C-C-C angles somewhat larger than those found in diamond. In n-octadecane, the marked librational motion of the carbon atoms in planes normal to the chain axis steadily increases from the center of the molecule extending toward either terminal carbon. This has the well known effect of bringing the mean positions of the carbon atoms inward towards the chain axis, thus shortening the C-C bonds and widening the C-C-C angles.

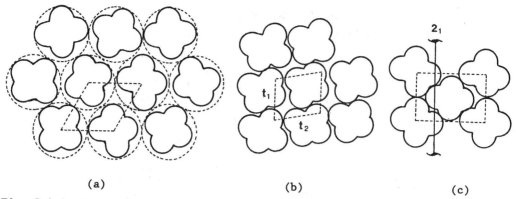

(a)   (b)   (c)

Fig. 5.1-8 Three possible types of close packing of chain molecules:
(a) hexagonal, (b) oblique, and (c) rectangular cell. (Adapted from ref. 17).

3.   Structure of compounds containing a long hydrocarbon chain

Compounds having a long hydrocarbon chain are very extensive, and their crystal structures share several common features: a) the carbon atoms in a long chain are nearly planar, b) the polar groups always assemble together, c) the long hydrocarbon chains are in close-packed parallel alignment, and d) the thermal vibration steadily increases from the polar groups or from the molecular center for nonpolar molecules. Some examples are discussed below.

a) Isostearic acid, $(CH_3)_2CH(CH_2)_{14}COOH$[22]

In this branched-chain fatty acid, the carbon chain is planar within 0.04Å. The zigzag plane forms an angle of 11.5° within the plane through the carboxyl group. This angle is 13° in 13-oxostearic acid. The molecules are held together by hydrogen bonds to form a centrosymmetric dimer. The O...O distance is 2.70Å, and the C-O...O angle is 115.7°.

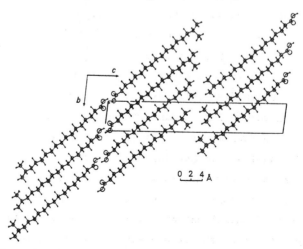

Fig. 5.1-9 Molecular packing of isostearic acid. (After ref. 22).

The packing of the dimers is shown in Fig. 5.1-9. The molecules slide along the chain axes relative to each other so that the methyl end of one chain just reaches the branch of another molecule. The branches are then accommodated between the methyl ends of the chains. This is one common way of accommodating the branch in monomethyl-substituted fatty acids. The chain axes form an angle of 44° with the end-group planes. The corresponding value for 13-oxoisostearic acid is 44° and for isopalmitic acid is 45°.

The chains pack laterally in the common triclinic packing mode. The subcell dimensions are: $a = 4.28$, $b = 5.37$, $c = 2.53$Å, $\alpha = 72.3$, $\beta = 108.8$, and $\gamma = 117.2°$.

b) *O-Palmitoyl benzophenone oxime* (I) and (*E*)-*O-palmitoyl phenyl 2-pyridyl ketone oxime* (II).[23]

The structures of these two molecules are shown in Fig. 5.1-10(a). The palmitoyl moieties of both molecules have the same extended conformation, and the C=N oxime part is nearly coplanar with the palmitoyl zigzag plane. The bond distances for the corresponding bonds of the two molecules are in good agreement.

The molecular packing modes of (I) and (II) have a bilayer arrangement, as shown in Fig. 5.1-10(b). The palmitoyl chains of (I) are packed in a typical triclinic $T^{||}$ arrangement. The subcell dimensions are $a = 2.54$, $b =$

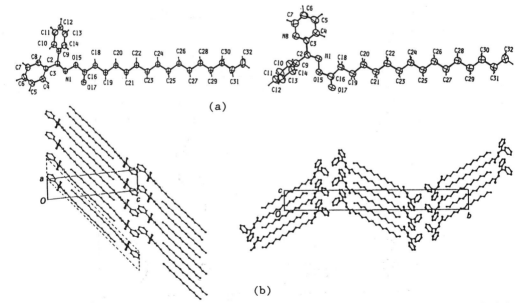

(a)

(b)

Fig. 5.1-10 (a) Molecular structure and numbering scheme of (I) (left), (II) (right). (b) Packing diagram of (I) projected down *b* and (II) projected down *a*. The broken lines indicate the conventional unit cell. (After ref. 23).

4.22, $c = 4.70\text{Å}$, $\alpha = 71.2$, $\beta = 86.4$, and $\gamma = 75.8°$. The palmitoyl chains of
(II) are arranged in a new type of triclinic subcell with unit-cell dimensions
$a = 8.38$, $b = 5.06$, $c = 2.544\text{Å}$, $\alpha = 93.4$, $\beta = 97.1$, and $\gamma = 117.8°$. This
subcell arises from hybridization of the $T^{||}$ and $M^{||}$ cells. The $T^{||}$ cell has
$a = 4.35$, $b = 5.06$, $c = 2.544\text{Å}$, $\alpha = 93.4$, $\beta = 109.6$, and $\gamma = 115.2°$; the $M^{||}$
cell has $a = 4.20$, $b = 5.06$, $c = 2.544\text{Å}$, $\alpha = 93.4$, $\beta = 94.9$, and $\gamma = 120.3°$.
These two cells contact each other in the bc plane. The head groups of the
long chains have close contacts between the molecules packed along the a axis
for (I) and the c axis for (II).

   c) Geranylamine hydrochloride, $C_{10}H_{20}NCl$

   This terpene derivative was the first structure determined with three-
dimensional Weissenberg data.[24] An interesting feature is that the thermal
vibration of the atoms is reflected by a gradual diminution in electron
density along the chain, starting from the N terminal.[25] Fig. 5.1-11 shows
a composite set of sections through the atoms. The approximate isotropic
temperature factors, B (in Å^2), for the atoms are:

Cl	N	C(1),C(2)	C(3),C(4)	C(5),C(6)	C(7),C(8)	C(9),C(10)
4.4	4.6	5.6	6.8	7.6	9.3	12.3

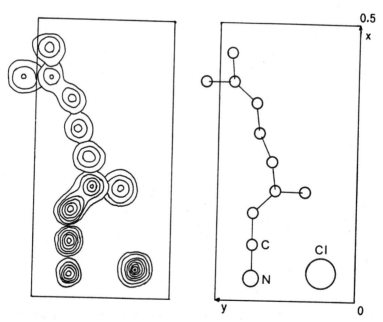

Fig. 5.1-11 Geranylamine hydrochloride. Left: increase in thermal
vibration along the chain. Right: atomic positions corresponding to the
electron density maxima. (After ref. 26).

d) Suberic acid, $HO_2C(CH_2)_6CO_2H$

The crystal structure of suberic acid was first determined in 1965,[27] and recently redetermined by neutron diffraction at 18, 75 and 123K and by X-ray diffraction at 123K.[28] Bond lengths for methylene C-C [1.526(2)Å] and C-H [1.086(4)Å] groups have been corrected for thermal motion, including internal and anharmonic vibrations. The most characteristic feature is that the atomic vibrational amplitudes are greater at the center of the $(CH_2)_6$ chain than at the terminal carboxylic acid groups.

References

[1] S. Abrahamsson, *Acta Crystallogr.* **16**, 409 (1963).

[2] P.B. Curtis-Prior (ed.), *Prostaglandins: Biology and Chemistry of Prostaglandins and Related Eicosanoids*, Churchill Livingstone, Edinburgh, 1988.

[3] S. Bergström, *Angew. Chem. Int. Ed. Engl.* **22**, 858 (1983). [Nobel Lecture].

[4] B. Samuelsson, *Angew. Chem. Int. Ed. Engl.* **22**, 805 (1983). [Nobel Lecture].

[5] J.R. Vane, *Angew. Chem. Int. Ed. Engl.* **22**, 741 (1983). [Nobel Lecture].

[6] N.H. Andersen, C.J. Hartzell and B. De in J.E. Pike and D.R. Morton, Jr. (eds.), *Advances in Prostaglandin, Thromboxane, and Leukotriene Research*, Vol. **14**, Raven Press, New York, 1985, p. 1.

[7] D.A. Langs, M. Erman and G.T. DeTitta, *Science (Washington)* **197**, 1003 (1977).

[8] G.T. DeTitta, D.A. Langs and M.G. Erman in B. Samuelsson, P.M. Ramwell and R. Paoletti (eds.), *Advances in Prostaglandin and Thromboxane Research*, Vol. **6**, Raven Press, New York, 1980, p. 381, and references cited therein.

[9] D.A. Langs, G.T. DeTitta, M.G. Erman and S. Fortier in B. Samuelsson, P.W. Ramwell and R. Paoletti (eds.), *Advances in Prostaglandin and Thromboxane Research*, Vol. **6**, Raven Press, New York, 1980, p. 477.

[10] G.T. DeTitta, D.A. Langs and J.W. Edmonds, *Biochemistry* **18**, 3387 (1979).

[11] J.D. Oliver and L.C. Strickland, *Acta Crystallogr., Sect. C* **41**, 1477 (1985).

[12] A.L. Spek, *Acta Crystallogr., Sect. B* **33**, 816 (1977).

[13] G.T. DeTitta, D.A. Langs, J.W. Edmonds and W.L. Duax, *Acta Crystallogr., Sect. B* **36**, 638 (1980).

[14] J.D. Oliver and L.C. Strickland, *Acta Crystallogr., Sect. C* **39**, 380 (1983).

[15] C.G. Chidester and D.J. Duchamp, *Abstr. Am. Crystallogr. Assoc.*, Vol. 2, Spring Meet., Berkeley, California, 1974, p. 34.

[16] J.W. Edmonds and W.L. Duax, *J. Am. Chem. Soc.* 97, 413 (1975).

[17] A.I. Kitaigorodsky, *Molecular Crystals and Molecules*, Academic Press, New York, 1973, chap. 1.

[18] A. Müller and W.B. Saville, J. Chem. Soc. 127, 599 (1925); A. Müller, *Proc. Roy. Soc. London* A120, 437 (1928); M.G. Broadhurst, *J. Res. Natl. Bur. Std.* A66, 241 (1962).

[19] H. Mathisen, N. Norman and B.F. Pedersen, *Acta Chem. Scand.* 21, 127 (1967).

[20] A.M. Merle, M. Lamotte, S. Risemberg, C. Hauw, J. Gaultier and J.P. Grivet, *Chem. Phys.* 22, 207 (1978).

[21] S.C. Nyburg and H. Lüth, *Acta Crystallogr., Sect. B* 28, 2992 (1972).

[22] S. Abrahamsson and B.-M. Lundén, *Acta Crystallogr., Sect. B* 28, 2562 (1972).

[23] T. Taga and T. Miyasaka, *Acta Crystallogr., Sect. C* 43, 748 (1987).

[24] G.A. Jeffrey, *Proc. Roy. Soc. London* A183, 388 (1945).

[25] D.W.J. Cruickshank and G.A. Jeffrey, *Acta Crystallogr.* 7, 646 (1954).

[26] S.C. Nyburg, *X-Ray Analysis of Organic Structures*, Academic Press, New York, 1961, chap. 6.

[27] J. Housty and M. Hospital, *Acta Crystallogr.* 18, 753 (1965).

[28] B.M. Craven, Q. Gao, H.P. Weber and R.K. McMullan, *Acta Crystallogr., Sect. A* 43, C-110 (1987).

Note The two alkyl chains in phospholipids and glycosylamides are nearly co-planar and run parallel to each other in an all-*trans* conformation with a spacing of 4.8Å. Glycolipids as immunomodulators are reviewed in O. Lockhoff, *Angew. Chem. Int. Ed. Engl.* 30, 1611 (1991).

The crystal structure of $C_{20}H_{42}$ is reported in S.C. Nyburg and A.R. Gerson, *Acta Crystallogr., Sect. B* 48, 103 (1992).

5.2 α-Glycine

$^{+}H_{3}N-CH_{2}-CO_{2}^{-}$

Crystal Data

Monoclinic, space group $P2_1/n$ (No. 14)

$a = 5.1054(6)$, $b = 11.9688(19)$, $c = 5.4645(9)$Å, $b = 111.70(1)°$,[1] $Z = 4$

Atom	x	y	z	Atom	x	y	z
C(1)	.07504	.12486	.06619	H(1)	.28972	.10036	-.45414
C(2)	.06474	.14485	-.21308	H(2)	.49450	.11929	-.13184
O(1)	.30494	.09439	.23539	H(3)	.29935	.00561	-.22613
O(2)	-.14722	.14150	.10708	H(4)	.07688	.23444	-.24322
N	.30116	.08984	-.25904	H(5)	-.13322	.11439	-.35718

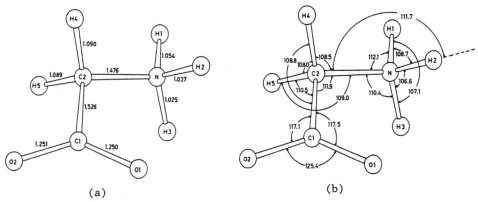

Fig. 5.2-1 Molecular structure of α-glycine. (a) Bond distances
and (b) angles. (After ref. 1).

Crystal Structure

α-Glycine was the first amino acid to be studied by X-ray
diffraction.[2] The structure was refined by neutron diffraction in 1972,[1]
and in 1980 a study of the charge density was carried out employing X-ray and
neutron diffraction data measured at 120K.[3] In the solid state the glycine
molecule takes the form of a zwitterion. The intramolecular bond distances
and angles are listed in Fig. 5.2-1. The atoms C(1), C(2), O(1) and O(2) are
nearly coplanar.

The hydrogen bonding system of α-glycine was described by Albrecht and
Corey[2] and by Marsh.[4] The neutron diffraction study provided more
accurate information on the geometry of the hydrogen bonds and of involvement
of the hydrogen atoms in other intermolecular contacts. The dipolar glycine
molecules are linked by two relatively short, nearly linear N-H...O hydrogen
bonds to forms layers parallel to the *ac* plane; the layers are connected, two
by two, by weaker N-H...O bonds, forming anti-parallel double layers. The

bonding situation about a glycine molecule is shown in Fig. 5.2-2, and the pertinent distances and angles are given in Table 5.2-1.

Table 5.2-1 Hydrogen bonds N-H...O and short H...O van der Waals contacts. [1]

X	H	O	X...O(Å)	X-H(Å)	H...O(Å)	X-H...O(°)	H...O-C(1)(°)
N	H(1)	O(1)	2.770(1)	1.054(2)	1.728(2)	169.3(2)	110.1(1)
N	H(2)	O(2)	2.855(1)	1.037(2)	1.832(2)	168.5(2)	125.6(1)
N	H(3)	O(2)	3.075(1)	1.025(2)	2.121(2)	154.0(1)	112.4(1)
N	H(3)	O(2)	2.955(1)	1.025(2)	2.365(2)	115.5(2)	135.0(1)
C(2)	H(4)	O(1)	3.277(1)	1.090(2)	2.390(2)	137.4(2)	148.1(1)
C(2)	H(4)	O(1)	3.362(1)	1.090(2)	2.453(2)	140.0(2)	98.7(1)

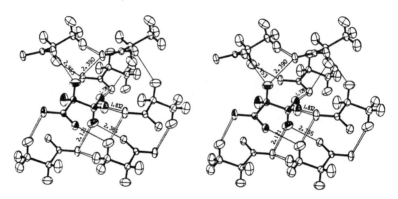

Fig. 5.2-2 Stereoscopic view of the hydrogen-bonding situation about a glycine molecule. All contacts where the H...O distances are less than 2.5Å are indicated by thin lines. (After ref. 1).

Albrecht and Corey proposed a "bifurcated hydrogen bond" for glycine, in which the hydrogen atom H(3) is shared between acceptor oxygen atoms O(1) and O(2). Their conclusions were based mainly on the N...O separations, but this is unsatisfactory in discussing the strength of nonlinear hydrogen bonds. The N...O(1) distance is appreciably shorter than the N...O(2) distance, and the H(3)...O(1) distance is considerably longer than the H(3)...O(2) distance of 2.121Å, and the N-H(3)...O(1) bond is much more bent than the N-H(3)...O(2) bond. Hamilton[5] has given an upper limit of 2.4Å for the H...O distance in a hydrogen bond. The H(3)...O(1) distance of 2.365Å is so close to this limit, and the angle N-H(3)...O(1) of 115.5° so unfavourable for hydrogen bond formation that this short contact is probably better classified as a short van der Waals contact rather than a very weak hydrogen bond.

Remarks

1. Amino acid residues occurring in proteins

In Table 5.2-2 are listed the 20 standard L-form α-amino acid residues occurring in proteins according to their relative abundance;[6] the commonly

used three-letter and one-letter abbreviations together with mnemonic aids are
also given. An amino acid residue refers to the -HN-CHR-CO- part of an amino
acid, $H_2N-C_\alpha HR-COOH$, occurring within a peptide chain; R denotes the side
chain. X-Ray diffraction data confirm that in polypeptides these α-amino acid
residues have the L-configuration at the C_α-atom; Fig. 5.2-3 shows the
absolute configuration of the corresponding L-amino acids. Fig. 5.2-4
illustrates the 20 common amino acid side chains R. A self-consistent set of
partial electronic charges for all atoms of the 20 naturally occurring amino
acids is given by Burley and Petsko.[32]

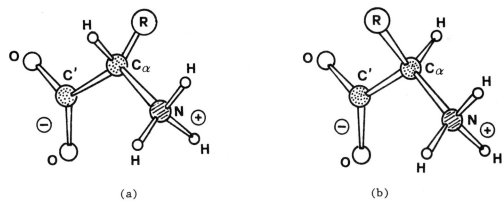

(a)   (b)

Fig. 5.2-3 The absolute configuration of (a) L-amino acids
and (b) D-amino acids. (After ref. 6).

2. Crystal structure of β- and γ-glycine

Three polymorphic forms of glycine have so far been reported:

a) Ordinary form α. In the crystal structure the hydrogen-bonded
double layers of glycine molecules are packed together by van der Waals
forces, as shown in Fig. 5.2-5(a).

b) An unstable form β. Crystal structure analysis[26] showed that
single molecular layers, whose internal arrangement is the same as in the α
form, are held together by hydrogen bonds to form a three-dimensional
framework, as shown in Fig. 5.2-5(b).

c) A third form γ. It is strongly piezoelectric and crystallizes with
trigonal hemihedral symmetry. X-Ray[27] and neutron[28] diffraction studies
showed that the structure consists of hydrogen-bonded molecular helices around
the crystallographic 3_2 screw axes, which are held together by lateral
hydrogen bonds forming a three-dimensional network, as shown in Fig. 5.2-6.

The crystal data and molecular geometry of the three polymorphic forms
of glycine are compared in Table 5.2-3.

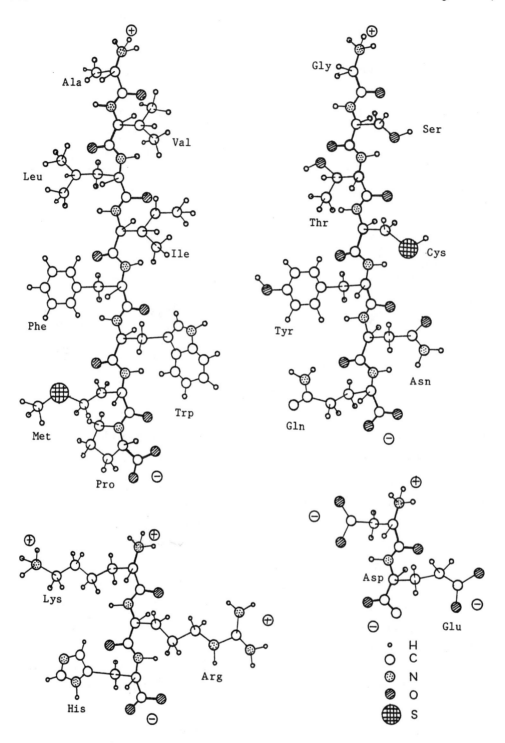

Fig. 5.2-4 The 20 standard amino acid side chains. All C$_\alpha$-atoms are represented by partially blackened circles. (Adapted from ref. 6).

Table 5.2-2 The 20 standard amino acid residues of proteins.[*]

Amino acid or residue thereof	Three-letter symbol	One-letter symbol	Mnemonic help for one-letter symbol	R	Relative abundance in E. Coli proteins (%)	Refs. for structure
Alanine	Ala	A	Alanine	Me	13.0	[7]
Glutamate	Glu	E	gluEtamic acid	$CH_2CH_2CO_2^-$	-	[8]
Glutamine	Gln	Q	Q-tamine	$CH_2CH_2CONH_2$	10.8	[9]
Aspartate	Asp	D	asparDic acid	$CH_2CO_2^-$	-	[10]
Asparagine	Asn	N	asparagiNe	CH_2CONH_2	9.9	[11]
Leucine	Leu	L	Leucine	CH_2CHMe_2	7.8	[12][a]
Glycine	Gly	G	Glycine	H	7.8	[1]
Lysine	Lys	K	befor L	$CH_2(CH_2)_3NH_3^+$	7.0	[13][b]
Serine	Ser	S	Serine	CH_2OH	6.0	[14]
Valine	Val	V	Valine	$CHMe_2$	6.0	[15]
Arginine	Arg	R	aRginine	$CH_2(CH_2)_2NHC\overset{\displaystyle NH_3^+}{\underset{\displaystyle \|NH}{\big|}}$	5.3	[16][c]
Threonine	Thr	T	Threonine	$CH(OH)Me$	4.6	[17]
Proline	Pro	P	Proline		4.6	[18]
Isoleucine	Ile	I	Isoleucine	$CHMeEt$	4.4	[19]
Methionine	Met	M	Methionine	CH_2CH_2SMe	3.8	[20]
Phenylalanine	Phe	F	Fenylalanine	CH_2	3.3	[21][d]
Tyrosine	Tyr	Y	tYrosine	CH_2OH	2.2	[22]
Cysteine	Cys	C	Cysteine	CH_2SH	1.8	[23]
Tryptophan	Trp	W	tWo rings	CH_2	1.0	[24][e]
Histidine	His	H	Histidine	CH_2	0.7	[25]

[*] The disulfide-bridged amino acid cystine occurring in some proteins, for which the three-letter symbol is Cys.Cys, has the formula $HO_2CCH(NH_2)CH_2SSCH_2CH(NH_2)CO_2H$. (a) L-leucine hydroiodide, (b) lysine hydrochloride dihydrate, (c) L-arginine dihydrate, (d) L-phenylalanine hydrochloride, (e) tryptophan hydrochloride.

Table 5.2-3   Crystal data and molecular geometry of glycine.

	α-Glycine	β—Glycine	γ-Glycine
space group	$P2_1/n$	$P2_1$	$P3_2$
a (Å)	5.1054(6)	5.0774	7.046(3)
b	11.9688(19)	6.2676	-
c	5.4645(9)	5.3799	5.491
$\beta(°)$	111.70(1)	113.2	-
Z	4	2	3
C(1)-O(1)	1.250(1)	1.233	1.25(8)
C(1)-O(2)	1.251(1)	1.257	1.247(2)
C(2)-N	1.476(1)	1.484	1.473(1)
C(1)-C(2)	1.526(1)	1.521	1.531(1)
C(2)-C(1)-O(1)	117.5(1)	117.8	117.7(1)
C(2)-C(1)-O(2)	117.1(1)	115.9	116.5(1)
O(1)-C(1)-O(2)	125.5(1)	126.2	125.8(1)
C(1)-C(2)-N	111.9(1)	110.8	111.5(1)

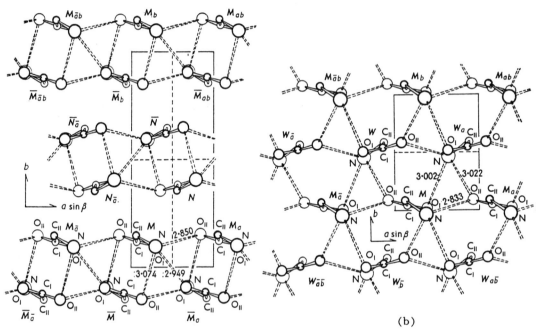

(b)

Fig. 5.2-5   Views of crystal structure of glycine parallel to the c axis: (a) α-glycine and (b) β-glycine.   (After ref. 26)

3.   Structure of L-arginine dihydrate

L-arginine, $^+(H_2N)_2CNH(CH_2)_3CH(NH_2)COO^-$, is one of the biologically important amino acid essential for animal growth.   It is also the most basic of the amino acids since, in addition to the α-amino group, it also contains a terminal guanidyl group.   The crystals are orthorhombic, space group $P2_12_12_1$, with $a = 5.68$, $b = 11.87$, $c = 15.74$Å, and $Z = 4$.   The crystal structure of L-arginine dihydrate was the first non-centrosymmetric structure to be determined by the direct method.[29]

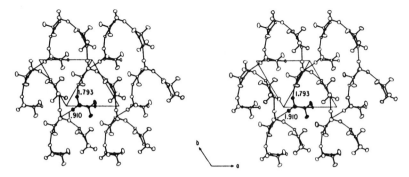

Fig. 5.2-6 A stereoscopic view of structure of γ-glycine viewed down the crystallographic *c* axis. (After ref. 28).

In the crystal, the arginine molecule is characterized by two planes, one through the carboxyl group and the other through the side chain that terminates with the guanidyl group. The molecule is a zwitterion with the guanidyl group, rather than the amino group, accepting a proton from the carboxylate group. The bond lengths and angles are shown in Fig. 5.2-7(a). The dimensions of the amino acid group are quite similar to those reported for glycine[1] and L-lysine[30].

The packing of the arginine molecules is characterized by a three-dimensional network of hydrogen bonds. [Fig. 5.2-7(b)]. The arginine molecules lie extended parallel to the *c* axis, forming infinite chains via hydrogen bonding between two guanidyl NH groups and the O atoms of the

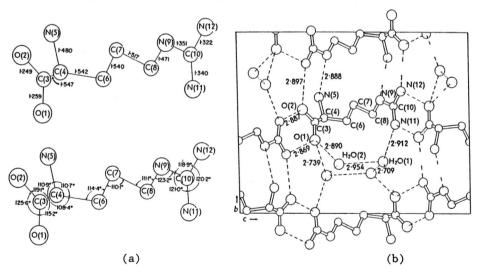

(a)   (b)

Fig. 5.2-7 Structure of L-arginine dihydrate. (a) Bond distances and angles. (b) Structure viewed along the *a* axis. The dashed lines denote the hydrogen bonds. (After ref. 29).

carboxyl group. Hydrogen bonding between arginine molecules also occurs parallel to the *b* axis, and O(2) from the carboxyl group is bonded to N(12')H(14') of the guanidyl group, and N(9')H(10') of the guanidyl group to N(5) of the amino group. The two hydrogen atoms, H(2) and H(3), in the amino group are not involved in hydrogen-bond formation.

4. Metal complexes of simple amino acids[31]

α-Amino acids with the general formula $H_3N^+CH(R)COO^-$ may have non-coordinating or coordinating side chains. The more common coordination mode of amino acids with non-coordinating side chains is as a bidentate chelate through the N and O atom (1), which gives rise to a thermodynamically stable five-membered ring for the α-amino acid-metal fragment:

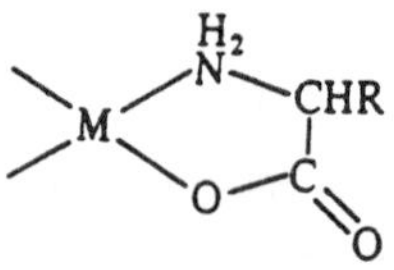

1

A number of amino acids shown in Table 5.2-2 have polar side groups and hence the potential of becoming tridentate. It is these third sites that are important in protein binding of metal ions. For example, the imidazole ring (2), as part of histidine and as its 5,6-dimethyl derivative, is an important ligand for transition metal ions in a number of biological systems. Tridentate chelation has been established in the solid state for [Co^{II}(L-His*O*)$_2$] (3), [Co^{II}(L-His*O*)(D-His*O*)] (4), in both isomeric forms for the bis complexes of Ni^{II}, Zn^{II} and Cd^{II}, for numerous complexes of Co^{III} and Cr^{III}, and is present in the Mo^{V} complex [Mo$_2$O$_4$(L-His*O*)$_2$] (5). A binuclear Ru^{III} complex in which the imidazole ring bridges the two metal centers as in (6) has also been reported.

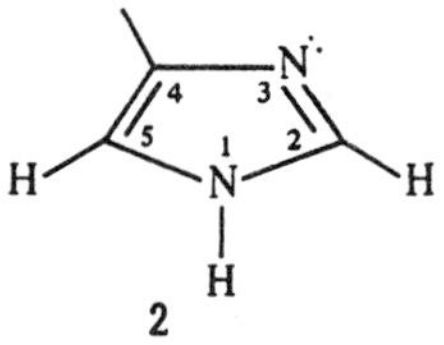

2 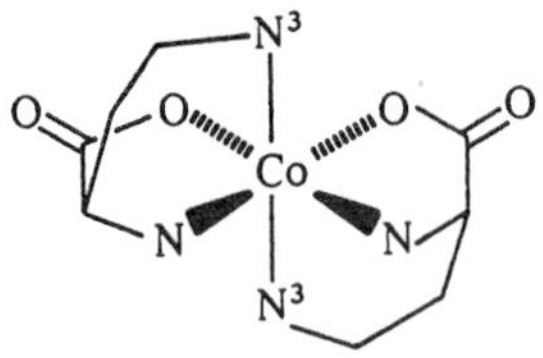   3 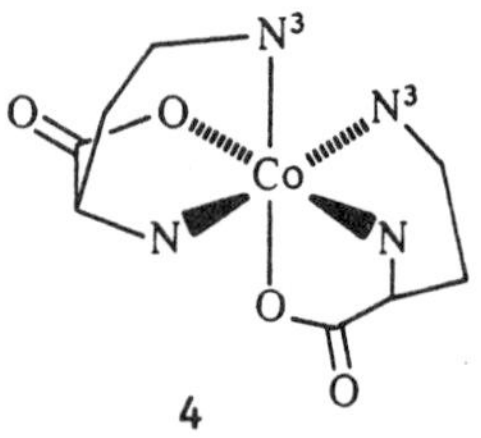   4

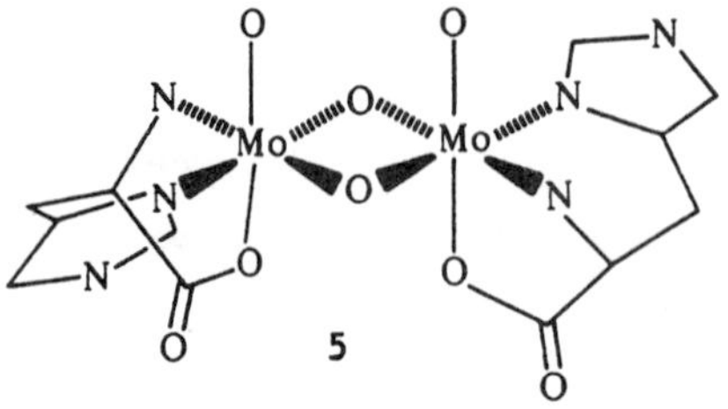

5 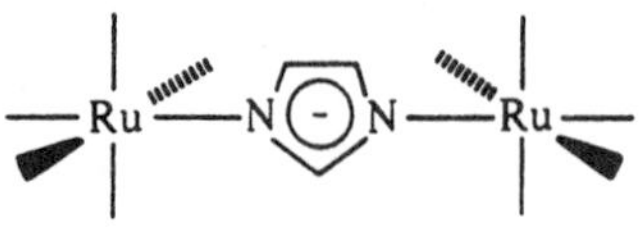   6

As with histidine, cysteine is a major metal-binding site in proteins.[23] Except the N and O atoms, the S atoms of this anion is also a very effective metal coordinate site. Sometimes the strong affinity of "soft" metal ions for ionized thiol groups gives rise to monodentate behaviour. Table 5.2-4 lists the binding modes of cysteine with metal ions.

Table 5.2-4   Observed modes of binding of cysteine showing its ambidentate nature.

Donor atoms	Metal ions
Tridentate (S,N,O)	Cd^{II}, Co^{II}, Co^{III}, Cr^{III}, In^{III}, Mo^{IV}, Mo^{V}, Pb^{II}, Sn^{II}
Bidentate (S,N)	Co^{II}, Co^{III}, Fe^{III}, Ni^{II}, Pd^{II}, Zn^{II}
Bidentate (S,O)	Cd^{II}, Co^{III}, Fe^{III}, In^{III}, Zn^{II}
Bidentate (N,O)	Co^{III}
Monodentate (S)	Ag^{I}, Cu^{I}, Hg^{II}, $MeHg^{+}$, Pt^{II}

References

[1] P.G. Jönsson and Å. Kvick, *Acta Crystallogr., Sect. B* **28**, 1827 (1972).

[2] G. Albrecht and R.B. Corey, *J. Am. Chem. Soc.* **61**, 1087 (1939).

[3] J.P. Legros and Å. Kvick, *Acta Crystallogr., Sect. B* **36** 3052 (1980).

[4] R.E. Marsh, *Acta Crystallogr.* **11**, 654 (1958).

[5] W.C. Hamilton in A. Rich and N. Davidson (eds.), *Structural Chemistry and Molecular Biology*, Freeman, San Francisco, 1968, p. 466.

[6] T.L. Blundell and L.N. Johnson, *Protein Crystallography*, Academic Press, New York, 1976.

[7] M.S. Lehmann, T.F. Koetzle and W.C. Hamilton, *J. Am. Chem. Soc.* **94**, 2657 (1972).

[8] N. Hirayama, K. Shirahata, Y. Ohashi and Y. Sasada, *Bull. Chem. Soc. Jpn.* **53**, 30 (1980); M.S. Lehmann and A.C. Nunes, *Acta Crystallogr., Sect. B* **36**, 1621 (1980).

[9] T.F. Koetzle, M.N. Frey, M.S. Lehmann and W.C. Hamilton, *Acta Crystallogr., Sect. B* **29**, 2571 (1973).

[10] J.L. Derissen, H.J. Endeman and A.F. Peerdeman, *Acta Crystallogr., Sect. B* **24**, 1349 (1968); B. Dawson, *Acta Crystallogr., Sect. B* **33**, 882 (1977).

[11] M.M. Harding and R.M. Howieson, *Acta Crystallogr., Sect. B* **32**, 633 (1976).

[12] M.O. Chaney, O. Seely and L.K. Steirauf, *Acta Crystallogr., Sect. B* **27**, 544 (1971).

[13] T.F. Koetzle, M.S. Lehmann, J.J. Verbist and W.C. Hamilton, *Acta Crystallogr.*, *Sect. B* **28**, 3207 (1972); R.R. Bugayong, A. Sequeira and R. Chidambaram, *Acta Crystallogr.*, *Sect. B* **28**, 3214 (1972).

[14] E. Benedetti, C. Pedone and A. Sirigu, *Gazz. Chim. Ital.* **103**, 555 (1973).

[15] K. Torii and Y. Iitaka, *Acta Crystallogr.*, *Sect. B* **26** , 1317 (1970).

[16] M.S. Lehmann, J.J. Verbist, W.C. Hamilton and T.F. Koetzle, *J. Chem. Soc.*, *Perkin Trans. II*, 133 (1973).

[17] D.P. Schoemaker, J. Donohue, V. Schomaker and R.B. Corey, *J. Am. Chem. Soc.* **72**, 2328 (1950)

[18] R.L. Kajusina and B.K. Vainstein, *Kristallografiya* **10**, 833 (1965).

[19] K. Torii and Y. Iitaka. *Acta Crystallogr.*, *Sect. B* **27**, 2237 (1971).

[20] K. Torii and Y. Iitaka, *Acta Crystallogr.*, *Sect. B* **29**, 2799 (1973).

[21] G.V. Gurskaja, *Kristallografiya* **9**, 839 (1964).

[22] M.N. Frey, T.F. Koetzle, M.S. Lehmann and W.C. Hamilton, *J. Chem. Phys.* **58**, 2547 (1973).

[23] K.A. Kerr and J.P. Ashmore, *Acta Crystallogr.*, *Sect. B* **29**, 2124 (1973).

[24] T. Takigawa, T. Ashida, Y. Sasada and M. Kakudo, *Bull. Chem. Soc. Jpn.* **39**, 2369 (1966).

[25] (I) J.J. Madden, E.L. McGandy and N.C. Seeman, *Acta Crystallogr.*, *Sect. B* **28**, 2377 (1972); (II) M.S. Lehmann, T.F. Koetzle and W.C. Hamilton, *Int. J. Peptide Protein Res.* **4**, 229 (1972); (III) J.J. Madden, E.L. McGandy, N.C. Seeman, M.M. Harding and A. Hoy, *Acta Crystallogr.*, *Sect. B* **27**, 2237 (1971).

[26] Y. Iitaka, *Acta Crystallogr.* **13**, 35 (1960).

[27] Y. Iitaka, *Acta Crystallogr.* **14**, 1 (1961).

[28] Å. Kvick, *Acta Crystallogr.*, *Sect. B* **36**, 115 (1980).

[29] I.L. Karle and J. Karle, *Acta Crystallogr.* **17**, 835 (1964).

[30] D.A. Wright and R.E. Marsh, *Acta Crystallogr.* **15**, 54 (1962)1.

[31] S.H. Laurie in G. Wilkinson, R.D. Gillard and J.A. McCleverty (eds.), *Comprehensive Coordination Chemistry*, Vol. **2**, Pergamon Press, Oxford, 1987, chap. 20.2, p. 739.

[32] S.K. Burley and G.A. Petsko, *Adv. Protein Chem.* **39**, 125 (1988).

5.3   Perdeuterio-α-Glycylglycine

$^+D_3N-CD_2-CO-ND-CD_2-CO_2^-$

Crystal Data

Monoclinic, space group $P2_1/c$ (No. 14)

$a = 8.1232(8)$, $b = 9.5589(8)$, $c = 7.8250(9)$Å, $\beta = 107.656(6)°$, [3] $Z = 4$

Atom	x	y	z	Atom	x	y	z
C(1)	-.3429	.0953	-.1034	D(1)	-.1927	.2748	-.0576
C(2)	-.4853	.1708	-.2428	D(2)	-.0873	.1335	.0390
C(3)	-.7665	.1553	-.4666	D(3)	-.1363	.1529	-.1877
C(4)	-.8616	.0536	-.6111	D(4)	-.3747	.0987	.0215
N(1)	-.1778	.1690	-.0769	D(5)	-.3291	-.0130	-.1396
N(2)	-.6179	.0932	-.3364	D(6)	-.6099	-.0130	-.3292
O(1)	-.4787	.2991	-.2612	D(7)	-.8564	.1946	-.4003
O(2)	-.8216	-.0735	-.5910	D(8)	-.7240	.2430	-.5311
O(3)	-.9734	.1065	-.7396				

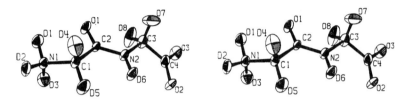

Fig. 5.3-1 Stereoscopic drawing of the molecular structure of fully deuterated α-glycylglycine. (After ref. 4).

Crystal Structure

Glycylglycine is an important amino acid because it contains the functional groups characteristic of all proteins. Of special interest are the electronic structure of the peptide bond, the amount and nature of the charge separation in the zwitterion, and the effect of hydrogen bonding on the electron distribution. The β form of glycylglycine was the first linear peptide to be subjected to X-ray analysis.[1] The crystal structure of the α form was first determined by X-ray diffraction[2] and later refined using neutron diffraction data on the perdeuterated compound.[3,4].

a) The peptide bond

In the molecule, the peptide group is appreciably distorted from planarity. A C-N single bond between sp^2 hybridized C and N atoms is about 1.49Å, while a double bond is about 1.27Å long. The length of the observed peptide bond, 1.326Å, is clearly intermediate. Bond peaks are apparent in the deformation density section through the N, C and O atoms; the maxima occur at 0.35 and 0.48 $e\text{Å}^{-3}$ in the C-N and C-O (1.238Å in length) bonds, respectively [Fig. 5.3-2]. The section of the deformation density perpendicular to the C-N bond through its midpoint clearly shows elongation perpendicular to the peptide plane, confirming the partial double-bond character.

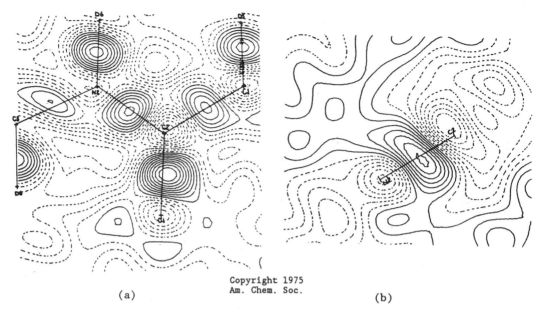

(a)   (b)

Fig. 5.3-2 ρ_{X-N} in (a) the peptide plane defined by C(2)-O(1)-N(2) and (b) section perpendicular to the peptide plane containing C(2)-N(2). Contours at 0.05 $e\text{Å}^{-3}$. (After ref. 3).

The peak height in the peptide oxygen-lone pair region is unusually small compared with values of 0.15-0.45 $e\text{Å}^{-3}$ observed in other studies. This effect would explain the displacement of the C-O overlap density maximum from the bond axis towards the NH_3^+ group.

b) The carboxyl group

The plane through the carboxyl group shows considerable asymmetry in the residual density in the two C-O bonds. The two bonds are similar in length, 1.255(4)Å and 1.240(4)Å, but have peak maxima of 0.48 $e\text{Å}^{-3}$ in C(4)-O(2) and 0.23 $e\text{Å}^{-3}$ in C(4)-O(3). This difference is consistent with an apparent absence of a second lone pair on the O(3) atom.

c) Hydrogen bonding

Each of the three deuterium atoms of the ND_3^+ group participates in a strong hydrogen bond to a carboxyl oxygen of three different carboxyl groups, while the peptide hydrogen is involved in a weaker bond with a peptide carbonyl, as listed in Table 5.3-1.

Table 5.3-1 Hydrogen bonds in α-glycylglycine.

Bonds	N-D/Å	N...O/Å	D...O/Å	N-D...O/deg.	D...O-C/deg.
N(1)-D(1)...O(2)	1.035(3)	2.753(3)	1.840(3)	150.9(3)	143.7(3)
N(1)-D(2)...O(3)	1.036(3)	2.724(3)	1.719(3)	162.5(3)	156.4(3)
N(1)-D(3)...O(2)	1.033(3)	2.790(3)	1.825(3)	147.5(3)	108.4(3)
N(2)-D(6)...O(1)	1.018(3)	2.956(3)	1.983(3)	158.9(3)	151.5(3)

There is no buildup of electron density near the midpoint of the line joining the deuterium atom and the oxygen acceptor, in accordance with an electrostatic model for the interaction. The N-D vectors tend to point toward the lone-pair density rather than to the center of the acceptor atom. The lone-pair density accumulates close to the line connecting the D and O atoms.

Remarks

1. Geometry and charge distribution of the peptide group

The geometry and average dimensions of the peptide group are shown in Fig. 5.3-3. These data have been derived by Pauling and co-workers.[5] from crystal structures of small molecules containing one or more peptide groups. Their most important result was that the C'-N bond is 0.15Å or 10% shorter than normal. Moreover, the C'=O double bond is 0.02Å longer than that known from aldehydes and ketones. The six atoms, C_i^α, C', O, N, H and C_{i+1}^α all lie in one plane.

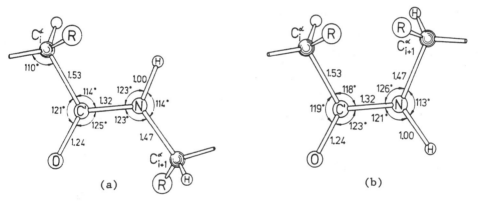

Fig. 5.3-3 Bond distances and bond angles of the peptide group in the (a) *trans*-conformation and (b) rare *cis*-conformation. (After ref. 6).

The partial charges of atoms in the peptide group are listed in Table 5.3-2; the calculated data are taken from the results of Momany and co-workers[7] who derived the partial charges of the twenty common amino

Table 5.3-2 Partial charges of atoms in the peptide group.

Atom	Calculated	Experimental[3]
C_α	+0.06	-0.10
C'	+0.45	+0.30
O	-0.38	-0.58
N	-0.36	-0.41
H_N	+0.18	+0.34
H_α	+0.02	+0.08

acid residues using the CNDO/2 method, and the experimental data are taken from ref. 3.

2. Characteristics of peptide conformations

Although the sequence of residues in a peptide can vary in many ways, the bond lengths and bond angles are nearly invariant in all peptide units, and the torsion angle about the C'-N bond, often called the peptide bond, is generally close to 180°(±5°). The angles of rotation about the remaining two types of bonds in a peptide backbone, N-C$^\alpha$ and C$^\alpha$-C', generally fall within a restricted range. The following questions arise. Do peptides exist in many conformations? Is there a preferred or stable conformation for a particular sequence? How do these flexible molecules fold? X-ray structure analyses to date have provided some answers.[8]

Cyclic hexaglycyl was the first cyclic polypeptide to be studied by X-ray analysis.[9] It crystallizes in the triclinic system with four quite different conformations of the backbone occurring side-by-side in the unit cell, as shown in Fig. 5.3-4. The four conformers occur in the ratio of 4:2:1:1. The most prevalent conformer has two intramolecular hydrogen bonds, whereas the remaining conformers participate only in intermolecular hydrogen bonding. The ease of forming several conformers is probably facilitated by the absence of any side-chains in this peptide which permits a greater latitude in the rotational ranges about the bonds in the backbone.

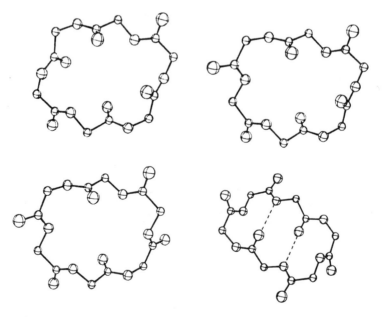

Fig. 5.3-4 Four conformers of cyclic hexaglycyl that occur in the same triclinic unit cell. (After ref. 8).

[Leu[5]]enkephalin, Tyr-Gly-Gly-Phe-Leu, a natural opiate found in the brain and gut, is a linear peptide that crystallizes with four different conformers coexisting in the same lattice. In all four conformers the backbones are fairly similar, but the side-chains assume a number of entirely different conformations, particularly Tyr and Leu (Fig. 5.3-5).

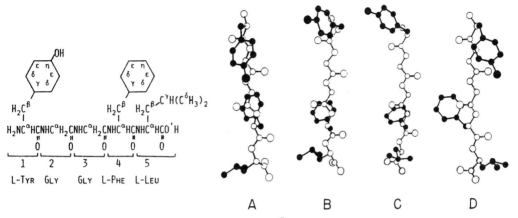

Fig. 5.3-5 Four conformers of [Leu[5]]enkephalin occurring in a crystal grown from dimethylformamide. Atoms in the side chains are distinguished by shading. (After ref. 8).

The larger peptides appear to have a unique folding pattern probably because there are many more intramolecular interactions, such as internal hydrogen bonds and hydrophobic attractions, as compared to lattice forces, such as intermolecular hydrogen bonds. For example, a unique conformation is maintained by the cyclic decapeptide antanamide in six different crystal

$$\text{antanamide} = \begin{array}{c} \text{Val}^1{-}\text{Pro}^2{-}\text{Pro}^3{-}\text{Ala}^4{-}\text{Phe}^5 \\ | \qquad\qquad\qquad\qquad\qquad | \\ \text{Phe}^{10}{-}\text{Phe}^9{-}\text{Pro}^8{-}\text{Pro}^7{-}\text{Phe}^6 \end{array}$$

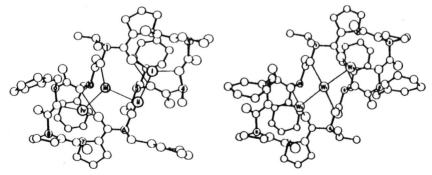

Fig. 5.3-6 Comparison of the conformations of antanamide with four intrinsic water molecules (left) and [Phe[4]Val[6]]antanamide with three intrinsic water molecules (right). (After ref. 8).

forms. Despite significant differences in regard to polarity of solvents,
co-crystallized solvent, amount of hydration, association between neighbouring
peptide molecules, extrinsic hydrogen bonding (if any), and even side chains
on residues 4 and 6, the conformation of the peptide in all the polymorphs
remains unchanged. The conformations of natural antamanide in polymorph I,
antamanide.$8H_2O$.X, and the active synthetic analogue in polymorph II,
[Phe^{4}Val^6]antamanide $3H_2O$.Y, are compared in Fig. 5.3-6.

The conformation of cyclic peptides undergoes profound changes by
complexation with ions. Uncomplexed peptides generally do not have a suitable
cavity already preformed, or a suitable polar region, to accept a metal ion
directly. A large amount of folding and twisting of the backbone occurs
during complexation.[10] Fig. 5.3-7(a,b) shows the change in conformation

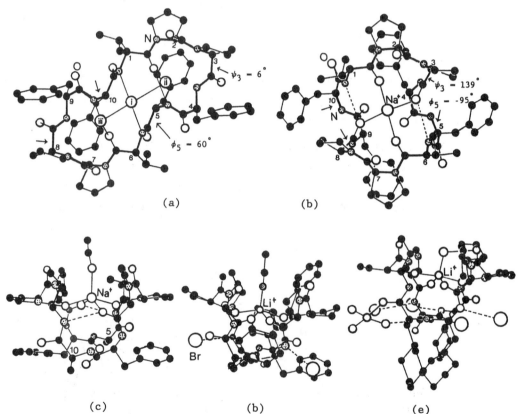

Fig. 5.3-7 The changes in conformation from uncomplexed [Phe^{4}Val^6]antamanide
(a) to the Na$^+$ complex (b). The arrows indicate bonds where twisting of the
backbone differs in these two cases; the largest rotations take place about
C_3^α-C_3' and C_8^α-C_8', as well as N_5-C_5^α and N_{10}-C_{10}^α. Comparison of the
conformations of (c) Na$^+$[Phe^{4}Val^6]antamanide.EtOH (phenyl rings belonging to
the Phe residues in postions 5 and 10 omitted), (d) Li$^+$.antamanide.CH_3CN.Br$^-$,
and (e) Li$^+$.perhydroantamanide.Li$^+$$4H_2O$.2Br$^-$. (Adapted from refs. 10 and 11).

from uncomplexed [Phe^{4}Val^6]antamanide to the Na$^+$ complex. A comparison of the conformations of the Na$^+$ and Li$^+$ complexes of some related antamanides is shown in Fig. 5.3-7(c-e).

3. Dihedral angles defining a polypeptide chain

The backbone of a polypeptide chain is described by two parameters per residue. With a stiff peptide bond and rather rigid bond lengths and bond angles, the conformation of the polypeptide chain is essentially specified by the dihedral angles ϕ and ψ at the C_α-atoms as defined in Fig. 5.3-8. The definitions follow the IUPAC-IUB recommendations of 1969.[12] The torsion angle ω is inserted although this rotation is inhibited. The zero positions of the dihedral angles are defined in Fig. 5.3-8. Thus the depicted main chain angles are $\phi_i = 180°$ and $\psi_i = 180°$. As indicated by hatching, normally the six atoms C_i^α, C', O_i, N_{i+1}, C_{i+1}^α and H_{i+1} all lie in one plane, so that ω equals 180°. The senses of rotation of ω, ψ, ϕ are taken as positive if the viewer is located on the atom at the N-terminal side of a bond, and the C-terminal side of this bond is rotated in the direction as indicated by the arrow (clockwise in Fig. 5.3-8). A corresponding definition applies to χ with the viewer placed on the side of the atom closer to the C_α-atom.

$\omega_i = 0$ for C_i^α-C_i' cis to N_{i+1}-C_{i+1}^α

$\psi_i = 0$ for C_i^α-N_i trans to C_i'-O_i

$\phi_i = 0$ for C_i^α-C_i' trans to N_i-H_i

$\chi_i^1 = 0$ for C_i^α-N_i cis to C_i^β-O_i^γ

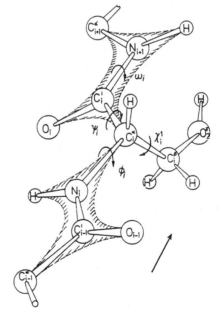

Fig. 5.3-8 Definition of dihedral angles in a polypeptide chain. (After ref. 6).

4. The helical configurations of the polypeptide chain[13]

On the basis of the information about interatomic distances, bond angles, and other configurational parameters from the crystal structures of amino acids, peptides, and other simple substances related to proteins, Pauling and his co-workers[5] succeeded in constructing two reasonable hydrogen-bonded helical configurations for the polypeptide chain. Besides the planarity of each peptide residue, it was assumed that each nitrogen atom forms a hydrogen bond with an oxygen atom of another residue, with N...O ≃ 2.72Å, and that the vector from the nitrogen atom to the hydrogen-bonded oxygen atom lies not more than 30° from the N-H direction.

Solution of this problem shows that there are five and only five configurations for the chain that satisfy the conditions other than that of the direction of the hydrogen bond relative to the N-H direction. These correspond to the values 165°, 120°, 108°, 97.2° and 70.1° for the rotational angle. In the first, third, and fifth of these structures the $>$C=O group is negatively and the $>$N-H group positively directed along the helical axis, taken as the direction corresponding to the sequence -CHR-CO-NH-CHR- of atoms in the peptide chain, and in the other two their directions are reversed. The first three of the structures are unsatisfactory, in that the N-H group does not extend in the direction of the oxygen atom at 2.72Å; the fourth and fifth are satisfactory, the angle between the N-H vector and N-O vector being about 10° and 25° respectively. The fourth structure has 3.69 amino acid residues per turn in the helix, and the fifth structure has 5.13 residues per turn. In the fourth structure each amide group is hydrogen bonded to the third amide group beyond it along the helix, and in the fifth structure each is bonded to the fifth amide group beyond it. Pauling and co-workers named these two structures the 3.7-residue structure, or α-helix, and the 5.1-residue structure, or γ-helix, respectively. The geometrical features of the α- and γ-helices are illustrated in Fig. 5.3-9 and Fig. 5.3-10, respectively.

For glycine both the 3.7-residue helix and the 5.1-residue helix could occur with either a positive or a negative rotational translation relative to the positive direction of the helical axis given by the sequence of atoms in the peptide chain. For other amino acids with the L configuration, the positive helix and the negative helix would differ in the position of the side chains, and it might well be expected that in each case one sense of the helix would be more stable than the other. An arbitrary assignment of the R groups has been made in these figures.

The translation along the helical axis in the α-helix is 1.47Å, and that in the γ-helix is 0.99Å. The values for one complete turn are 5.44Å and 5.03Å, respectively.

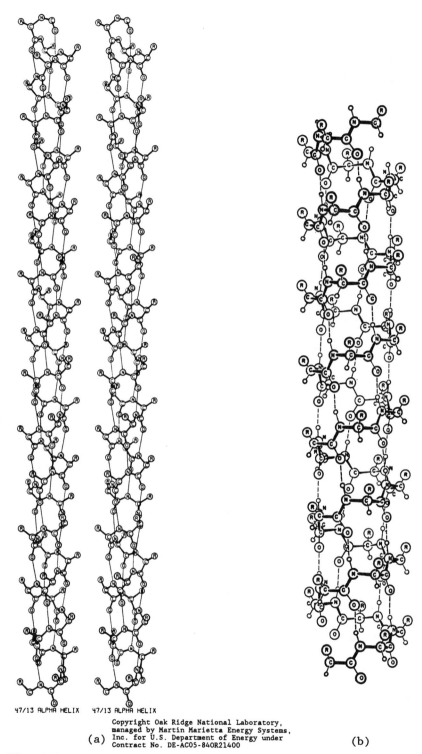

47/13 ALPHA HELIX   47/13 ALPHA HELIX

(a)   (b)

Fig. 5.3-9 The helical configurations of polypeptide chain. (a) The α-helix and (b) the γ-helix. (After refs. 14 and 15).

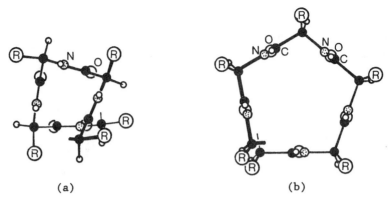

(a)   (b)

Fig. 5.3-10 Helical polypeptide chain viewed along its axis. (a) The
α-(3.7-residue)helix. (b) The γ-(5.1-residue)helix. (Adapted from ref. 15).

5. β-Pleated sheets

Extended polypeptide chains can associate by hydrogen bonding to form
sheet-like structures. Concomitantly with the postulation of the α-helix,
Pauling and Corey[16] also proposed the planar parallel and antiparallel β-
pleated sheets as suitable regular hydrogen-bonded structures for polypeptide
chains, as shown in Fig. 5.3-11.

In a β-sheet, the polypeptide chains lie side by side (either parallel
or antiparallel to each other) and form N-H...O=C hydrogen bonds in an
approximate plane on either side of the chain. The β-sheets have been
observed at high resolution in the structures of many different proteins.
Baker and Hubbard have listed 12 generalizations regarding the role of
hydrogen bonds in stabilizing the structures of proteins.[31]

6. Small peptides with antiparallel β-sheet structures

Some examples are discussed below.

a) Tripeptide L-Ala-L-Ala-L-Ala[18]

Tri-L-alanine crystallizes as a hemihydrate in space group *C*2. The
molecules are packed in rows in head-to-tail fashion (C-terminal opposite N-
terminal) with the methyl groups alternately above and below the planes
through the peptide chains, and are held together in sheets by hydrogen bonds
between all carbonyl oxygen and N-H groups of neighbouring molecules; this
corresponds to the antiparallel pleated-sheet arrangement for polypeptide
chains, as shown in Fig. 5.3-12. This arrangement lessens the steric
hindrance between side chains which would occur for fully extended ($\varphi = \psi = \omega$
$= 180°$) polypeptide chains. Tri-L-alanine provides the first precise and
direct experimental measurements of the conformational angles for an
antiparallel β-sheet arrangement.

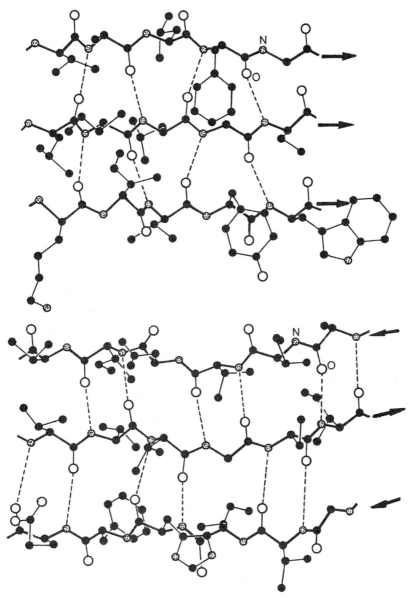

Fig. 5.3-11 Structure of β-pleated sheets: (Top) An example of a parallel β sheet from flavodoxin (residues 82-86, 49-53, and 2-6). The hydrogen bonds are evenly spaced but slanted in alternate directions. Since both sides of the sheet are covered by other main chain (as is almost always true for the parallel sheet), side groups pointing in both directions are predominantly hydrophobic except at the ends of the strands. (Bottom) An example of an antiparallel β sheet from Cu, Zn superoxide dismutase (residues 93-98, 28-33 and 16-21). Pairs of closely spaced hydrogen bonds alternate with widely spaced ones. The direction of view is from the solvent, so that side chains pointing up are predominantly hydrophilic and those pointing down are predominantly hydrophobic. (Adapted from ref. 17).

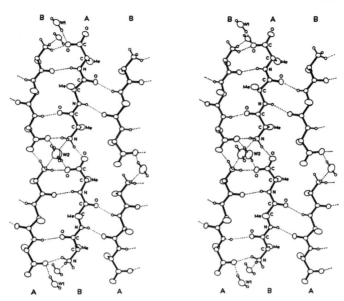

Fig. 5.3-12 Stereoview of the structure of tri-L-alanine with **b** pointing away from the viewer, **c** upward and **a** to the right. The oxygen atoms of $W(1)$ at the bottom of the figure lie on the twofold axis at $0,y,0$; those on $W(1)$ at the top are at $0,y,1$. (After ref. 18).

b) [Leu[5]]enkephalin[19]

[Leu[5]]enkephalin (Tyr-Gly-Gly-Phe-Leu) grown from N,N-dimethylformamide (DMF)/water crystallizes as a monoclinic crystal. The asymmetric unit includes four enkephalin molecules (Fig. 5.3-5), eight water molecules, eight

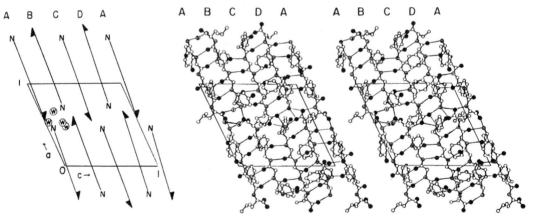

Fig. 5.3-13 Four independent molecules of enkephalin, labelled A, B, C and D, each with a different conformation, form an infinite antiparallel β-sheet in the crystal. The layer of hydrogen-bonded molecules near $y \sim 1/4$ is shown in the stereodiagram. The axial directions are a ↖ and c →. Four water molecules are labelled W, O atoms are indicated by filled circles, and N atoms by hatched circles. (After ref. 19).

DMF molecules, plus an unknown number of disordered solvent molecules: $4C_{28}H_{37}N_5O_7 \cdot 8H_2O \cdot 8C_3H_7NO \cdot X$. The four conformers with extended backbones form an infinite antiparallel β-sheet, as shown in Fig. 5.13-13. β-sheets related by the 2_1 axis are separated by a 12Å spacing. Side groups protruding above and below the β-sheets are entirely immersed in a thick layer of solvent occupying the volume between β-sheets.

    c) Cyclic cystine peptides

    The first example of a disulfide bridging across an antiparallel β-sheet was provided by the crystal structure of the cyclic cystine peptide Boc-Cys-Val-Aib-Ala-Leu-Cys-NHMe,* as shown in Fig. 5.3-14(a).[20] Hydrogen
underbrace: S—S
bonds between the two antiparallel portions of the chain occur between N(2)H...O(5), N(5)H...O(2) and N(7)H...O(0) with normal lengths. The S-S bridge from $C_β(1)$ to $C_β(6)$ fits comfortably from one β-strand to the antiparallel β-strand where the $C_α(1)...C_α(6)$ distance is 4.036Å. The S-S distance of 2.027(6)Å and the C-S-S angles of 106.3(5) and 105.3(4)°.

    The β-sheet conformation already present in the individual molecules is extended throughout the crystal by intermolecular hydrogen bonding. Fig. 5.3-14(a) depicts three neighbouring peptides along the *a* axis and the N(6)...O(1) hydrogen bonds that link these molecules into an infinite ribbon.

    The cyclic biscystine peptide [Boc-Cys-Ala-Cys-NHMe]$_2$ crystallizes in space group $C2$, with one-half formula unit of $C_{30}H_{52}N_8O_{10}S_8 \cdot 2(CH_3)_2SO$ per asymmetric unit. The molecule assumes an extended antiparallel β-sheet conformation with two disulfide bridges between the β-strands [Fig. 5.3-14(b)].[21] A crystallographic C_2 axis passes perpendicularly through the center of the molecule. The 22-membered ring contains two transannular NH...OC hydrogen bonds and two additional NH...OC bonds are formed at both ends of the molecule between the terminal $(CH_3)_3COCO$ and $NHCH_3$ groups. The antiparallel peptide strands are distorted from a regularly pleated sheet, caused mainly by the L-Ala residue in which $\phi = -155°$ and $\psi = 162°$. In the disulfide bridge S-S = 2.030Å and C-S-S = 107 and 105°.

7. Polypeptide chains in proteins

    The first protein structures to be successfully solved by X-ray analysis were those of myoglobin[22] and hemoglobin.[23,24] The principal function of these two proteins is the reversible binding of the oxygen molecule O_2. Hemoglobin is present in the erythrocytes of the blood and transports O_2 in the circulation, while myoglobin stores oxygen in the muscles. The molecular

---

* Aib is the residue of α-amino-isobutyric acid, and Boc stands for the (tert-butyloxy)carbonyl group

weight of myoglobin is about 18,000; it contains 153 amino acid residues. The
unit cell is monoclinic, space group $P2_1$, with a = 64.6, b = 31.1, c = 34.8Å,
β = 105.5° and Z = 2.

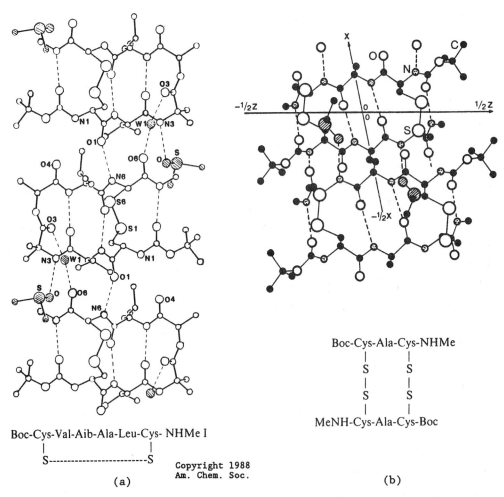

Boc-Cys-Ala-Cys-NHMe
      | |
      S   S
      | |
      S   S
      | |
MeNH-Cys-Ala-Cys-Boc

Boc-Cys-Val-Aib-Ala-Leu-Cys- NHMe I
    | |
    S------------------S

(a)   (b)

Fig. 5.3-14 (a) Extended β-sheet formed by a two-fold screw operating
parallel to the a axis and by hydrogen bonding N(6)H...O(1) between adjacent
molecules. Intermolecular distances N(3)H...O(6) and N(1)H...O(4) are 0.3-
0.4Å longer than the range of values usually observed for hydrogen bonds. The
solvent molecules DMSO and water (W1) are shown shaded. (b) Packing of the
two peptide molecules in the cell. Two-fold rotation axes occur at x = 0, z
= 0 and x = -1/2, z = 0. A two-fold screw axis occurs at x = -1/4, z = 0.
The lower molecule is translated from the upper molecule by -1/2 + x, 1/2 + y,
z. The peptide molecules, when viewed along the z-direction, are arranged in
a checkerboard pattern and do not form an extended β-sheet. Two DMSO
molecules are represented by shaded circles. (After refs. 20 and 21).

The molecule of myoglobin, as well as molecules of some other proteins, contain a "prosthetic group" attached to the polypeptide chain. It is a flat porphyrin molecule, the "haem" group, at whose center there is an iron atom binds an O_2 molecule. The polypeptide chain in myoglobin has a 75% α-helix configuration. The hemoglobin molecule is built up of four subunits, each similar to the myoglobin molecule as regards its tertiary structure. Fig. 5.3-15 shows a subunit of a hemoglobin molecule and the geometry around the Fe atom of a heme group.

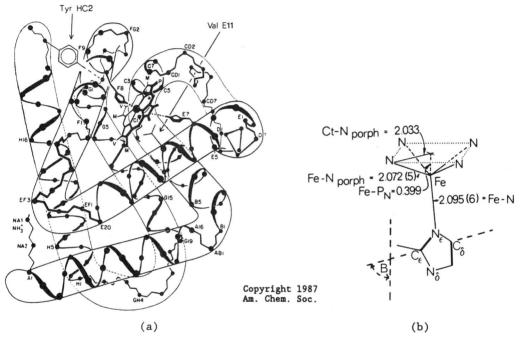

(a)   (b)

Fig. 5.3-15   (a) Structure of a subunit of the hemoglobin molecules. (b) The geometry around the Fe atom in 2-MeIm(TpivPP)Fe(II). (After ref. 24).

8. The structure of insulin[25,26]

Insulin, the most studied peptide hormone, is secreted by the beta islet cells of the pancreas in response to a rise in blood sugar. Its action is to lower the level of blood sugar by promoting peripheral utilization of glucose and limiting hepatic release. Inadequate insulin secretion generally results in excess sugar in the blood and urine, accompanied by progressive emaciation, extreme hunger and thirst, and eventually metabolic failure. The use of insulin, either natural or synthetic, as a drug in treating diabetes mellitus permits diabetic patients to lead normal lives.

All mammalian insulins are made up of 51 amino acid residues arranged in two chains, **A** and **B**, comprising 21 and 30 residues, respectively; the total molecular weight is about 5807. The chains are connected by two disulfide linkages, and a third disulfide bridge occurs within the shorter chain **A**. The complete amino acid sequence of insulin was determined in 1955 by Sanger (1958 Nobel Laureate in Chemistry) and his colleagues.[27] The structural formula of porcine (pig) insulin is shown below.

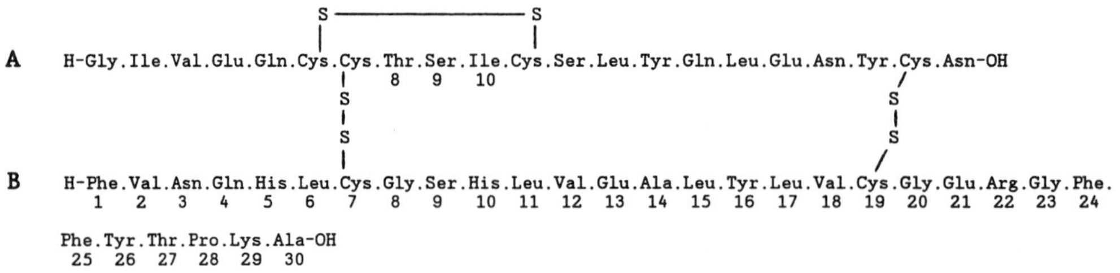

```
              S ————————————— S
              |               |
A    H-Gly.Ile.Val.Glu.Gln.Cys.Cys.Thr.Ser.Ile.Cys.Ser.Leu.Tyr.Gln.Leu.Glu.Asn.Tyr.Cys.Asn-OH
                         |    8   9  10   |                                          /
                         S                                                           S
                         |                                                           |
                         S                                                           S
                         |                                                           /
B    H-Phe.Val.Asn.Gln.His.Leu.Cys.Gly.Ser.His.Leu.Val.Glu.Ala.Leu.Tyr.Leu.Val.Cys.Gly.Glu.Arg.Gly.Phe.
        1   2   3   4   5   6   7   8   9   10  11  12  13  14  15  16  17  18  19  20  21  22  23  24

     Phe.Tyr.Thr.Pro.Lys.Ala-OH
      25  26  27  28  29  30
```

Species differences in insulin are usually confined to variation in one or more of the residues at positions 8, 9 and 10 of the **A** chain and position 30 of the **B** chain. The difference in composition of a few common species is shown in Table 5.3-3.

Table 5.3-3   Variation in insulin structure.

Species	A chain			B chain
	8	9	10	30
Human	Thr	Ser	Ile	Thr
Porcine (pig)	Thr	Ser	Ile	Ala
Rabbit	Thr	Ser	Ile	Ser
Beef	Ala	Ser	Val	Ala

Residue number

Many crystalline modifications of insulin exist, but the rhombohedral form of the zinc-containing porcine insulin (usually referred to as 2Zn pig insulin) was the first to be successfully tackled by X-ray crystallography. Dorothy Crowfoot (later Hodgkin) had obtained single-crystal X-ray photographs of insulin as early as 1935,[28] but it took her Oxford group thirty-four years to solve the structure of 2Zn pig insulin by the multiple isomorphous replacement method.[29]   The structure was also independently worked out by the Peking Insulin Structure Research Group in China in 1971.[30]

Rhombohedral 2Zn pig insulin, MW ~ 36,000, crystallizes in space group $R3$ with $a = 82.46$, $c = 33.94$Å and $Z = 3$ in a triply-primitive hexagonal unit cell.   In the asymmetric unit is a dimer comprising two nearly identical insulin molecules related by an approximate (local) two-fold axis that is perpendicular to the crystallographic three-fold axis and intersects it.   Thus six insulin molecules are organized to form a hexamer of approximate symmetry 32, as illustrated in Fig. 5.3-16.[25] The two zinc ions are 15.4Å apart, being located on the three-fold axis and equally distant from the local axis.

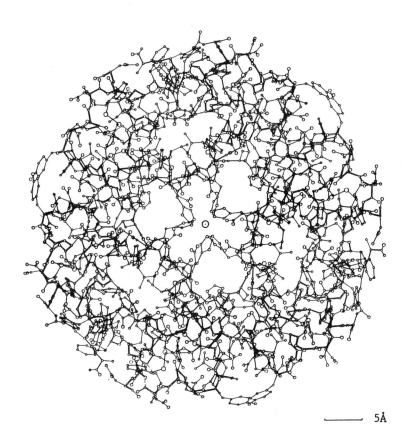

———— 5Å

Fig. 5.3-16 The structure of the insulin hexamer. (After ref. 25).

References

[1] E.W. Hughes and W.J. Moore, *J. Am. Chem. Soc.* **71**, 2618 (1949).

[2] A.B. Biswas, E.W. Hughes, B.D. Sharma and J.N. Wilson, *Acta Crystallogr., Sect. B* **24**, 40 (1968); E.W. Hughes, *Acta Crystallogr., Sect. B* **24**, 1128 (1968).

[3] J.F. Griffin and P. Coppens, *J. Am. Chem. Soc.* **97**, 3496 (1975).

[4] H.C. Freeman, G.L. Paul and T.M. Sabine, *Acta Crystallogr., Sect. B* **26** 925 (1970).

[5] L. Pauling, R.B. Corey and H.R. Branson, *Proc. Nat. Acad. Sci. USA* **37**, 205 (1951).

[6] G.E. Schulz and R.H. Schirmer, *Principles of Protein Structure*, Springer-Verlag, New York, 1979.

[7] F.A. Momany, R.F. McGuire, A.W. Burgess and H.A. Scheraga, *J. Phys. Chem.* **79**, 2361 (1975).

[8] I.L. Karle in J.J. Stezowski, J.-L. Huang and M.-C. Shao (eds.), *Molecular Structure: Chemical Reactivity and Biological Activity*, Oxford University Press, New York, 1988, pp. 46-59; I.L. Karle, J.L. Flippen-Anderson and T. Wieland, *Int. J. Peptide Protein Res.* **33**, 422 (1989).

[9] I.L. Karle and J. Karle, *Acta Crystallogr.* **16**, 969 (1963).

[10] I.L. Karle in R.H. Sarma and M.H. Sarma (eds.), *Biomolecular Stereodynamics III*, Adenine Press, Albany, New York, 1986, pp. 197-215.

[11] I.L. Karle in E. Benedetti, C. Pedone, T. Tancredi, P. Temussi and M. Goodman (eds.), *Biopolymers*, Vol. **28**, Wiley, New York, 1989, pp. 1-14, and references cited therein.

[12] Anon., IUPAC-IUB Commission on Biochemical Nomenclature, *J. Biol. Chem.* **245**, 6489 (1970).

[13] L. Pauling and R.B. Corey, *Fortschr. Chem. org. Naturstoffe* **11**, 180 (1954); *Proc. Nat. Acad. Sci. USA* **37**, 251, 729 (1951).

[14] C.K. Johnson, *ORTEP-II: A FORTRAN Thermal-Ellipsoid Plot Program for Crystal Structure Determination*, ORNL-5138, Oak Ridge National Laboratory, Oak Ridge, Tennessee, 1976.

[15] L. Pauling, R.B. Corey and H.R. Branson, *Proc. Natl. Acad. Sci. USA* **37**, 205 (1951).

[16] L. Pauling and R.B. Corey, *Proc. Nat. Acad. Sci. USA* **37**, 251 (1951).

[17] J.S. Richardson, *Adv. Protein Chem.* **34**, 168 (1981).

[18] J.K. Fawcett, N. Camerman and A. Camerman, *Acta Crystallogr., Sect. B* **31**, 658 (1975).

[19] I.L. Karle, J. Karle, D. Mastropaolo, A. Camerman and N. Camerman, *Acta Crystallogr., Sect. B* **39**, 625 (1983).

[20] I.L. Karle, R. Kishore, S. Raghothama and P. Balaram, *J. Am. Chem. Soc.* **110**, 1958 (1988).

[21] I.L. Karle, J.L. Flippen-Anderson, R. Kishore and P. Balaram, *Int. J. Peptide Protein Res.* **34**, 37 (1989).

[22] J.C. Kendrew, R.E. Dickerson, B.E. Strandberg, R.G. Hart, D.R. Davies, D.C. Phillips and V.C. Shore, *Nature (London)* **185**, 422 (1960).

[23] M.F. Perutz, H. Muirhead, J.M. Cox, L.C.G. Goaman, F.S. Mathews, E.L. McGandy and L.E. Webb, *Nature (London)* **219**, 29 (1968).

[24] M.F. Perutz, G. Fermi, B. Luisi, B. Schaanan and R.C. Liddington, *Acc. Chem. Res.* **20**, 309 (1987).

[25] T. Blundell, G. Dodson, D. Hodgkin and D. Mercola, *Adv. Protein Chem.* **26**, 279 (1972).

[26] G. Dodson, J.P. Glusker and D. Sayre (eds.), *Structural Studies on Molecules of Biological Interest*, Clarendon Press, Oxford, 1981.

[27] A.P. Ryle, F. Sanger, L.F. Smith and R. Kitai, *Biochem. J.* **60**, 541 (1955).

[28] D. Crowfoot, *Nature (London)* **135**, 591 (1935).

[29] M.J. Adams, T.L. Blundell, E.J. Dodson, G.G. Dodson, M. Vijayan, E.N. Baker, M.M. Harding, D.C. Hodgkin, B. Rimmer and S. Sheat, *Nature (London)* **224**, 491 (1969); E.J. Dodson, G.G. Dodson, D.C. Hodgkin and C.D. Reynolds, *Can. J. Biochem.* **57**, 469 (1979).

[30] Peking Insulin Structure Research Group (1971), *Peking Rev.* **40**, 11 (1973); *Scientia Sinica* **16**, 136 (1973); *ibid.* **17**, 752 (1974); *ibid.* **19**, 358 (1976).

[31] E.N. Baker and R.E. Hubbard, *Prog. Biophys. Mol. Biol.* **44**, 97 (1984).

Note The structure of the Zn-free pig insulin dimer in the cubic crystal (space group $I2_13$ with $a = 78.9$Å and $Z = 24$) is reported in J. Badger, M.R. Harris, C.D. Reynolds, A.C. Evans, E.J. Dodgson, G.G. Dodgson and A.C.T. North, *Acta Crystallogr., Sect. B* **47**, 127 (1991). The water structure is described in J. Badger and D.L.D. Caspar, *Proc. Nat. Acad. Sci. U.S.A.* **88**, 622 (1991).

Readers interested in protein structure and enzyme catalysis are referred to the following books: D. Voet and J.G. Voet, *Biochemistry*, Wiley, New York, 1990; R.E. Dickerson and E. Geis, *Haemoglobin: Structure, Function, Evolution, Pathology*, Benjamin/Cummings, Menlo Park, 1983; M. Perutz, *Mechanisms of Cooperativity and Allosteric Regulation in Proteins*, Cambridge University Press, Cambridge, 1989; M.I. Page and A. Williams (eds.), *Enzyme Mechanisms*, Royal Society of Chemistry, London, 1987; A. Cooper, J.L. Houben and L.C. Chien (eds.), *The Enzyme Catalysis Process: Energetics, Mechanism, and Dynamics* (NATO ASI Series Vol. **178**), Plenum Press, New York, 1989.

5.4 Potassium Dihydrogen Isocitrate
$K(C_6H_7O_7)$

Crystal Data

Orthorhombic, space group $P2_12_12_1$ (No. 19)

$a = 12.013(6)$, $b = 13.145(7)$, $c = 5.159(3)$Å, $Z = 4$

Atom	x	y	z	Atom	x	y	z
K	.3830	.0055	.6781	O(5)	.5096	.1970	.7832
C(1)	.1799	.1877	.7151	O(6)	.3977	.1596	.8995
C(2)	.2464	.2382	.4987	O(7)	.3000	.1877	.3582
C(3)	.3276	.3197	.6014	H(1)	.199	.294	.364
C(4)	.3714	.3861	.3821	H(2)	.277	.361	.740
C(5)	.3994	.4933	.4656	H(3)	.341	.401	.238
C(6)	.4192	.2751	.7698	H(4)	.442	.366	.288
O(1)	.1295	.0950	.8658	H(5)	.330	.172	.234
O(2)	.1716	.5441	.7347	H(6)	.088	.204	1.031
O(3)	.4551	.5296	.2935	H(7)	.452	.604	.257
O(4)	.3727	.3229	.6696				

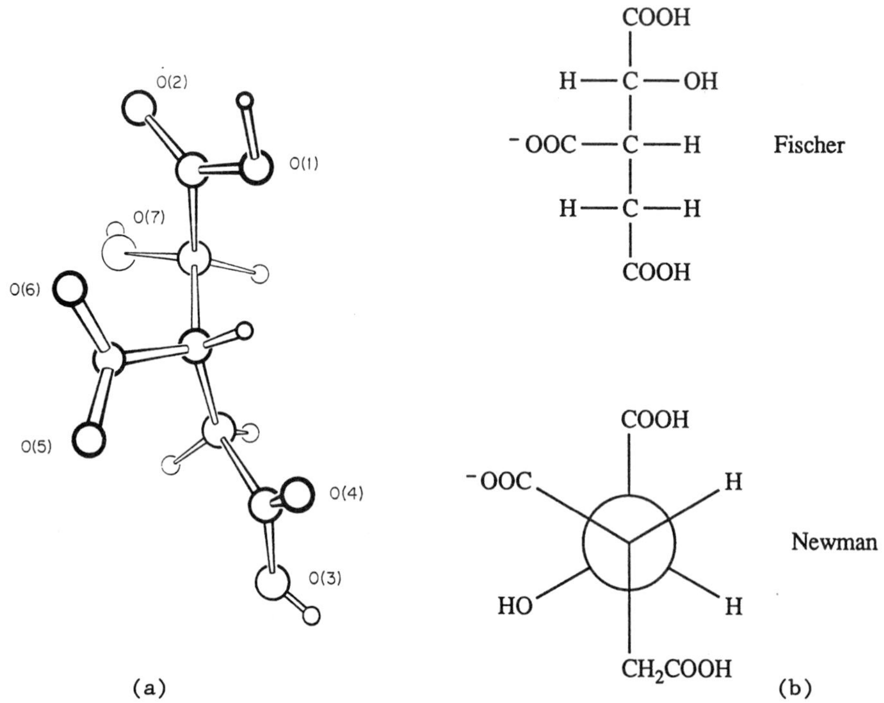

(a)   (b)

Fig. 5.4-1 Structure and absolute configuration of the isocitrate ion. (a) Viewed along the plane through C(1), C(2) and C(3). (b) Projection formulae representing the absolute configuration determined. (After ref. 1).

Crystal Structure

The absolute configuration of the biologically active isomer, (+)-isocitric acid, has been determined from anomalous dispersion data. The correct configurational formula for (+)-isocitric acid (Fig. 5.4-1) is

(1R,2S)-1-hydroxy-1,2,3-propanetricarboxylic acid. The bond lengths and angles in the isocitrate ion are given in Fig. 5.4-2.

All carbon-carbon bonds lie in the range 1.513-1.535Å. The O...O distance between the carboxyl group and the α-hydroxy group is short, 2.622Å, as found for other citrates, and the hydroxyl oxygen atom is 0.071Å out of the plane of the terminal carboxyl group. The C(1)-C(2)-C(3)-C(4)-C(5) carbon backbone of the anion is extended, with C(1) and C(5) displaced by 0.332Å and 0.698Å, respectively, from the plane through C(2)-C(3)-C(4). The central carboxyl group is slightly inclined to this plane.

There is no intramolecular hydrogen bonding, all three available hydrogen atoms being involved in hydrogen bonds of the type: O(7)-H(5)...O(6) = 2.687Å, O(1)-H(6)...O(5) = 2.516Å, and O(3)-H(7)...O(1) = 2.884Å.

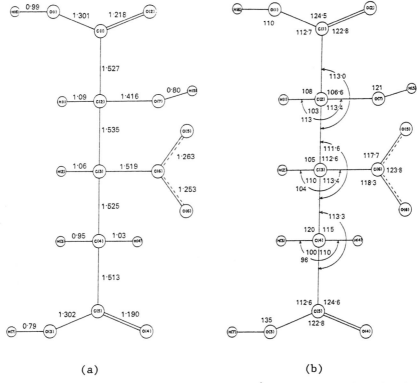

(a)   (b)

Fig. 5.4-2 (a) Covalent bond lengths (Å) and (b) angles (deg) for the isocitrate ion. (After ref. 1).

The arrangement of the ions in the crystal is shown in Fig. 5.4-3. The structure consists of columns of pairs of potassium ions surrounded by oxygen atoms; the six closest K...O contacts are in the range 2.725-3.058Å.

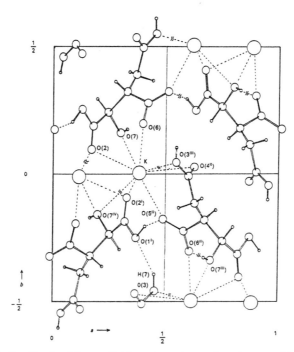

Fig. 5.4-3 Projection of the structure in the -c direction showing
the coordination of oxygen atoms round the K$^+$ and the hydrogen bond
system. (After ref. 1).

Remarks

1. Absolute configuration

The absolute configuration of a crystal or molecule refers to its
structure expressed in an absolute reference frame. For any compound
containing one asymmetric carbon or nitrogen atom that has four different
substituents arranged in tetrahedral fashion, there are two isomers of
opposite chirality, called enantiomers or enantiomorphs, for which three-
dimensional representations of their structures are illustrated in Fig. 5.4-4.
Absolute configuration can often be correlated with chiroptical phenomena,
such as the direction of rotation of the plane of polarization of light by a
solution of the compound. Solutions of two enantiomers rotate the plane of
polarized light in opposite directions.

L. Pasteur[2] in 1848 showed that crystals of sodium ammonium tartrate
bounded by a set of asymmetrically located "hemihedral faces" rotated the
plane of polarization of light clockwise, while crystals with similar faces in
mirror-image orientations rotated the plane of polarization counter-clockwise.
The external form of the crystals illustrated in Fig. 5.4-5 could be used to
separate enantiomers by hand.

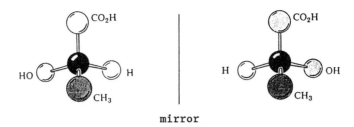

Fig. 5.4-4 Two chiral isomers related to each other by a mirror plane.

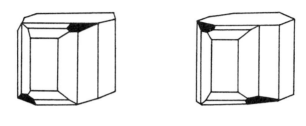

Dextrorotatory   Levorotatory

Fig. 5.4-5 Hemihedral faces on an enantiomorphic pair of sodium ammonium tartrate crystals (used by Pasteur to differentiate dextro- from levo-rotatory forms).

2. Determination of absolute configuration by comparing the intensities of Friedel pairs.

The possibility of utilizing the breakdown of Friedel's Law [3] (Section 2.9) to provide a bridge between macroscopic and molecular chirality was first proposed by J.M. Bijvoet in 1949. [4] At that time the configurations of opticaly active compounds were determined relative to (+)-glyceraldehyde as a standard, *arbitrarily* assigned the configuration I by Emil Fisher [5] and conventionally represented by the Fisher projection formula II. Within this system, for example, the configuration of (+)-tartaric acid was known to be III, and that of the naturally occurring L-amino acids to be **IV**, but there was no way to establish the actual configuration of the reference

       I   II   III   IV

molecule. Within a couple of years, Bijvoet's proposal had been put into
practice in a classical study on sodium rubidium $(2R,3R)$-(+)-tartrate
tetrahydrate.[6,7]

For monochromatized ZrKα radiation ($K\alpha_1$ = 0.78588Å), the imaginary part
of the anomalous scattering of Rb (K absorption edge = 0.81549Å) amounts to
$\Delta f''_{Rb}$ = 3.2. Fig. 5.4-6 shows how the introduction of a "phase lag" in the
scattering of the Rb atom destroys the Friedel equivalence of the $hk\ell$ and $\overline{hk\ell}$
reflections.

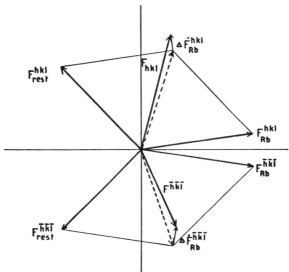

Fig. 5.4-6 The vectors F_{Rb} and F_{rest} are combined for the reflections
$hk\ell$ and $\overline{hk\ell}$ respectively, neglecting the imaginary part of the structure
factor of Rb. The resultant (dotted) amplitudes are of equal modulus
but opposite phase. The introduction of the term $\Delta F''_{Rb}$ with phase
increment $\pi/2$ with respect to F_{Rb} is seen to destory this equality of
the resultant modulus. (After ref. 6).

In Table 5.4-1 the experimental and calculated values of $I(hk\ell)$ and
$I(hk\overline{\ell})$ are compared, the latter by symmetry being equal to $I(\overline{hk\ell})$. The visual
estimations of the last column were made by several persons independently.
The calculated values are based on a structure model corresponding to the
absolute configuration assignment of Emil Fischer.

Table 5.4-1 Comparison of $I(hk\ell)$ and $I(hk\overline{\ell})$ in sodium rubidium tartrate.[11]

| | | | Calculated | | | | | | Calculated | | |
h	k	ℓ	$I(hk\ell)$	$I(hk\overline{\ell})$	Observed	h	k	ℓ	$I(hk\ell)$	$I(hk\overline{\ell})$	Observed
1	4	1	361	377	?	2	6	1	828	817	>
1	5	1	337	313	?	2	7	1	18	8	>
1	6	1	313	241	>	2	8	1	763	716	>
1	7	1	65	78	<	2	9	1	170	166	?
1	8	1	185	148	>	2	10	1	200	239	<
1	9	1	65	46	>	2	11	1	159	149	?
1	10	1	248	208	>	2	12	1	324	353	<
1	11	1	27	41	<	2	13	1	18	35	<

The concordance of inequality signs in the observed and calculated intensities proves that the chemical convention (Fig. 5.4-7) fortuitiously matches with reality, as the odds against this is 1:1

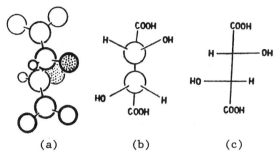

(a)   (b)   (c)

Fig. 5.4-7 Absolute configuration of natural dextrorotatory tartaric acid. (a) As determined by X-rays in sodium rubidium tartrate. (b) In normalized configuration by rotating around single bonds. (c) In projection. (After ref. 7).

3. Absolute configuration and absolute structure

The general term "absolute structure" has been advocated by Jones to signify a non-centrosymmetric structure successfully distinguished from its inverse using anomalous scattering techniques.[8] The structural features determined, other than bond distances and valence angles, in an X-ray analysis of a non-centrosymmetric crystal are summarized in Table 5.4-2.

Table 5.4-2 X-Ray analysis of non-centrosymmetric crystals.

Molecule	Non-centrosymmetric space group	Structural feature determined
chiral	chiral	absolute configuration
achiral	chiral	absolute conformation (i.e. the sign of all torsion angles); inverting the structure changes the sign of the torsion angles.
achiral	chiral or achiral, containing a polar axis (e.g. Pna2$_1$ and P2$_1$)	polar axis direction
achiral	achiral, lacking a polar axis (e.g. I$\bar{4}$)	absolute assignment of x and y axes; changing x to -y and y to x converts the set of symmetry operators to their inverse.
chiral	enantiomorphous pair of space groups (e.g. P3$_1$21 and P3$_2$21)	simultaneous determination of space group and absolute configuration
achiral	enantiomorphous pair	simultaneous determination of space group and absolute conformation

4. Determination of absolute structure

The classical method based on careful measurement of selected Bijvoet pairs is infrequently attempted nowadays. New techniques incorporated into standard least-squares refinement procedures have come into general use. A popular method, Hamilton's R ratio test, involves least-squares refinement of both the structure model (X_i, Y_i, Z_i) and its enantiomorph $(-X_i, -Y_i, -Z_i)$ followed by a statistical comparison of the resultant values of the generalized weighted R factor, namely

$$R_g = [\Sigma w(|F_o| - k^{-1}|F_c|)^2/\Sigma w|F_o|^2]^{\frac{1}{2}}$$

where the summations are taken over the set of unique reflections.[9]

An attempt to provide a more reliable method was made by Rogers,[10] who suggested refining a parameter η as a factor multiplying all imaginary components $\Delta f''$ of the anomalous dispersion terms of the atomic scattering factors; η should then refine to values of +1 or -1, corresponding to the correct or incorrect model, respectively. For a wide range of structures containing medium-to-strong anomalous scatterers, η refinement has been shown to be an effective and robust method.[8,11]

From theoretical considerations and computer simulations, Flack has shown that the η refinement technique has some shortcomings that might lead to a false least-squares minimum and hence unreliable estimated standard deviations.[12] As an alternative method, he suggested refining an "absolute-structure parameter" x, which describes the crystal as an "inversion twin"; $1-x$ and x are then respectively the fractions of the structure and its inverse in the macroscopic sample. The relevant structure-factor equation:

$$|F(h,k,\ell,x)|^2 = (1-x)|F(h,k,\ell)|^2 + x|F(-h,-k,-\ell)|^2$$

has been efficiently implemented in a least-squares program package which deals with non-centrosymmetric crystal structures in a very general way.[13]

The application of anomalous dispersion of X-rays to the determination of absolute configurations of organic molecules[14] and coordination compounds has been reviewed.[15]

References

[1] D. van der Helm, J.P. Glusker, C.K. Johnson, J.A. Minkin, N.E. Brown and A.L. Patterson, *Acta Crystallogr., Sect B* **24**, 578 (1968).

[2] L. Pasteur, *Ann. Chim. Phys.* [3] **24**, 442 (1848).

[3] J.P. Glusker and K.N. Trueblood, *Crystal Structure Analysis*, 2nd ed., Oxford University Press, Oxford, 1985.

[4] J.M. Bijvoet, *Proc. Kon. Ned. Akad. Weten.* **B52**, 313 (1949).

[5] E. Fisher, *Ber. Deut. Chem. Gesell.* **23**, 2114 (1890); **27**, 3189 (1894).

[6] A.F. Peerdeman, A.J. van Bommel and J.M. Bijvoet, *Proc. Kon. Ned. Akad. Weten.* **B54**, 16 (1951).

[7] J.M. Bijvoet, A.F. Peerdeman and A.J. van Bommel, *Nature (London)* **168**, 271 (1951).

[8] P.G. Jones, *Acta Crystallogr., Sect. A* **40**, 660 (1984).

[9] W.C. Hamilton, *Statistics in Physical Science*, Ronald Press, New York, 1964; *Acta Crystallogr.* **18**, 502 (1965).

[10] D. Rogers, *Acta crystallogr., Sect. A* **37**, 734 (1981).

[11] P.G. Jones and K. Meyer-Bäse, *Acta Crystallogr., Sect. A* **43**, 79 (1987).

[12] H.D. Flack, *Acta Crystallogr., Sect. A* **39**, 876 (1983).

[13] G. Bernardinelli and H.D. Flack, *Acta Crystallogr., Sect. A* **41**, 500 (1985).

[14] R. Parthasarathy in H.B. Kagan (ed.), *Stereochemistry: Fundamentals and Methods.* Vol. 1. *Determination of Configurations by Spectrometric Methods*, Georg Thieme, Stuttgart, 1977, p. 181.

[15] Y. Saito, *Inorganic Molecular Dissymmetry*, Springer-Verlag, Berlin, 1979.

Note Crystallization of $NaClO_3$ (rectangular prisms, space group $P2_13$) from a stirred solution yields almost exclusively only one of the enantiomorphic forms, as reported in D.K. Kondepudi, R.J. Kaufman and N. Singh, *Science (Washington)* **250**, 975 (1990); J.M. McBride and R.L. Carter, *Angew. Chem. Int. Ed. Engl.* **30**, 293 (1991).

5.5   Methyl *p*-Bromocinnamate

$$BrC_6H_4CH=CHCOOCH_3$$

Crystal Data

Monoclinic, space group $P2_1/n$ (No. 14)

$a = 8.485$, $b = 20.703$, $c = 5.764$Å, $\beta = 92.2°$, $Z = 4$

Atom	x	y	z	Atom	x	y	z
Br	-.0913	.2353	.0186	C(5)	.1517	.4068	.4304
O(1)	.2526	.5592	1.0461	C(6)	.2084	.3820	.2146
O(2)	.4218	.5562	.7647	C(7)	.1410	.3315	.0918
C(1)	.3559	.6077	1.1542	C(8)	.0052	.3053	.1836
C(2)	.3031	.5381	.8468	C(9)	-.0576	.3278	.3806
C(3)	.1961	.4874	.7326	C(10)	.0143	.3775	.5007
C(4)	.2359	.4582	.5498				

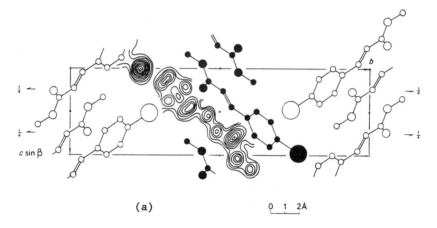

(a)   0 1 2Å

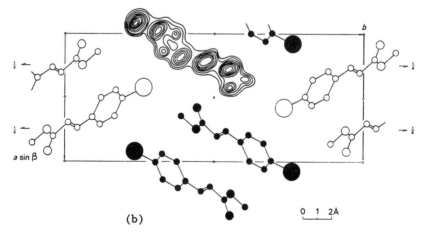

(b)   0 1 2Å

Fig. 5.5-1 Packing diagram and electron-density projection of the methyl *p*-bromocinnamate structure along (a) [100] and (b) [001]. Contours are drawn at intervals of 1eÅ^{-2} except at the bromine atom; lowest contour at 2eÅ^{-2}.   (After ref. 1).

Crystal Structure

The molecular dimensions of methyl *p*-bromocinnamate as determined by X-ray diffraction are shown in Fig. 5.5-2.

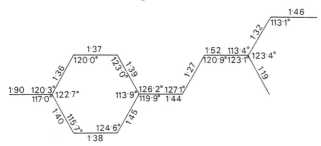

Fig. 5.5-2 Experimental bond lengths (Å) and angles of methyl *p*-bromocinnamate. (After ref. 1).

In the compound, the benzene ring, the ethylenic system and the C-COOCH₃ system are coplanar; the benzene ring and carboxyl group adopt the *trans* configuration.

The molecule displays the packing arrangement of the α-type in the cinnamic acid series since contacts between nearest-neighbour 〉C=C〈 groups occur across centers of symmetry. The center-to-center distance across the inversion centre at (0,1/2,1/2) is 4.11Å. Photochemical interaction across the center at (0,1/2,1/2) would be expected to lead to formation of the centrosymmetric dimer, dimethyl 4,4'-dibromo-α-truxillate.

Remarks

1. Solid reactions of *trans*-cinnamic acids

Although the results of X-ray structure determinations are static views of molecules or ions that have become aligned to give a minimum energy arrangement on crystallization, it is sometimes possible to see evidence of a tendency for reaction between adjacent molecules or ions.

Schmidt and co-workers[1-4] correlated the crystal structure of *trans*-cinnamic acids and their derivatives with the photo-reactivity and steric configuration of the products. The results established the "topochemical principle" which states that a solid-state reaction tends to occur with a minimum of atomic and molecular motion. In the *trans*-cinnamic acids the environment of the olefin double bonds conforms to one of three principal types: the α-type crystal, in which the double bonds of neighbouring molecules make contact at a separation of about 3.7Å across a center of symmetry; the β-type, characterized by a lattice having one axial length of 4.0 ± 0.2Å between translationally-related molecules; and the γ-type, in which double bonds of neighbouring molecules are more than 4.7Å apart. On photo-irradiation of cinnamic acid derivatives, an α-type crystal gives a centrosymmetric dimer related to α-truxillic acid ($\bar{1}$ dimer) and a β-type crystal a dimer of mirror symmetry related to β-truxinic acid (m-dimer); a γ-type crystal is photo-stable (Fig. 5.5-3).

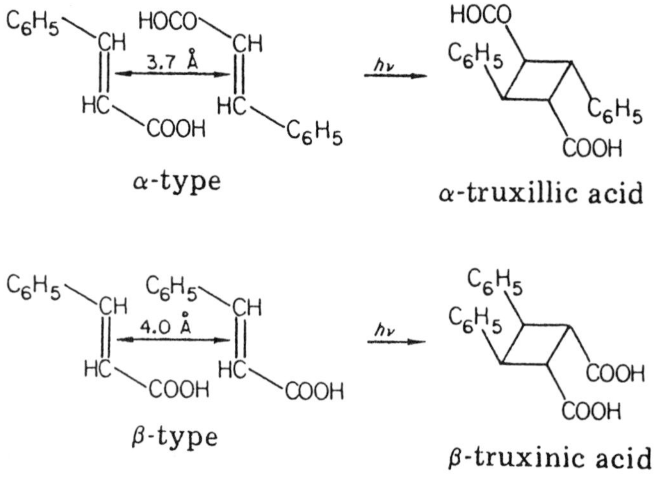

Fig. 5.5-3 Lattice-controlled photodimerization of *trans*-cinnamic acid to stereospecific products.

2. Photopolymerization of diolefin crystals[5]

When a brilliant-yellow crystal of 2,5-distyrylpyrazine (DSP) is exposed to sunlight, it turns into an insoluble polymeric substance. Investigation of this phenomenon has shown that the DSP crystal is converted by the action of sunlight into a linear, high-molecular-weight polymer crystal with cyclobutane units in the main chain. This new type of polymerization is now referred to as "four-center type photopolymerization".[6]

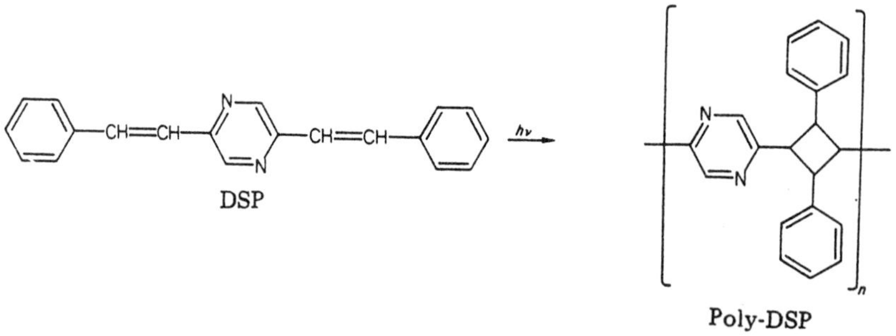

A large number of diolefin crystals have been found to photopolymerize to linear high-molecular-weight polymers. The four-center type photo-polymerization in the crystalline state is a general term for the reactions in which conjugated diolefin crystals are photochemically converted into crystals of linear polymers containing cyclobutane and aromatic groups alternating in

the main chain. On the basis of mechanistic and crystallographic results it
was concluded that the four-center type photopolymerization is a typical
topochemical reaction in which the quantitative transformation into the
polymer crystal is performed under a thermally diffusionless process, with
retention of the space group of the starting diolefin crystal.[7]

    In all the photopolymerizable crystals, nearly planar molecules are
piled up and displaced in the direction of the molecular longitudinal axis by
about half a molecule to form a parallel plane-to-plane stack. Fig. 5.5-4
shows the crystal structure of DSP(α).

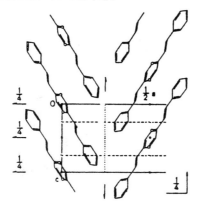

Fig. 5.5-4 Crystal structure of DSP(α). [Space group *Pbca*, *a* =
20.638, *b* = 9.599, *c* = 7.655Å, *Z* = 4]. (After ref. 5).

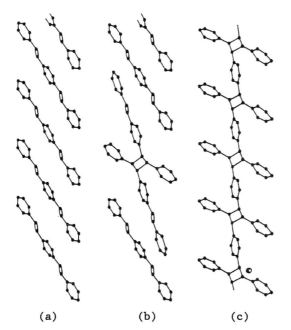

   (a)   (b)   (c)

Fig. 5.5-5 Schematic illustration of the conversion of (a) DSP(α) into
(b) dimer and (c) polymer. (After ref. 5).

The starting crystal of DSP(α) has pairs of double bonds each related by an inversion center, and thus affords initially a dimer of symmetry 1, and finally a polymer having a cyclobutane with a 1,3-*trans* configuration in the main chain, as shown in Fig. 5.5-5.

The polymer crystals prepared by a topochemical process are powdery in appearance and are aggregates of tiny crystallites with an extremely high degree of crystallinity.

In all the examples, monomer and polymer crystals are closely correlated in respect of unit cell dimensions, and as a result all the elementary processes including initiation, growth, and crystallization of the polymer in the four-center type photopolymerization are controlled by the symmetry of the crystal lattice. This is in remarkable contrast to the polymerization of cyclic ethers or vinyl-type compounds in the crystalline state.

References

[1] L. Leiserowitz and G.M.J. Schmidt, *Acta Crystallogr.* **18**, 1058 (1965).

[2] G.M.J. Schmidt, *J. Chem. Soc.*, 2014 (1964).

[3] M.D. Cohen, *Angew. Chem. Int. Ed. Engl.* **14**, 386 (1975).

[4] G.M.J. Schmidt, *Pure Appl. Chem.* **27**, 647 (1971).

[5] M. Hasegawa, *Chem. Rev.* **83**, 507 (1983).

[6] M. Hasegawa and Y. Suzuki, *J. Polym. Sci.* **B5**, 813 (1967).

[7] H. Nakanishi, M. Hasegawa and Y. Sasada, *J. Polym. Sci., Polym. Phys. Ed.* **15**, 173 (1977).

Note An overview of organic reactions in the solid state is presented by R. Perrin and R. Lamartine in Pierrot (eds.), *Structure and Properties of Molecular Crystals*, Elsevier, Amsterdam, 1990, p. 107.

5.6 Cyclooctatetraene (COT)
C₈H₈

Crystal Data

Orthorhombic, space group *Aba*2 (No. 41)

$a = 7.664(6)$, $b = 7.650(3)$, $c = 10.688(6)$Å, (129 K), $Z = 4$

Atom	x	y	z	Atom	x	y	z
C(1)	.0171	.0945	.0806	H(1)	-.017	.154	.157
C(2)	.1021	.1845	-.0069	H(2)	.127	.309	.010
C(3)	.1771	.1150	-.1228	H(3)	.303	.148	-.136
C(4)	.0929	.0245	-.2105	H(4)	.156	-.003	-.293

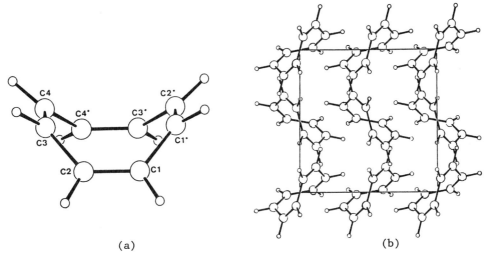

(a)   (b)

Fig. 5.6-1 (a) Molecular structure of C₈H₈ and (b) molecular packing viewed down the *c* axis. (After ref. 1).

Crystal Structure

In the crystal the tub-shaped COT molecule occupies a special position with a crystallographic two-fold axis bisecting the atom pairs C(1)-C(1*) and C(4)-C(4*) [Fig. 5.6-1(a)]. Selected interatomic distances (Å), bond angles (°) and torsion angles (°) are listed in Table 5.6-1. The measured C-C bond distances are as expected for a non-planar compound with isolated double bonds and no significant π-orbital overlap; the bond angles average to 126.7(2)°. Whereas adjacent C-H bonds at C-C double bonds are only slightly twisted out of the C-C bonding plane, the C-H bonds enclosing the C-C single bonds are twisted by 46 and 41°, respectively. The transannular distance between the centers of the opposite double bonds, characteristic for the tub conformation, is 3.085Å. The observed geometry is in striking accordance with the results of an earlier electron diffraction work.

Table 5.6-1   Selected interatomic distances (Å), bond angles (°) and
              torsion angles (°) of COT

C(1)-C(1*)	1.470(2)	C(1*)-C(1)-C(2)	126.6(1)
C(1)-C(2)	1.333(2)	C(3)-C(2)-C(1)	126.7(1)
C(2)-C(3)	1.465(3)	C(4)-C(3)-C(2)	126.4(1)
C(3)-C(4)	1.333(2)	C(4*)-C(4)-C(3)	126.9(2)
C(4)-C(4*)	1.473(2)		
C(1)-C(2)-C(3)-C(4)	55.4(3)	H(1)-C(1)-C(1*)-H(1*)	-45.5(21)
C(2)-C(3)-C(4)-C(4*)	0.3(5)	H(2)-C(2)-C(3)-H(3)	46.4(17)
C(1*)-C(1)-C(2)-C(3)	1.1(3)	H(1)-C(1)-C(2)-H(2)	0.2(19)
C(2*)-C(1*)-C(1)-C(2)	-57.9(3)	H(3)-C(3)-C(4)-H(4)	-4.3(19)
C(3)-C(4)-C(4*)-C(3*)	-56.6(3)	H(4)-C(4)-C(4*)-H(4*)	-40.6(20)

Remarks

1. Valence isomers of COT

The valence isomers of annulene comprise a group of compounds that
undergo many interesting photochemical, thermal, and catalytic
interconversions. The numbers of isomers and structures for a particular
annulene formula, and the reactivity relationships among these isomers are

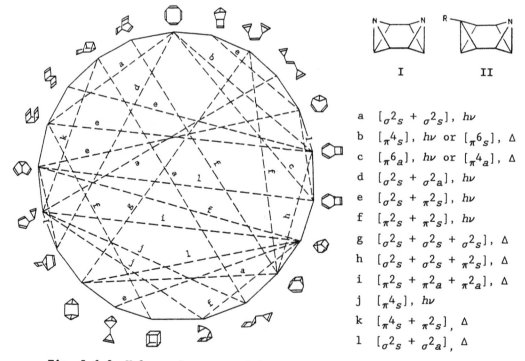

a $[_\sigma 2_s + _\sigma 2_s]$, $h\nu$

b $[_\pi 4_s]$, $h\nu$ or $[_\pi 6_s]$, Δ

c $[_\pi 6_a]$, $h\nu$ or $[_\pi 4_a]$, Δ

d $[_\sigma 2_s + _\sigma 2_a]$, $h\nu$

e $[_\sigma 2_s + _\pi 2_s]$, $h\nu$

f $[_\pi 2_s + _\pi 2_s]$, $h\nu$

g $[_\sigma 2_s + _\sigma 2_s + _\sigma 2_s]$, Δ

h $[_\sigma 2_s + _\sigma 2_s + _\pi 2_s]$, Δ

i $[_\pi 2_s + _\pi 2_a + _\pi 2_a]$, Δ

j $[_\pi 4_s]$, $h\nu$

k $[_\pi 4_s + _\pi 2_s]$, Δ

l $[_\sigma 2_s + _\sigma 2_a]$, Δ

Fig. 5.6-2   Valence isomers and interconvertive schemes for COT. Two
aza-analogues of octabisvalene are shown on the upper right corner.
(After ref. 2).

topics that are most usefully discussed using a systematic approach.

From the 17 constitutional formulae of the planar graphs (CH)$_8$, one may derive 21 possible isomers[2] as shown in Fig. 5.6-2, in which the criss-crossing lines indicate some conceivable conversions. The highly strained azapolycycles Z-diazaoctabisvalene **I** and 7-substituted azaoctabisvalene **II** are accessible by efficient synthesis.[3] Fig. 5.6-3 summarizes the known thermal (a), photochemical (b), and metal-ion catalyzed (c) interconversions.[4]

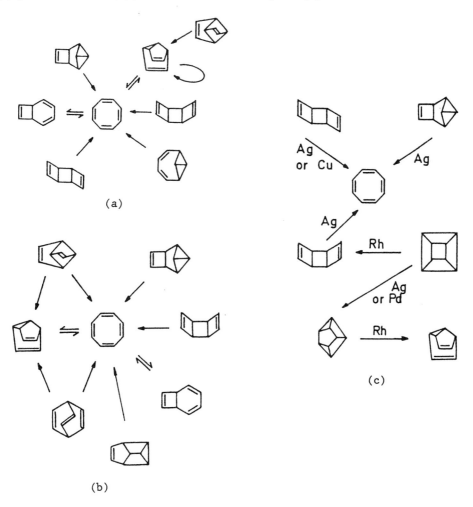

(a)

(b)

(c)

Fig. 5.6-3 (a) Thermal, (b) photochemical, and (c) metal-ion catalyzed interconversions of known isomers of COT. (After ref. 4).

2. Benzannelated cyclooctatetraenes[5]

All five possible benzannelated derivatives of cyclooctatetraene, **6A**, namely benzocyclooctatetraene, **1**, dibenzo[a,c]cyclooctatetraene, **2**,[6] dibenzo[a,e]cyclooctatetraene, **3**,[7] tribenzo[a,c,e]cyclooctatetraene, **4**,[6] and tetrabenzo[a,c,e,g]cyclooctatetraene (commonly known as tetraphenylene),

5[7] are known. The X-ray structural results of these compounds are listed in Table 5.6-1.

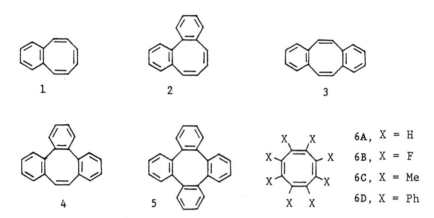

| 1 | 2 | 3 |

| 4 | 5 | 6A, X = H
6B, X = F
6C, X = Me
6D, X = Ph |

Table 5.6-1 Bond lengths (Å) and fold angles $\alpha(°)^*$ of benzannelated cyclooctatetraenes and related compounds.[5]

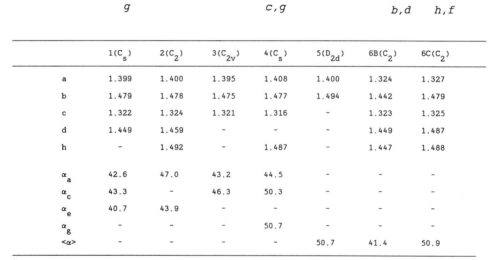

	1(C$_s$)	2(C$_2$)	3(C$_{2v}$)	4(C$_s$)	5(D$_{2d}$)	6B(C$_2$)	6C(C$_2$)
a	1.399	1.400	1.395	1.408	1.400	1.324	1.327
b	1.479	1.478	1.475	1.477	1.494	1.442	1.479
c	1.322	1.324	1.321	1.316	-	1.323	1.325
d	1.449	1.459	-	-	-	1.449	1.487
h	-	1.492	-	1.487	-	1.447	1.488
α_a	42.6	47.0	43.2	44.5	-	-	-
α_c	43.3	-	46.3	50.3	-	-	-
α_e	40.7	43.9	-	-	-	-	-
α_g	-	-	-	50.7	-	-	-
$<\alpha>$	-	-	-	-	50.7	41.4	50.9

* In the illustrated example, α_a is the dihedral angle between the plane containing bonds b, a and h, and the plane containing bonds c and g.

3. Planar unsaturated eight-membered carbocycles

     The study of planar cyclooctatetraenes was pioneered by Krebs, who reported the fugitive existence of 1,2-didehydrocyclooctatetraene, 7, in 1965.[8] The relatively more stable dibenzo derivative 8, which decomposes

readily at room temperature, was shown by low-temperature X-ray
crystallography to be planar within 0.100Å[9] The stable derivative 9, in
which the methyl substituents effectively shield the otherwise rather exposed
strained triple bond and kinetically stabilize it against dimerization and/or
air oxidation, has a "butterfly" conformation intermediate between those of
planar 8 and tub-shaped 10.[10]

7   8   9   10

11   F   F   12   13

The 5,6,11,12-tetradehydrodibenzo[*a,e*]cyclooctene molecule, 11,[11] in
the crystalline state is substantially planar, the slight deviation from
planarity being probably due to intermolecular effects. Three structural
features are worthy of note: (a) the average value of the angles at carbon
atoms involved in triple bonds is 155.8°, (b) the lengths of triple bonds are
1.195 and 1.199Å; and the distance between the triple bonds is 2.61Å, and (c)
atoms of the eight-membered ring and the two outer benzene rings are essential
coplanar.

MINDO/3 optimization indicates that a four-membered ring fused to
cyclooctatetraene helps to flatten the eight-membered ring skeleton, exact
planarity being reached with two cyclobutene rings annelated at positions
1,2,5 and 6.[12] Similarly, the annelation of cyclooctatetraene with one or
more three-membered rings converts the tub to a planar configuration.[13]
These findings are consistent with the planar carbon skeletons found for
12[14] and 13[15] by X-ray diffraction.

4. Cyclooctatetraenes as neutral π-ligands in silver(I) complexes [5]

Cyclooctatetraene, 6A, reacts with aqueous silver nitrate to form three
crystalline compounds of compositions $2C_8H_8 \cdot AgNO_3$, $C_8H_8 \cdot AgNO_3$ and

$2C_8H_8 \cdot 3AgNO_3$. The last compound has a complicated, three-dimensional polymeric structure in which two C_8H_8 ligands partition three Ag(I) ions in the asymmetric unit.[16]

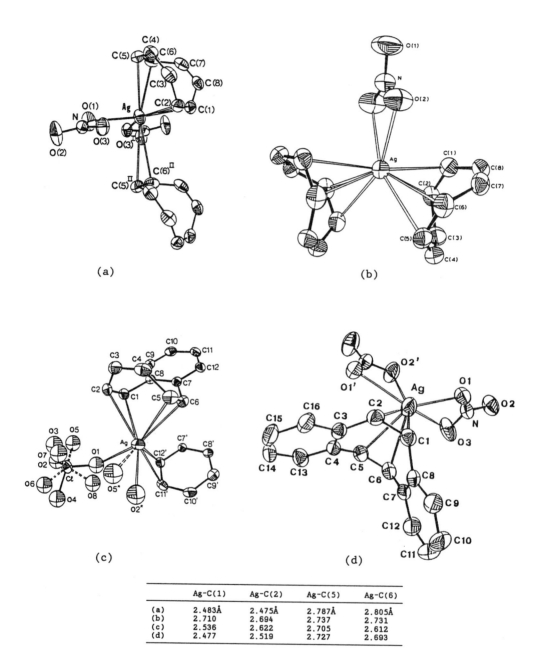

	Ag–C(1)	Ag–C(2)	Ag–C(5)	Ag–C(6)
(a)	2.483Å	2.475Å	2.787Å	2.805Å
(b)	2.710	2.694	2.737	2.731
(c)	2.536	2.622	2.705	2.612
(d)	2.477	2.519	2.727	2.693

Fig. 5.6-4 Perspective view of the coordination geometry about the Ag(I) ion in (a) $C_8H_8 \cdot AgNO_3$, (b) $(C_8H_8)_2 \cdot AgNO_3$, (c) $C_{12}H_{10} \cdot AgClO_4$ and (d) $C_{16}H_{12} \cdot AgNO_3$. (After ref. 5).

The crystal structure of $C_8H_8.AgNO_3$[17] consists of strongly-bonded Ag-
C_8H_8 units linked by weaker metal-ligand interactions to form infinite chains
parallel to the c axis; neighboring chains are cross-linked by bridging
nitrato groups to form corrugated sheets normal to the a axis. The
configuration of the ligands about Ag may be described as distorted trigonal
bipyramidal, with two metal-oxygen bonds (to different nitrato groups) and the
strongest metal-olefin bond lying in the equatorial plane, and the weakest π
ligand in an axial position [Fig. 5.6-4(a)].

In the crystal, each $(C_8H_8)_2.AgNO_3$ molecule lies on a crystallographic
diad, with the Ag(I) ion coordinated by a symmetrical bidentate nitrato group
and four double bonds, two from each C_8H_8 moiety in an unsymmetrical fashion
[Fig. 5.6-4(b)].[18] If the bidentate nitrato group is considered to occupy
one position in the coordination sphere, the configuration about Ag can be
described as approximately trigonal bipyramidal: the nitrato group and the
centers of the C(5)-C(6) and C(5)-C(6') double bonds lie in the equatorial
plane, and the pair of C(1)-C(2) and C(1')-C(2') double bonds occupy the axial
positions.

The 1:1 adduct of 1 $(C_{12}H_{10})$ with $AgClO_4$[19] provided the first example
of simultaneous donor π-bonding from aromatic and olefinic ligands to a
silver(I) ion [Fig. 5.6-4(c)]. With the disordered perchlorate group in its
principal orientation, the configuration of the ligands about the metal center
may be described as distorted trigonal bipyramidal. The silver ion is
coordinated by two non-adjacent double bonds of a benzocyclooctatetraene
molecule, one aromatic C-C bond from a neighbouring benzocyclooctatetraene
molecule, and two oxygen atoms belonging to different perchlorate groups; atom
O(1) and the center of the C(5)-C(6) olefinic ligand occupy the axial
positions of the trigonal bipyramid. The silver-olefin and silver-aromatic
interactions are approximately equal in magnitude.

In the 1:1 adduct of 3 $(C_{16}H_{12})$ and $AgNO_3$, the silver atom is
tetrahedrally coordinated by two non-adjacent double bonds of 3, a bidentate
nitrato group, and another bidentate nitrato group generated from the first by
a symmetry operation [Fig. 5.6-4(d)].

5. Structure of bicyclic non-benzenoid systems[20,21]

Bicyclic conjugated π-electron systems with two odd-numbered rings, such
as pentalene,[22] azulene or heptalene, can be viewed as annulenes bridged by
a single bond. Even when an even-numbered ring is inserted between the two
odd-numbered rings, the bridging bonds in all Kekulé structures remain single,
as substantiated by several crystal structure determinations.

a) Pentalene

Fig. 5.6-5 shows the structures of 5-*t*-butyl-1,3-bis(dimethylamino)-2-azapentalene (**14**) and 1,3-bis(dimethylamino)pentalene (**15**). The data reflect striking differences in the electronic structures of these two systems. While structure **14** exhibits localized bonds of the bicyclic 8π-electron system, the dimensions of **15** suggest that it should be described by the dipolar structure with an equilibration of bond lengths. This can be rationalized by the perturbation of the π-system by the nitrogen in (2)-position of **14**, which stabilizes the localized pentalene structure and destabilizes a dipolar structure with a cyclopentadienide moiety.

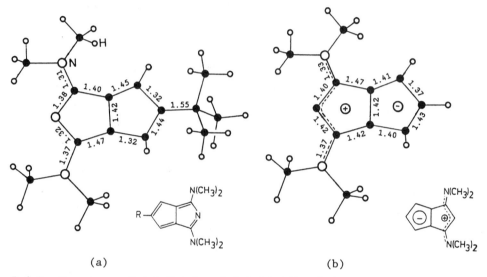

(a)   (b)

Fig. 5.6-5 Structure of (a) 5-*t*-butyl-1,3-bis(dimethylamino)-2-azapentalene (**14**) and (b) 1,3-bis-(dimethylamino)pentalene (**15**). (After ref. 20).

The first analogue of pentalene with a phosphorus atom in each ring, shown in Fig. 5.6-6, has a high melting point (194°C) and is stable towards air and hydrolysis.[23] The molecule is centrosymmetric and the important bond lengths are: P1-C1 1.758(1), C1-C1' 1.466(2), C1-C2 1.379(2), C2-C3 1.423(2), P1-C3' 1.759(1)Å; C1-C2-C3 115.1(1), C2-C3-P1' 109.4(1), C1-P1-C3' 93.5(1), P1-C1-C1' 108.7(1), C1'-C1-C2 111.8(1)°. The measured dimensions are consistent with a significant contribution of ylidic structures to the stability of the diphosphapentalene, which has the molecular peak (m/z 546) as the base peak in its mass spectrum.

b) Azulene[22]

Azulene (**16**), the prototype of non-benzenoid aromatic hydrocarbons, crystallizes in space group $P2_1/a$, with $a = 7.884(8)$, $b = 5.988(8)$, $c = 7.840(8)$Å, $\beta = 101.6(3)°$ and $Z = 2$.[24] The structure is the classical

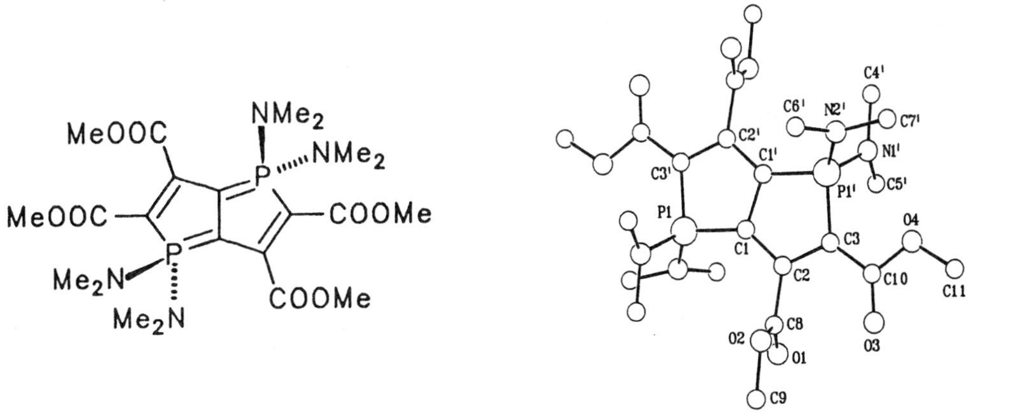

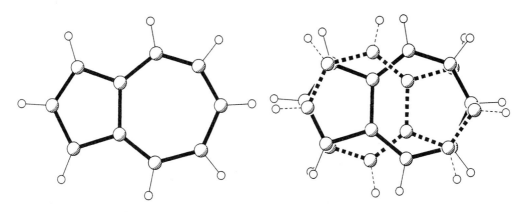

(a)   (b)

Fig. 5.6-6 Structural formula and molecular structure of a derivative
of $1\lambda^5,4\lambda^5$ -diphosphapentalene. (After ref. 23).

example of statistical disorder, which involves the occupation of a given site
by a molecule or its centric equivalent (Fig. 5.6-7). The molecule is planar
and the bond lengths are within the range expected for aromatic bonds; the
bridge bond of 1.48Å is suggestive of a single bond between sp^2 hybridized
carbon atoms. Bond angles are as expected, averaging 108° in the five-
membered ring and 128.6° in the seven-membered ring. Intermolecular contacts
are normal with a shortest approach of 3.54Å.

Fig. 5.6-7 Azulene molecule (16, $C_{10}H_8$) and superposition of two
azulene molecules in the crystal.

c) Heptalene

In contrast to cyclobutadiene and pentalene, their higher homologues
cyclooctatetraene and heptalene[25] are characterized by a nonplanar
structure. X-Ray structure analysis of heptalene derivatives 17 and 18

confirms a C_2 geometry for the bicyclic system with the two annelated seven-membered rings in boat conformations, as shown in Fig. 5.6-8. In these two structures, the lengths of the perimeter bonds follow a long-short alternating sequence.

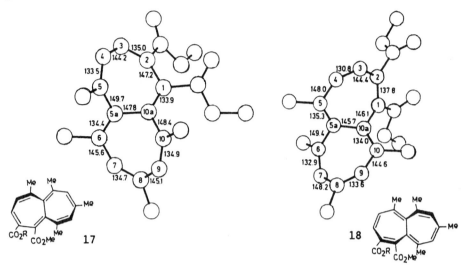

Fig. 5.6-8 Structure of heptalene derivatives. (After ref. 25).

d) Tricyclic non-benzenoid systems

Fig. 5.6-9 shows the structures of 2,6-di-t-butyl-4,8-bis(dimethyl-amino)-s-indacene (**19**), 2,4,6-tri-t-butyl-8-dimethylamino-s-indacene (**20**), and 2,7-di-*tert*-butyl-dicyclopenta[a,e]cyclooctene (**21**). A similar influence of dimethylamino groups on the perimeter is observed in the s-indacene system. Two dimethylamino groups in (4)- and (8)-position effect a delocalization in the perimeter (Fig. 5.6-9, molecule 19) whereas only one dimethylamino group in one of these positions (Fig. 5.6-9, molecule 20) leads to a structure with bond fixation showing fulvenoid units. These findings show clearly that electron-donating substituents exert an important influence on the bonding nature of 4n π-electron systems.[20]

Molecule 21 possesses a planar ring system with an inversion center. The maximum distance of the ring-C atoms from the mean plane is ±0.015Å. The perimeter of the molecule, with C-C bond lengths of about 1.40Å, exhibits substantial bond equivalence, whereas the two bridges are noticeably longer (1.49Å) with predominantly single-bond character. Compound 21 is unusual in that it belongs to the few hydrocarbons with a planar eight-membered ring, and is also a completely planar 14π-electron system that has no further cyclic conjugated subunits. The molecular structure and spectroscopic properties justify its classification as a non-benzenoid aromatic hydrocarbon.

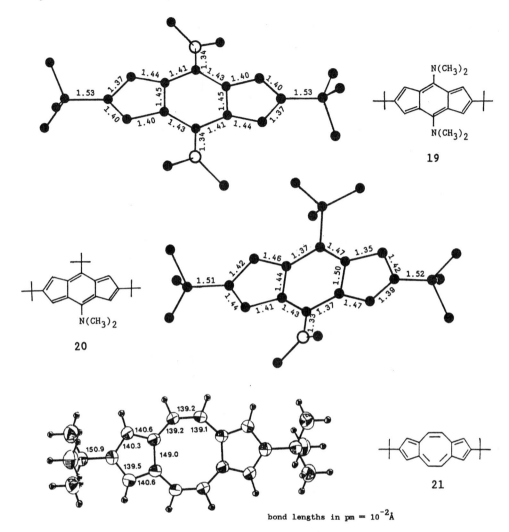

bond lengths in pm = 10^{-2}Å

Fig. 5.6-9 Structure of some tricyclic non-benzenoid systems.
(After refs. 20 and 21).

References

[1] K.H. Claus and C. Krüger, *Acta Crystallogr., Sect. C* **44**, 1632 (1988).

[2] L.R. Smith, *J. Chem. Educ.* **55**, 569 (1978).

[3] B. Trupp, H. Fritz and H. Prinzbach, *Angew. Chem. Int. Ed. Engl.* **28**, 1345 (1989).

[4] A.T. Balaban and M. Banciu, *J. Chem. Educ.* **61**, 766 (1984).

[5] T.C.W. Mak in J.J. Stezowski, J.L. Huang and M.C. Shao (eds.), *Molecular Structure: Chemical Reactivity and Biological Activity*, Oxford University Press, New York, 1988, pp. 325-344.

[6] N.Z. Huang and T.C.W. Mak, *J. Mol. Struct.* **101**, 135 (1983).

[7] H. Irngartinger and W.R.K. Reibel, *Acta Crystallogr., Sect. B* **37**, 1724 (1981).

[8] A. Krebs, *Angew. Chem. Int. Ed. Engl.* **4**, 953 (1965).

[9] R.A.G. de Graaff, S. Gorter, C. Romers, H.N.C. Wong, and F. Sondheimer, *J. Chem. Soc., Perkin Trans. II*, 478 (1981).

[10] T.-L. Chan, T.C.W. Mak, C.-D. Poon, H.N.C. Wong, J.H. Jia and L.L. Wang, *Tetrahedron* **42**, 655 (1986).

[11] R. Destro, T. Pilati and M. Simonetta, *J. Am. Chem. Soc.* **97**, 658 (1975).

[12] T.C.W. Mak and W.-K. Li, *J. Mol. Struct. (Theochem.)* **89**, 281 (1982).

[13] H.N.C. Wong and W.-K. Li, *J. Chem. Res. (S)*, 302 (1984).

[14] F.W.B. Einstein, A.C. Willis, W.R. Cullen and R.L. Soulen, *J. Chem. Soc., Chem. Commun.*, 526 (1981).

[15] H. Dürr, G. Klauck, K. Peters and H.G. von Schnering, *Angew. Chem. Int. Ed. Engl.* **22**, 332 (1983).

[16] T.C.W. Mak, *J. Organometal. Chem.* **246**, 331 (1983).

[17] W.C. Ho and T.C.W. Mak, *J. Organomet. Chem.* **241**, 131 (1983).

[18] T.C.W. Mak and W.C. Ho, *J. Organomet. Chem.* **243**, 233 (1983).

[19] T.C.W. Mak and W.C. Ho and N.Z. Huang, *J. Organomet. Chem.* **251**, 413 (1983).

[20] K. Hafner, *Pure Appl. Chem.* **54**, 939 (1982).

[21] K. Hafner, G.F. Thiele and C. Mink, *Angew. Chem. Int. Ed. Engl.* **27**, 1191 (1988).

[22] K. Hafner, *Nachr. Chem. Tech. Lab.* **28**, 222 (1980).

[23] J. Silberzahn, H. Pritzkow and H.P. Latscha, *Angew. Chem. Int. Ed. Engl.* **29**, 799 (1990).

[24] J.M. Robertson, H.M.M. Shearer, G.A. Sim and D.G. Watson, *Acta Crystallogr.* **15**, 1 (1962).

[25] K. Hafner, G.L. Knaup, H.J. Lindner and H.-C. Flöter, *Angew. Chem. Int. Ed. Engl.* **24**, 212 (1985).

Note The 14 isomeric strained $(CH)_8$ hydrocarbons are reviewed in K. Hassenrüch, H.-D. Martin and R. Walsh, *Chem. Rev.* **89**, 1125 (1989). 2,4,6,8-Tetrakis(morpholino)-1,3,5,7-tetrazocine has a boat-shaped configuration that resembles C_8Me_8 but somewhat flatter than C_8H_8, as reported in S. Ehrenberg, R. Gompper, K. Polborn and H.-U. Wagner, *Angew. Chem. Int. Ed. Engl.* **30**, 334 (1991).

Novel pentafulvenes are surveyed in K. Hafner, *Pure Appl. Chem.* **62**, 531 (1990). Novel azulenic π-electronic compounds are described in T. Asao, *Pure Appl. Chem.* **62**, 507 (1990).

5.7 Polyethylene Adipate (PEA)

$$[-CO(CH_2)_4CO.O(CH_2)_2O-]_n$$

Crystal Data

Monoclinic, space group $P2_1/a$ (No. 14)

$a = 5.47(3)$, $b = 7.23(2)$, $c = 11.72(4)$Å, $\beta = 113.5(2)°$, $Z = 2$

Atom	x	y	z	Atom	x	y	z
C(1)	.059	.069	.468	H(1)	.248	.125	.535
C(2)	-.060	.025	.258	H(2)	-.081	.182	.426
C(3)	.068	-.050	.173	H(3)	.281	-.014	.214
C(4)	-.063	.035	.044	H(4)	.044	-.202	.167
O(1)	.102	-.035	.374	H(5)	-.043	.186	.051
O(2)	-.280	.105	.230	H(6)	-.276	0	.004

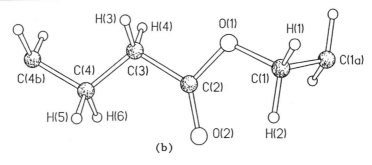

(b)

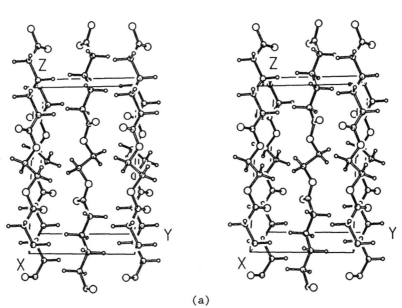

(a)

Fig. 5.7-1 Crystal structure of polyethylene adipate: (a) asymmetric unit with atom labeling and (b) stereoview showing the packing of polymeric chains.

Crystal Structure

The structure of the crystalline regions of cold-drawn fibres and highly oriented, stretched rubber-like specimens of polyethylene adipate (PEA) has been determined.[1] Fig. 5.7-1 shows the arrangement of the molecules viewed down the fiber axis (a), and the *b*-projection of the structure with hydrogen atoms omitted (b). Fig. 5.7-2 shows the configuration of a single molecule viewed roughly perpendicular and parallel to the planar chain of the adipate portion.

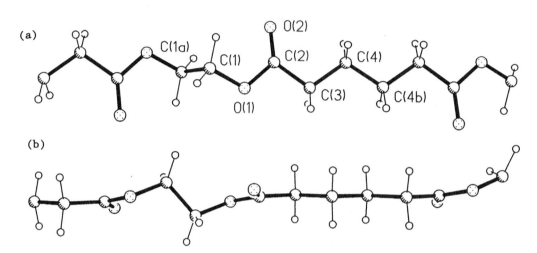

Fig. 5.7-2 PEA chain configuration viewed (a) perpendicular, (b) parallel to the plane of the adipate chain.

The bond lengths are: C-C all 1.53Å, C_1-O_1 1.43, C_2-O_1 1.37, and C_2=O_2 1.24Å. The planar hydrocarbon chain lies at a mean angle of 40° to the *ac* plane and is tilted to the *c* axis by a very small angle . The arrangement of the chains side by side is very similar to that found in polyethylene and other straight-chain monomeric hydrocarbons. The contacts between chain atoms of neighbouring molecules are of the van der Waals type.

Remarks

1. Configuration and conformation of chain polymers[2]

The polymers of monosubstituted olefins (-CH_2-C^*HR-)$_n$ should contain asymmetric carbon atoms (denoted by an asterisk) along the chain. The structure of stereoregular polymers is discussed below and some examples are listed in Table 5.7-1.

An isotactic (*it*) polymer is one in which the chemical structural unit possesses at least one main-chain carbon atom with two different substituent atoms or groups, and the configurations around the carbon atoms are the same along the chain. If an *it*-polymer is shown in Fisher projection, the substituents of each type in successive units all appear on the same side of the line representing the main chain.

Table 5.7-1 Configuration of chain polymers.[2]

Polymer	Configuration	Fischer and Newman projections
it-polypropylene [-CH$_2$CH(CH$_3$)-)$_n$	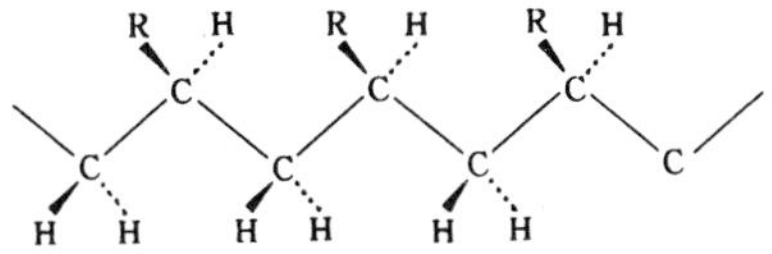	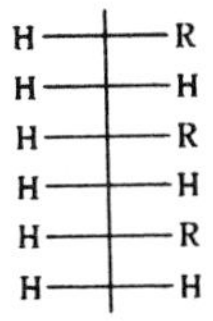
it-poly(propylene oxide) [-CH$_2$CH(CH$_3$)O-]$_n$	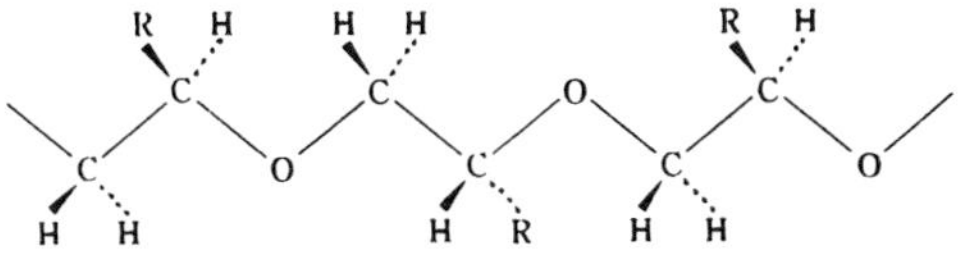	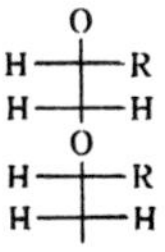
st-polypropylene [-CH$_2$CH(CH$_3$)-]$_n$	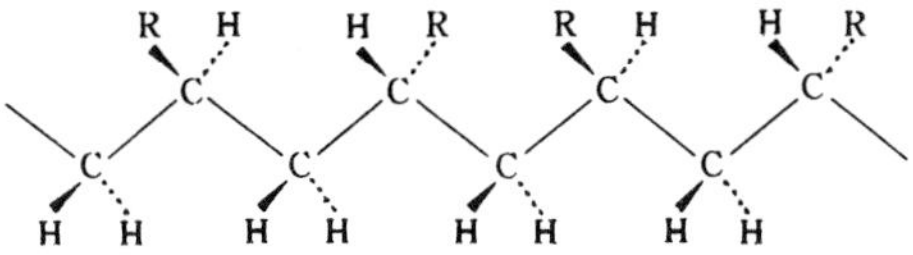	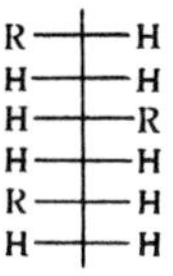
threo-*dit*-polymer*	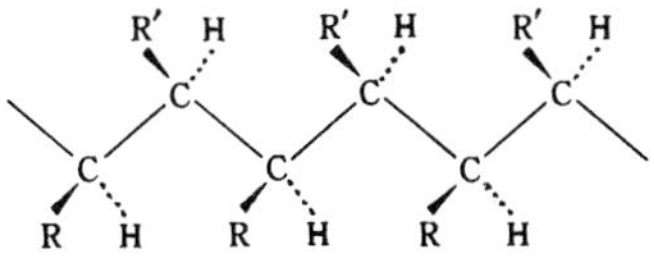	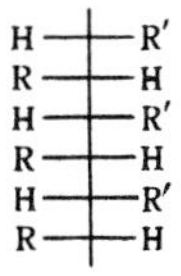 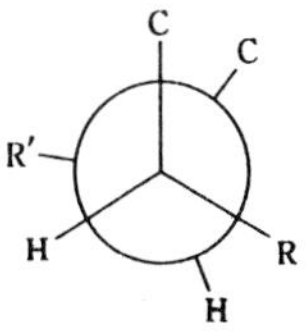
erythro-*dit*-polymer**	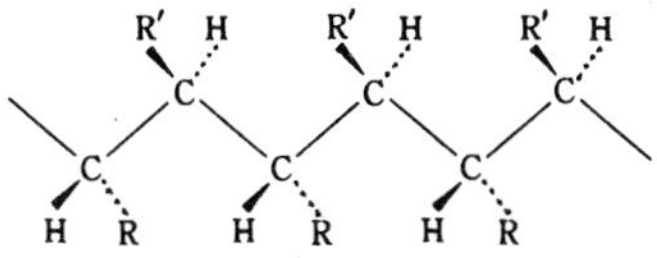	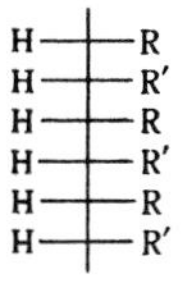 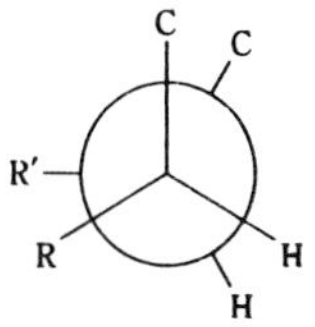
dst-polymer	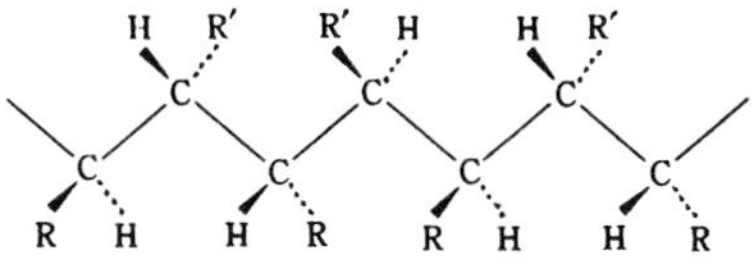	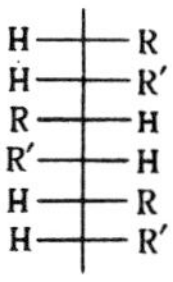

*D-threose

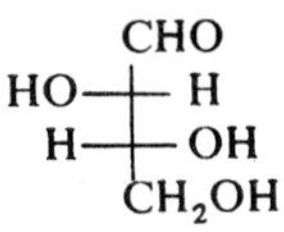

**D-erythrose

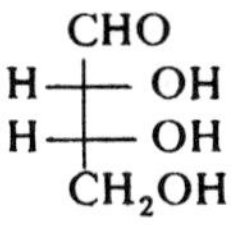

A syndiotactic (*st*) polymer is a polymer in which the chemical structural unit possesses at least one main-chain carbon atom with two different substituents, and the configurations around the carbon atoms are opposite in successive structural units.

An atactic (*at*) polymer is one in which there is complete randomness with regard to the configurations at all the main-chain sites of steric isomerism.

In a diisotactic (*dit*) or disyndiotactic (*dst*) polymer, the structural unit possesses two carbon atoms, each having two different substituents, with configurations around each carbon atom in successive units making the molecule

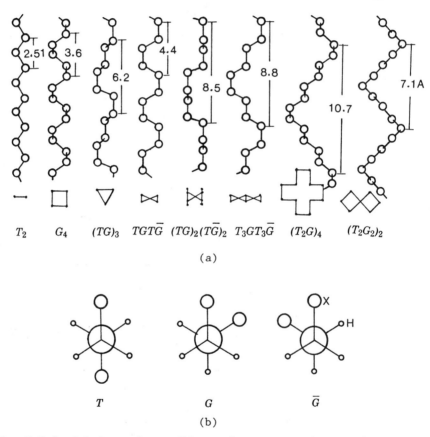

T₂   *G₄*   *(TG)₃*   *TGTḠ*   *(TG)₂(TḠ)₂*   *T₃GT₃Ḡ*   *(T₂G)₄*   *(T₂G₂)₂*

(a)

T   *G*   *Ḡ*

(b)

Fig. 5.7-3 (a) Several possible conformations of a single-bonded carbon chain. (b) The notations of conformation: *T*, *G*, and *Ḡ* stand for *trans*, *gauche*, and minus *G*, respectively. (After ref. 2).

isotactic or syndiotactic, respectively. There are two types of *dit*-polymers: erythro-*dit*- and threo-*dit*-polymers in which the configurations at the two main-chain atoms in the structural unit are alike or opposite, respectively. The threo-erythro terminology originates from the names of D-threose and D-erythrose.

Chain polymer molecules can adopt different conformations in the solid state, depending on the conditions of crystallization, which can result in various crystal modifications. Fig. 5.7-3(a) shows several possible conformations of single-bonded carbon chains. The notations T, G and G are represented in Fig. 5.7-3(b). For example, $(TG)_3$ denotes a helical chain containing three alternations of T and G in the fiber identity period. This gives rise to a left-handed helix, while $(TG)_3$ denotes a right-handed helix.

2. Structures of crystalline polymers

The structures of many crystalline polymers have been determined, and summaries of the crystallographic data and reviews are available.[2-5] Some examples are discussed below:

a) Poly(p-phenylene oxide), $(-O-C_6H_4-)_n$, orthorhombic, *Pbcn*, $a = 8.07$, $b = 5.54$, $c = 9.72$Å, $Z = 4$.[6]

The structure exhibits a zigzag chain with linear $O-C_6H_4$ elements and a bond angle of 124° at the oxygen atom. Potential energy calculations predict the existence of minima at ±40° from a planar configuration, which is in good agreement with the observed alternating 50° rotations in the crystal (Fig. 5.7-4).

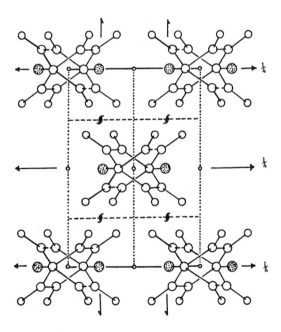

Fig. 5.7-4 Crystal structure of poly(p-phenylene oxide). The shaded circles represent oxygen atoms that are located on the two-fold axes at $z = 1/4$ and $3/4$. (After ref. 4).

b) Isotactic poly(3-methyl-1-butene), $[-CH_2-C(CHMe_2)H-]_n$, monoclinic, $P2_1/b$, $a = 9.55$, $b = 17.08$, $c = 6.84$Å, $\gamma = 116.5°$, $Z = 2$.

In the crystal structure of *it*-poly(3-methyl-1-butene), (4/1) helical molecules are packed in a monoclinic lattice (Fig. 5.7-5).[7]   Natta and co-workers developed the theory of close-packing as applied to helical molecules.[8]   When D_1 and D_2 are the helix radii of the main chain and the side group, respectively, and the ratio D_1/D_2 lies between 0.3 and 0.8, the most efficient way of packing is with each right-handed helix surrounded by four left-handed helices and vice versa.

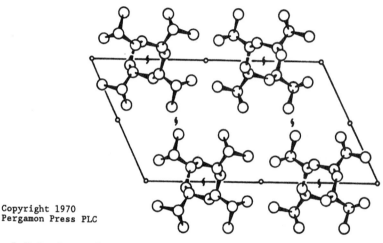

Fig. 5.7-5 Crystal structure of *it*-poly(3-methyl-1-butene). (After ref. 7).

c) Polyisobutylene, $[-CH_2-C(CH_3)_2-]_n$, orthorhombic, $P2_12_12_1$, $a = 6.88$, $b = 11.91$, $c = 18.60$Å, $Z = 2$.

The crystal structure of polyisobutylene is shown in Fig. 5.7-6.[9]   The helical molecule has a fiber period consisting of two asymmetric units each containing four monomeric units. The helix axis coincides with the 2_1 screw axis in the lattice. The averaged bond angle $C-C(CH_3)_2-C$ is about 110°, but the angle $C-CH_2-C$ is much larger (about 128°).

d) *cis*-1,4-Polyisoprene, monoclinic, $P2_1/a$, $a = 12.46$, $b = 8.89$, $c = 8.10$Å, $\beta = 92°$, $Z = 4$.

Natural rubber is *cis*-1,4-polyisoprene. Through developments in stereo-regular polymerization, polyisoprene consisting of almost entirely *cis*-1,4-bonding, "synthetic natural rubber", has been prepared. Nyburg[10] proposed a statistically-disordered crystal structure in which the molecular chain having the conformation denoted by the solid line in Fig. 5.7-7 and its mirror image, denoted by the broken line, exist with equal probability. Natta and Corrandini[11] reported the crystal structure shown in Fig. 5.7-7(c), where,

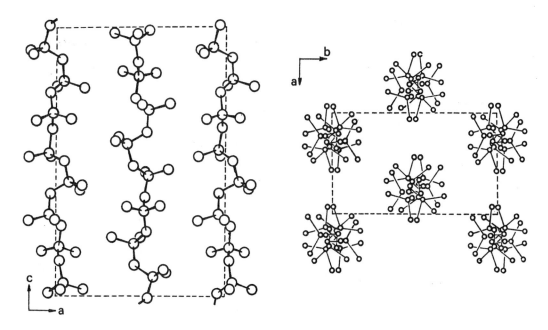

Fig. 5.7-6 Crystal structure of polyisobutylene. (Adapted from ref. 9).

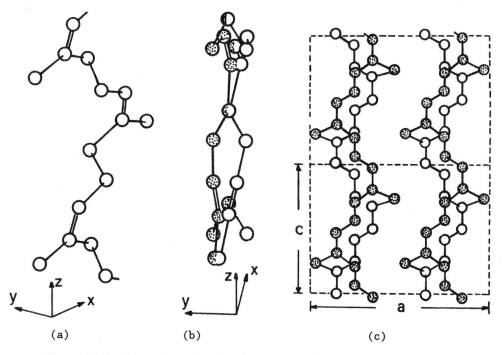

(a)   (b)   (c)

Fig. 5.7-7 Crystal and molecular structure of natural rubber.
(a) A single molecule, (b) a pair of chains that are mirror images,
(c) crystal structure. (Adapted from ref. 2).

for example, A and A' indicate the molecular chains that are mirror images. They concluded that alignment of the A-B-A'-B'... type does not occur, but that either the C-D-C-D... or the C'-D'-C'-D'-... type will pack adjacent to the A-B-A-B... type; statistical disorder occurs only along the *a* axis.

e) Nylon 66, $[-NH-(CH_2)_6-NHCO-(CH_2)_4-CO-]_n$, α-form, triclinic, $P\bar{1}$, $a = 4.9$, $b = 5.4$, $c = 17.2$Å, $\alpha = 48.5$, $\beta = 77$, $\gamma = 63.5°$, $Z = 1$.

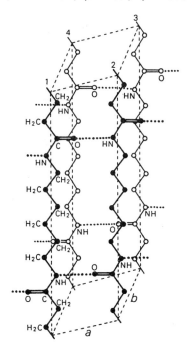

Fig. 5.7-8 Crystal structure of the α-form of nylon 66. (After ref. 2).

Fig. 5.7-8 shows the crystal structure of the α-form of nylon 66.[12] Planar zigzag chains form a sheet by hydrogen bonding, and such sheets stack parallel to the *ac* plane.

3. Factors governing the steric structure of polymers[2]

a) Molecular conformations in polymers are primarily determined by internal rotation potentials and nonbonded interactions. For a free molecular chain $T(180°)$, $G(60°)$, and $\bar{G}(-60°)$ are the predominant conformations resulting from the internal rotation potentials. However, nonbonded interactions are equally important, and conformations deviating markedly from T, G and $\bar{G}$ have been found for many polymers. Hydrogen bonding, if possible, also plays an important role, but the contribution of electrostatic interactions seems relatively unimportant.

b) The molecular conformation can be predicted to some extent by considering only intramolecular interactions, although the potential functions and parameters for energy calculations are not yet well established. In a favorable case such as poly(ethylene oxybenzoate), the calculated internal rotation angles based on intramolecular interactions are changed by at most 28° when intermolecular interactions are taken into account.[13]

    c) For chain packing in the crystalline regions, energy minimization through optimal contact between nearest neighbours appears to be the most dominant factor. Packing modes satisfying three-dimensional crystal symmetries may actually appear.

    d) Strictly speaking, a comparison of the stability of crystal modifications should be made in terms of the free energy rather than the static potential energy. For instance, theoretical calculations of the vibrational free energies predict that orthorhombic polyethylene is more stable than monoclinic polyethylene.[14]

4. Relation of structure to physical properties and synthesis[2]

    a) *Electric properties*

    According to X-ray analysis, poly(vinylidene fluoride) forms I and III, the β form of poly[3,3-bis(chloromethyl) oxacyclobutane], poly(trimethylene sulfide), and poly(m-phenylene isophthalamine) have polar crystal structures. These crystalline polymers are potentially useful as piezoelectric and pyroelectric elements.

    b) *High modulus, high tenacity, and high thermal stability*

    These distinguished properties of poly(p-phenylene terephthalamide) may be interpreted in terms of the all-*trans* extended-type molecular conformation and the very low molecular flexibility.

    c) *Synthesis*

    The following polymerization mechanisms have been clarified from structural studies: the *cis*-opening mechanism of coordinated anionic polymerization (ethylene and propylene) and cationic poymerization (β-substituted vinyl ethers), the inversion-opening polymerization mechanism of ethylene oxide and 1-butene oxide, and the stereoselective polymerization mechanism of monomers containing asymmetric carbon atoms, such as *tert*-butylethylene oxide, isopropylethylene oxide, propylene sulfide, and β-substituted-β-propiolactones.

    d) *Study of biopolymers*

    The accumulation of reliable data and the development of structure-analysis methods for synthetic polymers provide useful information for the study of biopolymers, which may be regarded generally as copolymers having side chains with complicated sequences.

References

[1]   A. Turner-Jones and C.W. Bunn, *Acta Crystallgr.* **15**, 105 (1962).

[2]   H. Tadokoro, *Structure of Crystalline Polymers*, Wiley, New York, (1979).

[3]   R.L. Miller in J. Brandrup and E.H. Immergut (eds.), *Polymers Handbook*, 2nd ed., Wiley, New York, 1975, pp. III-3 to III-137.

[4]   B. Wunderlich, *Macromolecular Physics*, Vol. 1, Academic Press, New York, 1973, chap. 2.

[5]   G. Natta and G. Allegra, *Tetrahedron*, **30**, 1987 (1974).

[6]   J. Boon and E.P. Magré, *Makromol. Chem.* **126**, 130 (1969).

[7]   P. Corradini, P. Ganis and V. Petraccone, *Eur. Polym. J.* **6**, 281 (1970).

[8]   G. Natta, P. Corradini, I.W. Bassi and G. Fagherazzi, *Eur. Polym. J.* **4**, 297 (1968).

[9]   T. Tanaka, Y. Chatani and H. Tadokoro, *J. Polym. Sci. Polym. Phys. Ed.* **12**, 515 (1974).

[10] S.C. Nyburg, *Acta Crystallogr.* **7**, 385 (1954).

[11] G. Natta and P. Corradini, *Angew. Chem.* **68**, 615 (1956).

[12] C.W. Bunn and E.V. Garner, *Proc. Roy. Soc. London* **A189**, 39 (1947).

[13] H. Kusanagi, H. Tadokoro, Y. Chatani and K. Suehiro, *Macromolecules* **10**, 405 (1977).

[14] M. Kobayashi and H. Tadokoro, *Macromolecules* **8**, 897 (1975).

5.8 8,8-Dichlorotricyclo[3.2.1.0^{1,5}]octane

C$_8$H$_{10}$Cl$_2$

Crystal Data

Monoclinic, space group $P2_1/m$ (No. 11)

$a = 6.796(24)$, $b = 8.721(35)$, $c = 6.903(25)$Å, $\beta = 102.62(7)°$ (-40°C), $Z = 2$

Atom	x	y	z	Atom	x	y	z
Cl(1)	.2234	1/4	.5552	H(1)	-.468	.432	.211
Cl(2)	.0949	1/4	.1318	H(2)	-.243	.447	.090
C(1)	-.1683	.3401	.3595	H(3)	-.472	1/4	.004
C(2)	-.3018	.3934	.1721	H(4)	-.351	1/4	-.095
C(3)	-.3668	1/4	.0469	H(9)	-.324	.397	.581
C(7)	-.2089	.3367	.5682	H(10)	-.118	.392	.664
C(8)	.0117	1/4	.3566				

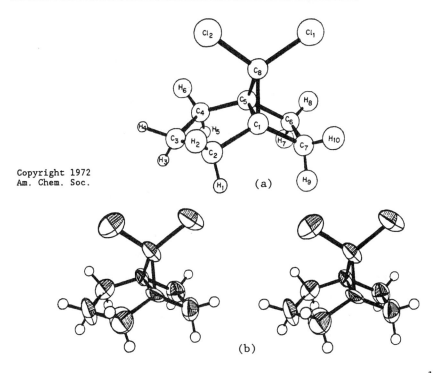

(a)

(b)

Fig. 5.8-1 Molecular structure of 8,8-dichlorotricyclo[3.2.1.0^{1,5}]-octane: (a) atom designation [the crystallographic mirror plane contains the atoms Cl(1), Cl(2), C(8), C(3), H(4) and H(3)]; (b) stereoscopic view. (After ref. 1).

Crystal Structure

The bond lengths and bond angles of the title molecule are listed in Table 5.8-1.[1] The most interesting aspect of the structure is the geometry at the bridgehead. The bridgehead carbon C(1) is displaced 0.093Å out of the plane formed by C(2), C(7) and C(8) and on the opposite side of the plane from C(5). Thus the four interatomic vectors from C(1) (i.e., the lines connecting the nuclear centers) all lie on one side of a plane. The length of the C(1)-C(5) bond is 1.572(15)Å.

The crystallographically imposed mirror plane requires the cyclobutane ring to be planar. Although the nonplanar conformation is considerably more prevalent, several cyclobutane rings, in simple derivatives, have been found to adopt this planar geometry. The cyclopentane ring is nearly planar despite the resulting increase in torsional strain. The planarity of the ring increases the angles at the bridgehead and this must lead to a decrease in strain which is greater than the increase in torsional strain. Within the cyclopropane ring there are two unusually short (mirror related) bridge-bridgehead bonds of 1.458(12)Å.

Table 5.8-1 Bond lengths (Å) and bond angles (°) of 8,8-dichlorotricyclo-[3.2.1.0^{1,5}]octane.

C(8)-Cl(1)	1.757	Cl(1)-C(8)-Cl(2)	108.7
C(8)-Cl(2)	1.764	C(1)-C(8)-C(5)	65.3
C(1)-C(5)	1.572	C(5)-C(1)-C(8)	57.4
C(1)-C(8)	1.458	C(2)-C(1)-C(5)	108.2
C(5)-C(6)	1.526	C(1)-C(2)-C(3)	106.4
C(1)-C(2)	1.483	C(2)-C(3)-C(4)	109.7
C(2)-C(3)	1.529		

Remarks

1. Structure of 1-cyanotetracyclo[3.3.1^{3,7}.0^{3,7}]decane

A single crystal X-ray analysis of a cyano- derivative of 1,3-dehydro-adamantane (Fig. 5.8-2)[2] showed that the bridgehead atoms C(3) and C(7) each has all four bonds directed to one side of the relevant atom, and the central C(3)-C(7) bond distance is 1.643(4)Å. Thus these atoms have "inverted" bond configurations. The side bonds (1.476-1.517Å) appear to correspond closely to the value expected for C(sp^2)-C(sp^3) bonds, giving some support for the assumption of sp^2 hybridization at the bridgehead, with pσ overlap for the central bond.

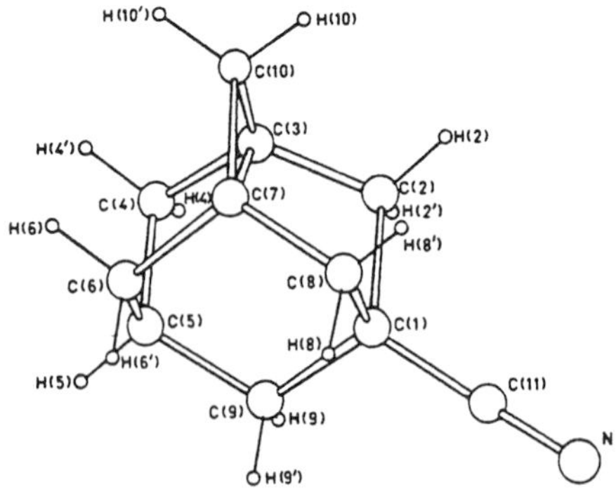

Fig. 5.8-2 Molecular structure of 1-cyanotetracyclo[3.3.1^{3,7}.0^{3,7}]-decane. (After ref. 2).

2. Fenestranes and the flattening of tetrahedral carbon[3]

The name fenestrane [fenestra(L.), window], originally coined specifically for tetracyclononane 1, has been conveniently extended to [m.n.p.q]fenestrane for the class of tetracyclic compounds represented by 2 and [m.n.p]fenestrane for the corresponding tricyclic systems. Hydrocarbon 1 then becomes [4.4.4.4]fenestrane, and 3 is named [4.4.4]fenestrane.

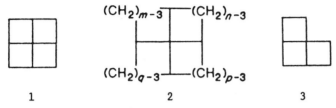

| 1 | 2 | 3 |

In fenestrane molecules the geometry of the central quaternary carbon atom is a somewhat flattened tetrahedron. The molecules 4 to 7 have been studied by X-ray analysis, and Fig. 5.8-3 shows their molecular structures.

| | | | |
| 4 | 5 | 6 | 7 |

In molecule 4 the internal cyclobutane ring [C(1)-C(4)-C(5)-C(6)] is essentially planar with no atom deviating from the least-squares plane by more than 0.003Å.[4] The two external rings are slightly puckered, with dihedral angles of 10.8° [C(1)-C(2)-C(3)-C(4)] and 12.6° [C(1)-C(6)-C(7)-C(8)]. The two shortest carbon-carbon bonds are associated with C(1); C(1)-C(2) is 1.515Å, and C(1)-C(8) is 1.524Å. Bonds C(1)-C(4) and C(1)-C(6) are 1.552 and 1.571Å, respectively, slightly shorter than the long (1.577Å) internal carbon-carbon bond of bicyclo[2.2.0]hexane.

There are several noteworthy features in the structure of 5.[5] The angles C(1)-C(10)-C(6) and C(3)-C(10)-C(8), which reflect the enforced flattening at C(10), are 128.3° and 129.2°, respectively. Two of the bond distances involving C(10) are quite short, both C(3)-C(10) and C(6)-C(10) being only 1.49Å. The perimeter bonds for the cyclobutane rings are correspondingly lengthened to an average of 1.574Å. These values are in quite good agreement with those given by MNDO calculations.

The structure of [4.4.5.5]fenestrane keto amide **6** reveals considerable
flattening at C(11); angle C(1)-C(11)-C(6) and C(3)-C(11)-C(9) are 128.2 and
123.0°, respectively. Once again these observed values are in excellent
agreement with MNDO calculations.[6]

The structure of the [5.5.5.5]fenestrane derivative **7** shows less
flattening at the central atom C(1) than in the fenestranes with four-membered
rings. Angle C(5)-C(1)-C(11) is 117.5, and C(2)-C(1)-C(8) is 115.1°; bond
lengths at C(1) range from 1.527Å [C(1)-C(11)] to 1.568Å [C(1)-C(8)].

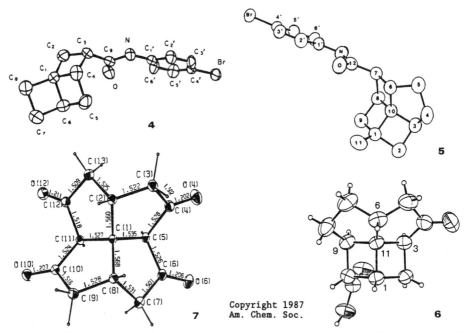

Copyright 1987
Am. Chem. Soc.

Fig. 5.8-3 Molecular structure of some fenestranes. (After ref. 3).

3. Structure of bullvalene

Doering and Roth[7] in 1963 proposed that bullvalene, $C_{10}H_{10}$, should
undergo degenerate Cope rearrangements that would make each of the ten CH
groups equivalent, some of which are illustrated in Fig. 5.8-4. If each of
the ten CH groups were individually labeled, there would be 10!/3 = 1,209,600
different ways of arranging them in the structure of bullvalene. Doering and
Roth pointed out that observation of a single line in the proton magnetic
resonance spectrum would mean that all these structures were simultaneously
present and rapidly interconverting. Later in the same year Schröder
announced the synthesis of bullvalene and confirmed that the proton magnetic
resonance spectrum at 100°C is indeed a single sharp line that broadens on
cooling and divides into two bands of area ratio (high-field: low-field peak)
4:6 at -25°C.[8]

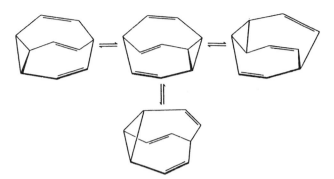

Fig. 5.8-4 A few Cope-rearranged isomers of bullvalene.

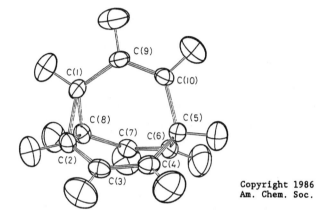

Fig. 5.8-5 Molecular structure of bullvalene. (After ref. 9).

Table 5.8-2 Molecular dimensions of bullvalene.

	Wing A C(8)-C(7)-C(6)-C(5)	Wing B C(1)-C(9)-C(10)-C(5)	Wing C C(2)-C(3)-C(4)-C(5)
C$_r$-C	1.471Å	1.474Å	1.473Å
C=C	1.343	1.342	1.342
C-C(5)	1.516	1.517	1.516
C$_r$-C=C	126.6°	126.6°	126.4°
C=C-C	123.6	123.9	123.7
C$_r$-C=C-C	-1.3	-0.4	+0.8
C$_r$-C=C-H	-179.5	179.8	179.9
H-C=C-C(5)	177.8	180.0	178.2
H-C=C-H	0.5	0.2	0.9

inter-wing angles

A:B   118.4°   A:C   121.8°   B:C   119.8°

cyclopropane ring

C(1)-C(8) 1.536Å   C(1)-C(2) 1.533Å   C(2)-C(8) 1.530Å

The molecular structure of bullvalene has been determined by neutron diffraction at 110K.[9] It consists of a cyclopropane ring with three HC-CH=CH-C wings tied to a common junction, which are all planar and significantly bent away from the ideal 120° relative orientations (Fig. 5.8-5). The molecule is distorted only very slightly from the C_{3v} symmetry expected for the idealized case, although the crystal symmetry is monoclinic. The geometrical parameters of the molecule are listed in Table 5.8-2.

4. Strain in organic chemistry[10]

Systematic investigation of small-ring compounds had a sensational beginning in 1883 with Perkin's synthesis of cyclobutanecarboxylic acid.[11] Next year he reported the first directed syntheses of cyclopropane. In 1885 A. von Baeyer proposed his strain theory[12] and correctly deduced that compounds with three- and four-membered rings would be less stable because of the deviation in bond angles from the normal tetrahedral values.

The concept of strain has greatly expanded from von Baeyer's original idea and is now discussed in terms of bond length and bond angle distortions as well as torsional deformation, nonbonded interactions, and rehybridization.[13] Strain energies have proved to be very valuable quantities for examining organic compounds having unusual geometries because of intramolecular interactions.

The strain energy of a molecule is derived from its heat of formation from the elements (ΔH_f) by comparison with a hypothetical strain-free model. For example, cyclohexane is generally taken to be a standard. Its heat of formation is -29.5 kcal mol^{-1}, or -4.92 kcal mol^{-1} per methylene group. Cyclopropane, with three methylene groups, would then be expected to have ΔH_f = -14.75 kcal mol^{-1}, whereas the observed ΔH_f is +12.73 kcal mol^{-1}. The difference between these values, 27.5 kcal mol^{-1}, is the strain energy. Table 5.8-3 gives the heats of formation and strain energies for a variety of cyclic hydrocarbons.

One might reasonably expect that increased strain would lead to increased reactivity. This is, however, not always the case. Factors that must be considered included the location of the activated complex along the reaction coordinate and the strain energies of the compounds that are formed as products or reaction intermediates. For example, the strain energies of [1.1.1]-, [2.1.1]- and [2.2.1]propellanes are essentially the same and, therefore, these compounds might be expected to have similar reactivities. However, whereas [1.1.1]propellane[14] is thermally stable to *ca.* 100°C, the other propellanes polymerize at 50K. The reason for the difference in

Table 5.8-3 Experimentally determined heats of formation (ΔH_f), calculated
             strain energies (SE), and entropies (S°) of cycloalkanes,
             methylenecycloalkanes and cycloalkenes. For methylenecyclo-
             alkanes and cycloalkenes, an olefinic strain (OS) is also
             given.[10]

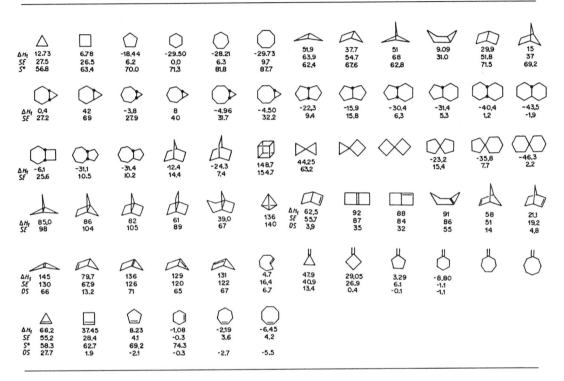

reactivity is easily seen by considering the energy required to "dissociate"
the central bonds:

$\Delta H = 5$ kcal.mol^{-1} $\Delta H = 30$ kcal.mol^{-1} $\Delta H = 65$ kcal.mol^{-1}

The differences in energy are mainly a result of the differences in strain
between the bicycloalkanes from which the diradicals are derived.
Bicyclo[1.1.1]pentane has two small rings, which, along with extra distortion,
lead to a large strain energy, whereas norbornane with no small rings has a
relatively low strain energy.

5. Structural homology

        Comparison of the molecular geometry of a bicyclo[1.1.1]pentasilane
derivative, I [Fig. 5.8-6(a)],[15] with that of a substituted carbon analogue

II[16] reveals a high degree of structural homology. In I the ratio of the "non-bonding distance" between apical atoms (NBD, Si1...Si3 2.98Å) to the average length of the single bond (BL, Si-Si 2.35Å) is 1.27, which is close to the corresponding ratio of 1.23 in II (Table 5.8-4). This *approximate* homology rule also applies to other series based on cyclobutane, bicyclo[2.2.1]heptane, and bicyclo[3.3.1]nonane.[15] Its validity even extends to the pentastanna[1.1.1]propellane derivative III shown in Fig. 5.8-6(b). The distance of 3.367(1)Å between the bridgehead Sn1 and Sn3 atoms far exceeds the longest known Sn-Sn single bond length of 3.052(1)Å in hexakis(2,6-diethylphenyl)distannane, thus pointing to significant singlet diradical character for III.[17] Compound III reacts quantitatively with CH_3I to yield the bicyclo[1.1.1]pentastannane IV whose Sn_5 framework has virtually the same dimensions [Fig. 5.8-6(c)].[18]

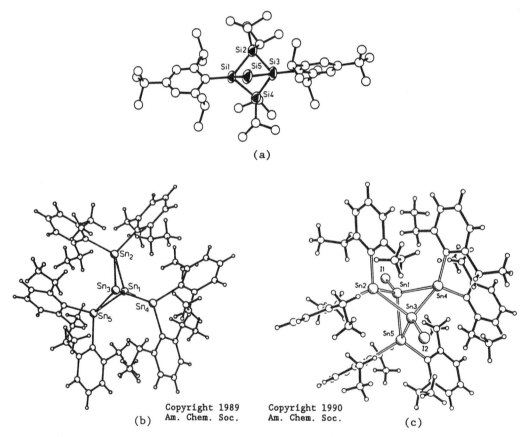

(a)

(b)   (c)

Fig. 5.8-6 Structures of (a) 1,3-bis(4-tert-butyl-2,6-diisopropylphenyl)-2,2,4,4-tetraisopropylbicyclo[1.1.1]pentasilane, (b) 2,2,4,4,5,5-hexakis(2,6-diethylphenyl)pentastanna[1.1.1]propellane and (c) 2,2,4,4,5,5-hexakis(2,6-diethylphenyl)-1-iodo-3-methylbicyclo[1.1.1]pentastannane (I_1 and I_2 represent statistically disordered CH_3 and I ligands). (After refs. 15, 17 and 18).

Table 5.8-4 Structural data illustrating the homology rule.

Compound	NBD/Å	BL/Å	NBD/BL	Fig.	Ref.
I	2.89	2.35	1.27	5.8-6(a)	[15]
II	1.89	1.54	1.23		[16]
III	3.367	2.856	1.18	5.8-6(b)	[17]
IV	3.361	2.833	1.19	5.8-6(c)	[18]

References

[1] K.B. Wiberg, G.J. Burgmaier, K.-W. Shen, S.J. La Placa, W.C. Hamilton and M.D. Newton, *J. Am. Chem. Soc.* **94**, 7402 (1972).

[2] C.S. Gibbons and J. Trotter, *Can. J. Chem.* **51**, 87 (1973).

[3] B.R. Venepalli and W.C. Agosta, *Chem. Rev.* **87**, 399 (1987).

[4] K.B. Wiberg, L.K. Olli, N. Golembeski and R.D. Adams, *J. Am. Chem. Soc.* **102**, 7467 (1980).

[5] V.B. Rao, C.F. George, S. Wolff and W.C. Agosta, *J. Am. Chem. Soc.* **107**, 5732 (1985).

[6] R. Mitschka, J. Oehldrich, K. Takahashi, J.M. Cook, U. Weiss and J.V. Silverton, *Tetrahedron*, **37**, 4521 (1981).

[7] W. von E. Doering and W.R. Roth, *Tetrahedron*, **19**, 715 (1963).

[8] G. Schröder, *Angew. Chem. Int. Ed. Engl.* **2**, 481 (1963).

[9] P. Luger, J. Buschmann, R.K. McMullan, J.R. Ruble, P. Matias and G.A. Jeffrey, *J. Am. Chem. Soc.* **108**, 7825 (1986).

[10] K.B. Wiberg, *Angew. Chem. Int. Ed. Engl.* **25**, 312 (1986); D. Cremer and E. Kraka in J.F. Liebman and A. Greenberg (eds.), *Structure and Reactivity*, VCH publishers, Weinheim, 1988, chap. 3.

[11] W.H. Perkin, *Ber. Deut. Chem. Gesell.* **16**, 1787 (1883).

[12] A. von Baeyer, *Ber. Deut. Chem. Gesell.* **18**, 2278 (1885).

[13] A. de Meijere, *Angew. Chem. Int. Ed. Engl.* **18**, 809 (1979).

[14] K.B. Wiberg and F.H. Walker, *J. Am. Chem. Soc.* **104**, 5239 (1982).

[15] Y. Kabe, T. Kawase, J. Okada, O. Yamashita, M. Goto and S. Masamune, *Angew. Chem. Int. Ed. Engl.* **29**, 794 (1990).

[16] A. Padwa, E. Shefter and E. Alexander, *J. Am. Chem. Soc.* **90**, 3717 (1968).

[17] L.R. Sita and R.D. Bickerstaff, *J. Am. Chem. Soc.* **111**, 6454 (1989).

[18] L.R. Sita and I. Kinoshita, *J. Am. Chem. Soc.* **112**, 8839 (1990).

Note For compounds containing a planar tetracoordinate carbon atom stabilized by α σ-donating/π-accepting metal, see G. Erker, R. Zwettler, C. Krüger, R. Noe and S. Werner, *J. Am. Chem. Soc.* **112**, 9620 (1990). To cite a recent example, Cp$_2$Zr(μ-C≡CPh)(μ-CPh=CMe)AlMe$_2$ exhibits a carbon atom of this type within the central metallacyclic ring, as reported in G. Erker, M. Albrecht, C. Krüger and S. Werner, *Organometallics* **10**, 3791 (1991).

Structural data for [3]rotane and higher homologues are given in R. Boese, T. Miebach and A. de Meijere, *J. Am. Chem. Soc.* **113**, 1743 (1991).

The structure of bicyclopropylidene is reported in M. Traetteberg, A. Simon, E.-M. Peters and A. de Meijere, *J. Mol. Struct.* **118**, 333 (1984). The synthesis and crystal structure of perspirocyclopropanated bicyclopropylidene and four-fold spiropropanated [3]rotane are reported in S. Zöllner, H. Buchholtz, R. Boese, R. Gleiter and A. de Meijere, *Angew. Chem. Int. Ed. Engl.* **30**, 1518 (1991).

Strained-ring and double-bond systems consisting of the elements Si, Ge, and Sn are reviewed in T. Tsumuraya, S.A. Batcheller and S. Masamune, *Angew. Chem. Int. Ed. Engl.* **30**, 902 (1991).

5.9   Cubane

$$C_8H_8$$

Crystal Data

Trigonal, space group $R\bar{3}$ (No. 148)

$a = 5.340(2)$Å, $\alpha = 72.26(5)°$, $Z = 1$

Atom	x	y	z
C(1)	-.1871(6)	.1952(6)	.1071(6)
C(2)	.1155(4)	.1155(4)	.1155(4)
H(1)	-.325(9)	.347(8)	.185(8)
H(2)	.210(6)	.210(6)	.210(6)

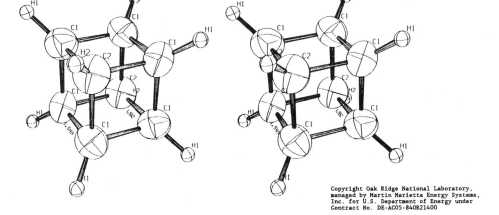

Copyright Oak Ridge National Laboratory,
managed by Martin Marietta Energy Systems,
Inc. for U.S. Department of Energy under
Contract No. DE-AC05-84OR21400

Fig. 5.9-1   Stereoview of the molecular structure of cubane.

Crystal Structure

Cubane was first synthesized by Eaton in 1964.[1] In the crystal, the cubane molecule has $\bar{3}$ symmetry, two body diagonal carbon atoms being located at (x,x,x) and $(-x,-x,-x)$.[2] The hydrogens attached to these carbons also lie on the $\bar{3}$ axis. The other carbons and hydrogens, six of each, occupy the general positions of the space group. The C-C bond lengths are 1.553(3) and 1.549(3)Å, the C-C-C bond angles are 89.3°, 89.6°, and 90.5°, all conforming nicely to the values expected for a regular cube. Fig. 5.9-1 shows a stereoview of the molecular configuration of cubane.

Remarks

1. C-C Bond lengths in cubanes[3]

Several cubane structures (a to f) have been determined by X-ray crystallography; the measured CH-CH edge distances in the cubane cage are listed below:

a	cubane[1]	1.551(3)Å	N = 2
b	cubylcubane[3]	1.553(8)Å	N = 9
c	2-*tert*-butylcubylcubane[3]	1.556(6)Å	N = 16
d	1,4-dinitrocubane[4]	1.564(2)Å	N = 3
e	1,3-dinitrocubane	1.556(5)Å	N = 6
f	1,4-dicarboxycubane[5]	1.560(4)Å	N = 6

The estimated standard deviations enclosed in parentheses are derived from the variation in each sample, not from the X-ray refinements. The number of independently determined bond lengths in each structure is denoted by N. The three cubanes, d, e and f with electron-withdrawing substituents have the slightly longer bond lengths, <C-C> = 1.559(1)Å for N = 15. When the values from cubanes a, b and c are pooled, <C-C> = 1.555(1)Å with N = 27. If the CH-CH edges from all the structures are considered together, they form a normal distribution with <C-C> = 1.556Å and N = 42.

Fig. 5.9-2 shows the structure of cubylcubane and 2-*tert*-butylcubylcubane. The intercage C-C bond lengths are significantly shorter than any recorded to date between quaternary carbon atoms. This phenomenon can be understood by considering a simple bonding picture for cubanes. The

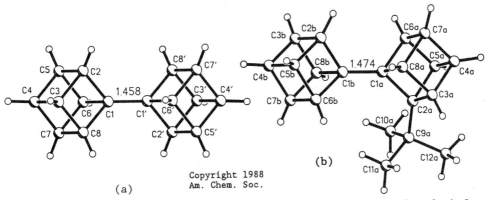

(a)

(b)

Fig. 5.9-2 Structure of (a) cubylcubane and (b) 2-*tert*-butylcubyl-cubane. The intercage bond lengths given are in Å. (After ref. 3).

cage geometry is better accommodated by endocyclic orbitals richer in p character than the sp^3 hybrid of typical tetracoordinated carbon compounds. (Consider that pure p orbitals, like the internuclear connections in cubane, are orthogonal to one another). Correspondingly, the exocyclic orbitals on cubane are richer in s character and form shorter, tighter bonds to substituents than would their sp^3 counterparts. Alternatively, the carbon atoms of cubanes are considered to be pyramidalized past tetrahedral geometry and "pulled back" away from the substituent. Accordingly there is less steric crowding in cubylcubane than would be found in a hexasubstituted ethane. In

1-adamantyladamantane, the intercage bond formed from pure (or nearly pure) sp^3-hybridized carbons is 1.578(2)Å long,[6] 0.12Å longer than in cubylcubane.

For the expected length of a sp^3-sp^3 bond, Allen obtained a value of 1.538(1)Å from an analysis of 1798 crystal structure determinations of various types of compounds chosen to exclude errors due to heavy atoms.[7] However, the "normal" distance between two carbons that have no hydrogen substituents might well be longer. Gilardi and co-workers.[3] searched the Cambridge Crystallographic Database for all bond lengths between two connected carbons each substituted by three other carbons and found that <C-C> = 1.572(35)Å for a sample of 138 distances.

2. Tetrahedrane

The $(CH)_n$ hydrocarbons constitute an interesting class of molecules whose carbon skeletons form convex polyhedra. A subset of these, the regular polyhedral hydrocarbons based upon the perfect solids of antiquity, is composed of three potential members: tetrahedrane ($n = 4$), cubane ($n = 8$) and dodecahedrane ($n = 20$). The remaining two regular polyhedra, the octahedra and the icosahedra, are not suitable skeletons for neutral hydrocarbons as they imply pentavalent or hexavalent carbons.

Tetrahedrane is the most highly strained of the formally saturated hydrocarbons. An unquestionable proof of the existence of tetrahedrane is still wanting.[8] However, tetra-*tert*-butyltetrahedrane (Fig. 5.9-3) has been successfully synthesized and confirmed by crystal-structure analysis.[9] Apart from the intermolecular effect of mutual shielding, the bulky substituents in the tetra-*tert*-butyltetrahedrane molecule play a second role

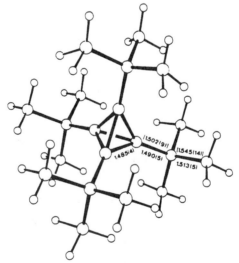

Fig. 5.9-3 Configuration of tetra-*tert*-butyltetrahedrane. (After ref. 9).

of decisive importance for its stability. The intramolecular repulsion of the four *tert*-butyl groups is at a minimum when their mutual distance is at a maximum. This condition is fulfilled by the T_d symmetry of the tetrahedron. Any other imaginable arrangement causes the substituents to be closer to each other, and the breaking of a skeletal bond in the tetra-*tert*-butyltetrahedrane molecule must equally lead to strong steric strain. In other words, the favored spherical arrangement of the four peripheral groups in the compound imposes the tetrahedral geometry on the molecule.

The crystal structure and X-X deformation densities of tetra-*tert*-butyltetrahedrane have been determined from low temperature data (103K) of the argon clathrate.[10] Two sections shown in Fig. 5.9-4 demonstrate the electron density distributions of the C-C bonds in the tetrahedrane skeleton. The density maxima of the tetrahedrane bonds are outwardly shifted by 0.37Å from the bond axes, which is equivalent to a bending of 26°. The density maxima of the exocyclic bonds from the tetrahedrane to the *tert*-butyl groups are localized exactly on the bond axes. The height of the density is more pronounced in these exocyclic bonds compared to the tetrahedrane bonds that are weakened by bending and strain.

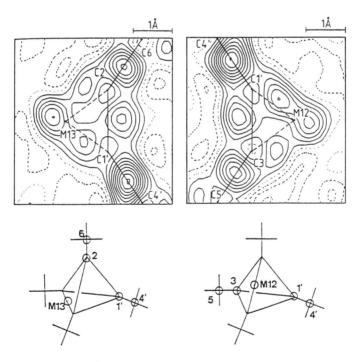

Fig. 5.9-4 Deformation density of tetra-*tert*-butyltetrahedrane. Section parallel to one bond and perpendicular to the opposite bond of the tetrahedrane. Contour interval 0.05 eÅ^{-3}. (After ref. 10).

In the present context, it is worth noting that tetra-*tert*-butyltetra-
hedrane is transformed at 135°C to the valence isomer tetra-*tert*-butylcyclo-
butadiene. X-Ray analysis of the latter compound at -150°C[11] revealed a
disordered structure modeled by the superposition of two mutually
perpendicular rectangular rings of sides approximately 1.60 and 1.34Å (chosen
to fit values found in other substituted cyclobutadienes) and with one
orientation several times more probable than the other.[12]

Renewed efforts have led to the successful synthesis of tri-*tert*-butyl-
(trimethylsilyl)tetrahedrane, which melts at *ca.* 162°C and rearranges into the
respective cyclobutadiene at *ca.* 180°C.[13]

3. Dodecahedrane

Dodecahedrane was successfully synthesized by Paquette and co-workers in
1982.[14] It is an almost spherical hydrocarbon molecule, which arises from
its exceptional icosahedral (I_h) symmetry, regular polyhedral nature, and
aesthetically appealing topology.

In the crystal structure of dodecahedrane, the molecule has
crystallographically imposed T_h symmetry, with the asymmetric unit containing
two carbon atoms and two hydrogen atoms.[15] The geometry of the molecule
obtained from X-ray analysis does not deviate significantly from I_h symmetry
(Fig. 5.9-5). The framework bonds, at C(1)-C(2) = 1.541(2)Å and C(2)-C(2) =
1.535(5)Å, are somewhat shorter than the 1.546Å value determined for
cyclopentane.[16] The exterior C-C-C bond angles [C(2)-C(1)-C(2) =
108.1(1)°, C(1)-C(2)-C(1) = 107.7(2)°, and C(1)-C(2)-C(2) = 107.9(4)°]
conform nicely to the value expected for perfect dodecahedral symmetry (108°).

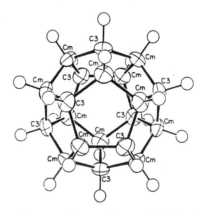

Fig. 5.9-5 Molecular configuration of dodecahedrane. The C(1) position
symmetry is 3, and C(2) is *m*. (After ref. 15).

Intracavity distances across the center of the molecule are 4.310(3)Å
for C(1)...C(1), and 4.317(5)Å for C(2)...C(2). If the generally accepted

value for the van der Waals radius of carbon (1.70Å) is taken into account, the transcavity diameter is only 0.91-0.93Å, too small for encapsulation of any but the smallest ions. Since the structure is face-centered cubic and the centers of the dodecahedrane molecules occupy the corners and the centers of the faces of the unit cell, the molecular packing is cubic closest packing. The shortest intermolecular contacts are 3.98Å for C...C and 2.33Å for H...H.

The crystal structures of C_{2v}-dimethylmonosecododecahedrane (1) and D_{3d}-dimethyldodecahedrane (2) have been determined.[17] Fig. 5.9-6 shows their

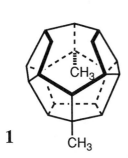 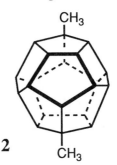

structures. In contrast to the very slight deviations from idealized dodecahedral symmetry present in 2, the topology of 1 is severely distorted. The high connectivity of the molecule constrains the two methylene groups to be closely juxtaposed. The enforced approach of the hydrogen atoms carried by these methylene carbons results in considerable flexing of the violated dodecahedral framework as it attempts to alleviate the close contact. The H(13)A...H(14)B distance is 1.95Å, *ca.* 0.47Å smaller than twice the van der Waals radius for hydrogen. Furthermore, the C(13)...C(14) distance is 3.03Å, nearly 1.5Å greater than the normal C-C single bond distance. This expansion in the C(13)...C(14) direction draws the two other sides of the framework toward each other, forming a somewhat flattened pouch.

Allinger and co-workers[18] have determined the crystal structures of 1,16-dimethyldodecahedrane and benzylmonosecododecahedrane at low temperature and calculated their geometries by the MM3 molecular-mechanics program. The two sets of results are in good agreement. Fig. 5.9-7 shows the structure of benzylmonosecododecahedrane, the framework of which, minus the benzyl group, can be described approximately in terms of C_{2v} symmetry.

X-Ray crystallographic analysis of methoxycarbonyldodecahedrane and 21-phenylcyclopropadodecahedrane (Fig. 5.9-8) has shown that their frameworks are modestly distorted by the substituents, the intracavity distance being more extended along the axis that contains the pendant group(s).[19] For the annulated dodecahedrane the cyclopropane ring is strikingly undistorted with

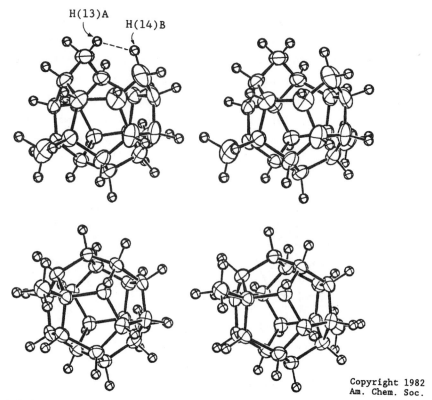

Fig. 5.9-6 Stereoview of dimethylmonosecododecahedrane (top) and 1,16-dimethyldodecahedrane (bottom). (After ref. 17).

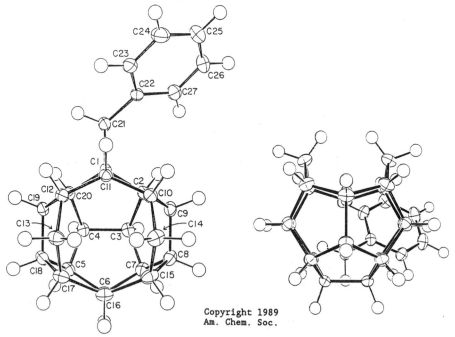

Fig. 5.9-7 Two views of the structure of benzylmonosecododecahedrane. (After ref. 18).

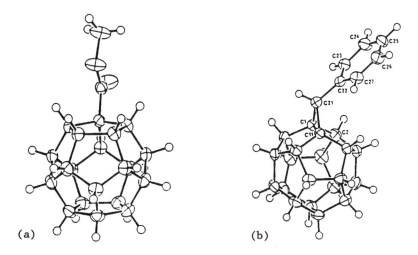

Fig. 5.9-8 Structures of (a) methoxycarbonyldodecahedrane and (b) 21-phenylcyclopropadodecahedrane. (After ref. 19).

C-C bond lengths in the range 1.503-1.509(4)Å, and those dodecahedrane bonds in the immediate vicinity of the ring fusion are extensively perturbed.

The energy-rich $C_{20}H_{20}$ hydrocarbon [1.1.1]pagodane **A** (X = Y = H) can be converted into dodecahedrane **B** (X = Y = Z = H) catalytically as well as in a stepwise manner via intermediates **C**, **D** and **E**. An elegant and elaborate "pagodane route" to four-, six-, and eight-fold functionalized dodecahedranes and dodecahedrenes of the types shown in the scheme below has been reported by Prinzbach and co-workers.[20]

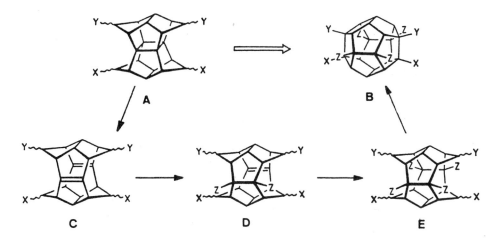

References

[1] P.E. Eaton and T. Cole, Jr., *J. Am. Chem. Soc.* **86**, 3157 (1964).

[2] E.B. Fleischer, *J. Am. Chem. Soc.* **86**, 3899 (1964).

[3] R. Gilardi, M. Maggini and P.E. Eaton, *J. Am. Chem. Soc.* **110**, 7232 (1988).

[4] P.E. Eaton, B.K. Ravi Shankar, G.D. Price, J.J. Pluth, E.E. Gilbert, J. Alster and O. Sandus, *J. Org. Chem.* **49**, 185 (1984).

[5] E.W. Della, P.T. Hine and H.K. Patney, *J. Org. Chem.* **42**, 2940 (1977).

[6] R.A. Alden, J. Kraut and T.G. Traylor, *J. Am. Chem. Soc.* **90**, 74 (1968).

[7] F.H. Allen, *Acta Crystallogr.*, *Sect. B* **37**, 890 (1981).

[8] P. Politzer and J.S. Murray in A. Greenberg and J.F. Liebman (eds.), *Structure and Reactivity*, VCH Publishers, Weinheim, 1988, chap. 1.

[9] H. Irngartinger, A. Goldmann, R. Jahn, M. Nixdorf, H. Rodewald, G. Maier, K.-D. Malsch and R. Emrich, *Angew. Chem. Int. Ed. Engl.* **23**, 993 (1984).

[10] H. Irngartinger, R. Jahn, G. Maier and R. Emrich, *Angew. Chem. Int. Ed. Engl.* **26**, 356 (1987).

[11] H. Irngartinger, M. Nixdorf, N.H. Riegler, A. Krebs, H. Kimling, J. Pocklington, G. Maier, K.-D. Malsch and K.-A. Schneider, *Chem. Ber.* **121**, 673 (1988); H. Irngartinger and M. Nixdorf, *Chem. Ber.* **121**, 679 (1988).

[12] J.D. Dunitz, C. Krüger, H. Irngartinger, E.F. Maverick, Y. Wang and M. Nixdorf, *Angew. Chem. Int. Ed. Engl.* **27**, 387 (1988).

[13] G. Maier and D. Born, *Angew. Chem. Int. Ed. Engl.* **28**, 1050 (1989).

[14] R.J. Ternansky, D.W. Balogh and L.A. Paquette, *J. Am. Chem. Soc.* **104**, 4503 (1982); L.A. Paquette, R.J. Ternansky, D.W. Balogh and G. Kentgen, *J. Am. Chem. Soc.* **105**, 5446 (1983).

[15] J.C. Gallucci, C.W. Doecke and L.A. Paquette, *J. Am. Chem. Soc.* **108**, 1343 (1986).

[16] W.J. Adams, H.J. Geise and L.S. Bartell, *J. Am. Chem. Soc.* **92**, 5013 (1970).

[17] G.G. Christoph, P. Engel, R. Usha, D.W. Balogh and L.A. Paquette, *J. Am. Chem. Soc.* **104**, 784 (1982).

[18] N.L. Allinger, H.J. Geise, W. Pyckhout, L.A. Paquette and J.C. Gallucci, *J. Am. Chem. Soc.* **111**, 1106 (1989).

[19] J.C. Gallucci, R.T. Taylor, T. Kobayashi, J.C. Weber, J. Krause and L.A. Paquette, *Acta Crystallogr.*, *Sect. C* **45**, 893 (1989).

[20] G. Lutz, D. Hunkler, G. Rihs and H. Prinzbach, *Angew. Chem. Int. Ed. Engl.* **28**, 298 (1989); and the four following papers in this series.

Note The structure of tetra-*tert*-butyltetraboratetrahedrane is described in T. Mennekes, P. Paetzold, R. Boese and D. Bläser, *Angew. Chem. Int. Ed. Engl.* **30**, 173 (1991).

The chemistry of "polyquinanes", hydrocarbons with multiply fused cyclopentane rings, is described in L.A. Paquette and A.M. Doherty, *Polyquinane Chemistry: Synthesis and Reactions*, Springer-Verlag, Berlin, 1987. "Cyclopolyindans", benzannelated polyquinanes with a central quaternary carbon atom, are reviewed by D. Kuck in I. Hargittai (ed.), *Quasicrystals, Networks, and Molecules of Fivefold Symmetry*, VCH Publishers, New York, 1990, chap. 19; notable examples are fenestridan (S_4 molecular symmetry) and centrohexaindan (T_d) as reported, respectively, in D. Kuck and H. Bögge, *J. Am. Chem. Soc.* **108**, 8107 (1986) and D. Kuck and A. Schuster, *Angew. Chem. Int. Ed. Engl.* **29**, 1192 (1988).

5.10 1',8':3,5-Naphtho[5.2.2]propella-3,8,10-triene
$C_{18}H_{14}$

Crystal Data

Orthorhombic, space group *Pnma* (No. 62)

$a = 6.815(1)$, $b = 12.827(2)$, $c = 13.938(2)$Å, $Z = 4$

Atom	x	y	z	Atom	x	y	z
C(1)	.83141	1/4	.57892	C(6)	.78543	.14999	.53756
C(2)	.92916	1/4	.66980	C(7)	.68178	.13539	.44259
C(3)	.97753	.15508	.71556	C(8)	.48539	.18900	.43585
C(4)	.93360	.06265	.67459	C(9)	.38593	.19889	.33855
C(5)	.83754	.06115	.58631	C(10)	.35189	.19859	.52312

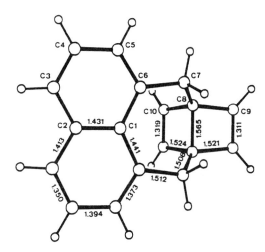

Fig. 5.10-1 Molecular structure of 1',8':3,5-naphtho[5.2.2]propella-3,8,10-triene. (After ref. 1).

Crystal Structure

The structure of the Dewar benzene derivative, 1',8':3,5-naphtho[5.2.2]-propella-3,8,10-triene (Fig. 5.10-1), has been determined by X-ray diffraction. In the crystal, the molecule lies on the mirror plane passing through the atoms C(1) and C(2) and the midpoints of the C(8)-C(8'), C(9)-C(9'), and C(10)-C(10') bonds. The bond lengths(Å) are shown in Fig. 5.10-1, and the bond angles(°) are as follows:

C(7)-C(8)-C(8')	117.1(2)	C(8')-C(8)-C(10)	85.4(1)
C(7)-C(8)-C(9)	119.3(2)	C(9)-C(8)-C(10)	116.0(2)
C(7)-C(8)-C(10)	121.1(1)	C(8)-C(9)-C(9')	94.8(2)
C(8')-C(8)-C(9)	85.2(1)	C(8)-C(10)-C(10')	94.6(2)

The dihedral angle between the cyclobutene rings of the Dewar benzene system is 116.6(1)°, which is near the value 117.3(6)° from an electron diffraction study of Dewar benzene itself.[2]

Remarks

1. Valence isomers of benzene[3,4]

Benzene $(CH)_6$, as a [6]annulene, has six valence isomers **1-6**, as shown below:

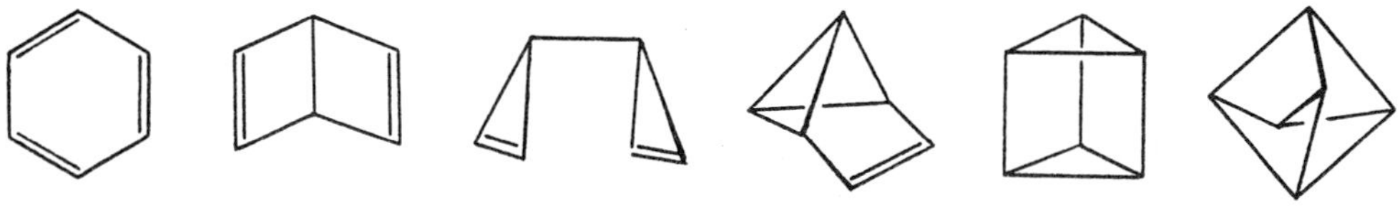

1	**2**	**3**	**4**	**5**	**6**
benzene	Dewar benzene	bicyclo-pro-2-enyl	benzvalene	triprismane (Ladenburg benzene)	

The first synthesis of valence bond isomers of aromatic compounds was carried out in 1962,[5] through the isolation of a stable Dewar-type benzene (**2**) by irradiation of tri-*t*-butylbenzene.

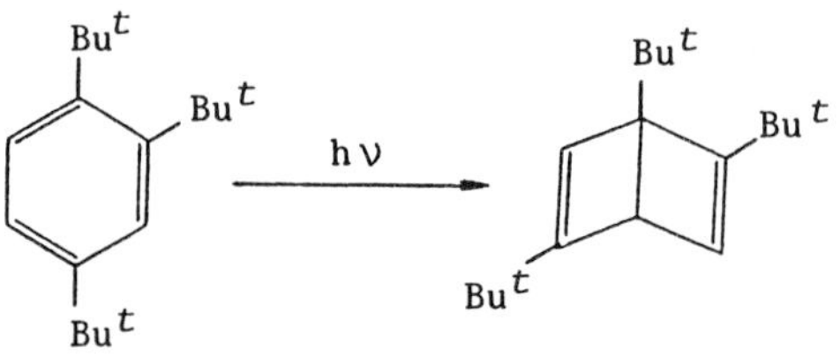

The guiding idea was that the bulkiness of the *t*-butyl groups would destabilize the flat structure of benzene, and steric repulsion would be relieved by conversion to the non-planar Dewar structure. Photolysis of hexafluorobenzene gives the Dewar form in a high yield. Also, hexakis(trifluoromethyl)benzene gives hexakis(trifluoromethyl)benzvalene upon irradiation by a high-pressure mercury lamp, or hexakis(trifluoromethyl) Dewar benzene and triprismane with a low-pressure mercury lamp.

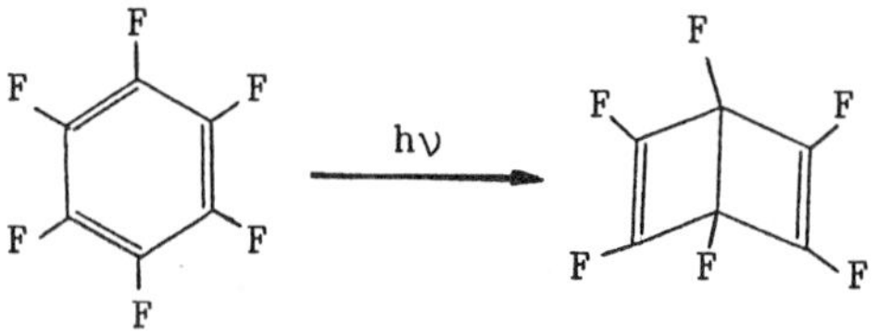

Bicyclopro-2-enyl, calculated to be the $(CH)_6$ species of highest energy,[6] has eluded synthesis until recently.[7]

Benzvalene 4 is the most extensively studied valence isomer of
benzene.[8] The structural parameters of benzvalene have been established
using microwave spectroscopy and electron diffraction methods. The dihedral

angle between its two three-membered rings is 106°, and the length of the
double bond is *ca.* 1.45Å. Benzvalene is one of the most reactive olefins
toward electron-deficient substrates; it is also bifunctional, since after
addition to the π system the ring strain of the σ system provides the driving
force for rearrangement or further addition reactions.

2. Structures and deformation densities of Dewar benzene derivatives[9]

In the Dewar benzene derivatives 7-10, the bond lengths between the
bridgehead carbon atoms vary from 1.562 to 1.612Å depending on the steric
influence of the substituents.

From the low temperature data of 10, the deformation densities have been
determined. The central bond between the bridgehead carbon atoms is bent
strongly by 28°, as shown in Fig. 5.10-2. The remaining bonds of the four
membered rings are deformed to a lesser extent.

The crystal structures of the Dewar pyridine derivative 11[10] and the
Dewar 2-phosphabenzene 12[11] have been determined. The F atoms in 11 have
extraordinarily large thermal parameters, and the bridge-head C-N bond (1.56Å)
is much longer than the other two C-N bonds (1.50Å). Fig. 5.10-3 shows the
molecular structure of 12 with selected bond distances and angles given in the
legend.

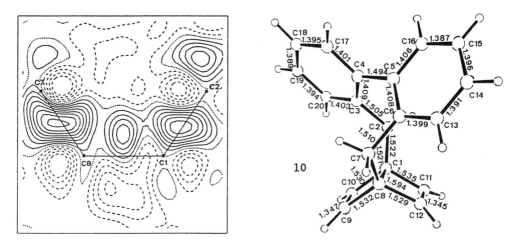

Fig. 5.10-2 Deformation density of **10** in a section along the central bond C(1)-C(8) including the methylene carbon atoms C(2) and C(7).   (After ref. 9).

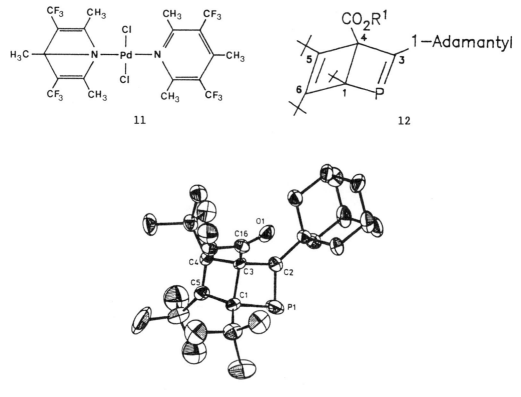

Fig. 5.10-3 Molecular structure of Dewar 2-phosphabenzene **12**.   Selected bond lengths (Å) and angles (deg): P1-C1 = 1.89(1), P1-C2 = 1.68(1), C1-C3 = 1.59(1), C1-C5 = 1.53(1), C2-C3 = 1.54(1), C3-C4 = 1.54(1), C4-C5 = 1.35(1); C1-P1-C2 = 79.5(4), P1-C1-C3 = 88.2(5), C3-C1-C5 = 85.3(6), P1-C2-C3 = 98.3(6), C1-C3-C2 = 93.9(6), C1-C3-C4 = 85.8(6), C3-C4-C5 = 93.9(6), C1-C5-C4 = 95.0(7).   (After ref. 11).

3. [n]Prismanes

The [n]prismanes are a class of saturated hydrocarbons having the structural formula $(CH)_{2n}$ and a semiregular, prismatic carbon skeleton composed of two parallel n-membered rings conjoined by n 4-membered rings. Besides tetraprismane (cubane), triprismane (Ladenburg benzene)[12] and pentaprismane (housane)[13] have succumbed to the unrelenting efforts of synthetic chemists. Some properties of the three known [n]prismanes are given in Table 5.10-1.[13]

X-Ray analysis of triprismanes 13,[14] 14,[15] and 15[16] have shown that the strained bonds are very sensitive to electronic effects of the

Table 5.10-1 Some properties of the known parent prismanes.

	Triprismane	Tetraprismane	Pentaprismane
point group	D_{3h}	O_h	D_{5h}
mp,°C	liquid	130-131	127.5-128.5
IR m, cm^{-1} a	3066, 1765, 1640	2992, 1235,	2973, 1273, 1231
	1233, 950, 881,	852	1069, 875, 768
	798, 733, 670		
d^{1}H	2.28	4.04	3.48
d^{13}C	30.6	47.3	48.6
J^{13}C-H (Hz)	180	155	148
~%s character(C-H) b	36	31	30

a In the region 3800-625 cm^{-1}. b Calculated as J(^{13}C-H)/5.

substituents and steric repulsion between the *tert*-butyl groups. In 13 and 15, the carbonyl moiety in the favorable bisecting orientation with respect to the three-membered ring gives rise to shortening of the distal bonds by 0.030-0.036Å. Repulsive interaction between the *tert*-butyl substituents stretches the prismane bonds up to 1.594Å in 15.

The synthetic conquest of hexaprismane (**16**), formally a face-to-face dimer of benzene, poses a formidable challenge and present progress has reached the stage exemplified by the synthesis of secohexaprismane (**17**)[17] and D_{2h}-bishomohexaprismane ("garudane" **18**).[18]

Ab initio calculations of the optimized geometries, vibrational frequencies, and heats of formation of the prismanes ($n = 3$ to 9) in D_{nh} symmetry have yielded results which offer encouragement in the quest for hexaprismane in that (i) its stability is similar to that of pentaprismane on

16   17   18

a per methine basis, and (ii) the vibrational frequencies of all 66 normal modes are real.[19] Equally interesting is the suggestion that [9]prismane is slightly less strained than the known [3]prismane.

The synthesis of prismanes consisting of framework atoms other than carbon also offers an intriguing synthetic challenge. The first germanium triprismane analogue, $[(Me_3Si)_2CHGe]_6$, has been successfully synthesized and unequivocally established by X-ray analysis.[20] The hexagermaprismane skeleton is constructed from two regular triangles and three regular squares of edges 2.58 and 2.52Å, respectively. These measured values are at variance with the trend predicted theoretically for C and Si prismanes[21] and confirmed in the X-ray structures of the carbon analogues.

4. Strained fused benzenes[22]

A series of strained fused benzenes have been studied by X-ray analysis, and the X-X deformation electron densities were determined for illustrating their bond properties.

a) Structure of 1H-cyclopropabenzene 19 and its 1,1-bis(triisopropyl-
   silyl) derivative 20

Fig. 5.10-4(a) and (b) show the structures of 19 and 20, respectively. In both structures the cyclopropabenzene moiety deviates from a strictly planar conformation. In the case of 19, the mm2 symmetry is reduced to m, the six-membered ring and the three-membered ring being tilted at an angle of 2.4° to each other. In 20, there is a reduction in symmetry from mm2 to 2, corresponding to a torsion angle $(C_2-C_1-C_6-C_7)$ of 1.9°. In both cases the angles in the benzene ring are deformed in the same manner by the annelation of the three-membered ring, and the C-C bond lengths are consistent with aromatic π delocalization. The C1-C2, C2-C3 and C3-C4 bonds are all normal (1.36-1.40Å), but the C1-C6 bond is significantly shortened [19, 1.334(4); 20, 1.336(3)Å]. In the cyclopropene ring, the C1-C7 bond lengths in the two compounds differ [**19**, 1.498(3); **20**, 1.541(1)Å] as a consequence of the nature of the substituents at the sp^3-hybridized carbon atom.

b) Structure of 3,4-dihydro-1*H*-cyclobuta[*a*]cyclopropa[*d*]benzene, **21**[23]

Fig. 5.10-4(c) shows the structure of **21**, which ranks among the most strained fused benzenes known. In **21**, the bonds of the cyclopropene ring are 0.015-0.022Å longer than those in **19**. The bond lengths in the six-membered ring are within 1.39-1.40Å; other bond lengths are C3-C4 1.522Å, C4-C4′ 1.579Å, and C1-C5 1.508Å.

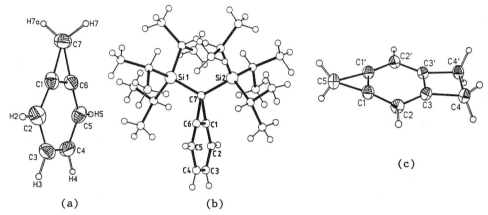

(a)   (b)   (c)

Fig. 5.10-4 Molecular structures of fused benzenes (a) **19**, (b) **20**, and (c) **21**. (After refs. 22 and 23).

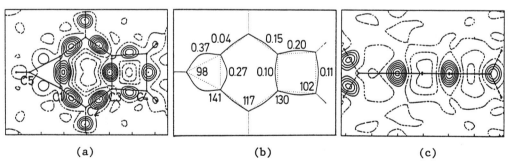

(a)   (b)   (c)

Fig. 5.10-5 (a) Contour lines of the X-X deformation electron densities in the molecular plane of **21**. Unbroken lines: positive values and zero, interval $0.05e\text{Å}^3$; broken lines: negative values, interval $0.1e\text{Å}^{-3}$. (b) Electron density maxima shifts (decimal points in Å) of the internuclear distances (dotted); the unbroken lines lead in every case from nucleus to nucleus via the electron density maximum. The quoted angles (°) refer to the direct electron density maximum-nucleus-electron density maximum connections. (c) Contour lines of the X-X deformation electron density lines perpendicular to the molecular plane in the C_2-axis of the molecule (edge view; left cyclopropene part, right cyclobutene part). (After ref. 23).

The X-X deformation electron densities in the molecular plane are shown
in Fig. 5.10-5(a). Consistent with the assumption that the deviations of the
electron density maxima from the direct C-C connections is a measure of
strain, the cyclopropene ring pushes its electrons more strongly (0.27Å) into
the π-system than its cyclobutene counterpart (0.10Å) [Fig. 5.10-5(b)]. The
electron densities are moved into the exocyclic region with respect to the
bonds C1-C2 (by 0.04Å) and C2-C3 (by 0.15Å), a result of the intracyclic
shifts of the annelated bonds. The extent of this effect is also evident in
the bond angles that can be assigned to the electron density maxima. In the
cyclopropene ring the "outside" angle is 98° instead of 53.2° for the nucleus-
nucleus connection, and that associated with the annelating carbon 141°
instead of of 170.2°; in the case of the four-membered ring the latter value
is 130° instead of 142.2°. Even the smallest bond angle in the benzene
nucleus is widened from 109.2° to 117°.

The X-X deformation electron density map perpendicular to the molecular
plane in the C_2-axis reveals a π contribution by its elliptical shape along
the distances C1-C1' and C3-C3', respectively, compared to the radial
distribution along the single bond C4-C4' [Fig. 5.10-5(c)]. To counter the
effects of the doubly induced strain, the system undergoes expansion, thus
avoiding an even greater displacement of the electron density maxima from the
direct internuclear connections.

    c) Structure of 1,2-dihydrocyclobutabenzene, **22**, and 1,2,4,5-tetra-
       hydrodicyclobuta[*a,d*]benzene, **23**[24]

Molecule **22** has *mm*2 symmetry, while **23** has *mmm* symmetry. Bond lengths
for the two molecules are as follows:

Molecule	Fig. 5.10-6	C3-C4	C4-C5	C5-C6	C6-C1	C6-C7	C7-C8
22	(a)	1.399	1.400	1.385	1.391	1.518	1.576
	(c)	1.40	1.41	1.40	1.41	1.56	1.65
		C1-C2			C2-C3	C2-C5	C5-C4
23	(d)	1.394			1.399	1.521	1.575
	(f)	1.40			1.41	1.53	1.60

The values of (a) and (d) are the distances between the nuclei while (c)
and (f) are calulated linearly from the maxima of the deformation electron
densities. These values are consistent with the concept of bent bonds; the
shifts of the bonding electron density maxima are clearly recognizable in Fig.
5.10-6(b) and (e).

If the bond angles are calculated with the assumption of bent bonds,
i.e. bonds which do not correspond to the nucleus-nucleus bond axes but to
lines joining the maxima of the deformation densities [Fig. 5.10-6(c) and

(f)], the resulting angles compare far better with those expected for sp² and sp³ hybridized C-atoms.

From Fig. 5.10-6(b) and (e), it is clear that the ring strain of the cyclobutene ring is transferred to the annelated bond, whose bonding electron density shifts to the center of the benzene ring. This causes an outward shift of the electron densities in the neighbouring bonds.

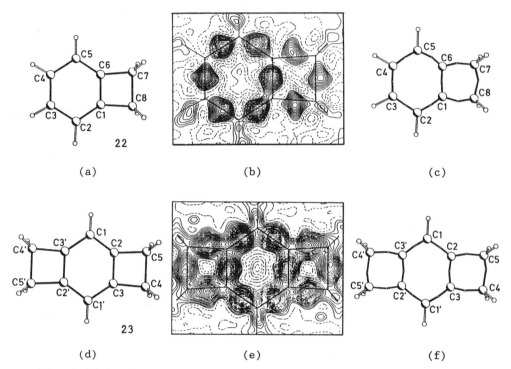

Fig. 5.10-6 (a) and (d) Formulae of **22** and **23** with atomic labeling; (b) and (e) X-X deformation electron densities of **22** and **23**; (c) and (f) lines joining the maxima of the deformation densities. (After ref. 24).

References

[1] K. Weinges, J. Klein, W. Sipos, P. Guenther, U. Huber-Patz, H. Rodewald, J. Deuter and H. Irngartinger, *Chem. Ber.* **119**, 1540 (1986).

[2] E.A. McNeill and F.R. Scholer, *J. Mol. Struct.* **31**, 65 (1976).

[3] Y. Kobayashi and I. Kumadaki, *Acc. Chem. Res.* **14**, 76 (1981).

[4] A.T. Balaban and M. Banciu, *J. Chem. Educ.* **61**, 766 (1984).

[5] E.E. van Tamelen and S.P. Pappas, *J. Am. Chem. Soc.* **84**, 3789 (1962); E.E. van Tamelen, S.P. Pappas and K.L. Kirk, *ibid.* **93**, 6092 (1971).

[6] A. Greenberg and J.F. Liebman, *J. Am. Chem. Soc.* **103**, 44 (1981).

[7] W.E. Billups and M.M. Haley, *Angew. Chem. Int. Ed. Engl.* **28**, 1711 (1989).

[8] M. Christl, *Angew. Chem. Int. Ed. Engl.* **20**, 529 (1981).

[9] H. Irngartinger and J. Deuter, *Chem. Ber.* **123**, 341 (1990);
H. Irngartinger, J. Deuter, H. Wingert and M. Regitz, *ibid.*, p. 345.

[10] Y. Kobayashi, A. Ohsawa and Y. Iitaka, *Tetrahedron Lett.*, 2643 (1973).

[11] J. Fink, W. Rösch, U.-J. Vogelbacher and M. Regitz, *Angew. Chem. Int. Ed. Engl.* **25**, 280 (1986).

[12] T.J. Katz and N. Acton, *J. Am. Chem. Soc.* **95**, 2738 (1973).

[13] P.E. Eaton, Y.S. Or and S.J. Branca, *J. Am. Chem. Soc.* **103**, 2134 (1981);
P.E. Eaton, Y.S. Or, S.J. Branca and B.K.R. Shankar, *Tetrahedron* **42**, 1621 (1986).

[14] H. Irngartinger, D. Kallfass, E. Litterst and R. Gleiter, *Acta Crystallogr., Sect. C* **43**, 266 (1987).

[15] G. Maier, I. Bauer, U. Huber-Patz, R. Jahn, D. Kallfass, H. Rodewald and H. Irngartinger, *Chem. Ber.* **119**, 1111 (1986).

[16] H. Wingert, H. Irngartinger, D. Kallfass and M. Regitz, *Chem. Ber.* **120**, 825 (1987).

[17] G. Mehta and S. Padma, *J. Am. Chem. Soc.* **109**, 2212 (1987).

[18] G. Mehta and S. Padma, *J. Am. Chem. Soc.* **109**, 7230 (1987).

[19] R.L. Disch and J.M. Schulman, *J. Am. Chem. Soc.* **110**, 2102 (1988).

[20] A. Sekiguchi, C. Kabuto and H. Sakurai, *Angew. Chem. Int. Ed. Engl.* **28**, 55 (1989).

[21] A.F. Sax and R. Janoschek, *Angew. Chem. Int. Ed. Engl.* **25**, 651 (1986); S. Nagase, M. Nakano and T. Kudo, *J. Chem. Soc., Chem. Commun.*, 60 (1987); S. Nagase, H. Teramae and T. Kudo, *J. Chem. Phys.* **86**, 4513 (1987); A.F. Sax, J. Kalcher and R. Janoschek, *J. Comput. Chem.* **9**, 564 (1988).

[22] R. Neidlein, D. Christen, V. Poignée, R. Boese, D. Bläser, A. Gieren, C. Ruiz-Pérez and T. Hübner, *Angew. Chem. Int. Ed. Engl.* **27**, 294 (1988).

[23] D. Bläser, R. Boese, W.A. Brett, P. Rademacher, H. Schwager, A. Stanger and K.P.C. Vollhardt, *Angew. Chem. Int. Ed. Engl.* **28**, 206 (1989).

[24] R. Boese and D. Bläser, *Angew. Chem. Int. Ed. Engl.* **27**, 304 (1988).

Note The structures of 1,4-dihydrodicyclopropa[b,g]naphthalene and cyclopropa[b]naphthalene are reported in B. Halton, R. Boese, D. Bläser and Q. Li, *Aust. J. Chem.* **44**, 265 (1991).
The following [5]prismane and [4]prismanes are known: $(RSn)_{10}$ [R = 2,6-diethylphenyl] in L.R. Sita and I. Kinoshita, *J. Am. Chem. Soc.* **113**, 1856 (1991); $(RSn)_8$ in *Organometallics* **9**, 2865 (1990); $(R'Si)_8$ [R = Bu^tMe_2Si] in H. Matsumoto, K. Higuchi, Y. Hoshino, H. Koike, Y. Naoi and Y. Nagai, *J. Chem. Soc., Chem. Commun.*, 1083 (1988).

5.11 [2.2]Paracyclophane
$C_{16}H_{16}$

Crystal Data

Tetragonal, space group $P4_2/mnm$ (No. 136)

$a = 7.781(1)$, $c = 9.290(2)$Å, $Z = 2$

Atom*	x	y	z
C(1)	.12618	.12618	.15060
C(2)	.24773	.03326	.07459
H(2)	.3238	-.0503	.1264
C(7)	.07098	.07098	.29926
H(71)	.1923	-.0165	.3391
H(72)	.0621	.1566	.3688

* The "half" H atoms attached to C(7) are numbered H(71) and H(72).

The atom numbering has been changed to conform to that of Fig. 5.11-2.

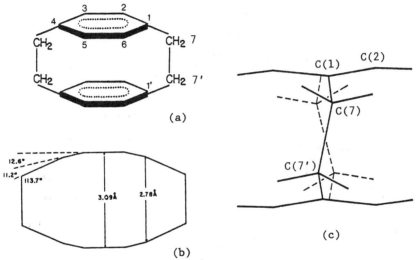

Fig. 5.11-1 Structure of [2.2]paracyclophane. (a) Structural formula, (b) molecular profile of the carbon skeleton, and (c) end view of the molecule, illustrating the effect of the disorder and the bending of the aromatic hydrogen atoms below the ring plane. (After ref. 1).

Crystal Structure

The crystal structure of [2.2]paracyclophane was first determined by Brown[2] and later reinvestigated by Lonsdale, Milledge and Rao.[3] More recently Hope and co-workers[1] determined the structure of octafluoro[2.2]paracyclophane, refined the structure of [2.2]paracyclophane, and compared the two sets of molecular parameters.

The carbon skeletons of the two related compounds are very similar. In the unsubstituted paracyclophane, the crystallographically-imposed point symmetry is *mmm*, and the bond lengths are: C-C bridge, 1.569; C_7-C_1, 1.514;

C_1-C_2, 1.394; and C_2-C_3, 1.394Å. The planes of the four unsubstituted C atoms of each ring are 3.09Å apart, the *para*-carbon atoms are bent about 12° out of the plane of the other four atoms of the aromatic ring, and C_1...$C_{1'}$ is 2.78Å, as shown in Fig. 5.11-1(b). Fig. 5.11-1(c) shows the disordered ethylene bridge in the structure.

Remarks

1. Historical importance

    The structural features of [2.2]paracyclophane prompted the early development of cyclophane chemistry.[4] The study of stereochemistry and electronic effects, including bent benzene rings, the face-to-face contact of aromatic nuclei, charge transfer effects and non-benzenoid π-systems, has continued to advance at a rapid pace.

    The bridging of molecules through long chains or short clamps has since been usefully applied in many branches of chemistry. New synthetic procedures for medium and large rings are now well developed, which lead to cyclophanes in excellent yields without using complicated techniques. In the design of novel compounds such as the multi-bridged, multi-layered, multi-stepped, and helically-wound cyclophanes, the static and dynamic stereochemistry, steric interactions, and electronic effects can be directionally adjusted and varied by using different bridge lengths and interior ring substituents.

2. Symbols used for the description of cyclophane structures[5]

    There are three general types of cyclophane structures: *para*-, *meta*- and *ortho*-type, as shown in Fig. 5.11-2.

    In all cyclophane types, the first bond in the bridge away from the aromatic ring is denoted by *a*, the second by *b*, the third by *c*, and the fourth by *d*. Letter *e*, *f*, *g*, *i* and *j* specifically denote the bond lengths in the aromatic rings. For the *para*-type ring [Fig. 5.11-2(a)] *e* describes the 1-2 (1-6) bonds (1 being the atom bound to the bridge) and, when related by symmetry, the 3-4 (4-5) bonds. If the latter bonds are unrelated to the former because of dissymmetry the value is denoted by the letter *g*. The 2-3 (5-6) bonds are denoted by *f*. In the *meta*- and *ortho*-type ring, the letter designations are shown in Fig. 5.11-2(b) and (c).

    Letters *p* and *q* denote the nonbonded distances as described in Fig. 5.11-2(b). For the *para*-type structure *p* represents the distance between the 1 and 1' (4 and 4') atoms, and *q* the distance between the 2 and 2' or 6 and 6' (3 and 3' or 5 and 5') atoms. For the *meta*-type structure *p* represents the distance between the 1 and 1' (3 and 3') atoms, and *q* denotes the distance between 2 and 2' atoms in either *syn* or *anti* form.

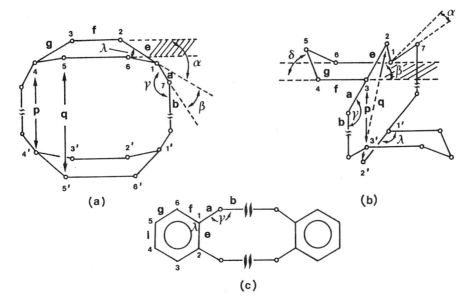

Fig. 5.11-2 Geometrical parameters used in the description of the three general types of cyclophane structures: (a) *para*, (b) *meta*, and (c) *ortho*. (After ref. 5).

The important bond angles are denoted by γ and λ as pictured in Fig. 5.11-2. Distortions associated with the aromatic rings are described by angles α, β and δ.

3. General structural features and properties of cyclophanes

Keehn[5] has collated the known X-ray crystal structure determinations of cyclophanes (up to the early part of 1982) and summarized their structural features:

a) Bond lengths

The bond lengths do not deviate dramatically from normal values, but some changes in the aromatic ring around the atom bound to the bridge and in the bridges are observed. The bridging bond lengths [*b*, in Fig. 5.11-2(a)] are generally the longest and in some instances have been found to be near or greater than 1.6Å.

b) Bond angles

The molecular distortion in cyclophanes is reflected largely in the deviation of the bond angles from the normal values of 109.5, 120, and 180° for sp³-, sp²-, and sp-hybridized atoms, respectively. These deviations are observed primarily in the internal angle of the bridging atom in the aromatic ring [λ, in Fig. 5.11-2(a)] and in the angle of the bridge at the atom once removed from the aromatic ring. The former values (sp^2) have been observed to be as low as 115°, whereas the latter (sp^3) can be as high as 118°. Angle γ also deviates from 120° when the bridging atom is sp^2-hybridized.

c) Nonplanarity of aromatic rings and displacement of appended atoms from the ring planes

In meta- and paracyclophanes with fewer than four atoms in each bridge, the aromatic ring atoms are distorted into boat shapes. In the paracyclophanes the 1 and 4 atoms of the aromatic rings are displaced out of the plane of the other four atoms toward the cyclophane cavity. In the metacyclophanes the 2 and 5 atoms of the aromatic ring are displaced from the plane of the other four atoms away from the cyclophane cavity. The 2 atom of the metacyclophane aromatic ring is generally more affected by the transannular π-π interaction, and thus angle β is usually larger than δ. In some instances, as in the multibridged and multilayered cyclophanes, chair geometries of the aromatic rings are also observed.

d) Proximity of nonbonded atoms and relationships of interplanar angles

The ideal nonbonded contact distance between aromatic rings in crystals containing these groups is 3.4Å, as observed in graphite. In the lower homologs of [$m.n$]para- and [$m.n$]metacyclophanes, mean intramolecular aromatic ring separations are near 3Å, and specific C...C nonbonded distances of the corresponding 1,4 and 1,3 atoms, which eclipse one another transannularly, are frequently less than 2.8Å. These close proximities cause distortions in the geometry of the ring sp^2 atoms such that the π density grows on the outer face of the aromatic ring. This rehybridization characteristic is observed crystallographically as the "turning-in" of the H atoms bound to the aromatic ring and has been observed in most structures in which the H atoms were found during the structure refinement.

Another aspect of intra-annular π-π interactions sometimes found in cyclophane structures is that the mean aromatic planes are not parallel but are inclined with respect to one another.

These structural features of cyclophanes have a direct bearing on their physical properties and chemical behavior. Some experimental facts are listed below:[6]

a) The deformed benzene rings exhibit NMR proton signals at chemical shift values (δ) of 6.30 ppm, well outside of the usual 6.5-8.0 ppm range.

b) When a carboxyl group substitutes a H atom of the benzene ring, the product is optically active as the carboxyphenyl group cannot rotate to give rise to the enantiomer.

c) The rate of monoacetylation of the [m,n]paracyclophanes increases as

m and n decrease but the rate of introducing an acetyl group into the second ring decreases. Thus the proximity of the second ring increases the availability of electrons for electrophilic substitution into the first ring, but the electron-withdrawing acetyl group in the first ring decreases the electron density not only in that ring but also in the neighboring second ring.

d) When paracyclophane forms a charge-transfer complex with an acceptor [A], the transannular effect makes the complex more stable. The charge-transfer complexes have a large polarizability perpendicular to the plane of the components as revealed by the fact that their absorption of polarized light is greatest when the electric vibrations are normal to the parallel planes of the two components. The delocalization of positive charge on the benzene ring is facilitated by the presence of a second adjacent benzene ring parallel to that attached to [A].

Accordingly, with a given [A], the λ_{max} of 1 occurs at longer wavelengths than that of 2, which implies on the basis of empirical observations that 2 is the more stable. This transannular effect was estimated to range from 2 to 14.4 kcal mole^{-1}.

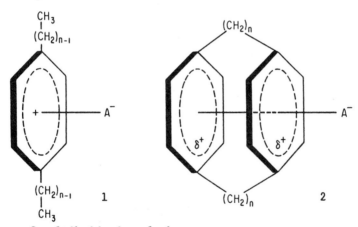

4. Structure of multibridged cyclophanes

[2.2]Paracyclophane is regarded as the parent molecule of the $[2_n]$ cyclophanes, the entire series beginning with [2.2]cyclophane and ending with superphane has been prepared, and their structural, spectroscopic, and chemical properties have been investigated. Fig. 5.11-3 shows some examples of multibridged cyclophanes.

X-Ray analysis revealed that bridge bonds in the $[2_n]$cyclophanes are universally lengthened. The only exception is the middle bridge bond of (c), which is of normal length because of the high tilt of the benzene ring (dihedral angle 42°).

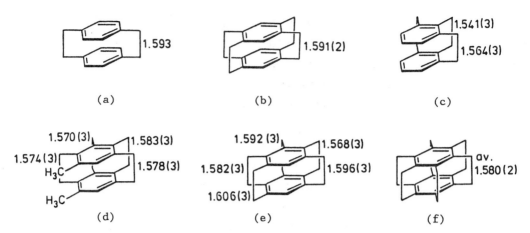

Fig. 5.11-3 Structure of multibridged cyclophanes. (After ref. 7).

5. Multilayered paracyclophanes[8]

    Compounds 3 and 4 (isolated as a mixture) were the first multilayered
cyclophanes prepared by the method of Hofmann elimination of quaternary
ammnonium hydroxides. The structure of tetramethyl quadruple-layered

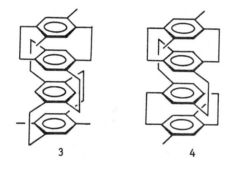

[2.2]paracyclophane, 4, the centrosymmetric isomer, has been determined [Fig.
5.11-4(a)].[9] The triple-layered bromo [2.2]paracyclophane, 5, has also been
prepared and characterized by crystal-structure analysis [Fig. 5.11-4(b)].[10]
Camparison of the bending angles (Table 5.11-1) shows that the multilayered
paracyclophanes have a highly strained configuration in their inner benzene
rings. Such highly strained molecular structures as those of the triple- and
quadruple-layered [2.2]paracyclophanes are suggestive of higher strain energy
than that of [2.2]paracyclophane.

    The detailed molecular structure of orthocyclophane 6 was determined by
X-ray analysis, as shown in Fig. 5.11-5.[11] The important structural
characteristics of 6 are that all three benzene rings are bent into a boat
form (with larger bending angles for the inner benzene ring), their mean
planes being stacked partially and nearly in parallel.

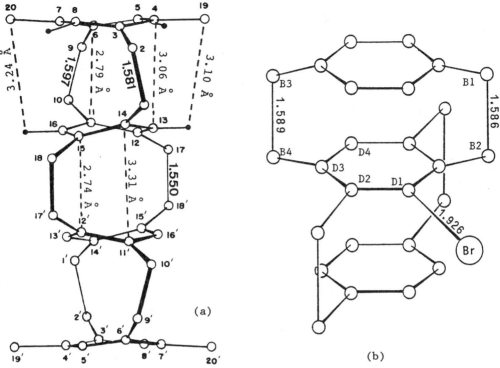

Fig. 5.11-4 Molecular structure of (a) tetramethyl quadruple-layered [2.2]paracyclophane centrosymmetric isomer 4, and (b) triple-layered bromo [2.2]paracyclophane 5. (After refs. 9 and 8).

Table 5.11-1 Bending angles of benzene rings in multilayered paracylophanes.

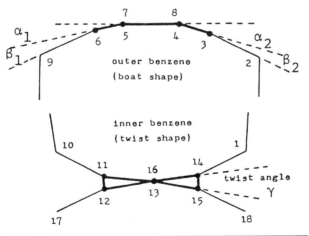

Compound	Outer benzene (deg)		Inner benzene(deg)
	α_1, α_2	β_1, β_2	γ
[2.2]paracyclophane	12.6	11.2	-
4	12.5	10.8	13.4
	12.9	11.8	
5	12.1	10.9	13.6
	12.3	11.5	

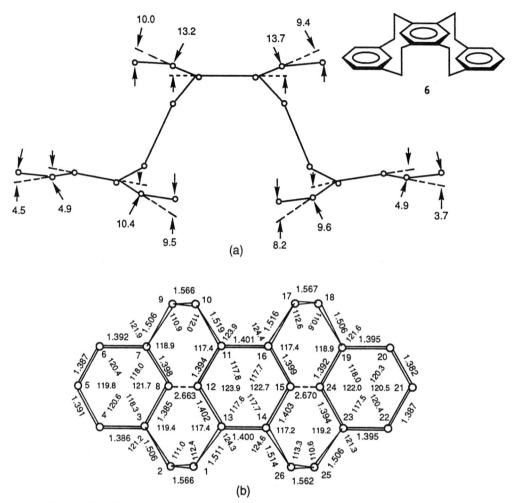

Fig. 5.11-5 Molecular projection of **6**: (a) side view and (b) bond lengths and bond angles. (After ref. 8).

6. Wiberg's "mini-superphane" and columnenes[12]

The structure of tricyclo[4.2.2.2^{2,5}]dodeca-1,5-diene ($C_{12}H_{16}$, "mini-superphane") has been determined by X-ray crystallography. The molecule is centrosymmetrical and consists of two olefinic groups linked by four bridging -CH_2CH_2- groups [Fig. 5.11-6(a)]. The experimental structural parameters and calculated values from *ab initio* optimization are listed in Table 5.11-2.

The intramolecular distance between the two double bonds is 2.395Å, which is about 1.0Å less than the π-π C...C van der Waals contact of 3.4Å found in simple aromatic compounds. The inherent strain appears primarily in the form of two molecular distortions: a) the elongation of the C-C single bonds in the bridging CH_2-CH_2 groups to 1.596Å, and b) the nonplanarity of the double bonds and their substituents (angle between the C_1-C_2 vector and the C_2-C_4-C_6 plane is 27.3°, and C_1 and C_2 lie 0.40Å out of the plane defined by the atoms C_3, C_4, C_5 and C_6).

Table 5.11-2   Comparison of experimental and calculated geometries of "mini-superphane".

Parameters	Exptal.	Ab initio	Parameters	Exptal.	Ab initio
C_1-C_2	1.354	1.328	C_1-C_2-C_6	120.6°	120.6°
C_2-C_4	1.517	1.535	C_4-C_6-$C_{5'}$	105.3°	105.3°
C_3-$C_{4'}$	1.596	1.589	C_4-C_2-C_6	110.0°	110.0°
C_1-$C_{2'}$	2.395	2.399	C_1-C_2-C_4-C_6	27.3°	27.4°

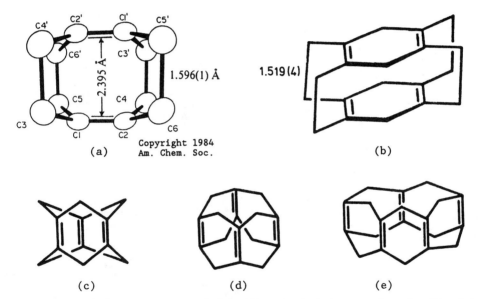

(a)

Copyright 1984
Am. Chem. Soc.

(b)

(c)   (d)   (e)

Fig. 5.11-6 Structures of (a) Wiberg's "mini-superphane", (b) the reduction product of [2.2.2.2](1,2,4,5)cyclophane, and (c)-(e) three columnenes. (After refs. 7, 12 and 14).

The structure of the reduction product of [2.2.2.2](1,2,4,5)cyclophane ($C_{20}H_{24}$) is shown in Fig. 5.11-6(b). The intramolecular distance between the six-membered rings is 2.809Å.[13]

Ab initio theoretical studies on three columnenes, illustrate in Fig. 5.11-6(c-e), have led to the prediction that these should have significantly pyramidalized π bonds.[14] The pyramidalization angles were estimated to be 47.3, 29.3 and 18.2° for compounds c, d, and e, respectively, based on STO-2G geometry optimization. These compounds are predicted to have strong through-space π-π interactions.

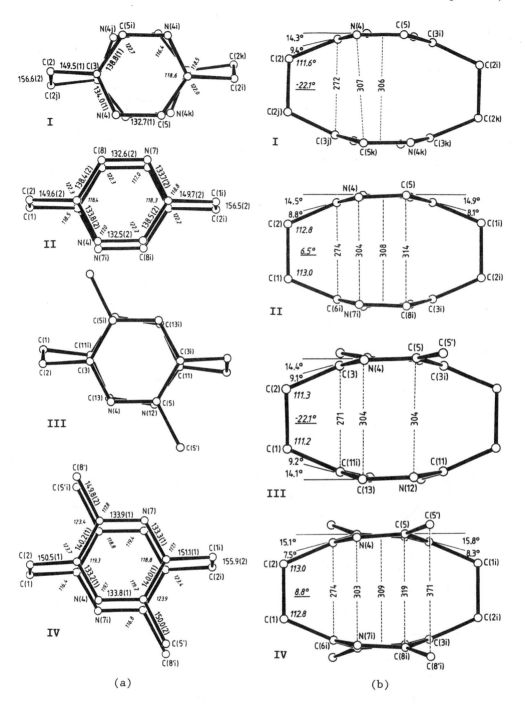

(a)   (b)

Fig. 5.11-7 Molecular structures of some [2.2](2,5)pyrazinophanes shown in top-views (a) and side-views (b). The bond lengths and transannular distances are in pm (100 pm = 1Å), and all angles are in degrees. (After ref. 15).

7. Isomeric [2.2](2,5)pyrazinophanes

The isomeric [2.2](2,5)pyrazinophanes **I** (pseudo-*ortho*) and **II** (pseudo-geminal), and their methyl-substituted derivatives **III** and **IV** have been synthesized and characterized by X-ray crystallography.[15]

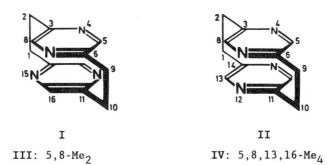

I   **II**

III: 5,8-Me₂   **IV**: 5,8,13,16-Me₄

Fig. 5.11-7(a) shows the structures each viewed along the normal to the least-squares plane through the four non-bridgehead atoms of a pyrazine moiety. In both pseudo-*ortho* N-heterocyclic[2.2]paracyclophanes **I** and **III**, one ring is rotated with respect to the other by a twist angle of *ca.* 10°. On the other hand, the pseudo-geminal phanes have centrosymmetric structures, and exact superposition of the two pyrazine rings is avoided by a slight parallel displacement of *ca.* 0.25Å.

Side-views of the molecular structures, as shown in Fig. 5.11-7(b), clearly show the twist boat-like deformation of the pyrazine rings, which take up most of the inherent strain. As compared to the aromatic rings of [2.2]paracyclophane (bending angles α = 12.6, β = 11.2°), the pyrazinophane rings are significantly more deformed (α angles lie in the range 14.3-15.8°) but the exocyclic bonds at the bridgehead carbons are bent to a lesser degree (β angles 9.4-7.9°). For **IV** the additional steric interaction of the "eclipsed" methyl groups forces the pyrazine rings to adopt a more distinct twist boat-like conformation.

8. Cyclophanes bridged by group IV elements

Tetrasila[2.2]paracyclophane (**V**) and hexasila[2.2.2](1,3,5)cyclophane (**VI**), novel cyclophanes bridged by two and three Si-Si bonds, respectively, exhibit interesting electronic properties resulting from strong through-bond interaction between the high-lying Si-Si σ orbitals and the benzene pπ system, which are parallel to one another.[16,17] Tetragerma[2.2]paracyclophane (**VII**), recently synthesized by the reductive coupling reaction of 1,4-bis(chlorodimethylgermyl)benzene with metallic sodium, is very similar to **V** in regard to its physical and chemical properties.[18]

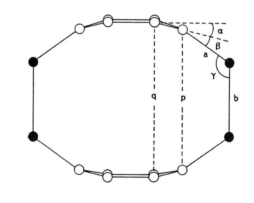

| V | VI | VII |

Molecule VII, as characterized by X-ray crystallography, is centrosymmetric with almost ideal *mmm* symmetry. The geometrical parameters of the isostructural series [2.2]paracyclophane, V and VII are compared in Table 5.11-3.[18]

Table 5.11-3 Bridging bond lengths, interplanar distances, and strain angles
            of paracyclophanes bridged by M-M (M = C, Si and Ge) bonds.

	C	Si	Ge
a(Å)	1.551	1.883	1.944
b(Å)	1.593	2.376	2.438
p(Å)	2.778	3.347	3.382
q(Å)	3.093	3.458	3.496
α(°)	12.6	4.3	4.5
β(°)	11.2	10.6	9.6
γ(°)	113.7	105.0	104.1

References

[1] H. Hope, J. Bernstein and K.N. Trueblood, *Acta Crystallogr.*, *Sect. B* **28**, 1733 (1972).

[2] C.J. Brown, *J. Chem. Soc.*, 3265 (1953).

[3] K. Lonsdale, H.J. Milledge and K.V.K. Rao, *Proc. Roy. Soc. London*, **A255**, 82 (1960).

[4] F. Vögtle (ed.), *Topics in Current Chemistry*, Vols. 113 and 115, Springer-Verlag, Berlin, 1983.

[5] P.M. Keehn in P.M. Keehn and S.M. Rosenfeld (eds.), *Cyclophanes*, Vol I, Academic Press, New York, 1983, chap. 3.

[6] L.N. Ferguson, *The Modern Structural Theory of Organic Chemistry*, Prentice-Hall, New York, 1963.

[7] E. Osawa and K. Kanematsu in J.F. Liebman and A. Greenberg (eds.), *Molecular Structure and Energeties*, Vol. 3, *Studies of Organic Molecules*, VCH Publishers, Weinheim, 1986, chap. 7.

[8] S. Misumi in P.M. Keehn and S.M. Rosenfeld (eds.), *Cyclophanes*, Vol. II, Academic Press, New York, 1983, chap. 10.

[9] H. Mizuno, K. Nishiguchi, T. Toyoda, T. Otsubo, S. Misumi and N. Morimoto, *Acta Crystallogr.*, *Sect. B* 33, 329 (1977).

[10] Y. Koizumi, T. Toyoda, H. Horita and S. Misumi, *32nd Nat. Meet. Chem. Soc. Jpn. Abstract I*, 1975, p. 184.

[11] Y. Kai, N. Yasuoka and N. Kasai, *Acta Crystallogr.*, *Sect. B* 33, 754 (1977).

[12] K.B. Wiberg, R.D. Adams, P.J. Okarma, M.G. Matturro and B. Segmuller, *J. Am. Chem. Soc.* 106, 2200 (1984).

[13] A.W. Hanson, *Acta Crystallogr.*, *Sect. B* 33, 2003 (1977).

[14] R.P. Johnson in J.F. Liebman and A. Greenberg (eds.), *Molecular Structure and Energetics*, Vol. 3, *Studies of Organic Molecules*, VCH Publishers, Weinheim, 1986, chap. 3.

[15] U. Eiermann, C. Krieger, F.A. Neugebauer and H.A. Staab, *Chem. Ber.* 123, 523 (1990).

[16] H. Sakurai, S. Hoshi, A. Kamiya, A. Hosomi and C. Kabuto, *Chem. Lett.*, 1781 (1986).

[17] A. Sekiguchi, T. Yatabe, C. Kabuto and H. Sakurai, *Angew. Chem. Int. Ed. Engl.* 28, 757 (1989).

[18] A. Sekiguchi, T. Yatabe, C. Kabuto and H. Sakurai, *J. Organomet. Chem.* 390, C27 (1990).

Note Meta- and paracyclophanes having short bridges ($n < 8$) and boat-shaped benzene rings are reviewed in F. Bickelhaupt, *Pure Appl. Chem.* 62, 373 (1990).

Cyclophanes form a variety of supramolecular complexes with neutral and charged organic species in the liquid state. The role of cyclophanes as synthetic receptors in molecular recognition is reviewed in F.N. Diederich, *Cyclophanes*, Royal Society of Chemistry, London, 1991.

5.12 1,1'-Binaphthyl

$C_{20}H_{14}$

Crystal Data

Tetragonal, space group $P4_12_12$ or $P4_32_12$ (No. 92 or 96)

$a = 7.181(2)$, $c = 27.681(10)$Å, $Z = 4$[1]

Atom	x	y	z	Atom*	x	y	z
C(1)	.4855	.3742	.25830	H(2)	.548	.181	.2060
C(2)	.4665	.2080	.23451	H(3)	.324	-.045	.2310
C(3)	.3354	.0743	.24926	H(4)	.132	.011	.2983
C(4)	.2249	.1067	.28800	H(5)	.038	.216	.3665
C(5)	.1308	.3115	.35551	H(6)	.072	.494	.4104
C(6)	.1488	.4725	.38076	H(7)	.285	.729	.3837
C(7)	.2748	.6093	.36530	H(8)	.472	.680	.3147
C(8)	.3832	.5810	.32535				
C(9)	.3706	.4118	.29876				
C(10)	.2398	.2757	.31407				

* The H atoms were given the numbers of the atoms to which they were attached.

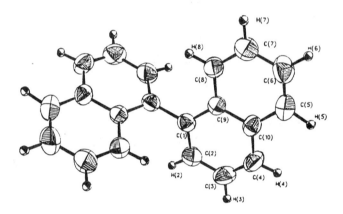

Fig. 5.12-1 Molecular structure of 1,1'-binaphthyl. (After ref. 3).

Crystal Structure

1,1'-Binaphthyl exists in its stable conformation as a chiral molecule.
In the crystalline state it exists in two forms: the higher melting (159°C)
and the lower melting (145°C). The latter form, the structure of which was
determined by Kerr and Robertson[4] following earlier work,[5] consists of
racemic crystals (space group $C2/c$). The structure of the chiral higher
melting form was determined independently by different research groups in 1980
to 1983.[1-3]

In the crystal, the asymmetric unit is one half of the molecule, the
other half being generated by rotation about a crystallographic two-fold axis
that bisects the C(1)-C(1') bond (Fig. 5.12-1). The molecule has a *trans*
arrangement about the central C(1)-C(1') bond with a dihedral angle of 103.1°;
both phenyl rings show slight distortions from strict planarity. The best

planes through the atoms of the two six-membered rings make an angle of 1.7°
with each other. The packing of the *trans* form is shown in Fig. 5.12-2.

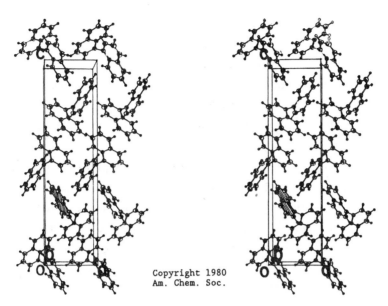

Fig. 5.12-2 Stereoscopic view of the packing in the crystal of the
trans form of binaphthyl. The aromatic rings of the reference half-
molecule are shaded. (After ref. 1).

Five binaphthyls have had their structures determined: two racemates
(binaphthyl and the 4,4'-dimethyl derivative[6]) and three optically-active
forms (binaphthyl, and the 4,4'-dimethyl[4] and 4,4'-diamino derivatives[3]).
The molecules in the racemic forms have dihedral angles between the two
naphthalene residues of about 68°; in the optically active forms this dihedral
angle is in the range 80-103°. The bond distances in the five structures
generally do not differ much from the lengths of the corresponding bonds in
naphthalene. The shortest bonds are those of type C(1)-C(2) (and 3-4, 5-6, 7-
8) with mean values in the range 1.360-1.369Å, the value in naphthalene being
1.357Å. Mean values for the other types of bond are all in the range 1.40-
1.44Å. The central C(1)-C(1') bonds are in the range 1.481-1.511Å, the
differences probably not being significant.

Remarks

1. Molecular complexation and crown ethers[7]

        Molecular complexation is a precondition receptor function such as
substrate selection, substrate transportation, isomeric differentiation, and
stereoselective catalysis. Although the investigation of such functions with
synthetically derived compounds is a relatively new development in chemistry,

they are well known and extensively studied functions in the biological domain. Evolution, gene expression, cell division, DNA replication, protein synthesis, immunological response, hormonal control, ion transportation, and enzymic catalysis are only some of the many examples where molecular complexation is a prerequisite for observing a biological process.

The recognition of a receptor function is the practical manifestation of successful receptor design. It follows that synthetic molecular receptors, in common with their natural counterparts, must be capable of recognizing other chemical species by achieving complementarity of steric and electronic shape as well as size.

The noncovalent bonds that hold molecular complexes together are largely electrostatic in origin. They include the following interactions: pole-pole, pole-dipole, dipole-dipole, dipole-induced dipole, and induced dipole-induced dipole, that is, dispersion forces of the van der Waals-London type. Particular structural features and conditions lead to the recognition of important special cases such as hydrophobic interactions, hydrogen bonding, and charge transfer stabilization. They are all potentially available to the synthetic chemist for exploitation. The binding forces that give rise to molecular complexes are no different from those present (i) in crystal lattices in the solid state and (ii) in solvating media in the solution state.

Chirality was recognized as a necessary and indispensible part of synthetic receptor molecular design and function, and a variety of chiral crown ethers have been synthesized and investigated.

The most extensively studied of the crown ethers is the parent macrocycle 1,4,7,10,13,16-hexaoxacyclooctadecane, or 18-crown-6. It forms molecular complexes with a wide range of substrate species including:

a) Alkali metal (e.g., K^+), alkaline earth metal (e.g., Ba^{2+}), as well as the harder transition metal and post-transition metal cations.

b) Nonmetal inorganic cations such as H_3O^+, NH_4^+, $H_2NNH_3^+$ and $HONH_3^+$.

c) Neutral inorganic complexes such as BF_3NH_3 and BH_3NH_3.

d) Transition metal complexes containing NH, OH, and CN acidic ligands (e.g., NH_3, H_2O and CH_3CN).

e) Organic cations such as PhN_2^+, $MeNH_3^+$, $PhCH_2NH_3^+$ and $Ph_3PCH_3^+$.

f) Neutral organic molecules containing polar N-H and C-H bonds.

The X-ray crystal structures of 18-crown-6 in the uncomplexed form, and its molecular complexes with KSCN, NH_4Cl, BF_3NH_3, *trans*-$Me_3PPtCl_2NH_3$, $PhCOCH_2NH_3PF_6$, and $CH_3SO_2CH_3$ are shown in Fig. 5.12-3.

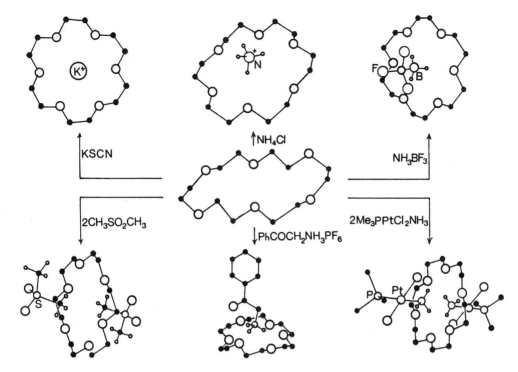

Fig. 5.12-3 Framework structures of 18-crown-6 and its complexes with a wide range of substrates. (After ref. 7).

2. Binaphthyl crown ethers

The first chiral crown ethers to exhibit chiral recognition towards enantiomeric substrates were described in 1973.[8] They were prepared from optically pure 2,2'-dihydroxyl-1,1'-binaphthyl, which can be obtained after

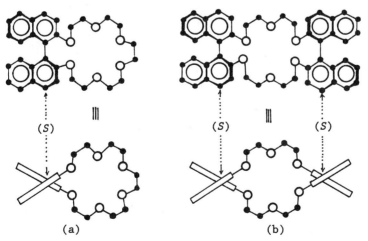

Fig. 5.12-4 Molecular structure of (a) binaphthyl 20-crown-6(*s*) and (b) bisbinaphthyl 22-crown-6(*ss*). (After ref. 7).

resolution of the binaphthol by a number of different methods. The isolation
of optically-pure enantiomers of this diol reflects its axis of chirality
resulting from hindered rotation about its naphthalene-naphthalene bond. Both
the binaphthyl-20-crown-6(s) and the bisbinaphthyl-22-crown-6(ss) [Fig. 5.12-
4] derivatives bear some structural resemblance to 18-crown-6.

Many binaphthol derivatives and unsaturated dihydroxy chiral compounds
(Fig. 5.12-5) have been incorporated into fused-ring all-oxygen crown
compounds.

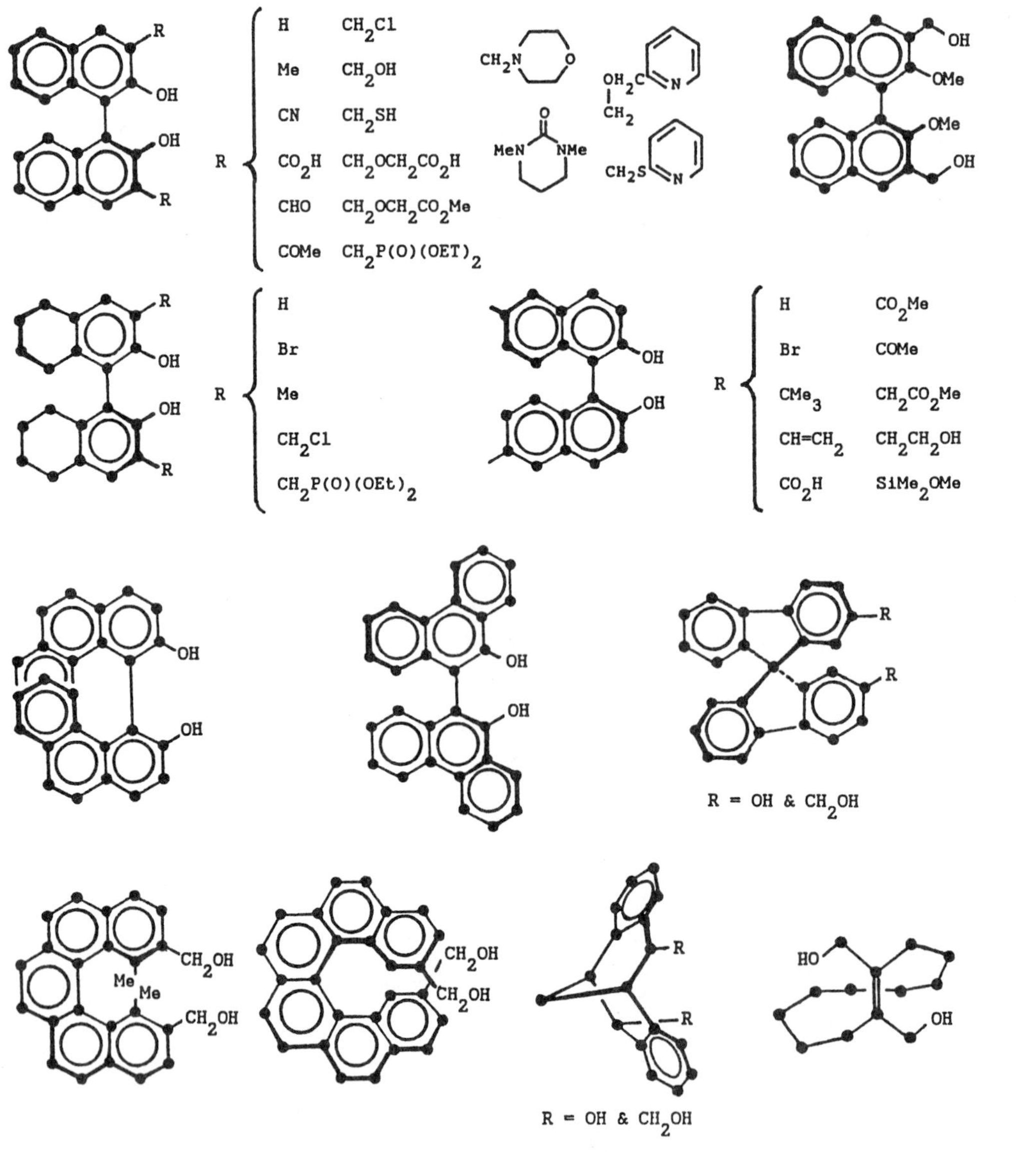

Fig. 5.12-5 Chiral binaphthol crown-ether compounds. (After ref. 7).

The monobenzhydryl derivative of (*S*)-binaphthol has played an important role, not only in the synthesis of chiral bisbinaphthyl crown ether derivatives, for example, compound (*SS*)-**A** (Fig. 5.12-6) containing two different bridges between the two binaphthyl units, but also in the provision of an entry into the constitutionally isomeric derivative (*SS*)-**B**, (Fig. 5.12-6). Rational stepwise synthetic schemes for macrocycles containing three binaphthyl units have been devised and applied to the synthesis of (*SSS*)-**C** and (*RSS*)-**D** (Fig. 5.12-6). Clearly, in all these procedures the C_2 symmetry of the chiral building block [(*S*)-binaphthol] restricts the number of products and defines the symmetries of the macrocycles formed.

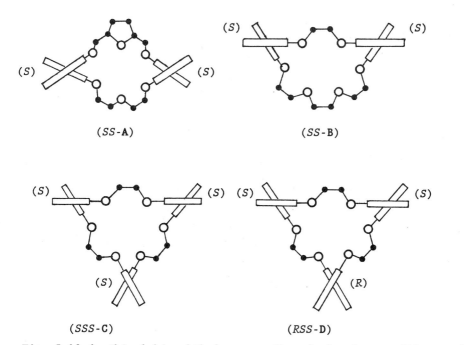

(*SS*-**A**)   (*SS*-**B**)

(*SSS*-**C**)   (*RSS*-**D**)

Fig. 5.12-6 Chiral binaphthyl crown ether derivatives. (After ref. 7).

The crystal structure of a complex of the 22-crown-6 derivative (*SS*)-**E** incorporating (*S*)-binaphthyl units, with the (*R*)-PhCH(CO$_2$Me)NH$_3^+$ cation has been determined.[9]

(*SS*)-**E**

3. Binaphthyl macropolycyclic compounds

The first optically active macrobicycles and macropolycycles were reported in 1974.[10] Employing the synthetic methodology developed for the preparation of cryptands, chiral molecular receptors, such as (*S*)-F and (*S*)-G, have been isolated and characterized.

(*S*)-F   (*S*)-G

The crystal structure of the free macropolycyclic receptor (*SS*)-H, consisting of two 1,1'-tetralyl units bridged by four polyether chains between their 2 and 3 positions, have been determined.[11]

(*SS*)-H

4. Homogeneous asymmetric catalysis using binaphthyl metal complexes

The efficient synthesis of optically active organic molecules from prochiral compounds by enantioselective catalysis has advanced rapidly in the last two decades. Chelating ligands possessing the binaphthyl skeleton have been employed in the design of new synthetic reagents for "multiplication of chirality".[12]

For example, the chiral reducing reagent BINAL-H exhibits exceptionally high enantioface-differentiating ability in the stoichiometric reduction of prochiral ketones bearing an aromatic, olefinic, or acetylenic substituent.[13] The general binaphthol/carbinol configurational relationship (*S/S* or *R/R*) is independent of the relative bulkiness of unsaturated and alkyl groups flanking the carbonyl function. This is rationalized by the proposal that in the (*S*)-BINAL-H reduction, the *S*-generating structure is favoured over the diastereomeric *R*-generating structure because the latter is destabilized by repulsive n/π interaction between a binaphthoxyl oxygen and the unsaturated moiety (*Scheme 1*).

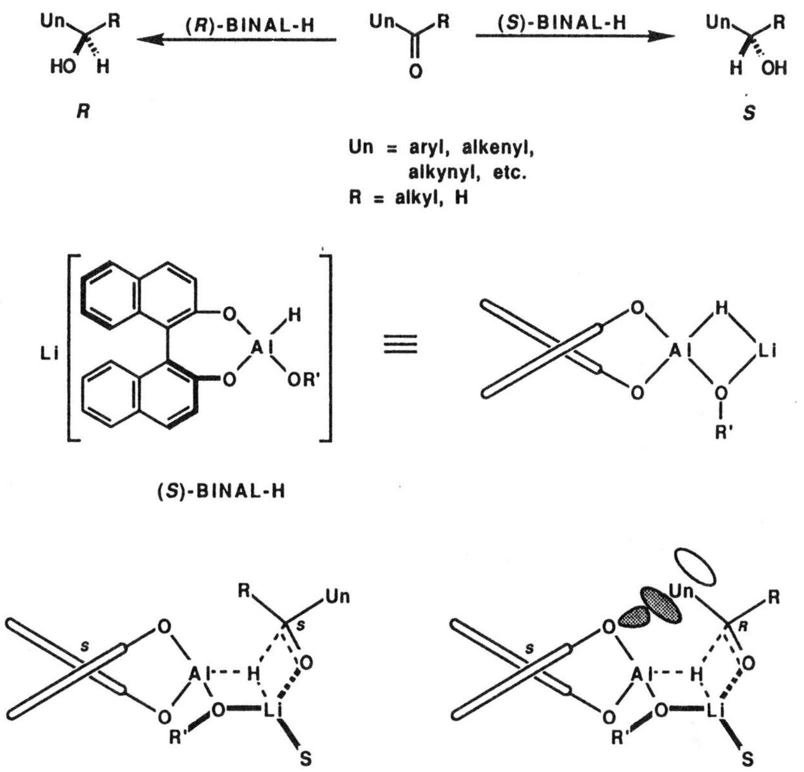

Un = aryl, alkenyl,
alkynyl, etc.
R = alkyl, H

(*S*)-BINAL-H

Scheme 1

A large-scale industrial process employing asymmetric catalysis is
illustrated by the production of (-)-menthol. The key step is the Rh-BINAP
catalyzed enantioselective isomerization of diethylgeranylamine to citronellal
diethylenamine in 96-99% optical yield (*Scheme 2*).[14] The atropisomeric
BINAP ligand is conveniently prepared by resolution of its dioxide with
camphorsulphonic acid or *O*-dibenzoyltartaric acid followed by reduction with
trichlorosilane.[15]

The newly designed Ru-BINAP dicarboxylate complexes[16] have proved to
be extremely versatile reagents for homogeneous catalytic hydrogenation in
stereoselective organic synthesis. A range of optically active compounds of
either chirality sense are obtainable by this clean, operationally simple and
economical method. For instance, the Ru-catalyzed hydrogenation of the
prochiral allylic alcohols geraniol and nerol gives (*S*)- or (*R*)-citronellol in
96-99% ee.[17] Either natural or unnatural forms can be made flexibly by
changing the chirality of the catalyst or geometry of the olefin substrates
(*Scheme 3*).

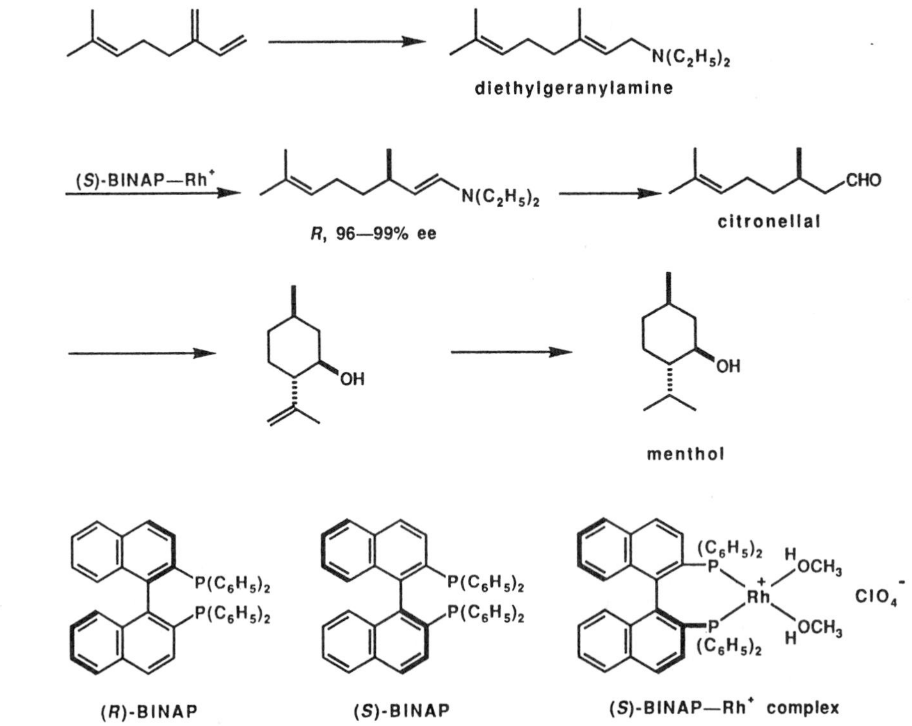
N(C2H5)2
diethylgeranylamine
(S)-BINAP—Rh+
N(C2H5)2
R, 96—99% ee
CHO
citronellal
OH
OH
menthol
P(C6H5)2
P(C6H5)2
(R)-BINAP
P(C6H5)2
P(C6H5)2
(S)-BINAP
(C6H5)2
P
Rh
H
OCH3
P
(C6H5)2
H
OCH3
ClO4-
(S)-BINAP—Rh+ complex

Scheme 2

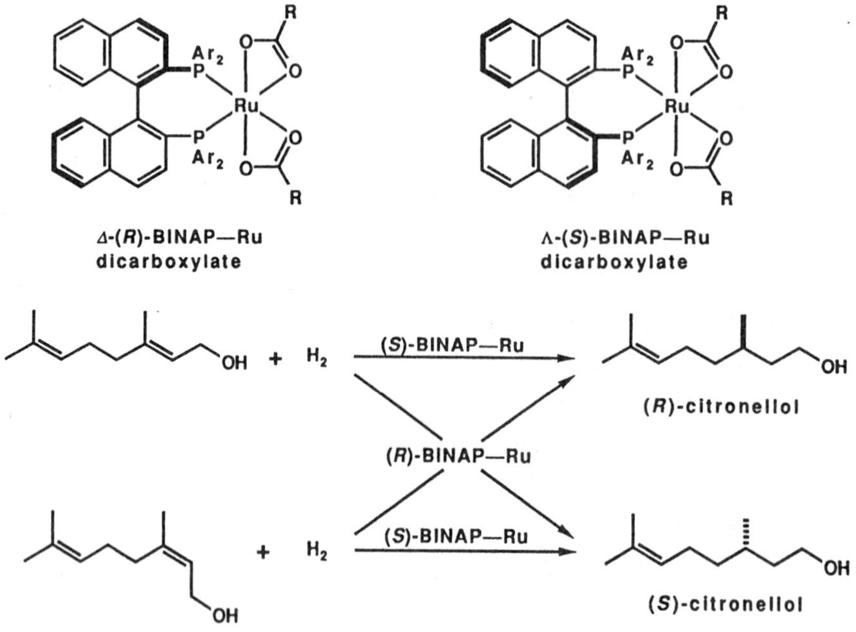
R
O
O
Ar2
P
Ru
P
Ar2
O
O
R
Δ-(R)-BINAP—Ru
dicarboxylate
Λ-(S)-BINAP—Ru
dicarboxylate
OH
+ H2
(S)-BINAP—Ru
OH
(R)-citronellol
(R)-BINAP—Ru
+ H2
(S)-BINAP—Ru
OH
OH
(S)-citronellol

Scheme 3

The molecular structure of the Ru-(S)-BINAP dipivalate complex has been determined by X-ray analysis.[16] As shown in Fig. 5.12-7, the dissymmetry of (S)-BINAP fixes the delta conformation of the seven-membered chelate ring containing the diphosphine and Ru, which in turn determines the chiral disposition of the four phenyl rings. The two equatorial phenyl rings exert profound steric influence on the other equatorial coordination sites so that the bidentate ligation of the pivalate moieties to Ru occurs stereo-selectively, leading to exclusive formation of the Λ diastereomer.

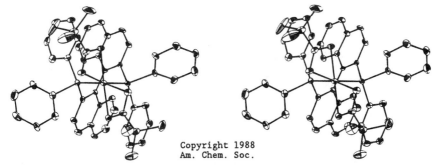

Copyright 1988
Am. Chem. Soc.

Fig. 5.12-7 Stereoview of the molecular structure of Λ-Ru[(S)-BINAP](Me₃CCO₂)₂. (After ref. 16).

References

[1] R.B. Kress, E.N. Duesler, M.C. Etter, I.C. Paul and D.Y. Curtin, *J. Am. Chem. Soc.* **102**, 7709 (1980).

[2] R. Kuroda and S.F. Mason, *J. Chem. Soc., Perkin Trans. II*, 167 (1981).

[3] R.A. Pauptit and J. Trotter, *Can. J. Chem.* **61**, 69 (1983).

[4] K.A. Kerr and J.M. Robertson, *J. Chem. Soc. (B)*, 1146 (1969).

[5] W.A.C. Brown, J. Trotter and J.M. Robertson, *Proc. Chem. Soc. (London)*, 115 (1961).

[6] R.A. Pauptit and J. Trotter, *Can. J. Chem.* **59**, 1149 (1981).

[7] J.F. Stoddart, *Top. Stereochem.* **17**, 207 (1987).

[8] E.P. Kyba, M.G. Siegel, L.R. Sousa, G.D.Y. Sogah and D.J. Cram, *J. Am. Chem. Soc.* **95**, 2691 (1973); E.P. Kyba, K. Koga, L.R. Sousa, M.G. Siegel and D.J. Cram, *J. Am. Chem. Soc.* **95**, 2692 (1973).

[9] I. Goldberg, *J. Am. Chem. Soc.* **99**, 6049 (1977).

[10] B. Dietrich, J.-M. Lehn, and J. Simon, *Angew. Chem. Int. Ed. Engl.* **13**, 406 (1974).

[11] I. Goldberg, *Acta Crystallogr., Sect. B* **36**, 2104 (1980).

[12] R. Noyori, *Chem. Soc. Rev.* **18**, 187 (1989); R. Noyori and M. Kitamura in R. Scheffold (ed.), *Modern Synthetic Methods 1989*, Springer-Verlag, Berlin, 1989, p.115.

[13] R. Noyori, I. Tomino, Y. Tanimoto and M. Nishizawa, *J. Am. Chem. Soc.*
      106, 6709 (1984); R. Noyori, I. Tomino, M. Yamada and M. Nishizawa,
      J. Am. Chem. Soc. **106**, 6717 (1984).

[14] K. Tani, T. Yamagata, S. Akutagawa, H. Kumobayashi, T. Taketomi,
      H. Takaya, A. Miyashita, R. Noyori and S. Otsuka, *J. Am. Chem. Soc.* **106**,
      5208 (1984).

[15] H. Takaya, S. Akutagawa and R. Noyori, *Org. Synth.* **67**, 20 (1988).

[16] T. Ohta, H. Takaya and R. Noyori, *Inorg. Chem.* **27**, 566 (1988).

[17] H. Takaya, T. Ohta, N. Sayo, H. Kumobayashi, S. Akutagawa, S. Inoue,
      I. Kasahara and R. Noyori, *J. Am. Chem. Soc.* **109**, 1596, 4129 (1987).

Note Inclusion compounds formed by "scissor-type" hosts such as 2,2'-
dihydroxy-1,1'-binaphthyl and 1,1'-binaphthyl-2,2'-dicarboxylic acid are
reviewed by E. Weber in J.L. Atwood, J.E.D. Davies and D.D. MacNicol
(eds.), *Inclusion Compounds*, Vol. **5**, Oxford University Press, New York,
1991, chap. 5.

"Lariat ethers", in which a sidearm bearing a donor group is attached to
a pivot atom in the macrocycle, are reviewed by G.W. Gokel in
J.L. Atwood, J.E.D. Davies and D.D. MacNicol (eds.), *Inclusion Compounds*,
Vol. **5**, Oxford University Press, New York, 1991, chap. 7; G.W. Gokel and
J.E. Trafton in Y. Inoue and G.W. Gokel (eds.), *Cation Binding by
Macrocycles*, Marcel Dekker, New York, 1990, p. 253. The crystallographic
data for cationic complexes of lariat ethers are covered by F.R. Fronzcek
and R.D. Candour, *ibid.*, p. 311.

Crown thioethers and their complexes with both transition and p-block
metal ions are reviewed in S.R. Cooper and S.C. Rawle, *Structure and
Bonding* **72**, 1 (1990); A.J. Blake and M. Schröder, *Adv. Inorg. Chem.*
35, 1 (1990).

The conformation of multidentate macrocyclic ligands in transition-metal
complexes is reviewed by J.C.A. Boeyens and S.M. Dobson in I. Bernal
(ed.), *Stereochemical and Stereophysical Behaviour of Macrocycles*,
Elsevier, Amsterdam, 1987.

5.13   Kekulene (Cyclo[d.e.d.e.d.e.d.e.d.e.d.e]dodecakisbenzene)
$$C_{48}H_{24}$$

Crystal Data

Monoclinic, space group $C2/c$ (No. 15)

$a = 27.951(4)$, $b = 4.579(1)$, $c = 22.680(2)$Å, $\beta = 109.64(1)°$, $Z = 4$

Atom	x	y	z	Atom	x	y	z
C(1)	.17207	1.0897	.27390	C(13)	.18988	-.4944	.63015
C(2)	.12601	.9874	.27051	C(14)	.23430	-.5920	.67536
C(3)	.11987	.7740	.31410	C(15)	.28224	-.4911	.67861
C(4)	.07298	.6605	.31217	C(16)	.21391	.7853	.36621
C(5)	.06793	.4529	.35486	C(17)	.16411	.6740	.36212
C(6)	.01930	.3388	.35304	C(18)	.15855	.4680	.40406
C(7)	.01570	.1407	.39552	C(19)	.11232	.3523	.40228
C(8)	.05953	.0314	.44481	C(20)	.10788	.1367	.44716
C(9)	.05637	-.1726	.48900	C(21)	.15044	.0287	.49298
C(10)	.09979	-.2765	.53532	C(22)	.14836	-.1763	.53720
C(11)	.09739	-.4850	.58167	C(23)	.19369	-.2899	.58519
C(12)	.13933	-.5874	.62609	C(24)	.24161	-.1934	.58821
H(1)	.1751	1.248	.2429	H(11)	.0630	-.552	.5002
H(2)	.0949	1.059	.2378	H(12)	.1383	-.726	.6592
H(4)	.0422	.738	.2803	H(14)	.2310	-.738	.7067
H(6)	-.0123	.411	.3203	H(18)	.1889	.397	.4371
H(7)	-.0179	.070	.3943	H(21)	.1852	.094	.4939
H(9)	.0215	-.249	.4875	H(24)	.2449	-.050	.5565

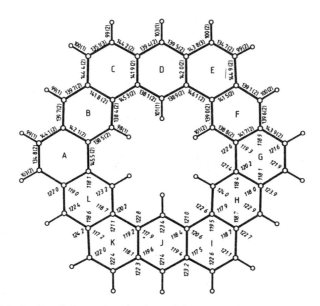

Fig. 5.13-1 Bond lengths (pm) and bond angles (°) in kekulene. (After ref. 1).

Crystal Structure

Kekulene, the synthesis of which was accomplished by Diederich and Staab in 1978,[2] is the first example of a class of aromatic hydrocarbons known as "cycloarenes", in which the circular annellation of arene units results in a macrocyclic structure enclosing a cavity lined with hydrogen atoms.

X-Ray analysis has shown that the kekulene molecule has almost perfectly planar D_{6h} symmetry.[1] The mean deviation of the carbon atoms from the least-squares plane through all 48 carbon atoms amounts to only 0.03Å; the maximum deviation from this plane is 0.07Å. The close conformity to planarity includes even the six internal hydrogens (maximum deviation 0.09Å) although the non-bonding distances of 1.92(2)Å between adjacent hydrogens in the central cavity are unusually small.

With regard to the electronic structure of kekulene the carbon-carbon bond lengths are of special interest (Fig. 5.13-1). For the group of equivalent benzene units (B,D,F,H,J,L) "normal" arene bond lengths are observed; for these rings the mean value of the carbon-carbon bonds in the outer perimeter amounts to 1.395(2)Å, that of the bonds in the inner perimeter to 1.386(3)Å, whereas the radial bonds are slightly stretched to a mean value of 1.418(2)Å. Drastic differences in bond lengths are found for the second group of (A,C,E,G,I,K) rings in kekulene. The six peripheral carbon-carbon bonds of these rings, with a mean bond length of 1.349(2)Å, have almost the length of normal carbon-carbon double bonds. For the bonds linking these "double bonds" to the "aromatic" rings (B,D,...) a mean length of 1.442(4)Å is found, indicating a relatively high single-bond character. An even stronger extension to a mean bond length of 1.456(4)Å is observed for the bonds which connect the "aromatic" rings in the inner perimeter; these bonds approach the type of single bonds found in polyphenyl systems.

On the basis of the X-ray analysis, the valence-bond structural formula shown in Fig. 5.13-2(a) is undoubtedly the best representation of the actual bonding state of kekulene.

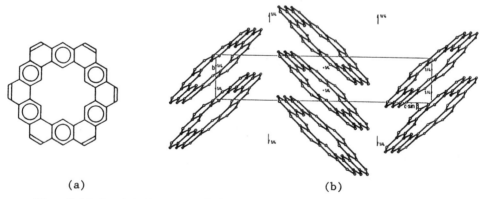

| (a) | (b) |

Fig. 5.13-2 (a) Structural formula of kekulene. (b) Molecular packing of kekulene in projection along *a* (for the sake of clarity only stacks centered at (1/4)*a* are drawn; corresponding stacks at (3/4)*a* which result from *C*-centering are omitted). (After ref. 1).

In the crystal, the kekulene molecules are stacked along the b axis with the stacking axis making an angle of 42.9° with the molecular planes. Within such a stack of equidistant molecules the interplanar distance is 3.35Å; neighbouring molecules are parallel-shifted by 3.12Å resulting in appreciable π overlap. The molecular planes in adjacent stacks are inclined to each other by 86°, leading to a typical "herringbone pattern" of molecular packing [Fig. 5.13-2(b)].

Remarks

1. Structure of circulenes

A [m]circulene is formed from m-condensed aromatic rings arranged in a closed macrocycle.[3] [5]Circulene (corannulene) has a bowl-shaped molecular structure,[4] as shown in Fig. 2.5-5. [6]Circulene, given the classical name "coronene", has a beautifully simple crystal structure in which the molecule lies at a center of symmetry.[5] Mean lengths for the four types of bond present are: central ring, 1.422(4)Å; radial, 1.430(3)Å; second outermost, 1.415(3)Å; and outermost, 1.346(4)Å. (These values are corrected for thermal libration; corrections are 0.002 to 0.003 Å). The molecule is slightly but significantly nonplanar, pairs of outer atoms lying alternately above and below the mean molecular plane. The maximum deviation is 0.029 Å.

Perchlorocoronene is expected to be nonplanar owing to severe steric interaction between the peri-related chlorine atoms. Of the many chiral and achiral conformations possible, X-ray analysis has established that perchlorocoronene adopts the **aabbaabbaabb** conformation with pairs of chlorine atoms situated alternately above and below the mean molecular plane.[6] The displacements are in the range ±0.94Å for the chlorine atoms and ±0.32Å for

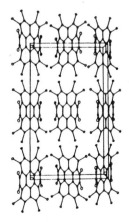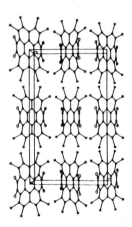

Fig. 5.13-3 A stereoview, along the b axis, illustrating the molecular packing of perchlorocoronene. (After ref. 6).

their parent carbon atoms. The shape of the molecule, which displays two
concave faces, can be seen from the stereo packing diagram shown in Fig.
5.13-3. The molecule is located on a site of 2/m symmetry and its geometry
quite closely approximates to D_{3d}. The mean bond lengths of the central ring,
radial, second-outermost, and outermost carbon-carbon bonds are 1.425(3),
1.420(3), 1.425(2), and 1.360(4)Å, respectively.

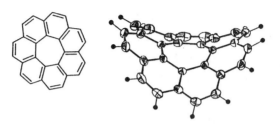

Fig. 5.13-4 Molecular structure of [7]circulene. (After ref. 7).

In [7]circulene (Fig. 5.13-4), the central seven-membered ring assumes
a boat conformation, and the whole molecular shape closely approximates a
saddle form.[7] The C-C bonds can be classified into four groups: central
core bonds with bond distances of 1.447-1.466Å, average 1.457Å; spoke bonds
with bond distances of 1.433-1.435Å, average 1.434Å; the next outer bonds with
bond distances of 1.407-1.423Å, average 1.414Å; and the outermost bonds with
bond distances of 1.327-1.344Å, average 1.338Å.

2. Structure of helicenes[8]

Helicenes are benzologues of phenanthrene in which a regular cylindrical
helix is formed through an all-*ortho* annellation of the aromatic rings. The
helical structure is a consequence of the repulsive steric overlap of the
terminal aromatic nuclei. In 1956, Newman and Lednicer[9] successfully
synthesized the first helicene, namely hexahelicene or [6]helicene.

The helicene molecules usually have a C_2 axis, which is not necessarily
coincident with the crystallographic axis. The structure of the helicenes can
be described generally in a simplified way by distributing the atoms of the
molecular skeleton on three spirals: one inner helix having (n+1) C atoms, one
outer helix composed of 2n C atoms and one middle helix having (n+1) C atoms,
where n is equal to the number of benzene rings.[10]

Generally, the pitch of the inner helix is smaller than that of the
middle helix, which is in turn smaller than that of the outer helix. The
former is relatively independent of the value of n, whereas the latter two
decrease with increasing n. Replacing hydrogen atoms of the terminal benzene
rings by larger substituents leads to increased steric repulsion. With

increasing n, the two outer pitches only change a little. The radius of the inner helix also remains approximately constant, whereas those of the middle and outer helices increase with increasing n. In relation to benzene, the bond lengths of helicenes are shorter in the inner part of the helix, whereas in the periphery they are more elongated. The strain is not equally distributed inside the helicene molecule: the torsion angle between the inner bonds in the molecule are large, *ca.* $25\pm3°$. The inner rings are markedly nonplanar, whereas the terminal rings are nearly completely planar with bond lengths and angles similar to those of the phenanthrenes.

In general all helicenes can be constructed from trigonal planar C_4 units each consisting of one inner, one median, and two outer carbon atoms. This characteristic has been used in an elaborate "refined staircase model" which gives an accurate description of the actual helicene geometry.[10]

[6]Helicene, the historically first and structural prototype of the helicenes, crystallizes in space group $P2_12_12_1$.[11] The interplanar angle between the terminal benzene rings is remarkably large at 58.5°. This is due to the repulsion of the facing terminal benzene rings at both ends of the helix [Fig. 5.13-5(a)].

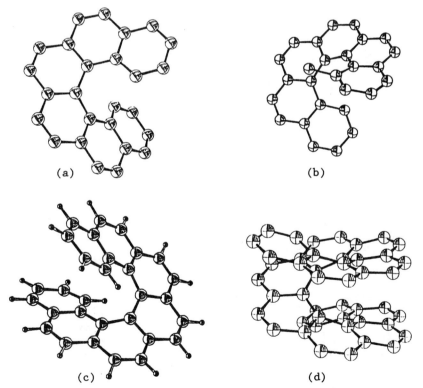

(a)   (b)

(c)   (d)

Fig. 5.13-5 Molecular structure of some helicenes: (a) [6]helicene, (b) 1-methyl[6]helicene, (c) [7]helicene, (d) [11]helicene. (After ref. 8).

Introducing substituents into [6]helicene sometimes leads to changes in the molecular skeleton and the space group. The changes are easily recognized from a comparison of the interplanar angles of the terminal bezene rings: [6]helicene: 58.5°, 2-methyl[6]helicene: 54.8°, 1-methyl[6]helicene: 42.4° [Fig. 5.13-5(b)], 1-formyl[6]helicene: 44.5°, 1-acetyl[6]helicene: 42.5°, 1,16-dimethyl[6]helicene: 29.6°.

[7]Helicene exists in two crystalline modifications: one in space group $P2_1$ with an interplanar angle of 30.7°,[12] and the other in space group $P2_1/c$ with an angle of 32.3°[13] [Fig. 5.13-5(c)].

[11]Helicene crystallizes in space group $P2_1$.[14] The interplanar angle between the facing benzene rings A and H is 4°, as shown in Fig. 5.13-5(d).

The helicenes are unique with respect to their very high optical rotation values, which are a consequence of the inherent chiral chromophore enclosing the whole molecular skeleton. On the other hand, they undergo thermal racemization by a conformational process with ralatively low potential barriers, since the necessary molecular deformations are distributed over the entire helicene skeleton.[15]

3. Structure of ovalene

The crystal structure of ovalene, or octabenzonaphthalene, $C_{32}H_{14}$, was solved in 1954[16] and refined in 1973.[17] The structure of the planar molecule, which occupies a centrosymmetric site in the crystal, is shown in Fig. 5.13-6. The C-C bond lengths, which have been corrected for the effects of libration, are in good agreement with the theoretical values computed by the PPP-SCFMO method.[18] The results are displayed in Table 5.13-1.

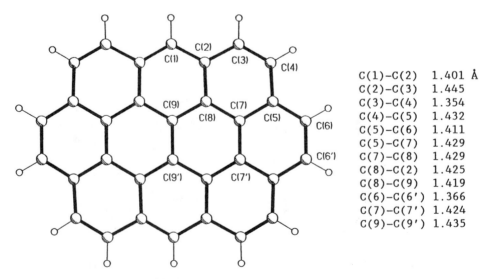

C(1)–C(2)	1.401 Å
C(2)–C(3)	1.445
C(3)–C(4)	1.354
C(4)–C(5)	1.432
C(5)–C(6)	1.411
C(5)–C(7)	1.429
C(7)–C(8)	1.429
C(8)–C(2)	1.425
C(8)–C(9)	1.419
C(6)–C(6')	1.366
C(7)–C(7')	1.424
C(9)–C(9')	1.435

Fig. 5.13-6 Molecular structure of ovalene.

Table 5.13-1 Bond lengths of ovalene.

Bond	Observed[17]	Calculated	Bond	Observed[17]	Calculated
C(2)-C(3)	1.445	1.442	C(2)-C(8)	1.425	1.424
C(3)-C(4)	1.354	1.367	C(8)-C(9)	1.419	1.419
C(4)-C(5)	1.432	1.440	C(9)-C(9')	1.435	1.441
C(5)-C(6)	1.411	1.423	C(8)-C(7)	1.429	1.434
C(6)-C(6')	1.366	1.383	C(7)-C(7')	1.424	1.429
C(1)-C(2)	1.401	1.408	C(7)-C(5)	1.429	1.414

4. Crystal structures of polynuclear aromatic hydrocarbons[19]

Polynuclear aromatic hydrocarbons crystallize in four basic structural types that can be clearly differentiated by energetic and geometrical criteria. The simplest pattern that can be identified is the herringbone motif. In this structural type, C...C nonbonded interactions are between non-parallel nearest neighbour molecules. Naphthalene belongs to this type. In the second type, called "sandwich herringbone" or sandwich, the herringbone motif is made up of sandwich-type diads. Pyrene belongs to this type. In the third type, the main C...C interactions are between parallel translated molecules; a sort of flattened-out herringbone, called γ, can be defined. Coronene belongs to this type. All these structures are also stabilized by C...H interactions. A fourth type is a layered structure made up of "graphitic" planes. This is labelled β and is characterized by strong C...C interactions without much contribution from C...H contacts. Tribenzopyrene belongs to this type. Fig. 5.13-7 shows the four exemplary crystal structures.

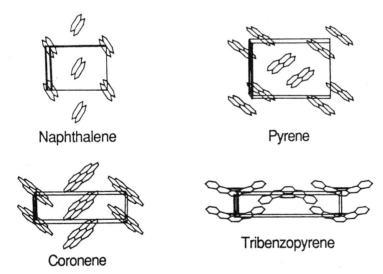

Naphthalene   Pyrene

Coronene   Tribenzopyrene

Fig. 5.13-7 The four basic types of molecular packing for planar aromatic hydrocarbons: naphthalene (herringbone), pyrene (sandwich herringbone), coronene (γ) and tribenzopyrene (β); in each case the short axis is normal to the plane of the paper. (After ref. 19).

The key parameters in distinguishing the four structural types are the shortest unit-cell axis and the interplanar angle, defined as the angle between the mean plane of one molecule and that of its nearest neighbours. A study of 32 representative hydrocarbons shows that the adoption of one or the other structural type depends on the relative importance of C...C and C...H interactions and therefore on the number and positioning of C and H atoms in the molecule.[19]

References

[1]   H.A. Staab, F. Diederich, C. Krieger and D. Schweitzer, *Chem. Ber.* **116**, 3504 (1983).

[2]   F. Diederich and H.A. Staab, *Angew. Chem. Int. Ed. Engl.* **17**, 372 (1978).

[3]   E. Clar, *Polycyclic Hydrocarbons*, Academic Press, New York, 1964.

[4]   W.E. Barth and R.G. Lawton, *J. Am. Chem. Soc.* **93**, 1730 (1971).

[5]   J.M. Robertson and J.G. White, *J. Chem. Soc.*, 607 (1945); J.K. Fawcett and J. Trotter, *Proc. Roy. Soc. London* **A289**, 366 (1965).

[6]   T. Baird, J.H. Gall, D.D. MacNicol, P.R. Mallinson and C.R. Michie, *J. Chem. Soc., Chem. Commun.*, 1471 (1988).

[7]   K. Yamamoto, T. Harada and M. Nakazaki, *J. Am. Chem. Soc.* **105**, 7171 (1983).

[8]   K.P. Meurer and F. Vögtle, *Top. Curr. Chem.* **127**, 1 (1985).

[9]   M.S. Newman and D. Lednicer, *J. Am. Chem. Soc.* **78**, 4765 (1956).

[10] J. Navaza, G. Tsoucaris, G. Le Bas, A. Navaza and C. de Rango, *Bull. Soc. Chim. Belg.* **88**, 863 (1979).

[11] C. de Rango, G. Tsoucaris, J.P. Declercq, G. Germain and J.P. Putzeys, *Cryst. Struct. Commun.* **2**, 189 (1973).

[12] P.T. Beurskens, G. Beurskens and T.E.M. van den Hark, *Cryst. Struct. Commun.* **5**, 241 (1976).

[13] T.E.M. van den Hark and P.T. Beurskens, *Cryst. Struct. Commun.* **5**, 247 (1976).

[14] G. Le Bas, A. Navaza, M. Knossow and C. de Rango, *Cryst. Struct. Commun.* **5**, 713 (1976).

[15] R.H. Martin, *Angew. Chem. Int. Ed. Engl.* **13**, 649 (1974).

[16] D.M. Donaldson and J.M. Robertson, *Proc. Roy. Soc. London* **A220** 157 (1954).

[17] R.G. Hazell and G.S. Pawley, *Z. Kristallogr.* **137**, 159 (1973).

[18] D.H. Lo and M.A. Whitehead, *Can. J. Chem.* **46**, 2027 (1968).

[19] G.R. Desiraju and A. Gavezzotti, *Acta Crystallogr., Sect. B* **45**, 473 (1989).

5.14   *s*-Triazine

$C_3H_3N_3$

Crystal Data

Rhombohedral, space group $R\bar{3}c$ (No. 167)

$a = 9.647(3)$, $c = 7.281(3)$Å, $Z = 6$

Atom	MoKα			Neutron		
	x	y	z	x	y	z
C	.1317	0	1/4	.1328	0	1/4
N	-.1408	0	1/4	-.1400	0	1/4
H	.2334	0	1/4	.2411	0	1/4

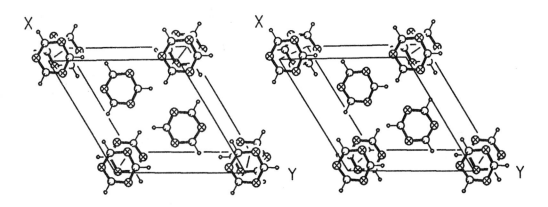

Fig. 5.14-1 Stereodiagram of the crystal structure of *s*-triazine viewed approximately along the *c* axis.

Crystal Structure

The crystal structure of *s*-triazine was first solved by Wheatley.[1] The atomic charge distribution in the heterocyclic molecule was later determined from X-ray and neutron diffraction data by Coppens.[2]

In the crystal, the molecule is planar but deviates quite markedly from a regular hexagon. From the more accurate neutron study, the C-N and C-H bond lengths are 1.315(2)Å and 0.93(4)Å, respectively, and the N-C-N angle is 125.2(10)°. Fig. 5.14-1 shows the packing of the molecules in the unit cell. The molecules are arranged in columns parallel to *c* with a regular spacing of *c*/2. Besides the pair of neighbours above and below it, each molecule is surrounded by six others at a center-to-center separation of 5.7Å.

Crystalline *s*-triazine undergoes a phase transition from a rhombohedral structure (space group $R\bar{3}c$) at room temperature to a monoclinic structure (space group $C2/c$) below 198K. The unit cell parameters of the low-temperature modification are:

$5K^{[3]}$: $a = 6.719(1)$, $b = 9.528(1)$, $c = 7.030(1)$Å, $\beta = 125.28(1)°$;

$150K^{[4]}$: $a = 6.884(8)$, $b = 9.569(2)$, $c = 7.093(3)$Å, $\beta = 126.61(3)°$.

s-Triazine was selected for a study of the deviations from spherical symmetry of the atomic charge distributions because (i) it is a simple molecule containing only first-row atoms, so that bonding effects are relatively important; (ii) it contains electron lone pairs on the nitrogen, aromatic C-N bonds, and C-H bonds, all of which are features of interest; and (iii) the number of independent structural parameters is small since the molecule occupies a site of symmetry 3/*m* in space group $R\bar{3}c$.

The difference-density maps (Fig. 5.14-2) show that density has migrated from the atoms into the bonding region and the lone-pair region of the nitrogen atom, and only a small maximum is observed in the region of the C-H bond.

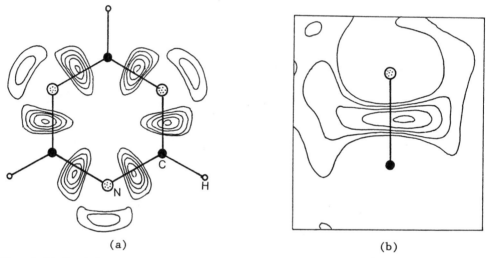

(a)   (b)

Fig. 5.14-2   (a) Section at $z = 1/4$ through the molecular plane of the function $\Delta\rho_{X-N}$; contours at interval of 0.05 eÅ^{-3}.   (b) Section through the C-N bond perpendicular to the molecular plane; contours as in (a).   Note that the extension of the bond peak in section (b) is compatible with π-bonding in the aromatic ring.   (Adapted from ref. 2).

Remarks

1. Deformation density features in organic molecules[5]

"Deformation density" is defined as the difference between the observed density from the X-ray experiment and the superposition of spherical atoms each located at the position determined by an independent neutron diffraction experiment.   Because of the combined use of X-ray and neutron data, this method is referred to as the X-N method.   Neutron data are used because

X-ray atomic positions obtained by least-squares refinement are based on the assumption that atoms are spherical. X-Ray positions are therefore in error by small amounts, usually less than 0.01Å, except in the case of hydrogen atoms for which the bias is much larger (~ 0.1-0.2Å), because the single electron of the hydrogen atom is also its valence electron and therefore perturbed by chemical bonding. The use of neutron diffraction data eliminates

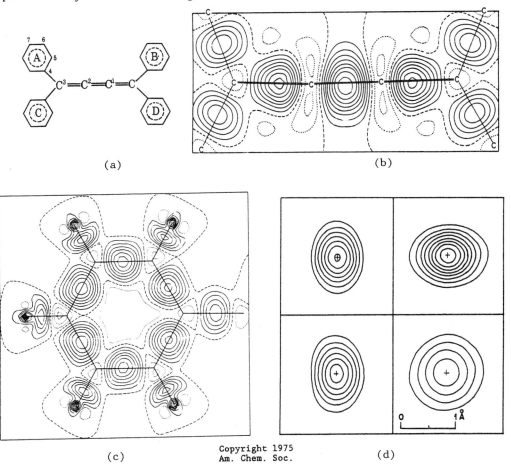

(a)   (b)

(c)   (d)

Fig. 5.14-3 (a) Structural formula and numbering of tetraphenylbutatriene. (b) Deformation density in the butatriene plane. Contour interval 0.1 eÅ^{-3}, zero contour broken, negative contours dotted. (c) Deformation density section through the plane of the phenyl ring. (d) Deformation density sections perpendicular to the various C-C bonds through their centers: (top left) the outer $C^2=C^3$ bond, (top right) the inner $C^1=C^2$ bond, (bottom left) an aromatic bond, (bottom right) the exocyclic C^3-C^4 bond. The vertical direction is perpendicular to the butatriene plane for top two, and to the phenyl plane for bottom two. The center of the bond is marked +. (After ref. 6).

this complication since neutrons are scattered by the nuclei of the atoms that are not deformed at all on the scale of the wavelengths 0.7-1.5Å. In the X-X method, which has been applied extensively in more recent studies, the neutron data are replaced by high-order X-ray data (i.e., at large Bragg angles). These high-order X-ray data have much smaller contributions from the valence electrons and are therefore less affected by chemical bonding.

Deformation-density studies provide a vivid picture of the nature of chemical bonding in a diverse range of chemical compounds. For simple organic molecules, single, double and triple bonds can be readily distinguished. When sections through the midpoint of the bonds perpendicular to the bonds are plotted, single and triple bonds show up as being circular within the uncertainty of the experiment, which is typically about $0.1e\text{Å}^{-3}$, while double bonds have elliptical shapes. That the bonds are extended in the direction of the p_π orbital axis is very clearly demonstrated in cumulenes. Adjacent double bonds in cumulenes, such as tetraphenylbutatriene, must use p_x or p_y orbitals on the common atom, where the z axis is taken as the bond direction. The deformation electron density in adjacent bonds is therefore elongated in mutually perpendicular directions. Experimental results confirming this theoretical prediction are shown in Fig. 5.14-3.[6]

2. Single-ring nitrogen heterocycles

Nitrogen is the only atom from the first row of the Periodic Table which can replace a CH group of benzene and give an uncharged aromatic heterocycle. There are altogether eight heterocyclic systems known which have one or more sp^2-hybridized nitrogen atoms in the six-membered ring:[7]

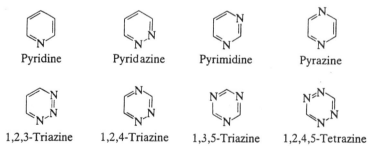

Pyridine   Pyridazine   Pyrimidine   Pyrazine

1,2,3-Triazine   1,2,4-Triazine   1,3,5-Triazine   1,2,4,5-Tetrazine

Many heterocyclic nitrogen compounds occur naturally and their functions are often of fundamental importance to living systems in biological processes. A large proportion of modern pharmaceuticals is derived from nitrogen heterocycles. The structures of many single-ring nitrogen heterocycles and their derivatives have been determined. The bond lengths in some heteroaromatic nitrogen compounds are shown in Fig. 5.14-4.

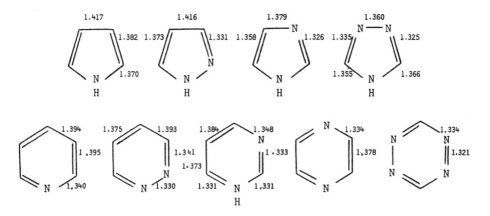

Fig. 5.14-4 Bond lengths in some single-ring nitrogen heterocycles.
(Adapted from refs. 7-10).

3. Structure of imidazole

Imidazole is a constituent of histamine and of the histidine residues of
proteins. The crystal structure of imidazole has been determined by X-ray and
neutron diffraction at room temperature and -150°C.[11] In the crystal, the
ring atoms are coplanar with the H atoms all slightly displaced from this
plane on the same side. The molecular shape, bond lengths, and bond angles
are shown in Fig. 5.14-5(a) and (b). The crystal structure is shown in Fig.
5.14-5(c). It is noteworthy that the C(4)-H(4) and C(5)-H(5) bonds are
subject to considerable in-plane bending away from the direction of the
bisector of the ring angles at C(4) and C(5). The H-C-C angles (130°) are

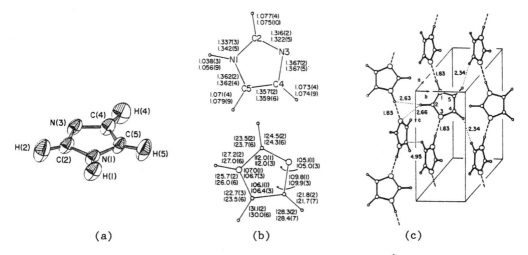

(a)   (b)   (c)

Fig. 5.14-5 (a) Molecular shape, (b) bond lengths (Å) and angles (°)
and (c) crystal structure of imidazole. (After ref. 11).

greater than H-C-N angles (123°). A similar effect also occurs in the
imidazolium ion (in imidazolium phosphate).

The presence of the NH bond at atom N(1) but not at N(3) gives rise to a
perturbation from point symmetry 2mm in the rest of the imidazole molecule.
The internal ring angles at N(1) and N(3) differ by about 2°, the internal
ring angles at C(4) and C(5) by almost 4°, and the N(1)-C(2) and N(3)-C(2)
bond lengths by 0.02Å. This perturbation is absent in the imidazolium ion in
which both N atoms from NH bonds. The most significant differences between
imidazole and the imidazolium ion occur in the C(2)-N bond lengths [1.337(2),
1.316(2)Å in imidazole, and 1.324(4)Å in the imidazolium ion] and in the
N(1)-C(2)-N(3) angle [112.0(1)° in imidazole and 108.6(1)° in the imidazolium
ion]. These comparisons suggest that proton transfer from N(1) to N(3) or
proton addition at N(3) does not involve major changes in electronic
structure, particularly when N(1) and N(3) are already engaged in strong
hydrogen bonds such as those in the crystal structures of imidazole and
imidazolium salts. In this respect, the imidazole moiety of histidine would
appear to be well suited for mediating proton-transfer reactions in biological
systems.

4. Structure of 2,4,6-triamino-1,3,5-triazine (melamine)

The crystal structure of melamine was first determined, from three axial
projections, by Hughes[12] who introduced a new least-squares method for
refining atomic parameters in conjunction with Fourier synthesis to locate the
atoms. A later reinvestigation, using both X-ray and neutron diffraction
data, was undertaken to obtain more accurate structural information.[13]

Melamine crystallizes in space group $P2_1/a$ with a = 10.601(1), b =
7.495(1), c = 7.295(1)Å, β = 112.26(2)°, and Z = 4. The molecule is
significantly nonplanar. The ring has a small amount of boat character and
the amine groups deviate from the least-squares plane through the six ring
atoms by up to 0.10Å. The three C-N bonds to each of the ring C atoms are
approximately coplanar with the ring, but in each case the H atoms deviate
significantly from the ring plane. The best planes through the amine groups
are inclined at angles between 0.7° and 3.0° to the ring plane.

Bond lengths and angles derived from X-ray and neutron diffraction are
shown in Fig. 5.14-6. The results reveal that the amine groups at N(4) and
N(5) are close to trigonal, but the group at N(6) has a large amount of
pyramidal character; the mean angle subtended at N(6) is 114°, and the H atoms
deviate from the plane through N(6), C(3), N(7) and N(8) by about 0.33Å.

The bond lengths of the C-NH$_2$ bonds appear to depend on the

configuration of the amine groups. The two C-N bonds associated with N(4) and
N(5) are similar, while the C-N distance involving the pyrimidal group N(6) is
0.02Å longer. Within the ring the two C-N bonds adjacent to C(3) are shorter
than the other four ring bonds in the X-ray structure. This is consistent
with larger contributions from quinonoid valence structures having a double
bond on C(3), which is directly bonded to the amine group N(6).

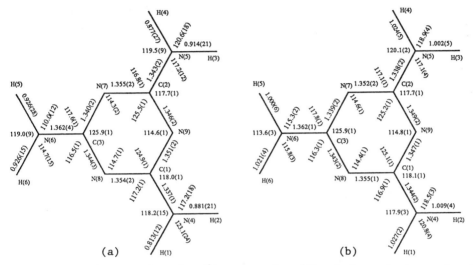

Fig. 5.14-6 Bond lengths (Å) and angles (°) of the melamine molecule:
(a) X-ray and (b) neutron. (After ref. 13).

5. Structure of s-triazine dihydrofolate reductase inhibitors[14]

The enzyme dihydrofolate reductase (DHFR), a necessary component for
cell growth, is responsible for maintaining intracellular folate pools in
their biochemically-active reduced state. This enzyme is also strongly and
specifically inhibited by certain substrate analogues with binding affinities
so great that they are not readily displaced by the natural folic acid
substrates. Furthermore, many of these antifolates have an inhibitory action
that is specific for different species of DHFR.

Structure-activity studies have shown that a 2,4-diaminopyridine, s-
triazine, quinazoline, or pteridine ring structure is sufficient for tight
binding to DHFR. In order to provide potent inhibitors with improved
capacities for cellular uptake, a series of 1,2-dihydro-2,2-dimethyl-4,6-
diamino-s-triazines (Fig. 5.14-7) have been investigated, leading to the
conclusion that the hydrophobic character of the N(1)-phenyl substituent is
the dominant factor in their relative potency as mammalian DHFR
inhibitors.[15] The available structural data[14,16-18] show that: (i) the s-
triazine ring has greater flexibility than the ring systems of the other

antifolate inhibitors; the *s*-triazine ring adopts a twist-sofa conformation
with C(2) nearly 0.5Å above the plane made by N(3), C(4), N(5), C(6) and N(1),
and the 2,2-dimethyl groups are equatorial-axial; (ii) the conformation and
electronic properties of the *s*-triazine ring are sensitive to the nature of
the N(1)-substituents; (iii) the orientation of the N(1)-phenyl ring is nearly
perpendicular to the *s*-triazine ring, in agreement with minimum energy
calculations and with antifolate binding in the active site of chicken liver
dihydrofolate reductase; and (iv) the major structural difference between *s*-
triazines and 2,4-diaminopyrimidines lies in the C(2) and C(21) positions, and
the equatorial C(22) lies in a position analogous to the 6-methyl position of
2,4-diaminopyrimidine antifolates.

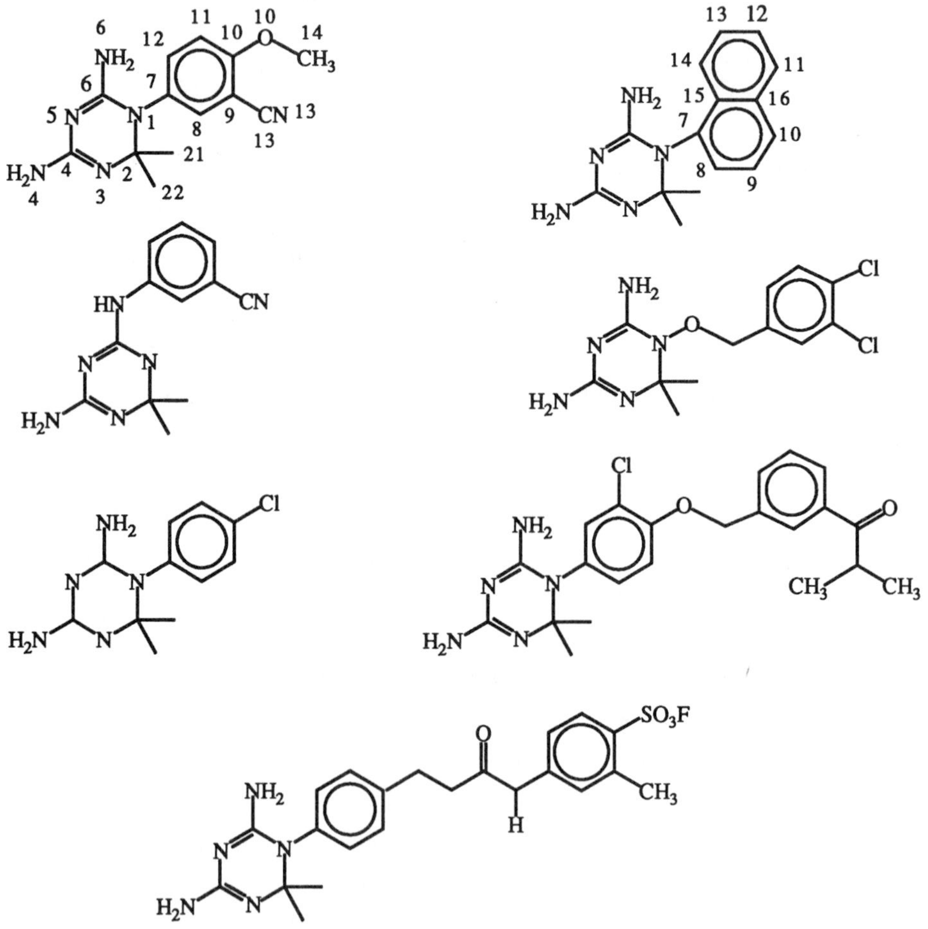

Fig. 5.14-7 1,2-Dihydro-2,2-dimethyl-4,6-diamino-*s*-triazines for which
crystal data have been reported. (After ref. 14).

References

[1] P.J. Wheatley, *Acta Crystallogr.* **8**, 224 (1955).

[2] P. Coppens, *Science (Washington)* **158**, 1577 (1967).

[3] S.M. Prasad, A.I.M. Rae, A.W. Hewat and G.S. Pawley, *J. Phys. C* **14**, L929 (1981).

[4] J.H. Smith and A.I.M. Rae, *J. Phys. C* **11**, 1761 (1978).

[5] P. Coppens, *J. Chem. Educ.* **61**, 761 (1984), and references therein; C.L. Klein and E.D. Stevens in J.F. Liebman and A. Greenberg (eds.), *Structure and Reactivity*, VCH Publishers, Weinheim, 1988, chap. 2; A.A. Low and M.B. Hall in Z.B. Maksic (ed.), *Theoretical Models of Chemical Bonding*, Part 2: *The Concept of the Chemical Bond*, Springer-Verlag, 1990, p.543.

[6] Z. Berkovitch-Yellin, and L. Leiserowitz, *J. Am. Chem. Soc.* **97**, 5627 (1975).

[7] T.L. Gilchrist, *Heterocyclic Chemistry*, Pitman, London, 1985.

[8] F. Bertinotti, G. Giacomello and A.M. Liquori, *Acta Crystallogr.* **9**, 510 (1956).

[9] G.A. Jeffrey, J.R. Ruble and J.H. Yates, *Acta Crystallogr., Sect. B* **39**, 388 (1983).

[10] G. de With, S. Harkema and D. Feil, *Acta Crystallogr., Sect. B* **32**, 3178 (1976).

[11] B.M. Craven, R.K. McMullan, J.D. Bell and H.C. Freeman, *Acta Crystallogr., Sect. B* **33**, 2585 (1977).

[12] E.W. Hughes, *J. Am. Chem. Soc.* **63**, 1737 (1941).

[13] J.N. Varghese, A.M. O'Connell and E.N. Maslen, *Acta Crystallogr., Sect. B* **33**, 2102 (1977).

[14] V. Cody and P.A. Sutton, *Anti-Cancer Drug Design* **2**, 253 (1987).

[15] J.M. Blaney, C. Hansch, C. Silipo and A. Vittoria, *Chem. Rev.* **84**, 333 (1984).

[16] H.L. Ammon and L.A. Plastas, *Acta Crystallogr., Sect. B* **35**, 3106 (1979).

[17] A. Camerman, H.W. Smith and N. Camerman, *Acta Crystallogr., Sect. B* **35**, 2113 (1979).

[18] W.E. Hunt, C.H. Schwalbe, K. Bird and P.D. Mallinson, *Biochem. J.* **187**, 533 (1980).

Note The crystal structure of pyridazine at 100K is reported in A.J. Blake and D.W.H. Rankin, *Acta Crystallogr., Sect. C* **47**, 1933 (1991).

5.15 Di(2,3,6,7-tetramethyl-1,4,5,8-tetraselenafulvalenium)
Hexafluorophosphate
$$(C_{10}H_{12}Se_4)_2PF_6$$

Crystal Data

Triclinic, Space group $P\bar{1}$ (No. 2)

$a = 7.297(1)$, $b = 7.711(1)$, $c = 13.522(2)$Å, $\alpha = 83.39(1)°$,

$\beta = 86.27(1)°$, $\gamma = 71.01(1)°$, $Z = 1$

Atom	x	y	z	Atom	x	y	z
Se(1)	.2925	.3391	.6147	C(2)	.1670	.7093	.6516
Se(2)	.1681	.7458	.5116	C(3)	.2524	.4927	.4943
Se(11)	.3672	.1746	.3880	C(4)	.1030	.8839	.7034
Se(12)	.2343	.5819	.2839	C(5)	.2282	.4744	.8085
P	0	0	0	C(11)	.3653	.2138	.2463
F(1)	.0362	-.1074	.1066	C(12)	.3127	.3856	.2033
F(2)	.2119	-.0984	-.0266	C(13)	.2796	.4278	.4026
F(3)	.0442	.1644	.0380	C(14)	.3052	.4353	.0932
C(1)	.2212	.5377	.6959	C(15)	.4351	.0411	.1960

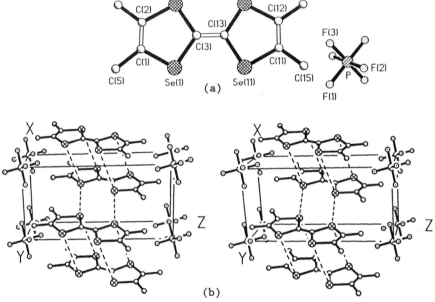

Fig. 5.15-1 (a) Asymmetric unit with atom labeling; formal double bonds are shown as open lines. (b) Stereoview of the crystal structure of (TMTSF)$_2$PF$_6$; intermolecular Se...Se contacts under 3.95Å are indicated by broken lines.

Crystal Structure

The first superconducting organic solid to be discovered, (TMTSF)$_2$PF$_6$ or $(C_{10}H_{12}Se_4)_2PF_6$, comprises the two molecular constituents illustrated in Fig. 5.15-1(a). This compound belongs to a series of organic mixed-valence cation-radical salts (TMTSF)$_2$X, X = AsF$_6^-$, PF$_6^-$, SbF$_6^-$, BF$_4^-$, NO$_3^-$ etc., which were found to conduct appreciably even at ambient pressure below 20K.

The bond lengths and angles are given in Fig. 5.15-2. The values for the TMTSF moiety are in good agreement with those found in the charge-transfer salts (1:1 compounds) with TCNQ.[2] In neutral TMTSF[3] the formal double bonds corresponding to C(3)-C(13) and C(1)-C(2) [or C(11)-C(12)] are 1.352 and 1.315Å, respectively, slightly shorter than in the present compound. This is also to be expected because of the larger extent of electron delocalization in a charge-transfer salt with a formal charge of +0.5 on the donor molecule.

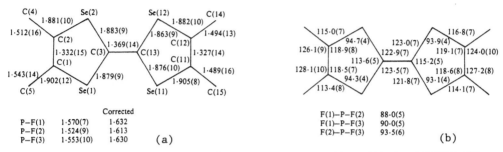

Fig. 5.15-2 (a) Bond lengths (Å) and (b) bond angles (°) in (TMTSF)₂SF₆. The P-F distances have been corrected for thermal motion. (After ref. 1).

The TMTSF molecule as a whole deviates insignificantly from planarity. The TMTSF units are stacked along the short *a* axis with two molecules per unit cell [Fig. 5.15-1(b)]. Within a stack the units repeat by an inversion leading to overlap displacements alternating in the direction of the long molecular axis. The two molecules on each side of a given molecule are related by a unit translation along *a*, but since this axis makes an angle of 1.1° with the normal to the molecular plane, the stacking is slightly irregular as reflected in the intermolecular Se...Se distances. As seen in Fig. 5.15-1(b), each TMTSF moleccule makes twice as many short Se...Se contacts (3.874 and 3.927Å) on one side as on the other (3.934Å).

Remarks

1. Organic superconductors and organic metals[4,5]

   The majority of known organic materials are electrical insulators, e.g. anthracene has an electrical conductivity, σ, less than 10^{-22} $\Omega^{-1}cm^{-1}$. A few organic systems exhibit metallic properties. The electrical conductivity of these substances increases with decreasing temperature, and they comprise a class of materials known as "organic metals" ($\sigma \approx 10\text{-}10^4$ $\Omega^{-1}cm^{-1}$). The first organic metal, reported in 1954, was a perylene-bromine complex (perylene.Br₂, $\sigma \approx 1$ $\Omega^{-1}cm^{-1}$), but intensive research on organic conductors proceeded only after the discoveries of the powerful π-molecular acceptor tetracyano-*p*-quinodimethane (TCNQ) in 1960,[6] and the sulfur-based tetrathiofulvalene

(TTF) in 1970,[7] and finally in 1972 their combination to form the π-molecular donor-acceptor complex TTF-TCNQ.[8]

TTF-TCNQ has high metallic conductivity, rising to $\sim 10^4$ $\Omega^{-1} cm^{-1}$ around 60K, at which point a metal-semiconductor transition occurs. This discovery prompted many chemists, physicists, and theoreticians to enter the field of organic metals. Until 1979, when the first organic superconductor was discovered,[9] research focussed on the synthesis of new TTF, TMTSF, and BEDT-TTF(ET) derivatives and other novel donor and acceptor systems. Fig. 5.15-3 shows some molecules that form conducting crystals.

The understanding of superconductivity in elemental metals underwent dramatic improvements with the development of the theory of Bardeen, Cooper, and Schrieffer.[10] According to this BCS theory, electrons form bound pairs (the Cooper pairs) as a result of their interactions with lattice vibrations (the phonons). The upper limit for the superconducting transition temperature is determined by the maximum phonon frequency, the Debye frequency. This understanding led Little[11] to propose that high transition temperatures could be achieved in organic metals (if these could be made) since the high frequencies of their internal vibrations might serve to play the same role phonons play in ordinary superconductors. Little's proposal sparked

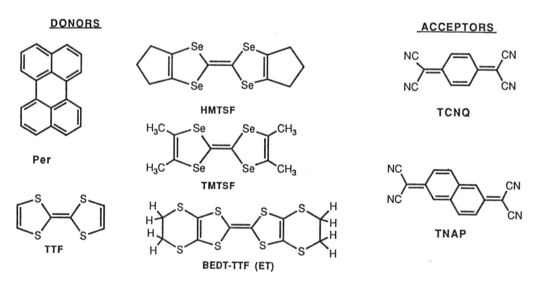

Fig. 5.15-3 Some molecules that form conducting crystals. The left column shows the donors: perylene (per), tetrathiofulvalene (TTF), hexamethylenetetraselenafulvalene (HMTSF), tetramethyltetraselena-fulvalene (TMTSF), and bis(ethylenedithio)-TTF (BEDT-TTF or ET). The right column are the acceptor molecules tetracyanoquinodimethane (TCNQ) and tetracyanonaphthalene (TNAP).

considerable interest and marked the onset of the search for superconductivity in organic materials.

2. Structural aspects of $(TMTSF)_2X$ conductors[12]

A knowledge of the crystal and molecular structures of the $(TMTSF)_2X$ [X = PF_6^-, AsF_6^-, BF_4^-, BrO_4^-, ClO_4^-, NO_3^-] conductors provides key information for understanding their electrical properties and gives vital insight required for the design of new conducting materials. A most unusual feature of these systems is that they are all isostructural, belonging to space group P1. In the crystal the nearly planar TMTSF molecules stack in a zigzag fashion forming a quasi-one-dimensional chain that runs parallel to the axis (*a*) of highest conductivity (see Fig. 5.15-4).

The lack of complete crystallographic regularity in the stacking motif is indicated in the varying intrastack Se...Se contacts in Fig. 5.15-4.

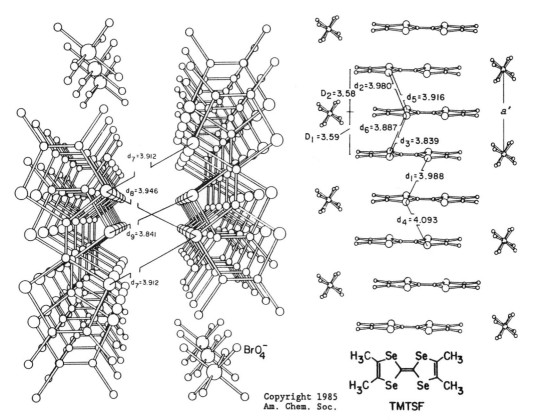

Copyright 1985
Am. Chem. Soc.

TMTSF

Fig. 5.15-4 Illustration of the crystal structure of the conductor $(TMTSF)_2BrO_4$ viewed down the stacking direction (left) and showing the zigzag stacking of TMTSF molecules (right-side view). Not all of the oxygen atom positions of the anion, which result from crystallographic disorder, are shown. Note the Se...Se contacts (d ≈ 3.9-4.0Å) and that there is no dimerization in the TMTSF stack. (After ref. 12).

The most noteworthy structural feature is the existence of Se...Se contact distances that are often considerably less than twice the van der Waals radius of Se. There exists an infinite "sheet network" of Se...Se interactions between TMTSF molecules extending in the *ab* plane, and within these sheets there are unusually short (d < 4.0Å) interstack Se...Se interactions (Fig. 5.15-5). However, the TMTSF molecules do not themselves form a complete three-dimensional network because the sheets are separated along *c* by the anions X. The anions appear to play a minor moderating role in the electrical conductivity, which arises from the network of Se...Se interactions.

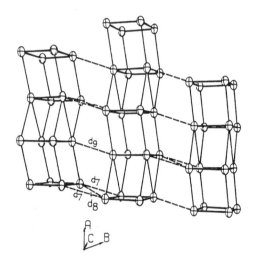

Fig. 5.15-5 A view of the "infinite sheet network" of short Se...Se interactions found in $(TMTSF)_2X$ conductors. The interstack distances d_7, d_8 and d_9 are also shown in Fig. 5.15-4. (After ref. 12).

The ramifications of the anisotropic structural changes in the interstack Se...Se network (see Fig. 5.15-4) and their relationship to pressure-induced superconductivity in some derivatives have been thoroughly discussed.[13] Although all Se...Se contact distances are about 3.9-4.0Å at room temperature, this is certainly not the case as the temperature is decreased. Surprisingly, as observed by use of the 125 K diffraction data, there is a marked change in interstack Se...Se distances such that they contract much more than the intrastack separations. At 125K the ratio of the decrease in interstack to intrastack Se...Se distances is nearly 2:1 when compared to the room-temperature distances. Thus the distances between the vertical "chains" decrease, on the average, by twice as much as the distances between the stacks of TMTSF molecules. This leads to considerably increased inter-chain electronic delocalization through the Se...Se network as the temperature is lowered if the orbital overlap is favorable.

The magnitude of the structural changes with temperature (298K → 125K) in terms of the interstack Se...Se distances (d_7, d_8, and d_9 in Fig. 5.15-4) reveals that some of these interatomic separations may be up to 0.3Å smaller than the van der Waals sum. Theoretical calculations have shown that considerable Se...Se bonding interactions are involved in the cases of these shortened separations.

The Se...Se sheet-network distances are also anion dependent and vary systematically with the anion size. It is most noteworthy that the incipient (pressure is required) superconductors [(TMTSF)$_2$X, X = TaF$_6^-$, SbF$_6^-$, PF$_6^-$, AsF$_6^-$, ReO$_4^-$] all have $d_{av} > d_{av}$ (X = ClO$_4^-$) [$d_{av} = (2d_7 + d_9)/3$]. From these structural findings the conclusion may be drawn that when the incipient superconductors are placed under pressure the Se...Se network contracts in a predictable fashion until its geometry closely approximates that in (TMTSF)$_2$ClO$_4$. At this point a superconducting transition becomes "structurally" favored.

3. Structural aspects of TTF-TCNQ and (ET)$_2$X conductors

Fig. 5.15-6 shows the crystal structure of TTF-TCNQ. This complex was the first molecular crystal found to exhibit genuine metallic behavior.[14]

TTF-TCNQ-like metals (segregated double stack, incomplete charge-transfer) possess an important structural feature. Because of the strong intrastack declocalization there are only weak interactions between the segregated stacks. This results in a highly anisotropic conductivity ($\sigma \approx 10^4$ Ω^{-1}cm^{-1} at 54K) in TTF-TCNQ. Without increased two- and three-dimensional electronic interactions between the segregated stacks, all of these materials

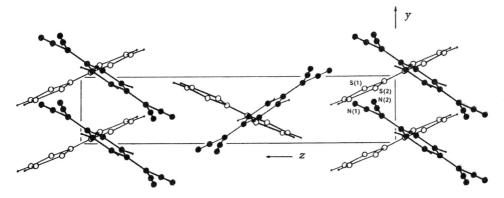

Fig. 5.15-6 A view down [100] of the crystal packing in TTF-TCNQ. The shaded molecules have their centroids at $x = 1/2$. (After ref. 14).

would likely behave as quasi-one-dimensional conductors and have a tendency to undergo insulating transitions at low temperature. Such is the case for TTF-TCNQ; this material is metallic for $T > 54K$, at which temperature a Peierls distortion[15] occurs. This transition involves a one-dimensional instability, a crystal lattice distortion, and a rapid decrease in electrical conductivity. It is known in TTF-TCNQ that a second transition occurs at lower temperature (38K) and that the two anomalies are associated with the two different types of stacks. These results were derived from X-ray diffuse scattering and elastic neutron scattering studies. A large number of studies of TTF-TCNQ-like (segregated stack) systems were conducted before the need for increasing interchain communication was fully appreciated and eventually fulfilled with the synthesis of a true molecular organic superconductor.

The system $(ET)_2X$ ($X = BrO_4^-$, ReO_4^-, ClO_4^- ...) has notable structural differences in comparison with $(TMTSF)_2X$: (i) whereas the TMTSF molecule is always found to be nearly planar, the ET moiety is decidedly non-planar with the $-CH_2$ groups protruding out of the molecular plane (Fig. 5.15-7); (ii) the ReO_4^- anion is ordered at ambient temperature; (iii) there exist mainly short *inter*stack S...S interactions, rather than short *intra*stack S...S separations. The "corrugated sheet network" of $(ET)_2BrO_4$ is shown in Fig. 5.15-8.[16] The S...S network is the main pathway for electrical conduction in all of the $(ET)_2X$ systems.

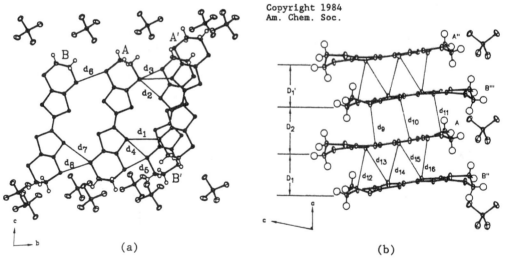

Copyright 1984
Am. Chem. Soc.

(a)   (b)

Fig. 5.15-7 Views showing the intermolecular S...S interactions in $(ET)_2BrO_4$ (at 125 K). (a) The interstack S...S contact distances less than 3.60Å (the van der Waals sum), $d_1 = 3.505$, $d_2 = 3.448$, $d_3 = 3.483$, $d_4 = 3.550$, $d_5 = 3.402$, $d_6 = 3.450$, $d_7 = 3.434$, $d_8 = 3.427$Å. (b) The S...S contact distances, d_9-d_{16}, all longer than 3.60Å. (After ref. 16).

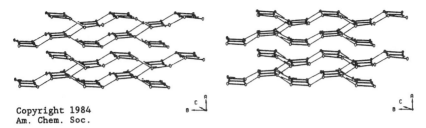

Fig. 5.15-8 Stereoview of the "corrugated sheet network" of short interstack S...S interactions (d < 3.60Å) observed in $(ET)_2BrO_4$. (After ref. 16).

Several β-phase $(ET)_2X$ superconductors with X = monovalent linear anions have a corrugated-sheet network structure that is remarkably similar to that of $(ET)_2BrO_4$, except that the loosely packed molecular stacks occur in the [110] direction rather than along the *a* axis. In the cation-radical ET salts containing IBr_2^-, AuI_2^- and I_3^-, the closest interstack S...S contacts and superconducting transition temperature (T_c) both increase concomitantly with linear anion length, the T_c values being 2.8, 4.98 and 8K(at 0.5 kbar), respectively.[5] By employing $Cu(SCN)_2^-$ as a counter anion, a stable ambient-pressure ET superconducting meterial with T_c = 10.4K has been obtained.[17] The T_c value can be raised further to 11.6K at ambient pressure for κ-$(ET)_2Cu[N(CN)_2]Br$[18] and to 12.8K (0.3 kbar) for the chloride analogue.[19] The crystal structure of κ-$(ET)_2Cu[N(CN)_2]Br$ exhibits the typical κ-type arrangement of mutually perpendicular face-to-face $(ET)_2$ dimers (Fig. 5.15-9), and the anion layer consists of zigzag polymeric copper dicyanamide chains with an extra bromide ligand on each copper(I) atom (Fig. 5.15-10).[18]

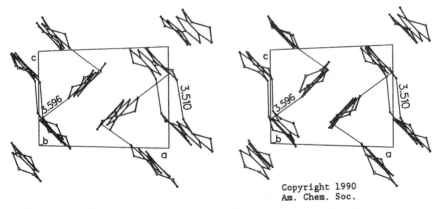

Fig. 5.15-9 Stereoview of the ET donor molecular layer in κ-$(ET)_2Cu[N(CN)_2]Br$ showing the κ-type arrangement of mutually perpendicular face-to-face dimers. The S...S contacts shorter than 3.60Å are indicated by thin lines. (After ref. 18).

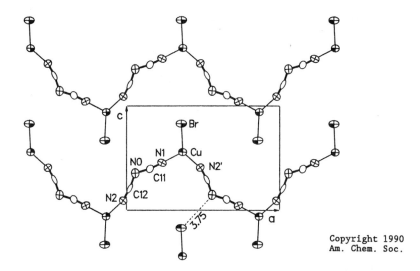

Fig. 5.15-10 Polymeric anion layer in κ-(ET)$_2$Cu[N(CN)$_2$]Br consisting of zigzag copper dicyanamide chains with an extra bromide ligand on each copper(I) atom. All atoms shown are located on a mirror plane. (After ref. 18).

4. DCNQI CT-complexes and radical anion salts

Well established rules have been formulated for the design of organic metals.[20] *N,N*-Dicyanoquinonediimines (abbreviated as DCNQI's) have been explored as a substitute for TCNQ in the synthesis of new charge transfer (CT) complexes.[21] Over two dozen complexes of TTF with DCNQI's bearing different ring substituents and CN configurations (*anti* or *syn*) have been synthesized. Most samples show powder conductivities of 10^{-2} to 10^{-1} Ω^{-1}cm^{-1}. In the "chess-board" structure of crystalline DCNQI.TTF.2H$_2$O [Fig. 5.15-11(a)], the water molecules closely match the additional cyano groups in the corresponding TCNQ.TTF complex. By contrast, the stacks of donor and acceptor molecules in DCNNQ.TTF are arranged in double layers parallel to the *bc* plane [Fig. 5.15-11(b)].

Many conducting DCNQI-radical anion salts with organic or metal counter cations have been prepared. In the crystal structure of (NMQ)$_2$(2,5-Br$_2$-DCNQI)$_3$, slightly shifted trimers of radical anions form stacks that are separated by stacks of equidistant *N*-methylquinolinium (NMQ) cations (Fig. 5.15-12).[21] The structure of (NMe$_4$)(2,5-Cl$_2$-DCNQI)$_2$ contains zigzag stacks of equidistant acceptor molecules with the cations occupying the holes in the crystal lattice.[22]

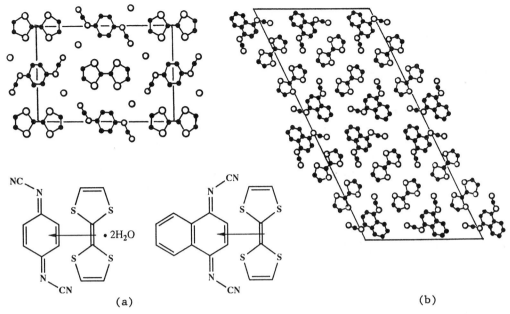

Fig. 5.15-11 Projection of the crystal structure of (a) DCNQI.TTF.$2H_2O$ along *a* (space group $P2_1/c$; $\sigma = 10\ \Omega^{-1}cm^{-1}$) and (b) DCNNQ.TTF along *b* (space group $C2/c$; $\sigma = 25\ \Omega^{-1}cm^{-1}$). (After ref. 21).

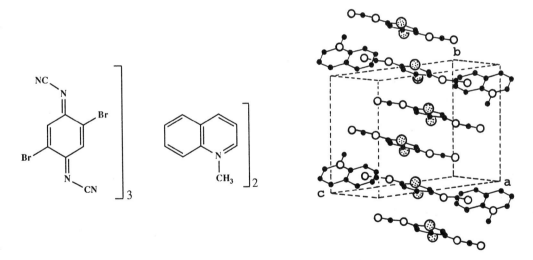

Fig. 5.15-12 Crystal structure of $(NMQ)_2(2,5\text{-}Br_2\text{-}DCNQI)_3$ (space group $P1$; $\sigma = 0.1\ \Omega^{-1}cm^{-1}$ for powder sample). (After ref. 21).

The radical anion salts $M(2\text{-}R,5\text{-}R'\text{-}DCNQI)_2$ (M = monovalent Cu, Ag, Tl, Li, Na, K, NH_4, Rb, Cu) are of interest as they have relatively high conductivities (10^{-2} to $10^3\ \Omega^{-1}cm^{-1}$) and possess the same crystal structure ($I4_1/a$ or a closely related space group). The highest conductivity is achieved with the copper(I) salts. In the crystal structure of $Cu(2,5-Me_2-DCNQI)_2$, each column of metal ions (separated by 3.78-3.97Å) are surrounded by four stacks of DCNQI ligands such that tetrahedral coordination occurs at each metal center (Fig. 5.15-13). Conductivity is directed principally through the DCNQI stacks parallel to *c*, but may also occur in the *a* and *b* directions if the metal ions can easily change their oxidation states at the appropriate potentials.

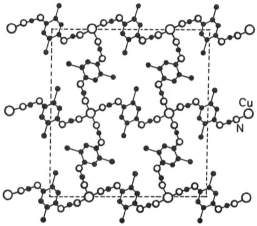

Fig. 5.15-13 Crystal structure of $Cu(2,5-Me_2-DCNQI)_2$ projected along *c*. (Adapted from ref. 21).

References

[1] N. Thorup, G. Rindorf, H. Soling and K. Bechgaard, *Acta Crystallogr.,
     Sect. B* **37**, 1236 (1981).

[2] K. Bechgaard, T.J. Kistenmacher, A.N. Bloch and D.O. Cowan, *Acta
     Crystallogr., Sect. B* **33**, 417 (1977).

[3] T.J. Kistenmacher, T.J. Emge, P. Shu and D.O. Cowan, *Acta Crystallogr.,
     Sect. B* **35**, 772 (1979).

[4] J.M. Williams and K. Carneiro, *Adv. Inorg. Chem. Radiochem.* **29**, 249
     (1985).

[5] J.M. Williams, H.H. Wang, T.J. Emge, U. Geiser, M.A. Beno, P.C.W. Leung,
     K.D. Carlson, R.J. Thorn, A.J. Schultz and M.-H. Whangbo, *Prog. Inorg.
     Chem.* **35**, 51 (1987).

[6] D.S. Acker, R.J. Harder, W.R. Hertler, W. Mahler, L.R. Melby, R.E. Benson
     and W.E. Mochel, *J. Am. Chem. Soc.* **82**, 6408 (1960).

[7] F. Wudl, G.M. Smith and E.J. Hufnagel, *J. Chem. Soc., Chem. Commun.*, 1453
     (1970).

[8] J. Ferraris, D.O. Cowan, V. Walatka, Jr. and J.H. Perlstein, *J. Am. Chem.
     Soc.* **95**, 948 (1973).

[9] D. Jérome, A. Mazaud, M. Ribault and K. Bechgaard, *J. Phys. Lett.* **41**, 95 (1980).

[10] J. Bardeen, L.N. Cooper and J.R. Schrieffer, *Phys. Rev.* **108**, 1175 (1957).

[11] W.A. Little, *Phys. Rev.* **A134**, 1416 (1964).

[12] J.M. Williams, M.A. Beno, H.H. Wang, P.C.W. Leung, T.J. Emge, U. Geiser and K.D. Carlson, *Acc. Chem. Res.* **18**, 261 (1985).

[13] J.M. Williams, M.A. Beno, J.C. Sullivan, L.M. Banovetz, J.M. Braam, G.S. Blackman, C.D. Carlson, D.L. Greer and D.M. Loesing, *J. Am. Chem. Soc.* **105**, 643 (1983).

[14] T.J. Kistenmacher, T.E. Phillips and D.O. Cowan, *Acta Crystallogr., Sect. B* **30**, 763 (1974).

[15] R.E. Peierls, *Quantum Chemistry of Solids*, Clarendon Press, Oxford, 1964.

[16] J.M. Williams, M.A. Beno, H.H. Wang, P.E. Reed, L.J. Azevedo and J.E. Schirber, *Inorg. Chem.* **23**, 1790 (1984).

[17] H. Urayama, H. Yamochi, G. Saito, K. Nozawa, T. Sugano, M. Kinoshita, S. Sato, K. Oshima, A. Kawamoto and J. Tanaka, *Chem. Lett.*, 53 (1988).

[18] A.M. Kini, U. Geiser, H.H. Wang, K.D. Carlson, J.M. Williams, W.K. Kwok, K.G. Vandervoot, J.E. Thompson, D.L. Stupka, D. Jung and M.-H. Whangbo, *Inorg. Chem.* **29**, 2555 (1990).

[19] J. Williams, A.M. Kini, H.H. Wang, K.D. Carlson, U. Geiser, L.K. Montgomery, G.J. Pyrka, D.M. Watkins, J.M. Kommers, S.J. Boryschuk, A.V.S. Crouch, W.K. Kwok, J.E. Schriber, D.L. Overmyer, D. Jung and M.-H. Whangbo, *Inorg. Chem.* **29**, 3272 (1990).

[20] J.M. Williams, H.H. Wang, T.J. Emge, U. Geiser, M.A. Beno, P.C.W. Leung, K.D. Carlson, R.J. Thorn, A.J. Schultz and M.-H. Whangbo, *Prog. Inorg. Chem.* **35**, 51 (1987).

[21] S. Hünig, *Pure Appl. Chem.* **62**, 395 (1990).

[22] S. Hünig, A. Aumüller, P. Erk, H. Meixner, J.V. von Schütz, H.-J. Groß, V. Langohr, H.-P. Werner, H.C. Wolf, C. Burschka, G. Klebe, K. Peters and H.-G. von Schnering, *Synth. Metals* **27**, B181 (1988).

Note Recent advances in organic CT salts are covered in M.R. Bryce, *Chem. Soc. Rev.* **20**, 355 (1991); J.Y. Becker, J. Bernstein, S. Bittner and S.S. Shaik, *Pure Appl. Chem.* **62**, 467 (1990); M.R. Bryce and A.J. Moore, *ibid.*, 473; K. Nakasuji, *ibid.*, 477; G.C. Papavassiliou, *ibid.*, 483. Molecular metals and superconductors derived from metal complexes of 1,3-dithiol-2-thione-4,5-dithiolate are reviewed in P. Cassoux, L. Valade, H. Kobayashi, A. Kobayashi, R.A. Clark and A.E. Underhill, *Coord. Chem. Rev.* **110**, 115 (1991).

5.16   Silenes (Silaethenes)

$$\text{I} \quad Me_2Si{=}C(SiMe_3)(SiBu^t{}_2Me)\cdot C_4H_8O$$

$$\text{II} \quad Me_2Si{=}C(SiMe_3)(SiBu^t{}_2Me)$$

Crystal Data

I   Orthorhombic, space group $P2_12_12_1$ (No. 19)

$a = 10.000(5)$, $b = 15.386(8)$, $c = 15.509(8)$Å, $(-45°C)$, $Z = 4$[1]

Atom	x	y	z	Atom	x	y	z
Si(1)	.1784	.1764	.2608	C(10)	.4306	.3931	.0400
Si(2)	.1828	.3424	.3647	C(11)	.3128	.2579	.0476
Si(3)	.3892	.3258	.2136	C(12)	.5638	.2744	.2262
C(1)	.2597	.2749	.2807	C(13)	.5910	.2615	.3167
C(2)	-.0038	.1754	.2403	C(14)	.5775	.1857	.1863
C(3)	.2469	.0886	.1914	C(15)	.6755	.3287	.1869
C(4)	.0791	.4354	.3236	C(16)	.0788	.0463	.3899
C(5)	.3038	.3884	.4460	C(17)	.1446	.0000	.4644
C(6)	.0587	.2841	.4371	C(18)	.2487	.0621	.4975
C(7)	.4194	.4430	.2486	C(19)	.2987	.1064	.4186
C(8)	.3370	.3382	.0938	O	.1777	.1123	.3639
C(9)	.2017	.3851	.0913				

II   Orthorhombic, space group $P2_12_12_1$ (No. 19)

$a = 8.109(1)$, $b = 14.886(2)$, $c = 16.470(2)$Å, $(-35°C)$, $Z = 4$[2]

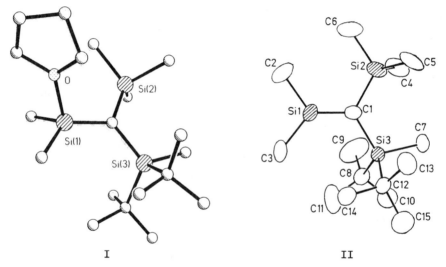

| I | II |

Fig. 5.16-1 Molecular structures of I and II. (After refs. 1 and 2).

Crystal Structure

Silenes (or silaethenes) are compounds which contain the Si=C double bond. The stable silaethene I is obtained in crystalline form as an 1:1 adduct with THF.[1] It can be converted into the solvent-free crystalline form II by azeotropic distillation with benzene.[2]

The molecular structures of I and II are shown in Fig. 5.16-1, and their important bond distances(Å) and bond angles(°) are listed in Table 5.16-1.

Table 5.16-1   Selected bond distances (Å) and angles (°) of **I** and **II**.

	I	II
Si(1)-C(1)	1.747	1.702
Si(1)-C(2)	1.850	1.851
Si(1)-C(3)	1.858	1.842
C(1)-Si(2)	1.835	1.865
C(1)-Si(3)	1.836	1.890
Si(1)-O	1.878	-
C(1)-Si(1)-C(2)	119.7	127.0
C(1)-Si(1)-C(3)	124.1	129.0
C(2)-Si(1)-C(3)	104.9	104.3
Si(1)-C(1)-Si(2)	114.9	119.5
Si(1)-C(1)-Si(3)	126.7	122.8
Si(2)-C(1)-Si(3)	117.2	117.7

In **I**, the most interesting feature is an essentially planar, sp^2 hybridized carbon atom, C(1), engaged in bonding with a tetracoordinate silicon atom Si(1), which is flanked by two methyl groups and closely approached by the oxygen atom of the coordinated THF molecule. The Si(1)-C(1) bond length of 1.747(5)Å is drastically shorter than Si-C single bonds with sp^3- or sp^2-hybridized C atoms which range typically from 1.87 to 1.94Å.[3] The planarity of C(1) is indicated by the sum of the bond angles around it (358.8°) and by its small deviation, only 0.11Å, from the plane of the three Si atoms. The Si(1)-O distance is 1.878(4)Å, much longer than 1.63-1.66Å found for Si-O bonds with two-coordinated O atoms.

Compound **II** shows an essentially planar C₂Si=CSi₂ skeleton. The deviations of the bond angles at the unsaturated centers Si(1) and C(1) from standard sp^2 values may be attributed to steric strain exerted by the SiMe₃ and especially the SiMe(*t*-Bu)₂ groups at C(1). The length of the Si=C bond is 1.702(5)Å, in excellent agreement with the range of 1.69-1.71Å predicted theoretically by *ab initio* calculations.[4,5] The observed double bond geometry of the silaethenes closely resembles that of ethenes and implies a similar bonding situation in both systems.

Remarks

1. Bonding in Me₂Si=C(SiMe₃)(SiBu$^t_2$Me).(C₄H₈O)

The bonding situation in **I** can be described in terms of the resonance formulation **Ia** ⟷ **Ib**:[1]

                      Ia   **Ib**

Formula **Ia** implies adduct formation between THF and the Si atom engaged in a Si=C double bond. Because of this interaction, the Si atom has to accommodate 10 valence electrons, which necessarily means d-orbital participation. Formula Ib involves a 1,3-charge separation as a consequence of strong adduct formation; the adduct can be regarded as structurally related to phosphorus ylid $R_3P\text{-}CR_2$. The rather long Si-O bond in I is consistent with a relatively weak adduct Ia, in which suitable d functions at Si permit overlap with O donor orbitals. Although at first glance this agreement seems to point to an essentially doubly bonded Si atom, as in Ia, it should also be noted that the short ylidic bond in Ib can be accounted for in terms of $p_\pi\text{-}d_\pi$ overlap of the lone pair at the ylidic C atom with suitable d orbitals at the Si atom, as well as electrostatic attraction between the localized partial charges.

The silicon-carbon double bond has attracted the attention of theoretical chemists who have performed a large number of high quality calculations on various properties of silenes. The Si=C bond in the hypothetical parent silene $H_2C=SiH_2$, of calculated length 1.71Å, is polarized so that the silicon and carbon atoms carry partial positive and negative charges, respectively. Substituents that increase this polarization and thus the degree of ionicity of the Si=C bond are expected to shorten it. Electronegative substituents at carbon reduces the bond ionicity and the bond length increases.[6,7]

2. Structure of $(Me_3Si)_2Si=C(OSiMe_3)(C_{10}H_{15})$[8]

This adamantyl derivative is the first silene to have its structure determined by X-ray analysis. Fig. 5.16-2 gives an ORTEP plot of the molecule and some of the bond lengths and bond angles that are of particular relevance.

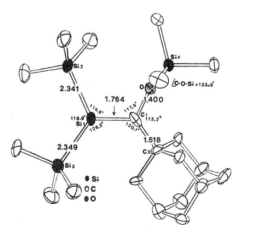

Fig. 5.16-2 Molecular structure of $(Me_3Si)_2Si=C(OSiMe_3)(C_{10}H_{15})$. (After ref. 8).

The molecule is actually twisted at -50°C, an angle of 14.6° existing between
the planes

$$\begin{matrix} Si \\ \diagdown \\ Si{=}C \\ \diagup \\ Si \end{matrix} \qquad \text{and} \qquad \begin{matrix} Si{=}C\diagup^{OSi} \\ \phantom{Si{=}C}\diagdown_{C} \end{matrix} \quad .$$

The maximum deviation from planarity of atoms in these planes is 0.01 and
0.03Å, respectively, indicating that the central Si and C atoms are truly sp^2
hybridized.

The Si-C(methyl) bonds range in length from 1.854 to 1.877Å, which are
consistent with the Si-C single bonds observed for $SiMe_4$ (1.875(2)Å from
electron diffraction)[9] and its Si/C inverse analogues $C(SiH_3)_4$ (1.875(7)Å
from electron diffraction) and $C(SiH_2Ph)_4$ (average 1.885(2)Å).[10]

3. Disilenes[11-16]

Disilenes are compounds which contain the Si=Si double bond.
Tetramesityldisilene, $(Mes)_2Si=Si(Mes)_2$, was the first to be prepared and,
together with *trans*-(Mes)(CMe₃)Si=Si(CMe₃)(Mes), had their structures
established by crystal structure analysis.[11] The tetramesityldisilene
molecule possesses a C_2 axis passing through the silicon-silicon bond (Fig.
5.16-3). The four aryl substituents are disposed about the silicon-silicon
framework in a roughly helical fashion in which two of the *cis*-related rings
are more nearly coplanar with the silicon-silicon bond axis. The Si=Si bond
length is 2.160(1)Å, and the independent Si-C bond lengths are 1.880(2) and
1.871(2)Å.

The *trans*-(Mes)(CMe₃)Si=Si(CMe₃)(Mes) molecule possesses a center of
symmetry midway between the two silicons. The Si=Si bond length is
2.143(1)Å. The Si-C bond lengths involving the *t*-butyl and mesityl groups are
1.904(3) and 1.884(2)Å, respectively.

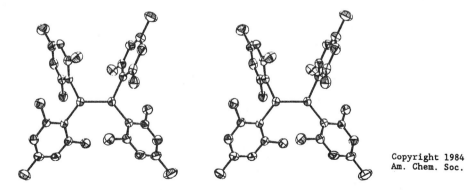

Fig. 5.16-3 Stereoscopic view of the structure of tetramesityldisilene.
(After ref. 11).

The Si=Si bond length in these two disilenes is about 0.20Å less than the normal Si-Si single bond length of 2.340(9)Å in hexamethyldisilane, Si_2Me_6. The bond length decrease is quite similar to the bond shortening observed on going from C-C to C=C, indicating the presence of significant π-bonding in the Si=Si double bond. The disilenes may exhibit deviation from the regular planar structure of olefins, because the Si=Si double bond is much more flexible toward distortion than C=C. The barrier to rotation about Si=Si is about 70% as large as that for comparable alkenes. In sharp contrast to alkenes, the disilenes are yellow or orange compounds, with electronic absorption bands between 400 and 470 nm. The visible spectra of disilenes are due to $3p\pi \rightarrow 3p\pi^*$ transitions; these have energies only about half of those for typical olefins, whose $2p\pi \rightarrow 2p\pi^*$ absorptions fall in the vacuum ultraviolet region. This difference is explicable in term of a crude molecular orbital model. As shown in Fig. 5.16-4, the π orbital for disilenes lies much higher than for alkenes, and the π^* level much lower.

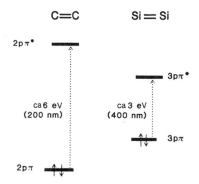

Fig. 5.16-4 Qualitive MO diagram showing relative energies of π and π^* levels for alkenes and disilenes. (After ref. 14).

Although the reactions of disilenes are known only very incompletely, they have led to several new classes of compounds. Some examples are shown as follow:

4. Metal silene complexes[17,18]

Silenes are usually reactive organosilicon intermediate whose formation can be established indirectly with trapping reactions. Recent interest in silenes has been stimulated by the synthesis of isolable examples that are stabilized by steric protection of the Si=C double bond. Given the well known ability of transition metals to stabilize reactive species, e.g. carbenes, carbynes, cyclobutadienes, by ligation, one can reasonably assume that stable transition-metal silene complexes could also be isolated and studied.

The crystal structure of $(\eta^5\text{-}C_5Me_5)[P(i\text{-}Pr)_3]Ru(H)(\eta^2\text{-}CH_2SiPh_2)$ has been determined.[17] Fig. 5.16-5(a) shows the molecular structure of one of two

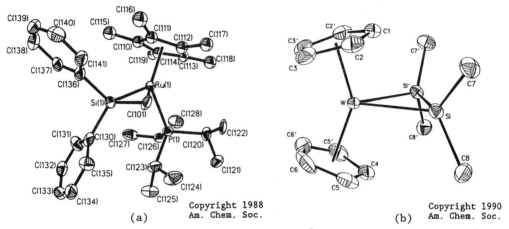

Copyright 1988
Am. Chem. Soc.

(a)

Copyright 1990
Am. Chem. Soc.

(b)

Fig. 5.16-5 Molecular structure of (a) $(\eta^5\text{-}C_5Me_5)[P(i\text{-}Pr)_3]Ru(H)(\eta^2\text{-}CH_2SiPh_2)$ and (b) $Cp_2W(SiMe_2)_2$. (After refs. 17 and 20).

enantiomers in the asymmetric unit. The bond angles about Ru indicate that the hydride ligand is *cis* to the ligated silicon atom. The $Si\text{-}CH_2$ distances in the silene ligand is 1.78(2) and 1.79(2)Å, which seem to reflect partial double bond character. The metal-ligand distances [Ru(1)-Si(1), 2.382 and 2.365Å; Ru(1)-C(101), 2.25 and 2.26Å] show that the compound contains M-η^2-silene bonds similar to those in transition metal η^2-alkene complexes.

The first transition-metal η^2-disilene complexes in which the Si=Si bond is formally bonded to a (diphos)Pt center has recently been synthesized and characterized, although a definitive crystal structure determination is not yet available.[19] The first successful X-ray study was accomplished with $Cp_2W(SiMe_2)_2$, in which the disilene-metal bonding exhibits properties intermediate between the "metal-olefin" and "metallacyclopropane" formalisms [Fig. 5.16-5(b)].[20] The molecule has crystallographic mirror symmetry, and the measured dimensions are: W-Si 2.606(2), Si-Si' 2.260(3), Si-C7 1.889(8), Si-C8 1.898(9)Å; Si-W-Si' 51.39(6), W-Si-Si' 64.30(7), Si'-Si-C7 124.0(3), Si'-Si-C8 120.3(3)°. The isomorphous Mo complex is also known.

5. Stable germaethenes and digermenes

Until recently, the existence of short-lived germaethenes could only be detected by trapping reactions. Stabilization of the Ge=C double bond has now been achieved in I by resonance with an ylide form,[21] and in II by charge transfer into a fluorenylidene system.[22]

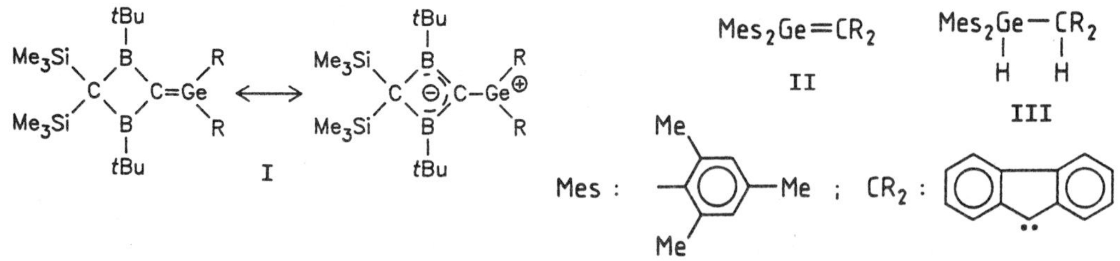

The molecular structures of germaethenes I, II and the related dihydride III are shown in Fig. 5.16-6. In I the Ge=C bond length is 1.827(4)Å, and the short C(2)-B distances [avg. 1.529(6)Å] support the significance of ylid resonance. The Ge atom is practically planar coordinated, the C(2) atom slightly pyramidalized, and the average twist angle about the Ge=C bond is 36°. Compound II has two crystallographically independent molecules which are virtually identical with trigonal-planar configurations at Ge and C(1); the average Ge=C bond length and twist angle about it are 1.803(4)Å and 5.9°, respectively. In III the mean Ge-C(mesityl) distance of 1.965(3)Å is similar to the Ge-C single bond (1.98Å) in $Ge(CH_3)_4$, but the Ge-C(fluorenyl) bond of 2.010(4)Å is markedly long owing to the increased steric encumberance.

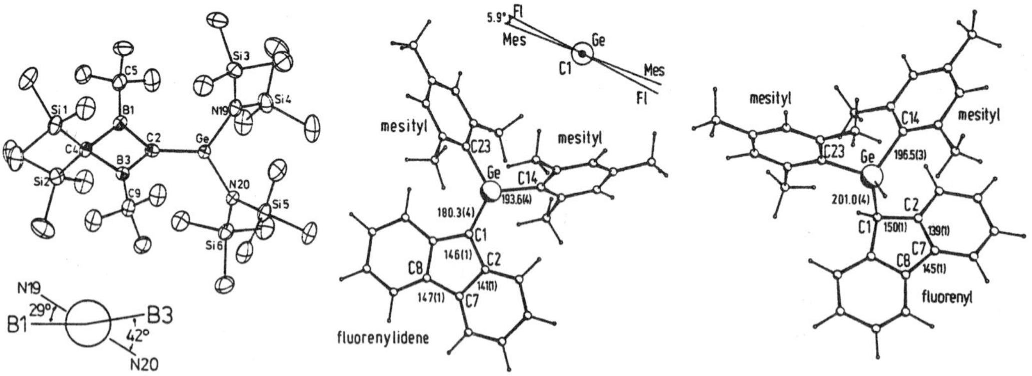

Fig. 5.16-6 Molecular structures of two stable germaethenes and a dihydro derivative. (After refs. 21 and 22).

Two digermenes of the type $R_2Ge=GeR_2$ [R = $(Me_3Si)_2CH$, **IV**; R = 2,6-$Et_2C_6H_3$, **V**] have been characterized by single-crystal X-ray analysis (Fig. 5.16-7). Digermene **IV** has a *trans*-folded C_{2h} framework with a fold angle of 32° at the germanium atom and a Ge=Ge bond length of 2.347(2)Å,[23] which is only 5% shorter than the standard Ge-Ge single bond (2.465Å in $(Ph_2Ge)_4$;[24] 2.463 and 2.457Å in $(Ph_2Ge)_6$[25]; 2.429Å in $Ph_3Ge-GeMe_2-GePh_3$[26]). Compound **V** has a C_2 axis normal to the Ge=Ge bond, with a twist angle of 11° about it and a fold angle of 15°.[27] The different Ge=Ge bond lengths in the two digermenes are in accord with the fact that **IV** disproportionates very easily in solution into two singlet germylenes while **V** retains its structural integrity. In the Z isomer of RR'Ge=GeRR' (R = 2,6-$Et_2C_6H_3$, R' = Mes), the fold angle is 32° and the Ge=Ge and Ge-C bond lengths are 2.301(1) and [1.972(5) and 1.988(5)]Å, respectively.[28] Recent developments on multiply bonded germanium derivatives have been reviewed.[29] The stannaethene $Me_2Sn=C(SiMe_3)_2$ exists as a short-lived species at -100°C.[30]

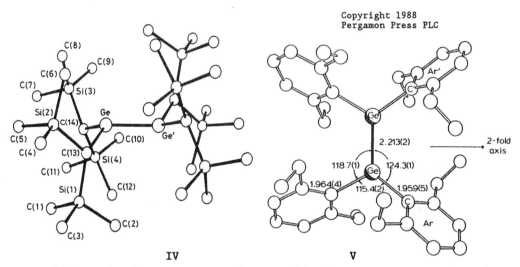

Copyright 1988
Pergamon Press PLC

IV   **V**

Fig. 5.16-7 Molecular structures of two stable digermenes. For compound **IV**, Ge-C(13) 2.048(9) and Ge-C(14) 1.979(9)Å. (After refs. 23 and 27).

References

[1] N. Wiberg, G. Wagner, G. Müller and J. Riede, *J. Organomet. Chem.* **271**, 381 (1984).

[2] N. Wiberg, G. Wagner and G. Müller, *Angew. Chem. Int. Ed. Engl.* **24**, 229 (1985).

[3] V. Bazant, V. Chvalovsky and J. Rathousky, *Organosilicon Compounds*, Academic Press, New York, 1965.

[4] H.F. Schaefer III, *Acc. Chem. Res.* **15**, 283 (1982).

[5] Y. Apeloig and M. Karni, *J. Chem. Soc., Chem. Commun.*, 768 (1984).

[6]   M.S. Gordon, *J. Am. Chem. Soc.* **104**, 4352 (1982).

[7]   Y. Apeloig and M. Karni, *J. Am. Chem. Soc.* **106**, 6676 (1984).

[8]   A.G. Brook, S.C. Nyburg, F. Abdesaken, B. Gutekunst, G. Gutekunst,
      R. Krishna, M.R. Kallury, Y.C. Poon, Y.-M. Chang and W. Wong-Ng, *J. Am.
      Chem. Soc.* **104**, 5667 (1982).

[9]   B. Beagley, J.J. Monaghan and T.G. Hewitt, *J. Mol. Struct.* **8**, 401
      (1971).

[10] R. Hager, O. Steigelmann, G. Müller, H. Schmidbaur, H.E. Robertson and
      D.W.H. Rankin, *Angew. Chem. Int. Ed. Engl.* **29**, 201 (1990).

[11] M.J. Fink, M.J. Michalczyk, K.J. Haller, J. Michl and R. West,
      Organometallics **3**, 793 (1984).

[12] S.G. Baxter, K. Mislow, and J.F. Blount, *Tetrahedron* **36**, 605 (1980).

[13] G. Raabe and J. Michl, *Chem. Rev.* **85**, 419 (1985).

[14] R. West, *Angew. Chem. Int. Ed. Engl.* **26**, 1201 (1987).

[15] K.S. Pitzer, *J. Am. Chem. Soc.* **70**, 2140 (1948).

[16] R.S. Mulliken, *J. Am. Chem. Soc.* **72**, 4493 (1950); **77**, 884 (1955).

[17] B.K. Campion, R.H. Heyn and T.D. Tilley, *J. Am. Chem. Soc.* **110**, 7558
      (1988).

[18] A.G. Brook and K.M. Baines, *Adv. Organomet. Chem.* **25**, 1 (1986).

[19] E.K. Pham and R. West, *J. Am. Chem. Soc.* **111**, 7667 (1989).

[20] D.H. Berry, J.H. Chey, H.S. Zipin and P.J. Carroll, *J. Am. Chem. Soc.*
      112, 452 (1990).

[21] H. Meyer, G. Baum, W. Massa and A. Berndt, *Angew. Chem. Int. Ed. Engl.*
      26, 798 (1987).

[22] M. Lazraq, J. Escudié, C. Couret, J. Satgé, M. Dräger and R. Dammel,
      Angew. Chem. Int. Ed. Engl. **27**, 828 (1988).

[23] P.B. Hitchcock, M.F. Lappert, S.J. Miles and A.J. Thorne, *J. Chem. Soc.,
      Chem. Commun.*, 480 (1984).

[24] L. Ross and M. Dräger, *J. Organomet. Chem.* **199**, 195 (1980).

[25] M. Dräger and L. Ross, *Z. Anorg. Allg. Chem.* **476**, 95 (1981).

[26] M. Dräger and D. Simon, *J. Organomet. Chem.* **306**, 183 (1986).

[27] J.T. Snow, S. Murakami, S. Masamune and D.J. Williams, *Tetrahedron Lett.*
      25, 4191 (1984).

[28] S.A. Batcheller, T. Tsumuraya, O. Tempkin, W.M. Davis and S. Masamune, *J.
      Am. Chem. Soc.* **112**, 9394 (1990).

[29] J. Barrau, J. Escudié and J. Satgé, *Chem. Rev.* **90**, 283 (1990).

[30] N. Wiberg and S.-K. Vasisht, *Angew. Chem. Int. Ed. Engl.* **30**, 93 (1991).

Note Disilenes, digermenes and distannenes are reviewed in T. Tsumuraya, S.A.
      Batcheller and S. Masamune, *Angew. Chem. Int. Ed. Engl.* **30**, 902 (1991).

5.17 Sucrose

$C_{12}H_{22}O_{11}$

Crystal Data

Monoclinic, space group $P2_1$ (No. 4)

$a = 10.8633(5)$, $b = 8.7050(4)$, $c = 7.7585(4)$Å, $\beta = 102.945(6)°$, $Z = 2$

Atom	x	y	z	Atom	x	y	z
C(1)	.29961	.35792	.48487	H(C1)	.3347	.2451	.5388
C(2)	.31253	.47474	.63600	H(C2)	.4116	.4693	.7117
C(3)	.28545	.63673	.56447	H(C3)	.1871	.6448	.4897
C(4)	.37404	.67095	.44198	H(C4)	.4717	.6681	.5218
C(5)	.35925	.55107	.29529	H(C5)	.2638	.5613	.2093
C(6)	.45754	.57083	.18455	H(C6)a	.4531	.6873	.1345
C(1')	.10301	.13110	.54380	H(C6)b	.4372	.4927	.0719
C(2')	.12446	.19262	.36895	H(C1')a	.0520	.0224	.5216
C(3')	.00718	.19075	.21485	H(C1')b	.1947	.1091	.6330
C(4')	.06478	.16653	.05476	H(C3')	-.0495	.0884	.2290
C(5')	.17635	.06133	.12864	H(C4')	.0984	.2772	.0155
C(6')	.28927	.08194	.04668	H(C5')	.1465	-.0599	.1133
O(1)	.17143	.34630	.39165	H(C6')a	.3672	.0119	.1195
O(2)	.22954	.43550	.74766	H(C6')b	.2651	.0399	-.0887
O(3)	.30801	.74770	.70279	H(O2)	.2719	.3724	.8466
O(4)	.34880	.81412	.35631	H(O3)	.2318	.7660	.7426
O(5)	.37719	.39878	.36864	H(O4)	.3407	.8913	.4325
O(6)	.58144	.54525	.28621	H(O6)	.6015	.4383	.2866
O(1')	.03017	.23548	.62119	H(O1')	.0845	.3247	.6542
O(2')	.21205	.09445	.31572	H(O3')	-.0310	.4089	.1761
O(3')	-.07367	.31776	.20452	H(O4')	-.0167	.1538	-.1961
O(4')	-.02123	.09734	-.08904	H(O6')	.3477	.2780	.1603
O(6')	.32644	.23799	.04035				

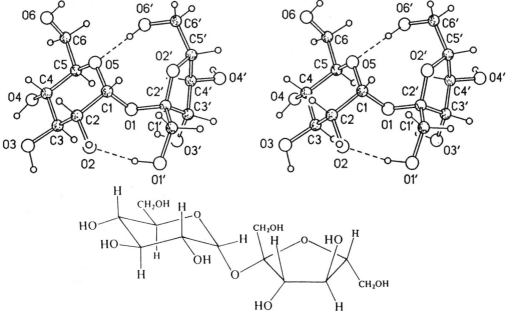

Fig. 5.17-1 Stereoview and structural formula of the sucrose molecule shown in its absolute configuration.

Crystal Structure

The structure of sucrose was solved by Beevers and co-workers.[3] in 1952, subsequently refined by Brown and Levy[4] in 1963, and finally subjected to highly precise refinement using neutron[1] and X-ray[2] diffraction data by two research groups in 1973. The glycosidic linkage was first established from the crystal structure analysis of sucrose.

Since sucrose is a non-reducing sugar, both glucose and fructose must be linked through their respective reducing groups. The structural data indicate that α-glucose is linked to β-fructose to give α-D-glucopyranosyl-β-D-frucotofuranoside. Sucrose is hydrolyzed by dilute acids or by the enzyme invertase to an equimolecular mixture of D(+)-glucose and D(-)-fructose. Thus glucose is present in the pyranose form and fructose in the furanose form.

Bond lengths, bond angles and torsion angles for the sucrose molecule are shown in Fig. 5.17-2.

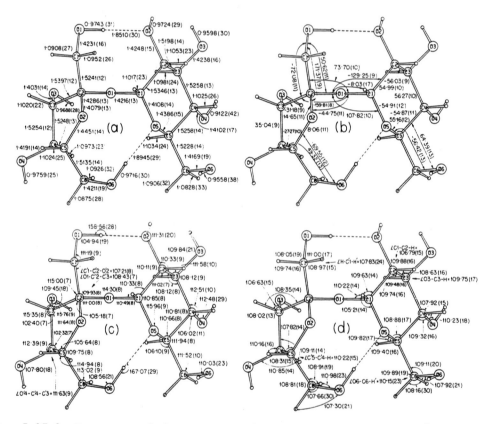

Fig. 5.17-2 Structure of the sucrose molecule: (a) bond lengths (Å); (b) selected torsion angles (°); (c) bond angles (°) not involving C-H bonds; (d) bond angles involving C-H bonds. All lengths and angles are uncorrected for the effects of thermal motion. (After ref. 1).

The distance and angle parameters describing the hydrogen bonds are given in Table 5.17-1. A figure showing the hydrogen bonds is shown in Fig. 5.17-3. There are seven crystallographically-distinct hydrogen bonds, excluding the weak interactions of H(O4). They involve the ring oxygen O(5) as an acceptor and each OH group except O(4)-H(O4) and O(6)-H(O6) as both donor and acceptor; O(6)-H(O6) acts as a donor only, whereas O(4)-H(O4) and ring oxygen O'(2) are not involved at all.

Table 5.17-1 Description of the hydrogen bonds in sucrose.

O-H...O	Distances/Å			Angle
	O-H	H...O	O...O	O-H...O(°)
Intramolecular				
O(1')-H...O(2)	0.974	1.851	2.781	158.6
O(6')-H...O(5)	0.972	1.895	2.850	167.1
Intermolecular				
O(2)-H...O(6')	0.972	1.892	2.855	170.2
O(3)-H...O(3')	0.959	1.907	2.862	172.8
O(6)-H...O(3)	0.956	1.921	2.848	162.9
O(3')-H...O(4')	0.969	1.908	2.864	168.5
O(4')-H...O(1')	0.976	1.760	2.716	165.4
Non-participating OH				
	O(2')	2.309	2.838	116.6
O(4)-H	O(3) 0.912	2.534	2.879	103.0
	O(6)	2.539	3.373	152.1

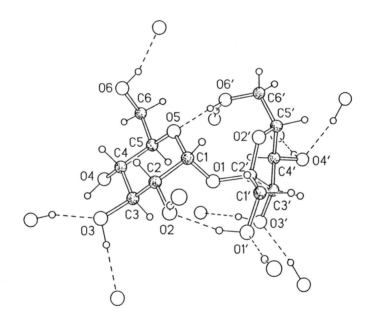

Fig. 5.17-3 Hydrogen-bonding environment of the sucrose molecule in the crystal structure, showing all donor and acceptor interactions with its neighbouring molecules.

Remarks

1. Carbohydrates

Carbohydrates[5] are polyhydroxy aldehydes, polyhydroxy ketones, or compounds that can be hydrolyzed to them. A carbohydrate that cannot be hydrolyzed to simpler compounds is called a monosaccharide. A carbohydrate that can be hydrolyzed to two monosaccharide molecules is called a disaccharide. A carbohydrate that can be hydrolyzed to many monosaccharide molecules is called a polysaccharide.

A monosaccharide may be further classified. If it contains an aldehyde group, it is known as an aldose; if it contains a keto group, it is known as a ketose. Depending upon the number of carbon atoms in it, a monosaccharide is known as a triose, tetrose, pentose, hexose, and so on. An aldohexose, for example, is a six-carbon monosaccharide containing an aldehyde group; a ketopentose is a five-carbon monosaccharide containing a keto group. Most naturally-occurring monosaccharides are pentoses or hexoses.

Investigations on carbohydrates have made important contributions to conformational analysis, the importance of which in many areas of chemistry and biochemistry was recognized by the award of the 1969 Nobel Prize in Chemistry to Barton[6] and Hassel.[7] A comprehensive "lead article" on the structural data for carbohydrates has been published by Jeffrey.[8]

2. Structure of D-glucose

According to the Fischer convention, D-sugars have the same absolute configuration at the asymmetric center farthest removed from their carbonyl group as does D-glyceraldehyde. Fig. 5.17-4 shows the molecular structure of α-D-glucose ($C_6H_{12}O_6$) as determined by neutron diffraction analysis.[9] The OH group attached to the anomeric carbon atom C(1) and the CH_2OH substitutent

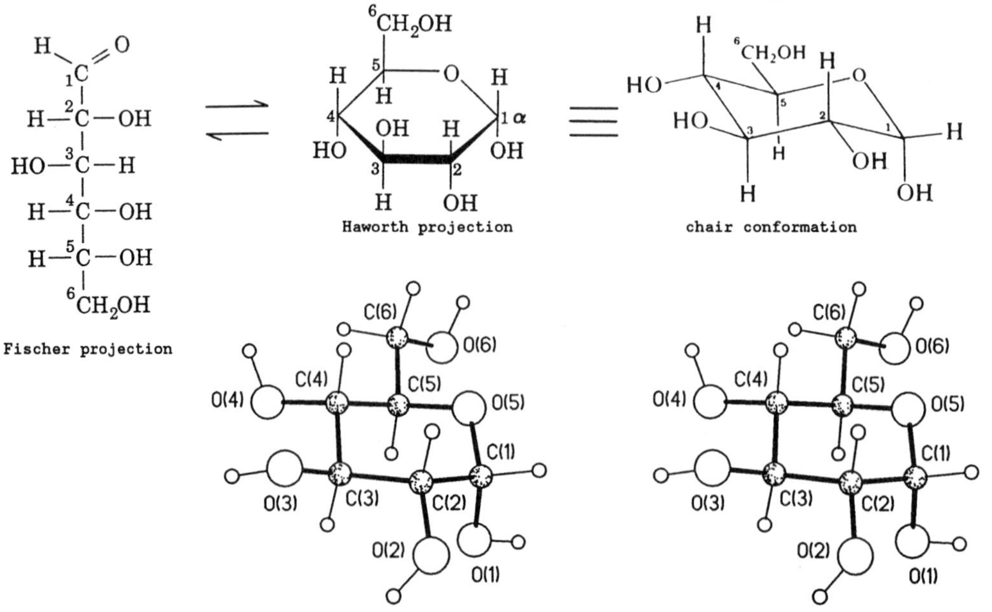

Fig. 5.17-4   Structural formula and stereoview of α-D-glucose.

on the chiral center C(5) lie on opposite sides of the sugar ring. The C-C, C-H and O-H bond lengths deviate only slightly from their means of 1.523, 1.098 and 0.968Å, respectively. The C(1)-O(1) bond length, 1.389Å, is significantly shorter than the mean value, 1.420Å, of the other C-O bonds. The valence angle at the ring oxygen is 113.8°.

The structure of the β anomer of D-glucose (Fig. 5.17-5) has also been determined[10] and refined.[11]   The mean C-C bond length is 1.520Å, and the mean C-O bond length, excluding ring oxygen O(5) and anomeric C(1), is 1.425Å.

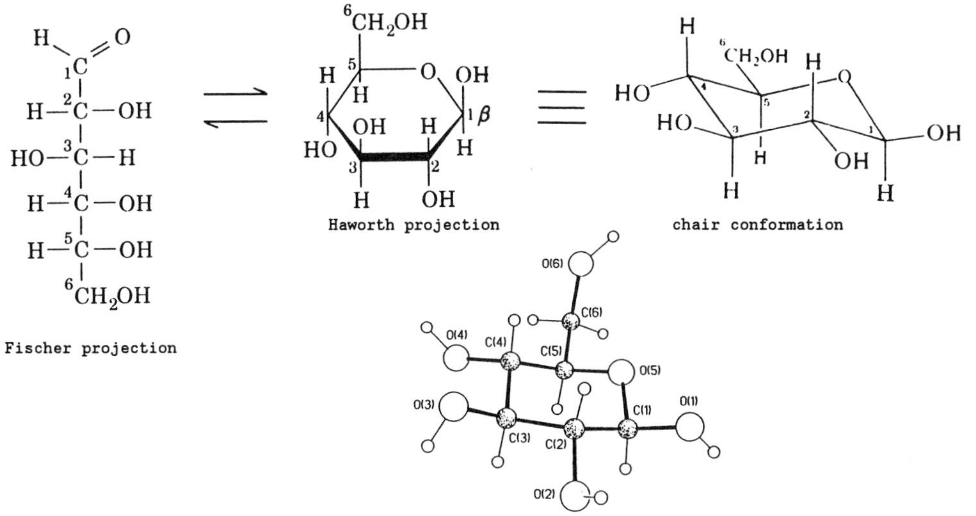

Fischer projection

Haworth projection

chair conformation

Fig. 5.17-5 Structural formula and conformation of the β-D-glucose molecule.

3. Structure of lactose, a disaccharide[13]

The α- and β-lactoses are disaccharides composed of a galactose and an α- or β-glucose moiety linked through a glycoside bond. They constitute a pair of anomers that interconvert spontaneously in solution.

α-Lactose crystallizes from water as a monohydrate in space group $P2_1$, with a = 7.982, b = 21.652, c = 4.824Å, β = 109.52°, and Z = 2.[12]   The molecular structure is illustrated in Fig. 5.17-6(a). The crystals display a characteristic wedge-shaped morphology with the apex at the -b end, which arises as a consequence of the molecular packing [Fig. 5.17-6(b)].

The crystals were observed to grow only in the +b direction, which was interpreted as an inhibition of growth along -b by the β-anomer present in solution. It is obvious that β-lactose can be adsorbed only at the -b end of the crystal by virtue of the unmodified galactose moiety. Once adsorbed, the modified β-glucose inhibits growth perpendicular to the {011} faces. Therefore, by fixation of the absolute polarity of the molecule inside the crystal, its absolute configuration can be assigned.

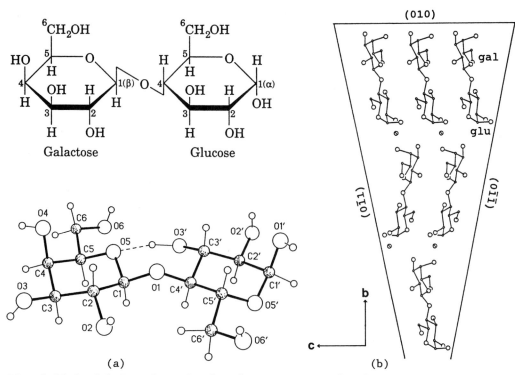

Fig. 5.17-6 (a) Formula and molecular structure of α-lactose. (b) Packing arrangement of α-lactose monohydrate delineated by the observed crystal faces, as viewed along the *a* axis. (Adapted from ref. 13).

β-lactose crystallizes in space group $P2_1$ with $a = 10.839(6)$, $b = 13.349(6)$, $c = 4.954(5)$Å, $β = 91.31(9)°$, and $Z = 2$.[14] In the crystal, the molecular conformation [Fig. 5.17-7] is nearly isostructural with that of β-cellobiose [Fig. 5.17-9] except for the axial O(4) atom. The exocyclic C(5)-C(6) (1.508Å) and C(5')-C(6') (1.510Å) bonds are significantly shorter than the other C-C bonds. The lengths of the two ring C-O bonds are unequal in the galactose unit and equal in the glucose unit. The C(1)-O(1) (1.402Å) and C(1')-O(1') (1.388Å) bond lengths are shorter by 0.015Å and 0.030Å than the mean exocyclic C-O length of 1.417Å. The structure contains an intramolecular hydrogen-bond [O(3')-H(3')...O(5)] and exhibits an unsymmetrical twist about the bridge bond. All the oxygen atoms, except the bridge oxygen atom, are involved in the hydrogen-bonding network, which comprises two terminating chains.

4. Structure of raffinose, a trisaccharide

Raffinose is a trisaccharide, galactosyl-glucosyl-fructose, that occurs naturally in great abundance. The crystal data for the pentahydrate are: space group $P2_12_12_1$, $a = 8.966(10)$, $b = 12.327(15)$, $c = 23.837(24)$Å, and

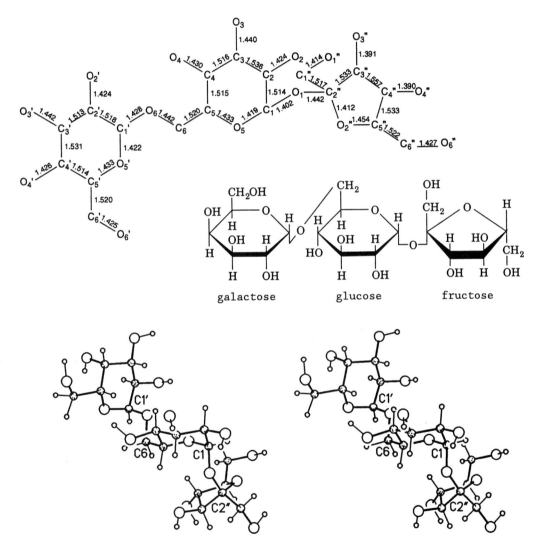

Fig. 5.17-7 Structural formula and molecular conformation of β-lactose.

Fig. 5.17-8 Structure formula, bond lengths, and stereoview of the conformation of raffinose.

$Z = 4$.[15]   Fig. 5.17-8 illustrates the configuration and conformation of the raffinose molecule with standard atom numbering.

The two pyranose rings are in the chair form and the furanose ring is puckered. The $C(1')-O(6)-C(6)$ angle (111°) is the smallest one reported thus far for a glycosidic link. The conformation angles of the pyranose rings range from 52.7 to 59.5° and 53.8 to 62.2° for the glucose and galactose units respectively, in comparison with 54.8 to 56.0 for sucrose.[4] The furanose ring has a geometry similar to that of the fructose moiety of sucrose. In sucrose the fructoside $O(1')$ and $O(6')$ are hydrogen-bond donors to the glucosidic oxygen atoms $O(2)$ and $O(5)$, whereas in raffinose the sucrose moiety is not involved in intramolecular hydrogen bonding.

5. Structure of cellulose

Cellulose, $[C_6H_{10}O_5]_n$, is the chief component of wood and plant fibers, and cotton is nearly pure cellulose. It is insoluble in water and is tasteless, and is a non-reducing carbohydrate. Complete hydrolysis by acid yields D-(+)-glucose as the only monosaccharide. Cellulose is made up of chains of D-glucose units each joined by a glycoside linkage to C-4 of the next, as follow:

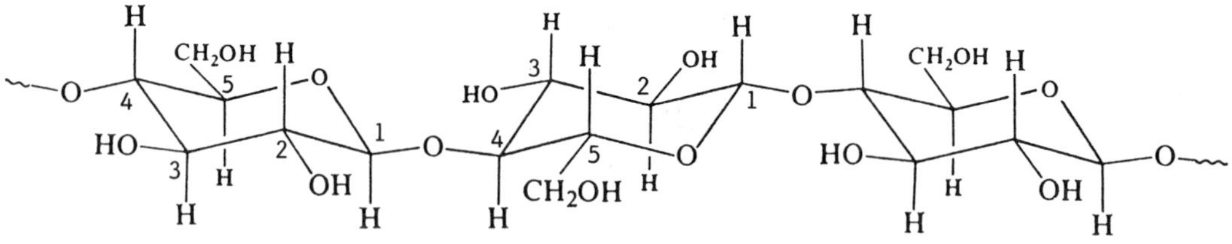

The molcular weight of cellulose ranges from 250,000 to 1,000,000 or even higher, with at least 1500 glucose units per molecule. Cellulose I occurs naturally in practically all native cellulose structures in terrestrial and aquatic plants, as well as in those produced by microorganisms. Cellulose II is the normal conversion product obtained by mercerization (treatment with NaOH) of cellulose I.

In the crystal structure of cellobiose,[11] the repeating disaccharide of cellulose, the $^4C_1$ pyranose ring shape of the D-glucose residue and the equatorial disposition of the $C(1)-O(1)$ bond corresponding to the β-D configuration were confirmed (Fig. 5.17-9). The disaccharide assumes the "Hermans conformation" with an intramolecular $O(3')...O(5)$ hydrogen bond, that is, from the reducing to the non-reducing residue.[16]

Recent investigations employing powerful new methods, particularly those which combine X-ray diffraction analysis with computer-based stereochemical

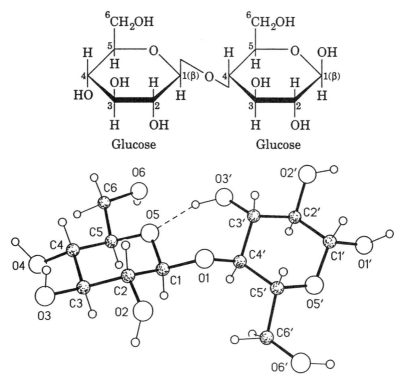

Fig. 5.17-9 Structural formula and molecular conformation of cellobiose.

modeling and cross polarization-magic angle spinning ^{13}C NMR, have yielded new information on the structure of cellulose.[17]

Cellulose I has a parallel chain structure [Fig. 5.17-10(a)], whereas cellulose II has an antiparallel structure and extensive three-dimensional hydrogen bonding [Fig. 5.17-10(b)], making it very stable. Cellulose I consists of three types of cellobiose units: (i) O(3)H...O(5'), (ii) O(2)H...O(6') intramolecular hydrogen bonds, and (iii) units without hydrogen bonds. Cellulose II contains four types of cellobiose units: (i) O(3)H...O(5'), (ii) O(2)H...O(6'), (iii) O(6)H...O(2') intramolecular hydrogen bonds, and (iv) units without intramolecular hydrogen bonds. This knowledge has led to the development of a commercial process for modifying and/or destroying the intramolecular hydrogen bonds in wood cellulose pulp.

Unlike cotton, the cellulose in wood is locked within a well-defined matrix made up of cellulose (42-48%), hemicellulose (27-38%) and lignin (19-29%) with varying proportions between species (Fig. 5.17-11). The cellulose microfibrils constitute the framework; the lignin-rich middle lamella is the bonding medium that interconnects the hemicellulose within the cell walls and imparts structural support. This composite structure is responsible for its toughness and high tensile strength.

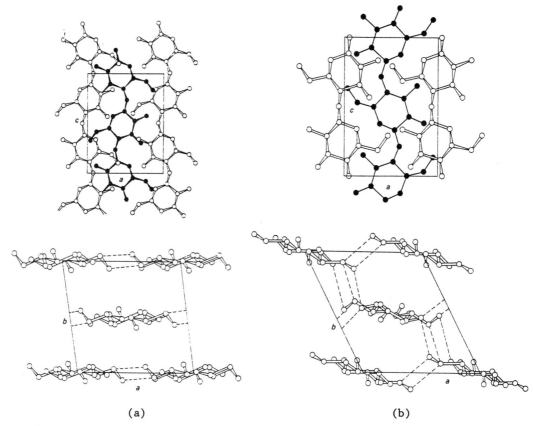

(a)   (b)

Fig. 5.17-10 The structure of cellulose I (a) and cellulose II (b).
(After ref. 18).

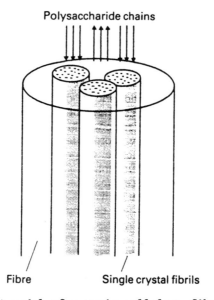

Fig. 5.17-11 A model of a ramie cellulose fiber. (After ref. 18).

6. Absolute configuration of raffinose adsorbed on a sucrose crystal[13]

A sucrose crystal is chiral as it contains a polar axis. It has been demonstrated that radiolabeled occluded reffinose is adsorbed through the {110} faces of a sucrose crystal, and that raffinose inhibits growth of sucrose at the +b pole of the crystal and also along its +c and -c directions. These observations can be taken advantage of for the direct assignment of absolute configuration. Fig. 5.17-12(a) depicts the packing arrangement of sucrose. The structure is delineated by the various faces of the crystal; the fructose moiety emerges at the {01̄1} faces, directed toward -b, whereas the glucose residue emerges at the {110} faces. The O-H vector of the CH_2OH group attached to C(6) in sucrose points toward -b, the C(6)-C-O-C linkage adopting a *gauche* conformation [Fig. 5.17-12(b)]. The faces of the sucrose crystal to which raffinose may adhere depend on the conformation of the trisaccharide.

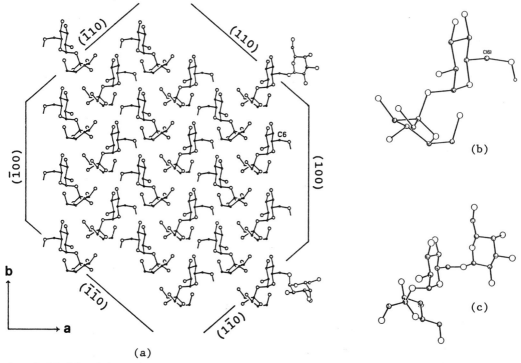

(a)

Fig. 5.17-12 (a) Packing arrangement of sucrose bounded by the observed crystal faces, as viewed along the c axis. The emerging galactosyl moiety of a raffinose inhibitor molecule has been inserted at the {110} face in the *exo* conformation. Note that the corresponding hydroxyl of sucrose adopts an *endo* conformation. The sugar ring inserted at the {1̄10} face indicates the position that would be taken by the galactosyl moiety of a raffinose inhibitor molecule in the *endo* conformation. (b) and (c) Conformations of sucrose and raffinose in their native crystals. (After ref. 13).

The raffinose molecule bears a sucrose moiety linked to a galactose unit through a glycoside bond at C(6) of glucose. In the fully extended *exo* conformation [i.e., C(6)-C-O-C (galactose) anti-planar as shown in Fig. 5.17-12(c)], raffinose may adhere only at the {110} faces, whereas in the *endo* conformation [with C(6)-C-O-C (galactose) *gauche*], it may adhere only at the {1$\bar{1}$0} faces and thus impede growth along -b [Fig. 5.17-12(a)]. The latter possibility would be compatible with the observed conformation of the CH$_2$OH group of sucrose [Fig. 5.17-12(b)]. From the experimental results one may infer that raffinose must assume the *exo* conformation (as in its parent crystal[19]) on the {110} surface of a growing sucrose crystal.

References

[1] G.M. Brown and H.A. Levy, *Acta Crystallogr., Sect. B* **29**, 790 (1973).

[2] J.C. Hanson, L.C. Sieker and L.H. Jensen, *Acta Crystallogr., Sect. B* **29**, 797 (1973).

[3] C.A. Beevers, T.R.R. McDonald, J.H. Robertson and F. Stern, *Acta Crystallogr.* **5**, 689 (1952).

[4] G.M. Brown and H.A. Levy, *Science (Washington)* **141**, 921 (1963).

[5] J.F. Stoddart, *Stereochemistry of Carbohydrates*, Wiley-Interscience, New York, 1971.

[6] D.H.R. Barton, *Experientia* **6**, 316 (1950); *J. Chem. Soc.*, 1027 (1953).

[7] O. Hassel, *Quart. Rev. (London)* **7**, 221 (1953).

[8] G.A. Jeffrey, *Acta Crystallogr., Sect. B* **46**, 89 (1990).

[9] G.M. Brown and H.A. Levy, *Science (Washington)* **147**, 1038 (1965); *Acta Crystallogr., Sect. B* **35**, 656 (1979).

[10] W.G. Ferrier, *Acta Crystallogr.* **16**, 1023 (1963).

[11] S.S.C. Chu and G.A. Jeffrey, *Acta Crystallogr., Sect. B* **24**, 830 (1968).

[12] D.C. Fries, S.T. Rao and M. Sundaralingam, *Acta Crystallogr., Sect. B* **27**, 995 (1971).

[13] L. Addadi, Z. Berkovitch-Yellin, I. Weissbach, M. Lahav and L. Leiserowitz, *Top. Stereochem.* **16**, 1 (1986).

[14] K. Hirotsu and A. Shimada, *Bull. Chem. Soc. Jpn.* **47**, 1872 (1974).

[15] H.M. Berman, *Acta Crystallogr., Sect. B* **26**, 290 (1970).

[16] R.H. Marchessault and A. Sarko, *Adv. Carbohydr. Chem.* **22**, 421 (1967).

[17] R.H. Atalla (ed.), *The Structure of Cellulose: Characterization of the Solid States*, Symposium Series 340, American Chemical Society, Washington, DC, 1987.

[18] G.O. Phillips, *Chem. Brit.* **25**, 1006 (1989).

[19] G.A. Jeffrey and Y.J. Park, *Acta Crystallogr., Sect. B* **28**, 257 (1972).

5.18 (+)-8-Bromocamphor
$$C_{10}H_{15}OBr$$

Crystal Data

Monoclinic, space group $P2_1$ (No. 4)

$a = 9.411(1)$, $b = 7.928(2)$, $c = 6.873(2)$Å, $\beta = 92.71(2)°$, $Z = 2$

Atom	x	y	z	Atom	x	y	z
Br	1.1274	.5000	.1812	C(5)	.6620	.6666	.3640
O	.6106	.3411	-.0782	C(6)	.5931	.4903	.3722
C(1)	.6885	.3705	.2586	C(7)	.8371	.4528	.3070
C(2)	.6668	.4216	.0459	C(8)	.9487	.3692	.1834
C(3)	.7318	.6004	.0333	C(9)	.8880	.4418	.5250
C(4)	.7894	.6296	.2390	C(10)	.6746	.1789	.2918

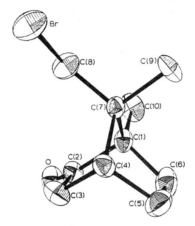

Fig. 5.18-1 Structure and absolute configuration of (+)-8-bromocamphor. (After ref. 1)

Crystal Structure

The molecular structure and absolute configuration of (+)-8-bromocamphor are shown in Fig. 5.18-1. The norbornane skeleton has normal geometry; the angles between the three-atom bridge plane [C(1), C(7) and C(4)] and the two four-atom planes of the boat-shaped six-membered ring are 124.5° and 124.2°. Bond lengths and angles (Table 5.18-1) are close to normal values, the bridgehead angle C(1)-C(7)-C(4) being 94°, and intermolecular distances

Table 5.18-1 Bond lengths (Å) and angles (°) of (+)-8-bromocamphor.

Br-C(8)	1.98	C(2)-C(1)-C(6)	106
O-C(2)	1.17	C(2)-C91)-C(7)	100
C(1)-C(2)	1.52	C(6)-C(1)-C(7)	100
C(1)-C(6)	1.54	C(1)-C(2)-C(3)	105
C(1)-C(7)	1.57	C(2)-C(3)-C(4)	102
C(1)-C(10)	1.54	C(3)-C(4)-C(5)	107
C(2)-C(3)	1.55	C(3)-C(4)-C(7)	103
C(3)-C(4)	1.51	C(5)-C(4)-C(7)	103
C(4)-C(5)	1.54	C(4)-C(5)-C(6)	101
C(4)-C(7)	1.54	C(1)-C(6)-C(5)	106
C(5)-C(6)	1.54	C(1)-C(7)-C(4)	94
C(4)-C(8)	1.53	C(1)-C(7)-C(8)	109
C(7)-C(9)	1.55	Br-C(8)-C(7)	112

correspond to van der Waals interactions.

It is worth noting that the X-ray study of d-α-Br, Cl, and CN-camphor by Wiebenga and Krom constituted one of the earliest applications of isomorphous replacement to the direct determination of crystal structures.[2] Historically this was preceded by the investigation of the metal phthalocyanines by Robertson (Section 4.10), and of glucosamine hydrochloride and hydrobromide by Cox and Jeffrey.[3]

Remarks

1. Terpenes and the isoprene rule

The terpenoids are perhaps the most numerous and varied class of organic compounds occurring in nature, embracing a wide range of substances from the simple monoterpenoids to the highly complex triterpenoids and carotenoids.[4] Most are found in plants but others are produced by microorganisms and as sex hormones by insects.

Camphor is a monoterpene occurring in essential oils, which are of great commercial importance in the perfume and flavoring industries. The chemical constituents of the essential oils have a large variety of structures, many of them containing one or more rings and/or double bonds. Most of these compounds are composed of even multiples of five-carbon isoprene (2-methyl-1,3-butadiene) fragments.

Terpenoids are classified according to the number of isoprene units in the molecule.

Number of carbon atoms	Class
10	Monoterpenoids
15	Sesquiterpenoids
20	Diterpenoids
25	Sesterterpenoids
30	Triterpenoids
40	Tetraterpenoids
>40	Polyterpenoids

Carotenes are plant pigments in the tetraterpenoid class, and polyterpenoids include rubber, gutta-percha and balata. Some examples of monoterpenes are shown below.

| Camphor | R-(+)-limonene from lemon and orange oils | (-)-menthol from peppermint oil | R-(-)-linaloöl (-)-from rose oil (+)-from orange oil |

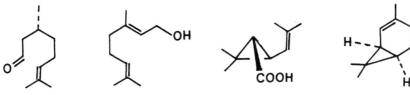

R-(+)-citro-   geraniol   (+)-*trans*-chrysanthemic   (+)-2-carene
nellal from   from Turkish   acid   from various
citronella oil   geranium oil   essential oils

The last two compounds shown are interesting from a historical perspective. In 1920, Ruzicka furnished a proof of the structure of chrysanthemic acid while the structure of 2-carene was determined by Simonsen. Ruzicka was struck by the structural resemblance between the two compounds and recognized the five-carbon isoprene skeleton as a common building block.

Ruzicka hypothesized that these plant constituents were synthesized in nature by a head-to-tail connection of isoprene units. He examined a large number of terpenoid compounds over many years and found that the structures of most of them showed this regularity, which he termed the "isoprene rule". In 1939, Ruzicka shared the Nobel Prize in Chemistry for his contributions to terpenoid chemistry with Butenandt who worked on sex hormones.

Given a choice of possible structures for a compound that contains a multiple of five carbon atoms, chemists generally favour the one that appears to be made of isoprene units connected in a head-to-tail fashion. The isoprene rule does not predict whether the compound contains double bonds or whether functional groups containing oxygen are present. A wide variety of ring patterns, unsaturation, and combination of alcohol, carbonyl, and carboxylic acid functional groups are found in natural terpenes.

2. Monoterpenoids

The crystal structures of many monoterpenes have been reported.[5] Some examples are listed in Table. 5.18-2.

3. Structure of qinghaosu, a sesquiterpene

Qinghaosu is an active principle of the Chinese medicinal herb *Artemisia annua* L., which has been shown to be an effective antimalarial of low toxicity. Its crystal structure was determined in 1980, and the absolute configuration [Fig. 5.18-3(a)] was established through careful measurement of the intensities of some enantiomer-sensitive Bijvoet pairs.[15]

The molecular skeleton of qinghaosu comprises four fused rings: A[C(1), C(2), C(3), C(4), C(5), C(6)], B[C(6), C(7), O(3), C(8), C(9), C(10), C(1)],

Table 5.18-2 Some monoterpenoids of known structure.

Monoterpene	Structural formula in Fig. 5.18-2	Structural feature	Ref.
Iridomyrmecin	(a)	The six-membered ring is in boat conformation with the cyclopentanoid ring endo to the six-membered ring.	[6]
(-)-Bromodihydro-unbellulone	(b)	The ring has a planar conformation.	[7]
trans-2,8-Dihydroxy-1(7)-p-menthene	(c)	The ring is in chair conformation with the ring -OH group axial and the hydroxyisopropyl group equatorial.	[8]
1-p-Menthene-3,6-diol	(d)	The cyclohexene ring is in the half-chair conformation. The two -OH groups are oriented syn with respect to one another, anti with respect to the isopropyl group.	[9]
3,4-Epoxy-1,2-naphthalenediol	(e)	The saturated ring is in a twist-boat conformation. The epoxide ring appears to be polarized toward the O atom.	[10]
Xylomollin	(f)	The ether ring has a chair conformation and the lactone ring a slightly flattened chair.	[11]
Cannabidiol	(g)	The cyclohexene ring is in the half-chair conformation.	[12]
Anhydrobromonitro-camphane	(h)	The isoxazoline ring is not planar.	[13]
Camphoric anhydride	(i)	The six-membered ring shows a half-boat comformation with marked non-planarity of the anhydride moiety.	[14]

$C[C(6), C(7), O(3), C(8), O(1), O(2)]$ and $D[C(5), C(6), C(7), O(4), C(12), C(11)]$. Ring A is chair-shaped while ring D is a δ-lactone that assumes the shape of a distorted chair. The B and C rings are both saturated oxo-

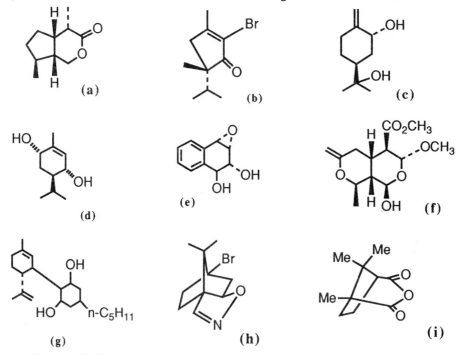

Fig. 5.18-2 Structural formulae of some monoterpenes.

heterocycles. The junctions **A/D**, **A/B**, and **C/D** are *cis* while the **D** and **B** rings
are fused in the *trans*-manner.

In the qinghaosu molecule, all five oxygen atoms crowd on the same
side of the molecule. In the carbon-oxygen chain O(5)-C(12)-O(4)-C(7)-O(3)-
C(8)-O(1)-O(2)-C(6) the C-O bond distances (Å) starting from C(12)-O(4) are in
the sequence short, long, short, long, ... as follows:

C(12)-O(4)	1.339(8),	O(4)-C(7)	1.451(8),
C(7)-O(3)	1.390(7),	O(3)-C(8)	1.437(7),
C(8)-O(1)	1.403(8),	O(1)-O(2)	1.478(6),
O(2)-C(6)	1.456(6).		

All these bond distances lie well within the range between a normal single
bond and a partial double bond. This implies that the lone pair electrons on
oxygen are no longer localized, and a variation of bond types tends to make
the entire molecule more stable. The result is in agreement with the fact
that qinghaosu is stable towards both heat and light. Probably bond-length
alternation and the presence of the peroxide linkage both contribute to the
chemotherapeutic activity of the qinghaosu molecule.

The arteether, a β-ethyl ether isomer of dihydroqinghaosu, is a novel
antimalarial.[16] The crystal structure of the ether dimer of

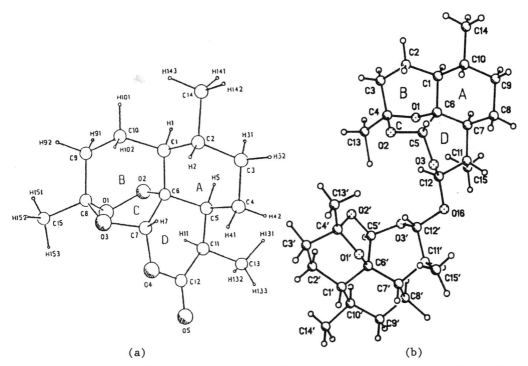

(a)   (b)

Fig. 5.18-3 Molecular structures of (a) qinghaosu and (b) a dimer of
deoxydihydroqinghaosu. (After refs. 15 and 17).

deoxydihydroqinghaosu, a potential metabolite of the antimalarial arteether, has been determined recently [Fig. 5.18-3(b)].[17] The dimer adopts a conformation such that the two deoxyarteether moieties are *cis* to each other. The X-ray structures of both the *cis*[16] and *trans*[18] [C(15) relative to O(16)] forms of the deoxyarteether monomer have been reported. In the dimer the conformation of the deoxyarteether moieties is the same as that found for the *trans* isomer. Ring A has a normal chair conformation while ring B has a slightly distorted chair conformation. Ring D assumes a somewhat distorted conformation, and the five-membered ring C is a normal envelope with O(1) as the out-of-plane atom. In the dimer, the methyl group on C(11) and the oxygen atom on C(12) are *gauche* with respect to each other.

4.   Structure of 3-O-acetylgibberellin A_3, a diterpene

The gibberellins are a large group of plant growth hormones isolated from culture filtrates of the fungus *Fibberella fujikuroi*. Among these active fungal metabolites, gibberellin A_3 (or gibberellic acid) is known to be the most important metabolite of the fungus and has been the subject of exhaustive chemical studies. All gibberellins are diterpenoid acids based on the tetracyclic gibberellane skeleton. The structure and absolute configuration of gibberellin A_3 have been determined from X-ray studies of some of its derivatives.[19,20]

The structure of 3-O-acetylgibberellin A_3 was carried out to characterize the compound fully.[21] The atom-numbering scheme and ring designations of the molecule are given in Fig. 5.18-4, and a stereoview is shown in Fig. 5.18-5. Bond distances and bond angles are within the expected ranges. Slight lengthening of bonds C(3)-C(4) = 1.557(6), C(6)-C(8) = 1.575(5) and C(8)-C(9) = 1.564(5)Å can be related to steric strain as substituents at all these atoms have the β configuration. In the lactone ring, the distances C(19)-O(4) = 1.337(5) and C(10)-O(4) = 1.509(4)Å agree with the corresponding values of 1.345 and 1.492Å observed in 3-dehydro-gibberellin A_3 methyl ester,[22] and 1.34 and 1.51Å in 3-dehydrogibberellin A_3.[23]

The gibberellane skeleton shows the conformational variety characteristic of many other gibberellin derivatives. Ring A is in a 5β-sofa conformation, the five-membered ring B has an intermediate conformation between a 10α-envelope and $10\alpha,5\beta$-half-chair, ring C is boat-shaped, ring D adopts an envelope conformation, and the γ-lactone ring is an almost ideal envelope with atom C(5) as the flap.

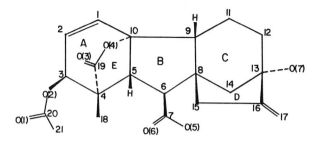

Fig. 5.18-4 Structural formula of 3-*O*-acetylgibberellin A$_3$ showing the atom numbering and the absolute configuration. (After ref. 21).

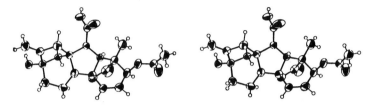

Fig. 5.18-5 A stereoview of the 3-*O*-acetyl-gibberellin A$_3$ molecule. (After ref. 21).

5. Triterpenoid saponins

The saponins are a group of plant constituents possessing characteristic properties such as hemolytic and surface activities, the ability to form molecular complexes with cholesterol, and toxicity towards fishes and amphibians. A saponin consists of a sugar and a sapogenin, which constitutes the aglucon moiety of the molecule. The sugar may be glucose, galactose, a pentose, or a methylpentose, and the sapogenin may be steroidal or terpenoid in nature. The saponins are widely distributed in plants; they foam easily when shaken with water, and the emulsions can act as protective colloids.

Abruslactone A, a new triterpenoid sapogenin isolated from the roots and vines of *Abrus precatorius* L., has been characterized by spectral and X-ray analysis as the γ-lactone of 3β,22α-dihydroxy-olean-12-en-29α-oic acid (Fig. 5.18-6).[24] The most notable feature in this structure is that ring E adopts a boat conformation, which differs markedly from the usual chair conformation observed in the basic olean-12-ene skeletons of hederagenin, sophoradiol, and soyasapogenol B. Examination of a molecular model shows that formation of the γ-lactone ring is facilitated by manipulating ring E into the observed boat conformation.

The crystal structures of cantoniensistriol and sophoradiol, two oleanane-type triterpenes from the roots of *Abrus cantoniensis* Hance, have been determined by X-ray analysis of their acetates.[25] The atom numbering

schemes and torsion angles (°) are shown in Fig. 5.18-7. In these two
sapogenol derivatives, the six-membered rings A/B and B/C are *trans*-fused and
D/E *cis*-fused; the presence of the double bond in ring C accounts for the
observed twist conformation. The basic olean-12-ene skeletons of these
compounds have essentially the same geometry as that found in hederagenin.[26]

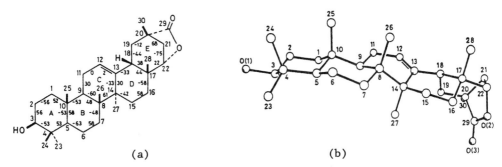

Fig. 5.18-6 (a) Molecular formula and (b) molecular structure of
abruslactone A. (After ref. 24).

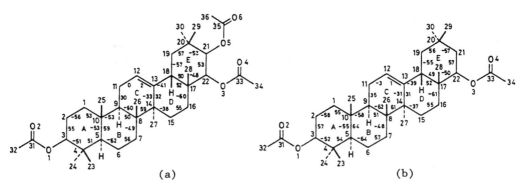

Fig. 5.18-7 Atomic numbering and torsion angles (°) for (a) cantoniensistriol
triacetate and (b) sophoradiol diacetate. (After ref. 25).

References

[1] C.A. Bear and J. Trotter, *Acta Crystallogr., Sect. B* **31**, 903 (1975).

[2] E.H. Wiebenga and C.J. Krom, *Recueil* **65**, 663 (1946).

[3] E.G. Cox and G.A. Jeffrey, *Nature (London)* **143**, 894 (1939).

[4] J.S. Glasby, *Encyclopaedia of the Terpenoids*, Vols. 1 and 2, Wiley-
     Interscience, Chichester, 1982.

[5] R.B. Yeats in J.R. Hanson (Senior Reporter), *Terpenoids and Steroids*,
     Vol. 9, The Chemical Society, London, 1979, p. 3.

[6] J.F. McConnell, A. McL. Mathieson and B.P. Schoenborn, *Acta
     Crystallogr.* **17**, 472 (1964).

[7] H.E. Smith, R.T. Gray, T.J. Shaffner and P.G. Lenhert, *J. Org. Chem.*
     34, 136 (1969).

[8] W.E. Scott and G.F. Richards, *J. Org. Chem.* **36**, 63 (1971).

[9] T.J. Delord, A.J. Malcolm, F.R. Fronczek, N.H. Fischer and S.F.
     Watkins, *Acta Crystallogr., Sect. C* 44, 765 (1988).

[10] C.L. Klein and E.D. Stevens, *Acta Crystallogr., Sect. B* **44**,
     50 (1988).

[11] M. Nakane, C.R. Hutchinson, D. VanEngen and J. Clardy, *J. Am. Chem.
     Soc.* **100**, 7079 (1978).

[12] P.G. Jones, L. Falvello, O. Kennard and G.M. Sheldrick, *Acta
     Crystallogr., Sect. B* **33**, 3211 (1977).

[13] G.L. Dwivedi and R.C. Srivastava, *Acta Crystallogr., Sect. B* **28**,
     2567 (1972).

[14] K. Wichmann, H. Bradaczek, Z. Dauter and T. Polonski, *Acta Crystallogr.,
     Sect. C* 43, 577 (1987).

[15] Liang Li, Dong Yicheng and Zhu Naijue, *Jiegou Huaxue (J. Struct. Chem.)*
     5, 73 (1986).

[16] A. Brossi, B. Venugopalan, D.L. Gerpe, H.J.C. Yeh, J.L. Flippen-
     Anderson, P. Buchs, X.D. Luo, W. Milhous and W. Peters, *J. Med.
     Chem.* **31**, 645 (1988).

[17] J.L. Flippen-Anderson, C. George, R. Gilardi, Q.-S. Yu,
     L. Dominguez and A. Brossi, *Acta Crystallogr., Sect. C* **45**, 292 (1989).

[18] D.L. Gerpe, H.J.C. Yeh, A. Brossi and J.L. Flippen-Anderson,
     Heterocycles **27**, 897 (1988).

[19] J.A. Hartsuck and W.N. Lipscomb, *J. Am. Chem. Soc.* **85**, 3414 (1963).

[20] F. McCapra, A.T. McPhail, A.I. Scott, G.A. Sim and D.W. Young, *J.
     Chem. Soc. (C)*, 1577 (1966).

[21] M.B. Hossain, D. van der Helm, R. Sanduja and M. Alam, *Acta
     Crystallogr., Sect. C* 44 1022 (1988).

[22] L. Kutschabsky, G. Reck, E. Hohne, B. Voigt and G. Adam, *Tetrahedron* **36**,
     3421 (1980).

[23] L. Kutschabsky, G. Reck and G. Adam, *Tetrahedron* **31**, 3065 (1975).

[24] H.-M. Chang, T.-C. Chiang and T.C.W. Mak, *J. Chem. Soc., Chem.
     Commun.*, 1197 (1982).

[25] T.C.W. Mak, T.-C. Chiang and H.-M. Chang, *J. Chem. Soc., Chem.
     Commun.*, 785 (1982).

[26] R. Roques, J.P. Declercq and G. Germain, *Acta Crystallogr., Sect. B*
     34, 1634 (1978).

Note The structures, reactions, and syntheses of arteannuin (Qinghaosu) and
related compounds are reviewed by W.-S. Zhou and X.-X. Xu in A. Rahman
(ed.), *Studies in Natural Products Chemistry*, Vol. 3: *Stereoselective
Synthesis, Part B*, Elsevier, Amsterdam, 1989.

5.19   Cholesterol Hemiethanolate

$$C_{27}H_{46}O \cdot \tfrac{1}{2}C_2H_5OH$$

Crystal Data

Triclinic, space group $P1$ (No. 1)

$a = 12.787(2)$, $b = 35.310(11)$, $c = 12.225(1)$Å

$\alpha = 97.80(2)$, $\beta = 100.40(2)$, $\gamma = 99.06(2)°$, $Z = 8$

Atom	x	y	z	Atom	x	y	z
Ethanol molecules							
C(1)E1	.4243	.4924	.0137	C(1)E2	1.0336	.5077	.4941
C(2)E1	.5047	.5126	-.0252	C(2)E2	.9618	.4907	.5292
OE1	.4480	.4981	.1361	OE2	.9866	.4950	.3651
C(1)E3	.4344	.4970	.5243	C(1)E4	1.0244	.5087	.9832
C(2)E3	.5119	.5038	.4751	C(2)E4	.9331	.4980	1.0315
OE3	.4637	.4928	.6423	OE4	1.0045	.5003	.8698
Cholesterol molecules							
O(3)A	.3067	.5222	.2536	O(3)B	.8039	.5182	.3056
C(1)A	.3441	.6266	.2163	C(1)B	.8513	.6212	.2546
C(2)A	.3421	.5831	.1865	C(2)B	.8547	.5781	.2344
C(3)A	.3200	.5648	.2906	C(3)B	.8071	.5596	.3268
C(4)A	.2039	.5694	.3112	C(4)B	.6954	.5688	.3350
C(5)A	.2011	.6142	.3242	C(5)B	.6914	.6101	.3419
C(6)A	.1658	.6305	.4139	C(6)B	.6491	.6288	.4205
C(7)A	.1478	.6714	.4320	C(7)B	.6339	.6690	.4353
C(8)A	.1599	.6924	.3314	C(8)B	.6573	.6893	.3381
C(9)A	.2569	.6824	.2862	C(9)B	.7551	.6759	.2953
C(10)A	.2381	.6376	.2379	C(10)B	.7370	.6317	.2582
C(11)A	.2812	.7072	.1937	C(11)B	.7918	.7016	.2075
C(12)A	.2978	.7513	.2365	C(12)B	.8073	.7444	.2426
C(13)A	.2021	.7607	.2787	C(13)B	.7012	.7566	.2768
C(14)A	.1832	.7374	.3708	C(14)B	.6776	.7323	.3701
C(15)A	.0976	.7533	.4262	C(15)B	.5939	.7527	.4162
C(16)A	.1252	.7971	.4171	C(16)B	.6209	.7954	.3980
C(17)A	.2137	.8014	.3481	C(17)B	.7158	.7971	.3373
C(18)A	.1012	.7528	.1800	C(18)B	.6107	.7484	.1751
C(19)A	.1513	.6259	.1284	C(19)B	.6605	.6171	.1415
C(20)A	.2186	.8367	.2871	C(20)B	.7161	.8326	.2686
C(21)A	.3133	.8426	.2277	C(21)B	.8198	.8358	.2120
C(22)A	.2124	.8742	.3619	C(22)B	.7164	.8695	.3445
C(23)A	.1964	.9100	.2991	C(23)B	.7214	.9069	.2935
C(24)A	.1037	.9062	.2112	C(24)B	.6069	.9028	.2155
C(25)A	.1137	.9445	.1496	C(25)B	.5940	.9417	.1776
C(26)A	.1218	.9706	.2149	C(26)B	.6056	.9406	.0717
C(27)A	-.0047	.9362	.0652	C(27)B	.4940	.9487	.1684
O(3)C	.1448	.4704	.2746	O(3)D	.1662	.4779	.7768
C(1)C	.1153	.3627	.2766	C(1)D	.1102	.3690	.7845
C(2)C	.0817	.4038	.2892	C(2)D	.0854	.4113	.7881
C(3)C	.1767	.4331	.2675	C(3)D	.1835	.4380	.7657
C(4)C	.1956	.4203	.1475	C(4)D	.2009	.4255	.6488
C(5)C	.2174	.3871	.1366	C(5)D	.2226	.3833	.6425
C(6)C	.3047	.3715	.0984	C(6)D	.3080	.3746	.6010
C(7)C	.3316	.3318	.0756	C(7)D	.3350	.3339	.5800
C(8)C	.2380	.2983	.0824	C(8)D	.2383	.3030	.5844
C(9)C	.1841	.3105	.1780	C(9)D	.1868	.3162	.6817
C(10)C	.1363	.3475	.1648	C(10)D	.1381	.3529	.6676
C(11)C	.1006	.2755	.1933	C(11)D	.0955	.2808	.6956
C(12)C	.1497	.2389	.2052	C(12)D	.1421	.2437	.7093
C(13)C	.2026	.2263	.1113	C(13)D	.1949	.2308	.6131
C(14)C	.2846	.2608	.1011	C(14)D	.2824	.2669	.6090
C(15)C	.3510	.2458	.0209	C(15)D	.3512	.2500	.5329
C(16)C	.3589	.2054	.0507	C(16)D	.3469	.2075	.5559
C(17)C	.2773	.1979	.1321	C(17)D	.2693	.2007	.6357
C(18)C	.1187	.2144	.0037	C(18)D	.1131	.2174	.5016
C(19)C	.0321	.3381	.0736	C(19)D	.0380	.3463	.5721
C(20)C	.2266	.1533	.1189	C(20)D	.2190	.1589	.6164
C(21)C	.1480	.1454	.2013	C(21)D	.1398	.1480	.6884
C(22)C	.3258	.1262	.1307	C(22)D	.3065	.1310	.6337
C(23)C	.2681	.0849	.0781	C(23)D	.2743	.0868	.5757
C(24)C	.3754	.0657	.0550	C(24)D	.3834	.0744	.5511
C(25)C	.4187	.0707	-.0602	C(25)D	.3952	.0620	.4245
C(26)C	.3427	.0521	-.1529	C(26)D	.3401	.0575	.3630
C(27)C	.4966	.0512	-.0463	C(27)D	.4816	.0535	.4566

O(3)E	.3286	.5288	.7375	O(3)F	.8324	.5239	.7598
C(1)E	.2439	.6254	.8157	C(1)F	.7391	.6180	.8138
C(2)E	.2596	.5837	.8247	C(2)F	.7569	.5771	.8308
C(3)E	.3068	.5676	.7305	C(3)F	.8121	.5631	.7371
C(4)E	.4124	.5920	.7278	C(4)F	.9238	.5878	.7500
C(5)E	.4055	.6355	.7258	C(5)F	.9015	.6283	.7329
C(6)E	.4395	.6539	.6483	C(6)F	.9436	.6475	.6580
C(7)E	.4393	.6952	.6414	C(7)F	.0373	.6886	.6419
C(8)E	.4071	.7173	.7447	C(8)F	.9040	.7105	.7422
C(9)E	.3147	.6916	.7773	C(9)F	.8136	.6860	.7840
C(10)E	.3498	.6537	.8131	C(10)F	.8417	.6485	.8181
C(11)E	.2693	.7143	.8701	C(11)F	.7654	.7103	.8740
C(12)E	.2425	.7532	.8434	C(12)F	.7403	.7481	.8416
C(13)E	.3382	.7781	.8120	C(13)F	.8343	.7737	.8091
C(14)E	.3667	.7537	.7164	C(14)F	.8681	.7469	.7168
C(15)E	.4439	.7832	.6695	C(15)F	.9445	.7770	.6686
C(16)E	.3958	.8208	.6882	C(16)F	.8990	.8154	.6761
C(17)E	.3085	.8131	.7541	C(17)F	.8074	.8073	.7488
C(18)E	.4302	.7920	.9164	C(18)F	.9286	.7887	.9155
C(19)E	.4277	.6612	.9283	C(19)F	.9182	.6559	.9350
C(20)E	.2981	.8512	.8292	C(20)F	.8023	.8468	.8160
C(21)E	.2208	.8449	.9064	C(21)F	.7253	.8428	.8960
C(22)E	.2529	.8792	.7379	C(22)F	.7524	.8736	.7287
C(23)E	.2437	.9178	.7813	C(23)F	.7571	.9132	.7733
C(24)E	.2137	.9426	.6911	C(24)F	.7095	.9350	.6795
C(25)E	.1016	.9366	.6368	C(25)F	.5767	.9274	.6479
C(26)E	.0457	.9478	.7238	C(26)F	.5752	.9615	.7483
C(27)E	.0929	.9731	.5761	C(27)F	.5706	.9482	.5488
O(3)G	.6244	.4681	.2048	O(3)H	.6486	.4714	.7236
C(1)G	.7315	.3737	.1684	C(1)H	.7337	.3736	.6704
C(2)G	.7263	.4175	.1741	C(2)H	.7404	.4194	.6806
C(3)G	.6187	.4274	.2067	C(3)H	.6388	.4308	.7114
C(4)G	.6090	.4162	.3125	C(4)H	.6206	.4172	.8188
C(5)G	.6161	.3726	.3166	C(5)H	.6252	.3749	.8213
C(6)G	.5421	.3511	.3543	C(6)H	.5453	.3503	.8529
C(7)G	.5403	.3094	.3627	C(7)H	.5412	.3084	.8611
C(8)G	.6468	.2960	.3494	C(8)H	.6498	.2961	.8455
C(9)G	.6912	.3140	.2570	C(9)H	.6934	.3140	.7552
C(10)G	.7123	.3594	.2801	C(10)H	.7190	.3585	.7814
C(11)G	.7845	.2969	.2268	C(11)H	.7967	.2975	.7335
C(12)G	.7667	.2526	.2068	C(12)H	.7729	.2525	.7082
C(13)G	.7331	.2361	.3049	C(13)H	.7348	.2356	.8063
C(14)G	.6277	.2521	.3226	C(14)H	.6309	.2522	.8159
C(15)G	.5791	.2276	.3973	C(15)H	.5745	.2263	.8890
C(16)G	.6068	.1869	.3705	C(16)H	.6136	.1869	.8666
C(17)G	.6886	.1918	.2833	C(17)H	.6871	.1916	.7833
C(18)G	.8257	.2461	.4095	C(18)H	.8227	.2468	.9126
C(19)G	.8139	.3771	.3746	C(19)H	.8191	.3747	.8746
C(20)G	.7666	.1640	.2886	C(20)H	.7648	.1628	.7849
C(21)G	.8419	.1677	.2064	C(21)H	.8448	.1665	.7041
C(22)G	.7016	.1217	.2708	C(22)H	.7010	.1199	.7666
C(23)G	.7794	.0895	.2775	C(23)H	.7629	.0874	.7685
C(24)G	.8184	.0872	.3925	C(24)H	.8305	.0915	.8881
C(25)G	.8795	.0527	.4326	C(25)H	.8617	.0472	.8991
C(26)G	.8018	.0199	.4564	C(26)H	.8701	.0415	1.0000
C(27)G	.9618	.0556	.3826	C(27)H	.9704	.0583	.8667

Crystal Structure

In the crystal, there are eight cholesterol molecules (labeled **A**, **B**, ..., **H**) in the asymmetric unit (Fig. 5.19-1). Molecular packing is a characteristic bilayer with alternating hydrophilic and hydrophobic regions. The long directions of the molecules are roughly parallel.

Cholesterol hemiethanolate crystallizes in a triclinic and a monoclinic form. The latter also has eight independent cholesterol molecules in the unit cell and a similar packing as the triclinic crystal, but there is a distinct kink at the hydrophobic contact. Bond distances and angles averaged over the 16 independent molecules in the triclinic and monoclinic hemiethanolate structures are shown in Fig. 5.19-1. The accuracy of these parameters for the side chains is low, especially at the ends where the thermal motion is very pronounced.

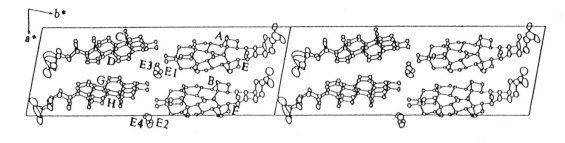

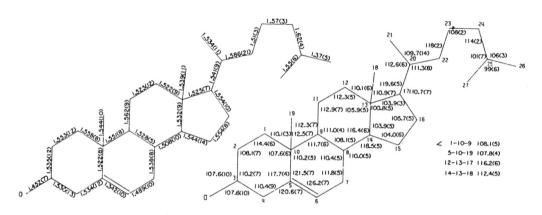

Fig. 5.19-1 Top: two triclinic unit cells of the cholesterol
hemiethanolate structure viewed in a direction 5° from the c axis,
showing the pseudosymmetry which relates molecules **C** and **G** to **D** and **H** by
a c/2 translation. The relationship of **A** and **E** to **B** and **F** by an a/2
translation can also be seen. **E1...E4** are the four ethanol molecules.
Bottom: bond distances (Å) and angles (°) given as means, and standard
deviations of the means of values from the 16 independent cholesterol
molecules in the triclinic and monoclinic hemiethanolate structures.
(After ref. 1).

In both crystal forms two chains of hydrogen bonds, running in the *a*
direction, link the cholesterol and ethanol molecules in the hydrophilic
region, as shown in Fig. 5.19-2(a). Fig. 5.19-2(b) shows part of the
triclinic hemiethanolate structure viewed along the pseudo-two-fold axes,
which are directed along [10$\bar{1}$].

Remarks

1. *Steroids and their functions*[2]

The naturally-occurring steroids and their synthetic derivatives all
contain a basic carbon skeleton or nucleus consisting of three six-membered
rings and one five-membered ring, as shown in Fig. 5.19-3.

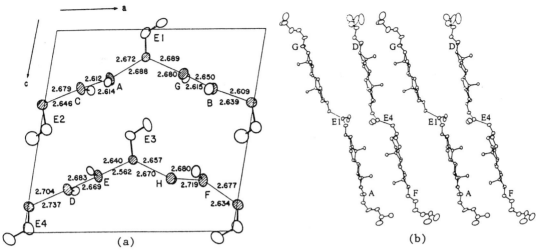

Fig. 5.19-2 (a) Hydrogen bonding in the cholesterol hemiethanolate structures. The O atoms (shaded) of the cholesterol molecules **A**, ..., **H** are shown with their adjacent C(3) atoms. *E*1, ..., *E*4 are ethanol molecules. Hydrogen-bond lengths in Å (e.s.d. 0.020Å) are given, for the triclinic (above) and monoclinic (below) form. (b) View of one half of the triclinic cholesterol hemiethanolate structure along the [10$\bar{1}$] axis. (After ref. 1).

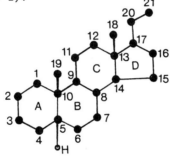

Fig. 5.19-3 Basic steroid skeleton with standard atomic numbering and ring designations.

The most important classes of steroids include: (i) the sex hormones, which are responsible for the development and maintenance of primary and secondary sex characteristics; (ii) the adrenal cortical hormones, which are essential for the maintenance of salt and water balance and carbohydrate metabolism; (iii) bile acids present in animal bile and used in chemical synthesis of anti-inflammatory agents; (iv) sterol constituents of cell membranes; and (v) cardiotonic agents useful in the treatment of congestive heart failure. In addition to the regulatory functions that steroids perform, they play a significant role in the body responses to emotional and physical stress and disease conditions.

The first complete X-ray crystal structure analysis of a steroid, cholesteryl iodide, was reported by Carlisle and Crowfoot (Hodgkin) in 1945.[3] The determination was made directly from the X-ray data and did not involve any chemical knowledge except in the last stages when an ambiguity imposed by the crystallographic symmetry made it necessary to choose between possible alternatives for certain coordinates of the atoms.

Nowadays a wealth of information concerning the molecular conformation and intermolecular interaction in steroids is available for the analysis of structure-function relationships. In the *Atlas of Steroid Structure*[4] pertinent crystallographic data have been collected and subjected to an exhaustive and uniform analysis in order to delineate subtle structural variations.

2. Crystallography of cholesterol polymorphs[5]

The importance of cholesterol is borne out by the fact that single crystal X-ray diffraction determinations of five different crystal forms have been reported. The crystallographic data are summarized in Table 5.19-1.

Table 5.19-1 Crystallographic data of cholesterol polymorphs.

Polymorph	Space group	Parameters of unit cell	Number of molecules in asymmetric unit	Long direction of molecule	Ref.
I (anhydrous)	P1	a = 14.172(7)Å, α = 94.64(4)° b = 34.209(18), β = 90.67(4) c = 10.481(5), γ = 96.32(4)	8	b	[6]
II (monohydrate)	P1	a = 12.39(3)Å, α = 91.9(1)° b = 12.41(3), β = 98.1(1) c = 34.36(6), γ = 100.8(1)	8	c	[7]
III (hemiethanolate)	P1	a = 12.787(2)Å, α = 97.80(2)° b = 35.310(11), β = 100.40(2) c = 12.225(1), γ = 99.06(2)	8	b	[1]
IV (hemiethanolate)	P2$_1$	a = 12.775(2)Å, b = 68.668(15), β = 100.43(1)° c = 12.213(2),	8	b	[1]
V (anhydrous 37°C)	P1	a = 27.565(10)Å, α = 94.25(3)° b = 34.594(15), β = 90.90(3) c = 10.748(4), γ = 96.55(3)	16	b	[8]

Some general structural features in these polymorphs are:

a) Eight or more independent molecules in the crystallographic asymmetric unit

The total number of independent molecules determined in the five crystal structures in 48; eight each in I, II, III, and IV, and 16 in V. Fig. 5.19-4 shows the molecular skeletons of the eight cholesterol molecules in an asymmetric unit in the anhydrous crystal form I.

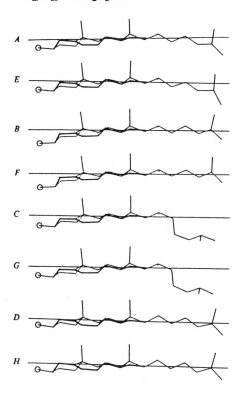

Fig. 5.19-4 The eight cholesterol molecules of I viewed parallel to the mean plane of the C and D rings. Pseudosymmetrically related molecules are shown next to one another. (After ref. 6).

b) Side chain conformation

The observed distribution of the seven-atom side-chain conformation in the 32 independent molecules of crystal forms I, II, III and IV is significantly different from the distribution in a sample of 123 molecules containing the fragment, particularly in the case of the solvated crystals. In the anhydrous crystal form I, six of the eight independent molecules are in the extended low-energy conformation. In the solvated forms where the cholesterol and solvent molecules form chains of hydrogen bonds on the polar surface, *none* of the 24 molecules is in the extended conformation. To maximize the hydrogen bonding at the A-ring end of each cholesterol molecule, conformational adjustments are made in the side chains to improve the van der Waals contacts between them.

c) Aggregation, solvation and bilayer formation

The aggregates found in the anhydrous forms of cholesterol differ from those found in the solvated forms. In the solvated forms the infinite bilayer results when the solvent links the aggregates together to generate well-defined polar surfaces in the bilayer [Fig. 5.19-5(a)]. In the anhydrous forms the 8-(and 16-) member "blocks" stack together in a slightly staggered fashion that disrupts the formation of infinite bilayers [Fig. 5.19-5(b)].

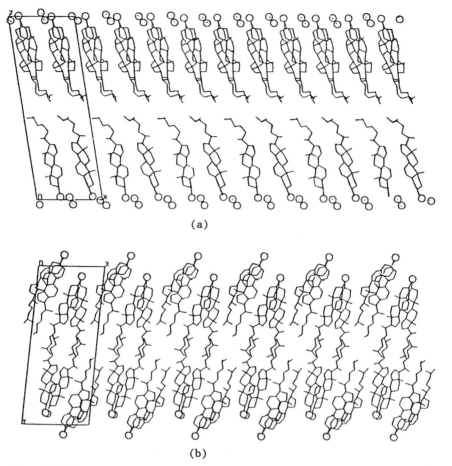

(a)

(b)

Fig. 5.19-5 (a) Bilayer structure found in the hydrated forms of cholesterol. (b) Aggregation in the anhydrous forms. (After ref. 5).

3. Ring conformation and structural features of steroids[2]

The six- and five-membered rings in a steroid molecule are all conformationally flexible, and a qualitative description of their symmetries is based on their approximation to idealized geometries. Fig. 5.19-6(a) illustrates three ideal conformations used to classify five-membered rings. The planar ring has the highest symmetry. The ideal envelope conformation has only a mirror plane passing through the out-of-plane atom, and the half-chair has only a two-fold axis bisecting the bond between the two out-of-plane atoms. Fig. 5.19-6(b) illustrates the symmetry elements that define the ideal forms of six-membered ring conformations.

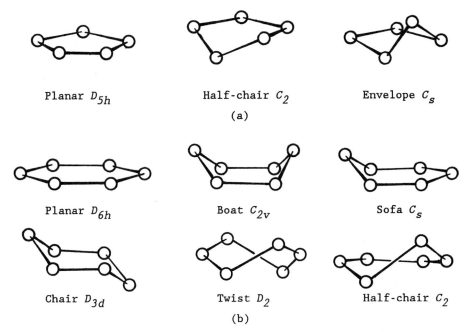

Planar D_{5h} Half-chair C_2 Envelope C_s

(a)

Planar D_{6h} Boat C_{2v} Sofa C_s

Chair D_{3d} Twist D_2 Half-chair C_2

(b)

Fig. 5.19-6 (a) Three ideal conformations of five-membered rings. (b) Six ideal conformations of six-membered rings. (Adapted from ref. 2).

The most important features brought to light by the comparative conformational analysis of steroids based upon X-ray crystallographic data have been summarized by Duax and co-workers[2]:

a) The conformations of steroids having flexible unsaturated rings and substituents are observed to have characteristic cluster patterns. These patterns of conformational isomerism are most apparent in 1,3,5(10)-estratriene structures: 17β-benzoate, 3β-acetate, and 3β-benzoate derivatives, and Δ^4-3-one structures having **B**- and **C**-ring substituents. In three observations estrone molecules have 7α,8β half-chair conformations that are indistinguishable, whereas a fourth molecule of estrone exists in the 8β sofa conformation. A 2:1 ratio of conformers is observed for three molecules of each of the 5α-androstane-3,17-dione and 17β-hydroxy-5α-androstane-3-one structures. This clustering of conformations of cocrystallization of significantly different conformers suggests equal population of the conformers in the crystallization medium and an appreciable barrier to interconversion.

b) Substitution will often remove, alter, or limit the flexibility of a steroid ring or side chain. Although the testosterone **A** ring ranges from 1α, 2β half-chair to 1α sofa conformations, most **B**- and **C**-ring substitutions produce a clustering about one or the other of these symmetric forms. For example, 9α-halosubstitution stabilizes the 1α,2β half-chair.

c) In contrast, the conformation of less flexible steroids is virtually unaffected by changes in crystal habit. The conformation of the **B** ring of estradiol is a distorted $7\alpha,8\beta$ half-chair in different adducts with water, propanol, and urea.

d) Specific substitutions are observed to have characteristic effects upon the conformations of flexible rings and side chains. The stabilization of a nearly perfect half-chair conformation brought about by 9α-halo substitution and the 12 degree shift in the relative orientation of the progesterone 17β side chain caused by 17α-hydroxy substitution are two of the best examples of these substituent effects that are clearly illustrated by the crystal structure data.

e) Correlations between specific substitutions and changes in bond lengths and angles at points distant from the substituent are discernible in the data. These conformational transmission or long-range effects are illustrated by the significant differences in bond lengths and angles throughout the steroid nucleus when contrasting 4-androsten-3-ones with 4-pregnen-3-ones.

f) Definite patterns are observed in the directionality of hydrogen bonds. Clearly, the *trans* orientations of the 11β-hydroxy and 17α-hydroxy hydrogen bonds in cortisol derivatives relative to the $C(9)$-$C(11)$ and $C(13)$-$C(17)$ bonds, respectively, are intramolecularly controlled. Consequently, details of intermolecular interaction in the solid may have parallels in the binding and reactive sites in a biological system.

g) Molecular packing patterns vary widely, suggesting that conformational features are largely controlled by intramolecular forces rather than crystal packing.

4. Some biologically important steroids

Some notable steroids of biological importance are listed in Table 5.19-2 and are discussed briefly below:

a) Sex hormones

Three groups of steroid sex hormones are produced in the genital glands: estrogens, progestins, and androgens.

Estradiol is necessary in the female for development of the secondary sexual characteristics, growth of the uterus at puberty, and for growth of the endometrium (uterine lining) during the first part of the menstrual cycle. As the most potent naturally occurring estrogen in mammals, estradiol is formed by the ovary, placenta, and testis. It has been isolated from the follicular liquor of sow ovaries and from the pregnancy urine of mares.

Progesterone, the human pregnancy hormone, is produced by the corpus luteum, adrenal glands, and placenta. It is the only naturally occurring progestin.

Testosterone is necessary in the male for spermatogenesis, proper functioning of the reproductive-tract glands, and maintenance of the secondary sexual characteristics; there are less apparent effects on a wide variety of tissues. Testosterone is the principal androgenic hormone produced by the interstitial cells of the testis.

b) Cortisone

Cortisone is one of the important adrenalcorticosteroid hormones which are vital to the maintenance of life. Both cortisone and cortisol (hydrocortisone) are potent anti-inflammatory agents, and their synthetic analogues are widely used in medicine.

c) Genins

Steroids occurring in digitalis plants, such as the purple foxglove, exist as conjugates with sugars; the conjugate is termed a glycoside, and the steroid portion is referred to as the aglycone or as a genin. In the pharmacological literature the glycosides are classified into two groups: the cardiac glycosides and the saponins. The former exhibit a characteristic stimulatory action on the activity of the mammalian heart; an overdosage can lead to heart stoppage. The steroids digitoxigenin, digoxigenin and gitoxigenin are cardiac aglycones.

The structures of the following digitalis-like glycosides have been determined.[18]

Triosides:	digoxin,
	gitoxin;
Biosides of:	digitoxigenin,
	gitoxigenin,
	digoxigenin;
Monosides of:	digoxigenin.

5. Receptor binding of steroid hormones[19]

The activities of steroid hormones depend on their binding to specific cytosolic protein receptors and the subsequent interaction of the steroid-receptor complex with chromatin. Steroids that bind well to the estrogen receptor have the phenolic A-ring as a common feature. The potent non-steroid estrogen agonist diethylstibestrol (DES) possesses two phenolic rings capable of mimicking the A-ring of estradiol; furthermore, the distance between the O atoms in either molecule is about 14.5Å.

Table 5.19-2 Some biologically important steroids.

Steroid	Molecular formula	Structural formula[a]	Reference for structure determination
Estradiol	$C_{18}H_{24}O_2$		$C_{18}H_{24}O_2 \cdot \frac{1}{2}H_2O$ [9]
			$C_{18}H_{24}O_2 \cdot C_3H_7OH$ [10]
			$C_{18}H_{24}O_2 \cdot CO(NH_2)_2$ [11]
Progesterone	$C_{21}H_{30}O_2$		$C_{21}H_{30}O_2$ [12]
Testosterone	$C_{19}H_{28}O_2$		$C_{19}H_{28}O_2$ (I) and (II) [13]
			$C_{19}H_{28}O_2 \cdot H_2O$ (I) [14]
			$C_{19}H_{28}O_2 \cdot H_2O$ (II) [15]
Cortisone	$C_{21}H_{28}O_5$		$C_{23}H_{30}O_6 \cdot CH_3COCH_3$[b] [16]
			$C_{23}H_{30}O_6 \cdot H_2O$ [17]
Digitoxigenin	$C_{23}H_{34}O_4$		$C_{35}H_{54}O_{10} \cdot CH_3COOC_2H_5$[c] [18]
Gitoxigenin	$C_{23}H_{34}O_5$		$C_{35}H_{54}O_{11} \cdot CH_3COOC_2H_5$[d] [18]
Digoxigenin	$C_{23}H_{34}O_5$		$C_{35}H_{54}O_{11} \cdot 4H_2O$[e] [18]
			$C_{35}H_{54}O_{11} \cdot CH_3COOC_2H_5$ [18]

[a] From M. Windholz, The Merck Index, 10th ed., Merck & Co., Inc. (1983). [b] $C_{23}H_{30}O_6$ is cortisone acetate.
[c] $C_{35}H_{54}O_{10}$ is bis-digitoxoside of digitoxigenin. [d] $C_{35}H_{54}O_{11}$ is bis-digitoxoside of gitoxigenin.
[e] $C_{35}H_{54}O_{11}$ is bis-digitoxoside of digoxigenin.

942

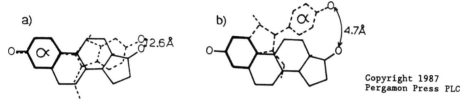

estradiol   diethylstilbestrol (DES)   indenestrol A (IA)

In the case of the racemates of indenestrol A (IA), only the C3-(*S*) enantiomer has high affinity for the estrogen receptor, and model studies based on X-ray structural data suggest that it is the α-ring that mimics the estrogen **A**-ring (Fig. 5.19-7).

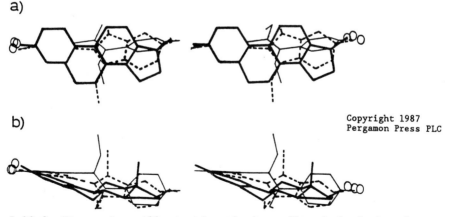

Fig. 5.19-7 Superposition of the structures of IA and estradiol, showing that the best fit is achieved when the α-ring matches the **A**-ring and the double bond in the β-ring overlaps the C(8)-C(9) bond in estradiol. (After ref. 20).

Fig. 5.19-8 is a superposition of the conformations of estradiol, IA, and the crystallographically observed form of DES with its two methyl groups on the same side of the double bond.

The *trans* or *Z* isomer of tamoxifen [Fig. 5.19-9(a)] is one of the most potent antiestrogens employed in cancer chemotherapy, whereas the *cis* or *E* isomer exhibits only weak estrogen properties. The antagonist utility of

Fig. 5.19-8 Stereoviews illustrating the best fit of the hydroxyl groups and the hydrophobic bulk of estradiol (dark solid lines), DES (light solid lines), and IA (dashed lines). (After ref. 20).

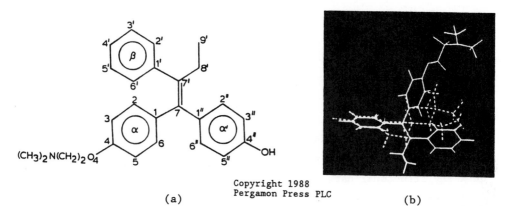

Copyright 1988
Pergamon Press PLC

(a)   (b)

Fig. 5.19-9 (a) Structural formula and ring designation of 4-hydroxy-*trans* tamoxifen. (b) Superposition of estradiol (dashed lines) and tamoxifen (solid lines). (After ref. 21).

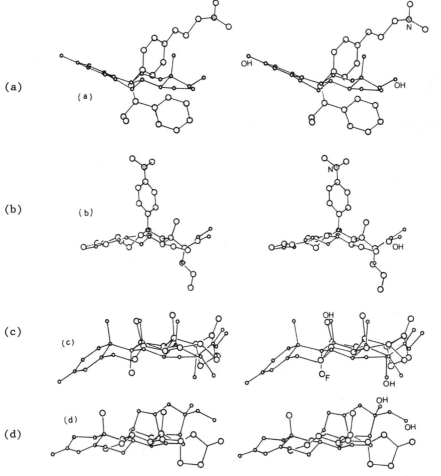

Fig. 5.19-10 Stereograms comparing the structures of potent agonists (small circles) and antagonists (large circles) that compete for binding to (a) estrogen, (b) progesterone, (c) glucocorticoid, and (d) mineralocorticoid receptors. (Adapted from ref. 19).

trans-tamoxifen is a consequence of the competition for receptor binding of its major metabolite, 4-monohydroxy tamoxifen. The skeletal fit achieved by overlapping the α-ring of tamoxifen and the A-ring of estradiol is illustrated in Fig. 5.19-9(b). The estrogen antagonist has a bulky substituent extending nearly perpendicular to the steroid skeleton that may interfere with a conformational change in the receptor needed for hormonal response.

Small-molecule crystal structure determinations have also yielded valuable information on the binding of a wide range of agonists and antagonists to the progesterone receptor, the androgen receptor, the glucocorticoid receptor, and the mineralocorticoid receptor (Fig. 5.19-10).[19,21] However, a thorough understanding of the mode of ligand binding and receptor conformation must await the successful preparation of single crystals of the steroid-protein complexes.

References

[1] H.-S. Shieh, L.G. Hoard and C.E. Nordman, *Acta Crystallogr., Sect. B* **38**, 2411 (1982).

[2] W.L. Duax, C.M. Weeks and D.C. Rohrer, *Top. Stereochem.* **9**, 271 (1976).

[3] C.H. Carlisle and D. Crowfoot, *Proc. Roy. Soc. London* **A184**, 64 (1945).

[4] W.L. Duax and D.A. Norton, *Atlas of Steroid Structure*, Plenum Press, New York, Vol. 1, 1975; Vol. 2, 1984.

[5] W.L. Duax, J.F. Griffin and C. Cheer in W.D. Nes and E.J. Parish (eds.), *Analysis of Sterols and Other Biologically Significant Steroids*, Academic Press, New York, 1989, p. 203.

[6] H.-S. Shieh, L.G. Hoard and C.E. Nordman, *Acta Crystallogr., Sect. B* **37**, 1538 (1981).

[7] B.M. Craven, *Acta Crystallogr., Sect. B* **35**, 1123 (1979).

[8] C.E. Nordman and L.-Y. Hsu, *Science (Washington)* **220**, 604 (1983).

[9] B. Busetta and M. Hospital, *Acta Crystallogr., Sect. B* **28**, 560 (1972).

[10] B. Busetta, C. Courseille, S. Geoffre and M. Hospital, *Acta Crystallogr., Sect. B* **28**, 1349 (1972).

[11] W.L. Duax, *Acta Crystallogr., Sect. B* **28**, 1864 (1972).

[12] P.B. Braun, J. Hornstra and J.I. Leenhouts, *Philips Res. Rep.* **24**, 427 (1969).

[13] P.J. Roberts, R.C. Pettersen, G.M. Sheldrick, N.W. Isaacs and O. Kennard, *J. Chem. Soc., Perkin Trans. II*, 1978 (1973).

[14] B. Busetta, C. Courseille, F. Leroy and M. Hospital, *Acta Crystallogr., Sect. B* **28**, 3293 (1972).

[15] G. Precigoux, M. Hospital and G. van den Bosche, *Cryst. Struct. Commun.* **3**, 435 (1973).

[16] V.J. van Geerestein and J.A. Kanters, *Acta Crystallogr., Sect. C* **43**, 136 (1987).

[17] V.J. van Geerestein and J.A. Kanters, *Acta Crystallogr., Sect. C* **43**, 936 (1987).

[18] K. Go and K.K. Bhandary, *Acta Crystallogr., Sect. B.* **45**, 306 (1989), and references cited therein.

[19] W.L. Duax and J.F. Griffin, *Adv. Drug Res.* **18**, 115 (1989).

[20] W.L. Duax and J.F. Griffin, *J. Steroid Biochem.* **27**, 271 (1987).

[21] W.L. Duax, J.F. Griffin, C.M. Weeks and Z. Wawrzak, *J. Steroid Biochem.* **31**, 481 (1988).

Note The parameters that define the degree of puckering of a n-membered ring are given in D. Cremer and J.A. Pople, *J. Am. Chem. Soc.* **97**, 1354 (1975); for a lucid discussion of the geometric constraints in cyclic molecules, see J.D. Dunitz, *X-Ray Analysis and the Structure of Organic Molecules*, Cornell University Press, Ithaca, New York, 1979, chap. 9.

The six-membered ring of hexakis(1,3-dithiol-2-ylidene)cyclohexane is in a twist D_2 conformation with torsion angles of $\pm63°$ and $\pm30°$(x2) around the ring, whereas hexakis(ethylidene)cyclohexane takes a slightly flattened chair conformation with torsion angles of $\pm46.2°$ as compared to $\pm54.9°$ in the parent cyclohexane. See T. Sugimoto, Y. Misaki, T. Kajita, Z.I. Yoshida, Y. Kai and N. Kasai, *J. Am. Chem. Soc.* **109**, 4106 (1987); W. Marsh and J.D. Dunitz, *Helv. Chim. Acta* **58**, 707 (1975).

A low-temperature neutron diffraction study of cholesteryl acetate is reported in H.-P. Weber, B.M. Craven, P. Sawzik and R.K. McMullan, *Acta Crystallogr., Sect. B* **47**, 116 (1991).

5.20 Codeine Hydrobromide Dihydrate
$$C_{18}H_{21}O_3N \cdot HBr \cdot 2H_2O$$

Crystal Data

Orthorhombic, space group $P2_12_12_1$ (No. 19)

$a = 13.089(10)$, $b = 20.825(15)$, $c = 6.808(5)$Å, $Z = 4$

Atom	x	y	z	Atom	x	y	z	Atom	x	y	z
C(1)	.1194	.4668	.4593	C(10)	-.0530	.4290	.3218	N	-.1549	.3249	.3504
C(2)	.2246	.4576	.4845	C(11)	.0561	.4199	.3813	O(1)	.3740	.3865	.4273
C(3)	.2699	.4017	.4229	C(12)	.1042	.3618	.3356	O(2)	.2386	.2973	.2619
C(4)	.2078	.3537	.3470	C(13)	.0570	.3037	.2409	OH	.2651	.2758	-.1329
C(5)	.1512	.2711	.1520	C(14)	-.0242	.3305	.0936	H₂O'	.4295	.3292	.0620
C(6)	.1625	.2837	-.0663	C(15)	.0021	.2607	.3905				
C(7)	.1188	.3484	-.1323	C(16)	-.0859	.2952	.4941	H₂O"	.8566	.4814	.8079
C(8)	.0307	.3687	-.0598	CH₃'	.4428	.4350	.5011				
C(9)	-.1018	.3719	.2131	CH₃"	-.2464	.3534	.4523	Br	.1403	.0954	.0489

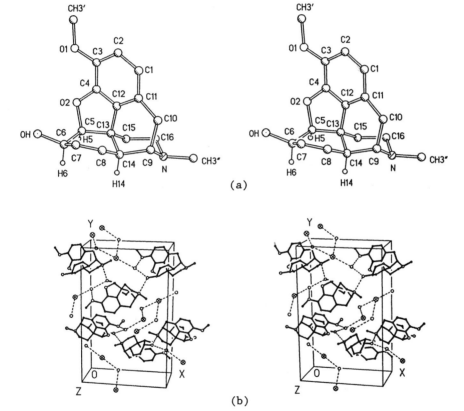

Fig. 5.20-1 Stereoviews of (a) absolute configuration of codeine and (b) crystal structure of codeine.HBr.2H₂O; the O...O and Br...O hydrogen bonds are shown as broken lines.

Crystal Structure

In 1931 Small[3] concluded that the structural formula of Gulland and Robinson[4] accounted best for all the facts of morphine chemistry available at that time. This formula was later confirmed by the total synthesis of codeine, the methyl ether of morphine, and of morphine. Following various attempts by others to establish an acceptable stereochemical configuration for morphine and related compounds, Stork[5] deduced a complete three-dimensional model that is shown (for codeine) in conventional form in Fig. 5.20-1(a).

The crystal structure analysis of codeine hydrobromide dihydrate[1,2] unambiguously confirmed the stereochemical configuration and established the absolute configuration of the codeine molecule on the basis of the anomalous scattering of CuKα radiation by the bromine atom, as shown in Fig. 5.20-1(b).

In the codeine molecule, the atoms O_1 and CH_3', the O-ring, and one of the carbocyclic rings (II) are very nearly in the same plane as that of the benzene ring (I), with the exception of C_5 and C_9. On the other hand the piperidine ring and the other carbocyclic ring (III) extend outwards from this plane on opposite sides to form the arms of a T-shaped molecule. Therefore, the five-ring system may be considered in terms of two planes: one through C_1, C_2, C_3, C_4, C_{11}, C_{12}, C_{10}, O_2, and the other through C_5, C_6, C_7, C_8, C_{13}, C_{14}, C_9, N, C_{16}, C_{15}. The angle between the two planes is 88.4°. It may be noted the carbocyclic ring (C_5, C_6, C_7, C_8, C_{14}, C_{13}) which contains the double bond $C_7{=}C_8$ adopts a boat form, and the piperidine ring takes a chair form.

The N atom of the piperidine ring forms an acceptor hydrogen bond (2.772Å) with the OH group on C_6 of an adjacent codeine molecule, which completes an approximately tetrahedral coordination around N (C_9-N-C_{16} 113.6°, C_{16}-N-CH_3" 110.4°, and C_9-N-CH_3" 113.1°). The configuration of the three hydrogen-bonds around OH, H_2O', and HBr is almost planar.

Remarks

1. Morphine alkaloids

Morphine is the principal alkaloid of opium (about 10% morphine, 0.5% codeine, 6% narcotine, 1% papaverine) and is very effective in eliminating pain, but it is dangerously addictive with repeated use. Its diacetyl derivative, heroin, is even more potent as a narcotic analgesic.

The molecular structure of morphine was determined from X-ray analysis of its hydroiodide dihydrate[6] and hydrochloride trihydrate.[7] The absolute configuration of the morphine skeleton was established from the crystal structure of codeine,[2] which is 3-methoxymorphine. Several opiate drugs (Fig. 5.20-2) related to morphine have been subjected to crystal structure analysis (Table 5.20-1).

The principal difference between the structures of morphine monohydrate and the hydrated hydrochloride salt is a shortening of the C-N bonds in the

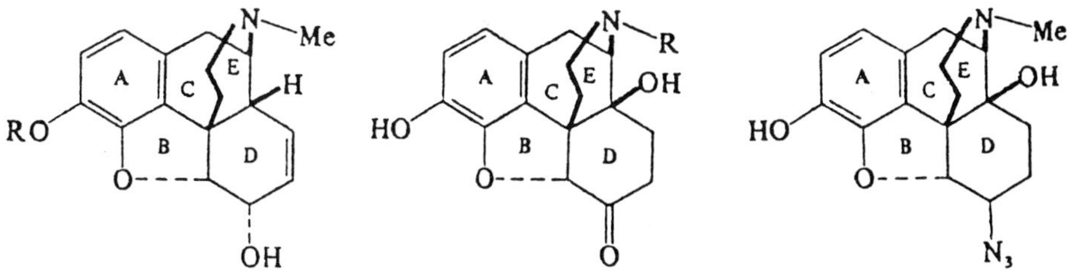

(a)   (b)   (c)

Fig. 5.20-2 Structural formulae of some opiates. (a) Morphine, R = H; codeine, R = Me. (b) Oxymorphone, R = Me; naloxone, R = CH_2-CH=CH_2. (c) 6-Deoxy-6-azido-14-hydroxydihydroisomorphine.

Table 5.20-1 Morphine and related opiates.

Alkaloid	Molecular formula	Structural formula in Fig. 5.20-2	Reference for crystal structure determination	
Morphine	$C_{17}H_{19}O_3N$	(a), R = H	$C_{17}H_{19}O_3N \cdot HI.2H_2O$	[6]
			$C_{17}H_{19}O_3N \cdot H_2O$	[8]
			$C_{17}H_{19}O_3N \cdot HCl.3H_2O$	[7]
Codeine	$C_{18}H_{21}O_3N$	(a), R = Me	$C_{18}H_{21}O_3N$	[9]
			$C_{18}H_{21}O_3N \cdot HBr.2H_2O$	[2]
Oxymorphone	$C_{17}H_{19}O_4N$	(b), R = Me	$C_{17}H_{19}O_4N \cdot H_2O$	[10]
Naloxone	$C_{19}H_{21}O_4N$	(b), R = CH_2-CH=CH_2	$C_{19}H_{21}O_4N$	[11]
Morphine derivative*	$C_{17}H_{18}O_3N_4$	(c)	$C_{17}H_{18}O_3N_4$	[12]

* 6-deoxy-6-azido-14-hydroxydihydroisomorphine

piperidine ring of the free base. A noteworthy feature of the structure of morphine monohydrate is the bridging water molecule that forms donor hydrogen bonds to the hydroxyl on C(6) and the phenolic oxygen atom O(1) on C(3). Oxymorphone, a narcotic agonist more potent than morphine, differs from the parent compound in three ways: C(7)-C(8) is hydrogenated, C(6) has a carbonyl function instead of a hydroxyl, and C(14) has a hydroxyl substituent. Oxymorphone differs from naloxone, the classic opiate antagonist, only in the R-group on nitrogen. When the carbonyl of oxymorphone is replaced by an azide group, improved pharmacological properties are obtained. Both oxymorphone and naloxone have an intramolecular hydrogen bond between the 14-hydroxy and the piperidine nitrogen.

2. Conformational factors in opiate receptor binding

Among the various classes of narcotic analgesics, the common pharmacophore interacting with the opiate receptor appears to be a 4-phenyl piperidine moiety having the heterocyclic ring in a chair conformation and the

phenyl ring in an axial orientation.   On the basis of studies on morphinan
analogues, Belleau and co-workers have demonstrated that the precise
orientation of the nitrogen lone pair may be of critical importance in
receptor binding.[13]   When the six-membered piperidine ring of the active
morphinan molecule I was contracted to a five-membered ring, they observed
that the resulting D-normorphinan II is devoid of receptor affinity.   The
explanation was provided by an X-ray analysis of the hydrobromide of II, which
showed that in this structure the nitrogen lone-pair projects towards the
phenyl ring, whereas in the classical morphinan derivatives this lone-pair
points away from the phenyl ring (Fig. 5.20-3).[14]

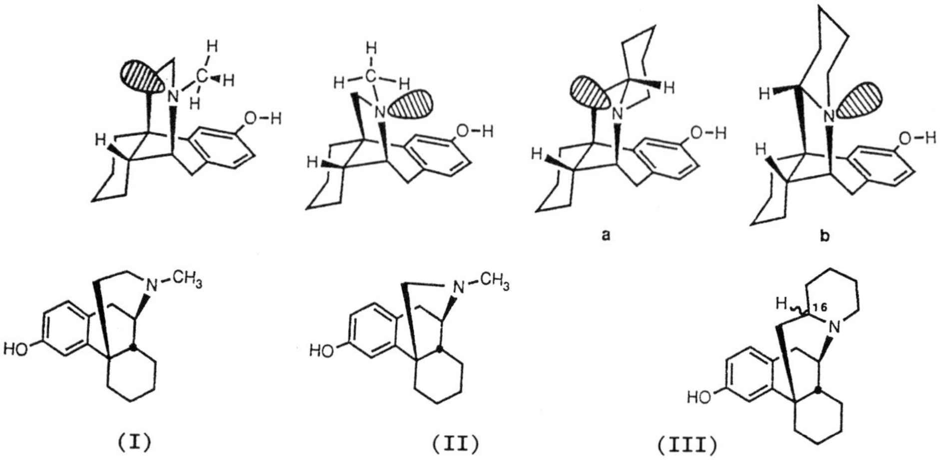

Fig. 5.20-3 Structural formulae and stable conformations (from left to
right) of morphinan I, D-normorphinan II, and the α and β epimers (a and
b, respectively) of 16,17-butanomorphinan III.   (After ref. 16).

Additional evidence supporting this model was provided by studies on
16,17-butanomorphinan III.[13]   When the configuration of the hydrogen atom at
C(16) is α the molecule exhibits agonist activity, whereas its β epimer has
neither agonist nor antagonist activity.   Subsequent X-ray studies have shown
that a chair conformation is possible for the piperidine ring D of the α
epimer, whereas owing to a combination of stereochemical and cyclic
constraints, this ring is forced into a boat conformation in the β epimer
(Fig. 5.20-3).[15]

The principal conclusions derived from studies on morphinan analogues
have been confirmed using the SCRIPT molecular modeling system.[16]   However,
the generality of the model cannot be considered as being firmly established
as there remain some examples that fail to meet this kind of stereo-electronic
requirements.

3. Cinchona alkaloids

This group of plant alkaloids can be divided into two skeletal types represented by quinine and quinotoxin, the latter possessing a secoquinuclidine ring.[17] The cinchona alkaloids with an indole nucleus are biogenetically related to these compounds.

Quinine and quinidine (a dextrorotatory stereoisomer of quinine) have opposite configurations at C_8 and C_9.[18] The structure, numbering, and absolute configuration of some of their derivatives are shown in Fig. 5.20-4.

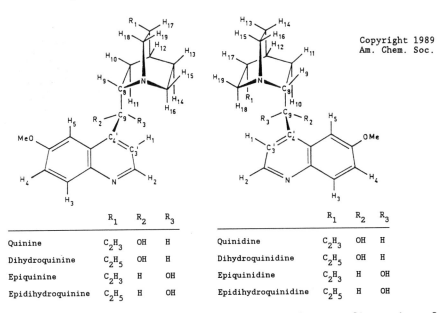

Copyright 1989
Am. Chem. Soc.

	R_1	R_2	R_3		R_1	R_2	R_3
Quinine	C_2H_3	OH	H	Quinidine	C_2H_3	OH	H
Dihydroquinine	C_2H_5	OH	H	Dihydroquinidine	C_2H_5	OH	H
Epiquinine	C_2H_3	H	OH	Epiquinidine	C_2H_3	H	OH
Epidihydroquinine	C_2H_5	H	OH	Epidihydroquinidine	C_2H_5	H	OH

Fig. 5.20-4 Structure, numbering and absolute configuration of eight cinchona alkaloids. (After ref. 18).

Quinine is a well known antimalarial and febrifuge, and quinidine is used to treat atrial and ventricular arrhythmias. In chemistry quinine finds extensive use as a chiral resolving agent.[19] The chromatographic separation of enantiomers employing quinine- and quinidine-impregnated supports is a recent significant development.[20] Wynberg has shown that catalytic amounts of quinine or quinidine are sufficient to induce pronounced asymmetry in a variety of synthetic transformations.[21] From the work of Sharpless it is known that cinchona alkaloid derivatives serve as excellent chiral ligands in the asymmetric dihydroxylation of olefins.[22]

The features of the ground-state conformations of (*p*-chlorobenzoyl)-dihydroquinidine and other alkaloids by a combined NMR, molecular mechanics, and X-ray crystallographic study have been reported.[18] The assignment by ^{1}H NMR of a "closed" conformation [Fig. 5.20-5] for (*p*-chlorobenzoyl)dihydroquinidine is confirmed by the X-ray analysis. The important torsion angles, C(3')-C(4')-C(9)-C(8) and H(8)-C(9)-C(8)-H(9), which are -86 and 170° in the crystal structure, closely resemble the corresponding angles of -90 and 155° for the alkaloid in solution as determined by difference NOE and NOESY NMR techniques.

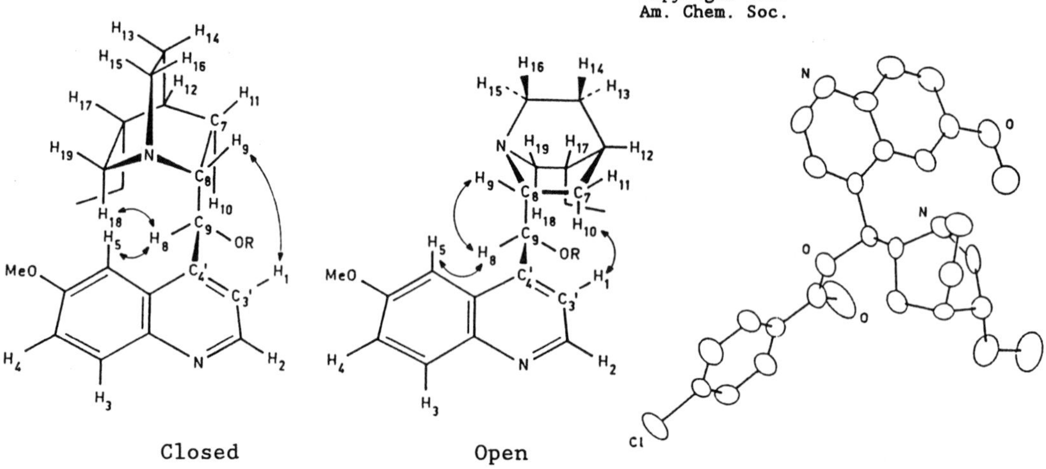

Closed   Open

R = *p*-ClBz   (a)   (b)

Fig. 5.20-5 (a) Schematic drawing showing the closed and the open conformation of the quinidine derivatives. (b) Perspective view of the molecular structure of (*p*-chlorobenzoyl)dihydroquinidine. (After ref. 18).

Table 5.20-2 Representative alkaloids of pharmaceutical importance.

Alkaloid	Structural formula in Fig. 5.20-6	Type	Pharmaceutical activity	Reference for crystal structure
Nicotine	(a)	pyridine	ganglionic depressant, depolarizing drug	[24]
Caffeine	(b)	purine	CNS cardiac and respiratory stimulant, diuretic	[25]
Cocaine	(c)	tropane	surface anesthetic	[26]
Lysergic acid diethylamide (LSD)	(d)	ergot	hallucinogen, psychotomimetic	[27]
Camptothecin	(e)	quinoline	antileukemic and antitumor	[28]
Reserpine	(f)	indole	antihypertensive, tranquilizer	[29]
Strychnine	(g)	indole	central stimulant, resolving agent	[30]
Vinblastine	(h), R^1 = CH_3 R^2 = CO_2CH_3	dimeric indole	antineoplastic	[31]
Yuehchukene	(i)	bis-indole	anti-implantation	[32]

4. Representative alkaloids of pharmaceutical importance

Roughly estimated, there are about ten thousand alkaloids of all structural types known at present.[23] The pharmaceutical uses of alkaloids have been known since the dawn of history, and many of them play prominant roles in vascular and neurophysiology, oncology, cell biology, and chemical ecology. Table 5.20-2 lists some alkaloids of pharmaceutical importance; their structural formulae are shown in Fig. 5.20-6. Useful data and literature references on individual alkaloids are to be found in Southon and Buckingham's *Dictionary of Alkaloids*[23] and Glasby's *Encyclopedia of the Alkaloids*.[33]

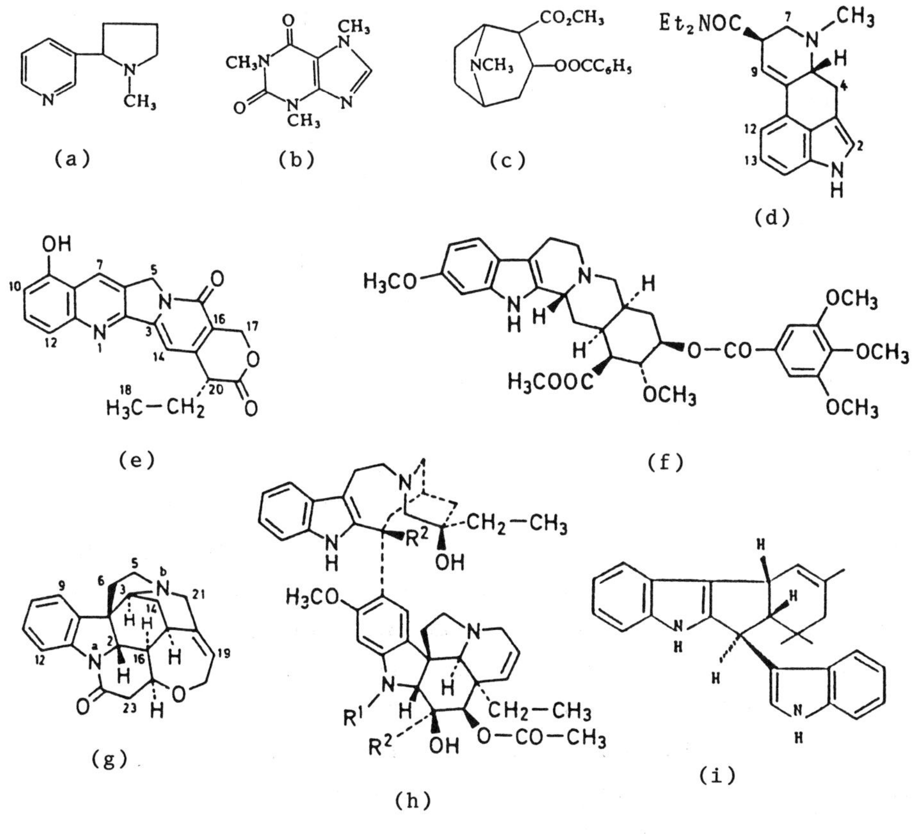

Fig. 5.20-6 Structural formulae of some alkaloids of pharmaceutical importance.

References

[1] J.M. Lindsey and W.H. Barnes, *Acta Crystallogr.* **8**, 227 (1955).

[2] G. Kartha, F.R. Ahmed and W.H. Barnes, *Acta Crystallogr.* **15**, 326 (1962).

[3]   L.F. Small, *Chemistry of the Opium Alkaloids*, Public Health Service, Washington, DC, 1932.

[4]   J.M. Gulland and R. Robinson, *Nature (London)* **115**, 625 (1925).

[5]   G. Stork in R.H.F. Manske and H.L. Holmes (eds.), *The Alkaloids*, Vol. II, Academic Press, New York, 1952.

[6]   M. Mackay and D.C. Hodgkin, *J. Chem. Soc.*, 3261 (1955).

[7]   L. Gylbert, *Acta Crystallogr.*, *Sect. B* **29**, 1630 (1973).

[8]   E. Bye, *Acta Chem. Scand. (B)* **30**, 549 (1976).

[9]   D.V. Canfield, J. Barrick and B.C. Giessen, *Acta Crystallogr.*, *Sect. C* **43**, 977 (1987).

[10] R.J. Sime, M. Dobler and R.L. Sime, *Acta Crystallogr.*, *Sect. B* **32**, 2937 (1976).

[11] I.L. Karle, *Acta Crystallogr.*, *Sect. B* **30**, 1682 (1974).

[12] A. Kalman, Z. Ignath, K. Simon, R. Bognar and S. Makleit, *Acta Crystallogr.*, *Sect. B* **32**, 2667 (1976).

[13] J. Dimaio, F.R. Ahmed, P. Schiller and B. Belleau in F. Gualtieri, M. Giannella and C. Melchiorre (eds.), *Recent Advances in Receptor Chemistry*, Elsevier, Amsterdam, 1979, p. 221.

[14] B. Belleau, T. Conway, F.R. Ahmed and A.D. Hardy, *J. Med. Chem.* **17**, 907 (1974).

[15] F.R. Ahmed, *Acta Crystallogr.*, *Sect. B* **37**, 188 (1981).

[16] N.C. Cohen in B. Testa (ed.), *Advances in Drug Research*, Vol. **14**, Academic Press, 1985, pp. 41-145.

[17] G. Grethe and M.R. Uskokovic in J.E. Saxton (ed.), *Heterocyclic Compounds*, Vol. **25**, Part 4, *The Monoterpenoid Indole Alkaloids*, Wiley, New York, 1983, chap. XII, p. 729, and references cited therein.

[18] G.D.H. Dijkstra, R.M. Kellogg, H. Wynberg, J.S. Svendsen, I. Marko and K.B. Sharpless, *J. Am. Chem. Soc.* **111**, 8069 (1989).

[19] J. Jacques, A. Collet and S.H. Wilen, *Enantiomers, Racemates and Resolution*, Wiley, New York, 1981, pp. 254, 257.

[20] P. Salvadori, C. Rosini, D. Pini, C. Bertucci, P. Altemura, G. Uccello-Barretta and A. Raffaelli, *Tetrahedron* **43**, 4969 (1987).

[21] H. Wynberg, *Top. Stereochem.* **16**, 87 (1986).

[22] J.S.M. Wai, I. Marko, J.S. Svendsen, M.G. Finn, E.N. Jacobsen and K.B. Sharpless, *J. Am. Chem. Soc.* **111**, 1123 (1989).

[23] I.W. Southon and J. Buckingham, *Dictionary of Alkaloids*, Chapman and Hall, London, 1989.

[24] C.H. Koo and S.H. Kim, *Daehan Hwahak Hwoejee (Bull. Kor. Chem. Soc.)* **9**, 134 (1965).

[25] D.J. Sutor, *Acta Crystallogr.* **11**, 453 (1958).

[26] R.J. Hrynchuk, R.J. Barton and B.E. Robertson, *Acta Crystallogr., Sect. A* **37**, C72 (1981), R.J. Hrynchuk, R.J. Barton and B.E. Robertson, *Can. J. Chem.* **61**, 481 (1983).

[27] R.W. Baker, C. Chothio, P. Pauling and H.P. Weber, *Mol. Pharmacol.* **9**, 23 (1973).

[28] M.E. Wall, M.C. Wani, C.E. Cook, K.H. Palmer, A.T. McPhail and G.A. Sim, *J. Am. Chem. Soc.* **88**, 3888 (1966).

[29] J.W. Moncrief and W.N. Lipscomb, *Acta Crystallogr.* **21**, 322 (1966).

[30] I.L. Karle and J. Karle, *Acta Crystallogr., Sect. B* **24**, 81 (1968).

[31] S.S.B. Glover, R.O. Gould and M.D. Walkinshaw, *Acta Crystallogr., Sect. C* **41**, 990 (1985).

[32] Y.-C. Kong, C.-N. Lam, K.-F. Cheng and T.C.W. Mak, *Jiegou Huaxue (J. Struct. Chem.)* **4**, 30 (1985).

[33] J.S. Glasby, *Encyclopedia of the Alkaloids*, Vol. 1, 1975, Vol. 2, 1975, Vol. 3, 1977, Vol. 4, 1983, Plenum Press, New York.

5.21   Sodium Benzylpenicillin (Sodium Penicillin G)

$$C_{16}H_{17}N_2O_4SNa$$

Crystal Data

Monoclinic, space group $P2_1$ (No. 4)

$a = 8.48$, $b = 6.33$, $c = 15.63$Å, $\beta = 94.2°$, $Z = 2$

Atom	x	y	z	Atom	x	y	z
S(1)	.384	.000	.228	O(13)	.174	.521	.054
C(2)	.373	.284	.212	N(14)	.723	-.107	.214
C(3)	.397	.336	.120	C(15)	.883	-.166	.223
N(4)	.484	.167	.090	O(16)	.958	-.189	.162
C(5)	.472	-.030	.127	C(17)	.952	-.183	.315
C(6)	.648	-.062	.127	C(18)	.864	-.107	.382
C(7)	.653	.167	.102	C(19)	.752	-.234	.424
O(8)	.757	.314	.099	C(20)	.666	-.161	.491
C(9)	.217	.355	.246	C(21)	.693	.039	.515
C(10)	.510	.358	.272	C(22)	.803	.164	.474
C(11)	.229	.336	.074	C(23)	.890	.093	.407
O(12)	.177	.156	.047	Na	.066	.834	.028

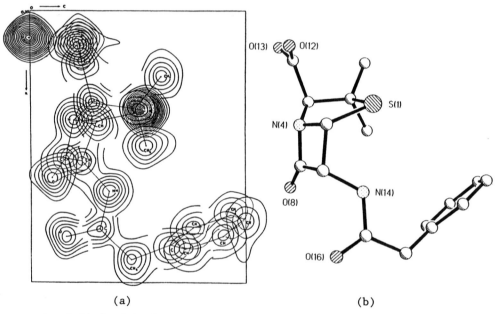

(a)   (b)

Fig. 5.21-1 (a) Electron-density map of sodium penicillin G. (After ref. 1). (b) Molecular structure of penicillin G.

Crystal structure

Penicillin is an antibiotic that has revolutionized the treatment of bacterial infections and saved countless lives otherwise lost to disease. It acts by interfering with cell-wall synthesis in bacteria.

Penicillin and vitamin B_{12} are the two principal structures of biological interest solved in the laboratory of Dorothy Crowfoot-Hodgkin in the 1940s and 1950s, and which convincingly demonstrated the power of X-ray crystallographic methods for the structural elucidation of compounds of unknown chemical formulae.

The X-ray analysis of sodium[1] and potassium benzylpenicillin[1-3]* showed that the penicillin anion has the thiazolidine-β-lactam core, i.e., it contains a thiazolidine five-membered ring condensed with a β-lactam four-membered ring; the benzamide residue is coupled to the latter at the N atom. The carboxylate group and β-lactam ring lie on opposite sides of the thiazolidine ring; the latter and the benzamide residue lie on one side of the β-lactam ring, as shown in Fig. 5.21-1(b). The three-dimensional electron density [Fig. 5.21-1(a)] was used to elucidate the structure of the sodium salt.

The thiazolidine ring is not planar, the CH group attached to the carboxyl group lying out of the plane of the other atoms. The carboxyl group is symmetrical and planar within the limits of experimental accuracy. In the β-lactam ring system the attached O atom appears to lie slightly out of the plane of the ring system towards the thiazolidine ring.

Remarks

1. Historical development of chemotherapy and antibiotics

Chemotherapy may be defined as the curative treatment of a bacterial infection with a chemical substance, either natural or synthetic, that produces a systematic anti-microbic action. Historically the first substances to be put to such use were natural products of plant origin, such as quinine (the active component in an extract of cinchona bark) for the treatment of malaria. The development of synthetic drugs dates from Ehrlich's discovery of salvarsan in 1909, and sulphonamides came into vogue in the mid 1930's. Table 5.21-1 presents the chronological discovery of early chemotherapeutic agents.

Table 5.21-1 Chronological discovery of early chemotherapeutic agents.[4]

Name	Date of discovery	Effective against
Salvarsan	1907	tripanosomes and syphilis
Prontosil	1935	bacterial infection
Sulfathiazole	1938	bacterial infection
Dapsone	1939	leprosy
p-Aminosalicylic acid (PAS)	1946	tuberculosis
Methothrexate	1949	acute leukemia
Thiotepa	1952	leukemia, malignant tumors
5-Fluorocytosine	1957	mycoses

The antibiotic era began with the discovery of penicillin by Fleming in 1929, and in 1940 Florey and co-workers. demonstrated that it was a chemotherapeutic agent of unprecedented potency.[5] The term "antibiotic" now

---

* The original structure[1] was twice refined in later X-ray studies.[2,3] The atomic coordinates given in refs. 1 and 2 lead to unreasonably close intermolecular contacts involving the phenyl rings, whereas those given in Ref. 3 even fail to generate a chemically reasonable molecule.

applies to substances produced by micro-organisms antagonistic to the life or growth of others in high dilution.

The early discoveries of the principal natural antibiotics are summarized in Table 5.21-2.[6]

Table 5.21-2 Chronological discovery and source of natural antibiotics.

Name	Date of discovery	Microbe
Gramicidin	1939	Bacillus brevis
Griseofulvin	1939	Penicillium griseofulvum Dierckx
	1945	Penicillium janczewski
Streptomycin	1944	Streptomyces griseus
Chloramphenicol	1947	Streptomyces venezuelae
Polymyxin	1947	Bacillus polymyxa
Chlortetracycline (aureomycin)	1948	Streptomyces aureofaciens
Cephalosporin C, N and P	1948	Cephalosporium sp.
Erythromycin	1952	Streptomyces erythreus

2. β-Lactam antibiotics

The research on penicillins (**A**), cephalosporins (**B**), and related β-lactam antibiotics is certainly among one of the most exciting therapeutic successes of this century. The fascinating structures of the chemical entities involved, the strict stereochemical requirements for the biological activities, as well as the understanding, in molecular terms, of their mechanisms of action continue to challenge the ingenuity of medicinal chemists and biologists.

A   **B**

Six different penicillins, designated as natural penicillins, were isolated and later identified as 6-acylamino-penicillanic acids, which differ only in the structure of the side chain, as listed in Table 5.21-3.

The crystal structures of two active Δ^3-cephalosporin antibiotics (cephaloridine.HCl.H_2O and cephaloglycine.HAc.H_2O) and a biologically inactive Δ^2-cephalosporin (phenoxymethyl-Δ^2-desacetoxyl cephalosporin) have been determined.[7] The most striking structural feature is the prominent pyramidal character of the β-lactam nitrogen atom in the penicillins and both active Δ^3-cephalosporin antibiotics, in contrast to the nearly planar lactam nitrogen in the inactive Δ^2-cephalosporin. The ease of base hydrolysis of the lactam amide bond in these antibiotics correlates with biological activity. This lability is rationalized as being due to decreased amide electron delocalization and is inferred from C-N and C-O bond lengths. Table 5.21-4 lists the structural parameters of some β-lactam compounds.

Table 5.21-3 Natural penicillins.[4]

$$R-CO-NH\overset{H\ H}{\underset{O\ \ \ N}{\overset{6\ \ 5\ \ S}{\underset{7\ \ \ 4\ 3}{\square}}}}\overset{1}{\underset{CH_3}{\overset{CH_3}{\underset{COOH}{2}}}}$$

R = side chain	Penicillin	Designation
$H_3C-CH_2-CH=CH-CH_2-$	2-Pentenyl	F
$H_3C-CH=CH-CH_2-CH_2-$	3-Pentenyl	Flavicidin
$H_3C-CH_2-CH_2-CH_2-CH_2-$	Pentyl	Dihydro F
$C_6H_5-CH_2-$	Benzyl	G
$H_3C-(CH_2)_6-$	Heptyl	K
$^-OOC-CH(NH_3^+)-(CH_2)_3-$	D-α-aminoadipyl	N
$C_6H_5-O-CH_2-$	Phenoxymethyl	V
$HO-C_6H_4-CH_2-$	p-Hydroxybenzyl	X

Table 5.21-4 Structural and other parameters of some β-lactam compounds.[7]

Compound	Sum of angles around nitrogen (°)	Distance of nitrogen from ring plane (Å)	C=O stretch (cm^{-1})	C=O bond length (Å)	C-N bond length (Å)	Antibiotic activity
Ampicillin	339(1)	-		1.198(7)	1.369(7)	Yes
Penicillin G	337(3)	0.40		1.17(4)	1.34(4)	Yes
Penicillins	-	-	1780-1770	-	-	Yes
Cephaloridine	350.7(8)	0.24		1.214(8)	1.382(8)	Yes
Cephaloglycine	353(6)	0.22		1.28(5)	1.48(5)	Yes
Δ^3-Cephalosporins	-	-	1776-1764	-	-	Yes
Δ^2-Cephalosporins V	359.3(8)	0.065		1.223(7)	1.339(7)	No
Δ^2-Cephalosporins	-	-	1760-1756	-	-	No
Unfused β-lactams	360	0	1760-1730	-	-	No

The discovery of the cephalosporin antibiotics (**B**) opened up new perspectives by showing that relatively important structural modifications of the penicillin nucleus (**A**) are possible without loss of antibacterial properties. Despite the extensive work undertaken in both biology and medicinal chemistry for several decades, the understanding of the molecular requirements necessary for desirable biological activities is still limited and does not permit the rational design of new molecular structures to be confidently envisaged. Practical synthetic work in medicinal chemistry has therefore been mainly restricted to the exploration of analogues by using various empirical working hypotheses.

Precise stereochemical requirements are necessary for the recognition of antibiotics by enzymes. The geometrical features of a set of representative active and inactive structures, as shown in the upper part of Fig. 5.21-2,

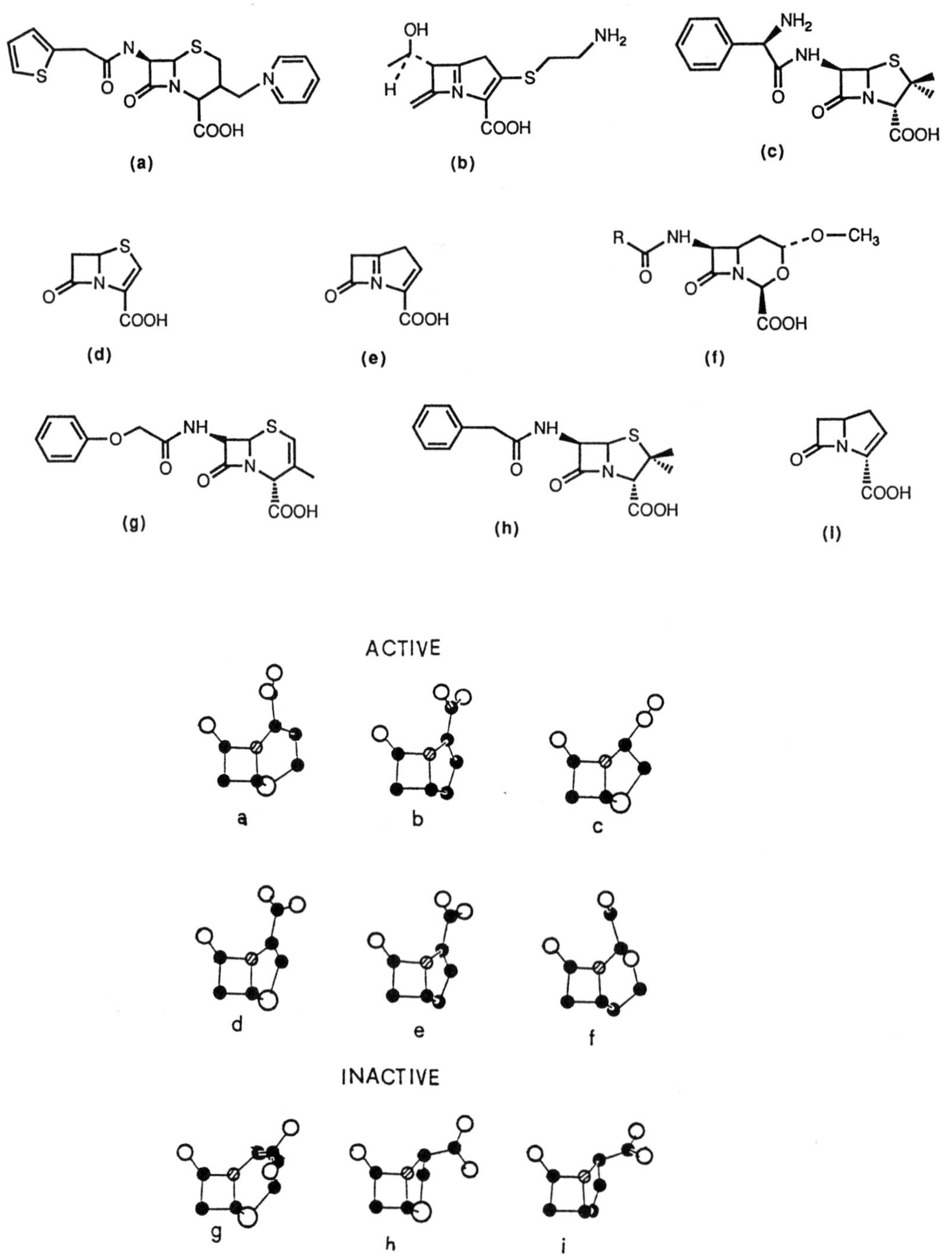

(a)

(b)

(c)

(d)

(e)

(f)

(g)

(h)

(I)

ACTIVE

a

b

c

d

e

f

INACTIVE

g

h

i

Fig. 5.21-2 Representative β-lactam molecules chosen for the comparison of stereochemical features (top) and their three-dimensional structures (bottom; hydrogen atoms have been omitted for clarity). (After ref. 8).

were analyzed.[8] The results demonstrate that active compounds have the carboxyl groups concentrated in the upper region of the drawings shown in Fig. 5.21-2, whereas the carboxyl groups of inactive molecules are almost perpendicular to an axis defining the active region. In this light, the active form of the penam nucleus of penicillins is that with the carboxyl function in a pseudoequatorial position (conformer **b** in Fig. 5.21-2).

3. Structure of tetracyclines

The tetracyclines are a family of closely related antibiotics possessing four fused rings. Table 5.21-5 lists some of the therapeutically important tetracyclines.

Table 5.21-5 Therapeutically important tetracyclines.[4]

Name	R^1	R^2	R^3	R^4	R^5	Refs. for structure
Tetracycline ("tetramycin")	H	H	CH_3	OH	H	[11]
7-Chlorotetracycline (chlortetracycline, "aureomycin")	H	H	CH_3	OH	Cl	[12, 13]
5-Hydroxytetracycline (oxytetracycline, "terramycin")	H	OH	CH_3	OH	H	[11, 12]
7-Chloro-6-demethyltetracycline (demethylchlortetracycline)	H	H	H	OH	Cl	[14]
6-Demethyl-6-deoxy-5-hydroxy-6-methylenetetracycline (methaycycline)	H	OH	$=CH_2$		H	
6-Deoxy-5-hydroxytetracycline (deoxycycline)	H	OH	CH_3	H	H	
7-Dimethylamino-6-demethyl-6-deoxy-tetracycline (minocycline)	H	H	H	H	$N(CH_3)_2$	
Pyrrolidinomethyltetracycline (rolitetracycline)	CH_2 (N-pyrrolidine)	H	CH_3	OH	H	

Woodward and co-workers in 1953 deduced the correct structure of "terramycin" (oxytetracycline) on the basis of chemical degradation studies. The chemical-structural properties and metal ion-ligand interaction for the tetracycline antibiotics have been discussed in detail by Stezowski and co-workers.[9,10] The principal structural features are summarized as follows.

a)   The structural data from naturally occurring, chemically modified,
or totally synthetic tetracyclines show that crystals grown under various
conditions exist in two different forms: a zwitterion (±) and a non-ionized
(0) form. This solvent-dependent equilibrium between the molecular species
has been demonstrated for several systems.

b)   The molecular entities in the hydrated and anhydrous crystals are
different in conformation and molecular structure. For example, the hydrated
crystals of oxytetracycline (OTC) are composed of zwitterionic molecules,
OTC(±), which display a general conformation similar to that of fully
protonated, biologically active tetracycline derivatives, whereas the
anhydrous oxytetracycline crystals contain unionized molecules, OTC(0). The
zwitterion is the important form in aqueous solutions in the pH range of their

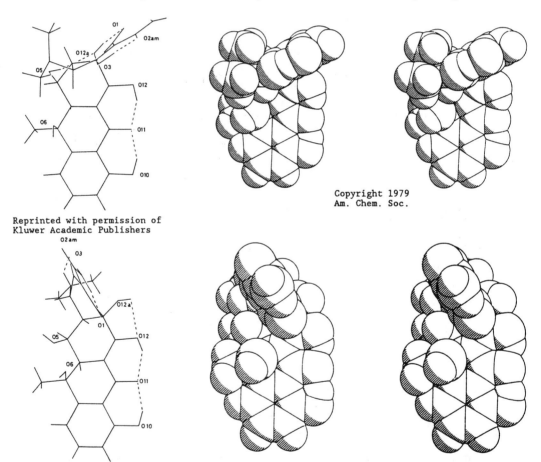

Copyright 1979
Am. Chem. Soc.

Reprinted with permission of
Kluwer Academic Publishers

Fig. 5.21-3 Structure of oxytetracycline (OTC): (top) non-ionized
OTC(0) and (bottom) zwitterion OTC(±). The left picture shows the
molecular structure and the right picture the stereoscopic space-filling
model. (After refs. 10 and 12).

optimum antimicrobial activity. The non-ionized molecule has considerably reduced polarity and exists in a conformation that displays extensive intramolecular hydrogen bonding. Fig. 5.21-3 shows the structure of the two forms of OTC.

c) The correlation between the conformation of non-ionized OTC(0) and the reported lipophilicities of its derivatives indicates that this form may play an important role in nearly anhydrous regions of biological systems, whereas the zwitterion OTC(±) must play an important role in the blood stream and in the aqueous rich regions of microbial cells. In the crystal structure of OTC(±), both charge centers are associated with the A ring. The effect of the charge distribution on the bonding geometry of this

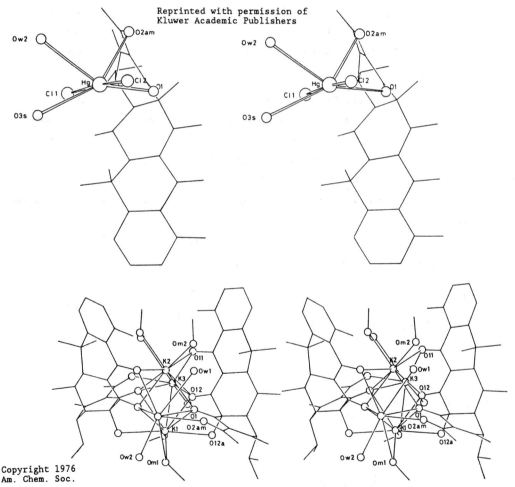

Fig. 5.21-4 Stereoscopic projections displaying metal-oxytetracycline interactions. Top: zwitterionic oxytetracycline mercuric chloride. Bottom: oxytetracycline dipotassium salt; K(2) and K(3) lie on a crystallographic two-fold axis relating the two divalent anions. (After refs. 10 and 15).

portion of the molecule can be clearly demonstrated by a comparison with average bond distances from the crystalline non-ionized derivatives. For example, the protonated amine group at C(4) displays average C-N bond distances that are 0.03Å longer than those of the nonionized derivatives.

   d) The tetracycline derivatives exhibit a broad spectrum of antimicrobial activity, which may in part result from their ability to form metal ion complexes. Complexes have been described in which the nature of the ligand ranges from the zwitterionic free base (MH_2L), a monovalent anion (MHL), a divalent anion (ML), two ligands per metal ion (ML_2), to more than one metal ion interacting with one ligand. Crystal structure analysis of a mercuric chloride complex of OTC showed metal-oxygen coordination involving only A-ring coordination sites of OTC($\pm$), whereas in the dipotassium salt the OTC moiety has two coordination sites utilizing both the BCD and A-ring chromophores (one through atoms O_1, O_{12} and O_{11} and the other through O_1, O_{12a} and O_{12}), as shown in Fig. 5.21-4.

4. Structures of some important chemotherapeutic agents and antibiotics[4]

   Nowadays more than one hundred antibiotics are used clinically, which constitute one of the most efficient weapons for the physician in the combat against infectious diseases. Medicine today also makes use of a large variety of chemotherapeutic agents. Figs. 5.21-5 and 5.21-6 show the structural formulae of some clinically important chemotherapeutic agents and antibiotics, respectively.

   Salvarsan was introduced into clinical medicine in 1910 and thus marked the beginning of modern chemotherapy. Sulfathiazine and sulfathiazole are

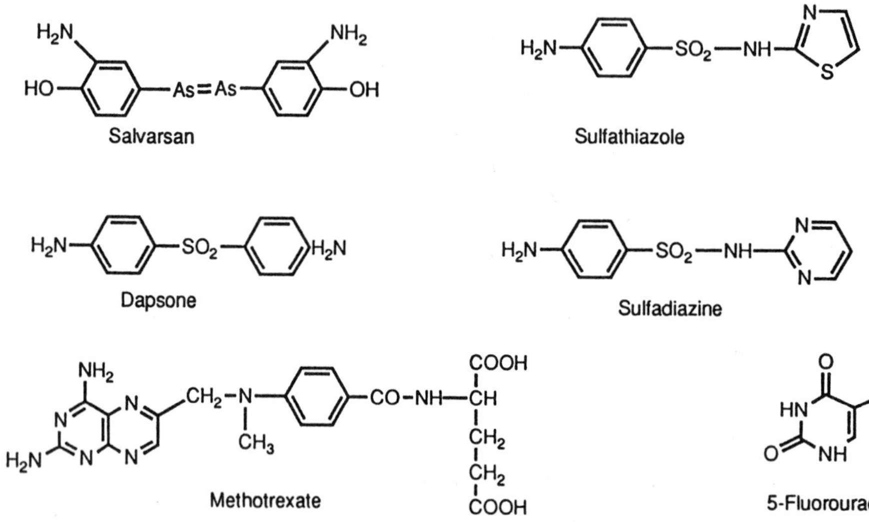

Salvarsan

Sulfathiazole

Dapsone

Sulfadiazine

Methotrexate

5-Fluorouracil

Fig. 5.21-5 Structural formulae of some chemotherapeutic agents.

both important sulfonamide drugs. In 1939, the sulfone dapsone was found to be highly active against leprosy. Dapsone acts like the sulfonamides as a p-aminobenzoic acid antagonist. Methotrexate, a diaminofolic acid analogue, is the most widely used folic acid antagonist, and 5-fluorouracil is a pyrimidine antagonist.

Chloramphenicol is an antibiotic derived from a simple amino acid for treating infections caused by Gram-positive and Gram-negative bacteria. Griseofulvin is orally effective against fungal infections. Streptomycin, an antibiotic derived from sugars, is effective against tuberculosis.

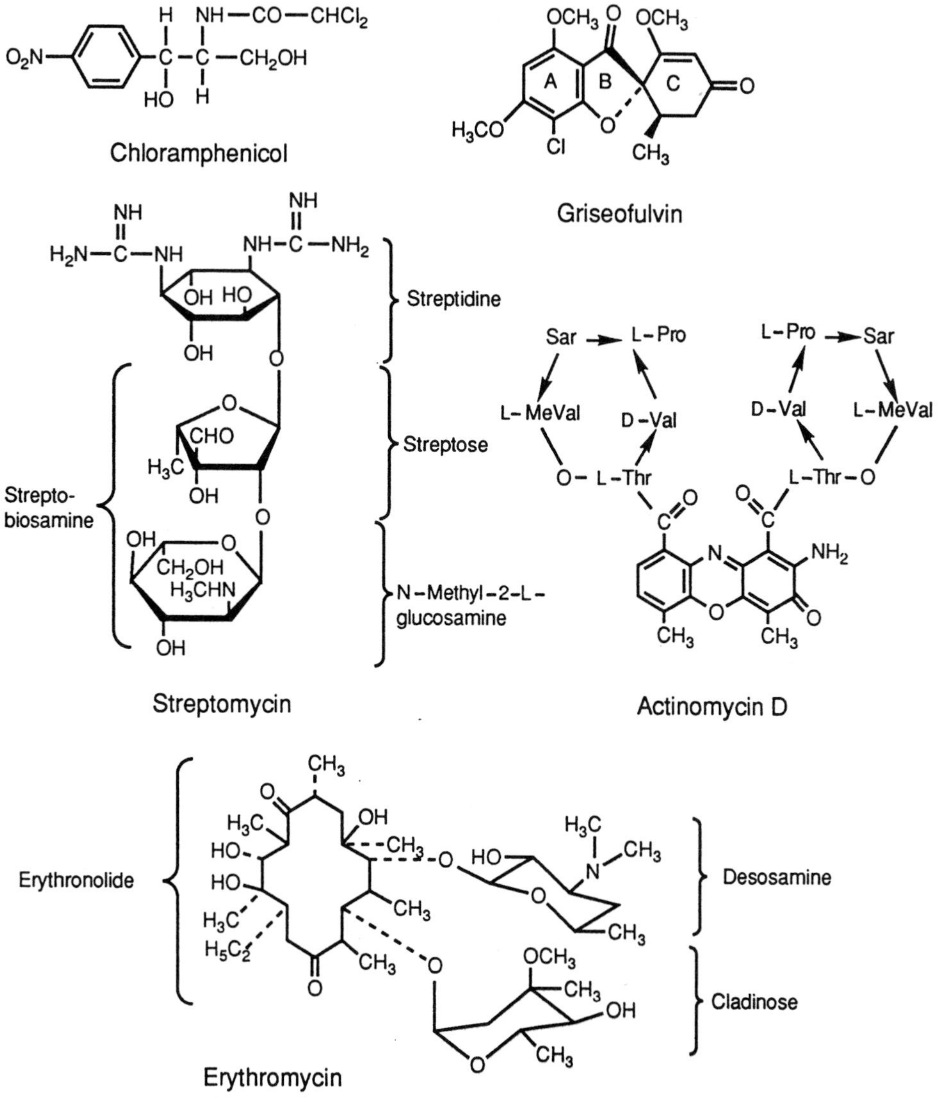

Fig. 5.21-6 Structural formulae of some antibiotics.

Actinomycin D is an antibiotic with cytostatic activity. The macrolide antibiotic erythromycin is the most active and important representative of its class against Gram-positive bacteria.

Structural data for sulfathiazine,[16] sulfathiazole form I,[17] sulfathiazole form II,[18] dapsone,[19] methothrexate tetrahydrate,[20] 5-fluorouracil,[21] chloramphenicol,[22] griseofulvin,[23] streptomycin,[24] erythromycin A,[25] actinomycin D bis(deoxyguanosine) dodecahydrate,[26] actinomycin D bis(deoxyguanylyl-(3'-5')-deoxycytidine).nH_2O ($n = 71.67$ at $0°C$),[27] and ampicillin (α-aminobenzylpenicillin)[28] are available in the crystallographic literature.

References

[1] D. Crowfoot, C.W. Bunn, B.W. Rogers-Low and A. Turner-Jones, *The Chemistry of Penicillin*, Princeton University Press, 1949, p. 310; *Structure Reports* 12, 424 (1949).

[2] G.J. Pitt, *Acta Crystallogr.* 5, 770 (1952).

[3] D.D. Dexter and J.M. van der Veen, *J. Chem. Soc., Perkin Trans I*, 185 (1978).

[4] R. Reiner in F. Korte and M. Goto (eds.), *Methodicum Chimicum*, Vol. 11, Part 2, Academic Press, 1977, chap. 1.

[5] H.W. Florey, *Antibiotics*, Oxford University Press, London, 1949.

[6] L.P. Garrod, H.P. Lambert and F. O'Grady, *Antibiotic and Chemotherapy*, 5th ed., Churchill Livingstone, Edinburgh, 1981.

[7] R.M. Sweet and L.F. Dahl, *J. Am. Chem. Soc.* 92, 5489 (1970).

[8] N.C. Cohen in B. Testa (ed.), *Advances in Drug Research*, Vol. 14, Academic Press, 1985, pp. 41-145; N.C. Cohen, *J. Med. Chem.* 26, 259 (1983).

[9] R. Prewo, J.J. Stezowski and R. Kirchlechner, *J. Am. Chem. Soc.* 102, 7021 (1980).

[10] J.J. Stezowski in B. Pullman and N. Goldblum (eds.), *Metal-Ligand Interactions in Organic Chemistry and Biochemistry*, Part 1, D. Reidel, Dordrecht-Holland, 1977, p. 375.

[11] J.J. Stezowski, *J. Am. Chem. Soc.* 98, 6012 (1976).

[12] R. Prewo and J.J. Stezowski, *J. Am. Chem. Soc.* 101, 7657 (1979); 102, 7015 (1980).

[13] J. Donohue, J.D. Dunitz, K.N. Trueblood and M.S. Webster, *J. Am. Chem. Soc.* 85, 851 (1963).

[14] G.J. Palenik and M. Mathew, *Acta Crystallogr., Sect. A* 28, S47 (1972).

[15] K.H. Jogun and J.J. Stezowski, *J. Am. Chem. Soc.* 98, 6018 (1976).

[16] H.S. Shin, G.S. Ihn, S.H. Kim and C.H. Koo, *Deehan Hwahak Hwoejee (Bull. Kor. Chem. Soc.)* **18**, 329 (1974).

[17] G.J. Kruger and G. Gafner, *Acta Crystallogr., Sect. B* **28**, 272 (1972).

[18] G.J. Kruger and G. Gafner, *Acta Crystallogr., Sect. B* **27**, 326 (1971).

[19] C. Dickinson, J.M. Stewart and H.L. Ammon, *J. Chem. Soc., Chem. Commun.*, 920 (1970).

[20] P.A. Sutton, V. Cody and G.D. Smith, *J. Am. Chem. Soc.* **108**, 4155 (1986).

[21] L. Fallon III, *Acta Crystallogr., Sect. B* **29**, 2549 (1973).

[22] J.D. Dunitz, *J. Am. Chem. Soc.* **74**, 995 (1952).

[23] G. Malmros, A. Wägner and L. Maron, *Cryst. Struct. Commun.* **6**, 463 (1977).

[24] S. Neidle, D. Rogers and M.B. Hursthouse, *Tetrahedron Lett.*, 4725 (1968).

[25] D.R. Harris, S.G. McGeachin and H.H. Mills, *Tetrahedron Lett.*, 679 (1965).

[26] S.C. Jain and H.M. Sobell, *J. Mol. Biol.* **68**, 1 (1972).

[27] F. Takusagawa, M. Dabrow, S. Neidle and H.M. Berman, *Nature (London)* **296**, 466 (1982).

[28] M.O. Boles and R.J. Girven, *Acta Crystallogr., Sect. B* **32**, 2279 (1976).

Note An up-to-date and comprehensive account of the chemistry of the major families of antibiotics is given in G. Lukacs and M. Ohno (eds.), *Recent Progress in the Chemical Synthesis of Antibiotics*, Springer-Verlag, Berlin, 1990.

Development of the new generation of fluoroquinolone antibiotics is described in K. Grohe, *Chem. Brit.* **28**, 34 (1992).

5.22   Vitamin B$_{12}$ Coenzyme
$C_{72}H_{100}CoN_{18}O_{17}P \cdot 17H_2O$

Crystal Data

Orthorhombic, space group $P2_12_12_1$ (No. 19)

$a = 27.701(7)$, $b = 21.608(6)$, $c = 15.351(4)$Å, $Z = 4$[*]

Atom	x	y	z	Atom	x	y	z
Co	.0503	.1827	.0192	C(56)	-.0770	.3669	.1328
C(1)	.0198	.2638	-.1199	C(57)	-.1087	.4004	.1925
C(2)	.0412	.2797	-.2138	O(58)	-.0991	.4064	.2717
C(3)	.0521	.2128	-.2507	N(59)	-.1477	.4286	.1603
C(4)	.0640	.1781	-.1669	C(60)	.0137	.4203	-.0701
C(5)	.0916	.1217	-.1659	C(61)	-.0271	.4651	-.0705
C(6)	.0921	.0857	-.0921	C(62)	-.0689	.4496	-.0835
C(7)	.1206	.0233	-.0801	N(63)	-.0148	.5243	-.0568
C(8)	.0921	-.0027	-.0012	C(Pr1)	-.1814	.4641	.2108
C(9)	.0765	.0556	.0452	C(Pr2)	-.2188	.4271	.2588
C(10)	.0719	.0607	.1339	C(Pr3)	-.2538	.4689	.3060
C(11)	.0611	.1135	.1794	P	-.2535	.3196	.2048
C(12)	.0629	.1190	.2766	O(P2)	-.2057	.2967	.1583
C(13)	.0280	.1764	.2913	O(P3)	-.2467	.3921	.1944
C(14)	.0343	.2089	.2050	O(P4)	-.2530	.3003	.2970
C(15)	.0224	.2705	.1916	O(P5)	-.2954	.3004	.1534
C(16)	.0230	.2979	.1055	C(R1)	-.1458	.1494	.1135
C(17)	.0134	.3666	.0827	C(R2)	-.1733	.2073	.0793
C(18)	.0040	.3614	-.0160	C(R3)	-.1920	.2325	.1667
C(19)	.0337	.3046	-.0424	C(R4)	-.1506	.2232	.2238
C(20)	-.0342	.2539	-.1193	C(R5)	-.1611	.2241	.3226
N(21)	.0456	.2047	-.0985	O(R6)	-.1332	.1624	.2008
N(22)	.0719	.1022	-.0133	O(R7)	-.1421	.2506	.0394
N(23)	.0480	.1681	.1415	O(R8)	-.2037	.1939	.3444
N(24)	.0325	.2660	.0369	N(B1)	-.1015	.1370	.0652
C(25)	.0071	.3167	-.2700	C(B2)	-.0572	.1640	.0779
C(26)	.0906	.3134	-.2055	N(B3)	-.0257	.1477	.0180
C(27)	.1185	.3171	-.2891	C(B4)	-.0360	.0787	-.1126
O(28)	.1055	.3497	-.3504	C(B5)	-.0674	.0428	-.1582
N(29)	.1584	.2808	-.2949	C(B6)	-.1161	.0383	-.1315
C(30)	.0132	.1780	-.3025	C(B7)	-.1323	.0686	-.0586
C(31)	.0107	.1951	-.3990	C(B8)	-.0989	.1015	-.0093
C(32)	.0572	.1851	-.4468	C(B9)	-.0499	.1083	-.0379
O(33)	.0760	.1339	-.4500	C(B10)	-.0514	.0070	-.2390
N(34)	.0770	.2332	-.4861	C(B11)	-.1518	-.0005	-.1824
C(35)	.1194	.1057	-.2468	N(A1)	.1981	.2691	.5161
C(36)	.1191	-.0202	-.1590	C(A2)	.1761	.3127	.4681
C(37)	.1711	.0423	-.0534	N(A3)	.1724	.3191	.3831
C(38)	.2078	-.0093	-.0323	C(A4)	.1949	.2735	.3432
O(39)	.1947	-.0654	-.0261	C(A5)	.2200	.2247	.3812
N(40)	.2524	.0100	-.0238	C(A6)	.2216	.2230	.4726
C(41)	.0480	-.0424	-.0262	N(A7)	.2408	.1867	.3197
C(42)	.0154	-.0599	.0501	C(A8)	.2286	.2117	.2450
C(43)	-.0234	-.1033	.0149	N(A9)	.2004	.2648	.2547
O(44)	-.0143	-.1593	.0068	N(A10)	.2423	.1759	.5203
N(45)	-.0628	-.0789	-.0143	C(A11)	.1765	.2993	.1864
C(46)	.1138	.1374	.3006	C(A12)	.2088	.3152	.1099
C(47)	.0494	.0596	.3234	C(A13)	.1951	.2710	.0404
C(48)	-.0242	.1582	.3026	C(A14)	.1408	.2607	.0584
C(49)	-.0367	.1401	.3971	C(A15)	.1207	.2028	.0233
C(50)	-.0750	.0899	.4118	O(A16)	.1381	.2642	.1530
O(51)	-.0758	.0618	.4801	O(A17)	.1971	.3782	.0839
N(52)	-.1050	.0774	.3482	O(A18)	.2042	.2922	-.0442
C(53)	.0103	.3090	.2722				
C(54)	.0612	.4021	.1044				
C(55)	-.0282	.4007	.1283				

[*] X-ray data; parameters of solvent molecules have been omitted.

Crystal Structure

The vitamin B$_{12}$ coenzyme molecule consists of a corrin nucleus with a Co(III) atom situated at the center forming bonds to four inner N atoms (see Fig. 5.22-1). Several side chains of biological interest in the context of water-protein and water-DNA interactions are attached to the outer atoms of

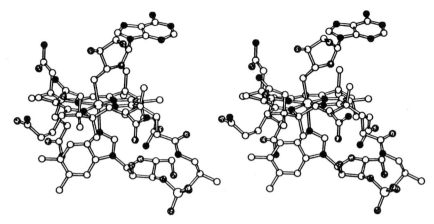

Fig. 5.22-1 Stereoview of vitamin B_{12} coenzyme, adenosylcobalamin. (After ref. 4).

the nucleus. These include three acetamides, three propionamides, one propionic acid residue and eight methyl groups. An isopropanolamine group links the propionic acid side-chain to the phosphate group of a nucleotide containing a benzimidazole base. This latter group is linked through the N(B3) position to the Co atom. On the opposite side, a 5'-deoxyadenosine nucleoside is situated with the C(5') position of the ribose moiety forming a direct link to the Co atom.

The crystal structure of hydrated vitamin B_{12} coenzyme provides an excellent system of intermediate size in which the distribution of the solvent molecules and their general interactions at the interface can be studied to atomic resolution using diffraction methods. The original structure determination[2-3] was later updated by a detailed neutron structural analysis (assuming exchange of 19 coenzyme H atoms by D atoms and 17 D_2O solvent molecules) to below 1.0Å resolution; to assist in the interpretation, a parallel X-ray study was also undertaken.[1]

One of the side chains of the corrin ring is disordered between two extreme positions. All the H- and D-atom positions for the coenzyme molecule and approximately 65% of the solvent D atoms were located from neutron difference Fourier maps. Of the eleven methyl groups present, six are well ordered and five disordered. An acetone molecule (with partial occupancy) was located in the solvent regions in both the neutron and X-ray analyses.

Remarks

1. Vitamin B_{12} and its coenzyme[4]

Vitamin B_{12} is a red crystalline compound that can be isolated from the liver. It has a functional role in preventing pernicious anaemia and also serves as a coenzyme in hydrogen and methyl transfer reactions (Co appears to be the only metal capable of catalyzing C transfer reactions *in vivo*; O and N transfer reactions are more common). Vitamin B_{12} is also a growth-promoting factor for several microorganisms.

The structure elucidation of vitamin B_{12} was one of the major triumphs of X-ray crystallography in the early 1950s,[5] as its constitution was established from the interpretation of electron-density maps with meager prior chemical information. The analysis revealed a porphyrin-like "corrin" ring with similar side chain arrangements - short, long; short, long; short, long; long, short. The existence of a cobalt-carbon bond aroused great interest since this bond is broken and made when vitamin B_{12} functions as a coenzyme in biochemical reactions. In recognition of her X-ray studies of vitamin B_{12}, penicillin and many other biologically important compounds, Dorothy Crowfoot Hodgkin was awarded the Nobel Prize in chemistry in 1964.

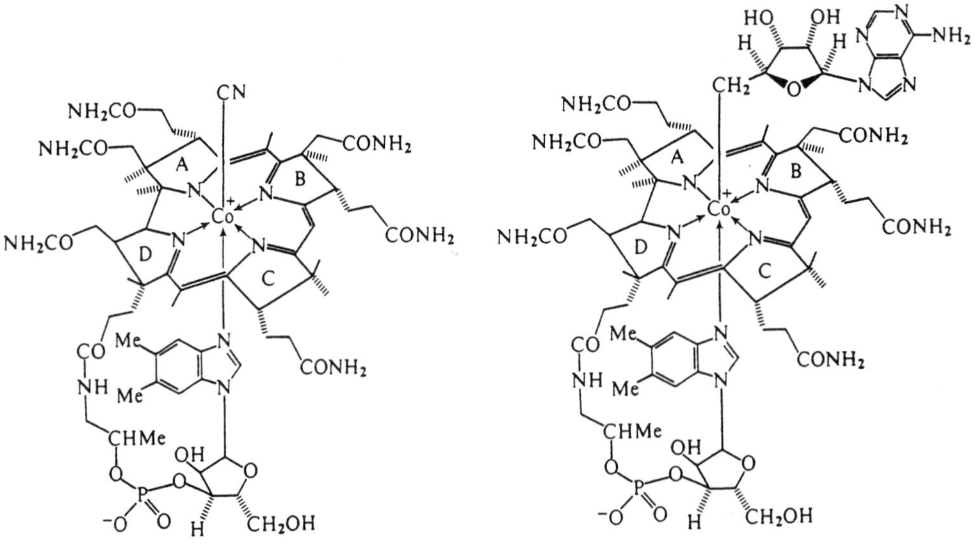

Fig. 5.22-2 Structural formula of (a) vitamin B_{12} and (b) vitamin B_{12} coenzyme. (After ref. 7).

Vitamin B_{12}, usually called cyanocobalamin, is a macrocyclic cobalt(III) complex as shown in Fig. 5.22-2(a). Four of the six ligand positions of the octahedron are occupied by the nitrogens of the corrin ring. The fifth position carries a cyano group, and the remaining one is coordinated by the nitrogen at position 3 of the 5,6-dimethylbenzimidazole moiety. The corrin ring consists of four pyrrole rings which are jointed by methene bridges, with the exception of a direct linkage between the two α-carbons of rings A and D, and the six conjugated double bonds constitute a unique resonating system. On the corrin ring of natural corrinoids, seven methyl, three carboxymethyl (acetic) and four carboxyethyl (popularly called propionic) groups are attached as side chains in the same order as in uro-porphyrin III. According to the results of X-ray analysis, the corrin ring is almost, but not completely, planar and all propionic side chains project toward the nucleotide side (usually called lower side) and all acetic groups project towards the other, upper side. Six of the carboxyl groups are simply amidated but the remaining one is bound by an amide linkage to D-1-amino-2-propanol, (isopropanol-amine), which is esterified to the phosphate of 3'-mononucleotide of 5,6-dimethyl-benzimidazole. The sugar is D-ribofuranose and the glycoside linkage is α, unlike the β-linkage in nucleic acids.

The active species in the body is not vitamin B_{12} but a coenzyme derived from it, which acts as a cofactor, that is, an activator essential for the action of certain enzymes. There are two B_{12} coenzymes, namely adenosylcobalamin and methylcobalamin; the former is often referred to as the B_{12} coenzyme, as shown in Fig. 5.22-2(b). The crystal structure of methylcobalamin has also been determined.[6]

Vitamin B_{12} shows three types of interactions with proteins. One is the ability of its coenzymes to act as the cofactors for a variety of enzymatic reactions. In this capacity the coenzyme is bound and controlled by the enzyme. Second is its relationship with the cobalamin-binding proteins, such as intrinsic factor and transport proteins, to insure a continuing supply to the organism. In this capacity the vitamin or coenzyme is bound with high affinity. Third, it acts as a substrate for the enzymes that add the adenosyl or methyl group to the vitamin to form the biologically active coenzymes.

In humans and other mammals the only two reactions proven to be B_{12}-mediated are methylmalonate-succinate isomerization with methylmalonyl CoA mutase and methylation of homocysteine to methionine with methyl-transferase. The first reaction is mediated by adenosylcobalamin and is part of the pathway for the degradation of propionyl CoA produced by the catabolism of isoleucine or valine. The second reaction is mediated by methylcobalamin.

The structural changes that occur during the Co-C bond homolysis of vitamin B_{12} coenzyme, which involves a transition from a hexacoordinate cobalt(III) corrin to a pentacoordinate cobalt(II) corrin, have been investigated by an X-ray analysis of cob(II)alamin.[8] The structure of this low-spin Co(II) corrin is strikingly similar to that of the corrin moiety of coenzyme B_{12} and of methylcobalamin, the "upward folding"[4] of the macrocyclic ligand being 16.3°, 13.3° and 15.8°, respectively. The respective bond distances(Å) in the three related corrins are: equatorial Co-N 1.89,

1.90, 1.92; axial Co-N 2.13, 2.24, 2.19. Position and orientation of the
benzimidazole base with respect to the corrin nucleus are virtually unchanged
in B_{12} coenzyme and its Co(II) corrinoid homolysis fragment. Significant
differences in ligand structure are only evident for ring D and its
substituents, and conformational changes occur in the ribophosphate segment of
the nucleotide loop.

     Cobaloximes, which are cobalt(III) complexes of the monoanions of
dimethylglyoxime, can serve as simple models for cobalamins. Two
dimethylglyoxime anions chelate the metal and also form hydrogen bonds to each
other, resulting in a planar coordination environment about the cobalt atom.
It is interesting that the Co-N(benzimidazole) bond is longer in the
cobalamins than in the cobaloxime model compounds, which suggests that the
strength of the bond *trans* to the Co-C bond is weaker for the coenzymes than
for the cobaloximes. The structures and properties of cobaloximes are
discussed in section **6.16**.

2. Vitamins

     Vitamins, which are organic compounds necessary for the life, growth and
health of human and other animals in trace amounts (micrograms to milligrams
per day), cannot be produced by the body and hence must be supplied. However,
vitamin D may be obtained from food or produced in the skin by irradiation
(ultraviolet light) of sterols. Each newly discovered vitamin is designated
by a letter of the alphabet, but once its structure has been established, the
vitamin is generally renamed.

     The vitamins have been classified into the "lipid-soluble group" and the
"water-soluble group", as listed in Tables 5.22-1 and 5.22-2; their structural
formulae are shown in Figs. 5.22-3 and 5.22-4, respectively.

Table 5.22-1 Lipid-soluble vitamins.[9]

Vitamin	Chemical name	Main natural source	Function	Deficiency disorders	Ref. for structure
A_1	retinol	animal liver, fish liver oil,	synthesis of rhodopsin	night blindness (nyctolopia)	[10]
A_2	3-dehydroretinol	carrot, spinach			
D_2	ergocalciferol	fish liver oil	absorption of calcium and	rickets in the young and osteomalacia in	[11]
D_3	cholecalciferol		phosphorus from intestine into blood	adults	[12]
E	d-α-tocopherols	corn oil, peanut oil	antioxidant	not fully established	[13]
K_1	phylloquinones	alfalfa, cabbage, cauliflower,	synthesis of prothrombin	easy bleeding and hemorrhage	[14]
K_2	menaquinones	soybean, spinach, beef liver, pork liver			

R = CH_2OH for retinols, CHO for retinals, CO_2H for retinoic acids:

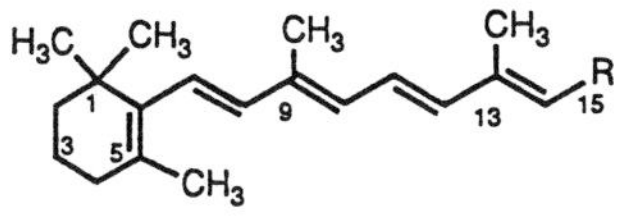

A_1 type

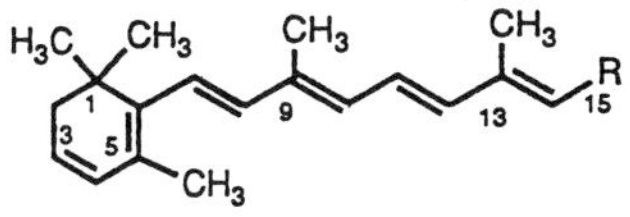

A_2 type

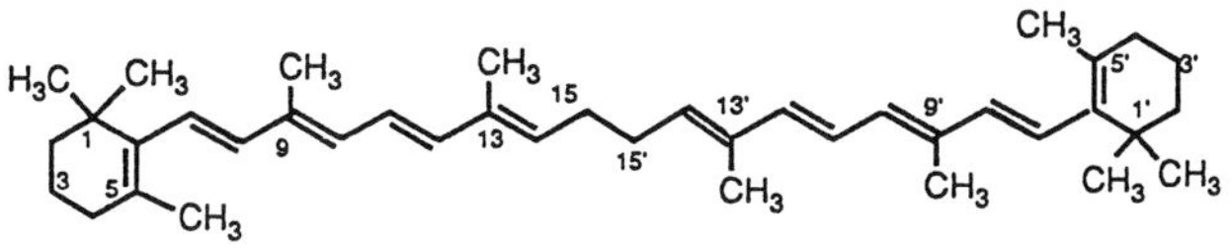

Carotene (provitamin A)

Structures of vitamin A and provitamin A

R = 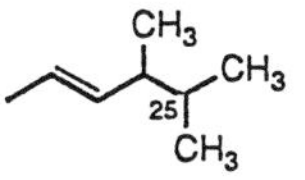 for D_2 series 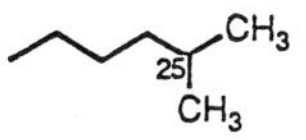 for D_3 series

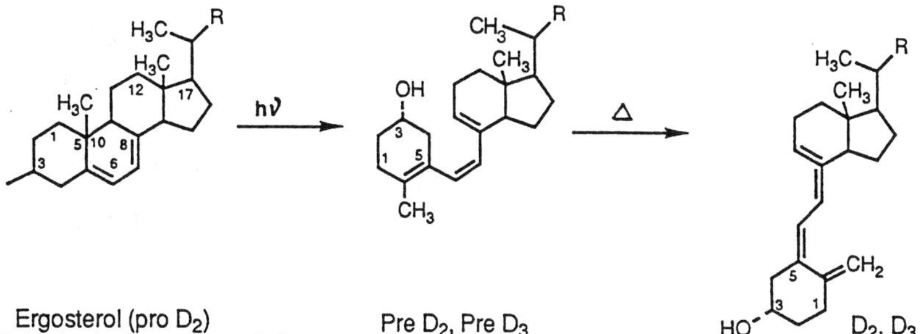

Ergosterol (pro D_2)
7-Dehydrocholesterol (pro D_3)

Pre D_2, Pre D_3

D_2, D_3

Conversion of previtamin forms of D to vitamin D

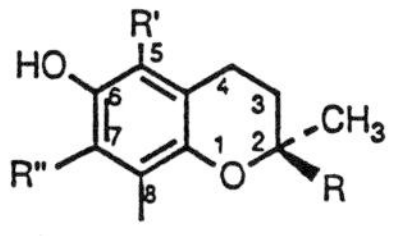

R', R" = CH_3, CH_3 for α
CH_3, H for β
H, CH_3 for γ
H, H for δ

R = 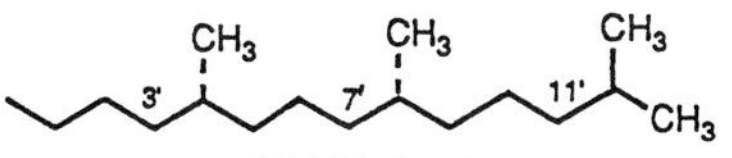

for tocopherols

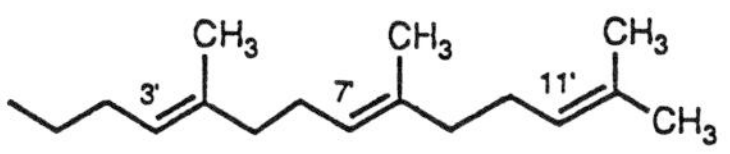

for tocotrienols

Structures of Vitamin E and related compounds

K_1 type
(Phylloquinones)

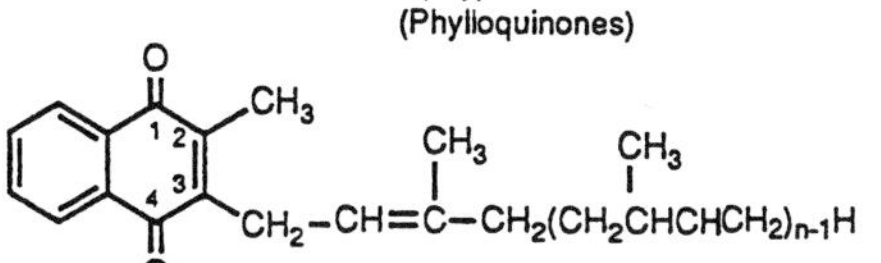

e.g., $K_{1(20)}$ when n = 4

K_2 type
(menaquinones)

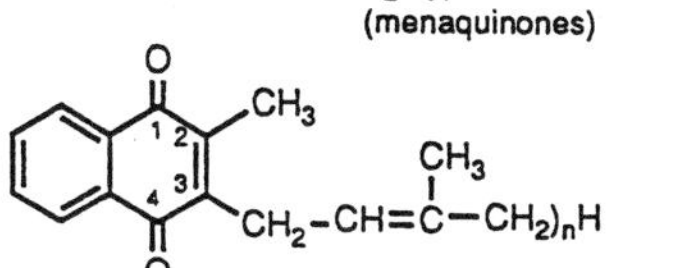

e.g., $K_{2(35)}$ when n = 7

Fig. 5.22-3 Structural formulae of lipid-souble vitamins. (Adapted from ref. 9).

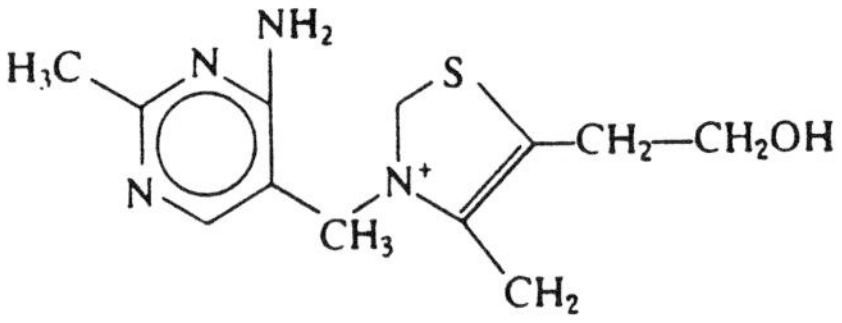

Thiamine

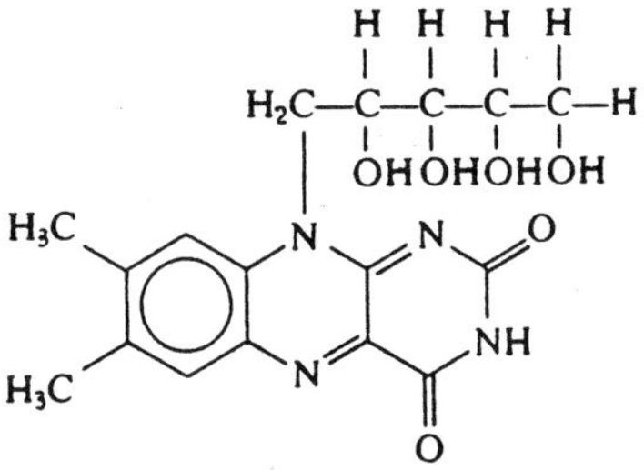

Riboflavin

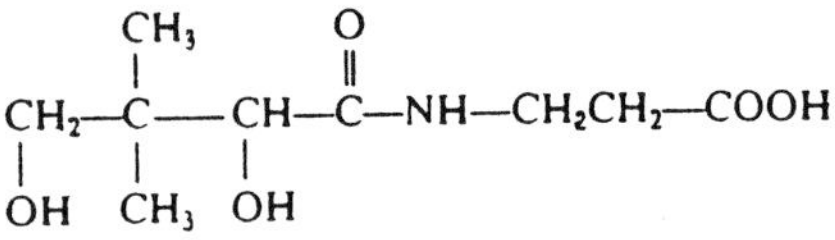

Pantothenic acid

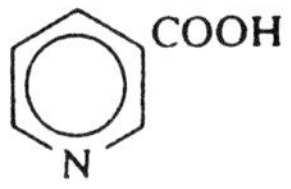

Nicotinic acid

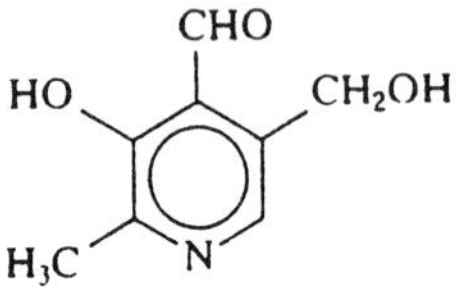

Pyridoxal

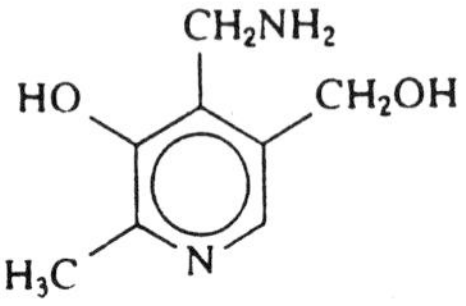

Pyridoxamine

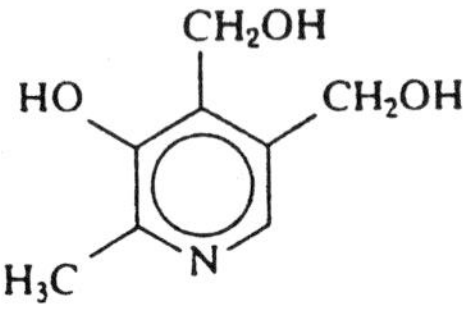

Pyridoxine

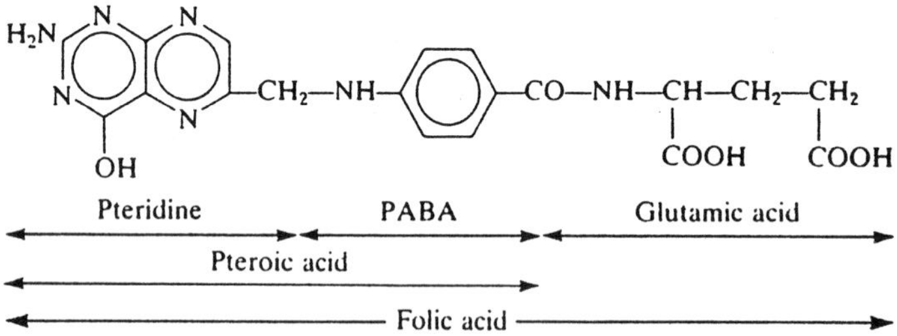

Pteridine   PABA   Glutamic acid

Pteroic acid

Folic acid

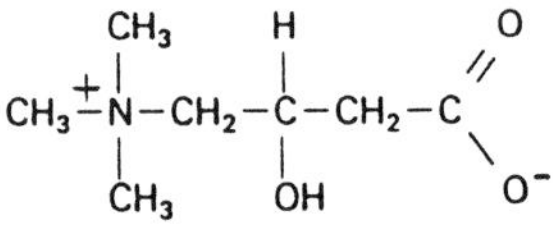

Carnitine

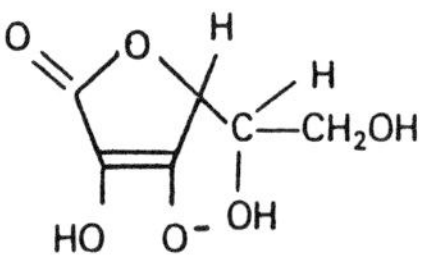

Ascorbate

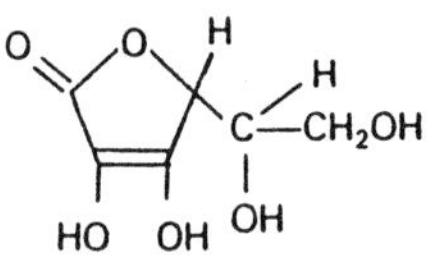

Ascorbic acid

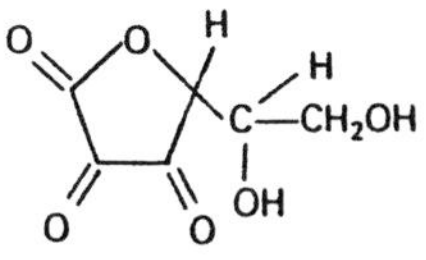

Dehydroascorbic acid

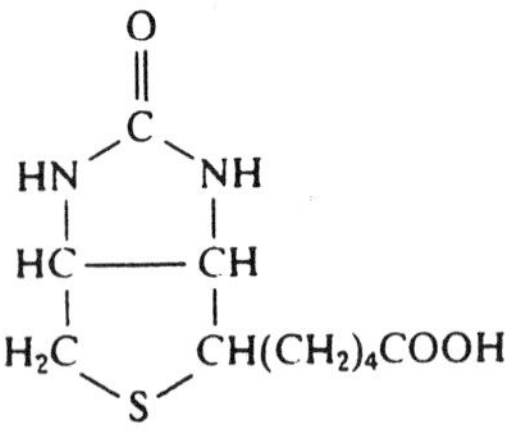

Biotin

$[(CH_3)_2S^+CH_2CH_2CH(NH_2)COOH]Cl^-$

Methylmethioninesulfonium chloride

Fig. 5.22-4 Structural formulae of water-soluble vitamins.

From the point of view of chemical structure, there is very little resemblance among various vitamins. However, many of the water-soluble vitamins have one feature in common, and that is their ability to take part in reversible oxidation-reduction processes. Thus they form a part of various coenzymes.

Table 5.22-2   Water-soluble vitamins.[15]

Vitamin	Chemical name	Main natural source	Deficiency disorders	Ref. for structure
B_1	thiamin	wheat germ, soybean, yeast, ham	beriberi	[16]
B_2 (or G)	riboflavin	yeast, animal liver and kidney	inflammation of tongue and skin, corneal vascularization	[17]
B_3 (or B_5)	pantothenic acid	yeast, liver, egg	not established in humans	
B_4 (or PP)	nicotinic acid, nicotinamide	roasted peanut, liver, yeast, white meat	pellegra	[18]
B_6	pyridoxine	yeast, walnut peanut, liver	cutaneous lesions, anemia	[19]
B_{12}	cobalamin, cyanocobalamin	kidney, liver	retarded growth, pernicious anemia	[5]
B_C (or M)	folic acid	yeast, liver	megaloblastic anemia, sprue	[20]
B_T	L-carnitine	meat extract, liver	hyperlipoproteinemic	[21]
C	L-ascorbic acid	broccoli, green pepper, black currant, guava	scurvy	[22]
H	biotin	royal jelly, yeast, liver	loss of weight and hair, dermatitis	[23]
U	S-methyl-L-methionine chloride (methylmethionine sulfonium chloride)		ulcer and certain hepatic disorders	[24]

3. Structure of retinal, vitamin A aldehyde

The two natural forms of vitamin A, retinol (A_1) and 3-dehydroretinol (A_2), are C_{15} isoprenoid alcohols containing substituted β-ionone and 3-dehydro-β-ionone rings, respectively (Fig. 5.22-3). These yellowish oils are sensitive to oxygen and ultraviolet light, which induces a greenish fluorescence. The human body derives retinal from intestinal dioxygenase-catalyzed cleavage of plant-based carotenes which serve as provitamins. Retinal is then reversibly reduced by pyridine nucleotide-dependent enzymes to retinol.

Provitamin A precursors comprise over fifty carotenoids and apocarotenoid compounds, the most common and effective being β-carotene (Fig. 5.22-3).

The crystal structure of 13-*cis*-retinal,[10] $C_{20}H_{28}O$, discloses the presence of two conformers. In the 6-*s-trans* conformer, the double bond in the cyclohexene ring lies nearly in the plane of the polyene chain of the molecule, while in the 6-*s-cis* conformer, the ring is rotated 110° from this orientation, as shown in Fig. 5.22-5. The 6-*s-cis* conformer consists of atoms C(1) to O(21), with a C(5)-C(6)-C(7)-C(8) torsion angle (henceforth denoted ϕ_{6-7}) of -65.4(6)°, and the nearly planar 6-*s-trans* conformer comprises atoms C(22) to O(42), with a corresponding C(26)-C(27)-C(27)-C(28)-C(29) torsion angle of -174.9(4)°.

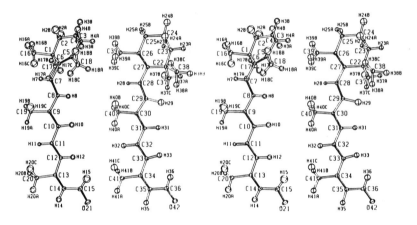

Fig. 5.22-5 Stereoview of the 6-*s-cis* and 6-*s-trans* conformers of 13-*cis*-retinal which coexist in the crystal structure. The two conformers are shown with similar orientations of the C(11)-C(12) and C(32)-C(33) bonds to facilitate comparison. (After ref. 10).

Either the twisted 6-*s-cis* (40°<$|\phi_{6-7}|$<65°) or the nearly planar 6-*s-trans* (165°<$|\phi_{6-7}|$<175°) conformers are found in vitamin-A-related compounds and carotenoids (Table 5.22-3).[25] Although the greater extent of the conjugated system in the 6-*s-trans* conformer indicates that it has more resonance stabilization than the 6-*s-cis* conformer, its intramolecular contacts and distortions show that it also has a higher steric energy. For the two conformers to coexist in the crystal, these two energy differences must be nearly balanced.

The molecule of all-*trans*-retinal$_2$ consists of a trimethylcyclo-hexadienyl ring connected to an all-*trans*-polyene side chain which terminates in an aldehydic group. Aside from the presence of two ring double bonds in all-*trans*-retinal$_2$, and only one in all-*trans*-retinal, they are nearly isomorphous.[25]

Table 5.22-3   Torsion angle C(5)-C(6)-C(7)-C(8) (°) for some retinals
               and related compounds.[25]

Compound	C(5)-C(6)-C(7)-C(8)
(a) 6-s-cis conformers	
11-cis-retinal (β-cis form)	40.0(1)
all-trans-retinoic acid* (6-s-cis conformer)	41.2(5)
11-cis-retinal (α-cis form)	41.4(7)
all-trans-retinal$_2$	55.6(5)
all-trans-retinal	58.3(6)
13-cis-retinal (6-s-cis conformer)	65.4(6)
19,19,19-trifluoro-9-cis-retinal	67.9(4)
9-cis-retinal	76.0(4)
(b) 6-s-trans conformers	
all-trans-retinoic acid* (6-s-trans conformer)	165.8(3)
13-cis-retinal (6-s-trans conformer)	174.9(4)

* Vitamin A acid.

4. Structure of vitamin B$_6$[19]

The B$_6$ vitamers contain a phenolic group and one or more basic sites
thus allowing the possibility of several tautomeric forms for each vitamer.
The crystal structure of pyridoxine (PN), pyridoxal (PL) and pyridoxamine
dihydrate (PM.2H$_2$O) have been determined.[19]

Pyridoxamine is a zwitterion in the PM.2H$_2$O crystal, where a proton has
been transferred from the phenolic group to the aminomethyl group [Fig. 5.22-
6(a)]. Bond distances and angles are similar to those in other vitamin B$_6$
derivatives. The differences which do occur are characteristic of the
particular zwitterion; e.g. the C(1)-O(1) distance is shorter in structures

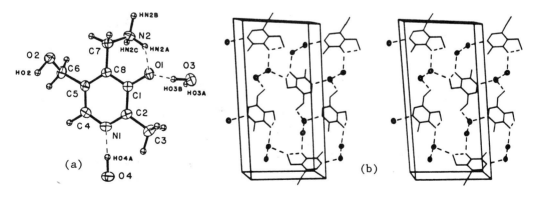

Fig. 5.22-6   (a) Molecular structure and numbering scheme for PM.2H$_2$O; dotted
lines indicate hydrogen bonds.   (b) Packing diagram for PM.2H$_2$O.   The *x* axis
is horizontal; *z* is vertical.   (After ref. 19).

containing an ionized phenolic group than in derivatives having an non-ionized phenolic group. The C(1)-O(1) distances in PL and PM.2H$_2$O are 1.290(1) and 1.327(2)Å, respectively, whereas this distance in non-dipolar pyridoxine is 1.374(4)Å.

There are two basic packing arrangements in vitamin B$_6$ structures. In the first, translationally related molecules are stacked and the *b* axis is about 4.6Å long. In the second, glide-related molecules are stacked along the *b* axis, which is about 9.1Å long. PM.2H$_2$O adopts the first arrangement, whereas PL adopts the second type of packing. Fig. 5.22-6(b) shows the molecular packing in PM.2H$_2$O.

5. Structure of vitamin C (L-ascorbic acid)

The crystal structure of L-ascorbic acid has been determined by X-ray and neutron diffraction.[22] The four molecules in the unit cell are related in pairs by pseudo screw axes, and each molecule consists of an almost planar five-membered ring plus a side chain; bond lengths and angles are shown in Fig. 5.22-7(a).

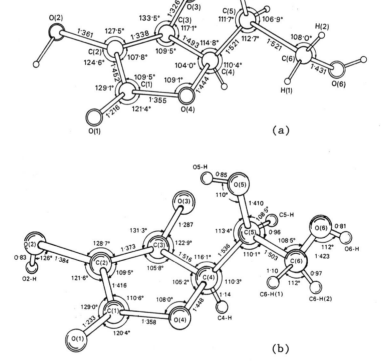

Fig. 5.22-7 The bond lengths and angles of (a) L-ascorbic acid and (b) the ascorbate anion in sodium ascorbate. (After ref. 22).

Comprehensive studies of ascorbic acid have provided empirical rules for the atomic arrangements necessary for the antiscorbutic effect. The presence of a furan ring which includes an enediol group next to a carbonyl group is imperative, and this configuration is also responsible for the acidity and reducing power of the substance. Moreover, the proper configuration at the asymmetric carbon atom C(4) and a side chain of at least two carbon atoms is necessary for the biological effect.

The crystal structure of sodium ascorbate has also been determined.[22] The bond lengths and angles of the ascorbate anion are shown in Fig. 5.22-7(b). Examination of the ring system reveals that the enediol group after dissociation of the hydrogen atom at O(3) is no longer planar as in ascorbic acid. Moreover, the lactone group which was moderately corrugated in the acid is now planar.

Isoascorbic acid[26] and sodium D-isoascorbate monohydrate[27] have been studied. An interesting deduction from these results is the small but significant differences between the dimensions of the acid molecules and the anions. The removal of a proton from the acid molecule, which belongs to the hydroxyl group attached to C(3), not only changes C(3)-OH to C(3)=O but also causes a redistribution of electron density throughout the five-membered ring. Within the ring the resulting bond length changes from acid molecule to anion agree for both isomers (Fig. 5.22-8). However, the ionization appears to affect the bond lengths in the conformationally different glycol side-chains of the two acid/ion pairs, as shown in Fig. 5.22-8.

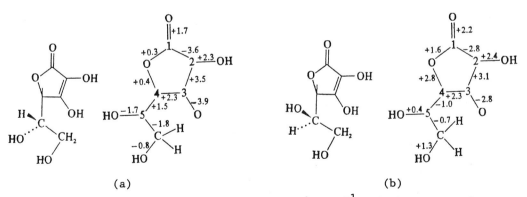

(a)   (b)

Fig. 5.22-8 Bond lengths differences (Å x 10^{-1}) of the anion relative to the neutral molecule of (a) ascorbic acid and (b) isoascorbic acid. (After ref. 28).

6. Structure of biotin, vitamin H

Biotin is an important coenzyme for many vital processes in plants and animals such as gluconeogenesis, biosynthesis of fatty acids and amino acid metabolism.

Early crystallographic studies of biotin by Traub established the relative stereochemistry at the asymmetric carbons.[29] The crystal structure of D-(+)-biotin has been determined to high precision by DeTitta and co-workers.[23] The molecular structure and crystal packing are shown in Fig. 5.22-9.

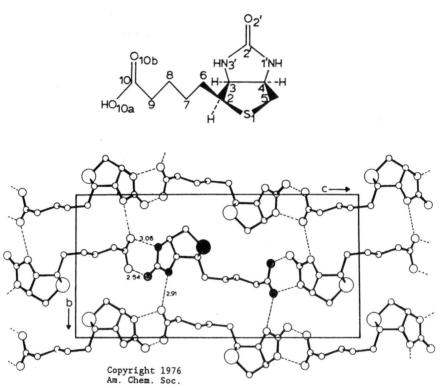

Fig. 5.22-9 (Top) Structural formula and absolute configuration of D-(+)-biotin. (Bottom) Crystal packing and hydrogen bonding viewed along the *a* axis. Bold lines are covalent bonds; dashed lines are hydrogen bonds. (After ref. 23).

The ureido ring, including the carbonyl oxygen, is essentially planar. The tetrahydrothiophene ring is envelope-shaped with the sulfur atom 0.87Å out of the least-squares plane fitted to the four carbon atoms. The bicyclic ring system possesses nearly perfect mirror symmetry about a plane passing through the sulfur and carbonyl carbon and bisecting the C(3)-C(4) bond. The dihedral angle of 122° describes the intersection of the six atoms of the ureido portion and the four carbon atoms of the tetrahydrothiophene ring, atoms C(3) and C(4) being common to both. The sulfur atom assumes the *endo* configuration with respect to the bicyclic ring system.

All potential hydrogen bond donors are utilized in the packing arrangement. The ureido N(1') and O(2') and a neighbouring carboxylic acid group form a {...O(2')-C(2')-N(1')-H...O(10b)=C(10)-O(10a)-H...} hydrogen-bonded ring. The carboxylic carbonyl oxygen O(10b) accepts a second hydrogen bond from another neighbouring N(3')H group. The sulfur atom, known to be protonated in strong acids, does not participate in the hydrogen bonding. Donor-acceptor relationships and distances are indicated in Fig. 5.22-9. The O(2')...H-O(10a) hydrogen bond distance, 2.54Å, is indicative of a strong hydrogen bond.

Biotin participates as a coenzyme in a variety of carboxylase, decarboxylase, and transcarboxylase systems. Its role is to fix CO_2 for eventual transfer, and it accomplishes its task in a two-step fashion during which a labile enzyme-biotin-CO_2 complex is formed. The recognition of bicarbonate by biotin may involve hydrogen bonding to the O(2') and N(1') atoms of the imidazolidone ring, facilitating the formation of a covalent bond between the bicarbonate C and N(1') atoms with the participation of ATP. Formation of an active enzyme-biotin-CO_2 complex requires that the biotin residue be correctly positioned on the protein; in addition to hydrogen bonding to O(2') and hydrophobic interaction with the side chain of biotin, the protein also uses a metal atom to form a coordinate bond with S(1).

Comparison of the structure of biotin with that of 1'-*N*-carboxy-biotin, as observed in its bis-*p*-bromoanilide derivative,[30] gives an assessment of the effects of carboxylation upon the ureido ring. Upon carboxylation the ureido ring remains planar, whereas the carbonyl bond shortens and the N-carbonyl C bonds lengthen to values more closely approximating those found in the barbiturates. This suggests that the planarity of the ureido ring in biotin may be a consequence of its incorporation into the bicyclic 2'-keto-3,4-imidazolidotetrahydrothiophene ring system rather than the effect of a charge delocalization.

References

[1] H.F.J. Savage, P.F. Lindley, J.L. Finney, and P.A. Timmins, *Acta Crystallogr., Sect. B* **43**, 280 (1987).

[2] P.G. Lenhert and D.C. Hodgkin, *Nature (London)* **192**, 937 (1961).

[3] P.G. Lenhert, *Proc. Roy. Soc. London* **A303**, 45 (1968).

[4] M. Rossi and J.P. Glusker in J.F. Liebman and A. Greenberg (eds.), *Environmental Influences and Recognition in Enzyme Chemistry*, VCH Publishers, Weinheim, 1988, p. 1.

[5] D.C. Hodgkin, J. Kamper, J. Lindsey, M. Mackay, J. Pickworth, J.H. Robertson, C.B. Shoemaker, J.G. White, R.J. Prosen and K.N. Trueblood, *Proc. Roy. Soc. London* A242, 228(1957); D.C. Hodgkin, J. Pickworth, J.H. Robertson, K.N. Trueblood, R.J. Prosen and J.G. White, *Nature (London)* 176, 325 (1955); D.C. Hodgkin, J. Kamper, M. Mackay, J. Pickworth, K.N. Trueblood and J.G. White, *Nature (London)* 178, 64 (1956).

[6] M. Rossi, J.P. Glusker, L. Randaccio, M.F. Summers, P.J. Toscano and L.G. Marzilli, *J. Am. Chem. Soc.* 107, 1729 (1985).

[7] R.D.W. Kemmitt and D.R. Russell in G. Wilkinson, F.G.A. Stone and E.W. Abel (eds.), *Comprehensive Organometallic Chemistry*, Vol. 5, Pergamon Press, Oxford, 1982, chap. 34.3.

[8] B. Kräutler, W. Keller and C. Kratky, *J. Am. Chem. Soc.* 111, 8936 (1989).

[9] L.J. Elsas and D.B. McCormick, *Vitamins and Hormones* 43, 103 (1986); W. Friedrich, *Vitamins*, Walter de Gruyter, Berlin, 1988.

[10] C.J. Simmons, R.S.H. Liu, M. Denny and K. Seff, *Acta Crystallogr.*, *Sect. B* 37, 2197 (1981).

[11] S.E. Hull, I. Leban, P. Main, P.S. White and M.M. Woolfson, *Acta Crystallogr.*, *Sect. B* 32, 2374 (1976).

[12] Trinh-Toan, H.F. DeLuca and L.F. Dahl, *J. Org. Chem.* 41, 3476 (1976).

[13] S. Krishnamurthy, *J. Chem. Educ* 60, 465 (1983); H. Mayer and O. Isler, *Methods Enzymol.* 18C, 241 (1971).

[14] H. Mayer, U. Gloor, O. Isler, R. Rüegg and O. Wiss, *Helv. Chim. Acta* 47, 221 (1964); L.M. Jackman, R. Rügg, G. Ryser, C. von Planta, U. Gloor, H. Mayer, P. Schudel, M. Kofler and O. Isler, *Helv. Chim. Acta* 48, 1332 (1965); O. Isler, R. Rüegg, L.H. Chopard-dit-Jean, A. Winterstein and O. Wiss , *Helv. Chim. Acta* 41, 786 (1958).

[15] A.S.V. Burgen and J.F. Mitchell, *Gaddum's Pharmacology*, 9th ed., Oxford University Press, New York, 1985, p. 137.

[16] R.E. Cramer, R.B. Maynard and J.A. Ibers, *J. Am. Chem. Soc.* 103, 76 (1981); J. Pletcher, M. Sax, A. Turano and C.H. Chang, *Ann. N.Y. Acad. Sci.* 378, 454 (1982).

[17] R.M. Burnett, G.D. Darling, D.S. Kendall, M.E. LeQuesne, S.G. Mayhew, W.W. Smith and M.L. Ludwig, *J. Biol. Chem.* 249, 4383 (1974); S. Fujii, K. Kawasaki, A. Sato, T. Fujiwara and K.-I. Tomita, *Arch. Biochem. Biophys.* 181, 363 (1977).

[18] W.B. Wright and G.S.D. King, *Acta Crystallogr.* **6**, 305 (1953); **7**, 473 (1954).

[19] J. Longo and M.F. Richardson, *Acta Crystallogr., Sect. B* **38**, 2721 (1982); C.L. MacLaurin and M.F. Richardson, *Acta Crystallogr., Sect. C* **41**, 261 (1985), and references cited therein.

[20] A. Camerman, D. Mastropaolo and N. Camerman in J.F. Griffin and W.L. Duax (eds.), *Molecular Structure and Biological Activity*, Elsevier Biochemical, New York, 1982, p. 1.

[21] R. Destro and A. Heyda, *Acta Crystallogr., Sect. B* **33**, 504 (1977).

[22] J. Hvoslef, *Acta Crystallogr., Sect. B* **24**, 23 (1968); **24**, 1431 (1968); **25**, 2214 (1969); P.A. Seib and B.M. Tolbert (eds.), *Ascorbic Acid: Chemistry, Metabolism and Uses*, American Chemical Society, Washington, D.C., 1982.

[23] G.T. DeTitta, J.W. Edmonds, W. Stallings and J. Donohue, *J. Am. Chem. Soc.* **98**, 1920 (1976); G.T. DeTitta, R.H. Blessing and G. Moss, *Acta Crystallogr., Sect. A* Supplement **37**, C128 (1981).

[24] G. Del Re, E. Gavuzzo, E. Giglio, F. Lelj, F. Mazza and V. Zappia, *Acta Crystallogr., Sect. B* **33**, 3289 (1977).

[25] C.J. Simmons, A.E. Asato and R.S.H. Liu, *Acta Crystallogr., Sect. C* **42**, 711 (1986).

[26] N. Azarnia, H.M. Berman and R.D. Rosenstein, *Acta Crystallogr., Sect. B* **28**, 2157 (1972).

[27] J.A. Kanters, G. Roelofsen and B.P. Alblas, *Acta Crystallogr., Sect. B* **33**, 1906 (1977).

[28] G.A. Jeffrey and A.D. French in L.E. Sutton and M.R. Truter (Senior Reporters), *Molecular Structure by Diffraction Methods*, Vol. **6**, The Chemical Society, London, 1978, p. 183.

[29] W. Traub, *Nature (London)* **178**, 649 (1956); Science *(Washington)* **129**, 210 (1959).

[30] C. Bonnemere, J.A. Hamilton, L.K. Steinrauf and J. Knappe, *Biochemistry* **4**, 240 (1965).

Note The structure of (adeninylpropyl)cobalamin, a coenzyme B₁₂ analogue in which a propylene chain substitutes for the ribose moiety of the adenosyl group, is described in T.G. Pagano, L.G. Marzilli, M.M. Flocco, C. Tsai, H.L. Carrell and J.P. Glusker, *J. Am. Chem. Coc.* **113**, 531 (1991).

5.23 Valinomycin

$$C_{54}H_{90}N_6O_{18} \cdot 3(CH_3)_2SO$$

Crystal Data

Orthorhombic, space group $P2_12_12_1$ (No. 19)

$a = 16.406(6)$, $b = 25.723(6)$, $c = 18.712(5)$Å, $Z = 1$

Atom*	x	y	z	Atom	x	y	z
S(1)	.3436	.9510	1.1078	O'(6)	.5080	.9575	.5241
O(1S1)	.4307	.9584	1.1356	C^α(6)	.4363	.9873	.5085
C(1S1)	.3134	.8907	1.1203	C'(6)	.4294	1.0324	.5520
C(2S1)	.3399	.9563	1.0239	O(6)	.3727	1.0671	.5498
S(2)	.4752	.8060	.7638	C^β(6)	.4325	1.0034	.4312
O(1S2)	.5604	.8193	.7436	$C^{\gamma 2}$(6)	.4217	.9560	.3857
C(1S2)	.4595	.8134	.8505	$C^{\gamma 1}$(6)	.4969	1.0330	.4028
C(2S2)	.4717	.7525	.7557	N(7)	.4920	1.0456	.5996
S(3)	.3013	1.0076	.7735	C^α(7)	.4990	1.0884	.6488
O(1S3)	.3354	1.0571	.7606	C'(7)	.5112	1.0689	.7251
C(1S3)	.3448	.9633	.7168	O(7)	.5316	1.0283	.7416
C(2S3)	.3226	.9792	.8310	C^β(7)	.5526	1.1316	.6264
N(1)	.5862	.9158	1.1020	$C^{\gamma 1}$(7)	.6435	1.1142	.6269
C^α(1)	.5946	.8650	1.1334	$C^{\gamma 2}$(7)	.5328	1.1528	.5527
C'(1)	.5767	.8220	1.0787	O'(8)	.5074	1.1115	.7665
O(1)	.5110	.8195	1.0518	C^α(8)	.5213	1.1041	.8398
C^β(1)	.5486	.8600	1.2015	C'(8)	.4588	1.1396	.8840
$C^{\gamma 1}$(1)	.5497	.8054	1.2291	O(8)	.4787	1.1633	.9358
$C^{\gamma 2}$(2)	.5715	.8927	1.2564	C^β(8)	.6065	1.1239	.8543
O'(2)	.6391	.7887	1.0686	N(9)	.3830	1.1309	.8624
C^α(2)	.6265	.7492	1.0170	C^α(9)	.3132	1.1607	.8904
C'(2)	.6570	.7678	.9396	C'(9)	.2920	1.1323	.9614
O(2)	.6486	.7309	.8966	O(9)	.2640	1.0887	.9639
C^β(2)	.6702	.6994	1.0458	C^β(9)	.2491	1.1651	.8389
$C^{\gamma 1}$(2)	.7593	.7099	1.0602	$C^{\gamma 2}$(9)	.2723	1.1934	.7714
$C^{\gamma 2}$(2)	.6179	.6810	1.1118	$C^{\gamma 1}$(9)	.1806	1.1924	.8715
N(3)	.6827	.8130	.9294	O'(10)	.3017	1.1626	1.0193
C^α(3)	.7062	.8260	.8561	C^α(10)	.2792	1.1424	1.0868
C'(3)	.6677	.8785	.8399	C'(10)	.3477	1.1112	1.1183
O(3)	.6189	.8999	.8759	O(10)	.3354	1.0910	1.1778
C^β(3)	.8014	.8284	.8552	C^β(10)	.2466	1.1830	1.1396
$C^{\gamma 2}$(3)	.8442	.7807	.8709	$C^{\gamma 1}$(10)	.2994	1.2207	1.1442
$C^{\gamma 1}$(3)	.8347	.8739	.8975	$C^{\gamma 2}$(10)	.1841	1.2145	1.0966
O'(4)	.6918	.8916	.7758	N(11)	.4233	1.1116	1.0862
C^α(4)	.6655	.9429	.7524	C^α(11)	.4892	1.0773	1.1113
C'(4)	.6454	.9374	.6695	C'(11)	.5182	1.0485	1.0540
O(4)	.6526	.9778	.6336	O(11)	.4967	1.0463	.9910
C^β(4)	.7270	.9821	.7659	C^β(11)	.5547	1.1101	1.1484
N(5)	.6089	.8974	.6473	$C^{\gamma 1}$(11)	.5239	1.1433	1.2109
C^α(5)	.5816	.8918	.5724	$C^{\gamma 2}$(11)	.5948	1.1459	1.1036
C'(5)	.5007	.9161	.5647	O'(12)	.5801	1.0165	1.0728
O(5)	.4422	.9037	.5939	C^α(12)	.6199	.9850	1.0230
C^β(5)	.5796	.8373	.5545	O(12)	.6947	.9056	1.0309
$C^{\gamma 1}$(5)	.5473	.8282	.4782	C'(12)	.6351	.9314	1.0496
$C^{\gamma 2}$(5)	.6586	.8115	.5589	C^β(12)	.6962	1.0105	.9980

* The first 12 entries are the atoms in the three co-crystallized $(CH_3)_2SO$ molecules.

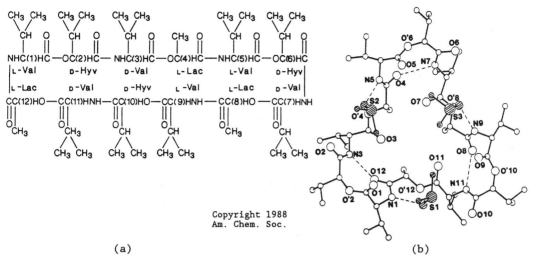

(a)   (b)

Fig. 5.23-1 (a) Structural formula of valinomycin. (b) Molecular structure of valinomycin.3DMSO, which contains three type 4 → 1 hydrogen bonds. (After ref. 1).

Crystal Structure

Valinomycin is an antibiotic that selectively transports potassium ions across natural and synthetic membranes. It is a cyclic dodecadepsipeptide built from a three-fold repetition of the alternating amino acid and hydroxy acid residues L-valine, D-α-hydroxyvaleric acid, D-valine, and L-lactic acid, (L-Val-D-Hyv-D-Val-L-Lac)$_3$ [Fig. 5.23-1(a)].

In the crystal structure of the valinomycin.3DMSO solvate,[1] the molecule assumes near C_3 symmetry for the backbone ring, the side chains, and even the S=O groups of the three solvent molecules that make hydrogen bonds with the three NH groups of the L-Val residues [Fig. 5.23-1(b)]. Three type II β-bends are formed encompassing the three repeating L-Val-D-Hyv sequences. The lengths of the hydrogen bonds (type 4 → 1, four C_α atoms are spanned) N(3)...O(12), N(7)...O(4) and N(11)...O(8) are 3.05, 3.22 and 3.24Å, respectively. These values are significantly larger than those observed in uncomplexed valinomycin (flattened bracelet conformation) from nonpolar solvents (Table 5.23-1), and in the K$^+$ complex, ~2.93Å. The other three N atoms, N(1), N(5) and N(9), are separated from O(10), O(2) and O(6), the atoms that participate in hydrogen bonding in the bracelet conformation, by 6.26, 6.37 and 6.08Å, respectively. Instead these NH groups form strong hydrogen bonds with the three DMSO molecules at N...O distances of 2.85, 2.81 and 2.80Å, respectively.

The conformation of the molecule is in the form of a shallow dish with

all the nonpolar side chains lying on the exterior. The outer diameter is
~15.6Å and the height is ~5.0Å. The dish could fit within a sphere of radius
~8.6Å. The bottom of the dish rests on the three $C^{\beta}H_3$ side chains from L-
Lac[4], L-Lac[8], and L-Lac[12] and forms a hydrophobic exterior surface. The inner
surface of the dish is covered with protruding carbonyl oxygens.

Remarks

1. Conformation of uncomplexed valinomycin

In 1975 Smith and co-workers[2] determined the structures of monoclinic
modification A (2:1 solvate crystallized from *n*-octane) and triclinic
modification B (unsolvated, from aqueous ethanol), while Karle[3]
independently solved the structure of modification B (crystallizing from
either *n*-octane or acetone). The molecular parameters of valinomycin (two
independent molecules in modification B) from these two studies are nearly
identical. The conformation can be considered as a "tennis-ball seam" or a
"flattened bracelet". All the hydrophobic side chains are on the surface of
the molecule, whereas the interior is lined with ester carbonyl oxygen atoms

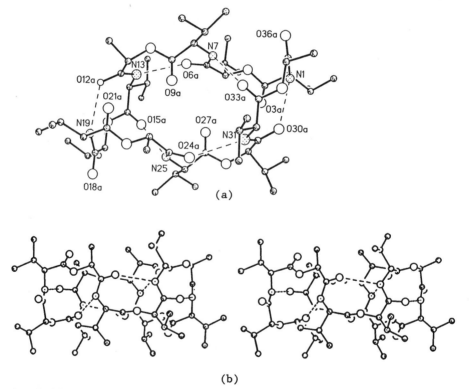

(a)

(b)

Fig. 5.23-2 (a) Molecular conformation of uncomplexed valinomycin
showing four 4 → 1 hydrogen bonds and two 5 → 1 hydrogen bonds.
(b) Stereoveiw showing the hydrophobic groups pointing up and down
around the bracelet-like structure.

[Fig. 5.23-2(a)]. The top part of the molecule is lined by three methyl groups and three isopropyl groups and the bottom part by six isopropyl groups [Fig. 5.23-2(b)], thus rendering the top less hydrophobic.

Table 5.23-2 Hydrogen bonds (average values in Å) in uncomplexed valinomycin.

Type 4 → 1	N(1)...O(30A)	L-Val-D-HyV	2.83
	N(13)...O(6A)	L-Val-D-HyV	2.98
	N(19)...O(12A)	D-Val-L-Lac	2.88
	N(31)...O(24A)	D-Val-L-Lac	3.07
Type 5 → 1	N(7)...O(33A)	D-Val-D-Val	3.11
	N(25)...O(15A)	L-Val-L-Val	3.02

Of the six intramolecular hydrogen bonds (Table 5.23-2), four are type 4 → 1 that link donor amide NH groups to neighbouring amide carbonyl oxygens. For the two remaining weaker hydrogen bonds, the acceptor in each case is a carbonyl oxygen belonging to an amino acid residue (type 5 → 1, five C_α atoms are spanned) [Fig. 5.23-2(a)]. Of the six carbonyl groups that do not partake in hydrogen bonding, four carboxylate oxygen atoms (O3A, O9A, O21A, and O27A) point inward whereas two amide oxygens (O18A and O36A) are quite exposed at the outer surface of the molecule.

2. Valinomycin as an ionophore

Ionophores are substances with the ability to promote the transfer of ions from an aqueous medium into a hydrophobic phase.[4] The hydrophobic phase can be a nonpolar solvent in contact with a polar one, such as a two-phase system made from chloroform and water. It can also be a biological membrane such as the one enclosing mitochondria, an artificial membrane such as a lipid bilayer, or an inert bulk membrane that supports the ionophorous materials.

There are two mechanisms by which ionophores can effect the transfer of ions from a polar into or across a nonpolar medium: the ion carrier and the channel-forming modes. A carrier ionophore forms a complex of well-defined stoichiometry with the ion, "carrying" the hydrophilic ion into or across the hydrophobic region, whereas channel-forming ionophores form hydrophilic pathways for the ions, spanning a lipid membrane barrier.

If isolated but functioning mitochondria are suspended in an aqueous medium containing potassium ions, the addition of catalytically small amounts of valinomycin can effect an active exchange of protons from inside the mitochondria against potassium. As a consequence of the uptake of potassium ions, the mitochondria will start to swell. It has been postulated that cyclic antibiotics like valinomycin can adopt a conformation with outward-directed hydrophobic groups, rendering the molecules lipid soluble, and with inward-directed polar groups contacting the solvation shell of the cations.

Formation of pores through the membrane by stacking antibiotic molecules was
suggested as the mode of action.[5]

3. The K^+ complex of valinomycin

Potassium iodo salts and valinomycin can form complexes in a variety of
modifications owing to the ease with which variable amounts of solvent
molecules may enter the crystals. X-Ray analysis of a crystalline KI_3-KI_5
complex[6] yielded the structure depicted in Fig. 5.23-3.

The K^+ complex has an approximate threefold rotation axis, corresponding
to the repetition of units in the chemical formula [Fig. 5.23-1(a)].
Approximate S_6 symmetry of the ring skeleton is extended reasonably well to
the side chains, with the exception of the D-Val isopropyl groups. All side
chains are either approximately parallel or perpendicular to the pseudo C_3
axis. Within experimental error the torsion angles about the C_α-C_β bonds of
all isopropyl groups are virtually identical with those found in uncomplexed
valinomycin. The K^+ ion is coordinated by six carbonyl oxygen atoms of ester
groups, forming a nearly regular octahedron with K^+...O distances in the range
of 2.72 to 2.79Å.

The six hydrogen bonds forming a part of the belt that embrace the
molecule are similar in lengths to the type 4 → 1 hydrogen bonds in
uncomplexed valinomycin.

4. Model of cation transport by valinomycin

Valinomycin adopts a compact conformation in nonpolar solvents, changing
to open conformations with increasing solvent polarity. It seems reasonable
to assume that metal complexation is favored by open and flexible structures.
Based on the crystal structures of valinomycin and its K^+ complex, a mechanism

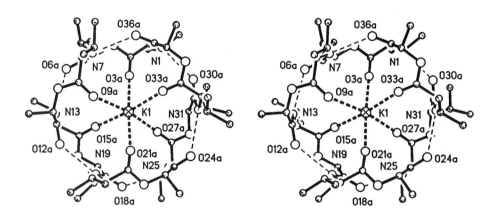

Fig. 5.23-3   Stereoview of the K^+ complex of valinomycin.

for the transport of cations has been suggested.[4] A valinomycin molecule in
its solid-state conformation lies at the membrane surface. Four carbonyl
oxygen atoms (0-3A, 0-9A, 0-21A, and 0-27A) are exposed on the surface of the
molecule, two on top and two at the bottom. A hydrated potassium ion may make
a loose contact with one of these oxygens, possibly stripping one of its water
molecules in exchange. Then the rather weak type 5 → 1 hydrogen bonds break
up and the carbonyl oxygens 0-15A and 0-33A become free to coordinate the K^+
ion. This process in a stepwise fashion entraps the cation, and the
previously free oxygens 0-18A and 0-36A finish up in the belt of type 4 → 1
hydrogen bonds of the complex. This process takes place with quite small
changes of torsion angles. The cation is now buried inside the valinomycin
molecule, with its six carbonyl oxygens replacing the water molecules of the
hydrated ion (Fig. 5.23-4).[7] The whole entity has a hydrophobic exterior
like a droplet of fat, rendering it soluble in the membrane. After transport
of the cation to the other side by driving forces, a similar reversed process
could release the cation, and thus the valinomycin could drift back, closing
the transport cycle.

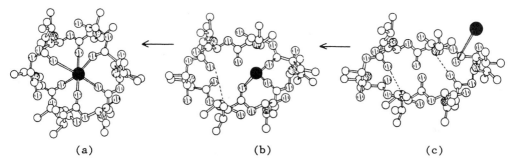

(a)   (b)   (c)

Fig. 5.23-4 Proposed model for the complexation of potassium ion by
valinomycin: (a) uncomplexed valinomycin (K^+ attached to a carbonyl
oxygen atom), (b) modelled by changes in torsion angles, and (c)
complexed valinomycin. The K^+ ion is depicted darkened, oxygen
stripped, and nitrogen stippled. (After ref. 7).

5. Structural studies on monensin A[4,8,9]

Monensin A [Fig. 5.23-5(a)] is a member of the family of monocarboxylic
acid, polycyclic, polyether antibiotics. It induces monovalent cation
transport across the mitochrondrial membrane in the reverse direction as
compared to valinomycin and exhibits a selectivity for Na^+ over K^+.

Historically the structure determination of monensin A in 1967, by X-ray
analysis of its silver salt dihydrate,[10] greatly stimulated the
investigation of carboxylic ionophores. Monensin and its monovalent metal

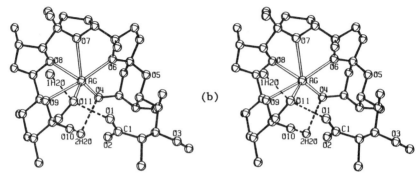

Fig. 5.23-5 (a) Chemical structure, numbering scheme, and ring identification of monensin. (b) Stereoview of the structure of the Ag$^+$ complex of monensin. For clarity the hydrogen bond O(2)...O(10) and two other hydrogen bonds to neighbouring molecules have been omitted. (Adapted from ref. 9).

salts crystallize in two different forms. The anhydrous sodium salt[11] and Form I dihydrates of sodium and silver salts[10,12] occur as "Type Na" in space group $P2_1$ with $Z = 2$. Dihydrates of the sodium (Form II),[11] potassium,[13] silver (Form II), and thallium salts, as well as a NaBr complex of monensin,[14] belong to "Type K" in space group $P2_12_12_1$ with $Z = 4$. The a and c axes of the two types are comparable in length, but the b axis is virtually doubled in going from the monoclinic to the orthorhombic form. A monohydrate of the free acid have very different unit cell dimensions in space group $P2_12_12_1$.[15]

All the cation complexes of monensin have nearly the same molecular structure. In the silver salt (Form I),[12] as in other complexes, the anion is wrapped around the Ag$^+$ ion which is coordinated by two hydroxyl, O(4) and O(11), and four ether oxygens, O(6), O(7), O(8) and O(9) [Fig. 5.23-5(b)]. The circular conformation of the ionophore is stabilized by two head-to-tail hydrogen bonds, O(1)...O(11) and O(2)...O(10), involving the carboxyl oxygens and two hydroxyl groups at the opposite end of the molecule.

The neutral monensin molecule in the monohydrate has a rather similar conformation to the cation complex, with the water molecule held in the center by hydrogen bonds with O(1), O(7) and O(11). Consequently there is only one head-to-tail hydrogen bond between O(2) and O(11).

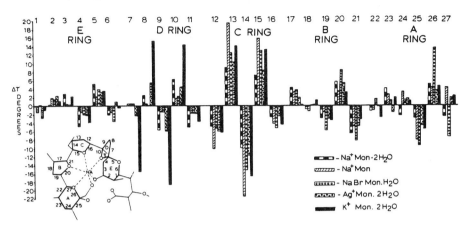

Fig. 5.23-6 The difference in the intra-ring torsion angles of the metal complexes of monensin A from those of the free acid. (After ref. 8).

Fig. 5.23-6 illustrates the magnitudes of the deviation of the individual ring torsion angles of the cation complexes from those of the free acid.[8] The conformations of all the complexes except the K^+ complex differ from the free acid in a very similar fashion. The significant torsion angle changes in ring D of the K^+ complex correspond to a pseudorotation from one envelope form to another, which acts as a hinge to expand the coordination sphere relative to that found in the Na^+ complex to accommodate the larger K^+ ion.

6. Classification of ionophores

     According to Dobler,[4] the ionophores are classified into four classes: (1) natural neutral ionophores; (2) natural carboxylic ionophores; (3) synthetic ionophores; and (4) quasi-ionophores.

     The natural neutral ionophores comprise two groups of antibiotics. The first are depsipeptides, molecules built from alternating amino acid and hydroxy acid residues as typified by valinomycin. The second group are the macrotetrolides, cyclic esters isolated from *actinomyces* strains with a high degree of biological activity. The first member of this group, nonactin, is a cyclic molecule built from four nonactic acid units with alternating chirality, resulting in a 32-membered ring of 20 carbon and 12 oxygen atoms. Fig. 5.23-7 shows the structure of its crystalline KCNS complex.[16]

     The carboxylic ionophores are a class of antibiotics produced by various *streptomyces* cultures. Their biological action differs from that of the

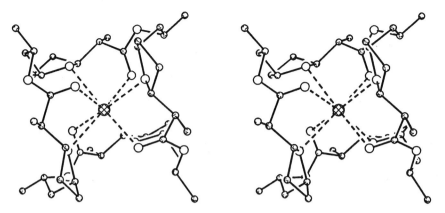

Fig. 5.23-7 Stereoview of the nonactin-K$^+$ complex along an *approximate* S_4 axis.

valinomycin-type antibiotics because they reverse the transport of alkali-metal cations across mitochrondrial membranes induced by neutral ionophores. Typical members of this class are monensin and nigericin,[17] and all carboxylic ionophores have the same basic skeleton composed of rings A to E [Fig. 5.23-5(a)] with very similar conformations in the free and complexed forms.

Five groups of compounds are prominent in synthetic ionophores: (i) macrocyclic polyethers ("crown" ethers); (ii) linear polyethers; (iii) macropolycyclic polyethers containing tertiary amino groups at the bridge-heads ("cryptands"); (iv) noncyclic ligands for alkaline earth cations; and (v) chirality-recognizing polyethers. The most widely studied variety are the compounds related to 18-crown-6 (Fig. 5.23-8).

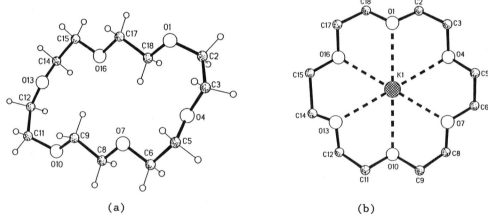

(a)   (b)

Fig. 5.23-8 The molecular structures of (a) uncomplexed 18-crown-6 and (b) [K$^+$.18-crown-6] in the KNCS complex.

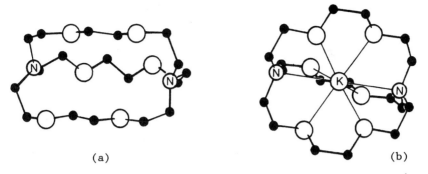

Fig. 5.23-9 Structures of (a) uncomplexed C(222) and (b) [K$^+$.C(222)] in the KI complex.

The cryptand C(222) [see Section 3.6] shows the highest selectivity for potassium ions. The structures of uncomplexed C(222) and its KI complex are shown in Fig. 5.23-9. The chemistry of crown ethers and analogues, including modern aspects of host-guest interaction with detailed structural information, is reviewed in a recent volume covering the literature to 1987-88.[18]

7. Ion channels in gramicidin A

Many substances, called quasi-ionophores, possess the ability to transport ions across lipid membranes in a novel way. Instead of transporting the ions as carriers, they form channels across membranes, possibly lined with polar functions on the inner surface.[19] Among the channel-forming ionophores, the gramicidins are the most investigated in terms of their conductance properties. The gramicidin family consists of a group of linear polypeptides built from fifteen amino acids terminated by a formyl group at one end and an ethanolamide group at the other. The structural formula of gramicidin A is:

HCO-L-Val-Gly-L-Ala-D-Leu-L-Ala-D-Val-L-Val-D-Val-L-Trp-D-Leu-L-Trp-D-Leu-L-Trp-D-Leu-L-Trp-NHCH$_2$CH$_2$OH.

The crystal structure of the gramicidin A/CsCl complex (at 1.8Å resolution)[20,21] contains two independent, but very similar, tube-like dimers approximately 26Å in length (Fig. 5.23-10). The peptide backbones of each dimer constitute a left-handed, antiparallel, double-stranded helix with a $\beta^{7.2}$ sheet hydrogen bonding pattern at 6.4 residues per turn. The solvent-filled channel contains two cesium and three chloride ions, arranged alternately and separated by more than the sum of their ionic radii. The hydrophilic side chains form a relatively uniform outer surface of diameter ~16Å, and there are two cesium sites (occupancy 0.6 and 0.4) in the region between the dimers.

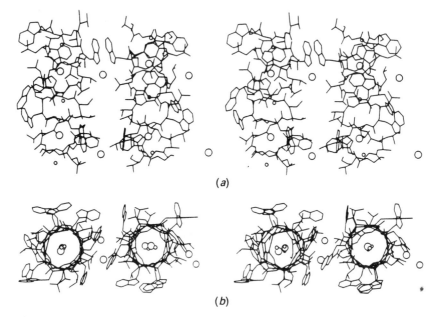

(a)

(b)

Fig. 5.23-10 Stereograms of the gramicidin A/cesium chloride complex viewed perpendicular (a) and along (b) the helical axes. The two independent dimers in the asymmetric unit are shown. Cesium and chlorine sites are represented by large and small circles, respectively. (After ref. 21).

The crystal structure of the uncomplexed form of gramicidine A (at 120K and 0.86Å resolution)[22,23] contains a helical dimer similar to those in the complexed form, but the double-stranded β ribbon is wound more tightly to form a longer channel (31Å) with a smaller diameter (average 4.85Å) at 5.6 residues per turn (Fig. 5.23-11). The tighter coiling shifts the hydrogen bonds joining the edges of the β ribbon to analogous interaction sites two peptide units away (Fig. 5.23-12).

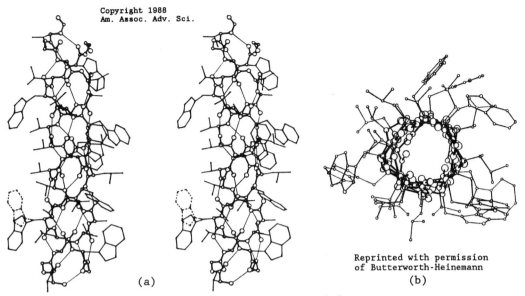

(a)   (b)

Fig. 5.23-11 The antiparallel double-stranded $\beta^{5.6}$ helical dimer in the crystal structure of uncomplexed gramicidin A. (a) Stereoview showing side chains and hydrogen bonds (lighter lines). One tryptophan is disordered as shown by dotted lines. (b) A view down the double helical channel. (After refs. 23 and 22).

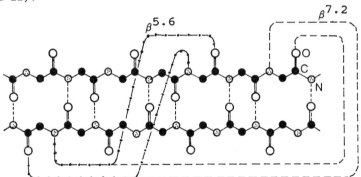

Fig. 5.23-12 Schematic diagram showing the hydrogen-bonding patterns for the uncomplexed $\beta^{5.6}$ (dot-dashed lines) and cesium-complexed $\beta^{7.2}$ (dashed lines) molecular models of the gramicidin A channel helices. (Adapted form ref. 23).

    The channel is devoid of solvent molecules but does contain three pockets with diameters that exceed 5.25Å. A potassium ion entering any one of these pockets cannot avoid severe weakening or disruption of some of the interstrand hydrogen bonds. It is conceivable that these potential binding sites can be induced to travel the length of the channel by a cooperative peristaltic flexing of the peptide backbone, but the mechanism remains to be elucidated.

References

[1] I.L. Karle and J.L. Flippen-Anderson, *J. Am. Chem. Soc.* **110**, 3253 (1988).

[2] G.D. Smith, W.L. Duax, D.A. Langs, G.T. DeTitta, J.W. Edmonds, D.C. Rohrer and C.M. Weeks, *J. Am. Chem. Soc.* **97**, 7242 (1975).

[3] I.L. Karle, *J. Am. Chem. Soc.* **97**, 4379 (1975).

[4] M. Dobler, *Ionophores and Their Structures*, Wiley-Interscience, New York, 1981.

[5] P. Mueller and D.O. Rudin, *Biochem. Biophys. Res. Commun.* **26**, 398 (1967).

[6] K. Neupert-Laves and M. Dobler, *Helv. Chim. Acta* **58**, 432 (1975).

[7] W.L. Duax, *J. Chem. Educ.* **65**, 502 (1988).

[8] W.L. Duax, J.F. Griffin and G.D. Smith, in J.J. Stezowski, J.-L. Huang and M.-C. Shao (eds.), *Molecular Structure: Chemical Reactivity and Biological Activity*, Oxford University Press, New York, 1988, p. 1.

[9] E.N. Duesler and I.C. Paul in J.W. Westley (ed.), *Polyether Antibiotics*, Vol. 2, Marcel Dekker, New York, 1983, chap. 3.

[10] A. Agtarap, J.W. Chamberlin, M. Pinkerton and L.K. Steinrauf, *J. Am. Chem. Soc.* **89**, 5737 (1967).

[11] W.L. Duax, G.D. Smith and P.D. Strong, *J. Am. Chem. Soc.* **102**, 6725 (1980).

[12] M. Pinkerton and L.K. Steinrauf, *J. Mol. Biol.* **49**, 533 (1970).

[13] W.L. Duax, W.A. Pangborn and D.A. Langs, *Am. Crystallogr. Assoc. Meet.*, Hamilton, Ontario, Canada, Abstr. No. PB56 (1986).

[14] D.L. Ward, K.T. Wei, J.G. Hoogerheide and A.I. Popov, *Acta Crystallogr., Sect. B* **34**, 110 (1978).

[15] W.K. Lutz, F.K. Winkler and J.D. Dunitz, *Helv. Chim. Acta* **54**, 1103 (1971).

[16] B.T. Kilbourn, J.D. Dunitz, L.A.R. Pioda and W. Simon, *J. Mol. Biol.* **30**, 559 (1967).

[17] M. Shiro and H. Koyama, *J. Chem. Soc. (B)*, 243 (1970).

[18] E. Weber, J.L. Toner, I. Goldberg, F. Vögtle, D.A. Laidler, J.F. Stoddart, R.A. Bartsch and C.L. Liotta (S. Patai and Z. Rappoport, eds.), *Crown Ethers and Analogs*, Wiley, New York, 1989.

[19] P. Läuger, *Angew. Chem. Int. Ed. Engl.* **24**, 905 (1985).

[20] B.A. Wallace and K. Ravikumar, *Science (Washington)* **241**, 182 (1988).

[21] B.A. Wallace, W.A. Hendrickson and K. Ravikumar, *Acta Crystallogr., Sect. B* **46**, 440 (1990).

[22] W.L. Duax, D.A. Langs, W.A. Pangborn, G.D. Smith, V.Z. Pletnev and V.T. Ivanov, *J. Mol. Graphics* **7**, 82 (1989).

[23] D.A. Langs, *Science (Washington)* **241**, 188 (1988).

Note The double-stranded, antiparallel, left-handed $\beta^{5.6}$-helix in a new monoclinic form of gramicidin A, as compared to that in the known orthorhombic form, is more uniform as a consequence of π interactions between dimers in which the tryptophan side chains are oriented normal to the helical axis to relieve distortion by lateral crystal packing forces. See D.A. Langs, G.D. Smith, C. Courseille, G. Précigoux and M. Hospital, *Proc. Natl. Acad. Sci. USA* **88**, 5345 (1991).

Chapter 6

Organometallic Compounds

6.1. *Methyllithium* $(CH_3Li)_4$ — 998

6.2. *Ethylmagnesium Bromide Dietherate (Grignard Reagent)*
$C_2H_5MgBr \cdot 2(C_2H_5)_2O$ — 1012

6.3. *Trimethylaluminum* $Al_2(CH_3)_6$ — 1023

6.4. *Zeise's Salt* $K[(H_2C=CH_2)PtCl_3] \cdot H_2O$ — 1032

6.5. *Bis(triphenylphosphine)hexafluorobut-2-yneplatinum(0)*
$(Ph_3P)_2(CF_3C\equiv CCF_3)Pt$ — 1038

6.6. *[1,2-Bis(dimethylphosphino)ethane](neopentylidyne)-*
(neopentylidene)(neopentyl)tungsten(IV)
$W(\equiv CCMe_3)(=CHCMe_3)(CH_2CMe_3)(dmpe)$ — 1051

6.7. *Dimanganese Decacarbonyl* $Mn_2(CO)_{10}$ — 1061

6.8. *Bis(tetraethyl)ammonium μ_6-Carbido-penta-μ-carbonyl-*
octacarbonyl-octahedro-hexacobaltate $[NEt_4]_2[Co_6C(CO)_{13}]$ — 1074

6.9. *Bis(cyclopentadienyl)iron (Ferrocene)* $Fe(C_5H_5)_2$ — 1092

6.10. *Bis(cyclopentadienyl)beryllium (Beryllocene)* $Be(C_5H_5)_2$ — 1106

6.11. *μ-Dinitrogen-bis[bis(pentamethylcyclopentadienyl)titanium]*
$\{(\eta^5\text{-}C_5Me_5)_2Ti\}_2N_2$ — 1119

6.12. *Tetraphenylcyclobutadiene Iron Tricarbonyl* $Fe(CO)_3(C_4Ph_4)$ — 1131

6.13. *Bis(cyclooctatetraenyl)uranium(IV) (Uranocene)* $U(\eta^8\text{-}C_8H_8)_2$ — 1139

6.14. *Tribenzo[b,e,h][1,4,7]trimercuronin (o-Phenylenemercury Trimer)*
$(C_6H_4Hg)_3$ — 1146

6.15. *Trimethyltin(IV) Chloride* $(CH_3)_3SnCl$ — 1152

6.16. *(R-1-Cyanoethyl)(3-methylpyridine)cobaloxime* $C_{17}H_{25}N_6O_4Co$ — 1164

6.1 Methyllithium
$(CH_3Li)_4$

Crystal Data

Cubic, space group $I\bar{4}3m$ (No. 217)

$a = 7.24(1)$Å, $Z = 2$

Atom	x	y	z
Li	.131	.131	.131
C	.320	.320	.320
H	.351	.351	.192

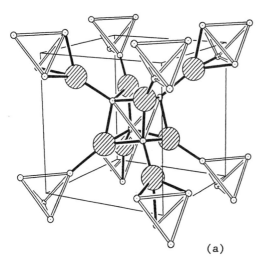 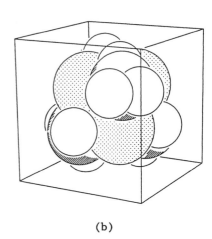

(a)   (b)

Fig. 6.1-1 (a) Crystal structure of $(CH_3Li)_4$. (b) Space-filling model of the $(CH_3Li)_4$ molecule.

Crystal Structure

The crystal structure of methyllithium, firstly determined in 1964[2] and refined in 1970 using powder diffraction data,[1] consists of tetrameric units of $(CH_3Li)_4$ with site symmetry T_d (Fig. 6.1-1). The molecule may be described as a tetrahedral array of Li atoms with a methyl C atom occupying each face of the tetrahedron or, alternately, as a distorted cubic arrangement of Li atoms and CH_3 groups. The C...C distances are much longer than the Li-Li bonds, and the Li-C-Li angle has the very low value of 68.3°.

The measured bond distances are:

within each tetramer	Li-Li	2.68(5)Å
	Li-C	2.31(5)Å
	C-H	0.96(5)Å
between two tetramers	C...C	3.68(5), 3.89(5)Å
	Li-C	2.36(5)Å

Remarks

1. Organometallic compounds

In an organometallic compound the organic entity forms carbon-metal bond(s) which can be σ- or π-type, but "ionic" bonding such as that in NaCN is excluded. The important organic ligands are listed in Table 6.1-1; the sections in this chapter are arranged roughly according to the entries in this table.

Table 6.1-1 Important ligands in organometallic compounds.

Ligand	Examples
1. alkyl	$-R$, $-CH_3$, $-C_2H_5$
aryl	$-Ar$, $-C_6H_5$
2. alkene	$\pi-C_2H_4$ derivatives
alkyne	$\pi-C_2H_2$ derivatives
allyl	$\pi-C_3H_5$ derivatives
conjugated diene	butadiene derivatives
3. carbene	$=CR_2$, $=C\diagup^{OR}_{\diagdown R'}$, $=C\diagup^{OR}_{\diagdown NHR'}$, $=C$(cyclic)
carbyne	$\equiv CR$, $\equiv CAr$
4. carbonyl	$C\equiv O$
5. carbido	C
6. aromatic carbocycle	cyclobutadiene (CBD,C_4H_4) derivatives
	cyclopentadienyl (Cp,C_5H_5) derivatives
	benzene (C_6H_6) derivatives
	cyclooctatetracene (COT,C_8H_8) derivatives

The discovery of the first organometallic compound, $K[(C_2H_4)PtCl_3]$, was by Zeiss in 1825 (Section 6.4). The first metal carbonyl complex, $[PtCl_2(CO)]_2$, was described in 1868 by Schutzenberger.[4] Mond's (1890) preparation of $Ni(CO)_4$,[5] the first binary metal carbonyl, had greater impact since this discovery led to a commercial process for the refining of nickel. The work on $Ni(CO)_4$ was soon followed by the discovery of $Fe(CO)_5$.[6] Grignard succeeded in synthesizing the alkylmagnesium halides that soon found wide application in organic synthesis (Section 6.2), and for this accomplishment he was awarded the Nobel Prize in 1912.

Modern research in organometallic chemistry is marked by the year 1951, when the nature of the metal-carbonyl bond, the prototype of π-backbonding, was proposed and ferrocene (Section 6.9) accidentally prepared. In the following year the correct "sandwich" structure of ferrocene was advanced by Wilkinson and Woodward,[7] while Fischer[8] described the isoelectronic

cationic Co(III) compound. The discovery of these novel compounds and their
chemical reactions greatly stimulated research in this field. Fischer and
Wilkinson were awarded the Nobel Prize in 1973.

The year 1955 witnessed the announcement of the organoaluminum compounds
of Ziegler and the mixed Al/Ti catalysts of Natta, which made possible the
low-pressure polymerization of olefins such as ethylene and propene. This
event of great technological significance initiated large-scale industrial
heterogeneous catalytic polymerization of alkenes.[9] Ziegler together with
Natta were honored with the Nobel Prize in 1963.

The universal application of X-ray crystallography has led to the
structural characterization of numerous organometallic compounds. The study
of compounds containing metal-carbon bonds has blurred the traditional
demarcation between inorganic and organic chemistry, and organometallic
chemistry serves as a bridge of the two disciplines as expounded by Hoffmann
in his Nobel Lecture of 1982.[10]

2. Structural chemistry of organolithium[3]

The first X-ray investigation of an organolithium compound, the
tetrahedral ethyllithium tetramer, was reported by Dietrich in 1963.[11] This
was followed a year later by Weiss's similar tetrameric methyllithium
structure.[2] Both investigators continue to be principal contributors in
this area. Stucky[12] also made many notable advances during the 1970's.

The most important organometallic compounds of the alkali metals are
those of lithium, which are perhaps the most widely used organometallic
synthetic intermediates.[13-15] The Li-C bond is a strongly polarized
covalent bond, so that organolithium compounds serve as sources of anionic
carbon.

The typical "inorganic" descriptions of lithium chemistry do not reflect
its importance. Bonds to lithium have a high degree of ionic character; this
leads to large energies of association. Dimeric, tetrameric, and other
oligomeric structures are common. Similarly, two or more lithium atoms often
bridge the same set of atoms simultaneously; this is the intramolecular
equivalent of aggregation. In addition, crystal structures of lithium
compounds often show interactions with electron-rich molecules. Lithium often
interacts with hydrocarbon π systems (olefins, arenes, or acetylenes) at many
different sites simultaneously; lithium tends to engage in multicenter
covalent bonding, but the ionic character still dominates. The propensity for
aggregation makes it particularly difficult to predict the structures of
organolithium compounds.

While lithium prefers four-coordination in a tetrahedral arrangement, exceptions are common. The coordination sphere around the lithium atom seems to be governed largely by steric effects.

3. Organolithium compounds containing lithium tetrahedra

When steric hindrance is minimal, organolithium compounds tend to form tetrameric aggregates. The simple alkyllithium compounds (MeLi)$_4$, (MeLi)$_4 \cdot 2[(CH_3)_2NCH_2CH_2N(CH_3)_2]$, and (EtLi)$_4$ adopt such tetrahedral structures (Fig. 6.1-2). The Li-C bond lengths range from 2.19 to 2.53Å, with an average value of 2.29Å, while the Li-Li distance lies in the range 2.42-2.63Å (average 2.56Å). These Li-C bond lengths are slightly longer than those for the terminally bonded organolithium compounds; this arises from the multicenter bonding in the tetrameric structures.

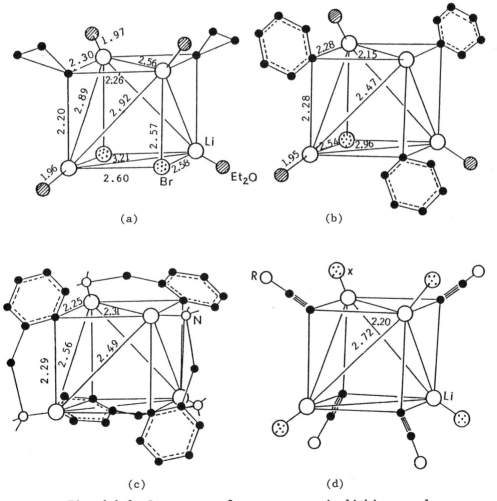

Fig. 6.1-2 Structures of some tetrameric lithium complexes.

The complex $(LiBr)_2 \cdot (CH_2CH_2CH\text{-}Li)_2 \cdot 4Et_2O$ [(Fig. 6.1-2(a)], the lithium bromide complex of phenyllithium monoetherate, [(PhLi$\cdot$Et$_2$O)$_3 \cdot$LiBr] [Fig. 6.1-2(b)], 1-lithio-2-dimethylaminomethylbenzene, [$C_6H_4CH_2N(CH_3)_2Li$]$_4$ [Fig. 6.1-2(c)], and two substituted lithium acetylides, (PhC≡CLi)$_4$(TMHDA)$_2$ and (t-BuC≡CLi)$_4$(THF)$_4$, [Fig. 6.1-2(d)], adopt similar tetrameric structures.

4. Organolithium compounds with bridging lithium atoms

Some dimeric organolithium compounds contain bridging lithium atoms in the solid state. The coordination sphere of each lithium atom is often completed by other ligands.

The dimer of lithiodimethylsulfide [Fig. 6.1-3(a)] illustrates the bridging sp^3-hybridized carbon atoms, while phenyllithium [Fig. 6.1-3(b)]

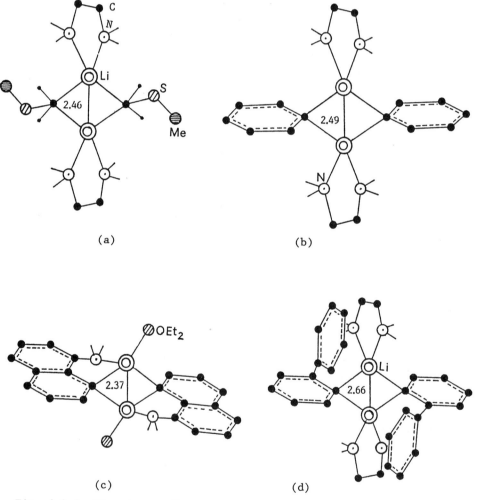

(a)   (b)

(c)   (d)

Fig. 6.1-3 Structures of some dimeric organolithium compounds with bridging lithium atoms.

exemplifies bridging between sp^2 carbon atoms. The dimers 8-(dimethylamino)-1-lithionaphthalene [Fig. 6.1-3(c)] and 2-lithiobiphenyl [Fig. 6.1-3(d)] also feature bridging sp^2 carbon atoms.

5. π Complexes of lithium[16]

Many examples of the interaction between lithium and diverse π systems are documented in the literature. In the first review in this field, Stucky[12] described the structure and bonding in π complexes of *N*-chelated lithium units. The synthesis and structure of organolithium compounds, including those containing unsaturated organic systems, have been reviewed by Wardell.[3] A collection of X-ray crystal structural data of organolithium species exhibiting lithium π interactions has been presented by Setzer and Schleyer in a comprehensive review covering the literature up to 1983.[13] The structures of some representative π complexes are schematically represented in Fig. 6.1-4.

The crystal structure of complex I shows symmetrical π bonding between the lithium atoms and allylic fragments. Each lithium atom interacts with two allylic π systems and further with the oxygen atom of a diethyl ether molecule.

The crystal structure of complex II shows that the metal is unusually coordinated by a η^5-Cp group and only one other monodentate ligand. The distance of the lithium atom to the centroid of the planar Cp ring (1.79Å) is considerably shorter (~0.2Å) than that in the corresponding Li-TMEDA complex; the same is true of the Li-N distance (1.99Å) in comparison with ~2.2Å in the TMEDA complex.

π-Allyl interactions are present in the crystal structures of some hydrocarbon π complexes with LiN$_2$ units (III-VI).

Metallation of indenofluorene with butyllithium yields ruby-red crystals (VII) of the first organolithium sandwich, having two lithium atoms each in linear coordination with two η^6 six-membered ring ligands.[17] The molecular geometry in this metallocenophane is consistent with both a simple electrostatic and a molecular orbital model.

Complexes VIII-XIII are dilithium compounds. In VIII-X, the two lithium atoms are not situated directly opposite to each other. In XI-XIII, "inverse sandwich" structures have been found, with two lithium atoms coordinating to opposite sides of a π system.

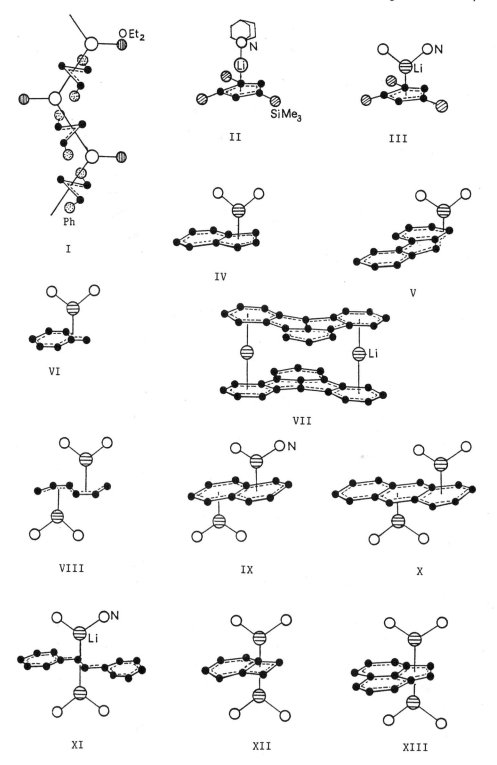

Fig. 6.1-4 Schematic structures of some π complexes of lithium. (Adapted from ref. 16).

6. Methyl derivatives of the heavier alkali metals

Methylsodium has been prepared by adding sodium *tert*-butoxide (dissolved in diethyl ether/hexane) to freshly prepared methyllithium.[18] Depending on the reaction conditions the products invariably contain variable amounts of methyllithium (Na:Li ratio *ca.* 36:1 to 3:1). Their common crystal structure (space group $F\bar{4}3c$, $a = 20.20$Å) has been determined by powder diffraction. The unit cell contains 24 tetrahedral $(CH_3Na)_4$ units, with distances Na-Na 3.12, 3.18Å; Na-C 2.58, 2.64Å; and Na-C 2.76Å between adjacent tetramers. Packing of the $(CH_3Na)_4$ units generates 8 large cavities, which can accommodate the smaller isostructural $(CH_3Li)_4$ units up to a maximum ratio of $(CH_3Na)_4$:$(CH_3Li)_4$ = 3:1 without noticeable lattice expansion. The crystal structures of cubic methylsodium and $3(CH_3Na)_4 \cdot (CH_3Li)_4$ are illustrated in Figs. 6.1-5 and 6.1-6, respectively.[18]

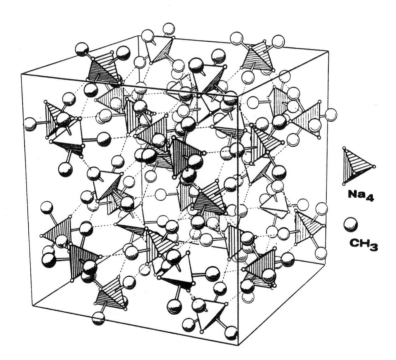

Fig. 6.1-5 Packing of $(CH_3Na)_4$ units in the unit cell of cubic methylsodium. (After ref. 18).

An orthorhombic form of methylsodium having a more compact structure results from the addition of dissolved LiCH₃ to previously prepared NaOtBu.[19] A combined neutron and synchrotron-radiation powder diffraction study of deuterated methylsodium, CD₃Na, at 1.5K yielded $a = 6.7686$, $b = 18.6016$, $c = 6.5762$Å, space group $I222$ and $Z = 16$.[20] A stereoview of the

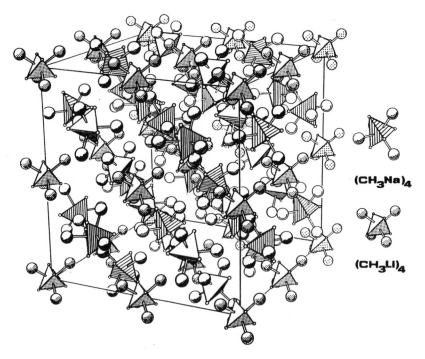

Fig. 6.1-6 Crystal structure of $3(CH_3Na)_4 \cdot (CH_3Li)_4$. (After ref. 18).

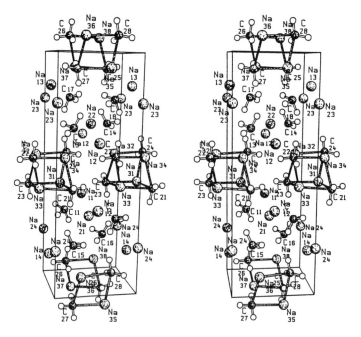

Fig. 6.1-7 Stereoview of the crystal structure of orthorhombic CD_3Na with atom numbering (symmetry-related atoms differ in the value of the second digit). Tetramers are highlighted by showing their heterocubane framework. (After ref. 19).

crystal structure of this *true methylsodium* is shown in Fig. 6.1-7. Half of the ions (eight Na3 and eight C2) constitute $(CD_3Na)_4$ units which are arranged in rows parallel to the x and z axes. The central Na_4C_4 core of the tetrameric unit is a strongly distorted cube (Na-Na 2.97-3.17, Na-C 2.57-2.64Å). The remaining Na^+ and CD_3^- ions (Na1, Na2 and C1) lie between these rows; the Na^+ ions form zigzag chains parallel to the z axis and are interconnected laterally by the CD_3^- ions. The Na-C distances involving these eight Na^+ and eight CD_3^- ions vary over a wider range than those within the tetramers. They are short (2.528 and 2.569Å) for direct contact between Na^+ and the lone pair of the methide ion and markedly longer (2.834 and 2.911Å) when Na^+ is located on the other side opposite the lone pair.

CH_3K, CH_3Rb, and CH_3Cs have been prepared by reacting methyllithium with potassium *tert*-butoxide, rubidium *tert*-butoxide, and cesium 2-methyl-2-pentanoate, respectively. The compounds were previously shown by X-ray powder diffraction to be isomorphous, possessing a hexagonal structure of the NiAs type with isolated methyl anions and alkali cations.[21,22] A recent neutron

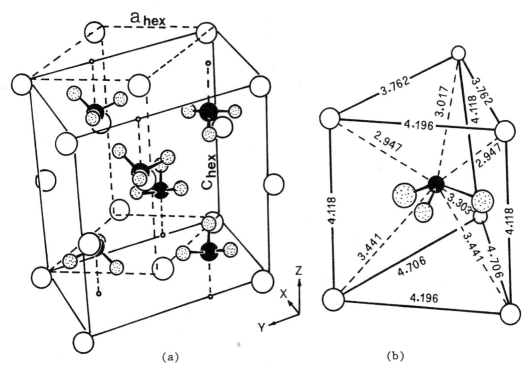

(a)   (b)

Fig. 6.1-8 (a) Crystal structure of methylpotassium. The relationship between the two choices of the unit cell are indicated. (b) The coordination environment of a CD_3^+ group in CD_3K. The distances shown are in Å and measured at 1.35K. (After ref. 23).

diffraction study on powder samples of CD_3K at 1.35 and 290K[23] revealed an
orthorhombic unit cell (space group *Pmcn*, a = 4.1956(2), b = 7.3073(3), c =
8.1644(3)Å at 1.35K, Z = 4), in contrast to the smaller hexagonal cell
($P6_3/mmc$, a = 4.27_8, c = 8.28_3Å, Z = 2) previously assigned to CH_3K.[21] The
two unit cells are related as shown in Fig. 6.1-8(a). In the crystal the
pyramidal methyl ions (bond angle 105° at 1.35K, 109° at 290K) have
alternating orientations, each being coordinated by six K^+ ions in a distorted
trigonal-prismatic array [Fig. 6.1-8(b)]. The three K^+ ions close to the sp^3
lone electron pair have short K...C contacts (2 x 2.95, 1 x 3.02Å), whereas
those close to the H atoms have longer K...C distances (2 x 3.44, 1 x 3.30Å).

7. π Complexes of sodium and potassium

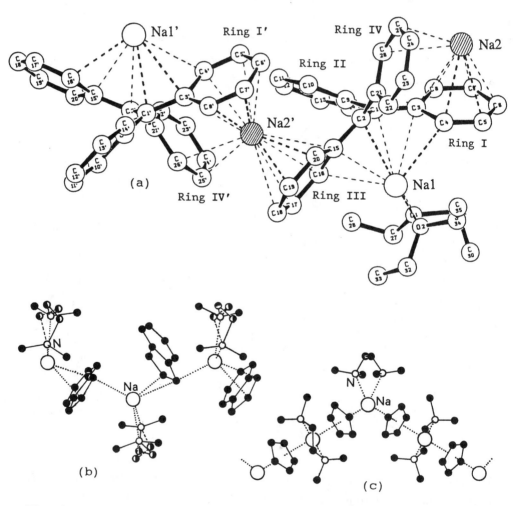

Fig. 6.1-9 Band structures of (a) $Na[(Ph_2C-CPh_2)^{2-}Na^+(OEt_2)_2]$,
(b) indenylsodium.tmeda and (c) cyclopentadienylsodium.tmeda. (After
refs. 25, 27 and 28).

In the band-like crystal structure of tetraphenylethylenedisodium. 2Et₂O [Fig. 6.1-9(a)],[25] the two halves of the $(Ph_2C-CPh_2)^{2-}$ dianion twist through 56° relative to each other, and the central C(1)-C(2) bond is lengthened to 1.49Å as compared to 1.36Å in $Ph_2C=CPh_2$.[26] The phenyl rings I-IV are twisted out of the C(1)-C(3)-C(9) and C(2)-C(15)-C(21) planes by 10-40°. The coordination sphere of Na(1) includes the negatively charged, trigonal-planar C(1) and C(2) atoms, one π bond from each of the phenyl rings I and III, and two Et₂O ligands. The unsolvated Na(2) center is sandwiched between phenyl rings III (binding from the backside) and I' (from an adjacent dianion), and further links to a π bond in ring IV'. The Na(1)-O distance is 2.32Å, and the Na(1)-C and Na(2)-C distances lie in the ranges 2.70-2.82 and 2.76-3.09Å, respectively.

In indenylsodium.tmeda the Na atom bridges two indenyl units in η^1 and η^2 modes to form a zigzag chain arrangement [Fig. 6.1-9(b)].[27,28] A similar infinitely aggregated chain structure is found in cyclopentadienylsodium.tmeda [Fig. 6.1-9(c)].

The crystal structure of $(Me_3Si)C_5H_4K$ features a polydecker sandwich as shown in Fig. 6.1-10(a); the zigzag bending is caused by weak interactions between neighbouring chains.[24] The organization of these chains to form double layers and the structure-determining influence of the trimethylsilyl groups are illustrated in Fig. 6.1-10(b).

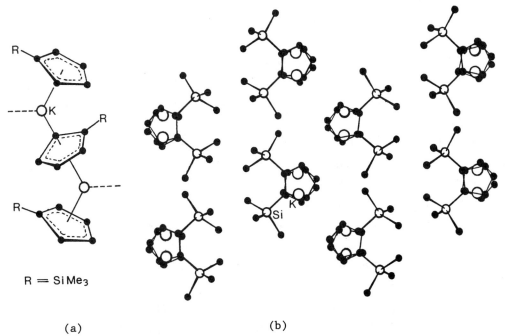

(a)   (b)

Fig. 6.1-10 Crystal structure of $(Me_3Si)C_5H_4K$. (Adapted from ref. 30).

References

[1]   E. Weiss and G. Hencken, *J. Organomet. Chem.* **21**, 265 (1970).

[2]   E. Weiss and E.A.C. Lucken, *J. Organomet. Chem.* **2**, 197 (1964).

[3]   J.L. Wardell in G. Wilkinson, F.G.A. Stone and E.W. Abel (eds.), *Comprehensive Organometallic Chemistry*, Pergamon Press, Oxford, 1982, Vol. 1, chap. 2, p. 43.

[4]   M.P. Schutzenberger, *Annalen* **15**, 100 (1868).

[5]   L. Mond, C. Langer and F. Quinke, *J. Chem. Soc.* **57**, 749 (1890); reprinted in *J. Organomet. Chem.* **383** (Special Volume: *Metal Carbonyl Mond Centenary*), 1 (1990).

[6]   L. Mond and C. Langer, *J. Chem. Soc.* **59**, 1090 (1891); M. Berthelot, *C.R. Acad. Sci.* **112**, 1343 (1891).

[7]   G. Wilkinson, M. Rosenblum, M.C. Whiting and R.B. Woodward, *J. Am. Chem. Soc.* **74**, 2125 (1952).

[8]   E.O. Fischer and W. Pfab, *Z. Naturforsch., Teil B* **7**, 377 (1952).

[9]   K. Ziegler, *Adv. Organomet. Chem.* **6**, 1 (1968); G. Natta, *Sci. Am.* **205**, 33 (1961); H. Sinn and W. Kaminsky, *Adv. Organomet. Chem.* **18**, 99 (1980).

[10] R. Hoffmann, *Angew. Chem. Int. Ed. Engl.* **21**, 711 (1982).

[11] H. Dietrich, *Acta Crystallogr.* **16**, 681 (1963).

[12] G.D. Stucky in A.W. Langer (ed.), *Polyamine-Chelated Alkali Metal Compounds*, American Chemical Society, Washington, D.C., 1974, chap. 3.

[13] W.N. Setzer and P.v.R. Schleyer, *Adv. Organomet. Chem.* **24**, 353 (1985).

[14] J.P. Collman, L.S. Hegedus, J.R. Norton and R.G. Finke, *Principles and Applications of Organotransition Metal Chemistry*, 2nd edition, University Science Books, Mill Valley, California, 1987.

[15] *Top. Curr. Chem.* **138** (1986). [Covers polylithiated aliphatic hydrocarbons, lithiation reactions, and electrochemistry of solvated electrons.]

[16] P. Jutzi, *Adv. Organomet. Chem.* **26**, 217 (1986).

[17] D. Bladauski, W. Broser, H.J. Hecht, D. Rewicki and H. Dietrich, *Chem. Ber.* **112**, 1380 (1979).

[18] E. Weiss, G. Sauermann and G. Thirase, *Chem. Ber.* **116**, 74 (1983).

[19] E. Weiss, S. Corbelin, J.K. Cockcroft and A.N. Fitch, *Angew. Chem. Int. Ed. Engl.* **29**, 650 (1990).

[20] E. Weiss, S. Corbelin, J.K. Cockcroft and A.N. Fitch, *Chem. Ber.* **123**, 1629 (1990).

[21] E. Weiss and G. Sauermann, *Chem. Ber.* **103**, 265 (1970).

[22] E. Weiss and H. Köster, *Chem. Ber.* **110**, 717 (1977).

[23] E. Weiss, T. Lambertsen, B. Schubert and J.K. Cockroft, *J. Organomet. Chem.* **358**, 1 (1988).

[24] P. Jutzi, W. Leffers, B. Hampel, S. Pohl and W. Saak, *Angew. Chem. Int. Ed. Engl.* **26**, 583 (1987).

[25] H. Bock, K. Ruppert and D. Fenske, *Angew. Chem. Int. Ed. Engl.* **28**, 1685 (1989).

[26] A. Hoekstra and A. Vos, *Acta Crystallogr., Sect B* **31**, 1722 (1975).

[27] C. Schade, P.v.R. Schleyer, G. Gregory, H. Dietrich and W. Mahdi, *J. Organomet. Chem.* **341**, 19 (1988).

[28] C. Schade and P.v.R. Schleyer, *Adv. Organomet. Chem.* **27**, 169 (1987).

[29] T. Aoyagi, H.M.M. Shearer, K. Wade and G. Whitehead, *J. Organomet. Chem.* **175**, 21 (1979).

[30] P. Jutzi, *Pure Appl. Chem.* **61**, 1731 (1989).

Note A comprehensive review of the crystallographic data for coordination compounds of lithium is given in U. Olsher, R.M. Izatt, J.S. Bradshaw and N.K. Dalley, *Chem. Rev.* **91**, 137 (1991).

The structures of organonitrogenlithium compounds featuring $(NLi)_n$ ring-stacking and ring-laddering are reviewed in R.E. Mulvey, *Chem. Soc. Rev.* **20**, 167 (1991).

The crystal structures of several organosodium compounds containing the carbanions benzyl, diphenylmethyl, *o*-xylyl and 1-phenylethyl are described in S. Corbelin, N.P. Lorenzen, J. Kopf and E. Weiss, *J. Organomet. Chem.* **415**, 293 (1991).

6.2 Ethylmagnesium Bromide Dietherate (Grignard Reagent)
$C_2H_5MgBr \cdot 2(C_2H_5)_2O$

Crystal Data

Monoclinic, space group $P2_1/c$ (No. 14)

$a = 13.18(3)$, $b = 10.27(3)$, $c = 11.42(3)$Å, $\beta = 103.3(3)°$, $Z = 4$

Atom	x	y	z	Atom	x	y	z
Br	.1413	-.0212	.2113	C(4)	.0930	.3284	.0313
Mg	.2762	.0185	.0962	C(5)	.2989	.2665	-.0336
O(1)	.2247	.1824	.0023	C(6)	.3342	.2163	-.1388
O(2)	.2453	-.1195	-.0374	C(7)	.1414	-.1473	-.1071
C(1)	.4410	.0261	.1726	C(8)	.1303	-.0886	-.2335
C(2)	.4759	-.0529	.2797	C(9)	.3224	-.2111	-.0550
C(3)	.1147	.2258	-.0480	C(10)	.3239	-.3286	.0312

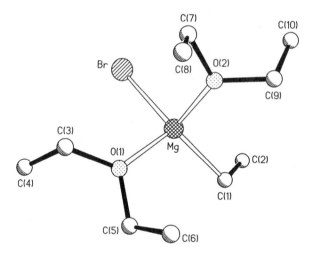

Fig. 6.2-1 Molecular structure of $C_2H_5MgBr \cdot 2(C_2H_5)_2O$.

Crystal Structure

The crystal structure of $C_2H_5MgBr \cdot 2(C_2H_5)_2O$ consists of a packing of discrete monomeric units, with one ethyl group, one bromine atom, and two ether molecules forming a distorted tetrahedron around the sp^3 magnesium atom (Fig. 6.2-1).[1] The Mg-O bond distance, 2.04Å (average), is among the shortest Mg-O distances known. The Mg-Br bond distance is 2.48Å. The next shortest Mg...Br contact is 5.81Å, which is too long to correspond to any conceivable chemical bond and hence rules out all dimeric structures involving bridging bromine atoms.

Since various structures proposed for the Grignard reagent have been based partly or wholly on evidence obtained from experiments on the ethyl Grignard reagent, a definitive result is provided by the X-ray analysis of $C_2H_5MgBr \cdot 2(C_2H_5)_2O$.

Remarks

1. Grignard reagents

The term "Grignard reagent" refers to the product (empirical formula RMgX) of a reaction between metallic magnesium and an appropriate organic halide in an ether solvent. The organic moiety can be alkyl, allyl, or aryl, and the halogen may be Cl, Br, or I. (Arylmagnesium chlorides must be made in the cyclic ether tetrahydrofuran instead of ethyl ether). These versatile reagents (etherated organomagnesium halides) are generally used in the preparation of alcohols, aldehydes, ketones, carboxylic acids, esters and amides; they are named after Victor Grignard, who received the Nobel Prize in 1912 for pioneering their utility in organic synthesis. In a Grignard compound the carbon atom bonded to the electropositive magnesium atom carries a partial negative charge as shown below:

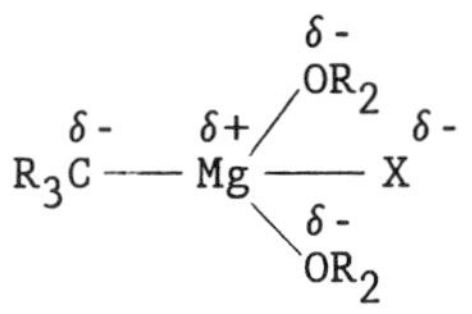

$$
\begin{array}{c}
\overset{\delta-}{OR_2} \\
R_3\overset{\delta-}{C} \!\!-\!\! \overset{\delta+}{Mg} \!\! \overset{\diagup}{\underset{\diagdown}{}} \!\! \overset{\delta-}{X} \\
\underset{OR_2}{\overset{\delta-}{}}
\end{array}
$$

Consequently a Grignard reagent is an extremely strong base, with its carbanion-like alkyl or aryl portion acting as a nucleophile.

2. The constitution of Grignard reagents

The composition of the Grignard reagent in ether as solvent is one of the most fascinating and fundamental problems.[2] A different but closely related problem involves the mechanisms by which Grignard reagents react with organic functional groups.[3,4] Much of the controversy over the nature of the Grignard reagent is concerned with the chemical species in solution. It now seems well-established that the principal equilibria can be represented in the following scheme (solvation of the various species have been omitted for simplicity):

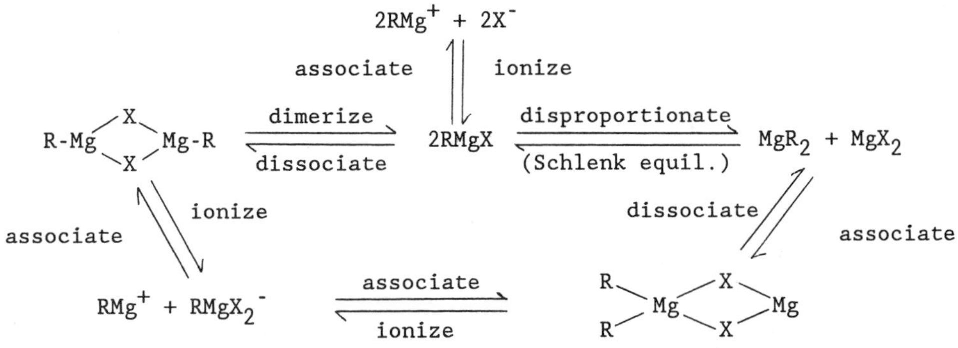

$$
\begin{array}{c}
2RMg^+ + 2X^- \\
\text{associate} \big\Updownarrow \text{ionize} \\
R\text{-}Mg\!\!\overset{X}{\underset{X}{\diagup\diagdown}}\!\!Mg\text{-}R \underset{\text{dissociate}}{\overset{\text{dimerize}}{\rightleftharpoons}} 2RMgX \underset{\text{(Schlenk equil.)}}{\overset{\text{disproportionate}}{\rightleftharpoons}} MgR_2 + MgX_2
\end{array}
$$

$$
\text{associate} \big\Updownarrow \text{ionize} \qquad\qquad \text{dissociate} \big\Updownarrow \text{associate}
$$

$$
RMg^+ + RMgX_2^- \underset{\text{ionize}}{\overset{\text{associate}}{\rightleftharpoons}} \quad R\!\!\overset{}{\underset{R}{\diagup}}\!\!Mg\!\!\overset{X}{\underset{X}{\diagup\diagdown}}\!\!Mg
$$

The structures of several pertinent Grignard compounds have been elucidated by X-ray crystallography (Fig. 6.2-2).

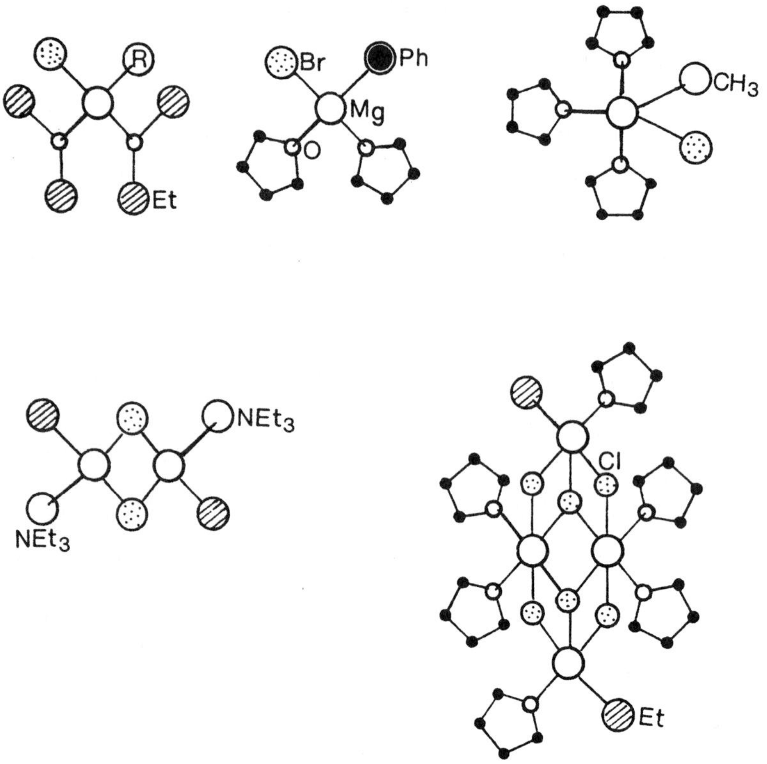

Fig. 6.2-2 Structures of some organomagnesium halide solvates. (After ref. 5).

3. Unsolvated alkylmagnesium compounds

Table 6.2-1 Selected structural data for some crystalline magnesium alkyls.

Compound[a]	Mg-C/Å	ΣCMgC/°	CN of Mg
$(MgR_2)_\infty$	2.105(4)	351.2[b]	3[b]
	2.117(4)		
	2.535(4)		
$[MgR(\mu\text{-Cl})(OEt_2)]_2$	2.131(8)	–	4
$MgR_2(OEt_2)$	2.17(1)	(360)[c]	3
$(MgR_2)_2(\mu\text{-dioxan-p})$	2.118(5)	(360)[c]	3
$[Mg(CH_2Bu^t)_3]^-$	2.22(1)	(360)[c]	3
$Mg(CH_2Bu^t)_2$ (gas)[d]	2.126(5)	180	2
$Mg[C(SiMe_3)_3]_2$	2.16(2)	180	2

[a] $R = CH(SiMe_3)_2$. [b] On the basis of the unit $Mg^a(R_2)_2(...CH_3)$.
[c] This is the sum of CMgC + 2CMgO angles. [d] Electron diffraction data.

The isolation and single-crystal X-ray diffraction study of unsolvated dialkylmagnesium compounds has been accomplished only recently.[7,8] The available structural data are summarized in Table 6.2-1.[8]

The isomorphous dialkyl compounds $M\{C(SiMe_3)_3\}_2$ (M = Mg,[7] Hg,[9] Mn[10]) are monomeric and the magnesium derivative is the first example of two-coordination for the metal in the solid state. The $Mg\{C(SiMe_3)_3\}_2$ molecule is centrosymmetric with interlocking $(Me_3Si)_3C$ groups [Fig. 6.2-3(a)]. On the other hand, $(MgR_2)_\infty$ [R = $CH(SiMe_3)_2$] has been shown by X-ray

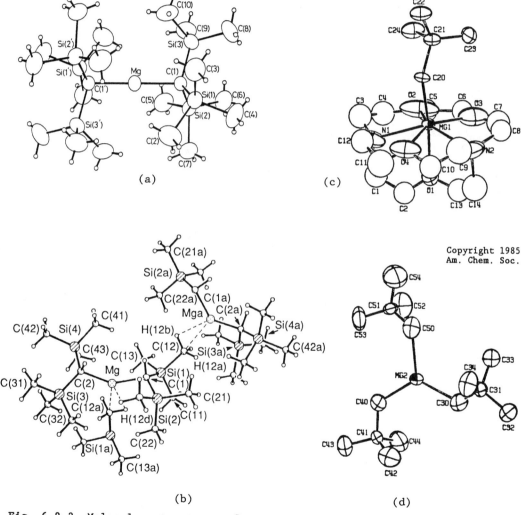

Copyright 1985
Am. Chem. Soc.

(a)   (c)

(b)   (d)

Fig. 6.2-3 Molecular structures of some magnesium alkyls. (a) $Mg\{C(SiMe_3)_3\}_2$, (b) $(MgR_2)_\infty$ [R = $CH(SiMe_3)_2$] (atom E^a is related to atom E by the symmetry operator 2_1), the cation (c) and anion (d) in crystalline $[Mg(CH_2Bu^t)(2,1,1\text{-cryptand})][Mg(CH_2Bu^t)_3]$. (After refs. 7, 8 and 11).

(190K) and neutron (15K) diffraction to have a polymeric structure featuring unprecedented weak (agostic) interaction between each magnesium atom with a γ-methyl group of a neighbouring MgR_2 unit [Fig. 6.2-3(b)].[8] The notable geometrical features are: (i) the $C(12)...Mg^a$ distance of 2.535(4)Å is much shorter than the sum of the van der Waals radii (3.4Å); (ii) the Si-$C(12)...Mg^a$ angle is 172.3(2)°; (iii) the Si-$C(12)$ bond [1.915(8)Å] is significantly longer than the Si-$C(11)$ and Si-$C(13)$ bonds [1.876(6) and 1.879(5), respectively]; (iv) the intermolecular $Mg^a...H$ contacts [2.333(4), 2.414(5) and 2.516(4) Å] are short; and (v) the H-$C(12)$-H angles (av. 110.5°) are opened out compared to other H-C-H angles (av. 107.1°).

The anion [Fig. 6.2-3(d)] of $[Mg(CH_2Bu^t)(2,1,1\text{-cryptand})][Mg(CH_2Bu^t)_3]$ has trigonal planar geometry which is rare for magnesium, whereas the coordination geometry of the cation is that of a pentagonal bipyramid with bonds to the neopentyl group and to all six heteroatoms of the cryptand [Fig. 6.2-3(c)].[11]

The crystal structure of a 1:1 adduct of dineopentylmagnesium and neopentylmagnesium bromide features a polymeric chain involving an alternating pattern of two Np_2Mg ($Np = Me_3CCH_2$) and two $NpMgBr$ fragments connected by bridging neopentyl and bromide groups, in which all metal atoms are four-coordinate (Fig. 6.2-4).[12] The four-membered $Br_2Mg(1)_2$ and $C_2Mg(2)_2$ rings are each located at a crystallographic inversion center. The Mg-Np-Mg bridges are remarkably asymmetric, having a short (2.20-2.23Å) and a long (2.41-2.42Å) Mg-C(Np) bond; the Mg-Br bonds are very long (2.808-2.818Å) as compared to the corresponding bond distance of 2.58Å in dimeric ethylmagnesium bromide.[13]

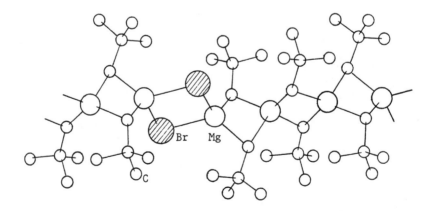

Fig. 6.2-4 Polymeric structure of $[(Np_2Mg)_2.(NpMgBr)_2]_n$. The short Mg-C bonds are indicated by solid lines, and long Mg-C bonds by open lines. (After ref. 12).

The "threaded" complex (18-crown-6).Et_2Mg, like its Zn analogue, has an exactly linear Et_2M (M = Mg, Zn) unit inserted through a centrosymmetric macrocycle such that the ether oxygen atoms surround the M atom in a quasi-equatorial manner.[14]

3. Alkylmagnesium alkoxides

This class of compounds can be made by the reaction of MgR_2 with an alcohol or ketone, or by reacting the metal directly with the appropriate alcohol and alkyl chloride in methylcyclohexane. NMR and infrared studies of the series MeMgOR (R = Pr^n, Pr^i, Bu^t, $CMePh_2$) in THF, Et_2O, and benzene have shown that the existence of the following structural types depends on the coordinating ability of the solvent L and the steric bulk of the organic group (Fig. 6.2-5).[6]

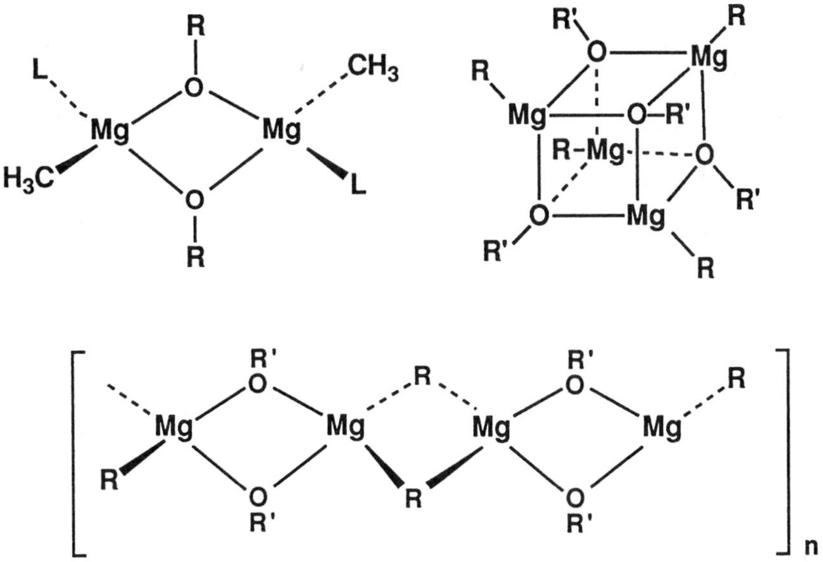

Fig. 6.2-5 Structural types of alkylmagnesium alkoxides. (After ref. 6).

4. Diarylmagnesium complexes

As depicted in the following scheme involving solvent L (e.g. THF), various aggregates of diarylmagnesium species exist in equilibrium in solution and can often be isolated under different crystallization conditions.

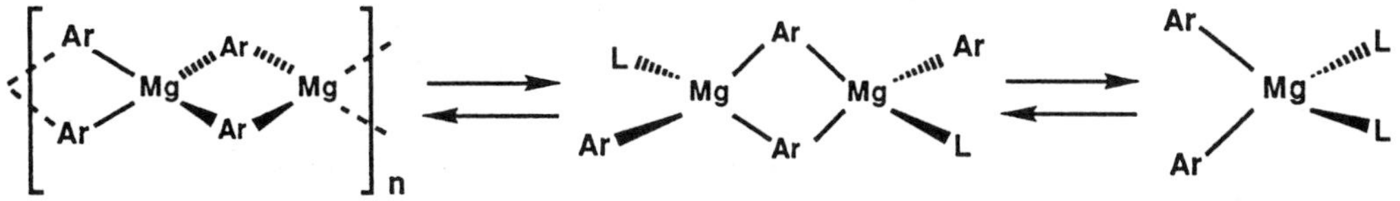

$[Ar_2Mg]_n$ $[Ar_2Mg.THF]_2$ $Ar_2Mg.(THF)_2$

In polymeric diphenylmagnesium, $[Ph_2Mg]_n$, the four-coordinate magnesium atoms are arranged in a linear chain and connected by symmetrically bridging phenyl groups [Fig. 6.2-6(a)].[15]   The Mg_2C_2 rings are centrosymmetric with Mg-C = 2.261(2)Å and Mg-C-Mg = 77.73(9)°.

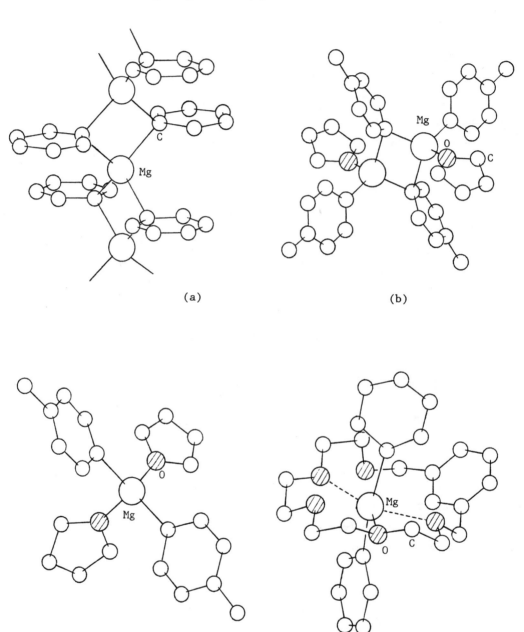

(a)   (b)

(c)   (d)

Fig. 6.2-6 Structures of (a) polymeric $[Ph_2Mg]_n$, (b) dimeric $[(p\text{-}tolyl)_2Mg.THF]_2$, (c) monomeric $(p\text{-}tolyl)_2Mg.(THF)_2$ and (d) "threaded" (1,3-xylyl-18-crown-5)diphenylmagnesium.

The crystallization of bis(p-tolyl)magnesium in tetrahydrofuran yielded a remarkable structure containing dimeric [(p-tolyl)$_2$Mg.THF]$_2$ [Fig. 6.2-6(b)] and monomeric (p-tolyl)$_2$Mg.(THF)$_2$ [Fig. 6.2-6(c)] in a 1:2 molar ratio.[15] The dimer is centrosymmetric, and the bridging p-tolyl group is linked unsymmetrically to the magnesium atoms at Mg(1)-C(1) = 2.245(7) and Mg(1)-C(1)' = 2.323(7)Å; the Mg-C and Mg-O bond distances involving the terminal ligands are 2.130(7) and 2.020(5)Å, respectively. In the monomeric complex the metal-ligand bond distances are Mg-C = 2.126(7)-2.132(8) and Mg-O = 2.031(6)-2.050(5)Å.

The complex (1,3-xylyl-18-crown-5).Ph$_2$Mg has a remarkable rotaxane or "threaded" structure in which the nearly linear Ph$_2$Mg group is equatorially surrounded by four of the five ether oxygen atoms [Fig. 6.2-6(d)].[16] the magnesium atom is tightly bound to O(2) and O(3) at Mg-O = 2.204(3) and 2.222(4)Å, respectively, and much less so to O(1) and O(4) at 2.516(4) and 2.520(4)Å, respectively. This type of intramolecular coordination differs from that in "threaded" (18-crown-6).Et$_2$Mg, whose Mg atom is weakly bound to the ether oxygen atoms at Mg-O distances in the range 2.767(1)-2.792(1)Å. In the Ph$_2$Mg unit the metal atom is bonded equally to both phenyl groups at Mg-C = 2.190(5)Å, and the C-Mg-C angle is 163.8(2)°.

5. Magnesium-anthracene derivatives

The reaction of anthracene and of 9,10-bis(trimethylsilyl)anthracene with metallic magnesium yields tetrahydrofuran (THF) solvates of 9,10-dihydro-9,10-anthrylenemagnesium (I) and of 9,10-bis(trimethylsilyl)-9,10-anthrylenemagnesium (II), respectively.[17]

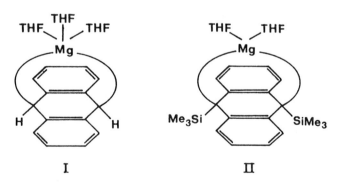

I   II

The structural analysis of (II) revealed that the Mg atom forms an intramolecular bridge between the 9- and 10- positions of the anthrylene system, the Mg-C distances being 2.233 and 2.234Å. The pseudo tetrahedral coordination sphere of the Mg atom is completed by the oxygens of the two THF solvate molecules; the Mg-O distances are 1.997 and 2.021Å.[18]

The structure of the 1,4-dimethyl derivative of (I) has also been determined.[19] It contains two crystallographically independent molecules of very similar geometry, in which the Mg atom forms an intramolecular bridge (average Mg-C = 2.32Å) between the 9- and 10- positions of the anthrylene system. Pentacoordination about each Mg atom is completed by the oxygen atoms of three THF ligand molecules.

The role of the magnesium-anthracene complexes such as (I) and (II) in the catalytic formation of MgH_2 has been rationalized according to the following reactions:[20]

$$\text{anthracene} \quad + \quad \text{Mg} \quad \xrightarrow[\text{THF}]{20\text{-}60°C} \quad \text{(I)}$$

$$\text{(I)} + CrCl_3(TiCl_4) \quad \xrightarrow[\text{THF}]{20\text{-}30°C} \quad \text{Cr(Ti)-catalyst} + \text{anthracene}$$

$$\text{(I)} + \text{Cr(Ti)-catalyst} + H_2 \quad \xrightarrow[\text{THF}]{20\text{-}30°C} \quad MgH_2 + \text{anthracene}$$

Samples of MgH_2 so produced are highly reactive and can be used both as a hydrogenating agent and a high-temperature hydrogen storage material.[20]

6. π Complexes of magnesium[21]

The structures of some π complexes of magnesium are shown in Fig. 6.2-7. Magnesocene, dicyclopentadienylmagnesium, can be prepared by the thermal decomposition of $(C_5H_5)MgBr$ as a colorless and very air-sensitive, crystalline compound. X-Ray diffraction yielded a typical sandwich structure [Fig. 6.2-7(a)] with Mg-C and C-C distances of 2.304(8) and 1.39(2)Å, respectively.[22] In the crystalline state the two parallel rings have a staggered D_{5d} conformation, whereas the electron-scattering pattern is consistent with an eclipsed D_{5h} comformation [Fig. 6.2-7(b)] in the gas phase.[23]

Hexakis(trimethylsilyl)magnesocene, $[(Me_3Si)_3C_5H_2]_2Mg$, crystallizes as a bent-sandwich molecule with a Cp centroid-Mg-Cp centroid angle of 171.1° which helps to minimize the steric repulsion between the bulky silyl groups [Fig. 6.2-7(j)].[24]

Mono-π-cyclopentadienylmagnesium compounds have so far only been isolated in the form of adducts with oxygen- or nitrogen-containing bases [Fig. 6.2-7(c-f)], including two Grignard reagents. The cyclopentadienyl-magnesium bromide tetraethyl-ethylenediamine adduct [Fig. 6.2-7(e)] was crystallized from a solution of C_5H_5MgBr in diethyl ether by slow addition of

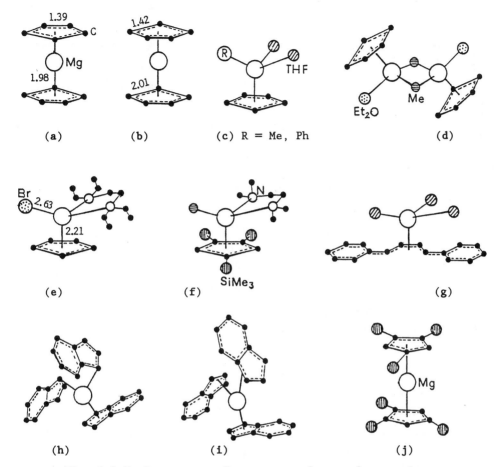

Fig. 6.2-7 Structures of some π complexes of magnesium.
(Adapted from refs. 21 and 24).

base.[25] The closest approach of the magnesium atom to the plane of the Cp group is 2.21Å, with an average Mg-C distance of 2.55Å. The terminal Mg-Br distance (2.63Å) is the same as for the six-coordinate magnesium in MgBr₂.4THF, but significantly longer than that for the four-coordinate magnesium in EtMgBr.2Et₂O.

References

[1] L.J. Guggenberger and R.E. Rundle, *J. Am. Chem. Soc.* **86**, 5344 (1964); **90**, 5375 (1968).

[2] B.J. Wakefield, *Organomet. Chem. Rev.* **1**, 131 (1966).

[3] E.C. Ashby, *Quart. Rev. (London)* **21**, 259 (1967).

[4] H.R. Rogers, C.L. Hill, Y. Fujiwara, R.J. Rogers, H.L. Mitchell and G.M. Whitesides, *J. Am. Chem. Soc.* **102**, 217 (1980), and the three following papers.

[5] I. Haiduc and J.J. Zuckerman, *Basic Organometallic Chemistry*, Walter de Gruyter, Berlin, 1985, p. 65.

[6] E.C. Ashby, J. Nackashi and G.E. Parris, *J. Am. Chem. Soc.* **97**, 3162 (1975).

[7] S.S. Al-Juaid, C. Eaborn, P.B. Hitchcock, C.A. McGeary and J.D. Smith, *J. Chem. Soc., Chem. Commun.*, 273 (1989).

[8] P.B. Hitchcock, J.A.K. Howard, M.F. Lappert, W.-P. Leung and S.A. Mason, *J. Chem. Soc., Chem. Commun.*, 847 (1990).

[9] F. Glockling, N.S. Hosmane, V.B. Mahale, J.J. Swindall, L. Magos and T.J. King, *J. Chem. Res. (S)*, 116; *(M)*, 117 (1977).

[10] N.H. Buttrus, C. Eaborn, P.B. Hitchcock, J.D. Smith and A.C. Sullivan, *J. Chem. Soc., Chem. Commun.*, 1380 (1985).

[11] E.P. Squiller, R.R. Whittle and H.G. Richey, Jr., *J. Am. Chem. Soc.* **107**, 432 (1985).

[12] P.R. Markies, G. Schat, O.S. Akkerman, F. Bickelhaupt, W.J.J. Smeets, A.J.M. Duisenberg and A.L. Spek, *J. Organomet. Chem.* **375**, 11 (1989).

[13] A.L. Spek, P. Voorbergen, G. Schat, C. Blomberg and F. Bickelhaupt, *J. Organomet. Chem.* **77**, 147 (1974).

[14] A.D. Pajerski, G.L. BergStresser, M. Parvez and H.G. Richey, Jr., *J. Am. Chem. Soc.* **110**, 4844 (1988).

[15] P.R. Markies, G. Schat, O.S. Akkerman, F. Bickelhaupt, W.J.J. Smeets, P. van der Sluis and A.L. Spek, *J. Organomet. Chem.* **393**, 315 (1990).

[16] P.R. Markies, T. Nomoto, O.S. Akkerman and F. Bickelhaupt, *J. Am. Chem. Soc.* **110**, 4845 (1988).

[17] H. Lehmkuhl, A. Shakoor, K. Mehler, C. Krüger and Y.-H. Tsay, *Z. Naturforsch.* **B40**, 1504 (1985).

[18] H. Lehmkuhl, A. Shakoor, K. Mehler, C. Krüger, K. Angermund and Y.-H. Tsay, *Chem. Ber.* **118**, 4239 (1985).

[19] B. Bogdanovic, N. Janke, C. Krüger, R. Mynott, K. Schlichte and U. Westeppe, *Angew. Chem. Int. Ed. Engl.* **24**, 960 (1985).

[20] B. Bogdanovic, *Angew. Chem. Int. Ed. Engl.* **24**, 262 (1985).

[21] P. Jutzi, *Adv. Organomet. Chem.* **26**, 217 (1986).

[22] W. Bünder and E. Weiss, *J. Organomet. Chem.* **92**, 1 (1975).

[23] J.L. Robbins, N. Edelstein, B. Spencer and J.C. Smart, *J. Am. Chem. Soc.* **104**, 1882 (1982).

[24] C.P. Morley, P. Jutzi, C. Krüger and J.M. Wallis, *Organometallics* **6**, 1084 (1987).

[25] C. Johnson, J. Toney and G.D. Stucky, *J. Organomet. Chem.* **40**, C11 (1972).

6.3   Trimethylaluminum

$$Al_2(CH_3)_6$$

Crystal Data

Monoclinic, space group $C2/c$ (No. 15)

$a = 12.74(2)$, $b = 6.96(1)$, $c = 14.63(2)$Å, $\beta = 123.67(3)°$, $Z = 4$

Atom	x	y	z	Atom	x	y	z
Al	.4708	.5747	.4073	H(2A)	.260	.466	.234
C(1)	.6221	.3814	.5084	H(2B)	.366	.300	.300
C(2)	.3518	.4325	.2701	H(2C)	.384	.517	.234
C(3)	.5520	.8152	.4098	H(3A)	.133	.300	.416
H(1A)	.133	.183	.083	H(3B)	.017	.400	.434
H(1B)	.133	.183	-.050	H(3C)	.000	.383	.350
H(1C)	.217	.067	.058				

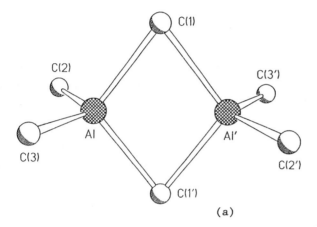

(a)

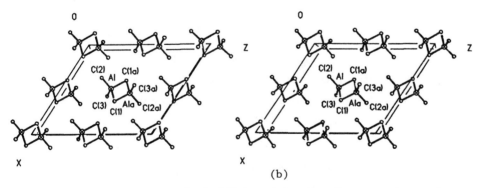

(b)

Fig. 6.3-1 Structure of $Al_2(CH_3)_6$: (a) molecular geometry and (b) arrangement of the dimeric molecules in the unit cell.

Crystal Structure

In the crystal structure of trimethylaluminum [Fig. 6.3-1(a)], two monomeric units related by a center of symmetry to form a dimer with bridging methyl groups. These dimers are then packed together at van der Waals

separations to form a molecular crystal [Fig. 6.3-1(b)]. The bond lengths and
bond angles are listed in Table 6.3-1.

Table 6.3-1 Bond lengths (Å) and bond angles (deg).

Bridge	Al-C(1)	2.134	Al-C(1)-Al'	74.7
	Al-C(1')	2.153	C(1)-Al-C(1')	105.3
	C(1)-H	1.07(av.)	H-C(1)-H	102(av.)
Terminal	Al-C(2)	1.983	C(3)-Al-C(2)	123.1
	Al-C(3)	1.958	C(2)-Al-C(1)	106.1
	Al...Al'	2.600	C(2)-Al-C(1')	107.2
	C(2)-H	1.02(av.)	C(1')-Al-C(3)	108.7
	C(3)-H	0.95	Al...Al-C(2)	118.2

The aluminum and terminal carbon atoms in the dimeric molecule are not
coplanar. A separation of 0.12Å is found between the parallel planes defined
by C(2)AlC(3) and C(2')Al'C(3'). This distortion can also be seen from the
angle of 2.6(3)° between the Al...Al' line and the line passing through Al and
the bisector of C(2)...C(3). Since the dihedral angle between the planes
defined by C(2)AlC(3) and C(1')AlC(1) is 89.2(5)°, the molecular symmetry
(exclusive of hydrogen) is C_{2h} with the twofold rotation axis bisecting the
Al...Al' line and perpendicular to the AlC(1)Al'C(1') plane.

According to Lewis and Rundle,[2] four principal factors govern the
stability of dimers held together by alkyl bridges. Association is favored by
(i) a large difference in electronegativity between the metal and carbon; (ii)
a low value for the energy required to promote an electron from a s orbital to
a p orbital in the valence shell of the metal; (iii) a large bond energy for a
normal single bond between the metal and carbon; and (iv) a minimal amount of
inner shell repulsion between the two metal atoms separated by an internuclear
distance dictated by the geometry of the dimer.

Remarks

1. Structures of carbon-bridged organoaluminum compounds

The crystal structures of many dimeric aluminum trialkyls and triaryls
have been determined. Fig. 6.3-2 shows several symmetrical organoaluminum
compounds, some of which have important implications with regard to electron-
deficient bonding.[3,4] In dimeric tricyclopropylaluminum, the cyclopropyl
rings are oriented normal to the Al...Al axis so that the π orbital on the
bridging carbon atom may interact with the appropriate vacant orbitals on the
metal atoms.[5] A similar situation prevails in the aromatic derivatives

Al$_2$(Ph)$_6$ and Al$_2$(Ph)$_2$(Me)$_4$. The structural data of these compounds are listed in Table 6.3-2.

Table 6.3-2 Bond distances and angles found in some carbon-bridged organoaluminum compounds.

Compound	Distances/Å			Angles/degrees		
	Al-C$_b$	Al-C$_t$	Al...Al	Al-C-Al	C-Al-C (internal)	C-Al-C (external)
Al$_2$Me$_6$ [1]	2.13 2.15	1.96 1.98	2.60	74.7	105.3	123.1
Al$_2$(cyclo-C$_3$H$_5$)$_6$ [5]*	2.062 2.098	1.90 1.95	2.618	78.2	96.7	115.2
Al$_2$Ph$_6$ [6]	2.184 2.180	1.960 1.956	2.70	76.5	103.5	115.4
Al$_2$Ph$_2$Me$_4$ [7]	2.129 2.152	1.965 1.968	2.690	77.5	101.3	121.5
Al$_2$(CH=CH-t-Bu)$_2$(i-Bu)$_4$ [8]	2.10 2.12	1.965 1.999	2.684	79	100.9	128.5
Al$_2$(C≡CPh)$_2$Ph$_4$ [9]	1.992 2.184	1.940	2.99	91.7	88.3	-

* The four-membered bridge ring is puckered with a dihedral angle of 31.9° between the two planar Al-C-Al moieties.

2. Bonding in the dimers

To account for the structures of the dimers of trimethylaluminum and related organoaluminum compounds, the electron-deficient bonding has been described in two limiting ways. First, one can assume that the bridge bond arises from the overlap of tetrahedral (sp^3) orbitals from the two aluminum atoms with the tetrahedral orbital of the bridging group.[2] Such a linear combination of atomic orbitals will yield a bonding, a non-bonding, and an antibonding molecular orbital; a three-center bridge bond would then result from the two electrons occupying the bonding molecular orbital. This view of the bonding is helpful in understanding the unusually acute Al-C-Al bridge angle, for a better orbital overlap is achieved if the aluminum orbital is not aligned along the Al-C axis, but is directed at the carbon orbital. In the second bonding picture, trigonal (sp^2) orbitals from the two aluminum atoms interact with the non-bridging methyl groups to form a tetramethylethylene-like σ-skeleton, and the p-orbitals on the aluminums form a π-orbital which can interact with the bridging groups.[10] The advantage of this model is that the relatively short Al...Al distance is explicable as an Al-Al σ bond and the relatively wide external C-Al-C angle as a consequence of sp^2 hybridization of aluminum.

The actual bonding situation in the dimers is probably intermediate between the two extremes.

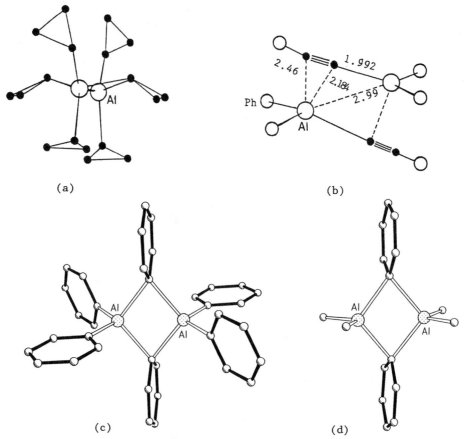

Fig. 6.3-2 Structures of (a) $Al_2(C_3H_5)_6$, (b) $Al_2(C{\equiv}CPh)_2Ph_4$, (c) Al_2Ph_6 and (d) $Al_2Ph_2Me_4$. (After refs. 3 and 5).

3. π Complexes of aluminum[11]

The X-ray crystal structure of dimeric diphenyl(phenylethynyl)aluminum [Fig. 6.3-3(a)] shows an interesting mode of phenylethynyl bridging, which suggests that the alkynyl group is largely σ bonded to one aluminum and π bonded to the other.[9] The bridging aluminum and carbon atoms form a rectangular array where the aluminum atoms are π bonded only to the α-acetylenic carbon atoms. The unusual π bonding in the dimer can be explained by overlap between one carbon $2p\pi$ orbital and the aluminum $3p_z$ orbital. A similar bonding situation has been revealed in dimeric dimethyl(1-propynyl)-aluminum by gas phase electron diffraction data [Fig. 6.3-3(b)].[12]

Dimethyl(cyclopentadienyl)aluminum [Fig. 6.3-3(c)] has been prepared by the direct reaction of Al_2Me_6 and excess cyclopentadiene. In the solid state the compound consists of infinite chains composed of dimethylaluminum groups bridged by cyclopentadienyl rings. The bridging units are unsymmetrically

bound to the metal and show significant distortion within the ring. One
bridge-bond distance has the length anticipated for an electron-deficient bond
while the other is significantly longer.[13]   In the gas phase only the
monomeric form has been observed; from electron diffraction data this is best
described as containing a cyclopentadienyl ring π bonded in an η^2 fashion.[14]
The structure of chloro(pentamethylcyclopentadienyl)methylaluminum [Fig.
6.3-3(d)] reveals a dimeric unit in which each aluminum atom is η^3 bonded to a
Me₅C₅ ring.[15]

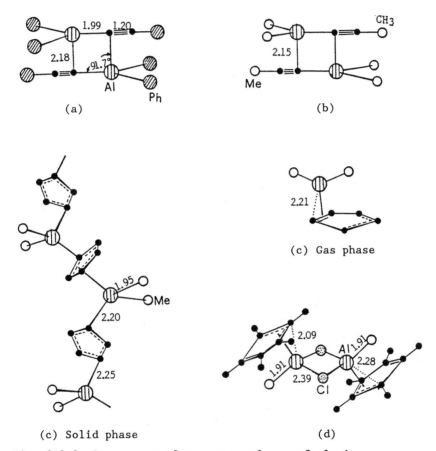

Fig. 6.3-3   Structures of some π complexes of aluminum.

4. Alkylaluminum compounds containing multidentate open-chain and macrocyclic
   amines

       The reaction of trimethylaluminum with the tetradentate open-chain amine
N,N'-bis(3-aminopropyl)ethylenediamine afforded the crystalline product
[AlMe][C₈H₁₉N₄][AlMe₂].[16]   In the molecule, Al(2) is tetrahedrally four-
coordinate, whereas Al(1) is clearly five-coordinate, being bonded to all four
nitrogen atoms of the amine as well as one methyl carbon atom (Fig. 6.3-4).

The Al(1) atom is coplanar with N(1), N(3) and C(9), its coordination sphere being completed by the remaining two nitrogen atoms situated at axial positions. The bond angle N(2)-Al(1)-N(4) is 169.2(1)°, so the coordination geometry of Al(1) may be described as trigonal bipyramidal. The bond lengths of Al(1)-N(2) and Al(1)-N(4) are 2.135(2) and 2.055(2)Å, respectively, longer than those of Al(1)-N(3) and Al(1)-N(1), which are 1.826 and 1.987Å, respectively.

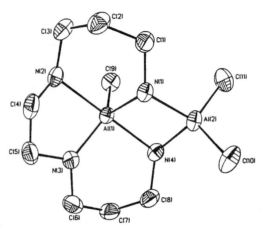

Fig. 6.3-4 Structure of the [AlMe][$C_8H_{19}N_4$][AlMe$_2$] molecule. Hydrogen atoms have been omitted. (After ref. 16).

Reaction of 3,3'-iminobispropylamine with trimethylaluminum yielded [AlMe]$_2$[$C_{12}H_{28}N_6$][AlMe$_2$]$_2$, which has been subjected to crystal structure analysis.[17] The centrosymmetric molecule comprises two AlMe and two AlMe$_2$ fragments bridged by three Al$_2$N$_2$ rings (Fig. 6.3-5). The Al(2) atom is in the usual tetrahedral four-coordinate environment. The Al(1) atom is in a trigonal-bipyramidal environment, being coplanar with C(7), N(1) and N(3), and the N(2)-Al(1)-N(3a) angle is 169.0(2)°. Both H atoms of N(3) and one H atom of N(1) are utilized in alkane elimination, whereas N(2) retains its single H atom.

In the analogous [AlMe]$_2$[$C_8H_{20}N_6$][AlMe$_2$]$_2$ complex, where the amine ligand is diethylenetriamine, both independent aluminum atoms are in square-pyramidal environments and each N atom loses one H atom in alkane elimination.[18] It has been suggested that trigonal-bipyramidal coordination is the favored configuration for five-coordinate organoaluminum compounds, and square-pyramidal geometry is observed only if the associated ligand(s) is sufficiently rigid.

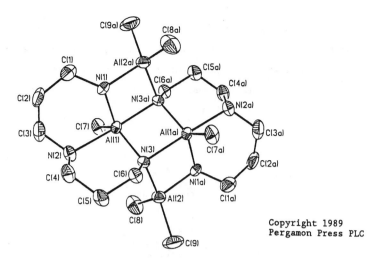

Fig. 6.3-5 Structure of the $[AlMe]_2[C_{12}H_{28}N_6][AlMe_2]_2$ molecule. Selected bond distances (Å): Al(1)-N(1) 1.967(4); Al(1)-N(2) 2.193(4); Al(1)-N(3) 1.878(4); Al(1)-N(3a) 2.065(4); Al(1)-C(7) 1.988(6); Al(2)-N(3) 1.891(4); Al(2)-N(1a) 1.944(4); Al(2)-C(8) 1.974(7); Al(2)-C(9) 1.987(7). (After ref. 17).

Investigation of the organoaluminum chemistry of the macrocyclic tetradentate secondary amine 1,4,8,11-tetraazacyclotetradecane, commonly known as cyclam, has yielded the complexes $[AlMe]_2[cyclam][AlMe_3]_2$[19] and $[AlEt]_2[cyclam][AlEtCl_2]_2$,[20] each having a centrosymmetric structure with a central planar Al_2N_2 ring and tetrahedral coordination about the Al atoms (Fig. 6.3-6).

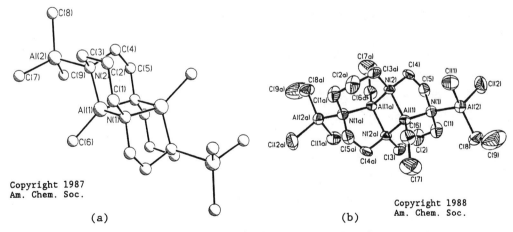

(a)   (b)

Fig. 6.3-6 Structure of (a) $[AlMe]_2[cyclam][AlMe_3]_2$ and (b) $[AlEt]_2[cyclam][AlEtCl_2]_2$. (After refs. 19 and 20).

The reaction of cyclam with trimethylaluminum and $ZrCl_4$ yielded the di-μ-chloro-stabilized $[AlMe_2Cl]_2[Al(cyclam)][AlMe_2]$ complex which contains a six-coordinate aluminum atom in a distorted octahedral environment (Fig. 6.3-7).[21] Also noteworthy is the fact that the macrocyclic ligand has been forced into a folded conformation.

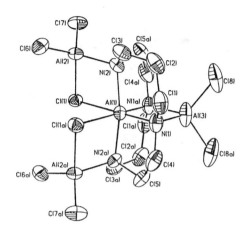

Fig. 6.3-7   Structure of $[AlMe_2Cl]_2[Al(cyclam)][AlMe_2]$, which has crystallographically imposed symmetry C_2.   (After ref. 21).

References

[1]   R.G. Vranka and E.L. Amma, *J. Am. Chem. Soc.* **89**, 3121 (1967).

[2]   P.H. Lewis and R.E. Rundle, *J. Chem. Phys.* **21**, 986 (1953).

[3]   J.P. Oliver, *Adv. Organomet. Chem.* **15**, 235 (1977).

[4]   J.J. Eisch in G. Wilkinson, F.G.A. Stone and E.W. Abel (eds.), *Comprehensive Organometallic Chemistry*, Pergamon Press, Oxford, 1982, Vol. 1, p. 555.

[5]   J.W. Moore, D.A. Sanders, P.A. Scherr, M.D. Glick and J.P. Oliver, *J. Am. Chem. Soc.* **93**, 1035 (1971).

[6]   J.F. Malone and W.S. McDonald, *J. Chem. Soc., Chem. Commun.*, 444 (1967); *J. Chem. Soc., Dalton Trans.* 2646 (1972).

[7]   J.F. Malone and W.S. McDonald, *J. Chem. Soc., Dalton Trans.*, 2649 (1972).

[8]   M.J. Albright, W.M. Butler, T.J. Anderson, M.D. Glick and J.P. Oliver, *J. Am. Chem. Soc.* **98**, 3995 (1976).

[9]   G.D. Stucky, A.M. McPherson, W.E. Rhine, J.J. Eisch and J.L. Considine, *J. Am. Chem. Soc.* **96**, 1941 (1974).

[10] A.H. Cowley and W.D. White, *J. Am. Chem. Soc.* **91**, 34 (1969).

[11] P. Jutzi, *Adv. Organomet. Chem.* **26**, 217 (1986).

[12] A. Almenningen, L. Fernholt and A. Haaland, *J. Organomet. Chem.* **155**, 245

[13] B. Teclé, W.R. Corfield and J.P. Oliver, *Inorg. Chem.* **21**, 458 (1982).

[14] D.A. Drew and A. Haaland, *Acta Chem. Scand.* **27**, 3735 (1973).

[15] P.R. Schonberg, R.T. Paine and C.F. Campana, *J. Am. Chem. Soc.* **101**, 7726 (1979).

[16] G.H. Robinson, S.A. Sangokoya, F. Moise and W.T. Pennington, *Organometallics* **7**, 1887 (1988).

[17] G.H. Robinson, F. Moise, W.T. Pennington and S.A. Sangokoya, *Polyhedron* **8**, 1279 (1989).

[18] G.H. Robinson and S.A. Sangokoya, *J. Am. Chem. Soc.* **109**, 6852 (1987).

[19] G.H. Robinson, A.D. Rae, C.F. Campana and S.K. Byram, *Organometallics* **6**, 1227 (1987).

[20] G.H. Robinson and S.A. Sangokoya, *Organometallics* **7**, 1453 (1988).

[21] G.H. Robinson, M.F. Self, S.A. Sangokoya and W.T. Pennington, *J. Am. Chem. Soc.* **111**, 1520 (1989).

Note The tetrameric aluminum(I) complex $[Al(\eta^5\text{-}C_5Me_5)]_4$ has a tetrahedral cluster structure with average Al-Al and Al-C bond distances of 2.769 and 2.334Å, respectively. See C. Dohmeier, C. Robl, M. Tacke and H. Schnöckel, *Angew. Chem. Int. Ed. Engl.* **30**, 564 (1991).

In the x-ray structures of both $[M(CH_3)_3]_2[14]aneN_4$ (M = Al, Ga) and $[Al(CH_3)_3]_2[14]aneN_4[Ga(CH_3)_3]_2$ the aza crown ligand assumes an unusual "inverted basket" conformation. See G.H. Robinson, W.T. Pennington, B. Lee, M.F. Self and D.C. Hrncir, *Inorg. Chem.* **30**, 809 (1991).

The aza crown ether-stabilized square-pyramidal coordination of Al in the $[(EtAl)_2 \cdot \text{diaza-18-crown-6}]^{2+}$ cation is described in M.F. Self, W.T. Pennington, J.A. Laske and G.H. Robinson, *Organometallics* **10**, 36 (1991).

6.4 Zeise's Salt
K[(H₂C=CH₂)PtCl₃]·H₂O

Crystal Data

Monoclinic, space group $P2_1/c$ (No. 14)

$a = 11.212(3)$, $b = 8.424(6)$, $c = 9.696(6)$Å, $\beta = 107.52(4)°$, $Z = 4$

Atom	x	y	z	Atom	x	y	z
Pt	.2862	.2505	.0595	H(2)	.4358	.1698	-.0966
Cl(1)	.0763	.2374	.0537	H(3)	.5173	.3588	.1896
Cl(2)	.2848	.5209	.0926	H(4)	.5224	.1440	.1704
Cl(3)	.2917	.9788	.0364	K	.9394	.0452	.2525
C(1)	.4390	.2715	-.0266	O	.8528	.3537	.2309
C(2)	.4856	.2566	.1212	H(5)	.7945	.3819	.1372
H(1)	.4304	.3874	-.0801	H(6)	.8124	.3889	.2973

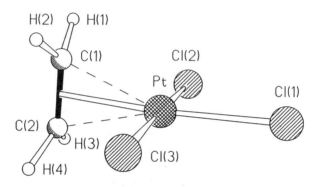

Fig. 6.4-1 Molecular structure of [H₂C=CH₂)PtCl₃]⁻.

Crystal Structure

The X-ray crystal structure of Zeise's salt was established in the correct space group by Jarvis *et al.*[2] Love *et al.* used neutron diffraction data to refine the structure and obtained improved positional parameters of the C and H atoms.[1]

The structure of the [(H₂C=CH₂)PtCl₃]⁻ anion is shown in Fig. 6.4-1. Some bond lengths and angles are listed follow:

Pt-Cl(1)	2.340Å		C(1)-C(2)	1.375Å
Pt-Cl(2)	2.302		C(1)-H(1)	1.096
Pt-Cl(3)	2.303		C(1)-H(1)	1.087
Pt-C(1)	2.128		C(2)-H(3)	1.079
Pt-C(2)	2.135		C(2)-H(4)	1.086
Cl(1)-Pt-Cl(2)	90.05°		Cl(1)-Pt-Cl(3)	90.43°
Cl(2)-Pt-Cl(3)	177.65°		C(1)-Pt-C(2)	37.6°

The distance from the midpoint of the C=C bond to the Pt atom is 2.022Å. The Pt-Cl bond *trans* to the ethylene (2.340Å) is longer than the average of the two *cis* Pt-Cl bonds (2.302Å). The hydrogen atoms of the ethylene group are symmetrically bent away from the central Pt atom, and the angle between the normals to the H-C-H planes is 32.5°. The distances of carbon atoms C(1) and C(2) from the plane of the four hydrogen atoms are 0.158 and 0.171Å, respectively, and the Pt atom is 2.185Å from this plane.

The chloride ligands of the $[(C_2H_4)PtCl_3]^-$ anion form a distorted trigonal prism around the K^+ ion. The water molecule caps two rectangular faces of the trigonal prism, and is in an approximate tetrahedral environment with two equivalent K^+ neighbours and two donor hydrogen bonds to nonequivalent Cl atoms.

Remarks

1. Structural features of metal-olefin complexes

In 1825, the Danish chemist W.C. Zeise[3] set into reflux a mixture of $PtCl_4$ and $PtCl_2$ in ethyl alcohol, treated the resulting black solid with KCl and HCl, and isolated cream-lemon crystals of composition $K[PtCl_3(C_2H_4)].H_2O$. This platinum-ethylene complex, later referred to as "Zeise's salt" in the literature, was the first organometallic compound to be isolated in pure form. This discovery spawned a tremendous growth in organometallic chemistry and still serves as the simplest example of transition metal-olefin complexation.

Accurate structural parameters for Zeise's salt and a number of metal-olefin complexes have led to the following generalizations.

a) Coordination of the olefin corresponds to an increase in the C-C bond length. The observed increase, which can be as large as 0.2Å, is sensitive to the oxidation state of the metal and the electron-donating properties of the other ligands.

b) The olefin substituents invariably bend away from the metal on coordination so as to give a quasi-tetrahedral character to the ligating C atoms.

c) The coordinated olefins show definite conformational preferences with respect to the coordination plane defined by the metal and the other ligands. Thus, in Zeise's salt and other Pt(II) complexes, the olefin carbon-carbon bond axis is orthogonal to the platinum-ancillary ligand plane. In the trigonal d^{10} complexes and trigonal bipyramidal d^8 complexes, the olefinic axis is coplanar to within a few degrees with the idealized equatorial coordination plane.

d) There is a manifestation of the *trans*-effect (Section 4.20). In Zeise's salt the Pt-Cl distance *trans* to olefin is longer than the two *cis* Pt-Cl distances.

2. Bonding in Zeise's salt

Elucidation of the bonding in metal-olefin π-complexes by Dewar[4] and by Chatt and Duncanson[5] provided a theoretical framework for the rapid growth of organometallic π-complex chemistry. This synergistic bonding model, now generally used to rationalize many trends in the structural, chemical, and spectroscopic properties of metal-olefin complexes, is considered to arise from two interdependent components as illustrated schematically in Fig. 6.4-2. In the first part, σ overlap between the filled π orbital of ethylene and a suitably directed vacant hybrid metal orbital forms the electron-pair donor

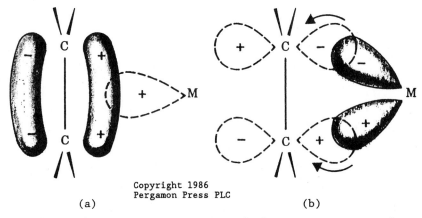

(a)   (b)

Fig. 6.4-2 Schematic representation of the two components of an ethylene-metal bond: (a) σ donation from filled π orbital of ethylene into vacant metal hybrid orbital; (b) π back donation from a filled metal d orbital into the vacant antibonding orbital of ethylene. (After ref. 6).

bond. This is reinforced by the second component involving the overlap of a filled metal d orbital with the vacant antibonding orbital of ethylene; these orbitals have π symmetry with respect to the bonding axis and allow $M{\rightarrow}C_2$ π back bonding to synergically assist the σ $C_2{\rightarrow}M$ bond.

  Ab initio molecular orbital calculations on Zeise's salt[7] have provided a very detailed picture of the metal-olefin bond and confirmed the essential features of the Dewar-Chatt-Duncanson model.

3. Reversible ethylene coordination in $(C_2H_4)Pt[P(C_6H_{11})_3]_2$[8]

  The complex $(\eta^2\text{-}C_2H_4)Pt[P(C_6H_{11})_3]_2$ exists in solution in which the equilibrium

$$\begin{array}{c}(C_6H_{11})_3P\\(C_6H_{11})_3P\end{array}{>}Pt\text{---}\underset{CH_2}{\overset{CH_2}{\parallel}} \;\rightleftharpoons\; \begin{array}{c}(C_6H_{11})_3P\\(C_6H_{11})_3P\end{array}{>}Pt^0 \;+\; C_2H_4$$

involves the reversible coordination of ethylene to the platinum(0) center. The crystal structure of $(\eta^2\text{-}C_2H_4)Pt[P(C_6H_{11})_3]_2 \cdot 1/3(C_6H_{14})$ provides evidence for weak coordination of the ethylene ligand in the solid state, as shown in Fig. 6.4-3. The molecule lies on a crystallographic two-fold axis which passes through Pt and the mid-point of the ethylene C=C bond. The ethylene ligand essentially lies in the PtP_2 coordination plane, and the $C(1)\text{-}C(1)^*$ bond distance of 1.440Å is much longer than that of free ethylene (1.335Å).

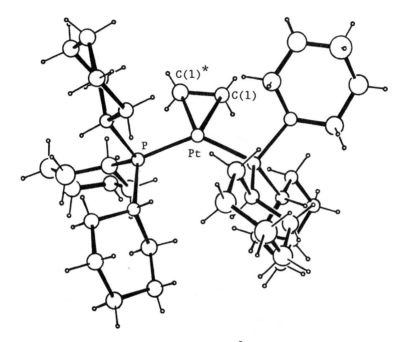

Fig. 6.4-3 Molecular structure of $(\eta^2\text{-}C_2H_4)Pt[P(C_6H_{11})_3]_2$. Atom $C1^*$ is related to $C1$ by a crystallographic two-fold axis. Relevant bond lengths (Å): Pt-P 2.284(1), Pt-C(1) 2.137(1), $C(1)\text{-}C(1)^*$ 1.440(7), P-C(11) 1.881(7), P-C(21) 1.861(6), P-C(31) 1.868(8); bond angles (°): $P\text{-}Pt\text{-}P^*$ 116.33(7), P-Pt-C(1) 141.5(2), $P\text{-}Pt\text{-}C(1)^*$ 102.2(2), $C(1)\text{-}Pt\text{-}C(1)^*$ 39.4(2), $Pt\text{-}C(1)\text{-}C(1)^*$ 70.3(4). (Adapted from ref. 8).

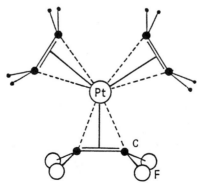

Fig. 6.4-4 Molecular structure of $Pt(C_2H_4)_2(C_2F_4)$. (Adapted from ref. 10).

$(\eta^2\text{-}C_2H_4)Pt(PPh_3)_2$ has a similar coordination geometry, in which the C-C bond distance in the ethylene ligand is 1.434Å.[9] $Pt(C_2H_4)_2(C_2F_4)$ has been shown to possess the planar conformation shown in Fig. 6.4-4.[10]

In trigonal bipyramidal d^8 complexes, for example $Fe(CO)_4(C_2H_4)$[11] and $[IrCl(CO)(PR_3)_2(olefin)]$, the olefin axis is coplanar to within a few degrees with the idealized trigonal coordination plane.

4. Allyl and conjugated dienes complexes

The allyl group, $CH_2=CH\text{-}CH_2\text{-}$, can act as an η^3 or η^2 π ligand. A great variety of both homoleptic and mixed-ligand η^3-allyl complexes is known.

The detailed geometry of *trans*-bis(η^3-allyl)nickel has been determined by a neutron diffraction study at 100K.[12] The molecule is centrosymmetric, and the C(1)-C(2)-C(3) plane is tilted by 110° from one containing Ni, C(1), C(3), C(1*) and C(3*); the *syn* and *meso* hydrogen atoms are bent by 8.9(4) and 15.8(1)°, respectively, towards the metal whereas the *anti* hydrogen atoms are bent 29.4(7)° away [Fig. 6.4-5(a)]. The deformation electron density map through Ni, C(2) and C(2*) is shown in Fig. 6.4-5(b).[13]

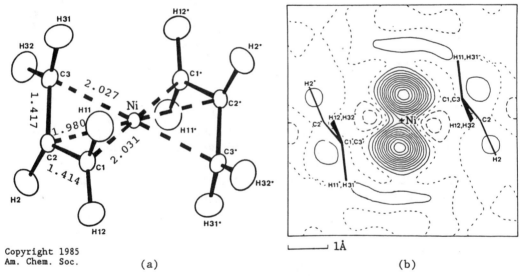

(a)   (b)

Fig. 6.4-5 Molecular structure (a) and deformation electron density map (b) of *trans*-bis(η^3-allyl)nickel. (After refs. 12 and 13).

The bonding in η^3-allyl complexes involves two ligand-to-metal donor bonds (one σ and the other π) and a metal-to-ligand $d\pi$-$p\pi$ back bond. The π-bonded allyl anion is a four-electron donor which can have some π-acidic character. The allyl group can also form bridging η^3-allyl complexes and σ-allyl derivatives as exemplified by the following two structures:

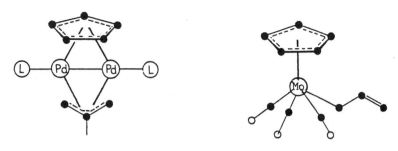

Conjugated dienes such as butadiene and its open-chain analogues can act as η^4 ligands. No new principles are involved in describing the bonding in these complexes, and appropriate combinations of the four p_z orbitals on the diene system can be used to construct MO's with the metal-based orbitals for donation and back donation of electron density. The C-C distances in diene complexes vary and the central C-C distance is often less than the two outer ones.

References

[1] R.A. Love, T.F. Koetzle, G.J.B. Williams, L.C. Andrews and R. Bau, *Inorg. Chem.* **14**, 2653 (1975).

[2] J.A.J. Jarvis, B.T. Kilbourn and P.G. Owston, *Acta crystallogr., Sect. B* **26**, 876 (1970); **B27**, 366 (1971).

[3] W.C. Zeise, *Overs. K. Dan. Vidensk. Selskabs. Forh.* 13 (1825); *Pogg. Ann. Phys.* **9**, 632 (1827).

[4] M.J.S. Dewar, *Bull. Soc. Chim. Fr.* **18**, C71 (1951).

[5] J. Chatt and L.A. Duncanson, *J. Chem. Soc.*, 2939 (1953).

[6] N.N. Greenwood and A. Earnshaw, *Chemistry of the Elements*, Pergamon Press, Oxford, 1986, chap. 8.

[7] N. Rösch, R.P. Messmer and K.H. Johnson, *J. Am. Chem. Soc.* **96**, 3855 (1974).

[8] H.C. Clark, G. Ferguson, M.J. Hampden-Smith, B. Kaitner and H. Ruegger, *Polyhedron* **7**, 1349 (1988).

[9] P.-T. Cheng, C.D. Cook, S.C. Nyburg and K.Y. Wan, *Inorg. Chem.* **10**, 2210 (1971); P.-T. Cheng and S.C. Nyburg, *Can. J. Chem.* **50**, 912 (1972).

[10] M. Green, J.A.K. Howard, J.L. Spencer and F.G.A. Stone, *J. Chem. Soc., Chem. Commun.*, 449 (1975).

[11] M.I. Davis and C.S. Speed, *J. Organomet. Chem.* **21**, 401 (1970).

[12] R. Goddard, C. Krüger, F. Mark, R. Stansfield and X. Zhang, *Organometallics* **4**, 285 (1985).

[13] G. Wilke in O. Vogl and E.H. Immergut (eds.), *Polymer Science in the Next Decade*, Wiley, New York, 1987, p. 89.

6.5   Bis(triphenylphosphine)hexafluorobut-2-yneplatinum(0)
$(Ph_3P)_2(CF_3C{\equiv}CCF_3)Pt$

Crystal Data

Triclinic, space group $P\bar{1}$ (No. 2)

$a = 11.799(2)$, $b = 16.062(3)$, $c = 9.723(1)$Å, $\alpha = 99.33(1)°$,

$\beta = 101.47(1)°$, $\gamma = 96.75(1)°$, $Z = 2$

Atom*	x	y	z	Atom*	x	y	z
Pt	.15611	.21612	.20163	F(5)	-.1017	.3294	.3018
P(1)	.31993	.15358	.24335	F(6)	-.1290	.3025	.0869
P(2)	.22725	.33254	.11948	C(1)	-.0347	.0768	.2980
F(1)	-.0533	.0054	.2016	C(2)	.0142	.1519	.2471
F(2)	-.1309	.0850	.3378	C(3)	-.0140	.2193	.2101
F(3)	.0373	.0599	.4099	C(4)	-.1136	.2654	.1966
F(4)	-.2130	.2205	.1848				

*Carbon atoms of the phenyl groups have been omitted.

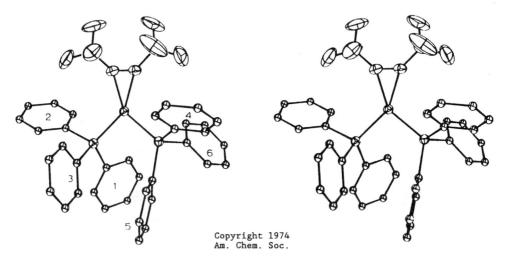

Fig. 6.5-1   Stereoview of the $(Ph_3P)_2(CF_3C{\equiv}CCF_3)Pt$ molecule.   (After ref. 1).

Crystal Structure

The crystal of the title compound consists of discrete molecular units. The coordination about the Pt atom is essentially planar (Fig. 6.5-2). The bond between the acetylenic C(2) and C(3) atoms makes an angle of 3.6(4)° with the plane of the Pt and the two P atoms. The Pt-P distances are 2.277 and 2.285Å, and the two P atoms subtend an angle of 100.17° at the Pt atom. The observed dimensions are consistent with a trigonal-planar geometry about the Pt atom, if the acetylene is viewed as a monodentate ligand.

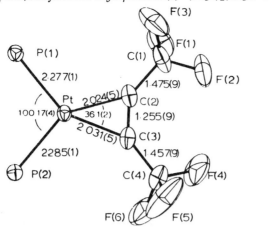

Fig. 6.5-2 Coordination about the Pt atom. (After ref. 1).

The geometry of the acetylenic ligand is perturbed considerably on coordination. The observed triple-bond length of 1.255(9)Å is larger than the average value of 1.204(2)Å in uncoordinated acetylenes. The acetylenic carbon atoms C(2) and C(3) subtend an angle of 36.1° at the Pt atom. The Pt-C distances to C(2) and C(3) are 2.024 and 2.031Å, respectively, shorter than the values observed in the corresponding olefin complexes. The acetylenic substituents are each bent back by an angle of 39.9°, and the mean $C-CF_3$ distance is 1.466Å.

Remarks

1. Structural types of acetylene complexes

Various types of acetylene complexes have been discussed in the context of the isolobal analogy,[2] and their interconversions have also been reviewed.[3] Some representative examples are as follows:

a) Pseudo-4-coordination

In the complex $Pt(\eta^2-C_2Bu^t_2)Cl_2(4\text{-toluidine})$,[4] the C≡C bond remains short and the alkyne group is normal to the plane of the $PtCl_2N$-group, as shown in Fig. 6.5-3(a).

A study of the geometrical effects of metal-carbon π bonding on complexed acetylene have been investigated using a model based largely on acetylene intramolecular interactions.[5]

b) Bridging ligand

The alkyne group can function as a bridging ligand as exemplified by the classic example $Co_2(CO)_6(C_2Ph_2)$,[6] which is formed by direct displacement of the two bridging carbonyls in $[Co_2(CO)_8]$ to give the structure shown in Fig. 6.5-3(b). The C-C group lies above and at right angles to the Co-Co vector;

the C-C bond length is 1.46Å, which has been taken to indicate extensive back donation from the two Co atoms. The Co-Co distance is 2.47Å as compared with 2.52Å in $Co_2(CO)_8$.

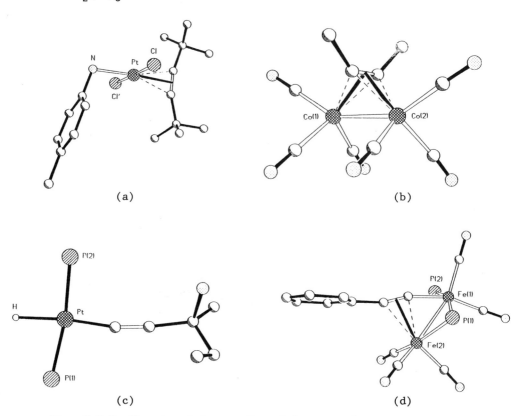

Fig. 6.5-3 Structural types of acetylene complexes.

c) Terminal ligand

A wide range of complexes of the type *trans*-HPt(C≡CR)(PPh$_3$)$_2$ were prepared by the reaction of the alkynes with the corresponding halide, and the complex with R = CEtMe(OH) [Fig. 6.5-3(c)] was characterized by X-ray crystallography.[7] In the complex $Fe_2(CO)_5(C_2Ph)(PPh_2)(PPh_3) \cdot C_6H_{12}$,[8] the structure as determined by X-ray diffraction contains an acetylide ligand σ-bonded to one Fe atom and π-bonded to the other. This σ-π acetylide has a C≡C acetylenic bond length of 1.225Å and a *trans* bent configuration, the C≡C-C angle being 164.7° [Fig. 6.5-3(d)].

2. Bonding in acetylene complexes[9]

The side-on coordination of acetylene may be conveniently described by the four orbital interactions shown in Fig. 6.5-4. The extent of overlap and the energy level difference between these interacting orbitals are important in determining the nature of bonding. The overlap decreases in the order:

(a)>(b)>(c)>(d). The energy levels are functions of the effective oxidation state of the metal, the nature of auxiliary ligands, and the substituents on the acetylenic carbons. Interaction (a) is usually bonding. Interaction (b) is in most cases bonding, since most transition metal ions or atoms have d electrons which may occupy $\pi_{||}$ orbitals. In early transition metal complexes having vacant $d\pi_{\perp}$ orbitals, such as $Cp_2W=O(RC\equiv CR)$ or $CpNb(CO)(PhC\equiv CPh)_2$ interaction (c) is bonding, but the contribution to the bond strength is less than that of (a) or (b). In later transition metal complexes, especially d^{10} complexes, the interaction should be repulsive and antibonding (c'). Interaction (d) can be neglected because of the poor overlap. The interactions (a), (b) and (c) or (c') are important in determining the overall bond strength.

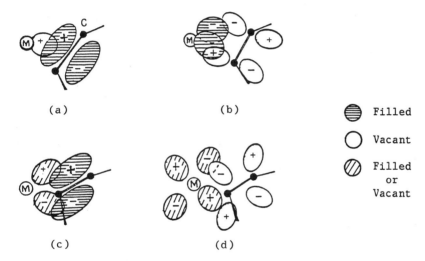

(a)   (b)   Filled

    Vacant

    Filled
    or
    Vacant

(c)   (d)

Fig. 6.5-4 Molecular orbital interactions for coordinated acetylene: (a) metal σ-orbital (s, pσ or dσ) and $\pi_{|| b}$; (b) metal $\pi_{||}$ orbital (pπ or d$\pi_{||}$) and $\pi_{||}^{*}$; (c) metal $\pi_{\perp}$ orbital (p$\pi_{\perp}$ or d$\pi_{\perp}$) and $\pi_{\perp b}$; and (d) metal δ orbital (dδ) and $\pi_{\perp}^{*}$.

The extent of C≡C bond lengthening and C≡C-C angle bending reflects a major contribution by π^{*} back donation [interaction (b) in Fig. 6.5-4] and a minor contribution by σ and π donation [interactions (a) and (c)]. Table 6.5-1 lists the C≡C bond lengths and C≡C-C angles in some acetylene complexes. The coordinated C≡C length of 1.24Å in a typical Pt(II) complex with an electron-donating acetylene is definitely shorter than the values in Pt(0) complexes with π-acidic acetylenes, implying the importance of back donation. The data in Table 6.5-1 also indicate that the majority lies in a relatively small range, namely 1.28 ± 0.02Å. The long C≡C bond (1.35Å) in an electron-deficient Nb complex may be accounted for by an effective π-donor interaction [Fig. 6.5-4(c)]. The C≡C-C angles also fall within a narrow range, 143° ± 5°. Larger values (162-165°) in $PtCl_2$(4-toluidine)($Bu^tC≡CBu^t$) suggest an extensive π-donor interaction (Fig. 6.5-4(a)].

Table 6.5-1 Observed C≡C bond lengths and C≡C-R angles in some acetylene complexes.

Acetylene complex	C≡C/Å	C≡C-R/°
Pt(cyclo-C_7H_{10})(PPh_3)$_2$	1.294	137,143
Pt(cyclo-C_6H_8)(PPh_3)$_2$	1.289	127,128
Pt($CF_3C≡CCF_3$)(PPh_3)$_2$	1.255	140.1
Pt(PhC≡CPh)(PPh_3)$_2$	1.32	140
Pt(NCC≡CCN)(PPh_3)$_2$	1.40	140
$PtCl_2$(4-toluidine)($Bu^tC≡CBu^t$)	1.24	162-165
Pd($CH_3O_2CC≡CCO_2CH_3$)(PPh_3)$_2$	1.28	145
[Pd($Ph_2PC≡CCF_3$)(PPh_3)]$_2$	1.28,1.29	138
Ni(PhC≡CPh)(Bu^tNC)$_2$	1.28	149
Nb(C_5H_5)(CO)(PhC≡CPh)$_2$	1.35	138
Nb(C_5H_5)(CO)(PhC≡CPh)(Ph_4C_4)	1.26	142
W(PhC≡CPh)$_2$(CO)	1.30	140

3. Molybdenum(II) and tungsten(II) alkyne complexes

Since 1980 much research has been done in transition metal alkyne complexes. The alkyne ligand is considered to be a "four-electron donor" when its filled $\pi_\perp$ orbital (orthogonal to the MC_2 plane) plays a significant role in forming a donor bond to the metal; the chemistry of monomeric Mo(II) and W(II) alkyne complexes of this type has been reviewed recently.[10]

The complex W(CO)(HC≡CH)(S_2CNEt_2)$_2$ contains the *cis*-M(CO)(RC≡CR) fragment common to all octahedral d^4 L_4M(CO)(RC≡CR) complexes characterized to date.[11] As shown in Fig. 6.5-5(a), the π-acid CO ligand preferentially stabilizes the d_{xz} and d_{yz} orbitals which accommodate the metal d^4 electrons, leaving the d_{xy} orbital vacant. The parallel alignment of the alkyne to the M-CO axis thus facilitates forward donation from $\pi_\perp$ to d_{xy} and back donation from d_{yz} to $\pi_{\parallel}^*$ (and simultaneously to π_y^* of CO to form a three-center, two-electron bond) [Fig. 6.5-5(b)].

In the mixed olefin alkyne complex W(MA)(PhC≡CH)(S_2CNEt_2)$_2$ the two unsaturated C_2 ligands, maleic anhydride and phenylacetylene, are *cis* and parallel to each other [Fig. 6.5-6(a)].[12] This class of octahedral d^4 complexes are characterized by a matching of each of the three metal $d\pi$ orbitals with a ligand π function [Fig. 6.5-6(b)].

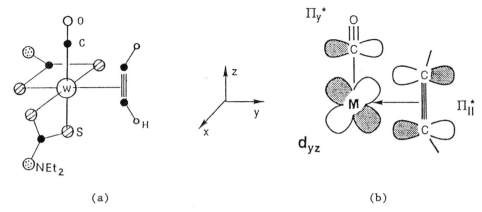

(a)   (b)

Fig. 6.5-5 (a) Geometry of the W(CO)(HC≡CH)(S₂CNEt₂)₂ complex. (b) Three-center, two-electron bond in L₄M(CO)(RC≡CR) complexes. (After ref. 10).

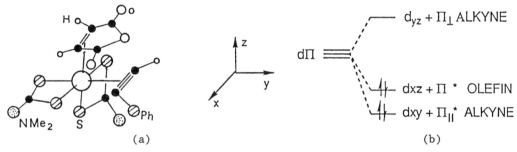

(a)   (b)

Fig. 6.5-6 (a) Geometry of the W(MA)(PhC≡CH)(S₂CNEt₂)₂ complex (MA = maleic anhydride). (b) Metal-ligand π interactions in an octahedral d^4 *cis*-(η^2-olefin)(η^2-alkyne)ML₄ complex. (After ref. 10).

In W(=CHPh)(PhC≡CPh)(PMe₃)₂Cl₂ the π-acid carbene ligand is *cis* to the alkyne and the plane of the metal benzylidene fragment is orthogonal to the *trans* P-W-P axis as anticipated from optimal $d\pi$ orbital utilization [Fig. 6.5-7(a)].[13] Alkyne orientation in pseudo-octahedral metal-alkyne complexes of

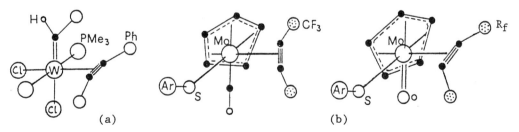

(a)   (b)

Fig. 6.5-7 (a) Geometry of the W(=CHPh)(PhC≡CPh)(PMe₃)₂Cl₂ molecule. (b) Different alkyne orientations in d^4 CpMo(CO)(CF₃C≡CCF₃)(SC₆H₅) and d^2 CpMo(O)(CF₃C≡CCF₃)(SC₆H₅). (After ref. 10).

the type CpML(RC≡CR)X depends on the nature of the two *cis* ancillary ligands.
A comparative study of the complexes CpMo(CO)(CF$_3$C≡CCF$_3$)(SC$_6$F$_5$) and
CpMo(O)(CF$_3$C≡CCF$_3$)(SC$_6$H$_5$) has revealed alkyne orientations differing by 90°
for the respective d^4 and d^2 configurations [Fig. 6.5-7(b)].[14]

Of the relatively few bisalkyne complexes which have been structurally
characterized, the two π ligands are invariably *cis* and parallel. Two
examples are illustrated in Fig. 6.5-8.[15,16]

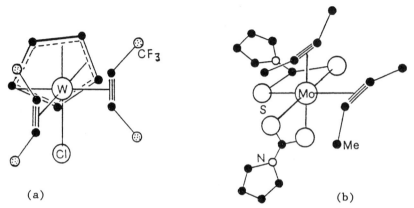

(a)   (b)

Fig. 6.5-8 Molecular geometry of (a) pseudo-octahedral CpW(CF$_3$C≡CCF$_3$)$_2$Cl and
(b) octahedral Mo(MeC≡CMe)$_2$(S$_2$CNC$_4$H$_4$)$_2$. (After ref. 10).

4. Mononuclear complexes containing benzyne and cycloalkyne ligands

Many metal complexes containing benzyne and cycloalkyne ligands have
been synthesized and characterized.[17,18] Some representative mononuclear
complexes are discussed below.

a) (C$_6$H$_{11}$)$_2$PCH$_2$CH$_2$P(C$_6$H$_{11}$)$_2$Ni(η^2-C$_6$H$_4$)

The molecule consists of a central Ni atom bonded to a symmetric η^2-
benzyne ligand [Ni-C(1) 1.870(4), Ni-C(2) 1.868(4)Å], the coordination
geometry being close to trigonal planar [Fig. 6.5-9(a)].[19] The C(1)-C(2)
bond length of 1.332(6)Å is significantly larger than those observed in alkyne
complexes of zero-valent Ni and Pt. The remaining C-C bonds of the six-
membered ring are almost equal in length (average 1.385Å), consistent with the
delocalized aromatic nature of the benzyne ligand.

b) (η^5-Cp)$_2$(PMe$_3$)Zr(η^2-C$_6$H$_4$)

The molecule consists of a central Zr atom coordinated tetrahedrally by
four ligand groups [Fig. 6.5-9(b)].[20] The η^2-benzyne ligand coordinates to
Zr with Zr-C(6) 2.267(5), Zr-C(7) 2.228(5)Å and C(6)-C(7) 1.364(8)Å. A high
level of back-bonding from Zr to the π*-orbital of benzyne is evidenced by the
fact that all the C-C bonds in the ring fragment are identical in length
within experimental error.

c) $(\eta^5\text{-Cp})_2(\text{PMe}_3)\text{Zr}(\eta^2\text{-C}_6\text{H}_8)$

The two independent molecules of this complex are nearly isostructural [Fig. 6.5-9(c)].[21] The average bond lengths are: Zr-C(1) = 2.17(1), Zr-C(2) = 2.24(1), and C(1)-C(2) = 1.30(2)Å in the cyclohexyne ligand. The C(1)-C(2)-C(3) and C(6)-C(1)-C(2) angles of 125.2(12)° and 126.0(12)° deviate only to a small extent from the usual value for simple olefins.

d) $(\eta^5\text{-Cp})_2(\text{PMe}_3)\text{Zr}[\eta^2\text{-C}_5\text{H}_4(\text{CH}_3)_2]$

The molecule consists of a central Zr atom coordinated tetrahedrally by four ligand groups [Fig. 6.5-9(d)].[22] The η^2-(3,3-dimethylcyclopentenyl) ligand coordinates to Zr with Zr-C(1) = 2.201(4), Zr-C(2) = 2.169(4)Å and C(1)-C(2) = 1.295(6)Å. The presence of the geminal dimethyl groups at C(3) does not cause a perceptible lengthening of the Zr-C(2) bond relative to the analogous bond in $(\eta^5\text{-Cp})_2(\text{PMe}_3)\text{Zr}(\eta^2\text{-C}_6\text{H}_8)$.

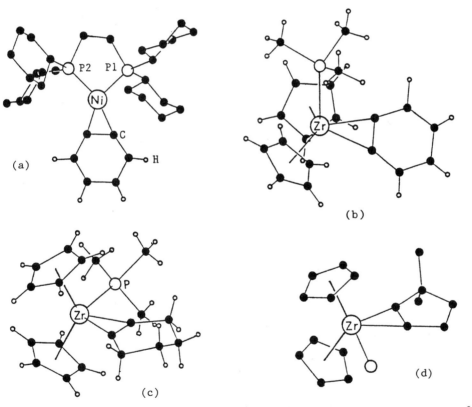

Fig. 6.5-9 Molecular structure of (a) $(C_6H_{11})_2PCH_2CH_2P(C_6H_{11})_2Ni(\eta^2\text{-}C_6H_4)$, (b) $(\eta^5\text{-Cp})_2(\text{PMe}_3)\text{Zr}(\eta^2\text{-C}_6H_4)$, (c) $(\eta^5\text{-Cp})_2(\text{PMe}_3)\text{Zr}(\eta^2\text{-C}_6H_8)$ and (d) $(\eta^5\text{-Cp})_2(\text{PMe}_3)\text{Zr}[\eta^2\text{-C}_5H_4(\text{CH}_3)_2]$. (After refs. 19-22).

5. Polynuclear benzyne complexes

The interaction of benzyne with a polynuclear metal framework can take a

variety of forms. A common coordination mode is μ_3-η^2, which has been found
in the trinuclear complex $[Ru_3(CO)_7(\mu$-$PPh_2)_2(\mu_3$-η^2-$C_6H_4)]$ (I).[23] The
hexanuclear cluster $[Ru_6(CO)_{12}(\mu_4$-$PMe)_2(\mu_3$-η^3-$C_6H_4)_2]$ (II) contains two
μ_3-benzyne ligands bound to the triruthenium faces of a Ru_6P_2 core.[24] A
variant of the μ_3-η^2 mode occurs in $[Os_3(\mu$-$H)(\mu_3$-$C_6H_4PMePh)(CO)_9]$ (III) in
which the benzyne ligand is a five-electron donor bonded through P to one Os
atom, by a σ Os-C bond to another, and by an η^2-C_6H_4 interaction to the
third.[25] The molecular structures of compounds II and III are shown in
Figs. 6.5-10(a) and (b), respectively.

I (R=Ph, R'=H)   II (M = Ru(CO)$_2$)   III (R = Me)

The trinickel complex $Ni_3(\mu_3$-$C_6H_4)(\mu_3$-C_6H_4-$C_6H_4)(P$-i-$Pr_3)_3$ contains both
μ_3-η^2-benzyne and a novel bridging 2,2'-biphenylyl group bound to a nearly
right-angled isosceles triangle of nickel atoms each bearing a terminal
triisopropylphosphine ligand [Fig. 6.5-10(c)].[26] The measured bond
distances are given in Fig. 6.5-10(d).

The μ_4-η^2, μ_4-η^4 and μ_5-η^6 coordination modes of benzyne have been
established by X-ray diffraction in the ruthenium complexes $[Ru_4(\mu$-
$H)_2(CO)_{11}(\mu$-$PR)(\mu_4$-η^2-$C_6H_4)]$ (R = ferrocenyl) (IV),[27] $[Ru_4(CO)_{10}(\mu$-$CO)(\mu_4$-
$PPh)(\mu_4$-η^4-$C_6H_4)]$ (V),[24] and $[Ru_5(CO)_{13}(\mu_4$-$PPh)(\mu_5$-η^6-$C_6H_4)]$ (VI),[24]
respectively. The molecular structures of complexes V and VI are illustrated

IV (R = C$_5$H$_4$FeC$_5$H$_5$)   V (ER = PPh)   VI (ER = PPh)

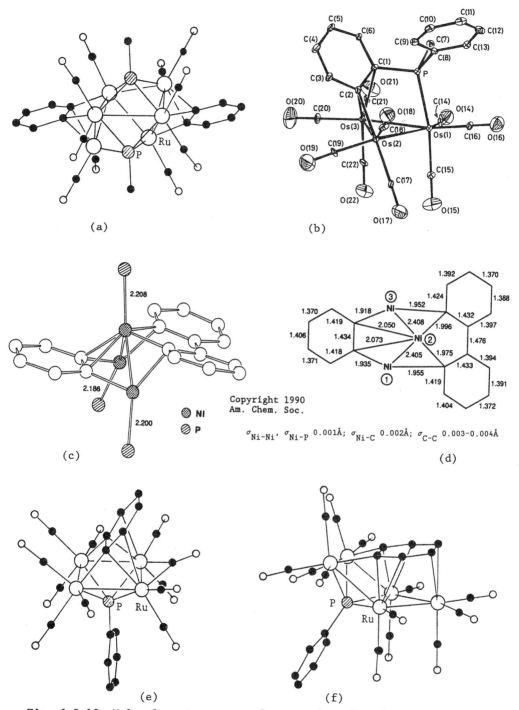

σ_{Ni-Ni}, σ_{Ni-P} 0.001Å; σ_{Ni-C} 0.002Å; σ_{C-C} 0.003–0.004Å

(a)   (b)   (c)   (d)   (e)   (f)

Fig. 6.5-10 Molecular structures of some polynuclear benzyne complexes:
(a) complex **II**, (b) complex **III**, (c) Ni₃(μ₃-C₆H₄)(μ₃-C₆H₄-C₆H₄)(P-i-Pr₃)₃
with the i-Pr groups omitted, (d) measured bond distances of the Ni₃
complex, (e) complex **V**, and (f) complex **VI**. (After refs. 24–26).

in Figs. 6.5-10(e) and (f), respectively. In each complex the benzyne ligand is bound to a Ru_4 square by Ru-C σ bonds to two adjacent ruthenium atoms, and by η^2-interactions to the other two; in VI there is also η^2 bonding to the fifth ruthenium atom, so that the overall coordination mode becomes μ_5-η^6.

6. Allene complexes

There are two coordination modes in allene complexes, one containing an allene double bond perpendicular to the principal molecular plane and the other containing the ligand in the plane (Fig. 6.5-11). In the former type, there are labile complexes, e.g., $RhI(PPh_3)_2(H_2C=C=CH_2)$, and fluxional ones, e.g., $Pt_2Cl_4(Me_2C=C=CMe_2)_2$. In the latter are found both labile complexes, e.g. $Pd(PPh_3)_2(H_2C=C=CH_2)$, and inert compounds, e.g. $Pt(PPh_3)_2(H_2C=C=CH_2)$. Table 6.5-2 lists the structural parameters of some transition metal-allene complexes.[9]

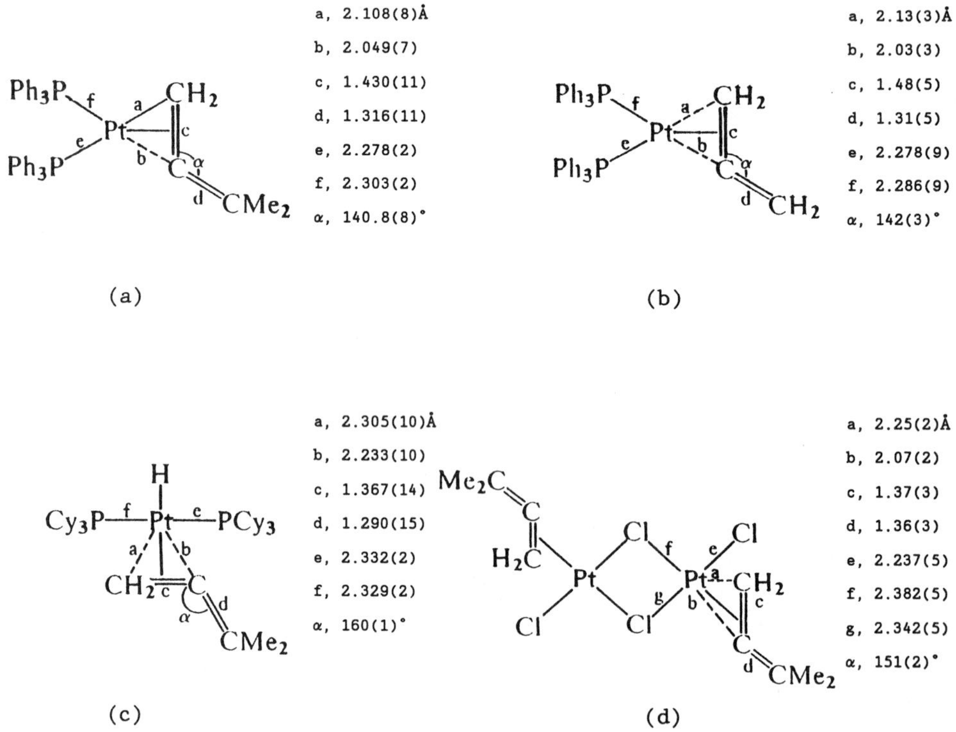

(a)
a, 2.108(8)Å
b, 2.049(7)
c, 1.430(11)
d, 1.316(11)
e, 2.278(2)
f, 2.303(2)
α, 140.8(8)°

(b)
a, 2.13(3)Å
b, 2.03(3)
c, 1.48(5)
d, 1.31(5)
e, 2.278(9)
f, 2.286(9)
α, 142(3)°

(c)
a, 2.305(10)Å
b, 2.233(10)
c, 1.367(14)
d, 1.290(15)
e, 2.332(2)
f, 2.329(2)
α, 160(1)°

(d)
a, 2.25(2)Å
b, 2.07(2)
c, 1.37(3)
d, 1.36(3)
e, 2.237(5)
f, 2.382(5)
g, 2.342(5)
α, 151(2)°

Fig. 6.5-11 Structures of some allene complexes. In (a) the planar allene ligand lies 9° out of the PtP_2-plane, whereas in (b) the dimethylallene ligand lies 7.8° out of this plane. In (c) and (d) the allene ligands are planar and make angles of 89.2(1)° and 85(1)° with the Pt(II) coordination plane, respectively. (After ref. 28).

Table 6.5-2   Structural parameters of some metal-allene complexes.

Compound	Structure*	$C(1)-$ $C(2)/Å$	$C(2)-$ $C(3)/Å$	$C(1)-C(2)-$ $C(3)/°$	$M-C(1)$ $/Å$	$M-C(2)$ $/Å$
$Rh(acac)(H_2C=C=CMe_2)_2$	perpendicular	1.40 1.41	1.30 1.29	153.3 152.6	2.13 2.13	2.07 2.06
$Rh(acac)(Me_2C=C=CMe_2)_2$	perpendicular	1.373 1.377	1.325 1.321	147.2 148.9	2.177 2.176	2.027 2.033
$Rh_2(acac)_2(CO)_2(H_2C=C=CH_2)$	perpendicular	1.37 1.41		144.5	2.12 2.14	2.05 2.06
$RhI(PPh_3)_2(H_2C=C=CH_2)$	perpendicular	1.35	1.34	158	2.17	2.04
$[PtCl_2(Me_2C=C=CMe_2)]_2$	perpendicular	1.37	1.36	151	2.25	2.07
$Pt(PPh_3)_2(H_2C=C=CH_2)$	in-plane	1.48	1.31	142	2.13	2.03
$Pt(PPh_3)_2(H_2C=C=CHMe)$	in-plane	1.44	1.32	146	2.12	2.05
$Pt(PPh_3)_2(H_2C=C=CMe_2)$	in-plane	1.430	1.316	140.8	2.107	2.049
$Pd(PPh_3)_2(H_2C=C=CH_2)$	in-plane	1.44	1.32	148	2.12	2.07

* "Perpendicular" refers to the molecular structure containing a coordinated double bond perpendicular to the principal molecular plane, and "in-plane" to that containing a double bond lying in the molecular plane.

References

[1] B.W. Davies and N.C. Payne, *Inorg. Chem.* **13**, 1848 (1974).

[2] D.M. Hoffman, R. Hoffmann and C.R. Fisel, *J. Am. Chem. Soc.* **104**, 3858 (1982).

[3] D.M. Hoffman and R. Hoffmann, *J. Chem. Soc., Dalton Trans.*, 1471 (1982).

[4] G.R. Davis, W. Hewertson, R.H.B. Mais, P.G. Owston and C.G. Patel, *J. Chem. Soc. (A)*, 1873 (1970).

[5] A.C. Blizzard and D.P. Santry, *J. Am. Chem. Soc.* **90**, 5749 (1968).

[6] W.G. Sly, *J. Am. Chem. Soc.* **81**, 18 (1959).

[7] A. Furlani, S. Licoccia, M.V. Russo, A.C. Villa and C. Guastini, *J. Chem. Soc., Dalton Trans.*, 2449 (1982).

[8] W.F. Smith, J. Yule, N.J. Taylor, H.N. Paik and A.J. Carty, *Inorg. Chem.* **16**, 1593 (1977).

[9] S. Otsuka and A. Nakamura, *Adv. Organomet. Chem.* **14**, 245 (1976).

[10] J.L. Templeton, *Adv. Organomet. Chem.* **29**, 1 (1989).

[11] L. Ricard, R. Weiss, W.E. Newton, G.J.-J. Chen and J.W. McDonald, *J. Am. Chem. Soc.* **100**, 1318 (1978).

[12] J.R. Morrow, T.L. Tonker and J.L. Templeton, *J. Am. Chem. Soc.* **107**, 6956 (1985).

[13] A. Mayr, K.S. Lee, M.A. Kjelsberg and D. Van Engen, *J. Am. Chem. Soc.* **108**, 6079 (1986).

[14] J.A.K. Howard, R.F.D. Stansfield and P. Woodward, *J. Chem. Soc., Dalton Trans.*, 246 (1976).

[15] J.L. Davidson, M. Green, D.W.A. Sharp, F.G.A. Stone and A.J. Welch, *J. Chem. Soc., Chem. Commun.*, 706 (1974).

[16] R.S. Herrick, S.J. Nieter-Burgmayer and J.L. Templeton, *Inorg. Chem.* **22**, 3275 (1983).

[17] S.L. Buchwald and R.B. Nielsen, *Chem. Rev.* **88**, 1047 (1988).

[18] M.A. Bennett and H.P. Schwemlein, *Angew. Chem. Int. Ed. Engl.* **28**, 1296 (1989).

[19] M.A. Bennett, T.W. Hambley, N.K. Roberts and G.B. Robertson, *Organometallics* 4, 1992 (1985).

[20] S.L. Buchwald, B.T. Watson and J.C. Huffman, *J. Am. Chem. Soc.* **108**, 7411 (1986).

[21] S.L. Buchwald, R.T. Lum and J.C. Dewan, *J. Am. Chem. Soc.* **108**, 7441 (1986).

[22] S.L. Buchwald, R.T. Lum, R.A. Fisher and W.M. Davis, *J. Am. Chem. Soc.* **111**, 9113 (1989).

[23] M.I. Bruce, J.M. Guss, R. Mason, B.W. Skelton and A.H. White, *J. Organomet. Chem.* **251**, 261 (1983).

[24] S.A.R. Knox, B.R. Lloyd, D.A.V. Morton, S.M. Nicholls, A.G. Orpen, J.M. Vinas, M. Weber and G.K. Williams, *J. Organomet. Chem.* **394**, 385 (1990).

[25] A.J. Deeming, S.E. Kabir, N.I. Powell, P.A. Bates and M.B. Hursthouse, *J. Chem. Soc., Dalton Trans.*, 1529 (1987).

[26] M.A. Bennett, K.D. Griffiths, T. Okano, V. Parthasarathi and G.B. Robertson, *J. Am. Chem. Soc.* **112**, 7047 (1990).

[27] See ref. 4a of W.R. Cullen, S.T. Chacon, M.I. Bruce, F.W.B. Einstein and R.H. Jones, *Organometallics* 7, 2273 (1988).

[28] F.R. Hartley in G. Wilkinson, F.G.A. Stone and E.W. Abel (eds.), *Comprehensive Organometallic Chemistry*, Pergamon Press, Oxford, 1982, Vol. **6**, chap. 39, p. 471.

Note Recent developments in the chemistry of phospha-alkynes, $RC{\equiv}P$, especially as building blocks in synthesizing novel organic, inorganic and organometallic compounds, including clusters such as $(Bu^tCP)_4$, $(Bu^tCP)_5$ and $P_6C_4Bu^t_4H_2$ are reviewed in J.F. Nixon, *Chem. Rev.* **88**, 1327 (1988); *Endeavour* **15**, 49 (1991); M. Regitz and P. Binger, *Angew. Chem. Int. Ed. Engl.* **27**, 1484 (1988); M. Regitz in E. Brock (ed.), *Heteroatom Chemistry*, VCH Publishers, Weinheim, 1990; M. Regitz and P. Binger in M. Regitz and O.J. Scherer (eds.), *Multiple Bonds and Low Coordination in Phosphorus Chemistry*, Thieme, Stuttgart, 1990.

The analogy between benzyne, C_6H_4, and ferrocyne, $Fe(\eta\text{-}C_5H_5)(\eta\text{-}C_5H_3)$, has been established through the synthesis and structural characterization of the ferrocyne complex $Os_3(CO)_9[\mu_3\text{-}(C_5H_3)Fe(C_5H_5)][\mu_3\text{-}P(C_5H_4)Fe(C_5H_5)]$ and the ferrodicyne complex $Os_3(H)_2(CO)_8(PPr^i_2C_5H_2)Fe(C_5H_2PPr^i_2)Os_3(H)_2(CO)_8$. See W.R. Cullen, S.J. Rettig and T.-c. Zheng, *Organometallics* **11**, 928 (1992).

6.6　[1,2-Bis(dimethylphosphino)ethane](neopentylidyne)(neopentylidene)(neopentyl)tungsten(VI)

W(≡CCMe₃)(=CHCMe₃)(CH₂CMe₃)(dmpe)

Crystal Data

Monoclinic, space group $P2_1/n$ (No. 14)

$a = 9.784(3)$, $b = 29.200(8)$, $c = 9.859(2)$Å, $\beta = 109.54(2)°$, $Z = 4$

Atom	x	y	z	Atom	x	y	z
W	-.00050	.11456	.10206	C(10)	-.332	.0841	-.287
P(1)	.2306	.10520	.3231	C(11)	-.0527	.0529	.2139
P(2)	.1827	.16057	.0453	C(12)	-.1999	.0321	.1783
C(1)	-.0711	.1645	.1593	C(13)	-.1901	-.0044	.2885
C(2)	-.1401	.2040	.2080	C(14)	-.2437	.0053	.0397
C(3)	-.2887	.1953	.1898	C(15)	-.3182	.0646	.1746
C(4)	-.1277	.2466	.1302	C(21)	.3247	.0507	.3577
C(5)	-.0574	.2165	.3567	C(22)	.2147	.1180	.4954
C(6)	-.1063	.1045	-.1008	C(23)	.3766	.1438	.3110
C(7)	-.2190	.1186	-.2376	C(24)	.3238	.1808	.2077
C(8)	-.2978	.1588	-.2294	C(25)	.1261	.2134	-.0534
C(9)	-.166	.1250	-.3490	C(26)	.2881	.1342	-.0548

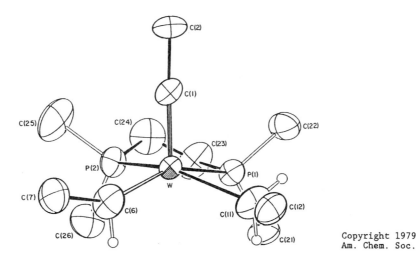

Copyright 1979
Am. Chem. Soc.

Fig. 6.6-1　Geometry of the W(≡CC)(=CHC)(CH₂C)(C₂PCCPC₂) core of the W(≡CCMe₃)(=CHCMe₃)(CH₂CMe₃)(dmpe) molecule.　Hydrogen atoms are placed in their estimated positions.　(After ref. 1).

Crystal Structure

The molecule in the crystal consists of a five-coordinate tungsten(VI) atom surrounded by a bidentate 1,2-bis(dimethylphosphino)ethane ligand (dmpe), a neopentylidyne ligand, a neopentylidene ligand, and a neopentyl ligand.　The coordination geometry about the W atom is distorted square-pyramidal.　The neopentylidyne ligand occupies the apical site, the apical-basal angles being C(1)-W-C(6) = 108.7°, C(1)-W-C(11) = 108.9°, C(1)-W-P(1) = 98.2°, and C(1)-W-P(2) = 90.8°.　The W and C(1) atoms lie 0.488 and 2.244Å, respectively, above the least-squares plane through the basally coordinated atoms [P(1), P(2), C(6), C(11)].

The structural interest lies in the comparison of the geometry of isoskeletal alkyl, alkylidene, and alkylidyne ligands coordinating to a common metal center. The W-C(α) distances for the neopentyl, neopentylidene, and neopentylidyne ligands are, respectively, W-C(11) = 2.258Å, W-C(6) = 1.942Å, and W-C(1) = 1.785Å; the successive decrements are 0.316 and 0.157Å for each unit increase in bond order. The W-C(α)-C(β) angles for the neopentyl, neopentylidene, and neopentylidyne ligands are, respectively, W-C(11)-C(12) = 124.5°, W-C(6)-C(7) = 150.4°, and W-C(1)-C(2) = 175.3°.

The dmpe ligand is linked to tungsten via the bonds W-P(1) = 2.577Å and W-P(2) = 2.450Å. The longer of these is for the phosphorus atom that is *trans* to C(6) of the alkylidene ligand, and the shorter is for the phosphorus atom *trans* to C(11) of the alkyl ligand.

The α-hydrogen of the neopentylidene ligand points towards the sixth coordination site. The unusually large W-C_α-C_β bond angle and low $J(C_\alpha$-$H_\alpha)$ value of 90 Hz are consistent with the formation of an α-agostic W $\leftarrow$ H-C bond.

Remarks

1. The first crystalline carbene and Fischer carbene complexes

Although carbenes :CR_2 have long been recognized as important reaction intermediates, the synthesis and structural characterization of the first crystalline carbene, namely 1,3-di-1-adamantylimidazol-2-ylidene, was achieved only recently.[2] The formula, structure, and selected molecular dimensions of this sterically and electronically stabilized carbene are shown in Fig. 6.6-2.

In 1964, Fischer and his co-workers reported the isolation of stable carbene complexes of the general formula L_nM=CR_2 (M = Cr, Mo, W) for the first time.[3] The first X-ray crystal structure determination was carried out on pentacarbonyl[methoxy(phenyl)carbene]chromium(0) (Fig. 6.6-3).[4] The O-$C_{carbene}$ distance of 1.33Å lies between the values for a single (1.43Å in diethyl ether) and a double (1.23Å in acetone) bond. According to Cotton,[6] a pure Cr-C σ-bond is expected to be 2.21Å long. Consequently, although the bond order of the Cr-$C_{carbene}$ bond (2.04Å) is distinctly less than that of the Cr-CO bonds (average 1.87Å) in the same complex, it is still greater than unity. The phenyl group is not significantly engaged in pπ-pπ bonding with the carbene carbon as it is twisted considerably from the plane containing the Cr, C, and O atoms. The double bond character of the O-$C_{carbene}$ bond is so substantial that *cis* and *trans* isomers can easily exist relative to this bond.

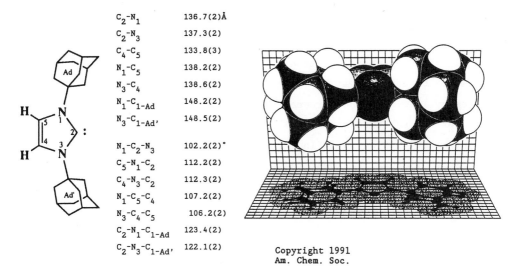

C_2-N_1	136.7(2)Å
C_2-N_3	137.3(2)
C_4-C_5	133.8(3)
N_1-C_5	138.2(2)
N_3-C_4	138.6(2)
N_1-C_{1-Ad}	148.2(2)
$N_3-C_{1-Ad'}$	148.5(2)
$N_1-C_2-N_3$	102.2(2)°
$C_5-N_1-C_2$	112.2(2)
$C_4-N_3-C_2$	112.3(2)
$N_1-C_5-C_4$	107.2(2)
$N_3-C_4-C_5$	106.2(2)
$C_2-N_1-C_{1-Ad}$	123.4(2)
$C_2-N_3-C_{1-Ad'}$	122.1(2)

Fig. 6.6-2 Structure and molecular dimensions of 1,3-di-1-adamantylimidazol-2-ylidene. (After ref. 2).

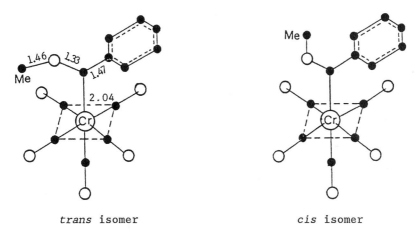

trans isomer   *cis* isomer

Fig. 6.6-3 Structure of $(CO)_5CrC(OMe)Ph$. (After ref. 5).

In the $(CO)_5CrC(OMe)Ph$ case, only molecules of the *trans* type are found in the crystal, but low-temperature ^{1}H NMR spectroscopy reveals the coexistence of the *cis* isomers in solution .

2. Bonding in Fischer metal carbene complexes[7,8]

Free singlet carbene :CR_2 is electron-deficient, with a bent configuration having a sp[2]-type electron lone pair within the CR_2 plane and an empty p orbital perpendicular to this plane. In a Fischer carbene complex the carbene unit donates electrons from its lone pair to an empty metal orbital, thus forming a metal-carbon σ-bond. The bonding interaction between both atoms is usually stabilized by back bonding, i.e., transfer of electron density from a filled d-like metal orbital into the empty carbene p orbtial. The degree of electron transfer via bonding and back bonding is determined by the separation of the energy levels of the orbitals involved. The energy levels of the carbene (CR_2) orbitals depend on the electronic nature of the substituents R, while the metal orbitals are influenced by the kind of metal, its coordination geometry and the ancillary carbonyl ligands.

In $(CO)_5CrC(XR)R'$ complexes (Table 6.6-1) replacement of an alkoxy group by an amino group, which is a much better π donor, results in a distinct

Table 6.6-1   Bond lengths (Å) in carbene complexes $(CO)_5CrC(XR)R'$.

XR	R'	Cr-C(carbene)	C(carbene)-X
OMe	Ph	2.04	1.33
OEt	C≡CPh	2.00	1.32
OEt	SiPh$_3$	2.00	1.33
SPh	Me	2.020	1.690
SCH=C(OMe)Ph	Ph	2.029	1.671
SCH$_2$SCH$_2$SCH=C(OMe)Ph	Ph	2.027	1.673
NHMe	Me	2.09	1.33
NEt$_2$	Me	2.16	1.31
NEt$_2$	H	2.084	1.300
NEt$_2$	Cl	2.110	1.299
NEt$_2$	SePh	2.172	1.28
NEt$_2$	OEt	2.133	1.328

lengthening of the Cr-C(carbene) distance from about 2.00-2.05Å to about 2.10-2.15Å, which is still stronger than a Cr-C single bond at 2.17-2.22Å.

The reaction of $(CO)_5W=C(OAc)Ph$ with the imine MeN=CHPh gives the *N*-aminal carbene complex that results from an unprecedented formal insertion of the imine into the heteroatom-carbene carbon bond [Fig. 6.6-4(a)].[9]   In general the carbene moiety in Fischer carbene complexes has a pronounced tendency to behave as a carbon nucleophile. Fischer carbenes are of considerable interest as reagents for cyclopropanation and other applications to organic synthesis.[10,11]

3. Schrock-type carbene (alkylidene) complexes[12]

In the alkylidene[*] complexes first described by Schrock, the carbene carbon behaves as a nucleophile.[8]   To date high-valent alkylidene compounds have been synthesized for the early transition metals Zr, Nb, Ta, Mo, W and

---

[*] In the older literature "alkylidene" is synonymous with "Schrock carbene".   IUPAC recommends that the term carbene be reserved for the free $:CR_2$ fragment, and accordingly carbene and carbyne (M≡CR) complexes should be termed alkylidene and alkylidyne complexes, respectively.

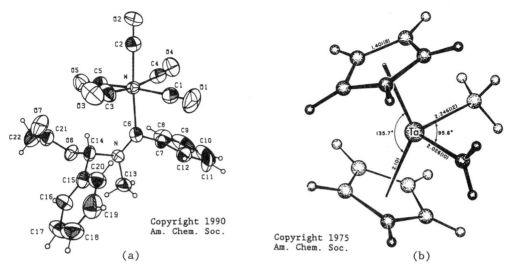

Fig. 6.6-4 Mollecular structures of (a) $(CO)_5W=CPh\{N(Me)CHPh(OAc)\}$ and (b) $(\eta^5\text{-}C_5H_5)_2TaMe(CH_2)$. (After refs. 9 and 16).

Re. The essential differences between Fischer and Schrock carbene complexes are summarized in Table 6.6-2.[13]

Table 6.6-2 Fischer and Shrock carbene complexes $L_nM=CR_2$.

Property	Fischer-type	Schrock-type
Reactivity of carbene carbon	electrophilic	nucleophilic
Typical metal	late transition metal in low oxidation state ($> d^2$)	early transition metal in high oxidation state
Typical R groups	good π donor (e.g. OR, NEt$_2$)	H, alkyl
Typical L Ligands	good π acceptor (e.g. CO, PPh$_3$)	η^5-C$_5$H$_5$, Cl, alkyl
Total number of electrons in valence shell of metal	18	≤ 18

In a Schrock-type carbene the $M=CR_2$ bond is best described as a metal-stabilized carbanion CR_2^{2-} acting as a σ and π donor to the metal. An alternative model describes the metal-carbon interaction as ethylene-like covalent bonding between a triplet metal center and a triplet carbene fragment. Both models adequately explain the nucleophilicity of the carbon center in Schrock-type carbene complexes.[14]

$(\eta^5$-C$_5$H$_5)_2$TaMe(CH$_2$) was the first transition metal methylene complex to be isolated.[15,16] X-Ray diffraction has shown that the molecule conforms to idealized C_s symmetry with its CH$_3$-Ta-CH$_2$ fragment lying in the mirror plane. The two eclipsed cyclopentadienyl rings are 2.10Å from Ta where their centroids subtend an angle of 135.7°; the C-Ta-C angle is 95.6°. Geometrical details of the CH_2 ligand and its bonding to Ta are as follows: (a) within experimental error the CH_2 plane is perpendicular to the C-Ta-C plane (88°) and the methylene carbon atom lies in the Ta-CH_2 plane; (b) the Ta-C bond distance is 2.026Å; and (c) the H-C-H angle is 107° [Fig. 6.6-4(b)].

Low-temperature neutron diffraction studies of $[Ta(CHCMe_3)(PMe_3)Cl_3]_2$ and $Ta(\eta^5-C_5Me_5)(CHCMe_3)(\eta^2-C_2H_4)(PMe_3)$ have established that the neopentylidene ligands are highly distorted because the metal draws electron density from the C_α-H_α bond.[17] The chloride bridges in the centrosymmetric dimer are markedly asymmetric, and the octahedral coordination geometry about each Ta atom is highly distorted with the neopentylidene ligand *trans* to the weakly bound bridging chloride [Fig. 6.6-5(a)]. The Ta=C_α bond is short and H_α is only 2.119(4)Å from Ta. In the first reported alkylidene-olefin complex shown in Fig. 6.6-5(b), the Ta...H(1) distance is 2.042(5)Å and the C=C bond distance in the ethylene ligand is much longer than that [1.375(4)Å, see Section 6.4] in Zeise's salt. The four ethylenic hydrogen atoms are bent away from the metal such that the angle between the normals to the two CH_2 planes is 68.5°.

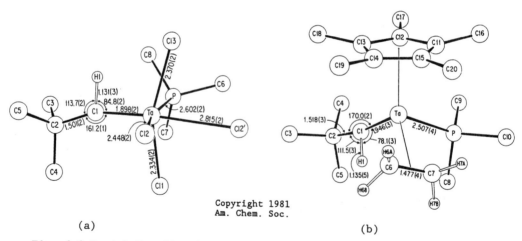

Copyright 1981
Am. Chem. Soc.

(a)   (b)

Fig. 6.6-5 (a) Coordination geometry around a metal center in dimeric $Ta(CHCMe_3)(PMe_3)Cl_3]_2$. (b) Molecular structure of $Ta(\eta^5-C_5H_5)(CHCMe_3)(\eta^2-C_2H_4)(PMe_3)$. (After ref. 17).

4. Structure of transition-metal dihalocarbene complexes[18]

X-Ray crystal structure determinations have provided the structural parameters listed in Table 6.6-3. Representative structures of a dichlorocarbene and a difluorocarbene complex are shown in Fig. 6.6-6.

Table 6.6-3 Structural parameters for monohalocarbene
and dihalocarbene complexes.

Compound	M=C/Å	C-X/Å	X-C-X/deg.	M-CO/Å
Mono- and dichlorocarbene complexes				
$[CpFe(=CCl_2)(CO)_2]BCl_4$	1.808(12)	1.694(12),1.731(11)	108.5(7)	1.796(12),1.805(10)
$Fe(TPP)(=CCl_2)(OH_2)$	1.83(3)	1.76(3)	-	-
$IrCl_3(=CCl_2)(PPh_3)_2$	1.872(7)	1.721(5)	107.5(4)	1.896(cis)
$Cr[=CCl(NEt_2)](CO)_5$	2.110(5)	1.780(6)	110.8(4)	1.858(trans)
$RuCl_2[=CCl(C_4H_4N)(CO)(PPh_3)_2$	1.949(12)	1.826(13)	104.4	1.857(16)
Mono- and difluorocarbene complexes				
$[CpFe(=CF_2)(CO)(PPh_3)]BF_4$	1.724(9)	1.334(10)	98.3(7)	1.798(10)
$Ru(=CF_2)(CO)_2(PPh_3)_2$	1.83(1)	1.37(1),1.36(1)	88.7(9)	1.87(1),1.91(2)
$Os(=CF_2)(CO)_2(PPh_3)_2$	1.915(15)	1.331(18),1.407(17)	101.4(12)	1.884(13), 1.938(15)
$OsCl(NO)(=CF_2)(PPh_3)_2$	1.967(6)	1.278(11),1.285(11)	105.0(7)	-
$Ir(CF_3)(=CF_2)(CO)(PPh_3)_2$	1.874(7)	1.30(2), 1.30(1)	89.5(8)	1.874(7)
$CpMn(=CFPh)(CO)_2$	1.830(5)	1.392(6)	106.1(4)	1.784(6),1.795(6)
$RuCl_2[CF(OCH_2CMe_3)](CO)(PPh_3)_2$	1.914(5)	1.307(6)	107.4(5)	1.910(7)

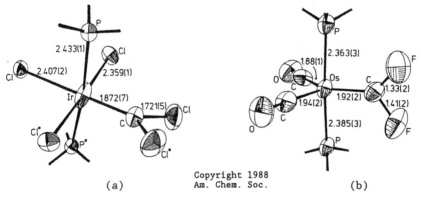

Copyright 1988
Am. Chem. Soc.

(a)   (b)

Fig. 6.6-6 Molecular structure of (a) $IrCl_3(=CCl_2)(PPh_3)_2$ and (b)
$Os(=CF_2)(CO)_2(PPh_3)_2$. (After ref. 18).

For each structure, the sum of angles around the carbene carbon atom is
very close to 360°, consistent with an sp^2-hybridized carbon atom. All
structures, except perhaps the $Cr[=CCl(NEt_2)](CO)_5$, exhibit short M=C bond
lengths as expected for some degree of metal-carbon multiple bonding. Each
compound containing both CX_2 and CO ligands has a metal-carbene distance which
is comparable to the metal-carbonyl distance. In some complexes the carbene
ligand exerts a pronounced structural *trans* influence, rendering the M-Cl bond
trans to it *ca.* 0.05Å longer than the *cis* M-Cl bond(s). The
$IrCl_3(=CCl_2)(PPh_3)_2$ molecule [Fig. 6.6-6(a)] lies on a crystallographically-
imposed twofold axis, the planar CCl_2 ligand being tilted at an angle of 24.4°

relative to the plane containing the three chloride ligands. The geometry of
$Os(=CF_2)(CO)_2(PPh_3)_2$ [Fig. 6.6-6(b)] illustrates the vertical orientation
(with respect to the equatorial plane) adopted by the carbene ligand in all
the d^8 structures. This is in contrast to the in-plane conformation observed
for olefin complexes of d^8 metal centers (Section 6.4).

5. Carbyne complexes[19]

Carbyne complexes were first made in 1973[20] by the reaction of
methoxycarbene complexes with boron trihalides:

$$(CO)_5MC{\overset{OMe}{\underset{R}{\diagdown}}} \quad + BX_3 \quad \xrightarrow[\text{low temp.}]{\text{pentane}} \quad (CO)_4(X)M{\equiv}CR + CO + BX_2(OMe)$$

$$M = Cr, Mo, W; \quad R = Me, Et, Ph; \quad X = Cl, Br, I.$$

The first successful X-ray study involved *trans*-(iodo)tetracarbonyl-
(phenylcarbyne)tungsten(0).[21] The analysis yielded an extremely short W≡C
distance of 1.90Å and a W≡C-Ph angle of 162°. In the structures of the
related complexes *trans*-(iodo)tetracarbonyl(methylcarbyne)chromium(0) and *mer*-
(bromo)tricarbonyl-(methylcarbyne)trimethylphosphinechromium(0) are found not
only the very short Cr≡C distance of 1.69Å, but also a linear arrangement of
the Cr≡C-C skeleton.[5]

In 1978 Schrock reported the first neutral d^0 alkylidyne complex, $(\eta^5-$
$C_5Me_5)(PMe_3)_2ClTa(\equiv CPh)$.[22] The coordination geometry around the Ta^V atom is
of the "four-legged piano stool" type; the Ta-C1 and C1-C2 bond lengths are
1.849(8) and 1.467(10)Å, respectively, and the Ta-C1-C2 angle is 171.8(6)°
[Fig. 6.6-7(a)]. The pseudo octahedral geometry of the tungsten alkylidyne-

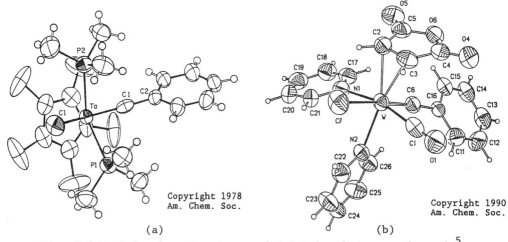

(a)   (b)

Fig. 6.6-7 Molecular structures of (a) Schrock-type carbyne $(\eta^5-$
$C_5Me_5)(PMe_3)_2ClTa(\equiv CPh)$ and (b) Fischer-type carbyne $[(\eta^2$-maleic
anhydride)Cl(CO)(py)_2W(\equiv CPh)]$. (After refs. 22 and 23).

alkene complex [(η^2-maleic anhydride)Cl(CO)(py)$_2$W($\equiv$CPh)] is typical of a low-valent monohalo group VI transition metal Fischer carbyne [Fig. 6.6-7(b)].[23] The W-C6 and C6-C16 bond distances are 1.801(6) and 1.448(8)Å, respectively, and the W-C6-C18 angle is 173.0(4)°. The dihedral angle between the mean plane of the η^2-maleic anhydride ligand and the W-C2-C3 plane is 109.5°.

6. Framework rearrangement of alkylidyne(carbaborane)tungsten complexes

Fig. 6.6-8 shows three related carbametallaborane complexes in which a tungsten atom functions as a vertex of an icosahedral cluster.[24] In anion I (PPh$_4$$^+$ salt) the metal atom is also bound to an alkylidyne ligand; the relevant W$\equiv$C and C-C bond distances are 1.942(7) and 1.166(9)Å, respectively, and the W$\equiv$C-C angle is 178.4(4)°.[25] Protonation of I results in migration and hydroboration of the alkylidyne group, as well as a framework rearrangment of the icosahedral cage, to form the anion II. Complexes of type II have non-adjacent CMe groups, and the one with Y = Cl has been characterized by X-ray diffraction. The reaction of II (PPh$_4$$^+$ salt) with MeC$\equiv$CPh followed by AgBF$_4$ gives the four-electron alkyne donor complex III, the structure of which was also confirmed by X-ray analysis.[24]

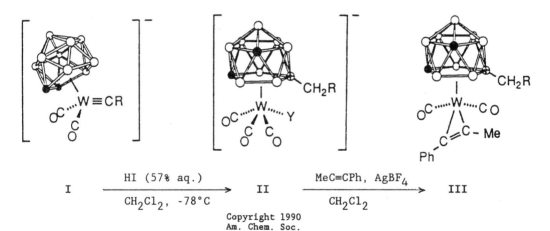

Fig. 6.6-8 Transformation of some related (carbaborane)tungsten complexes (R = C$_6$H$_4$Me-4). The solid, open, and crossed circles indicate CMe, BH, and B, respectively. (After ref. 24).

References

[1] M.R. Churchill and W.J. Youngs, *Inorg. Chem.* **18**, 2454 (1979).

[2] A.J. Arduengo III, R.L. Harlow and M. Kline, *J. Am. Chem. Soc.* **113**, 361 (1991).

[3] E.O. Fischer and A. Maasböl, *Angew. Chem. Int. Ed. Engl.* **3**, 580 (1964).

[4] O.S. Mills and A.D. Redhouse, *J. Chem. Soc. (A)*, 642 (1968).

[5]   E.O. Fischer, *Adv. Organomet. Chem.* **14**, 1 (1976).

[6]   F.A. Cotton and D.C. Richardson, *Inorg. Chem.* **5**, 1851 (1966).

[7]   U. Schubert, *Coord. Chem. Rev.* **55**, 261 (1984).

[8]   W. Kirmse, *Carbene Chemistry*, 2nd ed., Academic Press, New York, 1971.

[9]   C.K. Murray, B.P. Warner, V. Dragisich and W.D. Wulff, *Organometallics* **9**, 3142 (1990).

[10] D. Seyferth (ed.), *Transition Metal Carbene Complexes*, VCH Publishers, Weinheim, 1983.

[11] C.K. Murray, D.C. Yang and W.D. Wulff, *J. Am. Chem. Soc.* **112**, 5660 (1990).

[12] R.R. Schrock, *J. Am. Chem. Soc.* **96**, 6796 (1974); *Acc. Chem. Res.* **12**, 98 (1979); *J. Organomet. Chem.* **300**, 249 (1986).

[13] R.H. Crabtree, *The Organometallic Chemistry of the Transition Metals*, Wiley-Interscience, New York, 1988, chap. 11.

[14] T.E. Taylor and M.B. Hall, *J. Am. Chem. Soc.* **106**, 1576 (1984); E.A. Carter and W.A. Goddard III, *J. Am. Chem. Soc.* **108**, 2180, 4746 (1986).

[15] R.R. Schrock, *J. Am. Chem. Soc.* **97**, 6577 (1975).

[16] L.J. Guggenberger and R.R. Schrock, *J. Am. Chem. Soc.* **97**, 6578 (1975).

[17] A.J. Schultz, R.K. Brown, J.M. Williams and R.R. Schrock, *J. Am. Chem. Soc.* **103**, 169 (1981).

[18] P.J. Brothers and W.R. Roper, *Chem. Rev.* **88**, 1293 (1988).

[19] H. Fischer, P. Hofmann, F.R. Kreissl, R.R. Schrock, U. Schubert and K. Weiss (eds.), *Carbyne Complexes*, VCH Publishers, Weinheim, 1988.

[20] E.O. Fischer, G. Kreis, C.G. Kreiter, J. Müller, G. Huttner and H. Lorenz, *Angew. Chem. Int. Ed. Engl.* **12**, 564 (1973).

[21] G. Huttner, H. Lorenz and W. Gartzke, *Angew. Chem. Int. Ed. Engl.* **13**, 609 (1974).

[22] S.J. McLain, C.D. Wood, L.W. Messerle, R.R. Schrock, F.J. Hollander, W.J. Youngs and M.R. Churchill, *J. Am. Chem. Soc.* **100**, 5962 (1978).

[23] A. Mayr, A.M. Dorries, A.L. Rheingold and S.J. Geib, *Organometallics* **9**, 964 (1990).

[24] S.A. Brew, J.C. Jeffery, M.U. Pilotti and F.G.A. Stone, *J. Am. Chem. Soc.* **112**, 6148 (1990); F.G.A. Stone, *Adv. Organomet. Chem.* **31**, 53 (1990).

[25] O. Johnson, J.A.K. Howard, M. Kapon and G.M. Reisner, *J. Chem. Soc., Dalton Trans.*, 2903 (1988).

Note Silylenes, $:SiR_2$, are described in U. Boudjouk, R. Samaraweera, J. Sooriyakumaran, J. Chrusciel and K.R. Anderson, *Angew. Chem. Int. Ed. Engl.* **28**, 1355 (1988) [R = *t*-Bu] and D.H. Pae, M. Xiao, M.Y. Chiang and P.P. Gaspar, *J. Am. Chem. Soc.* **113**, 1281 (1991) [R = adamantyl].

6.7 Dimanganese Decacarbonyl
$Mn_2(CO)_{10}$

Crystal Data

Monoclinic, space group $I2/a$ (No. 15)

$a = 14.088(3)$, $b = 6.850(2)$, $c = 14.242(3)$Å, $\beta = 105.08(1)°$, (74K)

$Z = 4$

Atom	x	y	z	Atom	x	y	z
Mn	.34564	.23692	.06916	O(3)	.2375	.0456	.1996
C(1)	.4608	.2457	.1630	C(4)	.3736	-.0043	.0247
O(1)	.5322	.2518	.2239	O(4)	.3959	-.1493	-.0029
C(2)	.3931	.3461	-.0290	C(5)	.3091	.4844	.1016
O(2)	.4226	.4073	-.0899	O(5)	.2884	.6371	.1210
C(3)	.2774	.1189	.1490				

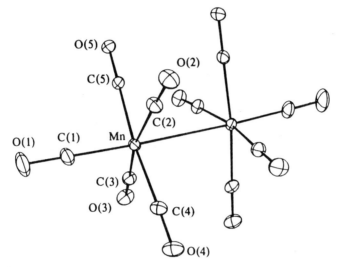

Fig. 6.7-1 Molecular structure of $Mn_2(CO)_{10}$ at 74K. (After ref. 1).

Crystal Structure

Dimanganese decacarbonyl has been the subject of many experimental and theoretical investigations. Its crystal structure was determined by X-ray diffraction in 1963,[2] 1981,[3] and 1982,[1] the latter being conducted at 74K. An electron diffraction study in the vapour phase was reported in 1969, and a number of quantum chemical calculations were performed.[4]

The molecule has approximate D_{4d} symmetry in the solid state (Fig. 6.7-1).[1] The Mn-Mn bond length is 2.895(1)Å, significantly shorter than 2.9038(6)Å from the room-temperature study.[3]. The axial M-C distance of 1.820(3)Å is 0.039Å shorter than the average equatorial M-C distance of 1.859(3)Å. This result is in accordance with the accepted model for M-CO bonding and arises from the competition for $d\pi$ electron density between mutually *trans* pairs of equatorial carbonyl ligands. The axial C-O distance

of 1.150Å is 0.010Å longer than the average distance of 1.140Å for equatorial
C-O bonds, which range from 1.136 to 1.144Å. The Mn-Mn-CO(axial) angle is
175.3°, and the average Mn-Mn-CO(equatorial) and OC(axial)-Mn-CO(equatorial)
angles are 86.1° and 93.9°, respectively. The torsion angle of the two
$Mn(CO)_5$ fragments about the molecular axis is 50.4°; at room temperature this
angle becomes 47.4°, much closer to 45° for the ideally staggered
conformation.

Remarks

1. The metal-metal bonding in $Mn_2(CO)_{10}$

$Mn_2(CO)_{10}$ is one of the simplest dimeric compounds containing a metal-
metal single bond unsupported by bridging ligands, and its diamagnetic
behaviour is accounted for in compliance with the 18-electron rule.

The X-X deformation density (total density minus free spherical atoms)
is shown in Fig. 6.7-2. The deformation density at the mid-point of the Mn-Mn

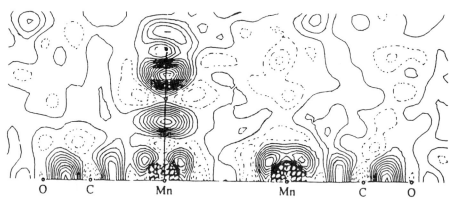

Fig. 6.7-2 Average deformation density in $Mn_2(CO)_{10}$. Resolution:
$Sin\theta/\lambda < 0.76Å^{-1}$. Contour interval: $0.05eÅ^{-3}$. Negative contours are
dashed. (After ref. 1).

bond is positive ($0.05eÅ^{-3}$) and not significant (e.s.d. is $0.07eÅ^{-3}$). The
slightly positive deformation density extends very far into the plane
perpendicular to the Mn-Mn bond at its mid-point. By integrating over all
points closer than 0.2Å to the plane, and within 1.5Å from the Mn-Mn axis, a
total excess of 0.037e is obtained, with an e.s.d. of 0.022e. Apart from the
small peak on the Mn-Mn bond, other maxima of $0.05eÅ^{-3}$ are observed away from
the bond, approximately between Mn' and the equatorial C atoms bonded to the
other Mn atom.

The configuration around the Mn atom is essentially octahedral, with
about 75% of the 3d electrons in the diagonal orbitals (d_{xy}, d_{xz}, d_{yz}) and the
remaining 25% in d_{z^2} and $d_{x^2-y^2}$. Atomic charges determined both by direct

integration and by least-squares refinement show a clear difference between axial and equatorial carbonyls, which is confirmed by comparing their electron density with that of free carbon monoxide: the differences are characteristic of a σ-bonding, π back-bonding mechanism, and are larger for the axial carbonyls than for the equatorial carbonyls (Fig. 6.7-3). This confirms the stronger bonding of the axial carbonyls, as reflected by the longer C-O bond and the shorter Mn-C bond. The directly integrated charge at Mn is slightly negative.

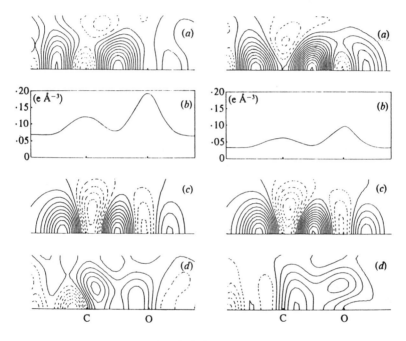

Fig. 6.7-3 Axial carbonyl (left) and average equatorial carbonyl (right). (a) Observed deformation density (cylindrical average); (b) e.s.d. of the deformation density along C-O axis; (c) calculated for carbon monoxide; (d) difference (a)-(c). (After ref. 1).

The Hartree-Fock-Slater (HFS-Xα) method[4] has shown that the interaction between the metal $3d_{xz}$ or $3d_{yz}$ orbitals and the carbonyl π^* orbitals is concentrated in the $11e_1$ and $11e_3$ molecular orbitals which are essentially the plus and minus combinations of the 11e orbital of Mn(CO)$_5$, so that the result is practically nonbonding as far as the interaction between the two Mn(CO)$_5$ fragments is concerned. The only direct cross-interaction is found in the HOMO, which accounts for the metal-metal bond: this orbital has about 60% metallic character, essentially $3d_{z^2}$ (27%) and $4p_z$ (24%), corresponding to a σ bond with the second metal. The $4p_z$ contribution enhances the density in the bond. However, there is also a 35% contribution of the π^* orbitals of the equatorial carbonyls which slightly overlap with the orbitals of both metals and tend to extend the density away from the bond axis.

2. The first polynuclear metal carbonyl

   Nonacarbonyldiiron, $Fe_2(CO)_9$, was the first polynuclear metal carbonyl [and only the third, after $Ni(CO)_4$ and $Fe(CO)_5$, of all metal carbonyls] to be discovered.[5] Its preparation constituted the first instance of photochemical generation of a polynuclear species from a simple mononuclear precursor. It was the first dinuclear metal carbonyl to be structurally characterized by X-ray crystallography, which also revealed the existence of the bridging carbonyl group.[6] An accurate crystal structure determination has been performed.[7] The space group is $P6_3/m$, with a = 6.436(1), c = 16.123(2)Å, and Z = 1. The molecular symmetry is C_{3h} ($\bar{6}$), but there is little deviation from the higher idealized symmetry D_{3h} ($\bar{6}m2$). The important molecular dimensions are: Fe-Fe 2.523(1)Å, Fe-C(terminal) 1.838(3)Å, Fe-C(bridge) 2.016(3)Å, C-O(term) 1.156(4)Å, C-O(brid) 1.176(5)Å, C(term)-Fe-C(term) 96.1(1)°, Fe-C(brid)-Fe 77.6(1)°, Fe-C-O(term) 177.1(3)° (Fig. 6.7-4).

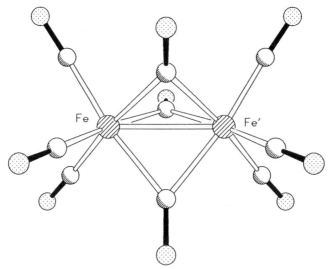

Fe   Fe'

Fig. 6.7-4 Molecular structure of $Fe_2(CO)_9$.

3. Dinuclear carbonyls related to $Mn_2(CO)_{10}$

   The crystal structures of $Tc_2(CO)_{10}$,[8] $Re_2(CO)_{10}$,[3] and $MnRe(CO)_{10}$,[9] which are isomorphous with $Mn_2(CO)_{10}$, have been determined. The measured M-M bond lengths are: Tc-Tc = 3.036(6)Å, Re-Re = 3.041(1)Å, Mn-Re = 2.909(1)Å.

   In $MnRe(CO)_{10}$ the average equatorial M-CO distance, 1.92(1)Å, is the average of the equatorial distances found in $Mn_2(CO)_{10}$ [1.856(7)Å] and

$Re_2(CO)_{10}$ [1.987(15)Å];[3] the axial M-CO distance in $MnRe(CO)_{10}$, 1.909(9)Å, is between those found in $Mn_2(CO)_{10}$ [1.811(3)Å] and $Re_2(CO)_{10}$ [1.929(7)Å].[3] The most striking feature is the short Re-Mn distance, 2.909(1)Å, versus Mn-Mn = 2.895(1)Å and Re-Re = 3.041(1)Å. Other Re-Mn distances reported for related structures are: $(CO)_5Re(H)Re(CO)_4Mn(CO)_5$, 2.960(3)Å; $(CO)_5MnRe(CO)_4(OCH_3)$, 2.972(1)Å; $ClMnRe(CO)_4(\mu\text{-}CO)(\mu\text{-}PhCCO)$, 2.817(3)Å. A bond order fractionally greater than one is proposed for the last entry.

X-Ray analysis has shown that the heterobimetallic complex anion in $PPN[HOsRe(CO)_8Br]$, where PPN is the bis(triphenylphosphine)iminium cation, comprises a $HOs(CO)_4$ fragment linked to a $Re(CO)_4Br$ fragment by an unsupported Os-Re single bond of length 2.995(1)Å.[10] The hydride and bromide ligands are in equatorial positions on osmium and rhenium, respectively, and the two parts of the complex anion are staggered with respect to each other.

4. Stable neutral binary carbonyls

The occurrence of stable neutral binary carbonyls is restricted to the central area of the d-block (Table 6.7-1), where there are low-lying vacant metal orbitals to accept σ-donated lone-pairs and also filled d orbitals for π back donation. Outside this area metal carbonyls are either very unstable (e.g. Cu, Ag), or anionic, or require additional ligands besides CO for stabilization.

Table 6.7-1   Known neutral binary metal carbonyls.

		$V(CO)_6$	$Cr(CO)_6$	$Mn_2(CO)_{10}$	$Fe(CO)_5$	$Co_2(CO)_8$	$Ni(CO)_4$	
Ti								Cu
				$Mn_4(CO)_{16}$	$Fe_2(CO)_9$	$Co_4(CO)_{12}$		
					$Fe_3(CO)_{12}$	$Co_6(CO)_{16}$		
Zr	Nb	$Mo(CO)_6$	$Tc_2(CO)_{10}$	$Ru(CO)_5$	$Rh_2(CO)_8$	Pd	Ag	
			$Tc_3(CO)_{12}$	$Ru_2(CO)_9$	$Rh_4(CO)_{12}$			
				$Ru_3(CO)_{12}$	$Rh_6(CO)_{16}$			
Hf	Ta	$W(CO)_6$	$Re_2(CO)_{10}$	$Os(CO)_5$	$Ir_2(CO)_8$	Pt	Au	
				$Os_2(CO)_9$	$Ir_4(CO)_{12}$			
				$Os_3(CO)_{12}$	$Ir_6(CO)_{16}$			

$Os_4(CO)_{13}$	$Os_5(CO)_{16}$	$Os_5(CO)_{19}$	$Os_6(CO)_{18}$	$Os_6(CO)_{20}$	$Os_7(CO)_{21}$	$Os_8(CO)_{23}$

5. Bonding in M-CO

CO is the most important and most widely studied of all organometallic ligands, and it is the prototype of the "π-acceptor ligands". The currently accepted view of the bonding is represented diagramatically in Fig. 6.7-5.

The top part (a) shows the formation of a σ bond by donation of the lone pair
into a suitably directed hybrid orbital on M. The lower part (b) shows the
accompanying back donation from a filled metal d orbital into the vacant
antibonding CO orbital having π symmetry with respect to the bonding axis.
This at once suggests why CO forms such strong complexes with transition
metals, since the drift of π-electron density from M to C tends to make the
ligand more negative and so enhances its σ-donor power. By implication, back
donation into antibonding CO orbitals weakens the CO bond and this is
manifested in the slight increase in interatomic distance from 1.128Å in free
CO to ~1.15Å in many carbonyl complexes. There is also a decrease in the C-O
force constant, and the drop in the infrared stretching frequency from 2143
cm^{-1} in free CO to 2125-1850 cm^{-1} for terminal CO's in neutral carbonyls has
been interpreted in the same way.

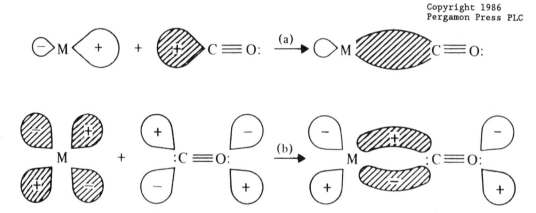

Fig. 6.7-5 Schematic representation of the orbital overlaps leading to
M-CO bonding: (a) σ overlap and donation from the lone pair on C into a
vacant metal hybrid orbital to form a σ M←C bond; (b) π overlap and
donation from a filled d_{xz} or d_{yz} orbital on M into a vacant antibonding
π_p^* orbital on CO to form a π M→C bond. (After ref. 11).

6. Coordination modes of the carbonyl ligand

In carbonyl metal clusters and carbidocarbonyl metal clusters, CO can
act as a terminal ligand [Fig. 6.7-6(a)], as an unsymmetrical or symmetrical
bridging ligand (μ_2-CO) [Fig. 6.7-6(b) and (c)], and as a triply bridging
ligand (μ_3-CO) [Fig. 6.7-6(e)]. In all these cases CO is η^1 but the
connectivity to metal increases from 1 to 3. It is notable that in the μ_2-
bridging carbonyls the angle M-C(O)-M is usually very acute (77-80°), whereas
in organic carbonyls the C-C(O)-C angle is typically 120-124°. This suggests
a fundamentally different bonding mode in the two cases and points to the
likelihood of a 2-electron, 3-center bond for the bridging metal carbonyls.

The hapticity can also be raised, and structural determinations afford examples in which one or both of the π^* orbitals in CO are thought to be involved in η^2 bonding to 1 or 2 metal atoms [Fig. 6.7-6(d), (f) and (g)]. A butterfly-type metallic skeleton with the carbonyl capping all four metal atoms has been observed, as shown in Fig. 6.7-6(h), (i) and (j). A bis-η^1-bridging mode has also been detected in an AlPh₃ adduct [Fig. 6.7-6(k)].

Fig. 6.7-7 shows some examples of unusual coordination modes of CO in metal carbonyl complexes.

a) The structure of the tetranuclear species [Ru₄(η^2-μ_3-CO)(CO)₉{μ-(RO)₂PN(Et)P(OR)₂}₂] has been determined.[12] The skeletal framework adopts a butterfly configuration in which the dihedral angle between the two triruthenium planes is 135°, and two opposite edges are bridged by the two diphosphazane ligands [Fig. 6.7-7(a)]. From the measured bond lengths [Ru(1)-C(5) 2.195(7), Ru(2)-C(5) 2.416(7), Ru(3)...C(5) 2.701(7), Ru(4)-C(5) 1.927(7) and Ru(2)-O(5) 2.159(4)Å], the unique carbonyl group belongs to the type shown in Fig. 6.7-6(i) and functions as a four-electron donor ligand.

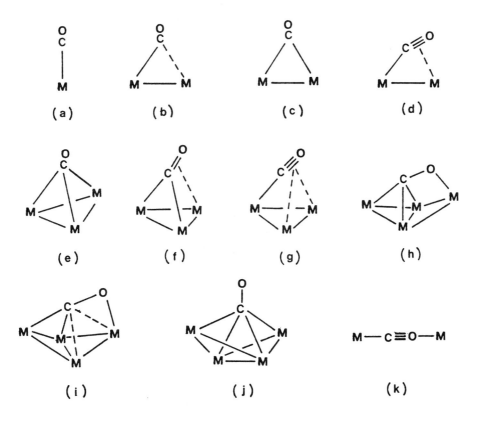

Fig. 6.7-6   Coordination modes of carbonyl group in transition metal complexes.

b) In $(\eta^5\text{-}Cp)_3Nb_3(CO)_6[\eta^2\text{-}\mu_3\text{-}C,\mu_2\text{-}O]$, both atoms of one of the carbonyl groups bond to metal centers, as shown in Fig. 6.7-7(b).[13] This coordination mode can be formulated as in Fig. 6.7-6(g).

c) In $Mo_2Ru_5(CO)_{14}(\eta^2\text{-}\mu_4\text{-}CO)_2Cp_2(\mu_4\text{-}S)$, two of the carbonyl groups each interacts with the metal core in the $(\mu_3\text{-}C,\mu\text{-}O)$ fashion, as shown in Fig. 6.7-7(c).[14] This coordination mode is represented in Fig. 6.7-6(h).

d) The new sulfido bimetallic cluster, $Cp_4Mo_2Ni_2S_2(CO)$, which contains an unprecedented $\eta^1,\mu_4\text{-}CO$ ligand, has been characterized by X-ray crystallographic analysis and other spectroscopic methods.[15] This coordination mode is shown in Fig. 6.7-6(j).

The geometrical arrangement of the core [see Fig. 6.7-7(d)] may be described either as a bicapped trigonal bipyramid with Ni atoms at apical positions and Mo and C atoms in equatorial positions (the carbonyl carbon is also counted as a vertex) or as a butterfly cluster with the carbonyl group bonded to the hinge Mo and wingtip Ni atoms. The Mo-Mo bond distance, 2.576Å, is shorter than those (2.64-2.67Å) found in geometrically similar clusters. The CO ligand is equidistant from both Ni atoms within experimental error. The $\eta^1\text{-}CO$ model is supported by the long Ni-O distances ($d_{Ni\text{-}O} > 3.03$Å) and the unbent CO: the angle between O, C, and the centroid of the Mo-Mo bond is approximately 180°.

The $\eta^1,\mu_4\text{-}CO$ assignment is indicated also by the extremely low CO stretching frequency in both the solid state (1654 cm^{-1} in a KBr pellet) and in solution (1653 cm^{-1} in THF) (cf. ν(CO) ~1850 cm^{-1} for $\mu_2\text{-}CO$ and ~1740-1720 cm^{-1} for $\mu_3\text{-}CO$). The long Ni-CO distances (average 2.39Å) suggest only a small bonding interaction between the Ni atoms and the carbonyl group.

e) The structure of the mixed-valent $\mu^2\text{-}\eta^2$-carbonyl complex $Cp^*_2V(\mu\text{-}CO)V(CO)_5$ is shown in Fig. 6.7-7(e).[16] The molecule possesses a linear V-C-O-V bond with V-O = 2.075(4)Å, C-O = 1.167(6)Å and V-C = 1.899(5)Å. This coordination mode is represented in Fig. 6.7-6(k). The complex is predominantly ionic, with a small covalent π back-donation from a t_{2g} orbital on $[V(CO)_6]^-$ into a partially filled π orbital on the $[Cp^*_2V]^+$ fragment.

f) A variant of the $\eta^2\text{-}\mu_4\text{-}CO$ coordination mode occurs in the alkoxide-supported tetranuclear tungsten cluster $W_4(OCH_2Pr^i)_{12}(\eta^2\text{-}\mu_4\text{-}CO)(CO)_2$ [Fig. 6.7-7(f)].[17] The metal atoms form a "spiked triangle" with W-W bond distances ranging from 2.658(2) to 2.806(3)Å. The $\eta^2\text{-}\mu_4\text{-}CO$ moiety may be viewed as a μ_3-oxyalkylidyne ligand and provides a model for partial reduction of carbon monoxide on a metal surface in the steps preceding reductive cleavage to carbide and oxide.

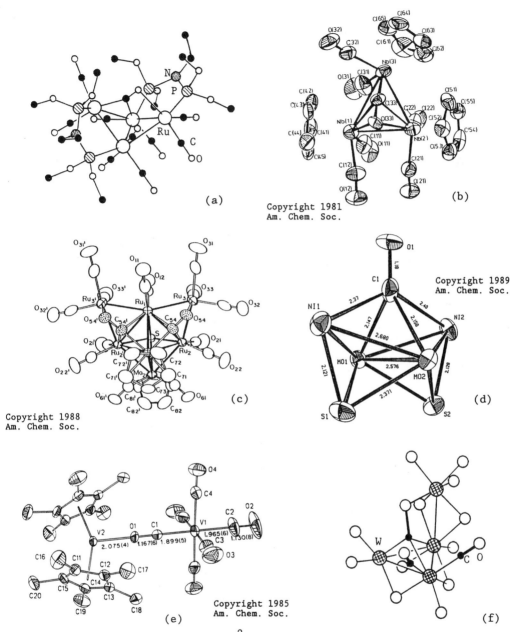

Fig. 6.7-7 Structure of (a) $Ru_4(\eta^2-\mu_3-CO)(CO)_9\{\mu-(MeO)_2PN(Et)P(OMe)_2\}_2$, (b) $(\eta^5-Cp)_3Nb_3(CO)_6(\eta^2-\mu_3-C,\mu_2-O)$, (c) $Mo_2Ru_5(CO)_{14}(\eta^2-\mu_4-CO)_2Cp_2(\mu_4-S)$, (d) $Cp_4Mo_2Ni_2S_2(\eta^1-\mu_4-CO)$, (e) $Cp^*_2V(\mu-CO)V(CO)_5$ and (f) $W_4(OCH_2Pr^i)_{12}(\eta^2-\mu_4-CO)(CO)_2$. (After refs. 12-17).

7. Metal complexes of some relatives of CO

The chemistry of ligands that are valence-isoelectronic with CO has attracted much attention.[18] The formal equivalence of NO^+ with CO is well established, and a great deal is known about the structure and properties of metal nitrosyl complexes.[19]   The $[RuCl(NO)_2(PPh_3)_2]^+$ ion is notable for having both linear and bent NO ligands [Fig. 6.7-8(a)].[20]   The analogy in reaction behaviour between coordinated CO and NO is demonstrated by the reaction of $[(\eta^5\text{-}C_5Me_5)Ru(NO)(Et)Ph]$ with PMe_2Ph to give a nitrosoalkane complex [Fig. 6.7-8(b)] *via* "NO insertion".[21]

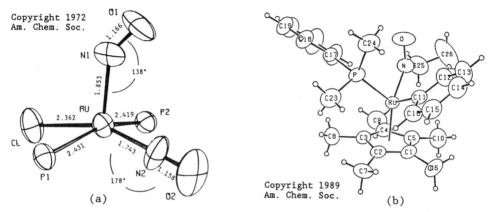

(a)   (b)

Fig. 6.7-8 Molecular structures of (a) $[RuCl(NO)_2(PPh_3)_2]^+$ and (b) $[(\eta^5\text{-}C_5Me_5)Ru\{N(O)Et\}(Ph)(PMe_2Ph)]$.   (After refs. 20 and 21).

A complete series of chalcocarbonyl complexes $[OsCl_2(CE)(CO)(PPh_3)_2]$ (E = O,S,Se,Te) have been characterized.[22]   In each case the osmium atom is octahedrally coordinated with the two PPh_3 ligands mutually *trans* and the CO and CE ligands mutually *cis*.   The bond lengths involving the terminal chalcocarbonyl ligands are: Os-CS 1.883(11), C-S 1.481(12); Os-CSe 1.913(5), C-Se 1.609(15); Os-CTe 1.813(12), C-Te 1.923(12)Å.   The *trans*-influence of the ligands increases in the order CO < CS < CSe < CTe.

The structures of three related chromium thiocarbonyl complexes are compared in Fig. 6.7-9.[23]   The dinuclear complex $(\eta^6\text{-}C_6H_5Me)(CO)_2CrCSCr(CO)_5$ contains an end-to-end bridging thiocarbonyl ligand with linear Cr-C≡S (bond angle 177.3°) and bent C≡S-Cr linkages.   The coordination of the S atom to the second chromium center leads to considerable lengthening of the C≡S bond distance relative to that in $Cr(\eta^6\text{-}C_6H_5COOMe)(CO)_2(CS)$, making it only slightly shorter than the C=S double bond in $Cr(SCMe_2)(CO)_5$.

The spectroscopic properties and chemical behaviour of the complexes $[M(CO)_5(CNH)]$ (M = Cr,W) indicate that the M-CNH bond is much stronger than the M-NCH bond in the corresponding isomers $[M(CO)_5(NCH)]$.   The structural chemistry of isocyanide complexes is well documented.[24]   The tetrahedral nickel clusters $Ni_4(CNR)_6$, $Ni_4(CNR)_7$, $Ni_4(CNR)_6L$, $Ni_4(CNR)_4(RC\equiv CR)_3$ and

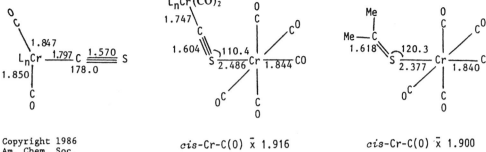

cis-Cr-C(0) $\bar{x}$ 1.916   *cis*-Cr-C(0) $\bar{x}$ 1.900

Fig. 6.7-9 Comparison of the structural parameters for three related chromium thiocarbonyl complexes. (After ref. 23).

$Ni_4(CNR)_6(RC{\equiv}CR)$ are all active precursors for the hydrogenation of acetylenes, and some are active for the hydrogenation of isocyanides and nitriles.[25] The structures of $Ni_4(CNCMe_3)_4(\mu_3-\eta^2-PhC{\equiv}CPh)_3$[25] and $Ni_4(CNCMe_3)_7$[26] are shown in Fig. 6.7-10(a) and (b), repectively.

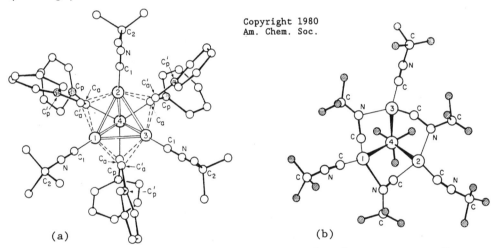

(a)   (b)

Fig. 6.7-10 Perspective drawing of the molecular structure of (a) $Ni_4(CNCMe_3)_4(PhC{\equiv}CPh)_3$ and (b) $Ni_4(CNCMe_3)_7$ viewed along the idealized C_3 axis through the apical Ni(4) atom. (After ref. 25).

The chemistry of vinylidene complexes (M=C=CRR') has been reviewed.[27] Examples of metal complexes containing a terminal allenylidene ligand (M=C=C=CRR')[28] and a μ_3-bridging vinylidene ligand[29] are shown in Fig. 6.7-11(a) and (b), respectively.

The first isolable silver carbonyl, $Ag(CO)B(OTeF_5)_4$, exhibits an unusually high CO stretching frequency at 2204 cm^{-1}.[30] The first complex containing PO, $[(Cp'Ni)_2W(CO)_4(\mu_3-PO)_2]$ ($Cp' = \eta^5-C_5iPr_4H$), has a tetragonal-pyramidal structure with a *trans*-P_2Ni_2 base.[31]

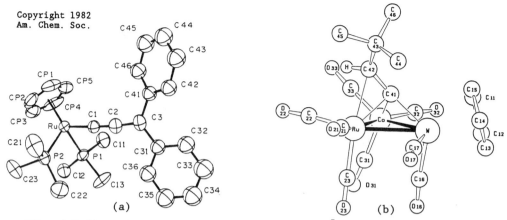

Fig. 6.7-11 Molecular structures of (a) $[Ru(\eta^5\text{-}C_5H_5)(PMe_3)_2(C_3Ph_2)]^+$ and (b) $RuCoW(\eta^5\text{-}C_5H_5)(CO)_8\{\mu_3\text{-}CCH(t\text{-}Bu)\}$. (After refs. 28 and 29).

References

[1] M. Martin, B. Rees and A. Mitschler, *Acta Crystallogr.*, *Sect. B* **38**, 6 (1982).

[2] L.F. Dahl and R.E. Rundle, *Acta Crystallogr.* **16**, 419 (1963).

[3] M.R. Churchill, K.N. Amoh and H.J. Wasserman, *Inorg. Chem.* **20**, 1609 (1981).

[4] W. Heijser, E.J. Baerends and P. Ros, *Discuss. Faraday Soc. (Symp.)* **14**, 211 (1980).

[5] J. Dewar and H.O. Jones, *Proc. Roy. Soc. (London)* **A76**, 558 (1905).

[6] H.M. Powell and R.V.G. Ewens, *J. Chem. Soc.*, 286 (1939).

[7] F.A. Cotton, and J.M. Troup, *J. Chem. Soc., Dalton Trans.*, 800 (1974).

[8] M.F. Bailey and L.F. Dahl, *Inorg. Chem.* **4**, 1140 (1965).

[9] A.L. Rheingold, W.K. Meckstroth and D.P. Ridge, *Inorg. Chem.* **25**, 3706 (1986).

[10] J.R. Moss, M.L. Niven and E.E. Sutton, *Inorg. Chim. Acta* **165**, 221 (1989).

[11] N.N. Greenwood and A. Earnshaw, *Chemistry of the Elements*, Pergamon Press, Oxford, 1986, chap. 8.

[12] J.S. Field, R.J. Haines and J.A. Jay, *J. Organomet. Chem.* **377**, C35 (1989).

[13] W.A. Herrmann, H. Biersack, M.L. Ziegler, K. Weidenhammer, R. Siegel and D. Rehder, *J. Am. Chem. Soc.* **103**, 1692 (1981).

[14] R.D. Adams, J.E. Babin and M. Tasi, *Inorg. Chem.* **27**, 2618 (1988).

[15] P. Li and M.D. Curtis, *J. Am. Chem. Soc.* **111**, 8279 (1989).

[16] J.H. Osborne, A.L. Rheingold and W.C. Trogler, *J. Am. Chem. Soc.* **107**, 6292 (1985).

[17] M.H. Chisholm, K. Folting, V.J. Johnson and C.E. Hammond, *J. Organomet. Chem.* **394**, 265 (1990).

[18] H. Werner, *Angew. Chem. Int. Ed. Engl.* **29**, 1077 (1990).

[19] R.D. Feltham and J.H. Enemark, *Top. Stereochem.* **12**, 155 (1981).

[20] C.G. Pierpont and R. Eisenberg, *Inorg. Chem.* **11**, 1088 (1972).

[21] J. Cheng, M.D. Seidler and R.G. Bergman, *J. Am. Chem. Soc.* **111**, 3258 (1989).

[22] G.R. Clark, K. Marsden, W.R. Roper and L.J. Wright, *J. Am. Chem. Soc.* **102**, 1206 (1980); G.R. Clark, K. Marsden, C.E.F. Richard, W.R. Roper and L.J. Wright, *J. Organomet. Chem.* **338**, 393 (1988).

[23] S. Lotz, R.R. Pille and P.H. Van Rooyen, *Inorg. Chem.* **25**, 3053 (1986).

[24] L. Malatesta and F. Bonati, *Isocyanide Complexes of Metals*, Wiley-Interscience, New York, 1969; E. Singleton and H.E. Oosthuizen, *Adv. Organomet. Chem.* **22**, 209 (1983).

[25] E.L. Muetterties, E. Band, A. Kokorin, W.R. Pretzer and M.G. Thomas, *Inorg. Chem.* **19**, 1552 (1980).

[26] V.W. Day, R.O. Day, J.S. Kristoff, F.J. Hirsekorn and E.L. Meutterties, *J. Am. Chem. Soc.* **97**, 2571 (1975).

[27] M.I. Bruce and A.G. Swincer, *Adv. Organomet. Chem.* **22**, 59 (1983); A.B. Antonova and A.A. Johansson, *Usp. Khim.* **58**, 1197 (1989); M.I. Bruce, *Chem. Rev.* **91**, 197 (1991).

[28] J.P. Selegue, *Organometallics* **1**, 217 (1982).

[29] T. Albiez, W. Bernhardt, C. von Schnering, E. Roland, H. Bantel and H. Vahrenkamp, *Chem. Ber.* **120**, 141 (1987).

[30] P.K. Hurlburt, O.P. Anderson and S.H. Strauss, *J. Am. Chem. Soc.* **113**, 6277 (1991).

[31] O.J. Scherer, J. Braun, P. Walther, G. Heckmann and G. Wolmershäuser, *Angew. Chem. Int. Ed. Engl.* **30**, 852 (1991).

Note The refined crystal structures of $Fe(CO)_5$ and $Mo(CO)_6$ are reported in R. Boese and D. Bläser, *Z. Kristallogr.* **193**, 289 (1990) and T.C.W. Mak, *ibid.* **166**, 277 (1984), respectively.

Ab initio MO studies of the stereochemistry and coordination modes of transition-metal complexes of N_2, CO, and CO_2 are reviewed by S. Sakaki in I. Bernal (ed.), *Stereochemistry of Organometallic and Inorganic Compounds, Vol. 4: Stereochemical Control, Bonding and Steric Rearrangements*, Elsevier, Amsterdam, 1990.

A comprehensive review of transition metal nitrosyl complexes is given in D.M.P. Mingos and D.J. Sherman, *Adv. Inorg. Chem.* **34**, 293 (1989).

Highly reduced metal carbonyl anions such as $[(Ph_3Sn)_4Zr(CO)_4]^{2-}$, which has a dodecahedral D_{2d} core, are discussed in J.E. Ellis, *Adv. Organomet. Chem.* **31**, 1 (1990); J.E. Ellis, K.-M. Chi, A.-J. DiMaio, S.R. Frerichs, J.R. Stenzel, A.L. Rheingold and B.S. Haggerty, *Angew. Chem. Int. Ed. Engl.* **30**, 194 (1991); W. Beck, *ibid.* **30**, 168 (1991).

6.8   Bis(tetraethyl)ammonium μ_6–Carbido-penta-μ-carbonyl-octacarbonyl-*octahedro*-hexacobaltate

$$[NEt_4]_2[Co_6C(CO)_{13}]$$

Crystal Data

Monoclinic, space group $C2$ (No.5)

$a = 20.750(2)$, $b = 11.470(2)$, $c = 16.097(2)$Å, $\beta = 91.968(3)°$, $Z = 4$

Atom	x	y	z	Atom	x	y	z
Co(1)	.30976	.62056	.82497	O(10)	.3485	.6912	.9941
Co(2)	.40949	.36996	.75247	C(11)	.3559	.2515	.7917
Co(3)	.32874	.48848	.67518	O(11)	.3518	.1496	.7946
Co(4)	.40744	.62443	.73976	C(12)	.3865	.3637	.6383
Co(5)	.40549	.51430	.88584	O(12)	.3967	.3163	.5763
Co(6)	.30616	.38978	.82700	C(13)	.3712	.6171	.6295
C(01)	.3616	.5005	.7842	O(13)	.3747	.6694	.5661
C(1)	.2297	.6642	.8398	N(1)	.3261	.4930	.3316
O(1)	.1770	.6874	.8472	C(14)	.3727	.5098	.4032
C(2)	.4851	.3014	.7609	C(15)	.4378	.4524	.3976
O(2)	.5332	.2577	.7688	C(16)	.2612	.5420	.3530
C(3)	.2654	.4780	.6015	C(17)	.2624	.6689	.3801
O(3)	.2235	.4714	.5542	C(18)	.3159	.3628	.3139
C(4)	.4761	.7007	.7127	C(19)	.2961	.2900	.3847
O(4)	.5226	.7449	.6906	C(20)	.3540	.5482	.2563
C(5)	.4816	.5807	.8817	C(21)	.3102	.5465	.1732
O(5)	.5327	.6212	.8879	N(2)	0	-.0135	0
C(6)	.2272	.3923	.7845	C(22)	.0002	.0876	-.0756
O(6)	.1735	.3898	.7634	C(23)	.0091	-.0244	-.1566
C(7)	.4172	.4279	.9777	C(24)	-.0594	.0664	-.0062
O(7)	.4251	.3699	1.0321	C(25)	-.1210	.0033	-.0132
C(8)	.2886	.3371	.9264	N(3)	0	1.01691	.5
O(8)	.2762	.3086	.9908	C(26)	.0459	.9401	.4549
C(9)	.3492	.7507	.7763	C(27)	.0948	1.0002	.4033
O(9)	.3455	.8500	.7669	C(28)	-.0360	1.0975	.4407
C(10)	.3497	.6379	.9326	C(29)	-.0742	1.0369	.3703

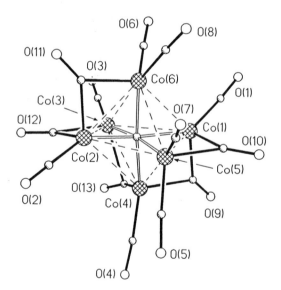

Fig. 6.8-1   Structure of $[Co_6C(CO)_{13}]^{2-}$.   Carbon atoms of the CO groups bear the same numbering as the O atoms.   Bonds to the carbide atom are indicated by open lines, and metal-metal bonds by broken lines.   The idealized two-fold axis passes through C(13)-O(13) and the midpoint of the Co(5)-Co(6) bond.

Crystal Structure

The cluster anion consists of a Co_6C core arranged in a distorted octahedron of metal atoms with the carbide atom in the cavity.[1] Of the 13 carbonyl ligands, five are bonded in edge-bridging mode and eight linearly. The bridged edges are on opposite sides of the octahedron equators and the edges of two opposite triangular faces are free of ligands. Therefore if these faces are taken as basal [see Fig. 6.8-1], the bridging ligands are found on five out of the six inter-basal edges. The cobalt atoms connected by the unbridged inter-basal edge [Co(5) and Co(6)] bear two terminal ligands and the idealized two-fold axis bisects this edge.

The Co-Co interactions are distinctly different for bridged and unbridged edges. The bridged Co-Co edges have normal lengths in the range 2.465(1)-2.505(1), average 2.483(1)Å. The unbridged edges show significantly longer and more scattered values in the range 2.468(1)-2.926(1), average 2.750(1)Å.

The Co-C(carbide) distances are in the range 1.86(1)-1.88(1), average 1.87(1)Å. The apparent radius of carbon turns out to be 0.55Å, almost equal to the value of 0.56Å found in $[Co_6C(CO)_{14}]^-$ and shorter than that in $[Rh_6C(CO)_{13}]^{2-}$, 0.60Å. These values cannot be considered true bond radii because they depend on the value chosen for the metal radius, i.e. one half of the average metal-metal bond in the compound under consideration.

$[Co_6C(CO)_{13}]^{2-}$ is not isostructural with the isoelectronic anion $[Rh_6C(CO)_{13}]^{2-}$, in which the ligands adopt an achiral arrangement with C_s idealized symmetry. The ligand arrangement adopted in the $[Co_6C(CO)_{13}]^{2-}$ is characterized by three-coordination of each of the cobalt atoms. In $[Rh_6C(CO)_{13}]^{2-}$, in contrast, the number of rhodium-carbonyl interactions is unevenly distributed and a uniform distribution of electrons achieved through asymmetric interactions of the bridging ligands and incipient bending of some terminal groups. In the cobalt species the ligand distribution can be considered more regular but the ligand-to-metal electron donation is unbalanced because Co(5) and Co(6), which bear two terminal ligands, are richer in electrons than the other metal atoms. In fact a compensation mechanism is set up through a system of unsymmetrical bridging ligands which bestows charge on atoms Co(1) and Co(2). Atoms Co(3) and Co(4) seem to remain deficient of electrons but some localization of anionic charge could help in equalizing the number of electrons around each cobalt atom. The ligand stereogeometry is mainly controlled by the need for an even distribution of the electrons provided to the metal core when repulsive inter-ligand interactions are not operative.

Remarks

1. Carbidocarbonyl metal clusters[2,3]

The complex $Fe_5C(CO)_{15}$ was the first member of this class of compounds to be discovered. The individual C atom fulfills predictable steric and electronic functions and a somewhat more systematic approach to the synthesis of such complexes is beginning to emerge. Table 6.8-1 lists known examples of these compounds. Fig. 6.8-2 shows some of the representative structures.

Table 6.8-1 Some carbidocarbonyl metal clusters.

$Fe_5C(CO)_{15}$	Black	$[Co_6C(CO)_{14}]^-$	Dark brown
$[Fe_5C(CO)_{14}]^{2-}$	Red-black	$[Co_6C(CO)_{15}]^{2-}$	Brown-red
$[Fe_6C(CO)_{16}]^{2-}$	Black	$[Co_8C(CO)_{18}]^{2-}$	Brown
$Ru_5C(CO)_{15}$	Red	$[Co_{13}(C)_2(CO)_{24}H]^{4-}$	Brown
$Ru_6C(CO)_{17}$	Deep-red	$[Rh_6C(CO)_{13}]^{2-}$	Red-Brown
$Ru_6C(CO)_{16}(L)$	Orange	$[Rh_6C(CO)_{15}]^{2-}$	Yellow
$Ru_6(arene)C(CO)_{14}$	Dark-red	$Rh_8C(CO)_{19}$	Black
$Os_5C(CO)_{15}$	Orange	$[Rh_{12}(C)_2(CO)_{24}]^{2-}$	Black
$Os_7C(CO)_{19}(H)_2$	Brown	$Rh_{12}(C_2)(CO)_{25}$	Black
$Os_8C(CO)_{21}$	Purple	$[Rh_{15}(C)_2(CO)_{28}]^-$	Brown
$[Co_6C(CO)_{13}]^{2-}$	Brown	$[Re_7C(CO)_{21}]^{2-}$	

The M-C(carbido) distances found in some carbidocarbonyl metal clusters are summarized in Table 6.8-2. It seems probable that in the smaller octahedral cavities the positive charge on the carbido atom will become higher to allow for the necessary contraction. Table 6.8-2 also compares the apparent radii and the coordination geometries of carbido atoms observed in the carbidocarbonyl metal clusters.

Table 6.8-2 Apparent radius and coordination of the carbon atom in some carbidocarbonyl metal clusters.[2]

Polyhedron	Species	Apparent radius from $\bar{d}(M-C) - \frac{1}{2}\bar{d}(M-M')/Å$
Square pyramid	$Fe_5C(CO)_{15}$	1.89 - 2.64/2 = 0.57
Octahedron	$[Fe_6C(CO)_{16}]^{2-}$	1.91 - 2.67/2 = 0.57
	$Ru_6C(CO)_{17}$	2.05 - 2.89/2 = 0.61
	$Ru_6C(CO)_{14}(C_9H_{12})$	2.04 - 2.88/2 = 0.60
	$[Ru_{15}(C)_2(CO)_{28}]^-$	2.04 - 2.87/2 = 0.60
Half trigonal prism	$Co_3C(CO)_9X$	1.92 - 2.48/2 = 0.68
Trigonal prism	$[Rh_6C(CO)_{15}]^{2-}$	2.13 - 2.79/2 = 0.74
	$Rh_8C(CO)_{19}$	2.13 - 2.81/2 = 0.72
Square antiprism (distorted)	$[Co_8C(CO)_{18}]^{2-}$	1.99 - 2.52/2 = 0.73

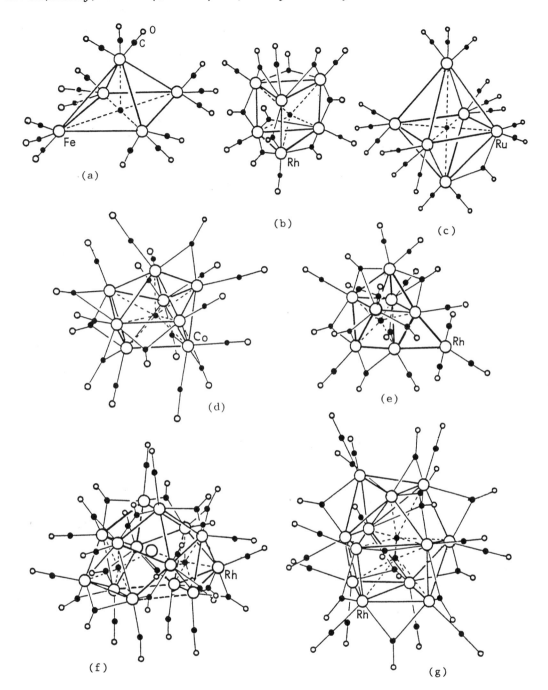

Fig. 6.8-2 Structures of some carbidocarbonyl metal clusters.

(a) $Fe_5C(CO)_{15}$, (b) $[Rh_6C(CO)_6(\mu_2\text{-}CO)_9]^{2-}$, (c) $Ru_6C(CO)_{16}(\mu_2\text{-}CO)$,

(d) $[Co_8C(CO)_{10}(\mu_2\text{-}CO)_8]^{2-}$, (e) $Rh_8C(CO)_{11}(\mu_2\text{-}CO)_6(\mu_3\text{-}CO)_2$,

(f) $[Rh_{15}(C)_2(CO)_{14}(\mu_2\text{-}CO)_{14}]^-$, (g) $Rh_{12}(C_2)(CO)_{14}(\mu_2\text{-}CO)_{10}(\mu_3\text{-}CO)$.

(Adapted from ref. 2).

Recently, an X-ray diffraction study has confirmed that reaction of $[Ru_5C(CO)_{15}]$ with excess pyridine gives an equimolar mixture of two isomers of $[HRu_5C(CO)_{14}(py)]$ (**a** and **b**) that differ only in the orientation of a bridging aromatic ligand [Fig 6.8-3], and provides the first example of this type of isomerism in the solid state.[4] The metal framework in each isomer may be described as a bridged "butterfly" or as *arachno*-pentagonal bipyramid. The hydride ligand bridges the Ru(1)-Ru(4) hinge bond in both compounds, and the Ru(4)...Ru(5) unbonded edge is bridged by the orthometallated pyridine via the nitrogen atom and the deprotonated carbon; in isomer **a** the N atom is bonded to the bridging metal [Ru(5)-N = 2.169(4)Å, whereas in isomer **b** it is bonded to the hinge ruthenium atom [Ru(4)-N = 2.134(12)Å]. Isomers **a** and **b** undergo quantitative thermal decarbonylation to give the same unstable product $[HRu_5C(CO)_{13}(py)]$, which can be recarbonylated quantitatively under mild conditions to regenerate equal proportions of the isomers.

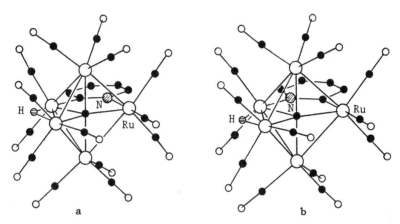

Fig. 6.8-3 Structures of the two isomers of $[HRu_5C(CO)_{14}(py)]$ **a** and **b**. (After ref. 4).

2. Mixed-metal carbido carbonyl clusters[5]

Many mixed-metal carbido carbonyl clusters have been prepared and characterized. Fig. 6.8-4 shows the structures of two examples. The carbido-carbonyl cluster species with a core $(M)_nRh_6C(CO)_{15}$, (M = Cu, Ag, Au), are all based on a trigonal-prismatic framework of Rh atoms capped by heteroatoms, which accommodates the interstitial C atom. In $[Rh_6C(CO)_{15}\{M(NCMe)\}_2]$, (M = Cu, Ag) and $[Rh_6C(CO)_{15}\{M(PPh_3)\}_2]$, (M = Cu, Ag, or Au), the trigonal-prismatic Rh_6 core is capped on both triangular faces by the heteroatoms. In the case of $[Rh_6C(CO)_{15}\{Au(PPh_3)\}]^-$ the Rh_6 core is monocapped by an $Au(PPh_3)$ fragment. The CO ligand stereogeometry is almost unaltered with respect to the parent dianion $[Rh_6C(CO)_{15}]^{2-}$, showing one CO terminally bound to each Rh

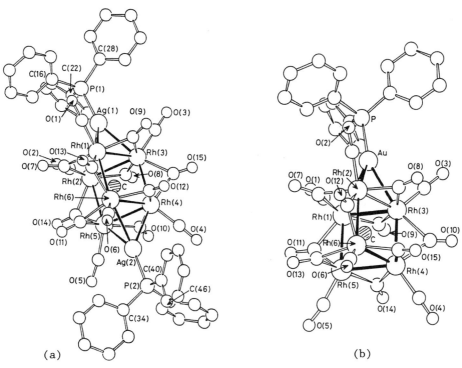

Fig. 6.8-4 Structure of (a) $[Rh_6C(CO)_{15}\{Ag(PPh_3)\}_2]$ and (b) $[Rh_6C(CO)_{15}\{Au(PPh_3)\}]^-$. (After ref. 5)

Table 6.8-3 Comparison of average bond lengths (Å) and angles (°) in some carbidocarbonyl Rh_6 clusters.

Compound	Rh-Rh				Rh-C-O	Rh-C-O bridge, basal	Rh-C-O bridge, interbasal	Rh-C-O terminal
	basal	interbasal	Rh-M	Rh-C	terminal			
$[Rh_6C(CO)_{15}]^{2-}$	2.776(3)	2.817(2)	-	2.134(6)	1.89(1)	2.12(1)	2.04(1)	178(1)
					1.13(1)	1.14(2)	1.17(2)	
$[Rh_6C(CO)_{15}\{Ag(NCMe)\}_2]$	2.775(1)	2.805(4)	2.822(1)	2.13(1)	1.91(1)	2.13(1)	2.05(1)	175(1)
					1.11(1)	1.16(1)	1.15(1)	
$[Rh_6C(CO)_{15}\{(AgPPh_3)\}_2]$	2.785(1)	2.808(1)	2.823(1)	2.13(1)	1.92(1)	2.14(1)	2.06(1)	176(1)
					1.11(1)	1.13(1)	1.13(1)	
$[Rh_6C(CO)_{15}\{Cu(PPh_3)\}_2]$	2.768(1)	2.813(1)	2.662(1)	2.129(6)	1.910(8)	2.133(6)	2.071(7)	176(1)
					1.13(1)	1.139(8)	1.146(8)	
$[Rh_6C(CO)_{15}\{Cu(NCMe)\}_2]$	2.765(1)	2.810(1)	2.660(1)	2.127(3)	1.88(1)	2.13(1)	2.05(1)	174(1)
					1.18(1)	1.17(1)	1.18(1)	
$[Rh_6C(CO)_{15}\{Au(PPh_3)\}]^-$	2.766(4);[a] 2.772(4)[b]	2.807(3)	2.820(3)	2.17(3) 2.09(3)	1.84(3)	2.06(3)	2.04(3)	177(1)
					1.20(4)	1.22(4)	1.18(4)	
$[Rh_6C(CO)_{15}\{Au(PPh_3)\}_2]$	2.780(1)	2.805(1)	2.824(1)	2.13(1)	1.91(1)	2.13(1)	2.07(1)	176(1)
					1.13(2)	1.16(2)	1.14(2)	

[a] Capped face. [b] Uncapped face.

atom and bridging CO groups on all prism edges (Fig. 6.8-4). A comparative analysis of the more relevant average structural parameters for the $[Rh_6C(CO)_{15}]^{2-}$ cluster and related species is shown in Table 6.8-3.

3. Sliding effect on CO ligand[6,7]

The term "sliding effect" refers to the apparent shift of the C atom of a bonded CO group towards the O atom, along the M-C-O vector, on simply passing from isotropic to anisotropic refinement. The differences in bond lengths for both terminal and bridging CO groups are of the order of 0.02-0.05Å and, though close to the experimental errors, they are systematic, invariably leading to underestimation of M-C and overestimation of C-O distances if isotropic refinement is used for the C and O atoms. This effect is illustrated by the data listed in Table 6.8-4 for $(NEt_4)_2[Co_6C(CO)_{15}]$ and $(PPh_4)_2[Rh_6C(CO)_{15}]$.

Table 6.8-4 Comparison of average M-C and C-O bond lengths (Å) determined using isotropic and anisotropic models.[6]

	(NEt$_4$)$_2$[Co$_6$C(CO)$_{15}$]							(PPh$_4$)$_2$[Rh$_6$C(CO)$_{15}$]					
$(\sin\theta/\lambda)_{max}$ (Å^{-1})	0.48			0.77			0.48-0.77	0.48			0.60		
$N_{obs}[I_o>2\sigma(I_o)]$	2853			5929			3076	3999			5582		
	Iso-tropic	Aniso-tropic	Δ	Iso-tropic	Aniso-tropic	Δ	Aniso-tropic	Iso-tropic	Aniso-tropic	Δ	Iso-tropic	Aniso-tropic	Δ
N_{var}	467	617		467	617		617	470	617		470	617	
R (%)	7.3	5.6		8.0	6.0		3.9	5.2	4.7		5.9	5.2	
M-C$_t$	1.70	1.75	.05	1.74	1.77	.03	1.77	1.86	1.89	.03	1.86	1.89	.03
C$_t$-O$_t$	1.19	1.14	-.05	1.17	1.13	-.04	1.14	1.14	1.13	-.01	1.14	1.13	-.01
M-C$_b$ triang. faces	1.93	1.97	.04	1.96	1.97	.01	1.97	2.09	2.10	.01	2.09	2.10	.01
C$_b$-O$_b$ triang.faces	1.21	1.16	-.05	1.18	1.16	-.02	1.16	1.18	1.17	-.01	1.18	1.16	-.02
M-C$_b$ sq. faces	1.88	1.90	.02	1.90	1.91	.01	1.91	2.00	2.02	.02	2.00	2.02	.02
C$_b$-O$_b$ sq. faces	1.21	1.18	-.03	1.17	1.17	.00	1.17	1.21	1.19	-.02	1.21	1.19	-.02

The following observations can be deduced from the available data:

a) There is a marked sliding effect on the terminal CO ligands which depends on the limiting value of $(\sin\theta/\lambda)_{max}$.

b) Bridging CO's in both $[Co_6C(CO)_{15}]^{2-}$ and $[Rh_6C(CO)_{15}]^{2-}$ show similar behaviour, yielding longer M-C and shorter C-O bond distances on passing from isotropic to anisotropic refinement (the M-C-M angle varies accordingly).

c) The two nonequivalent sets of bridging ligands (basal and interbasal ones) behave very much in the same way.

d) The effect decreases with increasing resolution of the data sets.

e) The sliding effect appears to be slightly less pronounced for M = Rh than M = Co although the former data set is smaller.

f) In the case of $[Co_6C(CO)_{15}]^{2-}$ a refinement carried out on high-order reflections only [rather arbitrarily selected within the $(\sin\theta/\lambda)$ range 0.48-0.77 Å^{-1}] yielded results which are almost identical to those obtained with anisotropic refinement using the full data set.

4. Effects of bridging CO ligands on M-M bond lengths[6,8]

A careful investigation of the structural features of the cluster compounds $(NEt_4)_2[Co_6C(CO)_{15}]$ and $(PPh_4)_2[Rh_6C(CO)_{15}]$ provides some insights into the relationships between the size of the metal-atom polyhedra (in terms of M-M bond lengths) and the surrounding ligand envelopes.

The $[M_6C(CO)_{15}]^{2-}$ cluster is rather exceptional in that all edges are spanned by bridging CO ligands, so that the effect of these ligands on the M-M bond distances should be more marked. Table 6.8-5 shows a comparison of the average values in the two prismatic dianions, $[Co_6C(CO)_{15}]^{2-}$ and $[Rh_6C(CO)_{15}]^{2-}$, and two octahedral dianions, $[Co_6C(CO)_{13}]^{2-[1]}$ and $[Rh_6C(CO)_{13}]^{2-}$.[9]

The Co-Co bond lengths are as expected shorter than the corresponding Rh-Rh ones, but the CO-bridged bonds in the related carbide species show Rh-Rh bonds of comparable length in $[Rh_6C(CO)_{15}]^{2-}$ and $[Rh_6C(CO)_{13}]^{2-}$, while Co-Co bonds appear to be shorter in $[Co_6C(CO)_{13}]^{2-}$ than in the prismatic core of $[Co_6C(CO)_{15}]^{2-}$.

Table 6.8-5 Comparison of relevant structural parameters (Å) between pairs of related prismatic and octahedral dianions.

	$[Co_6C(CO)_{15}]^{2-}$	$[Co_6C(CO)_{13}]^{2-}$	$[Rh_6C(CO)_{15}]^{2-}$	$[Rh_6C(CO)_{13}]^{2-}$
M-M (Bridged)	2.537(1)a 2.573(1)b	2.483(1)	2.772(1)a 2.819(1)b	2.773(1)
M-M (unbridged)		2.750(1)		3.039(10)
M-M (average)	2.549(1)	2.639(1)	2.790(1)	2.907(1)
M-C (carbide)	1.75(2)	1.87(2)	2.13(2)	2.05(2)
γ_C c	0.68	0.55	0.74	0.60
C...C d	2.64a 2.65b	3.00	2.91a 2.88b	3.07
O...O d	3.95a 3.83b	4.21	3.73a 3.55b	4.05

a Prism triangular faces.   b Prism square faces.

c Apparent C-atom radius, γ_C = (M-C) - (M-M)/2.

d Intramolecular next-neighbour contacts.

The reason for this discrepancy lies in the subtle interplay of several factors such as intramolecular-ligand, metal-ligand and metal-metal interactions. As can be seen in Table 6.8-5, the average next-neighbour CO...CO contact distances (calculated for both C...C and O...O contacts) reach a minimum in $[Co_6C(CO)_{15}]^{2-}$, being shorter than in the Rh prism and in both Co- and Rh-octahedral species. It has been shown that nonbonding interactions play a key role in establishing the more stable stereogeometry for the family of octahedral carbide species.

The interstitial C(carbide) atom appears to play a significant role in determining the optimum size of the surrounding metal cage. Although the apparent radius of C in the octahedral Co species, despite the presence of CO-bridged bonds of "normal" length, is much smaller (0.55Å) than within the prismatic cavity (0.68Å), the deformations of the former cage allow a longer average Co-C distance (1.87 *versus* 1.75Å).

X-Ray diffraction and ^{13}C NMR studies of $[N(PPh_3)_2]_2[Ru_{10}C(CO)_{24}].C_6H_{14}$ and $[N(PPh_3)_2][HRu_{10}C(CO)_{24}]$ revealed that the dianion $[Ru_{10}C(CO)_{24}]^{2-}$ and monoanion $[HRu_{10}C(CO)_{24}]^-$ are isostructural with their respective osmium analogues, all having a giant tetrahedral M_{10} cluster framework with an interstitial carbide at its center (Fig. 6.8-5).[10] In the decaruthenium monoanion the hydrido ligand resides in one of the tetrahedral cavities generated by the "Ru(CO)$_3$" fragments capping the central octahedron, resulting in a structure of C_{3v} symmetry.

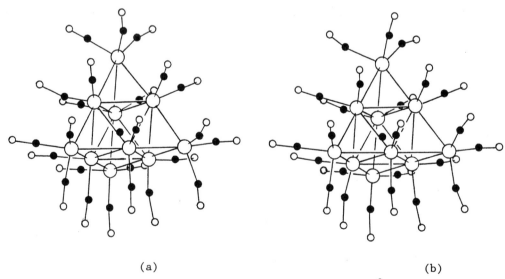

(a)   (b)

Fig. 6.8-5 Molecular structure of (a) $[Ru_{10}C(CO)_{24}]^{2-}$ and (b) $[HRu_{10}C(CO)_{24}]^-$. (After ref. 10).

5. Metal clusters containing encapsulated nitrogen atoms

Examples of this type of nitrido clusters include the trigonal prismatic species $[M_6N(CO)_{15}]^-$ (M = Co,Rh) and the related anions $[Rh_6MN(CO)_{15}]^{2-}$ (M = Co,Rh,Ir) whose metallic core consists of a trigonal prism capped on a square face. Fig. 6.8-6(a) shows that a nitride species occupies the prismatic cavity of $[Rh_7N(CO)_{15}]^{2-}$.[11] The cluster $[Rh_{14}N_2(CO)_{25}]^{2-}$ contains a metal-centered, three-layer core that can be described as a monocapped centered cubo-octahedron [the capping atom is Rh(14)] with one edge [Rh(12)...Rh(13)] broken [Fig. 6.8-6(b)].[12] As in other related high-nuclearity rhodium clusters, the interstitial N atoms are located close to surface square faces of the type $M_4(\mu\text{-}CO)_4(CO)_4$; each is in a square pyramidal environment with close contact to one metal atom and four more distant ones.

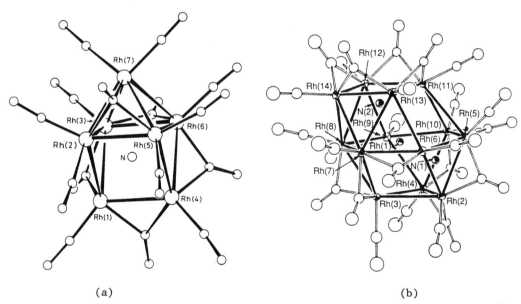

(a)   (b)

Fig. 6.8-6 Structures of nitrido carbonyl clusters (a) $[Rh_7N(CO)_{15}]^{2-}$ and (b) $[Rh_{14}N_2(CO)_{25}]^{2-}$. (After refs. 11 and 12).

6. Butterfly clusters containing an "exposed" carbon atom

In some carbido clusters, a carbon atom of "inverted" configuration bonds to four metal centers in an exposed position at the surface of the cluster. One of the first examples is $Fe_4C(CO)_{13}$ which comprises a butterfly configuration of four iron atoms, as shown in Fig. 6.8-7(a).[13] The dihedral angle between the two triangular planes is 101°. Each of the four iron atoms bears three terminal carbonyl ligands, and the remaining carbonyl [C(13)O(13)] bridges Fe(1) and Fe(4). The most remarkable feature of the molecule is the

μ_4-C atom: the Fe(2)-C(14)-Fe(3) angle is 175°, and the Fe-C(14) bond lengths
are shown in the bottom part of Fig. 6.8-7(a). The short Fe(2)-C(14) and
Fe(3)-C(14) bonds are suggestive of multiple bond character. The cluster
anion $[Fe_4C(CO)_{12}]^{2-}$ has the same butterfly core structure.[14]

    In the realm of metal alkoxide chemistry,[15] a cluster with an exposed
carbon atom is found in $W_4C(NMe)(OPr^i)_{12}$.[16] As shown in Fig. 6.8-7(b), the
complex exhibits a butterfly structure supported by a system of bridging
NMe^{2-}, OPr^i and carbido C ligands. For the wingtip tungsten atoms, the W-C-W
angle is 163.5(8)°, and the metal-carbide bond lengths are much shorter than
those involving the backbone tungsten atoms. The $W_4C(OCH_2-c-C_5H_9)_{14}$ cluster
has an analogous skeletal structure.[17]

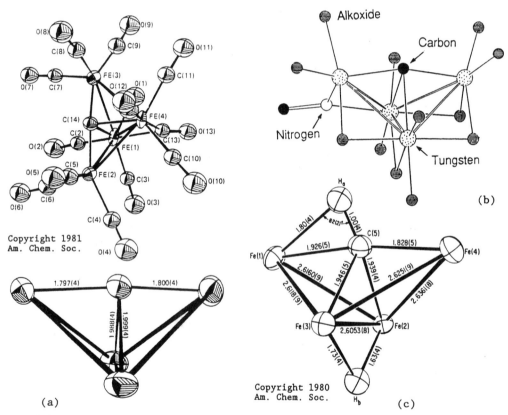

(a)

(b)

(c)

Fig. 6.8-7 Structure of (a) $Fe_4C(CO)_{13}$ (top) and core geometry (bottom),
(b) $W_4C(NMe)(OPr^i)_{12}$ and (c) $HFe_4(\eta^2-CH)(CO)_{12}$. (After refs. 13, 17 and 18).

    The $HFe_4(\eta^2-CH)(CO)_{12}$ cluster has a typical Fe_4C butterfly core in which
the η^2-methylidyne ligand forms an agostic hydrogen bond with one wingtip Fe
atom, and the pair of backbone Fe atoms are bridged by the hydridic hydrogen
atom [Fig. 6.8-7(c)].[18]

The nitrido cluster $HFe_4N(CO)_{12}$ also has a butterfly core with an exposed and four-coordinate nitrogen atom and a hydride bridging the two basal Fe atoms.[19] The Fe(apical)-N-Fe(apical) angle is 178.4(6)°, and the dihedral angle between the two Fe(apical)-2Fe(basal) planes is 101°. The cluster $HFe_5N(CO)_{14}$ is an isoelectronic analogue of $Fe_5C(CO)_{15}$ with the same square-pyramidal core geometry. The apical Fe atom is bonded to three terminal carbonyls, each of the basal Fe atoms carries two terminal carbonyls, three basal edges are each unsymmetrically bridged by one carbonyl, and the remaining edge spanned by a hydride.[19]

7. Metal clusters containing arene ligands

An arene may be attached to a transition-metal cluster complex in a conventional η^6 mode to one metal center, as in $[Ru_6C(CO)_{14}(\eta^6\text{-}C_6H_3Me)]$,[20] as a benzyne ligand as in $[Ru_4(CO)_{11}(\mu_4\text{-}PPh)(\mu_4\text{-}\eta^4\text{-}C_6H_4)]$ [Fig. 6.8-8(a)][21] and $[Os_3(\mu\text{-}H)(CO)_9(\mu_3\text{-}\eta^2\text{-}C_6H_4)(\mu\text{-}SMe)]$ [Fig. 6.8-8(b)],[22] or in a symmetric face-capping mode as in $[M_3(CO)_9(\mu_3\text{-}\eta^2:\eta^2:\eta^2\text{-}C_6H_6)]$ (M = Os,[23] Ru[24]) [Fig. 6.8-8(c)]. In the latter complex the benzene ligand exhibits C-C bond-length alternation with its H atoms bent away from the underlying M_3 triangle. The benzene-capped triosmium cluster has an almost exactly staggered configuration of approximate C_{3v} symmetry, whereas a twist of the benzene moiety in the triruthenium complex reduces the symmetry to C_3. In the related cyclohexa-1,3-diene complex $[HRu_3(CO)_9(C_6H_7)]$ the C-C bond distances are in accord with a 1,4-localization of the double bonds.[24] The C_6H_7 ligand is almost eclipsed with respect to the metal frame, but is shifted slightly to optimize the σ bonding between C(13) and Ru(3) [Fig. 6.8-8(d)].

Both η^6 terminal and $\mu_3\text{-}\eta^2:\eta^2:\eta^2$ face-capping modes of the benzene ligand are found in the cluster $[Ru_6C(CO)_{11}(C_6H_6)_2]$, as shown in Fig. 6.8-8(e)].[23] In the complex $[H_2Ru_6(CO)_{16}(\mu\text{-}C_6H_4O)]$ the ruthenium atoms are in a highly puckered "raft" arrangement, of which only Ru(3) is not coupled to the phenyl-olate ligand [Fig. 6.8-8(f)].[25] The measured dimensions of the C_6H_4O fragment correspond to those in free phenol, and the ligand binds five metal atoms in three modes: η^6 to Ru(4), ortho-metallated three-center bonding to Ru(5) and Ru(6), and $\mu\text{-}OR$ coordination to Ru(1) and Ru(2).

One important class of transition-metal arene complexes involves μ-phenyl ligands bridging two metal centers. In the heterobinuclear complex shown in Fig. 6.8-9(a), the metal centers are bridged by the μ-methylene carbon and the ipso carbon atom of the p-(N,N-dimethylamino)phenyl group.[26] The Ti and Rh atoms are in pseudo tetrahedral and psudo square-planar environments, respectively. Fig. 6.8-9(b) shows a dinuclear zirconocene

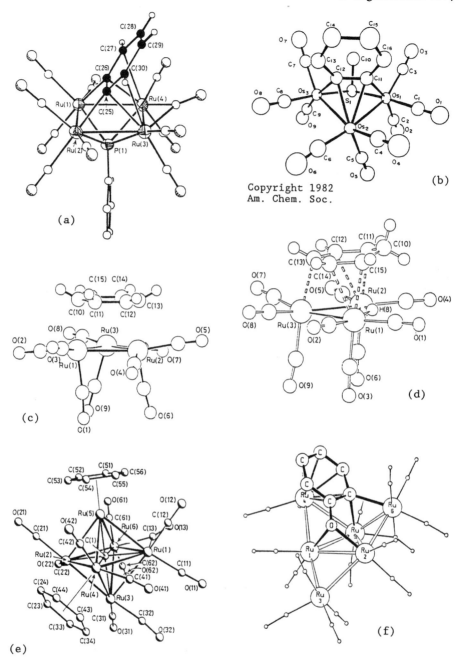

Fig. 6.8-8 Molecular structures of some metal clusters containing arene ligands: (a) $[Ru_4(CO)_{11}(\mu_4-PPh)(\mu_4-\eta^4-C_6H_4)]$, (b) $[Os_3(\mu-H)(CO)_9(\mu_3-\eta^2-C_6H_4)(\mu-SMe)]$, (c) $Ru_3(CO)_9(\mu_3-\eta^2:\eta^2:\eta^2-C_6H_6)]$, (d) $[HRu_3(CO)_9(C_6H_7)]$, (e) $[Ru_6C(CO)_{11}(\mu_3-\eta^2:\eta^2:\eta^2-C_6H_6)(\eta^6-C_6H_6)]$, and (f) $[H_2Ru_6(CO)_{16}(\mu-C_6H_4O)]$. (After refs. 21-25).

derivative that contains an in-plane bridging aryl ligand as well as a bridging methyl group.[27] The arene ring is completely planar, whereas the Zr1, Zr2 and methyl carbon atom are displaced from it by 0.130, 0.204 and 0.275Å, respectively.

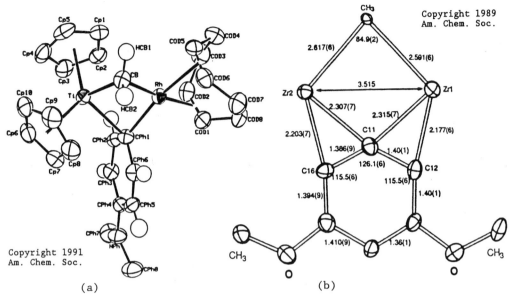

Copyright 1989
Am. Chem. Soc.

Copyright 1991
Am. Chem. Soc.

(a)   (b)

Fig. 6.8-9 Molecular structures of (a) $Cp_2Ti(\mu\text{-}CH_2)[\mu\text{-}p\text{-}C_6H_4Me_2]Rh(1,5\text{-}$cyclooctadiene) and (b) $(Cp_2Zr)_2(\mu\text{-}CH_3)[C_6H(OMe)_2]$ (for clarity the Cp rings and the aromatic H atom are omitted). (After refs. 26 and 27).

8. *μ*-Carbido-bridged complexes

The *μ*-carbido metalloporphyrin dimer [(TPP)Fe]$_2$C, an analogue of the well characterized [(TPP)Fe]$_2$O and [(TPP)Fe]$_2$N molecules, is the first reported compound possessing a formally dicarbenic carbon atom bridging two metal centers [Fig. 6.8-10(a)].[28] The structural parameters of this series of single-atom-bridged dimers are compared in Table 6.8-6.

Table 6.8-6 Structural parameters of [(TPP)Fe]$_2$X (X = C, N, O).

	μ-carbido*	*μ*-nitrido*	*μ*-oxo
Fe out-of-plane distance (Å)	0.26	0.32	0.50
Fe-X bond length (Å)	1.675	1.661	1.763
N-Fe-X angle (°)	97.5	99.2	103.7
Fe-X-Fe angle (°)	180	180	174
Mean separation between TPP planes (Å)	3.87	3.96	4.53

* The molecule is centrosymmetric.

The molecular structure of the first six-coordinate, *μ*-carbido-bridged Fe-phthalocyanine dimer containing coordinated 1-methylimidazole is shown in

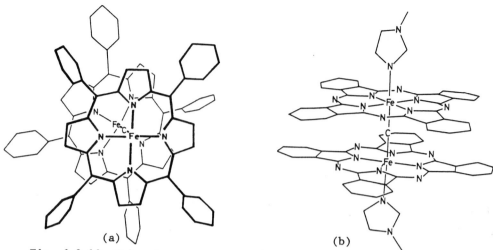

Fig. 6.8-10 Molecular structures of (a) [(TPP)Fe]$_2$C and (b) [(1-MeIm)PcFe]$_2$C. (After refs. 28 and 29).

Fig. 6.8-10(b). Although the X-ray analysis was affected by poor data ($R = 0.201$), it was sufficient to establish the essential structural features.[29]

In the heterodinuclear complex [(Me$_3$CO)$_3$W≡C-Ru(CO)$_2$(Cp)] the central carbon atom bridges the metal atoms in a linear fashion, the measured W-C-Ru angle being 177(2)°. The W-C and C-Ru bond lengths are 1.75(2) and 2.09(2)Å, respectively, and the relative conformation of the two coordinated metal fragments are "staggered" [Fig. 6.8-11(a)].[30] The trinuclear complex [(TPP)Fe=C=Re(CO)$_4$Re(CO)$_5$] exhibits an almost linear backbone with Fe-C-Re and C-Re-Re angles of 173.3(9) and 176.8(3)°, respectively. The important bond distances (Å) are: Fe-C 1.603(13), C-Re 1.957(12) and Re-Re 3.043(1)Å. The TPP ligand is severely twisted, and the equatorial CO ligands of the mutually staggered Re(CO)$_4$ groups are bent towards each other [Fig. 6.8-11(b)].[31]

9. μ-Dicarbido-bridged complexes

Some interesting metal complexes containing a C$_2$-bridging unit include dinuclear (CO)$_5$Re-C≡C-Re(CO)$_5$,[32] (t-BuO)$_3$W≡C-C≡W(O-t-Bu)$_3$,[33] (silox)$_3$Ta=C=C=Ta(silox)$_3$ (silox = t-Bu$_3$SiO),[34] and pentanuclear *trans*-Pt(PMe$_2$Ph)$_2$(C$_2$W$_2$(O-t-Bu)$_2$)$_2$.[35] In the σ,σ,-diacetylide bridged complex the two Re(CO)$_5$ groups are arranged in the eclipsed conformation. The ditantalum complex has a nearly linear cumulenic Ta(μ-C$_2$)Ta core; the centrosymmetric molecule has approximate D_{3d} symmetry with staggered (silox)$_3$Ta groups [Fig. 6.8-12(a)]. The heterometallic dicarbido complex is centrosymmetric, and its PtC$_2$W$_2$ core may be classified as a π-coordinated, dimetalla-substituted alkyne of the type {(μ-PtC≡CW$_\sigma$)=W$_\pi$) [Fig. 6.8-12(b)].

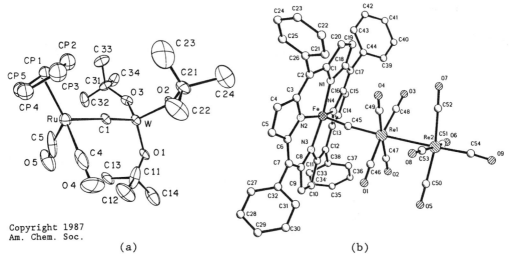

Fig. 6.8-11 Molecular structures of (a) [(Me₃CO)₃W≡C-Ru(CO)₂(Cp)] and (b) [(TPP)Fe=C=Re(CO)₄Re(CO)₅]. (After refs. 30 and 31).

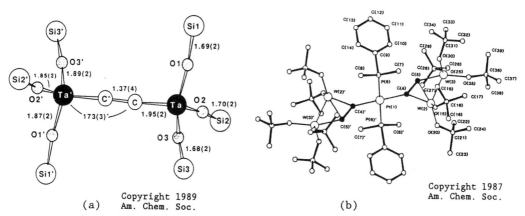

Fig. 6.8-12 Molecular structures of (a) (silox)₃Ta=C=C=Ta(silox)₃ (skeletal view) and (b) *trans*-Pt(PMe₂Ph)₂{C₂W₂(O-*t*-Bu)₂}₂. (After refs. 34 and 35).

References

[1] V.G. Albano, D. Braga and S. Martinengo, *J. Chem. Soc., Dalton Trans.*, 981 (1986), and references cited therein.

[2] P. Chini, G. Longoni and V.G. Albano, *Adv. Organomet. Chem.* **14**, 285 (1976)

[3] M. Tachikawa and E.L. Muetterties, *Prog. Inorg. Chem.* **28**, 203 (1981).

[4] G. Conole, M. McPartlin, H.R. Powell, T. Dutton, B.F.G. Johnson and J. Lewis, *J. Organomet. Chem.* **379**, C1 (1989).

[5] A. Fumagalli, S. Martinengo, V.G. Albano and D. Braga, *J. Chem. Soc., Dalton, Trans.*, 1237 (1988).

[6] V.G. Albano, D. Braga and F. Grepioni, *Acta Crystallogr., Sect. B* **45**, 60 (1989).

[7] D. Braga and T.F. Koetzle, *J. Chem. Soc., Chem. Commun.*, 144 (1987).

[8] B.F.G. Johnson and R.E. Benfield, *Inorg. Organomet. Stereochem.* **12**, 253 (1981).

[9] V.G. Albano, D. Braga and S. Martinengo, *J. Chem. Soc., Dalton Trans.*, 717 (1981).

[10] P.J. Bailey, B.F.G. Johnson, J. Lewis, M. McPartlin and H.R. Powell, *J. Organomet. Chem.* **377**, C17 (1989).

[11] S. Martinengo, G. Ciani and A. Sironi, *J. Chem. Soc., Chem. Commun.*, 1577 (1984).

[12] S. Martinengo, G. Ciani and A. Sironi, *J. Chem. Soc., Chem. Commun.*, 26 (1991).

[13] J.S. Bradley, G.B. Ansell, M.E. Leonowicz and E.W. Hill, *J. Am. Chem. Soc.* **103**, 4968 (1981).

[14] J.H. Davis, M.A. Beno, J.M. Williams, J. Zimmie, M. Tachikawa and E.L. Muetterties, *Proc. Natl. Acad. Sci. USA* **78**, 668 (1981).

[15] M. Chisholm, *J. Organomet. Chem.* **400**, 235 (1990).

[16] M.H. Chisholm, K. Folting, J.C. Huffman, J. Leonelli, N.S. Marchant, C.A. Smith and L.C.E Taylor, *J. Am. Chem. Soc.* **107**, 3722 (1985).

[17] M.H. Chisholm, C.E. Hammond, J.C. Huffman and V.J. Johnson, *J. Organomet. Chem.* **394**, C16 (1990).

[18] M.A. Beno, J.M. Williams, M. Tachikawa and E.L. Mutterties, *J. Am. Chem. Soc.* **102**, 4542 (1980).

[19] M. Tachiwara, J. Stein, E.L. Muetterties, R.G. Teller, M.A. Beno, E. Gebert and J.M. Williams, *J. Am. Chem. Soc.* **102**, 6649 (1980).

[20] R. Mason and W.R. Robinson, *J. Chem. Soc., Chem. Commun.*, 468 (1968).

[21] S.A.R. Knox, B.R. Lloyd, A.G. Orpen, J.M. Vinas and M. Weber, *J. Chem. Soc., Chem. Commun.*, 1498 (1987).

[22] R.D. Adams, D.A. Katahira and L.-W. Yang, *Organometallics* **1**, 235 (1982).

[23] M.P. Gomez-Sal, B.F.G. Johnson, J. Lewis, P.R. Raithby and A.H. Wright, *J. Chem. Soc., Chem. Commun.*, 1682 (1985).

[24] B.F.G. Johnson, J. Lewis, M. Martinelli, A.H. Wright, D. Braga and F. Grepioni, *J. Chem. Soc., Chem. Commun.*, 364 (1990).

[25] D.S. Bohle and H. Vahrenkamp, *Angew. Chem. Int. Ed. Engl.* **29**, 198 (1990).

[26] J.W. Park, L.M. Henling, W.P. Schaefer and R.H. Grubbs, *Organometallics* **10**, 171 (1991).

[27] S.L. Buchwald, E.A. Lucas and W.M. Davis, *J. Am. Chem. Soc.* **111**, 387 (1989).

[28] V.L. Goedken, M.R. Deakin and L.A. Bottomley, *J. Chem. Soc., Chem. Commun.*, 607 (1982).

[29] G. Rossi, V.L. Goedken and C. Ercolani, *J. Chem. Soc., Chem. Commun.*, 46 (1988).

[30] S.L. Latesky and J.P. Selegue, *J. Am. Chem. Soc.* **109**, 4731 (1987).

[31] W. Beck, W. Knauer and C. Robl, *Angew. Chem. Int. Ed. Engl.* **29**, 318 (1990).

[32] J. Heidrich, M. Steimann, M. Appel, W. Beck, J.R. Phillips and W.C. Trogler, *Organometallics* **9**, 1296 (1990).

[33] M.L. Listemann and R.R. Shrock, *Organometallics* **4**, 74 (1985).

[34] D.R. Neithamer, R.E. LaPointe, R.A. Wheeler, D.S. Richeson, G.D. Van Duyne and P.T. Wolczanski, *J. Am. Chem. Soc.* **111**, 9056 (1989).

[35] R.J. Blau, M.H. Chisholm, K. Folting and R.J. Wang, *J. Am. Chem. Soc.* **109**, 4552 (1987).

Note The classification of butterfly cluster geometries and complexes of the group VIII transition metals are reviewed in E. Sappa, A. Tiripicchio, A.J. Carty and G.E. Toogood, *Prog. Inorg. Chem.* **35**, 437 (1987).

$Ru_5(\mu_5-C_2)(\mu-SMe)_2(\mu-PPh_2)_2(CO)_{12}$ and $Ru_5(\mu_5-C_2)(\mu-SMe)_2(\mu-PPh_2)_2(CO)_{11}$ each contains a dicarbido unit attached to a bent open Ru_5 chain and a envelope-like five-membered Ru_5 ring, respectively. See C.J. Adams, M.I. Bruce, B.W. Skelton and A.H. White, *J. Chem. Soc., Chem. Commun.*, 26 (1992).

The molecular structure and interconversion of the benzene-coordinated pentanuclear carbido cluster isomers $[Ru_5C(CO)_{12}(\mu_3:\eta^2:\eta^2:\eta^2-C_6H_6)]$ and $[Ru_5C(CO)_{12}(\eta^6-C_6H_6)]$ are reported in P.J. Bailey, D. Braga, P.J. Dyson, F. Grepioni, B.F.G. Johnson, J. Lewis and P. Sabatino, *J. Chem. Soc., Chem. Commun.*, 177 (1992).

New azaalkane- and nitrene-bridged carbonyl metal clusters such as $Ru_4(EtN=NEt)(CO)_{12}$ and $Fe_4(\mu_4-NEt)_2(CO)_{11}$ are reported in B. Hansert, A.K. Powell and H. Vahrenkamp, *Chem. Ber.* *124*, 2697 (1991).

Sigma bonded homoleptic and related aryls of the transition metals are reviewed in S.U. Koschmieder and G. Wilkinson, *Polyhedron* **10**, 135 (1991).

6.9 Bis(cyclopentadienyl)iron (Ferrocene)

$$Fe(C_5H_5)_2$$

Crystal Data

Monoclinic, space group $P2_1/a$ (No. 14)

$a = 10.443(5)$, $b = 7.572(4)$, $c = 5.824(4)$Å, $\beta = 120.95(8)°$ (173K),[1]

$Z = 2$

Atom*	x	y	z	Atom	x	y	z
Fe	0	0	0	H(1)	-.062	.3543	-.053
C(1)	.0203	.2626	-.0317	H(2)	-.024	.193	-.414
C(2)	.0398	.1777	-.2201	H(3)	.201	-.024	-.179
C(3)	.1578	.0656	-.0943	H(4)	.3094	-.005	.321
C(4)	.2181	.0716	.1756	H(5)	.150	.234	.420
C(5)	.1327	.1977	.2283				

* The atomic coordinates shown are those resulting from conventional refinement with anisotropic thermal parameters.

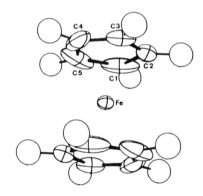

Fig. 6.9-1 Structure of ferrocene. (Afer ref. 2).

Crystal Structure

The early X-ray diffraction studies of ferrocene[3-5] established the sandwich structure and indicated that the cyclopentadienyl (henceforth abbreviated as Cp) ligands should be staggered, since the Fe atom lies at a center of inversion in space group $P2_1/a$. However, a variety of evidence from other experimental techniques strongly suggested that the crystal structure is disordered at room temperature. The heat-capacity data showed that a λ-point transition occurs at 164K,[6] corresponding to an onset of rotational disorder of the Cp rings. An electron diffraction study indicated that ferrocene in the gas phase has an eclipsed equilibrium configuration, and the barrier to internal rotation was estimated to be 3.8 kJ mol^{-1}.[7] A thorough reinvestigation of the structure of crystalline ferrocene by X-ray[2] and neutron diffraction[1] at temperatures of 173K and 298K have shed new light on this complicated problem.

In the ferrocene crystal, the apparently staggered arrangement of the cyclopentadienyl rings results from the presence of molecules in different orientations randomly distributed at both 173K and 298K. A model with

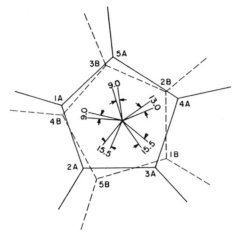

Fig. 6.9-2 A possible nearly-eclipsed configuration for ferrocene which is consistent with the disordered model. (After ref. 1).

disordered Cp ligands was introduced for refinement, as an alternative way of fitting the observed nuclear scattering density which exhibits very broad peaks. A possible configuration consistent with the model, with the two Cp ring rotated about 12° from the eclipsed position, is indicated in Fig. 6.9-2. (The rotation angle is 0° for the precisely eclipsed D_{5h} conformation, and 36° for the exactly staggered D_{5d} conformation).

The perpendicular distance between the rings is 3.25Å, and the mean interatomic distances are Fe-C = 2.033Å, C-C = 1.395Å, and C-H = 1.03Å. An important feature of the structure is that the H atoms on the Cp ligands are displaced significantly toward the Fe atom. At 173K, the mean displacement is 0.030(7)Å, corresponding to an inclination $\theta^* = 1.6(4)°$ of the C-H vector to the best least-squares plane through the five C atoms, which are coplanar to within 0.003Å. From gas-phase electron diffraction θ^* was found to be 3.7(9)°. The averaged geometrical parameters are shown in Fig. 6.9-3.

Remarks

1. Triclinic and orthorhombic ferrocene[8]

The low-temperature triclinic modification of ferrocene crystallizes in space group $F\bar{1}$ with a = 20.960(8), b = 15.019(6), c = 11.421(5)Å, α = 89.47(3), β = 119.93(3), γ = 90.62(3)° (at 101K), and Z = 16; this unconventional unit cell corresponds to doubling of all three axial lengths of the $P2_1/a$ unit cell for the room-temperature monoclinic modification. The two

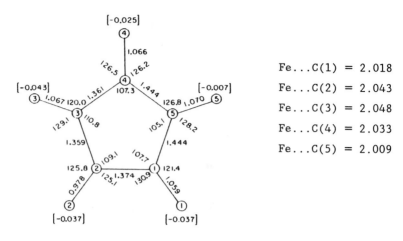

Fe...C(1) = 2.018

Fe...C(2) = 2.043

Fe...C(3) = 2.048

Fe...C(4) = 2.033

Fe...C(5) = 2.009

Fig. 6.9-3 Averaged bond distances (Å) and angles (°), and atomic displacements (Å) from the mean plane in ferrocene. (After ref. 1).

independent $Fe(C_5H_5)_2$ molecules show virtual D_5 symmetry (Fe-C = 2.052, C-C = 1.433Å), the rings being mutually rotated by about 9° from the eclipsed conformation.

The orthorhombic modification of ferrocene crystallizes in space group *Pnma*, with a = 6.987(6), b = 8.995(7), c = 12.196(5)Å, Z = 4 (98K). The ferrocene molecules are exactly eclipsed (molecular symmetry D_{5h}) with average bond distances Fe-C = 2.059 and C-C = 1.431Å, close to the values found for triclinic ferrocene at approximately the same temperature.

2. Simple ferrocene compounds

Ferrocene was first reported by Kealy and Pauson in 1951 as a product from the reaction of cyclopentadienyl magnesium bromide in benzene with anhydrous iron(III) chloride in ether.[10] The reaction was expected to lead to dicyclopentadienyl on a synthetic route towards fulvalene. They noted the exceptional stability of the orange crystalline compound $Fe(C_5H_5)_2$. An independent report of the same compound had been submitted for publication earlier by Miller, Tebboth and Tremaine but appeared later.[11] It was formed from the direct reaction of cyclopentadiene with iron in the presence of aluminum, potassium or molybdenum oxide at 300 °C. Wilkinson has given a personal account of the events leading to establishment of the true nature of this novel material.[12]

Fig. 6.9-4 shows some examples of substitution reactions of ferrocene, and Table 6.9-1 lists the structural data for some ferrocene and ferrocenium compounds.

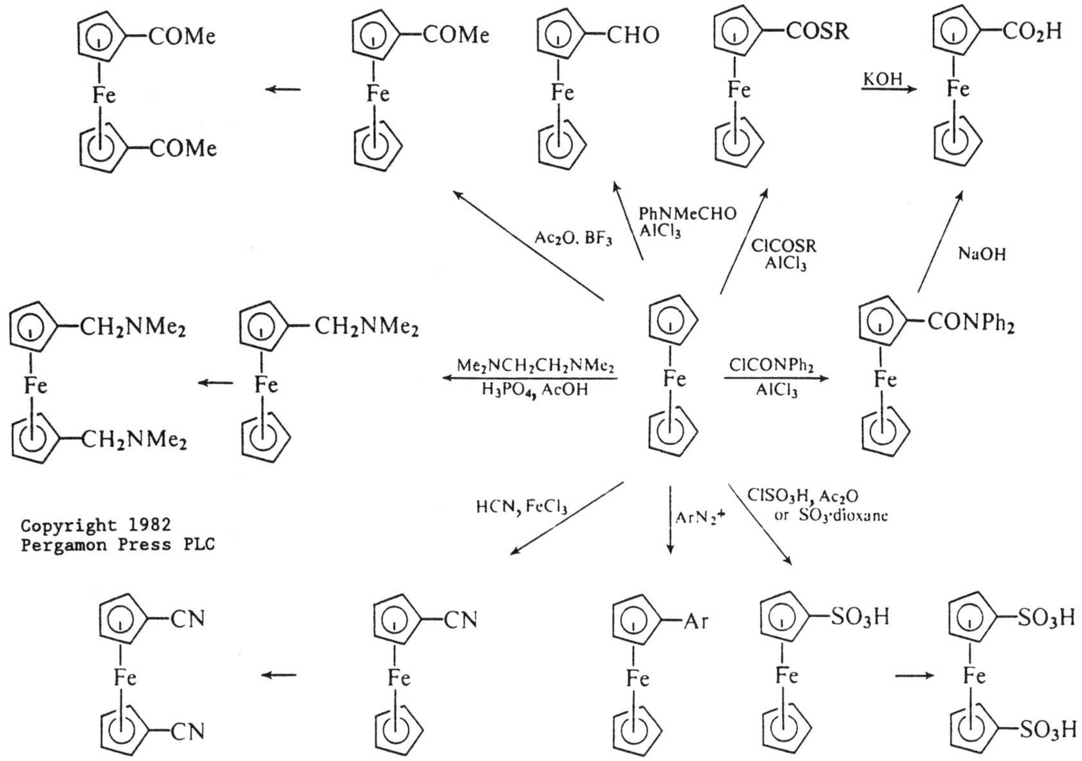

Copyright 1982
Pergamon Press PLC

Fig. 6.9-4 Some examples of substitution reactions of ferrocene.
(After ref. 13).

Table 6.9-1 Structural data for some ferrocene and ferrocenium
compounds.[13]

Compound	Fe-C/Å	C-C in ring/Å	Exocyclic C-C or C-H/Å	θ^*
$Fe(C_5H_5)_2$	2.033	1.395	1.03	+3.7°
$Fe(C_5Me_5)_2$, D_{5d}	2.050	1.419	1.502	-2.4°
$[Fe(C_5H_5)_2][FeCl_4]$	2.073	1.426		
$[Fe(C_5H_5)_2][BiCl_4]$	2.075	1.395		
$[Fe(C_5Me_5)_2][TCNQ](I)$	2.090	1.400	1.515	
$[Fe(C_5Me_5)_2][TCNQ](II)$	2.07	1.39	1.56	
$[Fe(C_5H_5)_2][As_4Cl_{10}O_2]$	2.067	1.380		
$[Fe(C_5H_4Me)_2][I_3]$	2.073	1.401		

3. Molecular orbital description of the bonding in ferrocene

The pπ atomic orbitals on the planar C_5H_5 group can be combined to give
five group orbitals: one combination has the full symmetry of the ring (a) and
there are two doubly degenerate combinations (e_1 and e_2) having respectively

one and two planar nodes at right angles to the plane of the ring. These five group orbitals can themselves be combined in pairs with a similar set from the second C_5H_5 ring. The combinations are labelled g or u (according to whether they are symmetric or antisymmetric with respect to inversion) and can in turn be matched with metal orbitals having similar symmetry properties. Each of the combinations (ligand orbitals + metal orbitals) shown in Fig. 6.9-5 leads in principle to a bonding MO of the molecule, provided that the energy of the two component sets is not very different. There are an equal number of antibonding combinations (not shown) corresponding to (ligand orbitals - metal orbitals). Note, however, that there are no metal orbitals of appropriate symmetry to combine with the degenerate ligand sets labelled e_{2u}.

The sequence of energy levels arising from these combinations is shown schematically in Fig. 6.9-6. The a_{1g} (s, z^2) bonding MO is mainly ligand-based with only a slight admixture of the Fe $4s$ and $3d_{z^2}$ orbitals, whereas $a_{1g}^*(z^2)$ is metal-based and only slightly antibonding, and is the HOMO in the ground state. The a_{2u} level has little, if any, admixture of the higher-lying Fe $4p_z$ orbital with which it is formally able to combine. The e_{1g} MOs arising from the bonding combination of the ligand e_{1g} with Fe $3d_{xz}$ and $3d_{yz}$ orbitals contribute mainly to the stability of the complex; the corresponding antibonding e_{1g}^* are unoccupied in the ground state but become involved in optical transitions. The e_{1u} bonding MOs are again mainly ligand-based but with some contribution from Fe $4p_x$ and $4p_y$. In the ground state there is room for just 18 electrons in bonding and nonbonding MOs and the antibonding MOs are unoccupied. In terms of electron counting the 18 electrons can be thought of as originating from the Fe atom ($8e$) and the two C_5H_5 groups ($2 \times 5e$), or alternatively from an Fe^{2+} ion ($6e$) and two $C_5H_5^-$ groups ($2 \times 6e$).

The X-ray crystal structures of several 19-electron sandwich complexes are known.[14] For example, in $CpFe^I(C_6Me_6)$ both rings are planar and parallel, and the Cp-Fe distance of 1.78Å is 0.1Å longer than in 18-electron complexes whereas the arene-Fe bond of 1.58Å is normal. The crystal structure of cobaltocene shows a lengthening of 0.05Å of the Co-C distances. The well-established sequence $d_{xz}, d_{yz}(e_1) > d_{z^2}(a_1) \geqslant d_{xy}, d_{x^2-y^2}(e_2)$ holds for the electronic structure of metallocenes, bis(arene)metal, and mixed CpM(arene) complexes. According to an extended Hückel study of $CpFe^I(C_6H_6)$, the ligand character of the e_1^* HOMO is 20-30%, so that the 19th electron mainly resides on the metal.

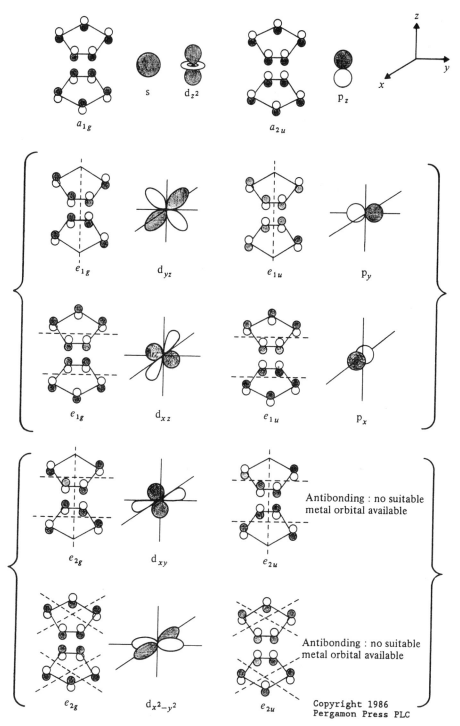

Fig. 6.9-5 Molecular orbitals of ferrocene from the interaction of ligand group orbitals with metal valence orbitals. (After ref. 9).

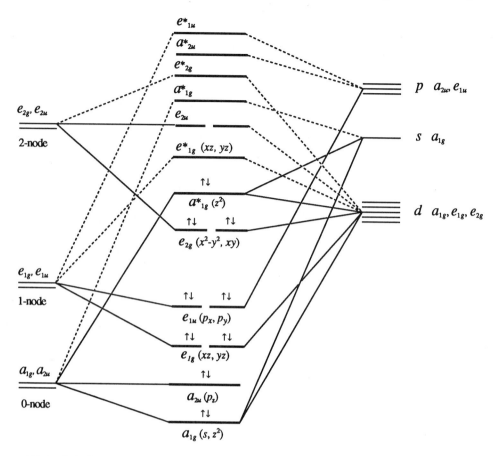

Fig. 6.9-6 A qualitative molecular orbital diagram for ferrocene.

4. Dibenzene chromium

The structure of dibenzene chromium (first described by F. Hein in 1919) is analogous to that of ferrocene, with a metal atom sandwiched between two parallel planar aromatic rings. A low-temperature (100K) X-ray study established that the molecule has D_{6h} symmetry (Fig. 6.9-7).[15] The C-C bond distance are all 1.42Å, as compared to 1.395Å for benzene in the vapour phase. The Cr-C bond distance is 2.15Å.

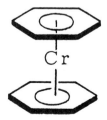

Fig. 6.9-7 Structure of dibenzene chromium.

The principal structural features of $(\eta^6\text{-arene})_2M$ complexes are as follows:[16] (i) the two C_6 rings are planar and parallel and have an eclipsed orientation for benzene and alkyl-benzene derivatives; unusual electron-withdrawing or -releasing substituents like fluorine may cause the C_6 rings to depart from planarity; (ii) the ring C-C distances are equal to within experimental accuracy for C_6H_6 or $C_6(CH_3)_6$ arene ligands, and not significantly different from those of the parent gaseous arene molecule; (iii) the metal-carbon(ring) distances for first-row transition metals in 18-electron complexes are ~2.15Å and the metal to C_6 ring-center separations range from about 1.6 to 1.7Å; and (iv) the ring substituents are slightly out of the C_6 ring plane with hydrogen substituents bent slightly toward the metal atom to achieve a more effective M-C overlap, but bulkier substituents like a methyl group tend to bend away from the metal atom for steric reasons. In first-row transition-metal complexes with more than 18 electrons, the additional electrons typically reside in the metal-carbon antibonding orbitals, resulting in an increase (~ 0.1Å) in the metal-carbon bond distance. Second-row (and presumably third-row) transition-metal complexes that have 20 electrons may behave differently.

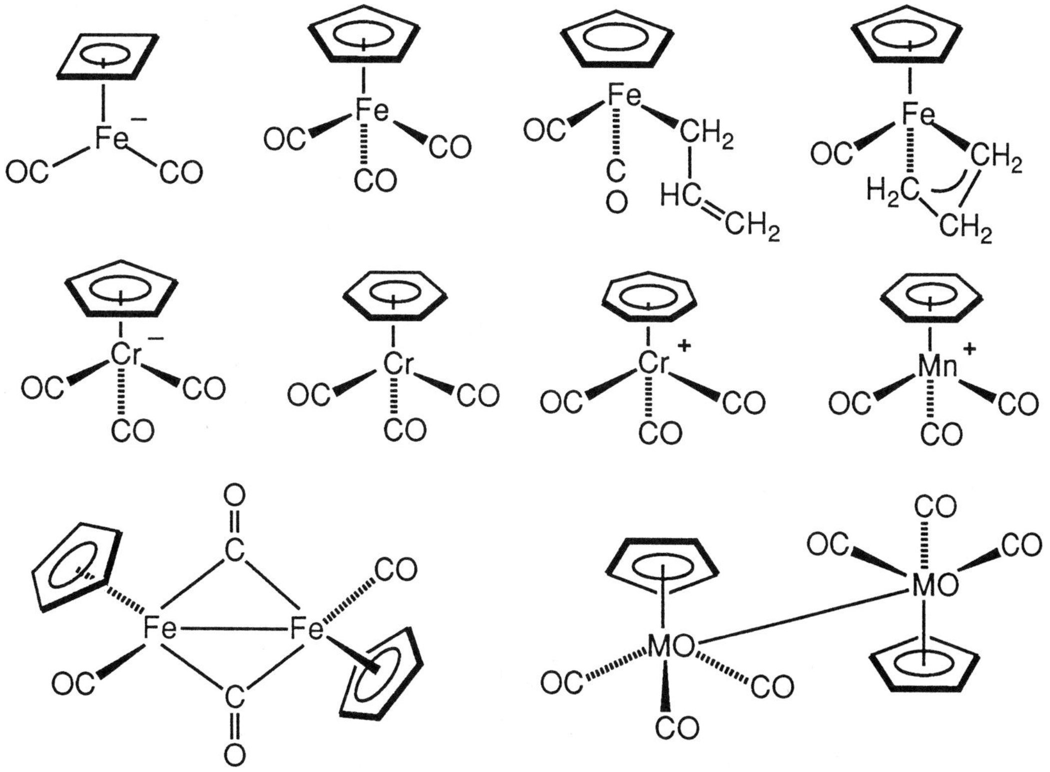

Fig. 6.9-8 Structures of some "half-sandwich" metallocenes.

5. Half-sandwich metallocenes and bent metallocenes[17]

Examples of "half-sandwich" metallocenes are shown in Fig. 6.9-8. Most of these substances conform to the eighteen-electron rule and in many cases the diamagnetism results from the formation of metal-metal bonds.

Some η^5-C_5H_5 complexes are "bent metallocenes", in which the two cyclopentadienyl rings are not parallel, with one to three additional ligands such as H, Cl, CO, R, etc., bonded to the metal (Fig. 6.9-9). The important Ti and Hf derivatives are covered in Section 6.11.

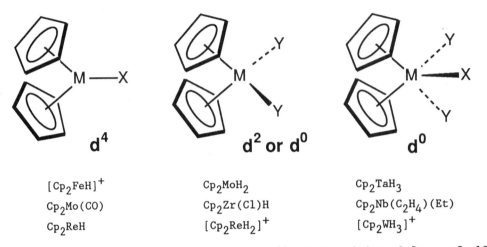

d^4	d^2 or d^0	d^0
$[Cp_2FeH]^+$	Cp_2MoH_2	Cp_2TaH_3
$Cp_2Mo(CO)$	$Cp_2Zr(Cl)H$	$Cp_2Nb(C_2H_4)(Et)$
Cp_2ReH	$[Cp_2ReH_2]^+$	$[Cp_2WH_3]^+$

Fig. 6.9-9 Structures of some bent metallocenes. (Adapted from ref. 18).

6. Derivatives of bis(pentamethylcyclopentadienyl)tungsten(IV)

Among the bent metallocenes, tungstenocene is notable for its wealth of chemical transformations including (i) the first example of the photochemical reductive elimination of dihydrogen, (ii) the insertion of $[Cp_2W]$ into sp^3 C-H bonds in alkene activation, (iii) the first evidence for α-H elimination for $[Cp_2W-CH_3]^+$, (iv) ring opening of a tungstenacyclobutane derivative to yield an alkylidene-olefin intermediate, leading to the Green-Rooney mechanism for olefin polymerization, and (v) the development of rules for predicting the regioselectivity of nucleophilic addition to organometallic cations.[19]

Parkin and Bercaw have described a "wet" entry into permethyltungstenocene chemistry through $Cp^*_2WCl_2$ ($Cp^* = \eta^5$-C_5Me_5) and synthesized a range of halide, hydride, alkyl, oxo and related derivatives.[20] The dioxo derivative, obtainable by the reaction of $Cp^*_2WH_2$ with H_2O_2(aq.), has the structure shown in Fig. 6.9-10(a).[21] In this (η^5-Cp^*)(η^1-Cp^*)WO_2 complex, the *monohapto* bonding mode represents the first example of Cp^* as a η^1-ligand. The other *pentahapto* Cp^* ligand is

asymmetrically bonded and displaced in the direction of the oxo ligands so
that the structure is more appropriately described as $(\eta^1,\eta^4\text{-Cp}^*)(\eta^1\text{-Cp}^*)WO_2$.

The highly reactive mono-oxo derivative $Cp_2^*W{=}O$ undergoes a remarkably
clean reaction with O_2 to form $Cp^*WO_2(OCp^*)$, which has the structure shown in
Fig. 6.9-10(b).[22] The $\eta^5\text{-Cp}^*$ ligand has an appreciable η^1 component as
carbon atom C_α is displaced from the mean plane of the other inner ring atoms
towards W by about 0.05Å, the $W{-}C_\alpha$ distance of 2.334(4)Å is 0.1-0.15Å shorter
than the other $W{-}C(Cp^*)$ distances, and the $C_\beta{-}C_\gamma$ bonds at 1.396(4)Å are
significantly shorter than the other ring bonds; its bonding mode is therefore
more aptly described as η^1,η^4.

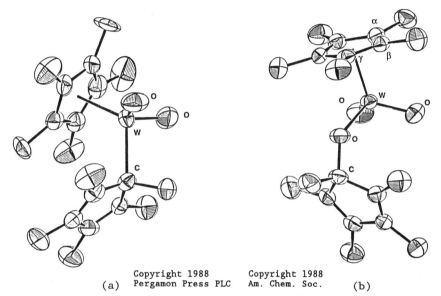

Copyright 1988   Copyright 1988
(a)   Pergamon Press PLC   Am. Chem. Soc.   (b)

Fig. 6.9-10 Molecular structure of (a) $(\eta^5\text{-Cp}^*)(\eta^1\text{-Cp}^*)WO_2$ and
(b) $(\eta^5\text{-Cp}^*)WO_2(OCp^*)$. (After refs. 20 and 22).

The asymmetry of the bonding of the $\eta^5\text{-Cp}^*$ ligand in $(\eta^5\text{-Cp}^*)(\eta^1\text{-Cp}^*)WO_2$
and $(\eta^5\text{-Cp}^*)WO_2(OCp^*)$ may be ascribed to the strong *trans* influence of the two
oxo ligands. The ligands *trans* to each oxo ligand may be considered to be the
double bond components of the Cp^* ligand. The strong W=O bonding weakens the
trans tungsten-ligand bonding, resulting in lengthening of the bonds to these
four carbon atoms.

7. Triple-decker sandwich compounds

Table 6.9-2 lists a number of "triple-decker sandwich" compounds with
measured M-M distances.[23]

The first structurally confirmed all-hydrocarbon triple-decker is
$(C_5H_5)Ni(C_5H_5)Ni(C_5H_5)BF_4$.[24] In the $[Ni_2(C_5H_5)_3]^+$ cation [Fig. 6.9-11(a)],

Table 6.9-2   Some triple-decker sandwich compounds.

Compound	M-M distance/Å	Compound	M-M distance/Å
$(C_5H_5)Ni(C_5H_5)Ni(C_5H_5)^+$	3.580	$(C_5H_5)Fe(C_2B_2SMe_2Et_2)Fe(C_5H_5)$	3.236
$(C_5H_5)V(C_6H_6)V(C_5H_5)$	3.403	$(C_5H_5)Cr(P_5)Cr(C_5H_5)$	2.728
$(C_6H_3Me_3)Cr(C_6H_3Me_3)Cr(C_6H_3Me_3)$	3.338	$(C_5Me_5)Cr(P_5)Cr(C_5Me_5)$	2.727
$(C_5H_5)Co(C_2B_3Me_5)Co(C_5H_5)$	3.140	$(C_5H_5)Fe(P_5)Fe(C_5Me_4Et)$	3.043
$(C_5H_5)Fe(C_3B_2HMe_2Et_2)Co(C_5H_5)$	3.204	$(C_5Me_4Et)Nb(P_6)Nb(C_5Me_4Et)$	2.791
$(C_5H_5)Ni(C_3B_2HMe_2Et_2)Co(C_5H_5)$	3.337	$(C_5Me_5)Mo(P_6)Mo(C_5Me_5)$	2.648
$(C_5H_5)Ni(C_3B_2HMe_2Et_2)Ni(C_5H_5)$	3.416	$(C_5H_4Me)Cr(As_5)Cr(C_5H_4Me)$	2.776
$(C_5H_5)Rh(C_4B_2H_4Me_2)Rh(C_5H_5)^{2+}$	3.440	$(C_5H_5)Mo(As_6)Mo(C_5H_5)$	2.762
$(C_5H_5)Ni(C_3B_2MeEt_4)Pt(C_3B_2MeEt_4)$	3.561	$(C_5Me_4Et)Mo(As_6)Mo(C_5Me_4Et)$	2.639

the average Ni-C distances involving the two terminal Cp rings are 2.09 and
2.08Å, respectively, whereas values of 2.13 and 2.16Å were found for the
average Ni-C distances to the bridging Cp ring; this difference accords well
with the picture obtained from the substitution reactions of the $[Ni_2(C_5H_5)_3]^+$

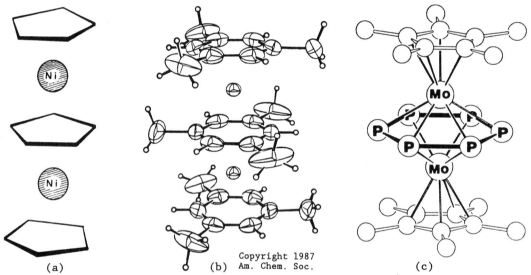

Copyright 1987
Am. Chem. Soc.

(a)   (b)   (c)

Fig. 6.9-11 Molecular structures of (a) $[Ni_2(C_5H_5)_3]^+$, (b)
$Cr_2(C_6H_3Me_3)_3$ and (c) $Mo_2(P_6)(C_5Me_5)_2$. (After refs. 24, 25 and 26).

ion with Lewis bases. The distances from the Ni atoms to the centers of the
rings are: outer rings 1.745 and 1.711Å; inner rings 1.771 and 1.805Å.

In $Cr_2(C_6H_3Me_3)_3$,[25] all three mesitylene rings are planar, parallel,
and nearly exactly eclipsed about the principal molecular axis. The relative
positions of the methyl groups on adjacent rings could not be determined owing
to crystallographic disorder [Fig. 6.9-11(b)].

The triple-decker sandwich compound $(C_5Me_5)Mo(P_6)Mo(C_5Me_5)$ with
hexaphosphabenzene as the central bridging ligand has been prepared and

characterized.[26] The molecule has a centrosymmetric structure in which both C_5 rings as well as the cyclo-P_6 ring (average edge length 2.170Å) are planar and parallel [Fig. 6.9-11(c)]. The Mo-P-Mo angle (mean value 62.8°) reveals an inter-relationship with P_4 and, in combination with the mean Mo-P distance of 2.541Å, gives 2.647Å as the length of the expected Mo-Mo bond. In the analogous complex $(C_5Me_4Et)Mo(As_6)Mo(C_5Me_4Et)$ with hexaarsabenzene as a middle deck, the mean As-As distance (2.35Å) is almost half-way between the values for single and double bonds, 2.44 and 2.24Å respectively. The mean Mo-As distance is 2.694Å, and the mean Mo-As-Mo angle of 58.7° reflects the increase in size of the cyclo-As_6 ring over the cyclo-P_6 ring.[27]

A noteworthy structural feature of the 26-valence-electron triple-decker complex $(C_5Me_5Et)Nb(P_6)Nb(C_5Me_5Et)$ is that its middle deck can be regarded formally as cyclo-P_6^{2-} built of two allyl-like P_3^- units, with two long [average 2.242(9)Å] and four short [average 2.157(9)Å] P-P bond distances.[28]

The pseudo-triple-decker $Cp_2Ru_2(\mu\text{-cyclo-}C_8H_8)$ [Fig. 6.9-12(a)] and the stable "flyover" complex $[Cp_2Ru_2(\mu\text{-cat-}C_8H_8)]^{2+}$ [Fig. 6.9-12(b)] are interconvertible by a two-electron process at ambient temperature.[29] In both isomeric molecules the bridging C_8H_8 ligand comprise two planar C_4 fragments, taking the form of a twisted ring in the neutral complex and an open (*catena*) chain with carbyne-like ends in the dication.

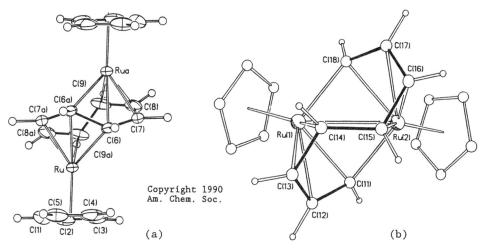

(a)   (b)

Fig. 6.9-12 Molecular structures of (a) $Cp_2Ru_2(\mu\text{-cyclo-}C_8H_8)$ and (b) the dication of $[Cp_2Ru_2(\mu\text{-cat-}C_8H_8)](PF_6)_2 \cdot 0.5C_6H_6$. (After ref. 29).

8. Open metallocenes and pseudo-metallocenes

In recent years bis(pentadienyl)metal complexes or "open metallocenes" and bis(cyclohexadienyl)metal complexes or "pseudo-metallocenes" have attracted considerable interest.[30] Both types of compounds generally adopt *anti* conformation. In the bis(2,3,4-trimethylpentadienyl)ruthenium molecule [Fig. 6.9-13(a)], with reference to the *cis*-eclipsed conformation, the twist angle between the two ligands is 52.5°.[31] In tris(2,4-dimethyl-pentadienyl)lutetium one of the π ligands bonds η^3 to the lanthanide while the remaining two interact with the metal in the normal η^5 fashion [Fig. 6.9-13(b)].[32] Pseudoferrocenes can be prepared in high yield by double addition of carbanions to $[(\eta^6\text{-arene})_2\text{Fe}]^{2+}$ in dichloromethane; in bis(6-*t*-butylcyclohexadienyl)iron [Fig. 6.9-13(c)] the twist angles are 59.5 and 57.5° for the two independent molecules in the crystal.[33]

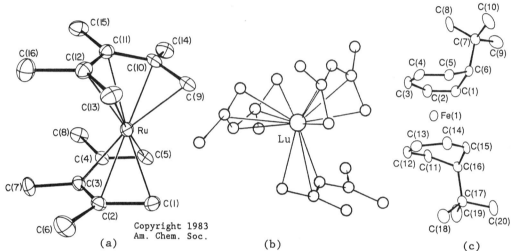

(a) Copyright 1983 Am. Chem. Soc.   (b)   (c)

Fig. 6.9-13 Molecular structures of some open and pseudo metallocenes. (After refs. 31-33).

References

[1] F. Takusagawa and T.F. Koetzle, *Acta Crystallogr., Sect. B* **35**, 1074 (1979).

[2] P. Seiler and J.D. Dunitz, *Acta Crystallogr., Sect. B* **35**, 1068 (1979).

[3] E.O. Fisher and W. Pfab, *Z. Naturforsch., Teil B* **7**, 377 (1952).

[4] P.F. Eiland and R. Pepinsky, *J. Am. Chem. Soc.* **74**, 4971 (1952).

[5] J.D. Dunitz and L.E. Orgel, *Nature (London)* **171**, 121 (1953).

[6] J.W. Edwards, G.L. Kington and R. Mason, *Trans. Faraday Soc.* **56**, 660 (1960).

[7] A. Haaland, *Top. Curr. Chem.* **53**, 1 (1975).

[8] P. Seiler and J.D. Dunitz, *Acta Crystallogr., Sect. B* **35**, 2020 (1979); **38**, 1741 (1982).

[9] N.N. Greenwood and A. Earnshaw, *Chemistry of the Elements*, Pergamon Press, Oxford, 1986, chap. 8.

[10] T.J. Kealy and P.L. Pauson, *Nature (London)* **168**, 1039 (1951).

[11] S.A. Miller, J.A. Tebboth and J.F. Tremaine, *J. Chem. Soc.*, 632 (1952).

[12] G. Wilkinson, *J. Organomet. Chem.* **100**, 273 (1975).

[13] A.T. Deeming in G. Wilkinson, F.G.A. Stone and E.W. Abel (eds.), *Comprehensive Organometallic Chemistry*, Pergamon Press, Oxford, 1982, Vol. 4, chap. 31.3, p. 377.

[14] D. Astruc, *Chem. Rev.* **88**, 1189 (1988), and references cited therein.

[15] E. Keulen and F. Jellinek, *J. Organomet. Chem.* **5**, 490 (1966).

[16] E.L. Mutterties, J.R. Bleeke, E.J. Wucherer and T.A. Albright, *Chem. Rev.* **82**, 499 (1982).

[17] J.W. Lauher and R. Hoffmann, *J. Am. Chem. Soc.* **98**, 1729 (1976).

[18] J.P. Collman, L.S. Hegedus, J.R. Norton and R.G. Finke, *Principles and Applications of Organotransition Metal Chemistry*, 2nd edition, University Science Books, California, 1987.

[19] M.L.H. Green, *Pure Appl. Chem.* **50**, 27 (1978).

[20] G. Parkin and J.E. Bercaw, *Polyhedron* **7**, 2053 (1988).

[21] W.P. Schaefer, G. Parkin and J.E. Bercaw, in preparation.

[22] G. Parkin, R.E. Marsh, W.P. Schaefer and J.E. Bercaw, *Inorg. Chem.* **27**, 3262 (1988).

[23] E.D. Jemmis and A.C. Reddy, *Organometallics* **7**, 1561, 2584 (1988).

[24] E. Dubler, M. Textor, H.-R. Oswald, and A. Salzer, *Angew. Chem. Int. Ed. Engl.* **13**, 135 (1974).

[25] W.M. Lamanna, W.B. Gleason and D. Britton, *Organometallics* **6**, 1583 (1987); W.M. Lamanna, *J. Am. Chem. Soc.* **108**, 2096 (1986).

[26] O.J. Scherer, H. Sitzmann and G. Wolmershäuser, *Angew. Chem. Int. Ed. Engl.* **24**, 351 (1985).

[27] O.J. Scherer, H. Sitzmann and G. Wolmershäuser, *ibid.* **28**, 212 (1989).

[28] O.J. Scherer, J. Vondung and G. Wolmerschäuser, *ibid.* **28**, 1355 (1989).

[29] W.E. Geiger, A. Salzer, J. Edwin, W. von Philipsborn, U. Piantini and A.L. Rheingold, *J. Am. Chem. Soc.* **112**, 7113 (1990).

[30] P. Powell, *Adv. Organomet. Chem.* **26**, 125 (1986); R.D. Ernst, *Chem. Rev.* **88**, 1255 (1988).

[31] L. Stahl and R.D. Ernst, *Organometallics* **2**, 1229 (1983).

[32] H. Schumann and A. Dietrich, *J. Organomet Chem.* **401**, C33 (1991).

[33] K.C. Sturge and M.J. Zaworotko, *J. Chem. Soc., Chem. Commun.*, 1244 (1990).

Note The triple dekker $(\eta^5\text{-}C_5Me_5)Co(\eta^5\text{-}C_5Me_5)Co(\eta^5\text{-}C_5Me_5)$ and dumbbell-like "dimetal sandwich" $[(\eta^5\text{-}C_5Me_5)_2Co_2]$ are described in J.J. Schneider, R. Goddard, S. Werner and C. Krüger, *Angew. Chem. Int. Ed. Engl.* **30**, 1124 (1991).

The isotypic diazametallocenes $(2,5\text{-}C_4H_2Bu_2{}^tN)_2M$ (M = Fe, Co) have C_2 molecular symmetry. See N. Kuhn, M. Köckerling, S. Stubenrauch, D. Bläser and R. Boese, *J. Chem. Soc., Chem. Commun.*, 1368 (1991).

For recent literature on derivatives of aza-, phospha-, arsa-, stiba-, and bismaferrocene, see A.J. Ashe, III, J.W. Kampf and S.M. Al-Taweel, *J. Am. Chem. Soc.* **114**, 372 (1992).

6.10   Bis(cyclopentadienyl)beryllium (Beryllocene)
$$Be(C_5H_5)_2$$

Crystal Data

Monoclinic, space group $P2_1/n$ (No. 14)

$a = 5.993(5)$, $b = 7.478(4)$, $c = 8.978(5)$Å, $\beta = 85.94(6)°$

$Z = 2$ (128K)

Atom	x	y	z	Atom	x	y	z
C(1)	.6045	.1745	.8279	H(1)	.762	.204	.812
C(2)	.4801	.0702	.7343	H(2)	.533	.011	.641
C(3)	.2566	.0571	.7988	H(3)	.136	-.016	.758
C(4)	.2410	.1523	.9316	H(4)	.099	.166	.001
C(5)	.4600	.2236	.9548	H(5)	.495	.306	.030
Be	.4694	-.0282	.9358				

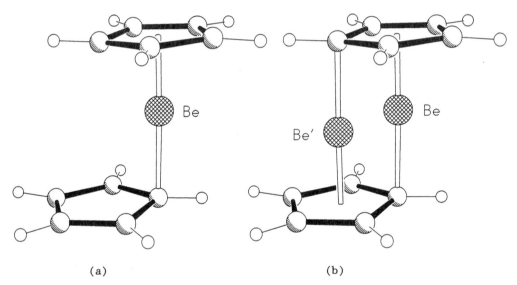

(a)   (b)

Fig. 6.10-1   (a) The ordered molecular configuration and (b) the disordered structure of $Be(C_5H_5)_2$ at 128K.

Crystal Structure

The crystal structure of bis(cyclopentadienyl)beryllium, $Be(C_5H_5)_2$, was first established from X-ray Weissenberg data taken at 153K[1] and later redetermined from diffractometer data at 128K.[2]

The $Be(C_5H_5)_2$ molecule has the "slip-sandwich" structure, as shown in Fig. 6.10-1(a). The beryllium atom is disordered over two equivalent sites, in each of which it is symmetrically bonded to one Cp ring (pentahapto, η^5) and peripherally bonded to the other (monohapto, η^1), being closest to a single carbon atom of that ring. Fig. 6.10-1(b) shows the disordered molecular structure of $Be(C_5H_5)_2$. In the solid the molecules are arranged in the "herringbone" fashion that is typical of many aromatic hydrocarbons (Section 5.13), and the parallel arrangement of the two Cp rings within each molecule does not seem to be determined by packing forces (Fig. 6.10-2).

The disorder of the beryllium atom between two crystallographically equivalent sites implies that the observed Cp ring is the mean between centrally bonded (η^5) and peripherally bonded (η^1) rings. The mean Be-C distances are 1.925Å and 1.826Å for the η^5 and η^1 rings, respectively. In order to arrive at a model for the ordered structure, the η^5 ring is taken to be a pentagon of edge length 1.41Å, which is the average of the five observed C-C distances. The geometrical parameters of the η^1 ring as calculated from twice the difference between the observed and the average C-C distances suggest that it is bonded to the beryllium atom through a largely sp^2 hybridized carbon. This indicates only a small perturbation of the delocalized π-electron structure of the η^1 ring and accounts for the reported Raman[3] and infrared[4] spectra of the compound.

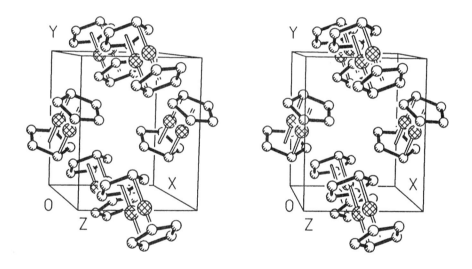

Fig. 6.10-2 Molecular packing in the crystal structure of Be(C₅H₅)₂.

Remarks

1. Monocyclopentadienyl complexes of beryllium[5]

During the last two decades a great number of beryllium compounds containing only one Cp ligand have been characterized (Fig. 6.10-3). Most of these have been synthesized by the reaction between beryllocene and the corresponding disubstituted beryllium compound BeR₂, leading to the mixed products (C₅H₅)BeR (see structure a-e in Fig. 6.10-3). The cyclopentadienyl

complexes **f-i** have been synthesized by the reaction of $(C_5H_5)BeCl$ with $LiBH_4$
(f) and KB_5H_8 (g), and the reaction of NaC_5H_5 with $B_5H_{10}BeBr$ (h) and $Be(B_5H_8)_2$
(i).

    All the complexes **a-i** are air-sensitive colorless solids or liquids.
Structural parameters are available mainly from X-ray structure analysis[6-7]
and from electron diffraction and microwave studies. The beryllium to ring
centroid distances are all similar to that to the nearest ring in beryllocene.
The $Be-CH_3$ bond distance in $(C_5H_5)BeCH_3$ (e) is not significantly different
from the Be-C bond distance in dimethylberyllium (1.698Å), but the Be-Cl bond
distance in $(C_5H_5)BeCl$ (b) is significantly longer both in the gas phase
(1.837Å) and the solid state (1.869Å) than in monomeric $BeCl_2$ (1.77Å). The

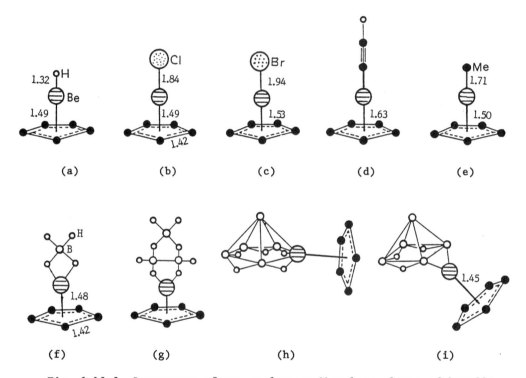

Fig. 6.10-3 Structures of monocyclopentadienyl complexes of beryllium.

hydrogen atoms of the cyclopentadienyl ring in $(C_5H_5)BeCl$ show a slight
tendency to bend toward the beryllium atom, in agreement with theoretical
predictions.[8] In complex (h) the beryllium atom appears to be incorporated
as a vertex in a six-atom *nido*-framework that is structurally and
electronically similar to that of the pentagonal pyramidal B_6H_{10} molecule. In
the π complex (i) the cyclopentadienylberyllium moiety resides in a nonvertex
bridging position between two adjacent basal boron atoms in a square-pyramidal
framework.

2. π Complexes of main group elements[5]

The term "metal π complex" is generally used in connection with compounds in which π ligands are polyhapto bonded to a central metal atom. In other sections we have discussed the π complexes of Li (**6.1**), Mg (**6.2**), Al (**6.3**) and Sn (**6.15**). Many π complexes of other main-group elements, such as Na, K, Ca, B, Ga, In, Tl, C, Si, Ge, Pb, P, As, Sb, S, and Br, have been prepared and investigated by X-ray analysis.

a) π Complexes of the heavier alkali metals

In Fig. 6.10-4 the simplified structures of some π complexes of the heavier alkali metals are illustrated.

Recently, the crystal structure of $K[C_5(CH_2Ph)_5] \cdot 3THF$ has been determined.[9] The potassium ion is bonded in an η^5 fashion to a planar cyclopentadienyl ring, the "ring slippage" being only 0.174Å; the K-Cp (ring center) distance is 2.788(3)Å, and the average K-C distance is 3.035Å. The three oxygen atoms of the THF molecules complete the coordination polyhedron around the central atom to give a coordination number of 8 and an average K-O distance of 2.735Å (Fig. 6.10-4, bottom center). For comparison, the K-C distances in $K[C_5H_4(SiMe_3)]$ is 3.00Å, and the K-O distances in $K[C_5(NO_2)_2Cl_3]$

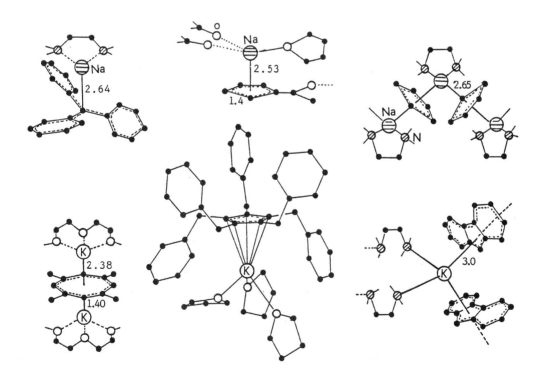

Fig. 6.10-4 Simplified structures of some π complexes of the heavier alkali metals. (Adapted from refs. 5 and 9).

and $K[C_5(CO_2Me)_5] \cdot MeOH$ are 2.58-2.94Å and 2.64-2.88Å, respectively.

   b) π Complexes of the heavier alkaline-earth metals

   The structure of polymeric $(C_5H_5)_2Ca$ shows interesting features [Fig. 6.10-5(a)]. In the crystal, each Ca atom is associated with four planar Cp rings which are bridging to varying degrees. Three of the rings are disposed about the Ca atom in a roughly trigonal manner, two being η^5 and one η^3

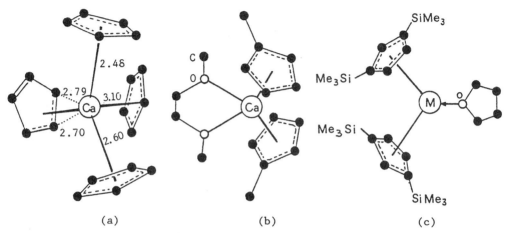

| (a) | (b) | (c) |

Fig. 6.10-5 Structures of (a) $(C_5H_5)_2Ca$ (After ref. 5), (b) $Ca(MeCp)_2 \cdot DME$ (After ref. 10) and (c) $M\{C_5H_3-1,3-(SiMe_3)_2\}_2(THF)$ (M = Ca, Sr). (Adapted from ref. 11).

bonded; the fourth ring is η^1 bonded, all the other carbon atoms in this ring being too far away for significant interaction. The bonding is presumably mainly ionic as evidenced by the long mean Ca-C distances.

   The crystal structure of $Ca(MeCp)_2 \cdot DME$[10] (MeCp = η^5-methylcyclo-pentadienyl, DME = 1,2-dimethoxyethane) shows that the calcium ion is peseudotetrahedrally coordinated by the centers of the two MeCp rings and the O atoms of the bidentate DME ligand [Fig. 6.10-5(b)]. This structure represents the first of the type $MCp_2 \cdot 2L$ or $MCp_2 \cdot \eta^2-L$ in the main group series.

   Calcium, strontium, and barium differ from their lighter group II congeners in that their bis(cyclopentadienyl) derivatives are isolable as Lewis base adducts, e.g., $Ca(C_5H_5)_2(THF)_2$, $M(C_5Me_5)_2(OEt_2)$ (M = Ca, Sr), $Ba(C_5Me_5)_2(THF)_2$, and $M\{C_5H_3-1,3-(SiMe_3)_2\}_2(THF)$ (M = Ca, Sr, Ba). X-ray analysis of latter two isostructural Ca and Sr adducts[11] showed that they comprise monomeric species of crystallographic C_2 symmetry (the important geometrical parameters are, respectively: M-O 2.310(9), 2.49(3)Å; M-ring centroid 2.397, 2.551Å; centroid-M-centroid angle 135.1, 134°) [Fig. 6.10-5(c)].

The solvent-free metallocenes M(Me$_5$C$_5$)$_2$ (M = Ca, Ba) have been obtained from distillation of a toluene solution of M(Me$_5$C$_5$)$_2$(OEt)$_2$ or vacuum sublimation of Ba(Me$_5$C$_5$)$_2$(THF)$_2$.[12] In the solid state both compounds contain two independent, bent molecules in the asymmetric unit. The ring centroid-M-ring centroid angles are (146.3, 147.7°) for the calcium metallocene and (130.9, 131.0°) for the barium analogue. In each crystal structure a third Me$_5$C$_5$ ring is in close proximity to the metal center of each metallocene molecule, as shown in Fig. 6.10-6.

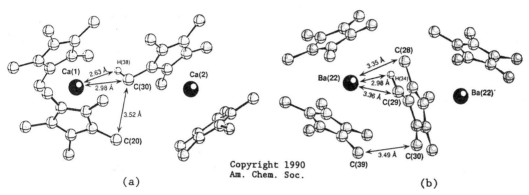

Copyright 1990
Am. Chem. Soc.

(a)   (b)

Fig. 6.10-6 The closest intermolecular contacts in (a) Ca(Me$_5$C$_5$)$_2$ showing the side-by-side approach of the two conformers, and (b) Ba(Me$_5$C$_5$)$_2$ between two symmetry-related molecules. (After ref. 12).

c) π Complexes of Ga, In and Tl

Some representative π complexes of Ga, In and Tl with tetrahedral coordination are shown in Fig. 6.10-7.

d) π Complexes of Si(II)

The first π complex with Si as the central atom, decamethylsilicocene, is a thermally stable but extremely air-sensitive, colorless compound.[13] X-ray analysis showed that two conformers are present in the unit cell in the ratio 1:2, as shown in Fig. 6.10-8.

Molecule **a** occupies a centrosymmetric site, and the two Cp rings are staggered and their planes strictly parallel; in molecule **b**, the two Cp rings are staggered and form an interplanar angle of 25.5°. In molecule **a**, the Si-C separations are equidistant with mean value 2.42Å; in molecule **b**, the Si-C distances lie between 2.324 and 2.541Å. In molecule **a**, the distance between the Si atom and the center of gravity of the Cp rings is 2.11Å; and in molecule **b** 2.12Å. The arrangement of the molecules in the crystal suggests

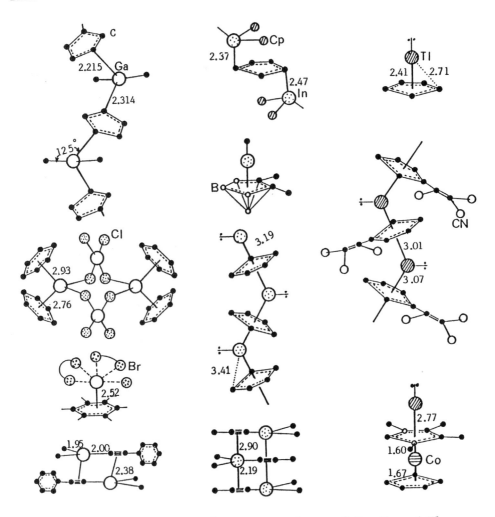

Fig. 6.10-7 Structures of some π complexes of Ga, In and Tl.

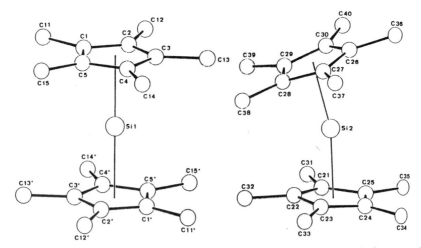

Fig. 6.10-8 Two conformers of decamethylsilicocene. (After ref. 12).

that the interplanar angle of molecule **b** is caused by intermolecular interactions and packing effects.

    e) π Complexes of Ge and Pb

    The structures of some π complexes of Ge and Pb are shown in Fig. 6.10-10. The common structure of Ge π complexes can be described as a more or less distorted sandwich accommodating the presence of a stereochemically active lone pair on the metal, as indicated in Fig. 6.10-9. The distortion can be

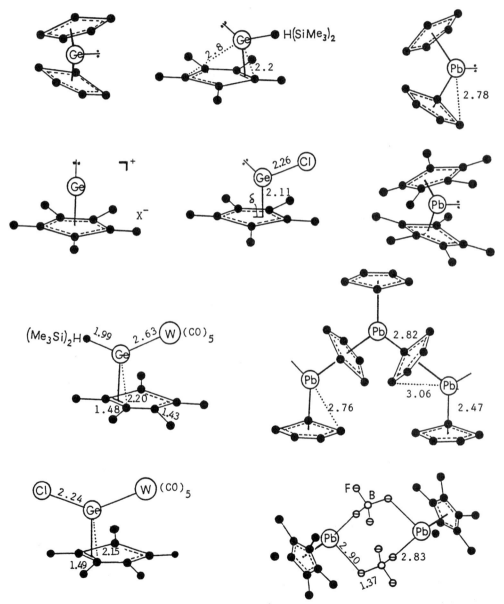

Fig. 6.10-9 Structures of some π complexes of Ge and Pb. (Adapted from 5 and 14).

regarded as a small bending of the Cp centroid-Ge-Cp centroid angle from linearity and as small tilts of the ring planes relative to the Ge-centroid vectors. As a consequence of these tilt the Ge-C distances are spread between 2.34 and 2.73Å, and the Ge atom is no longer required to reside over the center of the Cp rings.

Crystalline plumbocene has a polymeric structure, consisting of zigzag chains of Pb atoms separated by bridging Cp ligands lying perpendicular to the vector joining adjacent Pb centers. An additional Cp ligand is bound to each Pb atom in a η^5 fashion, as indicated in Fig. 6.10-9.

The structural parameters of the Group IVB $M(C_5Me_5)_2$ compounds are listed below:[14]

$M(C_5Me_5)_2$	Ge	Sn	Pb
α	22°	35.4°	37.1°
M-C(Å)	2.34-2.73	2.58-2.77	2.69-2.90
M-C(aver.,Å)	2.52	2.64	2.79

The structure of $Pb(C_5Me_5)BF_4$ determined from X-ray analysis is portrayed in Fig. 6.10-9.[15]

3. Cyclopentadienyl derivatives of group IIB metals

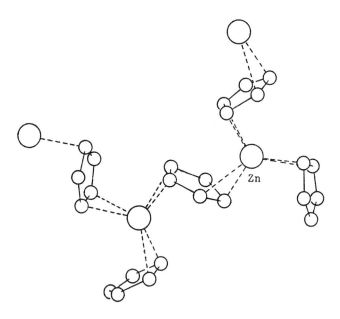

Fig. 6.10-10 Polymeric chain structure and coordination environment of the zinc atoms in $Zn(C_5H_5)_2$. Only one conformation of each of the disordered Cp rings C(11)-C(15) and C(51)-C(55) is shown, and broken lines indicate zinc-carbon distances under 2.50Å. (After ref. 16).

Cyclopentadienyl and isolobal five-membered ring systems serve as π-electron donor ligands to group IIB metals, but often both σ- and π-interactions contribute to the cyclopentadienyl-metal bonds. The cyclopentadienyl complexes of mercury are discussed in Section **6.14**.

The crystal structure of dicyclopentadienylzinc, $Zn(C_5H_5)_2$, comprises infinite zigzag chains of zinc atoms bridged by Cp groups; in addition, each metal atom carries a terminal Cp ligand (Fig. 6.10-10).[16] Two crystallographically independent zinc atoms occur in pairs along each chain running parallel to the *a* axis. The bridging Cp groups lying between zinc atoms of the same type are located at inversion centers and therefore disordered.

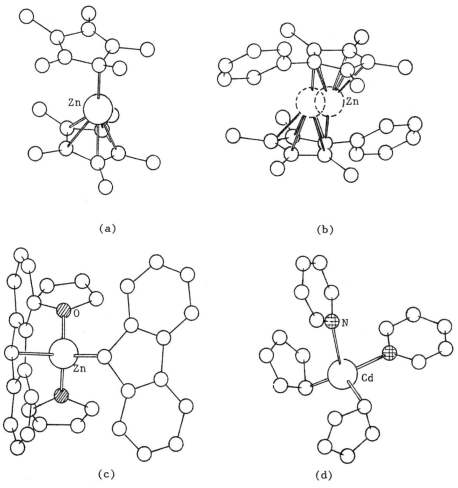

(a) (b)

(c) (d)

Fig. 6.10-11 Structures of (a) $Zn(C_5Me_5)_2$, (b) $Zn(C_5Me_4Ph)_2$, (c) $Zn(C_{13}H_9)_2 \cdot 2THF$ and (d) $Cd(C_5H_5)_2 \cdot 2py$. (After refs. 17-19).

Crystals of $Zn(C_5Me_5)_2$ and $Zn(C_5Me_4Ph)_2$[17] both comprise monomeric molecules with one η^5- and one η^1-bonded ring [Fig. 6.10-11(a) and (b)]. For $Zn(C_5Me_4Ph)_2$ the two parallel flat C_5Me_4Ph rings are centrosymmetrically related and enclose one Zn atom disordered over two crystallographically equivalent sites. The two phenyl groups are rotated 46.7(2)° with respect to the cyclopentadienyl ring plane.

In the crystals of the bis(tetrahydrofuran) complex of difluorenylzinc, $Zn(C_{13}H_9)_2 \cdot 2THF$,[18] and of the pyridine complex of dicyclopentadienylcadmium, $Cd(C_5H_5)_2 \cdot 2py$,[19] the metal atoms are all surrounded pseudotetrahedrally by four η^1-bonded ligands [Fig. 6.10-11(c) and (d)]. In $Zn(C_{13}H_9)_2 \cdot 2THF$, each Zn atom is coordinated by pairs of fluorenyl groups and THF molecules with C-Zn-C and O-Zn-O angles of 117.6(2) and 89.3(2)°, respectively. The Zn-O distances are normal for coordinating THF, being 2.114(5) and 2.095(4)Å. The Zn-C distances of 2.041(5) and 2.053(6)Å are normal for covalent Zn-C bonds. In $Cd(Cp)_2 \cdot 2py$, each Cd atom is coordinated by two η^1-Cp rings and two pyridine ligands with N-Cd-N and C-Cd-C angles of 96.4(1)° and 129.1(2)°, respectively. The Cd-C [2.307(5) and 2.353(5)Å] and Cd-N bonds [2.309(3) and 2.360(3)Å] are all sufficiently strong to favour the existence of a monomeric molecule.

The first zinc-nickel cluster compound $Zn_4Ni_2(C_5H_5)_6$ was synthesized by the reaction of dicyclopentadienylzinc with bis(1,4-cylooctadiene)nickel. Crystal-structure determination has shown that this heteronuclear compound has a *trans*-octahedral metal core, which is completely shielded by disordered η^5-C_5H_5 ligands (Fig. 6.10-12).[20]

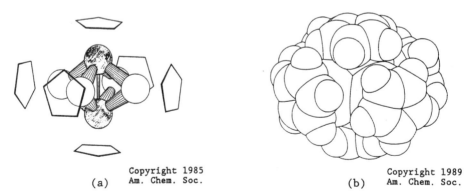

Fig. 6.10-12 Structure of the octahedral cluster $Zn_4Ni_2(C_5H_5)_6$: (a) perspective view showing only one orientation for each disordered η^5-C_5H_5 ring; (b) space-filling model showing the shielding of the metal core by the ligands. (After refs. 20 and 21).

The related mixed-metal, mixed-ligand cluster compound
$Zn_4Ni_2(C_5H_5)_4(C_5Me_5)_2$ was obtained by reacting equimolar amounts of $Zn(C_5H_5)_2$,
$Zn(C_5Me_5)_2$ and $Ni(COD)_2$. X-Ray analysis has established that the molecule is
a centrosymmetric cluster based on the same Zn_4Ni_2 core. The C_5Me_5 groups are
η^2-bonded to zinc, and the C_5H_5 rings coordinate to zinc and nickel in an
η^2-η^3 and η^5 mode, respectively (Fig. 6.10-13).[21]

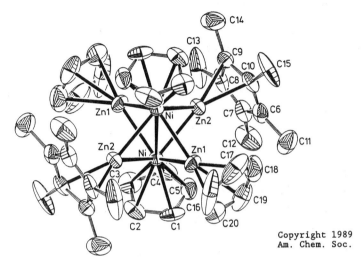

Fig. 6.10-13 Molecular structure of the $Zn_4Ni_2(C_5H_5)_4(C_5Me_5)_2$ molecule.
The bonding between Zn(1) and the C(16)-C(20) ring is intermediate
between η^2- and η^3-coordination. (After ref. 21).

References

[1] C.-H. Wong, T.-Y. Lee, K.-J. Chao and S. Lee, *Acta Crystallogr., Sect. B*
28, 1662 (1972).

[2] K.W. Nugent, J.K. Beattie, T.W. Hambley and M.R. Snow, *Aust. J. Chem.* **37**,
1601 (1984).

[3] J. Lusztyk and K.B. Starowieyski, *J. Organomet. Chem.* **170**, 293 (1979).

[4] K.W. Nugent and J.K. Beattie, *Inorg. Chem.* **27**, 4269 (1988).

[5] P. Jutzi, *Adv. Organomet. Chem.* **26**, 217 (1986); *J. Organomet. Chem.* **400**,
1 (1990).

[6] D.F. Gaines, J.L. Walsh, J.H. Morris and D.F. Hillenbrand, *Inorg. Chem.*
17, 1516 (1978).

[7] R. Goddard, J. Akhtar and K.B. Starowieyski, *J. Organomet. Chem.* **282**, 149
(1985).

[8] E.D. Jemmis, S. Alexandratos, P.v.R. Schleyer, A. Streitwieser Jr. and
H.F. Schaefer III, *J. Am. Chem. Soc.* **100**, 5695 (1978).

[9] J. Lorberth, S.-H. Shin, S. Wocadlo and W. Massa, *Angew. Chem. Int. Ed.
Engl.* **28**, 735 (1989).

[10] A. Hammel, W. Schwarz and J. Weidlein, *J. Organomet. Chem.* **378**, 347 (1989).

[11] L.M. Engelhardt, P.C. Junk, C.L. Ra ston and A.H. White, *J. Chem. Soc., Chem. Commun.*, 1500 (1988).

[12] R.A. Williams, T.P. Hanusa and J.C. Huffman, *Organometallics* **9**, 1128 (1990); T.P. Hanusa, *Polyhedron* **9**, 1345 (1990).

[13] P. Jutzi, D. Kanne and C. Krüger, *Angew. Chem. Int. Ed. Engl.* **25**, 164 (1986).

[14] P. Jutzi, *Pure Appl. Chem.* **61**, 1731 (1989).

[15] P. Jutzi, R. Dickbreder and H. Nöth, *Chem. Ber.* **122**, 865 (1989).

[16] P.H.M. Budzelaar, J. Boersma, G.J.M. Van Kerk, A.L. Spek and A.J.M. Duisenberg, *J. Organomet. Chem.* **281**, 123 (1985).

[17] B. Fischer, P. Wijkens, J. Boersma, G. van Koten, W.J.J. Smeets, A.L. Spek and P.H.M. Budzelaar, *J. Organomet. Chem.* **376**, 223 (1989).

[18] B. Fischer, J. Boersma, G. van Koten, W.J.J. Smeets and A.L. Spek, *Organometallics* **8**, 667 (1989).

[19] B. Fischer, G.P.M. van Mier, J. Boersma, W.J.J. Smeets and A.L. Spek, *J. Organomet. Chem.* **322**, C37 (1987).

[20] P.H.M. Budzelaar, J. Boersma, G.J.M. van der Kerk, A.L. Spek and A.J.M. Duisenberg, *Organometallics* **4**, 680 (1985).

[21] B. Fischer, H. Kleijn, J. Boersma, G. van Koten and A.L. Spek, *Organometallics* **8**, 920 (1989).

Note The polymeric structures of $(C_5H_5)SbCl_2$ (chiral chain containing bridging Cp rings) and $(C_5H_5)BiCl_2$ ($TcCl_4$-type chain with terminal Bi-Cp π bonding) are described, respectively, in W. Frank, *J. Organomet. Chem.* **406**, 331 (1991) and **386**, 177 (1990).

Main group I-IV metallocenes with substituted cyclopentadienyl ligands are reviewed in H. Schumann, *Pure Appl. Chem.* **63**, 813 (1991).

6.11 μ-Dinitrogen-bis[bis(pentamethylcyclopentadienyl)titanium] $\{(\eta^5\text{-}C_5Me_5)_2Ti\}_2N_2$

Crystal Data

Triclinic, space group $P\bar{1}$ (No. 2)

$a = 18.867(1)$, $b = 8.968(2)$, $c = 22.767(1)$Å

$\alpha = 98.22(1)°$, $\beta = 101.83(1)°$, $\gamma = 93.86(1)°$, $Z = 4$

Atom*	x	y	z
Ti(1)	.15393	.86362	.15287
Ti(2)	.34256	1.15462	.33908
Ti(3)	.33269	.64639	.84493
Ti(4)	.16398	.35278	.64669
N(1)	.22497	.98516	.22405
N(2)	.26792	1.04962	.26575
N(3)	.26318	.53812	.76793
N(4)	.22618	.47300	.72360

* The carbon and hydrogen atoms have been omitted.

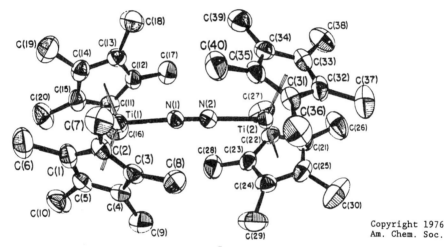

Copyright 1976
Am. Chem. Soc.

Fig. 6.11-1 Molecular configuration of $\{(\eta^5\text{-}C_5Me_5)_2Ti\}_2N_2$. (After ref. 1).

Crystal Structure

There are two independent molecules of the same structure in the crystal (Fig. 6.11-1).[1] The binuclear molecular skeleton consists of two $(\eta^5\text{-}C_5Me_5)_2Ti$ moieties bridged by the N_2 ligand in an essentially linear Ti-N≡N-Ti arrangement. The N-N distances are 1.165(14) and 1.155(14)Å for the two molecules in the assymmetric unit. This is a rare example of a trigonal bis(cyclopentadienyl)titanium derivative, for most titanocene derivatives have two additional ligand atoms occupying coordination sites about Ti. The average Cp centroid-Ti-Cp centroid angle is 145.7°. The Cp rings are closely planar and the average bond lengths are: $(C\text{-}C)_{ring}$ 1.404Å; $Ti\text{-}C_{(ring)}$ 2.387Å. The skeletal views of the two independent molecules are shown in Fig. 6.11-2.

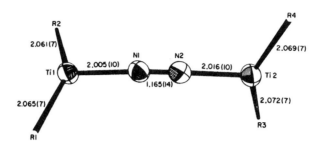

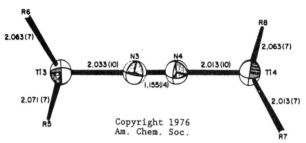

Fig. 6.11-2 Skeletal views of the two independent molecules of
$\{(\eta^5\text{-}C_5Me_5)_2Ti\}_2N_2$. The center of a C_5Me_5 ring is represented by R.
(After ref. 1).

Remarks

1. The quest for titanocene

Since 1956, many research groups have reported on the synthesis of
"titanocene" by various methods which produced complexes with different
colours and spectroscopic properties. A range of different materials was
isolated due to the very reactive nature of titanocene and its ability to
readily abstract hydrogens from the coordinated Cp ligands. None of the
complexes could be isolated in a crystalline form suitable for an X-ray
diffraction study. Neither $TiCp_2$ nor $[TiCp_2]_2$ has been isolated as a
crystalline solid.[2]

Reduction of $(\eta^5\text{-}C_5H_5)_2TiCl_2$ with potassium naphthalene in THF at low
temperature yielded a "bis(titanocene)". Single-crystal X-ray diffraction has
shown that its correct formulation is $(C_5H_5)_3(C_5H_4)Ti_2(THF)\cdot THF$, in which one
of the Cp ligands contains only four hydrogen atoms and serves to bridge the
two Ti centers in a η^1,η^5 bonding mode (Fig. 6.11-3).[3]

The most significant structural feature of this molecule is the high
degree of coordinative unsaturation about the Ti-Ti bond of length 3.336Å.
The THF ligand may readily be removed from the adduct by treatment with *n*-
octane under vacuum to yield very pure samples of $(\eta^5\text{-}C_5H_5)_2Ti\text{-}\mu\text{-}(\eta^1,\eta^5\text{-}C_5H_4)\text{-}$
$Ti(\eta^5\text{-}C_5H_5)$. This dimeric compound reacted with ammonia to produce a red

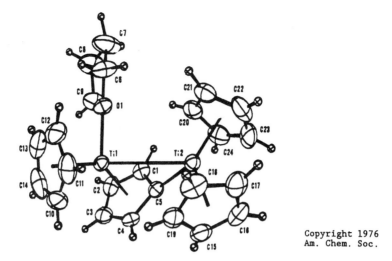

Fig. 6.11-3 Molecular structure of $(C_5H_5)_3(C_5H_4)Ti_2(THF)\cdot THF$ (the solvated THF molecule is not shown). (After ref. 3).

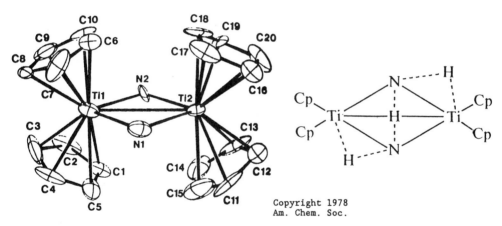

Fig. 6.11-4 Molecular geometry and structural formula of $(Cp_2TiNH)_2H$. (After ref. 4).

complex $(Cp_2TiNH)_2H$ and hydrogen. An X-ray study of $(Cp_2TiNH)_2H$ yielded Ti-Ti = 3.39Å, Ti-N = 2.2Å, Ti-C = 2.4Å, N...N = 2.9Å, N-Ti-N = 80°, and Ti-N-Ti = 98° (Fig. 6.11-4).[4] The hydrogen positions were deduced by IR spectroscopy and by reactions with hydride scavengers.

Decamethyltitanocene, $Ti(C_5Me_5)_2$, was isolated as an orange crystalline solid, but no structural information is available. On addition of N_2 to $Ti(C_5Me_5)_2$ at low temperatures, two complexes of stoichiometry $Ti_2N_2(C_5Me_5)_4$ (Fig. 6.11-1) and $Ti_2(N_2)_3(C_5Me_5)_4$ were formed.[5] The zirconium analogue of the latter has the structure shown in Fig. 6.11-5.[6]

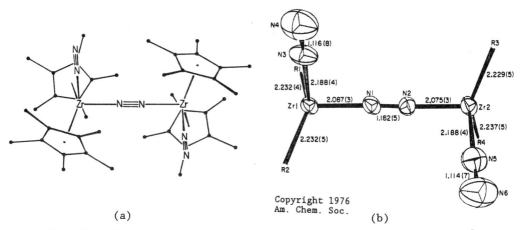

(a)   (b)

Fig. 6.11-5 (a) Structural formula and (b) skeletal view of $\{(\eta^5\text{-}$
$C_5Me_5)_2ZrN_2\}_2N_2$. (After ref. 6).

2. Structures of titanium cyclopentadienyl complexes

The organometallic chemistry of titanium is dominated by work on the
more stable Cp derivatives, many of which have been structurally
characterized.[14]

Ti(CO)$_2$Cp$_2$[7] possesses almost exact C_{2v} symmetry [Fig. 6.11-6(a)]. The
Ti-C(carbonyl) bond length of 2.03Å agrees with that predicted from metallic
radii considerations, while the Ti-Cp(centroid) distance of 2.025Å is shorter
than those in Ti(IV) complexes, indicating that the titanium d electrons are
in an orbital that is at least partially bonding with the Cp groups. In

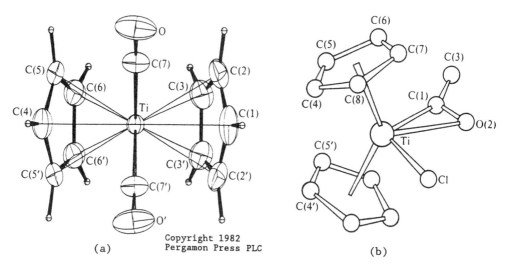

(a)   (b)

Fig. 6.11-6 Molecular structures of (a) Ti(CO)$_2$Cp$_2$ and
(b) TiCl(COMe)Cp$_2$. (After refs. 2 and 8).

TiCl(COMe)Cp$_2$,[8] the titanium-acyl bond is greatly distorted owing to a bonding interaction between the metal and the carbonyl oxygen, so that the

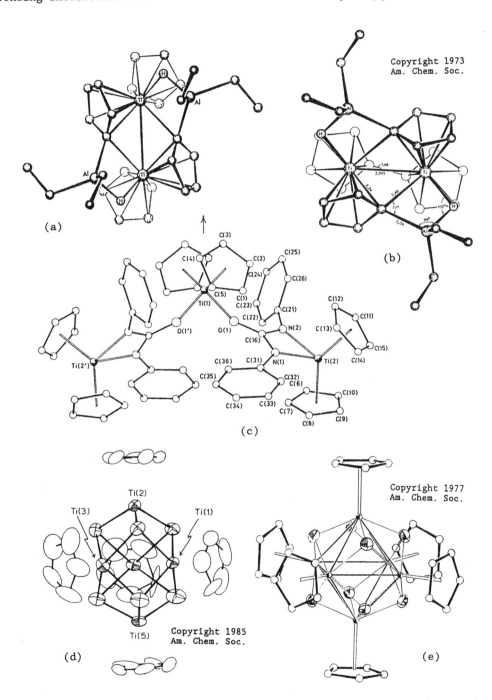

Fig. 6.11-7 Molecular structure of (a) {TiH(AlEt$_2$)(η-C_5H_4)Cp}$_2$, (b) {TiH(AlEt$_2$)(C_5H_4)}$_2$(η-$C_{10}H_8$), (c) Cp$_6$Ti$_3${μ-OC(NPh)$_2$}$_2$, (d) Cp$_5$Ti$_5$(μ_3-S)$_6$ and (e) Cp$_6$Ti$_6$(μ_3-O)$_8$. (After refs. 9-13).

acyl group may be regarded as a three-electron ligand [Fig. 6.11-6(b)].

In the complex {TiH(AlEt$_2$)(η-C$_5$H$_4$)Cp}$_2$ the titanium atoms are bridged by two aluminum atoms and hydrides.[9] The bridging hydrides were thought to be derived from hydrogen abstraction from an η-cyclopentadienyl ligand [Fig. 6.11-7(a)]. This complex was found to be a homogeneous alkene polymerization catalyst.

In the complex {TiH(AlEt$_2$)(C$_5$H$_4$)}$_2$(η-C$_{10}$H$_8$)[10] the fulvalene ligand remains intact but a hydrogen has been abstracted from each of the η-Cp groups that are linked to a titanium by an aluminum and hydride bridge [Fig. 6.11-7(b)].

The structure of Cp$_2$Ti$_3$(μ-OC(NPh)$_2$)$_2$ is composed of two Cp$_2$Ti(OC(NPh)$_2$) moieties symmetrically linked to a central Cp$_2$Ti unit through the carbonyl oxygen atoms of the diphenylureylene ligands, as shown in Fig. 6.11-7(c).[11] The complex has a magnetic moment of 1.72 BM per Ti atom at 296K. To account for the structural and magnetic properties of the compound, Ti(1) is assigned as Ti^{IV} while Ti(2) and Ti(2') are assigned as Ti^{III}.

The structure of Cp$_5$Ti$_5$(μ_3-S)$_6$ is composed of a distorted trigonal bipyramidal Ti$_5$ core.[12] The faces of the bipyramid are capped with triply-bridging thio ligands and each titanium atom is bonded to a Cp ligand, as shown in Fig. 6.11-7(d). The paramagnetism (μ_{eff} = 1.66 BM) and ESR properties (g = 1.993) of this compound are consistent with the presence of a single unpaired electron. The assignment of oxidation states III and IV to the equatorial and axial Ti atoms, respectively, is supported by the observed Ti-S distances [Ti(1)-S = 2.474, Ti(2)-S = 2.286, Ti(3)-S = 2.463, Ti(4)-S = 2.486, and Ti(5)-S = 2.272Å].

The structure of Cp$_6$Ti$_6$(μ_3-O)$_8$ has planar Cp rings located at the apices of a Ti$_6$ octahedron, each face being capped by a μ_3-O atom [Fig. 6.11-7(e)].[13] The high symmetry and diamagnetism of the complex precludes a localized Ti$_4^{IV}$Ti$_2^{III}$ description.

The green-black compound prepared from TiCl$_4$ and NaC$_5$H$_5$ has been formulated as Ti(η^1-Cp)$_2$(η^5-Cp)$_2$, which exhibits a single ^{1}H NMR resonance owing to two types of fluxional processess: rapid interchange of η^1 and η^5 rings and variation of the point of attachment of each η^1 ring over its five carbon atoms ("ring whizzing").

3. Zirconium and hafnium cyclopentadienyl complexes

X-ray analysis has shown that ZrCp$_4$ possesses one η^1 and three η^5 rings[15] whereas HfCp$_4$,[16] like its Ti analogue, contains two η^1 and two η^5 rings. These two compounds provide the first authenticated example of a

difference in the structural organometallic chemistry of zirconium and
hafnium.

Many zirconocene and hafnocene complexes have been prepared. In (s-cis-
η^4-1,3-diene)MCp$_2$ (M = Zr and Hf), the coordination of the double bonds of
the cis-1,3-diene ligands to the bent M(η^5-Cp)$_2$ fragment leads to a pseudo-
tetrahedral geometry, as shown in Fig. 6.11-8.[17] The conjugated diene
moiety (atoms C1, C2, C3, and C4) of the ligands remains essentially planar.
Fig. 6.11-9 shows the structures of (2,3-dimethylbutadiene)-ZrCp$_2$ and -HfCp$_2$.
The structural data of these two compounds are listed in Table 6.11-1.

Table 6.11-1 A comparison of selected structural data for
(s-cis-2,3-dimethylbutadiene)-ZrCp$_2$ and -HfCp$_2$.

	Zr	Hf
M-C1	2.300(3)	2.267(5)
M-C2	2.597(3)	2.641(5)
C1-C2	1.451(4)	1.472(8)
C2-C3	1.398(4)	1.378(8)
C1-M-C4	80.2(1)	80.3(2)
M-C1-C2	84.4(2)	87.3(3)
C1-C2-C3	122.7(3)	121.7(5)
C3-C2-R	121.8(3)	122.1(5)
α	123.9	124.3
β	112.8	116.5
γ	172.9	171.1

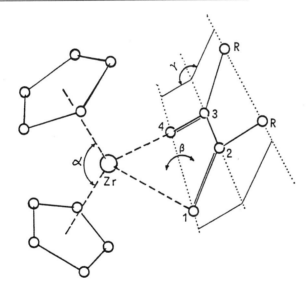

Fig. 6.11-8 Definition of some geometric parameters used for the
characterization of (s-cis-diene)metallocene complexes. (After ref. 17).

Differences in atomic radii between Zr(1.45Å) and Hf(1.44Å) suggest that
covalent bonds formed by these elements will be of comparable lengths with a
tendency for those of Hf to be somewhat shorter. The available structural
data may be summarized as follows:

a) Hf-C bonds are slightly shorter than Zr-C bonds;

b) (s-*cis*-η^4-diene)hafnocene complexes show the same bonding characteristics as do their zirconocene analogues;

c) the σ/π ratio in the diene bonding is shifted slightly towards a larger σ character in the case of Hf;

d) the ligand dependency of this σ/π ratio is the same for the hafnocene complexes examined so far as for the respective zirconocene complexes.

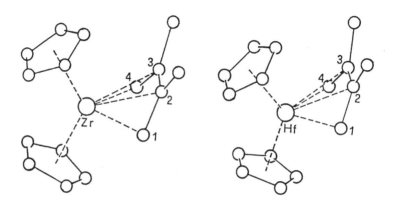

Fig. 6.11-9 The molecular geometries of (s-*cis*-2,3-dimethylbutadiene)-ZrCp$_2$ and -HfCp$_2$.

The s-*trans*-1,3-diene zirconocene complexes are also pseudotetrahedrally coordinated. The dienes are located in a plane roughly bisecting the Cp-Zr-Cp group. In contrast to the mirror symmetry of most s-*cis*-1,3-diene complexes, the molecular symmetry of s-*trans* complexes is approximately C_2. The bond length pattern of the Zr-C and C-C bonds is the reverse of that observed in the s-*cis* complexes. Thus the bonds of Zr to the terminal carbon atoms C1/C4 are longer than those to the internal ones C2/C3, the former coming close to the values of π-bonded Cp rings. In their short-long-short sequences the diene ligands more closely resemble those observed in free 1,3-dienes, as does their *trans* geometry. There is also a marked deviation from planarity.

4. Metallocene complexes as antitumor agents

Metallocene dichlorides of the early transition metals, $(\eta^5\text{-}C_5H_5)_2MX_2$, represent the first group of organometallic compounds found to exhibit cancerostatic properties. The *cis*-dichlorometal fragment, which is also present in the potent antitumor drug *cis*-diamminedichloroplatinum(II) (DDP, *cis*-platin, Section 4.20), apparently plays a critical role in the mechanism of action. Table 6.11-2 summarizes the relevant structural parameters for the metallocene dichlorides and the inorganic cytostatic drug DDP.[18]

Table 6.11-2 Structural parameters of DDP and some metallocene dichlorides.

Compound	M	M-Cl[a] (Å)	Cl-M-Cl (°)	Bite[b] (Å)
(C$_5$H$_5$)$_2$MoCl$_2$	Mo	2.471(5)	82.0(2)	3.242
(C$_5$H$_4$Me)$_2$VCl$_2$	V	2.398(2)	87.06(9)	3.303
cis-(NH$_3$)$_2$PtCl$_2$, DDP	Pt	2.330(1)	91.9(3)	3.349
(C$_5$H$_5$)$_2$NbCl$_2$	Nb	2.470(4)	85.6(1)	3.356
(C$_5$H$_5$)$_2$TiCl$_2$	Ti	2.364(3)	94.43(6)	3.470
[(CH$_2$)$_3$(C$_5$H$_4$)$_2$]HfCl$_2$	Hf	2.423(3)	95.87(8)	3.598
(C$_5$H$_5$)$_2$ZrCl$_2$	Zr	2.441(5)	97.1(2)	3.660

[a] Mean M-Cl distance. [b] Calculated non-bonded Cl...Cl distance.

For DDP the Cl...Cl bite of 3.35Å corresponds to the distance between two appropriate guanine-base N(7) atoms of a DNA helix, such that their coordination to the *cis*-Pt(NH$_3$)$_2$ fragment (after dissociation of the Cl ligands) makes the guanine rings tilt from their normal stacked positions, leading to disruption of the helix and to interference with the replication process.[19,20] All tumor-inhibiting metallocene dichlorides have bite distances within ±0.12Å of that of DDP. In contrast, the therapeutically ineffective zirconocene and hafnocene dichlorides have markedly larger bites which exceed the critical limit of 3.50-3.60Å for the formation of DNA-metal cross-links.

Studies on the structure-activity relation of metallocene dihalides led to the following generalizations:[18]

   a) There is a strong dependence of the antitumor action on the nature (especially size) of the metal atom. The complexes of Ti, V, Nb and Mo are most effective, but the activity declines markedly when Ta, W, Zr and Hf are used.

   b) Within the titanocene dihalide system, the Cl ligands can be replaced by other halide or pseudohalide ligands without reduction of antitumor potency.

   c) Chemical modification of the Cp rings, e.g. mono- or 1,1'-disubstitution with organic residues, leads to a diminution of cytostatic activity depending on the degree of substitution.

A variety of biological data from preclinical studies have established that cyclopentadienyl metal complexes of several structural types exhibit antiproliferative properties against many experimental tumors, e.g. Ehrlich ascites tumor, B16 melanoma, sarcoma 180, Lewis lung carcinosarcoma, colon 38 adenocarcinoma, leukemias L1210 and P388, as well as against various human

tumors heterotransplanted to athymic mice.[21] Apart from the well-documented
metallocene diacido complexes, antitumor activity has been detected for ionic
metallicenium salts of the electron-rich medium transition metals iron and
cobalt, such as $[(\eta^5-C_5H_5)_2Fe]^+FeCl_4^-$, and uncharged decasubstituted
metallocenes with tin(II) and germanium(II) as the central atom. X-Ray
analysis has shown that in decabenzylstannocene the Cp rings are inclined to
each other,[22] resulting in an "open sandwich" structure similar to that of
unsubstituted Cp_2Sn. On the other hand, the Cp rings adopt a parallel
arrangement as a consequence of the steric interaction between the phenyl
substituents in both decaphenylgermanocene[23] and decaphenylstannocene.[24]

5. Mononuclear titanium(III) cyclopentadienyl complexes

Only a few monomeric compounds of this class are known. In
$Cp_2TiO(2,6-Me_2C_6H_3)$[25] and $Cp_2TiO[2,6(Bu^t)_2-4-MeC_6H_2]$[26] dimerization is
hindered by the bulky residue that is σ-bonded to the Cp_2Ti group. In the
series $Cp_2^{**}MCl$ [Cp^{**} = 1,3-di(*tert*-butyl)cyclopentadienyl; M = Ti, Zr, Hf],
two bulky substituents are introduced into each Cp ring to achieve the same
purpose as well as to impart stability to the uncommon +3 oxidation state of
the metal.[27,28] X-Ray analysis of the Ti^{III} derivative (which lies on a
crystallographic C_2 axis)[26] and the Zr^{III} analogue[28] has shown that the
disubstituted Cp^{**} rings are oriented close to the idealized bent C_{2v}
"eclipsed" conformation as found in $Cp_2^{**}TiCl_2$ (Fig. 6.11-10). The important
structural parameters are compared with those of related compounds in Table
6.11-3.

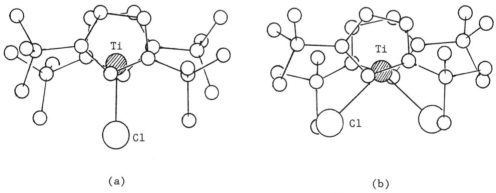

(a)   (b)

Fig. 6.11-10 Molecular structures of (a) $Cp_2^{**}TiCl$ and (b) $Cp_2^{**}TiCl_2$.
(After ref. 27).

Table 6.11-3   Important geometrical parameters in $Cp_2^{**}MCl$ (M = Ti, Zr) and some related bis-cyclopentadienyl complexes.

Compound	M-X /Å	M-C /Å	M-Cp[a] /Å	X-M-X /°	Cp-M-Cp /°
$Cp_2^{**}ZrCl$	2.423	2.505	2.200	–	133.3
$Cp_2^{**}TiCl_2$	2.349	2.499	2.124	93.1	121.0
$Cp_2^{**}TiCl$	2.337	2.386	2.066	–	134.8
$(C_5H_4Me)_2TiCl_2$	2.361	2.307	2.067	93.2	130.2
$[(CH_2(C_5H_4)_2]TiCl_2$	2.34	2.40	2.05	97.1	121.5
$[(C_5H_4Me)_2Ti]_2(\mu\text{-}Cl)_2$	2.547	2.376	2.062	79.2	130.9
$Cp_2Ti(2,6\text{-}Me_2C_6H_3)$	2.18	2.35	b	–	138.4
$Cp_2TiO(2,6\text{-}Bu^t_2\text{-}4\text{-}MeC_6H_2)$	1.892	2.362	b	–	135.5
$(C_5Me_5)_2TiCl$	2.363	b	2.06	–	143.6
Cp_2VCl	2.390	2.28	1.946	–	139.5

[a] The distance between the metal atom and the centroid of the Cp ring.   [b] Not calculated.

References

[1] R.D. Sanner, D.M. Duggan, T.C. McKenzie, R.E. Marsh and J.E. Bercaw, *J. Am. Chem. Soc.* **98**, 8358 (1976).

[2] M. Bottrill, P.D. Gavens and J. McMeeking in G. Wilkinson, F.G.A. Stone, and E.W. Abel (eds.), *Comprehensive Organometallic Chemistry*, Pergamon Press, Oxford, 1982, Vol 3, chap. 22.2, p. 281.

[3] G.P. Pez, *J. Am. Chem. Soc.* **98**, 8072 (1976).

[4] J.N. Armor, *Inorg. Chem.* **17**, 203 (1978).

[5] J.E. Bercaw, *J. Am. Chem. Soc.* **96**, 5087 (1974); J.M. Manriquez, D.R. McAlister, E. Rosenberg, A.M. Shiller, K.L. Williamson, S.I. Chan and J.E. Bercaw, *J. Am. Chem. Soc.* **100**, 3078 (1978).

[6] R.D. Sanner, J.M. Manriquez, R.E. Marsh and J.E. Bercaw, *J. Am. Chem. Soc.* **98**, 8351 (1976).

[7] J.L. Atwood, K.E. Stone, H.G. Alt, D.C. Hrncir and M.D. Rausch, *J. Organomet. Chem.* **132**, 367 (1977).

[8] G. Fachinetti, C. Floriani and H. Stoeckli-Evans, *J. Chem. Soc., Dalton Trans.*, 2297 (1977).

[9] F.N. Tebbe and L.J. Guggenberger, *J. Chem. Soc., Chem. Commun.*, 227 (1973).

[10] L.J. Guggenberger and F.N. Tebbe, *J. Am. Chem. Soc.* **95**, 7870 (1973).

[11] G. Fachinetti, C. Biran, C. Floriani, A.C. Villa and C. Guastini, *J. Chem. Soc., Dalton Trans.*, 792 (1979).

[12] F. Bottomley, G.O. Egharevba and P.S. White, *J. Am. Chem. Soc.* **107**, 4353 (1985).

[13] J.C. Huffman, J.G. Stone, W.C. Krusell and K.G. Caulton, *J. Am. Chem. Soc.* **99**, 5829 (1977).

[14] C.G. Young, *Coord. Chem. Rev.* **96**, 89 (1989).

[15] R.D. Rogers, R. Vann Bynum and J.L. Atwood, *J. Am. Chem. Soc.* **100**, 5238 (1978).

[16] R.D. Rogers, R. Vann Bynum and J.L. Atwood, *J. Am. Chem. Soc.* **103**, 692 (1981).

[17] G. Erker, C. Krüger and G. Müller, *Adv. Organomet. Chem.* **24**, 1 (1985).

[18] H. Köpf and P. Köpf-Maier in S.J. Lippard (ed.), *Platimum, Gold, and Other Metal Chemotherapeutic Agents*, American Chemical Society, Washington, D.C., 1983, chap. 16.

[19] S.E. Sherman and S.J. Lippard, *Chem. Rev.* **87**, 1153 (1987).

[20] J. Reedijk, A.M.J. Fichtinger-Schepman, A.T. van Oosterom and P. van de Putte, *Structure and Bonding* **67**, 53 (1987).

[21] P. Köpf-Maier and H. Köpf, *Structure and Bonding* **70**, 103 (1988).

[22] E.W. Neuse, R. Loonat and J.A.C. Boeyens, *Trans. Met. Chem.* **9**, 12 (1984).

[23] M.R. Churchill, A.G. Landers and A.L. Rheingold, *Inorg. Chem.* **20**, 849 (1981).

[24] M. Castagnola, B. Floris, G. Illuminati and G.J. Ortaggi, *J. Organomet. Chem.* **60**, C17 (1973).

[25] C.J. Olthof and F. van Bolhuis, *J. Organomet. Chem.* **122**, 47 (1976).

[26] B. Cetinkaya, P.B. Hitchcock, M.F. Lappert, S. Torroni, J.L. Atwood, W.E. Hunter and M.J. Zaworotko, *J. Organomet. Chem.* **188**, C31 (1980).

[27] I.F. Urazowski, V.I. Ponomaryov, O.G. Ellert, I.E. Nifant'ev and D.A. Lemenovskii, *J. Organomet. Chem.* **356**, 181 (1988).

[28] I.F. Urazowski, V.I. Ponomaryov, I.E. Nifant'ev and D.A. Lemenovskii, *J. Organomet. Chem.* **368**, 287 (1989).

6.12   Tetraphenylcyclobutadiene Iron Tricarbonyl
$Fe(CO)_3(C_4Ph_4)$

Crystal Data

Monoclinic, space group $P2_1/c$ (No. 14)

$a = 8.93$, $b = 18.72$, $c = 14.09$Å, $\beta = 92.7°$, $Z = 4$

Atom*	x	y	z
Fe	.0899	.1949	-.0119
O(1)	.1796	.0468	-.0476
O(2)	-.0852	.2231	-.1880
O(3)	.3748	.2582	-.0643
C(1)	.1428	.1074	-.0360
C(2)	-.0135	.2116	-.1179
C(3)	.2610	.2329	-.0435
C(4)	-.0395	.2631	.0636
C(5)	.1027	.2517	.1160
C(6)	.0653	.1777	.1316
C(7)	-.0804	.1884	.0808

* The carbon atoms of the phenyl groups have been omitted.

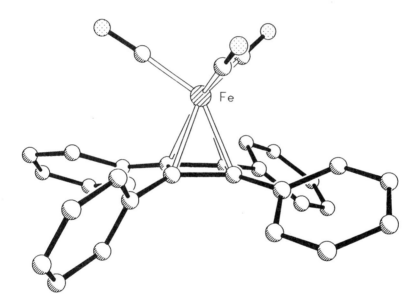

Fig. 6.12-1   Molecular structure of $Fe(CO)_3(C_4Ph_4)$.

Crystal Structure

The crystal consists of a packing of discrete $Fe(CO)_3(C_4Ph_4)$ molecules. In each molecule, the cyclobutadiene ring is square-planar to well within experimental accuracy, the C-C bond distances averaging 1.459Å [C(4)-C(5) 1.454Å, C(5)-C(6) 1.445Å, C(6)-C(7) 1.469Å, C(7)-C(4) 1.468Å].[1]

The substituent phenyl groups are bent out of the plane of the cyclobutadiene ring away from the iron atom by an average of 10.8°. In addition, the phenyl groups in each molecule are all twisted in the same sense.   Table 6.12-1 gives the bend and twist angles for the phenyl groups designated by the adjacent carbon atom of the cyclobutadiene ring.

Table 6.12-1   Bend and twist angles for the phenyl groups.

Phenyl group	Bend angle	Twist angle
at C(4)	6.9°	32.4°
at C(5)	9.6	36.4
at C(6)	10.1	28.6
at C(7)	16.7	60.8

Remarks

1. Bonding in metal cyclobutadienyl complexes

Cyclobutadiene (henceforth abbreviated as CBD) is a very unstable compound, having a very short life time at room temperature, but its instability was difficult to explain in terms of the valence bond theory.   On the basis of MO considerations, Longuet-Higgins and Orgel suggested in 1956 that a square-planar CBD molecule might be stabilized via coordination to a transition metal such as Ni, Pd and Pt.[2]   The molecular orbitals of CBD are illustrated in Fig. 6.12-2(a).   The non-bonding ψ_2 and ψ_3 orbitals are degenerate and have radical character, so that the CBD molecule is expected to

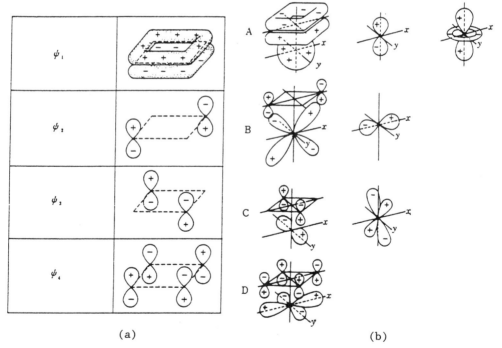

(a)   (b)

Fig. 6.12-2   Molecular orbitals of a metal CBD complex: (a) MO's of CBD; (b) the corresponding metal orbitals interacting with the MO's in (a). (Adapted from ref. 3).

be extremely reactive and ephemeral. The ψ_2 and ψ_3 orbitals can however overlap with the corresponding metal orbitals B and C to form covalent bonds, resulting in a stable complex [Fig. 6.12-2(b)].

In 1959, the first metal π-tetramethylcyclobutadiene complex, $Ni_2(C_4Me_4)_2Cl_4$, was prepared.[4] The CBD iron tricarbonyl $C_4H_4Fe(CO)_3$ was synthesized in 1965,[5] and the first CBD sandwich compound, $Ni(C_4Ph_4)_2$, in 1978.[6] The elusive CBD molecule was eventually made without the use of transition metals by Masamune.[7] Structural aspects of derivatives of CBD and its valence tautomer tetrahedrane are discussed in Section 5.9.

2. Structural features of metal cyclobutadienyl complexes

Structural parameters for some metal cyclobutadienyl complexes from X-ray diffraction studies are presented in Table 6.12-2.[8]

Table 6.12-2   Structural parameters for some complexes containing coordinated π-cyclobutadienyl (CBD) ligands.

Complex	CBD average parameters			Other noteworthy structural features
	M-C/Å	C-C/Å	C-C-C/deg.	
$Ph_4C_4Fe(CO)_3$	2.07	1.46	90.2	
$(OC)Fe(PhC_2C_6H_4C_2Ph)_2Fe(CO)_3$	2.05	1.46	89.6	Fe-Fe 2.494Å
$PhC^+[-C_4H_3Fe(CO)_3]_2BF_4^-$	2.05	1.44		C(exocyclic)-C(CBD) 1.40Å C(exocyclic)-C(Ph) 1.47Å
$[(1\text{-}Ph)(2\text{-}Ph)(3\text{-}Bu^t)(4\text{-}Bu^t)C_4]_2^-$ $Fe_2(CO)_3$	2.06	1.47		Fe-Fe 2.177Å
$[Me_4C_4NiCl_2]_2 \cdot C_6H_6$	2.02	1.43	90.0	
$Me_4C_4Ni(MeC_2Me)_2Fe(CO)_3$	2.02	1.46	90.0	Fe-Ni 2.449Å
$[Me_4C_4Pt(CF_3)(PMe_2Ph)_2]^+SbF_6^-$	2.22	1.47	90.0	
$(Ph_4C_4)_2Mo(CO)_2$	2.31	1.47	89.9	
$(OC)_2(Ph_4C_5O)Mo(Ph_2C_2)Mo(C_4Ph_4)(CO)$	2.30	1.47	90.0	Mo-Mo 2.772Å
$[Ph_4C_4Mo(CO)_2Br]_2$	2.25	1.46	90.0	Mo-Mo 2.954Å
$[(1\text{-}Ph)(2\text{-}Ph)(3\text{-}SiMe_3)(4\text{-}SiMe_3)C_4]^-$ $Co(C_5H_5)$	1.97	1.46	90.0	Co-C(Cp) 2.05Å
$C_4H_4CoC_5H_5$	1.96	1.44	90.1	Co-C(Cp) 2.04Å
$Ph_4C_4RhC_5H_5$	2.10	1.47	90.0	Rh-C(Cp) 2.22Å
$(Ph_4C_4)(C_5H_5)V(CO)_2$	2.26	1.47	90.0	V-C(Cp) 2.24Å
$(Ph_4C_4)(C_5H_5)(Ph_2C_2)Nb(CO)$	2.38	1.46	90.0	Nb-C(Cp) 2.44Å

a) Geometry of coordinated CBD rings

In the majority of the complexes listed in Table 6.12-2, the CBD ring exhibits a nearly perfect square-planar geometry which is not significantly affected by the ring substituents and/or the nature of the other ligands.

b) Metal-C(CBD) distances

In general, the metal atom is equidistant from all four carbon atoms of the CBD ring.

c) Comparisons of CBD and Cp rings

For complexes containing both CBD and Cp ring ligands, the average bond lengths conform to the order C(CBD)-C(CBD) > C(Cp)-C(Cp) and M-C(CBD) < M-C(Cp), indicating a stronger metal-cyclobutadiene than metal-cyclopentadienyl bonding interaction. This conclusion is consistent with the theoretical prediction by Longuet-Higgins and Orgel.[5]

d) Geometry of the substituents on the coordinated cyclobutadiene

Studies performed on $C_4H_4Fe(CO)_3$ (by ED and MW methods) yielded inconclusive results concerning the relative orientation of the C(CBH)-H bonds. For example, the average angle between the C-H bond and the planar CBD ring in this complex was reported by the various investigators to be as follows: 0°, 8.6±0.3° (endo to the Fe atom) and 6.5±0.5° (exo to the Fe atom). The C-C bond distances in the CBD ligands indicate a considerable sp^3 character for the ring carbon atoms; in the light of this observation, the exo C-H configuration in $C_4H_4Fe(CO)_3$ appears to be most plausible. Evidently, this conclusion is also consistent with the available structural data for the majority of the tetrasubstituted CBD metal complexes.

e) An unusual CBD Mo complexes

The main structural features of $(OC)_2(Ph_4C_5O)Mo(Ph_2C_2)Mo(C_4Ph_4)(CO)$ include the following: (i) a short Mo-Mo distance which has been assigned a double bond multiplicity; (ii) a nearly symmetrically bridging diphenylacetylene ligand; (iii) one of the Mo atoms being nearly symmetrically bonded to a Ph_4C_4 ligand whereas the other metal atom is coordinated to a tetraphenylcyclopentadienone ligand; (iv) the keto oxygen of the tetraphenyl-cyclopentadienone ligand being coordinated to the adjacent Mo atom; and (v) the terminal carbonyls being arranged to permit each of the Mo atoms to attain an eighteen-electron configuration.

Fig. 6.12-3 shows the structures of some metal cyclobutadienyl complexes.

The crystal structure of $PhC^+[C_4H_3Fe(CO)_3]_2BF_4^-$ [Fig. 6.12-3(a)][9] has several main features: i) The CBD rings are very close to being coplanar with the plane formed by the exocyclic carbon and the three carbon atoms to which it is attached, whereas the phenyl ring is twisted 43° from this plane. ii) The iron atom shows no large displacement from above the center of the CBD ring, and the Fe-C(11) distances are too large (2.85 and 2.94Å) to support any significant direct metal-exocyclic carbon interaction (the anchimeric effect).

The eight iron-ring carbon distances are close to being equal, suggesting the bond energy remains much the same in each case. iii) The C-C distances of the CBD ring remain close to being equal and similar to those in cyclobutadiene iron tricarbonyl itself.

In $Me_4C_4Ni(MeC_2Me)_2Fe(CO)_3$ [Fig. 6.12-3(b)[10]] the Ni atom is sandwiched between approximately coplanar CBD and ferracyclopentadiene rings. The nickel (CBD) fragment is coordinated to the five-membered ferracyclopentadiene ring both through a *cis*-butadiene-nickel interaction and through a metal-metal bond of length 2.449(3)Å.

In the structure of $[Me_4C_4Pt(CF_3)(PMe_2Ph)_2]^+SbF_6^-$ [Fig. 6.12-3(c)],[11] the angles P-Pt-P = 95.2(1)° and P-Pt-CF₃ = 98.7(5), 93.0(5)° indicate a distorted tetrahedral geometry about the Pt atom, taking the (Me_4C_4) group as a monodentate ligand. The fourfold axis of the CBD ring is approximately coincident with the pseudo-threefold axis of the Pt atom. The substituent methyl carbon atoms are bent out of the plane of the CBD ring away

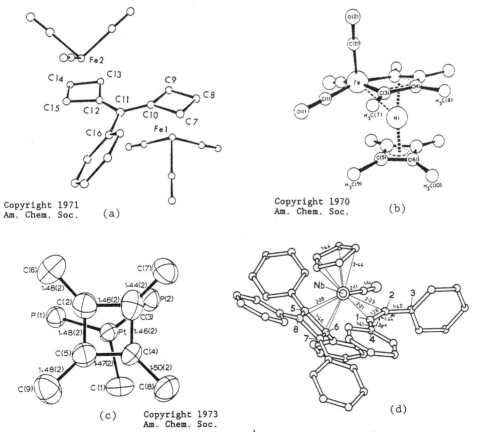

Fig. 6.12-3 Structure of (a) $PhC^+[C_4H_3Fe(CO)_3]_2BF_4^-$, (b) $Me_4C_4Ni(MeC_2Me)_2Fe(CO)_3$, (c) $[Me_4C_4Pt(CF_3)(PMe_2Ph)_2]^+SbF_6^-$ and (d) $(Ph_4C_4)Cp(Ph_2C_2)Nb(CO)$. (After refs. 9-12).

from the Pt atom by an average of 14(1)°, and the mean distance between the Pt atom and the four carbon atoms in the substituted CBD ring is 2.22(5)Å.

In $(Ph_4C_4)Cp(Ph_2C_2)Nb(CO)$ [Fig. 6.12-3(d)][12] the CBD ring is planar to within 0.02Å and makes an angle of 48° with the cyclopentadienyl ring. The average Nb-C(CDB) distance, 2.38Å, is distinctly shorter than the Nb-C(Cp) distance, 2.44Å. The coordinated triple bond C(1)-C(2) is roughly parallel to the C(6)-C(7) bond of the cyclobutadiene ligand [C(1)...C(7) 2.99, C(2)...C(6) 3.42Å]. Thus formation of hexaphenylbenzene by thermal decomposition of $(Ph_4C_4)Cp(Ph_2C_2)Nb(CO)$ is probably effected through an intra-complex addition of the tolane molecule to the C(6)-C(7) bond of cyclobutadiene, resulting in hexaphenyl-Dewar-benzene as an intermediate. This model for hexaphenylbenzene formation agrees well with the cyclobutadiene mechanism for formation of benzene derivatives by catalytic trimerization of acetylene on transition-metal complexes.

3. Cyclophanes with metal-stabilized cyclobutadiene stacks

Cyclobutadienophane bis(tricarbonyliron) is the smallest cyclophane containing two metalloaromatic (cyclobutadiene)iron tricarbonyl fragments held cofacially by two ethylene bridges [Fig. 6.12-4(a)].[13] The cyclobutadiene rings eclipse each other with no ring distortion, and there is a 20.6° out-of-plane bending of the aliphatic bridges away from the metal. The interstack separation of 2.704Å is the shortest known for cyclophane systems, and the

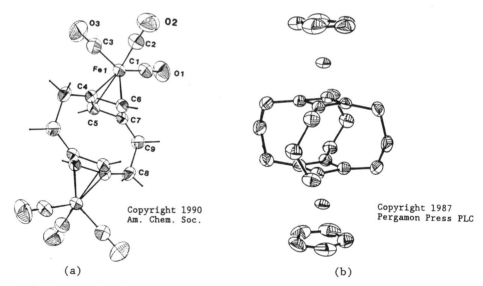

Copyright 1990
Am. Chem. Soc.

Copyright 1987
Pergamon Press PLC

(a)   (b)

Fig. 6.12-4 Molecular structures of (a) cyclobutadienophane bis(tricarbonyliron) and (b) [3_4](1.2.3.4)cyclobutadienophane bis(cyclopentadienylcobalt). (After refs. 13 and 14).

Fe(CO)$_3$ groups are in the staggered conformation. The related cobalt
[3.3.3.3]superphane has a sandwich structure with pairs of eclipsed four-
membered and staggered five-membered rings [Fig. 6.12-4(b)].[14] The
interdeck distance between the undistorted cyclobutadiene rings is 2.994Å.

4. Tetraphosphacyclobutadiene as a complex ligand

Structure **C** and **D** may be postulated for *cyclo*-P$_4$, the phosphorus
analogues of cyclobutadiene **A** and its dianion **B**.

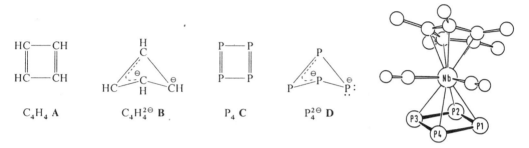

Fig. 6.12-5 Structures related to cyclobutadiene and molecular geometry
of Cp*(CO)$_2$Nb(η^4-P$_4$). (After ref. 15).

The complex Cp*(CO)$_2$Nb(η^4-P$_4$) has been prepared photochemically from
Cp*Nb(CO)$_4$ and white phosphorus.[15] X-Ray analysis shows that the molecule
has nearly C_s symmetry with four P coplanar atoms arranged in a slightly
distorted (kite-like) square (Fig. 6.12-5). The measured bond distances
(P1-P2 = 2.136(2), P2-P3 = 2.141(2), P3-P4 = 2.181(2), P4-P1 = 2.178(2)Å;
Nb-P2 = 2.607(1)Å, slightly shorter than the average of 2.635Å for the other
three Nb-P bonds) are consistent with either an allyl-like, folded *cyclo*-P$_4^{2-}$
ligand **D** (d^2-Nb) or a slightly distorted planar *cyclo*-P$_4$ ligand **C** (d^4-Nb).
The Wade-Mingos electron counting rules[16,17] give seven electron pairs for
the square pyramidal NbP$_4$ skeleton, in accordance with a *nido* structure.

References

[1] R.P. Dodge and V. Schomaker, *Acta Crystallogr.* **18**, 614 (1965).

[2] H.C. Longuet-Higgins and L.E. Orgel, *J. Chem. Soc.*, 1969 (1956).

[3] A. Yamamoto, *Organotransition Metal Chemistry*, Wiley, New York, 1986.

[4] R. Criegee and G. Schroder, *Liebigs Ann. Chem.* **623**, 1 (1959).

[5] G.F. Emerson, L. Watts, and R. Pettit, *J. Am. Chem. Soc.* **87**, 131 (1965).

[6] H. Hoberg, R. Krause-Göing and R. Mynott, *Angew. Chem. Int. Ed. Engl.* **17**,
     123 (1978).

[7] S. Masamune, *Pure Appl. Chem.* **44**, 861 (1975); T. Bally and S. Masamune,
     Tetrahedron **36**, 343 (1980).

[8] A. Efraty, *Chem. Rev.* **77**, 691 (1977).

[9] R.E. Davis, H.D. Simpson, N. Grice and R. Pettit, *J. Am. Chem. Soc.* **93**, 6688 (1971).

[10] E.F. Epstein and L.F. Dahl, *J. Am. Chem. Soc.* **92**, 502 (1970).

[11] D.B. Crump and N.C. Payne, *Inorg. Chem.* **12**, 1663 (1973).

[12] A.N. Nesmeyanov, A.I. Gusev, A.A. Pasynskii, K.N. Anisimov, N.E. Kolobova and Y.T. Struchkov, *J. Chem. Soc., Chem. Commun.*, 739 (1969).

[13] C.M. Adams and E.M. Holt, *Organometallics* **9**, 980 (1990).

[14] R. Gleiter, M. Karcher, M.L. Ziegler and B. Nuber, *Tetrahedron Lett.* **28**, 195 (1987).

[15] O.J. Scherer, J. Vondung and G. Wolmerschäuser, *Angew. Chem. Int. Ed. Engl.* **28**, 1355 (1989).

[16] K. Wade, *Adv. Inorg. Chem.* **18**, 1 (1976).

[17] D.M.P. Mingos, *Acc. Chem. Res.* **17**, 311 (1984).

Note The isolation of stable cyclobutadiene with a singlet ground state has been accomplished by synthesis carried out in the cavity ("inner space") of a hemicarcerand. See D.J. Cram, M.F. Tanner and R. Thomas, *Angew. Chem. Int. Ed. Engl.* **30**, 1024 (1991).

6.13   Bis(cyclooctatetraenyl)uranium(IV)   (Uranocene)
$$U(\eta^8\text{-}C_8H_8)_2$$

Crystal Data

Monoclinic, space group $P2_1/n$ (No. 14)

$a = 7.084(3)$, $b = 8.710(3)$, $c = 10.631(5)$Å, $\beta = 98.75(3)°$, $Z = 2$[2]

Atom	x	y	z
U	0	0	0
C(1)	.2655	.2049	.2070
C(2)	.3397	.0936	.1092
C(3)	.2817	.2025	.0221
C(4)	.1154	.2864	-.0115
C(5)	-.0646	.2951	.0336
C(6)	-.1440	.2184	.1296
C(7)	-.0853	.1106	.2162
C(8)	.0829	.0261	.2505

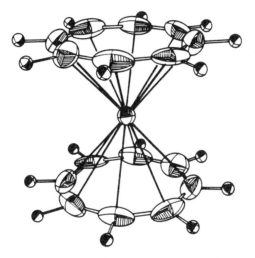

Fig. 6.13-1 Structure of $U(\eta^8\text{-}C_8H_8)_2$ and $Th(\eta^8\text{-}C_8H_8)_2$. (After ref. 3).

Crystal Structure

In the crystal structure of $U(\eta^8\text{-}C_8H_8)_2$, the U atoms occupy positions
$(0,0,0)$ and $(1/2,1/2,1/2)$ with site symmetry $\bar{1}$. The 10-π-electron
cyclooctatetraene (COT) dianions are planar octagons with an average C-C bond
length of 1.392(13)Å, and the molecule attains near perfect D_{8h} symmetry (Fig.
6.13-1). The average U-C bond distance is 2.647(4)Å, and the distance from
the U atom to the center of the dianion ring plane is 1.924Å.[1]

Thorocene has the same crystal structure as uranocene, with cell
parameters $a = 7.0581(11)$, $b = 8.8192(17)$, $c = 10.7042(18)$Å, and $\beta =$
98.44(3)°.[1] The mean Th-C bond distance is 2.701(4)Å, and the mean C-C bond
distance is 1.386(9)Å. The significant difference in M-C bond lengths in
these two compounds is explicable entirely on the basis of the different
ionic radii of the actinides.

Remarks

1. Bonding in uranocene[3]

     The bonding in uranocene and other $M(\eta^8\text{-}C_8H_8)_2$ organoactinides has been
the subject of much thorough investigation. In terms of orbital symmetry,
uranocene is an interesting f-orbital homologue of ferrocene.[4] The overlap
of the $C_8H_8^{2-}$ HOMO (e_{2u} in D_{8h}) with the symmetry-appropriate f_{xyz} and
$f_{z(x^2-y^2)}$ uranium orbitals is reminiscent of the interaction of the $C_5H_5^-$ e_{1g}
HOMO with the iron d_{xz} and d_{yz} orbitals. These relationships are illustrated
in Fig. 6.13-2. Electronic structure calculations carried out on actinide
$M(\eta^8\text{-}C_8H_8)_2$ compounds have confirmed the significance of such covalent
interactions.

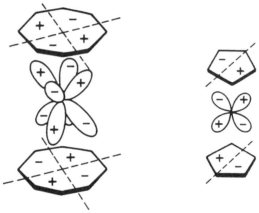

Copyright 1973
Am. Chem. Soc.

Fig. 6.13-2 Illustration of the cognate metal-ligand orbital
interactions in uranocene and ferrocene. (After ref. 4).

2. Structure of octaphenyluranocene, $U[\eta^8\text{-}C_8H_4(Ph)_4]_2$

     The crystal structure of octaphenyluranocene has been determined.[5]
The molecule is a sandwich compound with the COT rings rotated 6° from the
eclipsed configuration found in uranocene. The phenyl rings, attached to
alternate carbon atoms, are staggered; the four on one COT ring are tilted
like a right-handed propeller and those on the other ring in a left-handed
manner (Fig. 6.13-3). The dihedral angles between the planes of the COT ring
and the phenyl rings are in the range 43.8° to 40.3°.

     The plane-to-plane average distance between the COT rings is 3.793Å.
The C-C bond lengths average 1.42Å in the COT ring and 1.39Å in the phenyl
rings; the COT ring value is in agreement with the average 1.41Å found in
$U[\eta^8\text{-}C_8H_4(Me)_4]_2$[6] and with 1.407Å in a potassium salt of $[C_8H_4(Me)_4]^{2-}$.[7]

3. Uranocene half-sandwich compounds

     $Mono(\eta^8\text{-}C_8H_8)$uranium(IV) complexes, unlike their thorium(IV) analogues,
are difficult to synthesize from the metallocene in view of the low reactivity

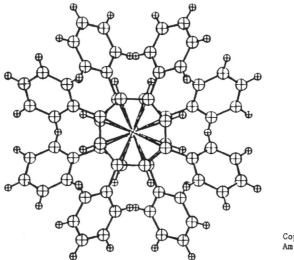

Fig. 6.13-3 Molecular structure of $U[(C_8H_4(Ph)_4]_2$. (After ref. 5).

of the uranium center with respect to ligand substitution. An early synthesis of $(C_8H_8)UCl_2$ has been achieved by the reduction of COT by UCl_3 or through the reaction of UCl_4, COT and NaH; its adducts $(COT)UCl_2(py)_2$ and $(COT)U(acac)_2$ have been isolated and characterized by X-ray crystallography.[8] Both have the four-legged piano-stool structure which is also adopted in the α and β (two independent molecules) forms of $(COT)ThCl_2$.[9]

4. Other metal complexes containing planar η^8-COT ligands

In general COT forms η^8-complexes with the early transition metals which require a large number of electrons to achieve a noble-gas electronic configuration.

The reaction of $Ti(OC_4H_9)_4$ with COT and $Al(C_2H_5)_3$ yielded either deep violet-red crystals of $Ti(COT)_2$ or yellow crystalline $Ti_2(COT)_3$, depending on the molar ratio of the reactants used.[10] X-Ray analysis has shown that the latter should be formulated as $[\{Ti(\eta^8-C_8H_8)\}_2(\mu-\eta^4,\eta^4-C_8H_8)]$, with a crystallographic C_2 axis passing through the centers of two opposite C-C bonds of the middle boat-shaped ring [Fig. 6.13-4(a)].[11] A similar coordination environment of the titanium atom exists in $Ti(C_8H_8)_2$, which contain one η^8- and one boat-shaped η^4-COT ring [Fig. 6.13-4(b)].[12]

η^8-COT complexes are formed by lanthanides in both the divalent and trivalent states, although the latter class is much more common; recent examples include $(C_8H_8)Yb(py)_3\cdot\frac{1}{2}py$[13] and $(C_8H_8)Lu(o-C_6H_4CH_2NMe_2)(THF)$.[14] The η^8-COT rings are exactly eclipsed in the D_{8h} dianion of $[K(dimethoxyethane)]_2[Yb(C_8H_8)_2]$,[15] staggered in the D_{8d} anion of $[K(diglyme)][Ce(C_8H_8)_2]$,[16] and nearly eclipsed but not parallel-planar in

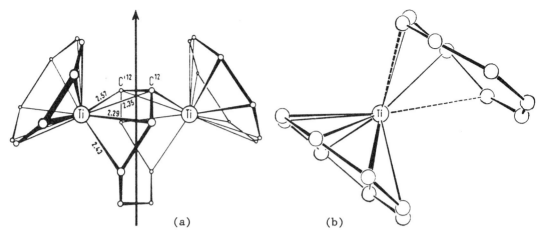

Fig. 6.13-4 Molecular structures of (a) $Ti_2(COT)_3$ and (b) $Ti(COT)_2$. (After refs. 11 and 12).

the anion of $[Nd(THF)_2(C_8H_8)][Nd(C_8H_8)_2]$.[17]

The structure of $Sc_2\{1,4(SiMe_3)_2C_8H_6\}Cl_2$.THF has been determined by single crystal X-ray diffraction,[18] and is shown in Fig. 6.13-5. The molecule is dimeric, the two metal centers being bridged by two chloro ligands (somewhat asymmetrically) and the THF ligand. Consideration of the two scandium-oxygen distances [Sc(1)-O 3.056(9), Sc(2)-O 2.324(7)Å] suggests that the THF molecule is best regarded as a "semi-bridging" ligand. The two η^8-$C_8H_6\{1,4(SiMe_3)_2\}^{2-}$ rings are both planar.

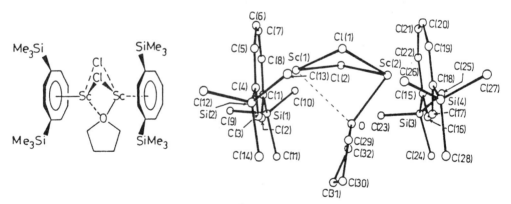

Fig. 6.13-5 Structural formula and molecular structure of $Sc_2[1,4(SiMe_3)_2C_8H_6]Cl_2(THF)$. (After ref. 18).

5. Structures of uranium tetracyclopentadienyl and its derivatives

The molecular structure of $U(C_5H_5)_4$ has overall S_4 symmetry[19] (Fig. 6.13-6(a)]. The coordination about the uranium atom is nearly tetrahedral with all ring centroid-U-ring centroid angles within 0.6° of 109°. The average U-C distance is 2.81Å, with a range of 2.79-2.83Å. The average C-C distance within a cyclopentadienyl ring is 1.39Å, with a range of 1.37-1.40Å. Other known actinide tetracyclopentadienyls include $Th(C_5H_5)_4$, $Pa(C_5H_5)_4$ and $Np(C_5H_5)_4$.

The $U(C_5H_5)_3X$ compounds may have tetrahedral [e.g., $UCp_3C\equiv CPh$][20] or trigonal bipyramidal [e.g., $UCp_3(NCS)(MeCN)$] coordination about the metal center.[21] Figs. 6.13-6(b) and (c) illustrate these structures. The structure of $UCp_3C\equiv CPh$ clearly reveals the σ-bonded nature of the phenylacetylene group.

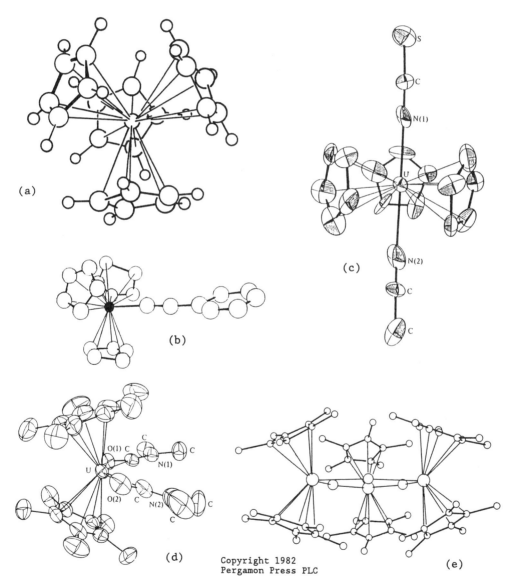

(a)

(b)

(c)

(d)

(e)

Fig. 6.13-6 Structures of (a) $U(C_5H_5)_4$, (b) $UCp_3C\equiv CPh$, (c) $UCp_3(NCS)(MeCN)$, (d) $U(Me_5C_5)_2(\eta^2\text{-}CONMe_2)_2$ and (e) $[U(Me_5C_5)_2Cl]_3$. (After refs. 3 and 20).

The compounds $U(Me_5C_5)_2(\eta^2\text{-}CONMe_2)_2$[22] and $[U(Me_5C_5)_2Cl]_3$[23] have tetrahedral geometries about the metal centers, their structures being shown in Fig. 6.13-6(d) and (e).

References

[1]   A. Avdeef, K.N. Raymond, K.O. Hodgson and A. Zalkin, *Inorg. Chem.* **11**, 1083 (1972).

[2]   A. Zalkin and K.N. Raymond, *J. Am. Chem. Soc.* **91**, 5667 (1969).

[3]   T.J. Marks and R.D. Ernst in G. Wilkinson, G.A. Stone and E.W. Abel (eds.), *Comprehensive Organometallic Chemistry*, Pergamon Press, Oxford, 1982, Vol. 3, chap. 21, p. 173; T.J. Marks and H.I.C. Fragal (eds.), *Fundamental and Technological Aspects of Organo-f-Element Chemistry*, D. Reidel, Drodrecht, 1985.

[4]   A. Streitwieser, Jr., U. Müller-Westerhoff, G. Sonnichsen, F. Mares, D.G. Morrell, K.O. Hodgson and C.A. Harmon, *J. Am. Chem. Soc.* **95**, 8644 (1973).

[5]   L.K. Templeton, D.H. Templeton and R. Walker, *Inorg. Chem.* **15**, 3000 (1976).

[6]   K.O. Hodgson and K.N. Raymond, *Inorg. Chem.* **12**, 458 (1973).

[7]   S.Z. Goldberg, K.N. Raymond, C.A. Harmon and D.H. Templeton, *J. Am. Chem. Soc.* **96**, 1348 (1974).

[8]   T.R. Boussie, R.M. Moore, Jr., A. Streitwieser, A. Zalkin, J. Brennan and K.A. Smith, *Organometallics* **9**, 2010 (1990).

[9]   A. Zalkin, D.H. Templeton, C. Le Vanda and A. Streitwieser Jr., *Inorg. Chem.* **19**, 2560 (1980).

[10] H. Breil and G. Wilke, *Angew. Chem. Int. Ed. Engl.* **5**, 898 (1966).

[11] H. Dietrich and H. Dierks, *Angew. Chem. Int. Ed. Engl.* **5**, 899 (1966); H. Dierks and H. Dietrich, *Acta Crystallogr., Sect. B* **24**, 58 (1968).

[12] H. Dietrich and M. Soltwisch, *Angew. Chem. Int. Ed. Engl.* **8**, 765 (1969).

[13] A.L. Wayda, I. Mukerji, J.L. Dye and R.D. Rogers, *Organometallics* **6**, 1328 (1987).

[14] A.L. Wayda and R.D. Rogers, *Organometallics* **4**, 1440 (1985).

[15] S.A. Kinsley, A. Streitwieser, Jr. and A. Zalkin, *Organometallics* **4**, 52 (1985).

[16] K.O. Hodgson and K.N. Raymond, *Inorg. Chem.* **11**, 3030 (1972).

[17] C.W. DeKock, S.R. Ely, T.E. Hopkins and M.A. Brault, *Inorg. Chem.* **17**, 625 (1978).

[18] N.C. Burton, F.G.N. Cloke, P.B. Hitchcock, H.C. de Lemos and A.A. Sameh, *J. Chem. Soc., Chem. Commun.*, 1462 (1989).

[19] J.H. Burns, *J. Organomet. Chem.* **69**, 225 (1974).

[20] J.L. Atwood, C.F. Hains, Jr., M. Tsutsui and A.E. Gebala, *J. Chem. Soc., Chem. Commun.*, 452 (1973).

[21] R.D. Fischer, E. Klähne and J. Kopf, *Z. Naturforsch., Teil B* **33**, 1393 (1978).

[22] P.J. Fagan, J.M. Manriquez, S.H. Vollmer, C.S. Day, V.W. Day and T.J. Marks, *J. Am. Chem. Soc.* **103**, 2206 (1981).

[23] P.J. Fagan, J.M. Manriquez, T.J. Marks, C.S. Day, S.H. Vollmer and V.W. Day, *Organometallics* **1**, 170 (1982).

Note The bonding and electronic structure of cyclopentadienyl-actinide complexes are reviewed in B.E. Bursten and R.J. Strittmatter, *Angew. Chem. Int. Ed. Engl* **30**, 1069 (1991).

The synthesis and X-ray structure of V(η_8-COT)(η_4-COT) is reported in D. Gourier, E. Samuel, B. Bachmann, F. Hahn and J. Heck, *Inorg. Chem.* **31**, 86 (1992).

6.14 Tribenzo[*b*,*e*,*h*][1,4,7]trimercuronin (*o*-Phenylenemercury Trimer) $(C_6H_4Hg)_3$

Crystal Data

Orthorhombic, space group $P2_12_12_1$ (No. 19)

a = 5.56(2), b = 23.36(2), c = 12.24(2)Å, Z = 4

Atom	x	y	z	Atom	x	y	z
Hg(1)	.5116	.0055	.5076	C(8)	-.0832	.1262	.3094
Hg(2)	-.0075	-.0032	.3317	C(9)	-.1942	.1790	.2771
Hg(3)	.2095	.1317	.4197	C(10)	-.4070	.1740	.2094
C(1)	.3725	-.0745	.4528	C(11)	-.4861	.1240	.1792
C(2)	.1819	-.0755	.3800	C(12)	-.3999	.0726	.2185
C(3)	.0675	-.1311	.3597	C(13)	.4965	.1347	.5269
C(4)	.1962	-.1821	.3957	C(14)	.6283	.0864	.5647
C(5)	.4109	-.1789	.4515	C(15)	.8470	.0891	.6346
C(6)	.4909	-.1264	.4878	C(16)	.9107	.1481	.6648
C(7)	-.1905	.0701	.2786	C(17)	.7855	.1976	.6248
				C(18)	.5824	.1896	.5603

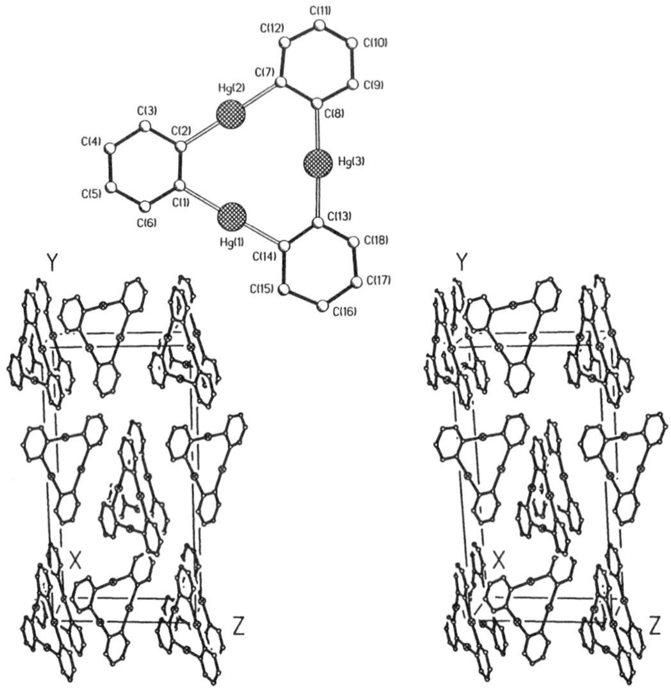

Fig. 6.14-1 Molecular and crystal structure of *o*-phenylenemercury trimer.

Crystal Structure

The crystal of the orthorhombic modification consists of trimeric *o*-phenylenemercury molecules (Fig. 6.14-1) packed together by van der Waals contacts. The Hg-C bond lengths fall in the normal range of 2.06-2.15Å for mercury-carbon compounds. The C-Hg-C bond angles do not differ significantly from 180° as expected for sp hybridization of the Hg orbitals. The intramolecular Hg...Hg distances are 3.61, 3.56, and 3.54Å.

Remarks

1. Monoclinic *o*-phenylenemercury trimer

Strain-free cyclic models can be constructed for "mercurials" of the formula $(C_6X_4Hg)_n$, where X = H or F and n = 3, 4, 6, 8, 10, etc. A monoclinic form, which had been assumed to be hexameric with Hg atoms at the corners of a hexagon,[2] was later shown to be a trimer (Fig. 6.14-2).[3] In this monoclinic modification of $(C_6H_4Hg)_3$, the Hg-C bond lengths fall in the range 2.04-2.16Å, and the C-Hg-C angles 175.7-177.1°.

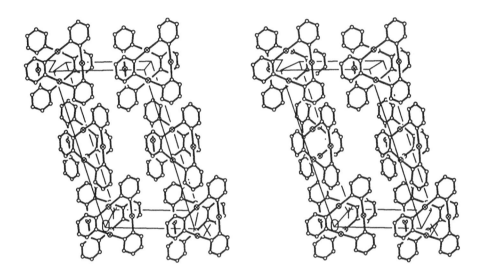

Fig. 6.14-2 Stereoview of the crystal structure of monoclinic $(C_6H_4Hg)_3$.

2. Structures of diorganomercury(II) compounds

The vast majority of diorganomercurials in the solid state exhibit the linear C-Hg-C geometry. The bonding is usually regarded as sp hybridization

Table 6.14-1 Structures of some diorganomercury(II) compounds.

Compound	Hg-C/Å	C-Hg-C/°	Additional features
$(CF_3)_2Hg$	2.109, 2.101	180	Discrete monomer: six intermolecular contacts of 3.181Å per Hg atom
$[(Bu^t\text{-}CO)_2CH]_2Hg$	2.13, 2.18	170	Four intermolecular contacts Hg...O 2.70Å
$(C_6H_5)_2Hg$	2.085	176.9	Coplanar rings, Hg bent out of plane of rings
$(2,4,6\text{-}Bu^t\text{-}C_6H_3)_2Hg$	2.077, 2.083	173.4	Angle between rings is 70.8°
$(o\text{-tolyl})_2Hg$	2.09	178.0	Angle between rings is 58.9°
$(p\text{-tolyl})_2Hg$	2.08	180	Coplanar rings
$(C_6F_5)_2Hg$	2.10	176.2	Angle between rings is 59.4°; Hg...F(ortho) 3.14, 3.22, 3.25, and 3.32Å
$(C_6H_5CH_2)_2Hg$	2.065	180	Angle Hg-C-C 117°, dihedral angle Hg-C-C-C 98°
$(2\text{-thienyl})_2Hg$	2.061	180	No Hg...S interaction, Hg bent out of plane of the aromatic rings

of Hg, but photoelectron spectra and calculations give little evidence of p character and a better description is three-center two-electron bonding.[4] Structural information from X-ray diffraction is summarized in Table 6.14-1.[5]

The predominant preference of mercury organometallics to achieve digonal σ coordination, if necessary by polymerization, is reflected in a reluctance to form π bonds, as exemplified by the X-ray structure of [(η^1-Ph$_3$PC$_5$H$_4$)-HgI]$_2$(μ-I)$_2$, in which the tetrahedrally coordinated mercury atoms are bridged by a pair of iodine atoms.[6]

3. Coordination of methylmercury(II) systems

The importance of methylmercury, MeHg^{II}, in the pollution of the environment became apparent following the surprising discovery that it constitutes the major fraction of mercury contaminants in fish, even though some of the fishes were taken from lakes and rivers into which no MeHg^{II} had been discharged.

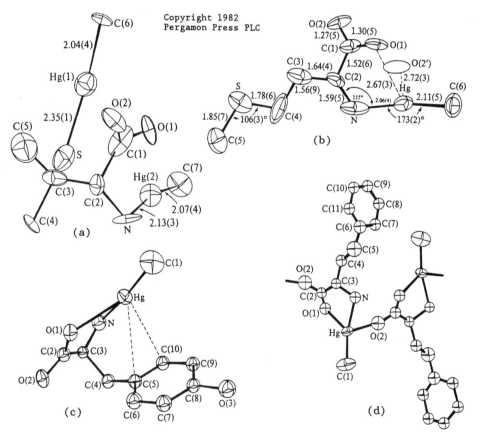

Fig. 6.14-3 Molecular structures of some methylmercury complexes. (After ref. 5).

Methylmercury has a rich coordination chemistry, forming complexes with a variety of organic and inorganic ligands. Although the mercury atom in MeHg^{II} has a strong tendency towards linear two-coordination, it does possess some residual Lewis acidity and takes part in secondary and weaker interactions with appropriate donor centers without basically disturbing the main collinear bond. Effective coordination numbers of 3 and 4 have been achieved by means of the secondary interactions in amino acid compounds, e.g. in μ-DL-penicillaminatobis(methylmercury), MeHgSCMe$_2$CH(CO$_2^-$)NH$_2$HgMe [Fig. 6.14-3(a)],[7] methyl(DL-methionine)mercury(II), MeHgNH$_2$CH(CO$_2^-$)CH$_2$CH$_2$SMe [Fig. 6.14-3(b)],[8] methyl(L-tyrosinato)mercury(II) monohydrate, MeHgNH$_2$CH(CO$_2^-$)-CH$_2$C$_6$H$_4$OH-p.H$_2$O [Fig. 6.14-3(c)],[9] and methyl[L-(2-amino-4-phenylbutanoato]mercury(II), MeHgNH$_2$CH(CO$_2^-$)CH$_2$CH$_2$Ph [Fig. 6.14-3(d)].[9] The complex of 2-methylpyridine with MeHg(CF$_3$COO) features a linear C-Hg-N entity, which may be significant in the understanding of MeHg^{II} binding in biological systems.[10]

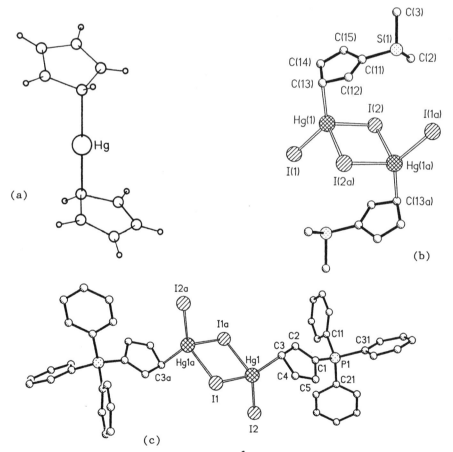

Fig. 6.14-4 Structures of (a) Hg(η^1-C$_5$H$_5$)$_2$, (b) [(CH$_3$)$_2$SC$_5$H$_4$HgI$_2$]$_2$, and (c) [Ph$_3$PC$_5$H$_4$HgI$_2$]$_2$. (After refs. 11-13).

4. Cyclopentadienyl complexes of mercury(II)

Cyclopentadienylmercury compounds generally contain a σ-bonded cyclopentadienyl ring. The structure of dicyclopentadienylmercury consists of monomeric $Hg(\eta^1\text{-}C_5H_5)_2$ units [Fig. 6.14-4(a)].[11] The C-Hg-C bond angle is 177.9(6)°, which is in line with sp-hybridization of the Hg center. The bond lengths are Hg-C(1) 2.12(1) and Hg-C(6) 2.15(1)Å.

Fig. 6.14-4 (b) and (c) show the structures of $[(CH_3)_2SC_5H_4HgI_2]_2$[12] and $[Ph_3PC_5H_4HgI_2]_2$,[13] respectively. Both compounds are iodine-bridged dimers located at inversion centers, with a tetrahedral coordination environment about the mercury atom. In $[(CH_3)_2SC_5H_4HgI_2]_2$, the bond lengths are: Hg-I (bridge) 2.896(1), 3.031(1)Å, Hg-I (terminal) 2.706(1)Å, and Hg-C(13) 2.196(14)Å. In $[Ph_3PC_5H_4HgI_2]_2$, the bond lengths are: Hg-I (bridge) 2.982(1), 2.937(1)Å, Hg-I (terminal) 2.681(1)Å, and Hg-C(3) 2.292(8)Å.

The reaction of KCp^* ($Cp^* = C_5Me_5$) with $HgCl_2$ at room temperature in ether gave $[Cp^*HgCl]_\infty$ in 74% yield.[14] The basic units of this novel polymeric structure consist of two independent, almost linear Cp^*-Hg-Cl "dumbbells" joined together via weak Hg-Cl bonds (3.10-3.24Å) to give a double-chain structure (Fig. 6.14-5). Each pair of antiparallel dumbbells, related by an inversion center, constitute a section of the "$[Hg_2Cl_2]_\infty$ battlement-like structure"; the folding angles at the Hg(1)-Cl(1)^a and Hg(2)-

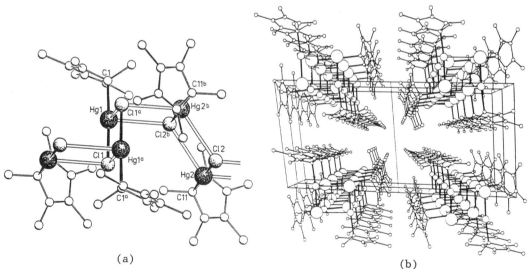

            (a)   (b)

Fig. 6.14-5 Polymeric structure of $[Cp^*HgCl]_\infty$. (a) Section viewed approximately along c. Selected bond lengths: Hg(1)-C(1) 2.08(2), Hg(1)-Cl(1) 2.361(5), Hg(1)-Cl(1)^a 3.098(5), Hg(1)-Cl(2)^b 3.238(7), Hg(2)-C(11) 2.10(2), Hg(2)-Cl(2) 2.350(6), Hg(2)-Cl(2)^b 3.177(6), Hg(2)^b-Cl(1)^a 3.232(6)Å. (b) Perspective view of the packing of the double chains along the a axis. (After ref. 14).

Cl(2) edges are 88 and 129°, respectively. The bulky Cp* rings bend away from the Hg-C bond axis at angles of about 113 and 108°, respectively, and form an organic protective covering of the polymeric chain.

References

[1] D.S. Brown, A.G. Massey and D.A. Wickens, *Acta Crystallogr., Sect. B* **34**, 1695 (1978).

[2] D. Grdenić, *Chem. Ber.* **92**, 231 (1959).

[3] D.S. Brown, A.G. Massey and D.A. Wickens, *Inorg. Chim. Acta* **44**, L193 (1980).

[4] R.L. DeKock, E.J. Baerends, P.M. Boerrigter and R. Hengelmolen, *J. Am. Chem. Soc.* **106**, 3387 (1984).

[5] J.L. Wardell in G. Wilkinson, F.G.A. Stone and E.W. Abel (eds.), *Comprehensive Organometallic Chemistry*, Pergamon Press, Oxford, 1982, Vol. 2, chap. 17, p. 863.

[6] N.L. Holy, N.C. Baenziger, R.M. Flynn and D.C. Swenson, *J. Am. Chem. Soc.* **98**, 7823 (1976).

[7] Y.-S. Wong, A.J. Carty and P.C. Chieh, *J. Chem. Soc., Dalton Trans.*, 1801 (1977).

[8] Y.-S. Wong, A.J. Carty and P.C. Chieh, *J. Chem. Soc., Dalton Trans.*, 1157 (1977).

[9] N.W. Alcock, P.A. Lampe and P. Moore, *J. Chem. Soc., Dalton Trans.*, 1324 (1978).

[10] R.D. Bach, H.B. Vardhan, A.F.M.M. Rahman and J.P. Oliver, *Organometallics* **4**, 846 (1985).

[11] B. Fischer, G.P.M. van Mier, J. Boersma, G. van Koten, W.J.J. Smeets and A.L. Spek, *Recueil* **107**, 259 (1988).

[12] N.C. Baenziger, R.M. Flynn and N.L. Holy, *Acta Crystallogr., Sect. B* **36**, 1642 (1980).

[13] N.C. Baenziger, R.M. Flynn, D.C. Swenson and N.L. Holy, *Acta Crystallogr., Sect. B* **34**, 2300 (1978).

[14] J. Lorberth, T.F. Berlitz and W. Massa, *Angew. Chem. Int. Ed. Engl.* **28**, 611 (1989).

Note The structural chemistry of mercury(II)-thiolate complexes is summarized in J.G. Wright, M.J. Natan, F.M. MacDonnell, D.M. Ralston and T.V. O'Halloran, *Prog. Inorg. Chem.* **38**, 323 (1990).

The cyclic tetrameric anion complex in $Li[(HgC_2B_{10}H_{10})_4Cl]$ possesses crystallographic C_{4h} symmetry with the four Hg atoms, the eight carborane C atoms, and eight B atoms located in the mirror plane, and the Cl⁻ ion on the C_4 axis is disordered over two equally populated sites at .383Å above and below the central cavity. See X. Yang, C.B. Knobler and M.F. Hawthorne, *Angew. Chem. Int. Ed. Engl.* **30**, 1507 (1991).

6.15   Trimethyltin(IV) Chloride
$(CH_3)_3SnCl$

Crystal Data

Monoclinic, space group $I2/c$ (No. 15)

$a = 12.541(8)$, $b = 9.618(11)$, $c = 11.015(11)$Å, $\beta = 92.62(7)°$,

(138K),[1] $Z = 8$

Atom	x	y	z
Sn	.70372	.48688	.35566
Cl	.6978	.4242	.1417
C(1)	.8699	.4709	.3984
C(2)	.5981	.3334	.4198
C(3)	.6406	.6899	.3438

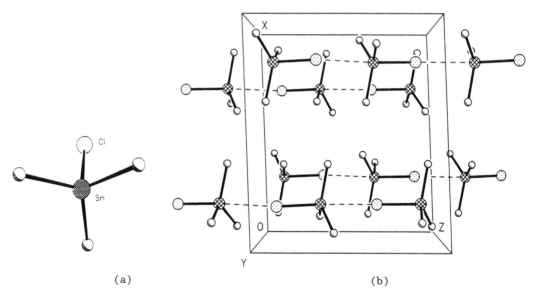

(a)   (b)

Fig. 6.15-1   (a) Asymmetric unit and (b) packing mode in crystalline $(CH_3)_3SnCl$.

Crystal Structure

The crystals are composed of polymeric chains of chlorine atoms bridging non-planar trimethyltin(IV) units at unequal (2.430 and 3.269Å) distances. The zigzag chains are bent at chlorine (angle Sn-Cl...Sn 150.30°), but nearly linear at tin (angle Cl-Sn...Cl 176.85°) to describe a trigonal bipyramidal geometry at tin with the trimethyltin groups eclipsed. The inter-chain Sn...Cl distances are greater than 4.1Å. The angles C-Sn-C 117.1°(mean) are larger than tetrahedral, while the angles C-Sn-Cl 99.9°(mean) are smaller, in accord with isovalent hybridization principles, but more severely distorted than in the gas-phase, monomeric structure. The Sn-Cl distance of 2.430Å is also longer than in the gas phase monomer, and the intermolecular contact of 3.269Å is shorter than in other organotin chloride bridged systems (sum of van der Waals radii 3.85Å).

Remarks

1. Coordination types of organotin compounds

Organotin compounds are finding increasing application as reagents in organic synthesis. A number of trialkyltin compounds are also used industrially in various biocidal applications, and some dialkyltin compounds are employed for catalyzing certain organic reactions, and as stabilizers for poly(vinyl chloride).

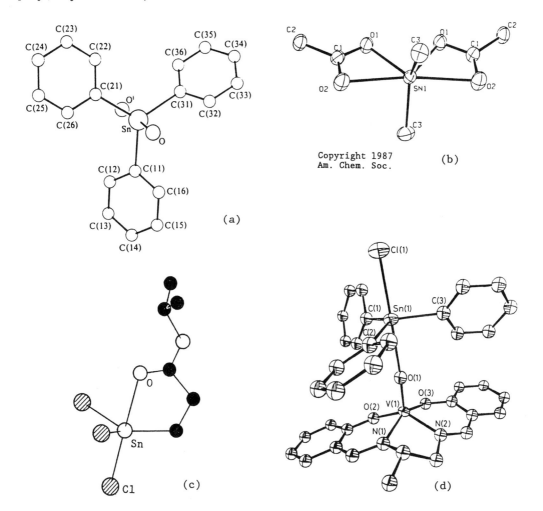

Copyright 1987
Am. Chem. Soc.

Fig. 6.15-2 Coordination types in organotin compounds.
(a) $Sn(C_6H_5)_3OH$, (b) $Sn(CH_3)_2(OAc)_2$, (c) $SnCl_3CH_2CH_2CO_2Pr\text{-}i$ and
(d) $Sn(C_6H_5)_3Cl\cdot VO(salpren)$. (After refs. 3-6).

In organotin compounds the tin atom can exhibit coordination numbers of two to seven in neutral, cationic and anionic oligomers and one-, two-, and three-dimensional polymeric arrays. Fig. 6.15-2 shows some representative coordination types. In 1978 Zubieta and Zuckerman[2] summarized the structural characteristics of organotin compounds in the following statements:

a) The preferred equilibrium, solid state molecular structures for organotin(IV) are tetrahedral, trigonal bipyramidal, octahedral and pentagonal bipyramidal for four- to seven-coordination.

b) The success of semi-quantitative models based on intramolecular nonbonded interactions suggests that the equilibrium geometry in the solid state is predominantly controlled by ligand repulsive forces and that the energy necessary to distort the molecules from ideal geometries is very small.

c) The observed angular distortions from perfect geometries are generally in the directions dictated by the isovalent hybridization model.

d) Electronegative atoms and groups tend to occupy the axial positions in trigonal bipyramidal complexes.

e) Diorganotin compounds tend to assume the *cis*-R_2Sn skeletal configuration in octahedral structures if this is sterically feasible. The role of the steric bulk of the ligands is crucial. The structures tend to adopt a *trans,trans,trans* arrangement in order to minimize ligand-ligand repulsions. Methyl groups are usually forced to be *trans*, except when the donor atoms are part of chelate rings whose bite is small. "Pointed" ligands, such as $O=ER_n$ in which the donor atom is free of further substitution, can take up *cis*-positions however.

f) Tetraorganotin compounds do not possess any Lewis acidity. However, the introduction of electronegative substituents increases the effective nuclear charge at tin, and hence the tendency for higher coordination. No authenticated hexacoordinated R_3Sn derivative is yet known.

g) A reduction in steric bulk of the anionic group allows donor atoms to bridge tin atoms to form dimers and oligomers and allows potentially bidentate ligands to chelate, raising the coordination number at tin; a further reduction in bulk allows oligomers and chelates to realign into associated, polymeric arrays in one-, two- or three-dimensions.

h) Organotin compounds are no different in general structural features than derivatives of other organo fourth-group elements below carbon, except that the detailed geometry and isomeric form, which depend on subtle steric interactions, will change with the size of the central atom.

Some of the above generalizations need to be revised in the light of more recent results. The first example of a six-coordinate triorganotin compound was found in trimethyltin tris(pyrazolyl)borate, which has crystallographically imposed C_3 symmetry.[7] More interestingly, *bis*[3-(2-pyridyl)-2-thienyl-*C*,*N*]diphenyltin is a six-coordinate tetraorganotin compound [Fig. 6.15-3(a)]. The pyridyl nitrogens are *cis* to each other [Sn-N = 2.560(2)Å, N-Sn-N' = 77.1(1)°], as are the phenyl groups [C-Sn-C' = 101.9(1)°], while the thienyl groups are in an approximately *trans* relationship [C-Sn-C' = 144.4(1)°].[8] Tribenzyltin 2-pyridinethiolate *N*-oxide adopts an unusual square-pyramidal geometry and provides the first example of polytopal dominance of this configuration in pentacoordinate organotin compounds [Fig. 6.15-3(b)].[9] In this molecule the basal plane is composed of the oxygen and sulfur atoms of the chelating 2-pyridinethiolate *N*-oxide ligand and the carbon atoms of two benzyl groups, and the C_{apical}-Sn-L_{basal} angles are in the range 100.1(1)-110.1(1)°.

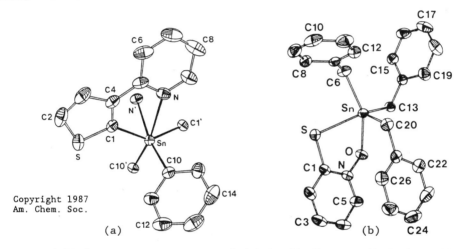

(a)   (b)

Fig. 6.15-3 Perspective views of (a) bis[3-(2-pyridyl)-2-thienyl]-diphenyltin (some atoms have been omitted for clarity, and the primed and unprimed atoms are related by two-fold axial symmetry) and (b) tribenzyltin 2-pyridinethiolate *N*-oxide, in which C(20) is the apical atom. (After refs. 8 and 9).

2. Reaction pathway for S_N2 substitution at tetrahedral tin[10]

     Investigation of the S_N2 reaction with inversion of configuration

$$Y + SnR_3X \rightleftharpoons [YSnR_3X] \rightleftharpoons YSnR_3 + X$$

by the "structure-correlation method" demonstrates that crystallography is capable of yielding valuable information about the dynamic aspects of molecular structure.[11,12] In applying this method, the structural parameters of relevant molecular fragments in the crystallographic literature are correlated with suitably chosen parameters to trace paths that approximate

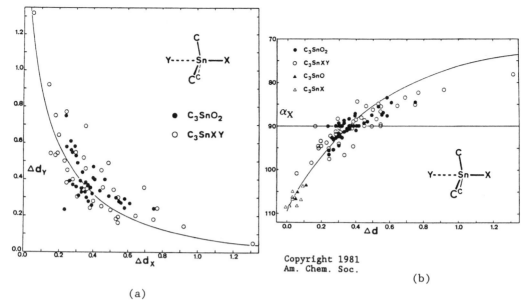

Fig. 6.15-4 Structural correlations for YSnR$_3$X ensembles: (a) plot of Δd_X versus Δd_Y (Å) and (b) C-Sn-X(α_X, in °) versus Δd_X [and C-Sn-Y(α_Y) versus Δd_Y]. (After ref. 10).

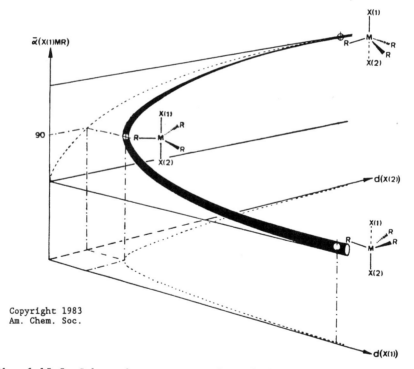

Fig. 6.15-5 Schematic representation of the reaction path for an S$_N$2-like inversion process in the three-dimensional parameter space spanned by the two X-M distances and an X-M-R angle. (After ref. 12).

the potential energy valleys in the parameter space. For $YSnR_3X$ ensembles two good correlations were found: as the Sn-Y distance d_Y becomes shorter, the Sn-X distance d_X tends to become longer and the C-Sn-X angles tend to become smaller. Fig. 6.15-4(a) illustrates the inverse relationship between Δd_X and Δd_Y ($\Delta d = 1.20 \log n$; $\Sigma n = 4$). The average C-Sn-X angle α_X shows a continuous change from obtuse through 90° to acute as the X-Sn bond is broken and the Sn-Y bond is formed [Fig. 6.15-4(b)]. The experimetal sample points for this associative ligand-exchange reaction thus provide an elegant mapping of the S_N2 pathway of the familiar Walden inversion type in three-dimensional parameter space (Fig. 6.15-5).

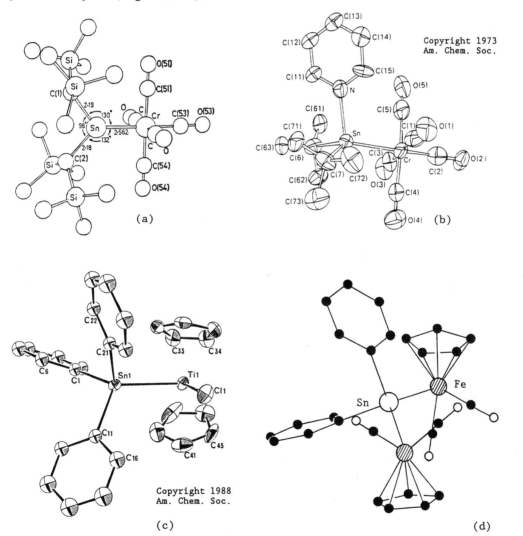

Fig. 6.15-6 Structures of some organotin-transition metal compounds.
(a) $[(Me_3Si)_2CH]_2Sn{:}Cr(CO)_5$, (b) $C_5H_5N.t{-}Bu_2Sn{:}Cr(CO)_5$,
(c) $Cp_2Ti(SnPh_3)Cl$, and (d) $[CpFe(CO)_2]_2SnPh_2$. (After refs. 13-16).

3. Structures of organotin-transition metal compounds

    Many organotin-transition metal compounds have had their structures
determined by X-ray analysis. The oxidation state of tin is II or IV; some
examples are shown in Fig. 6.15-6. With the +II oxidation state, the
organotin portion of the compound is formally a stannylene [R_2Sn], and as it
possesses a lone pair of electrons, a stannylene is also endowed with Lewis
basicity. For example, [$(Me_3Si)_2CH]_2Sn$:$Cr(CO)_5$ is a three-coordinate
molecule;[13] in the analogous di-t-butylstannylene compound
$C_5H_5N.t$-Bu_2Sn:$Cr(CO)_5$, the coordination number of tin is raised to four
because of the presence of a long dative N→Sn bond.[14] On the other hand,
where a formal σ bond exists between tin and the transition metal to give rise
to the +IV oxidation state, tin is invariably tetrahedral, as in
$Cp_2Ti(SnPh_3)Cl$[15] and [$CpFe(CO)_2]_2SnPh_2$.[16]

4. Structures of organotin compounds with Sn-Sn bonds.

    The structures of three organotin compounds with Sn-Sn bonds in the
solid state are shown in Fig. 6.15-7(a) to (c). In hexaphenylditin, the bond
angles at tin are close to tetrahedral and the Sn-Sn bond length is 2.77Å.[17]
Hexa(diphenyltin) adopts a chair conformation and the Sn-Sn bond length is
also 2.77Å.[18] In tetraphenyldiacetoxyditin, acetoxy groups bridge trigonal-
bipyramidal tin atoms and the Sn-Sn bond distance is reduced to 2.69Å (2.70Å

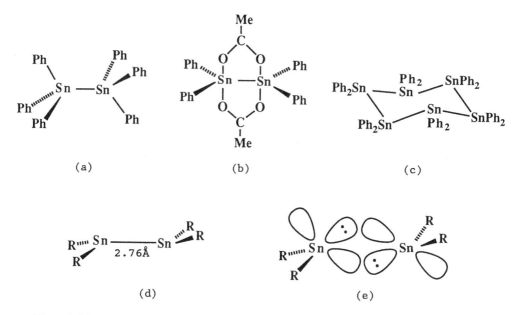

Fig. 6.15-7 Schematic structures of some organotin compounds with Sn-Sn
bonds. (a) Sn_2Ph_6, (b) $Sn_2Ph_4(OAc)_2$, (c) [$SnPh_2]_6$, (d) and (e)
$Sn_2[CH(SiMe_3)_2]_4$. (Adapted from ref. 21).

in the other independent molecule).[19] The tin atom in $Sn_2[CH(SiMe_3)_2]_4$ is best considered as being composed of two linked R_2Sn: units; the centrosymmetric demeric molecule is non-planar [Fig. 6.15-7(d)], the Sn-Sn contact being 2.76Å. The lone pairs of electrons are in the *anti* configuration, and it has been suggested that a weak double bond is formed between the tin atoms by overlap of the occupied sp^2 orbitals with the vacant p orbitals, as shown in Fig. 6.15-7(e).[20]

5. Structures of oligomeric organotin oxycarboxylates

Oligomeric organotin oxycarboxylates based on the compositions $[R'Sn(O)O_2CR]_6$ and $[(R'Sn(O)O_2CR)_2R'Sn(O_2CR)_3]_2$ have "drum" and "ladder" structures, respectively. The hexamer derivatives of this composition are drum-shaped molecules of idealized S_6 symmetry containing six-coordinate tin atoms. Fig. 6.15-8 shows a representative member, $[PhSn(O)O_2CC_6H_{11}]_6$, whose structure was determined by X-ray diffraction.[22]

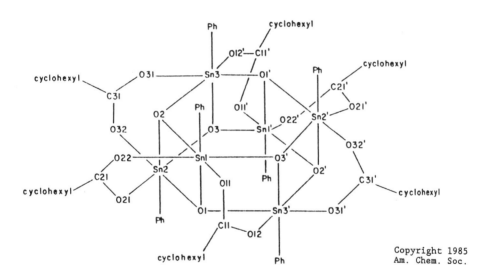

Fig. 6.15-8 Schematic representation of the drum structure of $[PhSn(O)O_2CC_6H_{11}]_6$. (After ref. 22).

The "stannoxane" framework of the molecule consists of two chair-shaped six-membered Sn_3O_3 rings, one lying directly on top of the other. Each Sn atom is thus bonded to three framework oxygen atoms, with Sn-O bond lengths ranging from 2.075(7) to 2.093(7)Å. Each of the four-membered rings of the core is spanned by a carboxylate group that forms a symmetrical bridge between two Sn atoms. The Sn-O bonds to the bridging carboxylate atoms are longer than the core bonds and range from 2.155(8) to 2.173(8)Å.

Fig. 6.15-9 illustrates the ladder structure of $[(n\text{-}BuSn(O)O_2CPh)_2.n\text{-}$

BuSn(Cl)(O$_2$CPh)$_2$]$_2$ which has crystallographic C_i symmetry.[23]   There are
three chemically inequivalent types of hexacoordinated Sn atoms in the
molecule.   The framework Sn-O bond length has the average value 2.063(3)Å.

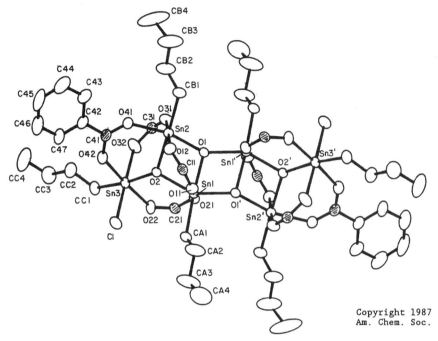

Fig. 6.15-9 Structure of [(n-Bu(Sn(O)O$_2$CPh)$_2$.n-BuSn(Cl)(O$_2$CPh)$_2$]$_2$.
Six of the eight phenyl groups and all hydrogen atoms have been omitted
for the sake of clarity.   (After ref. 23).

6. π Complexes of tin(II)[24]

   X-Ray crystallographic and gas phase diffraction data for some
stannocenes are shown in Fig. 6.15-10.   They reveal interesting information
about the steric effects of substituents at the cyclopentadienyl ring.
Stannocene[25] and most of its derivatives are bent-sandwich compounds with a
stereochemically active lone pair.   As expected the angle between the
cyclopentadienyl ring planes becomes smaller in going from the unsubstituted
to the perphenylated stannocene, in which the angle becomes zero.   This
complex is the first ferrocene-like sandwich with a group IV element as the
central atom.[26]   The perpendicular distance from tin to the cyclopentadienyl
ring planes is very similar in all stannocene structures.   Generally the tin
atom slips away from the ring centroid position; consequently very different
Sn-C distances of between 2.53 and 2.85Å are observed, and the angle between
the ring normals and the ring centroid-tin-ring centroid angle are different.

   The X-ray crystal structure of Me$_5$C$_5$Sn$^+$BF$_4^-$ confirmed the presence of
isolated cations and anions in the solid state.[27]   In the cation, the Me$_5$C$_5$

π system is η^5 bonded to the tin atom. The tin-ring centroid distance is considerably shorter than that in decamethylstannocene, and the methyl groups are bent away from the plane of the cyclopentadienyl ring.

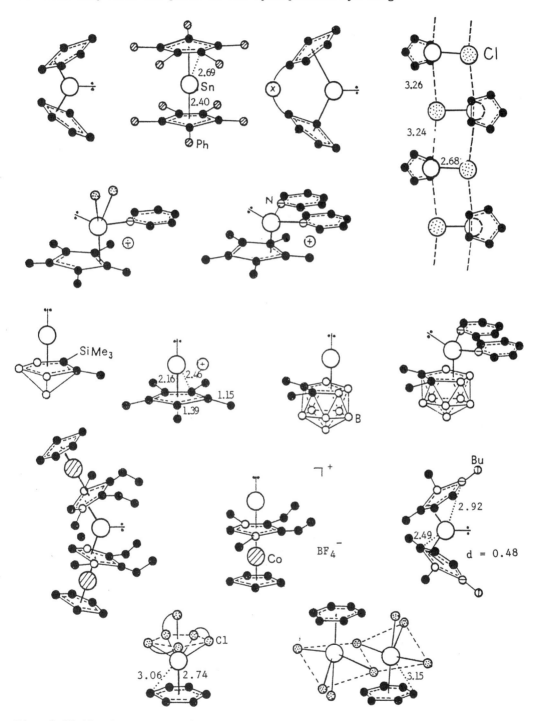

Fig. 6.15-10 Structures of some π complexes of tin(II). (After ref. 24).

References

[1] J.L. Lefferts, K.C. Molloy, M.B. Hossain, D. van der Helm and J.J. Zuckerman, *J. Organomet. Chem.* **240**, 349 (1982).

[2] J.A. Zubieta and J.J. Zuckerman, *Prog. Inorg. Chem.* **24**, 251 (1978).

[3] C. Glidewell and D.C. Liles, *Acta Crystallogr.*, *Sect. B* **34**, 129 (1978).

[4] T.P. Lockhart, J.C. Calabrese and F. Davidson, *Organometallics* **6**, 2479 (1987).

[5] R.A. Howie, E.S. Paterson, J.L. Wardell and J.W. Burley, *J. Organomet. Chem.* **304**, 301 (1986).

[6] B. Cashin, D. Cunningham, J.F. Gallagher, P. McArdle and T. Higgins, *J. Chem. Soc., Chem. Commun.*, 1445 (1989).

[7] B.K. Nicholson, *J. Organomet. Chem.* **265**, 153 (1984).

[8] V.G. Kumar Das, K.M. Lo, Chen Wei and T.C.W. Mak, *Organometallics* **6**, 10 (1987).

[9] S.W. Ng, Chen Wei, V.G. Kumar Das and T.C.W. Mak, *J. Organomet. Chem.* **334**, 283 (1987).

[10] D. Britton and J.D. Dunitz, *J. Am. Chem. Soc.* **103**, 2971 (1981).

[11] J.D. Dunitz, *X-Ray Analysis and the Structure of Organic Molecules*, Cornell University Press, Ithaca, New York, 1979, chap. 7.

[12] H. B. Bürgi and J.D. Dunitz, *Acc. Chem. Res.* **16**, 153 (1983).

[13] J.D. Cotton, P.J. Davison, D.E. Goldberg, M.F. Lappert and K.M. Thomas, *J. Chem. Soc., Chem. Commun.*, 893 (1974).

[14] M.D. Brice and F.A. Cotton, J. Am. Chem. Soc. **95**, 4529 (1973).

[15] W. Zheng and D.W. Stephan, *Inorg. Chem.* **27**, 2386 (1988).

[16] L. Pàrkànyi, K.H. Pannell and C. Hernandez, *J. Organomet. Chem.* **347**, 295 (1988).

[17] H. Preut, H.J. Haupt and F. Huber, *Z. Anorg. Allg. Chem.* **396**, 81 (1973).

[18] D.H. Olson and R.E. Rundle, *Inorg. Chem.* **2**, 1310 (1963).

[19] S. Adams, M. Dräger and B. Mathiasch, *J. Organomet. Chem.* **326**, 173 (1987).

[20] D.E. Goldberg, D.H. Harris, M.F. Lappert and K.M. Thomas, *J. Chem. Soc., Chem. Commun.*, 261 (1976); D.E. Goldberg, P.B. Hitchcock, M.F. Lappert, K.M. Thomas, A.J. Thorne, T. Fjeldberg, A. Haaland and B.E.R. Schilling, *J. Chem. Soc., Dalton Trans.*, 2387 (1986).

[21] A.G. Davies and P.J. Smith in G. Wilkinson, F.G.A. Stone and E.W. Abel (eds.), *Comprehensive Organometallic Chemistry*, Pergamon Press, Oxford, 1982, Vol. 2, chap.11, p. 519.

[22] V. Chandrasekhar, R.O. Day and R.R. Holmes, *Inorg. Chem.* **24**, 1970 (1985).

[23] V. Chandrasekhar, C.G. Schmid, S.D. Burton, J.M. Holmes, R.O. Day and R.R. Holmes, *Inorg. Chem.* **26**, 1050 (1987).

[24] P. Jutzi, *Adv. Organomet. Chem.* **26**, 217 (1986).

[25] A. Almenningen, A. Haaland and T. Motzfeldt, *J. Organomet. Chem.* **7**, 97 (1967).

[26] M.J. Heeg, C. Janiak and J.J. Zuckerman, *J. Am. Chem. Soc.* **106**, 4259 (1984).

[27] P. Jutzi, F. Kohl and C. Krüger, *Angew. Chem. Int. Ed. Engl.* **18**, 59 (1979).

Note The crystal structures of p-tol$_4$Pb, p-tol$_6$Sn$_2$ (two forms), p-tol$_6$SnPb, and p-tol$_6$Pb$_2$ are described in C. Schneider and M. Dräger, *J. Organomet. Chem.* **415**, 349 (1991).

Organometallic compounds of bivalent and tetravalent tin are reviewed in P.G. Harrison (ed.), *Chemistry of Tin*, Blackie, Glasgow and London, 1989.

6.16   (R-1-Cyanoethyl)(3-methylpyridine)cobaloxime
$C_{17}H_{25}N_6O_4Co$

Crystal Data

Orthorhombic, space group $P2_12_12_1$ (No. 19), $Z = 8$

Stage	$a/Å^*$	$b/Å$	$c/Å$
I	11.696(4)	38.34(2)	9.215(3)
II	11.644(3)	38.33(1)	9.262(2)
III	11.584(3)	38.29(1)	9.311(2)
IV	11.552(2)	38.22(1)	9.332(2)
V	11.548(2)	38.22(1)	9.334(1)
VI	11.537(2)	38.18(1)	9.347(1)

* At 293K

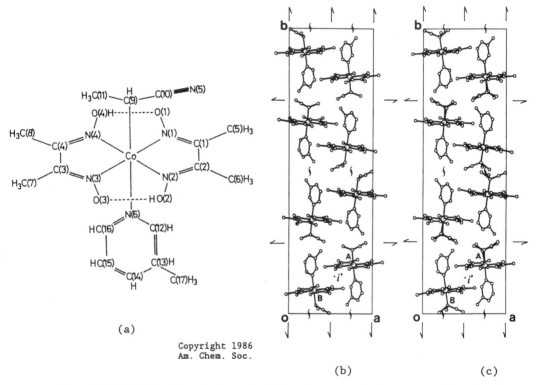

(a)

(b)   (c)

Fig. 6.16-1 Structural formula (a) and crystal structure viewed along the c axis at stage I (b) and at stage VI (c) of the cobaloxime complex. (After ref. 1).

The IUPAC name of the title complex is *trans*-(R-1-cyanoethyl)bis(dimethyl-glyoximato)(3-methylpyridine)cobalt(III).

Stage I

Atom	x	y	z	Atom	x	y	z
Co(A)	.27880	.32935	.01235	Co(B)	.30827	.08210	.54236
N(1A)	.1184	.3353	.0192	N(1B)	.1493	.0739	.5316
N(2A)	.2451	.3204	-.1844	N(2B)	.2726	.0885	.7401
N(3A)	.4386	.3237	.0042	N(3B)	.4676	.0880	.5550
N(4A)	.3092	.3368	.2113	N(4B)	.3434	.0755	.3446
O(1A)	.0622	.3437	.1406	O(1B)	.0967	.0661	.4058
O(2A)	.3253	.3109	-.2806	O(2B)	.3508	.0975	.8409
O(3A)	.4932	.3155	-.1184	O(3B)	.5211	.0944	.6803
O(4A)	.2275	.3431	.3091	O(4B)	.2639	.0677	.2430
C(1A)	.0626	.3317	-.1018	C(1B)	.0928	.0755	.6525
C(2A)	.1373	.3221	-.2212	C(2B)	.1654	.0842	.7756
C(3A)	.4931	.3270	.1269	C(3B)	.5239	.0860	.4335
C(4A)	.4175	.3347	.2474	C(4B)	.4509	.0781	.3089
C(5A)	-.0622	.3360	-.1129	C(5B)	-.0315	.0675	.6655
C(6A)	.0971	.3147	-.3706	C(6B)	.1235	.0859	.9289
C(7A)	.6179	.3210	.1422	C(7B)	.6514	.0891	.4258
C(8A)	.4526	.3370	.4030	C(8B)	.4931	.0720	.1581
C(9A)	.2634	.2766	.0463	C(9B)	.3303	.0294	.5771
C(10A)	.3542	.2638	.1400	C(10B)	.4160	.0147	.4813
C(11A)	.1519	.2637	.1041	C(11B)	.3556	.0181	.7296
N(5A)	.4196	.2527	.2165	N(5B)	.4870	.0031	.4136
N(6A)	.2980	.3813	-.0288	N(6B)	.2856	.1341	.5063
C(12A)	.2495	.4059	.0542	C(12B)	.3367	.1584	.5858
C(13A)	.2637	.4417	.0290	C(13B)	.3216	.1941	.5639
C(14A)	.3310	.4513	-.0840	C(14B)	.2526	.2044	.4522
C(15A)	.3799	.4269	-.1711	C(15B)	.1972	.1806	.3683
C(16A)	.3634	.3920	-.1404	C(16B)	.2168	.1457	.3987
C(17A)	.2057	.4673	.1285	C(17B)	.3849	.2201	.6600

Stage VI

Atom	x	y	z	Atom	x	y	z
Co(A)	.24769	.33073	.02332	Co(B)	.27847	.08204	.52654
N(1A)	.0851	.3352	.0305	N(1B)	.1161	.0744	.5136
N(2A)	.2143	.3224	-.1723	N(2B)	.2414	.0862	.7220
N(3A)	.4098	.3258	.0149	N(3B)	.4396	.0888	.5404
N(4A)	.2788	.3373	.2197	N(4B)	.3164	.0769	.3325
O(1A)	.0284	.3426	.1529	O(1B)	.0643	.0685	.3876
O(2A)	.2975	.3142	-.2689	O(2B)	.3206	.0929	.8234
O(3A)	.4659	.3186	-.1069	O(3B)	.4924	.0943	.6660
O(4A)	.1946	.3421	.3180	O(4B)	.2357	.0698	.2315
C(1A)	.0294	.3318	-.0890	C(1B)	.0607	.0740	.6350
C(2A)	.1057	.3238	-.2079	C(2B)	.1341	.0812	.7576
C(3A)	.4644	.3290	.1360	C(3B)	.4974	.0875	.4218
C(4A)	.3871	.3361	.2561	C(4B)	.4251	.0802	.2987
C(5A)	-.0984	.3342	-.0969	C(5B)	-.0640	.0653	.6447
C(6A)	.0621	.3159	-.3531	C(6B)	.0908	.0831	.9084
C(7A)	.5915	.3244	.1474	C(7B)	.6261	.0902	.4172
C(8A)	.4269	.3378	.4060	C(8B)	.4691	.0746	.1508
N(6A)	.2636	.3831	-.0121	C(9B)	.3035	.0291	.5581
C(12A)	.2113	.4067	.0716	C(10B)	.4090	.0170	.4873
C(13A)	.2272	.4427	.0529	N(5B)	.4892	.0068	.4334
C(14A)	.2944	.4539	-.0574	N(6B)	.2551	.1350	.4972
C(15A)	.3479	.4301	-.1425	C(12B)	.3085	.1587	.5795
C(16A)	.3321	.3946	-.1192	C(13B)	.2960	.1947	.5575
C(17A)	.1604	.4678	.1523	C(14B)	.2232	.2045	.4489
C(9A)	.2384	.2761	.0473	C(15B)	.1667	.1814	.3630
C(10A)	.3257	.2653	.1536	C(16B)	.1862	.1464	.3913
C(11A)	.1182	.2629	.0910	C(17B)	.3600	.2205	.6526
N(5A)	.3913	.2550	.2327	C(11B)	.2879	.0143	.7048
C(9'A)	.2192	.2798	.0749	C(11'B)	.2030	.0047	.5314
C(10'A)	.3229	.2624	.1288				
C(11'A)	.1450	.2570	-.0173				
N(5'A)	.4028	.2483	.1647				

Crystal Structure

The crystals of the cobaloxime complex with 3-methylpyridine and the chiral cyanoethyl group and 3-methyl pyridine as axial ligands are racemized by X-ray exposure without degradation of the crystallinity. The structures

were determined at several intermediate stages, which is essential for the
elucidation of the crystalline-state reaction mechanism.

The crystal has two molecules in the asymmetric unit. In the early
stages, both chiral cyanoethyl groups of the two molecules, **A** and **B**, were
gradually inverted into the opposite configuration. After 400h, the crystal
was completely racemized but the conversion of the two groups continued; the **B**
cyanoethyl group was further inverted whereas the **A** cyanoethyl group was
gradually restored to the original configuration.

At stage I, the crystal structure viewed along the c axis is shown in
Fig. 6.16-1(b). Except for the cyanoethyl group, the two crystallographically
independent molecules **A** and **B** are related by a local inversion center at
(0.5435, 0.1264, 0.2650). The cyanoethyl groups of the **A** molecules are
connected by a 2_1 axis to form a ribbon parallel to the a axis, whereas the **B**
cyanoethyl groups make a ribbon along another 2_1 axis in the c direction.
There are no unusually short contacts between the molecules.

At stage VI, the crystal structure viewed along the c axis is shown in
Fig. 6.16-1(c). Both of the **A** and **B** cyanoethyl groups are converted into the
disordered racemates. The local inversion center does not change to a
crystallographic one and the noncentrosymmetric space group $P2_12_12_1$ is
conserved. Fig. 6.16-2 shows the **A** and **B** molecules at stage VI viewed along
the normal to the cobaloxime plane. The newly appearing cyanoethyl group with
the S configuration has a different conformation in the two molecules. The
conformations of the cyanoethyl groups with the R configuration and those of
the 3-methylpyridine ligands are approximately the same as the corresponding
ones at the initial stage.

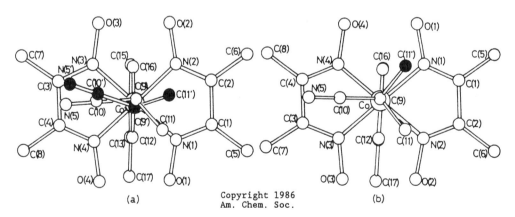

(a)   Copyright 1986   (b)
                       Am. Chem. Soc.

Fig. 6.16-2 The structures of (a) **A** and (b) **B** molecules viewed along
the normal to the mean plane of cobaloxime. The shaded atoms are the
newly appearing ones in the process of racemization. (After ref. 1).

Remarks

1. Dynamical structure analysis of a crystalline-state reaction[2]

In ordinary chemical reactions of solids, the initiation often occurs at defect or surface sites and the crystal lattice is destroyed as the reaction proceeds. A single crystal-to-single crystal transformation in the solid state is a very attractive system for study, since the reaction proceeds with retention of the crystalline form and the net movement of each atom can be presumed from the structures before and after the reaction. The structures at several intermediate stages are essential for elucidating the detailed reaction pathways.

In the course of structural studies on cobaloxime complexes aimed at interpreting their catalytic capability for asymmetric hydrogenation, Ohashi, Sasada and co-workers have found that the chiral cyanoethyl (cn) group in a cobaloxime complex crystal is racemized by X-ray exposure without degradation of crystallinity. The rate of racemization was so slow that several intermediate structures could be obtained by X-ray analyses. Moreover, they have observed various reaction pathways and obtained a quantitative relationship between the macroscopic reaction rate and the microscopic atomic arrangement in the crystal from intermediate structure analyses of a series of related crystals. Such a stepwise X-ray structure analysis is a very effective method in elucidating the reaction mechanism.

2. Racemization of the cn group in crystalline (R-1-cyanoethyl)bis(dimethyl-glyoximato)(S-α-methylbenzylamine)cobalt(III)[2]

The structure of this complex (Fig. 6.16-3) underwent a change in unit-cell dimensions without degradation of crystallinity when it was exposed to X-rays. The space group $P2_1$ remained unaltered.[3,4]

In order to follow the structural change, the unit-cell dimensions and intensity data were collected at the A to G stages indicated in Fig. 6.16-4. The structure analyses of the seven stages revealed that significant changes were observed only in the vicinity of the cn group. Fig. 6.16-5(a) shows the composite electron density map of the cn group viewed along the normal to the cobaloxime plane at stage A. The peaks of the cn substituents, especially the methyl group, were lowered and expanded at stage B. Fig. 6.16-5(b) shows a composite difference electron density map at stage B. This figure corresponds to the difference between the two electron density maps at the A and B stages; a trough appears at the position of the methyl group and a peak appears in the neighborhood of the methyl group. Fig. 6.16-5(c) is the corresponding composite difference electron density map at stage F; the new

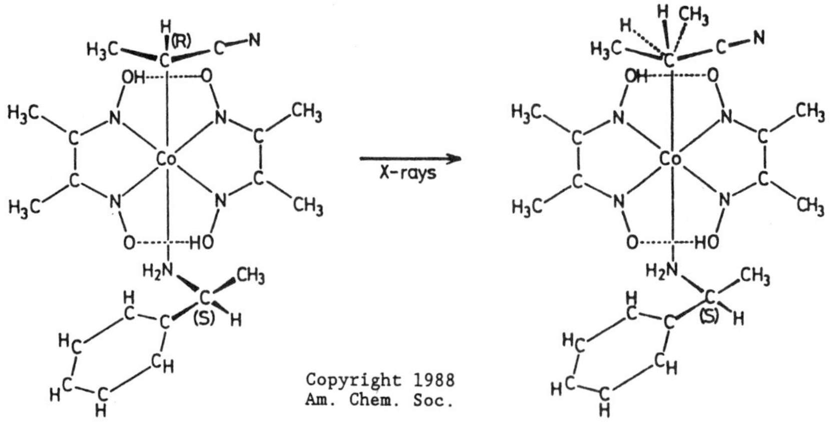

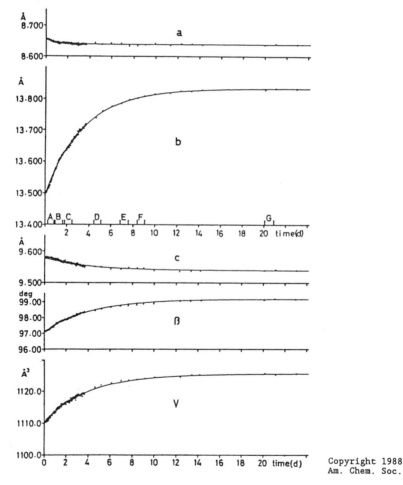

Fig. 6.16-3 Structural change in racemization of the cn group in a crystal of $Co(C_2H_4CN)(C_4H_7N_2O_2)_2(CH_3C_6H_4CH_2NH_2)$. (After ref. 2).

Fig. 6.16-4 Changes in unit-cell dimensions during racemization. At early stages the measurements were made every 30 min. (After ref. 2).

peak grows higher, and the trough becomes deeper. The composite electron density map at the final stage G is shown in Fig. 6.16-5(d). The height and position of the new peak correspond to those expected for a methyl substituent in a cn group of inverted configuration. The racemization process is thus mirrored by a gradual change of the electron density of the reactive group.

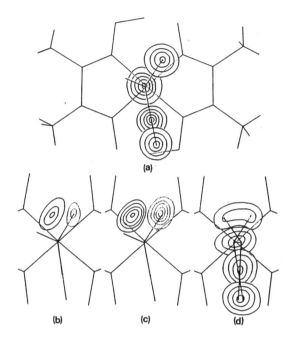

Fig. 6.16-5 (a) Composite electron density map of the cn group viewed along the normal to the cobaloxime plane at the initial A stage. The contour interval is 1.0 eÅ^{-3}. (b), (c) Composite difference electron-density maps at the B and F stages, respectively. The solid and dotted curves indicate positive and negative densities, respectively. The contour interval is 0.2 eÅ^{-3}. (d) Composite electron density map at the final stage G. The contour interval is the same as that of (a). (After ref. 2).

In this racemization, one or more of the four bonds around the chiral carbon atom, Co-C, NC-C, H$_3$C-C, and H-C, must have been cleaved by X-rays in the transition state. ESR spectra obtained after the crystal had been exposed to X-rays revealed that the Co-C bond was cleaved homolytically to produce the cn radical and the Co(II) complex.[5] The Co-C bond dissociation energy for these cobaloxime complexes has been determined to be 117-122 kJ mol^{-1}.[6] The racemization rate of the complex by X-rays is far lower than that by visible light, not only in the solid state but also in an aqueous methanol solution.

3. Structural characteristics of organocobaloximes

Organocobaloximes LCo(DH)$_2$R, where DH = monoanion of dimethylglyoxime, L = neutral ligand and R = alkyl group are useful model compounds of the vitamin B$_{12}$ system.[7,8] The study of the influence of steric and electronic features of various L and R groups on the cobalt-carbon bond has furnished support to the mechanism proposed for its cleavage, one of the vital steps in the enzymatic action of the B$_{12}$ coenzyme (Section 5.22).[9]

Table 6.16-1   Values of selected bond lengths (Å) and angles (°) for LCo(DH)$_2$R complexes.*

L	R	Co-C	Co-N	Co-N(ax)-C	Co-C-X	Orientation
Im	Me	1.985(3)	2.019(3)	129.7(1) 124.8(1)	-	B
1-MeIm	adamantyl	2.154(4)	2.065(4)	127.4(3) 127.8(3)	112.2(4)	B
	Me	2.009(7)	2.058(4)	127.4(3) 127.0(5)	-	A
	(CH$_2$)$_2$CN	1.989(5)	2.037(3)	125.4(2) 129.0(2)	127.9(5)	A
py	Me	1.998(5)	2.068(3)	122.3(3) 119.7(4)	-	B
	(CH$_2$)$_2$CN	2.002(7)	2.050(5)	121.9(5) 120.7(5)	123.6(6)	B
	CH$_2$NO$_2$	2.002(3)	2.028(3)	121.2(2) 121.2(2)	113.7(2)	B
Me$_3$Bzm	adamantyl	2.179(5)	2.137(4)	134.4(3) 120.8(4)	110.4(3)	B
	Me	1.989(2)	2.060(2)	133.6(1) 121.5(2)	-	B
	CH$_2$NO$_2$	1.988(5)	2.013(3)	132.7(3) 122.4(3)	115.7(4)	B
1,2-Me$_2$Im	Me	2.001(2)	2.086(1)	134.9(1) 119.1(1)	-	B
	(CH$_2$)$_3$CN	2.023(3)	2.083(2)	133.9(2) 120.7(2)	119.4(2)	B
	CH$_2$NO$_2$	1.999(3)	2.049(3)	132.2(2) 122.2(2)	114.5(1)	B

* Orientation of L with respect to the equatorial moiety is defined in Fig. 6.16-7; the mean values of the Co-C-X angles are reported for adamantyl derivatives.

X-Ray analyses of a series of complexes containing planar L ligands have yielded the principal structural results summarized in Table 6.16-1.[10]   For cobaloximes containing the same L ligand and different R groups, the electronic *trans*-influence is clearly shown in Fig. 6.16-6.[11]   The structural properties of cobaloximes in the solid state have been very useful in interpreting solution properties.

The available data allow the following conclusions to be drawn:

a)   For the same L ligand the Co-N(ax) distance increases with the σ-donor ability of the *trans* alkyl group.

b)   The L ligand in orientation **A** (see Fig. 6.16-7), which is found only in 1-MeIm derivatives with R = Me and (CH$_2$)$_2$CN, interacts more strongly with the equatorial macrocyclic ligand than in orientation **B**, leading to longer Co-N distances.

c)   For the same R and the same L orientation the Co-N(ax) distance increases in the order: Im ≃ 1-MeIm < py ≃ Me_3Bzm < $1,2-Me_2Im$ which follows the order of increasing bulkiness of L.

d)   The Co-C bond length and Co-C-X angle increase with the increasing bulkiness of R.

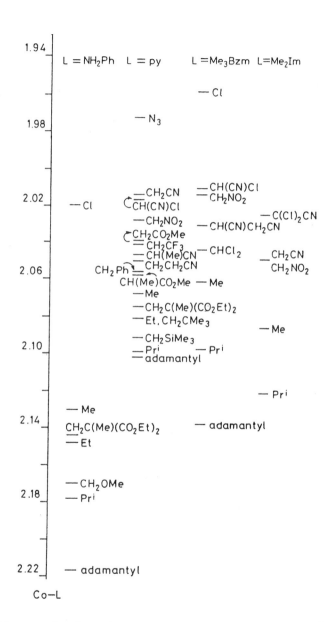

Fig. 6.16-6 Diagram showing the trend of the Co-L distances (Å) in cobaloximes for different alkyl groups *trans* to L = NH_2Ph, py, Me_3Bzm, Me_2Im. Available values for azido and chloro analogues are included for comparison. (After ref. 11).

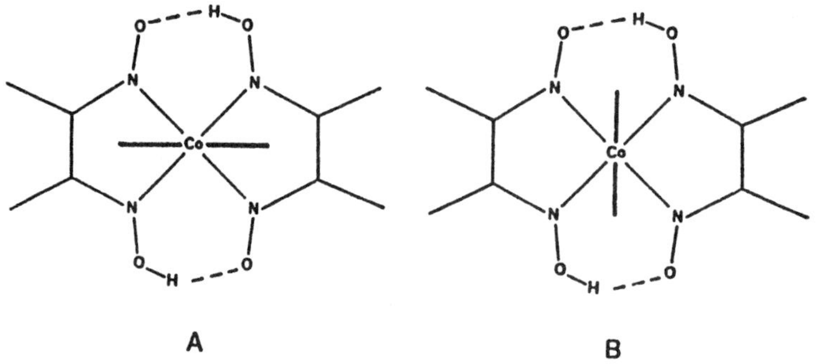

Fig. 6.16-7 The two limiting orientations **A** and **B** of the planar L ligand with respect to the equatorial macrocyclic ligand in an organocobaloxime complex. (After ref. 10).

References

[1] Y. Ohashi, Y. Tomotake, A. Uchida, and Y. Sasada, *J. Am. Chem. Soc.* **108**, 1196 (1986).

[2] Y. Ohashi, *Acc. Chem. Res.* **21**, 268 (1988).

[3] Y. Ohashi and Y. Sasada, *Nature (London)* **267**, 142 (1977).

[4] Y. Ohashi, K. Yanagi, T. Kurihara, Y. Sasada and Y. Ohgo, *J. Am. Chem. Soc.* **103**, 5805 (1981).

[5] C. Giannotti, G. Merle and J.R. Bolton, *J. Organomet. Chem.* **99**, 145 (1975).

[6] Y. Ohgo, K. Orisaku, E. Hasegawa and S. Takeuchi, *Chem. Lett.*, 27 (1986).

[7] N. Bresciani-Pahor, M. Forcolin, L.G. Marzilli, L. Randaccio, M.F. Summers and P.J. Toscano, *Coord. Chem. Rev.* **63**, 1 (1985).

[8] L.G. Marzilli, F. Bayo, M.F. Summers, L.B. Thomas, E. Zangrando, N. Bresciani-Pahor, M. Mari and L. Randaccio, *J. Am. Chem. Soc.* **109**, 6045 (1987).

[9] J. Halpern, *Science (Washington)* **227**, 869 (1985).

[10] N. Bresciani Pahor, W.M. Attia, S. Geremia, L. Randaccio and C. Lopez, *Acta Crystallogr.*, Sect. *C* **45**, 561 (1989).

[11] L. Randaccio, N. Bresciani Pahor, E. Zangrando and L.G. Marzilli, *Chem. Soc. Rev.* **18**, 225 (1989).

Chapter 7

Inclusion Compounds

7.1.   *Ethylene Oxide Deuterohydrate* $6(CH_2)_2O \cdot 46D_2O$ 1174

7.2.   *Tetra-n-Butyl Ammonium Benzoate Clathrate Hydrate*
       $(n\text{-}C_4H_9)_4N^+C_6H_5COO^- \cdot 39\frac{1}{2}H_2O$ 1191

7.3.   *β-Hydroquinone Hydrogen Sulfide Clathrate* $3C_6H_4(OH)_2 \cdot xH_2S$
       (x = 0.874)   1207

7.4.   *4-p-Hydroxyphenyl-cis-2,4-dimethylchroman Carbon Tetrachloride*
       Clathrate $6[C_{17}H_{18}O_2] \cdot CCl_4$ 1217

7.5.   *Hexakis(benzylthiomethyl)benzene–Dioxan 1:1 clathrate*
       $C_6(CH_2SCH_2C_6H_5)_6 \cdot C_4H_8O_2$ 1228

7.6.   *Tri-o-thymotide Chlorocyclohexane Clathrate*
       $2(C_{33}H_{36}O_6) \cdot C_6H_{11}Cl$ 1238

7.7.   *trans-Diamminemanganese catena-Tetra-μ-cyanonickelate-*
       Benzene (1/2) (Hofman-type Clathrate) $Mn(NH_3)_2Ni(CN)_4 \cdot 2C_6H_6$ 1248

7.8.   *(α-Cyclodextrin)$_2$ · Cd$_{0.5}$ · I$_5$ · 27H$_2$O* $(C_{36}H_{60}O_{30})_2 \cdot Cd_{0.5} \cdot I_5 \cdot 27H_2O$ 1258

7.9.   *1:1 Paraquat Bisparaphenylene-34-crown-10 Inclusion Compound*
       $[C_{12}H_{14}N_2 \cdot C_{28}H_{40}O_{12}][PF_6]_2 \cdot 2Me_2CO$ 1278

7.1 Ethylene Oxide Deuterohydrate

$6(CH_2)_2O.46D_2O$

Crystal Data

Cubic, space group $Pm3n$ (No. 223)

$a = 11.87(1)Å^{[1]}$ (80K), $Z = 1$

Atom*	x	y	z	Occupancy	Atom*	x	y	z	Occupancy
	Host atoms					Guest atoms			
O(i)	.18375	.18375	.18375	1.0	O(x)	.0314	.0314	.0314	.067
O(k)	0	.3082	.1173	1.0	OC(1)	.2782	.4621	.0313	.125
O(c)	0	1/2	1/4	1.0	OC(2)	.2514	.5669	-.0251	.125
D(ii)	.2323	.2323	.2323	0.5	OC(3)	.1622	.4994	.0237	.125
D(ck)	0	.4340	.2015	0.5	H(1)	.3182	.4680	.1128	.0833
D(kc)	0	.3787	.1588	0.5	H(2)	.3026	.3919	-.0212	.0833
D(kk)	0	.3161	.0357	0.5	H(3)	.2733	.6440	.0181	.0833
D(ki)	.0676	.2648	.1397	0.5	H(4)	.2576	.5679	-.1159	.0833
D(ik)	.1170	.2276	.1588	0.5	H(5)	.1237	.5306	.1000	.0833
					H(6)	.1080	.4544	-.0340	.0833

* D(ck) is a deuterium atom that is covalently bonded to O(c) and hydrogen-bonded to O(k). The small amount of entrapped air is represented by O(x). In the disordered ethylene oxide molecule, each OC atom is taken to be (O + ¼C)/3.

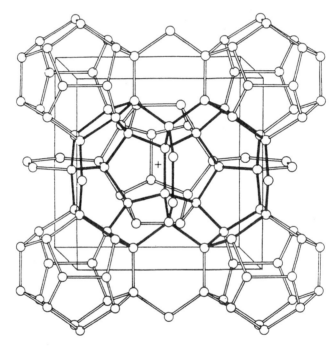

Fig. 7.1-1 Structure of the host framework of ethylene oxide deuterohydrate (a Type I clathrate hydrate). The deuterium atoms are omitted. (After ref. 2).

Crystal Structure

The initial X-ray crystal structure of $6(CH_2)_2O.46H_2O^{[2]}$ was later updated by a neutron diffraction study of $6(CH_2)_2O.46D_2O.^{[1]}$ The hydrogen-bonded D_2O framework structure as defined by the oxygen atoms is shown in Fig.

7.1-1.   The deuterium atoms are located precisely as disordered half-atoms, permitting a detailed description of the hydrogen bonding. Geometrical parameters for the four unique hydrogen-bonding interactions which form the edges of the polyhedra are listed in Table 7.1-1.

Table 7.1-1   Geometrical details of hydrogen bonding in $6(CH_2)_2O.46D_2O$.

	Bond length/Å			Bond angle/deg
	O...O	O-D	D...O	O-D...O
O(i), °D(ii)	2.724(2)	0.998(2)	1.726(2)	180
O(i), °D(ik)	2.750(9)	0.993(4)	1.761(4)	178.4(3)
		0.990(9)	1.757(9)	177.3(5)
O(k), °D(kc)	2.768(3)	0.971(5)	1.797(5)	173.6(4)
		0.972(4)	1.802(5)	177.5(4)
O(k), °D(kk)	2.784(4)	0.973(5)	1.818(5)	171.6(4)

The host framework can be described in terms of the close-packing of 12-hedra and 14-hedra that share faces in the ratio of 1 to 3. (In Fig. 7.1-1, the bold lines show two 14-hedra sharing a common face.) The centers of the 12-hedra are the two-fold positions (*a*), and the centers of the 14-hedra are the sixfold positions (*d*) of the space group *Pm3n*. Each 14-hedron shares its two hexagonal faces and eight of its twelve pentagonal faces with other 14-hedra. The remaining four faces are shared with the 12-hedra. There is no direct contact between the 12-hedra. The structure can be constructed from the vertices of 14-hedra arranged in columns with sharing of their opposite hexagonal faces. These columns are then placed in contact, so as to share a pentagonal face between each pair of 14-hedra, with the column axes directed along (*x*, 1/2, 0) of the cubic unit cell. The remaining space then comprises the 12-hedra in a pseudo body-centered cubic arrangement of their centers.

The 14-hedron encloses an approximately ellipsoidal volume with a free radius of 2.5 to 2.7Å normal to the hexagonal faces, and 3.0 to 3.15Å in the equatorial directions. For the ethylene oxide molecule within this void, there is an averaging of three orientations about a three-fold axis normal to the three-membered ring of the molecule. This model has 24-fold disorder: the electron density does not suggest a free rotor, but rather a dynamic disorder with small but significant barriers to reorientation. There is also non-stoichiometric occupancy of the 12-hedra by air molecules.

Remarks

1. Hydrate inclusion compounds

In the crystalline state an "inclusion compound" comprises two components, "host" and "guest", the association of which differs from that of other complexes or mixed crystals in two essential aspects:

a) Geometry: the guest molecules occupy voids (isolated cavities, channels, or interlayer regions) in a host lattice.

b) Bonding: construction of the host framework and the host/guest association involve van der Waals interaction and/or hydrogen bonding, but not conventional covalent or ionic bonding.

Hydrate inclusion compounds, which are characterized by a host lattice constructed wholly or principally from hydrogen-bonded water molecules, can be classified into four categories:

a) Clathrate hydrates, in which the neutral guest molecules are entrapped inside isolated cavities.

b) Alkylamine hydrates, of which the majority feature hydrogen-bonding interactions between the guest amine molecules and the water host lattice.

c) Peralkylated onium salt hydrates, in which the bulky hydrophobic cations occupy cages in a host framework built of water molecules and anions.

d) Layer-type hydrates, in which the generally puckered host layers are constructed from water molecules, functional groups (e.g. hydroxyl, amide, *N*-oxide, etc.) of organic molecules, and/or halide ions.

2. Clathrate hydrates[3-5]

Historically known as the "gas hydrates", these compounds have been subjected to extensive studies in regard to methods of preparation, non-stoichiometry, thermodynamical properties, crystal structure, motional behavior of the enclosed molecules, chemical separation, and other technical applications.

Nearly a hundred compounds ranging in size from argon to dioxan are known to form clathrate hydrates. The basic structural component is the $(H_2O)_{20}$ pentagonal dodecahedron [Fig. 7.1-2(a)], which has a volume of about 170Å^3. When these $[5^{12}]$ (12 faces each having 5 edges) polyhedra are associated to form a crystal lattice, the space-filling requirement necessarily generate other kinds of polyhedra. Other than the pentagonal dodecahedron, the important polyhedra found in the clathrate hydrates are the tetrakaidecahedron (14-hedron, $[5^{12}6^2]$), the pentakaidecahedron (15-hedron, $[5^{12}6^3]$) [Fig. 7.1-2(b)], and the hexakaidecahedron (16-hedron, $[5^{12}6^4]$; these larger polyhedra have volumes of about 230, 250 and 260 Å^3, respectively. Polyhedra containing fewer or more vertices are far less common. In the actual structures all polyhedra, including the pentagonal dodecahedron which can ideally assume I_h symmetry, generally have nonplanar faces with unequal edges and angles.

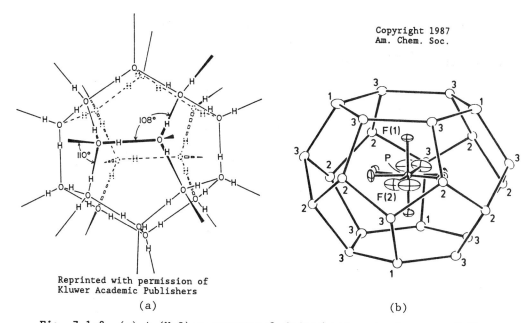

(a)   (b)

Fig. 7.1-2 (a) A $(H_2O)_{20}$ pentagonal dodecahedron showing one of the many possible ordered arrangements of the water protons. In the gas hydrates, the H atoms are twofold disordered across each O...O edge. (b) A tetrakaidecahedron enclosing a disordered PF_6^- guest species in the crystal structure of $HPF_6.7.67H_2O$. (After refs. 3 and 9).

3. Structural types of clathrate hydrates

The geometrical features and lattice parameters of the known or predicted crystal structures for the clathrate hydrates are given in Table 7.1-2. As shown in Table 7.1-3, the smaller molecules form the Type I structure and a few intermediate-sized molecules form both Type I and II hydrates. The larger molecules only form hydrates with the help of a smaller molecule such as H_2S or Xe. Bromine hydrate is believed to have the Type III structure. Dimethyl ether forms a Type II hydrate and a tetragonal hydrate believed to be Type III.

The acid hydrates $HEF_6.6H_2O$ (E = As, Sb) have a structure closely related to the Type IV structure with the proton incorporated into the cationic host lattice. Hitherto there is no example of a polyhedral clathrate hydrate based on the Type V structure. The Type VI structure is found in the hydrate of *tert*-butylamine, which is a true clathrate with no host/guest hydrogen bonding. In the Type VII structure adopted by the acid hydrates $HEF_6.5H_2O.HF$ (E = P, As, Sb), the cationic host lattice is constructed from water molecules, fluoride ions and protons.

In the Type I structure, when smaller guest species fully occupy both 14-hedra and 12-hedra, the ideal stoichiometry is, for example, $8CH_4.46H_2O$. However, crystallization of these hydrates does not require that the 12-hedra be filled and non-stoichiometry is expected for the smaller guest species over the range $6X.46H_2O$ to $8X.46H_2O$.

Interestingly, the compounds $Ge_{38}E_8I_8$ (E = P or As)[6] are structurally analogous to the Type I clathrate hydrates. In space group $P43n$ with $a =$

Table 7.1-2   Crystallographic structural data for clathrate hydrates.

Type	Cage types and number of cages per unit cell	Ideal unit cell formula	Space group	Example	Lattice parameters
I	$6[5^{12}6^2].2[5^{12}]$	$6X.2Y.46H_2O$	Pm3n	$6(CH_2)_2O.46D_2O$	a~11.87(1)Å, 80K
II	$8[5^{12}6^4].16[5^{12}]$	$8X.16Y.136H_2O$	Fd3m	$8C_4H_8O.16H_2S.136H_2O$	a~17.31(1)Å, 250K
III*	$16[5^{12}6^2].4[5^{12}6^3].10[5^{12}]$	$20X.10Y.172H_2O$	$P4_2/mmm$		a~23.5Å, c~12.5Å
IV**	$4[5^{12}6^2].4[5^{12}6^3].6[5^{12}]$	$8X.6Y.80H_2O$	P6/mmm		a~12.5Å, c~12.5Å
V***	$4[5^{12}6^4].8[5^{12}]$	$4X.8Y.68H_2O$	$P6_3/mmc$		a~12Å, c~19Å
VI	$16[4^35^96^27^3].12[4^45^4]$	$16X.156H_2O$	$I\bar{4}3d$	$16(CH_3)_3CNH_2.156H_2O$	a~18.81(2)Å, 250K
VII	$2[4^66^8]$	$2X.12H_2O$	Im3m	$HPF_6.5H_2O.HF$	a~7.678Å, 290K

*   Extrapolated from the framework structure of $(n-butyl)_4N^+C_6H_5COO^-.39°H_2O$.

**   Extrapolated from the framework structure of $(i-amyl)_4N^+F^-.38H_2O$.

*** Deduced from polyhedral packing related to the structure of $10(CH_3)_2CHNH_2.80H_2O$.

Table. 7.1-3   Neutral molecules reported to form clathrate hydrates.

Type I

Xe, CO, Cl_2, BrCl, H_2S, H_2Se, CO_2, N_2O, SO_2, ClO_2, PH_3, AsH_3 CH_4, C_2H_2, C_2H_4, C_2H_6, ethylene oxide, CH_3F, CH_3Cl, CH_2F_2, CH_2ClF, $CHClF_2$, CF_4, C_2H_5F, C_2H_3F, CH_3Br

Type I and II

COS, $cyclo-C_3H_6$, $(CH_3)_2O$, CH_3SH, CH_3CHF_2, $CH_2=CHF$, CH_3Br

Type II

Ar, Kr, O_2, N_2, air, SF_6, C_3H_8, $CH_3CH=CH_2$, $(CH_3)_3CH$, $cyclo-C_5H_{10}$, CH_3NO_2, CH_3I, CH_2Cl_2, $CHCl_2F$, $CHBrF_2$, $CHCl_3$, CCl_2F_2, $CBrF_3$, $CBrClF_2$, CCl_3F, CBr_2F_2, CH_3CHF_2, $CH_2=CHCl$, CH_3CH_2Cl, C_2F_4, CH_3CHCl_2, CH_3CH_2Br, CH_3CF_2Cl, $(CH_3)_3CF$, $CHF=CF_2$, CCl_3NO_2, CCl_3F, $(H_3C)_2C=O$, propylene oxide, $(CH_3)_2O$, $(C_2H_5)_2O$, furan, dihydrofuran, THF, cyclobutanone, 1,3- and 1,4-dioxolane, 1,3- and 1,4-dioxane

Type II with H_2S (help-gas)

I_2, CS_2, C_6H_6, CCl_4, C_2H_5I, CCl_3Br, C_2H_5I, C_3H_7Br, CCl_3NO_2, $(CH_3)_2S$, THF

Type III

Br_2, $(CH_3)_2O$

Type VI

$(CH_3)_3CNH_2$

10.507Å, the atoms of $Ge_{38}P_8I_8$ occupy the following positions: Ge(1) in 24(i) with $x = -0.0027$, $y = 0.1190$, and $z = 0.3069$; Ge(2) in 6(c); Ge(3) in 8(e) with $x = 0.1841$; P(1) in 8(e) with $x = 0.8148$; I(1) in 6(d) and I(2) in 2(a). The Ge and P atoms together constitute the host lattice with all cavities occupied by the I atoms. On the basis of their tetrahedral coordination and electron balance, the E atoms can be formally regarded as phosphonium or arsonium ions, and the two compounds represented by the structural formula $[Ge_{38}(E^+)_8]\cdot(I^-)_8$.

4. Structure of Type II clathrate hydrate

In the Type II structure, the host framework can be described in terms of 12-hedra and 16-hedra sharing faces in the ratio 2 to 1. Fig. 7.1-3 (a)

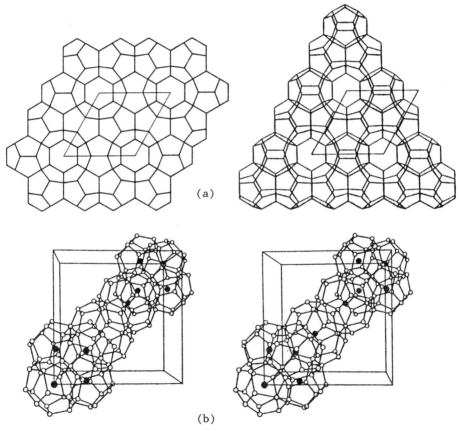

(a)

(b)

Fig. 7.1-3 (a) Layer formed by face-sharing 12-hedra: in projection (left) and in perspective (right). (b) Stereogram illustrating the Type II clathrate framework, with two 16-hedra centered at (3/8, 3/8, 3/8) and (5/8, 5/8, 5/8), and two cluster of 12-hedra centered at (1/8, 1/8, 1/8) and (7/8, 7/8, 7/8). The solid circles represent guest species occupying the voids. (After refs. 7, 18 and 19).

shows the association of face-sharing 12-hedra into a layer of hexagonal
symmetry, the residual voids being the 16-hedra.

There are two possible ways of arranging the layers to form a three-
dimensional framework: ABABAB... and ABCABC.... For the repeating sequence
ABCABC..., the normal to the layers corresponds to the [111] axis of the 17Å
cubic unit cell of the Type II structure. The centers of the 16-hedra form a
diamond lattice [Fig. 7.1-3(b)]. The structure generated by ABABAB...
stacking of the hexagonal layers has space group symmetry $P6_3/mmc$ and the same
stoichiometry as the 17Å cubic structure. A distortion of this hypothetical
Type V gas hydrate structure is represented by a hydrate of *iso*-propylamine,
$(CH_3)_2CHNH_2.8H_2O$.

Formation of Type II hydrates is facilitated by passing in H_2S or H_2Se
as a "help gas" during the preparation. The hydrates of COS, CH_3SH, CH_3CHF_2
are reported to change from Type I to Type II in the presence of H_2S. The H_2S
molecules occupy the small voids which cannot accommodate larger guest
species. Since other small gas molecules, such as air, O_2, N_2, O_3, Xe and Kr
may be trapped in the 12-hedra, all Type II hydrates usually contain non-
stoichiometric amounts of air unless special precautions are taken to avoid
their inclusion as guest species.

Crystal structure analyses have been reported for the tetrahydrofuran-
H_2S hydrate ($8C_4H_8O.16H_2S.136H_2O$) at 250K,[8] and the carbon disulphide-H_2S
hydrate ($8CS_2.16H_2S.136H_2O$) at 140K.[3] In the Type II hydrates, the radius
of the pentagonal dodecahedron is given by $\sqrt{3}a/8$ and is equal to 3.65Å for a
= 17.00Å.

Recently, an accurate neutron diffraction study of the Type II clathrate
hydrate $8CCl_4.nXe.136D_2O$ (n = 3.5) has been conducted at 13K and 100K.[7]
The D_2O molecules are disordered in six hydrogen-bonded orientations of equal
statistical weights, as in ice Ih. There are six non-equivalent hydrogen-bond
interactions with O...D distances (at 13K) between 1.738(3)Å and 1.802(1)Å and
O...D-O angles between 174.8(1)° and 180°. The covalent O-D lengths
(uncorrected for thermal motion) are in the range 0.986(1)Å to 1.001(2)Å and
are decreased significantly from 13K to 100K. The Xe atoms occupy
statistically 22% of the dodecahedra and vibrate about the cage centers with
r.m.s. displacements of 0.137Å at 13K and 0.179Å at 100K. The CCl_4 molecules
exhibit large amplitude libration motion about the the C atom located at the
center of the hexakaidecahedron. There are seven preferred molecular
orientations with C-Cl bonds directed toward the O vertices of the $(D_2O)_{28}$
hexakaidecahedron. The C-Cl bond length from the radius parameter of the

spherical distribution is 1.765(2)Å at 13K and 1.762(3)Å at 13K, as compared to the gas phase value of 1.766(3)Å.

5. Clathrate hydrates of strong monobasic acids

The structural parameters of clathrate hydrates of several strong monobasic acids are listed in Table 7.1-4.

The compounds $HPF_6 \cdot 7.67H_2O$, $HBF_4 \cdot 5.75H_2O$ and $HClO_4 \cdot 5.5H_2O$ are the first known examples of Type I clathrate hydrates of strong acids.[9] Owing to disorder in the crystals an accurate structure determination was successful

Table 7.1-4 Structural parameters of clathrate hydrates of strong monobasic acids and related compounds.

Compound	Structure type	Space group	Cell parameters (Å)	Figure showing structure	Ref. for structure
$HPF_6 \cdot 7.67H_2O$	I	Pm3n	a = 11.774(5) (111K)	7.1-1, 7.1-2(b)	[9]
$HBF_4 \cdot 5.75H_2O$	I	-	a = 11.744(5) (108K)	-	[9]
$HClO_4 \cdot 5.5H_2O$	I	-	a = 11.861(3) (193K)	-	[9]
$HAsF_6 \cdot 6H_2O$	IV	P6/mmm	a = 23.428(4) (148K) c = 13.841(3)	7.1-4	[10]
$HSbF_6 \cdot 6H_2O$	IV	P6/mmm	a = 23.680(4) (108K) c = 13.900(4)	-	[10]
$HPF_6 \cdot 5H_2O \cdot HF$	VII	Im3m	a = 7.544(2) (107K)	7.1-5	[11]
$HAsF_6 \cdot 5H_2O \cdot HF$	VII	Im3m	a = 7.665(3) (98K)	-	[11]
$HSbF_6 \cdot 5H_2O \cdot HF$	VII	Im3m	a = 7.777(4) (108K)	-	[11]

only for the HPF_6 hydrate. The smaller pentagonal dodecahedral voids of the hydrogen-bonded Type I host structure $2[5^{12}] \cdot 6[5^{12}6^2] \cdot 46H_2O$ were found to be vacant. The larger tetrakaidecahedral $[5^{12}6^2]$ voids are centered by octahedral PF_6^- species as anionic guests, each being orientationally twofold disordered about a F-P-F axis perpendicular to the hexagonal faces, as shown in Fig. 7.1-2(b). Disorder is also present in the cationic host structure, where the H atoms are distributed over two half occupied positions on every polyhedron edge, and statistical H_2O/HF substitution toward the ideal composition $HPF_6 \cdot 6.67H_2O \cdot HF$ must be assumed to account for the positive charge. In the HBF_4 and $HClO_4$ hydrates, the host structure may have vacant sites, thereby enabling the BF_4^- and ClO_4^- anions to partially fill the smaller $[5^{12}]$ voids.

The isomorphous compounds $HAsF_6 \cdot 6H_2O$ and $HSbF_6 \cdot 6H_2O$ are polyhedral clathrate hydrates.[10] The host structure is related to a three-dimensional network of hydrogen-bonded water molecules with the unit-cell formula $1[4^66^2] \cdot 3[4^25^86^4] \cdot 2[5^{12}6^2] \cdot 2[5^{12}6^3] \cdot 46H_2O$, which in turn is related to the fully four-connected hypothetical Type IV clathrate hydrate structure, as shown in Fig. 7.1-4. The positions of the H atoms could not be determined, but the host structure has to be assumed as protonated, i.e. cationic, and the guest species as deprotonated. The resulting anions, AsF_6^- or SbF_6^-, were found to occupy all but the smallest $[4^6 6^2]$ polyhedra with orientational as well as some positional disorder.

The polyhedral clathrate hydrates $HEF_6 \cdot 5H_2O \cdot HF$ (E = P, As, Sb) are based on a cubic cationic host lattice built of water molecules, fluoride ions, hydrogen fluoride and protons.[11] There is only one kind of polyhedral void in this Type VII hydrate, namely a truncated octahedron $[4^6 6^8]$. In these voids the deprotonated acid molecules are enclosed as anionic guests (PF_6^-,

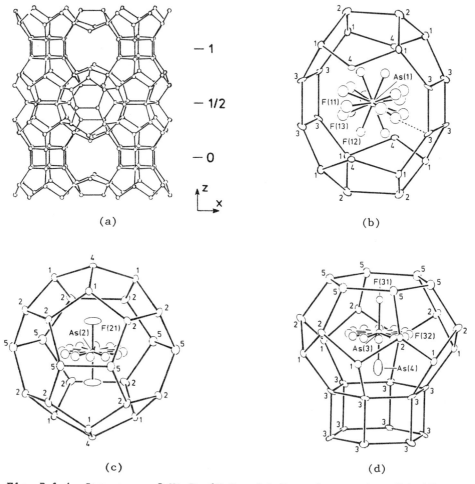

(a)   (b)   (c)   (d)

Fig. 7.1-4 Structure of $HAsF_6 \cdot 6H_2O$. (a) Host framework. (b)-(d) Orientations of disordered anions within cavities. (After ref. 10).

AsF$_6^-$ or SbF$_6^-$), as shown in Fig. 7.1-5. The orientational disorder of the
anions is described as being four-fold in all three compounds.

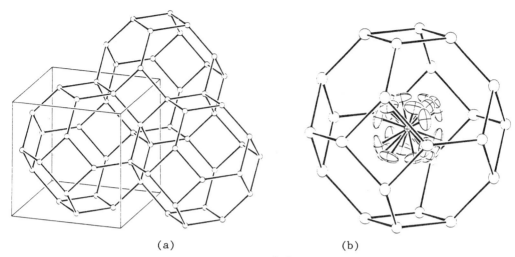

(a)   (b)

Fig. 7.1-5 (a) The framework of [$4^6 6^8$] polyhedra in HPF$_6$.5H$_2$O.HF. (b)
The orientations of the disordered anion in a cavity. (After ref. 11).

6. Clathrate hydrates of tetramethylammonium hydroxide

The structural parameters of crystalline hydrates of Me$_4$NOH are
summarized in Table 7.1-5.[12,13]

Table 7.1-5 Structural parameters of Me$_4$NOH.nH$_2$O.

Crystalline phase	Space group	Unit-cell parameters	Anionic host framework	Ref. for structure
α-Me$_4$NOH.5H$_2$O	Cmcm	a = 12.57(1), b = 10.96(1) c = 7.91(1)Å, Z = 4	2[$4^4(4)^2 6^6(6)^2$]	[12]
β-Me$_4$NOH.5H$_2$O	Im$\bar{3}$m	a = 8.146(3)Å (50°C), Z = 2	2[$4^6 6^8$]	[13]
α-Me$_4$NOH.7.5H$_2$O	P$\bar{1}$	a = 8.269(2), b = 11.813(2) c = 14.853(3)Å, α = 92.67(2) β = 103.86(2), γ = 106.08(2)° (-160°), Z = 4	2[$(4)^1 5^9(5)^1 6^4(6)^1$]. 2[$4^2 5^6(5)^2 6^3(6)^2$].2[$4^2(4)^1 5^5(5)^1$]	[13]
β-Me$_4$NOH.7.5H$_2$O	I4/mcm	a = 14.960(6), c = 12.447(6)Å (-5°C), Z = 4	8[$5^{12} 6^3$].4[$4^2 5^8$]	[13]
Me$_4$NOH.10H$_2$O	Pnma	a = 16.031(6), b = 8.453(3), c = 12.360(6)Å (-40°C), Z = 4	4[$4^1 5^{10} 6^6$].4[$4^3 5^6$]	[13]

The crystal structure of α-(CH$_3$)$_4$N$^+$.OH$^-$.5H$_2$O[12] resembles that found in
HPF$_6$.5H$_2$O.HF. The OH$^-$ ions and H$_2$O molecules form a hydrogen-bonded anionic
framework based on the space-filling arrangement of truncated octahedra, each
having three missing edges (Fig. 7.1-6). With the improper n-gonal faces
shown in parentheses, the host structure may be designated as 2[$4^4(4)^2 6^6(6)^2$].
Equivalent (CH$_3$)$_4$N$^+$ ions occupy the four available polyhedral cages in the
unit cell, and their bulkiness is responsible for a reduction of the symmetry

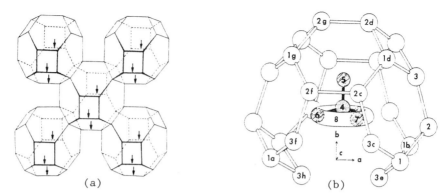

(a)   (b)

Fig. 7.1-6 (a) Idealized framework structure with closest-packed
truncated octahedra. The arrows indicate edges which are expanded in
the distorted structure found in α-$(CH_3)_4N^+OH^-\cdot 5H_2O$. (b) A distorted
truncated octahedron containing an axially disordered $(CH_3)_4N^+$ ion.
(After ref. 12).

of the oxygen host lattice from cubic to orthorhombic. Three carbon atoms of
each cation are disordered, giving rise to a toroidal distribution of electron
density about the fourth N-C bond axis, which suggests that the cations are
behaving as slightly hindered axial rotors.

The anionic host framework of the high-temperature β form of $Me_4NOH\cdot 5H_2O$
is isostructural to the cationic one of $HPF_6\cdot 5H_2O\cdot HF$.

The host structure of the decahydrate contains two new kinds of cages: a
17-hedron accommodating the cation and a vacant 9-hedron. Each 17-hedron
shares its hexagonal and four of its pentagonal faces with neighbouring
polyhedra of the same kind. In the resulting polyedral ensemble open zigzag
channels parallel to *b* are generated by the 9-hedra, each sharing two of its
tetragonal faces with two neighbours (Fig. 7.1-7).

Reprinted with permission of
Kluwer Academic Publishers

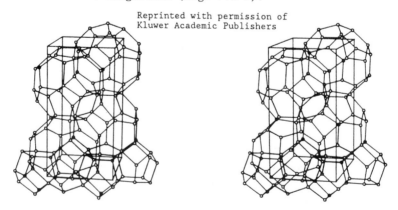

Fig. 7.1-7 Stereoview of the $4[4^15^{10}6^6]\cdot 4[4^35^6]$ host structure of
$Me_4NOH\cdot 10H_2O$ viewed approximately along *c*. (After ref. 13).

The high-temperature β form of the 7.5-hydrate has a host structure [Fig. 7.1-8(a)] constructed from well-known 15-hedra and vacant decahedra, which is isostructural with the hypothetical "tetragonal II" clathrate hydrate derived from the crystal structure of $(i-C_5H_{11})_4PBr.32H_2O$.[14] In the low-temperature α form the 15-hedra are modified to a 16-hedron and a 15-hedron, both with missing edges, and the vacant 9-hedron becomes distorted [Fig. 7.1-8(b)].

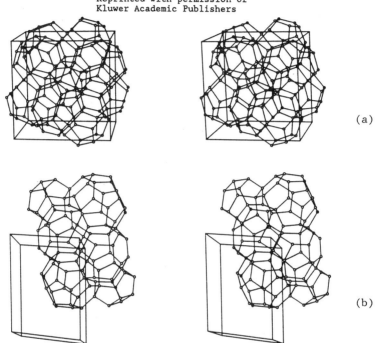

(a)

(b)

Fig. 7.1-8 Stereoview of the host structure of $Me_4NOH.7.5H_2O$: (a) β form viewed approximately along c; and (b) α form viewed approximately along a. (After ref. 13).

The phases $Me_4NOH.10H_2O$, $\beta-Me_4NOH.7.5H_2O$ and $\beta-Me_4NOH.5H_2O$ are the first examples of polyhedral clathrate hydrates with *proton-deficient*, yet fully four-connected water-hydroxide host structures. The proton distribution is disordered in these higher hydrates.

6. Clathrasils

The clathrasils all possess, as a common feature, a three-dimensional silicate framework containing cage-like voids occupied by guest species.[15] They are clearly distinct from the better known zeolites, which are characterized by channel-like pores in which the guest species can more or less readily migrate (Section 3.21).

A variety of clathrasils have been prepared and structurally characterized by Liebau and Gies. Table 7.1-6 gives a summary of the available structural data.

Table 7.1-6 Structural data on different clathrasils. [16]

Clathrasil family and cell formula*	Space group	Cell dimensions (Å)	Cage types and number of cages per unit cell	Structurally analogous to clathrate hydrate	Ref. for structure
Melanophlogites (Mel) $46SiO_2 \cdot 2M^{12} \cdot 6M^{14}$	Pm3n	a = 13.436	$2[5^{12}] \cdot 6[5^{12}6^2]$	Type I	[17]
Dodecasils 3C (D3C) $136SiO_2 \cdot 16M^{12} \cdot 8M^{16}$	Fd3m	a = 19.402	$16[5^{12}] \cdot 8[5^{12}6^4]$	Type II	[18]
Dodecasils 1H (D1H) $34SiO_2 \cdot 3M^{12} \cdot 2M'^{12} \cdot 1M^{20}$	P6/mmm	a = 13.783 c = 11.190	$3[5^{12}] \cdot 2[4^35^66^3] \cdot 1[5^{12}6^8]$	-	[19]
Deca-dodecasils 3R (DD3R) $120SiO_2 \cdot 6M^{10} \cdot 9M^{12} \cdot 6M^{19}$	R$\bar{3}$m	a = 13.860 c = 40.891	$6[4^35^66^1] \cdot 9[5^{12}] \cdot 6[4^35^{12}6^18^3]$	-	[20]
Nonasils (Non) $88SiO_2 \cdot 8M^8 \cdot 8M^9 \cdot 4M^{20}$	Fmmm	a = 22.232 b = 15.058 c = 13.627	$8[5^46^4] \cdot 8[4^15^8] \cdot 4[5^86^{12}]$	-	[21]
Deca-dodecasils 3H (DD3H) $120SiO_2 \cdot 6M^{10} \cdot 9M^{12} \cdot 1M^{15} \cdot 4M^{19} \cdot 1M^{23}$	**	a = 13.887 c = 40.989	$6[4^35^66^1] \cdot 9[5^{12}] \cdot 1[4^65^68^3] \cdot 4[4^35^{12}6^18^3] \cdot 1[5^{18}6^28^3]$	-	[16, 22]
Silica-sodalites (Sod) $12SiO_2 \cdot 2M^{14}$	Im3m	a = 8.836	$2[4^66^8]$	Type VII ($HPF_6 \cdot 5H_2O \cdot HF$)	[23]

* M^f: guest molecule located in a cage that has f faces.

** Space group R3, R$\bar{3}$, R32, R3m or R$\bar{3}$m.

Table 7.1-7 Size and maximum topological symmetry of the cages observed in clathrasils. [16]

Cage	Number of faces (f)	Free volume (Å³)	Max. topological symmetry	Number of cages per unit cell in						
				Non	Sod	Mel	D3C	D1H	DD3R	DD3H
$[5^46^4]$	8	25	222	8	-	-	-	-	-	-
$[4^15^8]$	9	30	mm	8	-	-	-	-	-	-
$[4^35^66^1]$	10	35	3m	-	-	-	-	-	6	6
$[5^{12}]$	12	80	m3	-	-	2	16	3	9	9
$[4^35^66^3]$	12	100	$\bar{6}$m2	-	-	-	-	2	-	-
$[4^66^8]$	14	130	m3m	-	2	-	-	-	-	-
$[5^{12}6^2]$	14	160	$\bar{4}$2m	-	-	6	-	-	-	-
$[4^65^68^3]$	15	200	$\bar{6}$m2	-	-	-	-	-	-	1
$[5^{12}6^4]$	16	250	m3	-	-	-	8	-	-	-
$[4^35^{12}6^18^3]$	19	350	3m	-	-	-	-	-	6	4
$[5^86^{12}]$	20	290	mmm	4	-	-	-	-	-	-
$[5^{12}6^8]$	20	430	6/mmm	-	-	-	-	1	-	-
$[5^{18}6^28^3]$	23	540	$\bar{6}$m2	-	-	-	-	-	-	1

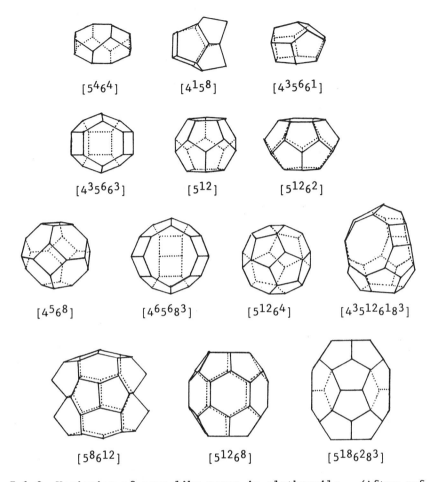

Fig. 7.1-9 Varieties of cage-like pores in clathrasils. (After ref. 16).

Most of the clathrasils have been synthesized with one or more combinations of guest molecules. In Fig. 7.1-9 and Table 7.1-7, a survey of the polyhedral cavities occurring in clathrasils is given. They vary considerably in size and maximum topological symmetry. The size of the guest molecules successfully incorporated ranges from that of Kr, N_2, CO_2 and CH_4 to that of 1-aminomethyladamantane, $C_{11}NH_{19}$.

Melanophlogite, $46SiO_2.2M^{12}.6M^{14}$ [space group $Pm3n$, a = 13.436(3)Å at 200°C] is a very rare mineral which is isostructural with the Type I gas hydrates, the tetrahedral [SiO₄] units being corner-linked in generating a 3-dimensional 4-connected host network (Fig. 7.1-10).[17] The [5^{12}] and [$5^{12}6^2$] cages are occupied by mixed (CH_4, N_2) and (CO_2, N_2) guest components, respectively. The observed Si-O distances (mean 1.576Å) and Si-O-Si angles (148.3° to 180.0°) are very different from those (mean values 1.608Å and 144°) found in silica frameworks.

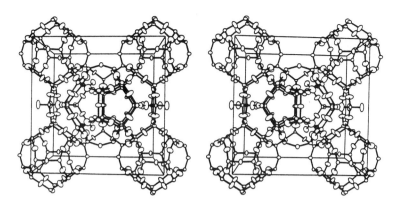

Fig. 7.1-10 Stereoview of the host framework of melanophlogite. (After ref. 17).

Dodecasil 3C is isostructural with Type II clathrate hydrates, with trimethylamine or dimethylamine occupying the $[5^{12}6^4]$ cages.[18] Synthetic dodecasil 1H, $34SiO_2.3M^{12}.2M'^{12}.1M^{20}$, has a silica host framework based on the same hexagonal layer arrangement of face-sharing $[5^{12}]$ polyhedra as in the Type II and V clathrate hydrates [Fig. 7.1-3(a)]. When the layers are stacked in the sequence $\overline{A}A\overline{A}A...$, new kinds of $[4^35^66^3]$ and $[5^{12}6^8]$ cages **(Fig. 7.1-9)** are generated. The large voids can accommodate large guest molecules such as piperidine or 1-aminoadamantane, and N_2 molecules are entrapped in the two types of small cages.[19]

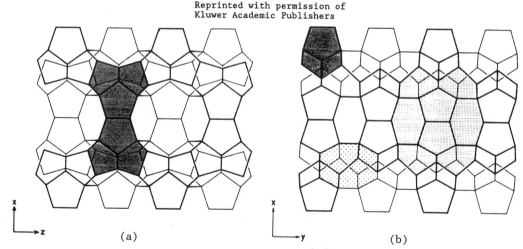

(a)   (b)

Fig. 7.1-11 Host structure of nonasil-$[4^15^8]$. (a) A double unit consisting of two face-sharing fundamental cages is highlighted by hatching. The two layers at different heights are distinguished by heavy and thin lines. (b) The three types of cage are emphasized by different shading. (After ref. 21).

Nonasil-$[4^15^8]$, $88SiO_2.8M^8.8M^9.4M^{20}$, has a silica framework which can be completely constructed by the linkage of $[4^15^8]$ polyhedra, which play the same fundamental role as the $[5^{12}]$ polyhedra in the clathrate hydrates and clathrasils 3C and 1H. Two $[4^15^8]$ "fundamental cages" share their tetragonal faces to form a "double unit", and each such unit shares edges with four others to generate a B-centered layer parallel to (010), as shown in Fig. 7.1-11(a). Linking such layers through Si-O-Si bridges in a stacking sequence in accordance with F-centering produces two other types of voids: a small $[5^46^4]$ cage and a large $[5^86^{12}]$ cage [Fig. 7.1-11(b)], the latter accommodating guest molecules such as 2-methylpyridine, 1,2-diaminocyclohexane, 2-aminopentane and hexamethyleneimine.[21]

7. Construction of polyhedral models in structural chemistry

Three-dimensional scaled models constructed from polyhedral units are particularly effective in the structural description and visual comprehension of clathrate hydrates, isopoly- and heteropolyanions, intermetallic compounds, minerals, and many non-molecular crystalline solids. Several papers and books have given detailed instructions for constructing various kinds of polyhedra from ball-and-spoke components, plastic tubing with either flexible connectors or rigid "valence clusters", cocktail picks, metal wire, cardboard, acrylic plastic sheets, and rigid polyurethane.[24] For the clathrate hydrates, cardboard models of the more common $[5^{12}]$, $[5^{12}6^2]$, $[5^{12}6^3]$, $[5^{12}6^4]$ and $[4^35^96^27^3]$ polyhedra can be constructed with the aid of cut-out plans, and the framework structure can then be assembled by glueing together the polyhedral faces into the appropriate face-sharing arrangements.[3] A versatile method for constructing all kinds of polyhedral molecular and crystal models makes use of drinking straws, and the finished models are cheap but precise, light yet durable, and functional as well as aesthetically pleasing.[24]

References

[1] F. Hollander and G.A. Jeffrey, *J. Chem. Phys.* **66**, 4699 (1977).

[2] R.K. McMullan and G.A. Jefferey, *J. Chem. Phys.* **42**, 2725 (1965).

[3] G.A. Jefferey in J.A. Atwood, J.E.D. Davies and D.D. MacNicol, *Inclusion Compounds*, Vol. 1, Academic Press, London, 1984, p. 135; G.A. Jeffrey, *J. Incl. Phenom.* **1**, 211 (1984).

[4] G. Tsoucaris in G.R. Desiraju (ed.), *Organic Solid State Chemistry*, Elsevier, Amsterdam, 1987, chap. 7.

[5] E.D. Sloan, Jr., *Clathrate Hydrates of Natural Gases*, Marcel Dekker, New York, 1989.

[6] H.-G. von Schnering and H. Menke, *Angew. Chem. Int. Ed. Engl.* **11**, 43 (1972).

[7] R.K. McMullan and Å. Kvick, *Acta Crystallogr.*, Sect. B **46**, 390 (1990).

[8] T.C.W. Mak and R.K. McMullan, *J. Chem. Phys.* **42**, 2732 (1965).

[9] D. Mootz, E.-J. Oellers and M. Wiebcke, *J. Am. Chem. Soc.* **109**, 1200 (1987).

[10] M. Wiebcke and D. Mootz, *Z. Kristallogr.* **183**, 1 (1988).

[11] M. Wiebcke and D. Mootz, *Z. Kristallogr.* **177**, 291 (1986).

[12] R.K. McMullan, T.C.W. Mak and G.A. Jeffrey, *J. Chem. Phys.* **44**, 2338 (1966).

[13] D. Mootz and R. Seidel, *J. Incl. Phenom.* **8**, 139 (1990).

[14] S.F. Solodovnikov, T.M. Polyanskaya, V.I. Alekseev, L.S. Alodko, Y.A. Dyadin and V.V. Bakakin, *Kristallografiya* **27**, 247 (1982); *Sov. Phys. Crystallogr.* (Engl. Transl.) **27**, 151 (1982).

[15] R.P. Gunawardane, H. Gies and F. Liebau, *Z. Anorg. Allg. Chem.* **546**, 189 (1987).

[16] F. Liebau in E.R. Corey, J.Y. Corey and P.P. Gasper (eds.), *Silicon Chemistry*, Ellis Horwood, Chichester, 1988, chap. 29.

[17] H. Gies, *Z. Kristallogr.* **164**, 247 (1983).

[18] H. Gies, F. Liebau and H. Gerke, *Angew. Chem. Int. Ed. Engl.* **21**, 206 (1982); H. Gies, *Z. Kristallogr.* **167**, 73 (1984).

[19] H. Gerke and H. Gies, *Z. Kristallogr.* **166**, 11 (1984); H. Gies, *J. Incl. Phenom.* **4**, 85 (1986).

[20] H. Gies, *Z. Kristallogr.* **175**, 93 (1986).

[21] B. Marler, N. Dehnbostel, H.-H. Eulert, H. Gies and F. Liebau, *J. Incl. Phenom.* **4**, 339 (1986).

[22] H. Gies, *J. Incl. Phenom.* **2**, 275 (1984).

[23] D.M. Bibby and M.P. Dale, *Nature (London)* **317**, 157 (1985).

[24] T.C.W. Mak, C.N. Lam and O.W. Lau, *J. Chem. Educ.* **54**, 438 (1977), and references cited therein.

Note Ar, Kr, O_2, N_2 and air preferentially form gas hydrates of Type II whereas CO forms the Type I structure as expected. See J.S. Tse, Y.P. Handa, C.I. Ratcliffe and B.M. Powell, *J. Incl. Phenom.* **4**, 235 (1986); D.W. Davidson, M.A. Desandro, S.R. Gough, Y.P. Handa, C.I. Ratcliffe and J.S. Tse, *ibid.* **5**, 219 (1987); M.A. Desandro, Y.P. Handa, R.E. Hawkins, C.I. Ratcliffe and J.A. Ripmeester, *ibid.* **8**, 3 (1990).

Porosil (clathrasil and zeosil) structure types are reviewed recently by H. Giese in J.L. Atwood, J.E.D. Davies and D.D. MacNicol (eds.), *Inclusion Compounds*, Vol. 5, Oxford University Press, New York, 1991, chap. 1.

Estimates of molecular size, especially the volume per molecule for various aliphatics and aromatics, are discussed in A.Y. Meyer, *Struct. Chem.* **1**, 265 (1990).

An exposition of the relationships between molecular volumes and properties is given in J.C. McGowan and A. Mellors, *Molecular Volumes in Chemistry and Biology: Applications including Partitioning and Toxicity*, Ellis Horwood, Chichester, 1986.

7.2   Tetra-n-Butyl Ammonium Benzoate Clathrate Hydrate
$(n-C_4H_9)_4N^+C_6H_5COO^-\cdot39\frac{1}{2}H_2O$

Crystal Data

Tetragonal, space group $P4_2/mnm$ (No. 136)

$a = 23.57(4)$, $c = 12.45(2)$Å, $(-30°C)$, $Z = 4$

Atom*	x	y	z	Occupancy	Atom	x	y	z	Occupancy
O(1)	.2301	.9955	.1912	1	C(21)	.373	.383	.533	1/4
O(2)	.3646	.1322	.1737	1	C(22)	.424	.412	.463	1/4
O(3)	.1315	.0022	.3151	1	C(23)	.340	.193	.460	1/4
O(4)	.3180	.0491	.3105	1	C(24)	.397	.181	.520	1/4
O(5)	.4749	.0963	.1174	1	C(25)	.431	.133	.460	1/4
O(6a)	.3823	.2338	.2923	1/2	C(26)	.492	.132	.501	1/4
O(6b)	.3951	.2433	.2179	1/2	C(27)	.256	.246	.410	1/4
O(7)	.0423	.0423	.1915	1	C(28)	.233	.191	.355	1/4
O(8)	.3155	.3155	.1657	1	C(29)	.188	.211	.267	1/4
O(9)	.3955	.3955	.1110	1	C(30)	.171	.150	.230	1/4
O(10)	.2055	.0551	0	1	C(31)	.271	.204	.590	1/4
O(11)	.2896	.1426	0	1	C(32)	.214	.223	.640	1/4
O(12)	.6487	.4345	0	1	C(33)	.212	.188	.733	1/4
O(13a)	.7648	.4054	0	1/2	C(34)	.151	.172	.769	1/4
O(13b)	.7639	.4351	0	1/2	C(35)	.242	.442	.315	1/4
N(14)	.7996	.2656	0	1/4	C(36)	.197	.445	.415	1/4
O(15)	.0887	.0887	0	1	C(37)	.173	.497	.428	1/4
O(16)	.2597	.2597	0	1	C(38)	.181	.395	.467	1/4
O(17)	1/2	0	1/4	1	C(39)	.133	.508	.483	1/4
O(18)	.4419	.2495	0	1/4	C(40)	.130	.408	.525	1/4
C(19)	.324	.284	.544	1/4	C(41)	.111	.458	.567	1/4
C(20)	.357	.327	.473	1/4					

* O(1) and O(6a) include the oxygen atoms of the benzoate group. C(19) to C(34) are alkyl carbon atoms, and C(35) to C(41) are benzyl carbon atoms.

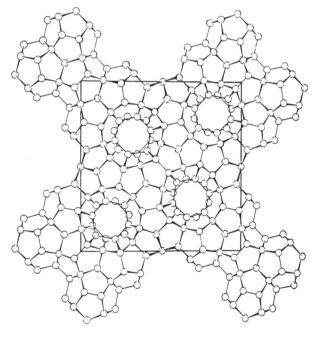

Fig. 7.2-1 Polyhedral host framework of $(n-C_4H_9)_4N^+C_6H_5COO\cdot39\frac{1}{2}H_2O$. (After ref. 1).

Crystal Structure

The polyhedral framework structure of this quaternary ammonium salt hydrate consists of face-sharing pentagonal dodecahedra (12-hedron, $[5^{12}]$), tetrakaidecahedra (14-hedron, $[5^{12}6^2]$) and pentakaidecahedra (15-hedron, $[5^{12}6^3]$) (Fig. 7.2-1). Fig. 7.2-2(a) shows the repeat unit of five 12-hedra, one 14-hedron and two 15-hedra and (b) shows the layer of face-sharing polyhedra at $z = 0$ and (c) at $z = 1/2$. The layer at $z = 1/2$ face-shares with that at $z = 0$ by superimposing the hexagonal A faces.

There are two five-fold groups of dodecahedra per unit cell, one centered at the origin and the other at the body center. They are related by the 4_2 symmetry axes at $(0,1/2)$ and $(1/2,0)$. These 10 dodecahedra account for 13 of the 17 vertices that form the asymmetric unit of the framework structure. The complete framework also contains 16 tetrakaidecahedra and 4 pentakaidecahedra per unit cell. The centers of the groups of five face-sharing pentagonal dodecahedra are the corners of the square base of the unit cell in Fig. 7.2-1. Adjoining them are the tetrakaidecahedra (A) with hexagonal faces parallel to (001) and an equal number (B) with their hexagonal faces perpendicular to (001). The polyhedra (C) are the pentakaidecahedra.

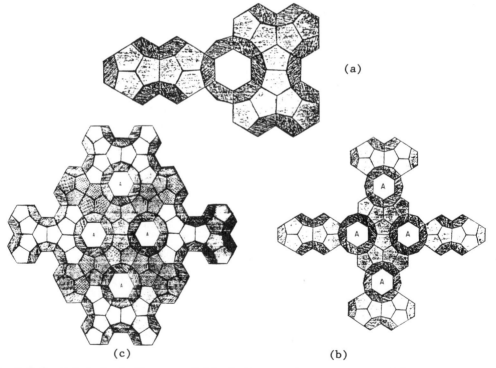

(a)

(c)   (b)

Fig. 7.2-2 Polyhedral diagram of ideal framework structure. (a) Arrangement of five face-sharing dodecahedra with one tetrakaidecahedron and two penta-kaidecahedra. (b) Layer at $z = 0$. (c) Layer at $z = 1/2$. (After ref. 1).

The midpoint of their shared hexagonal face at (1/2,1/2,0) is directly below the center of the group of five pentagonal dodecahedra at (1/2,1/2,1/2), as shown in Fig. 7.2-2(c). Here the letters A correspond to the top faces of the same tetrakaidecahedra as in Fig. 7.2-2(b), but the tetrakaidecahedra of type B have been omitted in order to make the arrangement of the polyhedra more obvious.

In the unit cell there are 172 vertices (framework sites) which form the 10 pentagonal dodecahedra, 16 tetrakaidecahedra, and 4 pentakaidecahedra. The ideal framework structure is modified in two respects in order to accommodate the ions of the quaternary ammonium salt (Fig. 7.2-3).

a) The central N atom of each cation occupies a vertex which is common to four of the larger polyhedra, so that the four n-butyl groups can be accommodated within them.

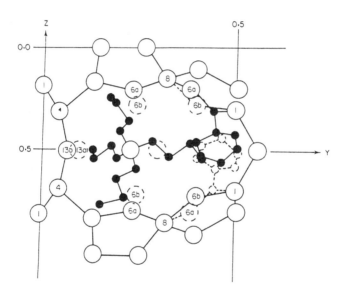

Fig. 7.2-3 A $(n-C_4H_9)N^+$ cation and a $C_6H_5COO^-$ anion included in the water framework. O(1) and O(6a) are carboxylate oxygen sites. O(13a) and O(13b) are alternate water sites depending upon the occupancy of the void. The O...O distances range from 2.6 to 2.9Å. (After ref. 1).

b) The benzoate anions have their carboxylate oxygens included in the hydrogen-bonded framework. They are disordered over two locations (full lines and dotted lines). In any one unit cell the carboxylate oxygens replace four of each of the 16-fold position O(1) and O(6), further reducing the number of framework water molecules per cell to 160. When the cation is in the void, the anions are absent and vice versa.

Remarks

1. Some polyhedral voids in alkylammonium salt clathrate hydrates[2,3]

    In this class of clathrate hydrates, the anions are incorporated by hydrogen-bonding into the water host framework and are located in the large voids. The cations are also located in large voids, which can be formed by fusion and distortion of small polyhedra. The large polyhedral voids observed in the alkylammonium salt hydrates are illustrated in Fig. 7.2-4(c-e). They can best be described by reference to the 14-hedra and 15-hedra of the gas hydrate structures, as shown in Fig. 7.2-4(a-b).

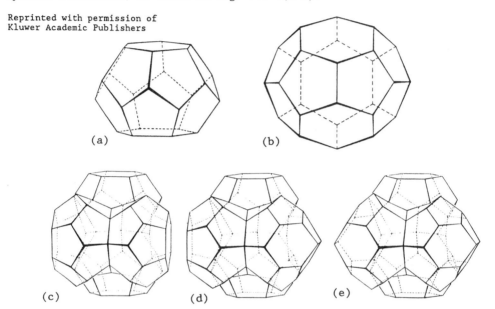

Fig. 7.2-4 Polyhedra. (a) The 14-hedron, $[5^{12}6^2]$; (b) the 15-hedron, $[5^{12}6^3]$; (c) 44-hedron, $[5^{40}6^4]$, formed from four 14-hedra; (d) 45-hedron, $[5^{40}6^5]$, formed from three 14-hedra and one 15-hedra; and (e) 46-hedron, $[5^{40}6^6]$, formed from two 14-hedra and two 15-hedra. (After ref. 3).

    The simplest is the 44-hedron formed by bringing together four 14-hedra so that two pairs of 14-hedra, sharing common hexagonal faces, fit together by sharing four pentagonal faces which have a common vertex. The numbers of faces (F), vertices (V), and edges (E) are related by Euler's formula:

$$44(F) + 70(V) = 112(E) + 2.$$

The common vertex is the site of the central nitrogen of the tetraalkyl-ammonium cation, so that the alkyl chains extend into the four voids of the original 14-hedra. The 44-hedron is then distorted to accommodate the central atoms of the cation at van der Waals separations.

    In the same way, a 45-hedron is constructed from three 14-hedra and one

15-hedron, and a 46-hedron is constructed from two 14-hedra and two 15-hedra giving:

$$45(F) + 72(V) = 115(E) + 2 \text{ and}$$
$$46(F) + 74(V) = 118(E) + 2, \text{ respectively.}$$

2. Structural types of alkyl-onium salt clathrate hydrates

The alkyl-onium salt hydrate structures fall into three main types as listed in Table 7.2-1. Type I is based on the cubic 12Å gas hydrate structure. Type III is a distortion of the tetragonal clathrate hydrate structure postulated for bromine hydrate which is, in fact, extrapolated from the structures of the tetragonal tetra-*n*-butyl ammonium benzoate and tetra-*n*-

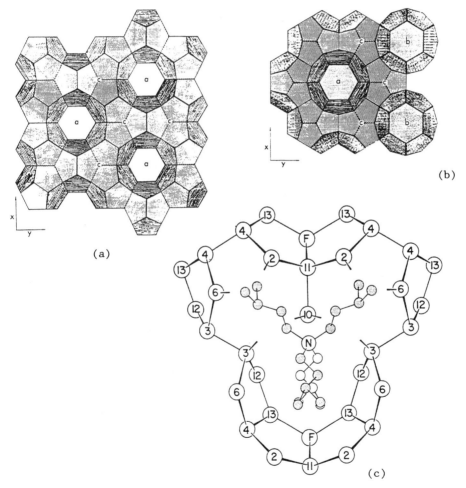

(a)

(b)

(c)

Fig. 7.2-5 Structure of $(i\text{-}C_5H_{11})_4NF.38H_2O$. (a) A layer of face-sharing polyhedra at $z = 0$ consisting of 12-hedra and 14-hedra. (b) A layer of face-sharing polyhedra at $z = \frac{1}{2}$ consisting of 14-hedra and 15-hedra. (c) A tetra-*iso*-amyl ammonium cation included in the distorted 46-hedron. (After ref. 5).

butyl ammonium fluoride hydrates.[4] Type IV is similarly a possible
clathrate hydrate structure extrapolated from the framework structure of the
orthorhombic tetra-*iso*-amyl ammonium fluoride hydrate, as shown in Fig. 7.2-5.

There is also the monoclinic structure, which is a further distorted
variant of the type IV orthorhombic structure, that exists in the second phase
of tri-*n*-butyl sulfonium fluoride hydrate. The type I, 24Å cubic structure is
believed to be a superstructure of the 12Å structure, but no experimental data
are available to confirm this.

Table 7.2-1 Polyhedral alkyl-onium salt clathrate hydrates
($R = n\text{-}C_4H_9$, $R' = iso\text{-}C_5H_{11}$).[2]

Type I		Type III		Type IV	
Cations	Anions	Cations	Anions	Cations	Anions
Cubic (12Å) Pm3n		Tetragonal P4$_2$/mmm		Hexagonal P6/mmm	
R_3S^+	F^-	R_4N^+	$C_6H_5CO_2^-$	-	-
R_4N^+	$n\text{-}C_3H_7CO_2^-$		$m\text{-}ClC_6H_4CO_2^-$	Orthorhombic	Pbmm
R_3NH^+	$n\text{-}C_4H_9CO_2^-$		$p\text{-}FC_6H_4CO_2^-$	R'_4N^+	F^-,Cl^-,OH^-
			$n\text{-}C_3H_7CO_2^-$		CrO_4^{2-}, $C_6H_5CO_2^-$
			$n\text{-}C_4H_9CO_2^-$		WO_4^{2-}, HCO_2^-
			$i\text{-}C_5H_{11}CO_2^-$	R'_3S^+	F^-
				$R'_2R_2N^+$	F^-
Cubic (24Å) Pm3n		Tetragonal P4$_2$/m		Monoclinic	P2$_1$/m
R_4N^+	$C_2O_4^{2-}$	R_4N^+	F^-, Cl^-, Br^-	R_3S^+	F^-
	WO_4^{2-}		CrO_4^{2-}, WO_4^{2-}, $C_2O_4^{2-}$	R_4P^+	$C_2O_4^{2-}$
R_4P^+	$C_2O_4^{2-}$		HCO_2^-, CH_3CO^-, OH^-	R'_4N^+	$C_6H_5CO_2^-$
R_3S^+	F^-		NO_3^-, HPO_4^{2-}		
			$p\text{-}ClC_6H_4CO_2^-$		
			$m\text{-}ClC_6H_4CO_2^-$		
		R_4P^+	F^-, $C_2O_4^{2-}$		
		$R_3R'N^+$	F^-		

3. Structure of alkylamine hydrates

The available data for the alkylamine hydrates are summarized in Table
7.2-2.[2] Some examples are discussed below.

a) Structure of *tert*-butylamine hydrate[6]

The tertiary butylamine hydrate $16(CH_3)_3CNH_2.156H_2O$, adoptindg the type
VI structure, is a true clathrate without hydrogen-bonding interactions
between the amine molecule and the water framework. The amine molecule
occupies a complex 17-hedron [$4^35^96^27^3$] with its C-N bond oriented along the
polar axis of the polyhedron (Fig. 7.2-6). These 17-hedra pack by sharing
their heptagonal and hexagonal faces, leaving small octahedral voids bounded
by four pentagonal and four square faces, which are presumably unoccupied.

Table 7.2-2   Crystallographic and structural data for
some alkylamine hydrates.[2]

Amine Guest G	Formula	Space group	parameters(Å)	Description of Structure
Dimethylamine $(CH_3)_2NH$	$6G.52H_2O$ $7G.49H_2O$ $8G.46H_2O$	P23 or Pm3	$a = 12.55(1)$	Structure unknown
Ethylamine $C_2H_5NH_2$	$6G.46H_2O$ $8G.40H_2O$	P$\bar{4}$3n or Pm3n	$a = 12.17(1)$	Structure unknown
Trimethylamine $(CH_3)_3N$	$4G.41H_2O$	P6/mmm	$a = 12.378(6)$ $c = 12.480(4)$	Structure is a distortion of the type IV gas hydrate structure. The voids are $[5^{12}]$, $[5^{12}6^3]$ and $[5^24 6^2]$ formed from two fused 14-hedra. Amine molecules are hydrogen-bonded with 15-hedra and 26-hedra.
Diethylamine $(C_2H_5)_2NH$	$12G.104H_2O$	Pbcn	$a = 13.44(1)$ $b = 11.77(1)$ $c = 27.91(2)$	Structure contains 18-hedra $[5^{12}6^6]$ and irregular voids $[4^35^86^6]$. Amine molecules are hydrogen-bonded within both voids. There are no 12-hedra.
Diethylamine $(C_2H_5)_2NH$	$4G.28H_2O$	P2$_1$/c	$a = 13.86(1)$ $b = 8.44(1)$ $c = 10.93(1)$ $\beta = 97.5(3)°$	Structure unknown
tert-Butylamine $(CH_3)_3CNH_2$	$16G.156H_2O$	I$\bar{4}$3d	$a = 18.81(2)$	Structure is a true clathrate. Amine is non-bonded in 17-hedra $[4^35^96^27^3]$, with vacant 8-hedra $[4^45^4]$.
n-Propylamine $(C_3H_7)NH_2$	$4G.40H_2O$ $4G.38H_2O$	P6/mmm	$a = 12.20(1)$ $c = 12.38(1)$	Structure unknown
n-Propylamine $(C_3H_7)NH_2$	$4G.26H_2O$	P2$_1$/n	$a = 12.43(2)$ $b = 20.73(3)$ $c = 17.28(2)$ $\beta = 89.3(3)°$	Structure contains 11-hedra $[4^25^86^1]$ and large irregular voids formed from fused 14-hedra $[5^{12}6^2]$ and 16-hedra $[5^{12}6^4]$. Amine is hydrogen-bonded in the large voids.
iso-Propylamine $(CH_3)_2CHNH_2$	$10G.80H_2O$	P6$_3$/mmc	$a = 12.30(1)$ $c = 24.85(2)$	Structure contains $[4^66^2]$, $[5^{12}]$, $[4^25^86^4]$ and $[5^{12}6^4]$. Amines are hydrogen-bonded within the 14- and 16-hedra.

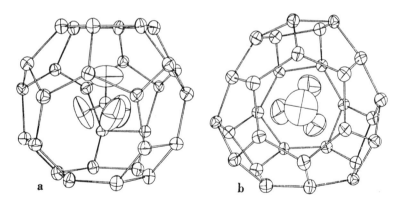

a   b

Fig. 7.2-6 Guest molcule trapped inside a cage in $16(CH_3)_3CNH_2.156H_2O$
viewed (a) normal to and (b) down C-N bond and threefold axis.
(After ref. 6).

b) Structure of *iso*-propylamine hydrate[7]

The structure of *iso*-propylamine hydrate is not a true clathrate because
the amine molecules are hydrogen-bonded to the water framework.   The hydrogen

bonds from the NH$_2$ group bridge one of the edges of the 16-hedron [Fig. 7.2-7(a)]. The amine molecules also occupy 14-hedra with the NH$_2$ group hydrogen-bonded in three different modes [Fig. 7.2-7(b-c)]. These 14-hedra can be generated by adding four additional vertices to two pentagonal dodecahedra, as shown in Fig. 7.2-7(d). As a result, additional small voids are generated in the structure which are hexagonal prisms, 8-hedra [4^{6}6^2]. The idealized amine hydrate stoichiometry of 10(CH$_3$)$_2$CHNH$_2$.80H$_2$O assumes that the pentagonal dodecahedra and 8-hedra are unoccupied.

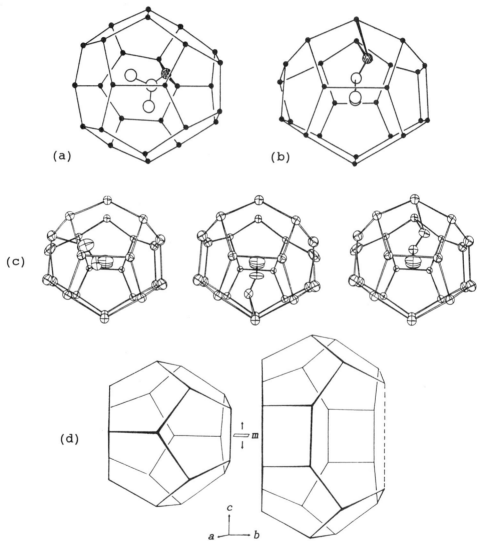

Fig. 7.2-7 Polyhedral cages in *iso*-propylamine hydrate. (a) The 16-hedron (5^{12}6^4) and encaged amine. (b) The 14-hedron (4^{2}5^{8}6^4) and encaged amine. (c) The disordered hydrogen-bonding over the three possible modes. (d) Relationship of the 14-hedron in (b) to the 12-hedron. (After ref. 7).

c) Structure of diethylamine hydrate[8]

Diethylamine hydrate, $12(C_2H_5)_2NH.104H_2O$, has a framework structure built entirely from 18-hedra $[5^{12}6^6]$ and 19-hedra $[4^35^86^67^2]$, to which the secondary nitrogen is bonded by donating and accepting a hydrogen bond as shown in Fig. 7.2-8. The 18-hedra is reasonably regular, but the polyhedral description of the 19-hedra stretches the imagination.

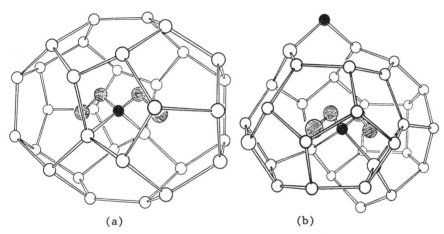

(a)	(b)

Fig. 7.2-8 Cages accommodating guest molecules in diethylamine hydrate. (a) Hydrogen-bonded in the 18-hedron $[5^{12}6^6]$. (b) Hydrogen-bonded in the 19-hedron $[4^35^86^67^3]$. (After ref. 8).

4. Quaternary ammonium salt hydrates of lower hydration number

X-Ray crystallographic studies have revealed that tetraalkylammonium salt hydrates of the general formula $R_4N^+X^-.nH_2O$ ($n \leq 4$) exist in a wide variety of stoichiometries and structures; their crystal data are presented in Table 7.2-3.

a) Structure of $Me_4NF.4H_2O$

In this crystalline hydrate, the tetrahedral Me_4N^+ ions are in voids within the hydrogen-bonded fluoride/water framework, which contains a buckled network of chair-shaped $(H_2O)_4F_2$ hexagons linked by four-fold screw spirals, as shown in Fig. 7.2-9(a). The N and F atoms lie alternately along the 4 axis at intervals of $c/2$. The fluoride ions are hydrogen bonded to four water oxygens in a configuration that is closer to square planar than to tetrahedral [Fig. 7.2-9(b)]. The water molecules are in an almost trigonal planar arrangement with two O-H...O hydrogen bonds of 2.732Å and one O-H...F bond of 2.630Å. The oxygen is 0.31Å out of the plane of its three bonded neighbours.

b) Structure of $Me_4NOH.4H_2O$

In this completely ordered crystal structure, the water molecules and

Table 7.2-3   Crystal data of $R_4N^+X^- \cdot nH_2O$ ($n \leq 4$) and related compounds.

Compound	Crystal data	Ref. for structure
$Me_4NF \cdot 4H_2O$	$I4_1/a$, a = 10.853, c = 8.065Å (-26°C), Z = 4	[9]
$Me_4NOH \cdot 4H_2O$	$P3_1$, a = 11.166, c = 6.671Å (-150°C), Z = 3	[10]
$4Et_4NF \cdot 11H_2O$	$Pna2_1$, a = 16.130, b = 16.949, c = 17.493Å, Z = 4	[11]
$Et_4NCl \cdot 4H_2O$	$Ccca$, a = 19.104, b = 23.084, c = 13.330Å, Z = 16	[12]
$(Me_4N)_2SO_4 \cdot 4H_2O$	$Pnam$, a = 15.438, b = 13.832, c = 7.923Å, Z = 4	[13]
$Et_4NOAc \cdot 4H_2O$	$P\bar{1}$, a = 12.327, b = 17.196, c = 8.753Å, α = 94.28, β = 91.09, γ = 120.30°, Z = 4	[14]
α-$Me_4NOH \cdot 2H_2O$	$Pnma$, a = 8.174, b = 8.777, c = 11.014Å (-150°C), Z = 4	[10]
β-$Me_4NOH \cdot 2H_2O$	$Cmcm$, a = 8.848, b = 11.145, c = 8.328Å (20°C), Z = 4	[10]
$Et_4NCl \cdot H_2O$	$C2/c$, a = 13.684, b = 14.144, c = 12.372Å, β = 110.75°, Z = 8	[15]
$[(CH_2)_6N_4CH_3]Br \cdot H_2O$	$P2_1/c$, a = 6.840, b = 6.898, c = 22.597Å, β = 94.91°, Z = 4	[16]
$[Zn(en)_3]F_2 \cdot 2H_2O$	$Pbcn$, a = 11.018, b = 14.805, c = 8.663Å, Z = 4	[17]

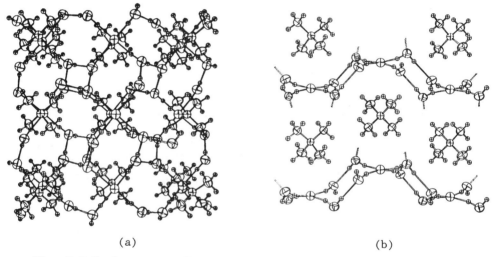

(a)   (b)

Fig. 7.2-9   Structure of $Me_4NF \cdot 4H_2O$ viewed (a) down the *c* axis and (b) in the *a* direction, showing the cations and a section of the hydrogen bonded anion/water framework. (After ref. 9).

hydroxide ions constitute fused channels which accommodate the cations [Fig. 7.2-10(a)]. Two of the independent water molecules are four-connected and the other two three-connected, each of the latter forming two donor and one acceptor hydrogen bonds. The hydroxide ion forms a weak donor hydrogen bond and three strong acceptor bonds. The host framework of this tetrahedrate can be derived from that of β-$Me_4NOH \cdot 5H_2O$ (Section 7.1) by selective removal of some water molecules [Fig. 7.2-10(b)].

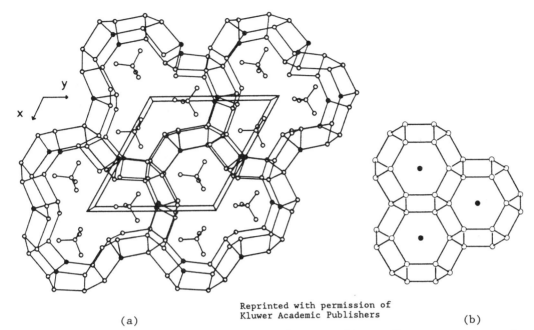

x y

(a)   (b)

Fig. 7.2-10 (a) Crystal structure of $Me_4NOH.4H_2O$. The oxygen atoms of H_2O and OH^- are represented by open and filled circles, respectively. (b) Structure of $\beta\text{-}Me_4NOH.5H_2O$ projected along [111]. The black circles represent the N atoms of the enclosed Me_4N^+ ions. (After ref. 10).

c) Structure of $4Et_4NF.11H_2O$

In $4Et_4NF.11H_2O$, the four independent fluoride ions are each tetrahedrally coordinated, and the resulting $(H_2O)_4F$ tetrahedra share pairs of opposite edges to form $[(H_2O)_2F]_\infty$ infinite chains, running through the structure in the direction of the a axis, as shown in Fig. 7.2-11(a). These chains are cross linked through O(5), O(9) and O(11) to give a three-dimensional water-anion framework. Of the eleven crystallographically-distinct water molecules, six are three-coordinate and the rest, including the three that serve bridging functions, are two-coordinate; consequently all water protons are fully utilized in consolidating the hydrogen-bonded host framework. The ordered Et_4N^+ cations occupy the voids in two open channel systems running in the b and c directions. Fig. 7.2-11(b) shows the structure of $4Et_4NF.11H_2O$ viewed parallel to c.

d) Structure of $Et_4NCl.4H_2O$

In the crystal structure of $Et_4NCl.4H_2O$, each chloride ion is tetrahedrally coordinated by water molecules with O-H...Cl bonds in the range 3.177-3.244Å. All $(H_2O)_4Cl$ units are stacked in columns in the direction of the c axis. Within each column, adjacent $(H_2O)_4Cl$ tetrahedra have different

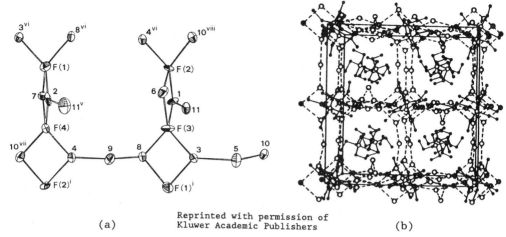

(a)   (b)

Fig. 7.2-11 Structure of $4Et_4NF.11H_2O$ viewed parallel to c showing (a) edge sharing $(H_2O)_4F$ tetrahedra and bridging water molecules, and (b) accommodation of the cations in channels. (After ref. 11).

orientations, and their vertices are linked pairwise by O-H...O hydrogen bonds to form an infinite chain of twisted six-membered $Cl(H_2O)_4Cl$ rings joined in a junction-sharing fashion, as shown in Fig. 7.2-12(a). The neighbouring columns are further interconnected by lateral O-H...O hydrogen bonds to generate a three-dimensional network featuring the occurrence of virtually planar, almost square $(H_2O)_4$ quadrilaterals. The water protons are fully utilized in an ordered scheme of hydrogen bonding. All four independent water

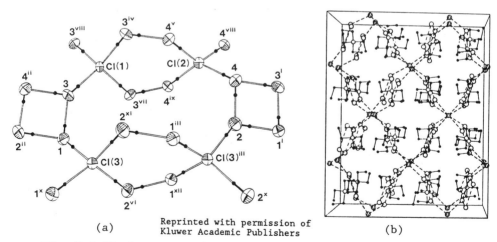

(a)   (b)

Fig. 7.2-12 Structure of $Et_4NCl.4H_2O$: (a) linkage of $(H_2O)_4Cl$ tetrahedra to generate twisted $(H_2O)_4Cl_2$ rings and nearly planar $(H_2O)_4$ quadrilaterals; (b) perspective view parallel to c showing channels filled by the cations. (After ref. 12).

molecules are three-coordinated, and their bonding configurations vary from almost planar to pyramidal.

The three-dimensional assembly of $(H_2O)_4Cl$ tetrahedra, interlinked by O-H...O hydrogen bonds between vertices, gives rise to two crisscross, planar channel systems normal to the c axis, which accommodate the ordered cations. [Fig. 7.2-12(b)].

e) Structure of $(Me_4N)_2SO_4\cdot4H_2O$

In the crystal structure, the water molecules are hydrogen-bonded into puckered hexagons which form infinite chains by sharing pairs of opposite edges. These chains run through the structure in the direction of the c axis and are cross-linked through oxygen atoms by interaction with the sulfate ions. When viewed down the c axis, as shown in Fig. 7.2-13(a), the structure is observed to possess definite channels which contain the cations [Fig. 7.2-13(b)].

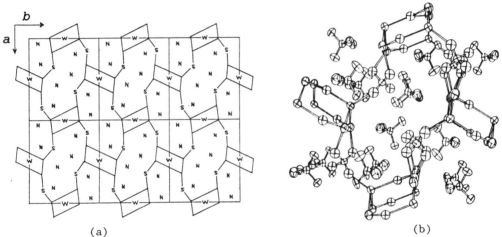

(a)   (b)

Fig. 7.2-13 Structure of $[Me_4N]_2SO_4\cdot4H_2O$ viewed down the c axis: (a) schematic pattern: W, water chains; S, sulfate ions; N, tetramethylammonium ions, and (b) water-sulfate interaction (for simplicity only one orientation of the disordered sulfate ion is depicted). (After ref. 13).

f) Structure of $Et_4NOAc\cdot4H_2O$

In the crystal structure, ordered Et_4N^+ cations are sandwiched between puckered layers of hydrogen-bonded water molecules and acetate anions. Each water-anion layer is constructed from an edge-sharing assembly of six independent, irregular, and non-planar polygons (two nonagons, two hexagons and two pentagons), as shown in Fig. 7.2-14. All protons in the scheme of hydrogen bonding are uniquely located. The water-anion layer structure of $(C_2H_5)_4N^+CH_3COO^-\cdot4H_2O$ bears a close resemblance to the two-dimensional arrays of puckered edge-sharing pentagons formed by water and hydroxyl groups in the tetrahydrates of 2,5-dimethyl-2,5-hexanediol[18] and 2,7-dimethyl-2,7-octanediol.[19]

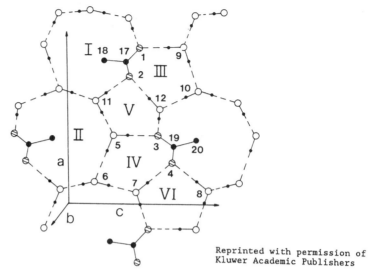

Reprinted with permission of
Kluwer Academic Publishers

Fig. 7.2-14 Structure of hydrogen-bonded water molecules and acetate anions in a puckered layer. The water oxygen, acetate oxygen, and carbon atoms are represented by open, half-shaded, and blackened circles, respectively. (After ref. 14).

It is of interest to compare the acetate ion with the hydroxide, fluoride, chloride, and sulfate ions as an ingredient for the construction of the water-anion lattice in a tetraalkylammonium salt hydrate. The preference of a monatomic anion for tetrahedral coordination is well manifested in the open framework structures of α-$(CH_3)_4NOH.5H_2O$,[20] $(CH_3)_4NOH.4H_2O$, $(CH_3)_4NF.4H_2O$, $4(C_2H_5)_4NF.11H_2O$, and $(C_2H_5)_4NCl.4H_2O$. In $[(CH_3)_4N]_2SO_4.4H_2O$, the highly symmetric molecular anion is disordered such that it has no well-defined hydrogen bonding interaction with neighbouring chains of edge-sharing puckered $(H_2O)_6$ hexagons. In contrast to this, the essentially planar acetate anion plays a dominating role in generating a net-like assembly of polygons in $(C_2H_5)_4NOAc.4H_2O$. The size, shape, and hydrogen bonding (acceptor) capability of the counter anion are thus important factors which dictate the crystal structure of a given tetraalkylammonium salt hydrate.

g) Structures of α- and β-$Me_4NOH.2H_2O$

Both forms (transition temperature at -85 °C) have essentially the same crystal structure except that the cations are ordered in the α form and disordered in the β form. Spiral chains of the type $^1_\infty[HO^-(HOH)_{4/2}]$ are generated by donor hydrogen bonds from the H_2O molecules to the OH^- ions (Fig. 7.2-15).

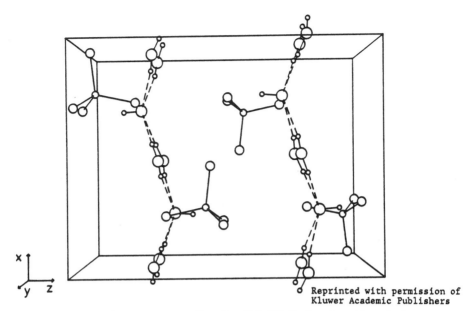

Fig. 7.2-15 Crystal structure of $\alpha\text{-}Me_4NOH \cdot 2H_2O$ showing the chains
parallel to the *a* axis. In the analogous structure of the β form, the
chains are parallel to the *c* axis. (After ref. 10).

h) Structures of $Et_4NCl \cdot H_2O$, $[(CH_2)_6N_4CH_3]Br \cdot H_2O$ and $[Zn(en)_3]F_2 \cdot 2H_2O$

In these three compounds, the water molecules and halide ions all form
planar, cyclic hydrogen-bonded dimers of the type $(H_2O \cdot X^-)_2$, which are
stabilized by the bulky cations in the solid state.[21] Fig. 7.2-16 shows the
structure of $(H_2O \cdot F^-)_2$ and the molecular packing in $[(CH_2)_6N_4CH_3]Br \cdot H_2O$. The
structural parameters for known $(H_2O \cdot X^-)_2$ dimers are listed in Table 7.2-4.

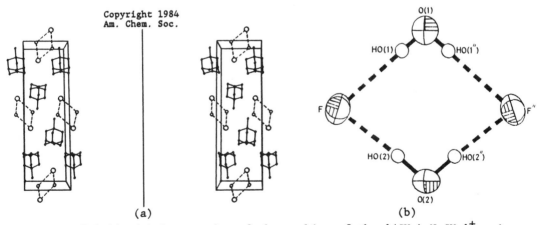

(a)   (b)

Fig. 7.2-16 (a) Stereo view of the packing of the $[(CH_2)_6N_4CH_3]^+$ and
$(H_2O \cdot Br^-)_2$ species in the crystal structure of $[(CH_2)_6N_4CH_3]Br \cdot H_2O$. (b)
Structure of the $(H_2O \cdot F^-)_2$ system in $[Zn(en)_3]F_2 \cdot 2H_2O$. (After refs. 16
and 17).

Table 7.2-4   Structural parameters for $(H_2O.X^-)_2$.

Complex	O...X(Å)	O...X'(Å)	X...O...X'($°$)	Ref.
$(H_2O.F^-)_2$	2.586	2.679	100.9	[17]
$(H_2O.Cl^-)_2$	3.204(4)	3.247(4)	103.8(1)	[15]
$(H_2O.Br^-)_2$	3.338(5)	3.386(5)	105.7(4)	[16]

References

[1] M. Bonamico, G.A. Jeffrey and R.K. McMullan, *J. Chem. Phys.* **37**, 2219 (1962).

[2] G.A. Jeffrey in J.L. Atwood, J.E.D. Davies and D.D. MacNicol (eds.), *Inclusion Compounds*, Vol. 1, Academic Press, London, 1984, p. 135.

[3] G.A. Jeffrey, *J. Incl. Phenom.* **1**, 211 (1984).

[4] R.K. McMullan, M. Bonamico and G.A. Jeffrey, *J. Chem. Phys.* **39**, 3295 (1963).

[5] D. Feil and G.A. Jeffrey, *J. Chem. Phys.* **35**, 1863 (1961).

[6] R.K. McMullan, G.A. Jeffrey and T.H. Jordan, *J. Chem. Phys.* **47**, 1229 (1967).

[7] R.K. McMullan, G.A. Jeffrey and D. Panke, *J. Chem. Phys.* **53**, 3568 (1970).

[8] T.H. Jordan and T.C.W. Mak, *J. Chem. Phys.* **47**, 1222 (1967).

[9] W.J. McLean and G.A. Jeffrey, *J. Chem. Phys.* **47**, 414 (1967).

[10] D. Mootz and R. Seidel, *J. Incl. Phenom.* **8**, 139 (1990).

[11] T.C.W. Mak, *J. Incl. Phenom.* **3**, 347 (1985).

[12] T.C.W. Mak, H.J.B. Slot and P.T. Beurskens, *J. Incl. Phenom.* **4**, 295 (1986).

[13] W.J. McLean and G.A. Jeffrey, *J. Chem. Phys.* **49**, 4556 (1968).

[14] T.C.W. Mak, *J. Incl. Phenom.* **4**, 273 (1986).

[15] J.H. Loehlin and Å. Kvick, *Acta Crystallogr., Sect. B* **34**, 3488 (1978).

[16] T.C.W. Mak, *Inorg. Chem.* **23**, 620 (1984).

[17] J. Emsley, M. Arif, P.A. Bates and M.B. Hursthouse, *J. Chem. Soc., Chem. Commun.*, 738 (1989); *Inorg. Chim. Acta* **165**, 191 (1989).

[18] G.A. Jeffrey and M.S. Shen, *J. Chem. Phys.* **57**, 56 (1972).

[19] G.A. Jeffrey and D. Mastropaolo, *Acta Crystallogr., Sect. B* **34**, 552 (1978).

[20] R.K. McMullan, T.C.W. Mak and G.A. Jeffrey, *J. Chem. Phys.* **44**, 2338 (1966).

[21] T.C.W. Mak, S.P. So, C. Chieh and K.S. Jasim, *J. Mol. Structure* **127**, 375 (1985).

Note A new polyhedral clathrate hydrate of composition $(i\text{-}C_5H_{11})_4NF.26.7H_2O$ is reported in J. Lipkowski, K. Suwińska, K. Udachin, T. Rodionova and Yu. Dyadin, *J. Incl. Phenom.* **9**, 275 (1990).

7.3 β-Hydroquinone Hydrogen Sulfide Clathrate
$3C_6H_4(OH)_2 \cdot xH_2S$ (x = 0.874)

Crystal Data

Rhombohedral, space group $R\bar{3}$ (No. 148)

$a = 16.616(3)$, $c = 5.489(1)$Å, [1] $Z = 3$

Atom	x	y	z	Atom	x	y	z
S	0	0	1/2	H(1)	.1316	-.0234	.024
O	.09751	-.08973	.0062	H(2)	.0720	-.2532	.007
C(1)	.13390	-.12590	.1716	H(3)	.2051	-.0113	.038
C(2)	.11093	-.21777	.1410	Group of atoms			
C(3)	.18941	-.07479	.3642	[OH]$_6$ ring	0	0	0
				Hydroquinone	1/6	-1/6	1/3

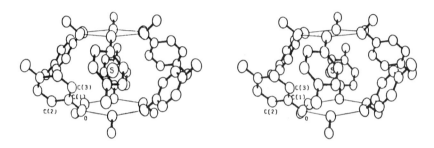

Fig. 7.3-1 Stereoview showing a H_2S guest molecule trapped inside a cage in the β-hydroquinone host lattice. (After ref. 2).

Crystal structure

β-Hydroquinone (quinol) clathrates are molecular inclusion compounds of the general formula $3C_6H_4(OH)_2 \cdot xG$, where G is the encaged guest species and x is a site-occupancy factor. The nature of molecular association in these compounds, and consequently the general concept of clathration, were clarified by Palin and Powell in their investigation of the β-hydroquinone sulfur dioxide molecular adduct.[3]

In order to define accurately the β-hydroquinone host lattice, H_2S was selected as the guest species to minimize complications due to disorder.[1,2] The hydroquinone molecule, with the exception of the hydroxy-hydrogen, is planar. The normal to the plane makes an angle of 55.2° with the c axis. The long axis of the hydroquinone molecule, which passes through the two C-O bonds, is inclined at an angle of 40.2° to the (001) plane. The bond length of C-O is 1.384(2)Å and the C-C, C-H, O-H bond lengths are normal. The β-hydroquinone host structure is composed of two interlocking three-dimensional hydrogen-bonded networks of rhombohedral symmetry. The oxygen atoms are grouped into hexagonal [OH]$_6$ rings normal to the c axis, and the guest molecule is sandwiched between two such rings and surrounded by six C_6H_4 groups, three of which originate from the top [OH]$_6$ ring, and the remaining three from the bottom [Fig. 7.3-1]. The clathration cavity is roughly spherical with a "free diameter" of *ca*. 4.8Å. In the [OH]$_6$ ring, which deviates slightly but significantly from exact planarity, the hydrogen-bond O...O distance is 2.696(1)Å, and the O...O...O angle near 120.°.

The H$_2$S molecule can be treated essentially as a sulfur atom in structure-factor calculations. The unusually large vibrational amplitude in the z direction is a consequence of the fact that the H$_2$S molecule has more freedom of movement in the direction of the open centers of the [OH]$_6$ rings.

Remarks

1. Historical significance[3-5]

The history of inclusion compounds dates back to 1823 when Michael Faraday reported the preparation of the clathrate hydrate of chlorine. Other early observations include the preparation of graphite intercalates in 1841, the β-hydroquinone H$_2$S clathrate in 1849, cyclodextrin inclusion compounds in 1891, the tri-*o*-thymotide benzene inclusion compound in 1909, clathrates of Dianin's compound in 1914, phenol clathrates in 1935, and urea adducts in 1940. In the late 1940s, the pioneering X-ray studies of Powell and his co-workers on the β-hydroquinone clathrates[3,4] firmly established the true nature of these systems. A new term "clathrate", which is derived from the Latin word "clathratus" meaning "enclosed by the bars of a grating",[4] was coined to describe inclusion compounds possessing cages-like host structures for the accommodation of guest species.

2. β-Hydroquinone clathrates

Table 7.3-1 lists the crystal data for β-hydroquinone clathrates.[6] All three crystallographically-distinguishable Types I, II, and III have the same general formula 3C$_6$H$_4$(OH)$_2$.xG, where G represents the encaged guest molecule and x is a site occupancy factor (e.g. Xe, x = 0.866; H$_2$S, x = 0.874). In the table, the first entry is "empty" β-hydroquinone (Fig. 7.3-2), whose cell dimensions are indeed the smallest in the series.[7]

a) Structure of Type I clathrates

In Type I clathrates, cavities having $\bar{3}$ (C_{3i}) symmetry are present. Fig. 7.3-2(a) shows a stereoview of such a centrosymmetric cage of the unsolvated form. The top and bottom of the void are formed by hexagons of hydrogen-bonded oxygen atoms; an ordered arrangment of hydrogen atoms is

Table 7.3-1 Crystal data for β-hydroquinone clathrates.[6]

Type	space group	Lattice parameters* (Å)		Guest	Hexamer dimensions O...O (Å)
		a	c		
I	$R\bar{3}$	16.613(3)	5.4746(5)	none	2.678(3)
		16.610(3)	5.524(1)	Xe	2.705(2)
		16.616(3)	5.489(1)	H_2S	2.696(1)
II	R3	16.31(5)	5.821(1)	SO_2	2.727(6), 2.733(6)
		16.621(2)	5.562(1)	MeOH	2.653(5), 2.779(5)
		16.650(1)	5.453(1)	HCl	2.61(1), 2.77(1)
		15.946(2)	6.348(2)	CH_3NC	2.779(6), 2.800(6)
III	P3	16.003(2)	6.245(2)	CH_3CN	2.778(mean)

* For $R\bar{3}$ and R3, the values of a and c given are referred to a hexagonal unit cell with Z = 3.

apparent in the $[OH]_6$ rings, and the host molecules point alternately above and below the mean plane of the (nearly planar) six oxygen atoms. The hexameric units forming the ceiling and floor of a given cage, as may be seen from Fig. 7.3-2(b), belong to two identical, but displaced, three-dimensional interlocking networks. The remarkably low packing coefficient,[8] 0.62, may be compared with the normal range, 0.65-0.77, for most organic molecular crystals, and demonstrates the realization of an "open" structure with unfilled cavities stabilized by an extended system of hydrogen bonds.

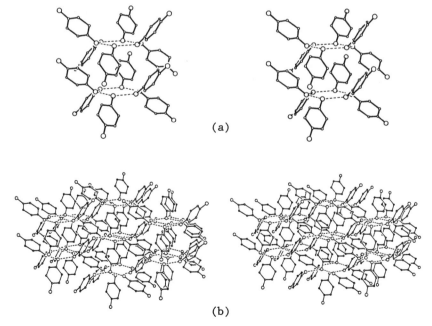

(a)

(b)

Fig. 7.3-2 Stereoviews illustrating (a) the construction of a single cage in the unsolvated form of β-hydroquinone and (b) more extended portions of the two identical, but displaced, three-dimensional networks from which cages are constructed.

In the Xe and H₂S clathrates, the encaged Xe atom and H_2S molecule
occupy an approximately spherical cavity of mean free diameter *ca.* 4.8Å.   The
H_2S guest molecule undergoes pronounced thermal motion, particularly in the
direction of the centers of the $[OH]_6$ rings.   The results are consistent with
rotational disorder of the guest molecule.

b)   Structure of Type II clathrates

In Type II clathrates, the space group symmetry is lowered from $R\bar{3}$ to
*R*3, and guest accommodation is provided in cages which are still trigonal,
though no longer centrosymmetric.   For the relatively long guest molecule
methyl isocyanide the cage length, corresponding to the *c*-spacing, is markedly
increased compared with the Type I systems.   Fig. 7.3-3 illustrates the
alignment of the CH_3NC molecule along the *c* axis in its Type II clathrate.[9]
In the MeOH clathrate,[10] the encaged MeOH molecule is located in three
preferred orientations related by three-fold rotation about the *c* axis; in
each orientation, the C-O bond is tilted by 35° from *c* to facilitate
interaction of the hydroxyl group with three phenolic oxygen atoms of the
adjacent $[OH]_6$ ring (Fig. 7.3-4).   In the MeOH clathrate, host-guest

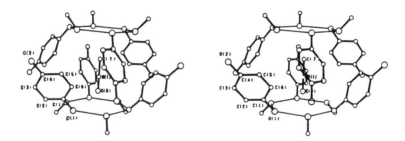

Fig. 7.3-3   Stereoview showing a CH_3NC guest molecule trapped inside a
cage in the structure of Type II β-hydroquinone clathrate.   For clarity
all hydrogen atoms have been omitted.   (After ref. 9).

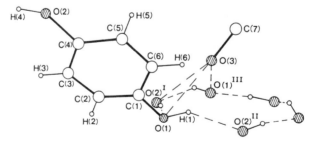

Fig. 7.3-4   Host-guest interactions in the methanol clathrate of β-
hydroquinone.   The O(3)-C(7) bond of the CH_3OH guest molecule is inclined at
an angle of 35° to the *c* axis of the crystal.   (After ref. 10).

interaction is reflected in unequal O-H...O hydrogen bonds in the $[OH]_6$ ring, a feature also found in the HCl clathrate.

c)   Structure of Type III clathrate

In the acetonitrile clathrate of hydroquinone, the only authenticated Type III system, a further reduction in symmetry from $R3$ of Type II to $P3$ occurs.   There are now three distinct types of clathration cavity in the shape of prolate spheroids.   The three symmetry-independent acetonitrile molecules fit snugly inside these cavities, one guest molecule being aligned in the opposite sense to the other two.   Fig. 7.3-5 shows electron density sections through the guest molecules; although molecule (c), in the opposite orientation from molecules (a) and (b), appears to be displaced from its "idealized" position along the z axis, the disposition of this molecule with respect to the top $[OH]_6$ ring of its cage is virtually the same as that of the other molecules with respect to their bottom rings.

3.   Structure of α- and γ-hydroquinone

α-Hydroquinone, the stable form at room temperature, belongs to trigonal space group $R\bar{3}$, with $a = 38.46(2)$, $c = 5.650(3)$Å, and 54 molecules of hydroquinone in the hexagonal unit cell.[11]   As shown in Fig. 7.3-6(a), of

----

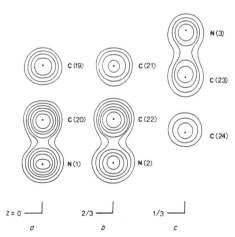

z = 0 ⌐   2/3 ⌐   1/3 ⌐
      a   *b*   *c*

Fig. 7.3-5 Electron-density sections through the CH_3CN guest molecules (a) at $(0, 0, z_a)$, (b) at $(1/3, 2/3, z_b \approx 2/3 + z_a)$, and (c) at $(2/3, 1/3, z_c)$ in the hydroquinone clathrate.   Contours are drawn at $1e\text{Å}^{-3}$ intervals starting at $2e\text{Å}^{-3}$.   For each cage, the dotted line represents the mean plane of its top $[OH]_6$ ring.   The disposition of guest molecule (c) relative to the top $[OH]_6$ ring of its cage is approximately the same as that of molecules (a) and (b) with respect to their bottom rings. (After ref. 9).

the three crystallographically independent hydroquinone molecules in the asymmetric unit, two are involved in forming two interpenetrating, open, hydrogen-bonded cageworks similar to those in β-hydroquinone and capable of enclathrating small molecules, whereas the third forms double helices consisting of hydrogen-bonded chains of molecules about a three-fold screw axis [Fig. 7.3-6(a)]. The cageworks and helices are hydrogen-bonded together in such a way that the interpenetrating cageworks are connected (unlike β-hydroquinone), as are two strands of the double helix. With three cages similar in size to those of the β-form per unit cell, the maximum host/guest mole ratio expected for an α-hydroquinone clathrate is 18:1.

The γ form of hydroquinone can be produced by sublimation or by rapid evaporation of a solution in ether. The crystals are monoclinic, space group $P2_1/c$, with $a = 8.07$, $b = 5.20$, $c = 13.20$Å, $\beta = 107°$, and $Z = 4$.[12] The

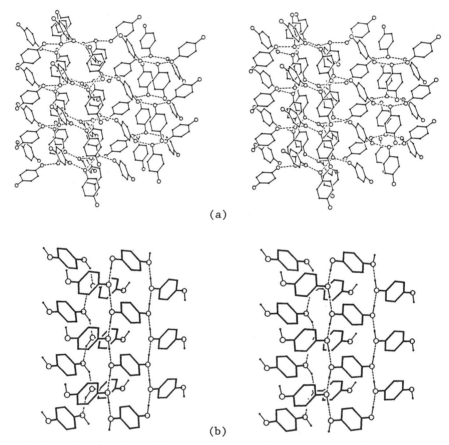

(a)

(b)

Fig. 7.3-6 Stereoviews illustrating (a) the structure of α-hydroquinone; two crystallographically distinct molecules form (unoccupied) cages on the right, while a third type forms double helices around a three-fold screw axis on the left; and (b) the structure of γ-hydroquinone built up from sheets of hydrogen-bonded molecules.

structure is built up from sheets of hydrogen-bonded hydroquinone molecules [Fig. 7.3-6(b)]. In accord with the instability and pronounced cleavage of the γ-form, these sheets are apparently held together only by weak van der Waals forces.

4.   Structure of hexamethylenetetramine hexahydrate

    Hexamethylenetetramine, $(CH_2)_6N_4$, was the first organic compound to have its crystal structure elucidated by X-ray analysis (see Section 2.24). The compound is rather unusual in that its solubility in water decreases with increasing temperature. This property is explicable in terms of increasing association with water at low temperatures. In fact, a hexahydrate of hexamethylenetetramine is easily obtained by slowly cooling a saturated aqueous solution towards 0°C. The hexahydrate crystallizes in space group $R3m$ with a = 7.30(1)Å, and α = 105.4(2)°.[13] The clathrate framework structure of this hydrate is shown in Fig. 7.3-7. The basic structural unit of the cagework is a slightly puckered six-membered ring of hydrogen-bonded water molecules, with its mean plane parallel to (0001). Hexagonal columns are formed by stacking these $[H_2O]_6$ rings, one on top of the other and separated by the c axis spacing. Neighbouring columns are staggered with respect to one another in the [0001] direction, giving rise to the required rhombohedral symmetry. The host water framework is then completed by the cross-lingkage of neighbouring columns through hydrogen bonds. This arrangement is analogous to that of a single open β-hydroquinone network in which each benzene ring with its two *para* links are replaced by a hydrogen bond.

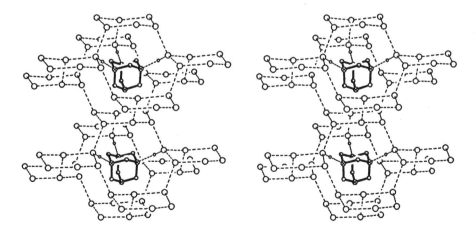

Fig. 7.3-7   Stereoview of the crystal structure of $(CH_2)_6N_4 \cdot 6H_2O$. Eight $[H_2O]_6$ rings form the immediate surroundings of each guest molecule; the N...H-O hydrogen bonds are also shown.

The hexamethylenetetramine molecules occupy cavities in the water framework. Each guest molecule is surrounded by eight $[H_2O]_6$ rings, as shown in Fig. 7.3-7; it is hydrogen bonded to three of these so as to hang "chandelier-like" to the upper walls of the cavity. The three nitrogen atoms at the top of the molecule in the figure are hydrogen-bond acceptors from the three nearest water molecules. The fourth nitrogen atom at the bottom is at van der waals distances from its neighbouring water molecules and is not involved in hydrogen bonding. This structure provided the first example of hydrogen bonding interaction between an encaged guest species and the host framework.

5. Phenol clathrates

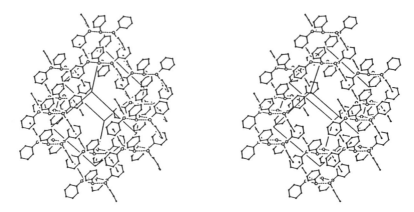

Fig. 7.3-8 Stereoview of the host structure of phenol clathrate; two pairs of hexameric units around the closest and most remotc rhombohedral lattice points as seen by the viewer have been omitted in order to reveal the elongated cage more clearly.

Phenol itself forms clathrates, which feature a basic structural unit comprising six phenol molecules linked by hydrogen bonds to form a $[OH]_6$ hexagon, with the phenyl groups pointing alternately above and below it (Fig. 7.3-8).[14] These sextets are arranged in a rhombohedral lattice, space group $R\bar{3}$, such that two types of centrosymmetric cavities are formed: a large elongated cage having an effective length of about 15Å and a cross-section of diameter 4-4.5Å, centered at (1/2, 1/2, 1/2), and a small spheroidal cage with a free diameter of *ca.* 4.5Å at the origin (0, 0, 0), as shown in Fig. 7.3-9. Both cages are capable of including suitably-sized guest molecules, and the limiting compositions[15] are:

$12C_6H_5OH.5G_1$,   where G_1 is a molecule of the size of HCl or HBr (four
                                    such molecules can be accommodated in the large cage and
                                    one in the small cage);

$12C_6H_5OH.4G_2$,   where G_2 is a molecule of the size of SO_2 (one and three guest molecules in the small and large cages, respectively);

$12C_6H_5OH.2G_3$,   where G_3 is a molecule of the size of CS_2 (two CS_2 molecules located in the large cavity and none in the small cavity);

$12C_6H_5OH.2G_3.G_1$, for a double clathrate (formed, for example, with CS_2 and air as guests).

The crystal structures of substituted phenols and their hydrogen-bonded molecular adducts have been reviewed.[16]

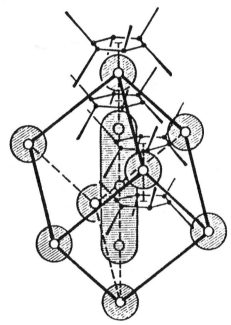

Fig. 7.3-9 The two types of cages in phenol clathrates. (After ref. 17).

References

[1] W.C. Ho and T.C.W. Mak, *Z. Kristallogr.* **161**, 87 (1982).

[2] T.C.W. Mak, J.S. Tse, C.-S. Tse, K.-S. Lee and Y.-H. Chong, *J. Chem. Soc., Perkin Trans. II*, 1169 (1976).

[3] D.E. Palin and H.M. Powell, *J. Chem. Soc.*, 208 (1947).

[4] H.M. Powell, *J. Chem. Soc.*, 61 (1948); J.E.D. Davis, W. Kemula, H.M. Powell and N.O. Smith, *J. Incl. Phenom.* **1**, 3 (1983).

[5] D.D. MacNicol in J.L. Atwood, J.E.D. Davies and D.D. MacNicol (eds.), *Inclusion Compounds*, Vol. **2**, Academic Press, London, 1984, chap. 1.

[6] G. Tsoucaris in G.R. Desiraju (ed.), *Organic Solid State Chemistry*, Elsevier, Amsterdam, 1987, chap. 7; T. Birchall, C.S. Frampton, G.J. Schrobilgen and J. Valsdóttir, *Acta Crystallogr., Sect. C* **45**, 944 (1989).

[7]   S.V. Lindemann, V.E. Shklover and Yu.T. Struchkov, *Cryst. Struct. Commun.* **10**, 1173 (1981).

[8]   A.I. Kitaigorodsky, *Molecular Crystals and Molecules*, Academic Press, New York, 1973, p. 18.

[9]   T.-L. Chan and T.C.W. Mak, *J. Chem. Soc., Perkin Trans. II*, 777 (1983).

[10] T.C.W. Mak, *J. Chem. Soc., Perkin Trans. II*, 1435 (1982).

[11] S.C. Wallwork and H.M. Powell, *J. Chem. Soc., Perkin Trans. II*, 641 (1980); H.M. Powell in J.L. Atwood, J.E.D. Davis and D.D. MacNicol (eds.), *Inclusion Compounds*, Vol. 1, Academic Press, London, 1984, chap. 1.

[12] K. Maartmaan-Moe, *Acta Crystallogr.* **21**, 979 (1966).

[13] T.C.W. Mak, *J. Chem. Phys.* **43**, 2799 (1965); G.A. Jeffrey and T.C.W. Mak, *Science (Washington)* **149**, 178 (1965).

[14] M. von Stackelberg, A. Hoverath and C. Scheringer, *Z. Elektrochem.* **62**, 123 (1958).

[15] L. Mandelcorn, *Chem. Rev.* **59**, 827 (1959).

[16] R. Perrin, R. Lamartin, M. Perrin and A. Thozet in G.R. Desiraju (ed.), *Organic Solid State Chemistry*, Elsevier, Amsterdam, 1987, chap. 8.

[17] D.W. Davidson, S.K. Garg, S.R. Gough, R.E. Hawkins and J.A. Ripmeester, *Studies of Molecular Motion in Clathrates and of Clathrate Structures by NMR and Dielectric Techniques*, Report No. C1093-82S, Division of Chemistry, National Research Council of Canada, Ottawa, 1982.

Note The structural types of simple phenols in the crystalline state are reviewed by R. Perrin, R. Lamartine, M. Perrin and A. Thozet in G.R. Desiraju (ed.), *Organic Solid State Chemistry*, Elsevier, Amsterdam, 1987, p. 271.

Inclusion compounds formed by alicyclic diol hosts are described by R. Bishop and I.G. Dance in J.L.Atwood, J.E.D. Davies and D.D. MacNicol (eds.), *Inclusion Compounds*, Vol. 5, Oxford University Press, New York, 1991, chap.1.

7.4 4-p-Hydroxyphenyl-cis-2,4-dimethylchroman Carbon Tetrachloride Clathrate

$$6[C_{17}H_{18}O_2]\cdot CCl_4$$

Crystal Data

Rhombohedral, space group $R\bar{3}$ (No. 146)

$a = 26.936(6)$, $c = 10.796(1)$Å (hexagonal unit cell, obverse setting),

$Z = 3$

Atom	x	y	z	Atom	x	y	z
Host molecule							
O(1)	.9172	.7439	1.1204	C(11)	.8666	.7563	.7989
C(2)	.8916	.7777	1.0809	C(12)	.9204	.7743	.7466
C(3)	.8348	.7397	1.0212	C(13)	.9433	.8161	.6545
C(4)	.8411	.7119	.9027	C(14)	.9129	.8416	.6137
C(5)	.8794	.6441	.8595	C(15)	.8595	.8248	.6630
C(6)	.9124	.6197	.8887	C(16)	.8371	.7834	.7539
C(7)	.9461	.6373	.9937	C(17)	.7803	.6634	.8656
C(8)	.9464	.6782	1.0692	C(18)	.8863	.8064	1.1981
C(9)	.9135	.7033	1.0387	O(20)	.9358	.8837	.5239
C(10)	.8791	.6862	.9336				

Atom	x	y	z	Occupany
Chloroform guest molecule				
Orientation (1)				
C(111)	0	0	.0900	0.50
Cl(4)	0	0	.2470	0.25
Cl(1)	.9300	.9700	.0590	0.25
Orientation (2)				
C(111)	0	0	.0900	0.50
Cl(42)	0	0	-.0650	0.25
Cl(12)	.9300	.9700	.1210	0.25

Crystal Structure

Fig. 7.4-1(a) shows a perspective view of the host molecule (I), in which the heterocyclic ring has a distorted half-chair conformation. Fig. 7.4-1(b) shows the host structure of this clathrate, in which six molecules are linked by hydrogen bonds involving their hydroxyl groups such that the oxygen atoms form a distorted hexagon, molecules of opposite configuration lying alternately on opposite sides of its mean plane. Such hexameric units are stacked along the c axis so that their bulkier parts interlock to form cages of imposed $\bar{3}$ symmetry. The O...O distance in the hexamer is 2.767(3)Å, and the displacement of the oxygen atoms from the mean plane is ±0.26Å. In the crystal, two independent hydroxyl H positions have occupancies of 0.7 and 0.3; corresponding O...H distances and O-H...O angles are 1.94(7) and 1.93(7)Å and 165(7) and 171(7)°, respectively. The CCl_4 guest molecule is disordered over two independent orientations.

A section through the van der Waals surface of the cavity is illustrated in Fig. 7.4-2(a). The marked change in cavity shape, brought about by removal of the six inward-pointing methyl groups of the parent molecule, Dianin's compound (II),[3] can be appreciated from Fig. 7.4-2.

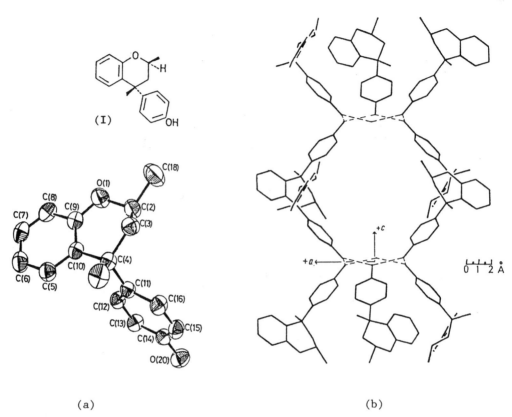

(a)   (b)

Fig. 7.4-1 Structure of 4-*p*-hydroxyphenyl-*cis*-2,4-dimethylchroman CCl₄ clathrate. (a) Molecular formula (I) and perspective view of the host molecule; hydrogen atoms are omitted for clarity. (b) Host framework; two host molecules which lie above and below the cavity as viewed in this direction have been excluded, apart from their hydroxy-oxygen atoms. The disordered guest molecules are not shown. (Adapted from refs. 1 and 2).

Remarks

1. Dianin's compound and related host molecules

The compound 4-*p*-hydroxyphenyl-2,2,4-trimethylchroman (II), widely known as Dianin's compound, was first prepared in 1914 by the gaseous HCl catalyzed condensation of phenol and mesityl oxide.[4] Subsequently this compound has been shown to be capable of including numerous diverse species, for example, argon, sulphur dioxide, iodine, ammonia, decalin, glycerol, sulphur hexafluoride, and di-*t*-butylnitroxide. The molecular structure of the host

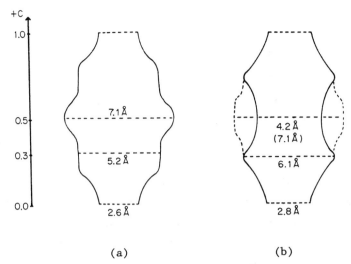

(a)   (b)

Fig. 7.4-2 Cavity shape. (a) Clathrate of compound (I) with CCl₄. (b) Clathrate of Dianin's compound (II) with CHCl₃; curved broken lines represent the effect of the formal removal of the "waist" methyl groups. (Adapted from refs. 1 and 2).

Table 7.4-1 Crystal data of clathrates of Dianin's compound and related molecules belonging to the space group $R\bar{3}$.[5]

| (II) | (III) | (IV) | (V) | (VI) | (VII) |

| Host formula | Lattice parameters (Å) | | Guest | Host/guest ratio |
	a	c		
Dianin (II)	26.94	10.94	none	–
	26.969	10.990	C₂H₅OH	3:1
	27.023	10.922	Xe	6:0.71[6]
	27.116	11.023	CHCl₃	6:1
	27.12	11.02	n-C₇H₁₅OH	6:1
(I)	26.936	10.796	CCl₄	6:1
S-Dianin (III)	27.81	10.90	C₂H₅OH	3:1
	27.91	10.99	Me₃CC≡CCMe₂OH	6:1
	28.00	11.08	di-t-butylacetylene	6:1
(IV)	29.22	10.82	cyclopentane	6:1
(V)	33.629	8.239	cyclooctane	4.5:1
(VI)	27.063	12.074	CCl₄	3:1[7]
Se-Dianin (VII)	28.225	10.859	n-hexane	6:1[8]

compound and the crystal structure of the clathrate were established in the mid-1950s.

The crystal data of some clathrates of Dianin's compound, its thia- and selena-analogues, and related molecules are given in Table 7.4-1.

2. Clathrate of S-Dianin's compound (III) with 2,5,5-trimethylhex-3-yn-2-ol

The clathrates formed by (III) are almost identical in structure to those of (II), and in each case guest accommodation is provided in extremely similar hour-glass shaped cavities [see Fig. 7.4-2(b)].[9] The 2,5,5-trimethylhex-3-yne-2-ol guest molecules adopt a staggered conformation (with a statistical disorder of the OH and Me groups to satisfy the imposed $\bar{3}$ symmetry of the cavity). As shown in Fig. 7.4-3(a), the acetylenic unit of the guest molecule is collinear with the c axis, the triple bond fitting neatly into the waist of the cavity, leaving a tetrahedral unit in the upper and lower halves

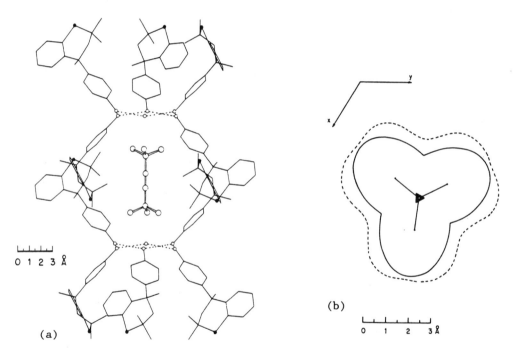

0 1 2 3 Å

(a)   (b)

0 1 2 3 Å

Fig. 7.4-3 (a) The structure of the clathrate of 4-p-hydroxyphenyl-2,2,4-trimethylthiachroman (3) with 2,5,5-trimethylhex-3-yn-2-ol as guest. Two host molecules have been excluded (apart from their hydroxyl oxygen atoms) to reveal the guest, which is accurately aligned along the c-axis; (b) The van der Waals contacts, as viewed along the c-axis, for a section at z = 0.26; the broken lines represent the van der Waals volumes of the atoms comprising the cage and the full lines the approximate van der Waals volume of the guest. (After ref. 9).

of the cavity. The match between host and guest is illustrated in Fig.
7.4-3(b), which shows van der Waals contacts for a section about a quarter way
up the cavity (at $z = 0.26$).

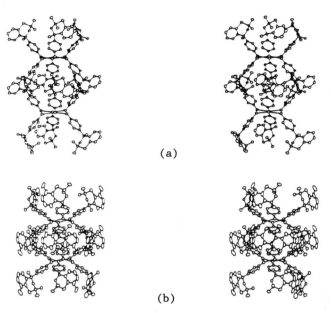

Fig. 7.4-4 Comparative stereoviews for (a) the 2,5,5-trimethylhex-3-yn-2-ol
clathrate of 4-p-hydroxyphenyl-2,2,4-trimethylthiachroman (III) and (b) the
cyclooctane clathrate of 4-p-hydroxyphenyl-2,2,4,8-tetramethylthiachroman (V).
The guest molecules are not shown. (After ref. 10).

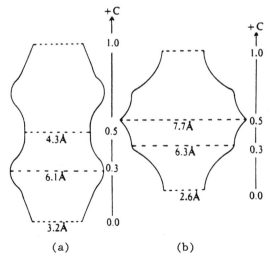

Fig. 7.4-5 Section through the van der Waals surface of the cavity for
(a) 4-p-hydroxyphenyl-2,2,4-trimethylthiachroman (III) and (b) 4-p-
hydroxyphenyl-2,2,4,8-tetramethylthiachroman (V), comparing the space
available for guest accommodation. (After ref. 11).

The cyclooctane clathrate of S-Dianin's compound (V) is isomorphous with those of parent (III) with a marked decrease in cavity length to 8.24Å, implying a fundamental change in cavity geometry (Table 7.4-1). As can be seen from the stereoviews in Fig. 7.4-4, the "legs" of the hexameric unit of (V) have "splayed out", compared with their disposition in (III), thus explaining the decrease in the *c* axial length and the corresponding increase in the *a* dimension, which reflects an increase in column width.[10] The dramatic change in cavity shape is illustrated in Fig. 7.4-5,[11] which shows that the initial hour-glass shaped cavity of (III) has been transformed into

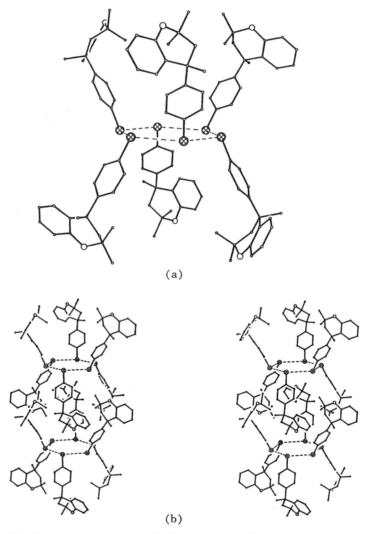

(a)

(b)

Fig. 7.4-6 (a) Perspective view of the hydrogen-bonded hexameric unit of 4-*p*-mercaptophenyl-2,2,4-trimethylchroman (VI) in the CCl₄ clathrate; and (b) stereogram illustrating the hexamer stacking in (VI), which is analogous to that found in Dianin's compound (II) itself.

the "Chinese-lantern" contour of (**V**). This change in cavity geometry is reflected in selective clathration properties. On recrystallization from an equimolar mixture of cyclopentane, cyclohexane, and cycloheptane, (**III**) greatly favours cyclopentane, whereas (**V**) favours the larger cycloparaffins, consistent with the view that van der Waals host-guest attractions are optimized during the crystallization process.

3. Clathrate of Dianin's compound (**VI**) with CCl₄

The relative values of lattice parameters for the 3:1 CCl₄ clathrate[7] of (**VI**) and clathrates of (**II**) are consistent with analogous packing modes for the two hosts, the increased *c* spacing for (**VI**) being in keeping with the longer C-S bond. The hydrogen-bonded hexameric units of six thiol molecules comprising the host structure in (**VI**) is shown in Fig. 7.4-6(a). The sulfur atoms form a near planar hexagon; the S-H...S hydrogen-bond is characterized by S...S and S...H distances of 3.76(1) and 2.67(9)Å respectively, and by the S-H...S angle of 164(5)°. Fig. 7.4-6(b) illustrates two hexameric units of (**VI**) stacked along the *c* axis such that their bulkier hydrophobic arms interlock to form a cage.

4. Inclusion compounds of calixarenes

Calixarenes, a class of synthetic macrocycles having phenolic residues in a cyclic array linked by methylene groups all *ortho* to the hydroxy groups, are exemplified by the molecules shown below:

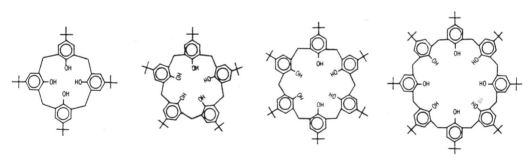

Calix[*n*]arenes (*n* = 4, 5, 6, 8)

These macrocyclic molecules exhibit a high degree of conformational flexibility.[12] For example the four isomeric forms of calix[4]arene shown in Fig. 7.4-7 are readily interconvertible in solution.

Many of the calixarenes form crystalline complexes with small molecules. For example, *p-t*-butylcalix[5]arene forms complexes with isopropyl alcohol and acetone; *p-t*-butylcalix[6]arene forms a complex containing chloroform and methanol; *p-t*-butylcalix[8]arene forms a complex with chloroform; and *p-t*-butyldihomooxacalix[4]arene forms a complex with methylene chloride. The most

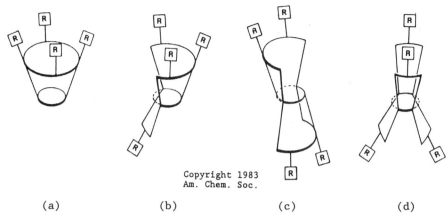

Copyright 1983
Am. Chem. Soc.

(a)   (b)   (c)   (d)

Fig. 7.4-7 Conformations of calix[4]arene: (a) cone, (b) partial cone, (c) 1,2-alternate, and (d) 1,3-alternate. (After ref. 12).

studied inclusion compounds are those of calix[4]arene and its derivatives (Table 7.4-2), in which the macrocyclic host invariably take the cone conformation. There are four types of inclusion compounds of calixarenes:[13]

a) Criptato-cavitate inclusion compounds have been observed in the case of p-t-butylcalix[4]arene and aromatic molecules such as toluene [Fig. 7.4-8(a)]. The interactions observed in the crystals have an intramolecular host-guest character and the guest is held in a chalice-like molecular cavity.[15]

b) Criptato-cavitate sandwich compounds have been observed in the case of p-t-butylcalix[4]arene with anisole [Fig. 7.4-8(b)], which shows 2:1 host-guest stoichiometry.[13]

c) Tubulato-clathrate inclusion compounds have been observed in the case of p-octylcalix[4]arene with toluene as guests (1:1 stoichiometry) [Fig. 7.4-8(c)].[14]

d) Intercalato-clathrate has been observed only in the orthorhombic phase of the 1:1 inclusion compound between calix[4]arene and acetone [Fig. 7.4-8(d)].[16]

Table 7.4-2 Calix[4]arene inclusion compounds.

R	Guest	Host/guest ratio	Space group	Z	Clathrate type	Structure in Fig. 7.4-8	Ref.
LOC	-	-	$P2_1/a$	4	-		[14]
LOC	toluene	1:1	Fmm2	4	tubulato	(c)	[14]
t-Bu	anisole	2:1	P4/n	2	closed cage (criptato-cavitate)	(b)	[13]
t-Bu	toluene	1:1	P4/n	2	half-cage	(a)	[15]
H	acetone	1:1	Pnma	4	intercalato	(d)	[16]
H	acetone	1:3	$P6_3/m$	6	tubulato		[16]

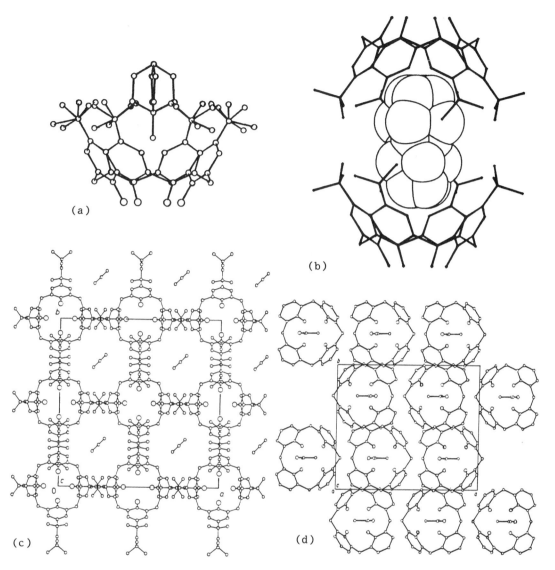

Fig. 7.4-8 Structural typies of inclusion compounds of calix[4]arenes. (Adapted from refs. 13-16).

5. Organic clays[17]

In this novel series of sulfonates, bilayers of anionic calixarenes in the cone configuration alternate with inorganic regions which contain the cations and the water molecules. The overall structures bear a close resemblance to those found for the clay minerals.

Na_5(calix[4]arene sulfonate).$12H_2O$, crystallizes in space group $P\bar{1}$ with a = 10.998(6), b = 13.582(5), c = 14.472(5)Å, α = 74.01(3), β = 89.09(4), γ = 86.50(4)°, and Z = 2. The structure may be divided into organic and

Calix[4]arene sulfonate =

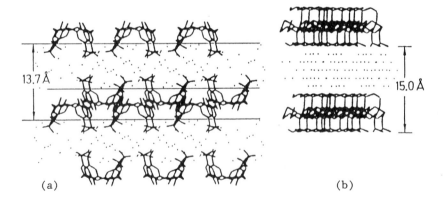

inorganic layers, as shown in Fig. 7.4-9(a). The organic layer is formed by
interlocking calixarenes in an up-down fashion, a sodium ion and a water
molecule interacting simultaneously with the phenolic oxygen atoms of one
calixarene and the sulfonate oxygen atoms of others. To accommodate this
cation, the superacidic proton of each calixarene has to be removed; the
resulting negatively charged oxygen atom forms a strong electrostatic bond
with the sodium ion. The inorganic layer is comprised of the remaining four
sodium ions and eleven water molecules. These exist in a vast hydrogen-bonded
array that includes the sulfonate head groups of the calixarenes. The highly
ordered layer arrangement is similar to the structure of clay, as shown in
Fig. 7.4-9(b).

13.7 Å

15.0 Å

(a)   (b)

Fig. 7.4-9 (a) Bilayer structure of Na_5(calix[4]arene sulfonate).$12H_2O$ in the
crystal; (b) layer structure of hydrated sodium vermiculite. (After ref. 18).

As with the clays, the sodium ions are exchangeable, and the
compositions of the hydrated layers can also be varied. The crystal
structures of the following "organic clay" compounds[17-19] have been
determined:

K_5(calix[4]arene sulfonate).$8H_2O$ Na_5(calix[4]arene sulfonate).$8H_2O.Me_2CO$

Rb_5(calix[4]arene sulfonate).$5H_2O$ $(NH_4)_6$(calix[4]arene sulfonate)$(MeOSO_3)$.$2H_2O$

Cs_5(calix[4]arene sulfonate).$4H_2O$

References

[1] J.H. Gall, A.D.U. Hardy, J.J. McKendrick and D.D. MacNicol, *J. Chem. Soc., Perkin Trans. II*, 376 (1979).

[2] A.D.U. Hardy, J.J. McKendrick and D.D. MacNicol, *J. Chem. Soc., Chem. Commun.*, 355 (1976).

[3] J.L. Flippen, J. Karle and I.L. Karle, *J. Am. Chem. Soc.* **92**, 3749 (1970).

[4] A.P. Dianin and *J. Russe, Phys. Chem. Soc.* **46**, 1310 (1914).

[5] D.D. MacNicol in J.L. Atwood, J.E.D. Davies and D.D. MacNicol (eds.), *Inclusion Compounds*, Vol. 2, Academic Press, London, 1984, chap. 1; G. Tsoucaris in G.R. Desiraju (ed.), *Organic Solid State Chemistry*, Elsevier, Amsterdam, 1987, chap. 7.

[6] F. Lee, E. Gabe, J.S. Tse and J.A. Ripmeester, *J. Am. Chem. Soc.* **110**, 6014 (1988).

[7] A.D.U. Hardy, J.J. McKendrick, D.D. MacNicol and D.R. Wilson, *J. Chem. Soc., Perkin Trans. II*, 729 (1979).

[8] D.D. MacNicol, P.R. Mallinson, R.A.B. Keates and F.B. Wilson, *J. Incl. Phenom.* **5**, 373 (1987).

[9] D.D. MacNicol and F.B. Wilson, *J. Chem. Soc., Chem. Commun.*, 786 (1971).

[10] A.D.U. Hardy, J.J. McKendrick and D.D. MacNicol, *J. Chem. Soc., Perkin Trans. II*, 1072 (1979).

[11] D.D. MacNicol, A.D.U. Hardy and J.J. McKendrick, *Nature (London)* **256**, 343 (1975).

[12] C.D. Gutsche, *Acc. Chem. Res.* **16**, 161 (1983); C.D. Gutsche, *Calixarenes*, Royal Society of Chemistry, Cambridge, 1989; C.D. Gutsche, J.S. Rogers, D. Stewart and K.-A. See, *Pure Appl. Chem.* **62**, 485 (1990).

[13] R. Ungaro, A. Pochini, G.D. Andreetti and P. Domiano, *J. Chem. Soc., Perkin Trans. II*, 197 (1985).

[14] G.D. Andreetti, A. Pochini and R. Ungaro, *J. Chem. Soc., Perkin Trans. II*, 1773 (1983).

[15] G.D. Andreetti, R. Ungaro and A. Pochin, *J. Chem. Soc., Chem. Commun.*, 1005 (1979).

[16] R. Ungaro, A. Pochini, G.D. Andreetti and V. Sangermano, *J. Chem. Soc., Perkin Trans. II*, 1979 (1984).

[17] J.L. Atwood, A.W. Coleman, H.-M Zhang and S.G. Bott, *J. Incl. Phenom.* **7**, 203 (1989).

[18] A.W. Coleman, S.G. Bott, S.D. Morley, C.M. Means, K.D. Robinson, H.-M. Zhang and J.L. Atwood, *Angew. Chem. Int. Ed. Engl.* **27**, 1361 (1988).

[19] S.G. Bott, A.W. Coleman and J.L. Atwood, *J. Am. Chem. Soc.* **110**, 610 (1988).

Note For recent reviews on the structural chemistry of calixarenes and their inclusion complexes, see J. Vicens and V. Böhmer (eds.), *Calixarenes: A Versatile Class of Macrocyclic Compounds*, Kluwer, Dordrecht, 1990; C.D. Gutsche in J.L. Atwood, J.E.D. Davies and D.D. MacNicol (eds.), *Inclusion Compounds*, Vol. 4, Oxford Univeristy Press, New York, 1991, chap. 2; G.D. Andreetti, F. Ugozzoli, K. Ungaro and A. Pochini, *ibid.*, chap. 3; G.D. Andreetti, F. Ugozzoli, Y. Nakamoto and S.-I. Ishida, *J. Incl. Phenom.* **10**, 241 (1991).

7.5   Hexakis(benzylthiomethyl)benzene-Dioxan 1:1 clathrate
$$C_6(CH_2SCH_2C_6H_5)_6 \cdot C_4H_8O_2$$

Crystal Data

Monoclinic, space group $P2_1/c$ (No. 14)

$a = 10.542$, $b = 20.863$, $c = 12.496$Å, $\beta = 95.48°$, $Z = 2$

Atom	x	y	z	Atom	x	y	z
Hexa-host molecule				C(18)	.1862	.2851	.3505
C(1)	-.0177	.0629	.0344	C(19)	.1779	.2289	.3897
C(2)	-.0340	.1315	.0693	C(20)	.2564	.1836	.3506
S(3)	-.1153	.1347	.1902	C(21)	-.1105	.0337	-.0368
C(4)	-.1307	.2210	.2032	C(22)	-.2326	.0678	-.0730
C(5)	-.2079	.2522	.1107	S(23)	-.2165	.1132	-.1947
C(6)	-.3291	.2321	.0776	C(24)	-.3805	.1397	-.2231
C(7)	-.4020	.2635	-.0048	C(25)	-.4735	.0861	-.2457
C(8)	-.3508	.3146	-.0541	C(26)	-.4858	.0553	-.3416
C(9)	-.2320	.3342	-.0241	C(27)	-.5704	.0056	-.3611
C(10)	-.1606	.3034	.0583	C(28)	-.6470	-.0131	-.2856
C(11)	.0904	.0291	.0731	C(29)	-.6357	.0168	-.1898
C(12)	.1857	.0592	.1557	C(30)	-.5494	.0667	-.1693
S(13)	.3058	.1035	.0929				
C(14)	.3866	.1413	.2124	1,4-Dioxan guest molecule			
C(15)	.3126	.1937	.2594	O(31)	.0028	-.0182	.3949
C(16)	.3030	.2527	.2068	C(32)	.1032	.0177	.4525
C(17)	.2320	.3014	.2533	C(33)	-.0502	-.0587	.4757

Crystal Structure

The center of the $C_6(CH_2SCH_2C_6H_5)_6$ host molecule, shown in the top part of Fig. 7.5-1, is located at a crystallographic center of inversion, and the sulfur atoms in the "legs" are situated alternatively above and below the plane of the central benzene ring. Consideration of the torsion angles C(1)-C(11)-C(12)-S(13), C(11)-C(1)-C(2)-S(3), and C(1)-C(21)-C(22)-S(23), which have the respective values 92, -90, and 91°, and also C(1)-C(2)-S(3)-C(4), C(11)-C(12)-S(13)-C(14), and C(21)-C(22)-S(23)-C(24) (-176, 172, and 174°), reveals an approximate three-fold "core" symmetry. Deviations from C_3 symmetry become more marked for the remainder of the molecule. In each of the three independent terminal phenyl groups, half the ring atoms lie above the central benzene ring plane, while the other half lie below. At the periphery of the molecule thermal-motion effects cause apparent bond shortening. The central benzene ring is planar within ±0.01Å, but the directly attached atoms C(2), C(12), and C(22) are displaced from this plane by -0.06, 0.09 and 0.09Å, respectively.

The packing arrangement in the crystal is illustrated in the bottom part of Fig. 7.5-1. The acute angle between the plane normal to the central benzene ring and the c axis is 36°. The 1,4-dioxan guest molecule has a chair conformation. There are no intermolecular contacts that are significantly shorter than the sum of the appropriate van der Waals radii.

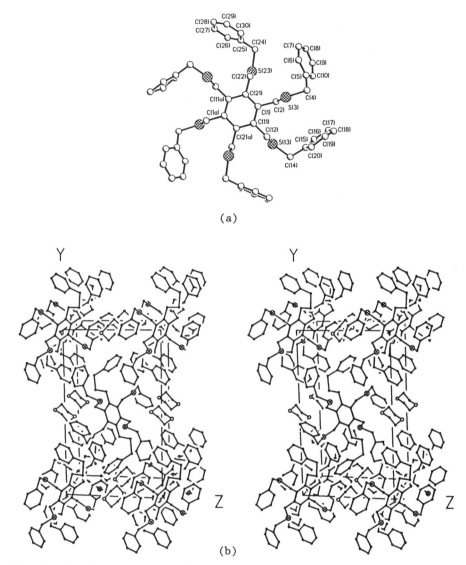

Fig. 7.5-1 (a) Structure of host molecule $C_6(CH_2SCH_2C_6H_5)_6$. (b) Stereoview of the molecular packing in the 1,4-dioxan clathrate of hexakis(benzylthiomethyl)benzene. (Adapted from ref. 1).

Remarks

1. The hexa-host geometric concept and host design[2-4]

A highly successful strategy to host design has been used by MacNicol and his coworkers[5] who noted the similarity between the geometry and dimensions of the hexagonal unit of hydrogen bonded O-H groups which is found in the β-hydroquinone, Dianin's compound, and related phenolic host frameworks with those of a hexa-substituted benzene ring, as shown in Fig. 7.5-2. Through proper choice of substituents it is therefore possible that the hexa-

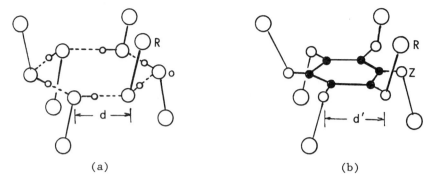

Fig. 7.5-2 Comparison of (a) hydrogen-bonded hexamer unit with
(b) hexa-substituted benzene analogue. (After ref. 6).

substituted benzenes may serve as prospective host frameworks. It has indeed
been found that benzenoid compounds with substituents such as SPh, CH_2OPh,
CH_2SPh and CH_2SCH_2Ph can act as "hexa-hosts". Fig. 7.5-3 shows a variety of
hexa-hosts and octa-hosts, many of which have been crystallized with guest
molecules to form clathrates.

Table 7.5-1   Selected crystal data for some inclusion compounds
formed by hexa-hosts.

Hexa-host[*]	Space group	Lattice parameters[**]	Guest	Mole ratio host:guest
1	R$\bar{3}$	a = 14.263, c = 20.717Å, Z = 3	CCl_4	1:2
3	R$\bar{3}$	a = 22.088, c = 12.232Å, Z = 6	pyridine, H_2O	1:6:1
5 [Ar = (p-t-Bu)Ph]	P$\bar{1}$	a = 14.710, b = 15.773, c = 20.417Å α = 107.40, β = 113.90, γ = 81.93°, Z = 2	squalene	2:1
7 (R = COCF$_3$)	P$\bar{1}$	a = 12.511, b = 12.781, c = 13.224Å α = 70.94, β = 66.04, γ = 75.27°, Z = 1	DMF	1:2
7 (R = COCH$_2$O- CH$_2$CH$_2$OCH$_3$)	P2/c	a = 15.249, b = 10.087, c = 30.328Å β = 95.10°, Z = 2	KSCN, H_2O	1:3:1
8 (Ar = Ph)	P2$_1$/c	a = 10.542, b = 20.863, c = 12.496Å β = 95.48°, Z = 2	1,4-dioxan	1:1
8 [Ar = (m-Me)Ph]	P2$_1$/c	a = 9.62, b = 15.45, c = 22.72Å β = 111.0°, Z = 2	p-o-xylene[***]	1:1
8 [Ar = (p-Me)Ph]	P2$_1$/c	a = 8.66, b = 23.69, c = 13.50Å β = 92.9°, Z = 2	none	-
8 [Ar = (p-Me)Ph]	Pcab	a = 18.67, b = 14.18, c = 23.22Å Z = 4	1,4-dioxan	1:1
9	P$\bar{1}$	a = 10.118, b = 15.172, c = 20.379Å α = 75.05, β = 82.64, γ = 68.45°, Z = 2	1,4-dioxan	1:1
10 (yellow form)	I2/c	a = 20.469, b = 10.567, c = 25.415Å β = 118.13°, Z = 4	none	-
10 (red form)	P$\bar{1}$	a = 9.149, b = 11.473, c = 12.466Å α = 101.42, β = 96.37, γ = 109.67°, Z = 1	none	-

[*] Hexa-hosts numbering as in Fig. 7.5-3.   [**] Z is equal to the number of host molecules.

[***] Recrystallization from an equimolar mixture of p- and o-xylene; the clathrate contains these two guests
in a 85:15 molar ratio.

Table 7.5-1 lists the crystal data for some inclusion compounds formed by hexa-hosts.[2,7]

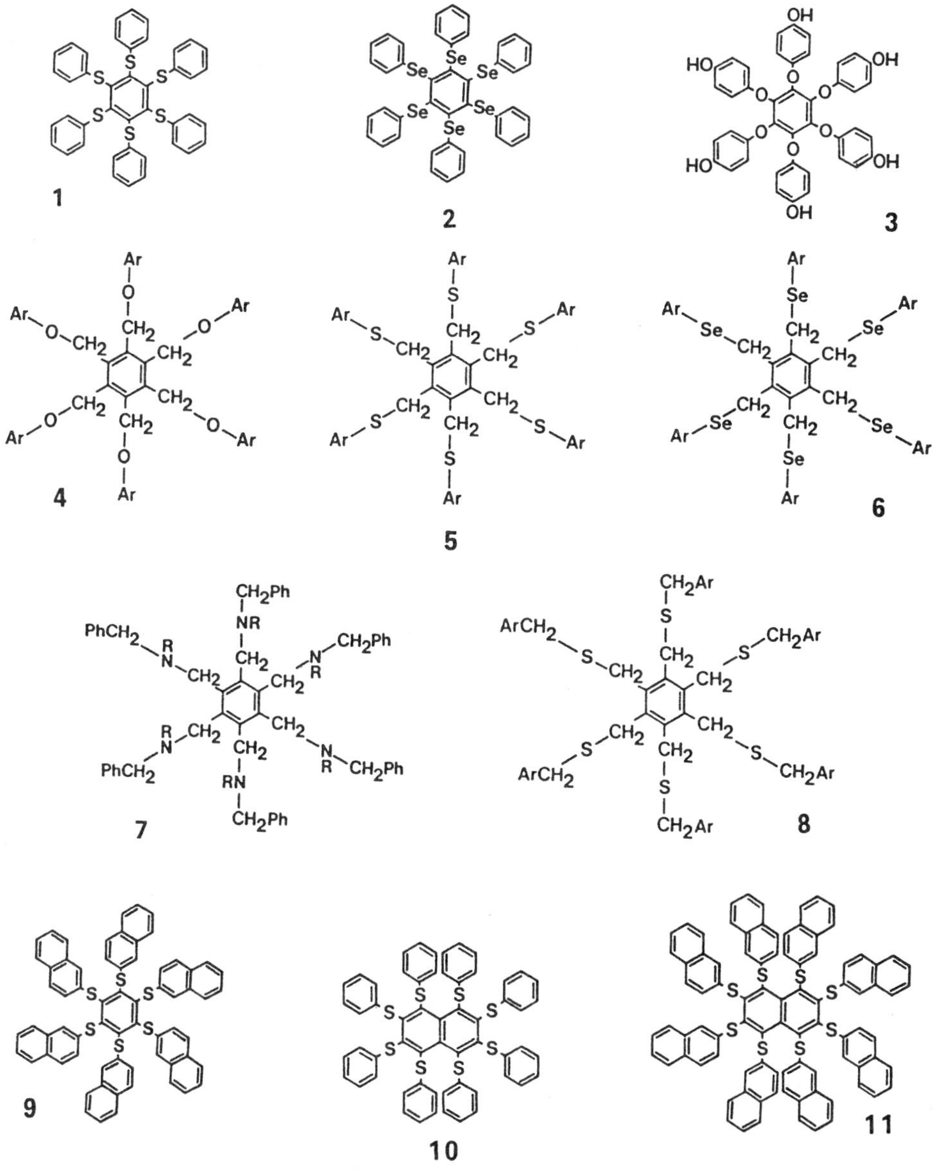

Fig. 7.5-3 Different hexa-hosts and octa-hosts.

2. Crystal structure of 1:2 inclusion compound of hexakis(phenylthio)benzene
   (1) with CCl_4[6]

The first inclusion compound of a hexa-host characterized by X-ray crystallography was the CCl_4 adduct of 1. The host molecule 1 is located at a site of 3 symmetry, so that the crystallographically equivalent "legs" point alternately above and below the plane of the central benzene ring. The

asymmetric unit consists of eight non-hydrogen atoms: C(1), S(2) and C(3)-
C(8). The relative disposition of the central and side-chain aromatic rings
may be described by the torsion angles C(1*)-C(1)-S(2)-C(3) and C(1)-S(2)-
C(3)-C(4), which are 56° and 28°, respectively. The true clathrate nature of
this adduct may be seen from the packing diagram shown in Fig. 7.5-4. In the
long closed cavity, of effective length *ca.* 17Å, two CCl$_4$ guest molecules are
accommodated such that a C-Cl bond of each is collinear with the *c* axis.

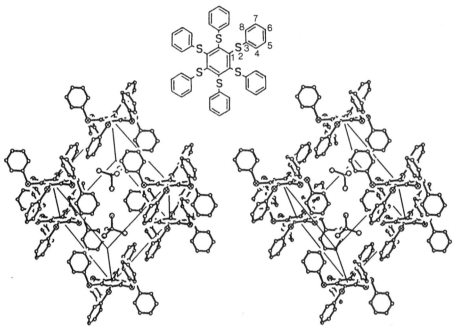

Fig. 7.5-4 Stereoview of the crystal structure of the CCl$_4$ clathrate
of **1**. Two of the eight host molecules that surround the cavity have
been excluded to show the pair of enclosed guest molecules more clearly.

3. Structure of 1:6 clathrate of hexakis(*p*-hydroxyphenyloxy)benzene (**3**) with
   pyridine[7]

The hexaphenol **3** has a functional array that suggests the possibility of
hexa-host character, in which case exactly half of the [OH]$_6$ rings of the
β-hydroquinone structure are replaced by permanent, hexa-oxygen-substituted
benzene units. Fig. 7.5-5(a) illustrates the structure of host molecule **3** in
its 1:6 clathrate with pyridine. Comparison of **3** with the hydrogen-bonded
hexameric unit of β-hydroquinone (empty cage form) [Fig. 7.5-5(b), also see
Fig. 7.3-2] reveals a close analogy. Both units are located on a site of
crystallographic 3̄ symmetry, with corresponding alternation of hydroxyl-
containing moieties above and below the central core. A significant
difference between the two units in that the torsion angle τ [denoted by the

dotted line in Fig. 7.5-5(b)], 58°, is significantly smaller in magnitude compared to the corresponding torsion angle $O(1^*)\text{-}O(1)\text{-}C(2)\text{-}C(7)$ of -105° for **3**. In contrast to the situation for β-hydroquinone where each hydroxyl group is involved in forming hydrogen-bonded hexamers, **3** has six pyridine molecules hydrogen-bonded to it.

Aggregates comprising one host molecule and six pyridine molecules are stacked along the c axis, parts of two neighbouring infinite columns being shown in the stereoview at right angles to the c axis in Fig. 7.5-6. Hydrogen bonds, length 2.71(2)Å, linking oxygen and nitrogen atoms are denoted by broken lines.

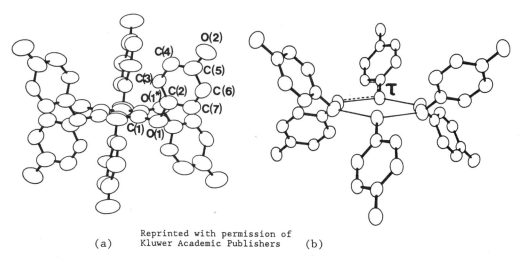

(a)   (b)

Fig. 7.5-5 A comparison of (a) the molecule of hexakis(p-hydroxyl-phenyloxy)benzene (**3**) in its pyridine clathrate with (b) the hydrogen-bonded hexameric unit of β-hydroquinone. (After ref. 7).

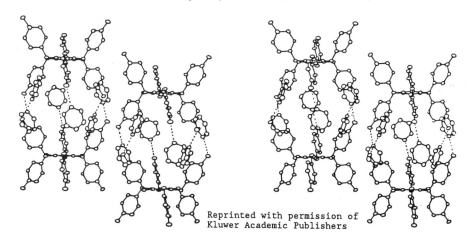

Fig. 7.5-6 A seteroview normal to the c axis illustrating the inter-column packing in the pyridine clathrate of host **3**. (After ref. 7).

4. Channel-type 2:1 inclusion compound of hexakis(*p-t*-butylphenylthiomethyl)-
   benzene (**5**) with squalene[8]

        This crystalline adduct of **5** with the $C_{30}H_{50}$ triterpene squalene was
   investigated because the exact 2:1 host-guest ratio suggested a possible high
   degree of order between host and guest components. Fig. 7.5-7(a) shows the
   host-guest packing arrangement, the squalene guests being accommodated in
   continuous channels running through the crystal. By virtue of location of the
   center of either of the double bonds adjacent to the molecular center of
   squalene at the crystallographic center (0,1/2,1/2), the disordered squalene
   exists as a pair of enantiomeric conformations belonging to the point group

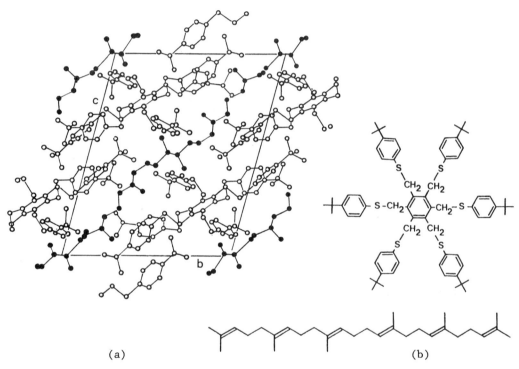

(a)   (b)

Fig. 7.5-7 (a) A view looking onto the *bc* plane illustrating the host-
guest packing in the adduct of **5** with squalene. (b) Molecular formulae
of **5** and squalene. (After ref. 8).

C_1, and appears as a continuous chain in Fig. 7.5-7. The successful
resolution of the guest disorder allows comparison of the guest squalene
conformation with that found in the molecular crystal of pure squalene itself
at -110°C.[9] As can be seen from Fig. 7.5-8, these conformations differ
markedly, the new squalene conformation being closely controlled by host-guest
interactions. This suggests the exciting possibility of specific reaction of
guests which have been subjected to "conformational selection". In the case

of squalene, because of low rotational energy barriers, the usual distribution of solution conformations is expected when the guest is released by dissolving the adduct.

(a)

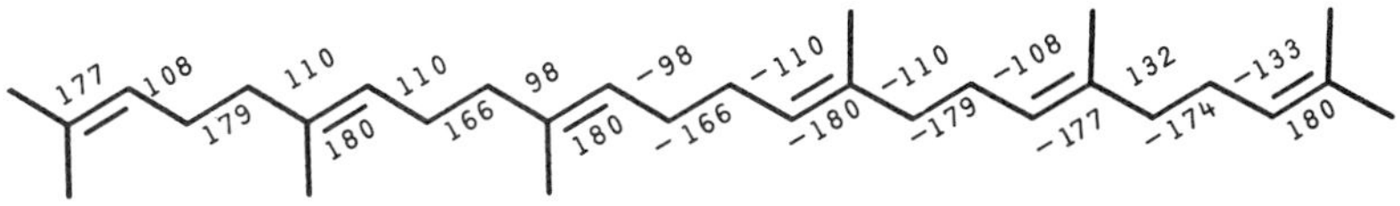

(b)

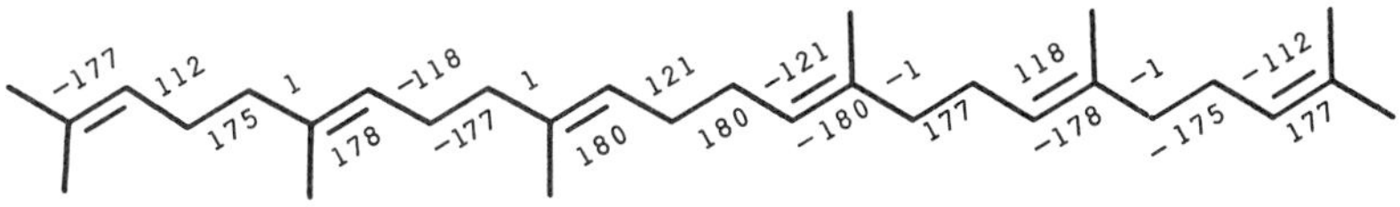

Fig. 7.5-8 A comparison of main chain torsion angles for (a) squalene in its inclusion compound with host **5** and (b) squalene in its molecular crystal at -110°C. (Adapted from ref. 8).

5. Structure of the 1:1 complex of **7** [$C_6(CH_2NRCH_2Ph)_6$, R = $COCH_2OCH_2CH_2OCH_3$] with 3KSCN.H_2O[10]

The X-ray study of this complex provided the first definitive structural information concerning a hexa-host inclusion compound containing an ionic guest. The host molecule **7** is centrosymmetric but deviates significantly from C_3 symmetry. In the complex, the K^+ ions occupy two independent sites. The K^+ ion in a general position is coordinated to four ether oxygens [K^+...O: 2.70(1), 2.81(1), 2.96(1), 3.08(1)Å] and two carbonyl oxygens [2.76(1), 2.84(1)Å] belonging to nonequivalent "legs" of adjacent host molecules, a K^+...N contact of 3.03(1)Å to a thiocyanate ion in a general position also being found. The K^+ ion on a two-fold rotation axis (coordinated to a water molecule) is situated between two equivalent host molecules, though here severe disorder of the two equivalent 2-methoxyethoxy units involved in coordination reduces the accuracy of the measured contact distances. The corresponding SCN^- ion, not coordinated to K^+, has its nitrogen on a two-fold rotation axis. The theoretically possible mode of binding of K^+ by three "legs" of a single host molecule is not employed at either site in this complex.

6. The Piedfort concept of inclusion compound design

The "Piedfort analogue" of a hexa-host consists of two *tri*-substituted aromatic rings, stacked in van der Waals contact, such that the resulting

alternating disposition of side-chain units and overall geometry are similar
(Fig. 7.5-9).[11]

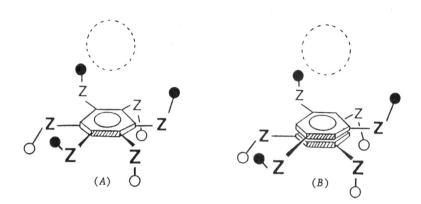

Fig. 7.5-9 Comparison of a hexa-host (*A*) with its Piedfort analogue
(*B*); Z represents a link atom or chain, the unfilled and filled circles
represent the outer groups, and the large dotted circle symbolizes the
projected guest region. (After ref. 11).

Two molecules of 2,4,6-tris[4-(2-phenylpropan-2-yl)phenoxy]-1,3,5-
triazine (**8**) constitute a Piedfort unit that forms an inclusion compound with
1,4-dioxane. The dimeric host units in this adduct nearly attains idealized 3
symmetry; two of the independent guest molecules are located on centers of
symmetry and the third is accommodated in the projected guest region (Fig.
7.5-10). The closest inter-ring (aromatic) contacts in the inclusion compound
(3.42-3.46Å) are significantly shorter than those (3.50-3.55Å) in the
unsolvated crystal.[11]

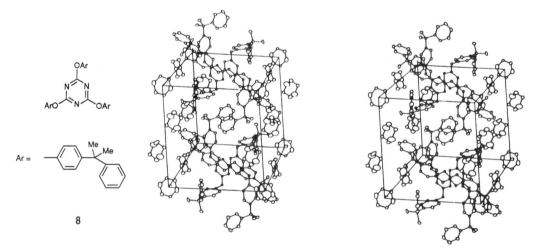

Fig. 7.5-10 Formula of host molecule **8** and stereo view of its 2:1
inclusion compounds with dioxan. (After ref. 11).

References

[1] A.D.U. Hardy, D.D. MacNicol, S. Swanson and D.R. Wilson, *J. Chem. Soc.,
     Perkin Trans. II*, 999 (1980).

[2] D.D. MacNicol in J.L. Atwood, J.E.D. Davies and D.D. MacNicol (eds.),
     Inclusion Compounds, Vol. **2**, Academic Press, London, 1984, chap. 5.

[3] J.E.D. Davies, W. Kemula, H.M. Powell and N.O. Smith, *J. Incl. Phenom.* **1**,
     3 (1983).

[4] G. Tsoucaris in G.R. Desiraju (ed.), *Organic Solid State Chemistry*,
     Elsevier, Amsterdam, 1987, chap 7.

[5] D.D. MacNicol, J.J. McKendric and D.R. Wilson, *Chem. Soc. Rev.* **7**, 65
     (1978).

[6] A.D.U. Hardy, D.D. MacNicol and D.R. Wilson, *J. Chem. Soc., Perkin Trans.
     II*, 1011 (1979).

[7] D.D. MacNicol, P.R. Mallinson, A. Murphy and C.D. Robertson, *J. Incl.
     Phenom.* **5**, 233 (1987).

[8] A.A. Freer, C.J. Gilmore, D.D. MacNicol and D.R. Wilson, *Tetrahedron
     Lett.*, 1159 (1980).

[9] J. Ernst and J.-H. Fuhrhop, *Liebigs Ann. Chem.* 1635 (1979).

[10] A.A. Freer, J.H. Gall and D.D. MacNicol, *J. Chem. Soc., Chem. Commun.*,
     674 (1982).

[11] A.S. Jessiman, D.D. MacNicol, P.R. Mallinson and I. Vallance, *J. Chem.
     Soc., Chem. Commun.*, 1619 (1990).

7.6   Tri-o-thymotide Chlorocyclohexane Clathrate
$2(C_{33}H_{36}O_6) \cdot C_6H_{11}Cl$

Crystal Data

Trigonal, space group $P3_121$ (No. 152)

$a = 13.604(1)$, $c = 30.605(1)$Å, (158K), $Z = 3$

Atom*	x	y	z	Atom	x	y	z
Host molecule							
O(1)	.9731	.3301	-.11008	C(22)	.7526	.4044	-.09396
O(2)	.8682	.4612	-.09595	C(23)	.7010	.4070	-.13658
O(3)	.8130	.3389	-.17409	C(24)	.7304	.3749	-.17587
O(11)	.8779	.2008	-.05718	C(25)	.6872	.3790	-.21575
O(22)	.7016	.3640	-.06087	C(26)	.6088	.4161	-.21575
O(33)	.6840	.1728	-.1463	C(27)	.5762	.4473	-.17759
C(01)	.8662	.1996	-.16731	C(28)	.6221	.4440	-.13825
C(02)	.9604	.2463	-.14061	C(29)	.5879	.4854	-.0976
C(03)	1.0474	.2218	-.14515	C(30)	.7212	.3421	-.25771
C(04)	1.0357	.1482	-.17951	C(31)	.7855	.4440	-.2882
C(05)	.9410	.1003	-.20552	C(32)	.6189	.2481	-.2814
C(06)	.8522	.1229	-.19999	C(33)	.7775	.2321	-.16100
C(07)	.7482	.0668	-.22917	Guest molecule			
C(08)	1.1512	.2728	-.1153				
C(09)	1.2635	.3186	-.1399	Cl(1)	.1896	-.0077	-.1332
C(10)	1.1376	.1834	-.0822	C(41)	.3072	-.0319	-.1272
C(11)	.9269	.2965	-.06907	C(42)	.2922	-.1297	-.1567
C(12)	.9583	.4009	-.04311	C(43)	.3073	-.0973	-.2054
C(13)	.9288	.4799	-.05659	C(44)	.4222	.0102	-.2135
C(14)	.9639	.5827	-.03481	C(45)	.4350	.1090	-.1852
C(15)	1.0322	.6018	.00231	C(46)	.4200	.0770	-.1366
C(16)	1.0622	.5239	.01583	Cl(2)	.1807	-.1297	-.1200
C(17)	1.0260	.4221	-.00564	C(51)	.3029	-.1019	-.1518
C(18)	1.0607	.3385	.01093	C(52)	.4108	-.0038	-.1315
C(19)	.9330	.6679	-.0501	C(53)	.4204	.1120	-.1396
C(20)	1.0374	.7834	-.0576	C(54)	.4014	.1307	-.1878
C(21)	.8510	.6773	-.0195	C(55)	.3847	.0323	-.2173
				C(56)	.2872	-.0806	-.2002

* Parameters of H atoms have been omitted.

Crystal Structure

Crystal structure analysis of the M-(-)-tri-o-thymotide [M-(-)-TOT] chlorocyclohexane clathrate established that the TOT host structure is of the cage type.[2,3] The clathration cavity is bounded by eight TOT molecules [Fig. 7.6-1(a)]: six form a cylindrical wall that is approximately parallel to the c axis, and the remaining two serve as the top and bottom lids, as shown in Fig. 7.6-1(c). The cage is located on a crystallographic two-fold axis, and there are three cages in each unit cell.

Two different guest conformations, axial-Cl chair and axial-Cl boat, are accommodated in the available cavities with an approximate occupancy ratio of 2:1 [Fig. 7.6-1(b)].

Remarks

1. Structural types of TOT clathrates[4,5]

TOT is quite a flexible chiral molecule. In solution, TOT exists primarily in a chiral, propeller-like conformation that undergoes rapid

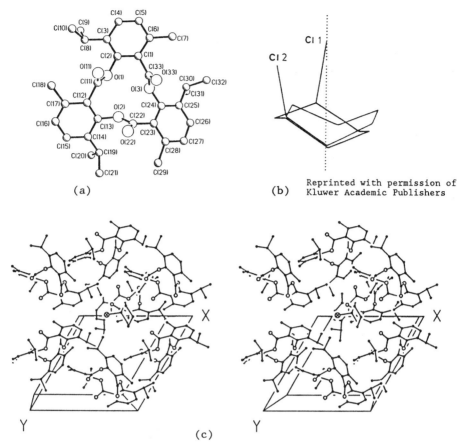

(a)   (b)   (c)

Fig. 7.6-1 (a) Tri-o-thymotide (TOT) molecular formula and atomic numbering. (b) Chlorocyclohexane molecular conformation. The dotted line denotes the crystallographic two-fold axis. (c) Stereoview, down the *z* axis, of the tri-o-thymotide molecules constituting the cage. For clarity the top host molecule and the minor guest conformer have been removed. The major guest conformer is shown single-positioned on a crystallographic two-fold axis. (Adapted from ref. 1).

interconversion between P (right-handed) and M (left-handed) forms, the activation energy for enantiomerization being 21 kcal mol^{-1} (half-life *ca.* 30 min. at 0°C). In its guest-free form, TOT crystallizes in the achiral space group *Pna*2_1.[6]

TOT clathrates crystallize in a variety of space groups [*P*$3_1$21 (No. 152), *P*6_1 (No. 169), and *P*1 (No. 2)] and display different modes of guest accommodation: cages, quasi-uniform channels, and channels of noticeably variable section. The *P*1 clathrate structure can be visualized as a distorted non-closest hexagonal packing of TOT molecules.[7] Table 7.6-1 lists various clathrates for which precise crystal structure data are available. Within each type, the lattice parameters may vary significantly (up to 0.35Å).

Table 7.6-1 TOT clathrates of known crystal·structure.

Space group	Guest (optically pure in initial solution)	Host/guest ratio
P3$_1$21	ethanol (achiral)	2:1
	pyridine (achiral)	2:1
	CH3CH2CHBrCH3 (racemic; 34% optically pure solution)	2:1
	trans-2,3-dimethyloxirane (optically pure)	2:1
	trans-2,3-dimethylthiirane (optically pure; racemic)	2:1
	CH3CH2SO(CH3) (racemic)	2:1
	chlorocyclohaxane (achiral)	2:1
P$\bar{1}$	trans-stilbene (achiral)	2:1
	cis-stilbene (achiral)	2:0.8
	benzene (achiral)	2:2.5
P6$_1$	cetyl alcohol (racemic)	6:1.3

The factors governing the "choice" of the crystal form include primarily the size of the guest and, to a lesser extent, the experimental conditions. It has been observed that guest molecules of length less than 9Å (including hydrocarbons, mono and dihalogenoalkanes, ethers, etc.) give rise to cage-type crystals, whereas those of greater length are accommodated in uniform-section channels. Larger molecules, e.g. *cis*- or *trans*-stilbene, induce crystallization in a form containing even larger cavities and channels of variable section.

An elegant illustration of the guest's influence on the crystal lattice was provided by Lawton and Powell[8]: within an isomorphous series of hexagonal channel clathrates, the length of the unit cell axis was shown to be correlated with the length of the included molecule.

2. Chiral discrimination

In its guest-free form, TOT is a racemic compound crystallizing in the achiral space group *Pna2*$_1$. But in the presence of solvent molecules of appropriate size, chiral clathrates are obtained, and each single crystal contains preferentially one of the guest enantiomers. The data in Table 7.6-2 represent the enantiomeric excess (e.e.) in crystals grown from a racemic solution of guest. Enantiomeric discrimination can be considerably enhanced when optically active solutions of guests are used; this results in amplification of guest optical purity after repeated crystallizations. For instance, after 3 cycles, the enantiomeric purity of 2-bromooctane reaches

Table 7.6-2 Enantiomeric excess of guest and correlation of guest and host chirality in P-(+)- TOT clathrate crystals.

Guest	Guest e.e.(%)	Guest configuration	Space group	Host/guest ratio
Cage-type clathrates				
2-chlorobutane	32 45	S-(+)	$P3_121$	2:1
2-bromobutane	34 35	S-(+)	$P3_121$	2:1
2-iodobutane	<1	-	$P3_121$	1:1
trans-2,3-dimethyloxirane	47	S,S-(-)	$P3_121$	2:1
trans-2,3-dimethylthiirane	30	S,S-(-)	$P3_121$	2:1
trans-2,4-dimethyloxetane	38		$P3_121$	2:1
trans-2,4-dimethylthietane	9		$P3_121$	2:1
propylene oxide	5	R-(+)	$P3_121$	2:1
2-methyltetrahydrofuran	2	S-(+)	$P3_121$	2:1
methyl methanesulphinate	14	R-(+)	$P3_121$	2:1
2,3,3-trimethyloxaziridine	7		$P3_121$	2:1
ethylmethylsulphoxide	83	S-(+)		
2-butanol	<5	S-(+)	$P3_121$	
2-aminobutane	<2	-		
Channel-type clathrate				
2-chlorooctane	4	S-(+)	$P6_1$	2.6:1
2-bromooctane	4	S-(+)	$P6_1$	2.6:1
3-bromooctane	4	S-(+)	$P6_1$	2.7:1
2-bromononane	5	S-(+)	$P3_1$	3:1
2-bromododecane	5	S-(+)	$P6_1$	3.8:1

80%, despite the very low enantiomeric purity (4%) resulting from the first crystallization.

The enantiomeric excess of guest depends on several factors:

a) Discrimination in channels is considerably weaker than in cages, presumably as a result of increased disorder.

b) Coincidence between cage symmetry (two-fold axis) and guest molecular symmetry could be expected to be a favorable factor.

c) The size of the guest is important.

d) Functional groups like -OH or $-NH_2$ do not seem to lead to a favorable effect.

The host/guest interactions are described in three levels in Table 7.6-3.

The chiral discrimination may be induced as follows:

a) In the presence of a racemic solvent, TOT undergoes spontaneous resolution and a conglomerate is obtained. Each single crystal of space group

Table 7.6-3 Host/guest interactions in the system: (guest)solution +
           (TOT)solution → clathrate.

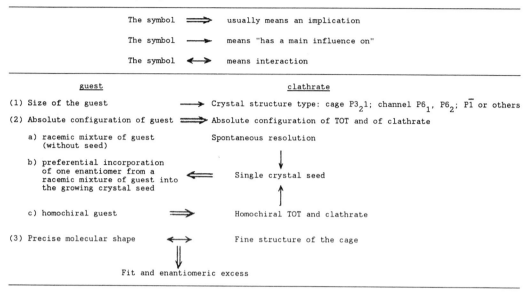

The symbol ⟹	usually means an implication
The symbol ⟶	means "has a main influence on"
The symbol ⟷	means interaction

guest | clathrate

(1) Size of the guest ⟶ Crystal structure type: cage $P3_21$; channel $P6_1$, $P6_2$; $P\bar{1}$ or others

(2) Absolute configuration of guest ⟹ Absolute configuration of TOT and of clathrate

   a) racemic mixture of guest (without seed) — Spontaneous resolution

   b) preferential incorporation of one enantiomer from a racemic mixture of guest into the growing crystal seed ⟸ Single crystal seed

   c) homochiral guest ⟹ Homochiral TOT and clathrate

(3) Precise molecular shape ⟷ Fine structure of the cage

Fit and enantiomeric excess

$P3_121$ contains TOT of the same chirality, but it includes both guest
enantiomers, the enantiomeric excess varying for different "host lattice/guest
molecule" combinations. Separation and identification of these crystals as
(+)- or (-)-TOT can be achieved by taking a chip from each crystal and then
measuring its rotation in solution at *ca.* 0°C.

   b) If a single crystal seed of given chirality is now introduced into
the solution of TOT in racemic guest, then a large single crystal of the same
chirality can be obtained by incorporating preferentially the same enantiomer
of guest and the same enantiomer of TOT.

   c) In the presence of optically-pure solvent, opitcally-pure TOT
clathrates are obtained: owing to the rapid P-M interconversion of TOT, the
guest chirality imposes itself on the whole clathrate, i.e. a given
configuration of guest always leads to single crystals or powder containing
the same configuration of TOT.

3. Determination of absolute configuration of TOT clathrate

   Powell first pointed out that TOT enclathration might be used to
establish the absolute configuration of guests. The absolute configuration of
TOT has been assigned by determing the configuration of TOT relative to guest,
for several guest molecules of known absolute configuration: (+)-TOT has a P-
propeller like configuration. Conversely, the absolute configuration of new
guest molecules can be determined by X-ray analysis of TOT clathrates. But

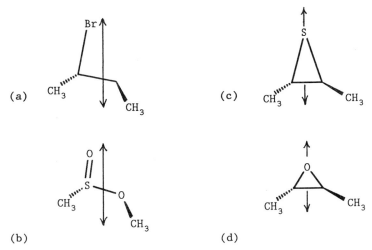

Fig. 7.6-2 Enantiomers of guest molecules (a) 2-bromobutane, (b) methyl methanesulfinate; (c) *trans*-2,3-dimethylthiirane, and (d) *trans*-2,3-dimethyloxirane, which are preferentially included in a clathrate crystal built of P-(+)-TOT molecules. The guests are shown in the orientations that they adopt relative to the two-fold axis of the cage cavity. (After ref. 5).

even without such an analysis a reasonable inference can be made about the absolute configuration of the preferred guest enantiomer. It has been noted that homologous atoms or groups of atoms in related molecules occupy similar positions in space, as shown in Fig. 7.6-2. The combined measurement of TOT and guest optical rotation would then lead to the determination of the absolute configuration of the guest.

Assignments of absolute configuration by means of TOT structures rely on the hypothesis that the correlation of the absolute configurations of the host and the preferred guest enantiomer is an intrinsic property of the complex and remains unchanged in every single crystal of a given clathrate.

4. Fine structural fit[4]

The characteristics of the cage are quite constant for several clathrates studied by X-ray analysis. The clathrate cavity is comprised of eight TOT molecules. The cage is located on a crystallographic two-fold axis, and there are three cages in each unit cell.

X-Ray analysis of several TOT clathrates has revealed some unexpected facts: a) the cage appears as a deformed ellipsoid and its capacity for enantiomeric discrimination is not obvious [Fig. 7.6-3]; b) enantiomeric disorder and crystal symmetry disorder are present; c) even for optically pure guest, two symmetry-related positions are located within the cage and are also disordered, as a result of the lack of spatial coincidence between the molecular and the crystallographic two-fold rotation axis.

Enantiomeric selectivity depends on several factors pertaining to the guest molecular properties relative to the host cage:

a) Molecular symmetry 2 seems to be a favourable factor, despite the disorder discussed above.

b) The size of the guest is an important factor.

c) The shape of the guest and its orientation in the cage apparently play an important role. It seems that there is a certain complementarity of guest and cavity shape as shown in Fig. 7.6-3. Moreover, small cage variations depending on the guest may be a way of "fine tuning" of the clathrate structure in order to achieve the best fit for the guest. Indeed, it seems that there is a correlation between the van der Waals volume of the guest, the volume of the van der Waals envelope of space accessible to the guest, and the volume of the unit cell.

d) Increased thermal movements of both guest atoms and TOT atoms at the cage internal surface result in the loss of chiral discrimination ability of the clathrate above a certain temperature.

5. Structure of tetra-1-naphthoide cyclohexanone 1:2 clathrate[9]

Tetra-1-naphthoide, a structural analogue of TOT, is able to enclathrate different guest molecules varying in shape and size from chloroform to naphthalene. All the clathrates of tetra-1-naphthoide which have been investigated so far are isomorphous and crystallize in space group $P\bar{1}$ with Z = 2. In the tetra-1-naphthoide cyclohexanone 1:2 clathrate crystal, the geometries of the host and guest molecules are shown in Fig. 7.6-4(a) and (b). The conformation of the host molecule differs considerably from that of guest-free tetra-1-naphthoide (space group $P4_2/n$, a = 13.142, c = 9.719Å and Z = 2)[10] which occupies a site of $\bar{4}$ symmetry. In the clathrate the dihedral angles subtended by the two pairs of opposite naphthalene moieties are 61.6 and 69.2°, as compared to the value 66.8° for the unsolvated host molecule. The packing of host molecules in the clathrate is characterized by channels running parallel to the a axis that accommodate two independent guest cyclohexanone molecules [Fig. 7.6-4(c)]. Guest molecules A are located close to the center of symmetry at (0,0,0). Molecules B are related through the center of symmetry at (1/2,0,0).

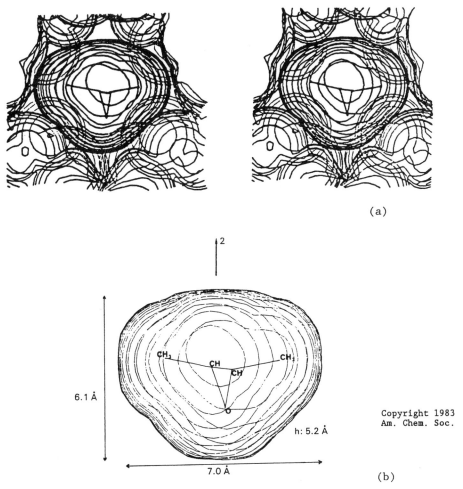

Fig. 7.6-3 (a) Stereoview of the contours of the van der Waals envelope of
TOT (light lines) and of the van der Waals envelope of the volume accessible
to guest atoms (heavy lines), viewed along the *c* axis. Sections are plotted
every 0.3Å, parallel to the *ab* plane, in a cube having 9.2Å sides and centered
on (0,1/3,1/6). The position of (+)-(*R,R*)-2,3-dimethyloxirane is shown. (b)
Dimensions of the van der Waals envelope of the space available to the guest;
sections and guest are drawn as in (a). (After ref. 2).

6. The trianthranilides

The trianthranilides, which are formally derived from TOT and related
trisalicylides by replacing the bridging ester O atoms by NR groups, comprise
a class of potential organic hosts.[11] *N,N'*-Dibenzyl and *N,N',N"*-
tribenzyltrianthranilide form 1:1 inclusion compounds with toluene and
ethanol, respectively. *N,N'*-Dimethyl-*N"*-benzyltri-3-methyltrianthranilide
(Fig. 7.6-5) shares with tri-*o*-thymotide the ability to undergo spontaneous

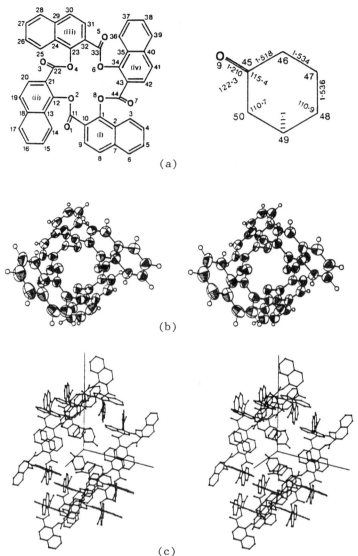

Fig. 7.6-4 (a) Atom-numbering scheme of tetra-1-naphthoide and "ideal" geometry (*MM2* force-field) of cyclohexanone. (b) Stereoview of the host molecule. (c) Stereoview of the channel along the *a* axis in the clathrate. (After ref. 9).

resolution, and X-ray analysis of its 1:1 channel inclusion compound with toluene showed that the host molecules adopt a helical conformation.[12] However, some important distinctions emerge in comparing the properties and assessing the potential of these two 12-membered ring compounds. Firstly, TOT crystallizes in chiral propeller conformations whereas *N,N'*-dimethyl-*N"*-benzyl-tri-3-methyltrianthranilide crystallizes in chiral helical conformations. Secondly, by changing the nature of the substituents on the nitrogen atoms during stepwise syntheses, a wide range of trianthranilide hosts incorporating desired functionalities becomes accessible for the entrapment of guest substrates in the solid state.

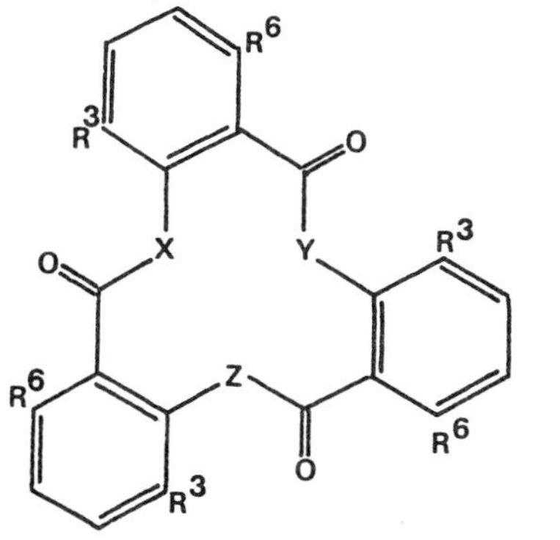

X = Y = NMe, Z = NCH$_2$Ph
R^3 = Me, R^6 = H

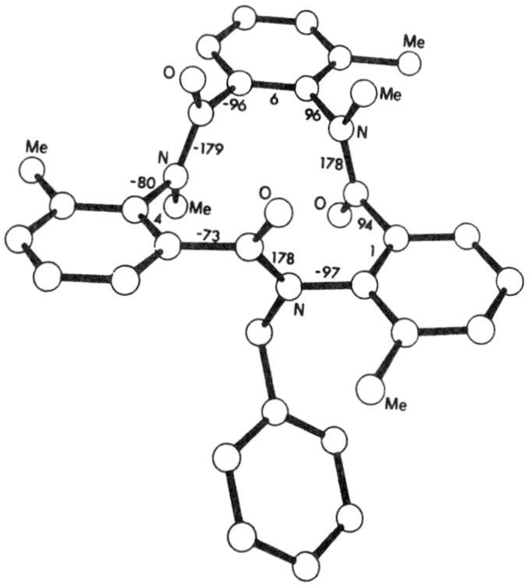

Fig. 7.6-5 Formula and molecular conformation of *N,N'*-dimethyl-*N"*-benzyltri-3-methyltrianthanilide; torsion angles shown in degrees. (After ref. 11).

References

[1]   R. Gerdil and A. Frew, *J. Incl. Phenom.* **3**, 335 (1985).

[2]   R. Arad-Yellin, B.S. Green, M. Knossow and G. Tsoucaris, *J. Am. Chem. Soc.* **105**, 4561 (1983).

[3]   S. Brunie, A. Navaza, G. Tsoucaris, J.P. Declercq and G. Germain, *Acta Crystallogr., Sect. B* **33**, 2645 (1977).

[4]   G. Tsoucaris in G.R. Desiraju (ed.), *Organic Solid State Chemistry*, Elsevier, Amsterdam, 1987, chap. 7.

[5]   R. Arad-Yellin, B.S. Green, M. Knossow, N. Rysanek and G. Tsoucaris, *J. Incl. Phenom.* **3**, 317 (1985).

[6]   D.J. Williams and D. Lawton, *Tetrahedron Lett.*, 111 (1975).

[7]   J. Allemand and R. Gerdil, *Acta Crystallogr., Sect. C* **39**, 260 (1983).

[8]   D. Lawton and H.M. Powell, *J. Chem. Soc.*, 2339 (1958).

[9]   K. Suwinska and R. Gerdil, *Acta Crystallogr., Sect. C* **43**, 898 (1987).

[10] R. Gerdil and G. Bernardinelli, *Acta Crystallogr., Sect. C* **41**, 1523 (1985).

[11] W.D. Ollis and J.F. Stoddart in J.L. Atwood, J.E.D. Davies and D.D. MacNicol, *Inclusion Compounds*, Vol. 2, Academic Press, London, 1984, chap. 6.

[12] S.J. Edge, W.D. Ollis, J.S. Stephanatou, J.F. Stoddart, D.J. Williams and K.A. Woode, *Tetrahedron Lett.* **22**, 2229 (1981).

7.7   trans-Diamminemanganese Catena-tetra-μ-cyano-nickelate-Benzene (1/2)(Hofmann-type Clathrate)

$$\mathbf{Mn(NH_3)_2Ni(CN)_4 \cdot 2C_6H_6}$$

Crystal Data

Tetragonal, space group $P4/m$ (No. 83)

$a = 7.432(6)$, $c = 8.335(5)$Å, $Z = 1$

Atom	x	y	z	Occupancy
Mn	0	0	0	1
Ni	1/2	1/2	0	1
N(1)	.2135	.2133	0	1
N(2)	0	0	.2725	1
C(1)	.3223	.3215	0	1
C(2)	1/2	0	.3331	1
C(3)A	.5670	.1474	.4158	0.4
C(3)B	.4353	.1523	.4120	0.6

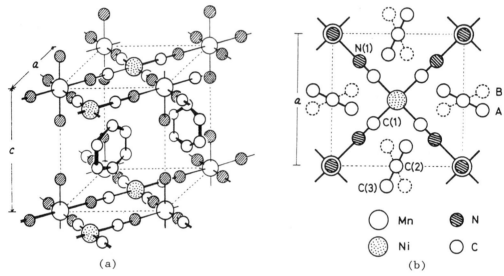

(a)   (b)

Fig. 7.7-1   (a) Crystal structure of $Mn(NH_3)_2Ni(CN)_4 \cdot 2C_6H_6$, and (b) a projection of the structure along the c axis with numbering of the atoms. (After ref. 1).

Crystal Structure

The host framework is isostructural with other Hofmann-type clathrates. The manganese atom is coordinated octahedrally by six nitrogen atoms, two from a pair of ammonia molecules in *trans* positions and four from adjacent cyano groups; the Mn-N distances are 2.24(1) and 2.27(1)Å, respectively. The nickel atom is surrounded by four carbon atoms of the cyano groups in square-planar coordination at Ni-C = 1.87(1)Å, thus constituting a tetracyanonickelate(II) moiety. The Mn and Ni atoms are thus bridged by cyano groups to form an infinite two-dimensional network, with the amine ligands protruding on both

sides. These layers are stacked parallel to each other along the c axis.
Benzene molecules are enclosed between the layers with their molecular planes
parallel to the c axis. Each benzene molecule has two orientations denoted as
A and B [Fig. 7.6-1(b)], which are mirror images of each other with respect to
(100). The population parameters for these two orientations were refined to
0.4 and 0.6, respectively.

Since the host framework possesses *P4/mmm* symmetry, the interaction
potential between a guest molecule and the host framework is equivalent in the
A and B orientations. Occurrence of the two orientations in one layer is
forbidden by intermolecular repulsion among them. On the other hand,
interactions bewteen guest molecules in adjacent layers are relatively weak,
so that within each layer the benzene molecules can take either orientation A
or B.

Remarks

1. Hofmann's clathrate and historical development[2]

The prototype clathrate, $Ni(NH_3)_2Ni(CN)_4.2C_6H_6$, was obtained in 1897 in
Hofmann's laboratory.[3] About a half century later, Powell and Raynor
determined its crystal structure,[4] which is isostructural with the Mn
complex shown in Fig. 7.7-1. There are thus two kinds of Ni(II) atom in the
host metal complex: one exhibits square-planar four-coordination in a
tetracyanonickelate(II) moiety, and the other involves an octahedral array of
nitrogen atoms about the metal.

Based on the crystal structure of $Ni(NH_3)_2Ni(CN)_4.2C_6H_6$, Schwarzenbach
and co-workers[5] and Iwamoto and co-workers[6] prepared a series of "Hofmann-
type" clathrates of the general formula $M(NH_3)_2M'(CN)_4.2G$ for M = Mn, Fe, Co,
Ni, Cu, Zn or Cd; M' = Ni, Pd, or Pt; and G = C_4H_5N, C_4H_4S, C_6H_6 and $C_6H_5NH_2$.

Several types of clathrates analogous to the Hofmann-type have been
developed by appropriate replacement of host moieties. The basic strategy of
host modification may be classified in three ways: a) replacement of a pair
of amine ligands confronting each other between adjacent sheets by an
ambidentate ligand such as ethylenediamine, triethylenediamine, etc.; b)
replacement of square-planar tetracyanometallate(II) by tetrahedral
tetracyanometallate(II); and c) introduction of bulky substituents into the
amine or diamine ligand. Fig. 7.7-2 shows some building blocks as
substituents for Hofmann-type clathrates.

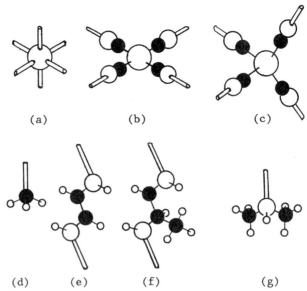

Fig. 7.7-2 "Building blocks" of Hofmann-type host structures. (a) Six-coordinate metal M, (b) square-planar $M'(CN)_4$, (c) tetrahedral $M'(CN)_4$, (d) NH_3, (e) $NH_2CH_2CH_2NH_2$ behaving as an ambidentate ligand, (f) $NH_2CH(CH_3)CH_2NH_2$ behaving as an ambidentate ligand, and (g) $NH(CH_3)_2$. (After ref. 2).

2.   Structural features of Hofmann-type inclusion compounds

The Hofmann type inclusion compounds, $M(NH_3)_2M'(CN)_4.2G$, are all tetragonal and their host complexes are isostructural to one another (Table 7.7-1).

Table 7.7-1   Structural data in a series of Hofmann-type clathrates, $M(NH_3)_2Ni(CN)_4.2G$.[2,15]

M	Mn	Ni	Ni	Cu	Cd	Cd
G	C_6H_6	C_6H_6	$C_{12}H_{10}$	C_6H_6	C_6H_6	$C_4H_8O_2$
Space group	P4/m	P4/m	I422	P4/mcc	P4/m	P4/m
a (Å)	7.432(6)	7.242(7)	7.240(3)	7.345(3)	7.542(2)	7.586(9)
c (Å)	8.335(5)	8.227(8)	25.301(1)	16.519(4)	8.308(4)	8.082(6)
Z	1	1	2	2	1	1
M-N(NH$_3$) (Å)	2.27(1)	2.08(6)	2.10(1)	2.05(2)	2.325(8)	2.32
M-N(CN) (Å)	2.24(1)	2.15(14)	2.095(5)	2.20(1)	2.334(4)	2.35
Ni-C (Å)	1.87(1)	1.78(17)	1.890(6)	1.88(1)	1.859(4)	1.81
C-N (Å)	1.14(1)	1.20(22)	1.135(8)	1.12(2)	1.140(6)	1.18
θ_{expt}*(°)	65.5° 66.9°	65.9°	71°	64.3°	66.5°	43.7°
θ_{calc}(°)	68	66		66	70	

*θ stands for the dihedral angle between the benzene ring and the plane (100).

In crystalline $Cu(NH_3)_2Ni(CN)_4 \cdot 2C_6H_6$ (space group $P4/mcc$ with a = 7.345 and c = 16.519Å),[7] the Cu(II) atom is coordinated by two NH_3 molecules and four nitrogen atoms of the cyanide anions (Fig. 7.7-3). The Cu-N distances are 2.05Å for NH_3 and 2.20Å for CN^-. The compressed tetragonal configuration of the ligands about Cu(II) appears to result from steric repulsion between the host framework and the guest molecule, since axial elongation is observed in the residual host framework, $Cu(NH_3)_2Ni(CN)_4$, that remains after the complete escape of the benzene guest component from the clathrate. The benzene molecules in the clathrate are arranged about the four-fold axis, the direction of inclination being alternately reversed in adjacent interlayer regions.

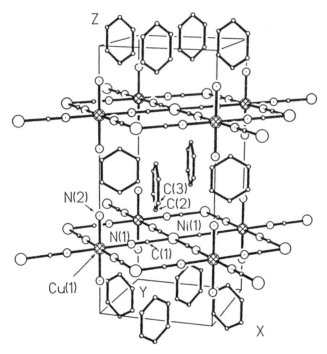

Fig. 7.7-3 Crystal structure of $Cu(NH_3)_2Ni(CN)_4 \cdot 2C_6H_6$. [The atoms listed in Table II, p. 99 of ref. 7 are in the order: Cu(1), Ni(1), C(1), N(1), N(2), C(2) and C(3).]

In the related $Ni(NH_3)_2Ni(CN)_4 \cdot C_{12}H_{10}$ crystal (space group $I422$ with a = 7.240 and c = 25.301Å),[8] shifting alternate layers of the host network by $(a + b)/2$ gives a body-centered unit cell (Fig. 7.7-4). Each of the phenyl rings of the biphenyl molecule is clamped by a pair of NH_3 molecules from the upper or lower sheet as in the case of the the benzene guest molecule.

The revised crystal structure of $Cd(NH_3)_2Ni(CN)_4 \cdot 2C_6H_6$ is shown in Fig. 7.7-5.[15] Molecular mechanics calculations were carried out to interpret the observed orientational angle (θ) of the benzene molecule enclathrated in this

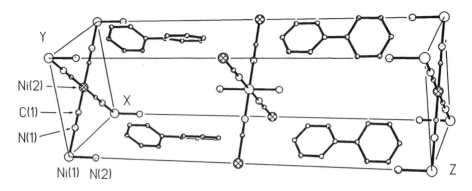

Fig. 7.7-4 Crystal structure of $Ni(NH_3)_2Ni(CN)_4 \cdot 2C_{12}H_{10}$.

inclusion compound and others. This angle is most influenced by the guest-to-guest contact in the interlayer space between the two-dimensional *catena*[metal(II) tetra-μ-cyanonickelate(II)] networks for the Hofmann-type series. The largest discrepancy between the calculated and the observed angles in each crystal structure does not exceed 3.5° (Table 7.7-1).

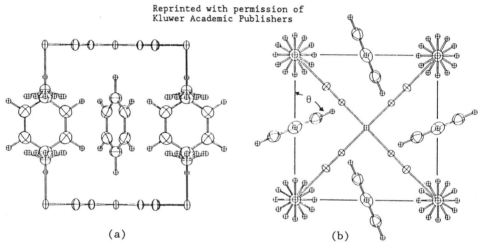

(a)   (b)

Fig. 7.7-5 Crystal structure of $Cd(NH_3)_2Ni(CN)_4 \cdot 2C_6H_6$. (a) Projection along the *a* axis; (b) projection along the *c* axis. Hydrogen atoms of the benzene molecule are located at the calculated positions with the C-H distance of 1.08Å; those of the ammine ligand, based on the structure of the ammonia molecule (N-H 1.011Å; H-N-H 106.7°), are distributed about the fourfold *c* axis. (After ref. 15).

3. Crystal structure of $Ni(H_2O)_2Ni(CN)_4 \cdot nH_2O$[9]

The compositions and structures of several hydrates of nickel cyanide have been determined. The structures of three characterized orthorhombic phases of $Ni(H_2O)_2Ni(CN)_4 \cdot nH_2O$ for $1 < n < 3$ (I), $3 < n$ (II), and $0 < n < 2$ (III) are shown in Fig. 7.7-6. In the metal cyanide network each octahedral

Ni(II) center is coordinated by two H_2O molecules. The excess water molecules are placed in cavities similar to those in $Ni(NH_3)_2Ni(CN)_4 \cdot 1.5H_2O$. Two phases of the hydrates, cubic $Ni(CN)_2 \cdot 2H_2O$ (a = 10.10Å) and orthorhombic $Ni(CN)_2 \cdot 1.5H_2O$ are the same as I and III, respectively.

4. Structure of modifications of Hofmann-type inclusion compounds[2]

Several series of inclusion compounds with the general formula $M(diam)M'(CN)_4 \cdot nG$ have been developed from Hofmann-type clathrates by appropriate selection of the *diam* (diammine, diamine, substituted amine, bis-amine) and $M'(CN)_4$ moieties. These are shown as building blocks with connectors in Fig. 7.7-2. A three-dimensional host structure is derived from the two-dimensional Hofmann-type host by replacing a pair of opposed NH_3 ligands between adjacent sheets by an ambidentate ligand such as ethylenediamine. Another three-dimensional host structure is derived by replacing the square-planar $M'(CN)_4$ moiety by a tetrahedral one such as $Cd(CN)_4$ or $Hg(CN)_4$; the structure can also be reinforced by bridges involving ambidentiate ligands. When a unidentate or bidentate amine with bulky substituents is introduced into the host structure, the substituents occupy the cavity partly or wholly in place of a guest molecule. Thus the numbers of guest molecules in the general formula may vary stepwise from 2, 3/2, 1 to 1/2 according to the number of cavities (0, 1, 2 and 3) occupied by substituents.

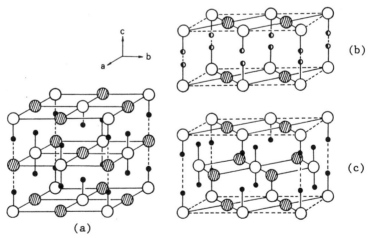

Fig. 7.7-6 Structures of $Ni(H_2O)_2Ni(CN)_4 \cdot nH_2O$. (a) Phase I with $1 < n < 3$, a = 10.15, b = 10.09, and c = 10.12Å; (b) phase II with $3 < n$, a = 7.09, b = 13.84, and c = 6.10Å; and (c) phase III with $0 < n < 2$, a = 7.26, b = 14.20, and c = 8.88Å. Solid circle, square-planar Ni; open circle, six-coordinate Ni; small open circle, coordinated water molecule. The cyanide linkages between the two kinds of Ni atoms are indicated by thick lines. (After ref. 9).

Some examples are discussed below.

    a) Cd(en)Cd(CN)$_4$.2C$_6$H$_6$

    This clathrate crystallizes in space group $P4_222$, with a = 8.265(1), c = 15.512(3)Å, and Z = 2.[10] In the crystal, the Cd(CN)$_4$ moiety is tetrahedral, and the en ligands are disordered in orientation. Two independent guest benzene molecules are centered in the α-cavity (0, 1/2, 0) and the β-cavity (0, 0, 1/4); the former is ordered, and in the latter a pair of benzene skeletons are distributed statistically. Fig. 7.7-7 shows the structure of this inclusion compound.

    b) Cd(NH$_2$CH$_2$CH$_2$OH)$_2$Ni(CN)$_4$.C$_4$H$_5$N

    This clathrate crystallized in space group $Pna2_1$, with a = 14.691(1), b = 15.881(1), c = 7.575(1)Å, and Z = 4.[11] The mea (NH$_2$CH$_2$CH$_2$OH) molecule behaves as a unidentate ligand similar to NH$_3$ in the Hofmann-type host network. As shown in Fig. 7.7-8, there are two kinds of mea molecules in the host structure. The hydroxyethyl group of one mea invades a cavity in the place of a pyrrole guest molecule; the other type of mea ligands are bridging bidentate, forming columns which partition the cavities occupied by the guests and by the tails of monodentate mea ligands. The metal complex network is considerably distorted from the coplanar structure of the Hofmann-type host and is shifted alternately in such a way that the six-coordinate Cd is located approximately above and below the four-coordinate Ni along the b axis of the

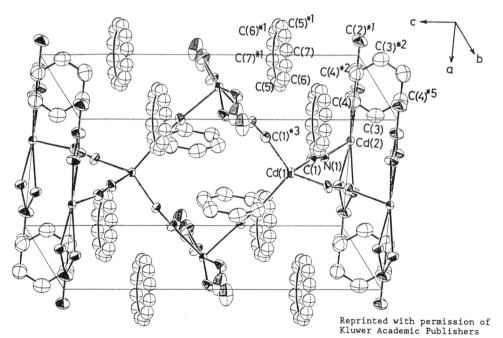

Fig. 7.7-7 Structure of Cd(en)Cd(CN)$_4$.2C$_6$H$_6$. (After ref. 10).

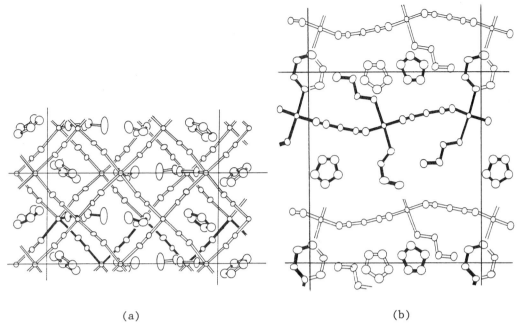

(a)   (b)

Fig. 7.7-8 Structure of $Cd(NH_2CH_2CH_2OH)_2Ni(CN)_4 \cdot C_4H_5N$. (a) Projection along the b axis and (b) along the c axis. Some atoms in the unit cell are not shown. (After ref. 11).

orthorhombic unit cell (which corresponds to the c axis of the Hofmann-type host).

   c) $Cd(NH_2(CH_2)_6NH_2)Ni(CN)_4 \cdot p\text{-}CH_3C_6H_4NH_2$

   The clathrate crystallizes in space group $P2/m$, with $a = 9.540(2)$, $b = 7.611(1)$, $c = 7.120(1)$Å, $\beta = 100.95(1)°$, and $Z = 1$.[12] In the host structure, the 1,6-diaminohexane ligand (dahxn) spans adjacent two-dimensional wavy cyanometal complex networks in an all-*trans* conformation of the aliphatic chain, which has normal bond distances and angles. The coordination structure of the $Ni(CN)_4$ moiety is little distorted from D_{4h} symmetry, and the Cd atom is in a compressed octahedral coordination with six nitrogen neighbours [Fig. 7.7-9(a)].

   The guest molecule is accommodated in a cavity roofed and floored by the networks of cyanometal complex, and columned and walled by the dehxn ligands. The openings of the cavity at both sides along the c axis are small enough to hinder the guest from moving to the next for m- and o-toluidine, but not for p-toluidine, as illustrated in Fig. 7.7-9(b).

   The crystal structures of $Cd[NH_2(CH_2)_4NH_2]Ni(CN)_4 \cdot (CH_3)_2C_6H_3NH_2$[13] and $Cd[NH_2(CH_2)_6NH_2]Ni(CN)_4 \cdot o\text{-}CH_3C_6H_4NH_2$[14] have also been determined.

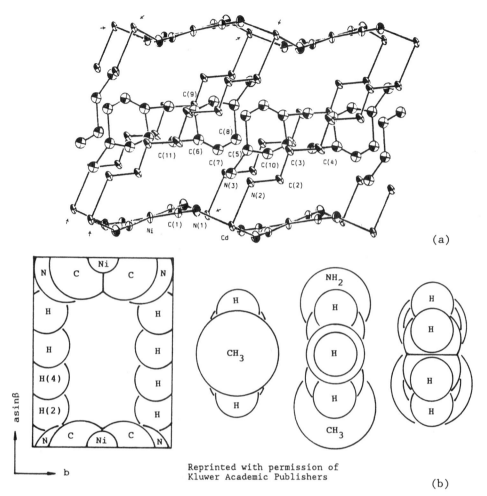

Reprinted with permission of
Kluwer Academic Publishers

(b)

Fig. 7.7-9 (a) ORTEP view of host structure of
$Cd[NH_2(CH_2)_6NH_2]Ni(CN)_4 \cdot p\text{-}CH_3C_6H_4NH_2$; 30% thermal ellipsoids are shown
for the host atoms, and hydrogen atoms are omitted. Each unit cell is
cornered by the Cd atom at the origin and seven other Cd atoms indicated
by arrows. (b) View of the opening of the cavity, and the projections
of *p*-, *m*-, and *o*-toludine molecules pictured with van der Waals radii: C
and N of CN group, 1.7Å; C, 1.5Å; H, 1.1Å; CH_3, 2.0Å; and NH_2, 1.8Å.
(After ref. 12).

References

[1] R. Kuroda and Y. Sasaki, *Acta Crystallogr., Sect. B* **30**, 687 (1974).

[2] T. Iwamoto in J.L. Atwood, J.E.D. Davies and D.D. MacNicol (eds.),
     Inclusion Compounds, Vol. 1, Academic Press, London, 1984, chap. 2.

[3] K.A. Hofmann and F.A. Küspert, *Z. Anorg. Allg. Chem.* **15**, 204 (1897).

[4] H.M. Powell and J.H. Raynor, *Nature (London)* **163**, 566 (1949); J.H. Raynor and H.M. Powell; *J. Chem. Soc.*, 319 (1952).

[5] R. Baur and G. Schwarzenbach, *Helv. Chim. Acta* **43**, 842 (1960).

[6] T. Iwamoto, T. Miyoshi, T. Miyamoto, Y. Sasaki and S. Fujiwara, *Bull. Chem. Soc. Jpn.* **40**, 1174 (1967).

[7] T. Miyoshi, T. Iwamoto and Y. Sasaki, *Inorg. Chim. Acta* **7**, 97 (1973).

[8] T. Iwamoto, T. Miyoshi and Y. Sasaki, *Acta Crystallogr.*, *Sect. B* **30**, 292 (1974).

[9] Y. Mathey and C. Mazieres, *Can. J. Chem.* **52**, 3637 (1974).

[10] S. Nishikiori and T. Iwamoto, *J. Incl. Phenom.* **3**, 283 (1985).

[11] S. Nishikiori and T. Iwamoto, *Chem. Lett.*, 1775 (1981).

[12] T. Hasegawa and T. Iwamoto, *J. Incl. Phenom.* **6**, 143 (1988).

[13] S. Nishikiori and T. Iwamoto, *J. Incl. Phenom.* **2**, 341 (1984).

[14] T. Hasegawa, S. Nishikiori and T. Iwamoto, *J. Incl. Phenom.* **2**, 351 (1984).

[15] S. Nishikiori, T. Kitazawa, R. Kuroda and T. Iwamoto, *J. Incl. Phenom.* **7**, 369 (1989).

Note Recent results from x-ray studies of Hofmann-type and analogous inclusion compounds are presented by I. Iwamoto in J.L. Atwood, J.E.D. Davies and D.D. MacNicol (eds.), *Inclusion Compounds*, Vol. 5, Oxford University Press, New York, 1991, chap. 6; S.-I. Nishikiori, T. Hasegawa and T. Iwamoto, *J. Incl. Phenom.* **11**, 137 (1991); M. Hashimoto, T. Kitazawa, T. Hasegawa and T. Iwamoto, *ibid.*, p. 153.

The host framework of *catena*-[(1,2-diaminopropane)cadmium(II) tetra-μ-cyanonickelate(II)] forms urea- and thiouea-like inclusion compounds of general formula $Cd(pn)Ni(CN)_4 \cdot xG$ ($x = 0.5$) with straight-chain [e.g. C_5H_{12}, C_6H_{14}, C_7H_{16} and $(C_2H_5)_2O$] and branched-chain [e.g. $Cl_2CHCHCl_2$, $CH_3CHClC_2H_5$, $ClCH_2CHClCH_3$ and $CH_3CHOHCH_3$] aliphatic guests, respectively. See K.-M. Park and T. Iwamoto, *J. Chem. Soc., Chem. Commun.*, 72 (1992).

The crystal structure of isomorphous $Zn(CN)_2$ and $Cd(CN)_2$ consists of two interpenetrating diamond-related frameworks of 3D-linked M-C≡N-M "molecular rods". $[NMe_4][Cu^I Zn^{II}(CN)_4]$ has a single diamond-related anionic framework with the oganic cations occupying half the adamantane-like cavities. $[Cu^I(4'4''4'''4''''$-tetracyanotetraphenylmethane)]BF_4 \cdot xC_6H_5NO_2$ is the first example of an engineered scaffolding-like 3D host structure based on $C-C_6H_4-C≡N-Cu$ rods that accommodates disordered $C_6H_5NO_2$ and BF_4^- guests. See B.F. Hoskins and R. Robson, *J. Am. Chem. Soc.* **112**, 1546 (1990).

7.8　(α-Cyclodextrin)$_2 \cdot$ Cd$_{0.5} \cdot$ I$_5 \cdot$ 27H$_2$O

(C$_{36}$H$_{60}$O$_{30}$)$_2 \cdot$ Cd$_{0.5} \cdot$ I$_5 \cdot$ 27H$_2$O

Crystal Data

Tetragonal, space group $P4_2 2_1 2$ (No. 94)

$a = 19.93$, $c = 30.88$Å, $Z = 4$

Atom	x	y	z	Atom	x	y	z
I(1)	.0000	.5000	.0876	C(2)2	-.2825	.5482	.2069
I(2)	.0000	.5000	.1835	C(3)2	-.2334	.4877	.2135
I(3)	.0283	.4847	.2832	C(4)2	-.2346	.4452	.1719
I(4)	.0000	.5000	.3840	C(5)2	-.2116	.4897	.1307
I(5)	.0000	.5000	.4801	C(6)2	-.2147	.4561	.0893
Cd	.5000	-.5000	.1248	C(1)3	-.0544	.7606	.1651
W(1)	.4655	.4405	.0624	C(2)3	-.1043	.7671	.2043
W(2)	.4643	.4336	.1828	C(3)3	-.1376	.7007	.2103
W(3)	.4027	.5699	.1173	C(4)3	-.1696	.6814	.1669
W(4)	.4556	-.3574	.0066	C(5)3	-.1183	.6748	.1327
W(5)	.3862	-.3998	.2086	C(6)3	-.1439	.6593	.0848
W(6)	.3438	.4205	.0429	O'(2)1	.0013	.8011	.3305
Cd*	.0000	.0000	.1169	O'(3)1	.1409	.7497	.3308
W(1)*	.0883	.0070	.1615	O'(4)1	.1573	.6502	.3930
W(2)*	.0840	.8888	.1108	O'(5)1	.0294	.7557	.4418
W(3)*	.0578	.0880	.0845	O'(6)1	.1452	.7445	.4955
W(4)*	.1196	.1848	.0362	O'(2)2	-.2519	.6434	.3292
W(5)*	.0950	-.0250	.0378	O'(3)2	-.1393	.7357	.3246
W(6)*	.1357	-.1352	.0000	O'(4)2	-.0447	.7118	.3901
W(7)*	.0957	-.1103	.2283	O'(5)2	-.2013	.6516	.4422
W(8)*	.2111	-.2568	.0396	O'(6)2	-.1274	.7460	.4942
W(9)*	.4018	-.4178	.2527	O'(2)3	-.2581	.3495	.3301
O(2)1	-.2099	.3099	.2453	O'(3)3	-.2757	.4950	.3286
O(3)1	-.0782	.2458	.2453	O'(4)3	-.2031	.5616	.3956
O(4)1	.0065	.2881	.1776	O'(5)3	-.2320	.3983	.4425
O(5)1	-.1631	.3100	.1306	O'(6)3	-.2716	.5021	.4980
O(6)1	-.0799	.2393	.0755	C'(1)1	-.0097	.7717	.4070
O(2)2	-.2787	.5909	.2443	C'(2)1	.0429	.7893	.3683
O(3)2	-.2618	.4445	.2459	C'(3)1	.0892	.7282	.3617
O(4)2	-.1821	.3973	.1788	C'(4)1	.1242	.7107	.4030
O(5)2	-.2585	.5480	.1287	C'(5)1	.0703	.6981	.4418
O(6)2	-.2789	.4280	.0820	C'(6)1	.1043	.6880	.4822
O(2)3	-.0676	.7845	.2432	C'(1)2	-.2300	.6236	.4043
O(3)3	-.1904	.7088	.2421	C'(2)2	-.2220	.6702	.3661
O(4)3	-.1920	.6099	.1751	C'(3)2	-.1450	.6809	.3590
O(5)3	-.0879	.7406	.1281	C'(4)2	-.1124	.7128	.3995
O(6)3	-.1918	.7058	.0711	C'(5)2	-.1269	.6588	.4380
C(1)1	-.1967	.3305	.1679	C'(6)2	-.1010	.6828	.4800
C(2)1	-.1743	.2871	.2065	C'(1)3	-.2281	.3577	.4075
C(3)1	-.1003	.2972	.2142	C'(2)3	-.2643	.3896	.3679
C(4)1	-.0669	.2717	.1685	C'(3)3	-.2347	.4598	.3597
C(5)1	-.0899	.3182	.1358	C'(4)3	-.2397	.5014	.4028
C(6)1	-.0649	.3028	.0872	C'(5)3	-.2009	.4625	.4386
C(1)2	-.2588	.5913	.1684	C'(6)3	-.2062	.4978	.4828

Atomic sites are fully occupied except for I(3), 0.5; W(5), 0.5; W(6)*, 0.5; W(9)*, 0.89; Cd, 0.18; Cd*, 0.07. Primed O and C atoms belong to α-CD molecule A; unprimed atoms to molecule B.

Crystal Structure

　　In this dark brown crystalline compound, the α-cyclodextrin (α-CD) molecules are in a head-to-head arrangement forming dimers. Adjacent α-CD molecules are rotated relative to each other by 13° such that their glucose moieties are staggered. The α-CD stacks and the resulting channel containing the polyiodide chain are strictly linear as required by the 4_2 symmetry axis [Fig. 7.8-1(a)]. The stacks in this tetragonal complex are arranged in a square array leaving large interstitial spaces to be filled by water molecules and Cd^{2+} cations, the latter occupying special positions on two-fold axes.

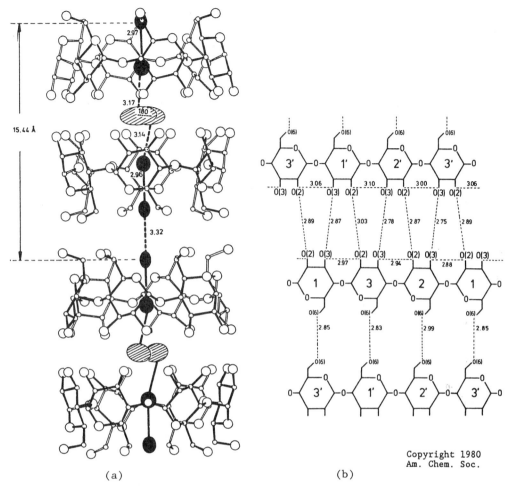

(a)   (b)

Fig. 7.8-1 Structure of $(\alpha\text{-CD})_2 \cdot \text{Cd}_{0.5} \cdot \text{I}_5 \cdot 27\text{H}_2\text{O}$. (a) Side view of the $(\text{I}_5^-)_\infty$ chain imbedded into the α-CD matrix. Disordered iodine sites are indicated by hatching. (b) Schematic representation of the hydrogen bonding between α-CD molecules. Additional intermolecular hydrogen bonds involving the water molecules are not indicated. (After ref. 1).

Of the five iodine atoms accommodated within one α-CD dimer, four lie on the 4_2 axis and the fifth is disordered over two sites that are only 0.9Å away from this axis. The repeat distance along the chain is $c/2 = 15.44$Å, and the bond distances (Å) are as follows:

$$\text{I(5)} \xrightarrow{3.32} \text{I(1)} \xrightarrow{2.96} \text{I(2)} \overset{3.14 \quad \text{--I(3)A} \sim 3.17}{\underset{3.14 \quad \text{I(3)B} - 3.17}{\rightrightarrows}} \text{I(4)} \xrightarrow{2.97} \text{I(5)} \text{ ---}$$

The polyiodide chain is interpreted as being composed of nearly linear I_5^-

units (with the disordered iodine as the central atom) that interact strongly
through I...I charge transfer.

The 13° rotation of opposing α-CD molecules allows the hydroxyl groups
to interlock by forming hydrogen bonds O(2)...O'(2) and O(3)...O'(3) [Fig.
7.8-1(b)]. From the O(6) side, opposing α-CD molecules give direct
O(6)...O'(6) hydrogen bonds. Hydrogen bonds continue from the O(6) side to
water molecules W(2)*, W(3), W(4), W(5)*, W(6) and W(8)*, and from the O(2),
O(3) side to W(1), W(2), W(5), W(7) and W(9).

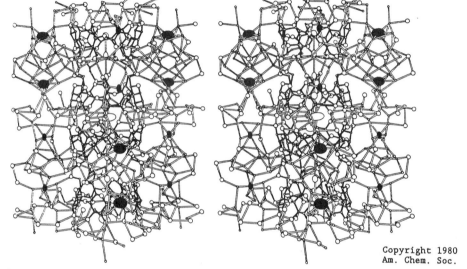

Fig. 7.8-2 Stereoview of the crystal structure of $(\alpha\text{-CD})_2 \cdot \text{Cd}_{0.5} \cdot \text{I}_5 \cdot 27\text{H}_2\text{O}$.
The iodine and cadmium atoms are indicated by unfilled and blackened
probability ellipsoids, respectively. (After ref. 1).

The two fractionally occupied cation sites Cd and Cd* are located on the
twofold axis parallel to c. The site Cd is octahedrally coordinated by water
molecules W(1), W(2), and W(3) while Cd* is surrounded irregularly by six
waters W(1)*, W(2)*, and W(3)*. A stereoview of the crystal structure of the
$(\alpha\text{-CD})_2 \cdot \text{Cd}_{0.5} \cdot \text{I}_5 \cdot 27\text{H}_2\text{O}$ complex is shown in Fig. 7.8-2.

Remarks

1. Structural features of cyclodextrin molecules[2,3]

Amylose can be degraded by glucosyl transferases into cyclic
oligosaccharides consisting of six, seven, or eight glucose units, called α-,
β-, and γ-cyclodextrin (α-, β-, and γ-CD), respectively. The three forms have
different properties as summarized in Table 7.8-1. In Fig. 7.8-3 and Fig.
7.8-4, the chemical structure of α-cyclodextrin and the overall molecular

shapes of α-, β- and γ-cyclodextrin are presented, respectively.

Table 7.8-1   Some properties of cyclodextrins.

Cyclodextrin	Number of glucose units	Diameter of cavity (Å)	Diameter of outer periphery (Å)	Height (Å)	Guest accommodated molecules
α	6	4.7-5.2	14.6±0.4	7.9-8.0	benzene, phenol, p-nitrophenol, iodides, benzaldehyde, etc.
β	7	6.0-6.4	15.4±0.4	7.9-8.0	benzophenone, diphenyl, naphthalene, etc.
γ	8	7.5-8.3	17.5±0.4	7.9-8.0	anthracene and analogues, vitamin K3, etc.

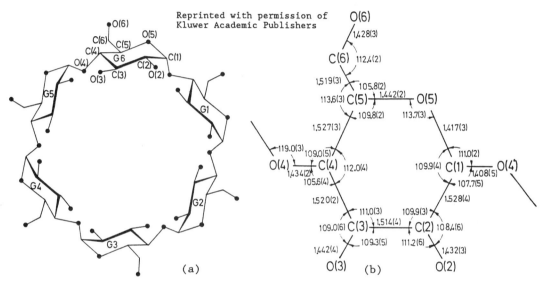

Reprinted with permission of Kluwer Academic Publishers

Fig. 7.8-3 (a) Chemical structure of α-cyclodextrin. (b) Average dimensions of the glucose unit taken from the data for α-CD.MeOH.5H₂O.[4] (After ref. 8).

As in starch, the glucose units in CDs are linked by α(1-4) bonds. The C(1) chair glucose rings are all aligned in register, so that the CD molecule takes the shape of a truncated cone with the secondary O(2)H/O(3)H hydroxyls pointing to the wider side and the primary O(6)H hydroxyls to the narrower side.   Thus hydrophilic groups occupy both rims of the cone and render the CD's soluble in aqueous solution. On the other hand, the inside of the cavity is hydrophobic in character because it is lined by C(3)-H and C(5)-H hydrogens and by the ether-like oxygen O(4). In solution, therefore, CDs are able to accommodate smaller guest molecules within their cavities, rather unspecifically and even only partially.

2. Channel-type structures of cyclodextrin inclusion compounds

In channel-type structures, CD molecules are stacked on top of one another like coins in a roll, the linearly aligned cavities producing channels

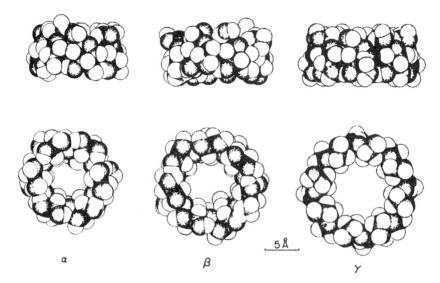

Fig. 7.8-4 Space filling plots of α-, β- and γ-CD viewed from the O(2)H/O(3)H and O(6)H rims and from the side. (After ref. 2).

in which guest molecules are embedded.

a) Linear polyiodide in α-CD channels as a model for the blue starch-iodine complex.

The Cd^{2+} and Li^+ polyiodide complexes of α-CD serve as model compounds for the well-known blue starch-iodine complex. In the $(\alpha\text{-CD})_2.LiI_3.I_2.8H_2O$ complex, the polyiodide chain is best described as alternating $I_3^- I_2 I_3^- I_2 \ldots$.[1]

$$--- I(1) \xrightarrow{2.882} I(2) \begin{array}{c} \xrightarrow{2.973} I(3)A \underset{}{\overset{3.693}{\dashrightarrow}} \\ \underset{4.005}{\dashrightarrow} I(3)B \xrightarrow{2.926} \end{array} I(4) \xrightarrow{2.832} I(5) \overset{3.486}{-----} I(1)$$

The deep colour of these complexes must be attributed to charge transfer along the polyiodide chain, facilitated by I...I distances in the range 2.9Å to 3.5Å (average 3.1Å as in starch iodine). These distances are much shorter than expected from the sum of the van der Waals radii, 4.3Å. There is no obvious interaction between iodine and the surrounding α-CD that could cause the colour.

b) Structure of β-CD.*p*-nitroacetanilide complex

The β-CD.*p*-nitroacetanilide complex[5] is a rarer case in which the guest is well ordered within the channel matrix. The reason might be that the asymmetric guest molecule wedges into the channel such that adjacent guests are in close contact, with nitro...nitro and acetyl...acetyl distances around 3.4Å, as shown in Fig. 7.8-5.

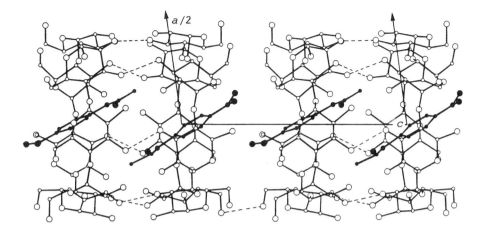

Fig. 7.8-5 Structure of β-CD.p-nitroacetanilide. (After ref. 5).

c) Structure of γ-CD.x(n-propanol).yH₂O complex

In γ-CD.x(n-propanol).yH₂O, a special situation is encountered because head-to-tail and head-to-head orientations alternate within the same stack.[6] In this crystal, channel axes coincide with four-fold rotation axes (space group P4), i.e. each γ-CD has four-fold symmetry and the crystal structure is composed of six quarter molecules labeled A-F in Fig. 7.8-6. Arrows indicate the orientations of γ-CDs, parallel orientation being associated with head-to-tail, and antiparallel with head-to-head interactions.

3. Cage-type structures of cyclodextrin inclusion compounds

In crystal structures belonging to the cage type, the cavity of one CD molecule is blocked off on both sides by adjacent CD's, thereby leading to isolated cavities. Two packing modes have been found: the herringbone-type of pattern most common in α-, β-, and γ-CD, and the brick-type so far observed for α-CD complexed with some para-disubstituted benzene derivatives, or with a dimethyl sulfoxide/methanol mixture.

a) Structure of α-CD.I₂.4H₂O[7]

In this reddish-brown crystal, the molecules are arranged in herringbone fashion, with the four water molecules as space-filling mediators (Fig. 7.8-7). The α-CD molecule assumes an almost hexagonal structure with one of the glucoses tilted a bit more towards the molecular axis than the other five. The I₂ and CD molecules are nearly coaxial, and the nearest separations between them in radial directions are in van der Waals contact at distances 3.75-4.17Å. In the axial directions, the alignment of the atoms:

$$
\begin{array}{ccccc}
& 3.07\text{Å} & & 2.677\text{Å} & & 3.32\text{Å} \\
O(42) & \text{-----} & I(1) & \text{———} & I(2) & \text{-----} & O(12)
\end{array}
$$

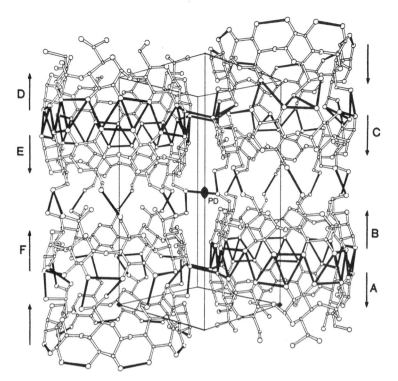

Fig. 7.8-6 Structure of γ-CD.x(n-propanol).yH$_2$O. Water molecules located between the γ-CD stacks have been omitted for clarity; n-propanol molecules within the cavity could not be located due to disorder; hydrogen bonds are drawn solid. (After ref. 6).

is roughly co-linear, the angular deviations of O(42) and O(12) from the I(1)-I(2) bond axis being within 13°. The difference between the I...O distances is significant, and the smaller value is appreciably less than the sum (3.55Å) of van der Waals radii, indicating attractive interactions.

b) Structure of α-CD.p-iodoaniline.3H$_2$O[9,10]

The structure is of the brick-cage type as shown in Fig. 7.8-8. Two kinds of cavities are present, one within the α-CD molecules and occupied by the p-iodoaniline (indicated by wide hatching), and the other between the α-CD molecules and filled with the three hydration water molecules (indicated by dense hatching). The p-iodoaniline molecule fits into the α-CD cavity such that the iodine atom is close to the O(6) side and the amino group protudes in brick-cage fashion about 1.5Å from the O(2),O(3) side. As the smallest dimension of the p-iodoaniline molecule, ~6.5Å, exceeds the 5.0Å wide diameter of the α-CD cavity, the α-CD molecule becomes elongated elliptically, as shown in Fig. 7.8-9.

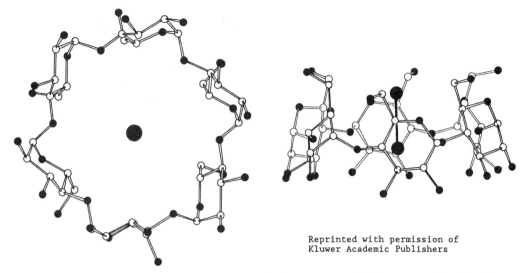

Fig. 7.8-7 Two different views of the α-CD.I$_2$.4H$_2$O compound with the hydration water molecules omitted for clarity. (After ref. 8).

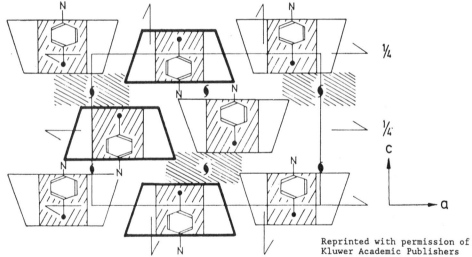

Fig. 7.8-8 Brick-cage type structure as found in the α-CD.p-iodoaniline.3H$_2$O complex. The α-CD molecules in the front are marked by heavy lines. (After ref. 8).

4. Hydrated cyclodextrins

When α-, β-, or γ-CD are crystallized from water, the following hydrates have been obtained and characterized by X-ray diffraction:

α-CD.6H$_2$O (Form I)   β-CD.12H$_2$O

α-CD.6H$_2$O (Form II)   γ-CD.17H$_2$O

α-CD.7.5H$_2$O (Form III)

In these crystals, cavities and interstices between CD molecules are filled by water molecules which are statistically disordered in the "round" α-CD.7.5H$_2$O

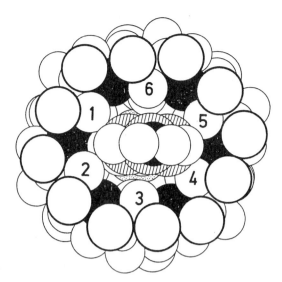

Fig. 7.8-9 Structure of the α-CD.p-iodoaniline.$3H_2O$ complex, showing elliptical distortion of the macrocycle. (After ref. 2).

and in β- and γ-CD, whereas they are in fixed positions in the "collapsed" α-CD.$6H_2O$.

a) Structure of α-CD.$6H_2O$ (Form I)[11]

α-CD.$(H_2O)_2$.$4H_2O$ is nearly isomorphous with α-Cd.I_2.$4H_2O$. The two water molecules included within the cavity are located approximately on the axis of the α-CD torus and are hydrogen bonded to each other. The water molecule close to the α-CD $O(6)$ side is further involved in two hydrogen bonds with $O(6)$ hydroxyl groups of the encaging α-CD molecule, as shown in Fig. 7.8-10(a), while the other water molecule forms hydrogen bonds with an adjacent, symmetry related α-CD.

The conformation of the α-CD.$(H_2O)_2$ adduct is no longer of nearly hexagonal symmetry as one of the six glucoses is rotated almost vertically with respect to the axis of the α-CD in order to facilitate hydrogen bond formation with one of the two included water molecules. The α-CD torus is partly "collapsed" by rotation of one glucose unit out of alignment with the other five. This movement disrupts two of the six $O(2)H...O(3)H$ hydrogen bonds, brings the $O(6)H$ "inside" to hydrogen bond to one of the included water molecules and, as a consequence, reduces the volume of the cavity [Fig. 7.8-10(a)].

b) Structure of β-CD.$12H_2O$[13,14]

In the crystal the β-CD cavity is filled by 6.5 water molecules distributed statistically over nine sites [Fig. 7.8-10(b)]. The remaining $5.5H_2O$ are spread over eight sites between β-CD molecules. The larger β-CD

molecule exists in β-CD-12H₂O in a "round" structure, and no conformational
change except rotation of C(6)-O(6) bonds is observed upon complex formation
with other guests.

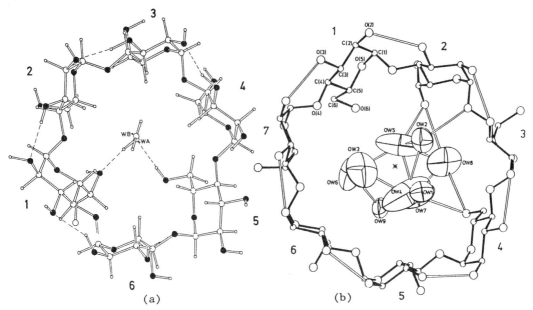

Fig. 7.8-10 (a) Results of a neutron diffraction study of α-CD.6H₂O (Form I).
All hydrogen atoms are drawn and hydrogen bonding interactions are indicated
by dashed lines. WA and WB mark enclosed water molecules, WA being hydrogen
bonded to two O(6)H groups whereas WB is bonded to WA and to an adjacent CD
molecule, not to the enclosing one.[12] (b) Structure of β-CD.12H₂O. The β-
CD torus exhibits a "round" structure with all O(2)H...O(3)H hydrogen bonds
formed. (After ref. 2).

c) Circular and flip-flop hydrogen bonds in hydrated cyclodextrins[2]

The term "circular hydrogen bonds" applies to a chain of O-H...O-H...O-H
hydrogen bonds arranged circularly to yield a 4, 5, 6, or higher membered ring
(Fig. 7.8-11). There are three kinds of circular hydrogen bonds: *homodromic*
if all hydrogen bonds run in the same direction, *antidromic* if two chains
emanate from one H₂O and converge at one oxygen, and *heterodromic* for randomly
oriented hydrogen bonds. The homodromic circles constitute a special case of
the frequently observed infinite chains of unidirectional hydrogen bonds and
both are energetically favoured. This is due to the cooperative effect which
leads to increased hydrogen bonding activity of an OH group if it is already
accepting or donating a hydrogen bond.

"Flip-flop hydrogen bonds" occur in β-CD.12H₂O, where 19 out of 58
hydrogen bonds are like those in ice Ih, which involve a dynamic system
O-H...O   O...H-O. Since several of these hydrogen bonds are linked, a

change in proton location requires cooperative, concerted change of all
involved hydrogen bonds.

Combination of circular and flip-flop hydrogen bonds leads to
oscillating systems which, if interconnected and fluctuating, can serve as
models for hydration of macromolecules and for "flickering clusters" in bulk
water. The flip-flop circles are energetically favourable not only because of
cooperative effects but also due to entropy contributions from the two
structurally and energetically equivalent flip-flop states.

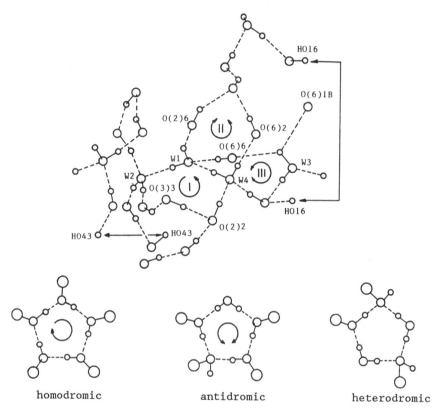

homodromic   antidromic   heterodromic

Fig. 7.8-11 Circular hydrogen bonds (top) observed in the crystal structure
of α-CD.6H$_2$O. Only a section of the asymmetric unit is shown, with chain-like
hydrogen bonds O-H...O-H...O-H indicated by arrows, and circular structures by
circular arrows. Circle I is homodromic and II, III are antidromic.
Nomenclature of circular hydrogen bonds (bottom). (Adapted from ref. 2).

5. Mechanism of inclusion compound formation[2]

Based on the crystal structures of α-CD.6H$_2$O [Fig. 7.8-10(a)], a
mechanism for inclusion compound formation of α-CD, reminiscent of the
induced-fit in enzyme-substrate interaction, has been proposed.[15] If a
substrate occupies the void, the ring of six O(2)H...O(3)H hydrogen bonds is

formed and α-CD adopts a "round" structure corresponding to minimum potential
energy, the "relaxed" state [Fig. 7.8-12]. In the "empty" α-CD, with water
enclosed in the cavity, the torus has collapsed and the ring of hydrogen bonds
is disrupted; the torsion angles about glycosyl C(4)-O(4) and O(4)-C(1'),
bonds indicate steric strain and therefore this state is called "tense".   The
inclusion of a substrate S can proceed *via* routes A, B, or C [Fig. 7.8-12].
In route A, cavity water is directly replaced by the substrate whereas B
involves first the formation of a "relaxed", round α-CD with enclosed
"activated" water, which is then expelled by the substrate.   The term

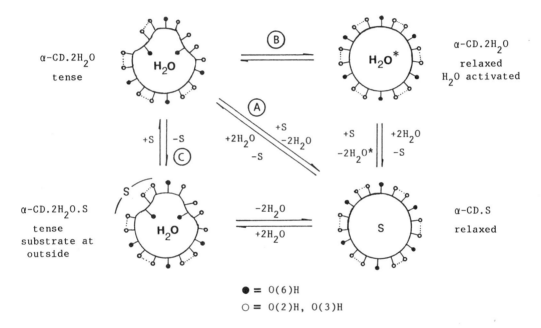

Fig. 7.8-12 Schematic representation of the "induced fit"-type complex
formation of α-CD.   S = substrate, H_2O^* = "active water". Hydrogen
bonds marked by broken lines.   (After ref. 15).

"activated water" was chosen because disordered water cannot satisfy its
hydrogen bonding potential and therefore exists in a higher energy state
compared with bulk water. In route C, the substrate first binds "outside" and
in a second step enters the cavity.

   This kind of mechanism is also supported by circular dichroism data
showing that α-CD undergoes conformational change when forming a complex. For
β-CD, such a change was not detected, in agreement with crystallographic
analysis which revealed "round" β-CD with $6.5H_2O$ molecules inside the cavity
disordered over 9 sites - the activated water [Fig. 7.8-10(b)].   It is

therefore assumed that β-CD does not change its conformation as drastically as α-CD when inclusion occurs and for γ-CD the same appears to hold. This "rigid" behaviour of β- and γ-CD, is probably caused by the ring of rather stable O(2)H...O(3)H hydrogen bonds.

6. Inclusion compounds containing inorganic and organometallic guests

In the crystal structure of the 1:1 complex of α-CD with Pt(NH$_3$)$_2$(cyclobutane-1,1-dicarboxylato), the hydrophobic cyclobutane ring of the organic ligand penetrates the host cavity through its wider aperture bearing the secondary hydroxyl groups.[16] Adduct formation between β-CD and *trans*-[Pt(PMe$_3$)Cl$_2$(NH$_3$)] yielded a 1:1 complex containing 5.5 water molecules of hydration. X-Ray analysis has shown that in 57% of the guest molecules the PMe$_3$ ligand is inserted into the narrower primary hydroxy-group-bearing face of the β-CD torus [Fig. 7.8-13(a)].[17] The remaining fraction of the guest molecules lies within a severely disordered surface forming a lid over the wider face of the receptor, and the water molecules, both ordered and disordered, occupy seven sites around the periphery of the β-CD belt.

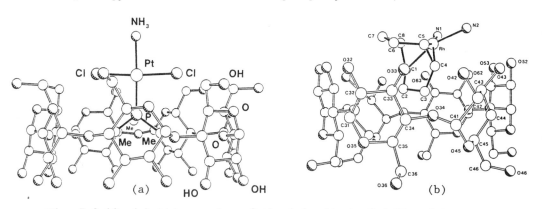

Fig. 7.8-13 (a) Side-on view of the 1:1 adduct of β-CD and *trans*-[Pt(PMe$_3$)Cl$_2$(NH$_3$)]; only the major component of the disordered guest molecule is shown. (After ref. 17). (b) Side-on view of the [α-CD.Rh(cod)(NH$_3$)$_2$]$^+$ complex. (After ref. 22).

Crystalline inclusion complexes of 1:1 stoichiometry have been prepared by treatment of β- and γ-CD with *o*-carborane, whereas α-CD forms both 1:1 and 2:1 complexes with this inorganic cage system.[18] Computer modeling suggests that *o*-carborane fits well into a β-CD cavity as shown in structure **1**, and structures **2a** and **2b** correspond to the 1:1 and 2:1 complexes of α-CD with *o*-carborane, respectively (Fig. 7.8-14).

Cyclodextrin inclusion compounds incorporating organometallic guests include 2β-CD.[Rh(L)Cl]$_2$ (L = norbornadiene or cycloocta-1,5-diene, cod) and β-CD.Pt(cod)X$_2$ (X = Cl, Br, I);[19] a wide range of 1:1 complexes of α-, β-,

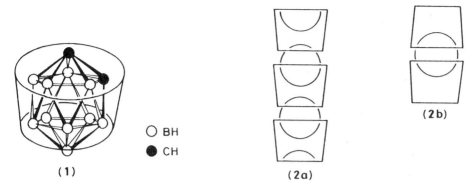

○ BH
● CH

(1)   (2a)   (2b)

Fig. 7.8-14 Proposed structures of inclusion complexes of *o*-carborane
with cyclodextrins. (After ref. 18).

and γ-CD with (π-arene)tricarbonylchromium;[20] and 2:1 α-CD, 1:1 β-CD, and
1:1 γ-CD complexes with ferrocene (F$_c$H) and its substituted derivatives.[21]
The proposed structures of these inclusion complexes are shown in Fig. 7.8-15.

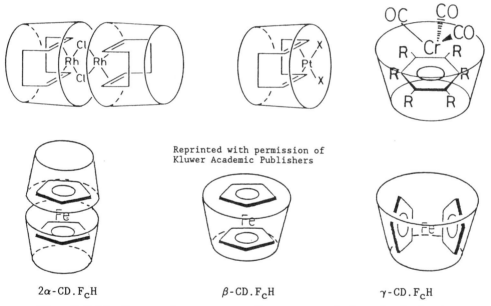

Reprinted with permission of
Kluwer Academic Publishers

2α-CD.F$_c$H   β-CD.F$_c$H   γ-CD.F$_c$H

Fig. 7.8-15 Proposed structures for some cyclodextrin inclusion
compounds of organometallics guests. (After refs. 19-21).

In the crystal structure of the ionic complex [α-CD.Rh(cod)(NH₃)₂]-
(PF₆).6H₂O,[22] the cod ligand is positioned almost directly over the center
of the α-CD torus on the wider side, with one ethylene group inserted into the
cavity [Fig. 7.8-13(b)].

The mixed-sandwich complex [(η⁵-C₅H₅)Fe(η⁶-C₆H₆)]PF₆ forms stable 2:1
inclusion compounds with both α-CD and β-CD. The crystalline hydrate

$(\alpha\text{-CD})_2 \cdot [(\eta^5\text{-}C_5H_5)Fe(\eta^6\text{-}C_6H_6)]PF_6 \cdot 8H_2O$ is the first example of a structurally characterized CD inclusion compound of an organometallic sandwich complex.[23] The sandwich cation is encapsulated within the cavity of a head-to-head α-CD dimer (cf. proposed structure for $2\alpha\text{-CD} \cdot F_cH$ in Fig. 7.8-15), but the C_5H_5 and C_6H_6 rings are tilted with respect to the mean planes of the α-CD molecules [Fig. 7.8-16]. The PF_6^- anion is located in a depression on the primary hydroxyl side of the α-CD molecule that includes the C_6H_6 ring. The dimers are stacked along the crystallographic c axis to form channels, with water molecules filling the intervening space between them. All water molecules are involved in at least four, and all hydroxy groups in at least two, hydrgogen bonds in a scheme which is similar to that in $(\alpha\text{-CD})_2 \cdot LiI_3 \cdot I_2 \cdot 8H_2O$.[1]

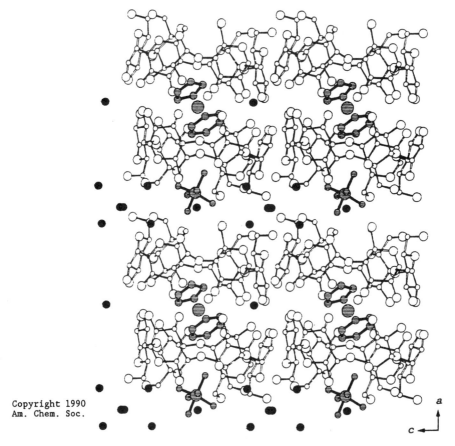

Fig. 7.8-16 Channel-type structure of $(\alpha\text{-CD})_2[(\eta^5\text{-}C_5H_5)Fe(\eta^6\text{-}C_6H_6)] \cdot PF_6 \cdot 8H_2O$; atoms of the guest molecules are shaded, and water molecules are represented by black circles. (After ref. 23).

7. A polymeric β-cyclodextrin inclusion compound

The monosubstituted β-CD derivative **1** is of interest since it has both host (cyclodextrin moiety) and guest (*tert*-butylthio group) character.[24] As

shown in Fig. 7.8-17, the molecules are arranged about a 2_1 screw axis to give
a polymeric structure where the SCMe₃ group of each molecule is inserted into
the hydrophobic cavity of an adjacent molecule. The cyclodextrin macrocycle
retains its approximate seven-fold axis and a round shape.

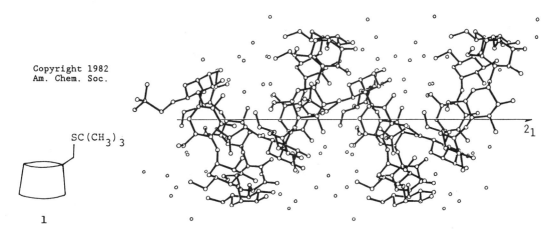

Fig. 7.8-17 Polymeric structure of the intermolecular inclusion complex
6-deoxy-6-(*tert*-butylthio)-β-cyclodextrin.22H₂0. The SBu^t group is
intermolecularly included in the hydrophobic cavity of the macrocycle.
The 2_1 screw axis is parallel to the c axis, and water oxygen atoms are
represented by smaller circles. (After ref. 24).

8. Cyclodextrin-based artificial enzymes[25,26]

    The cyclodextrin nucleus is a very attractive framework on which
catalytic groups may be grafted. It is possible to attach some functional
groups to a cyclodextrin so as to produce a cyclodextrin-based artifical
enzyme carrying more than one catalytic functionality and to permit catalyzed
reaction of ligands that are principally bound by the cyclodextrin cavity.
For example, Breslow and co-workers attached two imidazole rings to the β-CD
nucleus to produce a simple artifical enzyme, which can catalyse the selective
hydrolysis of a bound cyclic phosphate ester substrate via a bifunctional
mechanism similar to that used by ribonuclease.[27]

    Fig. 7.8-18(a) illustrates the selective reductive amination of keto
acids by a model catalyst containing a coenzyme (pyridoxamine) with a binding
group (β-CD). In competitive experiments with this artificial enzyme, the
conversion of phenylpyruvic acid to phenylalanine and indolepyruvic acid to
tryptophan was more rapid than that of pyruvic acid into alanine. The
artificial enzyme shown in Fig. 7.8-18(b) carries an asymmetrically mounted
basic group and can produce optically-active amino acids with ~6.7:1

preference. The artificial phosphatase shown in Fig. 7.8-18(c) is quite a good catalyst for phosphate ester hydrolysis.

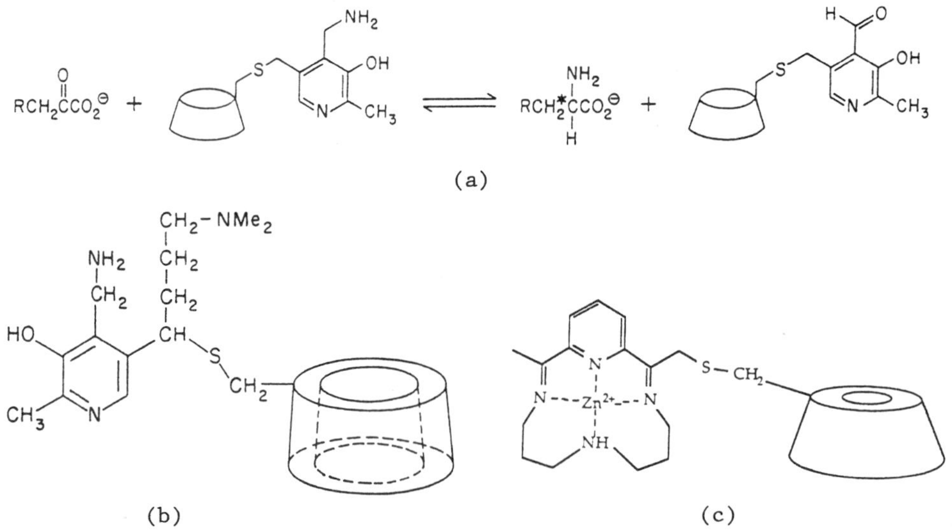

(a)

(b)   (c)

Fig. 7.8-18 (a) Attachment of pyridoxamine to cyclodextrin produces an artificial enzyme that can reductively aminate keto acids with substrate selectivity and some stereoselectivity in the product amino acids. (b) An artificial enzyme combining a coenzyme, binding group, and chirally mounted base catalyst. (c) An artificial phosphatase containing a binding group and a catalytic zinc complex. (After ref. 25).

9. Amylose inclusion compounds[28]

The reaction between starch and iodine (or iodine-iodide mixtures) in solution to form a deep blue inclusion complex is familar to all chemists through its applications in analytical chemistry. The host structure contains a helical arrangement of amylose, known as the V-form, that has an outer diameter of 13.0Å, an inner diameter of 5Å, and a pitch of 8.0Å with six glucose units per turn.[29]

For historical reasons V-amylose is subdivided into hydrated (V_h) and anhydrous (V_a) forms. Actually both forms have a significant water content and analogous crystal structures. In either case the amylose chains form left-handed helices packed in an antiparallel manner in a pseudohexagonal unit cell (space group $P2_12_12_1$), and the guest water molecules are located in interstitial sites between neighbouring helices. V_h-Amylose has 16 water molecules per unit cell (a = 13.65, b = 23.70, c = 8.05Å), half of which are present within the canals of sixfold helical conformation [Fig. 7.8-19(a)].[30] In V_a-amylose the number of water molecules per unit cell is reduced to 4, and the helical conformation has only twofold symmetry as the glucose hydroxymethyl groups adopt different orientations.[31]

V-Amylose forms inclusion compounds with a wide range of small organic molecules.[28] X-Ray crystallographic studies have been carried out on the alcohol[32] and DMSO[33] inclusion compounds. The latter compound contains DMSO molecules within the helical canals, and both DMSO and water molecules are present in the interstitial region. Although the space group is still $P2_12_12_1$, the unit cell is now pseudotetragonal owing to hydrogen bonding between the helical chains and the interstitial DMSO molecules.

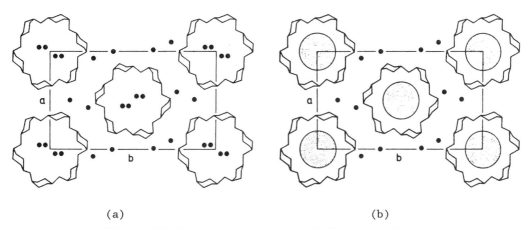

(a)   (b)

Fig. 7.8-19 Simplified representation of the crystal structures of (a) V_h-amylose and (b) the V_h-amylose polyiodide inclusion compound as projections along the c axis. The water molecules are shown as black dots, and the cross-sections of the polyiodide chains as stippled circles. (After ref. 28).

The V_h-amylose polyiodide inclusion compound crystallizes in space group $P2_1$ with $a = 13.60$, $b = 23.42$, and $c = 8.17$Å.[34] The iodine atoms are arranged in an almost linear manner (average I-I separation ~ 3.1Å) inside each helical amylose chain, but the exact nature is still uncertain; in addition, 8 water molecules are accommodated in the interstitial sites in the unit cell [Fig. 7.8-19(b)].

There exist other starch polymorphs which are inclusion compounds by virtue of their entrapment of water. The crystal structure of A-amylose is built of *right*-handed *double* helices which pack in an antiparallel manner.[35] The unit cell is orthorhombic and contains about 8 water molecules in the interstitial sites [Fig. 7.8-20(a)]. The B-polymorph has a related structure with water molecules (up to 3.5 per glucose residue) filling an open canal (diameter ~ 8Å) that is surrounded by a hexagonal array of the same type of double helices [Fig. 7.8-20(b)].[36] The C-polymorph has been found to exhibit structural features of both A-amylose and B-amylose.[37]

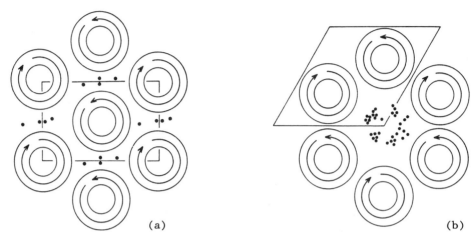

(a)   (b)

Fig. 7.8-20 Projection views of the crystal structures of (a) A-amylose and (b) B-amylose. The helical double strands of amylose are represented as rings, and the arrows indicate the antiparallel nature of their packing. The guest water molecules are shown as black dots. (After ref. 28).

References

[1]   M. Noltemeyer and W. Saenger, *J. Am. Chem. Soc.* **102**, 2710 (1980).

[2]   W. Saenger in J.L. Atwood, J.E.D. Davies and D.D. MacNicol (eds.), *Inclusion compounds*, Vol. 2, Academic Press, London, 1984, chap. 8.

[3]   W. Saenger, *Angew. Chem. Int. Ed. Engl.* **19**, 344 (1980).

[4]   B. Hingerty and W. Saenger, *J. Am. Chem. Soc.* **98**, 3357 (1976).

[5]   M.M. Harding, J.M. Maclennan and R.M. Paton, *Nature (London)* **274**, 621 (1978).

[6]   K. Lindner and W. Saenger, *Biochem. Biophys. Res. Commun.* **92**, 933, (1980).

[7]   R.K. McMullan, W. Saenger, J. Fayos and D. Mootz, *Carbohydrate Res.* **31**, 211 (1973).

[8]   W. Saenger, in B. Pullman (ed.), *Environmental Effects on Molecular Structure and Properties*, D. Reidel, Dordrecht, 1976, p. 265.

[9]   K. Harata, *Bull. Chem. Soc. Jpn.* **48**, 2409 (1975).

[10] W. Saenger, K. Beyer and P.C. Manor, *Acta Crystallogr., Sect. B* **32**, 120 (1976).

[11] P.C. Manor and W. Saenger, *J. Am. Chem. Soc.* **96**, 3630 (1974).

[12] B. Klar, B. Hingerty and W. Saenger, *Acta Crystallogr., Sect. B* **36**, 1154 (1980).

[13] K. Lindner and W. Saenger, *Angew. Chem. Int. Ed. Engl.* **17**, 694 (1978).

[14] K. Lindner and W. Saenger, *Carbohydr. Res.* **99**, 103 (1982).

[15] W. Saenger, M. Noltemeyer, P.C. Manor, B. Hingerty and B. Klar, *Bioorg. Chem.* **5**, 187 (1976).

[16] D.R. Alston, A.M.Z. Slawin, J.F. Stoddart and D.J. Williams, *J. Chem. Soc., Chem. Commun.*, 1602 (1985).

[17] D.R. Alston, A.M.Z. Slawin, J.F. Stoddart, D.J. Williams and R. Zarzycki, *Angew. Chem. Int. Ed. Engl.* **27**, 1184 (1988).

[18] A. Harada and S. Takahashi, *J. Chem. Soc., Chem. Commun.*, 1352 (1988).

[19] A. Harada and S. Takahashi, *J. Chem. Soc., Chem. Commun.*, 1229 (1986).

[20] A. Harada, K. Saeki and S. Takahashi, *Chem. Lett.*, 1157 (1985).

[21] A. Harada and S. Takahashi, *J. Incl. Phenom.* **2**, 791 (1984); Y. Odagaki, K. Hirotsu, T. Higuchi, A. Harada and S. Takahashi, *J. Chem. Soc., Dalton Trans 1*, 1230 (1990).

[22] D.R. Alston, A.M.Z. Slawin, J.F. Stoddart and D.J. Williams, *Angew. Chem. Int. Ed. Engl.* **24**, 786 (1985).

[23] B. Klingert and G. Rihs, *Organometallics* **9**, 1135 (1990).

[24] K. Hirotsu, T. Higuchi, K. Fujita, T. Ueda, A. Shinoda, T. Imoto and I. Tabushi, *J. Org. Chem.* **47**, 1143 (1982).

[25] R. Breslow, *Cold Spring Harbor Symposia on Quantitative Biology*, Vol. **LII**, 1987, p. 75, and references cited therein.

[26] I. Tabushi, *Acc. Chem. Res.* **15**, 66 (1982).

[27] R. Breslow, J. Doherty, G. Guillot and C. Lipsey, *J. Am. Chem. Soc.* **100**, 3227 (1978).

[28] R. Bishop and I.G. Dance, *Top. Curr. Chem.* **149**, 137 (1988).

[29] F.R. Senti and S.R. Erlander in L. Mandelcorn (ed.), *Non-Stoichiometric Compounds*, Academic Press, New York, 1964, p. 568.

[30] G. Rappenecker and P. Zugenmaier, *Carbohydr. Res.* **89**, 11 (1981).

[31] P. Zugenmaier and A. Sarko, *Biopolymers* **15**, 2121 (1976).

[32] R.M. Valletta, F.J. Germino, R.E. Lang and R.J. Moshy, *J. Polymer Sci. A* **2**, 1085, (1964).

[33] W.T. Winter and A. Sarko, *Biopolymers* **13**, 1461 (1974).

[34] T.L. Bluhm and P. Zugenmaier, *Carbohydr. Res.* **89**, 1 (1981).

[35] H.C.H. Wu and A. Sarko, *Carbohydr. Res.* **61**, 27 (1978).

[36] H.C.H. Wu and A. Sarko, *Carbohydr. Res.* **61**, 7 (1978).

[37] A. Sarko and H.C.H. Wu, *Starch* **30**, 73 (1978).

Note The crystal structures of (α-CD)₂.[(η⁵-C₅H₅)₂M)]PF₆ (M = Fe,Co,Rh) are described in B. Klingert and G Rihs, *J. Incl. Phenom.* **10**, 255 (1991).

A neutron and x-ray diffraction study of individual single crystals of partially deuterated β-CD.EtOH.8H₂O at 295K showed a significantly different arrangement of solvent molecules in the β-CD cavity. See T. Steiner, S.A. Mason and W. Saenger, *J. Am. Chem. Soc.* **113**, 5676 (1991).

A neutron diffraction study of partially deuterated γ-CD.15.7D₂O at 110K is reported in J. Ding, T. Steiner, V. Zabel, B.E. Hingerty, S.A. Mason and W. Saenger, *J. Am. Chem. Soc.* **113**, 8183 (1991).

Recent results on methylated CD complexes are summarized by K. Harata in J.L. Atwood, J.E.D. Davies and D.D. MacNicol (eds.), *Inclusion Compounds*, Vol. **5**, Oxford Univeristy Press, New York, 1991, chap. 9.

7.9   1:1 Paraquat Bisparaphenylene-34-crown-10 Inclusion Compound
$[C_{12}H_{14}N_2 \cdot C_{28}H_{40}O_{10}][PF_6]_2 \cdot 2Me_2CO$

Crystal Data

Triclinic, space group $P\bar{1}$ (No. 2)

$a = 10.204(2)$, $b = 11.562(2)$, $c = 13.835(5)$Å, $\alpha = 105.59(2)$, $\beta = 98.70(2)$, $\gamma = 115.47(1)°$, $Z = 1$

Atom	x	y	z	Atom	x	y	z
Bisparaphenylene-34-crown-10 molecule							
O(1)	.4530	.5708	-.3327	C(11)	.1420	-.1017	-.2391
C(2)	.2998	.5146	-.3944	C(12)	.1671	-.0046	-.1348
C(3)	.2748	.4123	-.4979	O(13)	.3215	.0529	-.0735
O(4)	.2896	.3034	-.4765	C(14)	.3719	.1489	.0262
C(5)	.2797	.2039	-.5638	C(15)	.2860	.1957	.0745
C(6)	.3428	.1222	-.5309	C(16)	.3467	.2876	.1759
O(7)	.2561	.0493	-.4751	C(17)	.5022	.6617	-.2317
C(8)	.3097	-.0342	-.4462	C(18)	.4131	.7053	-.1837
C(9)	.2115	-.1151	-.3949	C(19)	.5230	.2008	.0816
O(10)	.2232	-.0289	-.2967				
[Paraquat]²⁺ ion							
N(1)	.2055	.2798	-.2375	C(23)	.2912	.4297	-.0616
C(20)	.3458	.3029	-.2306	C(24)	.1787	.3440	-.1543
C(21)	.4637	.3898	-.1390	C(25)	.0789	.1800	-.3357
C(22)	.4390	.4547	-.0497				
Hexafluorophosphate ion							
P(1)	.8525	.4366	.7270	F(4)	.8594	.4088	.6182
F(1)	.8292	.2880	.7125	F(5)	1.0237	.5020	.7804
F(2)	.8799	.5843	.7423	F(6)	.6828	.3774	.6833
F(3)	.8286	.4578	.8359				
Acetone molecule							
O(26)	-.1247	.0788	-.1696	C(27)	-.3603	.0674	-.1836
C(26)	-.1973	.1248	-.1299	C(28)	-.1300	.2433	-.0265

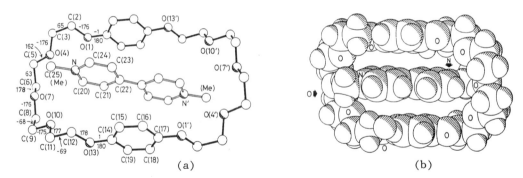

Fig. 7.9-1   (a) Molecular structure and (b) space-filling representation of [paraquat. BPP34C10]²⁺, **3**.   (After ref. 1).

Crystal Structure

Bisparaphenylene-34-crown-10 (BPP34C10), **2**, interacts with [paraquat]²⁺, **1**, to form a 1:1 molecular inclusion complex. The crystal structure consists of a packing of [paraquat. BPP34C10]²⁺, **3** (Fig. 7.9-1), PF_6^- ions, and acetone molecules, and reveals that the conformation of the receptor in **3** is virtually unchanged on complexation from that observed for the free macrocycle **2**,[2] as

Me—N⁺...N—Me

1

2 (j = k = 3)

3

shown in Fig. 7.9-2. The only significant torsion angle changes are 21° and 10°, respectively, about the O(1)-C(14′) and C(5)-C(6) bonds. In common with the X-ray crystal structure of free BPP34C10, that of its 1:1 complex with [paraquat]$^{2+}$ also possesses a crystallographic center of symmetry which here is coincident for the macrocycle and the dication. The binding of the planar [paraquat]$^{2+}$ dication can be interpreted as reflecting a balance between (i) charge transfer interactions involving the hydroquinol rings, (ii) electrostatic interactions involving the phenolic oxygen atoms, and (iii) C-H...O hydrogen bonds and van der Waals interactions involving the polyether chains of the receptor. There are some host-guest potential hydrogen bonding distances under 3.4Å: C(25)...O(4) 3.25Å, H(25)...O(4) 2.53Å, C(25)-H(25)...O(4) 132°; O(7)...C(25) 3.33Å, H(25)...O(7) 2.51Å, O(7)...H(25)-C(25) 144°.

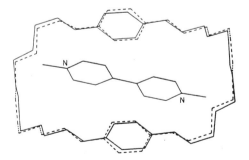

Fig. 7.9-2 Best least-squares fit of the conformation of free BPP34C10 (broken line) to that of [paraquat.BPP34C10]$^{2+}$ (full line). (After ref. 1).

Remarks

1. Structure of 1:1 complex **6** of paraquat **1** and 1,5-dihydroxynaphtho(38)-crown-10 ether **4**[3]

In crystalline [paraquat.1,5DN38C10][PF$_6$]$_2$, [**1.4**][PF$_6$]$_2$, the structure of [paraquat.1,5DN38C10]$^{2+}$ **6** possesses a crystallographic center of symmetry which is coincident for the macrocycle and the planar dication **1**, as shown in Fig. 7.9-3.

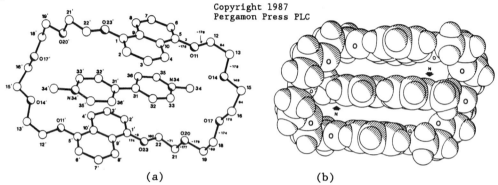

Copyright 1987
Pergamon Press PLC

                               (a)   (b)

Fig. 7.9-3   (a) Molecular structure and (b) space-filling representation of **6**.   (After ref. 3).

In this structure C-H...O hydrogen bonding and host-guest van der Waals interactions appear to be less important in stabilizing the 1:1 complex than charge transfer interactions between the π-electron deficient bipyridinium ring and the two π-electron-rich naphtho rings, and possibly also the electrostatic interactions between the C(32), C(33) and N(34) atoms in [paraquat]$^{2+}$ and the aryl oxygen atoms O(11) and O(23) in the receptor.   There is only one hydrogen bond C(34)-H(34)...O(14), in which C(34)...O(14) = 3.28Å, H(34)...O(14) = 2.35Å, and C(34)-H(34)...O(14) = 163°.

4 (m = 2), 5 (m = 3)

2. Structure of 2:1 complex **7** of paraquat **1** and 1,5-dihydroxynaphtho(44)-crown-12 ether **5**[4]

X-Ray analysis of [2paraquat.1,5DN44C12][PF$_6$]$_4$.Me$_2$CO reveals a continuously stacked π-donor/π-acceptor structure with paraquat dications alternately included within and sandwiched between adjacent macrocycles (Fig. 7.9-4). One of the two independent paraquat dications is positioned about a crystallographic inversion center and sandwiched within the macrocycle between the parallelly-aligned naphtho residues, and is oriented so as to maximize overlap between the two π-electron-rich naphtho units and each pyridinium ring of the π-electron deficient dication. The second paraquat dication occupies a centrosymmetric site between parallel lattice-translated complexes and is nearly perfectly co-aligned with its included counterparts. The interplanar separation between the naphtho units in **7** is 6.8Å, significantly shorter than the separation in **6**, with a consequent reduction in the receptor-substrate separation. The same interplanar separation is present between adjacent complexes and the intercalated paraquat dications.

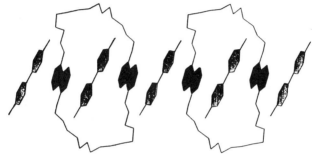

Fig. 7.9-4 Stacked π-donor/π-acceptor structure of **7** in the crystal. (After ref. 4).

3. Structure of cyclobis(paraquat-p-phenylene).4PF$_6$.3MeCN[5] and its inclusion compound with hydroquinone dimethyl ether[6]

    In the crystal structure of cyclobis(paraquat-p-phenylene).4PF$_6$.3MeCN, the [cyclobis(paraquat-p-phenylene)]$^{4+}$ cation **9** adopts a rigid centrosymmetric rectangular box-like conformation with the two paraquat units forming the

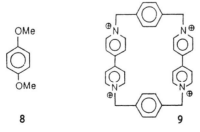

longer sides and the two *para*-xylylene residues the shorter ones, as shown in Fig. 7.9-5(a). There is a 19° twist angle between the two pyridinium rings of each paraquat unit; in addition, there are deformations of both the paraquat and *para*-xylylene components producing a bowing of the sides of the cyclophane. Thus the strain within the molecule is relieved by out-of-plane bending of the six aromatic rings and is distributed throughout the macrocycle with the maximum deviations associated with the exocyclic $C-CH_2$ bonds emanating from the *para*-phenylene residues: these bonds subtend an angle of 14° with respect to each other whilst the two N^+-CH_2 bonds associated with the paraquat units subtend an angle of 23°. There is also a concomitant reduction in the macrocyclic valence angles at C(7) and C(14), 108.3(7) and 108.5(7)° respectively, instead of *ca.* 111°. The overall dimensions of the macrocycle are 10.3Å between the centroids of the two *para*-phenylene rings and 6.8Å between C(1) and C(18). Inspection of the space-filling representation [Fig. 7.9-5(b)] shows a significant free pathway through the center of the macrocycle. Each macrocyclic ring is stacked almost directly above its neighbour in adjacent unit cells in the *a* direction; the small off-set is a result of the 109.36(1)° β angle. The region between unit-cell translated tetracations is occupied by four PF_6^- counterions and two MeCN molecules arranged in the form of a torus. The gross structure has continuous parallel open channels bounded by alternatively-charged entities. Measurements of the dimensions of the free pathways within these channels show these to be large enough to accommodate a wide range of different substrates. In the crystal, part of this void is filled by a MeCN molecule in two centrosymmetrically-related partially occupied sites.

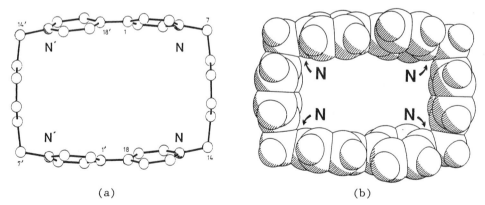

(a)   (b)

Fig. 7.9-5 (a) Molecular structure and (b) space-filling representation of **9** in **9**.4PF$_6$.3MeCN. (After ref. 5).

The X-ray structure analysis of **9**.4PF$_6$.2MeCN.**8** showed that the neutral hydroquinone dimethyl ether molecule **8** is inserted through the center of the centrosymmetric tetracationic macrocyclic receptor **9**, with the methyl groups of the methoxy substituents protruding above and below the rims of the

macrocyclic receptor [Fig. 7.9-6(a)]. The overall dimensions of the rigid
macrocyclic ring of **9** are virtually unchanged on complexation with **8**, the only
significant difference being a reduction in the twist angle between the two
pyridinium rings from 19° to 4°, as illustrated in Fig. 7.9-6(b). This
demonstrates the ability of **9** to flex so as to accommodate the π-electron-rich
substrate **8** within its π-electron-deficient cavity: this stereo-
electronically-induced fit does not perturb the overall channel structure of
the alternately-charged receptor stack. The toroidal arrangements of the four
interleaving PF_6^- ions and two MeCN molecules is unchanged from that found in
the substrate-free receptor. These are sandwiched in fixed positions between
the upper and lower rims of adjacent-stacked tetracations, indicating a strong
degree of stabilizing pole-pole and pole-dipole interactions.

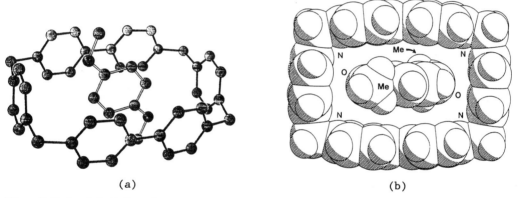

(a)   (b)

Fig. 7.9-6 (a) Molecular structure and (b) space-filling representation of **10**
= [**8.9**] in the crystal structure of **9**.4PF_6.2MeCN.**8**. (After ref. 6).

4. Structure of {[2]catenane of bisparaphenylene-34-crown-10 and
    cyclobis(paraquat-*p*-phenylene)}.4PF_6.5MeCN, (**13**.4PF_6.5MeCN)[7]

    The complex **3** is formed from **1** and **2** (Fig. 7.9-1), and complex **10** is
formed from **8** and **9** (Fig. 7.9-6). A continuous alternating electron donor-
acceptor stack is formed in crystal **7** (Fig. 7.9-4), which contains **5** and **2**(**1**).
From these structural data, a design concept to synthesize [2]catenane **13** has
been proposed.[7] The scheme in Fig. 7.9-7 shows a highly efficient one-step
template-directed synthesis of **13** from the bis(pyridinium) salt **11**.2PF_6, 1,4-

12

11

bis(bromomethyl)benzene **12**, and bisparaphenylene-34-crown-10 **2**.

Crystal structure analysis of (**13**.4PF$_6$.5MeCN) showed an elegantly ordered structural arrangement with the tetracationic cyclobis(paraquat-*p*-phenylene) intimately interlocked with the bisparaphenylene-34-crown-10 (Fig. 7.9-8). As anticipated, the dominant noncovalent bonding interactions are the electrostatic and dispersive forces associated with the mutual sandwiching of (i) a hydroquinol unit between parallelly-aligned bipyridinium moieties and of (ii) a bipyridinium moiety between parallelly-aligned hydroquinol units.

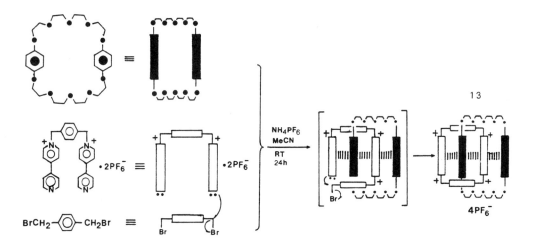

Fig. 7.9-7 Schematic representation of the synthesis of [2]catenane **13**. (Adapted from ref. 7).

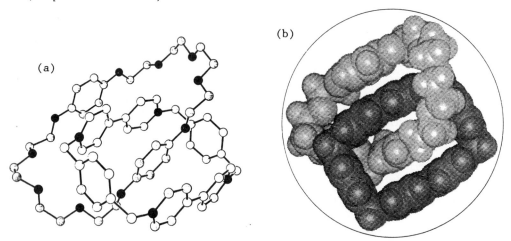

Fig. 7.9-8 (a) Molecular structure and (b) space-filling representation of [2]catenane **13**. (After ref. 7).

5. Structure of molecular belts, molecular collars, and molecular containers

Molecules with belt-, collar- and pocket-like structures (molecular belts, molecular collars and molecular containers) are of considerable

theoretical interest. An approach to the design and synthesis of these
molecules, which would offer a lot of flexibility in terms of structure, size,
electronic characteristics, functionality and properties, has been reported by
Stoddart *et al.*[8-12]

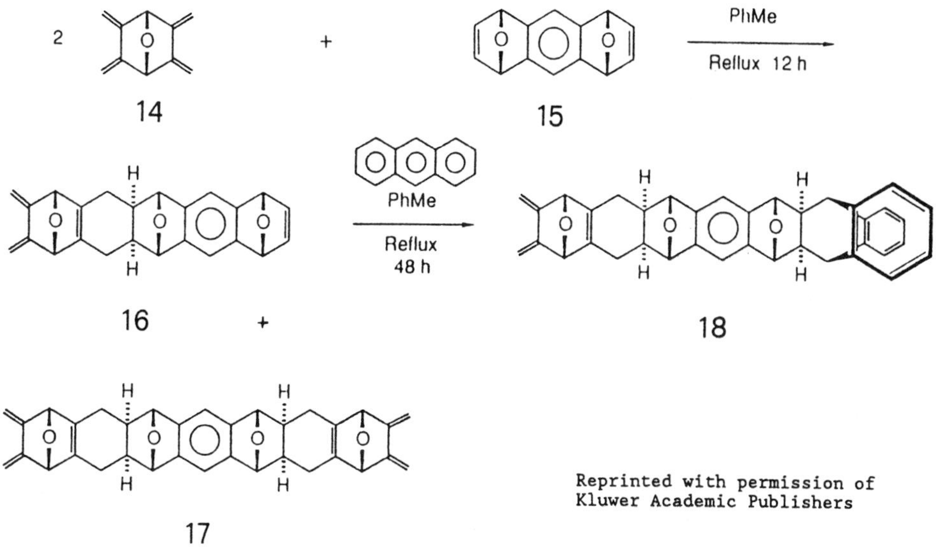

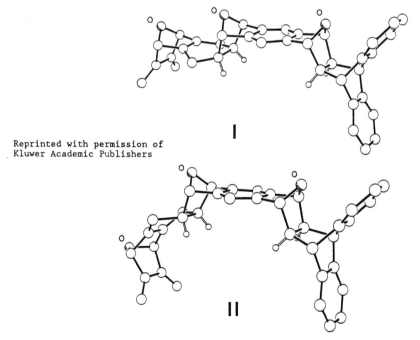

Reprinted with permission of
Kluwer Academic Publishers

Fig. 7.9-9   Schematic representation of synthesis of **18**.   (After ref. 8).

Reprinted with permission of
Kluwer Academic Publishers

Fig. 7.9-10   Structures of the two crystallographically independent
molecules (with the same configuration but differing in conformation) of
the anthracene adduct **18**.   (After ref. 8).

a) Molecular belts

Reaction of the bisdiene **14** with the bisdienophile **15** afforded the 1:1 adduct **16** and the 2:1 adduct **17** in yields of 24% and 61%, respectively, after chromatography (Fig. 7.9-9). The anthracene adduct **18** of **16** afforded good single crystals.

The crystal structure of **18** revealed two independent molecules, I and II, both of the same relative configuration consistent with the original 1:1 adduct having the *syn/endo*-H configuration **16**.[12] Molecules I and II differ mainly in the conformations of their substituted cyclohexene rings (Fig. 7.9-10). If appropriate functional groups are attached to the claws of pincer-like molecules of this type, the resulting expandable and contractible molecular grooves and clefts can be utilized to explore a wide range of molecular recognition phenomena.

b) Molecular collars

Cyclocoupling of **17** with **15** led to kohnkene **19**, which on deoxygenation smoothly afforded dideoxykohnkene **20**:

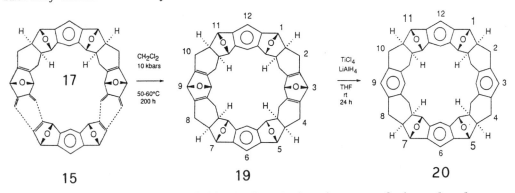

The crystal structure of **19** disclosed the elegance of the molecular architecture: six oxygen atoms are almost evenly spaced around the outside of an elliptical molecular belt (Fig. 7.9-11).[12] The interplanar separation between the two benzene rings at 6 and 12 o'clock positions is 7.9Å, leaving enough room (*ca.* 5Å between the van der Waals surfaces of the benzene rings) for a substituted benzene molecule to be accommodated orthogonally with respect to them. For a phenyl group (Ph) carrying an electron withdrawing substituent (X), the resulting edge-to face interaction with **19** could be sufficiently stabilising electrostatically to allow PhX molecules to enter inside its rigid molecular cavity.

The crystal structure of **20** revealed that the rigid cavity has an approximately square cross-section with distances between the mean planes of the parallel-disposed aromatic rings of 8.9Å (3 to 9 o'clock) and 9.6Å (6 to 12 o'clock).[10] The most striking feature of the structure is the Celtic

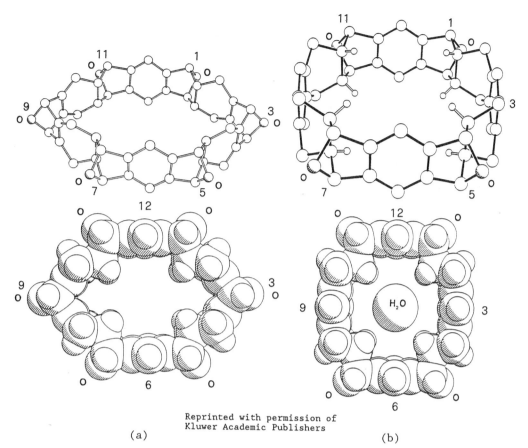

Reprinted with permission of
Kluwer Academic Publishers

(a)   (b)

Fig. 7.9-11 Ball-and-stick (top) and space-filling (bottom)
representations of the structure of (a) **19** and (b) **20**. (After ref. 8).

cross-like hydrophobic cavity within which a water molecule is insetted like a
gem. The water oxygen atom is > 3.5Å removed from any of the four pairs of
inward pointing methine hydrogen atoms in **20**. The individual hydrogen atoms,
which are > 2.7Å away from any potential interactive sites within the cavity,
are directed principally towards the pair of methine hydrogen atoms at 1:30
and 7:30.

c) Molecular container

The first closed molecular container compound, carcerand **21**, has been
designed and synthesized.[13] Its structural formula is shown in Fig. 7.9-
12(a).

The name "carcerand" originates from the Latin *carcer*, meaning prison.
The molecule is composed of two bowl-shaped cavitands coupled to each other at

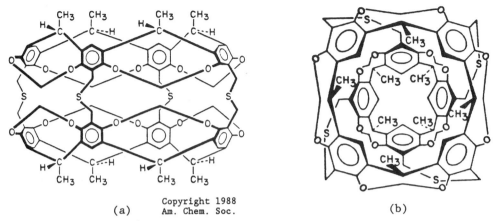

(a)   (b)

Fig. 7.9-12 (a) Structural formula of 21. (b) Looking through the portals along the four-fold axis of the carcerand. (After refs. 13 and 14).

their rims through four CH_2-S-CH_2 groups. The bridging carbons of the CH_2 groups are coplanar with their attached aryls, and the four S atoms are turned inward in a square-planar arrangement, with the orbitals of their unshared electron pairs facing one another across a substantial gap. Each CH_2-S-CH_2 bridge is flanked by two O-CH_2-O bridges in a complementary packing arrangement, leaving essentially no portal in this part of the shell of the molecular cell and little room for rotations about the Ar-CH_2 or CH_2-S bonds.

Molecule 21 belongs to point group D_{4h} and is shaped like an American football [Fig. 7.9-12(b)]. Two small openings in the shell occur at the top and bottom sides between the four methyl groups that protrude from the molecule to form a tunnel, constricted at the inner end by four aryl hydrogens. The carcerand is large enough to imprison behind its covalent bars small molecules of the solvents benzene, ether and chloroform, ions such as Cs^+ and Cl^-, or an Ar atom. Up to six water molcules can be accommodated altogether.

The construction of carcerands——hollow container-like molecules, which can imprison smaller molecules to form so-called carceplexes——raises the prospect of new organic materials with exciting properties and exotic applications, e.g. slow-release bio-degradable molecular encapsulants for pharmaceuticals and agrochemicals or metabolism-resistant molecular shells, containing radiation-emitting guests, attached to immunoproteins for use in cancer therapy. Incarcerated molecules inside soluble carcerands give very unusual NMR spectra, indicating that the inside of the carcerand is a new phase——the incarcerated molecules plus "vacuum"——whose properties are a function of the degree of occupancy of the interior of the carcerand by the imprisoned molecules.

6. Molecular recognition[15]

Molecular recognition has been defined as a process involving both *binding* and *selection* of substrate(s) by a given receptor molecule, as well as possibly a specific *function*. At the supramolecular level information may be stored in the architecture of the ligand, in its binding sites (nature, number, arrangement) and in the ligand layer surrounding bound substrate(s); it is read out at the rate of formation and dissociation of the supermolecule. Molecular recognition in the supermolecules formed by receptor-substrate binding or in the inclusion compounds formed by host-guest binding rests on the principles of molecular complementarity, as found in spherical and tetrahedral recognition, linear recognition by coreceptors metalloreceptors, amphiphilic receptors, and anion coordination. The chiral recognition in diastereomeric host-guest complexes in solution has attracted much attention in recent years.[16]

a) Spherical recognition

Cryptands **22-24** and **25**, as well as related compounds, display spherical recognition of appropriate cations and anions. Their complexation properties result from their macropolycyclic nature and define a cryptate effect characterized by high stability and selectivity, slow exchange rates, and efficient shielding of the bound substrate from the environment.

$22 = [2,1,1], m = 0, n = 1$
$23 = [2,2,1], m = 1, n = 0$
$24 = [2,2,2], m = 1, n = 1$

25

26

The macrobicyclic cryptands **22-24** form cryptates with a cation whose size is complementary to the size of the cavity, i.e., Li^+, Na^+ and K^+ for **22**, **23** and **24**, respectively.[17] The spherical macrotricyclic cryptand **25** binds strongly and selectively the larger spherical cations, giving a strong Cs^+ complex **26**.[18]

b) Tetrahedral recognition

Selective binding of a tetrahedral substrate is realized in the macrotricycle **25**, which contains four nitrogen and six oxygen binding sites located, respectively, at the corners of a tetrahedron and of an octahedron. It forms an exceptionally stable and selective cryptate **27** with the tetrahedral NH_4^+ cation, due to the high degree of structural and energetical complementarity.[19] The pKa of this NH_4^+ crypatate is about six units higher than that of free NH_4^+, indicating that similar effects exist in enzyme active sites and in biological receptor-substrate binding.

Other notable examples are **28** and **29**, which reflect the receptor-substrate binding complemetarity.

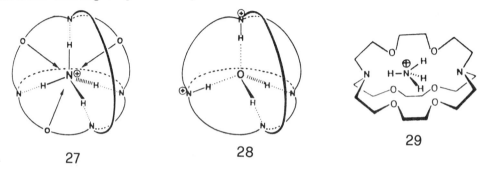

27   **28**   **29**

c) Recognition of anionic substrates

The hexaprotonated form of bis-tren, **30**, complexes various monoatomic and polyatomic anions. The crystal structures of four such anion cryptates

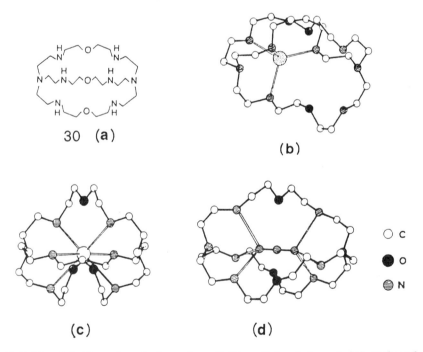

Fig. 7.9-13 (a) Structural formula of **30**. (b), (c) and (d) Molecular structures of the anion cryptates formed by **30** with F⁻, Cl⁻ (or Br⁻) and N_3^-, respectively. (After ref. 15).

(Fig. 7.9-13) provide a unique series of anion coordination patterns. The spherical halide ions are not complementary to the ellipsoidal receptor cavity and distort the structure, F⁻ being bound in a tetrahedral array of N-H...F⁻ bonds and Cl⁻ and Br⁻ having octahedral coordination. The linear triatomic anion N_3^- has a shape and size complementary to the cavity of **30** and is bound inside by a pyramidal array of three H-bonds to each terminal nitrogen forming the cryptate [**30**.N_3^-]. Thus **30** is a molecular receptor recognizing linear triatomic species such as N_3^-, which is indeed bound much more strongly than other singly charged anions.[20]

    d)   Linear recognition

       Receptor molecules possessing two binding subunits located at the two poles of the structure will complex preferentially substrates bearing two appropriate functional groups at a distance compatible with the separation of the subunits. This distance complementarity amounts to a recognition of molecular length of the substrate by the receptor. Such linear molecular recognition of dicationic and dianionic substrates corresponds to the binding modes illustrated by **31** and **32**.

       Incorporation of macrocyclic subunits that bind $-NH_3^+$ groups into cylindrical macrotricyclic[21] and macrotetracyclic[22] structures yields ditopic co-receptors that form molecular cryptates such as **33** with terminal diammonium cations $^+H_3N-(CH_2)_n-NH_3^+$. In the resulting supermolecules the substrate is located in the central molecular cavity and anchored by its two $-NH_3^+$ groups in the macrocyclic binding sites, as shown by the structure of **34**.[23] Changing the length of the bridges R in **33** modifies the binding selectivity in favor of the substrate of complementary length.

    e)   Multiple recognition in metalloreceptors[24]

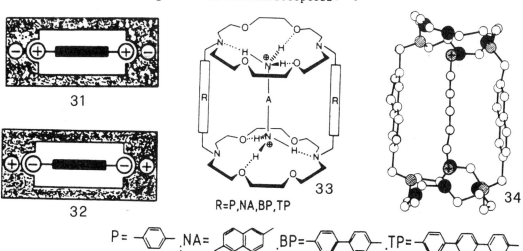

31

32

33

34

R=P,NA,BP,TP

P= ⬡— ; NA= ⬡⬡— ; BP=—⬡—⬡— ; TP=—⬡—⬡—⬡—

Metalloreceptors are heterotopic coreceptors that are able to bind both metal ions and organic molecules by means of substrate-specific units. Porphyrin and bipyridine groups have been introduced as metal-ion binding units in macropolycyclic coreceptors containing also macrocyclic sites for anchoring -NH₃⁺ groups. These receptors form mixed-substrate supermolecules by simultaneously binding metal ions and diammonium cations as shown in **35** and **36**. Metalloreceptors, and the supermolecules that they form, thus open up a vast area for the study of interactions and reactions between co-bound organic and inorganic species.

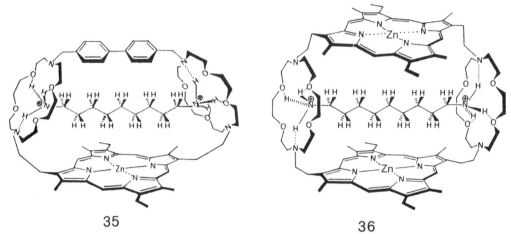

35   **36**

References

[1]   B.L. Allwood, N. Spencer, H. Shahriari-Zavareh, J.F. Stoddart and D.J. Williams, *J. Chem. Soc., Chem. Commun.*, 1064 (1987).

[2]   B.L. Allwood, H.M. Colquhoun, S.M. Doughty, F.H. Kohnke, A.M.Z. Slawin, J.F. Stoddart, D.J. Williams and R. Zarzycki, *J. Chem. Soc., Chem. Commun.*, 1054 (1987).

[3]   P.R. Ashton, E.J.T. Chrystal, J.P. Mathias, K.P. Parry, A.M.Z. Slawin, N. Spencer, J.F. Stoddart and D.J. Williams, *Tetrahedron Lett.* **28**, 6367 (1987).

[4]   J.-Y. Ortholand, A.M.Z. Slawin, N. Spencer, J.F. Stoddart and D.J. Williams, *Angew. Chem. Int. Ed. Engl.* **28**, 1394 (1989).

[5]   B. Odell, M.V. Reddington, A.M.Z. Slawin, N. Spencer, J.F. Stoddart and D.J. Williams, *Angew. Chem. Int. Ed. Engl.* **27**, 1547 (1988).

[6]   P.R. Ashton, B. Odell, M.V. Reddington, A.M.Z. Slawin, J.F. Stoddart and D.J. Williams, *Angew. Chem. Int. Ed. Engl.* **27**, 1550 (1988).

[7]   P.R. Ashton, T.T. Goodnow, A.E. Kaifer, M.V. Reddington, A.M.Z. Slawin, N. Spencer, J.F. Stoddart, C. Vicent and D.J. Williams, *Angew. Chem. Int. Ed. Engl.* **28**, 1396 (1989).

[8] J.F. Stoddart, *J. Incl. Phenom.* **7**, 227 (1989).

[9] F.H. Kohnke, J.P. Mathias and J.F. Stoddart, *Angew. Chem. Int. Ed. Engl.* **28**, 1129 (1989).

[10] P.R. Ashton, N.S. Isaacs, F.H. Kohnke, A.M.Z. Slawin, C.M. Spencer, J.F. Stoddart and D.J. Williams, *Angew. Chem. Int. Ed. Engl.* **27**, 966 (1988).

[11] F.H. Kohnke and J.F. Stoddart. *Pure Appl. Chem.* **61**, 1581 (1989).

[12] F.H. Kohnke, A.M.Z. Slawin, J.F. Stoddart and D.J. Williams, *Angew. Chem. Int. Ed. Engl.* **26**, 892 (1987).

[13] D.J. Cram, S. Karbach, Y.H. Kim, L. Baczynskyj, K. Marti, R.M. Sampson and G.W. Kalleymeyn, *J. Am. Chem. Soc.* **110**, 2554 (1988).

[14] F.H. Kohnke, J.P. Mathias and J.F. Stoddart, *Angew. Chem. Int. Ed. Engl.* **28**, 1103 (1989).

[15] J.-M. Lehn, *Angew. Chem. Int. Ed. Engl.* **27**, 89 (1988). [Nobel Lecture].

[16] F. Diederich, *Angew. Chem. Int. Ed. Engl.* **27**, 362 (1988).

[17] J.-M. Lehn and J.-P. Sauvage, *J. Am. Chem. Soc* **97**, 6700 (1975); B. Dietrich, J.-M. Lehn and J.-P. Sauvage, *Tetrahedron*, 2885, 2889 (1969).

[18] E. Graf and J.-M. Lehn, *J. Am. Chem. Soc.* **97**, 5022 (1975); *Helv. Chim. Acta* **64**, 1040 (1981).

[19] E. Graf, J.-M. Lehn and J. LeMoigne, *J. Am. Chem. Soc.* **104**, 1672 (1982).

[20] B. Dietrich, J. Guilhem, J.-M. Lehn, C. Pascard and E. Sonveaux, *Helv. Chim. Acta* **67**, 91 (1984).

[21] F. Kotzyba-Hibert, J.-M. Lehn and P. Vierling, *Tetrahedron Lett.* **21**, 941 (1980).

[22] F. Kotzyba-Hibert, J.-M. Lehn and K. Saigo, *J. Am. Chem. Soc* **103**, 4266 (1981).

[23] C. Pascard, C. Riche, M. Cesario, F. Kotzyba-Hibert and J.-M. Lehn, *J. Chem. Soc., Chem. Commun.*, 557 (1982).

[24] A.D. Hamilton, J.-M. Lehn and J.L. Sessler, *J. Chem. Soc., Chem. Commun.*, 311 (1984); *J. Am. Chem. Soc.* **108**, 5158 (1986).

Note Progress in understanding the non-covalent, selective binding of small molecules to biomolecules and unnatural macromolecules is covered in S.M. Roberts (ed.), *Molecular Recognition: Chemical and Biochemical Problems*, Royal Society of Chemistry, London, 1989.

A new series entitled *Monographs in Supramolecular Chemistry* under the editorship of J.F. Stoddart has been launched by The Royal Society of Chemistry to cover this rapidly developing field. Titles that have already published are:

C.D. Gutsche, *Calixarenes*, 1989

F.N. Diderich, *Cyclophanes*, 1991

G.W. Gokel, *Crown Ethers and Cryptands*, 1991

Novel host molecules belonging to the same class have been synthesized by attaching suitable spacers and "convergent functional groups" to the rigid skelton of Kemp's acid. The resulting open-chain "molecular clefts" serve as model receptors *via* selective binding to a wide range of acids, amines, amino acids, metal ions, heterocyclic compounds and nucleosides, and their applicability to problems in concerted catalysis has been extensively investigated. See J. Rebek, Jr., *Top. Curr. Chem.* **149**, 189 (1988); *Pure & Appl. Chem.* **61**, 1517 (1989); *Angew. Chem. Int. Ed. Engl.* **29**, 245 (1990). A recent study is reported in J.S. Nowick, Q. Feng, T. Tjivikua, P. Ballester and J. Rebek, Jr., *J. Am. Chem. Soc.* **113**, 8831 (1991).

The crystal structures of Kemp's acid and some of its derivatives are described in T.-L. Chan, Y.-X. Cui, T.C.W. Mak, R.-J. Wang and H.N.C. Wong, *J. Cryst. Spectrosc. Res.* **21**, 291 (1991).

For recent advances in the field of host-guest chemistry, the readers are referred to J.L. Atwood (ed.), *Inclusion Phenomena and Molecular Recognition*, Plenum Press, New York, 1990; V. Vögtle, *Supramolecular Chemistry: An Introduction*, Wiley, New York, 1991; V. Balzani and F. Scandola, *Supramolecular Photochemistry*, Ellis Horwood, Chichester, 1990; H.-J. Schneider and H. Dürr (eds.), *Frontiers in Supramolecular Organic Chemistry and Photochemistry*, VCH Publishers, Weinheim, 1991; *Host-Guest Molecular Interactions: From Chemistry to Biology* (Ciba Foundation Symposium 158; Chairman: I.O. Sutherland), Wiley-Interscience, Chichester, 1991.

Bibliography

Crystallography is a vast and interdisciplinary subject covering many related fields. A comprehensive **Crystallographic Booklist** [*J. Appl. Crystallogr.* **15**, 640-676 (1982)], edited by J.H. Robertson, annotates about 1200 titles published in the period 1970-1981 that are divided up into 29 categories. Most books and teaching materials of relevance to crystallography in the broadest sense can be ordered from Polycrystal Book Service, P.O. Box 3439, Dayton, OH 45401, U.S.A., which issues an updated catalogue annually. All publication of the **International Union of Crystallography** (IUCr) are now distributed by Kluwer Academic Publishers (formerly D. Reidel), Dordrecht, The Netherlands. Listed below is a compilation of selected books that are of interest to readers for reference and collateral reading; not included are continuing series such as *Structure and Bonding*, *Progress in Inorganic Chemistry*, and *Topics in Stereochemistry*.

Structural Chemistry of Solids

J. Donohue, *The Structures of the Elements*, Wiley, New York, 1974.

C.S. Barrett and T.B. Massalski, *Structure of Metals*, 3rd ed., Pergamon Press, Oxford, 1980.

W.B. Pearson, *The Crystal Chemistry and Physics of Metals and Alloys*, Wiley, New York, 1972.

F.R. DeBoer and D.G. Pettifor (eds.), *The Structures of Binary Compounds*, Elsevier, Amsterdam, 1990.

O. Muller and R. Roy, *The Major Ternary Structural Families*, Springer-Verlag, Berlin, 1974.

F.S. Galasso, *Structure and Properties of Inorganic Solids*, Pergamon Press, Oxford, 1970.

G.M. Clark, *The Structures of Non-Molecular Solids: A Coordinated Polyhedron Approach*, Applied Science Publishers, London, 1972.

D.M. Adams, *Inorganic Solids: An Introduction to Concepts in Solid-State Structural Chemistry*, Wiley, New York, 1974.

M. Krebs, *Fundamentals of Inorganic Crystal Chemistry*, McGraw-Hill, London, 1968.

I. Náray-Szabó, *Inorganic Crystal Chemistry*, Akadémiai Kiadó, Budapest, 1969.

B.G. Hyde and S. Andersson, *Inorganic Crystal Structures*, Wiley, New York, 1989.

H.D. Megaw, *Crystal Structures: A Working Approach*, Saunders, New York, 1973.

D.J. Vaughan and J.R. Craig, *Mineral Chemistry of Metal Sulfides*, Cambridge University Press, Cambridge, 1978.

F. Liebau, *Structural Chemistry of Silicates*, Springer-Verlag, Berlin, 1985.

D.T. Griffen, *Silicate Crystal Chemistry*, Oxford University Press, Oxford, 1991.

W.M. Meier and D.H. Olson, *Atlas of Zeolite Structure Types*, 2nd ed., Butterworths, London, 1987.

F. Lévy (ed.), *Crystallography and Crystal Chemistry of Materials with Layered Structures*, D. Reidel, Dordrecht, 1976.

D.B. Brown (ed.), *Mixed-Valence Compounds*, D. Reidel, Dordrecht, 1981.

R.J.D. Tilley, *Defect Crystal Chemistry and Its Applications*, Blackie, London, 1987.

F. Hulliger, *Structural Chemistry of Layer-Type Phases*, D. Reidel, Dordrecht, 1977.

G.W. Brindley and G. Brown (eds.), *Crystal Structures of Clay Minerals and Their X-Ray Identification*, Mineralogical Society, London, 1980.

A.S. Povarennykh, *Crystal Chemical Classification of Minerals* (Translated from the Russian edition by J.E.S. Bradley), Vols. 1 and 2, Plenum Press, New York, 1972.

周公度，《無機結構化學》，科學出版社，北京，1982。
[Zhou Gong-du, *Inorganic Structural Chemistry*, Science Press, Beijing, 1982.]

A.F. Wells, *Structural Inorganic Chemistry*, 5th ed., Oxford University Press, Oxford, 1984; reprinted with corrections, 1986.

M.F.C. Ladd, *Structure and Bonding in Solid State Chemistry*, Ellis Horwood, Chichester, 1979.

M. O'Keeffe and A. Novrotsky (eds.), *Structure and Bonding in Crystals*, Vols. I and II, Academic Press, New York, 1981.

J.D. Dunitz and J. Ibers (eds.) *Perspectives in Structural Chemistry*, Vols. I-IV, 1967-1971, Wiley, New York.

A.I. Kitaigorodskii, *Molecular Crystals and Molecules*, Academic Press, New York, 1973.

M. Pierrot (ed.), *Structure and Properties of Molecular Crystals*, Elsevier, Amsterdam, 1990.

G.R. Desiraju (ed.), *Organic Solid State Chemistry*, Elsevier, Amsterdam, 1987.

G.R. Desiraju, *Crystal Engineering: The Design of Organic Solids*, Elsevier, Amsterdam, 1989.

M. Dobler, *Ionophores and Their Structures*, Wiley, New York, 1981.

R.M. Izatt and J.J. Christensen, *Synthesis of Macrocycles: The Design of Selective Complexing Agents*, Wiley, New York, 1987.

E. Weber, J.L. Toner, I. Goldberg, F. Vögtle, D.A. Laidler, J.F. Stoddart, R.A. Bartsch and C.L. Liotta, *Crown Ethers and Analogs*, Wiley, Chichester, 1989.

M. Kakudo and N. Kasai, *X-Ray Diffraction by Polymers*, Kodansha, Tokyo, 1972.

H. Tadokoro, *Structure of Crystalline Polymers*, Wiley, New York, 1979.

W.L. Duax and D.A. Norton (eds.), *Atlas of Steroid Structure*, Vol. 1, 1975, Vol. 2 (J. Griffin, co-editor), 1983, Plenum Press, New York.

G.E. Schulz and R.H. Schirmer, *Principles of Protein Structure*, Spinger-Verlag, Berlin, 1979.

W. Saenger, *Principles of Nucleic Acid Structure*, Springer-Verlag, Berlin, 1983.

G. Dodson, J.P. Glusker and D. Sayre (eds.), *Structural Studies on Molecules of Biological Interest: A Volume in Honour of Dorothy Hodgkin*, Oxford University Press, Oxford, 1981.

J.P. Glusker (ed.), *Structural Crystallography in Chemistry and Biology* (Benchmark Papers in Physical Chemistry and Chemical Physics, Vol. 4), Hutchinson Ross, Stroudsburg, Pennsylvania, 1981. [Distributed by Academic Press, New York.]

V. Vögtle, *Supramolecular Chemistry: An Introduction*, Wiley, New York, 1991.

J.L. Atwood, J.E.D. Davis and D.D. MacNicol (eds.), *Inclusion Compounds*.
  Vol. 1: *Structural Aspects of Inclusion Compounds Formed by Inorganic and Organometallic Host Lattices*, Academic Press, London, 1984.
  Vol. 2: *Structural Aspects of Inclusion Compounds Formed by Organic Host Lattices*, Academic Press, London, 1984.
  Vol. 3: *Physical Properties and Applications*, Academic Press, London, 1984.
  Vol. 4: *Key Organic Host Systems*, Oxford University Press, Oxford, 1991.
  Vol. 5: *Inorganic and Physical Aspects of Inclusion*, Oxford University Press, Oxford, 1991.

H.-J. Schneider and H. Dürr (eds.), *Frontiers in Supramolecular Organic Chemistry and Photochemistry*, VCH Publishers, Weinheim, 1991.

Chemical Bonding and Molecular Structure

L. Pauling, *The Nature of the Chemical Bond*, 3rd ed., Cornell University Press, Ithaca, New York, 1960.

A. Zewail (ed.), *The Chemical Bond: Structure and Dynamics*, Academic Press, San Diego, 1992.

J.L.T. Waugh, *The Constitution of Inorganic Compounds*, Wiley, New York, 1972.

J.K. Burdett, *Molecular Shapes: Theoretical Models of Inorganic Stereochemistry*, Wiley, New York, 1980.

T.A. Albright, J.K. Burdett and M.H. Whangbo, *Orbital Interactions in Chemistry, Wiley, New York*, 1985.

B.E. Douglas and C.A. Hollingsworth, *Symmetry in Bonding and Spectra*, Academic Press, New York, 1985.

F.A. Cotton, *Chemical Applications of Group Theory*, 3rd ed., Wiley, New York, 1990.

R.J. Gillespie and I. Hargittai, *The VSEPR Model of Molecular Geometry*, Allyn and Bacon, Boston, 1991.

周公度，《結構化學基礎》，北京大學出版社， **1989**。
[Zhou Gong-du, *Fundamentals of Structural Chemistry*, Peking University Press, Beijing, 1989.]

徐光憲和王祥雲，《物質結構》第二版，高等教育出版社，上海， **1987**。
[Xu Guang-xian and Wang Xiang-yun, *Structure of Matter*, 2nd ed., Higher Education Press, Shanghai, 1987.]

I. Hargittai, *The Structure of Volatile Sulphur Compounds*, D. Reidel, Dordrecht, 1985.

J.F. Fergusson, *Stereochemistry and Bonding in Inorganic Chemistry*, Prentice-Hall, Englewood Cliffs, NJ, 1974.

D.L. Kepert, *Inorganic Stereochemistry*, Springer-Verlag, Berlin, 1982.

Y. Saito, *Inorganic Molecular Dissymmetry*, Springer-Verlag, Berlin, 1979.

R.S. Glass (ed.), *Conformational Analysis of Medium-Sized Heterocycles*, VCH Publishers, Weinheim, 1988.

P. Coppens and M.B. Hall (eds.), *Electron Distributions and the Chemical Bond*, Plenum Press, New York, 1982.

A. Domenicano and I. Hargittai, *Accurate Molecular Structures: Their Determination and Importance*, Oxford University Press, Oxford, 1991.

P. Schuster, G. Zundel and C. Sandorfy (eds.), *The Hydrogen Bond*, Vols. I-III, North-Holland, Amsterdam, 1976.

G.A. Jeffrey and W. Saenger, *Hydrogen Bonding in Biological Structures*, Springer-Verlag, Berlin, 1991.

W.C. Hamilton, *Statistics in Physical Science*, Ronald Press, New York, 1964.

Inorganic and Organometallic Chemistry

F.A. Cotton and G. Wilkinson, *Advanced Inorganic Chemistry*, 5th ed., Wiley, New York, 1989.

N.N. Greenwood and A. Earnshaw, *Chemistry of the Elements*, Pergamon Press, Oxford, 1984.

P.G. Harrison (ed.), *Chemistry of Tin*, Blackie, Glasgow, 1987.

G.A. Melson (ed.), *Coordination Chemistry of Macrocyclic Compounds*, Plenum Press, New York, 1979.

K. Nakamoto, *Infrared and Raman Spectra of Inorganic and Coordination Compounds*, 4th ed., Wiley, New York, 1986.

G. Wilkinson, R.D. Gillard and J.A. McCleverty (eds.), *Comprehensive Coordination Chemistry*, Vols. 1-7, Pergamon Press, Oxford, 1987.

F.A. Cotton and R.A. Walton, *Multiple Bonds between Metal Atoms*, Wiley, New York, 1982.

D.M.P. Mingos and D.J. Wales, *Introduction to Cluster Chemistry*, Prentice-Hall, Englewood Cliffs, NJ, 1991.

G.S. Hammond and V.J. Kuck (eds.), *Fullerenes: Synthesis, Properties and Chemistry of Large Carbon Clusters*, American Chemical Society, Washington, DC, 1992.

R.N. Grimes, *Metal Interactions with Boron Clusters*, Plenum Press, New York, 1982.

G.A. Olah, K. Wade and R.E. Williams (eds.), *Electron Deficient Boron and Carbon Clusters*, Wiley, New York, 1991.

E. Block (ed.), *Heteroatom Chemistry*, VCH Publishers, Weinheim, 1990.

H.W. Roesky (ed.), *Rings, Clusters and Polymers of Main Group and Transition Elements*, Elsevier, Amsterdam, 1989.

M. Moskovits (ed.), *Metal Clusters*, Wiley, New York, 1986.

B.F.G. Johnson (ed.), *Transition Metal Clusters*, Wiley, Chichester, 1980.

D.F. Shriver, H.D. Kaesz and R.D. Adams (eds.), *The Chemistry of Metal Cluster Complexes*, VCH Publishers, Weinheim, 1990.

J.P. Fackler, Jr. (ed.) *Metal-Metal Bonds and Clusters in Chemistry and Catalysis*, Plenum Press, New York, 1990.

W.A. Nugent and J.M. Mayer, *Metal-Ligand Multiple Bonds: The Chemistry of Transitional Metal Complexes Containing Oxo, Nitrido, Imido, Alkylidene, or Alkylidyne Ligands*, Wiley, New York, 1988.

M.F. Lappert, P.P. Power, A.R. Sanger and R.C. Srivastava, *Metal and Metalloid Amides*, Ellis Horwood, Chichester, 1980.

D.J. Cardin, M.F. Lappert and C.L. Raston, *Chemistry of Organo-Zirconium and -Hafnium Compounds*, Ellis Horwood, Chichester, 1986.

I. Haiduc and J.J. Zuckerman, *Basic Organometallic Chemistry*, Walter de Gruyter, Berlin, 1985.

Ch. Elschenbroich and A. Salzer, *Organometallics: A Concise Introduction*, VCH Publishers, Weinheim, 1989.

R.H. Crabtree, *The Organometallic Chemistry of the Transitional Metals*, Wiley, New York, 1988.

F.P. Pruchnik, *Organometallic Chemistry of the Transition Elements*, Plenum Press, New York, 1990.

A. Yamamoto, *Organotransition Metal Chemistry: Fundamental Concepts and Applications*, Wiley, New York, 1986.

J.P. Collman, L.S. Hegedus, J.R. Norton and R.G. Finke, *Principles and Applications of Organotransition Metal Chemistry*, 2nd ed., University Science Books, Mill Valley, California, 1987.

G. Wilkinson, F.G.A. Stone and E.W. Abel (eds.), *Comprehensive Organometallic Chemistry*, Vol. 1-9, Pergamon Press, Oxford, 1982.

N. Farrell, *Transition Metal Complexes as Drugs and Chemotherapeutic Agents*, Kluwer, Dordrecht, 1989.

Crystal Structure Analysis

J.P. Glusker and K.N. Trueblood, *Crystal Structure Analysis: A Primer*, 2nd ed., Oxford University Press, New York, 1985.

W.L. Bragg, *The Development of X-Ray Analysis*, Bell, London, 1975.

M.J. Buerger, *Crystal-Structure Analysis*, Wiley, New York, 1960.

周公度，《晶體結構測定》，科學出版社，北京，1981。
[Zhou Gong-du, *Determination of Crystal Structures*, Science Press, Beijing, 1981].

G.H. Stout and L.H. Jensen, *X-Ray Structure Determination: A Practical Guide*, 2nd ed., Wiley, New York, 1989.

M.F.C. Ladd and R.A. Palmer, *Structure Determination by X-Ray Crystallography*, 2nd ed., Plenum Press, New York, 1985.

P. Luger, *Modern X-Ray Analysis on Single Crystals*, Walter de Gruyter, Berlin, 1980.

J.V. Smith, *Geometrical and Structural Crystallography*, Wiley, New York, 1982.

J.M. Robertson, *Organic Crystals and Molecules*, Cornell University Press, Ithaca, New York, 1953.

J.D. Dunitz, *X-Ray Analysis and the Structure of Organic Molecules*, Cornell University Press, Ithaca, New York, 1979.

W.G. Laver and G. Air (eds.), *Use of X-Ray Crystallography in the Design of Antiviral Agents*, Academic Press, San Diego, 1990.

B.W. Rossiter and J.F. Hamilton (eds.), *Determination of Structural Features of Crystalline and Amorphous Solids (Physical Methods of Chemistry*, 2nd ed., Vol. 5), Wiley, New York, 1990.

D. Sherwood, *Crystals, X-Rays and Proteins*, Longman, London, 1976.

A. McPherson, *Preparation and Analysis of Protein Crystals*, Wiley, New York, 1982.

T.L. Blundell and L.N. Johnson, *Protein Crystallography*, Academic Press, New York, 1976.

H.W. Wyckoff, C.H.W. Hirs and S.N. Timasheff (eds.), *Methods in Enzymology*, Vol. 114 (Part A), Vol. 115 (Part B), Academic Press, New York, 1985.

A.M. Lesk, *Protein Architecture: A Practical Approach*, Oxford University Press, Oxford, 1991.

G.A. Jeffrey and J.F. Piniella (eds.), *The Application of Charge Density Research to Chemistry and Drug Design*, Plenum Press, New York, 1991.

M.F.C. Ladd and R.A. Palmer, *Theory and Practice of Direct Methods in Crystallography*, Plenum Press, New York, 1980.

G.E. Bacon, *Neutron Diffraction*, 3rd ed., Clarendon Press, Oxford, 1975.

M.A. Carrondo and G.A. Jeffrey (eds.), *Chemical Crystallography with Pulsed Neutrons and Synchrotron X-Rays*, D. Reidel, Dordrecht, 1988.

R.M. Hazen and L.W. Finger, *Comparative Crystal Chemistry: Temperature, Pressure, Composition and the Variation of Crystal Structure*, Wiley, New York, 1982.

R. Rudman, *Low Temperature X-Ray Diffraction*, Plenum Press, New York, 1976.

B.T.M. Willis and A.W. Pryor, *Thermal Vibrations in Crystallography*, Cambridge University Press, Cambridge, 1975.

General Works on Crystallography

International Tables for X-Ray Crystallography, revised ed., Kluwer, Dordrecht.
    Vol. I, N.F.M. Henry and K. Londsdale (eds.), *Symmetry Groups*, 1952.
    Vol. II, J.S. Kasper and K. Londsdale (eds.), *Mathematical Tables*, 3rd ed., 1972.
    Vol. III, C.H. MacGillavry, G.D. Rieck and K. Lonsdale (eds.), *Physical and Chemical Tables*, 2nd ed., 1983.
    Vol. IV, J.A. Ibers and W.C. Hamilton (eds.), *Revised and Supplementary Tables to Vols. II and III*, 1974.

International Tables for Crystallography, T. Hahn (ed.), Vol. A, *Space-Group Symmetry*, 2nd revised ed., Kluwer, Dordrecht, 1987.

J. Lima-de-Faria (ed.), *Historical Atlas of Crystallography*, Kluwer, Dordrecht, 1990.

P.P. Ewald (ed.), *Fifty Years of X-Ray Diffraction*, Kluwer, Dordrecht, 1962.

J.M. Bijvoet, W.G. Burgers and G. Hägg, *Early Papers on Diffraction of X-Rays by Crystals*, Vol. 1, 1969, Vol. 2, 1972, Kluwer, Dordrecht.

G.E. Bacon (ed.), *Fifty Years of Neutron Diffraction: The Advent of Neutron Scattering*, Kluwer, Dordrecht, 1987.

W.L. Bragg (ed.), *The Crystalline State*, Bell, London.
    Vol. I, W.L. Bragg, *A General Survey*, 1962.
    Vol. II, R.W. James, *The Optical Principles of the Diffraction of X-Rays*, 1962.
    Vol. III, H. Lipson and W. Cochran, *The Determination of Crystal Structures*, 3rd revised ed., 1966.
    Vol. IV, W.L. Bragg, G.F. Claringbull and W.H. Taylor, *Crystal Structures of Minerals*, 1965.

B.K. Vainshtein, A.A. Chernov and L.A. Shuvalov, *Modern Crystallography*, Springer-Verlag, Berlin.
Vol. **I**, B.K. Vainshtein, *Symmetry of Crystals. Methods of Structural Crystallography*, 1981.
Vol. **II**, B.K. Vainshtein, V.M. Fridkin and V.L. Indenbom, *Structure of Crystals*. 1982.
Vol. **III**, A.A. Chernov, *Crystal Growth*, 1984.
Vol. **IV**, L.A. Shuvalov, A.A. Urusovskaya, I.S. Zheludev, A.V. Zalesskii, B.N. Grechus, I.G. Chistyakov and S.A. Semiletov, *Physical Properties of Crystals*, 1984.

J.P. Glusker, B.K. Patterson and M. Rossi (eds.), *Patterson and Pattersons*, Oxford University Press, Oxford, 1987.

J.P. Steinhardt and S. Ostlund, *The Physics of Quasicrystals*, World Scientific, Singapore, 1987. [A collection of reprints.]

I. Hargittai (ed.), *Quasicrystals, Networks, and Molecules of Fivefold Symmetry*, VCH Publishers, New York, 1990.

Sources of Structural Data

W.B. Pearson, *Handbook of Lattice Spacings and Structures of Metals and Alloys*, Vol. 1, 1958, Vol. 2, 1967, Pergamon Press, New York.

G.W. Brindley and G. Brown (eds.), *Crystal Structures of Clay Minerals and Their X-Ray Identification*, Mineralogical Society, London, 1980.

Crystal Data: Determinative Tables, J.D.H. Donnay, H.M. Ondik and co-workers. (general editors), 3rd ed., National Bureau of Standards, Washington, DC.
Vol. 1, *Organic Compounds*, 1972.
Vol. 2, *Inorganic Compounds*, 1973.
Vol. 3, *Organic Compounds 1967-1974*, 1978.
Vol. 4, *Inorganic Compounds 1967-1969*, 1978.
Vol. 5, *Organic Compounds 1975-1978*, 1983.
Vol. 6, *Inorganic Compounds 1979-1981*, 1983.

Structurbericht, Band **I-VII** (1913-1914). [Reprint avaliable from Polycrystal.]

Structure Reports, J. Trotter and G. Ferguson (general editors), Vols. **8-56**, Kluwer, Dordrecht, 1942-1989. [Series **B** on Organic Compounds terminated with Vol. **52B** (1985); Series **A** on Metal/Inorganic Compounds continuing.]

Landolt-Börnstein, Numerical Data and Functional Relationships in Science and Technology, K.-H. Hellwege (editor-in-chief), New Series, Springer-Verlag, Berlin.
Group **III**: *Crystal and Solid Physics*.
Vol. 5, *Structure Data of Organic Crystals*, 1971.
Vol. 6, *Structure Data of Elements and Intermetallic Phases*, 1971.
Vol. 7, *Crystal Structure Data of Inorganic Compounds*, 1973-1985.
Vol. 10, *Structure Data of Organic Crystals, Supplement and Extension to Vol. III/5*, 1985.
Vol. 14, *Structure Data of Elements and Intermetallic Phases, Supplement to Vol. III/6*, 1986-1988.
Group **VII**: *Biophysics*, Vol. 1, *Nucleic Acids*.
Subvolume **a**, *Crystallographic and Structural Data I*, 1989.
Subvolume **b**, *Crystallographic and Structural Data II*, 1989.

R.W.G. Wyckoff, *Crystal Structures*, 2nd ed., Vols. **1-6**, Wiley, New York, 1963-1971.

O. Kennard, D.G. Watson and co-workers (eds.) *Molecular Structures and Dimensions: Bibliography*, Vols. **1-14**, Kluwer, Dordrecht, 1970-1983.

O. Kennard, F.H. Allen and D.G. Watson, *Molecular Structures and Dimensions: Guide to the Literature, 1935-1976*, Kluwer, Dordrecht, 1977.

Profiles, Pathways and Dreams

Beginning in 1990, a twenty-two volume set of the above title has been launched by The American Chemical Society under the editorship of J.I. Seeman. Each volume is written by an eminent *organic* chemist who recounts his career and achievements, and as a whole these autobiographies trace the progress of chemistry, with particular emphasis on organic materials, over the past century. The full list of titles are:

D.J. Cram, *From Design to Discovery*, 1990.

C. Djerassi, *Steriods Made It Possible*, 1990.

E.L. Eliel, *From Cologne to Chapel Hill*, 1990.

R.U. Lemieux, *Explorations with Sugar: How Sweet It Was,* 1990.

J.D. Roberts, *The Right Place at the Right Time*, 1990.

D.H.R. Barton, *Some Recollections of Gap Jumping*, 1991.

M.J.S. Dewar, *A Semiempirical Life*, 1991.

K. Nakanishi, *A Wandering Natural Products Chemist*, 1991.

T. Nozoe, *Seventy Years in Organic Chemistry*, 1991.

V. Prelog, *My 132 Semesters of Chemical Studies*, 1991.

M. Calvin, *Following the Trail of Light: A Scientific Odessey*, 1992.

E. Havinga, *Enjoying Organic Chemistry 1927-1987*, 1992.

R. Huisgen, *Mechanisms, Novel Reactions, Synthetic Principles*, 1992.

H. Mark, *From Small Molecules to Large: A Century of Progress*, 1992.

B. Merrifield, *The Concept and Development of Solid-Phase Peptide Synthesis,* 1992.

T. Mukaiyama, *To Catch the Interesting While Running*, 1992.

W.S. Johnson, *A Fifty-Year Love Affair with Organic Chemistry*, 1992.

F.G.A. Stone, *Organometallic Chemistry*, 1992.

C. Walling, *Fifty Years of Free Radicals*, 1992.

A.J. Birch, *To See the Obvious*, 1993.

P.v.R. Schleyer, *From the Ivy League into the Honey Pot,* 1993.

A. Streitwieser, Jr., *A Lifetime of Synergy with Theory and Experiment*, 1993.

Index

abruslactone A   929
absolute configuration   11, 788
absolute configuration, determination
  from Friedel pairs   789
absolute configuration, designation
  510
absolute structure   791
absolute structure, determination
  792
acetylene complexes, bonding   1040
acetylene complexes, C≡C bond
  parameters   1042
acetylene complexes, types   1039
3-O-acetylgibberellin A$_3$ 928
acetylides, heavy-metal   288
acid fluoride anions   200
acid hydrates   208, 1177
acid hydrates, table   208, 1181
actinide tetracyclopentadienyls
  1143, 1145
adamantane   39, 155
adamantane, homologues   40
adamantane-1,3,5,7-tetracarboxylic
  acid   42
agostic bond   553, 558
alkali metal anions, sizes of   227
alkali metal suboxides, bonding   235
alkali metal suboxides, table   234
alkalide   225
alkalides, mix-metal   230
alkaline earth halides   92
alkaloids, cinchona   951
alkaloids, morphine   948
alkaloids, pharmaceutically important
  952
n-alkanes   750
alkoxide clusters, of Mo and W   606
alkoxide and carbonyl cluster
  chemistry   608
alkylaluminum compounds   1027
alkylaluminum compounds, containing
  macrocyclic amines   1027, 1031
alkylaluminum compounds, containing
  multidentate open-chain amines
  1027
alkylamine hydrates, structural data
  1197
alkylidene aminoborane   271
alkylidene complexes   1054

alkylidyne (carbaborane) tungsten
  complexes   1059
alkylidyne complexes   1058
alkylmagnesium alkoxides, types
  1017
alkylmagnesium compounds, unsolvated
  1014
alkyne complexes of Mo(II) and W(II)
  1042
all-carbon molecules   49
allene complexes   1048
allene complexes, parameters   1049
allyl complexes   1036
alumatranes   327
alumazene   164
β"-alumina   274
β"'-alumina   276
β""-alumina   276
aluminosilicate catalysts   322
[Al(η5-C$_5$Me$_5$)]$_4$ 1031
Al$_5$CuLi$_3$ 444
(Al,Zn)$_{49}$Mg$_{32}$ cubic phase   443
aluminum, π complexes   1026
alums   276
alums, structural types   278
amalgams, alkali-metal   479
amino acids, absolute configuration
  of L-   759
amino acids, D- and L-   759
amino acids, metal complexes   764
amino acids, residues   758
amino acids, residues in proteins
  761
amino acids, side chains   760
ammonium hydrogen fluoride   197
amphiboles   142
amylose, inclusion compounds   1274
amylose, polymorphs   1275
anatase   94
Anderson anion   628
anomalous dispersion   74
antamanide   771
anthracene   171, 174
antibiotics   957, 964
antibiotics, discovery and sources
  958
antibiotics, structural formulae
  965

anticancer activity, of Pt complexes 699
anticancer activity, relation to molecular structure 699
anticancer agents 367
antifluorite structure 90
[Sb^{III}O$_4$] polyhedron 379
[Sb$_2$(d,ℓ-C$_4$H$_2$O$_6$)$_2$]$^{2-}$ ion 380
antimony, mixed-valence compounds 380
apophyllite 308
aragonite 147
aragonite structure, compounds with 148
arene ligands 1085, 1091
L-arginine dihydrate 762
arsa-capped cage molecules 520
arsenic, allotropes 356
arsenic, polyatomic fragments 356
artificial enzymes, cyclodextrin-based 1273
arylmagnesium complexes 1017
asbestos 307
atrane compounds 328
austenite 31
azasilatranes 317
azides, ionic 46
azulene 806
baddeleyite 96
Ba$_{0.62}$Mo$_4$O$_6$ 602
Ba$_{1.14}$Mo$_8$O$_{16}$ 602
BaTiO$_3$ 122
Ba$_2$YCu$_3$O$_6$ 126
Ba$_2$YCu$_3$O$_7$ 126
base-pairs in DNA 706
bases, in nucleic acid 704
bcc structure, elements with 35
bcc structure, interstices in 30
bcc structure, space occupied by spheres 30
benitoite 302
benzene, coordinated to metal cluster 1085, 1091
benzene, crystal structure 159
benzene, hexasubstituted 161
benzene, inorganic analogues 164
benzene, pyridine-like hetero 165
benzene, symmetry 159
benzene, triazatrimetalla 219
benzene, true equilibrium geometry
benzene, valence isomers 160, 840
benzenide 168
benzvalene 840
benzyne complexes, mononuclear 1044

benzyne complexes, polynuclear 1045
benzyne metal complexes 1044
beryl 301
beryllium, structural chemistry 238
beryllium acetate, basic 237
beryllium-oxide-carboxylates 238
beryllium-oxide-nitrate 239
beryllocene 1106
bicyclo[1.1.1]pentasilane 827
bicyclo[1.1.1]stannane 828
1,1'-binaphthyl 862
binaphthyl crown ethers 865
binaphthol crown ethers, chiral 866
binaphthyl macropolycyclic compounds 868
binaphthyl metal complexes 868
birefringence 308
bis(cyclohexadienyl)metal complexes 1104
bis(cyclopentadienyl)metal complexes, geometrical parameters 1129
bis(pentadienyl)metal complexes 1104
bis(titanocene) 1120
Bi(C$_6$H$_5$)$_5$ 385
Bi(C$_6$H$_5$)$_5$, dichroism 385
Bi$_2$(Sr,Ca)$_2$CuO$_{8-x}$ 594
Bi$_2$(Sr,Ca)$_3$Cu$_2$O$_{10-x}$ 594
Bi$_2$Sr$_{3-x}$Ca$_x$Cu$_2$O$_{8+y}$ 593
bismuth clusters, bonding 384
bismuth clusters, cationic 383
bismuth subchloride 382
body-centered cubic (bcc) 30
borabenzene 166
boranes 250
boranes, bonding 252
boranes, *styx* numbers 253
borates 267
borates, structural principles 267
borates, structural units 268
boratranes 327
borax 265
borazine 164
borazone 73
borides, icosahedral 245
borides, table 246
BBB bond, central 252
B-H-B bond, three-center two-electron 252
(B^{III})$_1$(A^{II})$_2$Ca$_{n-1}$Cu$_n$O$_{2n+2.5+\delta}$ 596
BC$_3$ 47
B$_9$C$_2$H$_{11}$$^{2-}$ ion 258
B$_{12}$ unit, icosahedral 241
B$_{12}$H$_{12}$$^{2-}$ analogues 255

$B_{12}H_{12}^{2-}$ anion 253
boron, allotropes 241
boron, α-rhombohedral 241
boron, α-tetragonal 243
boron, β-rhombohedral 244
boron arsenide 247
boron carbides 245
boron compounds, dicoordinate 269
boron halides, polyhedral 255
boron nitride 47
boron phosphide 247
boroxanes 269
Bragg equation 10
Bragg's law 65
brass, γ-form 440
bromocamphor 923
brookite 94
buckminsterfullerene 50, 53
buckyball 50
bullvalene 824
bullvalene, bonding arrangements 825
butterfly clusters 1083
butterfly clusters, containing carbon atom 1083
$(Bu^t)_4As_4S_4$ 923
$Cd(CN)_2$ 1257
$CdCl_2$ 112
CdI_2, polytypes 113
CdI_2 structure, compounds with 112
CdI_2 structure, layer sequence 111
$Cd(en)Cd(CN)_4.2C_6H_6$ 1254
$Cd(NH_2(CH_2)_6NH_2)Ni(CN)_4.p-CH_3C_6H_4NH_2$ 1255
$Cd(NH_2CH_2CH_2OH)_2Ni(CN)_4.C_4H_5N$ 1254
cadmium iodide 111
α-cage (truncated cubo-octahedron) 314
β-cage (cubo-octahedron) 314
caged metal ions 522
caged metal ions, LFSE 522
caged metal ions, redox properties 522
caged metal ions, strain effect 522
calcite 146
calcite structure, compounds with 148
CaBiN 221
$CaCO_3$, polymorphs 147
CaNiN 221
$CaTiO_3$ structure, compounds with 120

$CaTiO_3$ structure, conditions of formation 120
$CaTiO_3$ structure, derivative frameworks of 123
$CaTiO_3$ structure, distortion 122
calcium carbide 285
calcium carbide, isomorphous compounds 286
calcium carbide, modifications 285
calixarenes 1223, 1227
calixarenes, conformations 1224
calixarenes, inclusion compounds 1224
calomel 473
calthrasils 1185
calthrasils, cages 1186
calthrasils, structural data 1186
camphor, bromo 923
cantoniensistriol 929
carbametallaborane complexes 1059
carbene, first crystalline 1052
carbene complexes, bond lengths 1054
carbene complexes, dihalo 1056
carbene complexes, Fischer 1052
carbene complexes, Schrock 1054
carbide, scandium 289
carbide, thorium 286
carbides, C-C bond 287
μ-carbido-bridged complexes 1087
carbidocarbonyl metal clusters 1076, 1091
carbidocarbonyl metal clusters, radius of carbon atom 1076
carbidocarbonyl mixed-metal clusters 1078
carbidocarbonyl Rh_6 clusters, structural parameters 1079
carbocycles, eight-membered 802
carbohydrates 913
C_{60} 50, 53, 55, 182
C_{70} 53, 182
$[(n-C_4H_9)_4N]H_7[Si_8O_{20}](128/24)H_2O$ 321
$(C_5H_5)Fe(B_9C_2H_{11})$ 256
$(C_5H_5NH)Ag_5I_6$ 451
$\{(\eta^5-C_5Me_5)_2Ti\}_2N_2$ 1119
$C_6(CH_2NRCH_2Ph)_6.3KSCN.H_2O$ 1235
C_8K 46
$[C_{18}H_{36}O_6N_2K^+]_6Ge_9^{2-}\cdot Ge_9^{4-}.5/2(C_2H_8N_2)$ 330

C-H...O hydrogen bond 296
C-H activation, in coordinated
  catenand 528
carbon atom, "exposed", on carbido
  cluster 1083
carbon chain, conformations 814
carbon fibers 47, 48
carbon nitride, hypothetical
  structure 217
carbon oxides 50
carbon steels 34
carbon tubes 182
carbonate ion, absolute electron
  density 149
μ_6-carbonato ligand 485
carbonyl ligand, bridging effects on
  M-M bond 1081
carbonyl ligand, coordination modes
  1067
carboranes 257
carbyne complexes 1058
carcerand 1287
[2]catenane 1283
[2]catenane, synthesis 1284
catenands 525
catenanes 525
cellobiose 918
cellulose 918
cellulose fiber 920
cementite 32
ceside 225
$Cs^+(18C6)_2.Cs^-$ 226
$Cs^+(18C6)_2.e^-$ 229
$Cs^+(C222).Cs^-$ 226
Cs_7O 232
$Cs_{11}O_3$ 232
$Cs_{11}O_3$-Cs and $Cs_{11}O_3$-Rb, phase
  diagrams 234
CsCl structure, compounds with 70
CsCl structure, for intermetallics
  70
CsCl structure, volume occupied by
  spheres versus r_A/r_B 71
$CsF.Br_2$ 67
cesium chloride 237
cesium iron fluoride 501
cesium suboxide 232
chain polymers 812
chain polymers, configurations 813
chalcocarbonyl complexes 1070
chalcogenides, inter-alkali metal
  91
chalcogenides, pnictide 374

chemotherapeutic agents 964
chemotherapy 957
chiral discrimination 1240
chiral recognition 865
chirality, multiplication 868
$[Cl(HCl_4)_4]^-$ 203
chlorotris(triphenylphosphine)-
  rhodium (I) 534
cholesterol, hemiethanolate 932
cholesterol, polymorphs 936
cholesterol, structural features
  936
cholesteryl acetate 946
cholesteryl iodide 946
$[Cr(en)_3][Ni(CN)_5].1.5H_2O$ 514
chrysotile 307
circulenes 875
cis-1,4-polyisoprene 816
cisplatin 694
clathrasils 100, 1185
clathrasils, polyhedral cavities
  102
clathrate, hexakis(*p*-hydroxyphenyl-
  oxy)benzene with pyridine 1232
clathrate hydrate, tetra-*n*-butyl-
  ammonium benzoate 1191
clathrate hydrates 1176
clathrate hydrates, alkylammonium
  salt 1194
clathrate hydrates, alkyl-onium salt
  1195
clathrate hydrates, strong monobasic
  acids 1181
clathrate hydrates, structural data
  1178
clathrate hydrates, structural types
  1177
clathrate hydrates, tetramethylammonium hydroxide 1183
clathrate hydrates, type II 1179
clathrates, tetra-1-naphthoide
  cyclohexananone 1244
clathrochelating ligands 518
closest packed layer 22
cluster of clusters 468
cobalamine, adeninylpropyl 983
cobaloxime 1164
cobaloxime, (*R*-1-cyanoethyl)(3-
  methylpyridine) 1164
cobaloxime complexes, racemization
  1167
$[Co(C_{12}H_{30}N_8)]S_2O_6.H_2O$ 517
$[Co(en)_2(NO_2)]_2(H_3O_2)^{3+}$ 213
$(+)_{589}[Co(en)_3]_2Cl_6.NaCl.6H_2O$ 509

$[Co(NO_2)_6]^{3-}$, difference electron-density 514

cobalt inosine 5'-phosphate heptahydrate 702

cobalt sepulchrate complex 517

cobaltocenium salt 408

codeine 947

codeine, absolute configuration 947

codeine, hydrobromide dihydrate 947

coesite 100

color center 66

columbite 96

columnenes 856

conformational selection 1234

conjugate dienes, complexes 1036

coordination polyhedron, in ccp 23

Cope rearrangement 824

Cu_5Cd_8 441

Cu_9Al_4 441

copper 22

cooper(I) halides, polymorphs 448

copper oxide superconductors, local charge 598

copper phthalocyanine 582

copper sulfate pentahydrate 134

copper-aluminium alloy 443

copper-cadmium alloy 442

copper-zinc, alloys 441

copper-zinc, phase diagram 442

corannulene 51

cortisone 941

cristobalite 100

18-crown-6 864

18-crown-6, molecular complexes 864

18-crown-6, "threaded" complex 1017

crown ethers 863

crown thioethers 872

cryptand 224

cryptand(222) 993

crystal 2

crystal, optically anomalous 308

crystalline substances in Chinese alchemy texts 2

crystalline-state reaction 1166

crystalline-state reaction, dynamical structure analysis 1167

crystallization 2

cubane 830

cubanes, C-C bond lengths 830

cubic closest packing (ccp) 3, 23

ccp structure, elements with 25

ccp structure, holes in 24

ccp structure, layer sequence 23

ccp structure, space occupied by spheres 24

cuprous halides, polymorphs 448

1-cyanotetracyclo(3.3.1^{3,7}.0^{3,7})-decane 822

cyclic hexaglycyl 770

cyclic III-V compounds 283

cycloalkanes 40

cycloalkyne metal complexes 1045

cycloarenes 873

cyclobis(paraquat-*p*-phenylene) 4PF$_6$.3MeCN 1281

cyclobutadienophane bis(tri-carbonyliron) 1136

cyclobutadienophane bis(cyclo-pentadienylcobalt) 1136

α-cyclodextrin 1258, 1261

β-cyclodextrin 1260, 1261

β-cyclodextrin inclusion compound, polymeric 1272

γ-cyclodextrin 1260, 1261

cyclodextrin inclusion compounds, cage-type 1263

cyclodextrin inclusion compounds, channel-type 1261

cyclodextrin inclusion compounds, containing inorganic guests 1270

cyclodextrin inclusion compounds, containing organometallic guests 1270

(α-cyclodextrin)$_2$Cd$_{0.5}$.I$_5$.27H$_2$O 1258

(α-cyclodextrin)$_2$Cd$_{0.5}$I$_5$.27H$_2$O, polyiodide chain in 1259

cyclodextrins, hydrated 1265

cyclodextrins, hydrated, circular hydrogen bonds in 1267

cyclodextrins, hydrated, flip-flop hydrogen bonds in 1267

cyclodextrins, molecular shapes 1262

cyclodextrins, properties 1261

cyclodextrins, structural features 1260

cyclooctatetraene 799

cyclooctatetraene, benzannelated 801

cyclooctatetraene, valence isomers 800

cyclooctatetraene, vanadium complex 1145

cyclooctatetraenes, planar 802

cyclooctatetraenes, silver(I) complexes of 803

cyclopentadienyl derivatives of group IIB metals 1114

cyclopentadienyl-actinide complex 1145

cyclophanes 850

cyclophanes, historical importance 850

cyclophanes, intra-annular π-π interaction 852

cyclophanes, multibridged 853

cyclophanes, Si- and Ge-bridged 859

cyclophanes, structural description 850

cyclophanes, structural features and properties 851

cyclophanes, transannular π-π interaction 852

cyclophanes, with metal-stabilized cyclobutadiene stacks 1136

cyclophosphazanes, as anticancer agents 367

cyclophosphazanes, table 369

cyclopolyindans 756

cysteine, binding modes to metals 765

cystine peptide, cyclic 779

d-orbitals, electron populations 138

Dawson anion 626

DCNQI CT-complexes 898

DCNQI-radical anion salts 898

decaborane(14) 249

decagonal phase 446

decamethyltitanocene 1121

decamethylzirconcene 1121

decaphenylstannocene 342

decaselenium hexafluoroantimonate 414

deformation density 882

deformation density, in dimanganese decacarbonyl 1062

deformation density, in fused benzenes 845

deformation density, in hydrogen peroxide 389

deformation density, in metal porphyrin 670

deformation density, in organic molecules 882

deformation density, in oxalic acid 292

deformation density, in tetraphenyl-butatriene 883

deformation density, in *trans*-bis(η^3-allyl)nickel 1036

deformation density, in s-triazine 882

deformation density, quadruple bond in $Cr_2(mph)_4$ 721

deformation density and bonding 884

deoxyribonucleohelicate 532

deoxyribose, in nucleoside 705

D_2S, structural data 109

Dewar benzene, derivatives 839, 841

Dewar 2-phosphabenzene 842

Dewar pyridine, derivatives 841

Dewar-Chatt-Duncanson model 1033

dialkylmagnesium complexes 1014

diarylmagnesium complexes 1017

diamond 10, 37

diamond, hexagonal 39

diamond molecules 39

diamond structure, electron density 38

diamond structure, elements with 39

diamond structure, space occupied by spheres 38

diamondoid networks 42

Dianin's compounds 1218

Dianin's compounds, crystal data 1219

S-Dianin's compounds, cavity 1221

S-Dianin's compounds, clathrate of CCl_4 1223

S-Dianin's compounds, clathrate of 2,5,5-trimethylhex-3-yn-2-ol 1220

diarylmagnesium complexes 1017

diazametallocenes 1105

diazene (N_2R_2) complexes 640

diazenido (N_2R) complexes 640

dibenzene chromium 1098

μ-dicarbido-bridged complexes 1088, 1091

8,8-dichlorotricyclo(3.2.1.0^{1,5})-octane 821

dicyanoquinonediimines (DCNQI) 898

diene complexes, conjugated 1036

diethylamine hydrate 1199

digermenes 908, 910

dihydrofolate reductase 887

dimanganese decacarbonyl 1061

dimanganese decacarbonyl, Mn-Mn bonding in 1062

dimanganese decacarbonyl, other carbonyls related to 1064

dimetal sandwich 1105

dinitrogen complexes 1119

dinitrogen complexes, bonding nature 637

dinitrogen ligand, coordination modes 634

dinuclear complexes, di-bridged 726

dinuclear complexes, with S_n^{2-} ligands, 561

diopside 139

diorganomercury(II) compounds, table 1147

dioxygen complexes 662, 665

diplatinum complexes, tetra-bridged 725

dipotassium octachlorodirhenate(III) dihydrate 718

direct methods 9, 253

DNA 706

DNA, conformation 709

DNA-RNA hybrid 711

disilenes 905, 910

disilenes, MO diagram 906

disilenes, reaction products 906

distannenes 910

distyrylpyrazine 796

disulfur dintrogen, polymerization 410

dodecahedranes 834

dodecasil 3C 1188

dodecatungstophosphoric acid hexahydrate 618

double hcp structure 29

double helices, inorganic 530

double-helix, DNA 11, 706

electrides 228

electron compounds, table 443

electron density function 8

electron populations on d-orbitals 138, 582

electron-deficient compounds 12

electron-density map, anthracene 170

electron-density map, and bond length 171

electron-density map, naphthalene 170

electrostatic valence rule 11

enantiomeric excess 1241

enantiomeric selectivity 1244

enantiomorphs 788

κ-(ET)$_2$Cu[N(CN)$_2$]Br 897

ethylene oxide deuterohydrate 1174

ethylene oxide deuterohydrate, hydrogen bonding in 1175

(Et)$_2$X conductors 895

(Et)$_2$X conductors, corrugated sheet network 896

(Et)$_2$X superconductors, β-phase 930

(Et)$_2$X superconductors, κ-phase 930

ethylmagnesium bromide dietherate 1012

face-centered cubic (fcc) 23

feldspars 310

feldspars, geological evolution 311

fenestranes 823

fenestridan 756

ferredoxin (Fd) 646

ferrite 31

ferrocene, aza-, phospha-, arsa-, stiba- and bisma- 1105

ferrocene, bonding 1095

ferrocene, configuration 1092

ferrocene, MO diagram for 1098

ferrocene, open 1103

ferrocene, orthorhombic 1093

ferrocene, pseudo- 1104

ferrocene, structural data 1095

ferrocene, triclinic 1093

ferrocene compounds 1094

ferrocenium compounds 1095

ferrocyne complex 1050

ferrodicyne complex 1050

fluorite 89

fluorite-like compounds 89

fluorite-type oxides 91

fluoronium ions 202

fluoroquinolone 967

fluospar 89

forsterite 297

Fourier synthesis, first application to crystal structure 140

Frank-Kaspar phases 445

Friedel's law 75, 789

fullerene-60 50, 53, 55, 182

fullerene-70 53, 182

fullerene superconductors 182

furochlorophin 742

fused benzenes, strained 844

gallaborane 252

Ga, In and Tl halides, mixed-valence 281

gallium dichloride 280

gallium tetrachlorogallate 280

gas hydrates 1176

genins 941

geranylamine hydrochloride 744

germaethenes 908

germanazene 164

Ge=C bond 908

Ge=Ge bond 909

gibberellins 928

D-glucose 914

glycine, α- 757

glycine, β- and γ- 759

glycine, structural data 762

α-glycylglycine 767

Au$_{11}$I$_3$[P(p-C$_6$H$_4$F)$_3$]$_7$ 463

gold clusters 464
gold clusters, relativistic effects 470
gold complexes, linear chain 469
gold complexes, mixed-valent 468
gramicidin A, ion channels in 993
gramicidin A, nucomplexed 994
graphene 47, 48
graphite 43
graphite, comparison with diamond 44
graphite, icospiral particles 51
graphite, intercalation compounds 45
graphite, rhombohedral 44
graphite-like materials 46
Grignard reagent, contaminated 239
Grignard reagents 1012
Grignard reagents, constitution 1013
group IV atoms, high symmetry environments 341
group VA pentaphenyls 385
group VA pentaphenyls, stereo-chemistry 386
guest 1175
hafnium cyclopentadienyl complexes 1124
hafnocene complexes 1125
halocarbene complexes 1056, 1057
halogens 59
hcp structure 26
hcp structure, axial ratio 28
hcp structure, comparison with ccp 26
hcp structure, coordination polyhedron in 27
hcp structure, elements with 28
hcp structure, holes in 26
hcp structure, layer sequence 27
heavy-atom method 9, 579
helicate 530
helicenes 876
helicoidal complex 528
α-helix 11, 774, 775
γ-helix 774, 775
heme group 779, 781
heme proteins, oxygen binding 689
hemoglobin 781
hemoproteins, stereochemistry 688
heptalene 807
heteropolyanions, reduced Keggin-type 624
heteropolyanions, types of heteroatom 624

heteropolyanions, with As heteroatom 630
heteropolyanions, with Ce heteroatom 629
heteropolyanions, with octahedral heteroatom 628
heteropolyanions, with tetrahedral heteroatom 623
heteropolyatomic anions of Sn and Pb 332
hexa-host, examples 1231
hexa-host, geometric concept 1229
hexa-host, inclusion compounds 1230
hexaarsabenzene 1103
hexaarsenabenzene 166
hexaazabenzene 167
hexaazaoctadecahydrocoronene 162
hexabromoantimonate 380
hexachlorocyclophosphazene 364
hexachlorodigallium, electron diffraction 280
hexagermaprismane 844
hexagonal closest packing (hcp) 4, 24
hexakis(benzylthiomethyl)benzene-dioxan 1:1 clathrate 1228
hexamethylbenzene 10, 158
hexamethylenetetramine (urotropine) 10, 151
hexamethylenetetramine, historical significance 152
hexamethylenetetramine, in organic synthesis 152
hexamethylenetetramine, molecular adducts 154
hexamethylenetetramine, quaternized derivatives 155
hexamethylenetetramine, structural analogy 155
hexamethylenetetramine, structural chemistry 153
hexamethylenetetramine hexahydrate 1213
hexaphosphabenzene 166, 1102
hexatellurium tetrakis(hexafluoro-arsenate)-arsenic trifluoride (1/2) 423
high-albite 311
high-coordinate complexes, formation of 494
high-coordinate complexes, stereo-chemistry 495
high-potential iron proteins (HiPIP) 646
high-temperature supperconductors 125, 589

Hofmann-type clathrates 1248
Hofmann-type clathrates, building blocks 1250
Hofmann-type clathrates, history 1249
Hofmann-type clathrates, modification of 1253
Hofmann-type clathrates, structural data 1250
homogeneous asymmetric catalysis 868
host 1175
host design 1229
host-guest chemistry 13, 1293
hydrate inclusion compounds 1175
hydrate inclusion compounds, categories 1176
hydrates, alkylamine 1176, 1197
hydrates, layer-type 1176
hydrates, peralkylated onium salt 1176
hydrates, quarternary ammonium salt 1200
hydraton number 1199
hydrazido(2-) ($N_2R_2^{2-}$) complexes 640
hydrocarbon chain compounds 752
hydrochloric acid dihydrate 204
hydrochloric acid-water compounds 204
$[H_2C(CH_2)_5NLi]_6$ 220
$H_nF_{n+1}^-$ anions, 200
$H_3F_2^+$ 201
$(H_2O.X^-)_2$, structural parameters 1205
$H_3O_2^-$ bridging ligand 1217
$H_5O_2^+$ ion 206
$H_5O_2^+$ ion, compounds 206
hydrogen bonds, asymmetric strong 198
hydrogen bonds, S-H···S 110
hydrogen bonds, strongest 197
hydrogen bonds, trifurcated 201
hydrogen bonds in oxalic acid, deformation density 295
hydrogen difluoride anion 196
hydrogen difluoride anion, MO treatment 198
hydrogen oxide bridges 212
hydrogen oxide ion $H_3O_2^-$ 211
hydrogen peroxide 388

hydrogen peroxide, dihedral angles 390
hydrogen peroxide, hydrogen bonds 389
hydrogen peroxide, lone pairs 389
hydrogen peroxide, reactions 391
hydrogen sulfide, deuterated 296
hydrogen sulfide, ligand 110
hydrogen sulfide, solid phases 109
hydronium ion 206
hydroporphinoid Ni(II) complexes 679
hydroporphinoid Ni(II) complexes, ruffling of 680
α-hydroquinone 1211
γ-hydroquinone 1211
β-hydroquinone clathrates 1207
β-hydroquinone clathrates, types 1208
β-hydroquinone clathrates, with H_2S 1207
β-hydroquinone clathrates, with SO_2 1207
hydrosilylation 540
4-*p*-hydroxyphenyl-*cis*-2,4-dimethyl-chroman carbon tetrachloride clathrate 1217
hypofluorous acid 202
ice 185
ice, high-pressure polymorphs 185
ice, interpenetration framework 194
ice, phase diagram 186
ice, proton order/disorder 188
ice, structural features 194
ice I, ordered phase 108
ice Ic 107
ice Ih, distribution of H atoms 106
ice Ih, residual entropy 106
ice Ih 104
ice II 186
ice III 188
ice IV 189
ice V 190
ice VI 191
ice VII 192
ice VIII 193
ice IX 188
icosahedral phase 446
icospiral graphite particles 52
imidazole 885
iminoboranes 269
iminophosphanes 370
inclusion compound, formation mechanism 1268
inclusion compounds 1175
inclusion compounds, amylose 1274

inclusion compounds, containing inorganic and organometallic guests 1270

inclusion compounds, hexakis(phenyl-thio)benzene with CCl_4 1231

inclusion compounds, hexakis(*p-t*-butyl-phenylthiomethyl)benzene with squalene 1234

inclusion compounds, history 1208

inclusion compounds, thiourea-halide 182

inclusion compounds, urea-water-halide 182

In_4Se_3 281

indium-chlorine system 285

instrumentation, advances 14

insulin, dimeric pig 785

insulin, hexamer 783

insulin, rhombohedral 2Zn pig 782

insulin, structural formula 782

insulin, structure 781

insulin, variation in structure 782

inter-alkali metal chalcogenides 91

intercalation compound $(Te_2)_2(I_2)_x$ 61

interlocked macrocycles 525

interstices, in ccp 24

interstitial hydride complexes 550

iodine 57

ionic crystal fragment 67

ionic crystals, principles 11

ionophores 987

ionophores, classification 991

$IrCl(CO)(PPh_3)_2 \cdot Ag(B_{11}CH_{12})$ 667

$IrO_2Cl(CO)[P(C_6H_5)_3]_2$ 659

Fe atom, electron distribution in FePc 581

FeMo-cofactor, in nitrogenase 647

$Fe(O)_2(T_{piv}PP)(2\text{-MeIm}) \cdot EtOH$ 686

Fe_2S_2 rhombs 654

$[Fe_4S_4(\eta^5\text{-}C_5H_5)_4]^n$ 647

$[Fe_4S_4(SH)_4]^{2-}$ 647

$Fe_4S_4(SPh)_4^{2-}$ 644

Fe_6S_6 core and related compounds 648

$Fe_6S_6Mo_2$ core and related compounds 648

Fe_7S_6 core and related compounds 648

Fe_7S_8, structure 85

$[Fe_{18}S_{30}Na_2]^{8-}$ core 650

α-iron 30

iron, phase transition 31

iron porphyrins 669

iron(II) phthalocyanine 578

iron(II) porphyrin 688

iron(II) porphyrin, capped strapped 690

iron(II) porphyrin, CO complexes 692

iron(II)-O_2 porphyrin complexes 686

iron(II)-O_2 porphyrin complexes, picket fence 687

iron-sulfur clusters, cubane-type 651

iron-sulfur clusters, stereochemistry 654

iron-sulfur proteins 645

iron-sulfur tetrancuclear cluster 644

iron-sulfur tetranuclear clusters, dimensions in proteins 647

iron-sulfur tetranuclear clusters, energy-level diagram 648

iso-propylamine hydrate 1197

isoascorbic acid 979

isocitric acid 786

isolobal analogy 262

isomorphous replacement 9, 11, 579

isopolyanions, of molybdates 622

isopolyanions, of niobates 620

isopolyanions, of tantalates 620

isopolyanions, of tungstates 622

isopolyanions, of vanadates 620

isoprene rule 924

isostearic acid 752

Jahn-Teller distortion 137

kalide 229

keatite 101

Keggin structure 618

kekulene 873

Kemp's acid 1294

Kohnkene 1296

β-lactam compounds 958

β-lactam compounds, parameters 959

β-lactam compounds, stereochemical features 960

lactose 915

lactose, morphology 915

Ladenburg benzene 840

$La_{1.85}Sr_{0.15}CuO_4$ 126

lattice 3

Laue equation 5

lead oxides, table 342

lead(II) compounds, lone pair effects 341

lead(II) hexathiourea cation 345
lead(II) oxide 96, 341
lead(II) perchlorate hydrate, basic 339
[Leu^5]enkephalin 771, 778
LFSE effect 116
LiCaN 221
Li$_n$[FeN$_2$] (n = 3, 4) 223
Li[(HgC$_2$B$_{10}$H$_{10}$)$_4$Cl] 1151
Li(ND$_3$)$_4$ 220
lithium, coordination compounds 223, 1011
lithium, π complexes 1003
lithium nitride 214
lithium nitride, as solid ionic conductor 216
lithium nitride, difference density map 216
lithium nitride, phase transformation 215
lithium phosphorus(V) nitride 223
low-albite 309
lysozyme 11
M-H-M bridged compounds 545
M(NH$_3$)$_2$Ni(CN)$_4$.2G 1250
Mackay icosahedra 737
macromolecular crystallography 14
Mg$_2$Sn 10, 90
magnesium 26
magnesium, π complexes of 1020
magnesium alkyls 1015
magnesium aryls 1017
magnesium-anthracene complexes 1018
magnesocene 1020
Mn(NH$_3$)$_2$Ni(CN)$_4$.2C$_6$H$_6$ 1248
martensite 34
melamine 886
melanophlogite 1187
membrane protein, structure 14
mercurials 1147
Hg$_2$ compounds, table 475
Hg$_{3-\delta}$AsF$_6$ 478
mercury clusters, table 475
mercury cyclopentadienyl complexes 1150
mercury(I) compounds, evidence for Hg$_2^{2+}$ 474
mercury(II)-thiolate complexes 1151
metal carbonyl hydrides 545
metal carbonyls, binary 1065
metal carbonyls, coordination modes 1066
metal carbonyls, M-CO bonding 1065

metal clusters, butterfly 1083
metal clusters, containing arene ligands 1085
metal clusters, containing encapsulated nitrogen atoms 1083
metal clusters, electron counting 732
metal clusters, high nuclearity 731
metal clusters, containing selenide and phosphine ligands 738
metal clusters, ligand stabilized 738
metal clusters, table 734
metal clusters, with ccp cores 740
metal complexes containing agostic bonds 554
metal complexes, containing ligands related to CO 1069, 1073
metal complexes, containing planar η^8-COT ligands 1141
metal cyclobutadienyl complexes, bonding in 1132
metal cyclobutadienyl complexes, structural features 1133
metal cyclooctatetraene complexes 1140
metal disulfide complexes 566
metal hydrido complexes 548
metal nitrosyl complexes 1070
metal silene complexes 907
metal thiolate complexes, structural principles 574
metal thiolates 571, 573, 574
metal-metal bonding, in transition-metal fluorides 508
metal-metal bonds, non-integral order 724
metal-metal double bonds, reaction types 723
metal-metal double bonds, table 724
metal-metal multiple bonds 729
metal-metal quadruple bonds, bonding scheme 719
metal-metal quadruple bonds, reaction types 721
metal-metal quadruple bonds, table 719
metal-metal single bonds 724
metal-metal single bonds, table 724
metal-metal triple bonds 720
metal-metal triple bonds, reaction types 722
metal-metal triple bonds, table 721
metal-olefin complexes, structural features 1033
metal π complexes 13, 1109

metallicenium salts 1128
metalloboranes 1121
metallocarboranes 1122
metallocene, diaza 1105
metallocene complexes, as antitumor agents 1126
metallocene dichlorides, structural parameters 1127
metallocene dichlorides, structure-activity relation 1127
metallocenes 1099
metallocenes, bent 1100
metallocenes, half-sandwich 1100
metallocenes, open 1103
metallocenes, pseudo- 1103
metalloporphyrin complexes, cofacial 678
metalloporphyrins, five-coordinate 674
metalloporphyrins, nonplanar 676
metalloporphyrins, out-of-plane 675
metalloporphyrins, structural parameters 674
metalloreceptors 1292
methanetetraacetic acid 42
$[Me_2GaAs(t\text{-}Bu)_2]_2$ 283
$[Me_2InAsMe_2]_3$ 283
methyl derivatives of heavier alkali metals 1005
methyl p-bromocinnamate 794
methyllithium 998
methylmercury(II), pollution of the environment 1148
methylpotassium 1007
methylsodium, cubic 1005
methylsodium, deuterated orthorhombic 1005
mica groups 306
Miller indices 3
mini-superphane 856
mixed-valent compounds 457
mixed-valent compounds, of copper 458
mixed-valent compounds, of gold 460
mixed-valent compounds, of main group elements 457
molecular belts 1284
molecular biology 12
molecular clefts 1294
molecular collars 1284
molecular complexation 863
molecular compounds 13
molecular containers 1284
molecular hydrogen complexes 550
molecular knots 524

molecular knots, template sythesis 524
molecular nitrogen complexes 633
molecular oxygen carriers 661
molecular pentad 683
molecular recognition 524, 1289
molecular recognition, anionic substrate 1290
molecular recognition, linear 1291
molecular recognition, multiple 1291
molecular recognition, spherical 1298
molecular recognition, tetrahedral 1298
molecular rods 1294
molecular sieves 313
molecular size 1190
molecular volume 1190
$[Mo(CN)_8]^{4-}$ 494
Mo-O bond strength 602
Mo-S compounds 560
$[Mo_3O_2(C_2H_5COO)_6(H_2O)_2]_2(H_3O_2)Br_3 \cdot 6H_2O$ 210
molybdenum alkoxide clusters 606
molybdenum chalcogenides 605
molybdenum chalcogenides, ternary 604
molybdenum oxides, ternary 600
molybdoferredoxin 655
monensin A 989
monocyclopentadienyl complexes of beryllium 1107
mononuclear complexes, with S_n^{2-} ligands 560
monophosphazenes 370
monosaccharide 914
monoterpenoids, table 926
morphine 948
Moseley's law 234
muscovite 304
myoglobin 11, 779
naphthalene 169, 174
naphthalene bow 51
$1',8':3,5$-naphtho[5.2.2]propella-3,8,10-triene 839
natride 224
$Ni(H_2O)_2Ni(CN)_4 \cdot nH_2O$ 1252
Ni_2In, structure 258
$Ni_{34}Se_{22}(PPh_3)_{10}$ 738
NiAs structure, compounds with 82
nickel arsenide 81
nickel phthalocyanine 580

nitrides, covalent 217
nitrides, ionic 217
nitrides, ternary 221
nitrides, transition-metal 218
N_2H_x ligands 641
$(NEt_4)_2(Co_6C(CO)_{13})$ 1074
$(NH_4)_4[Mo_3S_2(NO)_3(S_2)_4(S_3NO)].3H_2O$ 559
$[NI_4]$ tetrahedra 346
$N_4(SF)_4$ 407
$(NO_2)ONa_3$ 128
$(NO)_n$ 413
nitrogen fixation 634, 655
nitrogen halides 349
nitrogen heterocycles, single ring 884
nitrogen triiodide, adduct with ammonia 346
nitrogen triiodide, amine adducts 348
nitrogenase 655
nitrogenase model compounds 641
nitrogenpentamminineruthenium(II) dichloride 633
Nobel lectures 20
Nobel prizes, the X-ray connection 6
non-benzenoid systems, bicyclic 805
non-benzenoid systems, tricyclic 808
non-centrosymmetric crystals, X-ray analysis 791
nonacarbonyldiiron 1064
nonactin 992
nonanuclear germanium anions 330
nonasil-$[4^{1}5^{8}]$ 1188
nonplanarity, aromatic rings 852
nucleic acid 703
nucleic acid, conformation 708
nucleic acid, interaction with metals 713
nucleoside 703
nucleotide 703
nylon 66 818
octa-host 1231
octahedral holes 24
octasulfur 396
olefin hydroformylation, mechanism 537
olefin hydrogenation, catalytic cycle 538
olefin hydrogenation, Halpern mechanism 538
olefin hydrogenation, mechanism 536
oligobipyridine ligands 530
olivine 297

one-dimensional electrical conductor 611
opiate drugs, conformation 949
opiate drugs, receptor binding 949
opiate drugs, table 949
optical data storage meterials 584
optically anomalous crystals 308
optically-active complexes 511
organic clays 1225
organic clays, layer structure 1226
organic metal 891, 901
organic pigments 579
organic superconductor 891
organoaluminum compounds 1024
organoaluminum compounds, carbon-bridged 1024
organocobaloximes 1170
organocyclophosphanes 362
organolithium chemistry 1000
organolithium compounds, containing Li_4 tetrahedra 1001
organolithium compounds, with bridging lithium atoms 1002
organomagnesium halide solvates 1014
organometallic compounds, discovery of 999
organometallic ligands, table 999
organonitrogenlithium compounds 1011
organosodium compounds 1011
organothiolate ligands 570
organotin compounds, coordination types 1153
organotin compounds, reaction pathway at tetrahedral tin 1155
organotin compounds, structural features 1154
organotin compounds, with Sn-Sn bond 1158
organotin oxycarboxylates 1159
organotin-transition metal compounds 1158, 1163
ovalene 878
oxalates, long C-C bond in 294
α-oxalic acid dihydrate 291
oxalic acid, hydrogen bond 291
oxalic acid, reducing properties in relation to C-C bond 294
oxide superconductors, factor influencing the T_c of 597
oxide superconductors, history of discovery 592
oxides, fluorite-type 90
oximes 753
oxo-bridged iron(III) complexes 680
O-H...O hydrogen bond 205

O-O bond, parameters 391
oxygenated cobalt complexes 663
oxytetracyclines, forms of 962
oxytetracyclines, interactions with
  metals 963
ozonides, table 392
pagodane 837
[2.2]paracyclophane 849
paracyclophanes, multilayered 854
paraquat bisparaphenylene-34-
  crown-10, inclusion compound 1278
paraquat 1,5-dihydroxynaphtho(38)-
  crown-10 ether 1280
paraquat 1,5-dihydroxynaphtho(44)-
  crown-12 ether 1280
Patterson function 8
penicillin 11, 956
penicillin, potassium benzyl- 957
penicillin, sodium benzyl- 956
penicillins, designation 959
penicillins, natural 959
pentacene 171
pentagonal dodecahedron 1176
pentagonal Frank-Kaspar phases 445
pentalene 806
pentalene, diphospha 807
pentaphenyls, group VA 385, 386
pentaphyrin, uranyl 685
pentastanna[1.1.1]propellane 828
peptide bond 767
peptide bond, ρ_{X-N} for 768
peptide group, charge distribution
  769
peptide group, conformations
  769, 770
peptide group, geometry 769
perchlorocoronene 875
perovskite 119
peroxides, oxygen-oxygen bond
  distance 392
phase inequalities 1106
phase problem 7
phenol clathrates 1214
o-phenylenemercury, modifications
  1147
o-phenylenemercury, trimer 1146
phospha-alkyne compounds 1050
phospha-capped cage molecules 520
phosphazanes 369
phosphazenes, bonding 366
phospholipid 756
phosphonitrilic halides, cyclic 365
$P_4N_{10}{}^{10-}$ 223
phosphorus, allotropes 352
phosphorus, cubic black 352

phosphorus, interconversion of
  various forms 353
phosphorus, monoclinic violet 354
phosphorus, orthorhombic black 352
phosphorus, polyatomic fragments 356
phosphorus, rhombohedral black 351
phosphorus oxides 375
phosphorus oxysulfides 375, 377
phosphorus sulfides 373
photopolymerization 796
photopolymerization, diolefin
  crystals 796
photopolymerization, lattice-
  controlled 796
phthalocyaninatopolymetalloxanes 586
phthalocyaninatopolymetalloxanes,
  crystallographic data 586
phthalocyanine 11
phthalocyanine, metal complexes 578
phthalocyanine sandwich compounds
  583
π complexes, Ca, Sr and Ba 1110
π complexes, Ga, In and Tl 1111
π complexes, Ge and Pb 1112
π complexes, heavier alkali metals
  1109
π complexes, mixed-metal 1116
π complexes, Si(II) 1111
Piedfort unit 1235
planar P_X and As_X fragments in
  complexes 359
plane groups 3
$Pt(Ph_3P)_2(CF_3C\equiv CCF_3)$ 1038
Pt-intercalator complexes 698
platinum complexes, of DNA
  constituents 697
pnictide chalcogenides 374
poly(3-methyl-1-butene), *it-* 816
poly(*p*-phenylene oxide) 815
polyarsenide clusters, heteroatomic
  358
polyatomic anions of Ge, Sn and Pb
  331
polyauriomethanes 468
polycarbons 49
polychalcogenide anions, chains 426
polychalcogenide anions, cyclic 426
polychalcogenide cations, cage
  structures 425
polyethylene adipate 811
polyhedral cages, in zeolites 316
polyhedral inclusion, principle
  732
polyhedral models in structural
  chemistry 1189
polyhydride complexes 552

polyiodide anions 59
polyiodide cations 61
polyisobutylene 816
polymer, atactic (*at*) 814
polymer, diisotactic (*dit*) 814
polymer, disyndiotactic (*dst*) 814
polymer, isotactic (*it*) 812
polymer, syndiotactic (*st*) 814
polymers, crystalline 815
polymers, factors governing steric structure 818
polymers, structure and properties 819
polymercury cations 475
polymorphs 78
polynuclear aromatic hydrocarbons 879
polynuclear clusters, with S_n^{2-} ligands 564
polyoxometallates 632
polypeptide 704
polypeptide chain, dihedral angles 773
polypeptide chain, helical configuration 773
polypeptide chain, in proteins 779
polyphosphazenes 368
polyphycene 685
polyquinanes 756, 838
polysaccharide 914
polyselenide ligands 427
polysulfide ligands 567
polytelluride ligands 427
polythiazyl, electrical conductivity 410
polythiazyl 409
polyvanadate chemistry 632
porosil 1190
porphinato-iron(II) complexes 672
porphinato-iron(II) complexes, spin state/stereochemical relationship 673
porphinato-iron(III) complexes 671
porphinato-iron(III) complexes, spin state/stereochemical relationship 672
porphinato-iron stereochemistry 671
porphyrin complexes, octaethyl 678
porphyrin core, ruffling of 676
porphyrin iron(III) complexes, oxo-bridged 680
porphyrin, steroid-capped 742
porphyrins, capped 689
porphyrins, picket-fence 689
porphyrins, strapped 691
porpyhrinogen tetraanion 685

KLiO, structure 92
$K^+(C222).e^-$ 229
$K_{1.75}[Pt(CN)_4].1.5H_2O$ 610
$K_2C_2O_4.H_2O_2$ 294
$K_2Mo_8O_{16}$ 600
$K_2Na[Co(NO_2)_6]$ 513
K_2NiF_4 122
K_2PtCl_4 structure, d-electron distribution 132
K_2PtCl_6 structure, compounds with 131
K_2PtCl_6 structure, d-electron distribution 130
K_2PtCl_6 structure, historical significance 129
potassium, π complexes 1008
potassium, polymeric complexes 1009
potassium antimony tartrate trihydrate 378
potassium dihydrogen isocitrate 786
potassium hexachloroplatinate(IV) 129
potassium hydrogon fluoride 196
potassium octacyanomolybdate(IV) dihydrate 493
potassium rhenium hydride 543
potassium tetrachloroplatinate(II) 132
potassium tetracyanoplatinate 610
POTCP (partially-oxidized tetracyano-platinate) 612
POTCP complexes 612
POTCP complexes, band theory 614
POTCP complexes, degree of partial oxidation 613
POTCP complexes, generalizations 615
POTCP complexes, MO treatment 615 oxidation 617
[*n*]prismanes 843, 848
[*n*]prismanes, table 843
propellane 826
prostacyclins 746
prostaglandin $PGF_{1\beta}$ 744
prostaglandins 745
prostaglandins, conformations 749
prostaglandins, types 746
prostanoids 748
proton-water complexes 208
Prussian blue 497
pure metals, structure 25
pyrazinophanes 858
pyridazine 889
pyridine-hydrogen fluorides 200

pyridine-like heterobenzenes 165
pyroxenes 142
qinghaosu 925
quadruple bond, deformation density in $Cr_2(mph)_4$ 721
quadruple bond, discovery 13
quartz 97
quasicrystal 446
quinidine 951
quinine 951
$R_4N^+X^-.nH_2O$, crystal data 1199
$(RO)_3M\equiv N$ compounds 639
racemization, electron density map 1169
racemization, unit-cell dimensions 1168
racemization process 1169
raffinose 916
rare-gas compounds, bonding 434
rare-gas compounds, discovery 431
rational indices, law of 3
realgar 375
refractivity, of calcite and aragonite 148
regioselectivity 697
retinal 975
reversible ethylene coordination 1034
Rh-BINAP complexes 869
Rh_{13} cluster 730
$[Rh_{13}H_2(CO)_{24}]^{3-}$ 730
ribose, in nucleoside 705
RNA 706
rotanes 829
rubber, natural 817
rubidide 230
Rb_9O_2 232
$RbAg_4I_5$ 449
$Rb_{15}Hg_{16}$ 480
Ru-BINAP complexes 869
rutile 93
rutile structure, charge density 95
rutile structure, compounds with 95
rutile structure, prediction 42
sandwich, dimetal 1105
sandwich compounds 1101
sandwich compounds, triple-decker 1102, 1105
saponins 929
sapphyrin 682
sarcophagine (sar) 518
Sayre equation 9
Sc_3C_4 287
Se-Dianin's compounds 1219

Se_8 400
Se_8^{2+} 418
Se_{10}^{2+} 414
selenium, polyatomic cations 416
selenium sulfides, cyclic 399
selenourea 356
sepulchrate (sep) 518
sequence of layers, in ccp 23
sesquichalcogenides of group IV elements, organo-substituted 156
sex hormones 940
β-sheets 776
β-sheets, antiparallel 776
silatranes 323, 329
silenes 902
silenes, bonding in 903
silenes, metal complexes 907
silicates, alumino 313
silicates, classification 298
silicates, double chain 140
silicates, double rings 303
silicates, general structural rule 311
silicates, layer 306
silicates, natural 311
silicates, replacement of cations in 299
silicates, single chain 324
silicates, single rings 303
silicates, types of chains 142
silicates, varieties of rings 302
$Si=C$ bond 902
$Si-O$ bond, nature in α-quartz 98
$Si-O$ bond, electron transfer 143
SiO_2, modifications 99, 101
SiO_2, transitions of modifications 99
$[SiO_4]$ tetrahedra 299
$[SiO_6]$ octahedra 299
$Si\leftarrow N$ bond, in silatranes 324
$Si=Si$ bond 906
silicon, five-coordination 323
silicon, lactam dervatives 326
silicon, six-coordination 300
silicon, stability of five-coordinate silatranes 325
silicon nitride Si_3N_4 217
$Ag(Ag_6O_8)NO_3$, electrical conductivity 456
$Ag(Ag_6O_8)NO_3$ 455
Ag_2HgI_4 452
AgI-type superionic conductors 453
$Ag_6(SC_6H_4Cl)_6(PPh_3)_5.2C_6H_5CH_3$ 569

silver acetylide, adduct with $AgNO_3$ 288

silver halides, polymorphs 448

α-silver iodide 448

α-silver iodide, as superionic conductor 449

silver nitrate oxide 455

silver(I) polyselenide complexes 429

silylenes 1060

sliding effect, on CO ligand 1080

sodide 224

$Na^+(C_{18}H_{36}N_2O_6)Na^-$ 224

NaAs structure, deficient and enriched phases 82

NaCl structure, compounds with 65

NaCl structure, historical significance 65

NaCl structure, incorporation of organic molecules 66

$NaClO_3$ 793

$NaMo_4O_6$ 602

Na_3NO_3 128

NaTl type, bonding 86

NaTl type, compounds with 87

NaTl type, physical properties 88

sodium, π complexes 1008

sodium β-alumina 272, 279

sodium β-alumina, as superionic conductor 274

sodium β-alumina, distribution of Na^+ ions 275

sodium ammonium tartrate 788

sodium chloride 10, 64

sodium chloride, electron density 64

sodium cryptate 224

sodium rubidium tartrate, intensities of Friedel pairs 790

sodium rubidium tartrate 790

sodium thallide 86

solid electrolytes 90, 448, 455

solid reaction 795

soot 52

sophoradiol 929

sp^3-sp^3 bond length 832

space groups 3

space lattice 3

spinel 114

spinel block 273

spinel structure, compounds with 115

spinels, crystallographic data 116

spinels, intermediate 115

spinels, inverse 115

spinels, magnetic 117

spinels, normal 115

stannoxane 1159

stereospecific products 796

steroid hormones 941

steroid hormones, receptor binding 941

steroids 934

steroids, basic skeleton 935

steroids, biologically important 940, 942

steroids, classes of 935

steroids, ring conformation 938

steroids, structural features 938

stishovite 100

strain, in organic chemistry 826

strain energies, small ring compounds 827

strain-free helical conformations 41

structural fit 1243

structural homology 827

suberic acid 755

sucrose 911

sucrose, absolute confuguration of crystal 921

sucrose, hydrogen bonds 913

S_4^{2+}, and related square-planar ions 3.36-4

S_8 molecule, *cyclo* 396

S_8^{2+} 418

S_{19}^{2+} 421

S_n^{2-} complexes, dinuclear 563

S_n^{2-} complexes, mononuclear 562

S_n complexes, polynuclear 564

$(SN)_x$ chain 409

$(SN)_x$ chain, brominated 412

$(SN)_x$, electric conductivity 410

S_2N_2, polymerization 410

α-sulfur, orthorhombic 394

α-sulfur, packing 395

β-sulfur 397

γ-sulfur 397

sulfur, allotropes 398

sulfur, modifications 394

sulfur, phase diagram 397

sulfur, polyatomic cations 416

sulfur, polymorphs 396

sulfur nitride, halogenated polymers 412

sulfur-nitrogen heterocycles and cages 403

superadamantane-5, $C_{35}H_{36}$ 42

superclusters 734

superclusters, building blocks 734

superclusters, diffraction data 735
superconductors, copper-oxygen based high-temperature 589
superconductors, coordination copper 597
superionic solids 90, 448, 455
supermolecules 742, 1292
supraclusters 466
supramolecular chemistry 1293
symmetry classes 3
synthetase (GlnRS) 16
tartar emetic 378
d-tartaric acid 11
tartaric acid, *d*-*ℓ* 791
$Te_2Se_4^{2+}$ 424
$Te_2Se_6^{2+}$ cation 420
$Te_2Se_8^{2+}$ 416
$Te_3S_3^{2+}$ 424
Te_6^{4+} 423
Te_8^{2+} cation 420
$(Te_2)_2(I_2)_x$ 61
tellurium, polyatomic cations 416
template effect 524
template synthesis 524, 632, 742
terpenes 924
terpenoids 924
tert-butylamine hydrate 1196
tetracene 173
tetraboratetrahedrane, tetra-*t*-butyl 756
tetracyanoquinodimethane (TCNQ) 891
tetracyclines 961
tetrahedral carbon, flattening 823
tetrahedral covalency of carbon 38
tetrahedral holes 24
tetrahedrane, tetrabora- 756
tetrahedranes 832
tetrahedranes, deformation densities 833
tetraindium triselenide 281
tetramethyl-tetraselenafulvalenium ion (TMTSF) 890
tetranitrogen tetra-(sulfurylfluoride) 407
tetraphenylcyclobutadiene iron tricarbonyl 1131
tetraphenylporphinato iron 669
tetraphosphacyclobutadiene 1137
tetraphosphorus decasulfide, reactions 375
tetraphosphorus trisulfide, iso-structural compounds 372
tetraphosphorus trisulfide 372

tetrasulfur tetraimide 407
tetrasulfur tetranitride 402
tetrasulfur tetranitride, adducts with transition metal halides 405
tetrasulfur tetranitride, as a tridentate ligand 406
tetrasulfur tetranitride, electron diffraction 403
tetrathallium trisulfide 282
tetrathiafulvalene (TTF) 892
tetrazocine, substituted 810
$(Tl,Pb)Sr_2Ca_2Cu_3O_9$ 593
$Tl_2Ba_2Ca_2Cu_3O_{10}$, coordination polyhedra 591
$Tl_2Ba_2Ca_2Cu_3O_{10}$, structure 589
$Tl_2Ba_2Ca_{n-1}Cu_nO_{4+2n}$ 589
Tl_4S_3 281
thiocarbonyl complexes 1070
thiolato complexes 571
thiourea, channel adducts 180
thiovanadyl complexes, five-coordinate 486
thorocene 1139
three-center, two-electron bonds 12
thromboxanes 746
$SnCl_3^-$ anion 336
Sn_8Tl^{3-} 333
Sn_9Tl^{3-} 333
tin, π complexes 1160
tin(II) chloride 336
tin(II) chloride dihydrate 335
tin(II) compounds, lone pair effects 337
tin(II) fluoride 337
tin(II) oxide 341
tin-sulfur clusters, cubane-type 652
TiO_2, polymorphs 94
TiO_2, predicted structure 42
titanium cyclopentadienyl complexes 1122
titanium(III) cyclopentadienyl complexes 1128
titanium-acyl bond 1123
titanocene 1120
titatranes 327
$(TMTSF)_2PF_6$ 890
$(TMTSF)_2X$ conductor 893
$(TMTSF)_2X$ conductor, molecular stack 893
$(TMTSF)_2X$ conductor, sheet network 894

topochemical principle 795
topochemical reaction 797
TOT cage 800
TOT clathrates, absolute
  configuration 1242
TOT clathrates, enantiomeric excess
  1241
TOT clathrates, structural types
  1238
TOT clathrates, table 1240
trans-cinnamic acid, solid reaction
  795
trans-cinnamic acid 795
trans-effect 701
transition-metal aryls 1091
transition-metal complexes containing
  μ-phenyl ligands 1085
transition-metal fluoro compounds,
  with multiple bridges 505
transition-metal fluoro compounds,
  with single bridges 501
transition-metal halocarbene
  complexes 1056, 1057
transition-metal hydride, types 544
transition-metal nitrides 218
tri-L-alanine 776
trianthranilides 1245
s-triazine 881
s-triazines, structures determined
  887
triazatrimetallabenzene 219
tridymite 100
trihydroxy acid 745
trimethylaluminum 1023
trimethylamine, adduct with hydrogen
  chloride 202
trimethylaluminum, bonding in dimeric
  1025
trimethyltin(IV) chloride 1152
trinuclear molybdenum-sulfur cluster
  559
tri-*o*-thymotide (TOT) 1238
tri-*o*-thymotide, chlorocyclohexane
  clathrate 1238
tris(ethylenediamine)cobalt(III)
  chloride 509
α-truxillic acid 796
β-truxinic acid 796
TTF-TCNQ conductors 895
$W(\equiv CCMe_3)(=CHCMe_3)(CH_2CMe_3)(dmpe)$
  1051
W-S compounds 560
tungsten alkoxide clusters 606
tungsten bronze structure 502
tungstenocene, derivatives 1100
Turnbull's blue 497

unit cell 3
unit cell dimensions, and molecular
  size 171
unit cell dimensions, anthracene
  171
unit cell dimensions, naphthalene
  171
uranium tetracyclopentadienyl 1142
uranocene 1139
uranocene, bonding 1140
uranocene half-sandwich compounds
  1140
uranocene, octaphenyl 1140
uranyl superphthalocyanine 586
urea 175
urea, acid adducts 177
urea, channel adducts 179
urea, hydrogen-bond framework 177
urea, inclusion compounds 178
urea, metal complexes 178
valence isomers, interconversions
  800
valence isomers 800
valence sum rule 11
valinomycin 984
valinomycin.3DMSO complex 984
valinomycin, as an ionophore 987
valinomycin, K^+ complex 988
valinomycin, model of cation
  transport by 988
valinomycin, uncomplexed 986
$VO(Q)_4L$ compounds 485
$VO(Q)_4L$ compounds, bonding 486
vanadium complexes, mononuclear
  488
vanadium complexes, dinuclear 490
vanadium complexes, mixed-valent 491
vanadyl bisacetylacetontate 484
vanadyl complexes 485
vanadylphthalocyanine 584
Vaska's compound 659
vinylidene complexes 1072
viruses 15
viruses, structures determined 16
vitamin A 975
vitamin B_6 977
vitamin B_{12} 969
vitamin B_{12} coenzyme 968
vitamin B_{12}, interactions with
  proteins 971
vitamin C (L-ascorbic acid) 978
vitamin H (biotin) 980
vitamins, structural formulae of
  lipid-soluble 973

vitamins, structural formulae of water-soluble 974
vitamins, table of lipid-soluble 972
vitamins, table of water-soluble 975
water, ligating 136
water, space-filling 141
Wilkinson's catalyst 534
Wilkinson's catalyst, mechanism 536
Wilkinson's catalyst, theoretical study 537
Wilkinson's catalyst, types 535
WORM (write once read many) 585
wurtzite 77
wurtzite structure, compounds with 79
X-ray crystallography, development 1, 4
X-ray diffraction by crystals 4
X-ray diffraction pattern, first observed 136
X-rays, applications 7
X-rays, discovery of 4
$XeF_2.XeF_4$ 431
XeF_5^+ ion 436
$Xe_2F_{11}^+$ ion 437
$[(XeOF_4)_3F]^-$ 436
$[MeCN-Xe-C_6F_5]^+$ 438
xenon compounds, table 432
xenon difluoride 433
xenon hexafluoride 434
xenon tetrafluoride 431
yeast *t*RNA 12
Zeise's salt 1032
Zeise's salt, bonding in 1033
zeolite A 314
zeolite 4A 313
zeolite ZSM-18 320
zeolites, cages and building units 319
zeolites, framework 316
zeolites, natural and synthetic 314
zeolites, structure-directing role of cations in synthesis of 318
zeolites, table 318
zeosil 1190
$Zn(CN)_2$ 1257
ZnS structure, compositions and properties 1109
ZnS structure, polytypes 1106, 1109
zinc blende 72

zinc blende structure, compounds with 73
zinc blende structure, multicomponent compounds with 73
Zintl phases 2.12-2
zirconium cyclopentadienyl complexes 87
porpyhrinogen tetraanion 685
zirconium(III) cyclopentadienyl complexes 1124
zirconocene complexes 1125

Note Added in Proof

Section 2.5

X-Ray analysis of a twinned crystal of C_{60} at 110K [space group $Pa\bar{3}$, $a = 14.052(5)$Å, $Z = 4$] led to 6:6 and 6:5 bond lengths of 1.355(9) and 1.467(21)Å, respectively. Se S. Liu, Y. Lu, M.M. Kappes and J.A. Ibers, *Science (Washington)* **254**, 408 (1991).

Crystalline $C_{60} \cdot 2I_2$ is the first example of an ordered fullerite intercalation compound with an alternating host-guest layer structure. See Q. Zhu, D.E. Cox, J.E. Fischer, K. Kniaz, A.R. McGhie and O. Zhou, *Nature (London)* **355**, 712 (1992).

The "rugby ball" structure of C_{70} has been determined from electron diffraction using a simulated-annealing technique. See D.R. McKenzie, C.A. Davis, D.J.H. Cockayne, D.A. Muller and A.M. Vassallo, *Nature (London)* **355**, 622 (1992).

Section 3.10

Microwave spectroscopy has demonstrated the existence of the clasical donor-acceptor complex $H_3N\text{-}BF_3$ in the gas phase with a N→B bond length of 1.59(3)Å, as reported in A.C. Legion and H.E. Warner, *J. Chem. Soc., Chem. Commun.*, 1397 (1991). However, the NBF_3 moiety in the gas phase adduct $CH_3CN\text{-}BF_3$ is a flattened trigonal pyramid with N-B = 2.011(7)Å and N-B-F = 95.6(6)°, which differ considerably from the corresponding values of 1.630(4)Å and 105.6(5)° measured for the crystalline solid. See M.A. Dvorak, R.S. Ford, R.D. Suenram, F.J. Lovas and K.P. Leopold, *J. Am. Chem. Soc.* **114**, 108 (1992).

Section 4.10

The first volume of a planned series that offers a comprehensive coverage of phthalocyanine chemistry is C.C. Leznoff and A.B.P. Lever (eds.), *Phthalocyanines: Properties and Applications*, VCH Publishers, Weinheim, 1989.

Section 5.6

The synthesis and characterization of the antiaromatic cyclic bi-, ter-, and quatercalicene π-systems are reported in T. Sugimoto, M. Shibata, S. Yoneda, Z. Yoshida, Y. Kai, K. Miki, N. Kasai and T. Kobayashi, *J. Am. Chem. Soc.* **108**, 7032 (1986); T. Sugimoto, M. Shibata, A. Sakai, H. Ando, Y. Arai, M. Sakaguchi, Z. Yoshida, Y. Kai, N. Kanehisa and N. Kasai, *Angew. Chem. Int. Ed. Engl.* **30**, 446 (1991).

Section 5.8

The rigid chair skeleton of hexaspiro[3.0.3.0.3.0.3.0.3.0.3.0]tetracosane (or [6.4]rotane) in the crystal persists in solution; the cyclobutane rings are almost ideally staggered and very slightly kite-shaped. See L. Fitjer, K. Justus, P. Puder, M. Dittmer, C. Hassler and M. Noltemeyer, *Angew. Chem. Int. Ed. Engl.* **30**, 436 (1991).